STRAMINIPILOUS FUNGI

Straminipilous Fungi

Systematics of the Peronosporomycetes Including Accounts of the Marine Straminipilous Protists, the Plasmodiophorids and Similar Organisms

by

Michael W. Dick
University of Reading

KLUWER ACADEMIC PUBLISHERS
DORDRECHT / BOSTON / LONDON

Library of Congress Cataloging-in-Publication Data

Dick, M. W.
Straminipilous fungi : systematics of the peronosporomycetes, including accounts of the marine straminipilous protests, the plasmodiophorids, and similar organisms / by Michael W. Dick.
p. cm.
Includes bibliographical references (p.).
ISBN 0-7923-6780-4 (hb : alk. paper)
1. Straminipilous fungi. I. Title.

QK604.2.S86 D53 2001
579.5'3--dc21

00-066290

ISBN 0-7923-6780-4

Published by Kluwer Academic Publishers,
P.O. Box 17, 3300 AA Dordrecht, The Netherlands.

Sold and distributed in North, Central and South America
by Kluwer Academic Publishers,
101 Philip Drive, Norwell, MA 02061, U.S.A.

In all other countries, sold and distributed
by Kluwer Academic Publishers,
P.O. Box 322, 3300 AH Dordrecht, The Netherlands.

Cover design:
Prof. Michael W. Dick

Printed on acid-free paper

PREFACE

Monographs of flagellate fungi have had their origins in the monographs of Schröter (1892-1893), Fischer (1892*a*, *b*) and von Minden (1911-1915). The second edition of *Aquatic Phycomycetes* (Sparrow, 1960) was in its final proof-stages when I was a post-doctoral instructor in Dr Sparrow's laboratory. Karling's monographs, on the lagenidiaceous fungi (1942*a*, 1981), the *Synchytriaceae* (1964*d*), the *Plasmodiophorales* (1942*b*, 1968*e*) and the chytrids (1977) are now also dated. Much has changed: new species have been described; new concepts of species and kingdom interrelationships are rapidly developing, and codes of nomenclature for plants and animals have also changed during the intervening years of the 20th century. The invaluable works of Karling and Sparrow had limitations: the later work of Karling was sadly marred by incorrect citations and inaccurate copies of original art-work. Sparrow gave limited coverage to the filamentous biflagellate taxa and endobiotic monads, so that his identification keys, based on family and genus concepts, could sometimes be misleading.

Many monographs have given abbreviated citations of the first descriptions of species without backup of the full reference (title and full pagination) in the bibliography. This practice may have saved paper but it greatly increases the length of time it takes to trace and obtain the original publications. To help future taxonomists, and particularly ***users*** of taxonomy, I have given entries with the authority unabbreviated and the journal title in full. In view of the often confused history of generic concepts, I have also given an extended synonymy under generic entries. In the past, the names of authorities for taxa have been more or less arbitrarily abbreviated. There are now comprehensive 'recommended' lists that can be followed, but they are still not universally recognized. For these reasons all citations are also listed alphabetically in the BIBLIOGRAPHY. APPENDIX 1 provides a 'binomials list' using the recommended authority abbreviation and APPENDIX 2 gives the recommended form for the authorities for all known straminipilous fungi. Journal abbreviations are not given because so many of the entries predate the *BIOSIS* files; librarians find it easier to trace full titles, and different journals follow different journal-abbreviation guide lines.

I believe that identification keys, rather than lengthy descriptions of species, provide the best ***initial*** introduction to the taxa. In most cases workers will then need to consult several publications in addition to the first publication. Comprehensive bibliographies for each species are included under each binomial. Illustrations are sometimes of such indifferent quality that I have preferred not to list them separately from these species references. Considerable artistic skill would be needed to provide recognizable portraits for the reduced morphology of the endobiotic parasites and this has not been attempted here; so many are known only from the original decriptions. The keys in this work are conceived from an ecological standpoint. I have therefore strayed outside the strict interpretation of straminipilous fungi, both in PART V: SYSTEMATICS and PART VI: KEYS (though to different extents) to indicate the relationships of organisms that at one time or another have been grouped, or perhaps confused, with the flagellate fungi. For practical identification of all of these small endobiotic flagellate fungi from natural habitats it is essential that the keys and systematic entries include the *Olpidiales* and part of the *Blastocladiales* of the *Chytridiomycetes*.

Preparation of the chapter entitled *Oomycota* (Dick, 1990*a*) for the *Handbook of Protoctista* (Margulis *et al.*,1990) made me realize the extent of the disarray in the taxonomy of the biflagellate endobiotic parasites, most of which have, at one time or another, been placed in the order *Lagenidiales*. At first it seemed that a simple reorganization of taxa would be sufficient, but it became increasingly obvious that a much more substantial theoretical reappraisal would be needed. I have had to question the basic tenets of the classification of flagellate fungi which have been accepted throughout the last half century. Foremost of these has been the reliance on the primacy of zoospore flagellar number. The importance of flagellar ornamentation was not initially recognized. The possession of tubular tripartite hairs is now considered to be of primary phylogenetic significance, and organisms possessing this feature are referred to as straminipiles (Patterson, 1989). Straminipilous ornamentation occurs on the anteriorly-directed flagellum of these fungi.

Assignation to a major taxon has often depended on observations by earlier workers without serious consideration of the existence of correlating secondary characters. Parallel evolution was invoked to account for patterns of parasitism and distribution, with the result that 'examples' of 'parallel evolution' are scattered throughout the classification. Seldom are the examples of parallel evolution accompanied by those minor modifications of ontogeny or form that would justify the hypothesis. The frequency of the use of such an explanation is not credible: the alternative is to question the premise.

More recently, the biochemical studies of the evolution of molecules at one extreme, and the recognition of coevolution of communities at the other extreme, have had their polemicists. All these approaches have their places, and each contributes to an understanding of the evolution of any group.

It is not enough to cast doubt on primary precepts without substituting another hypothesis. I have suggested that anisokont flagellation (flagella of different lengths), leading to heterokont flagellation (flagella of different surface morphology involving straminipilous ornamentation), may be more fundamental to that group of the *Protoctista* from which the Kingdom *Straminipila* may have evolved. This, in turn, provokes a reconsideration of the diagnostic criteria for, and thus the delimitation of, the *Straminipila*, a problem which is not yet fully resolved.

Incomplete loss of the flagellar apparatus, accepted as a probability by phycologists, may also have occurred in flagellate fungi. Thus, flagellar number could be of lesser significance than flagellar ornamentation or kinetosome organization: the inclusion of the *Hyphochytriomycetes* is therefore desirable.

This work has three main aims. The ***first aim*** is to review the diversity of morphology, morphogenesis and ultrastructure within the better-known biflagellate fungal taxa. PART I: CRITERIA FOR DIAGNOSES OF FLAGELLATE FUNGI thus also provides a source book for the general biology of the straminipilous fungi. The information at present available for the lagenidiaceous fungi is related to this framework. I hope that the discussion in PART II: CLASSIFICATIONS will help to bridge the divide between the zoological, phycological and mycological appproaches to evolutionary protistology by providing comprehensive coverage of the mycological references for non-mycologists.

The systematic treatments (PART III: THE BETTER-KNOWN PERONOSPOROMYCETES and PART IV: THE LAGENIDIACEOUS FUNGI) differ in style because of the very different stages that their systematics have reached. Whereas the hitherto stable generic classifications, based on morphology, of the *Peronosporales* and *Saprolegniales* are now being overturned by molecular studies to the extent that it will not be possible to present an overview for another two or three years, perhaps in a supplementary volume; there is little chance that sufficient numbers of the lagenidiaceous fungi will be studied with these techniques: it is still necessary to try to provide a morphological framework for a transitional, traditional classification for the latter group. For both groups of these straminipilous fungi a nomenclatural framework is essential; this is provided here.

The ***second objective*** is to present an hierarchical classification for as many as possible of the fungi left *incertae sedis* by the placement of the *Lagenidiaceae* in the *Pythiales* by Dick, Wong & Clark (1984). Brief reviews of the *Olpidiopsidaceae*, *Sirolpidiaceae* and other families are therefore relevant. The current generic integrity of *Lagenidium* is unsustainable. Nomenclatural changes are inevitable, but it would be unfortunate if such a change had to be made for *Lagenidium giganteum* which has been the subject of much recent research, so I have endeavoured to find a way around this problem.

The ***third aim*** is to provide a comprehensive, cross-referenced, source and introit into both the old and the more recent literature of both the straminipilous and other organisms. I have purposely broadened my remit to include marine, terrestrial and parasitic taxa that either may not otherwise be generally known or that are considered to be the 'province' of other specialists.

The European preference that italic should be used for taxonomic names at all hierarchical ranks can lead to disruptive typography in a text such as this, and also detracts from the emphasis intended for generic names and binomials. I have, therefore, in the general account, frequently adopted colloquial names for higher taxa.

Some readers may find this systematic scheme premature. It will serve its purpose if it alerts workers in other fields to potential problems in assuming that a well-established classificatory framework exists.

AUGUST 2000

M. W. DICK
University of Reading
Reading, U. K.

ACKNOWLEDGEMENTS

I owe an enormous debt of gratitude to three people who have influenced my work: first, to Victor Newby, who introduced me to water moulds in 1954, for his meticulous teaching of morphology and morphogenesis and for imbuing in me an ecological approach to these fungi in my postgraduate work; to Grace Waterhouse for her counsel, encouragement and caution since my postgraduate days; and to Fred Sparrow for teaching me precision and patience in observation.

I have great pleasure in thanking numerous colleagues for giving me their critical comments on early drafts of this work. I have tried to accommodate all their advice, but ultimately, all the opinions expressed are my own. Parts or all of these drafts have been seen by R. A. Andersen, D. J. S. Barr, C. M. Brasier, R. K. Brummitt, H. M. Canter, O. Constantinescu, J. C. David, M. S. Fuller, J. C. Green, J. B. Harborne, J. Hawkins, D. H. Jennings, C. B. Johnson, J. L. Kerwin, P. M. Kirk, G. R. Klassen, F. N. Martin, S. T. Moss, S. Y. Newell, J. D. Ross and B. C. Sutton.

I acknowledge with thanks the help that Andrew Wallace-Hadrill, University of Reading, Department of Classics, and John David, CABI Bioscience, have given me with the Latin descriptions.

I also thank colleagues at the Freshwater Biological Association, CABI Bioscience (formerly the International Mycological Institute), Egham, the Linnean Society of London, the Natural History Museum and the Royal Botanic Gardens, Kew, and especially Gina Douglas, the Librarian of the Linnean Society of London, and Malcolm Beasely, the Librarian of the Botany Department of the Natural History Museum, for helping me find the early literature and authorities for host taxa.

I am much indebted to my former research student, Sally L. Glockling, who provided me with the opportunity to look at first hand at the sporulation of some of the nematophagous lagenidiaceous fungi.

Finally, it gives me very great pleasure to acknowledge here the steadfast support and immeasurable help, mycologically and in innumerable other ways, given to me by my wife, Christine, throughout the long progress of this monograph.

CONTENTS

TABLES

FIGURES

INTRODUCTION

INTRODUCTION

The present monograph is intended to replace and be complementary to a large part of *Aquatic Phycomycetes, Second, Revised Edition* (Sparrow, 1960). I extend the scope of that work beyond Sparrow's restriction to the aquatic environment, but make no attempt to cover the chytridiomycetes extensively, although it is essential to make reference to these fungi for comparison in appropriate sections of the text. Similarly, the coverage embraces all of *Predominantly Holocarpic and Eucarpic Simple Biflagellate Phycomycetes* (Karling, 1981*a*), and part of *The Plasmodiophorales, Including a Complete Host Index, Bibliography, and a Description of Diseases Caused by Species of this Order* (Karling, 1968*e*).

This work is divided into six parts. PART I: CRITERIA FOR DIAGNOSES OF FLAGELLATE FUNGI reviews the characters available and used for classification throughout the flagellate fungi, providing a comprehensive survey of the early; seminal, and recent structural and molecular biological taxonomic literature. Reference is also made to cognate work on non-fungal organisms. This part may serve as a text for the structural biology, ecology and evolution of the straminipilous fungi. PART II: CLASSIFICATIONS provides an historical account of the development of systematic theory for the group. PART III: THE BETTER-KNOWN, MYCELIAR PERONOSPOROMYCETES discusses the systematics of the downy mildews and watermoulds. PART IV: THE LAGENIDIACEOUS FUNGI AND SIMILAR ORGANISMS deals with the systematics of other biflagellate organisms, some of which may belong to the *Peronosporomycetes*, while others are straminipilous protoctists. Organisms that may occur in similar habitats and which might be confused with straminipilous fungi are also discussed in this part. PART V: SYSTEMATICS presents the formal taxonomy of the straminipilous fungi, straminipilous *Labyrinthista*, plasmodiophorids and some other heterotrophs and PART VI: KEYS summarizes the morphological characteristics of these species in relation to their occurrence and identification. There are two APPENDICES, a GLOSSARY and an extensive BIBLIOGRAPHY.

The unifying structural feature of the biflagellate fungi; the chromophyte algae (which include diatoms, brown seaweeds, chrysophytes, yellow-green algae and other photosynthetic groups - see Preisig, 1999); the labyrinthulids and thraustochytrids; and free-living marine protists, is the possession of a distinctively ornamented flagellum, the **STRAMINIPILOUS FLAGELLUM** (see Dick, 1997*a*). By association, certain non-flagellate fungi, such as some downy mildews; a few uniflagellate fungi; and some biflagellate but non-straminipilous vertebrate gut commensals (Silberman *et al.*, 1996); are also to be included. Molecular sequencing has confirmed that this diverse group of organisms is monophyletic (Cavalier-Smith, 1998; Cavalier-Smith, Chao & Allsopp, 1995). The group certainly warrants kingdom status, being more deeply rooted within the eukaryotes than either the kingdoms *Animalia* or *Mycota*, but there is debate (discussed below) as to whether or not the photosythetic state is ancestral and therefore the kingdom may be referred to as the kingdom *Chromista* (photosynthetic endosymbiont ancestral) or the **KINGDOM *STRAMINIPILA*** (heterotrophy ancestral). The latter kingdom name is used hereafter.

The fundamental characteristic of fungi is that of nutrient assimilation by means of extracellular enzymes which are secreted through a cell wall, with the resultant digests being resorbed through the same cell wall. This physiological function has usually resulted in the familiar thallus morphology of a mycelium composed of hyphae. Whether filamentous or not, the heterotrophic, walled osmotrophs, endobiotic in living plants, animals and fungi, are **FUNGAL ORGANISMS** even though they are not monophyletic with fungi such as the *Ascomycetes* or *Basidiomycetes*.

Many of these endobionts are reduced, more or less holocarpic forms: they produce motile asexual spores with two, heterokont flagella. Such fungi, together with the much better-known filamentous fungi, including the downy mildews, phytophthoras and saprolegniaceous water moulds should now be known as '**STRAMINIPILOUS FUNGI**'. The filamentous straminipilous fungi are well-researched, but the term 'straminipilous fungi' should certainly embrace the uniflagellate hyphochytrids. The term 'straminipiles' is more appropriate when including thraustochytrids and labyrinthulids, which are, arguably, not fungi.

Straminipilous fungi are fungi which possess a zoospore with an ***anteriorly directed*** flagellum (the straminipilous flagellum) which ***pulls*** the zoospore through the water. The beat of this flagellum is initiated proximally to the base, but its hydrodynamic thrust is ***reversed*** because of two rows of stiff tubular tripartite hairs (the straminipilous ornamentation) held in the plane of the quasi-sinusoidal beat (for definitions, see *GLOSSARY* and for more detailed discussions of the straminipilous flagellum see PART I: CRITERIA FOR DIAGNOSES OF FLAGELLATE FUNGI and PART II: CLASSIFICATIONS). Cahill, Cope & Hardham (1996) have recently provided evidence, using antibodies, which strongly supports the hypothesis that these tubular tripartite hairs are responsible for this thrust reversal. A zoosporic stage with its straminipilous flagellum is not universal in straminipilous fungi: evolutionary loss of this stage has occurred several times within families or genera of the orders *Peronosporales*, *Pythiales*, *Saprolegniales*, *Sclerosporales* and *Leptomitales*.

Numerous small endobiotic parasites, ***without*** a well-developed mycelium, ***without*** proven heterokont flagellation and ***without*** oogamous sexual reproduction, have also been placed as 'oomycetes' (mostly in the order *Lagenidiales*). The genus *Lagenidium*, the family *Lagenidiaceae* and the order *Lagenidiales* all contain several disparate elements. These disparities have been recognized for many years, without attempts to resolve the taxonomic discrepancies (Bessey, 1950: 94; Sparrow, 1960, 1976; Barron, 1976*a*; Newell, Cefalu & Fell, 1977; Powell, Lehnen & Bortnick, 1985; Beakes, 1989; Dick, 1990*a*; Barr, 1992). The systematics of all of these holocarpic taxa is discussed in depth here. It is now clear that at least one species of *Lagenidium* (*L. giganteum* Couch) is biochemically closely related to *Pythium* and could even be regarded as congeneric (see Kerwin & Washino, 1983; Förster *et al.*, 1990; Belkhiri, Buchko & Klassen, 1992). This latter possibility requires that the status of the genera *Lagenidium* and *Myzocytium* be clarified with respect to *Pythium*, since their nomenclatural origins are contemporary. The problem lies not only with the order of priority of the generic names, but with the systematic and nomenclatural positions of the type and subsequent species referred to genera in the *Lagenidiaceae*.

The studies by Barron and co-workers (Davidson & Barron, 1973; Barron, 1973; Barron & Percy, 1975; Barron, 1976*a*, *b*, *c*, 1977*a*, 1978, 1980, 1981; Robb & Barron, 1982; Barron, 1985; Robb & Lee, 1986*a*, *b*; Barron, 1987, 1989b, 1990) have strongly suggested that some of the '***lagenidiaceous***' group of species may have affinity with the biflagellate, anisokont but non-straminipilous, protoctistan (non-fungal?) class, *Plasmodiophoromycetes*. In particular, the extent of similarities and the ultrastructural complexity of the infection apparatuses (extrusomes) in *Haptoglossa* (Barron, 1980; Glockling & Beakes, 2000*a*) with *Polymyxa* (Keskin & Fuchs, 1969) and *Plasmodiophora* (Aist & Williams, 1971) is such that these extrusomes are unlikely to have evolved independently. Recent studies (Beakes & Glockling, 1998; Glockling & Beakes, 2000*a*, *b*, *c*) have shown that the zoospore of *Haptoglossa*, while anisokont, is not straminipilous; if *Haptoglossa* should be placed with the *Plasmodiophoromycetes*, how many other 'lagenidiaceous' genera have a similar affinity?

A major problem is the incomplete descriptions for many of the 'lagenidiaceous' fungi, particularly the older descriptions amongst which will be found the taxonomic types. The opportunity for a comprehensive survey for fresh material of these organisms, especially the type species, most of which are relatively unimportant obligate parasites, is badly needed but unlikely to be funded. Earlier taxonomic dispositions therefore need to be assessed in an historical light: referal of a species to a genus is often dubious. Ultrastructural and biochemical features that are now recognized as having fundamental importance were unsuspected. On the other hand, more obvious features (now regarded as trivial differences) were accorded a prominence for the purposes of communication with contemporary colleagues. In addition to the host/parasite relationship itself, the thallus form in endobiotic, holocarpic, obligate parasites was sometimes the only feature described.

Given the diversity of habits, habitats and nutritional modes which are now coupled with one particular monophyletic origin, it is most important that cross-reference between work on monads, chromophyte algae and flagellate fungi should be facilitated. Many of the characters used today are ultrastructural or biochemical, thus they were not known when the first species were described. I have therefore attempted to integrate morphological and molecular approaches to systematics. For the next few decades classifications based on morphology and molecular biology must co-exist for the purposes of identification. I try to provide a middle way which will enable transition, but emphasise that biodiversity, and the fitness of particular species for particular environmental niches, depends on morphological and physiological attributes, not phylogeny.

The hierarchies above class level (superclass, subphylum, phylum or division, subkingdom) within the kingdom *Straminipila* are given in PART V: SYSTEMATICS and PART VI: KEYS: SYNOPTIC KEY, but their listing does not necessarily imply acceptance of all the hierarchical levels for the straminipilous fungi. According to the *International Code of Botanical Nomenclature* (*ICBN*), 1993, the rules of priority apply to the taxonomic rank of family and subordinate ranks. The situation for orders is less clear, although it is usual to apply the rule of priority. Higher taxa (kingdoms, phyla or divisions and classes) should be accompanied by a Latin diagnosis and be typified, but the rule of priority does not apply and the selection of a name is by personal preference. The adoption of plant, fungal, protistological or animal higher taxon suffixes is, or could be controversial, since some straminipiles are more akin to one or another of the fungal, algal or animal nutritional life-styles; species or genera, now grouped together, have been described under different codes. A pragmatic, perhaps illogical, scheme is preferable for the present, and so I use mixed suffixes for the higher ranks of the peronosporomycetes, labyrinthulids and thraustochytrids. Tables I: 1 - I: 4 provide alphabetic lists of the genera in the *Peronosporomycetes*, *Hyphochytriomycetes*, *Labyrinthista* and *Plasmodiophoromycetes* discussed below.

The fungal component of the *Straminipila* has been named as the sub-phylum (≡ sub-division) *Peronosporomycotina*, class *Peronosporomycetes*, using suffixes familiar to mycologists (Dick, 1995), but these suffixes may change in the future when a clearer hierarchical concept of the fungal and non-fungal components of the kingdom emerge (cf. *Labyrinthista* instead of *Labyrinthulomycetes*, *Labyrinthulales*, previously included within mycological works but now regarded as protoctists). The class *Peronosporomycetes* is divided into three sub-classes, one of which, the *Peronosporomycetidae*, at present contains two orders, the *Peronosporales* (dicotyledonous downy mildews) and the *Pythiales*. The other sub-classes are the *Saprolegniomycetidae* (including the order *Saprolegniales*) and the *Rhipidiomycetidae* (*Rhipidiales*). These four orders constitute what have been commonly regarded as 'the *Oomycetes*', usually referring only to the class *Peronosporomycetes* as defined here, but sometimes to the entire *Peronosporomycotina*. However, the systematic limits of the 'class *Oomycetes*', which includes 'lagenidiaceous' fungi, are extremely vague: I am unable to define the 'class *Oomycetes*'.

The stem 'oomyc-' (or oömyc-) as '*Oomycetes*' or '*Oomycota*', is "**an old-fashioned term, hoary with tradition ... *Oömycetes* should be put to rest!**" (Sparrow, 1976). The stem-name *Oosporae* was initially used as a class (Cohn, 1872) which included algae, and before 1879, as *Oomycetes* for the fungal component (Winter, 1879; and probably Brefeld, 1872-1912, quoted by Ramsbottom, 1916*c*, as 1881). Winter (1884) used the same name as an order which also included the *Monoblepharidaceae*. De Bary (1884) did not use the term oomycetes, but always referred to these fungi as Saprolegnieen (*Saprolegnieae*) and Peronosporeen (*Peronosporeae*), which he grouped together with other flagellate fungi and zygomycetes under the general term 'phycomycetes' within the *Ascomycetes*. It is now time to dispense with non-generic stem names and so I am following well-established precedents in selecting *Peronospora* Corda as the stem name to typify all the higher taxa. This leaves 'oomycetes' in its present day usage as a colloquial term, to be written with a lower case 'o'. To some extent this nomenclatural procedure circumvents the historical problem that older higher taxon names can be extremely difficult to date and determine status and rank. It also circumvents the use of the names of higher taxa of Cavalier-Smith (1986, 1989) which were given unfortunate stems. I do not and will not accept the taxonomic name *Pseudofungi* for any of the organisms discussed here because so much of the history of mycology and plant pathology has been closely concerned with *Pythium*, *Phytophthora* and the *Peronosporales*. These organisms are fungi (with a lower case 'f'); they are osmotrophic heterotrophs with a cell wall; they will continue to be studied by mycologists and plant pathologists, and approaching 10% of the current literature in plant pathology is concerned with these straminipilous fungi. Likewise, I will not accept higher taxon names apparently based on the genus *Pythium* (*Pythiista*, for example, although this could derive from the Greek stem for 'rotten') because the historic link and nomenclatural lineage is either with the *Peronosporales* (*Peronospora*) or *Saprolegniales* (*Saprolegnia*). Cavalier-Smith (1986, 1989) gave Latin diagnoses but the names were not typified. Cavalier-Smith (1986, 1989) has given various versions of higher taxon hierarchy, but I prefer to leave orders *incertae sedis* if this is desirable, rather than continuously change the breadth of higher taxa. All higher taxa named here are accompanied by Latin descriptions and are typified. I validate each name by description rather than diagnosis because the taxon concept depends on the totality of the characters available. At our present uneven level of understanding and information (especially with respect to flagellar insertion) the elimination of overlap between diagnostic criteria, and identification of cladistic nodes, is not always possible. The following diagnostics at the class level include: the type of ploidy cycle (i.e., haplomitotic-B which is diagnostic for *Peronosporomycetes*); type of oogamy. Diagnostics at the subclass level are: transitional zone structure, nutrient utilization, wall composition, sterol biochemistry, oosporogenesis, zoosporogenesis of the principal form zoospore, features of flagellar loss. Periplasmic oosporogenesis; ooplast structure; dimorphism; basal chromosome numbers; are of more value at the ordinal level. Appropriate information from molecular biology and thallus morphogenesis can be used at different hierarchical levels. The downy mildews may be polyphyletic within the straminipilous fungi (Dick, 1988; 2000*a*); the graminicolous downy mildews were placed in a different sub-class, the *Saprolegniomycetidae*, in the order *Sclerosporales* (Dick, Wong & Clark, 1984) from the dicotyledonous downy mildews. Molecular biological phylogenies (further discussed below) are placing the relationships among the families and genera of the *Peronosporomycetidae* in a state of flux.

Necrotrophic endobiotic parasitoids and biotrophic parasites may both show host specialization. Downy mildews and some related or comparable genera in the *Peronosporomycetidae* and *Saprolegniomycetidae* are necrotrophic to biotrophic obligate parasites. 'Biotrophic obligate parasitism' is not always fully developed, so that a limited range of host/parasite relationships may be covered by this phrase. Biotrophic obligate parasites have advanced genetic and biochemical attributes often, but sometimes unjustifiably, equated to an evolutionary status. Biotrophic obligate parasitism certainly requires a degree of specialization and a constraint to variation: there must be elements of genome protection or conservation in both partners. The basis for this harmony probably lies in unique pairings of 'metabolic packages', the principal components of which may differ from parasite to parasite, or host to host, or both (Dick, 1988, 2000*a*). Such 'pairings' are probable between the parasitic fungi and their hosts. Dependence might be based upon an 'empathy' between certain crucial metabolic pathways of host and parasite, so that the catabolism and anabolism of both are in accord, rather than there being a determining demand for a particular chemical. The stimuli for the development of secondary metabolic pathways may be sought in the exposure of angiosperms, which had evolved in sub-optimum light, to pressures for herbaceous development in open canopy. Here, photosynthetic activities would lead to excess photosynthate and high exposure would require UV protection. Straminipilous fungi, previously adapted to high protein/hydrocarbon/carbohydrate nutrition (perhaps primarily provided by animal substrata), might have been stimulated to colonize roots and crowns which had accumulated excess photosynthate. It should be noted that the straminipilous fungi have unique biochemical requirements and metabolic products, many of which are under-rated and some of which will be of significance to the establishment of parasitic relationships. The peronosporomycetes are, themselves, parasitized by a similarly limited range of parasites, including other peronosporomycetes, the hyphochytrids and plasmodiophorids. From the coevolutionary viewpoint, there are distinctions to be drawn between obligate parasitism, species-specific parasitism, and special-form relationships. A discussion on infra-specific differences could, in time, illuminate the processes of speciation compared with population diversity, but the data are too fragmentary at present. Whereas obligate parasitism merely requires the presence of a regular (but possibly periodic) and renewable (but possibly highly transient) nutrient availability from living protoplasm, species-specific parasitism implies a much more restricted range for potential complementary metabolisms. The concept of a 'tolerance range', probably much smaller *in planta* than *in vivo* and thus analogous to the ecological ranges of saprotrophs *in situ* in soils (Dick, 1992), might provide a better model than a search for a package of absolute metabolic requirements. The ***systematic*** range of hosts is both taxonomically diverse yet at the same time very limited. Angiosperms parasitized by downy mildews or *Saprolegniomycetidae* are mostly in highly specialized taxa, or in recently evolved families, or in taxa that may have a propensity to produce high levels of secondary metabolites. Different host pathways may be pre-eminent for different parasites. Because of these differences, individuals of a single host species may be infected by several parasites (e.g., Sansome & Sansome, 1974) and the parasites may, by the same token, also encompass different degrees of host specificity. The application of such principles to parasitism in algal or fungal hosts is distant.

It is now appropriate to place the major morphological emphasis on three criteria:

1. the detailed structure of the zoospore
2. the mechanism of zoosporogenesis, in so far as its precision has been determined
3. the ontogeny of sexual reproduction, particularly the position of the resting spore in the gametangium and the origins of the wall layers of the oospore

TABLE 1. *Alphabetic list of genera of* Peronosporomycetes, *with type species and approximate numbers of species*

Genus	Number of species	Type species
Achlya Nees (1823)	*ca* 50 species, polyphyletic	*A. prolifera* Nees
Albugo (Pers.) Roussel *ex* Gray (1821)	44 species	*A. candida* (J. F. Gmel.) Kuntze var. *candida*
Aphanodictyon M. W. Dick (1971)	1 species	*A. papillatum* M. W. Dick
Aphanomyces de Bary (1860)	*ca* 29 species	*A. stellatus* de Bary
Aphanomycopsis Scherff. (1925)	8 species	*A. bacillariacearum* Scherff.
Aplanes Humphrey (1893)	1 species	*A. androgynus* (W. Archer) Humphrey
Aplanopsis Höhnk (1952)	2 species, polyphyletic	*A. terrestris* Höhnk
Apodachlya Pringsh. (1883)	3 species	*A. brachynema* (Hildebr.) Pringsh.
Apodachlyella Indoh (1939)	1 species	*A. completa* (Humphrey) Indoh
Aqualinderella R. Emers. & W. Weston (1967)	1 species	*A. fermentans* R. Emers. & W. Weston
Araiospora (Cornu) Thaxt. (1896)	5 species	*A. spinosa* (Cornu) Thaxt.
Atkinsiella Vishniac (1958)	1 species	*A. dubia* (D. Atkins) Vishniac
Basidiophora Roze & Cornu (1869)	2 species	*B. entospora* Roze & Cornu
Benua Constant. (1998)	1 species	*B. kellermanii* (Sacc.) Constant.
Blastulidium Pérez (1903)	1 species	*B. paedophthorum* Pérez
Bremia Regel (1843)	1-20 species depending on species concepts	*B. lactucae* Regel var. *lactucae*
Bremiella G. W. Wilson (1914)	4 species	*B. megasperma* (Berl.) G. W. Wilson
Brevilegnia Coker & Couch (1927)	14 species	*B. subclavata* Couch
Brevilegniella M. W. Dick (1961)	1 species	*B. keratinophila* M. W. Dick
Calyptralegnia Coker (1927)	2 species	*C. achlyoides* (Coker & Couch) Coker
Chlamydomyzium M. W. Dick (1997)	6 species	*C. anomalum* (G. L. Barron) M. W. Dick
Ciliatomyces (I. Foissner & W. Foissner) I. Foissner & W. Foissner (1995)	1 species	*C. spectabilis* (I. Foissner & W. Foissner) I. Foissner & W. Foissner
Cornumyces M. W. Dick (2000)	8 species	*C. destruens* (Sparrow) M. W. Dick
Couchia W. W. Martin (1981)	1 species	*C. circumplexa* W. W. Martin
Crypticola Humber *et al.* (1989)	2 species	*C. clavulifera* Humber *et al.*
Cystosiphon Roze & Cornu (1869)	5 species	*C. pythioides* Roze & Cornu
Diasporangium Höhnk (1923)	1 species	*D. jonesianum* Höhnk
Dictyuchus Leitg. (1869)	7 species	*D. monosporus* Leitg.
Ducellieria Teiling (1957)	1 species	*D. chodatii* (F. Ducell.) Teiling
Ectrogella Zopf (1884)	6 species	*E. bacillariacearum* Zopf
Endosphaerium D'Eliscu (1977)	1 species	*E. funiculatum* D'Eliscu
Eurychasma Magnus (1905)	1 species	*E. dicksonii* (E. P. Wright) Magnus
Eurychasmidium Sparrow (1936)	3 species	*E. tumefaciens* (Magnus) Sparrow
Eurychasmopsis Canter & M. W. Dick (1994)	1 species	*E. multisecunda* Canter
Geolegnia Coker & J. V. Harv. (1925)	3 species	*G. inflata* Coker & J. V. Harv.
Gonimochaete Drechsler (1946)	4 species	*G. horridula* Drechsler
Gracea M. W. Dick (1997)	2 species	*G. gracilis* (E. J. Butler) M. W. Dick
Haliphthoros Vishniac (1958)]	4 species	*H. milfordensis* Vishniac
Halodaphnea M. W. Dick (1998)	5 species	*H. parasitica* (K. Nakam. & Hatai) M. W. Dick
Halophytophthora H. H. Ho & S. C. Jong (1990)	14 species	*H. vesicula* (Anastasiou & Churchl.) H. H. Ho & S. C. Jong
Hydatinophagus Valkanov (1931)	2 species	*H. apsteinii* Valkanov
Isoachlya Kauffman (1921)	9 species	*I. toruloides* Kauffman & Coker
Lagena Vanterp. & Ledingham (1930)	1 species	*L. radicicola* Vanterp. & Ledingham
Lagenidium Zopf (1859)	1 species	proposed: *L. giganteum* Couch
Lagenisma Drebes (1968)	1 species	*L. coscinodisci* Drebes
Leptolegnia de Bary (1888)	*ca* 3 species	*L. caudata* de Bary
Leptolegniella Huneycutt (1952)	5 species	*L. keratinophila* Huneycutt
Leptomitus (Roth) C. Agardh (1824)	1 species	*L. lacteus* (Roth) C. Agardh
Medusoides Voglmayr (1999)	1 species	*M. argyrocodium* Voglmayr
Mindeniella Kanouse (1927)	2 species	*M. spinospora* Kanouse
Myzocytiopsis M. W. Dick (1997)	17 species	*M. lenticularis* (G. L. Barron) M. W. Dick
Myzocytium Schenk (1858)	4 species	*M. proliferum* Schenk
Nellymyces A. Batko (1971)	1 species	*N. megaceros* A. Batko
Nematophthora Kerry & D. H. Crump (1980)	1 species	*N. gynophila* Kerry & D. H. Crump
Olpidiopsis Cornu (1872)	16 species	*O. saprolegniae* (A. Braun) Cornu var. *saprolegniae*
Pachymetra B. J. Croft & M. W. Dick (1989)	1 species	*P. chaunorhiza* B. J. Croft & M. W. Dick
Paraperonospora Constant. (1989)	8 species	*P. leptosperma* (de Bary) Constant.
Peronophythora W. H. Ko *et al.* (1978)	1 species	*P. litchi* W. H. Ko *et al.*
Peronosclerospora (S. Ito) Hara (1927)	11 species	*P. sacchari* (T. Miyake) Hara
Peronospora Corda (1837)	*ca* 460, perhaps only *ca* 75 species, over 800 binomials)	*P. rumicis* Corda
Petersenia Sparrow (1934)	3 species	*P. lobata* (H. E. Petersen) Sparrow

(Table 1, continued)

Phytophthora de Bary (1875)	65 species	*P. infestans* (Mont.) de Bary var. *infestans*
Plasmopara J. Schröt. (1886)	109 species	*P. pygmaea* (Unger) J. Schröt. forma *pygmaea*
Plectospira Drechsler (1927)	4 species	*P. myriandra* Drechsler
Pleocystidium C. Fisch (1884)	3 species	*P. parasiticum* C. Fisch
Plerogone M. W. Dick (1986)	1 species	*P. helodes* M. W. Dick
Pontisma H. E. Petersen (1905)	6 species	*P. lagenidioides* H. E. Petersen
Protoachlya Coker (1923)	4 species	*P. paradoxa* Coker
Pseudoperonospora Rostovzev (1903)	7 species	*P. cubensis* (Berk. & M. A. Curtis) Rostovzev
Pythiella Couch (1935)	2 species	*P. vernalis* Couch
Pythiogeton Minden (1916)	6 species	*P. utriforme* Minden
Pythiopsis de Bary (1888)	3 species	*P. cymosa* de Bary
Pythium Pringsh. (1858)	127 species	*P. monospermum* Pringsh.
Rhipidium Cornu (1872)	4 species	*R. interruptum* Cornu
Salilagenidium M. W. Dick (2000)	6 species	*S. callinectes* (Couch) M. W. Dick
Saprolegnia Nees (1823)	15 species	*S. ferax* (Gruith.) Thur.
Sapromyces Fritsch (1893)	3 species	*S. elongatus* (Cornu) Coker
Sclerophthora Thirum. *et al.* (1953)	8 species	*S. macrospora* (Sacc.) Thirum. *et al.*
Sclerospora J. Schröt. (1879)	3 species	*S. graminicola* (Sacc.) J. Schröt.
Scoliolegnia M. W. Dick (1969)	3 species	*S. asterophora* (de Bary) M. W. Dick*
Sirolpidium H. E. Petersen (1905)	7 species	*S. bryopsidis* (de Bruyne) H. E. Petersen
Sommerstorffia Arnautov (1923)	1 species	*S. spinosa* Arnautov
Syzygangia M. W. Dick (1997)	8 species	*S. marchaliana* (De Wild.) M. W. Dick
Thraustotheca Humphrey (1893)	3 species	*T. clavata* (de Bary) Humphrey*
Trachysphaera Tabor & Bunting (1923)	1 species	*T. fructigena* Tabor & Bunting
Verrucalvus P. Wong & M. W. Dick (1984)	1 species	*V. flavofaciens* P. Wong & M. W. Dick

TABLE 2. *Alphabetic list of genera of* Hyphochytriomycetes, *with type species and approximate numbers of species*

Canteromyces Sparrow (1960)	1 species	*C. stigeoclonii* (De Wild.) Sparrow
Cystochytrium Ivimey Cook (1932)	1 species	*C. radicale* Ivimey Cook
Hyphochytrium Zopf (1884)	6 species	*H. infestans* Zopf
Latrostium Zopf (1894)	1 species	*L. comprimens* Zopf
Reessia C. Fisch (1884)	2 species	*R. amoeboides* C. Fisch
Rhizidiomyces Zopf (1884)	12 species	*H. apophysatus* Zopf

TABLE 3. *Alphabetic list of genera of* Labyrinthista, *with type species and approximate numbers of species*

Althornia E. B. G. Jones & Alderman (1971)	1 species	*A. crouchii* E. B. G. Jones & Alderman
Aplanochytrium Bahnweg & Sparrow (1972)	1 species	*A. kerguelensis* Bahnweg & Sparrow
Developayella S. M. Tong (1995)	1 species	*D. elegans* S. M. Tong
Diplophrys J. Barker (1868)	1 species	*D. archeri* J. Barker
Elina N. J. Artemczuk (1972)	2 species	*E. marisalba* N. J. Artemczuk
Japonochytrium Kobayasi & Ôkubo (1953)	1 species	*J. marinum* Kobayasi & Ôkubo
Labyrinthula Cienk. (1867)	*ca* 12 species	*L. vitellina* Cienk.
Labyrinthuloides F. O. Perkins (1973)	5 species	*L. yorkensis* F. O. Perkins
Pyrrhosorus Juel (1901)	1 species	*P. marinus* Juel
Schizochytrium S. Goldst. & Belsky (1964)	5 species	*S. aggregatum* S. Goldst. & Belsky
Thraustochytrium Sparrow (1936)	15 species	*T. proliferum* Sparrow
Ulkenia A. Gaertn. (1977)	6 species	*U. visurgensis* (Ulken) A. Gaertn.

TABLE 4. *Alphabetic list of genera of* Plasmodiophoromycetes, *with type species and approximate numbers of species*

Genus	Species	Type species
Ectrogella Zopf (1884)	10 species, polyphyletic	*E. bacillariacearum* Zopf
Endemosarca L. S. Olive & Erdos (1971)	2 species	*E. hypsalyxis* Erdos & L. S. Olive
Haptoglossa Drechsler (1940)	10 species	*H. heterospora* Drechsler
Ligniera Maire & A. Tison (1911)	7 species	*L. verrucosa* Maire & A. Tison
Membranosorus Ostenf. & H. E. Petersen (1930)	1 species	*M. heterantherae* Ostenf. & H. E. Petersen
Octomyxa Couch *et al.* (1939)	2 species	*P. achlyae* Couch *et al.*
Plasmodiophora Woronin (1877)	5 species, probably polyphyletic	*P. brassicae* Woronin
Polymyxa Ledingham (1939)	2 species	*P. graminis* Ledingham
Sorodiscus Lagerh. & Winge (1912)	4 species	*S. callitrichis* Lagerh. & Winge
Sorosphaera J. Schröt. (1886)	2 species	*S. veronicae* (J. Schröt.) J. Schröt.
Spongospora Brunch. (1887)	2 species	*P. subterranea* (Wallroth) Lagerh.
Tetramyxa K. I. Goebel (1884)	1 species	*P. parasitica* K. I. Goebel
Woronina Cornu (1872)	7 species	*W. polycystis* Cornu

PART I

CRITERIA FOR DIAGNOSES OF FLAGELLATE FUNGI

THE ZOOSPORE

Asexual reproduction: definitions of sporangia, sporangiospores and zoospores

For most of the organisms discussed here, the normal method of asexual reproduction is by spores ***originating from*** a sporangium. Exogenously produced conidia or chlamydospores are either not known or of minor significance. In eucarpic organisms with hyphae, the sporangia are separated from the assimilative mycelia by septa or plugs. For holocarpic organisms, in which all nucleated protoplasm is utilized in spore production, the definition of a sporangium becomes difficult: at maturity, the assimilative thallus is converted into a sporangium, frequently with little morphological modification other than the formation of an exit tube and dehiscence papilla.

The normal definition of a **sporangium** is a spore container or vessel, but in many of these fungi the protoplasm is not in the form of spores when inside the container, and a looser definition is required. Cleavage of the protoplasm to form the spores can take place after the protoplasm of the sporangium has been expelled, and therefore does not necessarily occur in any recognizable, walled, morphological structure.

In much of the mycological literature the terms sporangium and zoosporangium are used synonymously. A flagellate asexual spore is known as a **zoospore**, and a sporangium ***containing zoospores*** is a **zoosporangium**. Strictly, a spore is a walled structure, and only a spore formed ***within*** a sporangium is a **sporangiospore** (cf. *Thraustotheca* or *Eurychasmopsis* - Canter & Dick, 1994; Canter, Heaney & Lund, 1990). The term **planont** is therefore useful when the term sporangiospore is inappropriate. The original definition of a planont (Sparrow, 1943) is of any ***motile*** (and therefore usually naked) cell; it should be noted that this definition specifically ***includes*** gametes and zygotes as well as asexual spores. Thus the term planont is usually interpreted as a ***flagellate*** cell: the term **aplanospore** is used to include any naked, amoeboid or non-amoeboid ***mobile*** cell as well as a walled sporangiospore.

The fungal zoospore has been defined by Lange & Olson (1983) as follows: "true zoospores are formed in a zoosporangium, by mitotic nuclear divisions and directly give rise to a vegetative thallus". Such a definition is untenable for straminipilous fungi. In *Pythium* the zoospores are differentiated ***outside*** the sporangium. In the *Peronosporomycetes* there are two ontogenetic patterns of sporangial development: in the *Saprolegniales* (Humphrey, 1893) and *Pythiales* (Hemmes, 1983) the zoosporangial protoplasm is delimited with its full complement of nuclei, but in the *Peronosporales* (*Peronospora*, Trigiano & Spurr, 1987; *Bremia*, Tommerup, 1989; *Plasmopara*, Burruano *et al.*, 1992) nuclear division occurs in the immature sporangium. In *Phytophthora* (Maltese, Conigliaro & Shaw, 1995) and *Albugo* (Khan, 1976) nuclear division and nuclear abortion occur in the immature sporangium. Doubt is thereby cast on whether *Phytophthora* should be classified in the *Pythiales* or the *Peronosporales*. In *Saprolegnia* polyplanetism means that the development of the subsequent vegetative thallus may be ***indirect***.

The **zoospore** must be more loosely defined as:

> a motile (and flagellate or potentially flagellate) naked cell or planont, ***normally*** uninucleate, and ***usually*** functioning as a dispersal agent for the same phase of the life-history from whence it was generated.

Oertel & Jelke (1986) reported the formation of abnormal, multinucleate zoospores in *Phytophthora*.

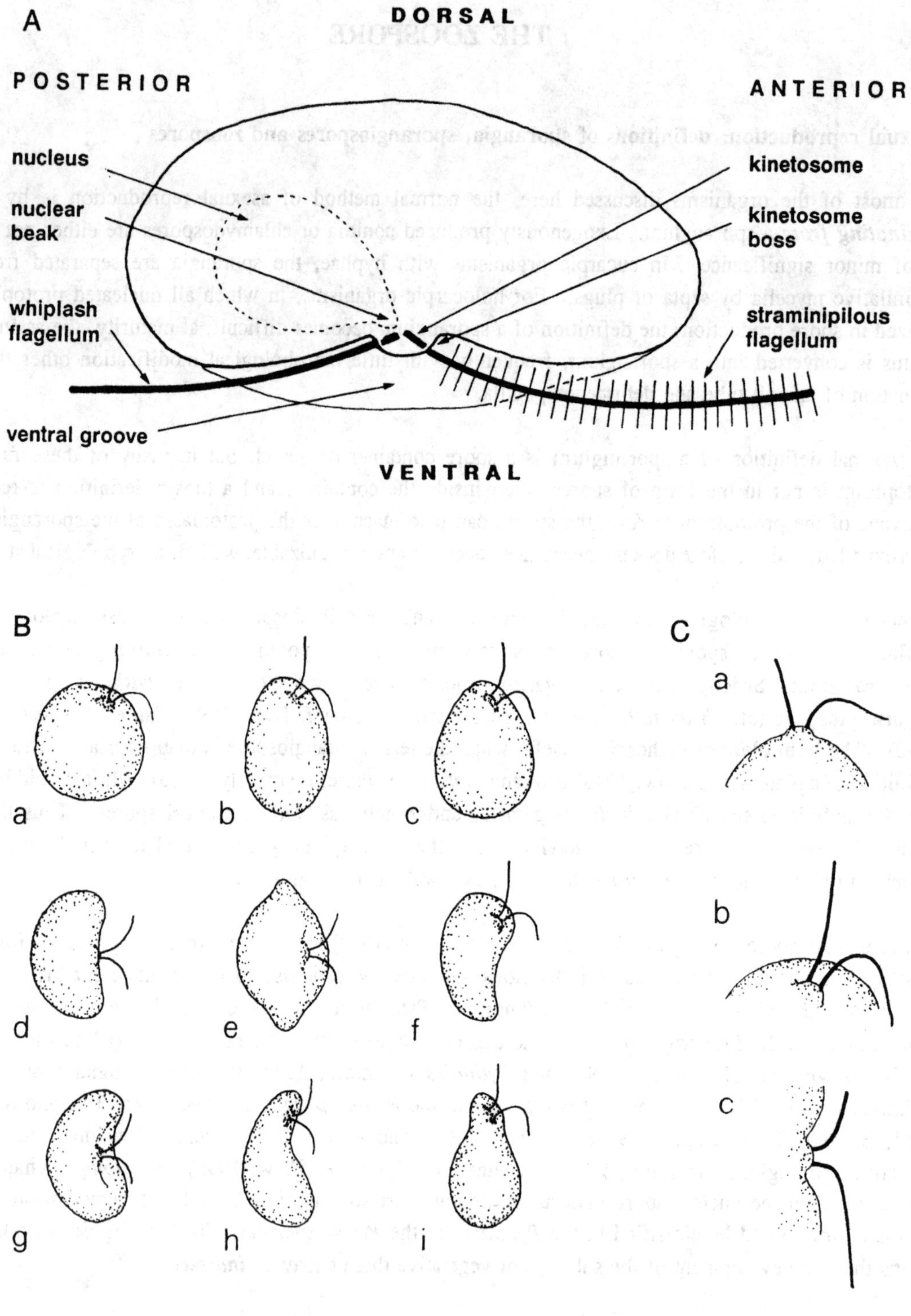

FIGURE I: 1. Zoospore orientation, terminology, shapes and flagellar insertion. A. Terminology; orientation of the principal-form zoospore. B. Zoospore shape, excluding flagellar insertion: a: spherical, anterior/posterior and dorsiventral axes equal; b: ellipsoid, anterior/posterior axis longer than the dorsiventral axis, symmetrical about the axes; c: ovoid, anterior/posterior axis longer than the dorsiventral axis which is greatest in the posterior half, symmetrical about the anterior/posterior axis; d: hemispherical, symmetrical about the dorsiventral axis; e: limoniform, ellipsoid with prolongation at the ends of the anterior/posterior axis; f: obovoid (or obpyriform), anterior/posterior axis longer than the dorsiventral axis which is greatest in the anterior half, asymmetrical; g: bean-shaped; h: grape-seed-shaped (arcuate pyriform); i: pyriform - ovoid, with a concave profile in the anterior half, symmetrical about the anterior/posterior axis. C. Flagellar insertion: a: apical; b: subapical; c: lateral. Fig. I: 1 A reproduced from Dick (1999: fig. 2 A) by permission of Academic Press.

A further modification of this definition from 'naked cell' to 'unwalled cell' would be necessary to encompass the thraustochytrids (but not the labyrinthulids) which have an investment of scales over the zoospore surface (*Althornia*, Jones & Alderman, 1971; *Thraustochytrium*, Kazama, 1974*a*; *Ulkenia*, Raghu Kumar, 1982*b*). In the *Chytridiomycetes*, at least, there is some evidence for a zoospore "cell coat" (Fuller, 1996: 168, and references therein). Compare also *Perkinsus* (*Labyrinthomyxa* of Perkins, 1976; regarded by Levine (1978) and Porter (1990) as an apicomplexan excluded from the *Labyrinthista*; (see PART V) and for molecular biology of this organism refer to Goggin & Barker, 1993; and Fong *et al.*, 1993.

Zoospore morphology and motile phases

Descriptions of zoospore morphology have become confused partly because of the definitions employed and partly because different structures or conditions have sometimes been described by the same words. Diagrams of zoospore symmetry and terminology are illustrated in Figure I: 1.

The phylogenetic significance of flagellar form and insertion was still debatable two or three decades ago, so it is not surprising that specific diagnoses and descriptions were imprecise regarding relative flagellar lengths and ornamentation. So few organisms have been studied by Transmission Electron Microscopy (TEM) and Scanning Electron Microscopy (SEM) that this systematic account must attempt an interpretation of the original descriptions in the light of reviews of ultrastructure, such as those by Barr (1981*c*), Barr & Désaulniers (1987*b*, 1989*a*) and Powell *et al.* (1985). The more structural/physiological/biochemical aspects, as opposed to the structural/taxonomic aspects, of zoospore organization are ably reviewed by Fuller (1996), with literature citations that hardly overlap with those given here.

The term 'flagellum' has received different definitions in recent years, sometimes being restricted to prokaryote appendages (Margulis *et al.*, 1990). The term 'cilium' has a long historical documentation, and has been preferred by some recent authors (e.g., Cavalier-Smith, 1991, 1993) but no consensus has been achieved. I prefer to retain 'flagellum' for cells with only one or two (perhaps a few) relatively long appendages for which the length of the quasi-sinusoidal flagellar beat is appreciably shorter (say, a third to a quarter of) the length of the flagellum. The term 'cilium' is more suitable when the appendage is replicated over the surface of the organism, often with interconnecting kinetids, and/or for those organisms in which the length of the quasi-sinusoidal wave is more nearly equal to or shorter than the length of the appendage (e.g., ciliates and *Chlamydomonas*). For straminipilous fungi and plasmodiophorids the term 'flagellum' is more appropriate.

It should be noted that substantial differences in zoospore volume and concomitant cytoskeletal complexity may influence zoospore shape and ultrastructure. Differences in zoospore functions between the life-histories of different but related species will also be reflected in biochemical biodiversity and complexity.

In the great majority of the peronosporomycetes there is only ***one*** zoospore form, this is the **principal zoospore** form, which is reniform or bean-shaped with flagella laterally inserted in a groove (Dick, 1973, 1990*a*, 2000*c*). The flagella are inserted on a protuberance, here termed the **kinetosome boss**, which may form a dyke bridging the flagellar groove (Dick, 1997*a*). The angle of divergence between the two flagella (and their subtending kinetosomes) is approximately 130°-150° (Barr, 1983). The principal zoospore exhibits polyplanetism (see p. 20) in both the *Saprolegniomycetidae* and *Peronosporomycetidae*. Whether the polyplanetic sequence involves mitosis is not established: if there is mitosis, the cyst is functioning as a microsporangium (Willoughby, 1977), but even if there is no mitosis the cyst may still be equivalent to a microsporangium (Pearson, Pike & Dick, 1991). In a small minority of taxa (*Saprolegnia*, *Protoachlya* and perhaps *Pythiopsis*) the more or less ovoid zoospore formed within the zoosporangium may develop flagella that are apically (or near-apically) inserted (Holloway & Heath, 1977*b*; Barr & Désaulniers, 1987*b*;

see also Money *et al.*, 1987). This zoospore, which has been termed an **auxiliary zoospore** (Dick, 1973, 1990*a*, 2000*c*), is a poor swimmer and there is no good evidence that it can ever show polyplanetism. It has often been asserted (e.g., Barr, 1983, Barr & Désaulniers, 1989*a*) that apical flagellar insertion represents the primitive state: there is no evidence to justify this statement (see p. 32). Certainly in *Saprolegnia* and *Protoachlya* any zoospore produced from a first-formed, auxiliary sporangiospore cyst is of the principal form, and the auxiliary zoospore may be merely an ecological adaptation.

The terms **primary zoospore** and **secondary zoospore** have different definitions for the plasmodiophorids compared with their use in much of the earlier literature on straminipilous fungi. These terms should be restricted to the occurrence of zoospores in different phases of the life-history, as in the plasmodiophorids, where the primary zoospore is derived from the resting cyst and the secondary zoospore is derived from the zoosporangium. A further source of confusion has recently been introduced by Glockling & Beakes (2000*a*) who have used the term 'secondary spore' as a synomym for the glossoid spore (gun cell) in *Haptoglossa*. The terms 'primary zoospore' and 'secondary zoospore' should not be used when referring to the *Peronosporomycetes*.

Of the lagenidiaceous fungi, *Lagenidium giganteum* has zoospores of the principal form (Domnas, Jaronski & Hanton, 1986; Brey & Lebrun, 1987). However, in *Salilagenidium callinectes* (Couch) M. W. Dick *loc. cit.* (Bland & Amerson, 1973*a*; Gotelli, 1974*b*), although the zoospore is described as bean-shaped, the flagellar attachment was stated to be sub-apical without any groove. Similarly, *Haliphthoros milfordensis* Vishniac (Overton, Tharp & Bland, 1983) has an ovoid zoospore with sub-apical flagellar insertion. The flagella appear to be inserted on a kinetosome boss, but again there is no indication of a flagellar groove. Schnepf, Deichgräber & Drebes (1978*d*) describe the zoospore of *Lagenisma* as bean-shaped, with lateral flagellar insertion in a "shallow sinus". The zoospore shape and flagellar insertion are not clearly depicted in the TEMs of a fourth marine genus, *Ectrogella* (S. Raghu Kumar, 1980*b*). In *Olpidiopsis saprolegniae* (A. Braun) Cornu var. *saprolegniae* (Bortnick, Powell & Bangert, 1985) the zoospore is elongate-ellipsoid (pyriform) with a narrower tip, the flagella are inserted sub-apically and there is a distinct, but small flagellar groove (M. J. Powell, pers. comm.). This, much smaller, zoospore apparently swims in a manner equivalent to that of the principal zoospore form of the *Peronosporomycetes* (Bortnick *et al.*, 1985). In *Lagena* the zoospore is ovoid or broadly pyriform with the flagella apically or sub-apically inserted (Barr & Désaulniers, 1987*b*) (after motility and prior to encystment the flagellar insertion readjusts to an apical position - D. J. S. Barr, pers. comm.). A TEM section of the zoospore of *Crypticola* (Frances, Sweeney & Humber, 1989) depicts an obpyriform zoospore with sub-apical flagellar insertion on a kinetosome boss. The zoospore of *Blastulidium* (Manier, 1976: fig. 3) is ellipsoid with sub-apical flagellar insertion. In contrast to these genera *Ducellieria* (Hesse, Kusel-Fetzmann & Carniel, 1989) has zoospores which are ellipsoid with distinctly lateral flagellar insertion in a groove.

In most other genera the shape of the zoospore and insertion of the flagella are less well established, relying on light microscope observations. When first-formed, the zoospore of *Syzygangia zygnematicola* M. W. Dick (Dick, 1997*b*) is apparently of the principal form, but it becomes more or less spherical when swimming (Ivimey Cook, 1935). Living material of dehiscing sporangia of *Myzocytiopsis lenticularis* (G. L. Barron) M. W. Dick (Dick, 1997*b*), a parasite of nematodes, also developed initially quiescent biflagellate zoospores which were nearly hemispherical with flagella inserted in the middle of the concave side (Glockling, 1994; Glockling & Dick, 2000). Glockling & Beakes (2000*b*) have described a range of zoospore forms and zoosporogeneses for *Myzocytiopsis*. In *Haptoglossa* (Barron, 1980) and *Myzocytiopsis humicola* (G. L. Barron & Percy) M. W. Dick (Barron & Percy, 1975; Dick, 1997*b*) the zoospores (photographs of living cells) are narrowly obpyriform with sub-apical flagellar insertion; in *Ciliatomyces* (Foissner & Foissner, 1986*b*, 1995) the zoospores (drawn) are also obpyriform with sub-apical flagellar insertion (***transverse*** sections ***below*** the region of flagellar insertion are described as reniform, due to a shallow groove). There are no TEMs of zoospores of the marine genera *Sirolpidium* or *Pontisma*, but

Sparrow (1934) has described the zoospores as narrowly pyriform, with a curved axis, and flagella inserted laterally on the attenuated anterior part.

In the plasmodiophorids the zoospore is subspherical and the flagella are apically inserted (Barr & Allen, 1982; Kole & Gielink, 1962). *Endesmosarca*, a genus possibly related to the plasmodiophorids, has an ellipsoidal zoospore with sub-apical flagellar insertion in a groove (Erdos & Olive, 1971; Olive, 1975). *Nephromyces* is a posteriorly biflagellate protoctist with some fungal characteristics (Saffo, 1981; Saffo & Fultz, 1986). Uniflagellate taxa tend to have flagella inserted apically (hyphochytrids - Cooney, Barr & Barstow, 1985) or posteriorly (chytrids - Lange & Olson, 1979) but there are exceptions such as *Rozella*, in which the flagellum is inserted at the base of a lateral, posteriorly directed canal (Held, 1975) (see p. 246).

Flagellar length is difficult to measure, and may be liable to misinterpretation because of partial withdrawal (Lange & Olson, 1983). The semi-conservative replication of flagella in heterokont biflagellates (Andersen *et al.*, 1991; Beech, Heimann & Melconian, 1991) presents a further complication, but its significance for zoosporic fungi is unclear. There may also be some lack of synchrony in the withdrawal of the two flagella (Holloway & Heath, 1974). Nevertheless, photographs of living cells of *Haptoglossa* and *Myzocytiopsis* indicate a very different proportionality in flagellar length with respect to body size, compared with *Ducellieria*, which has much longer flagella. Zoospores of other taxa are illustrated in Fuller & Jaworski (1987).

The ***shape*** of the zoospore, the ***precise pattern*** of flagellar insertion, and the ***relative lengths*** of the flagella to each other and to the zoospore all need to be documented, if possible, when describing new taxa.

Diagrams of representative ranges of the size and form of biflagellate zoospores are illustrated in Figures I: 2A and 2B on the following pages.

FIGURE I: 2. Zoospore outlines, drawn to the same scale. Composite drawings taken as far as possible from light microscope and TEM illustrations (principal sources indicated in brackets). Note the presence or absence of the flagellar groove and the kinetosome boss (which is relatively more prominent in the smaller zoospores). The presence and arrangement of tubular tripartite hairs is diagramatic, based on a shaft length of 1μm for the tripartite tubular hair, except where it is known to be longer (the straminipilous flagellum is interrupted by '?' where this feature has yet to be verified); the acronema has been indicated where known to occur. **Figure I: 2A**: a: *Saprolegnia* sp. principal form zoospore (Crump & Brandon, 1966); b: *Saprolegnia* sp. auxiliary form zoospore (Crump & Brandon, 1966); c: *Salilagenidium callinectes* (Bland & Amerson, 1973*a*); d: *Lagenidium giganteum* (Domnas *et al.*, 1986); e: *Ectrogella perforans* (C. Raghu Kumar, 1980*a*); f: *Haliphthoros milfordensis* (Overton *et al.*, 1983); g: *Myzocytium megastomum* (Canter, 1947); h: *Eurychasma dicksonii* (Sparrow, 1934); i: *Crypticola clavulifera* (Frances *et al.*, 1989); j: *Ducellieria chodatii* (Hesse *et al.*, 1989); k: *Gracea gracilis* (Dick, 1997*a*, 2000*d*); l: *Rhizidiomyces apophysatus* (Fuller & Reichle, 1965); m: *Sirolpidium bryopsidis* (Sparrow, 1934); n: *Pontisma lagenidioides* (Sparrow, 1934). **Fig. 2B**: a: *Lagena radicicola* (Barr & Désaulniers, 1990*a*); b: *Phytophthora palmivora* (Hemmes, 1983); c: *Haptoglossa mirabilis* (Barron, 1980); d: *Ciliatomyces spectabilis* (Foissner & Foissner, 1986*b*); e: *Woronina pythii* (Fuller & Jaworski, 1987) f: *Polymyxa graminis* (Barr & Allan, 1982); g: *Plasmodiophora brassicae* (Aist & Williams, 1971); h: *Endemosarca ubatubensis* (Olive, 1975); i: *Nematophthora gynophila* (Kerry & Crump, 1980); j: *Olpidiopsis saprolegniae* (Bortnick *et al.*, 1985); k: *Olpidiopsis varians* (Martin & Miller, 1986*b*); l: *Blastulidium paedophthorum* (Sigot, 1931); m: *Syzygangia zygnematicola* from zoosporangium (Ivimey Cook, 1935); n: *Syzygangia zygnematicola* from oospore (Ivimey Cook, 1935); o: *Rozella allomycis* (Held, 1975); p: *Dictyomorpha dioica* (Mullins, 1961); q: *Myzocytiopsis lenticularis* (*loc. cit.*); r: *Labyrinthula* sp. (Amon & Perkins, 1968); s: *Thraustochytrium* sp. (Kazama, 1974*a*); t: *Lagenisma coscinodisci* (Schnepf *et al.*, 1978*d*).

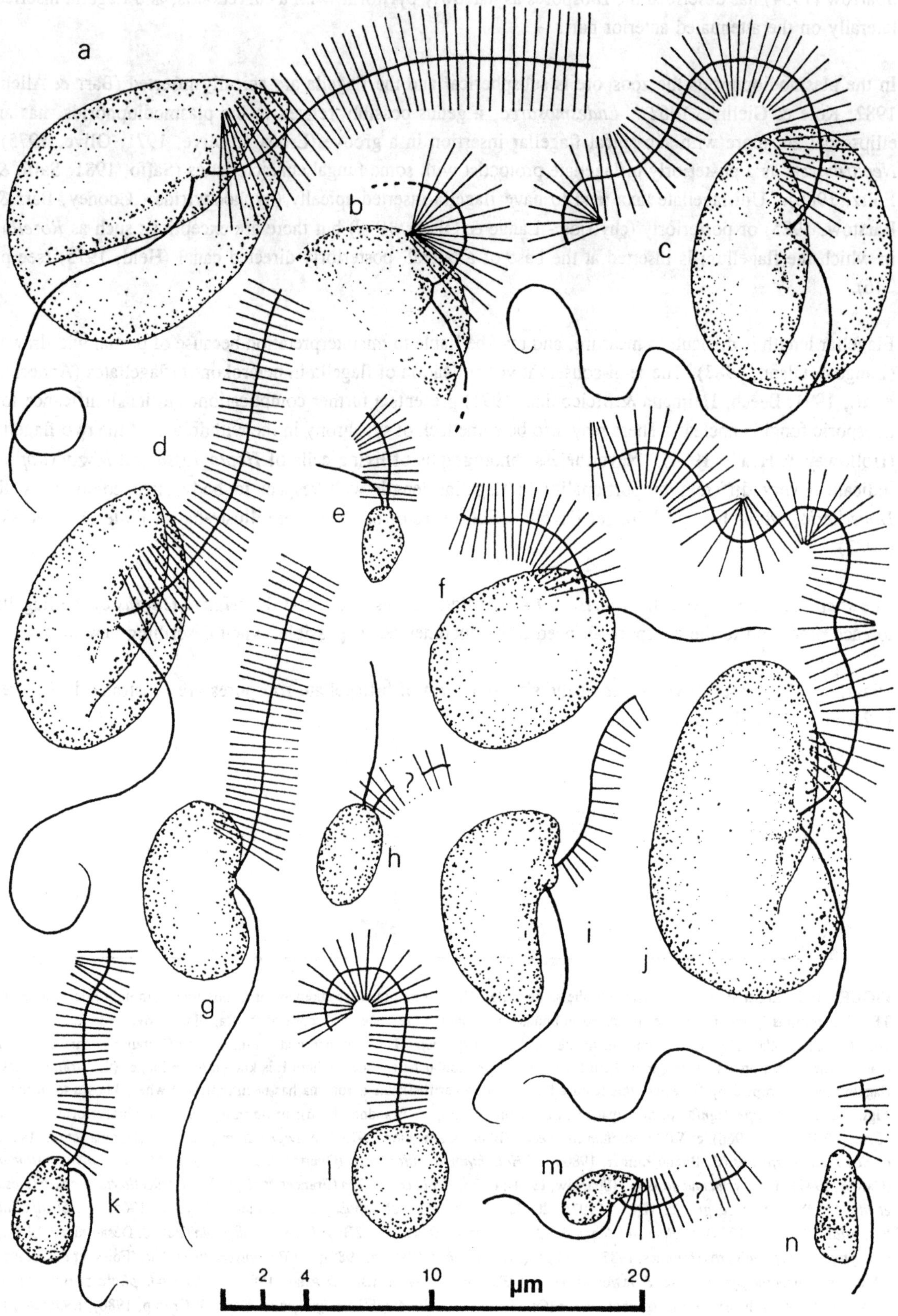

FIGURE I: 2A: a: *Saprolegnia* principal form; b: *Saprolegnia* auxiliary form; c: *Salilagenidium*; d: *Lagenidium*; e: *Ectrogella*; f: *Haliphthoros*; g: *Myzocytium*; h: *Eurychasma*; i: *Crypticola*; j: *Ducellieria*; k: *Gracea*; l: *Rhizidiomyces*; m: *Sirolpidium*; n: *Pontisma*.

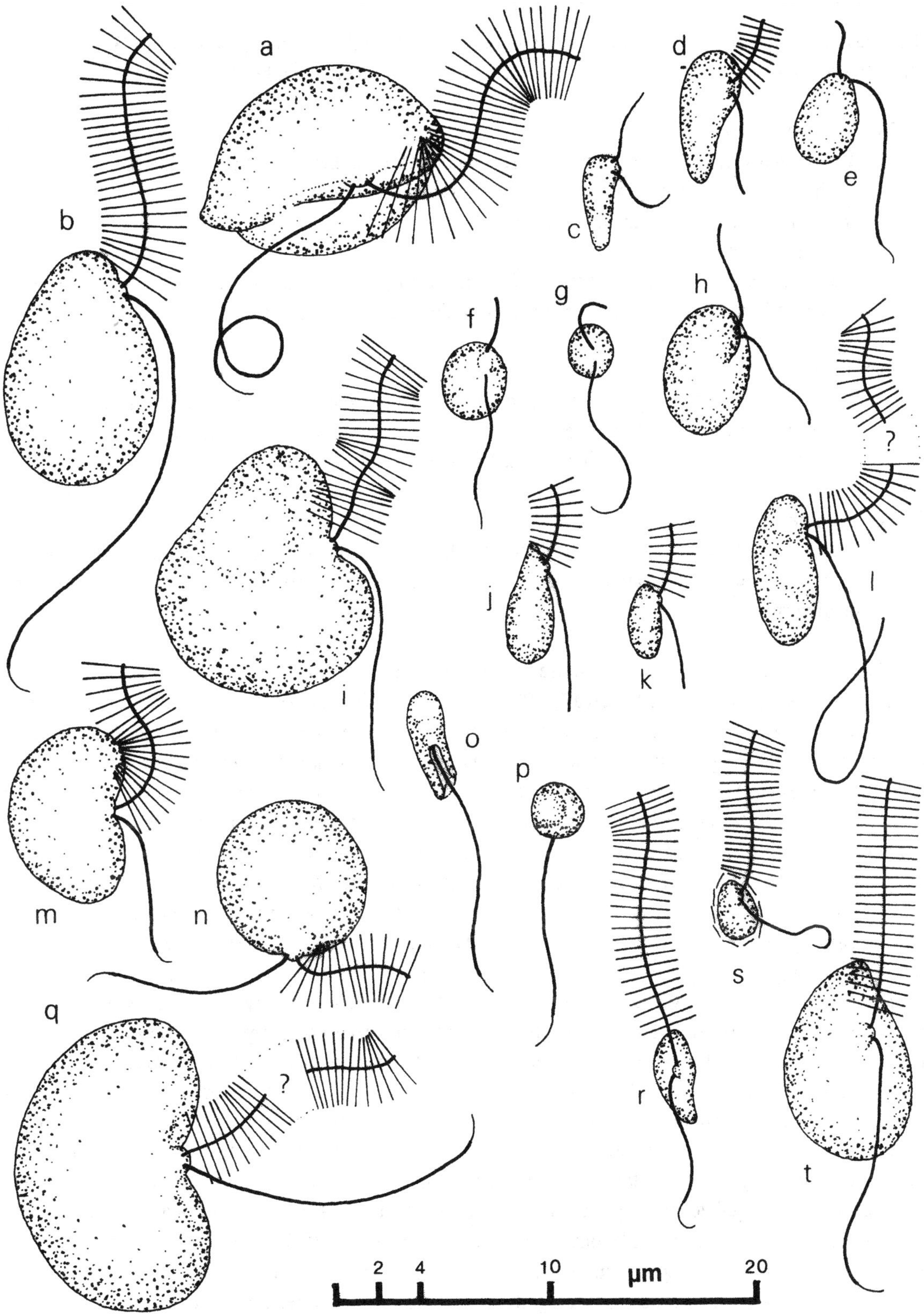

FIGURE I: 2B: a: *Lagena*; b: *Phytophthora*; c: *Haptoglossa*; d: *Ciliatomyces*; e: *Woronina*; f: *Polymyxa*; g: *Plasmodiophora*; h: *Endemosarca*; i: *Nematophthora*; j: *Olpidiopsis saprolegniae*; k: *O. varians*; l: *Blastulidium*; m: *Syzygangia* (zoosporangial); n: *Syzygangia* (oosporic); o: *Rozella*; p: *Dictyomorpha*; q: *Myzocytiopsis*; r: *Labyrinthula*; s: *Thraustochytrium*; t: *Lagenisma*.

Polymorphism, dimorphism and polyplanetism

There has been much debate about these terms in relation to fungal zoospores (Sparrow, 1958, 1960; Dick, 1973, 1990*a*; Martin, 1977; Ross, 1979: 61-62). They are defined again here:

Polymorphic spores
The existence of sporangiospores in more than one form, either co-existing during sporangiogenesis or showing a sequential progression after release, with or without quiescent phases. Spores which show amoeboid changes in shape are polymorphic.

Monomorphic spores
The existence of sporangiospores in only one form, but for the peronosporomycetes and other straminipilous fungi it is essential to qualify this term by stating whether the zoospore is of the principal or auxiliary type.

Dimorphic spores
The occurrence of zoospores with two morphologically and ultrastructurally distinct kinds of flagellar insertion. These types of zoospore are formed sequentially, each always followed by an encysted phase, and with the cysts formed from each kind of zoospore also often morphologically distinct.

Polyplanetism (diplanetism)
A sequence of two or more motile flagellate phases with interspersed mobile aplanosporic phases; the aplanosporic phase may be naked or as a walled cyst. The motile phases may be monomorphic or dimorphic.

Polyplanetism of the principal-form zoospore has been well documented for the *Saprolegniaceae* (Weston, 1919; Salvin, 1940; Cerenius & Söderhäll, 1984; Dieguez-Uribeondo, 1994) and all of these terms have been used principally in relation to this family, but polyplanetism also occurs in the *Pythiaceae* (Hallett & Dick, 1986). The principal-form zoospore of the *Saprolegniomycetidae* and *Peronosporomycetidae* is not known to be polymorphic. Polyplanetism is not documented for *Olpidiopsis saprolegniae*, but *Pleocystidium parasiticum* C. Fisch (≡ *Olpidiopsis schenkiana* Zopf) is reported to be polymorphic (Sparrow, 1960: 947). In other groups these criteria are seldom noted, but there are several references to polymorphic zoospores in lagenidiaceous fungi (e.g., Ivimey Cook, 1935 - *Syzygangia zygnematicola*, in which there is a change from an initially laterally flagellate condition to a pyriform-spherical shape when swimming; and Barron, 1976*c* - *Myzocytiopsis anomalum* (G. L. Barron) M. W. Dick (Dick, 1997*b*) in which the (naked-?) aplanospore changes slowly into a lens-shaped zoospore). Polyplanetism is recorded for *Myzocytiopsis parthenospora* (Karling) M. W. Dick (Karling, 1944*c*; Dick, 1997*b*) and *Syzygangia oedogonii* (Scherff.) M. W. Dick (Karling, 1981*a*; Dick, 1997*b*). In the plasmodiophorids polymorphic zoospores are frequently noted, but polyplanetism has not been reported for either the primary or the secondary zoospore. In some of the fungi of lagenidiaceous affinities the term diplanetic may have been used as if equivalent to dimorphism of the zoospore, even if the first naked aplanospores to be released are non-flagellate (cf.'achlyoid' discharge). The uninucleate protoplasts within the sporangium may or may not be flagellate; if flagellate, they may retract (?) their flagella upon emergence and become naked spheres or amoeboids; they may resume flagellate motility with or without an intervening encystment stage. In this respect comparisons with *Dictyomorpha* (Mullins, 1961) are relevant (see p. 252). Frequently, precise details are obscure: if '*Lagenidium*' zoospores encyst immediately upon emergence they may not do so in a hollow spherical cyst ball with the cysts cemented together as in true 'achlyoid' behaviour. Contrast the inferences for *Syzygangia oedogonii* (Karling, 1981*a*: 118) and *Myzocytiopsis parthenospora* (Karling, 1981*a*: 132). It is unclear whether flagella are present prior to 'achlyoid' encystment (cf. *Protoachlya*) or absent (cf. *Achlya sensu stricto*, see Money *et al.*, 1987).

In his earlier work Sparrow (1960) placed some reliance on descriptions of polyplanetism in the placement of peronosporomycete genera and families, but later (Sparrow, 1973*b*, 1976) he placed less emphasis on this character. I am in partial agreement with Sparrow (1976), in that I do not consider the descriptions to be sufficiently rigorous to justify the coupling of behaviour (within the definitions given above) with taxonomic hierarchies.

Body shape: reniform, bean-shaped, pip-shaped and grape-seed-shaped zoospores

All the terms describing body shape: reniform, bean-shaped, pip-shaped and grape-seed shaped zoospores, imply an asymmetry about the long axis of the zoospore. The descriptive problems arise because of:

1 the relationship of the asymmetry to the presence or absence of a groove parallel to the long axis
2 the extent to which the anterior and posterior parts of the zoospore have a different outline
3 the point of insertion of flagella

Asymmetry is primarily due to the construction of the cytoskeleton; that is, differences in the microtubular array from the kinetosomes. The possibility of morphological change, due to a change in the cytoskeleton, during maturation of the zoospore has not been considered. Such a change may account for the reniform to pyriform polymorphism noted in *Syzygangia zygnematicola* (Ivimey Cook, 1935), and revision of the characterization of this first-formed zoospore as a principal-form zoospore might then be necessary.

Interpretation of most descriptions involves a considerable amount of conjecture. 'Kidney-shaped' (reniform) and 'bean-shaped' are synonymous and commonly imply that there is a long axis groove; a similarity in shape of the anterior and posterior ends, and lateral insertion of flagella (in the groove). 'Pip-shaped' and 'grape-seed-shaped' commonly imply an axis of curvature, possibly with a shallow groove on the concave (ventral) side; a difference in shape between the anterior and posterior ends, and an apical or sub-apical insertion of flagella. However, different authors have attached different emphases to these terms. Sparrow (1960) tended to use 'bean-shaped' and 'grape-seed-shaped' as synonyms primarily to imply ***any*** asymmetry in contrast to the more or less spherical zoospore of the chytrids. Karling (1981*a*) tended to use the terms to distinguish sub-apical to lateral flagellar insertion from apical to sub-apical insertion. 'Lateral' insertion can be interpreted as 'sub-apical' if the zoospore is elongate. For example, see the attachment of flagella in *Olpidiopsis saprolegniae* (Bortnick *et al.*, 1985): either form of words could be used, yet the flagellation is typically heterokont on all ultrastructural criteria. Another example of the kind of contradictions that can occur is found with descriptions of *Sirolpidium*: Sparrow (1934: 10, plate 2Hb) "[flagella sub-apically] attached to the anterior, attenuated portion of the spore"; Sparrow (1960: 966, fig..77 I) "[flagella] anteriorly attached"; Karling (1981*a*: 57) "laterally inserted flagella". The ratio length:breadth is also a factor that has determined morphological description, the more elongate zoospores being described as 'grape-seed-shaped' or 'pyriform' (i.e., 'Conference' pear), while 'bean-shaped' and 'kidney-shaped' can indicate a variety of length:breadth ratios.

Since a degree of longitudinal asymmetry is present in both peronosporomycetous and plasmodiophoromycetous biflagellate zoospores, I attach more importance to dimensions, when they are given, than to light-microscope descriptions of body shape of these small (*ca* 5·5 μm diameter, $<75\ \mu m^3$) or very small (*ca* 3·5 μm diameter, $<25\ \mu m^3$) zoospores. Similarly, when descriptions have been based on light microscopy without accompanying photographic evidence, lesser weight is given to indications of point of flagellar attachment. Table I: 1 provides information on the diversity of zoospore/zoospore cyst volume found in straminipilous fungi, and Figure I: 2 provides an indication of the size ranges found in different taxa (often ***within*** a genus).

CRITERIA FOR DIAGNOSES

Inclusions

Vacuoles and refractive granules are occasionally mentioned in descriptions, but most descriptions make no mention of these inclusions and others are not readily interpretable. In light microscopy refractive granules are more likely to be recorded than vacuoles (note the photographs of *Haptoglossa* in Barron, 1980: fig. 10, and *Myzocytiopsis humicola* in Barron & Percy, 1975: fig. 11, which appear to show posterior refractive inclusions). Particular attention should be paid to this feature when mentioned, because refractive granules are not characteristic of peronosporomycete zoospores, though they are mentioned in descriptions of plasmodiophoromycete zoospores. The ultrastructural basis of these refractive granules has not been explored.

TABLE I: 1. *Calculations of mean zoospore/zoospore cyst volume for a miscellaneous range of taxa to give an indication of inter-generic variation (based on published cyst diameters - spheres,* $4/3\Pi r^3$ *- or mean zoospore dimensions assuming an ellipsoidal body -* $4/3\Pi r_1 r_2^2$*). Most members of the* Peronosporomycetidae *and the* Saprolegniomycetidae *have zoospore cyst diameters of 8-12µm; inter-specific differences have taxonomic value. Lagenidiaceous fungi (nomenclature as given in* PART V*) usually have zoospore cysts with much smaller diameters.*

	volume in µm³
Peronosporomycetidae	
Pythiogeton autossytum Drechsler	1767
Pythium anandrum Drechsler	1288
Phytophthora cinnamomi Rands	606
Pythium aquatile Höhnk	434
Pythium tracheiphilum Matta	203
Pythium angustatum Sparrow	113
Basidiophora entospora Roze & Cornu	29
Saprolegniomycetidae	
Saprolegnia anisospora de Bary (large cysts)	4905
Saprolegnia anisospora (medium cysts)	2399
Saprolegnia anisospora (small cysts)	1200
Verrucalvus flavofaciens P. Wong & M. W. Dick	755
Saprolegnia ferax (Gruith.) Thur.	668
Aphanomyces amphigynus Cutter	113
Leptolegniellaceae	
Aphanomycopsis bacillariacearum Scherff.	696
Leptolegniella exoospora W. D. Kane	243
Aphanomycopsis desmidiella Canter	179
Aphanomycopsis punctata Karling	179
Aphanodictyon papillatum Huneycutt *ex* M. W. Dick	170
Aphanomycopsis sexualis W. W. Martin	150
Brevilegniella keratinophila M. W. Dick	143
Leptolegniella keratinophila Huneycutt	130
Aphanomycopsis entophyta (Pringsh.) M. W. Dick	78
Aphanomycopsis saprophytica Karling	74

(Table I: 1, continued)

lagenidiaceous, hyphochytrid and plasmodiophorid species

Cystochytrium radicale Ivimey Cook	1288
Salilagenidium chthamalophilum (T. W. Johnson) M. W. Dick	573
Salilagenidium scyllae (Bian *et al.*) M. W. Dick	523
Cornumyces destruens (Sparrow) M. W. Dick	402
Cystosiphon pythioides Roze & Cornu	382
Salilagenidium callinectes (Couch) M. W. Dick	382
Haptoglossa heterospora Drechsler	382 and 75
Lagenidium giganteum Couch	350
Haliphthorus milfordensis Vishniac	283
Myzocytiopsis lenticularis (G. L. Barron) M. W. Dick	260
Lagena radicicola Vanterp. & Ledingham	256
Crypticola entomophaga (W. W. Martin) M. W. Dick	243
Atkinsiella dubia Vishniac	212
Pleocystidium parasiticum C. Fisch	205
Ducellieria chodatii (F. Ducell.) Teiling	180
Myzocytium proliferum Schenk	150
Crypticola clavulifera Frances *et al.*	143
Myzocytiopsis intermedia (G. L. Barron) M. W. Dick	143
Sorosphaera veronicae (J. Schröt.) J. Schröt. (2° zoospore)	130
Lagenisma coscinodisci Drebes	125
Cornumyces pygmaeus (Zopf) M. W. Dick	104
Halodaphnea parasitica (K. Nakamura & Hatai) M. W. Dick	87
Rozella septigena Cornu	83
Rozellopsis septigena Karling *ex* Cejp	83
Pleocystidium lundiae (Karling) M. W. Dick	78
Cornumyces muenscheri (Cutter) M. W. Dick	71
Haptoglossa mirabilis G. L. Barron	71
Halodaphnea hamanaensis (Bian & Egusa) M. W. Dick	65
Hyphochytrium saprobium (Karling) M. W. Dick	60
Polymyxa betae Woronin (1° zoospore)	56
Sorosphaera veronicae (J. Schröt.) J. Schröt. (1° zoospore)	44
Gracea gracilis (E. J. Butler) M. W. Dick	33
Plasmophagus deformans (Serbinow) M. W. Dick	33
Pleocystidium pygmaeoides (Karling) M. W. Dick	33
Rhizidiomyces apophysatus Zopf	33
Pythiella vernalis Couch	29
Olpidiopsis brevispinosa Whiffen	29
Ligniera junci (Schwartz) Maire & A. Tison (2° zoospore)	28
Endemosarca ubatubensis Erdos & L. S. Olive	24
Eurychasmidium tumefaciens (Magnus) Sparrow	24
Pseudosphaerita euglenae P. A. Dang.	23
Olpidiopsis achlyae D. A. McLarty	21
Endemosarca anomala Erdos	19
Phagomyxa algarum Karling	19
Rozellopsis simulans (A. Fisch.) Karling	18
Plasmophagus coleochaetis (Sparrow *et al.*) M. W. Dick	17
Rozella longicollis Karling	17
Sirolpidium bryopsidis (de Bruyne) H. E. Petersen	16
Dictyomorpha pythiensis (N. Sarkar & R. Dayal) M. W. Dick	15
Olpidiopsis saprolegniae (A. Braun) Cornu	14
Plasmodiophora brassicae Woronin (1° zoospore - Karling, 1968*e*)	14!
Sorodiscus radicicola (Cornu) A. Fisch. (2° zoospore)	14
Dictyomorpha dioica (Couch) Mullins	10
Pleotrachelus fulgens Zopf	9
Ectrogella bacillariacearum Scherff.	8
Plasmodiophora brassicae Woronin (2° zoospore - Karling, 1968*e*)	8!
Olpidiomorpha pseudosporae Scherff.	7
Hyphochytrium catenoides Karling	6
Woronina glomerata (Cornu) A. Fisch. (2° zoospore)	6
Ligniera plantaginis (Němec) Karling (2° zoospore)	2
Pseudosphaerita drylii Pérez Reyes	2

TABLE I: 2. *Variation in dimensions of tubular tripartite hairs and their density along the axoneme for various peronosporomycete zoospores (Hallett, 1975) and for other straminipiles (selected from Leadbeater, 1989 - density measurements were not given, base was included in shaft measurement, but given separately in parentheses).*

+* *Present, but in insufficient numbers for accurate mean measurements to be obtained.*

PERONOSPOROMYCETES	Density No.μm^{-1}	Length of tubular tripartite hairs (μm) Shaft + base	long hair	short hair	Ratio of long hair to shaft
Saprolegniaceae					
Achlya bisexualis Coker & Couch	6·8	2·10	0·69	0·16	1:3·0
Achlya heterosexualis Whiffen	13·0	1·50	0·50	+*	1:3·0
Achlya colorata Pringsh.	12·7	1·80	0·60	0·10	1:3·0
Dictyuchus monosporus Leitg.	22·9	2·00	0·50	0·20	1:4·0
Protoachlya paradoxa Coker	19·2	1·80	0·55	0·14	1:3·3
Pythiopsis cymosa de Bary	18·0	2·10	0·50	0·10	1:4·2
Saprolegnia parasitica Coker	14·6	1·80	0·40	+*	1:4·5
Saprolegnia ferax (Gruith.) Thur.		1·80 (0·30)	1·10		1:1·6
Leptolegniaceae					
Aphanomyces stellatus de Bary	10·1	1·70	0·62	0·20	1:2·7
Leptolegnia caudata de Bary	13·3	2·00	0·50	+*	1:3·9
Leptomitaceae					
Leptomitus lacteus Roth	18·8	1·30	0·75	+*	1:1·7
Apodachlya minima Cejp	13·6	1·90	0·50	0·19	1:3·8
Pythiaceae					
Pythium mamillatum Meurs	11·2	2·10	0·70	0·16	1:3·0
Pythium middletonii Sparrow	14·2	1·90	0·70	0·15	1:2·7
Pythium monospermum Pringsh.	8·9	1·60	0·46	+*	1:3·5
Pythium torulosum Coker & P. Patt.	16·3	1·90	0·45	0·16	1:4·2
Phytophthora gonapodyides (H. E. Petersen) Buisman	15·3	1·70	0·40	0·20	1:4·1
Phytophthora cryptogea Pethybr. & Laff.	6·6	1·30	0·40	0·15	1:3·3
Lagenidium giganteum Couch		1·16 (0·16)	0·27		1:4·3
Albuginaceae					
Albugo candida (Pers.) Kunze	24·0	0·60	0·40	0·06	1:1·8
Hyphochytriaceae					
Rhizidiomyces apophysatus Zopf		1·30 (0·20)	0·37		1:3·5
Hyphochytrium sp.	13·5	0·75	0·50	0·10	1:1·5
PHOTOSYNTHETIC STRAMINIPILES					
Bacillariophyceae					
Lithodesmium undulatum Ehrenb.		1·73 (0·23)	0·50		1:3·5
Chrysophyceae					
Ochromonas danica Pringsh.		1·25 (0·25)	0·50		1:2·5
Synura petersenii Korshikov		1·60 (0·20)	0·25		1:6·4
Eustigmatophyceae					
Polyedriella helvetica Vischer & Pascher		1·00	0·75		1:1·3
Fucophyceae					
Dictyota dichotoma (Huds.) J. V. Lamour		1·10	0·70		1:1·6
Raphidophyceae					
Heterosigma luteus		1·75 (0·25)	0·50		1:3·5
Tribophyceae					
Bumilleria sicula Borzi		2·00	0·50		1:4·0

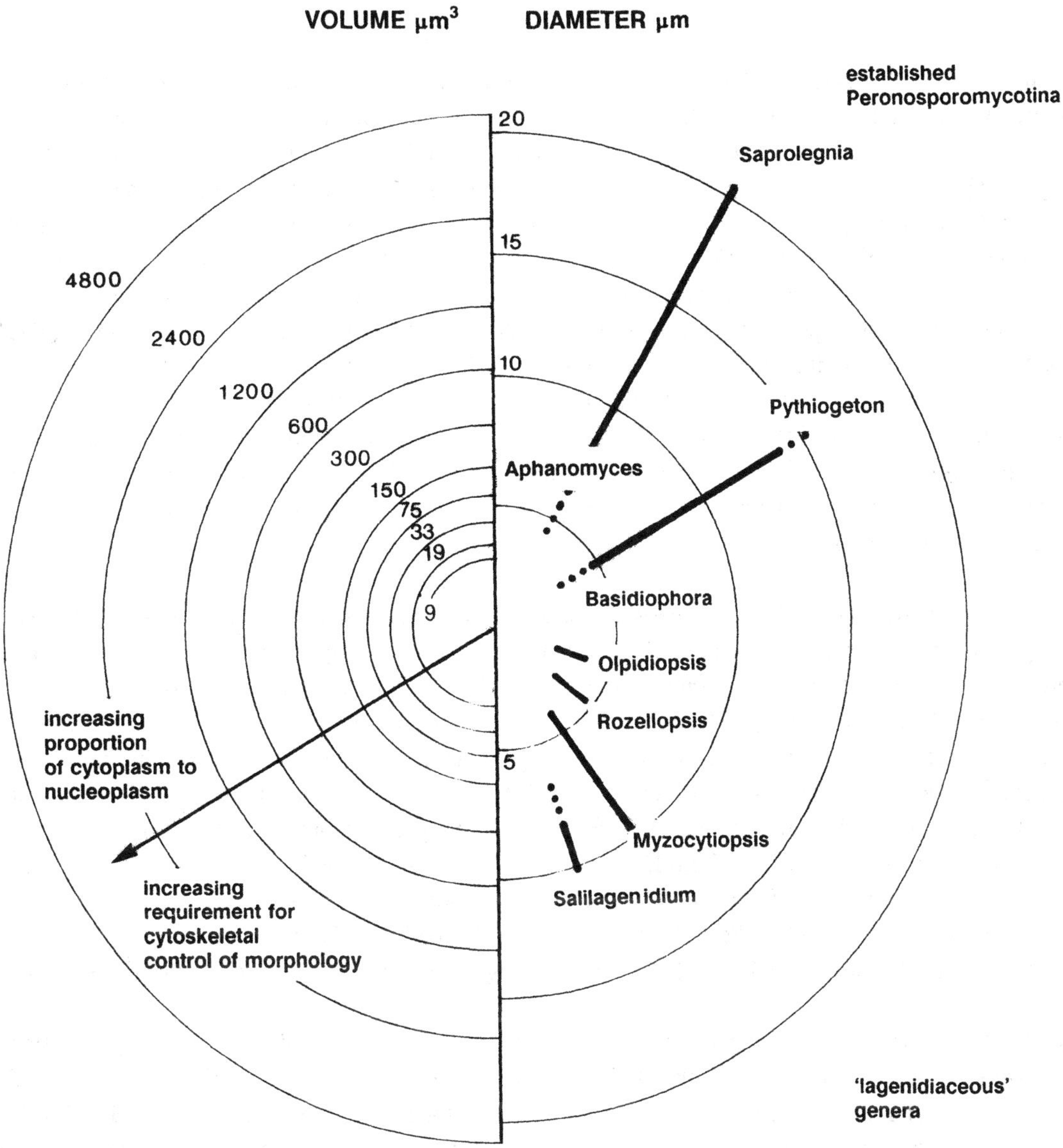

FIGURE I: 3. Inter- and intra-taxon variation in zoospore/zoospore cyst size; comparison of cyst diameters (right) with a geometric progression of volume equivalents (left) and ranges for a few selected taxa (see also Table I: 1).

Flagella, flagellar length and ornamentation, the straminipilous flagellum

Flagella are bounded by a flagellar plasmamembrane which is continuous with the cell plasmamembrane. The flagellar matrix (cytosol) is often devoid of structural detail except for the microtubules. There are normally two central and nine peripheral microtubules. The central pair are single microtubules, but the peripheral microtubules are usually doublets, composed of a complete microtubule of 13 subunits (A tubule) and an incomplete tubule of 11 subunits (B tubule), to which dynein arms may be attached (Smith & Sale, 1992). There may be ultrastructural diversity in both numbers and ultrastructural details of these microtubules.

CRITERIA FOR DIAGNOSES

Although Lange & Olson (1983) have stated that absolute estimates of flagellar length can be unreliable, relative lengths of flagella of zoospores fixed immediately after actively swimming are probably more reliable.

Sparrow (1976; cf. Sparrow, 1960: footnote, p. 913) and Karling (1981*a*) use the terms 'isokont' and 'heterokont' as antonyms, solely for differential length ratios, but there are different opinions for the terms used to describe differences in length and differences in morphology. In this text flagella of different lengths, whether of identical morphology or with ***non***-heterokont differences in morphology, are termed **anisokont** (in some algal groups, such as the *Pavlovales*, this term is used when there is non-heterokont difference in morphology and equal or unequal flagellar lengths). Most phycologists (e.g., Leedale, 1974: 269, the "Heterokont Kingdom"; Leadbeater, 1989) now restrict the term **heterokont** to the possession of two different kinds of flagella: the one ornamented with stiff **tubular tripartite (flagellar) hairs (TTHs)**, usually in two rows (the tinsel flagellum, flimmergeissel), the other unornamented or with a fibrillar surface coat (peitschengeissel). This restricted definition of the term **heterokont** is adopted here. The term 'mastigoneme' for this distinctive appendage is not now approved (Andersen *et al.*, 1991); a new adjective **straminipilous** is introduced for the possession of these tubular tripartite hairs, whether borne on flagella, naked cells or as cyst-wall ornaments (see *Glossary* for derivation) (the retroneme - Cavalier-Smith, 1989, is a tubular **bi**-partite hair).

The tubular tripartite hair (**TTH**) is composed of a cone-like base, a tubular shaft and usually two unequal, diverging terminal fibres. The number of the terminal fibres is not easy to determine; only one terminal fibre has been found in *Labyrinthula* (Kazama, 1974*a*), *Crypticola* (Frances *et al.*, 1989) and *Labyrinthuloides* (Bower 1987*a*). The most detailed description, for ***any*** member of the *Straminipila*, of the spiral organization of the tubular shaft of the **TTH** is given by Domnas *et al.* (1986) and in Figure I: 4 for *Lagenidium giganteum** (see also Fuller & Reichle, 1965); the relative densities per unit flagellum length along the axoneme, and the dimensions of **TTHs** differ within peronosporomycetes, but most have a shaft length of 1·0-2·0 μm (Table I: 1, from Hallett, 1975; Figs 3 & 4) (see Leadbeater, 1989: table 8.1 for equivalent algal data); their effect on motility has been discussed by Jahn, Landman & Fonseca (1964) using the theory of Taylor (1952). See Sleigh (1991) and Holwill (1982) for a more recent accounts of morphology and dynamics. An hypothésis on their arrangement is given in Figure I: 4 (after Dick, 1990*a*).

The **TTHs** are assembled in antiparallel fascicles enclosed in a vesicle, and released as a group (Heath, Greenwood & Griffiths, 1970; Bowerman & Domnas, 1997). In chromistan algae this assembly takes place in the chloroplast endoplasmic reticulum or the nuclear envelope (Andersen *et al.*, 1991). Bouck (1971), for *Ochromonas*, suggested a somewhat different morphogenesis, whereby a similarly antiparallel arrangement was initiated in perinuclear cisternae, but with the individual **TTHs** subsequently receiving further ornamentation from dictyosome-derived vesicles, before being released at the base of the flagellum as the axoneme elongated. In the fungi (and *Synura*) the origin of the **TTH** is a dictyosomal vesicle (Heath *et al.*, 1970). After release, the **TTHs** normally become attached to opposite sides of the flagellum approximately parallel to the plane of the two central microtubules (cf. Manton, 1956), each row being radially aligned with a peripheral microtubule. Manton (1956) suggested an alignment with microtubules 3 and 8 (see Figure I: 3), but whether the alignment is with microtubules 3 and 8 or microtubules 4 and 7 there will be a 20° divergence from a true plane, thereby producing a shallow trough along the axis of the

---

*FOOTNOTE: These data can be interpreted as a coil or stiff rope of 1 + 2 ply: in which case *Ochromonas* could be regarded as 1 + 3 ply. If one, two or all three of the more slender elements were spliced, the free ends along the length of the shaft could account for at least one of the two kinds of lateral hairs in *Ochromonas*, but compare accounts in Bouck (1971) and Deason (1971).

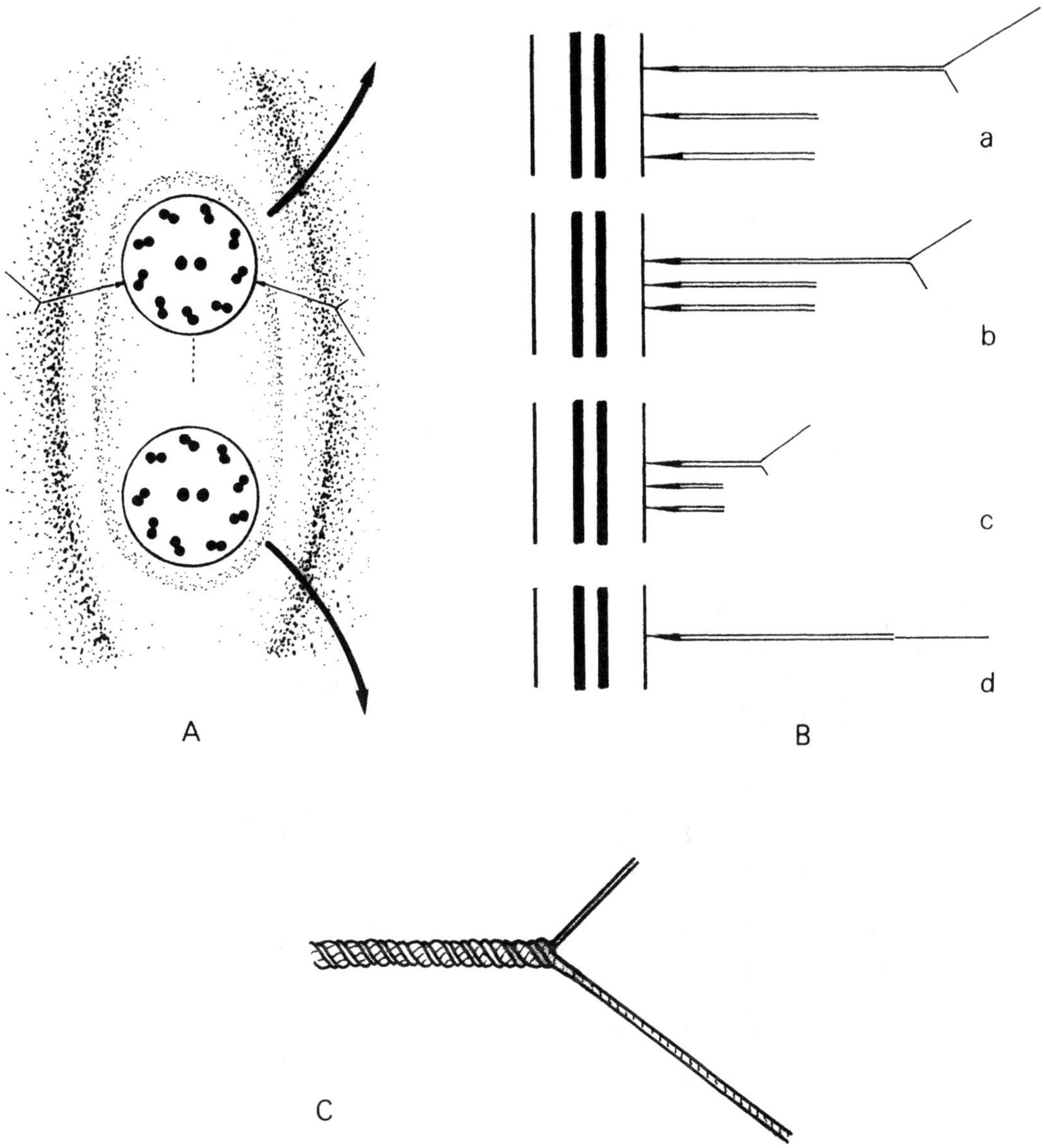

FIGURE I: 4. Flagellar structure and symmetry (after Manton, 1956; Dick, 1990*a*).

Fig. A: Hypothetical arrangement (freely adapted from Manton, 1956; Dick, 1990*a*) of straminipilous and whiplash flagella on the kinetosome boss (lightly stippled) within the longitudinal groove (heavily stippled). Note orientation of the two central microtubules and attachment of the tripartite tubular hairs to the plasmamembrane opposite peripheral doublet microtubules 3 and 8. The dotted line opposite doublet 1 represents the axis of the paraxonemal spine(s) or vane of fucophytes. The large arrows indicate the directions of the each flagellum; thus the peripheral doublet 1 is outward-facing or abaxial to the groove. The two ranks of tripartite tubular hairs are directed laterally so that the plane of the sine wave of the straminipilous flagellum is perpendicular to the dorsi-ventral axis.

Fig. B: generic variation in dimensions and densities of the tubular tripartite hairs: a: *Achlya bisexualis*: shaft length 2·1 μm, density 6·8 μm^{-1}; b: *Pythium torulosum*: shaft length 1·9 μm, density 16·3 μm^{-1}; c: *Albugo candida*: shaft length 0·6 μm, density 24·0 μm^{-1} (after Hallett, 1975; Dick, 1990*a*); d: *Crypticola clavulifera* single hair drawn to scale from Frances *et al.* (1989).

Fig. C: part of a tubular tripartite hair: tubular shaft composed of two spiral elements of different diameter, each terminating as one of the distal fibres (after Domnas *et al.*, 1986). Figure I: 4 reproduced from Dick (1999: fig. 2B-D) by permission of Academic Press.

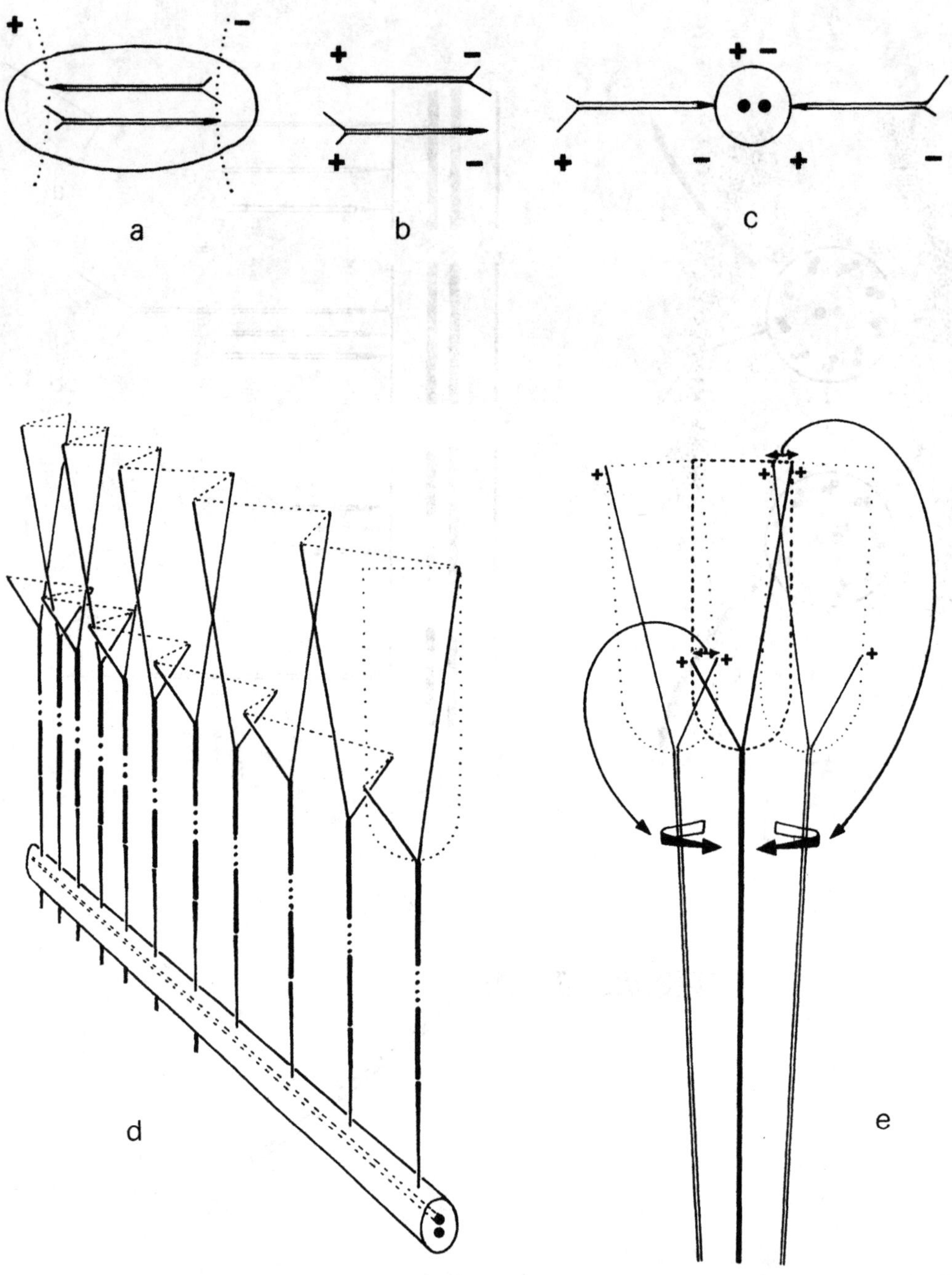

FIGURE I: 5. Hypothesis for the structural conservation and mode of function of the tubular tripartite hair (TTH). The theoretical problem is that, for the straminipilous flagellum to function in the way it has been observed to do, the TTHs must be held stiffly in the plane of the quasi-sinusoidal wave, yet they do not appear to have any structural mechanism to achieve this, particularly in view of the hydrodynamic stresses to which they are exposed. This hypothesis (after Dick, 1990*a*) is based on the antiparallel development of the TTHs, and depends on a resultant biochemical/electron polarity of the TTHs and the two central microtubules of the flagellar shaft; '+' and '-' are used loosely to indicate different polarity and 'like-repulsion'.

a: dictyosome-derived vesicle with antiparallel formation and array of TTHs (only 2 shown); b: TTHs of different polarity (each vesicle will produce populations of both); c: attachment of the two polarity classes of TTHs on the axoneme (possibly dependent on directional polarity of the central microtubules); d: perspective arrangement of one row of the TTHs: a common polarity of the solid bases will flagellum/**TTH** complex; such a trough may have fluid dynamic significance. In a few, non-fungal, straminipilous organisms (*Ochromonas danica* - Bouck, 1971; *Labyrinthuloides haliotidis* - Bower, 1988*a*) this pattern of alignment does not hold. The phylogenetic significance of straminipilous flagellar morphology and morphogenesis is discussed further on p. 104.

The range of variation in both anisokont and heterokont flagellation (Figures I: 1 & I: 5) is much greater than has been recognized by mycologists (e.g., Lange & Olson, 1983; Barr, 1983). Thus, descriptions must be recast using the terminology of Andersen *et al.* (1991); earlier transcripts and secondary sources cannot be relied upon.

There is no terminology in general mycological usage to distinguish an unornamented flagellum from one (e.g. the posterior flagellum of *Achlya*, *Dictyuchus*, *Pythiopsis*, *Saprolegnia*, *Pythium* and *Phytophthora* - Hallett, 1975; see Figure I: 5) clothed with fine flexuous hairs (the **fibrillar surface coat** as defined by Andersen *et al.*, 1991).

Flagellar shape is not usually mentioned, but sometimes the term 'whiplash' is used. Again this term has been used in morphological and functional contexts: it should be used to indicate a flagellum with a narrower extension to the flagellum tip (the **acronema**), due to the greater length of the central microtubules compared with the nine peripheral doublets (but cf. Manton, 1965). It has also been used as a general term for any unornamented flagellum, irrespective of whether it is abruptly truncate, sharply attenuated (acronemate) or with a gradually tapered tip due to progressive termination first of the central fibres, and secondly components of the longer peripheral doublets (Gibbons & Grimstone, 1960; Manton, 1965) (Figure I: 5). In the peronosporomycetes the straminipilous flagellum does not possess a whiplash extension, but in chrysophyte algae there may be a terminal tuft of non-tubular hairs on this flagellum (Bouck, 1971, 1972). In different fucophytes there may be a short (Manton & Clarke, 1951) or very long (Müller & Falk, 1973) acronema, while in *Gracea* the acronema varies in length from zoospore to zoospore (Dick, 1997*a*, 2000*d*). Additional flagellar features found in chromophyte algae, but not so far in straminipilous fungi, include paraxonemal bodies (PABs) such as the intra-plasmamembrane spines (*Himanthalia* - Manton, Clarke & Greenwood, 1953: *Xiphophora* - Manton, 1956; *Dictyota* - Manton, 1959, 1965) and the extra-plasmamembrane flagellar scales characteristic of certain prymnesiophytes. The orientation of the

---

(Figure I: 5, legend, continued)

space the TTHs at approximately regular intervals along the axoneme, the common polarity of the distal hairs will provide a mechanism of recoil and counter entanglement in the compressed phase of the sinusoidal wave; the dotted outline represents the 'face-view' of the 'paddle' created by the diverging terminal hairs, the opposed zig-zag dotted lines connect the long and short hairs respectively; e: interaction between a TTH (solid shaft) and its adjacent TTH (open shaft): as the adjacent TTH (left-hand) drifts *out of alignment* with the TTH (centre) the increasing strength of repulsion of the like polarity of the tips of the short terminal hairs (double-headed arrow) will correct alignment as indicated by the folded arrow (lower left); as the adjacent TTH (right-hand) drifts *out of alignment* with the TTH (centre) similar repulsion due to the like polarity of the tips of the long terminal hairs (double-headed arrow) will correct alignment as indicated by the folded arrow (lower right); the dimensions are such that the individual TTHs would stray from the palisade plane by less than 10°. These suggestions do not detract from the possible role of the terminal hairs in increasing viscous drag as they pass over the peak of the sinusoidal wave, thereby augmenting the action of the TTHs in reversing the thrust of the straminipilous flagellum.

The straminipilous flagellum 'paddles' the organism through the water. Using the paddle analogy, a 'skeleton blade' is represented by the long and short distal hairs. These hairs could be arranged in a plane perpendicular to the palisade of TTHs and also be perpendicular to the flagellar axis. Alternate TTHs are figured (d) with reciprocal arrangements of short and long hairs (possibly involving rotation at the point attachment of the TTH to the axoneme plasmamembrane).

The action of the straminipilous flagellum can be regarded as analogous to a rowing eight with fixed oars and a flexible keel (cf. the swimming action of nereid worms (Gray, 1939), and contrast both the 'sculls-action' of the pair of smooth flagella of *Chlamydomonas*, and the 'screw-action of a stern-rowlock single oar' of single, posteriorly directed flagella).

Fig. 5 d,e reproduced from Dick (1990*a*: fig. 3) by permission of Jones & Bartlett.

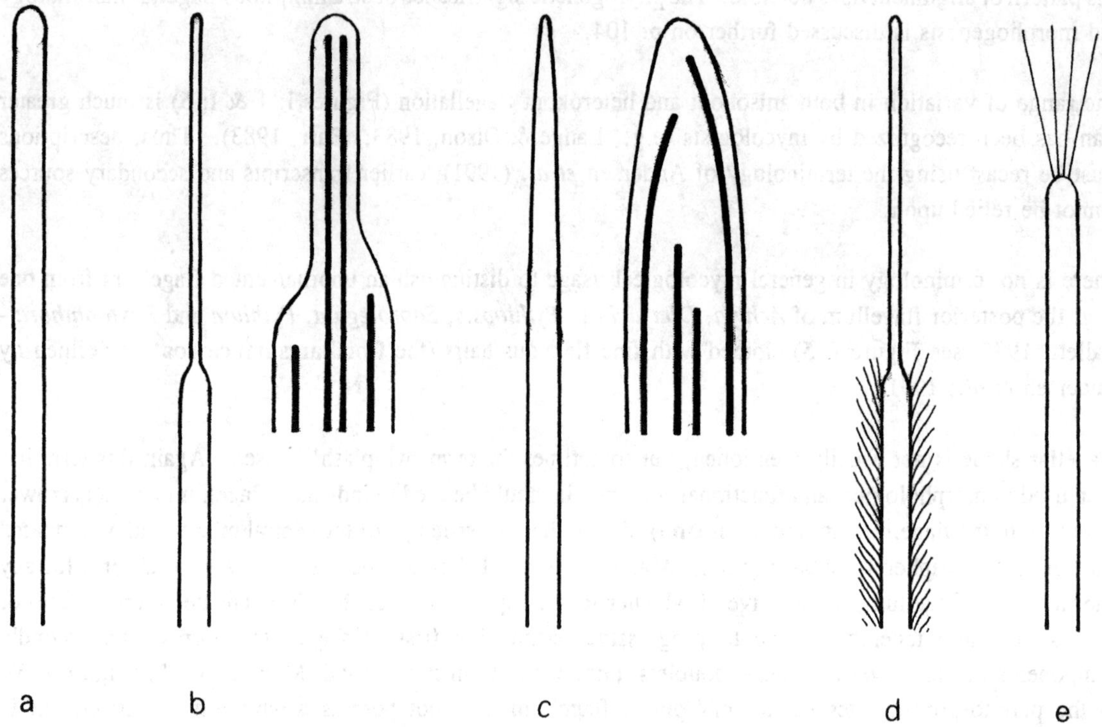

FIGURE I: 6. Diversity in flagellar tip morphology: a: truncate (central and peripheral microtubules approximately the same length); b: whiplash (central microtubules much longer than the peripheral microtubules); c: pseudo-whiplash (central microtubules shorter than the peripheral microtubules, which are also of different lengths); d: whiplash, with axoneme with a fibrillar surface coat; e: truncate with a terminal tuft of fine hairs (characteristic of some chrysophytes).

straminipilous flagellum relative to the whiplash flagellum and the spermatozoid body of fucophytes has been described by Manton (1956) and is adapted here in Figure I: 3, except that I am not adopting her numeration of the nine peripheral doublets as 0 (opposite the paraxonemal body) to 8 (with the tubular tripartite hairs associated with doublets 2 and 7), preferring to number the doublets 1 to 9. However, ***it must be emphasized that the orientation depicted in Figure I: 4A requires confirmation for straminipilous fungi*** with respect to the plane of the straminipilous flagellar beat (perpendicular to both the posterior- anterior and the dorsiventral axes); the direction of the arms on the peripheral doublets with respect to the transitional plate (dextrorotational or levorotational); whether this rotation on the whiplash flagellum is a mirror image or not, and whether the **TTH**s are associated with doublets 3 and 8 or 4 and 7.In the life-histories of the plasmodiophorids there are two zoosporic phases. The secondary zoospores, produced in the zoosporangia, possess two unornamented flagella of markedly different lengths, often in a ratio of 1:3 or 2:3. On this basis the flagellar morphology of the plasmodiophorids has been accepted as anisokont but not heterokont. However, there is still a very slight possibility of doubt. Kole & Gielink (1962) acknowledge that preparations of suspended cysts from soils, necessary for the production of primary zoospores, cannot be absolutely pure, but they provide quite convincing evidence (confirmed without presented data, by Aist & Williams, 1971) that the *Plasmodiophora* cyst produces a primary zoospore with a long whiplash flagellum and a short unornamented, often abruptly truncate flagellum. *Spongospora subterranea* (Wallr.) Lagerh. has two anisokont whiplash flagella (Kole & Gielink, 1961: fig. 3) and Glockling & Beakes (2000*a*, *b*) have shown that the two flagella of *Haptoglossa* are both acronemate. However, Kole & Gielink (1962) also reported that two other possibly contaminant zoospore morphologies occurred, albeit in very small numbers in their more highly purified preparations of *Plasmodiophora*: one was sub-anisokont with two unornamented whiplash flagella, the other was "entirely identical with" that figured for *Plasmodiophora* by Seaman, Larson & Walker (1961) from Wisconsin soils (Dick, 1998*a*, 2000*d*). Merz (1992) has also reported that heterokont and uniflagellate zoospore contaminants were "always" present in preparations of secondary zoospores of *Spongospora*, but no evidence of tubular tripartite hairs was presented, only SEM data of different flagellar insertion and flagellar base morphology. Merz (1992) also distinguished these contaminants by their different swimming characteristics as revealed by video recording (he also reported differences in swimming characteristics between the primary and secondary zoospores of *Spongospora*). Clay & Walsh (1997) have provided the most recent ultrastructural data for the flagella of *Spongospora*. The zoospore described by Seaman *et al.* (1961) had flagella of obviously different lengths, and the hairs on the tinsel flagellum were almost certainly tripartite, indicating a heterokont state. The straminipilous flagellum was the longer. Neither Kole & Gielink (1962) nor Merz (1992) gave measurements for the three kinds of zoospores or their flagella, but Seaman *et al.* (1961) gave dimensions of living zoospores as 3·0 × 3·0 × 3·9-4·8 μm with flagellar lengths of 6·0-8·0 μm and 2·0 μm respectively. It is remarkable that all three preparations from Wisconsin (Seaman *et al.*, 1961) and that from the Netherlands (Kole & Gielink, 1962) for *Plasmodiophora* (and perhaps the study by Merz (1992) on *Spongospora*) should yield a contaminant fungal zoospore type, with a longer straminipilous flagellum and a shorter whiplash flagellum on a broadly pyriform zoospore, that is otherwise confined, outside marine environments, to *Gracea gracilis* (E. J. Butler) M. W. Dick (Dick, 1997*b*, 2000*d*). (Note that *Tribonema* is a non-marine alga with a longer straminipilous flagellum - Massalski & Leedale, 1969.) It is regrettable, and in view of the interest in these fungi as possible virus vectors, surprising, that these 'contaminant' zoospores have not been subject to more thorough comparative video and ultrastructural study.

A much more controversial alternative explanation would be to suggest that the heterokont zoospore type is part of the plasmodiophoraceous life-history, in a way analogous to the different kinds of flagellation found in *Vaucheria*, in which the spermatozoids are heterokont and markedly anisokont while the coenocytic polyflagellate zoospores are not heterokont and only slightly anisokont (Koch, 1951; Møestrup, 1970; Ott & Brown, 1974).

In the *Peronosporomycotina* (Dick, 1976, the non-photosynthetic, fungal, part of the *Straminipila*) the heterokont flagellation may or may not also be anisokont. In most peronosporomycetes flagella are sub-equal with the straminipilous flagellum slightly shorter (in *Phytophthora* (Cho & Fuller, 1989: fig. 1) and *Plasmopara* (Beakes, 1989: fig. 17.2) this anisokont condition is very marked), while in the thraustochytrids the anterior straminipilous flagellum is the longer one and in the hyphochytrids it is the only extant flagellum. Vishniac (1955*d*) reported that the marine *Sirolpidium* was anisokont and heterokont, with the anterior straminipilous flagellum the shorter. Neither Martin & Miller (1986*b*) nor Bortnick *et al.* (1985) precisely described the anisokont condition in *Olpidiopsis*, but the straminipilous flagellum is presumed to be the shorter (*contra Gracea gracilis* studied by Pemberton *et al.* (1990) and Dick, 2000*d*). Similar variation between the relative lengths of the straminipilous anterior flagellum and the smooth posterior flagellum is found in the *Fucophyceae* and *Tribophyceae*.

For the biflagellate fungi under discussion, positively identified straminipilous flagellation (that is, with **TTHs** figured, or with antiparallel packets in the cytoplasm clearly identified in TEM) is limited to *Lagenidium giganteum* (Domnas *et al.*, 1986), *Salilagenidium callinectes* (Bland & Amerson, 1973*a*; Gotelli, 1974*b*, Fuller & Jaworski, 1987), *Haliphthoros milfordensis* (Overton *et al.*, 1983), *Lagenisma coscinodisci* Drebes (Schnepf *et al.*, 1978*d*), *Ectrogella perforans* H. E. Petersen (S. Raghu Kumar, 1980*b*), *Olpidiopsis saprolegniae* var. *saprolegniae* (Bortnick *et al.*, 1985), *Olpidiopsis varians* Shanor (Martin & Miller, 1986*b*), *Gracea gracilis* (Dick, 2000*d*) and *Lagena radicicola* Vanterp. & Ledingham (Barr & Désaulniers, 1987*b*). It is probable that *Ducellieria chodatii* (F. Ducell.) Teiling is also straminipilous (Hesse *et al.*, 1989), although the tripartite nature of the flagellar hair is not clear. *Ciliatomyces spectabilis* I. Foissner & W.

Foissner (Foissner & Foissner, 1986*b*, 1995) is stated to be heterokont, but without documentary illustrations, except for the packets of **TTHs**. Manier (1976: fig. 15) has described "corps fibreux" in *Blastulidium* which are probably packets of tubular tripartite hairs, but she did not give information on flagellar ornamentation.

The fungus described by Martin (1977) as *Atkinsiella entomophaga* W. W. Martin (≡ *Crypticola entomophaga* (W. W. Martin) M. W. Dick - Dick, 1998*a*), though stated to be heterokont, is not heterokont within the definitions used here, and the same may apply to *Syzygangia nodosa* (P. A. Dang.) M. W. Dick (≡ *Lagenidium nodosum* (P. A. Dang.) Ingold, Ingold, 1949; Dick, 1997*b*). The reference to heterokont flagellation for this taxon in Karling (1981*a*) is based on the work of Couch (1941) and the redisposition of *Resticularia* to *Lagenidium* by Ingold (1949). Couch (1941) did not identify the species, nor state the origin of his *Resticularia*. He wrote that this, and his likewise unidentified *Myzocytium* species, had a similar swimming motion. His figure is a drawing of two (*contra* Karling, 1981*a*: fig. 103) stained zoospores. From this pair of drawings it is probable that the anterior flagellum had only a single row of hairs but these hairs may not have been tubular or tripartite. Most recently, Frances *et al.* (1989) have provided an electron micrograph which suggests that *Crypticola* also has **TTHs** in a single row, the hairs are tripartite and tubular but with only a single fine distal fibre (Dr S. P. Frances, *in litt.* 29 Jun 1992). In other words, there may be a group of fungi which has a non-peronosporomycetous but straminipilous flagellar ornamentation comparable to the flagellum with uniseriate hairs in the cryptophytes. (Perkins & Menzel (1967) provided another report of a fungus, *Perkinsus* (≡ *Dermocystidium*) with an anterior flagellum possessing a single row of (tripartite?) hairs.) Bower (1987*a*) has figured straminipilous hairs assembled in an arc around the flagellar cross-section for *Labyrinthuloides haliotidis* S. M. Bower, while Tong (1995) has described, in *Developayella*, a straminipilous flagellum which is not directed anteriorly. It must be anticipated that further variants of the basic pattern may exist, but the extents to which phylogenetic significance is accorded to flagellar ornamentation are still likely to be reflected at appropriately high taxonomic levels.

The descriptions of flagellation by Barron must be reassessed, since both *Myzocytiopsis lenticularis* and *Pythium caudatum* (G. L. Barron) M. W. Dick *loc. cit.* (≡ *Lagenidium caudatum* G. L. Barron) are said to have anterior whiplash flagella (Barron, 1976*b*). If the anterior flagellum has **TTHs** it is unlikely to have a whiplash extension.

Flagellar base: kinetosome, its root organization and transitional zone

The flagellar bases are of phylogenetic importance in any discussion of the interrelationships within and between chromistans and other major taxa. The terminology of Andersen *et al.* (1991) is used in the following general account, but ultrastructural information is not available for most of the lagenidiaceous fungi. The flagellar base is composed of the kinetosome and the attached roots of microtubules. Between the kinetosome and the axoneme there is a transitional zone (for reviews see Grain, Mignot & Puytorac, 1988; Preisig, 1989, 1999; Barr, 1992; there is a useful summary diagram in Sleigh, 1989: fig. 5.33), and further structural features may be discernable at the base of the axoneme and in the core of the kinetosome.

The peronosporomycete root system is composed of six parts (Barr & Désaulniers, 1987*b*, 1989*a*) (Figure I: 6). Attached to the kinetosome of the anterior straminipilous flagellum there are two roots - **R1**: a triplet with ribbed microtubules directed anteriorly (three main microtubules with secondary microtubules extending laterally), with a posteriorly directed associated cord (**CD**) in *Phytophthora* and *Pythium* (Barr & Désaulniers, 1992), and **R2**: a doublet usually comprised of two simple microtubules. The ribbed triplet for *Phytophthora* is right-handed (Barr & Désaulniers, 1989*b*) not left-handed as presented in Barr & Allen (1985). I have followed the right-handed convention (Barr & Désaulniers, 1992) in Figure I: 6. Two parts

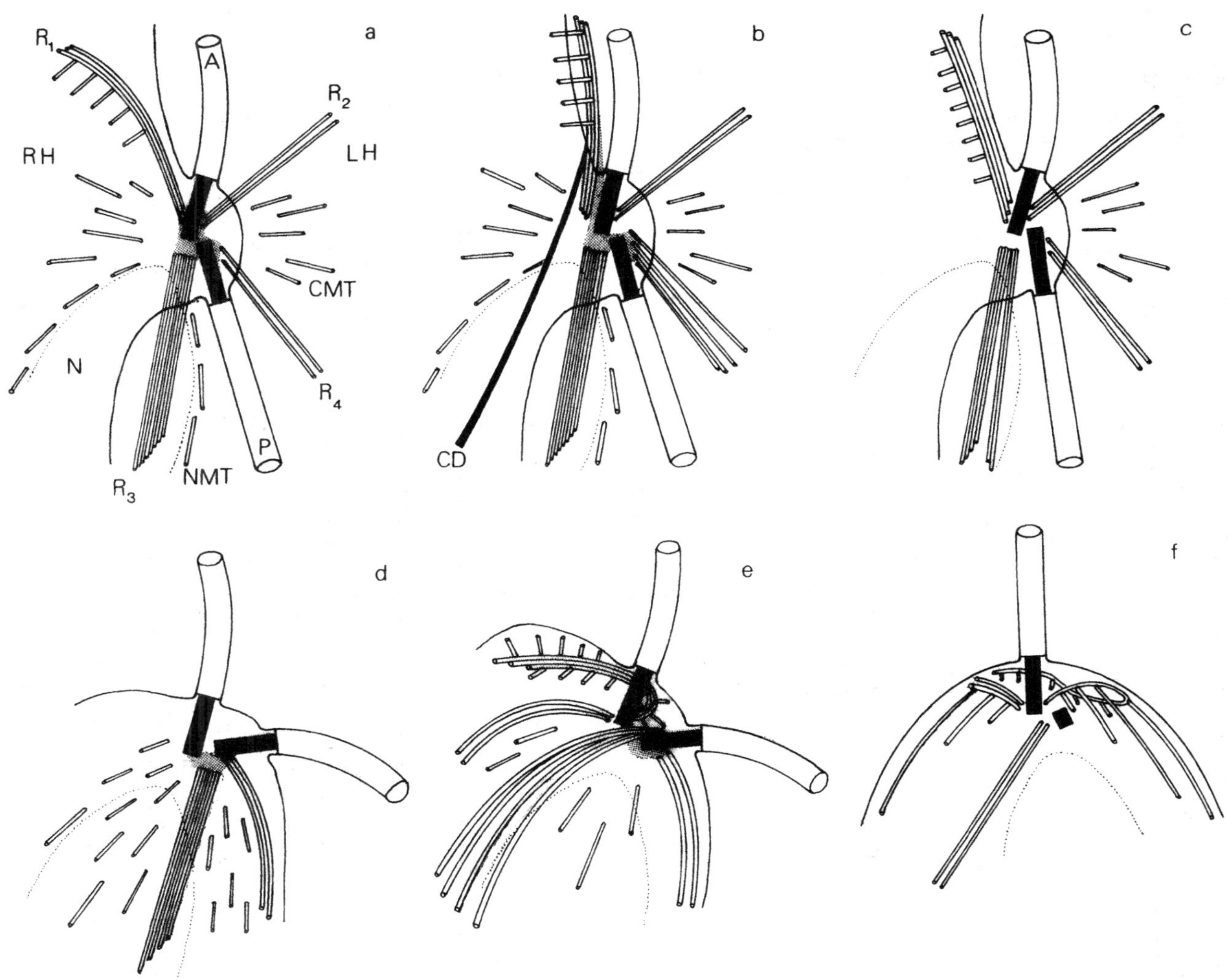

FIGURE I: 7. Simplified diagrams of kinetosome flagellar roots of zoospores, drawn from a slightly oblique ventral view: RH right-hand side of zoospore; LH left-hand side of zoospore; R1, R2 roots to kinetosome of the anterior flagellum; R3 (MS), R4 roots to kinetosome of the posterior flagellum; A anterior flagellum; P posterior flagellum; N nucleus; CMT cytoplasmic microtubules; NMT nuclear-associated microtubules; CD R1 cord. Kinetosomes shown as solid cylinders; electron-opaque microtubular organizing material stippled (after Barr, 1981, 1992; Barr & Allan, 1985; Barr & Désaulniers, 1987, 1989, 1992).

a: *Saprolegnia* principal-form zoospore; b: *Pythium*, note the R1 cord and the number of microtubules to R4; c: *Thraustochytrium*, note overlapping kinetosomes, divided R3 and sparce CMT; d: *Saprolegnia* auxiliary form, note the absence of R1 and R2, and density of Cmt; e: *Lagena*, note apparent rotation of each flagellum 'winding up' R1 and R3 and free R3 microtubules; f: *Rhizidiomyces*, note vestigial posterior kinetosome in same orientation as principal form zoospores, ribs on R2 and absence of R4.

Fig. 7 a,b reproduced from Dick (1999: fig. 3) by permission of Academic Press.

of the root complex are attached to the kinetosome of the posterior flagellum - **R3**: a posteriorly directed multistranded ('octet') band-shaped root, and **R4**: a counterbalancing posterior root of 3-6 microtubules; between the two kinetosomes is a microtubule organizing centre with microtubules radiating into the cytoplasm (**CMT**) and a group of specifically nucleus-associated microtubules (**NMT**). Fibrillar material connects the two kinetosomes, and in *Salilagenidium callinectes* (Bland & Amerson, 1973*a*) and *Haliphthoros milfordensis* (Overton *et al.*, 1983) this appears prominently striated. Although the overall pattern of roots is similar in the *Peronosporomycetes*, each root shows independent variation between taxa, and sometimes within a species (Dick, 1997*a*). **R1** possesses an electron-opaque (microtubule organizing ?) longitudinal rod or 'spine' giving attachment to the length of the kinetosome and which is extended posteriorly as a prominent cord in most of the *Pythiaceae* (all *Peronosporomycetidae* ?); it is confined to two microtubules in *Rhizidiomyces* and has two free ribbed microtubules in *Lagena*; it is absent (as is **R2**) from the auxiliary zoospore of *Saprolegnia*, suggesting that the straminipilous flagellum of this zoospore may have impaired function. **R2** is most constant in ultrastructure, but is reduced to a single ribbed microtubule in *Rhizidiomyces*. **R3** (multistranded root) varies in the number of microtubules involved, and the extent to which they are free (two groups in *Thraustochytrium*; reduced to two in *Rhizidiomyces* (Cooney *et al.*, 1985); absent in *Phytophthora infestans* (Mont.) de Bary (Barr & Désaulniers, 1992)). **R4** is also more conserved, although frequently of a pair of microtubules, it has more numerous components in the *Peronosporomycetidae*, with the greatest number of microtubules (5-6) in *Phytophthora infestans*. Variation in the frequencies of cytoplasmic and nuclear-associated microtubules is also reported. In *Lagena* each flagellum appears to be rotated clockwise between 90-180°, 'winding up' the R1 and R3 (MS) roots.

By way of contrast, a striated, fibrous root (sometimes termed a rhizoplast), is a prominent feature in *Rozella* (Held, 1975), and this structure is also found in *Chytridiomycetes* and the choanoflagellates. The structure and mechanism is distinct from the ultrastructural morphology described above.

Barr (1981) has stated that the primary function of the root system is to dissipate the forces of flagellar motion: it is reasonable to postulate that the powerful anterior flagellum with tripartite tubular hairs would require a stronger root system, and this may be manifest by the **R1** ribbed triplets (i.e., three microtubules with secondary microtubules extending laterally). It has also been suggested (M. J. Powell, pers. comm.) that the **R3** multistranded root (of two to ten parallel microtubules in the oomycetes) may be reduced in smaller zoospores. If there is homology, then *Olpidiopsis* has a quintet multistrand root, and *Lagena* 4-10 microtubules in its multistranded root. Zoospore volumes vary widely (Figure I: 1): the ***volume*** of the zoospore (or cyst) of *Saprolegnia* may be ***100 times*** greater than that of *Olpidiopsis* or *Haptoglossa*. It is conceivable that zoospores may show size-related differences in the degree of ultrastructural differentiation. It is difficult to determine whether reduction or elaboration from a primitive state is involved (Dick, 1997*a*).

The root complex must be assessed as a unit of several independent variables, and thus the data base is still much too small to make phylogenetic generalizations across the full range of straminipilous fungi. As with the diversity of flagellar construction, the extent of the diversity of root systems may be even greater than that already described. However, similarities between the *Saprolegniomycetidae* and the *Peronosporomycetidae*, and the differences ***within*** the *Peronosporomycetidae*, suggest that root organization and structure will be of taxonomic value within families and genera. There is at present no simple correlation or explanation for the variety of patterns between root organization and lateral or sub-apical flagellar insertion.

Barr & Désaulniers (1987*b*) distinguished four types of transitional zones (Figure I: 7). In each case there is a transitional plate, attached to the kinetosome at the end distal to the nucleus, which transects the axoneme core in a plane above that of the cell plasmamembrane and which extends to the flagellar plasmamembrane. At the centre of the transitional plate there is a bead-like projection. Towards the base

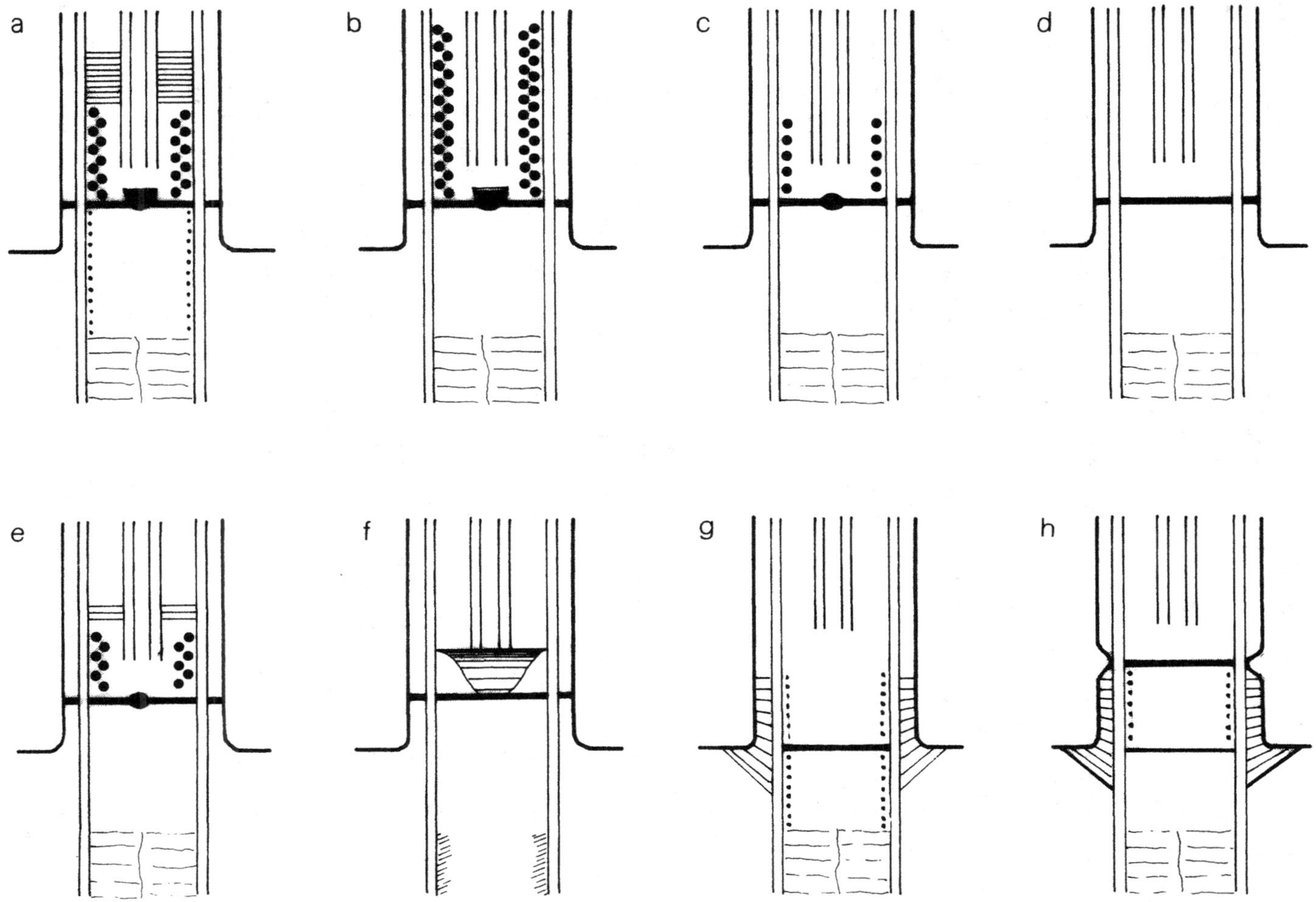

FIGURE I: 8. Simplified diagram of flagellar bases and transitional zones, see text for detailed descriptions (after Barr & Allen, 1985; Barr & Désaulniers, 1987, and Barr, 1992):
a: *Saprolegnia*
(and all *Saprolegniomycetidae* ?)
b: *Pythium*
(and all *Peronosporomycetidae* ?)
c: *Olpidiopsis*
d: *Lagena*
e: *Rhizidiomyces*
(and all *Hyphochtriomycetidae* ?)
f: *Thraustochytrium*
(and all *Labyrinthista* ?)
g: *Polymyxa*
(and all *Plasmodiophorales* ?)
h: a choanoflagellate.

Fig. I: 8 a,b reproduced from Dick (1999: fig. 5) by permission of Academic Press.

of the kinetosome, proximal to the nucleus, some indications of the 'cartwheel' of alternate triplet cross-linking (Manton, 1964, 1965) can usually be seen (see Glockling, 1994). However, the detailed serial transverse sections of this region necessary to describe its structure have not yet been made for flagellate fungi. It is possible that the detailed structure of this zone may differ between representatives of the straminipilous fungi.

Within the axoneme core (i.e., inside the cylinder of the nine doublets) and distal to the transitional plate there is an electron-opaque structure appearing in longitudinal section like a concertina or double helix of repeating units. In the *Pythiaceae* there may be up to 20 such units (Barr & Allen, 1985), but in the *Saprolegniaceae* (Heath & Greenwood, 1971; Holloway & Heath, 1977*a*, *b*; Hoch & Mitchell, 1972*b*), *Leptomitaceae* (Randolph & Powell, 1992), *Rhipidiaceae* (Gotelli & Hanson, 1987) and *Hyphochytriaceae* (Cooney *et al.*, 1985) and in *Petersenia palmariae* Van Der Meer & Pueschel (Pueschel & van der Meer, 1985), there is a short concertina (*ca* 6 units) with struts (struts unresolved for *Petersenia*) to the central doublet distally. Lange, Olson & Safeeulla (1984) described the flagellar base of *Sclerospora*, and Olson subsequently stated (L. W. Olson, pers. comm.) that the concertina of this material was also short and resembled that of *Saprolegnia*. *Olpidiopsis* (Bortnick *et al.*, 1985) also has a group of five or six rings arranged as a single helix. In *Salilagenidium callinectes* (Bland & Amerson, 1973*a*), *Haliphthoros milfordensis* (Overton *et al.*, 1983), *Ectrogella perforans* (Raghu Kumar, 1980) and *Crypticola clavulifera* Humber *et al.* (Frances *et al.*, 1989) the transitional plate is also above the plane of the plasmamembrane, but only in *H. milfordensis* has a concertina or short stack of about eight rings been described. *Lagena* lacks these electron-opaque units, and is unique in lacking a central bead to the transitional plate.

In the labyrinthulids the transitional zone is more complex. The flagellar base has been illustrated for *Thraustochytrium* (Kazama, 1972*a*) and *Schizochytrium* (Kazama, 1980), revealing a transitional plate situated above the plane of the plasmamembrane. Distal to this plate and dilating towards, and connecting to the peripheral doublets of the flagellum, was a cone-like structure, which might be homologous to a short concertina. The flagellar diameter was smaller proximal to the transitional plate. The middle third of longitudinal profiles of the kinetosome core also contained an ill-defined electron-opaque cylinder. Some of these details have been confirmed by Raghu Kumar (1982*b*) for *Ulkenia*.

Aist & Williams (1971), Barr & Allen (1982), Clay & Walsh (1990, 1997), Miller, Martin & Dylewski (1985) and Talley, Miller & Braselton (1978) have described the ultrastructure of the plasmodiophoraceous zoospore, and it is clear that the flagellar apparatus differs considerably from that of the peronosporomycetes. A fine transitional plate extends across the axoneme core at a point level with the cell plasmamembrane, but it does not extend to this membrane nor does it have a bead-like projection at its centre. Merz (1992) has used SEM to reveal ring-like swollen bases to the flagella at the junction with the plasmamembrane in *Spongospora*. The transitional zone also differs from that of the peronosporomycetes in lacking a concertina.

Schnepf (1994) did not compare his material of a *Phagomyxa*-like endoparasite with the *Thraustochytrium* of Kazama (1992*a*): there is some resemblance between the distally dilated cone on the transitional plate distal to the kinetosome (Schnepf, 1994: figs 14, 27; Kazama 1972*a*: figs 2, 4,,5). The absence of the dynein arms and the inner ring of electron-opaque material at the base of the axoneme are also similar. A somewhat similar construction, though less resolved, was noted by Barr & Allen (1982: fig. 1, 12) for *Polymyxa graminis*.

In common with many protists (cf. Hibberd, 1975), the kinetosome core of the plasmodiophorids contains two (?) rows of diffuse, slightly more electron-opaque spheres. Just within the triplets of the kinetosome there is a series of concentric fibres (visible in Barr & Allen, 1982: fig. 25 - D. J. S. Barr, pers. comm.) and this series of fibres continues throughout the transitional zone; in the peronosporomycetes a similar series of concentric fibres is also present in the kinetosome, but these fibres do not extend into the transitional zone (Dr D. J. S. Barr, pers. comm.). Evidence of this series of concentric fibres can be deduced for the kinetosomes of *Myzocytiopsis lenticularis* (Glockling, 1994). The kinetosome root system in the plasmodiophorids is simple and similar for each flagellum, each kinetosome having two sets of microtubules; a triplet and a doublet. Barr & Allen (1982) concluded that this root morphology is more similar to that for certain protists than to that of any of the described flagellate fungi or algae. (Could the triplet perhaps be equivalent to the **R1** ribbed triplet or **R3** octet of the peronosporomycete apparatus which are more prominent in these much larger zoospores?)

In the choanoflagellates (e.g., Leadbeater & Morton, 1974) and in the monoblepharids (Fuller, 1966) the transitional plate is complex with two electron-opaque zones, a very faint one level with the cell plasmamembrane, as in *Polymyxa* and many protists, and a second, more distinct zone a short way into the flagellum (Figure I: 7). The relationship of this construction with that of the thraustochytrids should be assessed. Cavalier-Smith (1987) uses kinetosome and rootlet structure, among other characters, to derive the *Chytridiomycetes* from a choanomastigote-like progenitor, but it should be noted that the range of morphological variation in the limited number of species studied makes relationships conjectural although they are supported, in general, by molecular analyses (Tehler *et al.*, 2000).

The kinetosome of the posterior flagellum butts on to the shaft of the kinetosome to the straminipilous flagellum in the oomycetes, *Olpidiopsis* (Bortnick *et al.*, 1985; Martin & Miller, 1986*b*) and the labyrinthulids (Kazama, 1972*a*, 1980; Porter, 1974, 1990). The abuttment may be less obvious when the base of the posterior kinetosome is approximately level with, but offset from, that of the anterior flagellum as in *Phytophthora* (Barr & Désaulniers, 1989*a*, *b*, 1990*b*) (Figure I: 8). (A similar kinetosome arrangement occurs in the *Pedinomonadales* -Melkonian, 1990.) In the plasmodiophorids the same base-butting-to-shaft arrangement is found, although the two flagella cannot be uneqivocally identified (*Sorosphaera*, Talley *et al.*, 1978; *Polymyxa*, Barr & Allen, 1982; *Ligniera*, Miller *et al.*, 1985; *Spongospora*, Clay & Walsh, 1990). The same appears to be true for *Blastulidium* (Manier, 1976). The abuttment may be related to the replication sequence of the centriole-kinetosome complex, the posterior kinetosome being derived from the 'older' centriole in motile photosynthetic unicells (cf. Beech *et al.*, 1991), irrespective of whether it has subtended a straminipilous or a smooth flagellum: it is inferred that centriole replication is semi-conservative, in which the maturation of the new centriole is not complete until the second mitosis. The angle between the kinetosomes of the plasmodiophorids is variable: within the zoosporangium prior to flagellar activity it approximates to 30° (Clay & Walsh, 1990), but in active, free-swimming zoospores the angle is 150° (Barr & Allen, 1982) or 180° (Merz, 1992).

In the uniflagellate fungi the non-functional centriole is located in different positions. In *Hyphochytrium* the non-functional centriole is offset, and at a different level from the functional kinetosome (Cooney *et al.*, 1985) but in the *Chytridiales* and *Monoblepharidales* it is approximately parallel to the kinetosome and their bases are at the same level (Lange & Olson, 1979; Montecillo, Bracker & Powell, 1980; Barr & Désaulniers, 1987*a*). In the *Spizellomycetales* the position of the non-functional centriole is variable, and in the *Blastocladiales* it is structurally diminished or absent, but if present it is at right angles to the functional kinetosome.

The correlations between the transitional zones and the root systems in non-hairy anisokont, and heterokont organisms have not been fully evaluated.

While the ultrastructural morphology of the flagella, the kinetosomes and root arrangement may show a certain amount of variation from organism to organism, there is a group of core characters that is common to the straminipilous fungi (Barr, 1981*c*; Barr & Allen, 1985; Barr & Désaulniers, 1987*b*, 1989*a*; Barr, 1992). Nevertheless, it should be restated that because flagellation is heterokont it is not necessarily peronosporomycetous; likewise, molecular sequences have shown that some flagellates without straminipilous ornamentation are related to the straminipiles (Silberman *et al.*, 1996). The flagellar ultrastructural characters that are ***diagnostic*** for the ***class Peronosporomycetes*** have ***not*** been established; it may be that no one character, on its own, will prove to be diagnostic. For the great majority of the lagenidiaceous organisms discussed here the information is either nonexistent, or far too fragmentary, to contribute to their classification.

ZOOSPOROGENESIS AND ZOOSPORE CYST GERMINATION

Asexual reproductive morphology

Most *Peronosporomycetes* have asexual reproduction but a few taxa are apparently incapable of producing any asexual stage. The diversity of asexual reproduction is considerable and can be summarized by noting the following points:

1. production of zoospores, aplanospores, or conidiosporangia
2. shapes and dimensions of the sporangia or conidiosporangia
3. morphogenesis and morphology of distinct sporangiophores or conidiosporangiophores
4. deciduous or caducous nature of the sporangia
5. ability of any sporangia to germinate indirectly by zoospores or directly by germ tubes
6. sporangial renewal: determinate or indeterminate (i.e., internal, cymose, basipetal, or percurrent) and/or accompanied by simultaneous or sequential sporangiogenesis
7. site and mechanism of protoplasmic cleavage and zoospore discharge including:
    - presence of a papilla or an operculum
    - presence of a discharge vesicle
8. site of encystment
9. ornamentations of the zoospore cyst wall

Asexual reproductive units may develop in terminal, lateral or intercalary positions on the assimilative thallus. The vegetative system then shows **eucarpic development**. Alternatively, the entire thallus may assume the role of a sporangium or, if septate, a series of sporangia (**holocarpic development**, irrespective of whether the sporangia mature simultaneously or sequentially). This transformation from an assimilative thallus to a sporangium (or sporangia) is frequently without morphological modification other than the formation of an exit tube and dehiscence papilla.

Most sporangia are of regular shape, frequently globose or fusiform, but *Pythium*, in particular, shows a wide intergrading range of sporangial shapes, from spherical, globose-clustered, through toruloid and digitate forms to otherwise-undifferentiated lengths of hyphae. Examples can be found of every one of these forms in terminal or intercalary positions. In the myceliar *Peronosporomycetes* the sporangium is usually terminal on a hypha and separated from the assimilative thallus by a **septum** (a wall deposit indistinguishable from and continuous with the thallus wall). In the *Rhipidiales* the sporangium is abstricted by a constriction and the remaining canal is blocked by a **plug** (material not confluent with the thallus wall) while in the *Leptomitales* the **cellulin granule** separates the sporangial protoplast from that of the other loments.

A zoosporangium usually has one or more apical papillae. The papilla normally protrudes from the general sporangial outline due to its thicker, often more hyaline wall. At maturity the papilla deliquesces to form the exit tube through which the planonts are extruded. In a few genera (e.g., *Calyptralegnia*, *Rhipidium*) the deliquescence is confined to a circumscissile ring to form a calyptra.

In the *Saprolegniales* the nuclei of the sporangial protoplast do not undergo further division, but in *Phytophthora* (Graham, 1954; Laviola, 1975; Maltese, Conigliaro & Shaw, 1995) and the *Peronosporaceae* (*Peronospora*, Trigiano & Spurr, 1987; *Bremia*, Tommerup, 1989; *Plasmopara*, Burruano *et al.*, 1992) the nuclei of the sporangioplasm undergo further mitoses. Khan (1976) described both nuclear degeneration and mitosis in the morphogenesis of sporangia of *Albugo*: in this genus the sporangia are formed in percurrent chains and so progressive morphogenetic stages could readily be traced. In taxa where the thallus becomes converted into the sporangium there cannot be different developmental pathways involving nuclear migration or nuclear mitosis, such as occur in the eucarpic *Peronosporomycetes*. At maturity the lagenidiaceous sporangia are multinucleate, but neither the sequence of mitoses leading up to and following thallus septation nor the pro-sporangial morphogenesis has been documented.

Descriptions of zoospore discharge by lagenidiaceous fungi are often difficult to interpret because they refer to a pattern typical for another genus without precisely conforming to every feature in that genus. This is true for references to 'achlyoid' discharge (see Dick, Wong & Clark, 1984) and for '*Pythium*' discharge, with which discharge by lagenidiaceous fungi is compared. In the descriptions for many species assigned in the past to the *Lagenidiales*, sporogenesis does not correspond strictly to that outlined above for *Pythium*. In the lagenidiaceous fungi there are no exact equivalents to the zoosporogenesis of *Pythiogeton* or *Phytophthora*.

There are three stages in zoosporogenesis that need to be assessed critically:

1 the process of cleavage into spore initials, its site, timing and ultrastructural morphogenesis
2 the nature of the vesicle
3 the behaviour of the expelled spore - as a naked protoplast; in relation to flagellar development; in encystment, and in possible re-emergence

Cleavage

Cleavage of zoospore initials occurs either within the zoosporangium (**intrasporangial zoosporogenesis**) or after discharge of the sporangial protoplasm (**extrasporangial zoosporogenesis**). Zoosporogenesis is intrasporangial and cleavage furrows developed from dictyosome-derived cisternae first become confluent with the prominent central tonoplast vacuole and eventually breach the plasmamembrane in *Achlya* and in *Saprolegnia* (Gay & Greenwood, 1966; Money *et al.*, 1987). There is a consequent ***loss of volume*** (ca. 10%) of the zoosporangium as turgor is lost. Zoospore discharge is achieved by imbibition of water through the sporangial cell wall in response to the release, within the confines of the zoosporangial wall, of water-soluble (osmotically active) ß-1,3-glucans from the zoospore initials and residues from the cleavage cisternae and central vacuole (Money & Webster, 1985, 1988; Money *et al.*, 1988). Intrasporangial cleavage may be followed by planont encystment before discharge (*Aphanodictyon, Calyptralègnia*), extrusion with subsequent encystment (*Achlya, Verrucalvus*), or the zoospore initials may develop flagella prior to extrusion (*Saprolegnia, Phytophthora*). In *Blastulidium*, *Eurychasma* and *Eurychasmopsis* planonts become parietally rearranged prior to intrasporangial encystment.

Pythium is characterized by the extrusion of uncleaved multinucleate protoplasm into an extraplasmamembranic, membranous, glucan-polymer vesicle which is formed simultaneously with discharge and is therefore initially in close contact with the sporangial protoplasm and confluent with the sporangium wall (extrasporangial zoosporogenesis). In *Phytophthora* (and *Pseudoperonospora* of the *Peronosporaceae*) cleavage takes place ***within*** a persistent zoosporangial plasmamembrane. Such a zoosporangial membrane, could account for the 'evanescent' vesicles also reported for the *Rhipidiomycetidae* if the entire plasmamembrane bound packet of zoospores were emergent.

Whereas in the *Saprolegniaceae* there is a prominent central vacuole up to zoospore delimitation, in *Phytophthora* the immature sporangium loses the central vesicle before the cleavage cisternae are in place and only small remnants of the initial vacuole occur among the cleavage cisternae (Hohl & Hammamoto, 1967; Williams & Webster, 1970). Since these earlier studies, Hardham & Mitchell (1998); Harold & Harold (1992); Heath & Harold (1992); Hyde, Gubler & Hardham (1991); Hyde & Hardham (1992, 1993) and Jackson & Hardham (1998) have shown that the actin fibres of the cytoskeleton play a vital role in the topographic organization of the nuclear spacing, cleavage cisternae and zoosporogenesis in *Phytophthora* and *Achlya*. Work of comparable detail is not available for other straminipilous fungi.

In the *Salilagenidiales* most of the cytoplasm is peripheral at the mid-cleavage stage, prior to discharge, and large vacuoles probably merge with the cleavage cisternae, so that the development is similar to, but not identical with, that of the *Saprolegniaceae*. Coincident with this zoosporogenesis is the presence or absence of a gelatinous matrix surrounding the planonts as they are discharged. A 'vesicle' develops from the exit tube apex as a gelatinous matrix ***prior to*** the extrusion of the protoplasm (see also Glockling & Dick, 1997: some *Myzocytiopsidales*). The protoplasm ***does not fill*** the clearly-defined 'vesicle'. At maturity this 'vesicle' becomes partially inverted and collapses down the outside of the exit tube during zoospore maturation, often remaining as a sleeve after discharge (Alderman, 1976). The boundary of a gelatinous matrix can often be distinct and it may be difficult to distinguish between such a boundary and the presence of a membranous vesicle. A fibrillar or amorphous vesicular 'membrane' could thus be formed as a precipitation reaction between such a colloidal matrix and the environment.

Studies of *Myzocytiopsis lenticularis* (Glockling, 1994) indicated that the vegetative thalloid segments do not possess a tonoplast, but that a few tonoplast vacuoles first develop early in the presporangial stage and persist while the cleavage cisternae reach an advanced stage of orientation. This developmental pattern is not saprolegniaceous (G. W. Beakes, pers. comm.) but neither can it be clearly equated with those of *Salilagenidium callinectes* or *Phytophthora*. Sparrow (1939*b*) also described a more or less *Saprolegnia*-like zoosporogenesis with a central vacuole for *Myzocytiopsis zoophthora* (Sparrow) M. W. Dick *sensu lato* (see Glockling & Dick, 1997; ≡ *Lagenidium oophilum* Sparrow). Photographs of living material (Canter-Lund & Lund, 1995: fig. 621) show that discharge in *Myzocytium megastomum* De Wild. and *Cystosiphon canterae* (Karling) M. W. Dick *loc. cit.* (≡ *Lagenidium canterae* Karling) is '*Pythium*'-like in every morphological detail. For many taxa assigned to the *Lagenidiales* there are clear statements that spore initials are fully preformed within the sporangium (but with extrusion of quiescent zoospores in *Myzocytiopsis lenticularis*, Glockling, 1994), or that the initials are to a greater or lesser extent delimited (*S. callinectes*, Bland & Amerson, 1973*a*), but with protoplasmic links similar to those described for *Aphanomyces* (Hoch & Mitchell, 1972*a*) and *Verrucalvus* (Dick *et al.*, 1984). In other species there are suggestions that both maturation and cleavage may be sequential, like *Schizochytrium*, with successive rather than simultaneous cleavage planes: for example the bipartitioning of the vegetative thallus of *Gonimochaete* (Drechsler, 1940) and chlamydospore formation in *Chlamydomyzium septatum* (Karling) M. W. Dick (≡ *Lagenidium septatum* Karling, Karling, 1969, Dick, 1997*b*). Cleavage sometimes occurs either within or outside the sporangium (*S. oedogonii*, *Cornumyces pygmaeus* (Zopf) M. W. Dick (≡ *L. pygmaeum* Zopf), Karling, 1981*a*: 118 and 126 respectively, Dick, 1998*b*). Another developmental pattern described for *Crypticola entomophaga* (≡ *Atkinsiella entomophaga*) (Martin, 1977), *Eurychasma dicksonii* (E. P. Wright) Magnus (Aleem, 1950*b*) and *Eurychasmopsis multisecunda* Canter (Canter & Dick, 1994 - the *Eurychasma* species described by Canter *et al.*, 1990)) is with the cleavage taking place within the sporangium but followed by sequential release of spores. Kinetosomes with their transitional plates are present in the uncleaved and as yet non-flagellate zoospore initials of *Myzocytiopsis* (Glockling, 1994).

How much variation in the morphogenesis of cleavage can be accepted within a single genus? Are these mechanisms of sporogenesis compatible with concepts for *Pythium*; the *Pythiaceae*, the *Pythiales* or the *Peronosporomycetes*? If these diverse patterns of sporogenesis are correlatable with other differences, there may be grounds for separation at a high taxonomic level.

Evidence for a vesicle

Pythium is characterized by the extrusion of uncleaved multinucleate protoplasm into an extra-plasmamembrane, membranous, glucan-polymer vesicle which is formed simultaneously with discharge. The vesicle membrane probably has a fibrillar nature since it has sufficient tensile strength to prevent escape of the vigorously motile zoospores for a perceptible length of time. Lunney & Bland (1976) have shown that the vesicle membrane is continuous with, and attached to the exit tube wall (≡ modified papilla) (cf. Hallett & Dick, 1986: fig. 11 for a *Pythium* 'microsporangium'). The dynamics of sporangial discharge are discussed by Money, Webster & Ennos (1988) for *Achlya* and Gisi & Zentmyer (1980) for *Pythium* and *Phytophthora*.

Within the *Pythiaceae* there are two other sequences which are derived conditions. In *Pythiogeton* the discharged and encapsulated sporangial protoplast becomes detached from the exit tube, but the membranous nature of the vesicle is preserved so that the vigorously motile zoospores are contained or restrained for a short time before rupture of the vesicle (M. W. Dick, unpublished obs.). In *Phytophthora* the inner wall layers of the discharge papilla extend down the inside of the sporangial walls for a considerable distance, rather as if a 'vesicle' had been retained within the sporangial cavity (Chapman & Vujičić, 1965; Gisi, Hemmes & Zentmyer, 1979).

Lange, Edén & Olson (1988) and Hyde, Gubler & Hardham (1991) have shown that cleavage takes place ***within*** a persistent zoosporangial plasmamembrane in *Pseudoperonospora* and *Phytophthora* respectively. Such a membrane, if emergent as it sometimes is, would be fragile and could account for the 'evanescent' vesicles described, for example, in the *Rhipidiomycetidae* (Sparrow, 1960). Hyde & Hardham (1992) and Jackson & Hardham (1998) have shown that the morphogenesis of cleavage is governed by the development and array of cytoskeletal filaments which, in turn, determine the nuclear spacing and cleavage planes. Inflation of the cleavage cisternae into vesicles may be an artefact.

Clay, Benhamou & Fuller (1991) have shown that the sporangial contents of *Rhizidiomyces apophysatus* Zopf are discharged as a single protoplast bounded by a thin, ephemeral, chitinous extension of the inner sporangial wall. The fate of this wall during cleavage is not established.

The range of patterns of discharge of 'lagenidiaceous' zoospores includes:

1. individual emergence of active, flagellate zoospores
2. extrusion of fully formed aplanospores
3. extrusion of fully formed quiescent (incompletely flagellate) zoospores
4. extrusion of partially delimited planonts
5. extrusion of a naked coenocytic protoplast

Coincident with this range is the presence or absence of a gelatinous matrix (Couch, 1942). The boundary of such a matrix will often be evident against the water of the environment (see Canter, 1979: plate 4A) and it may be difficult to distinguish between such a boundary and the presence of a membranous vesicle. Fibrillar or amorphous vesicular 'membranes' could also be formed as a precipitation reaction between a colloidal matrix and the environment (cf. Bland & Amerson, 1973*a*). An analogous situation is the mucilaginous condensation to form a wall-like membrane outside the oogonium in *Aplanopsis terrestris* Höhnk (Dick, 1969: plate 2, fig. 7). Two observations are important: the visible and ultrastructural nature of the junction of the vesicle with the discharge tube, and the sequence of cleavage and flagellar activity in relation to vesicle formation.

The developmental sequence for marine '*Lagenidium*' species has two features not found in the *Pythiaceae* and which may indicate a different vesicle morphogenesis. The vesicle develops from the exit tube apex as a gelatinous matrix ***prior to*** the extrusion of the protoplasm. The protoplasm ***does not*** fill the clearly-defined vesicle, nor is it initially spherical; also the vesicle becomes partially inverted and collapses down the outside of the discharge tube during zoospore maturation, often remaining as a sleeve after discharge (see Alderman, 1976). The zoosporogenesis of the *Myzocytiopsis* sp. described by Glockling & Dick (1997) is initially similar.

In contrast, *Lagenidium rabenhorstii* was described by Zopf (1884) as possessing a vesicle, but both Atkinson (1909) and Ivimey Cook (1935) state that the sporangial protoplast emerges naked and that cleavage takes place in the absence of a vesicle. There are several other instances (e.g., *Myzocytiopsis zoophthora* (Sparrow) M. W. Dick (≡ *Myzocytium zoophthorum* Sparrow, Sparrow, 1936*b*; Dick, 1997*b*; Glockling & Dick, 1997) in which cleavage and incipient flagellation appear to take place in the naked protoplast, so that the 'vesicle' is said to "deliquesce completely". Such 'vesicles' are probably not homologous with those of *Pythium*. Likewise, suggestions that the vesicle is developed ***subsequently*** to cleavage (*Myzocytiopsis humana* (Karling) M. W. Dick (≡ *Lagenidium humanum* Karling, Karling, 1947*b*; Dick, 1997*b*; the description for [*Sali*]*lagenidium* sp. (marine): Lightner & Fontaine, 1973) may also relate to an extra-spore and intra-sporangial secretion rather than to an essentially wall-like extra-plasmamembrane. Hesse *et al.* (1989) observed an initial interface between the protoplasm and the environment which developed into a "mucilaginous court" in *Ducellieria*.

A further possibility is that the vesicle may be derived from a persistent plasmamembrane as in *Phytophthora* and *Pseudoperonospora* (Hyde *et al.*, 1991; Lange *et al.*, 1988).

It thus appears that three major categories of 'vesicle' occur in the *Peronosporomycetes* and *Hyphochytriomycetes*:

Homohylic vesicle

An extra-plasmamembrane wall of fibrillar polyglucan or chitin continuous with one of the sporangial wall layers (e.g., *Pythium* and *Hyphochytrium* respectively) (ὁμο-homo - same ὑηλ-hyle - material).

Plasmamembranic vesicle

An evanescent plasmamembrane sac which was the sporangial plasmamembrane (e.g., *Phytophthora*, *Pseudoperonospora*).

Precipitative vesicle

A non-membranous (or initially gelatinous, non-membranous and non-fibrillar), fluid boundary between the peri-zoosporic oligoglucan colloid and the water of the environment (e.g., *Saprolegnia* and possibly *Salilagenidium callinectes*).

Data distinguishing this degree of precision are often totally lacking.

Zoospore taxis

Zoospores would inevitably have developed a recognition system and a mechanism for locating a suitable substratum over evolutionary time. The recognition system is not well understood, but the response has been documented more thoroughly. Tropic and taxic responses have been reviewed by Wynn (1981). Zoospore attraction to roots was first reported by Goode (1956). Directional movement, or **taxis** may have different stimuli, and chemotaxis (Khew & Zentmyer, 1973; Rai & Strobel, 1966; Zentmyer, 1979); electrotaxis (Khew & Zentmyer, 1974; Morris & Gow, 1993; Morris *et al.*, 1992; Troutman & Wills, 1964); negative geotaxis (Cameron & Carlile, 1977), and phototaxis (Muehlstein & Amon, 1987*b*) have all been described, usually for *Phytophthora*, *Aphanomyces* and other plant pathogenic fungi or *Peronosporomycetes* used for physiological and biochemical studies. Zoospore taxis has been recorded for parasites of plants (Mitchell & Deacon, 1986*a*); algae (Kerwin *et al.*, 1992); fish (Willoughby & Pickering, 1977) and nematodes (by *Catenaria*, see Jannson & Thiman, 1992).

The pattern of zoospore aggregation along roots has been reported by Dale & Erwin (1991), Halsall & Williams (1984), Hinch & Weste (1979), Ho & Hickman (1967*b*), Irwin (1976), Jones *et al.* (1991), Kraft *et al.* (1967), Kuan & Erwin (1980), Milholland (1975), Palloix *et al.* (1988*a*), Shishkoff (1989), Spencer & Cooper (1967), Tippett *et al.* (1976) and Zou & Paulitz (1993). Royle & Hickman (1964*a*, *b*) made a thorough study of aggregation to roots by *Pythium aphanidermatum*. Root surface topography has been implicated for zoospore aggregation by Deacon & Donaldson (1993).

In dense spore suspensions the phenomenon of autoaggregation may occur. The phenomenon was first noted by Lounsbury (1927) for *Isoachlya* and has occasionally been noted since then; Porter & Shaw (1978), working with *Phytophthora drechsleri*, viewed the autoaggregation as a taxis. Jones *et al.* (1991) have suggested that autoaggregation may be instigated soon after a few zoospores have clumped together. Reid, Morris & Gow (1995) have found that autoaggregation of zoospores is 'genus specific'. If the degree of specialization is confirmed for more genera, then this attribute could add to the criteria for diagnoses.

Flagellar retraction

The shedding or retraction of flagella at encystment and flagellar transformation (Beech *et al.*, 1991) are important, but rarely observed or recorded, criteria. For the chytrids, Koch (1968) has provided the most orderly account, defining five patterns of flagellar loss:

1. total detachment and loss
2. 'vesicular coiling', either from the flagellar tip or part-way along the flagellum, with eventual absorption of the vesicle into the zoospore body
3. 'straight-in', direct retraction
4. 'body-twisting', a rotation of the zoospore, acting as a spool for the axoneme
5. 'lash-around', an exceptional movement of the flagellum towards the zoospore body

The two last patterns have close similarities in that they result in the flagellar plasmamembrane making longitudinal contact with the zoospore body so that the flagellar axoneme temporarily resides tangentially within the encysting zoospore. There is no comparable review for straminipiles, although all of these patterns may occur.

Crump & Branton (1966) described a combination of vesicular coiling and body-twist for the auxiliary zoospore of *Saprolegnia*, and Holloway & Heath (1974) subsequently provided a more detailed account. There is non-synchronous shortening (presumed depolymerization) of the axoneme lengths of the two flagella, apparently unaccompanied by any change in kinetosome/nucleus orientation. When very short (<3 μm) the axoneme wraps against the zoospore plasmamembrane and becomes incorporated within the cell. A wrapping around of the straminipilous flagellum appears to be a feature of *Gracea gracilis* (Dick, 1998*a*). In the chrysophyte *Hibberdia*, Andersen (1989) also concluded that flagellar retraction was achieved by the flagellum wrapping round the body of the cell. Holloway & Heath (1974) were unable to distinguish between the straminipilous and smooth flagella, but it should be noted that a wrapping of the straminipilous axoneme around the body of the zoospore would be inconsistent with the presence of a tuft of tubular tripartite hairs on the resultant cyst wall (cf. Dick, 1990*a*). (Such tufts may be found during flagellar transformation and development in motile chrysophytes - R. A. Andersen, pers. comm., but this is not a strictly comparable situation.) Schnepf, Deichgräber & Drebes (1978*d*) reported that the flagella of the first-formed zoospores of *Lagenisma* were retracted by a movement of the whole kinetosome-axoneme unit towards the centre of the cell, which ought to account for a tuft of tubular tripartite hairs, but unfortunately these authors did not report the presence of any such tuft on the cysts of *Lagenisma*.

Whittick & South (1972) stated that, for *Pontisma antithamnionis* (Whittick & South) M. W. Dick *loc. cit.* (≡ *Olpidiopsis antithamnionis* Whittick & South), contact with the host is established by the longer posterior flagellum, followed by rapid activity (flagellar retraction?) resulting in the zoospore encysting at the host surface. Held (1973*a*) gave TEM evidence that the single smooth flagellum of *Rozella allomycis* F. K. Foust is retracted after making contact with the host. Equivalent information for *Rozellopsis* is more circumstantial: Prowse (1951) confirmed the observation by Fischer (1882) that attachment to the host is via the tip of the longer posterior whiplash flagellum, which then becomes retracted, drawing the encysting zoospore body to the host surface.

Ivimey Cook (1935: plate 4, fig. 1) photographed the attachment of a *Syzygangia zygnematicola* zoospore to a *Spirogyra* filament by the tip of the anterior flagellum. Similarly, Waterhouse (1940) described the attachment of *Gracea waterhouseae* (Karling) M. W. Dick (≡ *Rozellopsis waterhouseae* Karling, Karling, 1942*c*; Dick, 1997*b*) as by the longer anterior (straminipilous ?) flagellum. This would be the normal mode of attachment for cells with chemotactile-sensitive flagella, as has been demonstrated for fucophyte antherozoids (Müller & Falk, 1973).

For the plasmodiophorids, Fuchs (1966) has shown that the *Polymyxa* zoospore appears to make a "probe contact" with potential substrates by the tip of the shorter anterior flagellum, which is presumably chemotactile-sensitive. After contact, the anterior flagellum is retracted first. The suggestion has been made that virus particles, transmitted by plasmodiophorids, become bound to the flagellar membrane and become internalized following flagellar retraction (Temmink, Campbell & Smith, 1970; Stobbs, Cross & Monocha, 1982). Internalization of that part of the parasite plasmamembrane would not be necessary when the entire protoplast becomes ***inserted*** within the host cytoplasm. Presumably the mechanism would be different when *Lagena* is implicated as a vector.

Flagellar retraction has been presumed for hyphochytrids (Fuller & Reichle, 1965). Porter (1990) stated that it was not known whether the flagella of labyrinthulid (or thraustochytrid) zoospores are retracted or shed, but Bower (1987*a*) stated that the flagella of *Labyrinthuloides haliotidis* were shed. The planonts of species currently placed in *Olpidiopsis* appear to combine some characters of both the principal and the auxiliary zoospore types of *Saprolegnia*, but with additional, non-peronosporomycete features.

Flagellar detachment is characteristic for principal form zoospores of the *Peronosporomycetes* (Crump & Branton, 1966). The distinction between flagellar retraction or the shedding of flagella at encystment may be a more fundamental criterion than the point of flagellar insertion (cf. auxiliary and principal form zoospores).

Recognition of the substratum by the zoospore, or perhaps its straminipilous flagellum, is a prior requisite for encystment. Bimpong (1975) and Bimpong & Hickman (1975) described the metabolic reserves, enzyme activities and the time scales for encystment and germination. Paktitis, Grant & Lawrie (1986) noted surface changes which were consistent with stimulus-mediated secretion and there has been considerable work carried out on the calcium metabolism during these processes (see Broembsen & Deacon, 1996, 1997; Deacon & Donaldson, 1993; Donaldson & Deacon, 1992, 1993; Griffith, Iser & Grant, 1998; Grant, Griffith & Irving, 1986; Hill, Grayson & Deacon, 1998; Irving & Grant, 1984; Irving, Griffith & Grant, 1984; Iser *et al.*, 1989; Jackson & Hardham, 1996; Jackson & Heath, 1993; Reid, Morris & Gow, 1995; Warburton & Deacon, 1998 and references therein).

Encystment

The processes involved in the production and encystment of zoospores have been studied most intensively for *Phytophthora*, and to a lesser extent, *Achlya* (Hardham, 1987*b*, 1989; Hardham, Gubler & Duniek, 1990; Hardham & Mitchell, 1998; Hardham & Susaki, 1986; Hardham, Susaki & Perkin, 1985, 1986; Harold & Harold, 1992; Heath & Harold, 1992; Hyde, Gubler & Hardham, 1991; Hyde & Hardham, 1992, 1993, and Jackson & Hardham, 1998).

The zoosporangial and zoospore dictyosomes secrete three kinds of vesicles which are peripheral at the time of encystment. Two of these kinds of vesicle are located in the ventral groove of the swimming zoospore while the other is located under the remaining plasmamembrane. In *Phytophthora* the zoospore comes to rest on the substratum with the ventral groove against the substratum. The flagella are shed and the ventral vesicles attach the encysting zoospore to the substratum. The remaining vesicles secrete the cyst wall precursors which rapidly polymerize to form the cyst wall.

Cyst morphology

Cyst morphology has been described by Beakes (1983, 1987) and Dick (1990*a*) (Figure I: 9). For *Peronosporomycetes*, it is generally accepted that flagella of the auxiliary zoospore are retracted, except for the tubular tripartite hairs which are left in a tuft at the point of resorption (Dick, 1990*a*), while the flagella of the principal zoospore are shed (the membranes of these flagella may round up and bear scattered tubular tripartite hairs). Subsequent ornamentation of the cyst wall is derived from preformed structures in cytoplasmic vesicles (Beakes, 1983, 1987). References to cyst wall ornamentation are lacking in all but three of the lagenidiaceous taxa. In two of the exceptions spines are formed (*Haliphthoros milfordensis*, Overton *et al.*, 1983; and *Lagenisma coscinodisci*, Schnepf *et al.*, 1978*c*, *e*). In *Lagenisma*, as in *Leptomitus* (Hallett, 1975; Hallett in Dick, 1990*a*) and *Hyphochytrium* (Hallett, 1975) the spines are formed as finger-like protrusions, initially filled with cytoplasm. In *Gracea gracilis* the straminipilous flagellum is retracted by the 'lash-around' mechanism, leaving an equatorial ring of tubular tripartite hairs over the cyst surface (Dick, 2000*d*). The only other fungus known to have such hairs distributed over the cyst surface is the hyphochytrid, *Rhizidiomyces apophysatus* (Fuller & Reichle, 1965); in which the spiral banding of the tubular shafts was particularly prominent (cf. Domnas *et al.*, 1986, *contra* Beakes, 1987) (comparisons should also be made with the slopolinids - Patterson, 1989).

For a few lagenidiaceous species (see PART VI) the zoospore is reported to come to rest on a host, not by forming a spherical cyst as would be expected in any peronosporomycetous fungus, but by forming a hemispherical structure attached to the host over a large surface area. This again suggests that there is a greater tendency for the planont to revert to an amoeboid form than is normal for *Peronosporomycetes*. *Gracea*, *Rozella* and *Lagena* cysts undergo changes of shape during encystment but are spherical when attached to their hosts (Pemberton *et al.*, 1990; Held, 1973*a*; Barr & Désaulniers, 1987*b*).

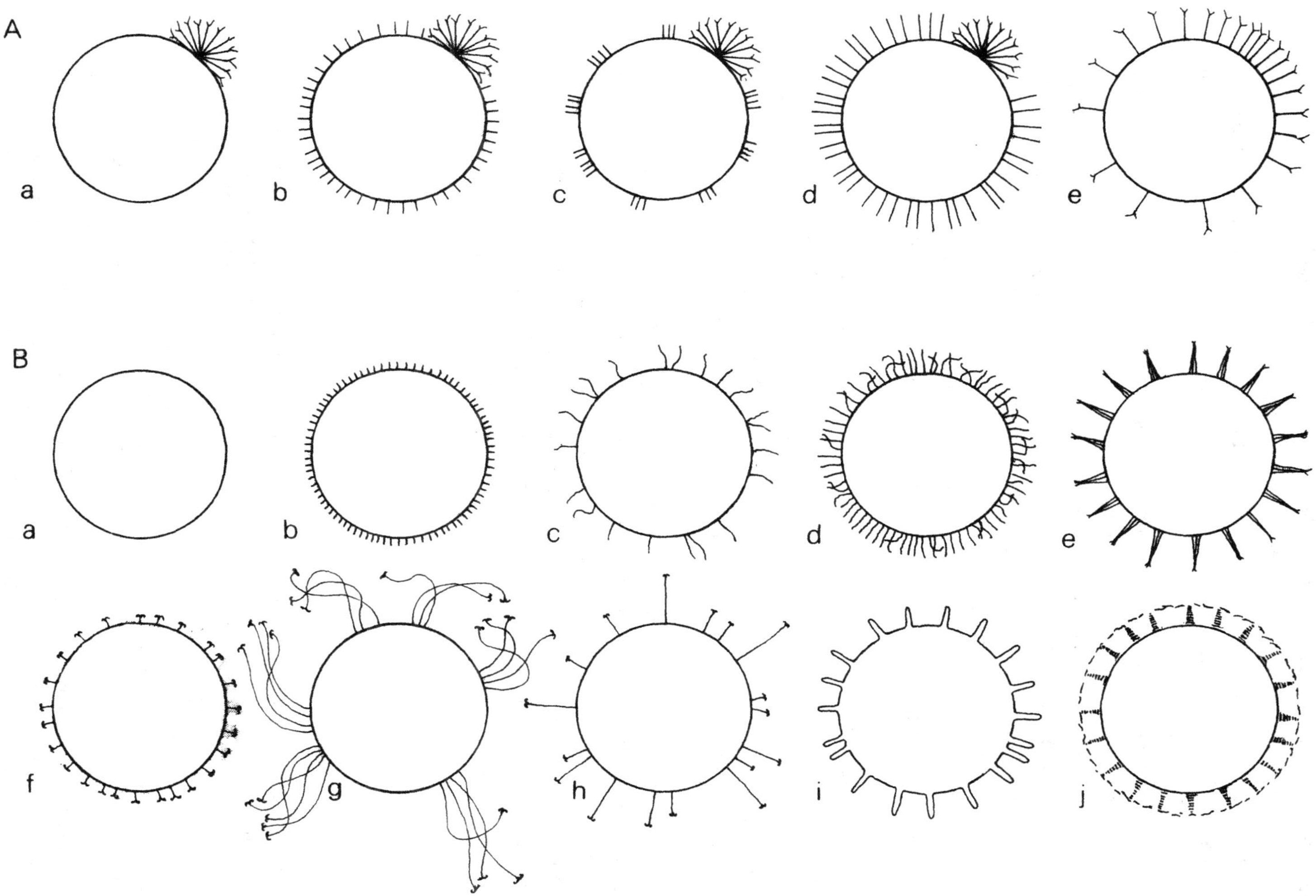

FIGURE I: 9. Zoospore cyst morphology (after Beakes, 1987; Dick, 1990*a*; Hallett, 1975).
Fig. A: cysts of auxiliary-form or pyriform zoospores (*Saprolegniomycetidae*) with a tuft of sloughed tripartite tubular hairs as straminipilous flagellum is resorbed: a: *Saprolegnia ferax*; b: *S. diclina*; c: *S. parasitica*; d: *S. megasperma*; e: *Rhizidiomyces apophysatus*; *Gracea gracilis* (this configuration is also found on membranes from shed straminipilous flagella).
Fig. B: cysts of principal-form zoospores (straminipilous flagellum shed): a: unornamented, *Saprolegnia turfosa*; *Pythium middletonii*; b: investment of short hairs - *Phytophthora cryptogea*; c: sparce investment of long hairs - *Pythium monospermum*; d: dense investment of long hairs - *Apodachlya minima*; e: regular distribution of tight fascicles of hairs, fimbriated at tips - *Dictyuchus sterile*; f: regularly spaced bifurcated hairs ('boat-hook' hairs - *Saprolegnia diclina*; g: tufts of long 'boat-hook' hairs - *S. parasitica*; h: 'boat-hook' hairs of varying lengths - *S. hypogyna*; i: hollow papillae - *Leptomitus lacteus*; *Hyphochytrium* sp.; *Lagenisma coscinodisci*; j: striated spines under a fibrillar layer - *Haliphthoros milfordensis*.
Fig. 9 Aa, Bb, Bh, Bi, reproduced from Dick (1999: fig. 4) by permission of Academic Press.

Germination

Cyst germination in the *Peronosporomycetes* can follow one of three basically different patterns of development, further elaborated in Figure I: 10):

1 in the non-filamentous *Rhipidiomycetidae* the germination is polar, with the production of a rhizoidal system followed by the enlargement of the original cyst to form an isodiametric or lobed basal cell, which subtends the sporulating structures. The 'mycelial' development of *Sapromyces* probably represents proliferations of potential reproductive units rather than a vegetative axis bearing sporangia

2 in the *Peronosporomycetidae* and the *Saprolegniomycetidae* (with the exception of the *Leptomitales*) two partially interchangeable patterns are known: either a very narrow (*ca* 1-3 μm diam.) germ-tube is produced which may maintain a branched vegetative existence as in the *Sclerosporales* (Dick *et al.*, 1984, 1989) or subsequently undergo rapid transition to a hyphal diameter of 5 μm or more; or a large diameter germ tube may be produced directly so that the spore cyst is scarcely discernible

3 in the *Leptomitales* the development is blastic and each segment or bud so produced is capable of further budding, or in *Leptomitus*, rhizoids may be produced from these secondary buds

The wide-diameter hyphal systems of the *Saprolegniales* may represent modifications from a basal-cell -type of organization, with isodiametric growth being replaced by tip growth accompanied by a shift of assimilatory function from a well-defined rhizoidal system to these broader hyphae (which, in *Achlya* particularly, may show continued growth in diameter). The 'vegetative' system of the *Saprolegniales* may thus be considered as an assimilative, growth-unlimited sporangio-gametangiophore.

There is probably a developmental relationship between the fine germ tube and the rhizoid. The development of the coralloid thallus of certain *Leptolegniellaceae* from an initial narrow diameter germ tube could be envisaged as an extension, with less coordinating control, of the blastic pattern (**3**). This same developmental pattern is found in *Haliphthoros* and in a number of other species which have been referred to the *Lagenidiales*. In most of the endoparasitic fungi which have been studied by TEM, the development of the fine penetration tube is accompanied by the expansion of a prominent distal vacuole in the zoospore cyst (Held, 1973*b*).

FIGURE I: 10. Diagrams of peronosporomycete thallus forms from zoospore cyst germination patterns to secondary developments: independent origins of the mycelial habit may be discerned. **Fig. A.** Tip growth of a narrow germ tube of limited length, with injection of protoplasmic contents to form a detached intra-host-protoplasmic non-mycelial thallus: *Olpidiopsis*, *Gracea* (*Olpidiopsidales*). **Fig. B.** Indeterminate tip growth and branching from a narrow germ tube (no vegetative expansion or intussusception) dilating only for reproductive function: *Verrucalvus*, *Sclerospora* (*Sclerosporales*), hyphae of *ca* 5 μm diam. **Fig. C.** Indeterminate tip growth and branching, presumably from a narrow germ tube, dilating periodically and producing secondary rhizoids (R_2), also dilating for reproductive function: *Medusoides* (*Pythiogetonaceae*), hyphae of *ca* 5 μm diam. **Fig. D.** Broad elongate growth from an expanded germ tube apex (narrow germ tube), allantoid or branched: *Myzocytiopsis* (*Myzocytiopsidaceae*). **Fig. E.** Broad elongate growth from an expanded germ tube apex (narrow germ tube), coralloid or branched: *Leptolegniella* (*Leptolegniellaceae*). **Fig. F.** Blastic with elongate segments (non-polar): *Apodachlya* (*Leptomitaceae*). **Fig. G.** Parablastic with elongate segments (non-polar), with or without secondary rhizoids (R_2): *Leptomitus* (*Leptomitaceae*). **Fig. H.** Cyst polar with primary rhizoids (R_1), cyst becoming swollen (forming the basal 'cell' by generalized intussusception) with proliferous tubular or clavate elements separated by thick-walled narrow isthmuses, eventually terminating in reproductive structures: *Sapromyces* (*Rhipidiaceae*). **Figs I, J.** Indeterminate tip growth and branching from a broad or very broad germ tube, with considerable later near-basal intussusception, with or without secondary rhizoids (R_2): *Achlya* (*Saprolegniaceae*), hyphae of 40-140 μm diam. **Fig. K.** Indeterminate tip growth and branching from a moderately broad germ tube, without intussusception, not dilating for reproduction, sometimes with retraction septa: *Pythium*, *Peronospora* (*Peronosporales*), hyphae of *ca* 10 μm diam.

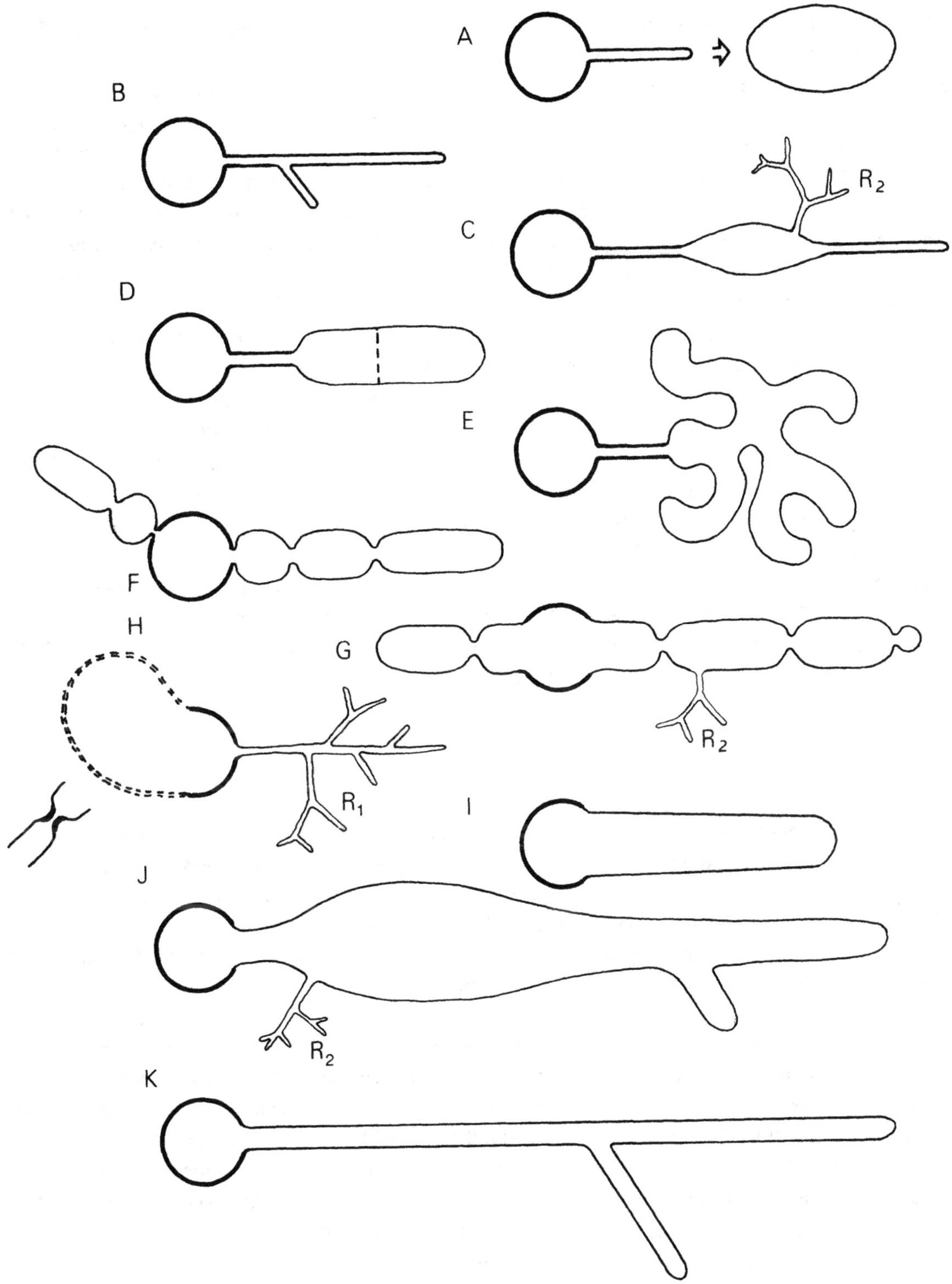
A
B
R2
C
D
E
F
H
G
R2
R1
I
J
R2
K

An additional feature of some of the lagenidiaceous fungi is the production of an ellipsoidal swelling part-way along the thin germ tube (*Cystosiphon canterae* (≡ *Lagenidium canterae* - Canter & Lund, 1969; *Pleocystidium lundiae* (Karling) M. W. Dick *loc. cit.* (≡ *Lagenidium lundiae* Karling) - Canter & Lund, 1969; *Lagenidium* sp. on *Oocystis* - Masters, 1971, and *Crypticola entomophaga* (≡ *Atkinsiella entomophaga*) - Martin, 1977). In *Gonimochaete* the cyst germinates on the surface of the nematode, producing an extremely fine penetration tube of *ca* 0·1-0·2 μm diameter (Saikawa & Anazawa, 1985). This tube subsequently enlarges to form the first unit of the disarticulating thallus. The discharged intermediary infection cell of *Haptoglossa mirabilis* (Barron, 1980) also has an ellipsoidal swelling in its tube-like extension analogous to those mentioned above.The *Peronosporomycetes* have a highly developed complex of extrusomes associated with encystment and germination. Exemplary studies by Hardham, Suzaki & Perkin (1985, 1986) and the overview by Hardham (1998) indicate that extra-plasmamembrane secretions in the region of the flagellar insertion are responsible for cyst attachments in *Phytophthora*. Donaldson & Deacon (1992) give further information on cyst adhesion for *Pythium* and Deacon & Saxena (1998) have extended the information to cover *Aphanomyces* and *Phytophthora*. In *Achlya* the encysting aplanospores become cemented together to form a hollow sphere of cysts. Secretory vesicles (mucosomes?, K-bodies - kinetosome-associated bodies - Lehnen & Powell, 1991) associated with this process may have a dictyosome origin. Further information on the distribution and ultrastructural differences of K_2 bodies, peripheral vesicles and encystment vesicles within the *Peronosporomycetes* (*Pythiaceae*, *Saprolegniaceae*, *Leptomitaceae* and *Rhipidiaceae*) is available in Sadowski & Powell (1990) and Randolph & Powell (1992). Held (1973*b*) has reviewed the more physiological aspects of germination in the endobiotic parasitic fungi. Further studies of 'lagenidiaceous' fungi since then have revealed a secretion/cementation of a different magnitude. An adhesive pad is apparently secreted from the cyst, whether that cyst has been formed from an aplanospore (*Myzocytiopsis subuliformis* M. W. Dick, Dick, 1997*a*) or from a zoospore (*Myzocytiopsis glutinospora* (G. L. Barron) M. W. Dick [≡ *Myzocytium glutinosporum* G. L. Barron], Dick, 1997*a*). Newell *et al.* (1977: 189) have postulated that the apparatus developed in *Haptoglossa* may merely be a further elaboration of such a secretion, but it is more likely that an intermediary infection cell, with its own adhesive pad (Barron, 1980) is developed from the cyst.

Extrusomes

Many protists possess organelles which extrude or evert their contents to the exterior, but which are not associated with generalized excretion or the development of extra-plasmamembrane investments such as walls or scales (Hausmann, 1978). Sleigh (1989: table 5:1) gives an indication of the distribution of extrusomes in flagellate protoctists; homologies have not been fully explored, but extrusomes are reported from diverse groups of chromistan algae - chrysophytes, raphidophytes, cryptophytes and other anisokont flagellates such as euglenoids and glaucocystophytes, and some green flagellates.

Extrusomes include vesicles with different levels of complexity, ranging from mucosomes to nematocysts. In the *Peronosporomycetes* mucosomes, or K_2-bodies, may be associated with specific secretions for wall modification (Lehnen & Powell, 1991). Hardham and coworkers (Gubler & Hardham, 1988, 1990; Gubler, Hardham & Duniec, 1989; Hardham, 1989, 1998; Hardham & Gubler, 1990; Hardham, Gubler, & Duniec, 1990; Hardham & Suzaki, 1986; Hardham *et al.*, 1985, 1986) have provided a detailed description of the three kinds of vesicle in *Phytophthora*. All are involved in encystment and attachment by cementation to the substrate. Prominent peripheral vesicles are figured for *Myzocytiopsis* (Glockling, 1994) and by Canter & Dick (1994: fig. 27) for zoospores of *Eurychasmopsis*.

The K_2-bodies, which have a tubular or structured matrix lumen, are regarded as having phylogenetic significance within the heterokont fungi and *Peronosporomycetes* in particular (Powell *et al.*, 1985; Powell & Blackwell, 1995; Randolph & Powell, 1992). They have been identified in several genera of the

Saprolegniaceae, where the ultrastructure is similar; in *Aphanomyces* there are some differences and Hoch & Mitchell (1972*b*) called them U-bodies. *Apodachlya* (*Leptomitaceae*) has similar K_2-bodies to the *Saprolegniaceae*. No comparable structures have been reported for the *Pythiaceae* or other *Peronosporomycetidae*. *Olpidiopsis saprolegniae* var. *saprolegniae* also has K_2-bodies, and these are compared with those of a tribophyte alga, as well as with other heterokont fungi (Powell *et al.*, 1985). These authors concluded that K_2-bodies were not discernable in micrographs published for *Ectrogella perforans*, *Salilagenidium callinectes*, *Lagenisma coscinodisci* or *Haliphthoros milfordensis*. The descriptions of *Ciliatomyces spectabilis* (Foissner & Foissner, 1986*a*, *b*) and *Ducellieria chodatii* (Hesse *et al.*, 1989) do not mention structures equatable with K_2-bodies.

Much more elaborate ultrastructure is possessed by the nematocysts of the dinoflagellates, the gullet extrusomes of the cryptophytes and the rohr/stachel/schlauch (glossoid cell/gun cell) complex of the plasmodiophorids and *Haptoglossa*. The relationship between the complex penetration apparatus of the plasmodiophorids and of *Haptoglossa* is reviewed in more detail in PART IV, pp. 202-213.

No comparable information is available for other lagenidiaceous fungi, either with respect to the simpler mucosomes or the more complex gun cells.

Dense-body vesicles

Dense-Body Vesicles (DBVs, also known as Finger-Print Vacuoles - FPVs - Hemmes, 1983) are often prominent in the reproductive structures of the *Peronosporomycetes*, although they can be found at all developmental stages. They are implicated in the production of the inter-protoplasmic colloids responsible for zoosporangial evacuation in the *Saprolegniaceae* (Money & Webster, 1985, 1988); in cleavage of ooplasm in oosporogenesis (Gay, Greenwood & Heath, 1971; Fletcher, 1979; Beakes, 1981*b*); in wall formation in oogonia of *Achlya radiosa* Maurizio (but not *Saprolegnia*) (Al-Rekabi, 1979), and are known to coalesce to form the ooplast in the mature oospores of all *Peronosporomycetes* (Howard & Moore, 1970; Fukutomi, Aki & Shiraishi, 1971; Hemmes & Bartnicki-Garcia, 1975; McKeen, 1975; Sargent, Ingram & Tommerup, 1977; Beakes & Gay, 1978*a*; Beakes, 1981*b*; Dick, 1990*a*, 1995). At different phases of the life-history and organ development they may show different and developmentally intergrading morphology, as revealed by TEM. At one extreme there is either a single central or one or more peripheral, more or less sharply defined electron opaque core or cores in an electron-translucent matrix. At the other extreme the matrix becomes almost completely filled with myelin-like configurations (hence the FPV connotation - Hemmes, 1983: figs 5 & 6) which initially take the form of layers joining the electron-opaque core to several parietal electron-translucent aggregates within the vesicle membrane (Al-Rekabi, 1979; Al-Rekabi in Dick, 1990*a*: fig. 2). These myelin configurations can also appear as 'exfoliations' from the electron-opaque core, and strata-like nonconformity between blocks of myelin configurations can also be found within the same vesicle. Wang & Bartnicki-Garcia (1973, 1974), Powell & Bracker (1977) and Bartnicki-Garcia & Wang (1983) have shown that the DBVs contain phosphorylated mycolaminarans, and such ionized glucans may be responsible for the myelin-like appearence in TEM.

The prominence and presumed importance of DBVs, and their involvement in glucan and phosphate metabolism and storage, might suggest an explanation for the phosphate/polyphosphate storage differences between *Eumycota* and oomycetes (absence of metachromatic granules as determined by toluidine blue O staining in the latter) that have been noted by Dietrich (1976) and Chilvers, Lapeyrie & Douglass (1985), and discussed by Niere, Griffith & Grant (1990) and Grant, Grant & Harris (1992).

The occurrence of DBVs in the plasmodiophorids has not been reported.

For the lagenidiaceous fungi there are no clear figures of the distinctive myelin-like form of DBVs, although TEMs may show possible cored DBVs, as in *Salilagenidium callinectes* (Gotelli, 1974*b*: fig. 9) and *Blastulidium* (Manier, 1976); the ooplast of *Ciliatomyces* (Foissner, 1987) also appears to have a DBV origin, although the electron-opaque regions are isolated and peripheral. Ultrastructural work (Glockling, 1994) has revealed the coexistence of two different sizes of DBVs with electron-opaque cores, and with possibly different functions, in species of *Myzocytiopsis* parasitic in nematodes. The larger class of DBVs appears to become terminalized and perhaps have a role in disarticulation and in providing templates for the restricted central swelling of the oogonial segments in *M. osiris* (Dick, 1995). Comparisons should be made between the large DBVs of *M. osiris*, and an illustration of a similar feature in *Crypticola clavulifera* (Frances *et al.*, 1989: fig. 3).

A careful search by electron microscopists for homologous DBVs in possibly related *Protoctista* and *Straminipila* (*Cafeteria*, *Developayella*, *Labyrinthuloides* spp.) may help to strengthen phylogenetic relationships indicated by molecular biological analyses (Leipe *et al.*, 1994, 1996; Tong, 1995; Silberman *et al.*, 1996). TEMs of *Haptoglossa* thalli (Glockling, 1994) reveal the presence of electron-opaque vesicles with peripheral electron-lucent 'bubbles' reminiscent of the coalescing DBVs of the pythiaceous ooplasts. However, fixation of *Haptoglossa* was consistently less satisfactory than for *Myzocytiopsis* (itself an interesting pointer to relatedness, but perhaps the thicker cell wall may have been a factor) although procedures have now been improved (S. L. Glockling, pers. comm.).

SEXUAL REPRODUCTION

Oogamy

The process of oogamous sexual reproduction is normally regarded as involving the production of an egg (oosphere or female gamete) in an oogonium (female gametangium) and an antherozoid in an antheridium (male gametangium) (see Kniep, 1928; Moewus, 1943). Different kinds of oogamy are known in lower plants as well as in animals, characterized for example, by life-history, or ploidy cycle (see Dick, 1987; GLOSSARY entries). In the *Peronosporomycetes* there are female gametes in the oogonia but the differentiation of antherozoids (discrete male gametes) does not occur. The antheridium merely contains donor gametangial nuclei. Gametangia of the *Peronosporomycetes* are **meiogametangia**, that is, diploid cells in which meiosis takes place. For instance, in some of the *Chytridiomycetes* the gametangia are haploid and the male gamete is motile. Gametangia may be developed in different positions on the thallus:

1 terminally, subterminally, or in an intercalary position on main or branch hyphae
2 as terminal or lateral appendages to a nonmycelial thallus
3 from the entire thallus

In the *Peronosporomycetes* synchronous meioses occur in the coenocytic gametangia. After meiosis, the contents of the receptor gametangium (oogonium) become separated as one or several uninucleate and initially unwalled gametes. For *Apodachlyella* Dick (1986) presented a deliberate ambiguity in descriptions of the sexual morphology. Preliminary work, which requires further research, indicated that the additional wall developed after meiosis; suggesting that in this genus walled female and male gametes are formed. The septate antheridia of *Eurychasmopsis* appear to be similar (Canter & Dick, 1994).

In the *Myzocytiopsidales* and a few *Pythiales* the thallus becomes septate and adjacent segments assume the function of gametangia. In such cases the thallus segment is the site of meiosis (**thalloid meiosis**), but the relationship with the eucarpic *Peronosporomycetes* is evident. Data are generally lacking on nuclear behaviour and ploidy cycles in *Lagenidium* and *Myzocytium*, so that the kind of oogamy cannot be determined. However, TEM profiles of *Myzocytiopsis* species (Glockling, 1994) suggest that meiosis precedes oosphere formation. Martin & Miller (1986*c*) have confirmed that the ploidy cycle of *Olpidiopsis varians* is haplomitotic B (gametangial meiosis).

The range of sexual morphology within a family such as the *Leptolegniellaceae* is also wide: from the eucarpic *Aphanodictyon*, which is homothallic with differentiated gametangia, through the intercalary gametangial development in *Brevilegniella*, to holocarpic genera such as *Leptolegniella* (coralloid thallus) and *Nematophthora* (ellipsoid thallus in which the whole thallus functions as an automictic gametangium, eventually containing multiple oospores).

The occurrence of multiple synchronous meioses in a gametangium makes it possible for karyogamy to take place between two haploid nuclei from adjacent meioses in the same gametangium (**automictic sexual reproduction**). Sexual reproduction can thus occur without a separate male gametangium or antheridium (Dick, 1995: fig. 2). The morphogenesis of the two kinds of gametangia and the sexual process (meiosis and karyogamy) should therefore be considered as separate criteria. The absence of a male gametangium does not necessarily indicate parthenogenetic (i.e., no meiosis, no karyogamy) development, therefore automictic sexual reproduction and parthenogenesis ***cannot be distinguished without cytological evidence*** (Dick, 1972, 1987*a*). There are few reports of the actual site of karyogamy; karyogamy is presumed to occur either in the haploid coenocyte of the oogonium after meiosis, or in the oosphere after fertilization. The distinctive cytoplasmic reorganization of the oospore, compared with that of the chlamydospore or resting sporangium, is indicative that functional or nonfunctional meioses precede oospore formation, as in *Aplanopsis* (Win-Tin & Dick, 1975).

The oospore protoplast (ooplast, lipid globules and nuclear spot) and oospore wall of the common *Peronosporomycetes* were clearly figured by de Bary (1887), and wall structure was already in use for taxonomic purposes by Schröter (1897) and Wilson (1907*b*) for the *Peronosporales*. The taxonomy of the *Peronosporomycetes* is based on oosporogenesis (vacuolar or periplasmic), oospore wall structure and oospore protoplasmic reorganization.

Sexuality and life-histories

The evolution of the *Peronosporomycetes* differs from that of eumycote fungi in that it is based on vegetative diploidy, not haploidy. The ploidy cycle is **haplomitotic B** (Dick, 1987), in which mitosis is confined to the diploid phase; haploid mitosis does not occur. Haplomitotic A or diplomitotic ploidy cycles are absent from this class, although they occur in the phylum and kingdom. For example, the life-history of certain labyrinthulids may involve pre-meiotic and post-meiotic mitosis (Perkins & Amon, 1969) so that the ploidy cycle would then be diplomitotic. Loss of sexuality has apparently occurred in several genera, leaving anamorphic diploids either with or without morphological structures homologous to oogonia and oospores. Diploidy has resulted in the ***oospore*** population functioning in ecological or population genetics in a way analogous to that of heterokaryotic ***anamorph spore*** populations in eumycote fungi. The functions of the sexual cycle may, in part, be for adverse condition survival and in part for control of genetic variability; the relative importance of these functions may differ between closely related taxa. Dick (1990*c*) has pointed out that in genera such as *Pythium* there may be congenitally foreshortened life-histories (Figure I: 11), in some of which sexuality has been lost.

Life-history studies in the *Peronosporomycetes* are relatively few. Dick (1970) showed that saprotrophic *Aphanomyces* species probably went through many asexual cycles in a summer season before a generation produced oogonia. A difference of emphasis is shown by *Plasmopara viticola* (Burruano *et al.*, 1994, 1995, 1997, 1999) which depends on the production, survival and germination of oospores for the annual production of conidiosporangia to initiate the new disease cycle in the spring.

Sexual systems have not been unequivocally described for the hyphochytrids and are unknown in thraustochytrids.

Homothallism and heterothallism: sex hormones and sterols

The majority of species of most of the genera of the *Peronosporomycetes* are homothallic; heterothallism may be secondarily derived. When present, the antheridial branch grows, under hormonal attraction, to the oogonium; the antheridium (donor or male gametangium) is differentiated after contact with the oogonium (the exception is the amphigynous antheridium of certain species of *Phytophthora* which is formed ***before*** the oogonium). The classic studies of heterothallism in *Achlya* initiated by Raper (1936, 1939*a*, *b*) have led to some understanding of the mating systems, morphogenesis of directional growth and penetration, and the identification of a C_{29} steroid sex hormone, antheridiol (Arsenault *et al.*, 1968; Barksdale, 1960, 1962*a*, *b*, 1963*a*, *b*, 1966, 1970; Barksdale & Lasure, 1973; Barksdale, Carlile & Machlis, 1965; McMorris, 1978; McMorris & Barksdale, 1967; Mullins (1968); Mullins & Raper (1965); Raper, 1951). This work was followed by research into heterothallism in *Phytophthora* (Galindo & Gallegly, 1960) and *Pythium* (Hendrix & Campbell, 1968; Hendrix, 1970). Leonian (1931) and De Bruyn (1935) reported heterothallism in *Peronospora*; Gustafsson & Arhammer (1983), Michelmore & Ingram (1980, 1981) and Michelmore &

Sansome (1982) have given accounts for *Bremia*, and Bishop (1940) described heterothallism in *Sapromyces*. Interpretation of the genetics underlying the differential sexuality in *Achlya* [six states, probably involving two independent genes, each with two alleles: homothallic (AABB); congenitally sterile (aabb); oogonial (AAbb); antheridial (aaBB); predominantly oogonial (#A>#B); predominantly antheridial (#A<#B)] was not possible until the kind of ploidy cycle had been resolved. Relative sexuality is also known in the *Peronosporales* but whether the mechanism depends on trisomy or lethals is disputed (Judelson et al. 1995; Van der Lee et al. 1997).

Various possible modes of action of sterols during sexual reproduction have been suggested, including induction (or suppression, Thomas & McMorris, 1987) of sexuality (with the indication that there may be substitution or analogue induction), directional growth of the gametangial axes (McMorris, 1978), localized stimulation of wall softening at the point of contact between gametangia (Gow & Gooday, 1987), and their effect on meiosis (Elliott & Sansome, 1977). Intertaxon induction of oosporogenesis (Brasier, 1975) and synergism (Sansome & Sansome, 1974) have also been suggested.

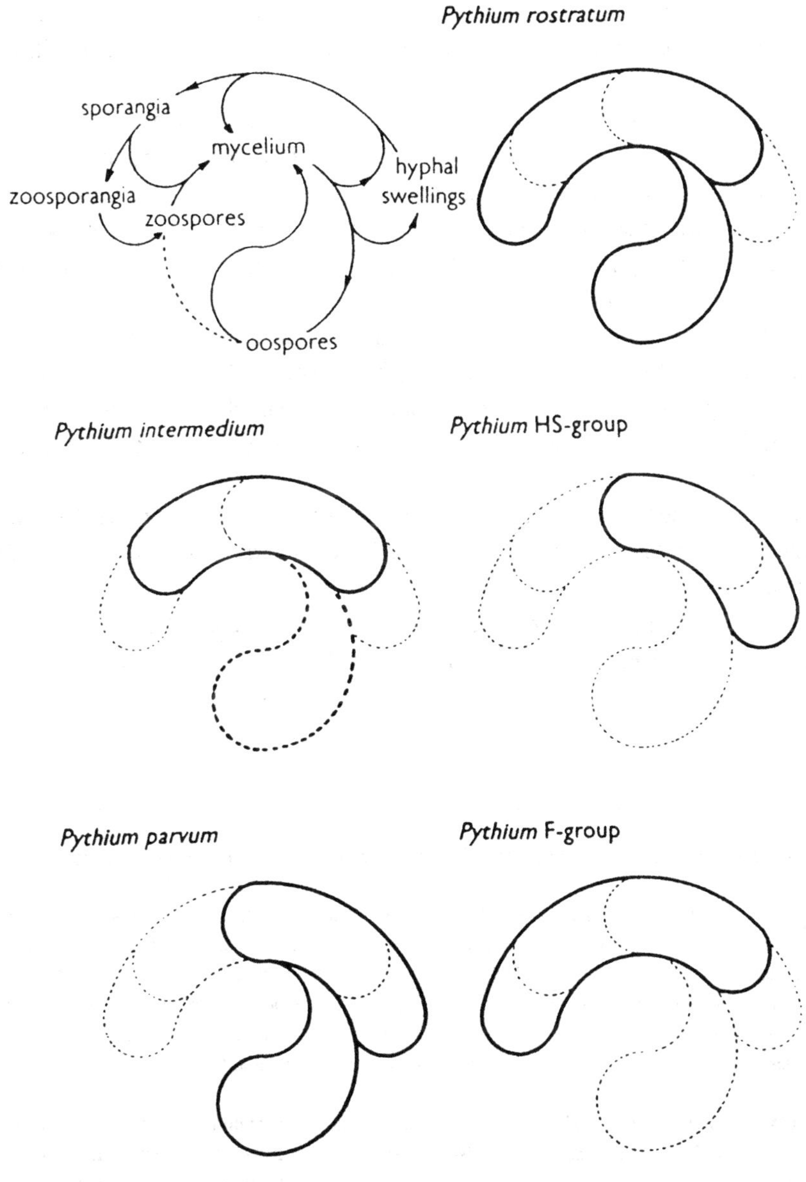

FIGURE I: 11. Diagrams of constitutionally foreshortened life-histories within the genus *Pythium* (from Dick, 1990*c*).

The oogonium

The oogonial initial develops as a swelling without septation or by a transformation of a gametangial segment. The first trigger for this morphogenetic change must be endogenous, from either a nutritional or biochemical stimulus.

The importance of intercalary initiation has not been sufficiently stressed (Dick, 1995). Intercalary differentiation is common in the *Pythiales* and the *Saprolegniales* and can occur in the *Leptomitales*. The oogonial initial is frequently subapical and terminal oogonia on main or lateral branch hyphae may have evolved by terminalization of the initial. Near-terminal oogonial differentiation leaving an apiculus is often seen in the *Saprolegniaceae* and *Pythiaceae*.

No single group within the *Peronosporomycetes* displays all the morphological diversity, morphogenetic patterns and wall layering that can be found in the class. Many species have oogonia with a smooth, more or less spherical outline, but many others are ornamented. The pattern of development of wall ornamentation depends on three criteria: initial expansion, secondary primordial initiation and wall deposition. Six examples of the sequential or simultaneous expression of these criteria will (Dick, 1995, 2000*c*) illustrate the diversity of form:

1 initial swelling accompanied by the generation of numerous new growing tips results in the formation of hollow papillae; further swelling and wall deposition result in the formation of a spherical oogonium with dense or scattered papillae (*Scoliolegnia subeccentrica*)

2 as above, but with the attenuation of these growing points before final swelling and wall deposition to form spines, the tips of which are sometimes completely occluded (*Achlya radiosa*, *Pythium echinulatum*)

3 as **1**, but with the apices of the papillae remaining as thin stretched membranes after secondary wall deposition to form truncate papillae (*Aphanomyces stellatus*, *Achlya recurva*)

4 initial swelling followed by the development of small circular areas under which there is no further wall deposition, resulting in the formation of simple pits (*Isoachlya toruloides*, *Saprolegnia ferax*)

5 secondary wall deposition proceeding simultaneously with swelling, resulting in plaques of thickened wall areas to form solid, U-shaped, vermiform or arched verrucae (*Pachymetra chaunorhiza*)

6 long outgrowths from the oogonial initial becoming almost completely occluded with oospore wall development extending into the outgrowths (*Medusoides argyrocodium*)

FIGURE I: 12. The biodiversity of oogonial and oosporic wall structure and oosporic cytoplasmic organization.

The diagram is arranged as three partially exploded wheels of sectors: the outermost sector is of oogonial wall structure; the middle sector is of oospore wall structure; the innermost, with the dotted line indicating the plasmamembrane of the zygote or oospore displays the protoplasmic reorganization in the resting spore. The diagram ignores the plerotic/aplerotic states of the oogonial/oospore relationship; it does not indicate the nuclear spot(s) and cannot show the fact that the ooplast is often not centric with respect to other organelles. The asterisk indicates a fluid layer. The three wheels of sectors may be partially, but not totally, independent of each other. The following notes relate to each wheel in turn, starting at the top and working anticlockwise.

Outermost wheel (oogonial wall structure): smooth, thin wall (*Pythium*); conic-papillate thin wall (*Pythium*); cylindric-papillate thin wall (*Pythium spinosum*); recurved-papillate, lanose wall (*Medusoides*); thick, acicular wall (*Aqualinderella*); thick spiny wall (*Olpidiopsis*); laminate veruccose wall (*Pachymetra*); thick wall with truncate papillae (apices thin) (*Achlya recurva*); thick wall with mamillate papillae (apices occluded) (*Achlya radiosa*); thick wall with hemispherical papillae (*Scoliolegnia*); smooth, pitted wall (*Saprolegnia*); smooth, mucilaginous wall (*Aplanopsis*).

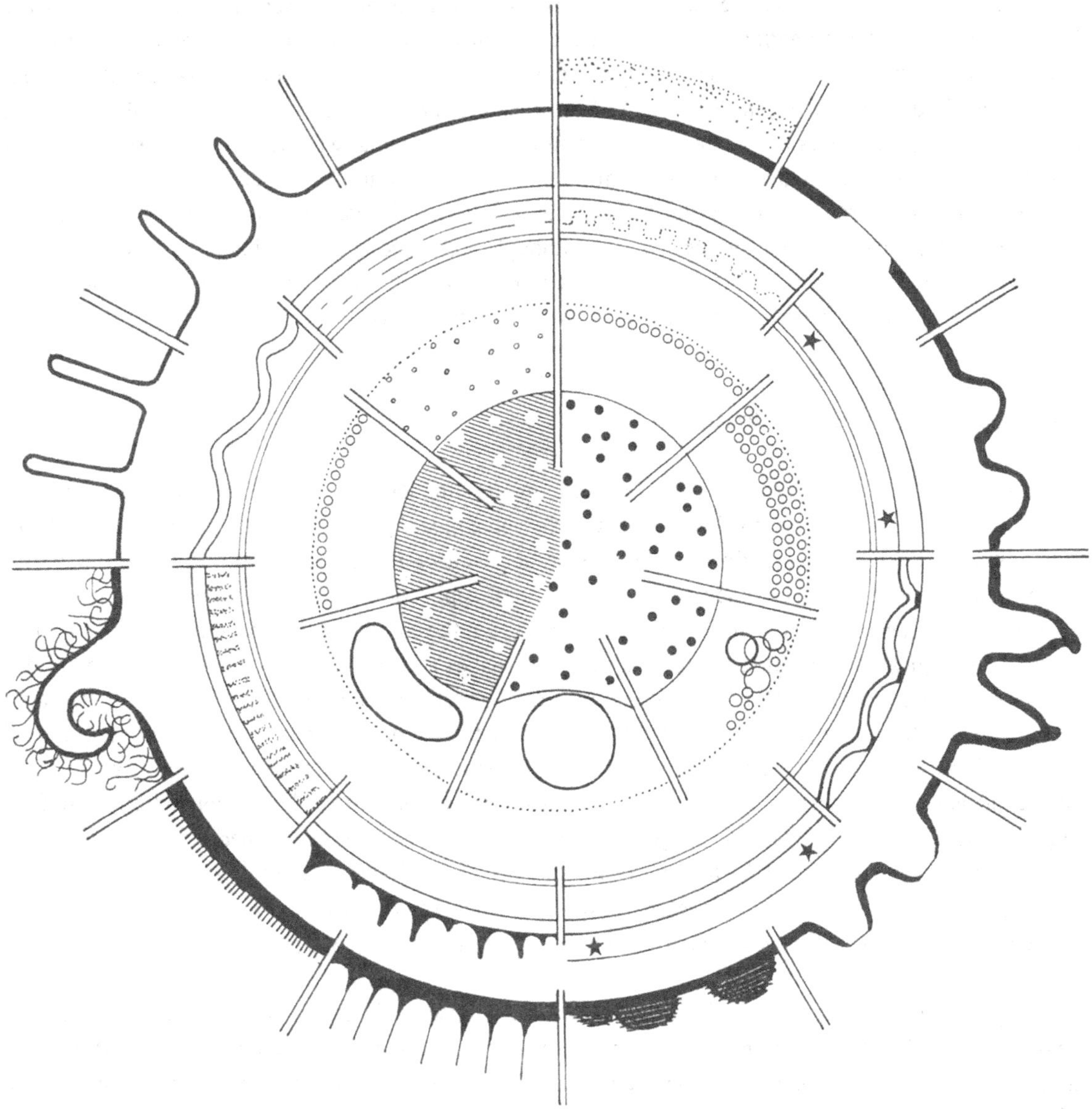

(Figure I: 12, legend, continued)

Middle wheel (oospore wall structure): outer epispore wall, thick, zoned endospore wall, inner boundary membrane; as before, but with a collapsed endospore wall giving rise to ridges/reticulations (*Pythium*, *Cystosiphon*); endospore formed of radial rods (*Medusoides*); as *Pythium*, but with a condensed outer exospore layer derived from oogonial periplasm (*Albugo*, *Peronospora*); female gamete membrane outside the epispore and endospore walls (*Apodachlyella*); reticulately ornamented inner wall layers between the smooth, outermost membrane and the smooth membrane bounding the protoplast (*Myzocytiopsis*); endospore with a fluid layer between the epispore and the boundary membrane (*Leptolegniella*); smooth epispore and endospore wall layers with some differentiation within the latter.

Innermost wheel (protoplasmic organization): solid ooplast with translucent zones, cytoplasm with little coalescence of lipids (Peronosporomycetidae); solid ooplast with translucent zones, cytoplasm with coalescence of lipids to form an outer shell (Leptolegniaceae); solid ooplast with translucent zones, cytoplasm with coalescence of lipids to form a single globule deformed by the ooplast (Leptolegniaceae); fluid ooplast with granules in Brownian motion, cytoplasm with coalescence of lipids to form a single globule deforming the ooplast (*Achlya*); fluid ooplast with granules in Brownian motion, cytoplasm with coalescence of lipids to form a single shell of uniformly sized globules (Saprolegniaceae); as previously, but with multiple shells of lipid globules (*Saprolegnia*); as previously, but with coalescence of lipid globules into globules of various diameters (*Scoliolegnia*).

The oogonial wall may be composed of up to three major layers, some of which can themselves be multilayered. The deposition of an internal layer or layers results in the thick oogonial walls characteristic of the *Saprolegniales*. The pits are not predetermined sites of antheridial attachment, but they are places through which the oospore germinates, either as a hypha or a zoosporangium. Some orders, such as the *Pythiales*, are normally characterized by having thin oogonial walls which can be contrasted with the thick oogonial walls of the *Saprolegniales*. Apart from the 'primary' and 'secondary' walls, there may be additional layers, sometimes as mucilage secretion through the oogonial wall (*Aplanopsis terrestris*). The mucilage may have an ecological function, cementing the oogonium to organic detritus. The size ranges of oogonia and oospores can be enormous: there is a 300-fold difference in oospore volume between *Calyptralegnia achlyoides* and *Pythium parvum* although both species have uniovulate oogonia and the oospores are presumably uninucleate. Oogonial morphology is thus useful for species identification, and some patterns of morphogenesis (e.g., pits) may have phylogenetic value.

The antheridium and the fertilization process

The morphology and morphogenesis of the antheridium are simpler than those of the oogonium, and the antheridial wall is seldom much thicker than that of the subtending hypha. A septum normally delimits the antheridium. When the antheridium is of regular occurrence, the shape, origin and position of the antheridium can be useful taxonomic criteria. The mode of application of the antheridium to the oogonium is distinctive for species and sometimes groups of species, but the continuous interspecific cline of variation between fully apical and completely lateral attachments makes accurate illustration essential. Antheridial origin can either be characteristic for a species, or highly variable within a species. Antheridia may arise from just below the oogonium, from the hypha subtending the oogonial branch or from a different hyphal system. These states are termed **closely monoclinous, monoclinous** and **diclinous** respectively. In a few species the antheridium may be delimited from a branch arising from the swollen oogonial initial itself (an **epigynous** antheridium). Where the oogonial initial grows ***through*** the ***preformed*** antheridial initial, a characteristic of some species of *Phytophthora*, the antheridial development is **amphigynous**. It is possible that the amphigynous condition of *Phytophthora capsici* (Stephenson & Erwin, 1972) may not be homologous with that for *Phytophthora megasperma* (Hemmes & Bartnicki-Garcia, 1975; Hemmes & Ribeiro, 1977).

The morphogenesis of the process of fertilization shows differences with respect to penetration and protoplast fusion (Dick 1995). When an antheridium is present, and fertilization occurs through **gametangial contact,** there is the injection of a small part of the antheridial protoplasm into the oogonium through a **fertilization tube**. This tube can be recognized as a walled structure within the oogonium ***after*** fertilization. In the polyoosporous *Saprolegniomycetidae* the fertilization tube may be branched, although there is no evidence that the number of branches ever equates with the number of apparently mature oospores, it is usually much less. In all ultrastructural studies of oogonial wall penetration by a fertilization tube there is a characteristic concentration of small vesicles on both sides of the penetration zone. Marchant (1968) claimed that there was no fertilization tube in *Pythium ultimum* Trow, and Win-Tin & Dick (1975) figured apparently naked antheridial protoplasm within the oogonium of *Achlya racemosa* Hildebr., but in both species walled fertilization tubes can be seen in mature oogonia. Laviola (1974*a*) has described the fertilization tube of *Phytophthora infestans*. It is significant that in the *Peronosporomycetes* the presence of this fertilization tube can be deduced even in strictly plerotic oospore/oogonium arrangements because of a deformation of the oospore wall (e.g., *Verrucalvus*, Dick *et al.*, 1984). In nearly plerotic arrangements, as in several *Pythium* species, the presence of the fertilization tube tends to push the oosphere to the opposite side of the oogonium (e.g., *P. lutarium* Ali-Shtayeh, Ali-Shtayeh & Dick, 1985: figs 41-45). In *Apodachlyella* and *Eurychasmopsis* a *fertilization hypha* develops when each cell of a chain of endogenous cells within the antheridium 'germinates' to cross the void between non-contiguous gametangia.

In *Lagenidium sensu stricto* and *Myzocytium sensu stricto* the thallus becomes septate and adjacent segments assume the function of gametangia. Thus, although the thallus is eventually holocarpic, the relationship with the eucarpic *Peronosporomycetes* is readily made. The range of morphology in the *Leptolegniellaceae* is greater, from the eucarpic *Aphanodictyon*, which is homothallic with differentiated gametangia, and *Brevilegniella*, in which swollen segments function as automictic gametangia, to holocarpic genera such as *Leptolegniella* and *Nematophthora*, in which the whole thallus functions as an automictic gametangium containing multiple oospores. In none of these taxa has the cytology been investigated, but it is presumed that meiosis occurs in the thallus/gametangium.

In *Lagenidium giganteum* the morphology of sexual reproduction has been illustrated by Willoughby (1971), Umphlett (1973), Kerwin & Washino (1983) and Brey (1985). These authors describe an extra-oogonial 'fertilization tube' which is not merely a fine tubular extension from an antheridial cell as in *Apodachlyella* or *Eurychasmopsis*, rather it resembles the structure originally figured for *Lagena*, and which Barr & Désaulniers (1990*a*) dismissed as an artifact. Although it would be possible to postulate that the antheridium is equivalent to a single antheridial cell of the kind found in *Apodachlyella* or *Eurychasmopsis*, the absence of an equivalent oosphere wall, such as occurs in these two genera, makes this unlikely. Also, given the many biochemical and morphological similarities between this particular species of *Lagenidium* and *Pythium*, I think it inconceivable that such a morphologically peculiar interpretation can be maintained. It is much more likely that the 'fertilization tube' is in fact the antheridium, which would then be epigynous (cf. *Achlya colorata* Pringsh.). A true fertilization tube may exist within the oogonium. The problem will only be resolved by a nuclear cytological investigation to reveal the site of meiosis. The proximity of the oospore to the antheridium may be due to the irregular shape of the oogonium which has its maximum diameter close to the point of antheridial contact.

Thalloid meiosis is also presumed in *Lagena*, *Ciliatomyces* and *Eurychasmopsis*; in *Lagena* no evidence for meiosis has been presented, but the organism is presumably automictic, given the significant protoplasmic reorganization. *Ciliatomyces* and *Eurychasmopsis* are heterothallic although it is not possible to decide whether heterothallism is genotypically or environmentally determined in these hologametangial fungi. Foissner (1987: fig. 9) figured a fertilization pore for *Ciliatomyces*, but in this species oosphere formation is periplasmic.

Thalloid meiosis has been demonstrated in *Lagenisma* (Schnepf *et al.*, 1978*b*, Schnepf & Heinzmann, 1980). Sexual reproduction (meiosis) appears to be induced in stressed environments (Schnepf & Drebes, 1977; Schnepf *et al.*, 1978*b*, *e*). The subsequent development, taken as a whole, is unique. However, it is possible to put parts of this development into a comparative sequence. Unlike any other peronosporomycete (with the possible exception of *Apodachlyella* and *Eurychasmopsis*, see previous page) the meiotic products of *Lagenisma* become discrete uninucleate cells. These flagellate meiospores are discharged like the zoospores. Aggregations of encysted meiospores occur and a functional 'male' forms a fertilization hypha which grows towards and then penetrates a functional 'female'. This growth could be compared with the fertilization hyphae of *Apodachlyella* and *Eurychasmopsis*. Plasmogamy results in the growth of a short zygotial hypha from the 'female' cyst. The binucleate protoplast then contracts to form a non-periplasmic zygote which secretes a wall. Karyogamy occurs in the developing zygospore. There is no evidence of haploid mitosis, so the ploidy cycle is still haplomitotic B. The morphological sequence opens up the possibility of deriving a diplomitotic cycle (cf. the *Blastocladiales*, which are freshwater not marine) from the haplomitotic B ploidy cycle.

In the *Olpidiopsidales* the donor gametangium is transformed from a contiguous smaller thallus, the **companion cell**. This smaller thallus (invariably of less than 10% of the volume of the receptor gametangium) is presumed to be derived from an independent infection, thus making the sexual union heterothallic. For *Olpidiopsis varians*, Martin & Miller (1986*c*) have figured a penetration tube from the male gametangium that differs little from the fertilization tube of a strictly plerotic oomycete, although their other figures relate to a narrow pore *ca* 0·4 μm in diameter.

In the *Myzocytiopsidales* the two gametangial initials are from contiguous thallus segments (homothallic) and of approximately the same size. The two gametic protoplasts condense towards a common pore in the contiguous walls of the gametangia and **gametangial copulation** occurs. This system should be contrasted with the above account of gametangial contact. The condensation is dependent on vesicle expansion distal to the point of union in both gametangia. The morphogenetic distinction between a fertilization tube which forces the oosphere away from the point of union and a conjugation pore, towards which both gametangial protoplasts converge, appear to be diametrically opposite systems. Although Zopf (1884) in his figures for *Lagenidium rabenhorstii* and *Myzocytium proliferum* Schenk, depicts a typical peronosporomycetous fertilization tube and describes the oogonial contents as aggregating at the tip of this tube, his account for *Myzocytiopsis vermicola* (Zopf) M. W. Dick (≡ *Myzocytium vermicola* Zopf, Dick, 1997*a*) is different. For *M. vermicola* the fertilization tube is shown as trumpet-shaped, dilating on to the sphere of the maturing resting spore: it is thus likely that the structure is cytoplasmic rather than extra-cytoplasmic. Newell *et al.* (1977) have provided a more detailed description of a marine parasite which they equate with this species. For their isolate they describe a septal papilla (deformed pore ?) which is not attached to the oosphere, but there is a tubular remnant cytoplasmic membrane which passes from the ***distal*** septum of the donor gametangium through the papilla to the developing zygote. Dangeard (1906), also working with *M. vermicola*, described the copulation as occurring between different thalli. All the protoplasm of both gametangia is involved (Barron & Percy, 1975). The aggregation of the protoplasm of both protoplasts against the common wall has been figured by many authors and is particularly well illustrated by Barron (1985: fig 12). Light micrographs and TEMs of the pore in the common wall for *Myzocytiopsis* have been published in Dick (1995). The question remains as to whether copulation involves wall dissolution or gametangial penetration. Some drawings (e.g., Karling 1981*a*: plate 25, fig. 61; plate 27, fig. 93) indicate that the pore is of sufficient width that wall dissolution is the more probable alternative. Gametangial copulation accompanied by aggregation and contraction of the combined protoplast in one of the gametangia is followed by the formation of a resting spore wall around that protoplast. Given the proximity of the gametangial protoplasts, it is a small step to the formation of that spore in the pore of the common gametangial wall. Such a position is described for *Myzocytiopsis subuliformis*, *Syzygangia sacculoides* (Karling, 1942*a*, 1981*a*; Dick 1997*b*) and possibly for *Syzygangia oedogonii* (Scherffel, 1925*a*; Dick, 1997*b*).

Drawings (Karling, 1942*a*, 1981*a*) of the morphogenesis and structure of the mature resting spore indicate that there is a contraction of the protoplasts away from the distal gametangial walls and descriptions sometimes indicate that there is no periplasm "as in *Pythium*". *Pythium* has minimal periplasm and the real distinction may concern the outline of the contracting protoplast. In *Pythium* the developing oosphere is always more or less spherical with a smooth plasmamembrane whereas the contracting protoplast of these lagenidiaceous fungi is more often figured as irregular, amoeboid, or with point attachments to the gametangial wall (e.g., *Myzocytiopsis humana* - Karling, 1947*b*; *M. microspora* (Karling) M. W. Dick (≡ *Lagenidium microsporum* Karling, Dick, 1997*b*) - Karling, 1944*c*; *Gonimochaete lignicola* G. L. Barron - Barron, 1985: fig. 12). This morphogenesis is confirmed by observations of living material (S. L. Glockling; M. W. Dick: Glockling, 1994): during mutual gametangial contraction one protoplast possesses a convex boundary to the intra-wall space while the other protoplast has a concave boundary to this space (which sometimes contains some large vesicular spheres). Ultrastructural studies (Glockling, 1994) have confirmed these observations but add a further complication. The 'pore' appears to be a deformed septum (or an incipient penetration tube?) formed from ***both*** protoplasts, protruding ***into*** the concave protoplast, which might have been presumed to be the 'antheridial' or donor protoplast because of the concavity. Furthermore, this 'tube' differs from that of *Olpidiopsis* in the involvement of both septal origins (cf. Martin

& Miller, 1986*c*) and in the apparent absence of the zone of small vesicles at the tip of the fertilization tube (Hemmes & Bartnicki-Garcia, 1975; Beakes & Gay, 1977; Al-Rekabi, 1979).

A further anomalous description of sexual development is provided by Ivimey Cook (1935) for *Syzygangia zygnematicola*, in which a walled oosphere is formed attached to the oogonium wall by a stalk. The stalk appears to be tubular (Ivimey Cook, 1935: plate 1, fig. 6) and although antheridial penetration is described as separate from this tube (Ivimey Cook, 1935: plate 1, figs 6, 8); there could have been confusion between a normal fertilization tube and supernumary antheridial cells.

Oosporogenesis: protoplasmic reorganization and oospore wall structure

The morphogenesis of the oosphere into the mature oospore is of major phylogenetic significance in the *Peronosporomycetes* in two ways:

1. the protoplasmic organization subsequent to meiosis follows one of two alternatives: **centripetal oosporogenesis** or **centrifugal oosporogenesis**
2. the extent to which there is exclusion of part of the nucleate or enucleate oogonioplasm from the oosphere(s) as **periplasm**

The kind of oosporogenesis is diagnostic at sub-class level. Cleavage in the polyoosporous taxa of the *Saprolegniales* is essentially centrifugal and similar to zoosporangial cleavage (Lohwag, 1926), with a tonoplast vacuole fusing with cleavage cisternae to form peripheral mounds of presumptive oosphere protoplasts. When the oogonial plasmamembrane is finally breached, the oospheres tumble to the centre of the oogonium. A similar development appears to take place in *Sclerospora* (McDonough, 1974*a*, *b*) although the single oosphere makes the interpretation more difficult. In the *Pythiales* the oogonioplasm does not contain a tonoplast vacuole, and the peripheral oogonioplasm gradually loses its organelles to define a centripetal oosphere. In the *Peronosporales* and *Rhipidiales* this peripheral protoplasm (periplasm) may be substantial, persistent and sometimes initially nucleate, eventually contributing the exospore layer (not a wall), often with distinctive reticulations. When the entire oogonial cavity is occupied by the oospore, the oospore is **plerotic**: very few taxa are truly plerotic: examples are confined to the *Saprolegniomycetidae*, *Olpidiopsidales* and perhaps the *Pythiogetonaceae* (Voglmayr *et al.*, 1999). In *Pythium* the concept of an **aplerotic index** has been proposed to aid taxonomic assessment of species differences (Dick, 1990b; Shahzad *et al.*, 1992).

The presence of nucleated periplasm has been demonstrated for *Ciliatomyces* (Foissner, 1987: fig. 5; Foissner & Foissner, 1986*a*) and *Lagena* (Barr & Désaulniers, 1990*a*). The full significance of this feature will not become apparent until the morphogenesis of the oosphere of the *Peronosporales* (nuclear persistence and the formation of the exospore?) and *Araiospora* (*Rhipidiales* - a possibly ***cellular*** periplasm?) has been reinvestigated.

Within the oospore, the protoplasmic contents become reorganized during maturation (Beakes, 1980*c*). Apart from the deposition, outside the plasmamembrane but inside the zygotic wall, of glucan polymers as the **endospore** (which is usually as concentric layers, Beakes, 1981; but rarely as radial rods, Voglmayr, *et al.*, 1999), there are two complementary processes of coalescence which proceed simultaneously:

1. the DBVs gradually coalesce to form a large single membrane-bound structure, the **ooplast** (Howard & Moore, 1970), which can be seen clearly in the oospores of almost all Peronosporomycetes
2. there may be a variable degree of coalescence of **lipid globules**

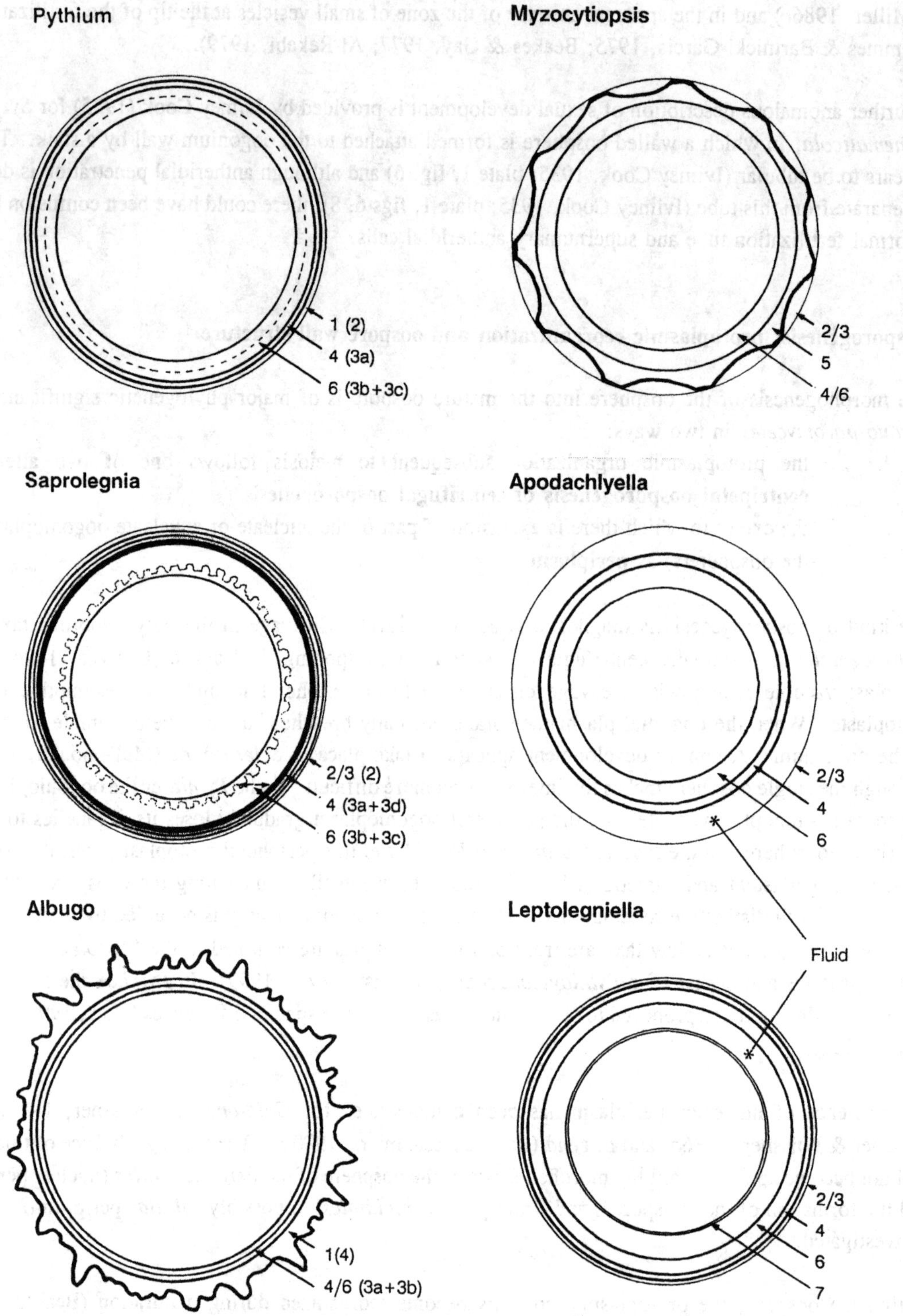

FIGURE I: 13. Oospore wall diversity: homologous and non-homologous wall layers as seen in light and TEM microscopy (oogonial wall not indicated). Hypothetical sequence of wall layers (arabic numerals) from the oogonial cavity to the oospore protoplast (after Beakes, 1981*b*; Dick, 1995): (1) periplasmically derived exospore deposit and ornamentation; (2) oosphere membrane (if present); (3) zygote membrane (continuous with (4)?); (4) smooth concentric epispore layer(s) (stratified ?); (5) reticulate epispore deposition with or without outer or inner smooth concentric layers; (6) endospore (resorbed on germination); (7) tertiary boundary membrane. Provisional identification of wall layers in different genera, Beakes' numbering in brackets. The asterisk marks a non-membranous discontinuity.

Both coalescence patterns contribute to the characteristic bubbly appearence of the partially mature oospore. One or more transparent ellipsoid zones in the mature oospore mark the position(s) of nuclei (**nuclear spots** or **pellucid spots**). At the most extreme state of coalescence, seen in some species of *Apodachlya* and *Achlya*, the mature oospore contains a single ooplast and a single lipid globule. The patterns of lipid coalescence and their distribution underneath the plasmamembrane (states described as **centric**; **subcentric**; **subeccentric**; and **eccentric**) have taxonomic value and possibly some ecological or phylogenetic significance.

Under both light and electron microscopy the ooplast has different characteristics within the class (Dick, 1971*a*, 1990*a*; Dick *et al.*, 1989; Howard, 1971). In the *Saprolegniaceae sensu stricto* the ooplast is fluid, with Brownian movement of granules (DBVs?), but in the *Leptolegniaceae* the granules are not in motion and are less numerous. In the *Pythiales* and *Leptomitales* the ooplast appears to be homogeneous under light microscopy. The *Pythiales* possess a distinctive ooplast (Dick, 1969; Howard & Moore, 1970; Dick *et al.*, 1984; Hemmes & Stasz, 1984) which is electron-opaque with a dispersed electron-lucent phase. The difference between a solid or fluid ooplast can be deduced from surface tension effects: in *Apodachlya* the ooplast bulges into the lipid globule; in *Achlya* the lipid globule causes deformation of the ooplast (Dick, 1969). *Ciliatomyces* (Foissner, 1987) again appears to be anomalous with respect to the discrete electron opaque cores. It is often difficult to interpret the oospore structure from the illustrations and light microscope descriptions of lagenidiaceous taxa; clear photographs, such as that for *Myzocytium megastomum* (Canter-Lund & Lund, 1995) or *Myzocytiopsis intermedia* (G. L. Barron) M. W. Dick (Dick, 1997*b*; Glockling, 1994), are rare. The information is usually limited to statements concerning the presence of vaguely defined fat globules. In a few cases (e.g., *Salilagenidium callinectes*, *S. chthamalophilum* (T. W. Johnson) M. W. Dick *loc. cit.* (≡ *Lagenidium chthamalophilum* T. W. Johnson) the descriptions are more detailed, and serve to emphasise a divergence from the typical morphology of the *Pythiales*, but with closer similarities to the *Saprolegniales*, rather than to other lagenidiaceous fungi. The ooplast was wrongly identified as the nucleus in *Pythium ultimum* by Marchant (1968).

The oospore wall itself is complex, possibly up to seven or more layers can be distinguished, but not all organisms have all layers (Dick, 1999, 2000*d*). It is not clear whether any exosporial deposit is universally present on aplerotic oospores, but it will be absent from ***strictly*** plerotic oospores. In the *Peronosporomycetes* as a whole, the wall layers of the resting spores are concentric and smooth. The wall layers as observed by light microscopy were redefined by Dick (1969) as:

1 **endospore**, which is resorbed on germination
2 **epispore**, representing the zygote wall (and oosphere wall if one is present)
3 **exospore**, which is deposited by periplasmic activity

Beakes (1981*b*) based his numbering system on ultrastructural and physiological criteria, while the numbers allocated to wall layers by Dick (1995) were topographic and morphogenetic. Oogonial wall layers were included by Beakes (1981*b*) as 1*a* and 1*b* (the ex-oogonial spines of *Olpidiopsis* were not considered), but I prefer to exclude oogonial wall layers from this account of the oospore walls (Figure I: 12). The sequence from the outside to the inside of the oospore, with Beakes' numbering in parentheses, is as follows:

1 periplasm-derived exospore (4*c*, 4*b*, 4*a* - numbered centrifugally)
2 primary oosphere wall (2 - not always present)
3 primary zygote wall; may not be ultrastructurally distinct from 2 (3*a* - 3*a*, 3*d*, 3*b*, 3*c* numbered centripetally (except for 3*d*)
4 epispore wall; not resorbed (3*d*?)
5 endospore/epispore wall; outer, collapsed layer, not resorbed (3*b*?, + 3*c*?)
6 endospore wall; resorbed (3*b*? + 3*c*?)
7 tertiary boundary membrane; resorbed? (3*c*?)

The ontogeny of these complex wall layers is not well understood (Beakes & Gay, 1978*a*, *b*: *Saprolegnia*; Haskins *et al.*, 1976: *Pythium*; Fukutomi *et al.* (1971) and McKeen (1975): *Phytophthora*; Lippman, Erwin & Bartnicki-Garcia (1974); Sargent *et al.* (1977): *Bremia*; Tewari & Skoropad, 1977: *Albugo*; and reviews by Beakes, 1981*b*; Dick, 1995), but seems to be evolutionally conserved and therefore of taxonomic value. It cannot be assumed that the component monomers or the degrees of hydration are the same between species, so numbering systems can only provide an approximate comparisons. Nevertheless, the degree(s) of condensation in the endospore and the ooplast are consistent for a species and will have a biological significance in maintaining the biodiversity in the community.

Foissner (1987) described the formation of an exospore in *Ciliatomyces* which was not periplasmic in origin and therefore not an exospore within the definition given above. Exceptional formations occur in other genera and these need to be studied in much more detail; the species concerned may need to be reclassified (note *Leptolegnia eccentrica* Coker & V. D. Matthews and *Cystosiphon dictyosporum* (Racib.) M. W. Dick *loc. cit.* (see pp. 189-190; PART V and PART VI). In the *Leptolegniellaceae* the endospore is also distinctive (Dick 1971*b*), although it is doubtful whether *Aphanomycopsis* (punctate?) has precisely the same structure as *Leptolegniella*, *Brevilegniella* and *Nematophthora*. There appear to be similarities between the wall ornamentation of the resting spores of *Aphanomycopsis* and *Latrostium* (Zopf, 1894), placed in the *Hyphochytriomycetes* by Karling (1977) and Fuller (1990).

Certain species of lagenidiaceous fungi have resting spores with tubercular or spiny ornamentation. Such ornamentation is only known elsewhere in the *Peronosporomycetes* in taxa whose taxonomic position is not well established. Ultrastructural work (Glockling & Beakes, 2000*b*) on *Myzocytiopsis* indicates that the prominently ridged oospore may develop within a thin endothallial membrane (layer 2 or 3?). Drawings and photographs of ornamented oospores of *Myzocytiopsis* (*Myzocytium*) (e.g., Barron, 1976*b*: figs 6b, 10, 11) would appear to support this morphology. Homology between the gametic (?) membranes of *Apodachlyella*, *Eurychasmopsis*, *Leptolegniella*, *Brevilegniella* and *Myzocytiopsis* is thus a possibility. Exo-oospore ornamentation (exospore wall), such as occurs in the *Peronosporales*, has not been suggested for *Lagenidium* or *Myzocytium*. Exo-oogonial ornamentation in *Olpidiopsis* (Martin & Miller, 1986*c*) and resting spore ornamentation in *Rozella* and *Rozellopsis* have not been examined in detail, and exo-oogonial ornamentation has no equivalent in the endobiotic lagenidiaceous fungi, although the *Aqualinderella* oogonium has a fibrillose outermost wall layer (Emerson & Weston, 1967).

Oospore germination

The oospores of many species of Peronosporomycetes are notoriously difficult to germinate, but success has been achieved for some species (see Dick, 1995; Lucas *et al.*, 1995). The method of germination may be either by zoospores, the oospore behaving as a zoosporangium, or by a germ tube. Constitutive dormancy may relate to wall composition or nuclear behaviour, such as delayed karyogamy. Germination may depend on the elution of various substances or layers of the oogonium and oospore wall, and light effects have also been implicated (for references see Dick 1995).

The germination of *Plasmopara viticola* and *Phytophthora* oospores have been studied most thoroughly from Gregory (1916) to Burruano, Ciofalo & Conigliaro (1995), Burruano, Conigliaro & Ciofalo (1997), Burruano, Conigliaro & Laviola (1999) and Burruano *et al.* (1992). In the spring the oospore of *Plasmopara* germinates, in the decayed detritus of its host, to produce a single conidiosporangiophore. The conidiosporangia are dispersed and set up the initial infection on young vine leaves. The germination of *Phytophthora* oospores has been studied primarily in relation to genetic studies (Banihashemi & Mitchell, 1976; Beakes, El-Hamalawi & Erwin, 1986; Blackwell (1943); Brasier & Brasier (1978); Brasier & Sansome (1975); El-Hamalawi & Erwin, 1986; Eriksson & Laane (1982); Erwin & McCormick, 1971;

Förster, Ribeiro & Erwin, 1981, 1983; Jiang *et al.*, 1989; Jimenez & Lockwood, 1982; Kaosiri, Zentmyer & Erwin, 1980; Klisiewicz, 1970; Meyer, 1975; Ribeiro, Erwin & Zentmyer, 1975; Salvatore, Gray & Hine, 1973; Shaw, 1967; Sney, Eye & Lockwood, 1981). One of the problems in genetic studies is the often poor percentage germination achieved. Jiang & Erwin (1990) have suggested the use of tetrazolium bromide as a criterion for determining percentage viability of *Phytophthora* oospores as an adjunct to the use of morphology and plasmolysis. *Albugo* oospore germination has been clearly illustrated by Verma & Petrie (1975) and *Pythium* oospore germination has been studied by Adams (1971) and Ruben & Stanghellini (1978). *Saprolegnia* oospore germination has been studied by Beakes (1980*a*, *b*, *c*, *d*, 1981*d*; Beakes *et al.*, 1986), that for *Aphanomyces* by Scharen (1960). Much more information will be needed before oospore germination can provide diagnostic criteria.

Oospore germination has only been reported once in the lagenidiaceous fungi, for *Syzygangia zygnematicola* by Ivimey Cook (1935). The oospore was reported to germinate to produce a single zoospore, slightly larger, but otherwise similar in morphology to the sporangial zoospores (see Figure I: 1 B m, n).

Morphometry of the oogonium and oospore

Descriptions of species have nearly always included measurements of reproductive structures. The science of measurement came much later and developed more gradually. Size ranges were included in descriptions, sometimes with extreme observed limits, sometimes with central tendencies, but no consensus regarding the presentation of these data was achieved. The number of measurements made for any one criterion was arbitrary. Latterly, statistics have usually been employed to assess central tendencies, but without critical evaluation of either the most appropriate statistic or the degree of accuracy that would be most useful for taxonomic purposes. **Continuous variation** should be expressed as a **mean** with standard variation; **discontinuous variation** should be expressed as a **modal value**. Shahzad, Coe & Dick (1992) were the first to consider the presentation of morphometric data for any straminipilous fungi. They found that measurements for 20 oogonia were sufficient to provide the central tendencies for oogonial and oospore parameters in *Pythium*.

Gäumann (1918*a*) was the first to assess morphometry of the conidia of *Peronospora* and applied his method in his monograph (Gäumann, 1923). However, his experimental approach was questioned by Gustavsson (1959*b*) because the field variation was as great as that found in experimental plots. Smith (1970) made further investigations into the effects of temperature and light on conidial size and concluded that the variation due to temperature was significant, but that variation between pathogenic strains under identical environmental conditions was also significant. Doubt was therefore cast on the value of dimensions for species determinations.

Hendrix & Campbell (1974) also dismissed the value of dimensions when assessing variation in *Pythium irregulare*. After assessing each parameter one by one, they concluded that the variation recorded was not of taxonomic use. Unfortunately they did not consider the total morphometric data. When Shahzad *et al.* (1992) reviewed the morphometric data obtained for a large number of species of *Pythium* they found that the size distributions for **single** characters within and between species formed a continuum. However, when these characters were combined to form derived character indices, there were distinct discontinuities between species (see also Dick, 1990*c*, 1992, 1995). Dick (1990*c*) used three indices:

1 **aplerotic index**, the ratio between oospore volume and oogonial volume
2 **wall index**, the volume of wall material as a function of the entire oospore volume
3 **ooplast index**, the ratio of ooplast volume to the entire oospore volume

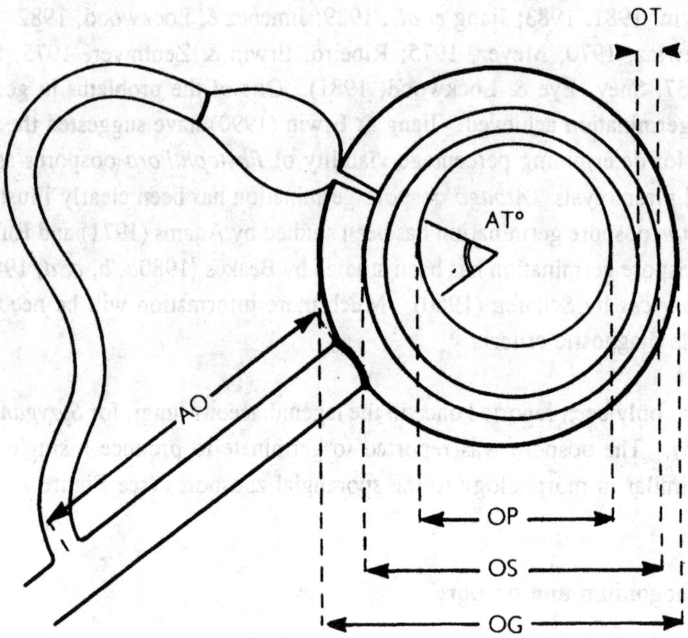

FIGURE I: 14. *Parameters for morphometry of* Pythium *oogonia and oospores.* OG *oogonial diameter;* OS *oospore diameter;* OP *ooplast diameter;* OT *oospore wall thickness;* AO *distance between antheridial origin and oogonial basal septum;* AT *angle of application of antheridial penetration. From Dick (1990c).*

FIGURE I: 15. *The use of ooplast index and oospore wall index to demonstrate species discontinuities for 39 species of* Pythium, *following canonical variate analysis. 79 isolates (1 to 3 isolates per species) were plotted as follows:* ○ hypogynum, okanoganense, lucens, middletonii, multisporum, opalinum, salpingiphorum, rostratum, vexans *(spp. with spherical sporangia);* aquatile, coloratum, pachycaule, tenue *(spp. with filamentous sporangia);* australe, parvum, pleroticum *(spp. with sporangia unknown);* ● aphanidermatum, arrhenomanes, dissimile, dissotocum, ultimum *var.* ultimum; ◆ iwayami, ultimum *var.* sporangiiferum; ◇ deliense, graminicola, torulosum; ▲ echinulatum, irregulare, mamillatum; △ nagae, paddicum; ▽ anandrum, debaryanum, heterothallicum; ★ acanthicum, hydnosporum, sinense; ☆ acanthophoron, periplocum. *The lines parallel to the axes represent 5% tolerance limits. See Shahzad* et al. *(1992) for data base, calculations of indices and further diagrams of principal component analyses.*

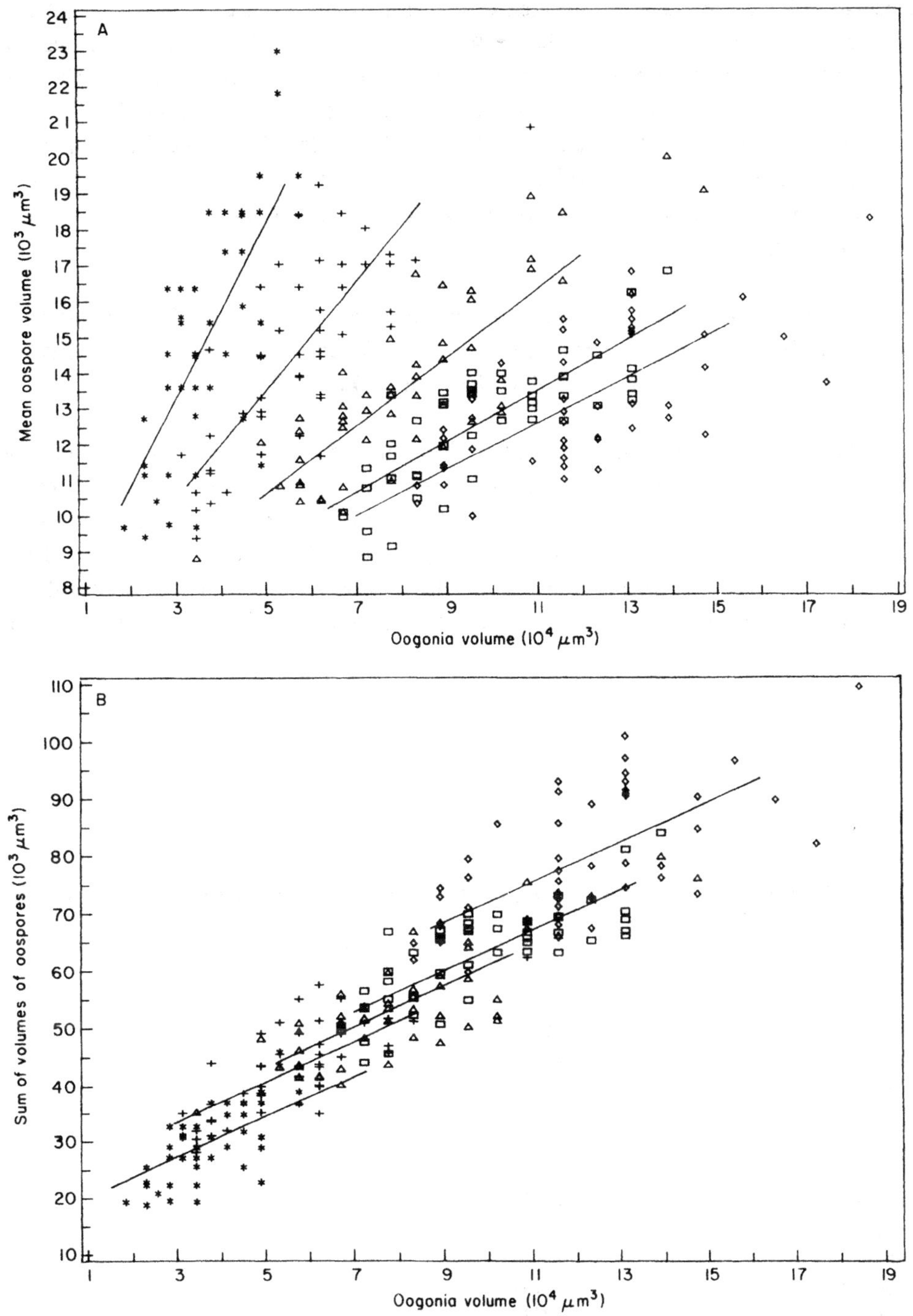

FIGURE I: 16 A,B. *Relationships between oospore size and oogonial size in pluriovulate oogonia of* Saprolegniaceae. Saprolegnia litoralis. *A. Mean oospore volume plotted against oospore volume for three modal classes of oospore number per oogonium. Computer-generated graph with common intercept lines fitted for each modal class. B. Sum oospore volume plotted against oogonial volume for the three modal classes of oospore number. Computer-generated graph with parallel lines fitted for each modal class (the slope of the parallel lines is distinctive for different species, see Table I: 3). Symbols for modal classes:* * *2 oospores per oogonium;* + *3 oospores per oogonium* △*4 oospores per oogonium;* □*5 oospores per oogonium;* ◇*6 oospores per oogonium. From Soumati & Dick (1989).*

The raw data of Shahzad *et al.* (1992) have been analysed by a completely independent statistic (statistical classification trees) which confirmed that of canonical variate analysis: the derived indices are of prime importance in determining discontinuities between species (P. A. Taylor & M. W. Dick, unpublished). With such a precise protocol for measurement it might be expected that variation would be introduced by different operators or different methods of presenting images for measurement. This has also been examined using Australian isolates of *Pythium*, although differences of means were recorded, the discontinuities beween isolates and species were unaffected (B. G. Hawke & M. W. Dick, unpublished data). The value of such indices has recently received independent confirmation from Møller & Hockenhull (2000), who used more isolates from fewer species. However, they preferred to use the linear measurements rather than calculated volumes. There will not be any difference in the relative differentials although the discontinuities will be less because the micrometer measure is raised to the third power in volumes. The purpose of using volumes is to express the actual endogenous reserves of the ooplast and endospore in oospores of one species compared with another.

Dick (1969) suggested that in the *Saprolegniaceae* and *Pythiaceae* there were 1:2:4 ratios of oospore volumes between different species of a genus. However, the relationship of such proportionality could not be based on any known factor such as polyploidy (Win-Tin & Dick, 1975).

In the multiovulate *Saprolegniaceae* Soumati & Dick (1989) investigated the morphometry of oogonia and oopores of 13 species. They showed that the number of oospores in an oogonium affects the mean oospore volume, but that the intercept slopes for each class of oospore number were the same for a species, although different species had different intercept slopes for the combined data (Figure I: 16; Table I: 3). The control of the range of oogonial size and shape is probably complex. Very careful selection of statistical treatments is required to handle the large intra-isolate variation that is concomitant with interspecific variation.

TABLE I: 3. *Ascending order of slopes of parallel lines regressions in 13 species of* Saprolegniaceae *(from Soumati & Dick, 1989)*

Species	**Slope of parallel lines regressions**
Thraustotheca clavata	0·1234
Achlya racemosa	0·1433
Achlya recurva	0·1746
Achlya hypogyna	0·1878
Achlya apiculata	0·1918
Achlya prolifera	0·1962
Isoachlya monilifera	0·2028
Achlya colorata	0·2085
Saprolegnia diclina	0·2116
Saprolegnia anisospora	0·2193
Saprolegnia furcata	0·2227
Saprolegnia terrestris	0·2796
Saprolegnia litoralis	0·3649

The higher the slope the more proportionally direct is the relationship between oogonial size and oospore size (i.e., the most proportionally direct relationship was for *S. litoralis*).

The functional significance of oospore morphology

The protoplasmic reorganization that takes place during the formation of the oospore is extensive and involves the redistribution of three kinds of reserve: the endospore (glucan polymers secreted outside the plasmamembrane), the ooplast (phosphate rich) and the lipid globules. The oospore is organized for the rapid mobilization of each of these reserves for specific aspects of growth and the proportions of each of these components is probably related to the function of the oospore in its life history. The endospore is mobilized rapidly at germination. Endospore mobilization may differ between taxa: extraplasmamembranic depolymerization may enable cell wall material to flow ***over*** the elongating germ tube plasmamembrane, while in other species pinocytotic ***resorption***, followed by secretion, may be involved. The endospore would thus be essentially potential wall material; the ooplast, membrane precursor; the lipid, an endogenous energy supply (Dick, 1995). Germ tube formation and rapid elongation will require relatively more wall glucan; zoospore production will require more plasmamembrane and DBVs for low polymer dehiscence mechanisms; zoospore production with extended periods of motility before any assimilative activity will require large lipid fuel reserves. These functional requirements are reflected in the proportionality of the reserves, which are specific rather than generic characteristics (Shahzad *et al.*, 1992).

The numerous (up to 40) oospores (*ca* 25 μm diam.) contained in single oogonia of aquatic species of *Saprolegnia* and the large populations of small (*ca* 12 μm diam.) unioosporus oogonia of some terrestrial species of *Pythium* reflect very different life-history strategies. The nutrient status of the pre-oogonial assimilative phase can affect the size of oogonia and number of oospores produced in multiovulate species. Depauperate oogonia can often be found in rough culture, leading to relatively large standard deviations about the mean but almost nothing is known about oogonial production in natural conditions (Dick, 1992).

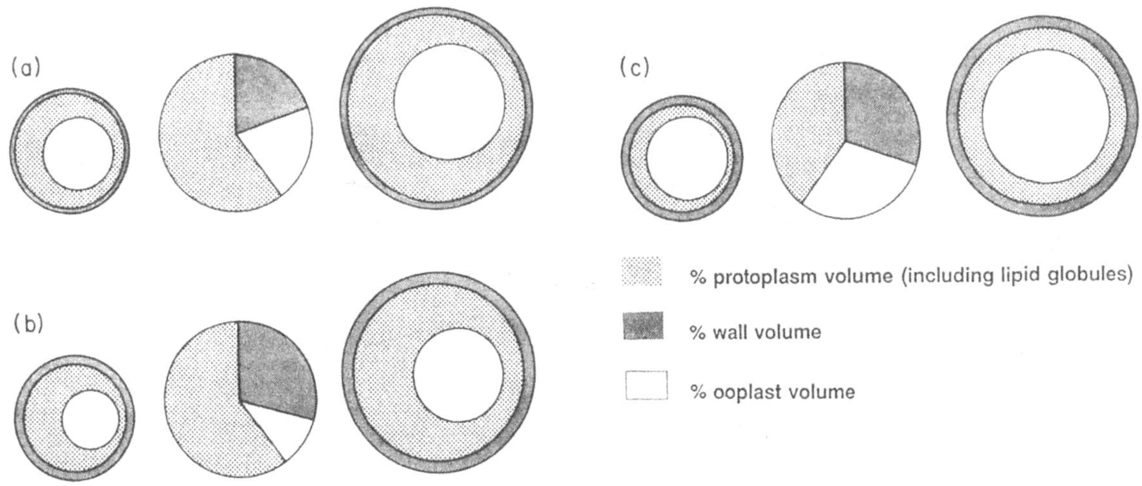

FIGURE I: 17 *Proportionality of the different oospore reserves in* Pythium. *Representative oospores with (1) different relative sizes and (2) different relative proportions of reserve materials. The large oospores have four times the volume of the small oospores. The pie-chart diagrams between the two oospore sizes provide the percentages by volume of endospore:ooplast (the lipid reserve can only be estimated separately in eccentric oospores with a single lipid droplet). Note that the protoplasm:ooplast plus endospore ratios are reciprocal between diagram series* b *and* c. *Note that the endospore wall thickness (measured thickness) is the same for the large oospore in* a *and the small oospore in* b. *See Dick (1990c), Shahzad* et al. *(1992); diagram rearranged from Dick (1995).*

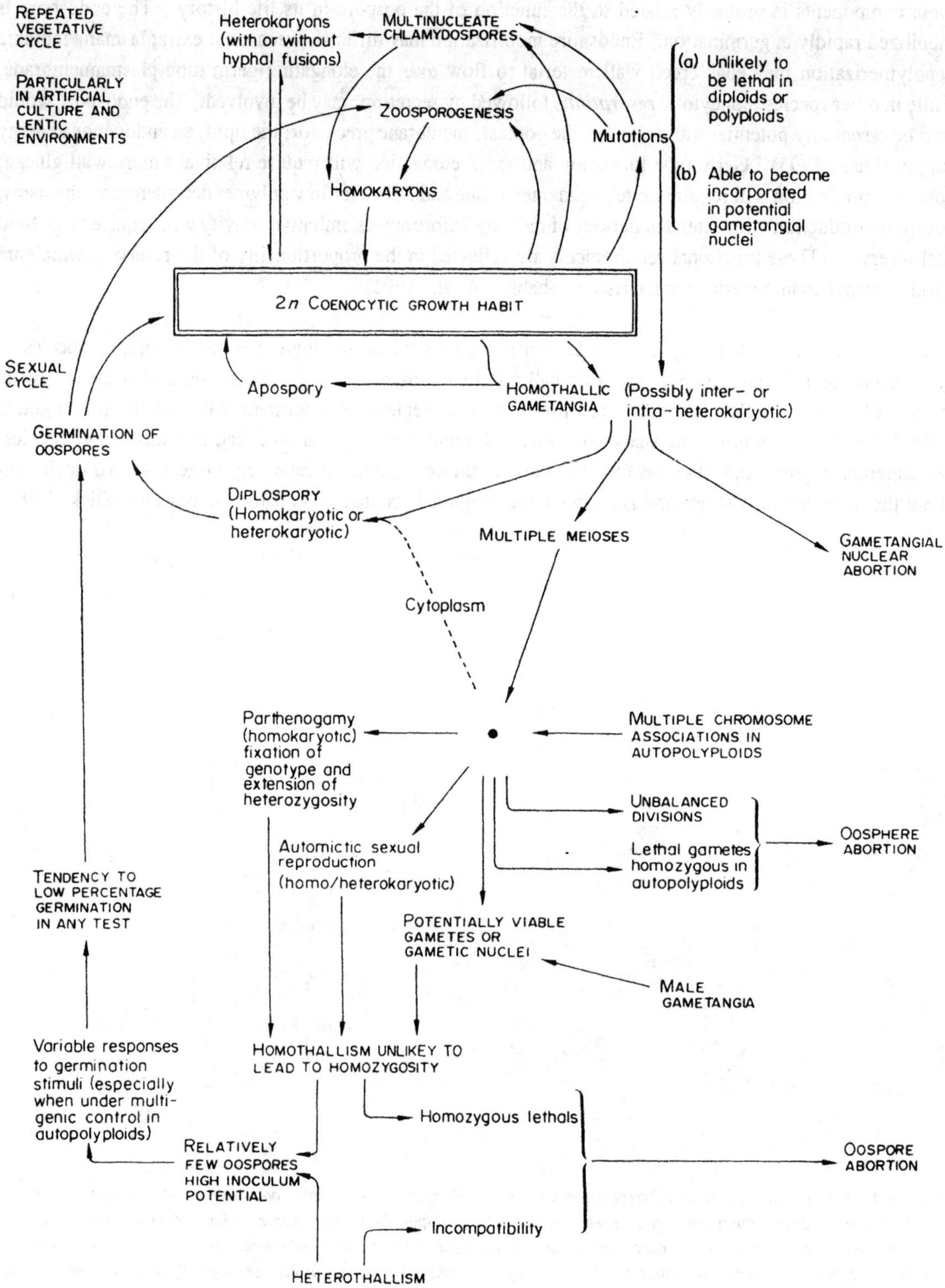

FIGURE I: 18. *Life-history of the* Peronosporomycetes *with emphasis on the nuclear developments, known events shown in upper case, unproven events in lower case. From Dick (1972: fig. 1)*

NUCLEAR CYTOLOGY: MITOSIS AND MEIOSIS

Mitosis

Early work on mitosis in the assimilative thallus was reviewed by Bakerspiegel (1960). Problems in resolving the irregularity of the spindle structure resulted in misinterpretations (binary fission) which were resolved by Heath (1974*a*, *b*, 1980*a*, *b*). Mitotic analysis is difficult and unlikely to be used as a diagnostic criterion.

Meiosis

Dick & Win-Tin (1973) reviewed the earlier accounts of nuclear behaviour in sexual reproduction of the *Peronosporomycetes* after Sansome (1961) had revived the earlier controversy regarding the site of meiosis. Despite some further controversy (Timmer *et al.*, 1970), it is now accepted that meiosis occurs in the gametangia of the *Peronosporomycetes* so that the ploidy cycle is haplomitotic B (Dick, 1987). The most recent and comprehensive overview of peronosporomycete nuclear cytology is that by Sansome (1987).

Much of the earlier work was based on paraffin wax sectioning, while more recent work has used squash techniques. The earlier work still has value because, although the interpretations of the division configurations may be obsolete, the spatial distribution and nuclear orientations will have been less disrupted than is the case after squash preparations. Table I: 3 provides a genus- and species-related summary of these earlier references from Dick & Win-Tin (1973) and Win-Tin & Dick (1975), with a few later citations.

Gametangia are meiogametangia in which meiosis occurs. Gametangia are coenocytic from the moment that the pro-gametangium is functionally predetermined; mitotic activity then ceases. All nuclei are diploid and apparently capable of undergoing meiosis. In paired gametangia the meioses are also either simultaneous, or nearly so, between the two protoplasts. The position of the meiotic nuclei in the gametangium is determined by the vacuolation of the protoplast in relation to centrifugal or centripetal oosporogenesis. Multiple concentric envelopes of ER may protect the functional nuclei (Dick, 1995). The numbers of nuclei entering the oogonial and antheridial initials are greater than the numbers of mature uninucleate haploid female gametes, which, in turn, are more numerous than the numbers of zygotes in the oogonium. A comparable reduction in the number of nuclei occurs in the antheridium (Dick, 1972). Figure I: 17, from Dick (1972) provides a chart of the probable patterns of nuclear abortion. Sterols have been implicated in relation to the patterns of nuclear abortion both prior to and after meiosis (Elliott & Sansome, 1977). Fertilization has been discussed on pp. 58-61; karyogamy may be precocious, as in *Achlya* (Win-Tin & Dick, 1965) or much delayed, as in *Phytophthora* (Jiang *et al.*, 1989).

All species of the *Peronosporomycetes* have a closed cruciform meiosis in which the nuclear membrane remains intact until the second telophase and the metaphase I and metaphase II spindle poles are in the same plane (Howard & Moore, 1970; Win-Tin & Dick, 1975). However, fenestration of the nuclear membrane, especially in the region of the poles, is a feature of *Olpidiopsis* (Bortnick *et al.*, 1985). Very few species have been examined by transmission electron microscopy for details of synaptonemal complexes (SCs). In *Achlya* the SCs are wider than in most other organisms, they tend to lack the central element, and fewer recombination nodules than would have been expected from the number of chromosomes present were found (Williamson & Yitzchak, 1991).

TABLE I: 4. *Cytology references from Dick & Win-Tin (1973) listed by species and augmented by later references*

Taxon	References
Peronosporomycetidae	
Peronosporales	
Albugo bliti	Stevens (1899, 1901?)
Albugo candida	Davis (1900); Krüger (1910); Wager (1896*a*, *b*)
Albugo evolvuli	Thirumalachar, Whitehead & Boyle (1949)
Albugo ipomoeae-panduranae	Stevens (1904)
Albugo lepigoni	Ruhland (1902, 1904)
Albugo mysorensis	Safeeulla & Thirumalachar (1956)
Albugo platensis	Damle (1943)
Peronospora ficariae	Krüger (1910)
Peronospora parasitica	Wager (1889, 1900); Sansome & Sansome (1974)
Peronospora spp.	Tsang (1929)
Plasmopara alpina	Rosenburg (1903)
Plasmopara halstedii	Nishimura (1926)
Plasmopara viticola	Arens (1929); Bosc (1946); Berlese (1898)
Pythiales	
Phytophthora cactorum	Blackwell (1943); Buddenhagen (1958); Erwin & McCormick (1971 - germination of oospores); Sansome (1965, 1966); Sansome & Harris (1962); Shaw & Elliott (1968)
Phytophthora cambivora	Allain (1934)
Phytophthora capsici	Timmer *et al.* (1970); Sansome (1976); Stephenson, Erwin & Leary (1974*a*, *b*)
Phytophthora cinnamomi	Brasier & Sansome (1975); Pratt *et al.* (1972)
Phytophthora drechsleri	Brasier & Sansome (1975); Sansome (1970)
Phytophthora erythroseptica	Murphy (1914, 1918); Sansome (1966)
Phytophthora fagi	Hartig (1880)
Phytophthora heveae	Bennett (1972)
Phytophthora himalayensis	Mundkur (1949)
Phytophthora infestans	Brasier & Sansome (1975); Gallegly (1968); Laviola (1969, 1971); Marks (1965); Sansome (1977); Sansome & Brasier (1973)
Phytophthora megasperma	Sansome & Brasier (1974)
Phytophthora megakarya	Brasier & Griffin (1979)
Phytophthora palmivora	Bennett (1972); Brasier & Medeiros (1978); Sansome (1977)
Phytophthora parasitica	Huguenin & Boccas (1970)
Phytophthora spp.	Elliott & McIntyre (1973); Galindo & Zentmyer (1967*a*, *b*); Gallegly (1970); Leal & Gomez-Miranda (1965); Sansome (1965, 1966); Sansome & Harris (1962); Shaw & Khaki (1971)
Pythium aphanidermatum	Seshadri & Payak (1970)
Pythium debaryanum	Miyake (1901); Sansome (1961, 1963*a*); Sansome & Harris (1962); Win-Tin & Dick (1975)
Pythium echinulatum	Win-Tin & Dick (1975)
Pythium multisporum	Win-Tin & Dick (1975)
Pythium torulosum	Patterson (1927*b*); Win-Tin & Dick (1975)
Pythium ultimum	Trow (1901); Win-Tin & Dick (1975)
Pythium spp.	Child & Haskins (1971); Marshall Ward (1883*b*); Saksena (1936)

(Table I: 4, continued)

Saprolegniomycetidae	
Saprolegniales	
Achlya ambisexualis	Raper (1940); Barksdale (1968); Ellzey (1974); Win-Tin & Dick (1975)
Achlya americana	Trow (1899)
Achlya apiculata	Win-Tin & Dick (1975)
Achlya benekei	Win-Tin & Dick (1975)
Achlya bisexualis	Raper (1936); Barksdale (1960, 1966); Mullins & Raper (1965)
Achlya caroliniana	Win-Tin & Dick (1975)
Achlya colorata	Patterson (1927*a*); Win-Tin & Dick (1975)
Achlya debaryana	Mäckel (1928)
Achlya dubia	Murdia (1938)
Achlya flagellata	Wolf (1938); Win-Tin & Dick (1975)
Achlya hypogyna	Clausz (1968); Cooper (1929*a*); Win-Tin & Dick (1975)
Achlya inflata	Win-Tin & Dick (1975)
Achlya klebsiana	Flanagan (1970)
Achlya polyandra	Mücke (1908)
Achlya prolifera	Pringsheim (1851); Mäckel (1928)
Achlya racemosa	Bakerspigel (1960); Carlson (1929); Win-Tin & Dick (1975)
Achlya radiosa	Win-Tin & Dick (1975)
Achlya recurva sensu Latham	Win-Tin & Dick (1975)
Achlya sparrowii	Win-Tin & Dick (1975)
Achlya treleasiana	Win-Tin & Dick (1975)
Achlya spp.	de Bary (1883); Sansome (1965, 1966); Sansome & Harris (1962)
Aplanopsis terrestris	Win-Tin & Dick (1975)
Brevilegnia diclina	Cooper (1929*b*)
Dictyuchus	Couch (1926)
Dictyuchus monosporus	Sherwood (1969, 1971)
Isoachlya anisospora var. *indica*	Bhargava (1946)
Pythiopsis cymosa	Win-Tin & Dick (1975)
Pythiopsis intermedia	Burton (1939)
Saprolegnia ferax	Mäckel (1928), Höhnk (1935), Bakerspigel (1960), Flanagan (1970), Win-Tin & Dick (1975)
Saprolegnia furcata	Beakes & Gay (1977); Win-Tin & Dick (1975)
Saprolegnia monoica	Reinke (1869)
Saprolegnia parasitica	Bakerspigel (1960)
Saprolegnia terrestris	Bryant & Howard (1969)
Saprolegnia spp.	Davis (1903, 1905); de Bary (1883); Hartog (1895, 1896, 1889, 1899); Marshall Ward (1883*a*); Moreau & Moreau (1935); Smith (1923); Slifkin (1967); Trow (1895, 1904, 1905); Ziegler (1953)
Thraustotheca clavata	Schrader (1937), Murdia (1938), Heath (1974*a*)
Apodachlya brachynema	Howard & Bryant (1971)
Aphanomyces laevis	Kasanowsky (1911)
Sclerosporales	
Sclerospora graminicola	McDonough (1937)
Sclerospora sorghi	Safeeulla & Thirumalachar (1955 - as *Sclerospora andropogonis-sorghi*); Sansome (1963, 1966)
Sclerophthora macrospora	McDonough (1946*a*, *b*); Tasugi (1953)
Sclerophthora cryophila	Safeeulla, Thirumalachar & Shaw (1963)
Peronosclerospora sorghi	Safeeulla & Thirumalachar (1955)
Peronosclerospora sorghi	Sansome (1963*b*)
Order *incertae sedis*	
Olpidiopsidales	
Olpidiopsis achlyae	McLarty (1941)

The prevalence of self-fertilization and automictic sexual reproduction would appear to deny an outbreeding function for sexual reproduction in this group. The lack of genetic opportunity resulting from karyogamy of gametic nuclei derived from meioses of recent-sister diploid nuclei may be offset by the development of polyploidy or polysomy and B chromosome development, but further work on genome synteny (see below) is essential. The populations will nevertheless be heterogeneous because of the non-selection of presumptive meiotic nuclei from very large diploid-mitotic populations.

A comparative study of the cytotaxonomy of the *Saprolegniaceae* and *Pythium* (Win-Tin & Dick, 1975) indicated probable base chromosome numbers of $x=3$ for *Saprolegniaceae*, $x=4$ for *Leptomitaceae* and $x=5$ for *Pythiaceae*, with polyploid series in both *Saprolegniaceae* and *Pythiaceae* (Win-Tin & Dick, 1975; Sansome, 1965). Base numbers of $x=9$ or $x=10$ are also common. Much more detailed cytology of *Phytophthora* has been carried out by Sansome, Brasier and co-workers (see Sansome, 1987, for a review of recent literature).

Electrophoretic karyotypification of chromosome-length pieces of DNA by means of Contour-clamped Homogenous Electric Field electrophoresis (**CHEF**) has been published for *Phytophthora* (Howlett, 1989; Mao & Tyler, 1991; Tooley & Carras, 1992; Judelson *et al.*, 1993), *Bremia* (Francis & Michelmore, 1993) and *Pythium* (Martin, 1995*a*, *b*). In *Phytophthora*, Judelson *et al.* (1993) found 6-8 chromosome-sized bands ranging in length from 2·9 - 10·0 Mb, which is comparable with the 10-13 chromosomes estimated by Sansome & Brasier (1974). In *Bremia*, Francis & Michelmore (1993) found a minimum of 7 chromosome-sized bands ranging in length from 3·0 - >8·0 Mb, corresponding to chromosome numbers estimated by light microscopy (Michelmore & Sansome (1982), but in addition they found variable numbers of up to 5 smaller components, identified as possible B chromosomes or large plasmids. In the wider spectrum of 18 species of *Pythium*, Martin (1993*a*) found 7-20 chromosome-sized bands ranging in length from 18·8 - 41·5 Mb, again comparable but not identical with the estimates of Win-Tin & Dick (1975). Mao & Tyler (1991) concluded that some of the bands that they obtained could represent doublets or triplets, thus parity between light microscopical methods and CHEF electrophoresis might be difficult to establish. Intraspecific differences for both the number and the lengths of the chromosome-sized bands were reported for *Pythium* (Martin, 1995*a*). The ranges of chromosome lengths could provide taxonomic information if and when the criterion is shown to be species-stable for particular genera, but this is by no means certain at the present time.

A different approach to phylogeny, using comparative DNA-based data (Feulgen Image Analysis) is that of Voglmayr & Greilhuber (1998) who have produced data suggesting that *Peronospora* and *Plasmopara* are probably not closely related (cf. Dick, 1988). Electrophoretic Karyotype (EK) polymorphisms (determined by CHEF analysis) occur in *Pythium* species (Martin, 1995*a*), but only one DM, *Bremia lactucae*, has been assessed (Francis & Michelmore, 1993) so again, evolutionary predictions would be premature. Analyses of EK whch show intra-specific heterozygosity are already known for several species of *Pythium* (F. N. Martin, pers. comm.), therefore, a large database will be necessary to establish whether chromosome size and number also show consistent inter-specific differences.

Light-microscope nuclear cytology of the plasmodiophorids has been reviewed by Karling (1968*e*), who concluded that the data indicated a base number of $x=2$; Harris, Braselton & Miller (1980) provided a later summary. The nuclei of all taxa are very small and TEM analyses of synaptonemal complexes (Harris *et al.*, 1980; Braselton, 1982, 1983, 1984, 1989*a*, *b*, 1990; Braselton & Short, 1985; Braselton & Dylewski, 1986) suggest that estimates of chromosome number from light microscopy are likely to be underestimates (Harris *et al.*, 1980; Williamson & Yitzchak, 1991).

The ultrastructure of nuclear division in the *Peronosporomycetes* was pioneered by Howard & Moore (1970) and has been reviewed by Beakes (1981*a*). Heath (1981) has provided a preliminary cluster analysis of nuclear cytological features of the *Peronosporomycetes* compared with other groups of protoctists. This analysis of mitosis was based on fourteen characters. Four of these diagnostic features are of relevance to this discussion:

1 the configuration of the centrioles
2 the persistence or reconstitution of the nuclear membrane
3 persistent telophase spindles (interzonal bridges)
4 the behaviour of the nucleolus

To these should be added two features of meiosis:

1 synaptonemal complexes
2 the arrangement of telophase II nuclei

Two patterns of centriole behaviour during mitosis have been described: in the *Peronosporomycetes* the spindle develops during polar separation of the centrioles, but in *Rhizidiomyces* and the chytrids polar migration is complete prior to spindle development. Perhaps related to this difference in synchrony, is the orientation of the two centrioles at the end of interphase. The centrioles are oriented end-to-end in the *Peronosporomycetes* (Berlin & Bowen, 1964: *Albugo*; Howard & Moore, 1970: *Saprolegnia*); the labyrinthulids (Perkins, 1970: *Labyrinthula*; Kazama, 1974*b*: *Thraustochytrium*), and the plasmodiophorids (Brazelton & Miller, 1973: *Sorosphaera*). Centrioles are approximately at right angles to each other in *Rhizidiomyces* (Fuller & Reichle, 1965), *Catenaria anguillulae* Sorokīn (Ichida & Fuller, 1968), *Blastulidium* (Manier, 1976) and some *Chytridiomycetes* (Whisler & Travland, 1973; McNitt, 1973).

In the *Peronosporomycetes* the nuclear membrane remains intact until the end of telophase in both meiosis and mitosis (Heath & Greenwood, 1968; Howard & Moore, 1970; Heath, 1974*a*; Ellzey, 1974). However, fenestration of the nuclear membrane, especially in the region of the poles, is a feature of *Olpidiopsis* (Martin & Miller, 1986*a*), *Rhizidiomyces* (Barstow, Freshour & Fuller, 1989), the labyrinthulids (Perkins & Amon, 1969), the plasmodiophorids (Braselton & Miller, 1973) and two orders of chytrids (*Spizellomycetales* (*Entophlyctis*): Powell,1975; *Chytridiales* (*Rhizophidium*): Powell, 1980) but not the *Blastocladiales* (*Catenaria anguillulae* - Ichida & Fuller, 1968). *Rhizidiomyces*, the plasmodiophorids and a few chytrids exhibit the type *iii* reconstitution behaviour (Heath, 1980*a*), whereby new nuclear envelopes are developed ***within*** the remnants of the original membrane (Garber & Aist, 1979*b*; Barstow *et al.*, 1989). In the thraustochytrid, *Ulkenia*, the nuclear membrane disappears completely at prophase (S. Raghu Kumar, 1982*b*).

Abstriction of the daughter nuclei may be achieved by nuclear membranes traversing the telophase spindle in two places rather than one, thus leaving a spindle remnant as an interzonal bridge (type *ii* reconstitution behaviour of Heath, 1980*a*). Such a pattern has been shown for *Olpidiopsis* (Martin & Miller, 1986*a*), *Thraustochytrium* (Kazama, 1975) and *Catenaria anguillulae* (Ichida & Fuller, 1968); it has also been reported for *Vaucheria* (Ott & Brown, 1972).

Pickett-Heaps (1970) has reviewed variations of nucleolar behaviour during mitosis. The nucleolus remains outside the spindle during mitosis in the oomycetes (Beakes, 1981*a*). In the plasmodiophorids the nucleolus also persists during mitosis, but has a more prominent role. The nuclei of the coenocyte undergo synchronous divisions prior to sporangial development (Karling, 1968*e*, Dylewski, 1990), and in these divisions there is an elongate nucleolus occupying the centre of the spindle. A comparable central structure is found in *Euglena* (Leedale, 1967) resulting from the fusion of numerous interphase nucleoli at prophase. In stained preparations the axile nucleolus of the plasmodiophorids gives the impression of a cruciform structure at metaphase, one axis being the nucleolus while the metaphase chromosomes form a plate perpendicular to the nucleolar axis (Miller, 1958*a*; Keskin, 1971; Braselton, Miller & Pechak, 1975; Dylewski, Braselton & Miller, 1978). The nucleolus is not persistent in meiosis and so cruciform configurations do not occur (Braselton & Miller, 1973). However, direct comparisons with *Euglena* are not possible because the arrangement of chromosomes in mitosis is different and meiosis is unknown.

Sparrow (1960, 1976) and Waterhouse (1973) suggested that, in the then present state of knowledge, the anisokont zoospore and the mitosis, with its persistent axile nucleolus at metaphase, should constitute the diagnosis for the *Plasmodiophoromycetes*. (Since the class and order are monotypic, this could be downgraded and interpreted as a family criterion.)

Synaptonemal complexes (SCs) have been repeatedly demonstrated in *Peronosporomycetes* (Howard & Moore (1970: *Saprolegnia*); Ellzey (1974) and Ellzey & Huizar (1977: *Achlya*); Hemmes & Ribiero (1977: *Phytophthora*); Beakes & Gay (1977: *Saprolegnia*); Gotelli (1979: *Sapromyces*); Traquair & McKeen (1980: *Aphanomyces*); Martin & Miller (1986*c*: *Olpidiopsis*); Williamson & Yitzchak (1991: *Achlya*). Williamson & Yitzchak (1991) regard the structure of the SC in oomycetes as "essentially unique" because they are wider and tend to lack a clearly defined central element, and although the SCs have recombination nodules, these were fewer than would have been expected from the chromosome number. The chromosome number determined by Williamson & Yitzchak (1991) was $n=15$, which is consistent with the base number proposed for the *Saprolegniaceae* by Win-Tin & Dick (1975). Martin & Miller (1986*c*) demonstrated the presence of SCs in the gametangia of *Olpidiopsis varians* but did not give an estimate of chromosome number.

Synaptonemal complexes have been more intensively studied in the plasmodiophorids (*Sorosphaera veronicae* (J. Schröt.) J. Schröt. - Harris *et al.*, 1980; *Plasmodiophora brassicae* - Braselton, 1982; *Polymyxa betae* - Braselton, 1983; *Polymyxa graminis* Ledingham - Braselton, 1984; *Plasmodiophora diplantherae* (Ferd. & Winge) Ivimey Cook - Braselton & Short, 1985; *Woronina pythii* Goldie-Sm. - Braselton & Dylewski, 1986; *Ligniera verrucosa* Maire & A. Tison - Braselton, 1989*a*; *Membranosorus heterantherae* Ostenf. & H. E. Petersen - Braselton, 1989*b*; *Tetramyxa parasitica* K. I. Goebel - Braselton, 1990; *Spongospora subterranea* - Braselton, 1992). Serial sections have revealed chromosome numbers of $n=9$ to $n=38$. Braselton (1989*a*, 1992) has noted that *Plasmodiophora brassicae* and *Woronina* have distinctly smaller nuclei and a shorter total length with less well-defined central zones to the SCs (cf. *Peronosporomycetes*, above) than the other plasmodiophorids.

For other straminipilous fungal groups, Moens & Perkins (1969) reported nine SCs in an unnamed species of *Labyrinthula*. The SCs of algal straminipiles have been much less intensively studied (see Berkaloff & Rousseau, 1979; Katsaros & Galatis, 1986; Markey & Wilce, 1976; Toth & Markey, 1973).

In the lagenidiaceous fungi, apart from *Olpidiopsis*, accounts of conventional nuclear cytology are limited to those on *Lagenisma* (Schnepf *et al.*, 1978*b*). The only comparative information relates to ultrastructural studies of *Salilagenidium callinectes* (Amerson & Bland, 1973) and *Lagenisma* (Schnepf *et al.*, 1978*b*). In the latter species the SCs appear normal, but in *S. callinectes* polycomplexes with a fine structure similar to synaptonemal complexes are described from nuclei of encysting [zoo]spores. These polycomplexes occurred in the presence of a nucleolus and the nuclear membrane appeared to be intact. Amerson & Bland (1973) did ***not*** claim that the polycomplexes established the site of meiosis. There must be some doubt as to whether these polycomplexes can be equated with synaptonemal complexes, although Beakes (1981*a*) and Williamson & Yizchak (1991) assume them to be synaptonemal complexes. Schnepf *et al.* (1978*b*) have also queried this interpretation. However, if the polycomplexes were to be considered equivalent to synaptonemal complexes, then meiosis would occur in the encysting [zoo]spore. The ploidy state of the life-history following germination, and the site of karyogamy, would need to be determined, but the life-history could hardly be considered peronosporomycetous. The lack of attachments between the polycomplexes and the nuclear membrane, and the absence of a regular series of true SCs, suggest that polycomplexes may be an unexplained chromosomal condensation which ***may not*** indicate the site of meiosis. Comparisons should be made with situations in which SCs are associated with mitotic ploidy reduction, as in endopolyploid oxymonads (Hollande & Carruette-Valentin, 1970) and radiolarians (Lécher, 1978), and in premeiotic mitoses (ascogenous hyphae - Zickler, 1973; pollen mother cells - MacQuade & Bassett, 1977).

The four telophase meiotic products can be arranged in three patterns depending on the orientations of the axes of the first and second metaphase spindles. If the spindles are all in the same orientation a linear array will be formed; if the spindles are in the same plane but with the second metaphase spindles at 90° to the first metaphase spindle a cruciform arrangement will result; if the two second metaphase spindle axes are at 90° to each other a tetrad will be formed. In the *Peronosporomycetes* the nuclear membrane persists until the end of the second telophase and the spindle axes are all in the same plane (Howard & Moore, 1970), resulting in the cruciform arrangement. This appears to be the case in *Myzocytiopsis* (Dick, 1995). In the plasmodiophorids the haploid nuclei are not bound by a common nuclear membrane and there is an interval between the two divisions so that a precise pattern is not established.

The diversity of nuclear cytology and the independence of some of the characters undoubtedly provides scope for the formulation of diagnostic criteria within the flagellate fungi. At present the number of organisms examined is insufficient to trace any evolutionary patterns.

Genome synteny

Genome synteny is the term which has been used to describe the total gene content of nuclei in relation to the chromosome number and the distribution of genes on the chromosomes. It is now viewed somewhat differently following the demonstration that while most of the genes in the genome are similar, they may be distributed differently between the chromosomes, so that, as in the grasses (Moore *et al.*, 1995), considerable differences in chromosome size and number conceal an underlying genome similarity. Genome synteny has been well established in animals and plants has not been documented or considered for eumycote fungi or straminipiles. The chromosomal distribution of genes may affect mode of action of the genes, with the possibility that a gene occurring in different positions may affect the biodiversity/speciation without much change in nucleotide sequence data. Polyploidy, as in *Achlya* (Win-Tin & Dick, 1975), *Pythium* (Win-Tin & Dick, 1975) and *Phytophthora* (Sansome, 1965) and polysomy, as in *Phytophthora* (Shaw, 1991, Shaw & Shattock, 1991) could affect gene function. When ploidy levels are different the breeding systems of the straminipilous fungi need to be taken into account, particularly when selfing and automictic sexual reproduction may be involved (Dick, 1972, 1987, 1995; Win-Tin & Dick, 1975). Brasier (1992) and Brasier & Hansen (1992) have reviewed the evolution of *Phytophthora* from a genetic standpoint.

It is also possible that differences in virulence could be attributed to Simple Sequence Repeats (SSRs ≡ microsatellite DNA). Infraspecific variation in *Pythium ultimum* is well-known, both at the varietal level (Drechsler, 1960) and within one of the varieties (Tojo *et al.*, 1998). Is the infraspecific variation in *Pythium ultimum*, with its varieties *P. ultimum* var. *ultimum* and *P. ultimum* var. *sporangiiferum* due to such changes in SSRs? Pathological distinctiveness (e.g., *formae speciales*) and interbreeding barriers, not be accompanied by morphological differences, may be the result of slight and recent genetic changes which will be of importance when determining species discontinuities. This possibility is relevant to molecular biological analyses which purport to determine conspecificity.

THE ASSIMILATIVE THALLUS

The assimilative thallus: morphology

The thallus is initiated from a uninucleate zoospore cyst or an aplanospore; a multinucleate asexual spore or propagule; or a uninucleate sexually produced oospore. The multinucleate asexual spores and oospores either germinate directly to give a thallus, hypha or conidiosporangiophore, or function as a zoosporangium. The thalli of members of the *Peronosporomycetes* may be filamentous, composed of hyphae forming a mycelium; or coralloid (eucarpic or holocarpic), allantoid or ellipsoid (holocarpic); or monocentric and eucarpic, having an assimilative system composed of branched rhizoids. Hyphae are ***analogous*** to hyphae of the *Eumycota* and show tip growth, first demonstrated in *Pythium* (Grove, 1970; Grove & Bracker, 1978; Grove, Bracker & Morre, 1970). Hill & Mullins (1979) used *Achlya* for a study of tip growth, and over a long period of time Heath and coworkers (e.g., Heath & Rethoret (1980) Gupta & Heath (1997) Kaminskyj, Garrill & Heath (1992) Heath & Steinberg (1999) have used *Saprolegnia ferax* as an experimental organism for investigating the fundamental processes of tip growth.

The protoplasm is coenocytic, with or without a conspicuous central tonoplast vacuole. Bidirectional cytoplasmic streaming along cytoplasmic strands can readily be seen in wide hyphae, where the cytoplasmic strands follow a long spiral. The nuclei are small, up to about 3 μm in diameter, but often deformed in hyphae. During active assimilative growth the nuclear cycle is short (about 36-76 min in *Saprolegniaceae* and 75-155 min in *Pythiaceae*, see Dick, 1990*a*, 2000*c*).

The mycelial habit has probably developed on several separate occasions (Dick, 1995, 2000*a*, *c*). Different views are held on the evolution of hyphal form (Barr, 1981; Dick, 1997*a*). The stout hyphae of the *Saprolegniaceae* have been regarded as a primitive character retained from an algal ancestor, but I regard this morphology as a more recent ecological specialization. The wide-diameter hyphal systems of the *Saprolegniaceae* may represent a modification from a monocentric kind of organization, with isodiametric growth being replaced by tip growth accompanied by a shift of assimilatory function from a well-defined rhizoidal system to these broader hyphae. This 'vegetative' system may thus be considered as an assimilative, growth-unlimited sporangio-gametangiophore. The origin of hyphae in the *Peronosporomycetes* was probably also recent, either from a sporangio-gametangiophore with indeterminate tip growth (wider hyphae), or from a narrow germ tube developed from an infection peg (narrower hyphae).

The possibility of different evolutionary origins of the myceliar habit might be supported from data on hyphal anastomoses. Wilde (1961) and Stephenson, Erwin & Leary (1974) have provided evidence for hyphal anastomoses in *Phytophthora*, but anastomoses have never been reported for the *Saprolegniomycetidae*. Stephenson *et al.* (1974) reported anastomoses between A^1 and A^2 sexual strains for three species of *Phytophthora*, but whether anastomosis is confined to union between hyphae differing ***only*** in the sexual mating type, or whether more general heterokaryotic union occurs has not been established. In eumycote fungi heterokaryon formation does ***not*** occur between sexually compatible strains so the mechanism and control must be different.

The assimilative thallus is bounded by a wall membrane at maturity, but initially may be naked in some endobiotic parasites. In mycelial forms, hyphae vary in diameter from 1·8-2·5 μm (*Pythiogeton, Verrucalvus*) up to 150 μm (after intussusception, in older hyphae of *Achlya*). Septa are normally present to delimit reproductive structures and sometimes develop as **retraction septa** in old mycelia of *Pythium*; plugs of wall material may replace septa at reproductive junctions in the *Rhipidiaceae*, *Peronosporaceae* and *Phytophthora*. In the *Saprolegniaceae* there is frequently excessive synthesis of wall material at the central closure of the septum, resulting in the development of an irregular peg or callus on one or both sides of the septum. Generalized intussusception of wall material occurs in the monocentric thalli of *Rhipidiaceae* and in the hyphae of some *Saprolegniaceae* as the structures increase in girth, but the mechanism of this synthesis is unresearched. Regular differential wall synthesis may result in ***thickened*** walls at the thallus

constrictions in the *Rhipidiaceae* (Dick 1973*b*). The entire wall of the basal cell of the *Rhipidiaceae* can become very thick, with exfoliation of wall material as the basal cell expands.

Septation of the mycelial thallus occurs in *Pythium*, and if *Lagenidium* is defined on the basis of differentiated gametangia, then there is hardly any discontinuity with *Pythium*. Similarly there is little distinction between a lobed thallus in an algal cell and the toruloid thallus typical of some saprotrophic members of the *Leptolegniellaceae*. Other taxa of *Leptolegniellaceae* (e.g., *Brevilegniella*) have a filamentous thallus with swollen intercalary reproductive segments. A thallus form such as that of "*Lagenidium* sp. on *Gracilaria*" (Johnson, 1957) is intermediate between that of *Brevilegniella* and the mycelium of certain species of *Pythium*. Diagram of diversity in thallus form are given in Figures I: 10 (p. 48-49) and I: 19 (p. 80).

It has long been recognized that experienced workers are able to suggest probable generic designatons for non-reproductive mycelia. This has been especially true for *Saprolegniaceae* (rounded or acute shape of hyphal tip and mean hyphal diameter distal from the tip), *Pythium* (sometimes 'fastigiate') and *Phytophthora* ('twiggy' and forming a 'chrysanthemum pattern'). Ho (1978) used nine parameters of hyphal branching systems and suggested that these characters might be useful in the delimitation of taxa. In addition, the modal angle of branching on agar is often useful. The water relations of *Achlya* on agar are such that extensive cracks develop in the agar well before the mycelium has reached the Petri dish circumference. With new programs for image analysis it should be possible to develop branching pattern as a diagnostic criterion, but it will be necessary to establish the reliability of these patterns under different cultural conditions.

Obligate parasites may be entirely confined ***within*** a single host protoplast (endobiotic), **intracellular** (some hyphae invading the protoplasts of a host thallus), or **intercellular** with specialized side branches (**haustoria**) that penetrate the cell walls, but not the protoplasts, of the host cells. In most of the endoparasitic fungi which have been studied by transmission electron microscopy, an extremely fine penetration tube of *ca* 0·1-0·2 μm diameter enters the host and subsequent tip expansion enables the formation of the first unit of the thallus within the protoplast of the host.

Thallus form has formerly been the principal determinant for placement in *Myzocytium* or *Lagenidium*. *Myzocytium* has a thallus which becomes articulated into swollen segments, often with the cross walls becoming thickened and hyaline, resembling *Catenaria* of the *Chytridiomycetes*. In some species these segments become disarticulated at maturity. In *Blastulidium* the irregularly loment thallus becomes septate and the segments become more or less spherical before disarticulating (Manier, 1976). On the other hand, *Lagenidium* has an allantoid, tubular branched or lobed thallus which may or may not be septate. If septate, the segments are not normally inflated. Obviously these forms intergrade, particularly in depauperate specimens or in small hosts, and reliance should not be placed on this criterion.

There is a fine distinction between the definition of a fungus as an osmotroph with a cell wall during the assimilative phase, and the naked intra-protoplasmic plasmodium of an endogenous parasite. This line of distinction appears to be crossed to different extents during the thallus development of species placed in the *Olpidiopsidales*, *Rozellopsidales* and *Plasmodiophorales*. The initial thallus development is plasmodial in *Olpidiopsis*, but a walled thallus, which continues expanding, soon becomes apparent. However, in *Rozellopsis* and *Rozella*, fungi which may or may not be related to either *Olpidiopsis* or the *Plasmodiophoromycetes*, more extensive plasmodial phases involving subdivision of the assimilative plasmodium into several independent units can occur (compare Held, 1980). The transition from a plasmodial to a walled assimilative state may be precocious or retarded in endoparasitic fungi, and thus not of great significance at high taxonomic hierarchies. Terminal elongation, with septation of the thallus and intercalary segments functioning as sporangia or gametangia, may be derived from a monocentric state and give rise to the 'hyphal' morphology.

Reliance on thallus/sporangial form, particularly when intermediate between a lobulate or saccate holocarpic form and a more or less minimalist ovoid (*Olpidium*- or *Olpidiopsis*-like) shape, has resulted in heterogeneous groups of organisms in genera such as *Olpidiopsis* and *Petersenia* (see pp. 215 and 228).

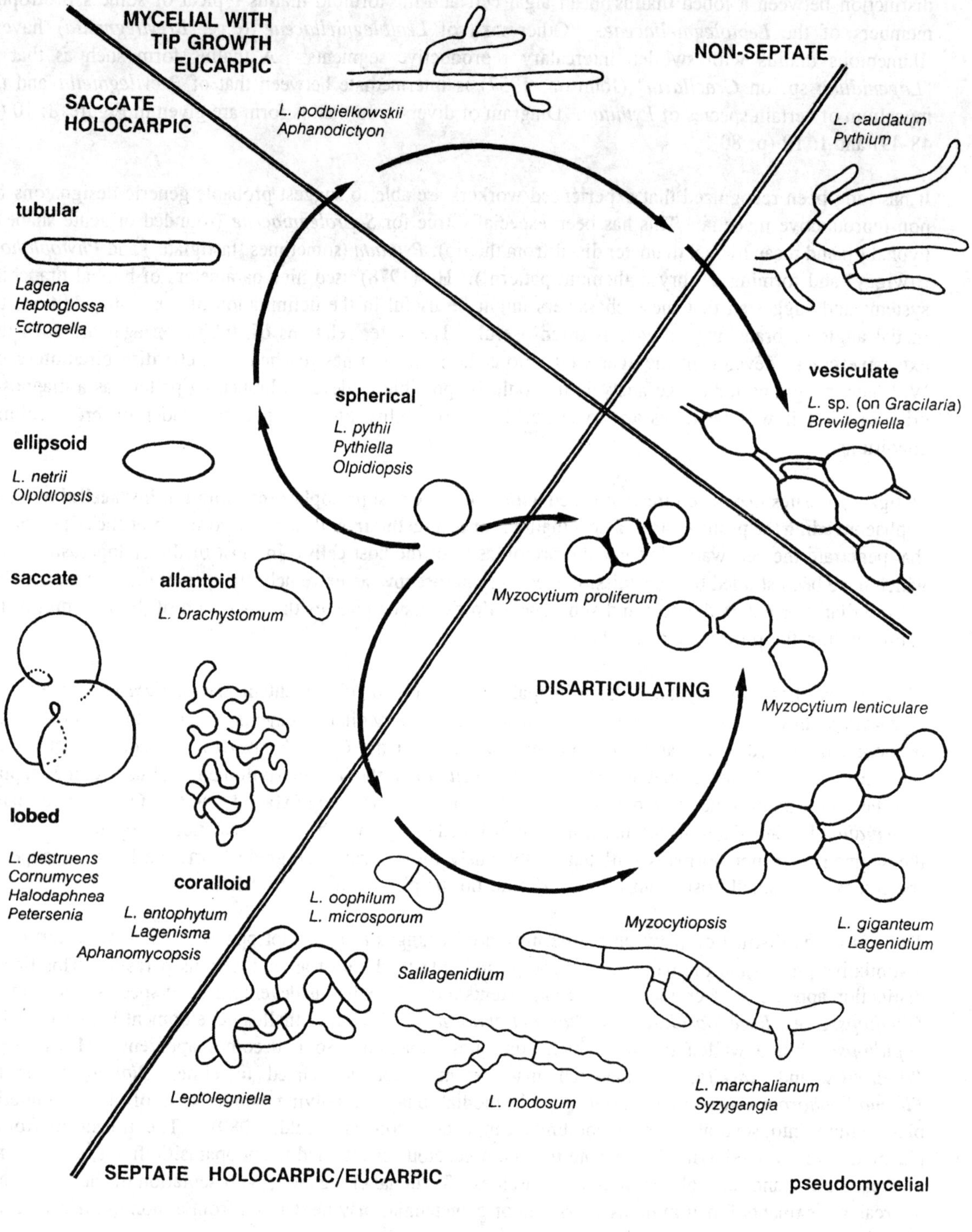

FIGURE I: 19. *Divergence and convergence in thallus form. Phylogenetic significance is not intended. The diagonal straight lines indicate major divisions with respect to holocarpy/eucarpy and mycelial tip growth/intercalary wall morphogenesis. Many states are represented by species of* Lagenidium sensu lato *(see PART IV), and by genera of the* Leptolegniellaceae.

While thallus form may be valuable at the species level, it cannot be accorded significance, on its own, at any higher taxonomic level until there is a series of much more precise, and rigorously applied, ontogenetic definitions.

Thallus form in the plasmodiophorids is initially plasmodial, and morphological distinctions between taxa only become evident when the plasmodium becomes sporogenous. After the mitotic phase a walled sporangium is formed which may take the form of an irregular chain of swollen segments, as in *Polymyxa betae*, or a convoluted branch system or reticulum, as in *Woronina pythii* (Dylewski, Miller & Braselton, 1978). The extent to which the protoplasm of these elements may be completely separated so that the elements function as individual sporangia is unclear. Ciafardini & Marotta (1988, 1989) have shown that the wall between contiguous parts of the sporangial complex of *Polymyxa betae* is more fragile than the other walls. In both of the above species there are resemblances to lagenidiaceous fungi (cf. Karling, 1968*e*: 95). Cystogenous plasmodia show cleavage to produce uninucleate segments of small volume *ca* 15-120 μm^3), each of which develops a smooth or spiny cyst wall. In *Octomyxa* these are commonly found in tetrads or octets, but larger groups occur in *Sorophaera* and *Sorodiscus*. Powell (1984) noted the probability of phagocytosis towards the end of the assimilative phase in *Rozella*, and there is a similar suggestion for the anisokont biflagellate parasite of protists, *Endemosarca* (Erdos & Olive, 1971, Erdos, 1973). Phagocytosis has not previously been regarded as an attribute of plasmodiophorids, but Clay & Walsh (1997) have provided evidence that it takes place in *Spongospora*.

The labyrinthulids are straminipilous heterotrophs, but they are not fungi because the assimilitive ectotrophic net, while osmotrophic, is never bounded by a cell wall. I therefore prefer to use the class name *Labyrinthista*, rather than the fungal hierarchical nomenclature *Labyrinthulomycota* (division) and *Labyrinthulomycetes* (class).

Intercellular hyphae of parasites

One of the striking features of the *Peronosporomycetes* (*Peronosporales*), the downy mildews (DMs), is the development of biotrophy from necrotrophy. Savile (1968, 1976) has suggested that the first step towards phytoparasitism would have been the development of ***systemic*** (whole plant) myceliar parasitism to protect the hyphae from desiccation (note the extremely narrow and vulnerable hyphae of the *Sclerosporaceae*), and that pathological lesions of limited mycelial extent would have evolved later. However, systemic infections are known to occur in the *Peronosporaceae* (Goosen & Sackston, 1968: *Plasmopara*; Heller, Rozynek & Spring, 1997, Ramsay, Smith & Wright, 1954, *Peronospora*; Marlatt, Lewis & McKittrick, 1962: *Bremia*) and *Albuginaceae* (Jacobson *et al.*, 1998). Infections caused by the DMs of panicoid grasses may also be systemic (Kenneth, 1981). Parasitism by the downy mildews must be contrasted with the parasitoidal associations of the *Myzocytiopsidaceae* with nematodes and algae (Dick, 1997*b*). These endobiotic parasites are always necrotrophic. Similarly, endobiotic *Saprolegniaceae* (*Aphanomyces parasiticus*), root-parasitic *Saprolegniaceae* (*Aphanomyces euteiches* and *A. cladogamus*) and *Pythiaceae* (many *Pythium* species) are necrotrophic.

Biotrophic obligate parasitism is not always fully developed, so that a limited range of host/parasite relationships may be covered by this phrase. Biotrophic obligate parasites, such as DMs, have advanced genetic and biochemical attributes often, but sometimes unjustifiably, equated to an evolutionary status. Significant cell damage is caused, for example, by *Peronospora tabacina* and *Plasmopara viticola* (Lafon & Bulit, 1981): the plasmamembranes of the host mesophyll cells become excessively leaky, resulting in a distinctive greasy or wet appearance to the infected part of the leaf. This is essentially a moderated manifestation of the symptoms associated with wet rots caused by certain species of *Phytophthora* (Keen & Yoshikawa, 1983) and probably resulting from a similar biochemical interaction. Biotrophic obligate parasitism certainly requires a degree of specialization and a constraint to variation: there must be elements of genome protection or conservation in both partners.

CRITERIA FOR DIAGNOSES

Haustoria

Another most important step would have been the development from mixed intra- and inter-cellular hyphae to mycelia solely of intercellular hyphae and haustoria (Fraymouth, 1956; Peyton & Bowen, 1963; Berlin & Bowen, 1964; Davison, 1968; Coffey, 1975). Parallel evolution of intercellular hyphae and haustoria (biotrophic parasitism) is manifest by the occurrence of these features in both the DMs and the phylogenetically unrelated *Uredinales* (*Puccinia*). Spencer-Phillips (Clark & Spencer-Phillips, 1993; Spencer-Phillips, 1997) has shown that the intercellular hyphae of the DMs retain the capacity for assimilation in the presence of haustoria. Differences could exist between the functions of haustoria in the nutrition of unrelated taxa. Thus, there is no reason to consider that this biotrophic development, even within the DMs, represents a monophyletic line. Indeed, the fact that the ***morphology*** of the haustoria is different in *Albugo* (small and spherical), *Peronospora* (frequently large and lobed), and *Sclerospora* (peg-like) could point to independent origins, each possibly with a characteristic physiology. Blackwell (1953) described haustoria from *Phytophthora infestans* and noted that about 15 other species formed haustoria although other species of the genus apparently lacked such a development.

Parasitism by the obligate, host-range specific, downy mildews

From the coevolutionary viewpoint, there are distinctions to be drawn between obligate parasitism, species-specific parasitism, and special-form relationships. A discussion on infra-specific differences could, in time, illuminate the processes of speciation compared with population diversity, but the data are too fragmentary at present. Whereas obligate parasitism merely requires the presence of a regular (but possibly periodic) and renewable (but possibly highly transient) nutrient availability from living protoplasm, species-specific parasitism implies a much more restricted range for potential complementary metabolisms. The concept of a 'tolerance range', probably much smaller *in planta* than *in vivo* and thus analogous to the ecological ranges of saprotrophs *in situ* in soils (Dick, 1992), might provide a better model than a search for a package of absolute metabolic requirements.

Potentially interacting organisms must be able to come into contact, and there must be sufficient compatibility for nutritional requirements to be satisfied. Frequently, this will be because new hosts are phylogenetically close to former hosts. Host populations at the frontiers of their realizable niches are more liable to become involved in new coevolutionary initiatives, but the development of a stable relationship will depend on the generation cycles of the parasite and its capacity for genetic change. The critical factors for the nutritional environment of the parasite, the pathways, or the specific metabolites produced, may occur in organisms of differing phylogeny; or, they may only become evident in certain populations because of environmental circumstances. Two facets interconnect: the coevolutionary reliance by the parasite on a host species, and the restrictive nature of this reliance to particular metabolitic pathways. The critical factors involved may require subtle definition. Obvious basic carbon and nitrogen sources are unlikely to be crucial, but sulphur and combined forms of carbon and nitrogen may be so for the DMs.

There must be physical or chemical similarities or analogues that enable an appropriate degree of association between previously separated populations. Too great a vulnerability will lead to an unstable and ephemeral (necrotrophic) relationship. The essence of coevolution is adaptive change in balanced relationships. It is possible that chance associations may lead to new relationships, as has been proposed by Baum & Savile (1985) for certain rusts. This may be more possible for parasites that produce a limited mycelium and for which physical rather than chemical environmental factors are more important. Chance associations leading to coevolution must be less likely for fungi that are essentially systemic, because there would be less likelihood that either host or parasite would survive long enough to reach reproductive maturity.

An obligate parasite that cannot be grown apart from its living host either requires particular metabolites that have not yet been identified, or the organism is intolerant of arbitrary levels of fluctuations in the concentrations and rates of supply of nutrients, or some other *in planta* factor is necessary. There are no suggestions that nutritional requirements are invariably linked to host range restriction in the DMs. The efficiency of waste removal may be a contributory factor. There is little evidence to support or refute any of these contentions. Moreover, extrapolations made from studies of related fungi that can be grown axenically could be misleading.

If parasite dependence is not based on a demand for particular chemical units, the dependence must have a different origin. I have suggested (Dick, 1988) that this could be based upon an 'empathy' between certain crucial metabolic pathways of host and parasite, so that the catabolism and anabolism were in harmony. Different host pathways may be pre-eminent for different parasites, whether these are taxonomically related or not. Thus, individuals of a single host species may be infected by several parasites. The most notable example for DMs is the suggested synergism between *Peronospora* and *Albugo* in the *Brassicaceae* (Sansome & Sansome, 1974). However, my hypothesis of critical pathway differences would not only explain the occurrence of simultaneous parasitism of a host by different, but systematically related biotrophic obligate parasites: it would also allow for the possibility that these parasites may have different degrees of host specificity. Biphasic culture has been achieved for several genera and species (Ingram, 1980; Lucas *et al.*, 1991, Lucas, Hayter & Crute, 1995), but none is yet in axenic culture.

Whatever the biochemistry underlying attraction to a particular host, and stimulation to germination and colonization by the parasite, there are well-documented examples of parasite-mediated modification of host physiology after establishment. Green ear hyperplasia of pearl millet caused by *Sclerophthora* (Williams, 1984), hypoplasia of sunflower by *Plasmopara* (Sackston, 1981), and the well-known hypertrophy of crucifer stems by *Albugo* are three of the clearest examples relating to growth substance induction. The precise mechanisms of the biochemical modifications have not been researched.

Symptomless occurrence of *Peronosporales* and *Pythiales* in angiosperms suggests that the evolution of parasitism has achieved the ultimate balance in some associations. Haustoria are not essential. *Pachymetra* in *Imperata cylindrica* var. *major* in Queensland (R. C. Magarey, pers. comm.), *Phytophthora* in roots of raspberry and strawberry in Scotland (J. M. Duncan, pers. comm.), and *Pythium* in grass and herbaceous roots are all good examples of such symptomless associations. Symptomless association does not imply a 'no yield loss' situation.

The boundaries between obligate parasitism, species-specific parasitism, and special-form relationships are unclear: more research and discussion (cf. Skalický, 1964; Skidmore & Ingram, 1985) should elucidate the processes of speciation as opposed to different levels of infraspecific (population) diversity. Species-specific parasitism implies a much more restricted range for potential complementary metabolisms. This can be viewed as a ***tolerance range*** rather than a package of absolute metabolic requirements. The breadth of this tolerance range may well be extremely narrow *in planta*, in much the same way that saprotrophic *Pythium* species may co-exist in soil, but have very different patterns of relative frequency of occurrence *in situ* than might be predicted from growth studies *in vitro* (Dick, 1992). The endpoint of this progression is the race concept of the special form for which biochemical compatibility is presumed to be the only apparent distinguishing feature. This may be merely the result of extremely narrow tolerance ranges for a number of factors. But it may be, as with race induction in response to resistance cultivar production, a gene-for-gene evolution that may function through a variety of biochemical, physiological or morphological requirements. An hypothesis for absolute metabolite requirement in the absence of strong selective pressure might require an improbably large number of genetic lesions to explain race-specific parasitism (*formae-speciales*) between related parasites and related hosts. From the systematic viewpoint, the above environmental/host distinctions of the parasite rest uneasily with the infra-specific categories of variety and form, together with *formae speciales* which are not governed by the rules of the *International Code of Botanical Nomenclature* (*ICBN*).

Discussions of single-gene host resistance in different systems of host resistance and pathogen virulence (e.g., Keen & Yoshikawa, 1983) ignore the attraction and stimulation that enables both species to coexist. It is unlikely that studies concentrating on intraspecific differences will reveal underlying coevolutionary factors. There is a long-standing inverse relationship between the outlook and research momentum for plant pathology and the quest by mycologists for an understanding of species-specific coevolution.

It should be noted here that the straminipilous fungi have unique biochemical requirements and metabolic products, many of which are under-rated and some of which will be of significance to the establishment of parasitic relationships. Phylogeny inferred from assimilative morphology is not acceptable, although subtle differences in morphogenesis might be invoked.

Conidiosporangia, chlamydospores, hyphal bodies and gemmae

When the coenocytic asexual reproductive initial has a distinct shape and size, but shows no further differentiation or development after septation and is disseminated by disarticulation, it is functionally a conidium. However, it may germinate directly by a germ tube or retain the capability of germinating as a zoosporangium; this propagule would then most correctly be termed a multinucleate **conidiosporangium** (Snell & Dick, 1957). In some *Peronosporomycetes* asexual propagules lack a clearly defined shape or size and are released from the mycelium by decay or autolysis of the hyphal system; such structures are termed **hyphal bodies** (*Pythium*) or **gemmae** (*Saprolegniaceae*). If such structures also develop a thick wall they may be termed *chlamydospores*, but this development is infrequent in the Peronosporomycetes.

Development from the thallus: conidiosporangiophores and sporangial regeneration

The patterns of growth and regeneration of the hyphae bearing sporangia are varied. When the axis is terminated by a sporangium the development is termed **determinate**, but when the sporangia are lateral or developed in intercalary positions development is termed **indeterminate**. Hyphal regrowth may take place through the sporangial septum (**internal renewal** in *Phytophthora*, *Saprolegnia*) or by a lateral branch (**cymose renewal** in *Achlya*). Sporangia may be produced in sequence on the same determinate axis (**basipetal development** in *Scoliolegnia*) or by limited internal renewal so that the successive sporangial septa are formed at approximately the same point on the axis (**percurrent development** in *Albugo*). The conidiosporangiophore may be differentiated from the assimilative hyphae to a greater (*Peronosporales*) or lesser (some *Phytophthora* species) extent; it may be swollen (*Basidiophora*, *Sclerospora*) with dichotomous (*Peronospora*), pseudodichotomous (*Sclerospora*) or more irregular branching (*Plasmopara*). In the *Peronosporales* the conidiosporangiophore is persistent and can be observed on herbarium material of the hosts, but in the *Sclerosporales* it is evanescent. The junction of the conidiosporangium on the conidiosporangiophore is pedicellate and plugged in the *Peronosporales* but delimited by a septum in the *Sclerosporales*. The evolution of sporangial form is also open to various interpretations which may depend on the evolution of thallus form. In *Pythium*, for example, there is a complete continuum from undifferentiated (and often intercalary) hyphal segments through coralloid branched sporangia, toruloid sporangia, sessile lateral inflated sporangia to spherical sporangia (Dick 1990*b*).

Mitochondrial morphology

Most mycology texts emphasise the difference between the tubular mitochondrial cristae of the biflagellate fungi and the plate-like cristae of the *Eumycota*. Cavalier-Smith (1981) distinguished beween plate-like cristae (true fungi); tubular cristae (chromistans); discoid cristae (euglenoids), and tubular - vesiculate cristae (protozoa) (Figure I: 20). However, this broad categorization masks minor but possibly significant differences between fungi, as noted by Dick & Abro (1990). Mitochondrial profiles can be observed in either the vegetative thallus or the planont. There has been no search for evidence of any difference in construction of the mitochondrion related to the phase of the life-history, even when the mitochondria occupy distinctive positions, as in chytrid zoospores (Lange & Olson, 1979) or in those cases (e.g., *Rozella*) where the mitochondria are uniquely positioned in relation to other organelles (Held, 1975).

In the *Peronosporomycetes* the profile of the mitochondrial crista is as a tube of more or less uniform diameter, and the cristae appear in TEMs in longitudinal sections with connection to the inner mitochondrial membrane and as transverse circular profiles (Figure I: 19).

Confusion has arisen because of the terminology used at different times. Mitochondrial cristae in the plasmodiophorids have been described as "with relatively few tubular cristae" (Williams & Yukawa, 1967) and "tubular" (Sleigh, 1989). They correspond to the vesicular category of Cavalier-Smith (1981) and a more accurate description would be: flattened finger-like lobes with ellipsoidal profiles in transverse section (e.g., *Polymyxa*, Barr & Allen, 1982: fig. 4, lowermost mitochondrion). Although by no means clear from the published TEMs, the mitochondria of *Gonimochaete* (Saikawa & Anazawa, 1985: figs 2, 7, 14) and *Haptoglossa* (Lee *et al.*, 1992) appear to have tubular/vesicular cristae; TEMs of *Haptoglossa* (Glockling, 1994) have similarly indefinite profiles, in contrast to those she obtained for *Myzocytiopsis*. Some micrographs of mitochondrial profiles of *Catenaria anguillulae* also show tubular/inflated finger-like cristae (Ichida & Fuller, 1968: figs 6, 8). Reference might also be made to the problematic protoctist *Nephromyces*, which is reported as having tubular mitochondrial cristae and chitin synthesis (Saffo, 1981; Saffo & Nelson, 1983; Saffo & Fultz, 1985).

Variations on the vesicular category of mitochondrial cristae can be seen in illustrations (e.g., *Rozella*, Powell, 1984), where the inflated lobes have a constricted attachment to the inner membrane, but such variations tend to be ignored in the two-category classification of tubular or plate-like cristal arrangement.

Mitochondrial profiles for many of the critical species of lagenidiaceous fungi are not available. However, Gotelli (1974*b*), Molina (1986) and Foissner (1987) have reported tubular cristae for mitochondria of *Salilagenidium callinectes*, *Petersenia* and *Ciliatomyces* respectively; Glockling (1994) obtained TEMs of mitochondria of *Myzocytiopsis* and *Chlamydomyzium* which have clearly defined tubular cristae.

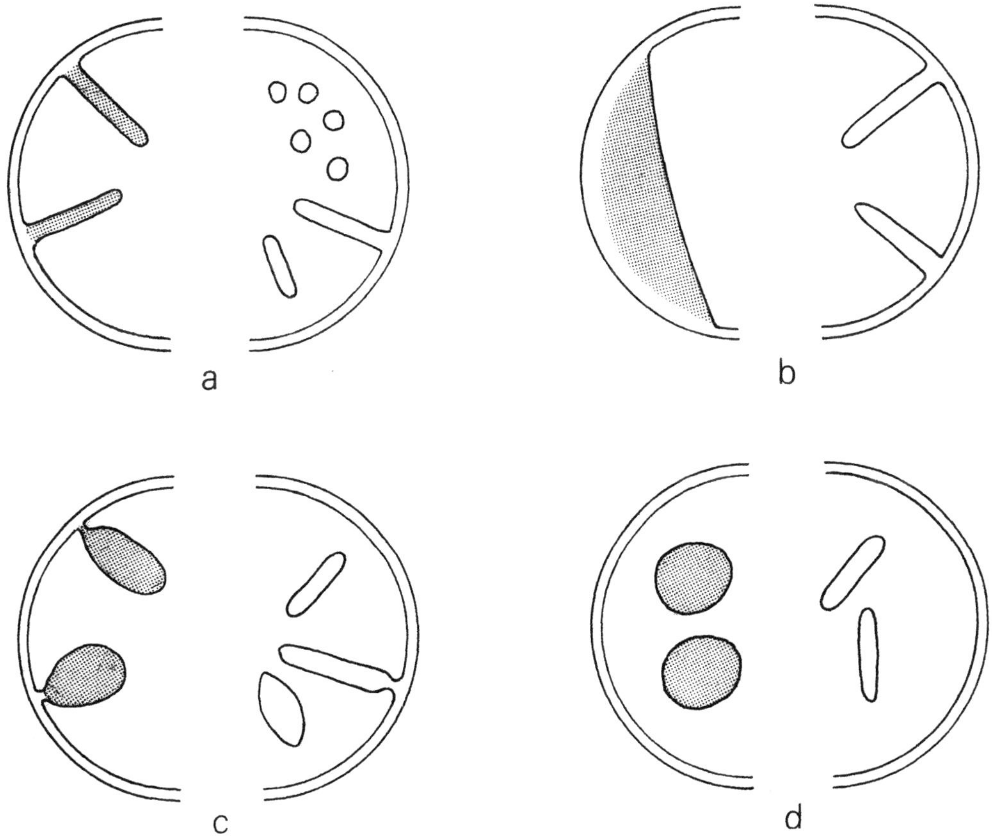

FIGURE I: 20. *Mitochondria: lateral diagrams (left) and observable ranges of transverse, longitudinal and oblique profiles (right) of cristae.*

a: tubular cristae; b: plate-like or shelf-like cristae; c: vesiculate or flattened, finger-like cristae; d: discoid cristae.

BIOCHEMISTRY

It is inevitable that the fungi used for biochemical investigations are those which are most readily culturable. Hence, the potentially powerful biochemical criteria are regrettably diminished to a role of providing confirmation of otherwise fairly obvious taxonomic affinities. Technical difficulties still need to be surmounted before these criteria can be employed where they are most needed, in the endobiotic obligate parasites.

The following sections relate only to those aspects of the biochemistry of the straminipilous fungi which have been accorded systematic significance. Different approaches have been used for these sections because of the different levels of phylogenetic significance and the extents of the research activity.

Cell walls

The physical and chemical composition of the walls of culturable peronosporomycete fungi has been extensively studied, with early work reviewed by Aronson (1965) and Bartnicki-Garcia (1966). The physical differences between peronosporomycete and chromophyte algal walls were highlighted by Parker, Preston & Fogg (1963) and these differences are probably only partly due to the different physiological requirements of heterotrophs and autotrophs. The amount of fibrillar material in the cell wall is much less in the fungi, and the cellulose (ß-1,4-glucan) tends to be masked by the much larger amounts of ß-1,3- and ß-1,6-glucans (reviewed by Cooper & Aronson, 1967, in relation to *Pythium*). Comparative studies (Novaes-Ledieu, Jiménez-Martínez & Villanueva, 1967; Sietsma, Eveleigh & Haskins, 1969; Dietrich, 1973, 1975; Vaziri-Tehrani & Dick, 1980*c*) indicate that there are differences in carbohydrate composition between species, at least in their characteristics after extraction.

The cellulosic nature of peronosporomycete cell walls has long been recognized and the use of chlor-zinc iodide histochemistry was in routine use for these fungi a century ago. This test is not always reliable; it sometimes gives different colour reactions which are difficult to interpret (see Couch, 1935*a*: *Pythium*: blue; *Olpidiopsis*: purplish; *Pythiella*: no reaction). Even for *Pythium ultimum*, a species which would be expected to have various glucans including cellulose, the stain reaction may be negative (Trow, 1901). The standard formulation of the cellulose stain (Zimmerman, 1901) was modified by Post & Laudermilk (1942) who also categorized some of the different colour reactions obtained when using this stain for cellulose from various plants. Until the techniques of gold-complexed lectins in association with glycosidic enzymes (Benhamou *et al.*, 1987) are shown to be applicable to foreign fungal walls within walled ***fungal*** protoplasts, the unsatisfactory chlor-zinc iodide test remains the only way of indicating the chemistry of the walls of endobiotic parasites.

Hegnauer & Hohl (1978) have distinguished two basic types of wall layer in the peronosporomycetes: the A layer, which is amorphous as perceived by TEM, and the FA layer, which is fibrillar with an amorphous component. The relative thicknesses and numbers of these alternating layers varies according to species and to wall origin, reaching an extreme development in the oogonial spines of *Pachymetra* (Dick *et al.*, 1989). The diversity of peronosporomycete zoospore cyst walls is well-known (Beakes, 1987; Dick, 1990*a*) but the chemistry of the ornamentations has not been established. Possibly similar alternating electron-opaque/electron lucent layers occur in the cyst spines of *Haliphthoros* (Overton *et al.*, 1983).

In the *Labyrinthista* the wall is composed of one or many layers of imbricated, dictyosome-derived, circular scales (Jones & Alderman, 1971; Perkins, 1973*a*, *b*; Darley, Porter & Fuller, 1973; Bahnweg & Jäckle,

1986; Moss, 1986). These walls contain 30-43% protein and variable amounts of carbohydrate, with the rare isomer L-galactose as the predominant component of the carbohydrate polymer, except for *Aplanochytrium*, in which fucose predominates (Bahnweg & Jäckle, 1986). The presence of sulphated polysaccharides, similar to those found in algae, has been demonstrated in *Thraustochytrium* (Chamberlain, 1980).

The unpolymerized precursor of chitin, *N*-acetyl-D-glucosamine occurs, and the presence of chitin has been inferred for saprolegniaceous fungi because of the effect of polyoxin D on hyphal diameters (Dietrich & Campos, 1978) and the occurrence of chitin has recently been confirmed by Bulone *et al.* (1992). Asiegbu, Lönneborg & Johansson (1996) have demonstrated the occurrence of relatively insignificant amounts of chitin in the middle layer of the hyphal wall of *Pythium*, masked by larger amounts of *N*-acetyl galactosamine, based on the use of fluorescine isothiocyanate and gold-labelled lectins. Chitin is known as a wall component for genera of the hyphochytrids (Nabel, 1939; Fuller, 1960; Fuller & Barshad, 1960; Clay *et al.*, 1991: *Hyphochytrium* and *Rhizidiomyces*) and *Leptomitaceae* (Lee & Aronson, 1975; Aronson & Lin, 1978; Huizar & Aronson, 1985, 1986; Bertke & Aronson, 1980: *Leptomitus, Apodachlya* and *Plerogone**). In the *Rhipidiaceae* glucosamine is absent (Bertke & Aronson, 1985: *Araiospora, Mindeniella*) or present only as a trace after digestion (Pau & Aronson, 1970: *Sapromyces*).

Clay *et al.* (1991) have demonstrated that *Hyphochytrium* and *Rhizidiomyces* contain both cellulose and chitin in their inner cell walls, and have shown that there are differences between these genera in the relative abundance (or ease with which they can be labelled) of these two wall components. The distribution of cellulose and chitin may differ in different walls (discharge tubes and septa) of the same organism. These authors stress that differences in wall composition may be of significance at low levels in the taxonomic hierarchy: such differences would thus have correspondingly less value at ordinal or class levels.

The walls of *Blastulidium* were merely noted as being "of polysaccharide" (Manier, 1976).

Proteins may be a large component of the dry weight of cell walls, being particularly high in the labyrinthulids (Darley *et al.*, 1973). Vaziri-Tehrani & Dick (1980*b*) reviewed the data then available for peronosporomycetes and other fungi, and produced a novel polygraphic display of relative amino acid ratios, which distinguished between the *Saprolegniomycetidae* (with *Sapromyces* of the *Rhipidiomycetidae*) and the *Peronosporomycetidae* (with *Atkinsiella dubia* (D. Atkins) Vishniac). One particular hydroxyproline-rich protein (HRP) is ubiquitous in plant walls and it has been suggested that HRP is involved in the control of growth (Lamport, 1970). Novaes-Ledieu *et al.* (1967) and Vaziri-Tehrani & Dick (1980*c*) have indicated that levels of HRP are greater in the *Pythiaceae* than in the *Saprolegniaceae*. Aronson & Fuller (1969) noted that the wall of *Atkinsiella dubia* was also rich in hydroxyproline. Takenaka & Kawasaki (1994) have used alanine-rich, hydroxyproline containing cell wall proteins for serological identification of *Pythium* species.

On the basis of negative results from histochemical tests using chlor-zinc iodide (Wisselingh, 1898; Maire & Tison, 1911*a*; Pendergrass, 1950; Goldie-Smith, 1951, 1954, 1956), Waterhouse (1973) claimed that plasmodiophorid walls were chitinous. Work on plasmodiophorid walls (Moxham & Buczacki, 1983; Moxham, Fraser & Buczacki, 1983; Buckzacki & Moxham, 1983) has shown that the spore-cyst wall is composed of four layers, the inner two containing chitin, which accounted for 25% of the dry weight.

---

*FOOTNOTE: As *Apodachlyella completa* (Humphrey) Indoh: the 1962 U.K. culture from M. W. Dick which was widely distributed in the U.S.A. by J. M. Aronson. This isolate was renamed by Dick (1986) on the rediscovery of Humphrey's fungus in the U.K. Cultures from American provenances of both *Apodachlyella* and *Plerogone* are now available (Longcore, Brooks & Homola, 1987).

Protein content of the entire wall was high, accounting for 33% of the dry weight. The outermost layer, and the next layer, which generated the spines, contained fibrillar protein. It is apparent that the range of variation within and between taxa is too high for generalizations to be established from the small number of species so far studied. See also Yano *et al.* (1994) for more recent data. Species comparisons using wall samples of different life-history origins need to be treated with caution.

Lysine synthesis

The two pathways for lysine synthesis are completely distinct, appear to be mutually exclusive and have major phylogenetic importance (Vogel, 1960, 1961, 1964). LéJohn (1972, 1974) has proposed that the DAP (α,ϵ-diaminopimelic acid) lysine synthesis pathway antedated the AAA (α-aminoadipic acid) lysine synthesis pathway, which correlates with chitinous cell walls, because the DAP pathway interferes with chitin biosynthesis. The control mechanisms of chitin biosynthesis and glutamate/glutamine biosynthesis appear to be connected, and the few organisms (notably the hyphochytrids) which possess both chitin and glucan polymers in their walls, lack a specific control of uridylates and UDP-amino sugar derivatives, which activate glutamic dehydrogenase. This would suggest potentially inefficient wall construction in the hyphochytrids. *Leptomitales* and *Saprolegniales* have yet to be tested for all these parameters. Rothschild & Heywood (1987) note another correlation, which might support LéJohn, namely that while AAA lysine synthesis is correlated with the presence of mitochondria with flat, plate-like cristae, DAP lysine synthesis may be associated with a range of mitochondrial types. The suggestion (Vaziri-Tehrani & Dick, 1980*a*) that high lysine:valine ratios, accompanied by generally lower but variable proline:valine ratios, may correlate with the DAP lysine synthesis pathway has been extrapolated by Buczacki (1983) to postulate that the plasmodiophorids have the AAA lysine synthesis pathway. However, the data of Vaziri-Tehrani & Dick relate to hyphal walls primarily of glucan polymers while the plasmodiophorid data were obtained from cyst walls, which are mostly protein (see above). There are no data to resolve these discrepancies, and there is no information for lagenidiaceous fungi with respect to combined lysine synthesis, wall composition and glutamine dehydrogenase characteristics. Paton & Jennings (1989) were unable to establish the presence of either pathway in *Thraustochytrium*. In the wider context it should be noted that euglenoids have the AAA pathway and mitochondria with discoid cristae, and the choanomastigotes have mitochondria with tubular cristae and produce investments of cellulose or chitin.

Lysine synthetic pathways cannot yet be used to resolve the classification of the lagenidiaceous fungi or the plasmodiophorids in the kingdoms *Straminipila* or *Protoctista*.

Sterol metabolism

Sterols fulfill two functions: as bulk sterols in membrane architecture where they act as stabilizing compounds, and as 'trigger' sterols for particular cellular activities (Leshem, 1992: 59). Bulk sterols will not be discussed here. In the *Peronosporomycetes* it was the 'trigger' function that was first given prominence by the study of the sterol requirements for sexual reproduction in *Pythium* (Haskins, Tulloch & Micetich, 1964). The same year Elliott *et al.* (1964), Harnish, Berg & Lilly (1964), Hendrix (1964) and Leal, Friend & Holliday (1964), and later, Elliott, Hendrie & Knights (1966) defined a steroid growth requirement for *Phytophthora*. Work on sterols was extended by Hendrix (1970, 1975), and Elliott & Sansome (1977) reviewed the role of sterol biosynthesis for meiosis. The interaction between sterol metabolism and progressive changes in phospholipases during the life-history of *Lagenidium giganteum* are under investigation (MacKichan, Tuininga & Kerwin, 1994).

Bulk sterols in membranes differ in the nature of the aliphatic side chain and in what groups are attached to the steroid skeleton, with sitosterol, campesterol and stigmasterol being commonly found in plants and cholesterol in animals, though cholesterol also occurs in plants. In a study which may be of relevance to sterol diversity and metabolism, Ellenbogen *et al.* (1969) pointed out that thraustochytrids, euglenoids and *Ochromonas* have retained a range of biochemical pathways for the biosynthesis of polyunsaturated fatty acids. These authors also pointed out the possible phylogenetic significance of particular polyunsaturated fatty acids, and since then the fatty acid 22:5w6 has been proposed as a signature compound for the thraustochytrids (Findlay *et al.*, 1986).

Detailed studies of sterol metabolism are confined to organisms which can be maintained in axenic culture. Nevertheless, information of taxonomic value is provided with respect to cycloartenol and lanosterol. Nes (1990, and references therein) has shown that lanosterol is formed from squalene oxide cyclization ***via*** cycloartenol in photosynthetic lineages, but ***directly*** in non-photosynthetic lineages. The *Peronosporomycetidae*, *Saprolegniomycetidae* and the hyphochytrids fall into the latter category, but differ in that fucosterol has not been detected in the hyphochytrids (Weete *et al.*, 1989). Nes *et al.* (1990) regard the different biosynthetic pathways as of a fundamental evolutionary significance equivalent to those for lysine synthesis. Some *Peronosporomycetes*, but not the *Pythiales*, can synthesize sterols from mevalonate. Domnas, Srebro & Hicks (1977) and Kerwin & Washino (1983, 1986*a*, *c*) have demonstrated the need for exogenous sterols for asexual and sexual reproduction by *Lagenidium giganteum*, but have also shown that this species differs, for example, from *Phytophthora* in being unable to use sterols with certain substituents. Dick (1988) suggested that there might be an interaction between the sterol requirements of obligate parasites such as the downy mildews and their host plant metabolisms. This suggestion was an extrapolation of the exogenous sterol requirements of *Phytophthora cactorum* (Lebert & Cohn) Schröt. (Nes, Saunders & Heftmann, 1982) in relation to the production of sitosterol by the host. The possible utilization of a principle based on reciprocity of sterol supply and demand for the understanding fungicide efficacy has been discussed in a perceptive review by Griffith, Davis & Grant (1992).

Warner, Sovocool & Domnas (1983*b*) have compared the sterol utilization by *Lagenidium giganteum* with that of *Salilagenidium callinectes*. Nes *et al.* (1986) and Berg & Patterson (1986) have taken this comparison further, showing that the similarity between *L. giganteum* and *Phytophthora cactorum* can be contrasted with the similarities between *S. callinectes* and *Achlya americana* Humphrey, *Dictyuchus monosporus* Leitg., *Saprolegnia ferax* (Gruith.) Thur. and *Plerogone helodes* M. W. Dick (see p. 88). The inference must be that while *L. giganteum* is closely related to the *Peronosporomycetidae*, *S. callinectes* is more likely to have affiliation with the *Saprolegniomycetidae*.

Some differences in sterol metabolism may prove to be of low evolutionary value, with loss of sterol anabolic pathways occurring at subgeneric levels, as may be inferred from sexual reproductive capacity within *Pythiaceae* (Kerwin & Duddles, 1989) and perhaps heterothallic and homothallic *Achlya* species. Vishniac (1955*b*, *c*, 1957) reported differences in steroid requirements between different isolates of *Labyrinthula*, with *L. minuta* possessing cholesterol. Bahnweg (1980) has provided data for *Haliphthoros* and *Halophytophthora epistomium* (Fell & Master) H. H. Ho & S. C. Jong on the stimulation of growth by exogenous sterols, but this provides little information of evolutionary or systematic significance.

MOLECULAR BIOLOGY

Molecular systematics, evolutionary origins and systematics

Molecular phylogeny is still in its infancy despite considerable research activity. The advantages and disadvantages of Linnaean classifications need to be evaluated, since an alternative system, based on molecular phylogenies, has been proposed which would challenge the nomenclatural hierarchy (Hibbett & Donoghue, 1998). Thus there exists and there will continue to be a tension between Linnaean/ICBN taxonomy and phylogenetic systematics (Brummitt, 1996; de Quiroz & Gauthier, 1994). Nevertheless, molecular phylogeny will provide information about relationships even if these relationships are not resolved into classifications. In addition to problems in translating molecular phylogeny into classifications (see PART II), diagnostic requirements raise several additional considerations summarized as follows.

To what extent should a clade node correspond to a 'classical' hierarchical level? Diversity within an ancient lineage may coexist with a more recently evolved, but fundamental attribute which so changes the evolutionary potential that the erection of a higher taxon is of practical value. Computer-generated similarity indices will reflect probable lineages, but these will not negate intra-subclass diversity in higher taxon concepts. Some higher taxa will encompass several nodes. Because of the progressively bifurcating nature of the cladogram, or lack of resolution for the origins of several lineages, phylogenetic approaches are not always best suited for establishing correlations (i.e., discontinuities) with currently recognized hierarchies in systematics. It is not always possible to distinguish between derived (apomorphous) and ancestral (plesiomorphous) character states. At ultimate branches of phylogenetic trees single cladistic characters may be insufficiently diagnostic, so that a 'suite' of characters is necessary for separation at species (and sometimes genus) level (see Donoghue, 1985). With finger-printing techniques separation proceeds through infraspecific taxa all the way to populations, clones and individuals (Lévesque *et al.*, 1994; Liew *et al.*, 1998; Panabières *et al.*, 1989).

The type concept is fundamental to systematics. Genera are defined by historically determined **type species**, irrespective of whether the type species is uncharacteristic of the taxa presently included in the genus. The type species is based upon a type, which again may deviate from the central tendency of the population from which it came. Although the type material may no longer be extant, or if extant no longer suitable for molecular analysis, it remains essential for the type species to be characterized before systematic changes can be justified. When the type material is not available, more recent isolates of the fungi (determined on morphological criteria) have to be used. These precepts are most pertinent to the systematics of the straminipilous fungi. The type species of *Plasmopara* (*Peronosporales*), *Phytophthora sensu lato* and *Pythium sensu lato* (both *Pythiales*) occupy extreme positions in the genera they characterize. Type species, characterized by their 18S rDNA data, which could be employed as outgroups now include: *Saprolegnia ferax*, *Leptolegnia caudata* and *Apodachlya brachynema* (*Saprolegniomycetidae*) and *Pythium monospermum* (*Peronosporomycetidae*) (Dick *et al.*, 1999).

There is no possibility of obtaining information from extinct taxa to qualify probabilities. In any systematic and phylogenetic (evolutionary) molecular reconstruction it is essential to recall that only relationships between ***extant*** species will be displayed. Taxa of extinct lines, which might have modified the cladograms, cannot be assessed.

The basis for phylogenetic placement and relationships within the straminipiles depends, very largely at present, on long sequences of nucleotides in the gene encoding for ribosomal RNA. It is possible that one part of one gene is sufficient to establish a robust cladistic framework, but justification and support is normally required (see Doyle, 1992). In angiosperm phylogeny ***three*** independent genes are being used (Soltis, Soltis, Chase, *et al.*, 1998*a, b*; Soltis, Soltis & Chase, 1999; The Angiosperm Phylogeny Group (APG), 1998). For entirely understandable reasons, the independent, endosymbiont genes most studied in straminipiles are ***either*** in the photoendobiont ***or*** in the mitochondrial endosymbiont (heterotrophs), so that comparability is lacking across the whole kingdom. Other genes have not yet been studied in sufficiently large samples of straminipilous fungi or other straminipiles to enable construction of a robust phylogenetic hypothesis, similar to that for angiosperms.

Another question relates to the number of informative, variable sites within the sequences that are necessary to give adequate characterization and separation within a particular group of related taxa: the region for data analysis must contain sufficient differences in sequences to allow closely related species to be separated; these differences should be the result of a single base change and be free of length mutations. Berbee *et al.* (1998) have shown, with ascomycetes, that while shorter sequences are sometimes adequate, there are some taxa for which much longer sequences are essential. It will be necessary to characterize the straminipiles in this respect.

Evolution is on-going. Species concepts (both real and postulated) vary widely, even in a single genus. Incipient speciation (sibling species) will occur. Isolation and/or modification of the gene pool may not initially be correlated with or represented by morphological attributes. Molecular biologists concerned with identification markers have not yet addressed the problems posed by ***recent*** speciation caused either by ploidy change (e.g., from diploid to tetraploid) or by change in genes other than the rDNA gene, which so affect the physiology of the organism that a new, but as yet cryptic species (i.e., a species lacking any ***recognized*** morphological and diagnosable discontinuity) has evolved. In neither of these scenarios would differences in any part (18S- 28S- or ITS) of the rDNA necessarily show change. The molecular biology must not be given disparate weight with respect to morphology and physiology. Intra-specific population diversity and formally defined infra-specific taxa (subspecies, varieties and forms) require reassessment; genetically controlled host/parasite associations which are characterized ***only*** by their hosts, are known as *formae speciales*. Similarity, even identity, in nucleotide sequences with respect to one gene may be yoked to variation in another gene which codes for such host-specific functional differences.

There is no absolute time-scale for rates of molecular evolution, but eventually the molecular phylogeny should be integrated with geological time. The molecular clock will be influenced by rates of base substitutions, life-histories and sexual systems. Van der Peer *et al.* (1996) have provided a more accurate formula for calculating 'substitution rate calibration' to avoid anomalies caused by the presence of extremely long branch sequences with high evolutionary rates. If a molecular clock cannot be determined, the apparent evolutionary distance, as represented by nucleotide sequence changes, will not necessarily be the same as the absolute evolutionary time-scale for all organisms (the 'sloppy' clock hypothesis of Doyle, 1992). For the straminipiles generation times and population sizes will have a profound effect on the absolute time scale of change. The diatoms and other marine straminipilous unicells have enormous populations, short generation times and sexual reproduction is rare. *Hyphochytrium* (and all described members of the *Hyphochytriales*) and *Halophytophthora* are apparently constitutionally anamorphic; almost nothing is known of chromosomal or genetic stability in these two genera. In *Aphanomyces* there may be many asexual generations between sexual events (Dick, 1970). Furthermore, the rates of evolution of mitochondrial genes and nuclear genes may differ by a factor of 10 in other organisms. For a well founded phylogeny of straminipiles, the factorial differences between the genes selected should be clarified.

The stability of the cladistic arrangement has yet to be established. The placement of some ordinal branches within the *Peronosporomycetes*, such as the *Leptomitales* (Dick, *et al.*, 1999; Riethmüller, Weiss & Oberwinkler, 1999; Hudspeth, Nadler & Hudspeth, 2000; Cook, Hudspeth & Hudspeth, 2000), is still equivocal, even after analysis of long (>1800) nucleotide sequences from 18S rDNA. A comparable situation holds for the photosynthetic straminipiles (Potter *et al.*, 1997). Association depends on the algorithm used. Additional, independent data are needed. Positioning of so few, deeply rooted taxa in cladograms can also be influenced by the size of the data base and the outgroups used. Divergent orders with very few known species, such as the *Leptomitales*, present problems when interpreting cladograms. It must also be recognized that the addition of new information may affect the branching of the cladogram. In all cases it is desirable to rationalize cladogram differences with structural features (the 'common sense' factor).

Diagnostic criteria based on molecular biology will become more firmly based when techniques have been developed and used for the extraction and sequencing of nucleic acids from unicellular endobiotic straminipilous parasites in straminipilous hosts. At present, three kinds of approach have been employed for readily cultured straminipilous fungi: the study of nuclear genomic DNA, initially as nucleotide ratios but now largely concerning the rDNA gene and its substructure; and mitochondrial DNA.

Nucleotide ratios

The first report of a DNA G:C ratio for a peronosporomycete was that for *Phytophthora infestans* (Clark *et al.*, 1968). Storck & Alexopoulos (1970) provided comparative data for other saprotrophic and facultatively parasitic species. Green & Dick (1972) gave a more detailed analysis for the DNA of other *Saprolegniaceae* and reported the presence of 'satellite' DNA, which was later shown to be mitochondrial DNA (Clark-Walker & Gleason, 1973; Neish & Green, 1976). Bahnweg & Jäckle (1986) have shown a wide range of G:C ratios for thraustochytrids. In CsCl-bisbenzimide gradients the rDNA repeating units produce a third band which facilitates separation (Belkhiri & Dick, 1988), and Martin (1990) has reported the presence, in some species, of a fourth band which he attributed to fragments derived from junctions between genomic DNA and rDNA. Belkhiri & Dick (1988) provided a basis for comparison between species of *Pythium* based on these gradients.

The rDNA gene

The rDNA gene has several regions which have been used for comparative sequencing: the small subunit (18S); the large subunit (28S, including the 5·8 segment); and the 5S and ITS1 and ITS2 regions. Most deep phylogeny still depends on the sequences of the small subunit (18S) of the rDNA gene. No other sequences can compare, in the numbers and diversity of organisms assessed, with 18S rDNA at this stage. Phylogenies based on 18S rDNA are well-established for straminipilous organisms and largely confirm prior taxonomic conclusions from kingdoms down to orders: relationships between families and genera are more open to debate.

Phylogenetic relationships based on the 18S rRNA analyses of a few members of the *Straminipila* have been made by Gunderson *et al.* (1987: *Achlya* and chrysophyte algae), Tan & Druehl (1993: *Fucophyceae*), Leipe *et al.* (1994: *Cafeteria* and *Labyrinthuloides*), Leipe *et al.* (1996: *Developayella*, *Labyrinthuloides* and *Proteromonas*) and Potter *et al.*, (1997: chromophytes). These data have been augmented by Förster *et al.* (1990: *Blastocladia*, *Phytophthora* and *Lagenidium*) and quoted in relation to other fungi by Illingworth *et al.* (1991) and Geber *et al.* (1992). Until recently the data for *Achlya*, *Phytophthora* and *Lagenidium* were the only 18S rDNA data available for deep phylogenetic analysis: note that the phylogenetic trees published by Illingworth *et al.* (1991) and Geber *et al.* (1992) showed different relative evolutionary distances for the same *Lagenidium* and *Phytophthora* data, perhaps because of the different algorithms used. A restriction map of the rDNA of *Lagenidium giganteum* shows that this species shares more sites with *Pythium* than it does with species of the *Saprolegniaceae* or *Leptomitaceae* (McNabb, 1989). Unpublished data (Lee &

Fuller, 1992) apparently provided information on the genetic relatedness of representatives of other major taxonomic groups of straminipiles, including *Salilagenidium*. The genetic distance between *Lagenidium* and *Phytophthora* is shorter than that between either and *Achlya*; the three *Peronosporomycetes* are more closely related to chrysophyte algae and diatoms than to other organisms. The phylogenetic value of genetic distance deduced by such analyses is not in dispute, but as Förster *et al.* (1990) have commented, the assessment of relatedness at particular genetic distances may depend on the appropriateness of the relative conservation (the numbers of informative sites) and not just the length of the subunit (number of nucleotides) that is selected. Förster *et al.* (1990) regard small sequences such as that for the 5S rRNA as inappropriate for extreme evolutionary distances, thus caution is necessary when interpreting data, for example, for the evolutionary position of the thraustochytrids (compare MacKay & Doolittle, 1982; Huysmans *et al.*, 1983; Hori, Lin & Osawa, 1985; Izzo, Lee & Porter, 1994).

Castlebury & Domier (1998) have presented 18S rDNA sequences for *Plasmodiophora brassicae*. More comparative data, based on about 800 bp-sequences for *Polymyxa* and *Plasmodiophora* has been provided by Ward & Adams (1998). They concluded that the wider relationships of the plasmodiophorids were unclear, but that the closest matches were with the straminipiles. A much larger range of outgroup organisms, a longer set of sequences for analysis and additional independent data will be necessary before a deep phylogeny is apparent.

The large subunit of the rDNA gene is now being used for phylogenetic analysis (Petersen & Rosendahl, 2000; Riethmüller *et al.*, 1999; Van der Auwera, Chapelle & De Wachter, 1994; Van der Auwera & De Wachter, 1997, 1998; Van der Auwera *et al.*, 1995). In general, the data support those from 18S rDNA, but the different ranges of organisms studied make comparisons difficult. The 28S rDNA data are particularly important in placing the hyphochytrids (Van der Auwera *et al.*, 1995). The relationships of the *Rhipidiales* and the *Leptomitales* with the *Saprolegniomycetidae* or *Peronosporomycetidae* are again inconclusive.

Restriction mapping has also revealed variability in the intergenic regions of the rDNA (Klassen, McNabb & Dick, 1987; Klassen & Buchko, 1990; Martin, 1990). Information (mainly restricted to shorter sequences) is known for straminipilous fungi, especially *Phytophthora* and *Pythium* (Briard *et al.*, 1995; Cooke *et al.*, 1996, 1999; Cooke *et al.*, 2000; Herrado & Klemsdal, 1998; Lévesque *et al.*, 1993, 1994, 1998; Moller, de Cock & Prell, 1993). The molecular data support, only in part, the hierarchical classification within the *Peronosporomycetes* (compare Grosjean, 1992 [pers. comm., J. M. Duncan, Scottish Crops Research Institute]; Panabières *et al.*, 1997; Ristaino *et al.*, 1998). The robustness of the cladogram branching order, as supported by Bootstrap and Jackknife procedures, is not always secure (values $<75\%$ should be viewed with caution; for the very much larger multigenic angiosperm database, this value could be set at $<50\%$, pers. comm., M. W. Chase, Royal Botanic Garden, Kew).

Genetic relatedness as assessed by ITS1 sequences may indicate centres of speciation but not necessarily the evolutionary phylogeny. The same may also apply to the position of the 5S rDNA relative to the NTS of the rDNA repeat. Shorter sequences such as pertain to the ITS region are frequently used for identification, but in *Pythium* there are length mutations in this region so that analysis becomes highly dependent on sequence editing. In spite of this complication, the ITS region can be effective in distinguishing between closely related species. Cock *et al.* (1996) have pointed out that heterogeneous base distribution can affect the efficacy of restriction enzymes and this should enable the optimal selection of restriction enzymes for determining similarity in restriction fragment patterns.

One of the problems in attempting to equate molecular identity with morphological criteria has been illustrated for the species pair *Pythium graminicola* and *P. aphanidermatum*. Chen (1990) originally concluded that the two species were conspecific, but later more detailed studies demonstrated their distinctiveness (Chen, Hoy & Schneider, 1992). Chen (1993) preferred to use antheridial number to distinguish these two species whereas Dick (1990*c*) used oogonial morphometry.

Klassen *et al.* (1997) are now using amplification of the 5kb intergenic spacer (the D2 region) in the rDNA repeat unit to identify *Pythium* species. In a study of 5S rRNA gene family organization it was found that species of *Pythium* with globose sporangia (or sporangia unknown) do not have a 5S gene in the NTS of the rDNA repeat unit, but that species of *Pythium* with filamentous sporangia have an inverted copy of the 5S gene in the NTS (Belkhiri, Buchko & Klassen, 1992). The latter arrangement is also found in *Lagenidium giganteum*, *Pachymetra* and *Verrucalvus*. In *Phytophthora cryptogea* Pethybr. & Laff. and *Achlya klebsiana* Pieters the 5S gene is also present in the NTS, but in the non-inverted orientation. There is thus variation within both subclasses *Peronosporomycetidae* and *Saprolegniomycetidae* with respect to this character, although the relationship between *L. giganteum* and certain species of *Pythium* is probably close. Alignments in other unresolved regions may be similarly problematic elsewhere in the entire rDNA gene. Although the D2 region of the 28S rDNA gene appears to be less suitable for phylogeny (pers. comm., F. N. Martin, U.S.D.A., Salinas, Ca.), Briard *et al.* (1995) have used the variable D2 domain of the 5' end of the 28S rRNA to display relationships (as "phylogeny") between certain species of two genera of the *Pythiaceae*. More caution should be exercised when using limited (i.e., <200) sequences from relatively few species for phylogenetic conclusions.

Infra-generic molecular taxonomy

Förster *et al.* (1988), Förster, Oudemans & Coffey (1990) and Moller *et al.* (1993) have used the genomic and mitochondrial DNA to discuss relatedness within the genus *Phytophthora*. Förster & Coffey (1990) have also used similar data to provide sexual markers in heterothallic crosses of *Phytophthora* and Wu, Mathur & Rimmer (1994) have applied similar techniques for crosses between races of *Albugo*. Numerous studies have reported on the use of molecular biology for understanding the genetics of *Phytophthora infestans*, but this topic is outside the present considerations.

Intergenic repeats have been studied by Klassen & Buchko (1990), who found that *Pythium ultimum* isolates showed variation in a region downstream of the 3' end of the large subunit. Each isolate showed a unique pattern of heterogeneity in the subrepeat. They suggested that the simple sequence repeats (SSRs ≡ microsatellite DNA) may have implications with regard to pathogenicity and virulence. These data are therefore outside the criteria for taxon diagnosis.

The reverse blot technique, Random Amplified Polymorphic DNA (RAPD) technique and other comparable short sequence analyses are now being used to indentify individual species and to separate closely related species. Ultimately, techniques will be applicable to *in vitro* and in field studies (e.g., Wiglesworth *et al.*, 1994). At present, research has mostly been confined to *Phytophthora* (Belkhiri & Dick, 1988; Briard *et al.*, 1995; Cooke *et al.*, 1996; Förster & Coffey, 1991: *Ph. parasitica*; Förster, Learn & Coffey, 1995; Förster *et al.*, 1998; Förster, Oudemans & Coffey, 1990; Liew, Maclean & Irwin, 1998: *Ph. medicaginis*; Möller, Cock & Prell, 1993; Panabières *et al.*, 1989; Shumard-Hudspeth & Hudspeth, 1990) and *Pythium* (Belkhiri & Dick, 1988; Briard *et al.*, 1995; Herraro & Klemsdal, 1998: *P. aphanidermatum*; Klassen, Balcerzak & Cock, 1996; Klassen *et al.*, 1997; Lévesque *et al.*, 1993; Lévesque, Vrain & de Boer, 1994: *P. ultimum*; Martin, 1990, 1991, 2000; Martin & Kistler, 1990; Matsumoto *et al.*, 1999; Matthew, Hawke & Pankhurst, 1995; Rafin, Brygoo & Tirilly, 1995; Shumard-Hudspeth & Hudspeth, 1990). At present there is no consensus as to the most appropriate protocols to be followed. However, the screening for distinctiveness against rarer species has not always been thoroughly explored so that the reliability of the procedure is not yet assured. In any case, the distinctiveness of a species in its natural substratum or habitat will depend on features other that DNA characteristics.

The use of restriction fragment length polymorphism (RFLP) and other techniques for assessing infraspecific variation are being employed routinely for *Phytophthora* and *Pythium* but rarely for other genera of the *Peronosporomycetes*. References to work with *Phytophthora* include: *Ph. capsici*: Hwang *et al.* (1991); *Ph. infestans*: Carter *et al.* (1990); Judelson, Spielman & Shattock (1995); Newhouse *et al.* (1992); Tooley & Therrien (1987); *Ph. megasperma*: Förster *et al.* (1999); *Ph. megasperma* var. *glycinea*: Judelson *et al.* (1993); Mao & Tyler (1991); *Ph. sojae*: Mao & Tyler (1996); *Ph. parasitica*: Förster & Coffey (1990). References to work with other genera include: *Achlya*: Hudspeth *et al.* (1983), and *Bremia lactucae*: Hulbert *et al.* (1988).

Genes other than rDNA

Other nuclear genes which might be suitable for providing data on the deeper phylogenies of the *Peronosporomycetidae* include the actin coding regions (Bhattacharya & Ehlting, 1995; Dudler, 1990; Hightower & Meagher, 1986; Uncles *et al.*, 1997) and ß tubulin (Cameron *et al.*, 1990; Sogin in Bhattacharya, Stickel & Sogin, 1991). Actin gene studies by Dudler (1990: *Phytophthora*) and Bhattacharya *et al.* (1991: *Achlya*) have confirmed the straminipile phylogeny, although the genetic distance between *Phytophthora* and *Achlya* appears to be greater than that between *Achlya* and the brown alga *Costaria* (*Laminariales*, *Fucophyceae*). The discrepancies noted by Bhattacharya *et al.* (1991) between the phylogenetic trees using actin or 18S/16S rRNA for protists/bacteria suggest that origins derived from these cladistics are still debatable. Complications arising from the use of actin gene sequences may arise because of gene duplication. In *Pythium irregulare* four 'copies' of the gene sequence occur. From sequence analyses, these 'copies' do not always fall in the same clade (one grouped with those for *Phytophthora* species, pers. comm., F. N. Martin, U.S.D.A., Salinas, Ca.). Another gene receiving attention, but which has not been used for straminipiles, is the DNA-dependent RNA polymerase (Klenk, Palm & Zillig, 1994; Klenk *et al.*, 1995; Palenik, 1992).

Mitochondrial DNA

The mitochondrial genome in straminipilous organisms is characteristically large. The length of the mitochondrial genome has been shown to vary between 36.4 and 73.0 kb (McNabb & Klassen, 1988). In some taxa Martin (1995*b*) found the mitochondrial genomes to be linear but he did not regard this feature as having phylogenetic importance. For straminipilous fungal genera such as *Pythium* and *Achlya* (i.e., in both subclasses), there is an inverted repeat in the mitochondrial genome (McNabb *et al.*, 1987). The length of this inverted repeat in *Pythium* (27-29 kb - kilobase pairs) is quite different from that in *Achlya* (10 kb) or mycote fungal mitochondria (4-5 kb), but it is very similar to that in chloroplasts (20-28 kb) (Whitfield & Bottomley, 1983). An inverted repeat of comparable size is present in the *Rhipidiales* (McNabb *et al.*, 1987; McNabb & Klassen, 1988) but *Phytophthora* (Shumard-Hudspeth & Hudspeth, 1990) has only a short (0·5-0·9 kb) inverted repeat. Shumard *et al.* (1986) concluded that the mere presence of this inverted repeat was not responsible for any evolutionary stability of this feature. The presence of an inverted repeat enables variation in the two unique regions which separate the repeat during replication.

Studies which have focused on the mitochondrial DNA include those of Boyd *et al.*, 1984; Cock *et al.*, 1992; Förster *et al.*, 1988; Hudspeth *et al.*, 1983; Hwang *et al.*, 1991; Klimczak & Prell, 1984; McNabb *et al.*, 1987; McNabb & Klassen, 1988; McNabb, Eros & Klassen, 1988; Möller, Cock & Prell, 1993; Shumard, Grossman & Hudspeth, 1986; Shumard-Hudspeth & Hudspeth, 1990). Alignments of the *Cox* II sequence from the mitochondrial genome have been prepared by Hudspeth *et al.* (2000) and Cook, Hudspeth & Hudspeth (2000) for a wide range of *Peronosporomycetes*, including *Pythiaceae*,

Saprolegniaceae, *Leptomitaceae*, *Rhipidiaceae* and *Lagenidium giganteum*. The most striking feature is the tripeptide indel sequence:

absent			from all *Saprolegniaceae* so far examined
present,	as	-Leu-Glu-Phe/Tyr-	in *Pythiaceae* and *Rhipidiaceae*
present,	as	-Tyr-Thr-Asp-	in *Leptomitaceae*
present,	as	-Leu-Glu-Tyr-	in *Lagenidium giganteum*

Further variations have been revealed by Cook *et al.* (2000) for some of the marine taxa:

present,	as	-Phe-Ser-Leu-	in *Atkinsiella dubia*
present,	as	-Thr-Asp-Leu-	in *Halodaphnea okinawensis*
present,	as	-Leu-Glu-Tyr-	in *Salilagenidium callinectes* (as for *L. giganteum*)

Lagenidium giganteum groups most closely with *Pythium ultimum* (D. S. S. Hudspeth, pers. comm., Northern Illinois University, DeKalb). According to D. S. S. Hudspeth (pers. comm.), the mitochondrially encoded cytochrome *c* oxidase subunit 2 locus (*cox2*) may be more informative of phylogeny where the 18S rRNA shows insufficient differences.

Summary

The deep phylogenetic divide between the *Peronosporomycetidae* and the *Saprolegniomycetidae* within the *Peronosporomycetes* (de Bary, 1866; Dick *et al.*, 1984) has now been confirmed with 18S rDNA data (Dick *et al.*, 1999) and 28S rDNA data (Riethmüller *et al.*, 1999; Petersen & Rosendahl, 2000). This divide is supported by the mitochondrially encoded cytochrome oxidase (*cox* II) data of Hudspeth *et al.* (2000), Martin (2000) and Cook *et al.* (2000). However, the placement of *Sapromyces* (*Rhipidiales*, *Rhipidiomycetidae*) may fall either in the *Saprolegniomycetidae* with the *Leptomitales* (Petersen & Rosendahl, 2000) or with the *Peronosporomycetidae* (Hudspeth *et al.*, 2000) depending on the molecular data used. Nevertheless, outgroups for phylogenetic analysis *within* the *Peronosporomycetidae* may be obtained from the *Saprolegniomycetidae* for the *Peronosporomycetidae* and *vice versa*. This will enable comparisons of longer nucleotide sequences, perhaps with additional variable and informative sites, than more distant outgroups taken from outside the *Peronosporomycetidae*.

Grosjean (1992), using ITS1, placed *Peronospora viciae* with *Phytophthora infestans*, and *Albugo candida* with *Pythium insidiosum* and *P. echinulatum*; both placements were at ultimate branches of their cladogram. Hudspeth (D. S. S. Hudspeth, pers. comm.), using mitochondrial *cox* II sequences, placed *Peronospora tabacina* and *P. nicotianae* with *Phytophthora megasperma*, also at ultimate branch points; in contrast, *Albugo candida* was basal to their phylogenetic tree, along with *Hyphochytrium* and *Sapromyces*. Cooke *et al.* (2000) again using ITS sequences, have placed a small sample of *Peronospora* species parasitic in the Rosids and Asterids (see PART III; Table III: 2 and Figure III: 1) with an intermediate branch of *Phytophthora* which includes *Ph. infestans* (the type species), *Ph. nicotianae* and *Ph. megakarya*. It is noteworthy that there is a measure of agreement with the molecular biological conclusions of Grosjean (1992) and Hudspeth (D. S. S. Hudspeth, pers. comm.) with respect to this particular group of *Peronospora* species with a similarly restricted group of *Phytophthora* species. There is no agreement concerning the placement of *Albugo*, although unpublished work generally places *Albugo* at some distance from *Phytophthora* and *Peronospora* (see PART III).

However, when *Peronospora rumicis* (the type species, Corda, 1837) has been studied and shown to belong to this group, then the genus *Phytophthora sensu stricto* (type species *Ph. infestans*) would become a synonym of *Peronospora*, heralding a nomenclatural nightmare! One solution might be to set the monophyletic generic concepts at a very low heirarchical level so that neither *Phytophthora* nor *Peronospora* would need to be abandoned, but this would necessitate the simultaneous erection of numerous other genera from *Phytophthora sensu lato* and *Pythium sensu lato*. The alternative is to accept that the genus is paraphyletic. A meticulous taxonomic reassessment of historic generic names, and their type species, which have been placed in synonymy with *Phytophthora sensu lato* would be required for monophyly (see PART III: Table III: 3). Stamps *et al.* (1990) separated *Phytophthora* into six morphological groups but molecular studies have not entirely endorsed this division (compare Cooke *et al.*, 1996, 1999; Ristaino *et al.*, 1998). Dick (in Klassen, McNabb & Dick, 1987; Dick, 1990*c*: Venn diagram) suggested that there were perhaps five major centres of speciation within *Pythium*, based on morphological criteria, and exemplified by (1) *P. monospermum*, *P. torulosum* and *P. diclinum*, (2) *P. anandrum*, (3) *P. ultimum*, (4) *P. irregulare* and (5) *P. ostracodes* and *P. oedochilum*: all of these groups are supported by deep clades in the ITS data of Grosjean (1992), but again, these data should not be regarded as sufficiently robust at this stage. Data from Cook *et al.* (2000) suggest that *Phytophthora sensu lato* is similarly nested within *Pythium sensu lato* with *Lagenidium*.

The genera *Phytophthora sensu lato* and *Pythium sensu lato* may need to be retained for non-systematists, even though they may be paraphyletic, until, and even after, a concensus of relationships has been established.

The symplesiomorphic trait in the *Phytophthora* line that gave rise to the apomorphies of the genera of the Peronosporales is yet to be defined. To put nomenclatural order into the phylogenetic classification of the Peronosporales, it will eventually be necessary to make a large number of name changes at genus and family levels using comprehensive molecular, biochemical and morphological criteria. For example, this approach will be necessary for *Phytophthora undulata* (Dick, 1989; Mugnier & Grosjean, 1995) which is neither a *Phytophthora sensu stricto nor* a *Pythium sensu stricto*, while the *Pythium vexans* group almost certainly belongs to *Phytophthora sensu lato*, (cf. Dick, 1990*b* - morphology; Panabières *et al.*, 1997 - elicitor characteristics).

Genetic distance, as determined by a single gene, however important, ***on its own*** is ***insufficient*** to determine the limits of higher taxa. The fundamental criterion (or seminal event) that defines the node of the cladogram must also be identified. A diverse, species-rich and relatively ancient lineage will encompass extant species widely spaced by genetic distances. Some of these species will have lesser genetic distances from other species belonging to a much younger, derived, but significantly distinct major taxon. Thus short genetic distances, if they bridge the critical cladogram node, may appparently contradict robust higher taxon concepts based on several independent criteria.

The precise pattern of branching in the unrooted phylogenetic tree at large genetic distances may still be unreliable when data for whole groups of higher taxa are not available. The greatest problem lies in the fact that workers unfamiliar with a higher taxon (genus, family or order) may infer that data for one or two species placed in that higher taxon will adequately represent the higher taxon.

COEVOLUTION

Some attention must also be given to the possible evolution of obligate parasites, particularly the downy mildews and the thalloid endobiotic parasites. The evolution and coevolution of the downy mildews is considered in detail in PART III. It is concluded that a cophylogenetic coevolution is improbable. Coevolution is much more likely to depend on the availability of host secondary metabolites, particularly in relation to the more highly evolved angiosperms. The evolution of the downy mildews was probably instigated and driven by climate change in warm-temperate climates in the Cretaceous. Any coevolutionary pattern is therefore recent. It is possible that there was diversity in the geographic provenances of the different groups of downy mildews. The importance of physiological and biochemical attributes cannot be overstressed. It is remarkable that host/parasite relationships show a common pattern from the downy mildews, through the parasitic *Pythiaceae* and *Aphanomyces*, to the plasmodiophorids. Physiological traits and patterns of mycoparasitism (Pemberton *et al.*, 1990; Dick, 2000*c*) could also provide further insight to the interface between biochemistry and systematics.

In contrast to the advanced angiosperms, the evolution of the hosts for the thalloid endobiotic straminipilous fungi has a long geological history. Endobiotic parasites, such as the *Olpidiopsidales* and *Myzocytiopsidales*, appear to have a world-wide distribution. For many, the host range appears to be restricted and often within the *Charophyta* (*Charales*, *Zygnematales*), *Chlorophyta* (*Oedogoniales*), *Bacillariophyta*, *Peronosporomycetes*, *Aschelminthes* (*Rotifera* and *Nematoda*) and *Crustacea*. Most of these hosts are freshwater or wet-terrestrial, but a few are marine-littoral, estuarine or oligohaline. No evidence for a stenohaline requirement has been presented. Disjunct relict distributions could be invoked, but there is little evidence that could support this hypothesis. Similarly, morphological barriers do not seem to offer an explanation for host/parasite specialization. It is more probable that genetic adaptations have evolved which confine vulnerability to, and/or virulence of a pathogen by means of restricted ranges of chemical attractants. Thus, the endobiotic parasites would have evolved in association with their hosts. This would be particularly relevant if the association is parasitoidal (Newell *et al.*, 1977). Such a concept has been epitomized by creating a new binomial for each parasite in a novel host/parasite relationship, but this approach is not supported by cross-infection studies for *Myzocytiopsis* and *Haptoglossa* between nematode and rotifer hosts (Barron, 1977*b*, 1989*b*).

Proximity within a common habitat will facilitate the evolution of specific adaptations, as could occur between communities of *Spirogyra*, *Oedogonium*, desmids, *Lemnaceae* and other aquatic angiosperms, and animals and fungi associated with such a community. Close evolutionary connections between species which would involve both widely differing hosts ***and*** different ecosystems are rather less probable. Observations of host specificity in genera such as *Olpidiopsis* and *Petersenia*, where restriction is to particular species or groups of species in host genera (Shanor, 1940; Slifkin, 1961; Sparrow, 1960; Whittick & South, 1971,1972) is in accord with systems of biochemical recognition. Such data are difficult to reconcile with implications that species of a genus may be found in widely separated hosts and habitats, which in *Petersenia* range from marine red algae to grass leaves (*P. panicicola* Thirum. & Lacy), and in *Olpidiopsis* from freshwater *Saprolegniales* to marine red algae. In these parasitic associations I prefer to conclude that the existing taxonomy is suspect. While diversity in host organisms may occur for a particular genus, it is more probable that the more closely related taxa will share a common community, a related group of hosts, or both.

PART II

CLASSIFICATIONS

CLASSIFICATION OF THE STRAMINIPILOUS FUNGI WITHIN THE EUKARYOTE KINGDOMS

In order to discuss the systematics of the mycelial and the endobiotic straminipilous fungi in detail, it is necessary to review the precepts for the primary categories of flagellate eukaryote organisms. This is particularly important when substantial realignments are being proposed.

Relationships between photosynthetic and walled or naked heterotrophic straminipiles

In recent decades at least two comprehensive arrangements of higher taxa have been published (Parker, 1982; Margulis *et al.*, 1990). In addition, Cavalier-Smith (1981, 1986, 1989) has discussed the concept of his Kingdom *Chromista*; a concept which can ultimately be derived from a suggestion made by Sachs (1882). Margulis *et al.* and Cavalier-Smith have debated the criteria for the classification of the composite eukaryote cell with its mitochondria and plastids and the problems inherent in phylogenies involving reticulate evolution of endosymbiont organelles. The endosymbiont theory has a long history, with its origins in the work of Schimper (1883). Altmann (1890) was the first to suggest that the mitochondria had an independent origin and the symbiosis of plastids with host cells was originally proposed by Fimintzin (1907). However, it is only since the application of molecular biological techniques that scientific proof for the endosymbiont theory has been presented. Taylor (1978) has considered diversity in mitochondrial morphology with particular reference to the lower eukaryotes. Rothschild & Heywood (1987) discussed problems relating to plastid incorporation, pointing out that while the evidence for photosynthetic prokaryote incorporation was strong, it was not, at the time of writing, conclusive. Margulis *et al.* (1990) were more dogmatic: "mitochondria ... originated ... by several independent acquisitions [there are no data cited to support this statement]. Therefore mitochondrial characteristics in different protoctist lineages cannot be used as the basis for classification until details of mitochondrial polyphyly are available".

Different Kingdom concepts have developed from Copeland (1956) and Whittaker (1969) (see Leedale, 1974, for a brief review of earlier theories). Two opposed concepts are now current: Kingdom status is accorded to each of a cluster of diverging phylogenetic lines (animals, green plants, true fungi and straminipiles), based on rDNA and other characters (Margulis *et al.*, 1990); a single kingdom encompasses all of these major lines, perhaps excluding some eukaryotic protists with extreme genetic distances as determined by rDNA analyses (Sogin, 1989).

Margulis *et al.* (1990) based the Kingdom *Protoctista*, which includes all protozoans, all algae and the flagellate fungi, on the "ultrastructure and sexual patterns of the cytoplasmic (heterotrophic) components ... exclusive of the mitochondria, plastids ...". Although the simple logic of this concept is unassailable, it is fundamentally flawed by its inconsistent treatment of multicellular organisms. Large chromophyte algae with advanced cellular differentiation were included in the *Protoctista* by Margulis *et al.* while liverworts and angiosperms, rotifers and chordates, yeasts and toadstools were excluded and placed in the Kingdoms *Plantae*, *Animalia* and *Fungi* (*Mycota*) respectively. Cavalier-Smith (1986: 332) has pointed out the same inconsistency in a more detailed critique.

At this point reference should be made to the use of the term **straminipiles** (cf. Patterson, 1989). Patterson (1989) made only passing reference to the heterokont fungi and apparently countenanced certain exclusions ("some or all oomycete fungi" - Patterson, 1989: 374) from the core of chromophyte algae plus heterotrophic protistan organisms with heterokont flagellation. He introduced the colloquial name "stramenopiles" (a spelling which has been used in many recent publications (see the *Glossary* for corrected derivation from the Latin, which is essential when the name is used in a formal taxonomic and nomenclatural context). While the concept of straminipiles, based on a single diagnostic criterion of a flagellum with row(s) of tubular tripartite hairs has merit, no one has made a formal proposal for a higher taxon, the Straminipila. The economic importance of heterokont fungi is such that they cannot be ignored, nor left without a formal higher taxon. I therefore retain the ***classificatory concept*** of the taxon *Chromista* (Cavalier-Smith, 1981), the name of which is derived from earlier groupings of those algae which lack chlorophyll *b* (e.g., Chadefaud, 1950; Manton, 1965), but rename the kingdom and change the ***diagnosis*** to exclude the endosymbiont moieties.

The reversal of direction of water movement by the straminipilous flagellum with its two rows of tubular tripartite hairs (TTHs) is so important (Gray, 1939; Holwill & Peters, 1974; Holwill & Sleigh, 1967; Holwill, 1982; Taylor, 1952) that convergent evolution would be favoured. The ontogeny and morphology of the TTHs are too elaborate and too constant to support an hypothesis of convergent evolution. The grouping of straminipilous organisms was first postulated by me in 1969 (Dick, 1969). A concensus on the phylogenetic significance of this flagellar morphology has been established Cavalier-Smith (1986), Leadbeater (1989), Patterson (1989) and (Dick, 1990*a*). The straminipilous flagellum is associated, through its kinetosome, with the centrioles which are intimately involved in cell division: some of the microtubules of the kinetosome roots are also nucleus-related. The flagellum - kinetid - nucleus complex has probably evolved as a unit. The diversity of organisms possessing a flagellum with TTHs is thus monophyletic, and the levels of cellular differentiation reached in some groups justifies the separation of the chromists from the Kingdom *Protoctista* by the same tokens that the Kingdom *Plantae* (*Viridiplantae*) is separated.

Cavalier-Smith (1986) based his Kingdom *Chromista* on "all protists having either tubular ciliary mastigonemes or chloroplast endoplasmic reticulum or both". One of these criteria would be acceptable to Margulis *et al.* (1990), but chloroplast endoplasmic reticulum (CER) requires the presence of the plastid, and would thus be an inadmissible criterion. Plastids in chromophyte algae have additional membranes compared with green plants. The evolution of these additional plastid membranes in chromophyte algae has been reviewed by Cavalier-Smith (1989): he refers to the inner membrane of the CER as the periplastid membrane (PPM) and the outer membrane, which is confluent with the cellular rough ER, as the periplastid rough ER (PRER).

I diverge from both Margulis *et al.* (1990) and Cavalier-Smith (1986) with respect to the values to be placed on the endosymbiont components, the mitochondria and plastids. Mitochondrial phylogeny must be separately researched, but while this endosymbiosis may have occurred on separate occasions to provide several evolving lines, separated in space and time, Gray (1989, 1994) has postulated a ***single*** mitochondrial endosymbiotic event and ***subsequent*** separate mitochondrial phylogenies. The evidence from cytochrome *c* oxidase gene analysis (Sachay *et al.*, 1993) indicates a common phylogenetic origin for the mitochondria of *Phytophthora* and green plants. No explanation for the conservation of diverse mitochondrial cristal forms has yet been proposed. Certain combinations of endosymbionts may have proved to be more successful because of enzymic integration between host and endosymbiont. Gene transfer between mitochondrial and nuclear genomes could have stabilized such relationships, so that the mitochondrial internal morphology may or may not reflect phylogeny. The fact that in complex multicellular animals the mitochondrial morphology may differ from tissue to tissue is probably irrelevant to the evolution of mitochondrial form in simpler thalli. It is therefore over cautious (cf. Margulis *et al.*, 1990) to deny the use of mitochondrial form in phylogeny.

It is not necessary to assume that the TTH and the chromophyte photo-endosymbiont were coevolutionarily locked together. The monophyletic development of the 3-thylakoid, chlorophyll *a* plus c_1 c_2, PPM-PRER-enclosed endosymbiont, is of equal standing to the stacked grana, chlorophyll *a* plus *b*, PPM-PRER-lacking endosymbiont of green plants, even though some authors exclude the *Prymnesiophyta* (haptophytes) from this monophyletic line, preferring the hypothesis of at least two independent photo-endosymbiotic events (Whatley, 1989). Preisig (1999) has provided the most recent comprehensive review of the systematics and evolution of all algal groups. Trophic flagellar activity in sedentary protists (Sleigh, 1981)* probably occurred contemporaneously with motility conferred by flagella (Cavalier-Smith, 1986) in relation to the development of photo-endosymbiont eukaryotes.

If mitochondrial morphology is used, then a narrow concept for the chromists, which excludes the cryptophytes must follow: if the chromophyte photoendosymbiont presence is used, a broader concept, in which at least two mitochondrial morphologies are allowed, is evident. Neither equates precisely with the possession of TTHs. Data for 18S rDNA also indicate divergence between cryptophytes and chromophytes (R. A. Andersen, pers. comm.).

Margulis *et al.* (1990) use the development of sexuality as a prime feature for the organization of the Kingdom *Protoctista*. Cavalier-Smith (1975) has proposed that the tubulin development for organizing genome division enabled mitosis and meiosis to evolve at the same time. Dick (1987) has pointed out that only three eukaryotic ploidy cycles are possible and that in most organisms other morphological constraints will reduce the likelihood of interconversion between these ploidy cycles. Heath (1981) provided some interesting cluster analysis associations based on mitotic characters, but was restricted by the range of organisms studied and the gaps in the data base. Nevertheless, there was a peronosporomycete association, and rather diverse and unpredicted links between other straminipilous and non-straminipilous flagellate fungi. Nuclear cytology coupled with life-history studies identifying meiosis and karyogamy are thus likely to be important phylogenetic aids.

A cluster of characters can be assembled to describe the core of the chromists. The brown algae and peronosporomycetes are both straminipilous; have mitochondria with tubular cristae; diploid mitosis and sexuality; DAP lysine synthesis; glucopyranoside polymers (ß-1,3- hexose polymers) as a major wall constituent. Other major groups conform to this core, but there are additional taxa which may or may not be monophyletically included within the kingdom and there is a third group of taxa which are circumjacent even to this broader concept. Both circumjacent and derived groups may never have had, or may have lost some of the attributes associated with the chromistan core, and may have accrued additional diagnostic criteria. Consider the placement by Silberman *et al.* (1996) of *Blastocystis* (which has few other diagnostic features) with *Proteromonas* (for an account of *Proteromonas* see Margulis *et al.*, 1990).

Deep-phylogenetic discussions have been based on protist and algal studies. Mycological investigations have been fewer, with the notable exception of the studies by Barr and co-workers (see Barr & Désaulniers, 1989*a*) on the kinetosome and root structure of the flagellum. In protist, algal and fungal studies flagellar base studies are more valuable below the kingdom level, except for the separation of the *Chytridiomycota*.

---

*FOOTNOTE: It is tempting to reverse the priorities implied by Sleigh (1981, 1991) for the evolution of flagella for movement and flagella for feeding (Dick, 1997*a*). In primaeval anaerobic environments heterotrophy would not have needed light. The evolution of oxidative as opposed to sulphur photobionts would have had a devastating effect on the anaerobic communities. Apart from the adjustments to an oxidative environment, the buoyancy conferred by oxidative photobionts would have lead to concentrations of organisms at the water/gaseous environmental interface, giving shade. Motility would thus confer advantages for both phototrophs and heterotrophs. Moreover, oxidative metabolism would be necessary for the energy for motility; thus flagellar motility and mitochondrial endosymbiosis would be integrated, but not necessarily with photoendosymbiont development.

There are two further factors which have not always been given prominence in phylogenetic debates, but which are relevant to fungi. These are the broad environmental conditions in which a major taxon evolved and the importance of heterotrophy and parasitism in this evolution.

Flagellate organisms must be presumed to have originated in wet environments. Geochemical data (Holland, 1984: 461 ff) indicate that biological productivity was initially concentrated in the shallow seas of continental shelves, and that ocean salinity has been relatively stable, varying from 35‰ and probably never higher than 45‰ over geological time (but individual ionic concentrations may have varied more widely). This salinity range is less than that experienced diurnally in an intertidal rock pool subject to high insolation or a tropical downpour. Present day salinity diverges from 35‰ in open seas to low values of 12‰ in the Baltic Sea and 17·5‰ in the Black Sea due to river inflow (although benthic salinity in the anoxic conditions of the Black Sea is 38·5‰) and up to ten times greater than the open sea values (335‰) in the Dead Sea. Such variation related to topography has probably always existed; the critical question is whether its duration was long enough to have had a major evolutionary impact. Changes, and the durations of these changes, in rainfall and sea levels would have been critical. Holland (1978) has suggested that continental drift probably had profound effects on patterns of rainfall distribution and the development of playa lakes, and Holland (1984) has concluded that sea levels may have fluctuated by 350m, which would have changed the extents of seas and lakes with different salinities. Although adaptations to different salinities have occurred frequently in minor components of most groups of organisms, for extant flagellate organisms, evolutionary development and adaptive radiation appears to have been in ***either*** marine ***or*** freshwater environments. Few, if any, well-researched groups are held to straddle this dichotomy from their inception. For example, diatoms, now ubiquitous, evolved in the marine environment (Round & Crawford, 1990), became prominent in the Cretaceous, but had only a 31% survival after the terminal Cretaceous event (compared with a 69% survival of higher plants) (Emiliani, Kraus & Shoemaker, 1981) and only subsequently became evident in freshwater in the Miocene. Most of the core straminipilous groups can be placed firmly in one or other environment, with secondary colonization of the alternate milieu (Table II: 1 - algal data from John, 1994).

While the Cretaceous terminal event was critical for the flora and fauna that exist today, the main evolutionary lines of the straminipilous organisms were probably determined much earlier. For example, Dick (1988; 2000*a*) has proposed that the *Phytophthora - Peronosporales* evolutionary line must have been established long before the late Cretaceous. Many geological time scales indicate a Silurian (450 m. y. B.P.) origin for the first land plants, but the origins of freshwater organisms are obscure. Did freshwater organisms evolve in estuarine lagoons, playa lakes or both?

I think it unwise, without strong evidence, to derive internal phylogenies which disregard this marine/freshwater divide (note the comment, which could be misconstrued, by Barr & Désaulniers (1987*b*) that *Lagena*, a parasite of grasses, has a flagellar transitional zone reminiscent of the brown algae in that it lacks the concertina or helix, presumably by independent loss). In the chromists the photoendosymbiont is thought to be derived from a red algal eukaryote (because of the multiple PPM - PRER enclosures). The origins of the principal chromist groups (*Fucophyceae*, *Bacillariophyceae* and *Chrysophyceae*) were therefore probably marine. The probability that *Salilagenidium* and *Halophytophthora* represent secondarily marine evolutionary developments requires assessment, as do the evolutionary positions and origins of the biflagellate families *Sirolpidiaceae*, *Pontismataceae and*, *Eurychasmataceae* and the *Anisolpidiaceae*, all of which are marine, but not yet proven to be straminipilous.

Heterotrophy and parasitism also pose problems for circumscribing the kingdoms. Heterotrophy may involve phagocytosis (animal-like) or fluid assimilation (osmophagy), normally across an intervening cell-wall (fungus-like). Phagocytosis can coexist with photosynthesis. In the chromists, according to Cavalier-Smith (1986), saprogenicity and parasitism involved the loss of any photosynthetic ability and the concommitant loss of associated morphology. Cavalier-Smith, Chao & Allsopp (1995) have provided further

TABLE II: 1. *The major taxa of the* Straminipila *and circumjacent taxa, listed according to their posssible original habitats. All groups now have extant representatives in both marine and freshwater. The approximate numbers of described species are indicated, but some of these groups may have many more undescribed species.*

Marine	*number of species*	**Freshwater**	*number of species*
CORE ALGAL AND FUNGAL GROUPS			
Algae			
Fucophyta (phaeophytes)	*ca* 1500		
Bacillariophyta	>10,000		
		Eustigmatophyta	*ca* 12
Dictyochophyceae	*ca* 10		
Pedinellales	*ca* 10		
Pelagophyceae	*ca* 10		
		Chrysophyceae	>1000
		Synurophyceae	*ca* 200
		Tribophyta (xanthophytes)	*ca* 600
Fungi			
Labyrinthulista	*ca* 40		
		Peronosporomycetes	*ca* 1500
		Hyphochytriomycetidae	*ca* 20
PERIPHERAL ALGAL AND FUNGAL GROUPS			
Algae			
Prymnesiophyta (haptophytes, including *Pavlovales*)	*ca* 500		
		Glaucocystophyta	*ca* 10-20
Cryptophyta	*ca* 200		
		Raphidophyta	*ca* 30
Fungi			
		Plasmodiophoromycetes	*ca* 48
CORE AND PERIPHERAL PROTISTAN GROUPS			
Choanomastigota	*ca* 120		
		Euglenida	*ca* 900
Bicosoecida	*ca* 45		
		Kinetoplastida (bodonids)	*ca* 550
Actinophryida (*Heliozoa* in part)	*ca* 100		
		Slopalinida (*Opalinidida* in part) *Proteromonadida*	*ca* 5
Opalinidida	*ca* 400		

support for the hypothesis that the heterotrophs have been derived from phototrophs in certain cases. To make this assumption for all heterotrophic straminipilous fungi, the fact is ignored that these fungi differ from freshwater photosynthetic chromists in many ways, for example: their coenocytic protoplast (except *Vaucheria*), wall chemistry (Parker *et al.*, 1963), haplomitotic B ploidy cycle and the absence of flagellate gametes.

There is no solid evidence to support a ***secondarily*** heterotrophic habit for the *Peronosporomycetes*; on the contrary, Berg & Patterson (1986) and Nes (1990) suggest that the evolution of the *Peronosporomycetes* from photosynthetic ancestors is ***unlikely*** because of the direct synthesis of lanosterol from squalene oxide in sterol synthesis, rather than via the cycloartenol pathway. Several heuristic cladograms of 18S rDNA relationships for the bicosoecids, slopalinids, labyrinthulids, hyphochytrids, photosynthetic straminipiles and the now much larger sample of *Peronosporomycetes* have been prepared (M. A. Spencer & M. W. Dick, unpublished)*. All the most robust of these cladograms place the straminipilous algae as a major clade subsidiary ***from*** the clade leading to the straminipilous fungi. Thus, the proposition for the evolution of the straminipilous fungi from the straminipilous algae by ***loss*** of photosynthetic ability is by no means proven.

The same is not the case for the thraustochytrids, where the pattern of biosynthesis of the polyunsaturated fatty acids suggests a phytoflagellate ancestry (Ellenbogen *et al.*, 1969). Biochemists have tended to assume a simple plant/animal divide which may not help in discussing heterotrophy within multikingdom frameworks. A wider comparative base is necessary. The relative lack of internal grouping within the 18S rDNA cladograms for the marine freeliving and gut comensal/parasitic heterotrophs also points to the need both for a larger body of data and for additional, independent supportive criteria. Leipe *et al.* (1994), in relation to *Labyrinthuloides* and *Cafeteria*, noted that colloquial names have value when the generic and specific names are '*sedis mutabilis*'. These authors also note that molecular biological data are equivocal with respect to plastid derived or non-plastid derived pleisiomorphy. In relation to the tribophyte algae, Potter *et al.* (1997) have pointed out that deep branching order is difficult to determine using only the 18S rDNA gene sequence comparisons. Additional data (e.g., from other genes) may be necessary before a consensus higher classification can be achieved.

The complex sequences of endosymbioses and membrane loss, accepted for the chromophyte algae because of their PPM-PRER vestiges of endosymbiosis, could lead to either the host or endosymbiont nucleus becoming dominant in the composite cell (Lee, 1989). Fundamental differences in membrane composition might reflect such alternative origins, but have not been sought or researched. Detachment of the nucleus from the PPM-PRER, as in diatoms, raphidophytes and *Vaucheria*, all of which have numerous chloroplasts, is presumably a further derived state, possibly involving a change between uninucleate and coenocytic protoplasts. Thus it might be possible to erect a seemingly improbable hypothesis involving more recent reticulate evolution between eukaryotes. Perhaps *Vaucheria* and the *Peronosporomycetes* do have a common ***cytoplasmic*** ancestor, but with the cytoplasm now governed by ***differently originating***, but still straminipilous (and therefore heterokont flagella-linked) nuclei!? Such an hypothesis would involve a progenitor which was phagocytic, at least in one phase of its life-history, or a progenitor with an elaborate extrusome armoury.

---

*FOOTNOTE: References, where available, or GenBank citation data, for 18S rDNA data for heterotrophs are as follows: *Blastocystis* (Silberman *et al.* 1996); *Cafeteria roenbergensis* (Fenchel & Patterson, 1988; Leipe *et al.* 1994); *Developayella elegans* (Tong, 1995; Leipe *et al.* 1996); *Labyrinthuloides haliotidis* (Bower 1987*a*; Leipe *et al.* 1996); *Labyrinthuloides minuta* (Watson & Raper 1957; Leipe *et al.* 1994); *Picophagus flagellatus* gen. et sp. nov. (*Heterokonta*) - Guillou, L., Chretiennot-Dinet, M. J., Boulben, S., Moon-van der Staay, S. Y. & Vaulot, D. (*Protist* - in press); *Proteromonas lacertae* (Leipe *et al.* 1996); *Protoopalina* sp. (*Slopalinida*) - Affa'a, F. M. & Hickey, D. A. (unpublished); *Rhizidiomyces apophysatus* (Hausner, G., Belkhiri, A. & Klassen, G. R. - unpublished); *Syluania monomastiga* gen. et sp. nov. (*Bicosoecida*) - Karpov, S. A., Kersanach, R. & Williams, D. M. (*European Journal of Protistology*); *Symbiomonas scintillans* gen. et sp. nov. (*Heterokonta*) - Guillou, L., Chretiennot-Dinet, M. J., Boulben, S., Moon-van der Staay, S. Y. & Vaulot, D. (*Protist* - in press); *Wobblia lunata* gen et sp. nov. (*incertae sedis*) - Moriya, M., Nakayama, T. & Inouye, I. (*European Journal of Protistology*) [see Tong, 1997: ≡ *Pendulomonas adriperis* gen. et sp. nov.]; labyrinthulids and thraustochytrids - Honda, D., Yokochi, T., Nakahara, T., Raghukumar, S., Nakagiri, A. & Schaumann, K. (*Journal of Eukaryotic Microbiology* - in press).

Parasitism often shows coevolutionary links and may provide supporting evidence of phylogenetic affinity (Dick, 1988). In the present context, the fact that many of the obligate parasites of the *Saprolegniales* and *Pythiales* are placed in the *Olpidiopsidaceae* and *Plasmodiophoraceae* may be significant. In fact, the *Plasmodiophoraceae* show a co-evolutionary pattern of host range similar to that of the *Peronosporomycetes* themselves.

Where is the line to be drawn which will mark the origin of the straminipiles from circumjacent groups, and how are circumjacent groups to be distinguished from derived offshoots of these straminipiles? The post-eocyte population that gave rise to the straminipiles must have been:

1 phagotrophic (to enable endosymbiont incorporation of mitochondrial and plastid procursors)
2 it must have evolved a two-flagellar kinetosome-centriole complex
3 its flagellation was probably anisokont
4 it possessed the capability of DAP lysine synthesis
5 it must have had a dictyosome-derived secretory system with the potential capability of producing tripartite tubular hairs
6 it would have had machinery for producing extracellular deposits of glucan polymers, chitin and silica either as loricas, scales, frustules or cell walls
7 it was probably marine

The energy requirements for the above attributes probably indicate that mitochondrial incorporation had been achieved prior to photoendosymbiosis*. This is in accord with the view of Cavalier-Smith (1991) that the first metakaryote was a non-photosynthetic phagotroph.

Cavalier-Smith (1986) includes in his kingdom the cryptophytes and the haptophytes (prymnesiophytes) because of their photo-endosymbionts. The kingdom concept is therefore very broad indeed, embracing organisms with pellicles (intra plasmamembrane "walls"), flattened finger-like mitochondrial cristae, and non-straminipilous flagellation. The circumjacent groups excluded, but equally worthy of consideration are:

1 choanoflagellates (phagotrophic, loricate forms with a lorica of cellulose, chitin or silica, a single flagellum but with a second, obliquely-placed centriole, flattened cristae, no sex)
2 euglenids and bodonids (phagotrophic, or photosynthetic by the incorporation of eukaryote chlorophyte-derived endosymbionts, endoplasmamembrane pellicle, anisokont flagellation, discoid cristae, no sex)
3 glaucocystophytes (photosynthetic by the incorporation of cyanobacteria-derived endosymbionts, naked or with a cellulosic or non-cellulosic cell wall, two differently directed flagella both with simple (?) hairs, flattened cristae, no sex)

If the broad kingdom concept is accepted, then there is a case for including the *Plasmodiophoromycetes* in this kingdom. There may be merit in this broad kingdom concept, but it makes the definition of the kingdom very difficult, with anisokont flagellation, ***not*** heterokont flagellation, being almost the only consistent character. The ***narrow*** kingdom concept (which I favour because it is more probably securely monophyletic) would need to exclude both the cryptophytes and the haptophytes, even though they have chromophyte endosymbionts. It is worth remembering that some dinoflagellates, excluded from Cavalier-Smith's *Chromista*, also have chromophyte endosymbionts. Although D. J. Patterson (pers. comm.) did not intend his "stramenopiles" to have formal taxonomic status, the stem is pre-eminently suitable for renaming the Kingdom *Straminipila* based on a more restricted diagnosis, which includes the more or less aniskont flagellation and the possession of straminipilous appendages, but excludes for the present, the characters of the endosymbiont organelles.

On the other hand, Cavalier-Smith's broad concept is pragmatically useful here, because by so doing, attention is drawn to the problems without presuming to make decisions on circumjacent groups outside the

---

*FOOTNOTE: Such a proposal would accommodate the *Pedinomonadales* (Melkonian, 1990) as a circumjacent group to the chromists despite the photoendobiont having chlorophylls *a* and *b*.

scope of this monograph. For example, the case for the inclusion of the *Chytridiomycetes sensu* Sparrow (1960) in the Kingdom *Mycota* is not as strong as is sometimes claimed (e.g., Cavalier-Smith, 1987; Barr, 1992; Bowman *et al.*, 1992; Tehler *et al.*, 2000). The full molecular and structural diversity that may eventually be shown for all the protist:fungi divides (including the endoparasitoids discussed here) has yet to be assessed. The relationship between the *Chytridiomycetes* on the one hand, and the *Zygomycotina*, *Ascomycotina* and *Basidiomycotina* on the other hand, depends on some shared biochemical criteria, which while important, are also found in association in the Kingdoms *Straminipila* and *Animalia*. The molecular biological data, based purely on genetic distances (Bowman *et al.*, 1992) undoubtedly reveals a clade including the *Chytridiomycetes*, *Ascomycotina* and *Basidiomycotina*. However, the placement of the *Chytridiomycetes* in the *Mycota* depends on whether the loss of flagella and the arrival of mycelial development should be regarded as the ***fundamental*** determinants of this ***fungal*** lineage, just as the lack of dikaryophase separates the *Zygomycotina* from the *Ascomycotina* and the *Basidiomycotina*, and the switchin the major and minor growth axes of the regenerating dikaryophase cell determines the divergence of the *Ascomycotina* and the *Basidiomycotina*. It could be argued that restricted concepts for the Kingdom *Straminipila* (core groups) and the Kingdom *Mycota* (excluding all the flagellate fungi) might be a more satisfactory arrangement, but this would also require a much more thorough and comprehensive reappraisal of the relationships between the *Chytridiomycetes* and the circumjacent protoctist groups discussed above. In his diagram, Scherffel (1925*a*: facing p. 39) derived both the straminipilous fungi and the chytrids from monad algal parasites based on *Aphelidium* and other genera (see p. 263). A reevaluation of the modern systematic position of these protists could be relevant.

Cladistic, phylogenetic and phenetic analyses and classifications

Cladistic analyses will be inefficient or inappropriate when reticulate evolution is involved and have necessarily been based on the use of a very few characters and incomplete coverage of all major groups. The addition of either a new character or a new group can and does change the parsimony. This is particularly true when cladistic analyses fail to include obligate, endobiotic fungal parasites. Until molecular biological techniques have been developed and become routine for endobiotic straminipilous fungal parasites of straminipilous fungi, reliance will still have to be placed on other criteria to characterize the endobiotic parasites, such as ultrastructure and biochemical pathways.

Equivalent taxonomic groups may be of very different breadths with respect to numbers of species and/or their biochemical and physiological diversity. Such differences may or may not be independent of evolutionary age. For example, the *Peronosporomycetidae* is species-rich, with relatively little biochemical diversity compared with the *Saprolegniomycetidae*. However, the latter subclass may be the derived lineage on the basis of 18S rDNA and 28S rDNA. Cladistic diagrams based on only one or two representatives of taxa of such different circumscription or biodiversity are then liable to provide a distorted view of relationships, especially when the cladogram is based on a single character, irrespective of the evolutionary importance of that character. At very high taxon levels there may be undue dependence on ultrastructural criteria such as the flagellar apparatus, the nucleus and its division, and the origins of vacuolar systems and their end products. At least some of these criteria may show diagnostic diversity at a slightly more subordinate hierarchical level.

The tacit acceptance (e.g., Sparrow, 1960; Karling, 1981) that obligately parasitic species with limited host ranges can be both closely related and have diverse hosts in widely disparate ecosystems must be rejected (Dick,2000*c*). Within the *Peronosporomycetidae* it is becoming apparent that long-accepted family and genus relationships are polyphyletic when viewed from a molecular biological perspective. To what extent should a working classification and nomenclature be based on phenetic or phylogenetic criteria?

In any systematic and phylogenetic (evolutionary) reconstruction four factors are paramount: it is essential to recall:

1. phylogenetic systematics is based largely on explaining relationships between extant species
2. molecular data may not provide diagnostic traits for particular nodes while cladistic classifications require that the clade nodes be diagnosable and taxonomically valid

3 taxonomically valid hierarchical systematic ranks will not necessarily be co-ordinate because several branch steps may be encompassed within one hierarchical category

4 the type species must be characterized and placed in any systematic study

The lineages defined by molecular data do not necessarily relate directly to prominent morphospecies attributes (dependent on different genes) of the groups of species sampled and it is rare for mycologists, using molecular biology, to attempt to correlate clade nodes determined by particular sequences within the gene(s) being studied with nodes based on morphology or biochemically significant functional attributes. Molecular and morphological data should, therefore, be complementary. In systematic classifications the taxa are grouped according to several or many shared (or partially shared) characters. Inevitably, there is a mismatch between these approaches to phylogeny and evolution. The probable phylogenies produced through molecular biology cannot usually take account of extinct taxa: the relationships presented are thus potentially flawed. Sister groups, often strongly supported by bootstrap and Jackknife procedures may be readily identified from the data available but it is not always possible to distinguish between (derived) apomorphous and (ancestral) plesiomorphous character states. At ultimate branches single cladistic characters may be insufficiently diagnostic, so that a suite of characters is necessary for separation at genus or species level. With finger-printing techniques separation proceeds through infraspecific taxa to populations, clones and individuals. Because of its progressively bifurcating nature, the molecular biology cladogram is not always best suited for establishing correlations (i.e., discontinuities) with currently recognized hierarchies in systematics.

The 18S rDNA genes are well-established tools for phylogenetic reconstruction and often confirm prior taxonomic conclusions from kingdoms down to orders: relationships between families and genera are more open to debate. Species concepts vary widely, even in a single genus. Diversity in an ancient lineage may coexist with a more recently evolved, but fundamental, attribute which so changes the evolutionary potential that the erection of a higher taxon is of practical value. Computer-generated similarity and evolutionary distance indices may reflect probable lineages, but these will not negate differing hierarchical systematic concepts. At present, it is not clear to what extent the molecular data actually conflict with non-molecular characters rather than with the essentially intuitive classifications based on such characters. Combining morphological, morphogenetic and biochemical evidence with that of DNA sequences ought to be a priority of future research.

Powell & Blackwell (1995, 1998) have discussed the phenetic characters used in the generic classification for the *Saprolegniaceae* with the view of supporting the traditional generic limits. However, recent molecular biological studies (Dick *et al.*, 1999; Riethmüller *et al.*, 1999) suggest that these basic tenets may be misplaced. Molecular phylogeny shows fairly close resemblance, at this stage, to the disposition of genera in Dick (1990*a*: fig 6). In particular, the divergence of *Leptolegnia*, *Aphanomyces* and *Plectospira* from *Saprolegnia* and *Achlya* was predicted on morphological criteria. A reappraisal of the morphological features of these three genera reveals a number of character states common to these genera and not found together in the Saprolegniaceae *sensu stricto*. The characters relate to: the ooplast; zoosporogenesis; zoosporangium differentiation; intussusception hyphal wall material; tendency to modal uniovulate oogonia. Traditional generic characters (such as cymose zoosporangial renewal - *Achlya*, *Pythiopsis* - and the formation of hollow balls of zoospore cysts at the zoosporangial orifice(s) - *Aphanomyces*, *Achlya*) are not supported by the phylogeny. The positions of *Achlya bisexualis* and *A. apiculata* suggest polyphyly in *Achlya*, a point made many years ago, on molecular data, by Green & Dick (1972). The molecular biological data of Molina, Jong & Ma (1995) are too meagre and too skewed to support the characterization of *Saprolegnia*.

Well-known 'phylogenetic' schemes for the *Peronosporales* (Shaw, 1981; Barr 1983) have been questioned by Dick *et al.* (1984) and Voglmayr & Greilhuber (1998). The simplistic linear evolutionary classification schemes of Shaw (1981) and Barr (1983), which have been largely based on subjectively selected morphological criteria, should be disregarded. The points of divergence may have been earlier than indicated by the familiar orders '*Peronosporales*' and '*Pythiales*' and families '*Peronosporaceae*', '*Albuginaceae*' and '*Pythiaceae*' of standard texts. The currently-used genera of the Peronosporaceae may still be paraphyletic: one group of species may have a larger number of 'primitive' characters as well as other, more conspicuous 'advanced' characters. The earlier suggestion (Skalický, 1966) that *Peronospora* and *Plasmopara* represent diverging lines rather than a progression along a single unbranching line from 'more primitive' to 'more advanced' characters is supported by their different spectra of hosts, geographic centres and climatic zones, different conidiophore morphology and Feulgen analyses (Dick, 2000*a*). *Albugo*, currently placed in its own family in the *Peronosporales*, is the first branch from the *Peronosporomycetidae* lineage (Riethmüller *et al.*, 1999). Dick (1988) proposed that the *Peronosporaceae* was polyphyletic, with separate geographic origins for the different genera of downy mildews on dicotyledonous families from different sections of *Phytophthora sensu lato*. If cladistic theory is employed, ***all*** that is missing is that the symplesiomorphic trait in the *Phytophthora* line that gave rise to the apomorphies of the genera of the *Peronosporales* has yet to be defined. A number of groups of both *Pythium* species and *Phytophthora* species will require reassessment (e.g., *Phytophthora undulata*, and the *Pythium vexans* group, which almost certainly belongs to *Phytophthora sensu lato*, cf. Dick, 1990*b*; Panabières *et al.*, 1997) on comprehensive molecular, biochemical (elicitor data) and morphological criteria. To put order into the phylogenetic classification of the Peronosporales in taxonomic terms, it will be necessary to make a large number of nomenclatural changes at genus and family levels. This should not be undertaken lightly when these fungi are so important and subject to numerous publications in fields that are not primarily systematic. Partial renaming will only confuse the nomenclature. The existing orders, families and genera have phenetic value and will continue to be useful, provided that their limitations are understood.

Hierarchical and non-hierarchical classifications

With the advent of an increasingly large data base of DNA sequences there is a school of thought which is advocating the abandonment of the traditional hierarchical classification (Hibbett & Donahgue, 1998; Pine, Hibbett & Donahgue, 1999). This would be a mistake for several reasons.

First and foremost, morphologically-based taxonomy will coexist with molecular phylogenies for the foreseeable future because of the small percentages of known species studied by molecular biological techniques and the estimates of percentages of undescribed species. DNA-based studies cannot disregard accepted taxonomic tenets: nomenclatural change based on molecular biological data can only be justified if type species have been studied. For most fungi the type material is not available and, as here, more recent isolates of the fungi (determined on morphological criteria) have to be used. Biodiversity at gene, species and ecosystem levels can all be demonstrated using molecular markers, provided that the precise aim of any project is predetermined: additional (i.e., phylogenetic) conclusions may need qualification (compare Hibbett & Donoghue, 1998). A synthesis between phylogenetic relationships, genome synteny, genetic similarities, gene loss and gene suppression and the progress of evolution through geologic time, aided by molecular biology, will provide an hypothesis for biological diversity in a given group of organisms can be generated.

Historical review of the previous classifications for the *Peronosporomycetes*

From PART I, two principal conclusions emerge. The first concerns the inter-relationships of the classes of straminipilous fungi; the second relates to the assignation of the lagenidiaceous fungi within this classification. The ordinal and family classification within the *Peronosporomycetes* will require revision to integrate molecular biological data with morphological data. The lagenidiaceous fungi are so diverse that some may have affinities outside the class boundary; some even outside the straminipilous fungi.

A brief review of the development of systematic theory for the *Peronosporomycetes* will provide the background for the changes to be made below. The traditional classification of the orders of flagellate fungi (Sparrow, 1960: 3) was: *Chytridiales*, *Blastocladiales*, *Monoblepharidales*, *Hyphochytriales*, *Plasmodiophorales*, *Saprolegniales*, *Leptomitales*, *Lagenidiales*, *Peronosporales*.

Flagellar number, first advocated by Fischer (1892*a*) and Butler (1907) as a taxonomic criterion, was used to separate the first four of these orders, and apart from the *Hyphochytriales*, may have fortuitously correlated with heterokont flagellation in the more easily studied organisms. Barr (1980*a*, 1992) has revised the classification of the first three (posteriorly uniflagellate) orders constituting the *Chytridiomycetes* and these are excluded from further consideration. I and colleagues (Dick *et al.*, 1984; Dick, 1988; Dick *et al.*, 1989; Dick, 1990*a*, 1995, 2000*c*; Voglmayr *et al.*, 1998; Dick *et al.*, 1999) have gradually changed the family relationships in the last four (peronosporomycetous) orders of Sparrow to the following:

Peronosporomycetes
- *Peronosporomycetidae*
  - *Pythiales*
    - *Pythiaceae* (including *Lagenidiaceae*)
    - *Pythiogetonaceae*
  - *Peronosporales*
    - *Peronosporaceae*
    - *Albuginaceae*
- *Rhipidiomycetidae*
  - *Rhipidiales*
    - *Rhipidiaceae*
- *Saprolegniomycetidae*
  - *Saprolegniales*
    - *Saprolegniaceae*
    - *Leptolegniaceae*
  - *Leptomitales*
    - *Leptomitaceae*
    - *Apodachlyellaceae*
    - *Leptolegniellaceae*
  - *Sclerosporales*
    - *Sclerosporaceae*
    - *Verrucalvaceae*

The following families listed by Sparrow (1976) were left *incertae sedis* by Dick (2000*c*): *Olpidiopsidaceae*, *Sirolpidiaceae*, *Haliphthoraceae*, *Ectrogellaceae*, *Atkinsiellaceae*, *Eurychasmaceae*. The *Atkinsiellaceae* was not named in accordance with the rules of the *International Code of Botanical Nomenclature*. Karling (1981*a*) did not retain the *Haliphthoraceae* (which was included in the *Sirolpidiaceae*) or the *Atkinsiellaceae* (which was included in the *Eurychasmaceae*).

The three commonly accepted classes of straminipilous fungi are the *Peronosporomycetes*, the *Hyphochytriomycetes* and the *Labyrinthulomycetes* (*Labyrinthista*). There can be no doubt that the most distant of these three classes is the *Labyrinthista*. It is distinct in the architecture of the flagellar transitional zone; in the dictyosome-derived investment of scales; in its longer, anterior straminipilous flagellum; in the unique sagenogenetosome-derived ectoplasmic net, and in its marine habitat. The *Labyrinthista*, which I do not consider to be fungi, clearly belong within the kingdom *Straminipila*. Papers on them are frequently presented in mycological journals and the species are indexed in the *Index of Fungi*. Nevertheless, because of their historic associations with the mycological literature, I have covered their systematics and identification in PART V and PART VI. There are no indications that any of the lagenidiaceous fungi could be associated with the *Labyrinthista*.

The separation of the *Hyphochytriomycetes* from the *Peronosporomycetes* is less well established, in spite of the conclusions of Fuller (Barstow *et al.*, 1989; Fuller, 1990). The principal claim for a separation at Class level rests on flagellar number; the behaviour of the nuclear membrane during mitosis, and the importance of chitin in the wall membranes. It must also be remembered that the occurrence of sexual reproduction has not been established in the hyphochytrids and that the number of described species is small, with few of these species having been rigorously studied. There are some features which suggest that there may be a link between the *Hyphochytriales* and the *Leptomitales*, such as the finger-like protrusions of the zoospore cysts of some species, and the presence of chitin. The limited molecular biological data at present available place the *Hyphochytriales* as a basal clade to the *Peronosporomycetes*. Whether the *Hyphochytriales* should be placed as the holotypic order to a Class or Subclass is a different question. Beakes (1989) concluded, on the basis of zoospore and zoospore cyst ultrastructure, that four major groups of *Peronosporomycetes* could be distinguished, but he did not address the systematic problems raised by his assessment. The anomalous separation of the two species of *Sapromyces* was not discussed, nor did he consider the hierarchical implications of grouping the Class *Hyphochytriomycetes* and genus *Olpidiopsis* with the Subclass *Saprolegniomycetidae*. Beakes (1989) isolated *Lagena* as the sole member of his fourth category. The possibility that the *Hyphochytriales* is an anamorph order sharing some characters with the *Leptomitales* and coequal with the Subclass *Saprolegniomycetidae* should be retained. The additional wall layer between the gametangial wall and the developing zygote wall in the *Apodachlyellaceae*; the wall layers of the resting spores of the *Leptolegniellaceae*; and the possibly punctate nature of the resting spore wall in *Aphanomycopsis* and *Latrostium*, add to the plexus of characters which may necessitate a reappraisal of the Class distinctions between the *Peronosporomycetes* and *Hyphochytriomycetes*.

The informal division of the class *Peronosporomycetes* into two groups, which goes back to de Bary (1866), was formally reinstated with the Subclasses *Saprolegniomycetidae* and *Peronosporomycetidae* by Dick *et al.* (1984). Dick *et al.* (1999), Riethmüller *et al.* (1999), Petersen & Rosendahl (2000) and Hudspeth, Nadler & Hudspeth (2000) have now demonstrated that this division is fully supported by 18S rDNA, 28S rDNA and mitochondrial cytochrome *c* oxidase (*cox II*) sequences respectively. The classification of Dick *et al.* (1984), as revised by Dick *et al.* (1989), was in accord with a supra-grouping suggested by Beakes (1989), except that Dick *et al.* (1989) placed the *Rhipidiales* in a third Subclass *Rhipidiomycetidae* (now supported by the data of Lee & Fuller, 1992, but not those of Randolph & Powell, 1992). The recent *cox II* data (Hudspeth *et al.*, 2000) support the separation of *Hyphochytrium catenoides*, the *Leptomitaceae*, three groups within the *Saprolegniaceae sensu lato*, and the *Pythiaceae* including *Lagenidium*. The separation of the *Rhipidiomycetidae* (as represented by *Sapromyces elongatus*) from the *Peronosporomycetidae* (*Pythiaceae*) remains uncertain. The status of the third Subclass, *Rhipidiomycetidae*, is still in question, requiring consolidation of data both from other species and from other gene sequences.

A majority of the species assigned to *Lagenidium* are undoubtedly peronosporomycetous, but, even for these, their affinities within the *Peronosporomycetes* are scattered; a few are very close to *Pythium* (*Peronosporomycetidae*), but the marine lagenidiaceous parasites of *Crustacea* may be better considered with the *Haliphthoraceae* in the *Saprolegniomycetidae*. In this respect I agree with Beakes (1989). Dick *et al.* (1984) and Dick, (1988) removed the *Olpidiopsidaceae* and *Sirolpidiaceae* from the *Lagenidiales* and placed them *incertae sedis*. The remaining family, *Lagenidiaceae*, containing three genera, *Lagenidium*, *Myzocytium* and *Lagena*, was placed in the *Pythiales* close to the *Pythiaceae*. It is apparent that other species of *Lagenidium* and species of *Myzocytium* do not have all the characters of these subclasses.

Groups of taxa which do not fit easily within the *Peronosporomycetes-Hyphochytriomycetes* concept should be identified. Principal among these are:

- group 1 *Olpidiopsis* species parasitic on *Peronosporomycetes*
- group 2 nematophagous *Myzocytium* and *Lagenidium* species, together with *Gonimochaete*
- group 3 phycophagic parasitoids
- group 4 *Lagena* and *Ciliatomyces* (both monotypic)
- group 5 nematophagous *Haptoglossa* species, together with *Ectrogella*
- group 6 *Rozellopsis* - *Rozella*
- group 7 parasites of other protoctists such as *Pseudosphaerita*

Groups 1-4 are straminipilous, or probably so, and have distant but distinct affinity with the *Peronosporomycetes*, but at hierarchical levels probably comparable, to those of the Subclasses. The few *Olpidiopsis* species which have affinity to the *Peronosporomycetes* possess: straminipilous and anisokont flagellation (relative lengths mostly undetermined); 'simplified' versions of the peronosporomycete kinetosome/transitional zone ultrastructure (one species studied); haplomitotic B ploidy cycle (one species studied). But they also show: gametangial copulation; plerotic resting spores. They are also parasitoidal endobionts. With respect to *Olpidiopsis*, *Lagena* and *Ciliatomyces* are comparably but not similarly reduced. These two genera and *Pythiella* have aplerotic resting spores and nucleated periplasm.

Group **2**, while not yet proven to be straminipilous, has many oomycetous characters, notably the mitochondria with tubular cristae and large numbers of typical DBVs, but the details of oospore production may indicate a more distant relationship intermediate between the *Leptomitales* and *Peronosporales* rather than between the *Saprolegniales* and *Pythiales*. The latest molecular data suggest that, even with the observed morphological similarities, the nematophagous species may be polyphyletic, though still within the *Peronosporomycetes*.

The majority of the lagenidiaceous taxa possess several characters that are common within this group of lagenidiums-myzocytiums. Group **5** is almost certainly to be allied with the plasmodiophorids. In addition, groups **6** and **7** may also be intermediate between the peronosporomycetes and the plasmodiophorids. A considerable rearrangement of species within existing and new genera is necessary to allow the diagnoses of supraspecific taxa. Apart from the underlying theme of biochemically mediated, host-range restriction of these obligately endobiotic organisms, the characters which contribute to these linkages are:

1 zoospore volume small
2 zoospores with refractive granules
3 anisokont flagellation
4 zoosporogenesis without a homohylic vesicle
5 site of zoospore cleavage indeterminate or variable
6 sexual reproduction ranging from gametangial copulation involving all of both protoplasts and resulting in aplerotic resting spores, to plerotic resting spores developed without any identified act of fusion

Before the following revision can be regarded as other than an interim, but I hope practical classification, there are several other important TEM and biochemical criteria which will need to be investigated in a wide range of carefully selected taxa. These criteria include:

1 heterokont flagellation
2 kinetosome/transitional zone organization
3 kinetosome/centriole orientation
4 nuclear cycle and ploidy state for mitosis
5 wall structure and chemical composition
6 sterol biochemistry
7 lysine synthesis pathway

Evolutionary origins of the *Straminipila*, including the *Peronosporomycetes*

It is possible that the straminipilous *fungi* themselves (as a component of the undoubtedly early-originating monophyletic line of straminipilous *organisms*) have an ancient origin (Pirozynski & Malloch, 1975; Stanghellini, 1974), but the fossil evidence is equivocal (Pirozynski, 1976*a*, *b*). No indubitable fossil *Peronosporomycetes* have been reported: the establishment of various fossil genera such as those of Duncan (1876: *Palaeoachlya*); Seward (1898: *Peronosporites*); Pampaloni (1902: *Peronosporites*, *Pythites*); Elias (1966: *Propythium*, *Ordovicimyces*); Douglas (1973: *Peronosporoides*); Stidd & Consentino (1975: *Albugo*-like oospores) can be discounted (Dick, 1988).

Hypotheses for the evolution of the straminipilous fungi must therefore be based on circumstantial evidence of structure, morphogenesis and biochemistry (including molecular biology) of extant taxa. The high energy requirements of the straminipilous flagellum and the bacterially-perpetuated anaerobic environment that would surround potential aquatic substrata appear to be mutually exclusive. The evolution of mechanisms for shifting the site of zoospore discharge from the site of zoosporogenesis would therefore have been beneficial and probably developed on more than one occasion. Similarly, substrata yielding readily available nutrients would also be favoured; concommitant tendencies to develop anaerobic metabolic pathways would follow, as shown by the *Rhipidiales* (Emerson & Held, 1969; Held, 1970) and *Pythiogetonaceae* (Winans in Emerson & Natvig, 1981; Voglmayr *et al.*, 1999). The sub-cuticular coenocytium of nematodes and the ecdysic fluids of aquatic arthropods are obviously also such nutrient-rich substrata, and these substrata would have existed in the Palaeozoic. The freshwater/terrestrial origin of heterotrophs was probably coevolutionarily linked to arthropods, nematodes and the animal food chain (cf. *Saprolegniales* and *Myzocytiopsidales*), with freshwater algae (green algae and certain straminipilous algae) being the primary producers. Freshwater green algae would have been well advanced, calcified charophytes being known from the upper Silurian (Feist & Grambast-Fessard, 1991; Kenrick & Crane, 1997). Relationships with angiosperms, must have occurred much later, probably in the late Cretaceous. Most extant saprotrophic straminipilous fungal species are associated with animals or seeds and fruits. Links with vascular plant substrata may have started with detrital decay in water by transfer from animal substrata to fruit and seed decay with fermentative metabolism. Twigs and leaves are less common substrata, but are utilized by *Phytophthora gonapodyides* and species of *Dictyuchus*, *Sapromyces* and *Apodachlya*. Leaf- and twig-decaying fungi are normally hyphomycetes which have a much older fossil history (Dilcher, 1965). It is noteworthy that evidence of associations of *Peronosporomycetes* with bryophytes, ferns, gymnosperms and early-divergent angiosperms (*Nymphaceae*, *Ceratophyllales*, *Laurales*, *Magnoliales* and *Piperales*) is all but non-existant (but see *Albugo tropica* and compare with *Phytophthora* on *Lauraceae*, Erwin & Ribeiro, 1996).

The straminipilous flagellum is unlikely to have evolved more than once (Dick, 1990*c*; Van Der Auwera & De Wachter, 1998); it is advanced (derived ***after*** the evolution of the standard 9 + 2 flagellar axoneme) and evolutionarily conserved (Leipe *et al.*, 1996), occupying a place as significant as the evolution of chlorophylls *b* and *c* (both derived from chlorophyll *à*). Flagellar loss, or partial flagellar loss (including loss or partial loss of straminipilous ornamentation), has probably occurred several times in the straminipilous fungi, as it has in the straminipilous algae (Leadbeater, 1989; Cavalier-Smith *et al.*, 1995) and 'straminipilous' protoctists (see Silberman *et al.*, 1996). Loss of the zoospore, and therefore flagellation, is a feature of both the *Peronosporales* and *Sclerosporales* and has minor phylogenetic significance.

Apart from the straminipilous flagellum, the kingdom *Straminipila* is also characterized by the possession of a mitochondrion with tubular cristae; a DAP (α, ϵ-diaminopimelic acid pathway) lysine synthesis pathway, and if photosynthetic, a plastid with a second chlorophyll, chlorophyll *c* (not chlorophyll *b* as in green plants). In addition to these straminipilous characters, the fungal Class *Peronosporomycetes* is characterized by a combination of five characters not found in any of the other major groups of straminipilous organisms (Dick, 2000*a*, *c*):

1 haplomitotic B ploidy cycle (mitosis confined to the diploid phase)
2 cruciform meiosis in a persistent nuclear membrane
3 multiple synchronous meioses in coenocytic (paired) gametangia (meiogametangia)
4 gametes without flagellation (donor gametes without cellular identity)
5 formation of zygotic resting spores (oospores) in oogonia (receptor gametangia)

The diversity, extant lineages of, and genetic distances between the straminipilous organisms are such that the Kingdom must have originated in the marine ecosystem, most probably in littoral and lagoon or estuarine environments (the photoendobiont was probably a red alga, strengthening the marine provenance; Potter *et al.*, 1997) and at an early geological epoch (Cambrian? Precambrian?). Nevertheless, all existing evidence points to a freshwater or terrestrial origin for the straminipilous fungi, or *Peronosporomycetes*. The warm temperate lagoon ecosystem would rapidly have become world-wide during the early tectonic movements and sea level changes of the Gondwanaland, Laurentian, European and Siberian plates, which were all equatorial and separate during the Precambrian and Cambrian epochs, but which became united to form the supercontinent Pangaea during the Permian (Tarling, 1980).

The principal questions remaining are:

1 Did the fungal organisms evolve from a ***freshwater heterotrophic*** ancestor (*Saprolegnia*- or *Pythium*-like) or from a ***marine heterotrophic*** ancestor (*Halophytophthora*-like)?
2 Did the freshwater fungal straminipilous organisms evolve from a *freshwater photosynthetic* ancestor, or from an ***originally and fundamentally heterotrophic*** ancestor in ***freshwater***?

In contrast to the oceanic margin ecosystems, the freshwater systems would not have been physically confluent, and therefore different communities could have evolved in isolation. Most species of the *Saprolegniaceae* and *Pythiaceae* are now cosmopolitan, evidence of provincialism is limited and this might be taken as evidence for an early origin. However, within the *Saprolegniaceae* the Atlantic Ocean has provided a barrier allowing evolution of separate species of *Aphanomyces* and the genus *Aplanopsis*. The possibility of very rare events of transcontinental movement, and opportunistic saprotrophism, would still allow the hypothesis that *most* of the extant *Peronosporomycetes* could have evolved in the very recent (late Tertiary) past. Thus, although the origins of the straminipiles (Kingdom *Straminipila*) were probably Cambrian or Precambian, the straminipilous fungi (Class *Peronosporomycetes*) might have evolved at any time between the Early Palaeozoic (Ordovician?) some 438-488 million years Before Present (m.y.B.P.). There remains an enormous gap between such a postulated early origin (whether photosynthetic or fundamentally heterotrophic) and the distributions and diversity of present-day taxa.

TABLE II: 2. *Selected characters of the* Peronosporomycetes *and* Plasmodiophoromycetes *with anomalous 'lagenidiaceous' and other biflagellate genera which do not fit one or other of the alternative criteria listed in the first two columns*

PERONOSPOROMYCETES	*PLASMODIOPHOROMYCETES*	REPRESENTATIVE ANOMALOUS 'LAGENIDIACEOUS' GENERA 'straminipilous'	 'non-straminipilous'
cytology			
haplomitotic B ploidy cycle	haplomitotic A ploidy cycle	*Lagenisma* [diplomitotic?], *Labyrinthula* [haplomitotic A]	
chromosome numbers low	chromosome numbers high	?	
synchronous coenocytic meioses	synchronous coenocytic meioses	?	
nucleolus persists outside mitotic spindle	nucleolus persistent in centre of mitotic spindle		
mitotic nuclear membrane intact until telophase	mitotic nuclear membrane reconstituted within remnants of previous membrane	*Olpidiopsis*, *Rhizidiomyces*	
centriole arrangement	centriole arrangement		
thallus			
assimilative thallus walled	assimilative thallus plasmodial	*Olpidiopsis*	*Rozellopsis*,
osmotrophic	osmotrophic		*Pseudosphaerita*, *Endemosarca*
thallus septation infrequent, without disarticulation	thallus division and dissociation occurs	*Myzocytium*, *Myzocytiopsis* *Haliphthoros*, *Sirolpidium*	*Rozella*, *Rozellopsis*
wall fibre of β-1,3- and β-1,6-glucans, with some β-1,4-glucan, rarely with glucosamines	wall proteinaceous, fibrous component of chitin (?)	*Hyphochytrium*	
mitochondrial cristae tubular	mitochondrial cristae vesiculate (flattened, finger-like)		*Haptoglossa*, *Rozella* *Catenaria*, *Nephromyces*
DBVs present, metachromatic granules absent	DBVs not known, phosphate storage system not identified		*Rozella* ?
DAP lysine synthesis	AAA lysine synthesis presumed	*Hyphochytrium* ?	

(Table II: 2, continued)

asexual reproduction			
zoosporogenesis: cleavage synchronous	zoosporogenesis: cleavage synchronous	*Schizochytrium*	*Pseudosphaerita*
zoospores biflagellate	zoospores biflagellate		*Rozella*, *Sphaerita*
flagellar insertion lateral longitudinal groove insertion	flagellar insertion subapical protuberance insertion	*Sirolpidium* ? *Ciliomyces*	*Haptoglossa*
slightly/strongly anisokont anterior flagellum shorter, straminipilous	strongly anisokont, anterior flagellum longer, unornamented	? *Crypticola*, *Gracea*, *Labyrinthula*, *Thraustochytrium*	
flagella of principal form shed on encystment	flagella retracted	*Gracea*	*Rozellopsis*
zoospore size $>150\mu m^3$, usually $>500\mu m^3$	zoospore size $<100\mu m^3$	*Myzocytiopsis*	*Ectrogella* (*pro parte* ?)
transitional plate with bead	transitional plate without bead	*Lagena*	
transitional plate above plane of plasmamembrane	transitional plate level with plane of plasmamembrane		
kinetosome concentric fibres not not extending into transitional zone	kinetosome concentric fibres extending into transitional zone		
sexual reproduction			
oogamous sexual reproduction	isogamous sexual reproduction	*Lagenisma*	
male gametes not delimited	naked gamete cells	*Lagenisma*, *Eurychasmopsis* ?	
female gametes enclosed	naked gamete cells	*Lagenisma*	
zygote cytoplasm almost entirely female	zygote cytoplasm derived from both gametes (?)	*Myzocytiopsis*	
oospore structure oogonial wall structure [not applicable]	[not applicable] [not applicable] cyst walls spiny or smooth		

THE CLASSES *PERONOSPOROMYCETES* AND *PLASMODIOPHOROMYCETES*

The separation of the *Straminipila* from the *Protoctista* is justified, but the circumscription of the straminipiles and chromists is still in doubt. The *Straminipila* could have at least two possible nodes of origin and divergence. Trophic and thallic diversity is a feature of the *Straminipila* and the requirement of a cell wall for osmophagy is dispensible for non-phagotrophic endobiont fungi. The plasmodiophorids are also non-phagotrophic heterotrophs with obscure relationships to other flagellates. However, Ward & Adams (1998), using an extensive but incomplete 18S rDNA sequence concluded that the nearest, but still very distant, phylogenetic links of the *Plasmodiophoromycetes* were with the *Straminipila*.

The wider Kingdom concept, which is necessary if all chromophytes and straminipiles are to be included, embraces more than one kind of mitochondrial morphology; it also requires the postulation of subsequent loss of tubular tripartite hairs (TTHs) on the anterior flagellum. Loss of the entire posterior flagellum has occurred in diverse core chromistan groups (diatoms, brown algae and hyphochytrids) but the possibility of loss of the straminipilous flagellum should be reconsidered. Ultrastructural data for *Labyrinthuloides haliotidis* (Bower, 1987*a-d*) and *Developayella elegans* (Tong, 1995) suggest that the diversity in straminipilous ornamentation is greater than has been recognized. The molecular biological relationship of *Blastocystis* and *Proteromonas* (Silberman *et al.*, 1996), which lack straminipilous ornamentation, to the straminipiles also points to a greater diversity.

Cavalier-Smith (1986) argued that although partial loss of TTHs has occurred in the chromists *sensu lato*, total loss of TTHs would be improbable because of the effect on direction of movement. The possibility of loss or change of straminipilous ornamentation from one of a ***markedly*** anisokont pair of flagella on ***very small*** motile units has not been considered from the mechanical viewpoint. The role of the cytoskeleton and kinetid roots will be a crucial consideration. The apical or subapical flagellar insertion coupled with a much longer posteriorly directed flagellum and small zoospore size in the plasmodiophorids needs to be incorporated in a general hypothesis of flagellar diversity and movement. If ***anisokont*** flagellation is regarded as a prerequisite for straminipilous flagellation, then it is possible that straminipilous ornamentation of the anterior flagellum was acquired after the inception of the monophyletic line. Anisokont flagellation could allow for the association of the plasmodiophorids within the broader Kingdom concept.

Ultrastructural details of flagellation must be the principal criterion for the classification of straminipilous fungi. Unfortunately not all straminipilous fungi produce zoospores, though in most cases secondary characters are sufficiently numerous to compensate for the absence of the primary criterion. In small, endobiont obligate parasites, secondary characters are fewer, and the acquisition of flagellar details is more difficult. For many of the lagenidiaceous fungi it is necessary to examine the classification of the straminipilous fungi before determining their correct placement. A resumé of the characters of the classes *Peronosporomycetes* (as represented by the sub-classes *Peronosporomycetidae* and *Saprolegniomycetidae* only) and *Plasmodiophoromycetes* respectively is given in Table II: 2. It is concluded that these two classes, while phylogenetically remote, might be very distantly related; also that a few of the lagenidiaceous and other biflagellate fungi, on the criteria that are presently available, show as close a relationship to the *Plasmodiophoromycetes* as they do to the *Peronosporomycetes* or any of the other classes of flagellate fungi. The proposed placements of the *Haptoglossaceae*, *Ectrogellaceae*, *Rozellopsidaceae* and *Endemosarcaceae* are provisional and cannot be justified without further supportive data.

PART III

MYCELIAR PERONOSPOROMYCETES:

DOWNY MILDEWS AND WATER MOULDS

INTRODUCTION

The fungi discussed here are those commonly known as the water moulds and downy mildews. These two groups of fungi have long characterized the two subclasses, *Saprolegniomycetidae* and *Peronosporomycetidae*, respectively. Traditionally, the saprotrophic water moulds have been regarded as the less 'advanced'. This perception may change in the light of molecular biological data now becoming available. Because they are so well known, the families and genera of both subclasses of these fungi have been regarded as stable. Molecular biological data is confounding this comfortable concept to such an extent that it is not possible, at this stage, to present a satisfactory systematic account with entries for each species. Therefore, extensive bibliographies are included here for genera because species names are not separately entered in PART V. Some of the problems are discussed in more detail below.

The species of the *Saprolegniaceae sensu lato* have stout mycelia; they are very common in most aquatic ecosystems; they are easily isolated, and valuable for demonstrating many of the cytological and physiological attributes of fungi (see PART I). Few species are parasitic, but those of economic importance are discussed below. The large diameter of the hyphae of *Saprolegnia* and *Achlya* has resulted in their being selected for numerous studies of fundamental attributes of the cytoplasm and cytoplasmic and wall morphogenesis. These studies have been noted in PART I under the appropriate sub-headings.

In contrast, many of the *Peronosporomycetidae* are obligate parasites: this subclass is comprised of two orders: the *Peronosporales* includes the downy mildews and white blister rusts; the *Pythiales* contains *Phytophthora* and *Pythium*. The fungi of the *Peronosporales* and *Phytophthora* are recognised as some of the most commercially important plant pathogens. Partly because of this importance, and partly because *Peronospora* Corda was one of the first genera to be described, I have adopted the generic stem '*Perono-*' for all the higher taxa to class level. The *Peronosporomycetidae* will therefore be discussed first.

THE *PERONOSPORALES*: THE DOWNY MILDEWS

What are the downy mildews?

The downy mildews (DMs) are parasitic in only a few groups of angiosperms. No evidence for parasitism of other vascular, but non-angiospermic, plants exists. The DMs are typically confined to the stem cortex and leaf mesophyll, but some species may be systemic, with the mycelium ramifying throughout the host plant (see p. 81). The long conidio-sporangiophores which emerge from stomata are responsible for the downy appearence of the mildew. Not all species are fully biotrophic. Significant cell damage is caused, for example, by *Peronospora tabacina* and *Plasmopara viticola* (Lafon & Bulit, 1981): the plasmamembranes of the host mesophyll cells become excessively leaky, resulting in a distinctive greasy or wet appearance to the infected part of the leaf. This is essentially a moderated manifestation of the symptoms associated with wet rots caused by certain species of *Phytophthora* (Keen & Yoshikawa, 1983) and probably resulting from similar biochemical interactions.

The DMs and *Phytophthora* have received attention because of the serious reduction in yields from crop plants. Together these organisms have been responsible for socio-economic-political change (Large, 1940;

Smith, 1884; Woodham-Smith, 1962); the birth of plant pathology (de Bary, 1863, 1876); and the first developments of the agricultural chemical industry instigated by Millardet (1885; see also Ainsworth, 1976; Schneiderhan, 1933). It is essential to acknowledge that distributions of many of the *Phytophthora* parasites were provincial until very recently, when trade movements disseminated these species, often with an eventually dramatic new pathogenic impact (Late Blight of potatoes and the Irish Famine).

The importance of the DMs in agriculture has only arisen within the last 500 years, brought about by significant intercontinental trade and movement in grain, root and fruit crops. There was almost certainly previous limited movement around the Mediterranean; across the Panamanian isthmus; and between the great south-east Asian river delta systems. But this trade did not, as far as is known, cause problems with pathogen introductions, nor did it impinge on hosts vulnerable to DMs. Despite this extremely recent movement (in evolutionary contexts), ploidy differences have arisen in *Phytophthora infestans* in Europe following the introductions of host and parasite, both of which originate from the Equador - Central American region (Boussingault, 1845; Daggett, Knighton & Therrien, 1995; Lucas *et al.*, 1991).

Taxonomic history of the downy mildews

Any discussion of the systematics and evolution of the downy mildews (DMs) must involve some related pathogens in the orders *Peronosporales* (*Albugo* in the monogeneric *Albuginaceae*) and *Pythiales* (particularly *Phytophthora* and *Pythium*). The white blister rusts (*Albugo* species) are also obligately biotrophic parasites of dicotyledons, commonly recorded from stems and leaves. Some of the obligate parasites in the *Peronosporaceae* have been maintained in biphasic culture, but none of the species of the *Peronosporaceae* or *Albuginaceae* has been obtained in axenic culture. In the *Pythiales*, the sister order to the *Peronosporales*, *Phytophthora* is known as a pathogen of a much wider range of woody and herbaceous angiosperms and conifers; different species may parasitize roots, hypocotylar regions, leaves or fruits. A few species are saprotrophic. *Phytophthora* species are culturable, but require more care than *Pythium* species. Whether primarily parasitic or primarily saprotrophic, almost all species of *Pythium* are readily culturable and possibly because of this, the extent of the truly saprotrophic habit of species in this genus is unknown. Table III: 1 provides a chronology for the development of taxonomic theory in the DMs and white blister rusts.

The white blister rust of crucifers, now known as *Albugo candida*, was the earliest of these plant pathogens to be recognized, dating back to the writings of Persoon (1801) and observations of Prévost (in Ainsworth, 1976). The species was listed, with two other species, by Gray (1821). However, *Albugo* was not recognized as being closely related to the DMs for another 40 years, until the comparative descriptions of sexual reproduction in the plant pathogenic genera by de Bary (1863). The generic name *Cystopus* was erected by Léveillé (1847) for the sexual (perfect) stage. The life-history of the fungus was established by de Bary (1863) and the combination *Cystopus candidus* was used by him and most other writers until (and after!) Wakefield (1927). Subsequent to the establishment of the principal genera, *Paraperonospora* has been erected by Constantinescu (1989) and *Benua* by Constantinescu (1998).

The first DM to be formally and acceptably described and diagnosed was *Peronospora rumicis* by Corda (1837) from Poland; Unger (1833) had earlier noted what was to become known as *Plasmopara pygmaea*. Descriptions of the genera *Bremia* (Regel, 1843) and *Basidiophora* (Roze & Cornu, 1869) followed. Note that *Phytophthora* was not formally described until 1876 (de Bary, 1876).

TABLE III: 1. *The early chronology of taxonomic and plant-pathogenic studies of the dicotyledonicolous downy mildews and related taxa up to Fitzpatrick (1930)*

date	fungus (only generic authorities given), citation and brief annotation
1801	*Uredo* section *Albugo* named and listed by Persoon (Persoon, 1801)
1806	*Albugo* named by Roussel (Roussel, 1806)
1807	*Albugo* zoospores described by Prévost (see Ainsworth, 1976: 62)
1821	*Albugo* (Pers.) Roussel [erstwhile (Pers.) Roussel: *ex* Gray] (Gray, 1921: lists type species *A. candida*, and 2 other species)
1833	*Botrytis pygmaea* [type species of *Plasmopara*] named (Unger, 1833)
1837	*Peronospora* Corda (Corda, 1837: type species *P. rumicis* [holotype]) in Amaranthaceae
1843	*Bremia* Regel (Regel, 1843: type species *B. lactucae* [holotype]) in Asteraceae
1845	*Botrytis infestans* Mont. (Montagne, 1845) described as the causal agent of Late Blight of Potatoes
1847	*Cystopus candidus* (Pers.) Lév. [new name for *Albugo candida*] (Léveillé, 1847)
1851	*Peronospora pygmaea* [type species of *Plasmopara*] described with other species of *Peronospora* (Unger, 1847)
1851	*Botrytis viticola* [*Plasmopara*] described (Berkeley, 1851)
1863	sexual reproduction in DMs described (de Bary, 1863): [modern nomenclature] *Bremia* (1 sp.), *Paraperonospora* (1 sp.), *Peronospora* (32 spp.), *Plasmopara* (5 spp.), *Albugo* (6 spp.), *Phytophthora* (1 sp) [the first synopsis of the classification of the DMs; still the most important comparative morphological account]
1866	the Peronosporeen and Saprolegnieen separated as the two major groups of biflagellate fungi (de Bary, 1866, 1887)
1869	*Basidiophora* Roze & Cornu described (Roze & Cornu, 1869; type species *B. entospora* [holotype])
1869	*Peronospora cubensis* [type species of *Pseudoperonospora*] described (Berkeley and Curtis, 1869: 363)
1876	*Phytophthora* de Bary described (de Bary, 1876; type species *P. infestans* [holotype] [group IV])
1876	*Plasmopara* [as *Peronospora*] found on vines (Farlow, 1876)
1878	*Plasmopara viticola* [as *Peronospora*] found in Europe (Millardet, 1885, in Schneiderhan, 1933)
1884	succinct accounts by Smith (1884) of plant pathology and morphology of DMs: [modern nomenclature] *Peronospora trifoliorum*, *P. destructor*, *P. parasitica*, *Plasmopara umbelliferarum*, *Bremia lactucae*, *Albugo candida*, *Phytophthora infestans*
1885	Bordeaux Mixture described, effective on *Plasmopara viticola* (Millardet, 1885, in Schneiderhan, 1933)
1886	*Plasmopara* Schroeter described (Schroeter, 1886; type species *P. pygmaea*)
1893	order Peronosporales described (Schröter, 1893)
1898	*Peronospora megasperma* [type species of *Bremiella*] described (Berlese, 1898)
1899-1901	cytology of DM oosporogenesis by Stevens and others (summarized in German, with illustrations, by Lotsy, 1907; see Dick & Win-Tin, 1973; Table I: 3, pp. 72-73)
1900 [**1901**]	*Plasmopara* subgenus *Peronoplasmopara* Berlese [type species *Plasmopara celtidis*] established by Berlese (1897-1902)
1902	subgenus *Peronoplasmopara* Berl. described (Berlese, 1897-1902, type species *P. cubensis*)
1903	Cucurbit DM disease in Europe
1903	*Pseudoperonospora* Rostovsev described (Rostowzow, 1903; type species *P. cubensis*)
1905	*Peronoplasmopara* (Berl.) G. P. Clinton described (Clinton, 1905; type species *P. cubensis*)
1907	*Rhysotheca* G. W. Wilson described (Wilson, 1907; type species *R. umbelliferarum*, *Plasmopara* remaining monotypic)
1914	*Bremiella* G. W. Wilson described (Wilson, 1914; type species *B. megasperma* [holotype])
1922	*Pseudoplasmopara* Sawada described (Sawada, 1922, type species *P. justiciae* [holotype])
1923	monograph on *Peronospora* by Gäumann (1923) - 243 species considered
1927	*Albugo* reestablished as the generic name for the white blister rusts [cf. *Cystopus*] by Wakefield (1927)

Note that the authority for the genus *Albugo* is '(Pers.) Roussel', not '(Pers.) Roussel *ex* Gray' as in Hawksworth *et al.* (1995) since it is no longer a requirement to rebase to Gray (1821). I am grateful to J. David (CABI Bioscience, pers. comm.) for this amendment.

The DMs have been classified within the family *Peronosporaceae* since de Bary (1866) first coined the family concepts "Saprolegnieen" and "Peronosporeen". Support for these subclass divisions is now available from molecular data (Dick *et al.*, 1999) and has been discussed in PART I (pp. 91-98). Evolutionary origins are summarized below.

The ***diagnostic*** criteria for the genera of the *Peronosporaceae* may require amendment as a result of molecular biological studies. Until then, the nomenclatural niceties (Skalický, 1990) regarding the validity of the genera *Pseudoperonospora* Rostovzev and *Peronoplasmopara* G. P. Clinton and the subgenera are largely irrelevant, pending convincing molecular biological data for ***all*** the type taxa involved. (The relevant subgenera are *Plasmopara* J. Schröt. subgenus *Peronoplasmopara* Berl.; type species: *Plasmopara celtidis* (Waite) Berl., and *Peronospora* Corda subgenus *Peronoplasmopara* (Berl.) Skalický; type species: *Peronospora celtidis* Waite.)

The taxonomy of *Peronospora* and *Plasmopara* is complicated by the species/specific concepts championed for the former genus by Gäumann (1923); later authors from eastern Europe (e.g., T. Rayss, A. Săvulescu, O. Săvulescu) western Asia (e.g., P. N. Glovin, N. P. Golvina, B. K. Kalymbetov, V. I. Ul'yanishchev) and eastern Asia (e.g., C. J. Li, T. R. Liu, C. K. Pai, Z. S. Yang, G. Y. Yin, Zhen Y. Zhao) have adopted this approach, with the result that the species lists are enormous (see APPENDIX I; Constantinescu, 1991*a*). The species concept has been variously interpreted, with workers sometimes resorting to *sensu lato* designations and sometimes to *sensu stricto* usage. Apart from the host identity and relationships (see Tables III: 2 and III: 3), simplistic morphological assessments are often all that are used. The use of the infra-specific categories of variety and form have taxonomic status governed by the rules of the *International Code of Botanical Nomenclature* (*ICBN*), but whether these infra-specific categories have been fully understood by authors proposing such taxa is less clear. From the systematic viewpoint, the environmental/host distinctions of the parasite rest uneasily with *formae speciales* which are not so governed. Infra-specific taxa have been ignored in Table III: 3, but their existence is indicated by the use of the type variety or form after the binomial in APPENDIX 1. Hall (1996) provided a more recent review of species concepts. Gustavsson (1959*a*, 1959*b*) and Constantinescu (1979, 1989, 1991*a*, 1991*b*, 1992, 1996*a*, 1996*b*, 1998) have adopted a more considered, but still morphological, approach to species delimitation.

TABLE III: 2. *Molecular phylogeny of the angiosperms (after Angiosperm Phylogeny Group (APG), 1998) and the occurrence of the dicotyledonicolous downy mildews and* Albugo. *A few families were not assigned to orders in APG (1998). The levels of Jackknife support, given in the original publication, have been omitted because they are not pertinent to this diagram. Unplaced families of no relevance to downy mildew parasitism have been excluded. Families from the Cretaceous include:* Caryophyllales*:* Amaranthaceae*;* Malpighiales*:* Euphorbiaceae*;* Myrtales*:* Onagraceae*; families from the Paleocene include:* Solanales*:* Convolvulaceae*;* Malvales*:* Malvaceae*;* Gentianales*:* Gentianaceae.

*The basal angiosperms, including the palaeoherbs (*Ceratophyllales, Piperales*),* Laurales *and* Magnoliales *have been omitted (but note that one species of* Albugo *occurs on* Piperales*). Monocotyledons have also been omitted, but one species of* Peronospora *occurs on* Asparagales*; one species of* Bremia *and two species of* Plasmopara *on* Poaceae. *In the absence of definitive check lists for DM genera, the numbers refer to binomials rather than taxa, so there may be some redundancy within and between columns.*

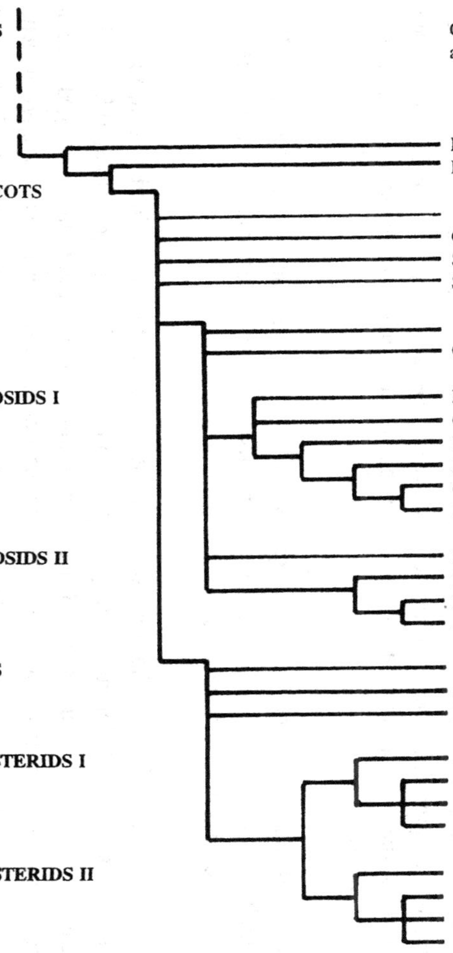

Clade	Orders of angiosperms	Number of angiosperm families	Number with *Peronospora* parasites (0* see *Plasmopara*)	Number of *Peronospora* binomials (total 464)	Number of *Plasmopara* binomials (total 109)	Number of *Albugo* binomials (total 44)	Other DMs (25 binomials)
ANGIOSPERMS							
EUDICOTS	Ranunculales	7	3	39	13	2	
	Proteales	3	0				
CORE EUDICOTS	Vitaceae	1	0*		3		
	Caryophyllales	26	7	90	1	15	
	Santalales	5	1	1			
	Saxifragales	13	1	4	1		
ROSIDS	Zygophyllaceae	1	1	1			
	Geraniales	6	1	3	4		
EUROSIDS I	Malpighiales	31	3	12	1		□
	Oxalidales	6	1	1			
	Fabales	4	1	65	1	1	
	Rosales	11	3	5	5	2	○
	Cucurbitales	7	0*		1		○
	Fagales	8	0				
EUROSIDS II	Myrtales	14	1	1	2		
	Brassicales	15	3	100	1	5	
	Malvales	10	1	2	4		¶
	Sapindales	9	0*		1		§¶
ASTERIDS	Boraginaceae	1	1	23	1	1	
	Cornales	5	0		1		
	Ericales	24	3	9	1		
EUASTERIDS I	Garryales	4	0				
	Gentianales	5	3	21	3	2	¶
	Lamiales	21	4	59	12	4	
	Solanales	5	2	7		7	
EUASTERIDS II	Aquifoliales	3	0				
	Apiales	7	0*		26		□
	Asterales	13	2	12	25	5	●
	Dipsacales	6	1	9	2		

● = *Basidiophora* (2 spp) , *Benua* (1 sp), *Bremia* (1 sp), *Paraperonospora* (9 spp); □ = *Bremiella* (3 spp); ○ = *Pseudoperonospora* (7 spp); § = *Peronophythora* (1 sp) in *Litchi* (Sapindaceae)
¶ = *Trachysphaera* (1 sp) in *Theobroma* (Malvaceae, formerly Sterculariaceae), *Citrus* (Rutaceae - Watson, 1971), *Coffea* (Rubiaceae) and *Persea* (Lauraceae - Spaulding, 1961) and *Musa* (Musaceae - Kader, 1999)

Evolution of the biotrophic habit: nutrition and biochemistry

Coevolution is often taken to mean co-phylogenetic evolution (e.g., Crute, 1981) but co-phylogenetic coevolution is seldom found. Reference to Table III: 2 shows that *Peronospora* has been found on relatively few of the angiosperm families that are now grouped into orders on the basis of molecular data (Angiosperm Phylogeny Group (APG), 1998). Even within the angiosperm orders there appear to be highly clustered centres of speciation, although it must be remembered that species-specific nomenclature has greatly exagerated this skewness. The fact that the widest spectrum of angiosperm hosts is parasitized by the morphologically 'advanced' (i.e., lacking zoospores) genus *Peronospora* is not incongruous if the 'attraction' is by biochemical stimuli. For species-specific parasitism the more intensively studied phenomena of suppression and inhibition are likely to be less important than the elements of attraction and growth stimulation.

The propensity of *Peronosporomycetes* for the products of angiosperm secondary metabolism may be the most important stimulus and the most important constraint for their evolution (Figure III: 1). Secondary metabolite production may have been stimulated by high carbohydrate levels or C:N ratios that would otherwise be deleterious to the angiosperm. Secondary biochemical pathways are thought by Ross & Sombrero (1991) to have evolved to remove excess photosynthate from the cell systems by producing osmotically inactive, secondary metabolites. Some of the secondary metabolites may have evolved initially to provide a UV screening function, others subsequently gained a protective function against animal predators, with the result that the web of angiosperm-insect-vertebrate coevolution became increasingly complex. It is immaterial for the fungus whether high insolation, stress, or root system associations induced the developments of the secondary metabolic pathways. The interaction between the physiological activities of leaves, shoots, and roots upon secondary metabolite production and translocation could have a marked effect in creating conditions suitable for disease establishment. In particular, the distribution of metabolites in clones of a host that have been grown in different communities, especially root rhizosphere and hyphosphere communities, has rarely been examined (but see Paxton, 1983). Mycorrhizal associations apart, the nitrogen-fixing potential of the *Fabaceae* would have led to a different secondary metabolism involving proteins and cyano-compounds. Distinct sulphur-related metabolic pathways in *Allium* could similarly explain this otherwise anomalous host of *Peronospora destructor*. From the coevolutionary point of view, genotypic variation is less important than the community structure. The equilibrium between the composition and stability of the community with environmental pressures to produce secondary metabolites, allowing their accumulation in potential target organs, would be crucial. The exogenous organic compounds essential for the growth and reproduction of DMs (and other straminipilous fungi) and the complementary integration of these substances with the unique biochemical pathways of the straminipilous fungi have yet to be fully investigated.

Although vitamin requirements by *Peronosporomycetidae* are well-known (e.g., Hohl, 1983), auxotrophs can usually survive on the low concentrations normally leaked from autotrophs. Vitamin requirements are unlikely to be determining factors for obligate parasitism.

The study of the physiology of host-parasite relationships has not so far yielded hints as to the coevolution of the species-specific relationship. The problem with nutritional studies (cf. Hohl, 1983) is that they have to take place in artificial environments. Conditions conducive to maximum and prolonged vegetative growth may not be the most advantageous (i.e., fittest or most competitive) in the natural environment. Even for *Phytophthora* no sound case has yet been presented to suggest that particular nutrient requirements are ***essential*** prerequisites for successful attack.

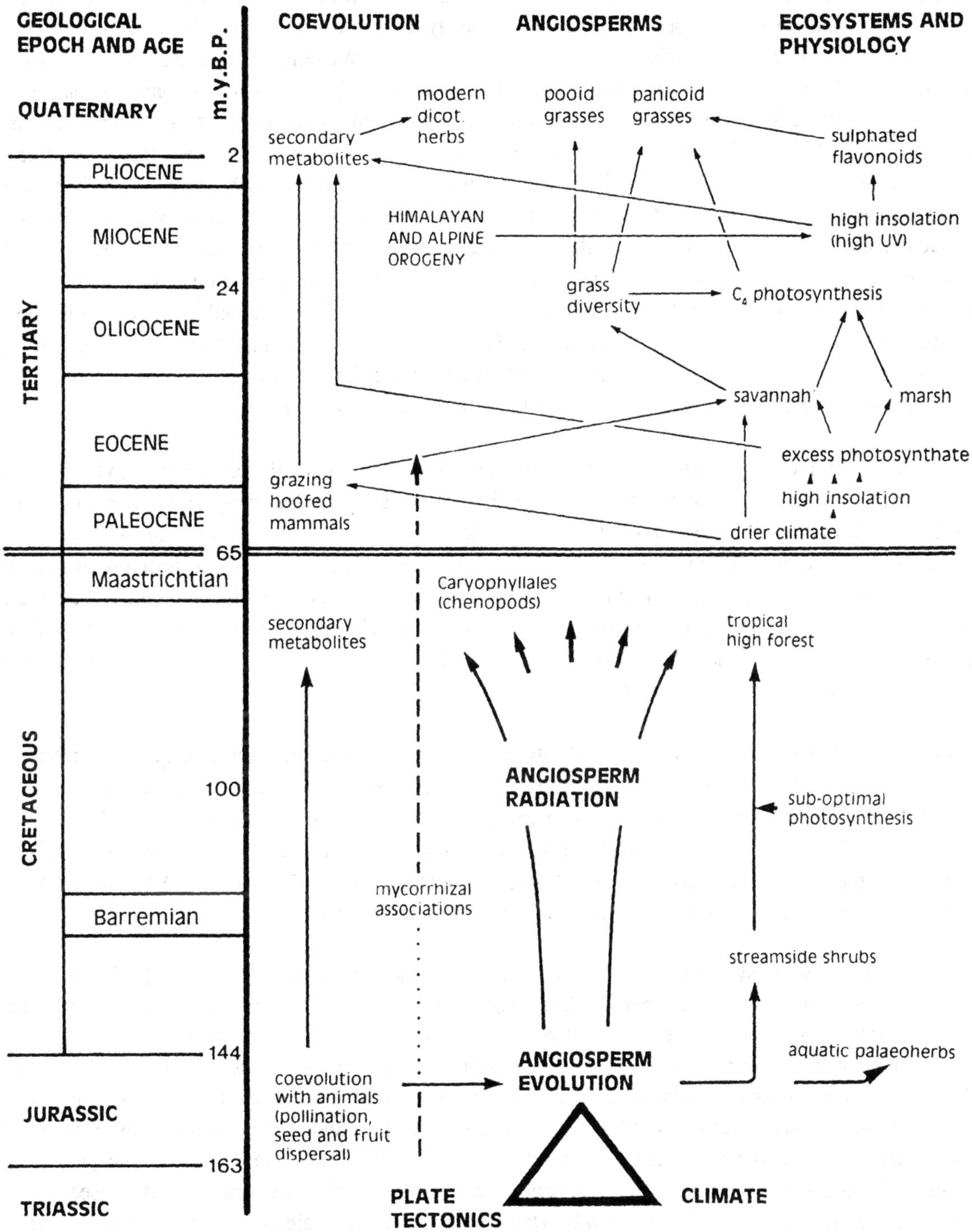

FIGURE III: 1. *Summary diagram of angiosperm evolution through geological periods (e.g., CRETACEOUS), epochs (e.g., EOCENE) and ages (e.g., Maastrichtian) with the nutritional, biochemical, climatic and geological influences for the development of the obligate biotrophic parasitic habit of the* Peronosporales *(from Dick, 2000a). For further explanation, see text.*

Most of the DMs are leaf and herbaceous stem parasites (exceptions are two root-parasitic species of *Plasmopara*, *P. halstedii* and *P. lactucae-radicis* - Stanghellini, Adaskaveg & Rasmussen, 1990; and *Peronospora parasitica* in radish roots - Ramsay, Smith & Wright, 1954). Photosynthesis may therefore be an important physiological influence. Diversifications from the C_3 metabolism (C_4 and Crassulacean Acid Metabolism, CAM) are common; they are assumed to have arisen frequently and independently in different angiosperm orders, but the charting of their systematic distributions (Moore, 1982; Harborne, 1988) is still in its infancy and does not appear to have been recently reviewed. It is probable that both carbon intermediates and ionic balances differ between C_4 and CAM. The availability of organic carbon to parasites may also be different. If the angiosperm ordinal distributions of C_4 and CAM are compared, and also compared with phytoalexin distribution (Dick, 1988: fig. 5), the only host orders producing phytoalexins, but for which no C_4 or CAM has been reported, are in the more advanced orders of the Eurosids (*Fabales*, *Rosales*, including *Urticaceae*; *Malvales*) and Euasterids (*Apiales*). The *Caryophyllales* and *Malpighiales* (*Euphorbiaceae*), and probably the *Asterales* and *Asparagales*, are orders in which C_4, CAM and phytoalexin production are common. A C_4 metabolic relationship has coevolutionary potential for *Albugo* and *Peronospora* which, on molecular phylogeny, appear to be widely divergent although their host family profiles are very similar (Table III: 2). However, the association of C_4 orders with *Plasmopara* is poor, and it is non-existent with *Pseudoperonospora*. For *Plasmopara* an association is possible with CAM in the Eurosids and, similarly, for *Pseudoperonospora* with the *Cucurbitales* and *Rosales*. All the circumstantial evidence points to a biochemical, rather than a co-phylogenetic coevolution.

It is generally accepted that genetic lesions resulting in loss of biological function are unlikely to be restored in full, or in the same way. This is particularly true for mutations, less so for transposons. Genetic lesions that lead to absolute requirements for particular organic chemicals are known to occur among races of a parasite (e.g., *Phytophthora infestans*; Hohl, 1983). Genotypic differences in the host production of secondary metabolites may contribute to the resistance or susceptibility of the individual, clone or cultivar. An example of such intraspecific differences has been shown in relation to catechol production by *Allium* and its effect on *Colletotrichum* (Walker & Stahmann, 1955) - perhaps a similar hypothesis could be tested for parasitism by *Peronospora destructor*.

Cantino (1955) proposed that the loss of the ability to utilize certain inorganic forms of sulphur and nitrogen was irreversible. The differences in mineral nutrition between the *Peronosporomycetidae* and the *Saprolegniomycetidae* correlate with other morphological criteria in classifications (see PART VI: SYNOPTIC KEY: *Peronosporomycetidae*: ability to use $SO_4^=$, ability to metabolize different inorganic N sources variable; *Saprolegniomycetidae*: inability to utilize $SO_4^=$, NO_3^-). Gleason (1976) and Hohl (1983) have also reviewed mineral nutritional aspects of the *Peronosporomycetidae*.

The pathways of lysine synthesis are regarded as of phylogenetic importance (see p. 89). Unlike most fungal parasite-angiosperm host systems, both the angiosperm hosts and their straminipilous parasites have the same (DAP) lysine synthesis pathway. Lysine is a precursor for many polyamines. Galston (1983) linked polyamine metabolism with the metabolism of growth substances such as gibberellins and cytokinins. The stunting (***hypoplasia***) of sunflowers caused by *Plasmopara* should be recalled (Sackston, 1981). While most attention has been given to secondary metabolites such as flavonoids and alkaloids, Galston regarded polyamines as neglected but probably important secondary metabolites. He noted that the polyamine putrescine is probably to be found in all cells: could this be the cause of the characteristic odour (see Smith, 1884: 281) of rot caused by *Phytophthora*? Other polyamines may help to control the nuclear cycle. Galston also reported a link between CAM and polyamines mediating cellular pH stasis. Perhaps the polyamines could be another component of the 'metabolic packages' for biotrophic parasitism? There are ample data to support the assumption that most of the *Peronosporomycetidae* have a dependence for the intermediates of sterol metabolism (Warner, Sovocool & Domnas, 1982; Elliott, 1983).

There seems to be a connection between phytoalexin phenomena, the sterol requirements of the parasitic species, and the secondary metabolites characteristic of the host order. The key could perhaps be an intense sterol-precursor requirement by the straminipilous fungus that can be satisfied by certain hosts which, for various reasons, have accumulations of a variety of secondary metabolites such as essential oils, saponins, and alkaloids. The kind of photosynthetic pathway, high insolation, or absence of a mycorrhizal carbon 'sink' could all individually or collectively give rise to this accumulation. The phytoalexin theory (for reviews see Dixon & Piava, 1995; Smith, 1996; and Ebel, 1998) originated from studies of resistance to *Phytophthora infestans* in potato. The parasite/host interaction, which results in the production of rishitin, involves an induced shift in the biochemical pathway normally leading to steroid glycoalkaloids in the potato (Friend & Threlfall, 1976). It could also involve diversion of sterols to the parasite, which probably requires an exogenous source of these substances. Investigations with *Phytophthora* (Hohl, 1983) have failed to reveal significant changes in sterol concentration or composition in relation to intra-specific variations in resistance or virulence. Stössel & Hohl (1981) found that high sterol ratios tend to correlate withsusceptibility; this would be expected if host sterols were crucial to the evolution of the parasitic relationship. If this is so, the biochemical diversity of the phytoalexins elicited in *Fabaceae* (isoflavonoids) and *Solanaceae* (terpenoids) may be explained. The phytoalexin phenomenon could be regarded as a by-product of parasite-induced diversion of host-steroid metabolism.

Clarke (1983) has also suggested that phytoalexins are of secondary importance in hypersensitive reactions. Discussions of differences in phytoalexin elicitation reflect abnormalities and deviations in the coevolutionary relationship. In natural populations, phytoalexin elicitation may confer a biological advantage ***to the parasite*** by limiting the extent of parasitism so that perennation of heterogeneous gene pools of both parasite and host is maintained. Only in artificial systems of clonal culture will this mechanism assume undue prominence. Paradoxically, the production of secondary metabolite pathways, evolved to alleviate physiological stress in the host, may have been evolutionarily conserved because they conferred a biological advantage (defence systems) by deterring fungal parasites as well as animal predators. It is noteworthy that many biochemical defence systems, including phytoalexins, are not as efficient as might have been expected, were they ***primarily*** evolved, under strong selective pressure, for defence. Sterol biosynthesis is linked through the terpenoids that relate to all these secondary metabolites.

A second steroid metabolism link is with the action of the polyene antibiotics (for early references, see Tsao, 1970). Polyene antibiotics act on *Mycota* but not on straminipilous fungi. Their selectivity suggests that there is a difference between these two groups of heterotrophs with respect to their membrane-bound sterols, as these antibiotics are thought to function by acting on membrane-bound sterols. The variety of sterols and their precursors which act as the messengers and recognition systems between the organism and its environment is not understood. In terms of response to the anti-straminipile fungicides Hymexazol and Metalaxyl (Kato *et al.*, 1990) the graminicolous DM relatives, the Verrucalvaceae, were relatively resistant to Metalaxyl but responded to Hymexazol similarly to *Pythium* and in contrast to *Phytophthora*. Ann & Ko (1992) surveyed antibiotic resistance and dependence in *Phytophthora*. Griffith, Davis & Grant (1992) have emphasised the importance of understanding the utilization of differences in metabolism and sterol production and requirements in assessing fungicide efficacy. The actions of sterol-inhibiting fungicides (Fletcher, 1987; Gisi, 2000) are also relevant to coevolution.

Synergism

Other possible biochemically-mediated organismal complexes have been envisaged, for example, the suggestion (Sansome & Sansome, 1974) that *Albugo* and *Peronospora* are synergistic on crucifers. The susceptibility of beans to *Pseudoperonospora cubensis* only after prior infection by *Uromyces* (Yarwood, 1977) may be another example of biochemical synergism.

Mycorrhizal associations

Most, but not all, angiosperms have mycorrhizal associations which enable more efficient absorption of phosphorus and/or nitrogen. From fossil evidence, vascular plant axis-fungal associations are known to have existed at least from the Devonian (*Asteroxylon-Glomus*-type inclusions; Pirozynski, 1976*a*, *b*). Such associations would have gained importance under savannah conditions because mycorrhizal associations are beneficial in climates producing increased water stress and depletion of nitrogen and phosphorus from the soil in storms and flash floods. For present-day angiosperms, mycorrhizal associations are most researched for grassland and forest. Information on mycorrhizal associations in wetland species is mainly confined to ericoid mycorrhiza, which may not be relevant to coevolution of the *Peronosporales*, since few *Ericales* are hosts to these fungi. Mycorrhizal associations will result in a concommitant redress to imbalance in C:N ratios, and possible reduction in secondary metabolite production. One interesting conjunction, which needs more substantiation, is that between ruderal herbs (e.g., *Brassicaceae* and *Caryophyllaceae*) and DMs. Ruderal herbs lack mycorrhizas (Gerdemann, 1968). It is interesting to speculate that perhaps the absence of mycorrhizal associations alters the flow of metabolites, and perhaps induces the manufacture of certain secondary metabolites, so that an accumulation of essential oils, alkaloids, and saponins is facilitated (Figure III: 2). For most organisms, including mycote fungi, these substances are antagonistic or repellent. For a few organisms (straminipilous fungi?) they may be attractants, or serve as undocumented intermediates to essential metabolism.

Symbiotic associations other than with fungi also evolved, such as *Rhizobium*-root nodules, and the increased nitrogenous uptake may have contributed to the cyano-compounds characteristic of the *Fabaceae*, a family that showed increasing diversity during the drier savannah conditions of the Eocene. Host-root associations (mycorrhizas or weak root parasites) might so affect the well-being of the host as to make a significant change (either way) to its vulnerability to parasitism of ***aerial*** parts. In experimental studies of biological control (Elad & Chet, 1987), induced root communities have been explored within root systems. Communities on and in roots do not appear to have been evaluated in connection with the metabolism of the whole angiosperm plant in different growing conditions or with DM parasitism of the aerial shoot. Lack of data for whole-plant pathological communities is a serious gap in knowledge.

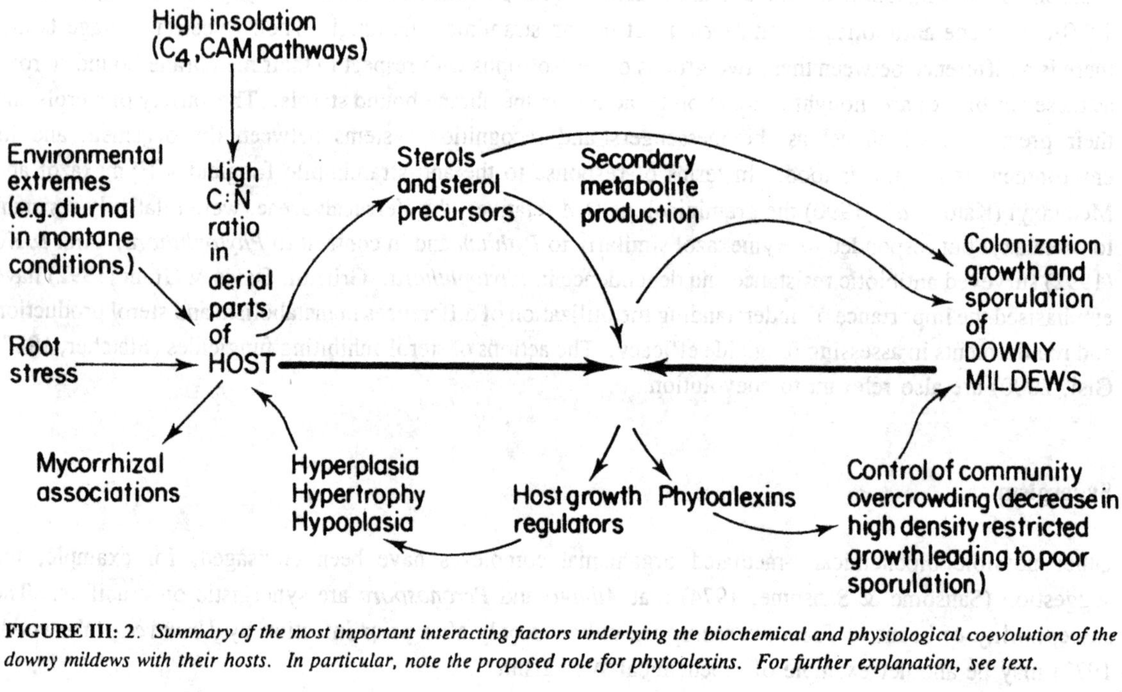

FIGURE III: 2. *Summary of the most important interacting factors underlying the biochemical and physiological coevolution of the downy mildews with their hosts. In particular, note the proposed role for phytoalexins. For further explanation, see text.*

Coevolution of the downy mildews with angiosperms in the Tertiary

A complex web between angiosperm/animal coevolution and secondary metabolite-mediated straminipile/angiosperm coevolution has evolved. It is impossible to countenance evolutionary hypotheses that disregard palaeo-geographic or climatic factors, or the way in which these factors have determined the evolution of their primogenitors. For biotrophic parasites the evolution of the hosts is as important as the evolution of their pathogenicity: the importance of high insolation might be inferred for tropical uplands (*Solanaceae*), temporary clearings (climbing plants), and montane and mediterranean habitats (the last two habitats would also be more xeric). Relatively extreme habitats are potentially physiologically stressful for the hosts. In such habitats the selection for, or from, non-zoosporic biotrophic *Peronomycetes* would be unrelated to their phylogeny, which should not therefore be based on zoosporic morphology and morphogenesis.

In the Tertiary the tropical/sub-tropical forest with a closed canopy gave way to more open vegetation. The circum-Mediterranean climate became 'mediterranean'. The mountain-building activity changed the topography and climate. The angiosperms, previously adapted to ***sub-optimal light*** conditions; ***high ambient*** temperatures and ***high*** humidity, were now exposed to ***greater insolation***; ***fluctuating diurnal*** temperatures and ***variable*** (extreme seasonal) humidity. Climatic pressures would have encouraged herbaceous evolution in open canopy ecosystems: exposure to high levels of UV irradiation would have been deleterious to under-canopy trees which had evolved in sub-optimum light; the high photosynthetic activities would have resulted in excess photosynthate. The consequent development of secondary metabolites from both of these causes would have produced further ramifications of the angiosperm/animal coevolution. Was this also a stimulus which enabled the straminipilous fungi, previously adapted to high protein/hydrocarbon/carbohydrate nutrition, to colonize roots? Was the climate change (hot days - cool nights - clear skies) sufficient to enable foliar colonization by *Phytophthora* species with caducous sporangia?

For coevolution with angiosperms to have taken place, there must be, or have been:

1 physical proximity between populations of the photosynthetic host and heterotrophic parasite taxa involved
2 sufficient similarity between biochemical pathways and/or products for nutritional requirements of the parasite to be satisfied

Sometimes, this will be because new hosts are phylogenetically close to former hosts. But phylogenetically distant or unrelated potential hosts may also provide the critical factors for the nutritional environment of the parasite. The factors involved are not necessarily the most evident and the major nutritional requirements, carbon and nitrogen sources, are unlikely to be critical. Harlan (1976) has discussed the influence of parasitism by diverse organisms on angiosperm evolution. Coevolution may embrace more than two taxa and the populations need not be large, stable, or of equivalent size. However, if the potential parasite is sexually reproducing, there is probably a minimum population size below which a coevolutionary relationship is unlikely to develop. This postulate could be argued on the out-breeding and colonization potentials of the parasite as well as the ability of the host population to develop resistance. Proximity may be of long standing, with a gradually established, balanced dependence. Alternatively, a conjunction may happen suddenly, either when a vulnerable alien host becomes introduced to an environment containing a previously stabilized host-parasite relationship, or when the spore population of a parasite is alien but able to exploit a susceptible but previously inaccessible host (Hijwegen, 1988). There must be physical or biochemical similarities or analogues that enable an appropriate degree of association between previously separated populations. Too great a vulnerability will lead to an unstable and ephemeral relationship. The essence of coevolution is adaptive change in balanced relationships. Chance associations leading to coevolution must be less likely for fungi that are essentially systemic, because there would be less likelihood that either host or parasite would survive long enough to reach reproductive maturity.

Despite the postulation of aquatic ancestral origins for both the angiosperms and the DMs, the host dicotyledonous angiosperm orders are not noted for their aquatic members. The habitats of extant hosts for DMs are diverse and an enormous morphological and anatomical range is encompassed. Coevolution is not, therefore, a development from closely associated ancestral stocks of hosts and DMs. High carbohydrate levels are conducive to the production of secondary metabolites (sterols, essential oils, flavonoids from phenylalanine; alkaloids from carotenoid precursors). Since the protoplasms of a haustorial obligate parasite and a host are never in contact, there must be a dynamic equilibrium between the components of the plasmamembranes to control the differential permeabilities. The metabolism of component phospholipids and sterols of the plasma membranes could obviously be involved (and could provide a role for haustoria; see Spencer-Phillips, 1997).

The evolutionary age; extent of speciation within; and genetic diversity of genera can be very different. If the various genera of the DMs arose from different 'pro-*Phytophthora*' ancestors, then the genetic distance between selected monophyletic DM genera and a particular group of species of *Phytophthora* may be less than the genetic distances within *Phytophthora sensu lato*. Comprehensive sequence data will be essential to assess whether the genetic distances between species of a more ancient genus (e.g., *Phytophthora*) may be greater than those between any recently evolved DM genera, even if these genera have arisen from different, though still relatively close, *Phytophthora* antecedents.

The link is not with the pattern of angiosperm phylogeny, as such, but with the highly evolved and greatly stressed angiosperms, characterized by their armoury of secondary metabolites functioning as deterrents to animal predators and mycote fungal pathogens. But these same metabolites rendered such angiosperms uniquely vulnerable to the straminipile heterotrophs. The phylogeny of the DMs is thus polyphyletic, with the dicotyledonous DMs arising from 'pro-*Phytophthora*' Cretaceous ancestors, at different ages of the Tertiary or even the Quaternary, and in different continents: independent origins, both in geographic location and geologic time, are postulated for the genera *Albugo*, *Plasmopara* and *Peronospora*. The evolutionary origins of the downy mildews are recent compared with the origins of their hosts.

Present-day geographical distributions are of little value in elucidating evolution. They merely confirm the opportunism and rapid physiological adaptation of these parasites. The susceptibility of high latitude Gondwanaland species to the sub-tropical Northern Hemisphere *Peronosporomycetidae* is exemplified by the susceptibility of Southern Hemisphere *Myrtaceae* and *Proteaceae* to the spread of *Phytophthora cinnamomi* on *Eucalyptus* (Shearer & Tippett, 1989; Weste, 1994) and *Banksia* (Shearer & Hill, 1989), which is probably now uncontainable less than a century after the introduction of the fungus. Appreciable intraspecific coevolution can nevertheless occur in relatively short time-scales: in this century accommodations between *Peronospora tabacina* on tobacco in North America and Europe have occurred (Welzien, 1981).

The outstanding characteristic of the distribution of the DM parasites of dicotyledons is that it is not primarily the more primitive or ancient orders of angiosperms that are affected. Angiosperms parasitized by DMs are mostly in ***highly evolved families***, ***OR*** in taxa that may have a propensity to produce ***high levels of secondary metabolites***. If the ordinal rather than family level (*contra* Palti & Kenneth, 1981) of the angiosperms is reviewed, a possible pattern emerges.

The majority of named species in *Peronospora* are parasitic in taxa of four clusters of angiosperm orders:

1 Eudicots - Core Eudicots (*Ranunculales*, *Caryophyllales*)

2 Eurosids I (*Fabales*, *Rosales* (including *Urticaceae*) and *Malpighiales* (*Euphorbiaceae* and *Violaceae*)

3 Eurosids II (*Brassicales*)

4 Asterids (but not *Asteraceae*)

TABLE III: 3. *Dicotyledonicolous downy mildews and white blister rusts arranged according to classification of hosts (Angiosperm Phylogeny Group (APG), 1998). This list is rearranged from APPENDIX I and lists in Dick (2000b). Type species are indicated. Subspecific taxa and recognised synonyms have been deleted, but the existence of subspecific taxa is indicated by use of the type subspecific rank. Synonymies within the genera* Peronospora *and* Plasmopara *have yet to be evaluated (see notes to APPENDIX 1).*

BASAL ANGIOSPERMS: dicots, omitting monocots

PIPERALES: Piperaceae

Albugo tropica Lagerh.

EUDICOTS: orders without supra-ordinal phylogeny

RANUNCULALES: Berberidaceae

Peronospora achlydis S. Ito & Tokun.

RANUNCULALES: Papaveraceae with *Fumariaceae*

Albugo eomeconis Z. Y. Zhang & Ying X. Wang
Albugo keeneri Solheim & Gilb.

Peronospora affinis Rossmann
Peronospora arborescens (Berk.) Casp.
Peronospora argemones Gaüm.
Peronospora bocconiae Syd.
Peronospora bulbocapni Beck
Peronospora chelidonii Jacz. & P. A. Jacz.
Peronospora corydalis de Bary
Peronospora corydalis-intermediae Gaüm.
Peronospora cristata Tranzschel
Peronospora dicentrae Syd.
Peronospora gaeumannii Mundk.
Peronospora glaucii Lobik
Peronospora hylomeconis S. Ito & Tokun.
Peronospora hypecoi Jacz. & P. A. Jacz.
Peronospora meconopsidis Mayor
Peronospora roemeriae Zaprom.

Plasmopara nakanoi S. Ito & Muray.

RANUNCULALES: Ranunculaceae

Peronospora aconiti Y. N. Yu
Peronospora alpicola Gaüm.
Peronospora apiospora G. Poirault
Peronospora consolidae Jacz. & P. A. Jacz.
Peronospora ficariae de Bary
Peronospora gigantea Gaüm.
Peronospora hellebori-purpurascentis Săvul. & Rayss
Peronospora hiemalis Gaüm.
Peronospora illyrica Gaüm.
Peronospora iwatensis S. Ito & Muray.
Peronospora leptopyri C. K. Pai
Peronospora myosuri Fuckel
Peronospora parvula Jacz. & P. A. Jacz.
Peronospora pennsylvanica Gaüm.
Peronospora pulveracea Fuckel
Peronospora ranunculi Gaüm.
Peronospora ranunculi-carpatici Săvul. & Rayss
Peronospora ranunculi-oxyspermi Jacz. & Sergeeva
Peronospora ranunculi-peduncularis Roiv.
Peronospora ranunculi-sardoi Săvul. & Rayss
Peronospora ranunculi-steveni Săvul. & Rayss
Peronospora yamadana Togashi

Plasmopara alpina (C. J. Johanson) A. Blytt
Plasmopara anemones-dichotomae Benua
Plasmopara anemones-nemorosae Săvul. & O. Săvul.
Plasmopara anemones-ranunculoidis Săvul. & O. Săvul.

Plasmopara baicalensis Jacz. & P. A. Jacz.

Plasmopara cimicifugae S. Ito & Tokun.

Plasmopara delphinii (Gapon.) Novot.
Plasmopara hellebori-purpurascentis Săvul. & O. Săvul.
Plasmopara isopyri Skalický
Plasmopara isopyri-thalictroidis (Săvul. & Rayss) Săvul. & O. Săvul.
Plasmopara pygmaea (Unger) J. Schröt. forma *pygmaea* **[type species]**
Plasmopara wartenweileri Skalický

CORE EUDICOTS: families and orders without supra-ordinal phylogeny

Vitaceae

Plasmopara chinensis Gorlenko
Plasmopara cissi Vienn.-Bourg.
Plasmopara viticola (Berk. & M. A. Curtis) Berl. & De Toni var. *viticola*

(Table III: 3, continued)

CARYOPHYLLALES: Aizoaceae

Albugo austro-africana Syd. & P. Syd.

Peronospora mesembryanthemi Verwoerd

Albugo trianthemae G. W. Wilson

CARYOPHYLLALES: Amaranthaceae

Albugo achyranthis (Henn.) Miyabe
Albugo amaranthi (Schwein.) Kuntze
Albugo eurotiae Tranzschel
Albugo bliti (Biv.) Kuntze
Albugo gomphrenae (Speg.) Cif. & Biga
Albugo occidentalis G. W. Wilson
Albugo portulacae (DC.) Kuntze

Peronospora amaranthi Gaüm.
Peronospora atriplicis-halimi Săvul. & Rayss
Peronospora atriplicis-hastatae Săvul. & Rayss
Peronospora atriplicis-hortensis Săvul. & Rayss
Peronospora atriplicis-tataricae Oescu & Rădul.
Peronospora axyridis C. Benois
Peronospora bohemica Gaüm.
Peronospora boni-henrici Gaüm.
Peronospora chenopodii Schltdl.
Peronospora chenopodii-ficifolii Sawada
Peronospora chenopodii-glauci Gaüm.
Peronospora chenopodii-opulifolii Săvul. & Rayss
Peronospora chenopodii-polyspermi Gaüm.
Peronospora chenopodii-rubri Gaüm.
Peronospora chenopodii-urbici Săvul. & Rayss
Peronospora chenopodii-vulvariae Săvul. & Rayss
Peronospora daturae Hulea
Peronospora echinopsili Tunkina
Peronospora effusa (Grev.) Tul.
Peronospora eurotiae Kalymb.
Peronospora farinosa (Fr.) Fr. *sensu lato*
Peronospora farinosa (Fr.) Fr. *sensu stricto*
Peronospora iolotanica Kolosch.
Peronospora kochiae Gaüm.
Peronospora kochiae-scopariae Kochman & T. Majewski
Peronospora litoralis Gaüm.
Peronospora minor (Casp.) Gäum.
Peronospora monolepidis Gaüm.
Peronospora muralis Gaüm.
Peronospora nitens Oescu & Rădul.
Peronospora obionis-verruciferae Săvul. & Rayss
Peronospora salicorniae Jenkina
Peronospora tatarica Săvul. & Rayss
Peronospora teloxydis Jacz. & P. A. Jacz.
Peronospora ussuriensis Jacz. & P. A. Jacz.
Peronospora variabilis Gäum.
Peronospora vistulensis Wróbl.

Plasmopara achyranthis J. F. Tao & Y. Qin

CARYOPHYLLALES: Caryophyllaceae

Albugo caryophyllacearum (Wallr.) Cif. & Biga

Albugo quadrata (Wallr.) S. D. Baker

Peronospora agrostemmatis Gaüm.
Peronospora alsinearum Casp.
Peronospora arenariae (Berk.) Tul.
Peronospora atlantica Gaüm.
Peronospora campestris Gaüm.
Peronospora cerastii-anomali Săvul. & Rayss
Peronospora cerastii-brachypetali Săvul. & Rayss
Peronospora cerastii-glandulosi S. Ito & Tokun.
Peronospora cucubali S. Ito & Tokun.
Peronospora dianthi de Bary
Peronospora fontana A. Gustavsson
Peronospora gypsophilae Jacz. & P. A. Jacz.
Peronospora helvetica Gaüm.
Peronospora herniariae de Bary
Peronospora holostei de Bary
Peronospora lepigoni Fuckel
Peronospora lychnitis Gaüm.
Peronospora media Gaüm.
Peronospora melandrii Gaüm.
Peronospora melandrii-noctiflori Săvul. & Rayss
Peronospora moenchiae C. Camara & Oliviera
Peronospora obovata Bonord.
Peronospora parva Gaüm.
Peronospora paula A. Gustavsson
Peronospora polycarpi Mayor & Vienn.-Bourg.
Peronospora pseudostellariae G. Y. Yin & Z. S. Yang
Peronospora septentrionalis Gaüm.
Peronospora scleranthi Rabenh.
Peronospora silenes G. W. Wilson
Peronospora stellariae-aquaticae Sawada
Peronospora stellariae-radiantis S. Ito & Tokun.
Peronospora stellariae-uliginosae Sawada
Peronospora tomentosa Fuckel
Peronospora tornensis Gaüm.
Peronospora trivialis Gäum.
Peronospora uralensis Jacz. & P. A. Jacz.
Peronospora vernalis Gaüm.

CARYOPHYLLALES: Molluginaceae

Albugo molluginis S. Ito

(Table III: 3, continued)

CARYOPHYLLALES: Nyctaginaceae

Albugo platensis (Speg.) Swingle

Peronospora oxybaphi Ellis & Kellerm.

CARYOPHYLLALES: Plumbaginaceae

Peronospora constantineanui Săvul. & Rayss
Peronospora limonii Simonyan
Peronospora statices Lobik

CARYOPHYLLALES: Polygonaceae

Albugo polygoni Z. D. Jiang & P. K. Chi

Peronospora americana Gaüm.
Peronospora ducometii Siemaszko & Jank.
Peronospora eriogoni H. Solheim & Gilb.
Peronospora fagopyri Jacz. & P. A. Jacz.
Peronospora jaapiana Magnus
Peronospora polygoni (Thüm.) A. Fisch.
Peronospora polygoni-convolvuli A. Gustavsson
Peronospora rumicis Corda **[type species]**
Peronospora rumicis-rosei Rayss
Peronospora sinensis D. Z. Tang

CARYOPHYLLALES: Portulacaceae

Albugo portulacearum (Schltdl.) Kochman & T. Majewski

Peronospora calindriniae Speg.
Peronospora claytoniae Farl.

SANTALALES: Santalaceae

Peronospora thesii Lagerh.

SAXIFRAGALES: Crassulaceae

Albugo tilleae (Lagerh.) Cif. & Biga

SAXIFRAGALES: Grossulariaceae

Plasmopara ribicola Davis

SAXIFRAGALES: Saxifragaceae

Peronospora chrysosplenii Fuckel
Peronospora minima G. W. Wilson
Peronospora saxifragae Bubák
Peronospora whippleae Ellis & Everh.

CORE EUDICOTS: ROSIDS: sister groups (families and orders) to Eurosids I and II

Zygophyllaceae

Peronospora tribulina Pass.

GERANIALES: Geraniaceae

Peronospora conglomerata Fuckel
Peronospora effusa-ciconia Becc.
Peronospora erodii Fuckel

Plasmopara geranii (Peck) Berl. & De Toni
Plasmopara geranii-pratensis Săvul. & O. Săvul.
Plasmopara geranii-sylvatici Săvul. & O. Săvul.
Plasmopara pusilla (de Bary) J. Schröt.

(Table III: 3, continued)

CORE EUDICOTS: ROSIDS: EUROSIDS I

CUCURBITALES: Cucurbitaceae

Plasmopara australis (Speg.) Swingle

Pseudoperonospora cubensis (Berk. & M. A. Curtis) Rostovzev **[type species]**

FABALES: Fabaceae (including *Caesalpinaceae*)

Albugo mauginii (Parisi) Cif. & Biga

Peronospora aestivalis Syd.
Peronospora astragali Syd.
Peronospora astragali-purpurei Mayor & Vienn.-Bourg.
Peronospora astragalina Syd.
Peronospora cilicia Bremer & Gäum.
Peronospora coronillae Gaüm.
Peronospora coronillicola C. Camara & Oliviera
Peronospora cytisi Rostr.
Peronospora desmodii S. Ito & Tokun.
Peronospora dipeltae Jacz. & P. A. Jacz.
Peronospora dorycnii Uljan.
Peronospora ervi A. Gustavsson
Peronospora esperaussensis Mayor
Peronospora fabae Jacz. & Sergeeva
Peronospora fulva Syd.
Peronospora galegae Săvul. & Rayss
Peronospora lagerheimii Gaüm.
Peronospora lathyri-aphacae Săvul. & Rayss
Peronospora lathyri-hirsuti Săvul. & Rayss
Peronospora lathyri-humilis C. Benois
Peronospora lathyri-maritimi Jermal.
Peronospora lathyri-palustris Gaüm.
Peronospora lathyri-pisiformis M. I. Nikol.
Peronospora lathyri-rosei Osipian
Peronospora lathyri-verni A. Gustavsson
Peronospora lathyri-versicolaris Săvul. & Rayss
Peronospora lathyrina Vienn.-Bourg.
Peronospora lentis Gaüm.
Peronospora lotorum Syd.
Peronospora manshurica (Naumov) Syd.
Peronospora mayorii Gaüm.
Peronospora medicaginis-minimae Gapon.
Peronospora medicaginis-orbicularis Rayss
Peronospora medicaginis-tianschanicae Gapon.
Peronospora meliloti Syd.
Peronospora melissiti Byzova & Dejeva
Peronospora moreaui Rayss
Peronospora narbonensis Gaüm.
Peronospora ononidis G. W. Wilson
Peronospora ornithopi Gaüm.
Peronospora orobi Gaüm.
Peronospora oxytropidis Gaüm.
Peronospora phacae Gäum.
Peronospora pisi Syd.
Peronospora pratensis Syd.
Peronospora romanica Săvul. & Rayss
Peronospora ruegeriae Gaüm.
Peronospora savulescui Rayss
Peronospora senneniana Gonz. Frag. & Sacc.
Peronospora tetragonolobi Gaüm.
Peronospora tetragonolobi-palestini Rayss
Peronospora trifolii-alpestris Gäum.
Peronospora trifolii-arvensis Syd.
Peronospora trifolii-formosi Rayss
Peronospora trifolii-hybridi Gaüm.
Peronospora trifolii-minoris Gaüm.
Peronospora trifolii-pratensis A. Gustavsson
Peronospora trifolii-repentis Syd.
Peronospora trifoliorum de Bary *sensu lato*
Peronospora trifoliorum de Bary *sensu stricto*
Peronospora trigonellae Gäum.
Peronospora viciae (Berk.) Casp. *sensu lato*
Peronospora viciae (Berk.) Casp. *sensu stricto*
Peronospora viciae-venosae Jacz. & P. A. Jacz.
Peronospora viennotii Mayor

Plasmopara cercidis C. G. Shaw

Pseudoperonospora cassiae Waterh.

MALPIGHIALES: Euphorbiaceae

Peronospora andina Speg.
Peronospora cyparissiae de Bary
Peronospora embergeri Mayor & Vienn.-Bourg.
Peronospora esulae Gäum.
Peronospora euphorbiae Fuckel
Peronospora euphorbii-glyptospermae Gäum.
Peronospora euphorbii-thymifoliae Sawada
Peronospora favargeri Mayor & Vienn.-Bourg.
Peronospora hypericifoliae Sinha & Mathur
Peronospora valesiaca Gäum.

Plasmopara acalyphae (G. W. Wilson) G. W. Wilson

MALPIGHIALES: Linaceae

Peronospora lini J. Schröt.

(Table III: 3, continued)

MALPIGHIALES: Violaceae

Bremiella megasperma (Berl.) G. W. Wilson **[type species]**

Peronospora violae J. Schröt.

OXALIDALES: Oxalidaceae

Peronospora oxalidis Verwoerd & du Plessis

ROSALES: Rosaceae

Peronospora potentillae de Bary
Peronospora sparsa Berk. *sensu lato*
Peronospora sparsa Berk. *sensu stricto*

Plasmopara sanguisorbae C. J. Li *et al.*

ROSALES: Ulmaceae

Pseudoperonospora celtidis (Waite) G. W. Wilson

ROSALES: Urticaceae

Albugo pileae J. F. Tao & Y. Qin

Peronospora boehmeriae G. Y. Yin & Z. S. Yang
Peronospora debaryi E. S. Salmon & Ware
Peronospora parietariae Vanev & E. G. Dimitrova

Plasmopara illinoensis (Farl.) Davis
Plasmopara elatostemmatis (Togashi & F. Onuma) S. Ito & Tokun.
Plasmopara miyakeana S. Ito & Tokun.
Plasmopara pileae (Gäum.) Jacz. & P. A. Jacz.

Pseudoperonospora cannabina (G. H. Otth) Curzi
Pseudoperonospora humuli (Miyabe & M. Takah.) G. W. Wilson
Pseudoperonospora urticae (Berk.) E. S. Salmon & Ware

CORE EUDICOTS: ROSIDS: EUROSIDS II

BRASSICALES: Brassicaceae

Albugo candida (J. F. Gmel.) Kuntze var. *candida* **[type species]**
Albugo capparidis (de Bary) Cif.
Albugo chardonii W. Weston
Albugo lepidii A. N. S. Rao

Peronospora alliariae-wasabi Gaüm.
Peronospora alyssi-calycini Gaüm.
Peronospora alyssi-incani Gaüm.
Peronospora alyssi-maritimi Kochman
Peronospora arabidis-alpinae Gaüm.
Peronospora arabidis-glabrae Gaüm.
Peronospora arabidis-hirsutae Gaüm.
Peronospora arabidis-oxyphyllae Gaüm.
Peronospora arabidis-strictae Jacz. & P. A. Jacz.
Peronospora arabidis-turritae Gaüm.
Peronospora arabidopsidis Gaüm.
Peronospora aubrietae Mayor
Peronospora barbareae Gaüm.
Peronospora berteroae Gaüm.
Peronospora biscutellae Gaüm.
Peronospora brassicae Gaüm.
Peronospora buniadis Gaüm.
Peronospora cakiles Savile
Peronospora calepinae Gaüm.
Peronospora hesperidis Gaüm.
Peronospora hornungiae A. Gustavsson
Peronospora hymenolobi Annal.
Peronospora iberidis Gaüm.
Peronospora iranica Petr. & Esfand.
Peronospora isatidis Gaüm.
Peronospora jordanovii R. Krusch.
Peronospora lepidii-perfoliati Săvul. & Rayss
Peronospora lepidii-sativi Gaüm.
Peronospora lepidii-virginici Gaüm.
Peronospora leptalei Kolosch.
Peronospora litwinowiae Kalymb.
Peronospora lobulariae Ubrizsy & Vörös
Peronospora lunulariae Gaüm.
Peronospora malcolmiae Lobik
Peronospora malyi Lindtner
Peronospora matthiolae Gaüm.
Peronospora maublancii Săvul. & Rayss
Peronospora menioci Kolosch.

(Table III: 3, continued)

Peronospora camelinae Gaüm.
Peronospora cardamines-laciniatae A. Gaüm.
Peronospora cardaminopsis A. Gustavsson
Peronospora cardariae-repentis N. P. Golovina
Peronospora carpoceratis Byzova
Peronospora chartomatis Golvin & Kalymb.
Peronospora cheiranthi Gaüm.
Peronospora chorisporae Gäum.
Peronospora cochleariae Gaüm.
Peronospora coringiae Gäum.
Peronospora coronopi Gaüm.
Peronospora crambes Jacz. & P. A. Jacz.
Peronospora cryptosporae Annal.
Peronospora dentariae Rabenh.
Peronospora dentariae-macrophyllae Gaüm.
Peronospora desertorum Jacz. & P. A. Jacz.
Peronospora diplotaxidis Gaüm.
Peronospora diptychocarpi Kalymb.
Peronospora drabae Gaüm.
Peronospora drabae-majusculae Lindtner
Peronospora eigii Rayss
Peronospora erophilae Gaüm.
Peronospora erucastri Gaüm.
Peronospora erysimi Gaüm.
Peronospora euclidii Săvul. & Rayss
Peronospora gaeumanniana Jaap
Peronospora galligena S. Blumer
Peronospora goldbachiae Jacz. & P. A. Jacz.
Peronospora golovinii Kalymb.
Peronospora heliophilae E. Müll. & Poelt
Peronospora myagri Mayor
Peronospora nasturtii-aquatici Gaüm.
Peronospora nasturtii-montani Gaüm.
Peronospora nasturtii-palustris S. Ito & Tokun.
Peronospora nesliae Gaüm.
Peronospora niessliana A. Berl.
Peronospora norvegica Gaüm.
Peronospora pachyphragmatis-macrophylli Savinceva
Peronospora parasitica (Fr.) Fr.
Peronospora rapistri Jacz. & Sergeeva
Peronospora rhaetica Gaüm.
Peronospora rorippae-islandicae Gaüm.
Peronospora simonianii Osipian
Peronospora sisymbrii-intermedii Gaüm.
Peronospora sisymbrii-loeselii Gaüm.
Peronospora sisymbrii-officinalis Gaüm.
Peronospora sisymbrii-orientalis Gaüm.
Peronospora sisymbrii-sophiae Gaüm.
Peronospora sophiae-pinnatae Gaüm.
Peronospora streptolomatis Golovin & Kalymb.
Peronospora syreniae-cuspidatae Oescu & Rădul.
Peronospora takahashii S. Ito
Peronospora taurica Jacz. & P. A. Jacz.
Peronospora tauscheriae Kalymb.
Peronospora teesdaliae Gaüm.
Peronospora thlaspeos-alpestris Gaüm.
Peronospora thlaspeos-arvensis Gaüm.
Peronospora thlaspeos-perfoliati Gaüm.
Peronospora turritidis Gäum.
Peronospora vvedenskyi Golovin

Plasmopara venezuelana Chardón

***BRASSICALES**: Limnanthaceae*

Peronospora floerkeae Kellerm.

***BRASSICALES**: Resedaceae*

Albugo resedae (Jacz.) Cif. & Biga

Peronospora crispula Fuckel

***MALVALES**: Malvaceae (Grewioideae)* (formerly *Tiliaceae*)

Plasmopara heliocarpi Lagerh.
Plasmopara satarensis P. B. Chavan & U. V. Kulkarni
Plasmopara skvortovii Miura
Plasmopara triumfettae A. D. Sharma & Munjal

***MALVALES**: Cistaceae*

Peronospora alpestris Gaüm.
Peronospora leptoclada Sacc.

***MYRTALES**: Onagraceae*

Peronospora arthurii Farl.

Plasmopara epilobii (G. H. Otth) Sacc. & P. Syd.
Plasmopara latifolii Savile

***SAPINDALES**: Meliaceae*

Plasmopara portoricensis (Lamkey) Waterh.

(Table III: 3, continued)

CORE EUDICOTS: ASTERIDS: sister groups to ASTERIDS I and II

ASTERIDS: EUASTERIDS I: family without ordinal grouping

Boraginaceae

Albugo cynoglossi (Unamuno) Cif. & Biga

Peronospora alkannae Vienn.-Bourg.
Peronospora anchusae Ziling
Peronospora arnebiae Golovin
Peronospora asperuginis J. Schröt.
Peronospora bothriospermi Sawada
Peronospora cerinthes Uljan.
Peronospora cynoglossi Swingle
Peronospora eritrichii S. Ito & Tokun.
Peronospora hydrophylli Waite
Peronospora lithospermi Gaüm.
Peronospora microulae Y. R. Meng & G. Y. Yin
Peronospora myosotidis de Bary
Peronospora nemophilae C. G. Shaw
Peronospora noneae Jacz. & Sergeeva
Peronospora omphalodis Gaüm.
Peronospora pulmonariae Gaüm.
Peronospora rocheliae Kalymb.
Peronospora rugosa Jacz. & P. A. Jacz.
Peronospora solenanthi Byzova
Peronospora symphyti Gaüm.
Peronospora thyrocarpi L. Ling & M. C. Tai
Peronospora trigonotidis S.Ito & Tokun.
Peronospora uljanishchevii Tunkina

Plasmopara myosotidis C. G. Shaw

ASTERIDS: EUASTERIDS I: orders without supra-ordinal phylogeny

CORNALES: Hydrangeaceae

Plasmopara harae S. Ito & Muray.

ERICALES: Balsaminaceae

Bremiella sphaerosperma Constant.

Plasmopara obducens (J. Schröt.) J. Schröt.

ERICALES: Polemoniaceae

Peronospora phlogina Dietel & Holw.

ERICALES: Primulaceae

Peronospora agrorum Gaüm.
Peronospora anagallidis J. Schröt.
Peronospora androsaces Niessl
Peronospora candida Fuckel
Peronospora cortusae Gaüm. & S. Blumer
Peronospora gregoriae S. Blumer
Peronospora mirabilis Jacz. & P. A. Jacz.
Peronospora oerteliana J. G. Kühn

ASTERIDS: EUASTERIDS I: orders and families

GENTIANALES: Apocynaceae

Peronospora vincae J. Schröt.

Plasmopara gonolobi (Lagerh.) Swingle
Plasmopara vincetoxici Ellis & Everh.

GENTIANALES: Gentianaceae

Albugo centaurii (Hansf.) Cif. & Biga
Albugo swertiae (Berl. & Kom.) G. W. Wilson

Peronospora canscorina Thite & M. S. Patil
Peronospora chlorae de Bary
Peronospora erythraeae Gaüm.
Peronospora gentianiae Rostr.

(Table III: 3, continued)

GENTIANALES: Rubiaceae

Peronospora borealis Gaüm.
Peronospora calotheca de Bary
Peronospora crucianellae Maire
Peronospora galii-anglici Uljan.
Peronospora galii-pedemontani Săvul. & Rayss
Peronospora galii-rubioides Săvul. & Rayss
Peronospora galii-trifidii S. Ito & Tokun.
Peronospora galii-veri Gäum.
Peronospora hiratsukae S. Ito & Tokun.
Peronospora hommae S. Ito & Tokun.
Peronospora insubrica Gaüm.
Peronospora rubiae Gäum.
Peronospora sakamotoi S. Ito & Tokun.
Peronospora seymourii Burrill in Underwood
Peronospora sherardiae Fuckel
Peronospora silvatica Gaüm.

Plasmopara borreriae (Lagerh.) Constant.

LAMIALES: Buddlejaceae

Peronospora hariotii Gaüm.

LAMIALES: Lamiaceae

Albugo evolvulae (Damle) Cif. & Biga

Peronospora amethysteae Jacz. & P. A. Jacz.
Peronospora calaminthae Fuckel
Peronospora chamaesphaci Kalymb.
Peronospora clinopodii Terui
Peronospora dracocephali C. J. Li & Zhen Y. Zhao
Peronospora davisii C. G. Shaw
Peronospora elsholtziae T. R. Liu & C. K. Pai
Peronospora galeopsidis Lobik
Peronospora glechomae Oescu & Rădul.
Peronospora hedeomatis Kellerm. & Swingle
Peronospora ibrahimovii T. M. Achundov
Peronospora lallemantiae Golovin & Kalymb.
Peronospora lamii A. Braun
Peronospora leonuri T. R. Liu & C. K. Pai
Peronospora lophanthi Farl.
Peronospora menthae X. Y. Cheng & H. C. Bai
Peronospora perillae Miyabe
Peronospora rossica Gaüm.
Peronospora saturejae-hortensis Osipian
Peronospora sideritidis Byzova
Peronospora scutellariae Bejlin
Peronospora stachydis Syd.
Peronospora stigmaticola Raunk.
Peronospora teucrii Gaüm.
Peronospora ziziphorae Byzova

Plasmopara calaminthae S. H. Ou
Plasmopara cephalophora Davis
Plasmopara densa (Rabenh.) J. Schröt.
Plasmopara elsholtziae J. F. Tao & Y. Qin
Plasmopara plectranthi L. Ling & M. C. Tai
Plasmopara satureiae F. L. Tai & C. T. Wei

Pseudoperonospora elsholtziae D. Z. Tang

LAMIALES: Plantaginaceae

Peronospora akatsukae S. Ito & Muray.
Peronospora alta Fuckel
Peronospora canescens C. Benois
Peronospora plantaginis Underw.
Peronospora plantaginis Burrill

Plasmopara plantaginicola T. R. Liu & C. K. Pai

LAMIALES: Acanthaceae

Plasmopara asystasiae Vienn.-Bourg.
Plasmopara justiciae (Sawada) Skalický
Plasmopara wildemaniana Henn. var. *wildemaniana*

LAMIALES: Scrophulariaceae

Albugo aechmantherae Z.-y. Zhang & Y.-x. Wang
Albugo evansii Syd.

Peronospora agrestis Gaüm.
Peronospora andicola Speg.
Peronospora antirrhini J. Schröt.
Peronospora aquatica Gaüm.
Peronospora arvensis Gaüm.
Peronospora canadensis Gaüm.
Peronospora melampyri (Bucholz) Davis
Peronospora melampyri-cristati Săvul. & Rayss
Peronospora palustris Gaüm.
Peronospora pedicularis Palm
Peronospora pocutica T. Majewski
Peronospora satarensis M. S. Patil

(Table III: 3, continued)

Peronospora celsiae Syd. & P. Syd.
Peronospora digitalis Gaüm.
Peronospora flava Gaüm.
Peronospora grisea (Unger) Unger
Peronospora indica Syd. & P. Syd.
Peronospora jacksonii C. G. Shaw
Peronospora lapponica Lagerh.
Peronospora linariae Fuckel
Peronospora linariae-genistifoliae Săvul. & Rayss
Peronospora saxatilis Gaüm.
Peronospora silvestris Gaüm.
Peronospora sordida Berk. & Broome
Peronospora tozziae S. Blumer
Peronospora tranzscheliana Bakhtin
Peronospora verbasci Gaüm.
Peronospora verna Gaüm.
Peronospora veronicae-cymbalariae Rayss

Plasmopara paulowniae C. C. Chen

LAMIALES: Phrymaceae (Verbenaceae)

Plasmopara phrymae S. Ito & Hara [also *Plasmopara halstedii* (Farl.) Berl. & De Toni - in *Verbena* as well as Asteraceae!]

SOLANALES: Convolvulaceae

Albugo hyoscyami Z. Y. Zhang *et al.*
Albugo ipomoeae-aquaticae Sawada
Albugo ipomoeae-hardwikii Sawada
Albugo ipomoeae-panduratae (Schwein.) Swingle var. *ipomoeae-panduratae*
Albugo ipomoeae-pes-caprae Cif.
Albugo minor (Speg.) Cif.
Albugo pes-tigridis Gharse

Peronospora convolvuli J. C. Lindq.
Peronospora fritzii J. Schröt.

SOLANALES: Solanaceae

Peronospora capsici J. F. Tao & T. J. Li
Peronospora dubia A. Berl.
Peronospora hyoscyami de Bary
Peronospora lycii L. Ling & M. C. Tai
Peronospora nicotianae Speg.

ASTERIDS: EUASTERIDS II: orders and families

APIALES: Apiaceae

Bremiella baudysii (Skalický) Constant. & Negrean

Plasmopara ammi Constant.
Plasmopara anethi Jermal.
Plasmopara angelicae (Casp.) Trotter
Plasmopara apii Săvul. & O. Săvul.
Plasmopara archangelicae Gapon.
Plasmopara carlottae Savile
Plasmopara caucalis Săvul. & O. Săvul.
Plasmopara chaerophyllii (Casp.) Trotter
Plasmopara cryptotaeniae J. F. Tao & Y. Qin
Plasmopara dahurici Benua
Plasmopara dauci Săvul. & O. Săvul.
Plasmopara laserpitii (Wartenw.) Săvul. & Rayss
Plasmopara mei-foeniculi Săvul. & O. Săvul.
Plasmopara oenanthes J. F. Tao & Y. Qin
Plasmopara pastinacae Săvul. & O. Săvul.
Plasmopara petroselini Săvul. & O. Săvul.
Plasmopara peucedani Nannf.
Plasmopara pimpinellae Săvul. & O. Săvul. var. *pimpinellae*
Plasmopara podagrariae (G. H. Otth) Nannf.
Plasmopara saniculae Săvul. & O. Săvul.
Plasmopara selini B. Wrońska
Plasmopara sii Gapon.
Plasmopara silai Săvul. & O. Săvul.
Plasmopara smyrnii Săvul. & M. Bechet
Plasmopara umbelliferanum (Casp.) Wartenw. var. *umbelliferarum*

APIALES: Araliaceae

Plasmopara panacis Bondartsev & Bunkina

ASTERALES: Asteraceae

Albugo chardiniae Bremer & Petr.
Albugo solivae J. Schröt.
Albugo solivarum (Speg.) Herter
Albugo spinulosus (de Bary) Herter
Albugo tragopogonis (Pers.) Gray var. *tragopogonis*

Basidiophora entospora Roze & Cornu **[type species]**
Basidiophora montana R. W. Barreto

(Table III: 3, continued)

Benua kellermanii (Sacc.) Constant. **[type species; monotypic]**

Bremia lactucae Regel var. *lactucae* **[type species]**

Paraperonospora apiculata Constant.
Paraperonospora artemisiae-annuae (L. Ling & M. C. Tai) Constant.
Paraperonospora artemisiae-biennis (Gäum.) Constant.
Paraperonospora chrysanthemi-coronarii (Sawada) Constant.
Paraperonospora leptosperma (de Bary) Constant. **[type species]**
Paraperonospora minor (Săvul. & Rayss) Constant.
Paraperonospora multiformis (J. F. Tao & Y. Qin) Constant.
Paraperonospora sulphurea (Gäum.) Constant.
Paraperonospora tanaceti (Gäum.) Constant.

Peronospora danica Gäum.
Peronospora radii de Bary

Plasmopara angustiterminalis Novot. forma *angustiterminalis*
Plasmopara asterea Novot. forma *asterea*
Plasmopara bidentis (Novot.) Novot.
Plasmopara callistephi (Novot.) Novot.
Plasmopara carthami Negru
Plasmopara cenolophii Jermal.
Plasmopara centaureae-mollis T. Majewski
Plasmopara galatellae (Novot.) Novot.
Plasmopara galinsogae L. Campb.
Plasmopara gnaphalii Novot.
Plasmopara halstedii (Farl.) Berl. & De Toni
Plasmopara helianthi Novot. forma *helianthi*
Plasmopara heteropappi (Novot.) Novot.
Plasmopara lactucae-radicis Stangh. & Gilb.
Plasmopara mikaniae Vienn.-Bourg.
Plasmopara palmae L. Campb.
Plasmopara petasitidis S. Ito & Tokun.
Plasmopara saussureae Novot.
Plasmopara savulescui Novot.
Plasmopara solidaginis Novot.
Plasmopara sphaerosperma Săvul.
Plasmopara spilanthicola Syd.
Plasmopara vernoniae-chinensis Sawada
Plasmopara yunnanensis J. F. Tao & Y. Qin

ASTERALES: Campanulaceae

Peronospora corollae Tranzschel
Peronospora erinicola Durrieu
Peronospora phyteumatis Fuckel
Peronospora speculariae Gaüm.

DIPSACALES: Dipsacaceae

Peronospora cephalariae Vincens
Peronospora cephalariae-laevigatae Săvul. & Rayss
Peronospora dipsaci de Bary
Peronospora karelii Bremer & Gaüm.
Peronospora knautiae J. Schröt.
Peronospora violacea Berk.

DIPSACALES: Caprifoliaceae

Plasmopara sambucinae Nelen
Plasmopara viburni Peck

DIPSACALES: Valerianaceae

Peronospora patriniae Kalymb.
Peronospora valerianae Trail
Peronospora valerianellae Fuckel

TABLE III: 4. Peronosporaceae, Phytophthora *species and* Sclerosporales *on monocotyledons arranged according to host classification Angiosperm Phylogeny Group (APG), 1999). Names are taken from the binomial lists in APPENDIX 1 and lists in Dick (2000b). Subspecific taxa and most recognised synonyms have been omitted. Synonymies within genera have yet to be evaluated.*

LILIOIDS

ASPARAGALES: Alliaceae

Peronospora destructor (Berk.) Casp.
Peronospora fugitai S. Ito & Tokun.

LILIALES: Liliaceae

Peronospora lilii Stenina [very dubious taxon]

(Table III: 4, continued)

ALISMATALES: Araceae

Phytophthora colocasiae Racib.
Phytophthora richardiae Buisman

COMMELINOIDS

POALES: Cyperaceae

Peronospora cyperi Ideta [*non P. cyperi* Miyabe & Ideta ≡ *Phytophthora cyperi*]

Phytophthora cyperi (Ideta) S. Ito
Phytophthora cyperi-bulbosi Seethal. & K. Ramakr.

POALES: Poaceae

Peronospora diplachnis Milovtz. [? = *Sclerospora graminicola*]
Peronospora graminicola (Sacc.) Sacc. [≡ *Sclerospora graminicola*]
Peronospora maydis Racib. [≡ *Peronosclerospora maydis*]
Peronospora setariae Pass. [= *Sclerospora graminicola*]

Bambusoideae (*Oryza*)

Phytophthora fragariae Hickman var. *oryzobladis* J. S. Wang & J. Y. Yu
Phytophthora japonica Waterh. (≡ *P. oryzae* (Brizi) K. Hara)

Paniceae

Plasmopara oplismeni Vienn.-Bourg.
Plasmopara penniseti R. G. Kenneth & J. Kranz

Bremia graminicola Naumov var. *graminicola*
Bremia graminicola Naumov var. *indica* Patel

Sclerophthora butleri W. Weston
Sclerospora graminicola (Sacc.) J. Schröt. **[type species]**
Sclerospora iseilmatis Thirum. & Naras.
Sclerophthora northii W. Weston
Sclerospora secalina Naumov

Peronosclerospora dichanthiicola (Thirum. & Naras.) C. G. Shaw
Peronosclerospora globosa Kubicek & R. G. Kenneth [*nom. invalid.*]
Peronosclerospora heteropogonis Siradhana *et al.*
Peronosclerospora maydis (Racib.) C. G. Shaw
Peronosclerospora miscanthi (T. Miyake) C. G. Shaw
Peronosclerospora noblei (W. Weston) C. G. Shaw
Peronosclerospora philippinensis (W. Weston) C. G. Shaw
Peronosclerospora sacchari (T. Miyake) Hara **[type species]**
Peronosclerospora sorghi (W. Weston & Uppal) C. G. Shaw
Peronosclerospora spontanea (W. Weston) C. G. Shaw
Peronosclerospora westonii (M. C. Sriniv. *et al.*) C. G. Shaw
Peronosclerospora zeae C.-L. Yao [*nom. invalid.*] (cited by Jeger *et al.*, 1998)

Pachymetra chaunorhiza B. J. Croft & M. W. Dick **[type species; monotypic genus]**

Sclerophthora cryophila W. Jones
Sclerophthora farlowii (Griffiths) R. G. Kenneth
Sclerophthora lolii R. G. Kenneth
Sclerophthora macrospora (Sacc.) Thirum. *et al.* **[type species]**
Sclerophthora rayssiae R. G. Kenneth *et al.* var. *rayssiae*
Sclerophthora rayssiae R. G. Kenneth *et al.* var. *zeae* Payak & Naras.

Verrucalvus flavofaciens P. Wong & M. W. Dick **[type species; monotypic genus]**

All are late-evolving angiosperm orders, of which many members are herbaceous and found in conditions of high insolation (montane, mediterranean, or both). Refer to Soltis, Soltis & Chase (1999); Soltis, Soltis & Chase *et al.* (1998*a, b*); APG (1998) and see Tables III: 2 and III: 3 for host systematics; Table III: 3 combines the occurrence of all the *Peronosporales* on dicotyledons. Table III: 4 lists the parasites of monocotyledons. Many of these angiosperms (*Brassicaceae*, *Caryophyllaceae* with *Amaranthaceae*, *Fabaceae* and Euasterids I) have a centre of speciation in the Armeno-Iranian floristic sub-zone of Takhtajan (1969). It is probably no coincidence that Takhtajan's floristic regions 2 and 8 within the sub-tropical zone correspond to the main centres of distribution of these orders and families. In particular he notes that the sub-region of group 8, the Armeno-Iranian region, has many endemics in the families *Amaranthaceae* (synonym *Chenopodiaceae*), *Caryophyllaceae*, *Brassicaceae*, *Rosaceae*, *Fabaceae*, *Zygophyllaceae*, *Scrophulariaceae*, *Lamiaceae*, *Campanulaceae*, *Asteraceae* and *Alliaceae*. Floristically this sub-region stretches from central Turkey, around the Caspian sea and skirts the northern flanks of the Himalayan massif; it continues as the southern border of the Euro-Siberian floristic zone, which extends westwards though the Balkan mountains, the Alps and the northern Appennines and eastwards to the Kuril Islands and the Kamschatka peninsula. At this most eastern end there is continuity from the Kuril islands into Hokkaido and the other Japanese islands. Takhtajan also draws attention to the floristic similarities between the Euro-Siberian and Canadian-Appalachian (4a) regions.

Peronospora is predominantly a Northern Hemisphere genus from the Old World north of *ca* 35° N. Exceptions from the Southern Hemisphere include: *P. andina* and *P. calindriniae* from Argentina and *P. ranunculi-peduncularis* from Tierra del Fuego (Roivaïnen, 1977) and *P. mesembryanthemi* and *P. oxalidis* from South Africa. Exceptions from the Northern Hemisphere New World include: *P. floerkeae* and *P. oxybaphi*. Exceptions from the south eastern flanks of the Himalayan massif include: *P. lycii* from S. W. China and *P. satarensis* from India. Species of the monocotyledonous genus *Allium* (*Alliaceae*, order *Asparagales*) which also shares the same centre of origin (Armeno-Iranian floristic sub-zone) and are noted for their sulphur-containing secondary metabolites, are hosts to the otherwise anomalous *P. destructor*. An appreciable part of this modern mediterranean flora is thought to have been derived from North Africa at different times, through Asia Minor, the Italian peninsula or Iberia depending upon proximities in different epochs. A mid- to late-Tertiary, peri-montane origin for *Peronospora* is suggested.

The most plausible origin of *Plasmopara* (note hosts in *Vitaceae* and Euasterids II (*Apiales*, *Asteraceae* and *Caprifoliaceae*) is from tropical South America in the mid- to late Palaeogene (Eocene - Oligocene): note that Wilson (1907) erected the genus *Rhysotheca* for all species ***other than*** the type species of *Plasmopara* (*P. pygmaea*) because these species produced zoospores (Table III: 1). Although there could have been early radiation through north-west Africa to Eurasia before the South Atlantic expanded, host origins are not entirely consistent with this suggestion. A tropical-host (not cold-temperature hardy?) element survives in *Plasmopara*, for example in *Mikania* (south American Asteraceae), *Plectranthus* (Lamiaceae), *Cissus* (west African Vitaceae) and the two anomalous parasites of *Oplismenus* and *Pennisetum* respectively (west African Poaceae). However, as the two American tectonic plates drew together (*ca* 50 m.y.B.P.), a later radiation through Central America was possible. It may be speculated that, unlike the South American flora, that of North America, due to the early decline of the Laurentian tropical element, would have been particularly vulnerable to parasites evolving from South American taxa of *Phytophthora* because these plants would have had little opportunity to develop genes for resistance to this genus. Alternatively, did *Ph. infestans* evolve from a precursor in south-east North America, becoming juxtaposed to *Solanum* from the northern Andean uplands?

The evolutionary pattern of *Basidiophora, Benua, Bremia* and *Paraperonospora* shows adaptive radiation in diverse *Asteraceae*. The species *Bremia graminicola* var. *graminicola* on the grass *Arthraxon ciliaris* (L.) P. Beauv. (Naoumoff, 1913) and *B. graminicola* var. *indica* on *A. lancifolius* Hoch. is anomalous: the drawings certainly suggest a placement in *Bremia*. The relationship with the *Asteraceae* is not primarily cophylogenetic: it is probably relatively recent after the host family had become cosmopolitan, mostly developing during the late Tertiary, less than 10 m.y.B.P., with *Basidiophora* and *Benua* evolving on the North and South American cordilleras. (The conspecificity of *Basidiophora montana* with *B. entospora* is not convincing: compare Barreto & Dick, 1991, with Constantinescu, 1998; consider the differences between

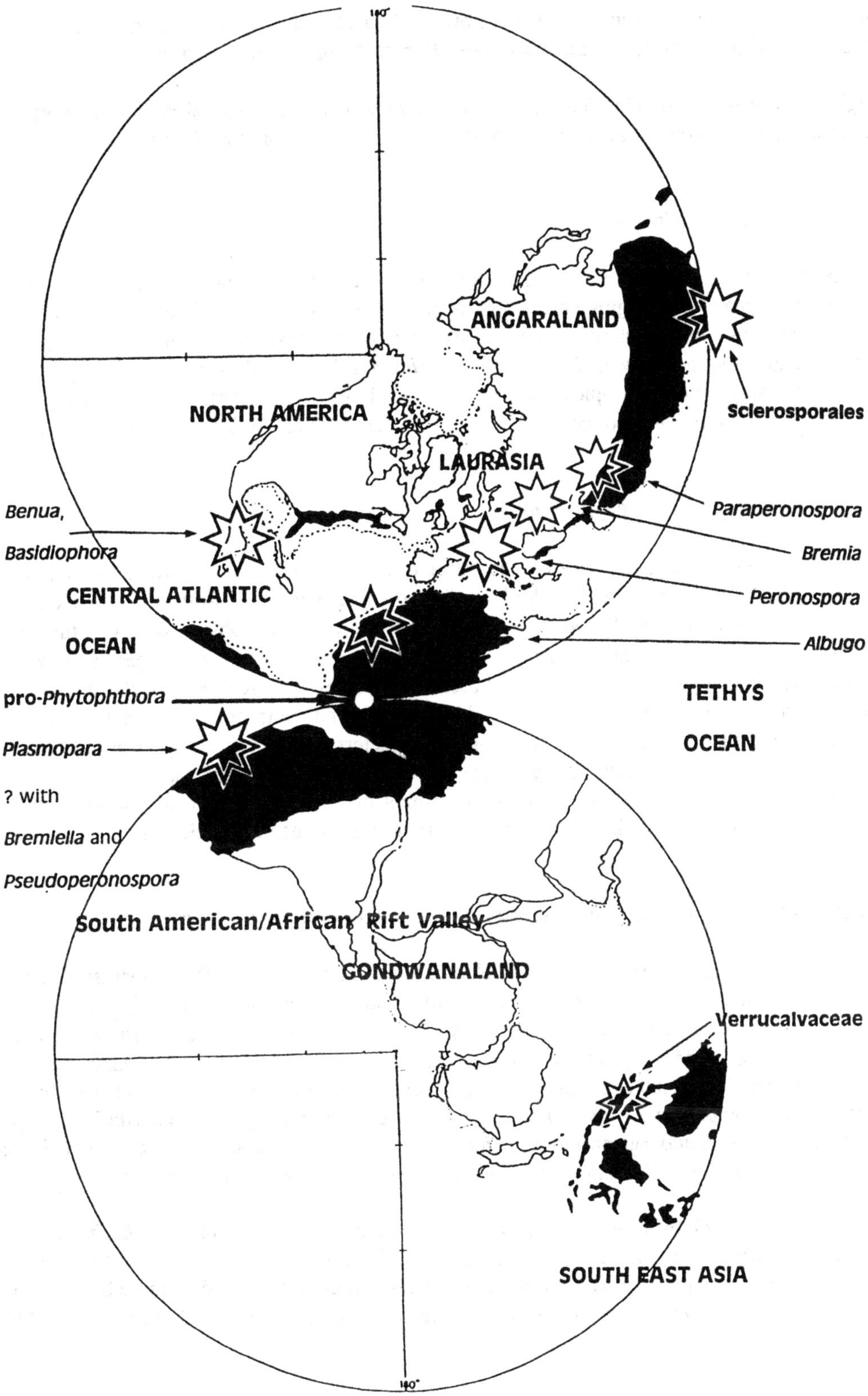

FIGURE III: 3. *Cretaceous drift map, 100 ± 10 m.y.B.P. with present-day continental outlines, adapted from Dick (1988: fig. 6), to show the probable limits of the tropical woody angiosperm flora (shaded black: note the island chains connecting Angaraland with southern Laurasia and North Africa respectively) and the proposed possible origins (open stars) of the dicotyledonous downy mildew genera* Peronospora *with* Paraperonospora *and* Bremia *(Miocene in Eurasia);* Plasmopara, *perhaps with* Pseudoperonospora *and* Bremiella, *(early Tertiary in South America);* Basidiophora *and* Benua *(Oligocene in Central America); the monocotyledonous downy mildews, order* Sclerosporales, *in Asia and the family* Verrucalvaceae *in south-east Asia (Pliocene/Quaternary), together with the family* Albuginaceae *(Maastrichtian/Paleocene in western North Africa). See text for further explanation.*

the cold-hardy *Erigeron* and tropical *Mikania* hosts.) *Bremia* and *Paraperonospora* evolved in the Old World, perhaps with *Bremia* in central Europe and *Paraperonospora* further to the east.

The origins of *Bremiella* (hosts in *Violaceae*, *Balsaminaceae* and *Apiaceae*) and *Pseudoperonospora* (hosts in *Urticaceae* and *Cucurbitaceae*) are less certain but may also be from the New World.

The white blister rusts: *Albugo*

Albugo may well be ancient, with a late Cretaceous origin in common with *Phytophthora*. *Albugo* and *Peronospora* have many host angiosperm orders in common (see Biga, 1955; Tables III: 2 and III: 3), but this may reflect only similar metabolite preferences. *Albugo* has hosts of evolutionarily early, occidental origin (*Piperales*, *Amaranthaceae*, *Convolvulaceae* (*Ipomoea*) in South America. *Albugo* shares with *Plasmopara* a warm-temperate/tropical element not found in *Peronospora*. The stimulus to parasitic diversification of *Albugo* may also have been the tectonic changes at the Cretaceous boundary.

Peronospora*, *Plasmopara* and *Phytophthora

It is becoming increasingly apparent that the separation of *Peronospora*, *Plasmopara* and *Phytophthora* into two orders, the *Peronosporales* and *Pythiales*, will not be tenable. Unfortunately, movement of the *Phytophthoraceae*, as defined by Pethybridge (1913), may not solve the problem because of the distinctions he made between *Phytophthora* and *Nozemia*. Molecular data indicates that some species of *Peronospora* and some species of *Phytophthora* appear to be closer to each other than are species of *Plasmopara* to either. Ordinal criteria will thus need reassessment. Host ranges and biotrophic obligate parasitism are not sufficient as diagnostic criteria. Other morphological characters are not sufficiently researched over a wide range of species; this is an inherent problem when some of the species are of major economic importance while other species are ignored. Suggestions for ordinal diagnostic criteria could be the occurrence of mitosis in the conidiosporangial initial and the formation of a plasmamembranic vesicle.

Dicotyledonicolous and graminicolous downy mildews: economic importance

It is not a coincidence nor is there a causal correlation that the occurrence of DM parasitism and the plants cultivated for food have depended on the production of secondary metabolites. The high levels of essential oils, mustard oils, and other secondary metabolites, that enhance the economic importance of crop plants in these taxa have rendered the plants vulnerable to the nutritional requirements and enzyme capacities of the DMs. Consider, for example, graminicolous grain crops (including flavour herbs such as Lemon Grass and sugar); pulses (plant proteins including enzymes); dicotyledonous grains (*Amaranthus*, *Fagopyrum*, *Helianthus*, *Linum*, *Raphanus*); green leaf (*Brassicaceae*, *Lactuca*, *Chenopodium*); tubers/roots (*Solanaceae*, *Convolvulaceae*, *Apiaceae*, *Alliaceae*); and fruits (*Solanaceae*, *Cucurbitaceae*, *Rosaceae*).

Some secondary metabolites may be toxic to man, or otherwise unpalatable (*Euphorbiaceae*), but all represent a high carbon (and nitrogen or sulphur) source. Comparative studies of the occurrence of downy mildews in non-crop and non-horticultural hosts is urgently required to provide a more balanced overview of the relationships between host secondary metabolites and the requirements of the obligately biotrophic DMs.

THE *PYTHIALES*

The order *Pythiales* is comprised of two families, the *Pythiaceae* and the *Pythiogetonaceae* (Dick, 1984; Voglmayr *et al.*, 1999). The *Peronophythoraceae* is placed in synonymy with the *Pythiaceae sensu lato*, together with its type species, which is affiliated to *Phytophthora sensu lato*. The two most important genera in this order are *Phytophthora* and *Pythium*. The former genus contains species which are primarily or obligately parasitic, although they can be cultured; the latter, while containing obligate parasites, has species which are facultatively parasitic or saprotrophic. *Lagenidium giganteum* has been placed with the *Pythiaceae* (Dick *et al.*, 1984; see PART IV).

The *Pythiaceae*: *Phytophthora*

Phytophthora has been well served by keys and monographs produced at regular intervals (Tucker, 1931; Leonian, 1934; Waterhouse, 1956, 1963, 1970; Newhook, Waterhouse & Stamps, 1978; Erwin *et al.*, 1983; Stamps *et al.*, 1990; Lucas *et al.*, 1991; Dowley *et al.*, 1995; Erwin & Ribeiro, 1996) and it is not appropriate to discuss this genus in detail here. *Phytophthora*, the type species epithet of which was given by Montagne (1845), was not formally described until 1876 (de Bary, 1876), although the symptoms of Late Blight of Potato had been observed since the end of the eighteenth century in Europe (Woodham-Smith, 1962) and South America (Boussingault, 1845), and the disease was particularly prevalent in Europe during the 1840s (Berkeley, 1846; Smith, 1884). These and other early taxonomic landmarks are summarized chronologically in Table III: 5.

Present day *Phytophthora* species embrace a wide range of habits, from root parasites to leaf and fruit pathogens; the monotypic genera *Peronophythora* (Ko *et al.*, 1978) on *Litchi* (*Sapindaceae*) and *Trachysphaera* (Tabor & Bunting, 1923) on *Theobroma* (*Malvaceae* formerly *Sterculiaceae*); *Citrus* (*Rutaceae*, Watson, 1971); *Coffea* (*Rubiaceae*); *Persea* (*Lauraceae*), cited by Spaulding (1961) and *Musa* (*Musaceae*, *Gingiberales* - Kader, 1999) are genera which have been separated from *Phytophthora sensu lato* on morphological criteria (branched conidiosporangiophore and spiny sporangia respectively) as well as their habit, restricted to fruits. Molecular data for *Peronophythora* suggests that this separation is untenable. *Diasporangium* (Höhnk, 1936), known only from records by Höhnk (1936; 1962) is also probably best placed with *Phytophthora sensu lato* (Voglmayr *et al.*, 1999).

Phytophthora and the dicotyledonicolous downy mildews (DMs) share a common ancestry but divergent groups of angiosperm hosts. *Phytophthora* possesses a range of species with different degrees of specialization (Erwin & Ribeiro, 1996). Any pattern of distribution with respect to the morphological (Roman numeral) groups of Stamps *et al.* (1990: Table III: 6) either with host-range or with restriction to a geographic region is difficult to discern. The widespread species of *Phytophthora*, in terms of both host range and geographic distribution (lists in Erwin & Ribeiro, 1996), include: *Ph. cactorum* [I]; *Ph. nicotianae* [II]; *Ph. citricola* [III]; *Ph. cinnamomi* [VI] and *Ph. drechsleri* [VI]; no such widespread representatives of groups IV or V are recognized. However, *Ph. infestans* [IV] and *Ph. palmivora* [II] are thought to have a central American provenance, thence widely distributed by mankind's activities. The angiosperm families *Lauraceae* and *Fagaceae*, prominent in the sub-tropical Eocene flora of Europe, include the genera *Cinnamomum* (much more widespread in the Tertiary) and *Quercus* respectively, with which one may bracket the parasites *Ph. cinnamomi* [VI] and *Ph. quercina* [I]. Several species are apparently more frequently recognized in particular regions of the world, such as India (*Ph. arecae* [II]; *Ph. cajani* [VI]; *Ph. cypari-bulbosi* [III]); south-east Asia (*Ph. colocasiae* [IV], *Ph. richardiae* [VI]); Australasia (*Ph. clandestina* [I]) and northern Europe (*Ph. fragariae* [V]; *Ph. idaei* [I]; *Ph. erythroseptica* [VI]).

TABLE III: 5. *The early chronology of taxonomic and plant-pathogenic studies of the* Pythiales

date	fungus (only generic authorities given), citation and brief annotation
1845	The Irish Famine - caused by Late Blight of Potatoes (symptoms also known as 'Potato Murrain'; 'Curl'; 'Rot') [after *ca* 50 years of similar symptom reports; also of well-known occurrence in rainy years in Bogota (Boussingault, 1845)]
1846	Berkeley: "The decay [caused by Late Blight of Potatoes] is the consequence of the mould, and not the mould of the decay." (*contra* Lindley - see Smith, 1884*a*; Large, 1940)
1858	*Pythium* Pringsh. described (Pringsheim, 1858; type species *P. monospermum* (filamentous-sporangiate), with other species, all, except the type species, have subsequently been transferred to other genera)
1863	sexual reproduction in DMs and *Phytophthora* described (de Bary, 1863): [modern nomenclature] *Bremia* (1 sp.), *Paraperonospora* (1 sp.), *Peronospora* (32 spp.), *Plasmopara* (5 spp.), *Albugo* (6 spp.), *Phytophthora* (1 sp) [the first synopsis of the classification of the DMs; still the most important comparative morphological account]
1866	the "Peronosporeen" and "Saprolegnieen" separated as the two major groups of biflagellate fungi (de Bary, 1866, 1887)
1869	*Cystosiphon* Roze & Cornu described (Roze & Cornu, 1869; type species *C. pythioides* [holotype]; the first valid generic name for spherical-sporangiate *Pythium* species in *Pythium s.l.*)
1876	*Phytophthora* de Bary described (de Bary, 1876; type species *P. infestans* [holotype] [group IV])
1884	succinct accounts by Smith (1884*a*) of plant pathology and morphology of DMs: [modern nomenclature] *Peronospora trifoliorum, P. destructor, P. parasitica, Plasmopara umbelliferarum, Bremia lactucae, Albugo candida, Phytophthora infestans*
1899-1901	cytology of DM oosporogenesis by Stevens and others (summarized, with facisimile illustrations, in Lotsy, 1907, see Dick & Win-Tin, 1973)
1903	*Kawakamia* Miyabe described (Miyabe & Kawakami, 1903; type species *K. [Phytophthora s.l.] cyperi* [holotype] [group III])
1906	*Phloeophthora* Kleb. described (Klebahn, 1905; type species *P. [Phytophthora s.l.] syringae* [holotype] [group III])
1909	*Pythiomorpha* H. E. Petersen described (Petersen, 1909; type species *P. [Phytophthora s.l.] gonapodyides* [holotype] [group VI])
1913	*Nozemia* Pethybr. described (Pethybridge, 1913; type species *Nozemia (Phytophthora s.l.) cactorum* [group I])
1913	family *Phytophthoraceae* described (Pethybridge, 1913)
1915	*Rheosporangium* Edson described (Edson, 1915; type species *R. aphanidermatum* [holotype])
1921	Jarrah (*Eucalyptus marginata*) dieback disease first noted (see Podger, 1972: causal agent *Phytophthora cinnamomi* [group VI])

Species with variously restricted host-ranges and geographic distributions can be recognized, perhaps usually confined to one angiosperm order (e.g., *Solanales*: *Ph. infestans*), or a genus, such as *Ph. inflata* [III] on *Ulmus* in North America; *Ph. quininea* [V] on *Chinchona* in north-western South America; *Ph. colocasiae* [IV] on *Colocasia esculenta*, or even a part of a plant, such as *Ph. megakarya* [II] on *Theobroma cacao* pods in west Africa.

Brasier, Cooke & Duncan (1999) have shown that the pathogen of *Alnus* is a *Phytophthora* of natural hybrid origin involving a *Phytophthora cambivora*-like species and an unknown taxon similar to *Phytophthora fragariae*, but the fungus has not yet been named.

Other restricted species are apparently anomalous, such as *Ph. ilicis*: in the literature supposedly restricted to the European *Ilex aquifolium* [IV] introduced and growing in North America, although it has now been isolated in Europe; and *Ph. lateralis* [II] isolated from native *Chamaecyparis* in western North America, but which is thought to have been introduced from the Mediterranean.

TABLE III: 6. *Characters of the Roman Numeral Categories of* Phytophthora *according to Stamps* et al. *(1990) and Erwin & Ribeiro (1996). Other characters (e.g., sporangial pedicel length) in the key of Stamps* et al. *(1990) are variable within each group.*

Category	Asexual reproduction: sporangial production, proliferation and exit pore	Sexual reproduction: oogonia and oospores	Sexual reproduction: antheridia
I	Abundant on solid media Usually caducous No internal proliferation Papillation: conspicuous Papilla: hemispherical, >4 μm high Exit pore: narrow (5-7 μm diam.)	Always formed in host Always formed in culture	All or most paragynous
II	Present, often aerial, on solid media Caducous or not caducous No internal proliferation Papillation: conspicuous Papilla: hemispherical, >4 μm high Exit pore: narrow (5-7 μm diam.)	Not readily formed in host Rarely formed in single-strain culture Formed in A1 × A2 crosses	Amphigynous
III	Present, sometimes aerial, on solid media Sporangiophores simple, not defined on agar Caducous or not caducous Proliferation: sympodial Papillation: semipapillate Papilla: hemispherical, <4 μm high Exit pore: narrow (5-7 μm diam.)	Always formed in host Always formed in single-strain culture Oogonial wall >2 μm thick	Predominantly paragynous
IV	Present, often aerial, on solid media Sporangiophores sympodial on host Usually deciduous No proliferation Papillation: semipapillate Papilla: hemispherical, <4 μm high Exit pore: narrow (5-7 μm diam.)	Variable in hosts Rarely formed in single-strain culture Formed in A1 × A2 crosses	Predominantlyamphigynous
V	Rarely produced on solid media Sporangiophores sympodial in aqueous systems Not deciduous Proliferation: internal in aqueous culture Non-papillate Scarcely detectable apical thickening Exit pore: broad (>12 μm diam.)	Always formed in host Always formed in culture Oogonia often large, (40 to more than 50 μm diam.)	All or most paragynous Subterminal or intercalary
VI	Rarely produced on solid media Sporangiophores sympodial in aqueous systems Not deciduous Proliferation: internal or sympodial in water Sporangial papilla: no papilla Apical thickening: scarcely detectable Exit pore: broad (>12 μm diam.)	Always formed in host Not always formed in culture Formed in A1 × A2 crosses in some species	All or most amphigynous

Without human intervention, *Phytophthora* would still be a mix of widespread-cosmopolitan taxa, with or without wide host ranges, and provincials adapted to much more restricted habitats. The DMs are presumed to be derived from the antecedents of such a gene pool.

Brasier (1992) and Brasier & Hansen (1992) have reviewed the evolution of *Phytophthora* from a genetic standpoint. Genome similarity should now be assumed between genera (see p. 77). The diversity of genome variation (see pp. 70-74) should be compared with that of angiosperms, where species-complexes in terms of chromosome sizes, chromosome numbers and genome size exist (see Vaughan, Taylor & Parker, 1996: *Scilla*). This is a factor which needs to be considered when reviewing DNA quantification (e.g., Martin, 1995*a*; Voglmayr & Greilhuber, 1998) and species based on a karyotype (*Phytophthora megakarya* - Brasier & Griffin, 1979).

The *Pythiaceae*: *Halophytophthora*

The position of the anamorphic marine genus, *Halophytophthora* (Ho & Jong, 1990), associated with mangroves, is rather less certain but may possibly be ancestral in some molecular phylogenies so that it should be included in any thorough phylogenetic molecular analyses. In view of this, it is worth noting that these estuarine and marine littoral vascular plants (*Avicennia* in the *Avicenniaceae*, *Lamiales*; and *Rhizophora* in the *Rhizophoraceae*, *Malpighiales*) can be traced back as far as the Paleocene (Plaziat, 1995). Fossils resembling, and from similar habitats to the mangrove palms (*Arecaceae* - *Nypa*) have earlier dates.

The *Pythiaceae*: *Pythium*

The genus *Pythium* (the generic name was conserved by Waterhouse, 1968*a*) has also received systematic treatments in recent decades (Waterhouse, 1967, 1968*b*; Plaats-Niterink, 1981; Dick, 1990*c*). However, the number of species described is much larger than for *Phytophthora* (see APPENDIX 1). Since the systematics of *Pythium* species is omitted from PART V, a complete list of taxonomic references, updated from Dick, 1990*c*) is given below:

Abad *et al.* (1995); Ahrens (1971); Ali-Shtayeh (1986); Ali-Shtayeh & Dick (1985); Apinis (1929); Atkins (1935); Atkinson (1895); Bakhariev (1984); Balakrishnan (1948*a*, *b*); Barron (1976*b*); Batko (1973); Berkeley (1845/8??); Blok & Plaats-Niterink (1978); Brandt (1881); Braun (1924, 1925); Buisman (1927); Butler (1907); Campbell & Hendrix (1967); Cejp (1955); Cheney (1932); Chesters & Hickman (1944); Coker & Patterson (1927); de Cock (1986); de Cock *et al.* (1987); Cornu (1872*b*); Dangeard (1886*a*, 1931); Dastur (1935); Deacon, Laing & Berry (1990); de Bary (1960, 1876, 1881*a*, *b*); De La Rue (1869); Deshpande & Deshpande (1965); Dick (1989, 1990*c*); Dick & Glockling (2000); Drechsler (1928, 1930, 1934, 1941, 1943, 1950, 1960); Duan (1985); Edson (1915); Fell & Master (1975); Fischer (1892*b*); Fitzpatrick (1923, 1930); Gobi (1899-1900); de Haan & Hoogkamer (1903); van Hall (1925); Hendrix & Campbell (1968, 1969, 1971); Hendrix & Papa (1974); Hesse (1874); Hirane (1960); Höhnk (1932, 1939, 1953); Ito, S. & Tokunaga (1933, 1935*a*); Ito, T. (1944); Jain (1951); Jokl (1918); Kanouse & Humphrey (1927); Kouyeas, H. & Theohari (1977); Kouyeas, V. & Kouyeas, H. (1963); Kröber (1985); Lechtova-Trnka (1931); Lifshitz, Stanghellini & Baker (1984); Lilienfeld (1911); Lindstedt (1872); Lipps (1980); Lohde (1874); Martin (1994); Matta (1965); Matthews (1928, 1931); Mendoza & Marin (1989); Mer & Khulbe (1983); Meurs (1928, 1934); Middleton (1943); Minden (1916 [1923]); Nicholas (1926); Park (1977); Paul (1986*a*, *b*, *c*, 1987*a*, *b*, 1988*a*, *b*, 1990, 1991, 1992*a*, *b*, *c*); Paul *et al.* (1998); Petersen (1909); Plaats-Niterink (1972, 1981); Poitras (1949); Pratt & Mitchell (1973); Preti (1936); Pringsheim (1858, 1873); Raciborski (1891); Rajagopalan & Ramakrishnan (1971); Ramsbottom (1916*a*); Rands (1930);Rangaswami (1962); Robertson (1979); Roldan (1932); Roze & Cornu (1869); Sadebeck (1874, 1876, 1887); Sawada & Chen (1926); Schenk (1859*b*); Schröter (1892-1897); Serbinow (1912); Shanor (1938); Shipton (1987); Sideris (1931, 1932); Smith (1876); Sorokin (1873); Sparrow (1931*b*, 1934, 1936*b*, 1960); Subramaniam (1919, 1928); Takahashi (1954); Takahashi, Ichitani & Sasaki (1977); Takahashi & Kawase (1965); Takahashi, Ohuchi & Alicbusan (1965); Takahashi *et al* (1972); Takimoto (1935, 1941); Tasugi & Siiono (1940); Tokunaga (1932); Tompkins (1975); Trow (1901); Vaartaja (1965); Vanterpool (1938); Vanterpool & Truscott (1932); Vuillemin (1888); Wahrlich (1887, 1889); Walz (1870); de Wildeman (1895*c*, 1896*c*, 1897*a*); Wittmack (1892); Waterhouse (1967, 1968*a*, 1968*b*; Yacevskie & Yacevskie (1931); Yu (1973); Yu & Ma (1989); Yu & Wang (1987).

The taxonomy of the genus has been regarded as 'difficult' and this can be attributed to the 'explosion' of new specific epithets introduced between the years 1925-1940 amounting to about one third of the names in current use (see Table III: 7, abstracted from Dick, 1990*c*, and excluding the earlier binomials and those introduced after 1948). Contrast this list of names with the number of species of *Phytophthora* recognised by Tucker (1931) and Stamps *et al.* (1990): 20 and 55, respectively. Discontinuities between species have still not been fully evaluated: the variability in oogonial origin and antheridial characters has resulted in lack of clarity in keys and descriptions, although the application of morphometric indices (see pp. 65-69) has helped to resolve some of the confusion; Venn diagram display of the sporangial characters is also useful (Dick, 1990*c*). Dick (in Klassen, McNabb & Dick, 1987; Dick, 1990*c*) suggested that there were five major centres of speciation within *Pythium*:

1. with filamentous sporangia and smooth oogonia (e.g., *Pythium monospermum* - the type species, *P. aquatile*)
2. with lobulate-toruloid sporangia and smooth oogonia (e.g., *Pythium torulosum*, *P. graminicola*)
3. with clustered globose sporangia and spiny oogonia (e.g., *Pythium anandrum*)
4. with small spherical sporangia and smooth or papillate oogonia (e.g., *Pythium irregulare*, *P. debaryanum*)
5. with large, internally proliferous spherical sporangia and smooth or papillate oogonia (e.g., *Pythium oedochilum*, *P. mastophorum*)

Molecular biological data (Martin, 2000; Matsumoto *et al.*, 1999; see also p. 95*ff*) to some extent supports these groupings, but the full range of species have yet to be examined. The molecular data may suggest synonymy between some of the species, but these data are not definitive. The continuing taxonomic debate between 'lumpers' and 'splitters' will not be settled until the ***biology*** of the taxa has been clarified; many of the species are cosmopolitan with slightly different morphololological characteristics in different continents. Isolates of '*Pythium irregulare*', for example, rarely produce sporangia in Europe (Plaats-Niterink, 1981) but those from Australia commonly do so.

Pythium, while commonly thought of as one of the causal agents of 'damping-off' of seedlings, has a much wider host range than *Phytophthora*, embracing invertebrate and vertebrate animals (Dick, 2000*b*); marine red and freshwater green algae; charophytes; and fungi. There is a very extensive bibliography on the plant pathology of *Pythium* species which cannot be covered in this volume: earlier bibliographic entries are listed in Tompkins (1975) and Plaats-Niterink (1981). Fungal parasitism by *Pythium* has been discussed by Deacon, Laing & Berry (1990), Laing & Deacon (1990, 1991), Foley & Deacon (1985, 1986) and Lifshitz, Stanghellini & Baker (1984). Parasitism *of Phytophthora* and *Pythium* by other flagellate fungi is documented in Table III: 8. The remaining genera of the *Pythiaceae* are listed in APPENDIX 1.

Perhaps the most notable vertebrate parasite is *Pythium insidiosum*. This fungal disease of horses was originally reported from India under the name 'bursattee', occurring when horses were grazing in or near stagnant water (Smith, F., 1884), although the fungal nature of the causal organism was first indicated by de Haan & Hoogkamer (1901). The disease has subsequently been described as 'phycomycosis of horses'; 'swamp cancer'; 'granular dermatitis', 'equine pythiosis', finally being named as 'pythiosis insidiosi'. The causal organism was initially reported as a non-sporulating mycelium, but zoospores were reported in culture by Austwick & Copland (1974) when the organism was regarded as a species of *Pythium*. The nomenclature of the *Pythium* species involved has depended on the interpretation of the earlier names using the specific epithet '*destruens*' (Bridges & Emmons, 1961; Dick, 1990*c*). However, following the description of sexual reproduction (Cock *et al.*, 1987), the nomenclature has stabilized as *Pythium insidiosum*. References to names such as '*Pythium gracile*' should be regarded as 'form names' not taxonomic entities, and much work still needs to be done before there is a fully reliable antigenic (Mendoza, Kaufman & Standard, 1986) or

Table III: 7. *Chronological list of binomials and basionyms in* Pythium*: Butler (1907) to Balakrishnan (1948), from Dick (1990c)*

Date	Name and citation
1907	*P. indigoferae* E. J. Butler, *Memoirs of the Department of Agriculture, India, Botanical Series* **1**: 73 (1907)
1907	*P. diacarpum* E. J. Butler, *Memoirs of the Department of Agriculture, India, Botanical Series* **1**: 80 (1907)
1907	*P. palmivorum* E. J. Butler, *Memoirs of the Department of Agriculture, India, Botanical Series* **1**: 82 (1907) [≡ *Phytophthora palmivora*]
1907	*P. rostratum* E. J. Butler, *Memoirs of the Department of Agriculture, India, Botanical Series* **1**: 84 (1907)
1909	*P. daphnidarum* H. E. Petersen, *Botanisk Tidsskrift* **29**: 392 (1909)
1909	*P. undulatum* H. E. Petersen, *Botanisk Tidsskrift* **29**: 394 (1909) [≡ *Phytophthora undulata*]
1909	*P. undulatum* H. E. Petersen var. *undulatum*, *Botanisk Tidsskrift* **29**: 394 (1909)
1911	*P. haplomitri* Lilienfeld, *Bulletin International. Académie Polonaise des Sciences et des Lettres. Classe des Sciences Mathematiques et Naturelles. Série B. Botanique, 1911* **(6B)**: 337 (1911)
1912	*P. perniciosum* Serbinow, *Scripta Botanica Horti Universitatis Imperialis Petropolitanae* **28**: 29 (1912)
1915	*Rheosporangium aphanidermatum* Edson, *Journal of Agricultural Research* **4**: 291 (1915)
1916	*P. pulchrum* Minden, *Mycologische Untersuchungen und Berichte von Dr Richard Falck* **[1]** 2(2): 224 (1916)
1916	*P. pythioides* (Roze & Cornu) Ramsbottom, *Transactions of the British Mycological Society* **5**: 315 (1916)
1918	*P. conidiophorum* Jokl, *Österreichische Botanische Zeitschrift* **67**: 33 (1918)
1919	*P. butleri* Subraman., *Memoirs of the Department of Agriculture, India, Botanical Series* **10**: 193 (1919)
1923	*P. aphanidermatum* (Edson) Fitzp., *Mycologia* **15**: 166 (1923)
1923	*Nematosporangium aphanidermatum* (Edson) Fitzp., *Mycologia* **15**: 168 (1923)
1923	*Nematosporangium aphanidermatum* (Edson) Fitzp. var. *aphanidermatum*, *Mycologia* **15**: 168 (1923)
1924	*P. complectens* Hans Braun, *Journal of Agricultural Research* **29**: 415 (1924)
1925	*P. splendens* Hans Braun, *Journal of Agricultural Research* **30**: 1043 (1925)
1925	*P. splendens* Hans Braun var. *splendens*, *Journal of Agricultural Research* **30**: 1043 (1925)
1925	*P. debaryanum* R. Hesse var. *pelargonii* Hans Braun, *Journal of Agricultural Research* **30**: 1060 (1925)
1925	*P. polyandrum* C. J. J. Hall [*nomen nudum*], *Mededeeling van het Instituut voor Plantenziekten, Buitenzorg* **67**: 46 (1925)
1926	*P. spinosum* Sawada, [*in* Sawada & Chen] *Journal of the Natural History Society Formosa* **16**: 199 (1926)
1926	*P. marchantiae* Nicolas, *Comptes Rendus hebdomadaires des Séances de l'Académie des Sciences, Paris* **182**: 82 (1926)
1927	*P. afertile* Kanouse & T. Humphrey, *Papers of the Michigan Academy of Science, Arts, and Letters* **8**: 137 (1927)
1927	*P. irregulare* Buisman, *Mededeelingen van het Phytopathologisch Laboratorium 'Willie Commelin Scholten'* **11**: 1 (1927)
1927	*P. irregulare* Buisman var. *irregulare*, *Mededeelingen van het Phytopathologisch Laboratorium 'Willie Commelin Scholten'* **11**: 1 (1927)
1927	*P. torulosum* Coker & P. Patt., *Journal of the Elisha Mitchell Scientific Society* **42**: 247 (1927)
1928	*P. mamillatum* Meurs, *Wortelrot, Veroorzaakt door Schimmels uit de Geslachten* Pythium *Pringsheim en* Aphanomyces *de Bary. Proefschrift ter Verkrijging van den Graad van Doctor... Rijks-universiteit te Utrecht...*, 1928: 44 (1928)
1928	*P. arrhenomanes* Drechsler, *Phytopathology* **18**: 873 (1928)
1928	*P. arrhenomanes* Drechsler var. *arrhenomanes*, *Phytopathology* **18**: 873 (1928)
1928	*P. graminicola* Subraman., *Bulletin of the Agricultural Research Institute, Pusa* **177**: 5 (1928)
1928	*P. graminicola* Subraman. var. *graminicola*, *Bulletin of the Agricultural Research Institute, Pusa* **177**: 5 (1928)
1928	*P. papillatum* V. D. Matthews, *Journal of the Elisha Mitchell Scientific Society* **43**: 231 (1928)
1929	*Pythiomorpha undulata* (H. E. Petersen) Apinis, *Acta Horti Botanici Universitatis Latviensis* **4**: 234 (1929)
1930	*P. dissotocum* Drechsler, *Journal of the Washington Academy of Sciences* **20**: 402 (1930)
1930	*P. periilum* Drechsler, *Journal of the Washington Academy of Sciences* **20**: 403 (1930)
1930	*P. myriotylum* Drechsler, *Journal of the Washington Academy of Sciences* **20**: 404 (1930)
1930	*P. periplocum* Drechsler, *Journal of the Washington Academy of Sciences* **20**: 405 (1930)
1930	*P. periplocum* Drechsler var. *periplocum*, *Journal of the Washington Academy of Sciences* **20**: 405 (1930)
1930	*P. paroecandrum* Drechsler, *Journal of the Washington Academy of Sciences* **20**: 406 (1930)
1930	*P. salpingophorum* Drechsler, *Journal of the Washington Academy of Sciences* **20**: 407 (1930)
1930	*P. acanthicum* Drechsler, *Journal of the Washington Academy of Sciences* **20**: 408 (1930)
1930	*P. oligandrum* Drechsler, *Journal of the Washington Academy of Sciences* **20**: 409 (1930)
1930	*P. anandrum* Drechsler, *Journal of the Washington Academy of Sciences* **20**: 410 (1930)
1930	*P. mastophorum* Drechsler, *Journal of the Washington Academy of Sciences* **20**: 411 (1930)
1930	*P. polymastum* Drechsler, *Journal of the Washington Academy of Sciences* **20**: 412 (1930)
1930	*P. helicoides* Drechsler, *Journal of the Washington Academy of Sciences* **20**: 413 (1930)
1930	*P. oedochilum* Drechsler, *Journal of the Washington Academy of Sciences* **20**: 414 (1930)
1930	*P. polytylum* Drechsler, *Journal of the Washington Academy of Sciences* **20**: 415 (1930)
1930	*P. palingenes* Drechsler, *Journal of the Washington Academy of Sciences* **20**: 416 (1930)
1930	*P. dactyliferum* Dreschler [*nomen nudum*], [*in* Rands] *Proceedings of the International Society of Sugar Cane Technologists* **3**: 122 (1930)
1931	*Nematosporangium arrhenomanes* (Drechsler) Sideris, *Mycologia* **23**: 272 (1931)
1931	*Nematosporangium arrhenomanes* (Drechsler) Sideris var. *arrhenomanes* Sideris, *Mycologia* **23**: 272 (1931)
1931	*Nematosporangium arrhenomanes* (Drechsler) Sideris var. *hawaiiense* Sideris, *Mycologia* **23**: 272 (1931)
1931	*Nematosporangium spaniogamon* Sideris, *Mycologia* **23**: 273 (1931)
1931	*Nematosporangium hyphalosticton* Sideris, *Mycologia* **23**: 275 (1931)
1931	*Nematosporangium polyandron* Sideris, *Mycologia* **23**: 276 (1931)
1931	*Nematosporangium thysanohyphalon* Sideris, *Mycologia* **23**: 277 (1931)
1931	*Nematosporangium rhizophthoron* Sideris, *Mycologia* **23**: 279 (1931)
1931	*Nematosporangium leiohyphon* Sideris, *Mycologia* **23**: 283 (1931)
1931	*Nematosporangium epiphanosporon* Sideris, *Mycologia* **23**: 283 (1931)
1931	*Nematosporangium aphanidermatum* (Edson) Fitzp. var. *hawaiiense* Sideris, *Mycologia* **23**: 286 (1931)
1931	*Nematosporangium leucosticton* Sideris, *Mycologia* **23**: 287 (1931)

(Table III: 7, continued)

1931 *Nematosporangium indigoferae* (E. J. Butler) Sideris, *Mycologia* **23**: 290 (1931)
1931 *Nematosporangium dictyospermum* (Racib.) Jacz. & P. A. Jacz., *Opredelitel' Gribov*: 96 (1931)
1931 *Nematosporangium gracile* (Schenk) Jacz. & P. A. Jacz., *Opredelitel' Gribov*: 96 (1931)
1931 *Nematosporangium tenue* (Gobi) Jacz. & P. A. Jacz., *Opredelitel' Gribov*: 96 (1931)
1931 *Nematosporangium subtile* (Wahrlich) Jacz. & P. A. Jacz., *Opredelitel' Gribov*: 97 (1931)
1931 *Nematosporangium perniciosum* (Serbinow) Jacz. & P. A. Jacz., *Opredelitel' Gribov*: 97 (1931)
1931 *Nematosporangium monospermum* (Pringsh.) Jacz. & P. A. Jacz., *Opredelitel' Gribov*: 97 (1931)
1931 *Nematosporangium aphanidermatum* (Edson) Jacz. & P. A. Jacz., *Opredelitel' Gribov*: 97 (1931)
1931 *Nematosporangium polysporum* (Sorokin) Jacz. & P. A. Jacz., *Opredelitel' Gribov*: 98 (1931)
1931 *P. vitis* Serbinow, [*in* Yačevskie (Jaczewski), A. A. & Yačevskie, P. A.] *Opredelitel' Gribov*: 98 (1931)
1931 *P. tuberculorum* (Vuill.) Lecht.-Trnka, *Étude sur les bactéries des légumineuses et observations sur quelques champignons parasites des nodosités*: 453 (1931)
1931 *P. inflatum* V. D. Matthews, *Studies on the Genus* Pythium: 45 (1931)
1931 *P. catenulatum* V. D. Matthews, *Studies on the Genus* Pythium: 47 (1931)
1931 *P. carolinianum* V. D. Matthews, *Studies on the Genus* Pythium: 71 (1931)
1931 *P. echinulatum* V. D. Matthews, *Studies on the Genus* Pythium: 101 (1931)
1931 *P. elongatum* V. D. Matthews, *Studies on the Genus* Pythium: 106 (1931)
1931 *P. angustatum* Sparrow, *Annals of Botany* **45**: 257 (1931)
1931 *P. adhaerens* Sparrow, *Annals of Botany* **45**: 272 (1931)
1931 *P. muscae* P. A. Dang., *Le Botaniste 22 Serie* **6**: 452 (1931)
1932 *P. allantocladon* Sideris, *Mycologia* **24**: 27 (1932)
1932 *P. ascophallon* Sideris, *Mycologia* **24**: 29 (1932)
1932 *P. polycladon* Sideris, *Mycologia* **24**: 32 (1932)
1932 *P. chamaehyphon* Sideris, *Mycologia* **24**: 33 (1932)
1932 *P. euthyhyphon* Sideris, *Mycologia* **24**: 34 (1932)
1932 *P. acanthophoron* Sideris, *Mycologia* **24**: 36 (1932)
1932 *P. araiosporon* Sideris, *Mycologia* **24**: 38 (1932)
1932 *P. splendens* Hans Braun var. *hawaiianum* Sideris, *Mycologia* **24**: 38 (1932)
1932 *P. teratosporon* Sideris, *Mycologia* **24**: 40 (1932)
1932 *P. irregulare* Buisman var. *hawaiiense* Sideris, *Mycologia* **24**: 40 (1932)
1932 *P. polymorphon* Sideris, *Mycologia* **24**: 43 (1932)
1932 *P. diameson* Sideris, *Mycologia* **24**: 47 (1932)
1932 *P. artotrogus* var. *macranthum* Sideris, *Mycologia* **24**: 47 (1932)
1932 *P. pleosporon* Sideris, *Mycologia* **24**: 50 (1932)
1932 *P. volutum* Vanterp. & Truscott, *Canadian Journal of Research, Section C*, **6**: 77 (1932)
1932 *P. arrhenomanes* Drechsler var. *canadense* Vanterp. & Truscott, *Canadian Journal of Research, Section C* **6**: (1932)
1932 *P. arrhenomanes* Drechsler var. *philippinense* Roldan, *Philippine Agriculturist* **21**: 165 (1932)
1932 *P. akenense* Tokun., *Transactions of the Sapporo Natural History Society* **12**: 122 (1932)
1932 *P. epigynum* Höhnk, *Mycologia* **24**: 505 (1932)
1932 *P. fabae* G. M. Cheney, *Australian Journal of Experimental Biology and Medical Science* **10**: 149 (1932)
1933 *P. oryzae* S. Ito & Tokun., *Journal of the Faculty of Agriculture, Hokkaido Imperial University* **32**: 208 (1933)
1933 *P. nagae* S. Ito & Tokun., *Journal of the Faculty of Agriculture, Hokkaido Imperial University* **32**: 209 (1933)
1933 *P. echinocarpum* S. Ito & Tokun., *Journal of the Faculty of Agriculture, Hokkaido University* **32**: 210 (1933)
1934 *P. marinum* Sparrow, *Dansk Botanisk Arkiv* **8**: 5 (1934)
1934 *P. deliense* Meurs, *Phytopathologische Zeitschrift* **7**: 179 (1934)
1934 *P. scleroteichum* Drechsler, *Journal of Agricultural Research* **49**: 887 (1934)
1935 *P. thalassium* Atkins, *Transactions of the British Mycological Society* **38**: 31 (1935)
1935 *P. dichotomum* Tokun., [*in* S. Ito & Tokunaga] *Transactions of the Sapporo Natural History Society* **14**: 12 (1935)
1935 *P. apleroticum* Tokun., [*in* S. Ito & Tokunaga] *Transactions of the Sapporo Natural History Society* **14**: 12 (1935)
1935 *P. diclinum* Tokun., [*in* S. Ito & Tokunaga] *Transactions of the Sapporo Natural History Society* **14**: 12 (1935)
1935 *P. iwayamai* S. Ito, [*in* S. Ito & Tokunaga] *Transactions of the Sapporo Natural History Society* **14**: 13 (1935)
1935 *Phytophthora fagopyri* S. Ito & Tokun., [Takimoto *ex* S. Ito & Tokunaga] *Transactions of the Sapporo Natural History Society* **14**: 15 (1935)
1935 *P. piperinum* Dastur, *Proceedings of the Indian Academy of Sciences* **1**: 803 (1935)
1936 *P. cactacearum* Preti, *Rivista di Patologia Vegetale, Padova, Serie* **26**: 349 (1936)
1936 *P. perigynosum* Sparrow, *Journal of the Linnean Society (Botany)* **50**: 464 (1936)
1938 *Phytophthora stellata* Shanor, *Journal of the Elisha Mitchell Scientific Society* **54**: 155 (1938) [no Latin diagnosis]
1938 *P. tardicrescens* Vanterp., *Annals of Applied Biology* **25**: 533 (1938)
1938 *P. aristosporum* Vanterp., *Annals of Applied Biology* **25**: 535 (1938)
1939 *P. maritimum* Höhnk, *Kieler Meeresforschungen* **3**: 347 (1939)
1940 *P. megalacanthum* de Bary var. *callistephi* Tasugi & Siino, *Annals of the Phytopathological Society of Japan* **10**: 292 (1940) [no Latin diagnosis]
1941 *P. cucurbitacearum* Takimoto, *Annals of the Phytopathological Society of Japan* **11**: 91 (1941) [no Latin diagnosis]
1941 *P. marsipium* Drechsler, *Phytopathology* **31**: 505 (1941)
1943 *P. hypogynum* Middleton, *Memoirs of the Torrey Botanical Club* **20**: 69 (1943)
1943 *P. ostracodes* Drechsler, *Phytopathology* **33**: 286 (1943)
1944 *P. violae* Chesters & Hickman, *Transactions of the British Mycological Society* **27**: 60 (1944)
1944 *P. helicum* T. Itô, *Journal of Japanese Botany* **20**: 56 (1944)
1944 *P. polypapillatum* T. Itô, *Journal of Japanese Botany* **20**: 58 (1944)
1944 *P. pleroticum* T. Itô, *Journal of Japanese Botany* **20**: 59 (1944)
1948 *P. indicum* Balakrishnan, *Proceedings of the Indian Academy Science, Section B* **27**: 171 (1948)
1948 *P. periplocum* Drechsler var. *coimbatorense* Balakrishnan, *Proceedings of the Indian Academy of Sciences, Section B* **28**: 31 (1948) [no Latin diagnosis]

molecular biological framework (Dick *et al.*, 1999) by which this pathogen can be placed in the genus *Pythium*.

The most recent, comprehensive review of the pathobiology and bibliography will be found in Mendoza, Ajello & McGinnis (1996), although their mycology is not always accurate. As indicated by the list of disease names above, there is a granuloma with an extensive lesion; this may (in horses) include the production of coralloid 'kunkers' which contain viable hyphae. The disease is found in cattle (Miller, Olcott & Archer, 1985) but is not confined to ungulates: a number of other mammals are known to be capable of infection (cats: Bissonette *et al.*, 1991; dogs: English & Frost, 1984; Fischer *et al.*, 1994; Foil *et al.*, 1984; Pavletic, Miller & Turnwald, 1983). Human infection is also known, notably from south east Asia (Chetchotisakd *et al.*, 1992; Imwidthaya, 1994; Parr, McKinney & Rogers, 1994; Sathapatayavongs *et al.*, 1989; Thianprasit, 1986, 1990; Virgile *et al.*, 1993). Mammal to mammal transfer has only been achieved using rabbits (Miller & Campbell, 1983).

The fungus is thermophilic, as would be expected for a parasite of mammals and may be keratinophilic (Miller, 1983). Its survival outside mammals presumably depends on existence in warm stagnant water, but extensive ecological surveys for its presence have not been carried out. It is therefore regarded as being restricted to the tropics and warm temperate zones, and is particularly prevalent in south east Asia (including Japan) and Australasia, and Central and South America. Surprisingly, it has not been reported from Africa. It is not regarded as indigenous in Europe.

For further details on this disease, see the following, and additional references therein:

Alfaro & Mendoza (1990); Amemiya (1982); Amemiya & Nishiyama (1966); Bridges & Emmons (1961); de Cock *et al.* (1987); de Haan & Hoogkamer (1903); Ichitani & Amemiya (1980); McMeekin & Mendoza (2000); Mendoza, Kaufman & Standard (1987); Mendoza & Marin (1989); Murdoch & Parr (1997); Shipton (1983, 1987); Shipton, Miller & Lee (1982); Thomas & Lewis (1998).

The affinity of *Phytophthora* for woody dicotyledonous hosts and the distinct, later origin of the pasture grasses makes it possible to suggest that, contrary to previously proposed phylogenies (Skaliký, 1966; Barr, 1983; Shaw, 1981), some sections of *Pythium sensu lato* may have evolved later than some species of *Phytophthora*.

The *Pythiogetonaceae*

The family *Pythiogetonaceae* was erected by Voglmayr, Bonner & Dick (1999) after consideration of the placement of the saprotrophic fungus *Medusoides argyrocodium*. They reviewed the taxonomy of the genus *Pythiogeton*, recognizing seven species. Since then, Jee, Ho & Cho (2000) have added another species, *P. zeae* parasitic in *Zea* (*Poaceae*), and provided a key to the genus. The family is distinguished by the very narrow hyphal diameter of the mycelium (1·0-3·5(7·0) μm diam.), a detachable homohylic vesicle and, in species for which oospores are known, oospores with very thick walls. The ultrastructure of the oospore wall of *M. argyrocodium* appears to be unique in having radial, electron-opaque rods in the endospore. The physiology of *Pythiogeton* has been investigated by Cantino (1949, 1951, and Natvig & Gleason, 1983) but needs further study. The metabolism of both genera may be facultatively or obligately fermentative (Emerson & Natvig, 1981; Natvig & Gleason, 1983; Voglmayr *et al.*, 1999).

Other literature on the family includes:

Batko (1971*a*); Beneke & Rogers (1970); Chaudhuri *et al.* (1947); Cornu (1872*b*); Czeczuga (1994, 1995*a*, 1995*b*, 1995*c*); Czeczuga, Chomutowska & Woronwicz (1990); Dayal (1968); Drechsler (1932); El-Hissy (1974); Gaertner (1954); Hamid (1942); Hsieh & Chang (1976); Ito & Nagai (1931); Ito & Tokunaga (1935*a*); Lund (1934); Minden (1916); Newell (1976); Shen & Siang (1948); Sparrow (1933*b*, 1952*c*, 1960); Watanabe (1974); Waterhouse (1942, 1968*b*, 1973); Wildeman (1896*c*); Wolf (1944); Zebrowska (1976).

TABLE III: 8. *Endobiotic flagellate parasites of* Pythium *and* Phytophthora. *The ectoparasites* Rhizophydium pythii *De Wild. (Karling, 1970) and* Solutoparies pythii *Whiffen (Whiffen, 1942) are excluded.*

Name	Type host	Distinguishing features
Cornumyces utriculobus (C. E. Mill.) M. W. Dick [≡ *Petersenia utriculoba* C. E. Mill.]	vegetative hyphae of *Pythium intermedium* from dry oat field soil	thallus lobed, up to 87 µm diam.; zoospores 4·0 × 5·5 µm; flagellar insertion lateral; oogonia pluriovulate, oospores 12-17 µm diam.
Gracea gracilis (E. J. Butler) M. W. Dick	*Pythium intermedium* de Bary	longer straminipilous flagellum; hyperplasia; zoosporangia 4-52 µm; zoospore pyriform, 3·0 × 5·0 µm; resting spores 12-27 µm, without companion cells, with or without extramural fibrils
Gracea waterhouseae (Karling) M. W. Dick	*Phytophthora cryptogea* Pethybr. & Laff. *Phytophthora megasperma* Dreschsler	longer straminipilous flagellum; zoosporangia up to 74 µm long; zoospores 5·0 × 8·0 µm; resting spores not known
Olpidiopsis brevispinosa Whiffen	a *Pythium* sp.	zoosporangia 11-68 µm diam; zoospores 2·0-3·0 × 3·0-4·5 µm, flagella ± isokont; oogonia 10-45 µm diam. with plerotic oospore and short extramural spines
Olpidiopsis curvispinosa Whiffen	a *Pythium* sp.	zoosporangia 12-68 µm diam.; zoospores elongate-reniform, dimensions not known; oogonia 17-24 µm diam. with plerotic oospore and curved extramural spines
Olpidiopsis indica G. C. Srivast.	*Pythium aphanidermatum* (Edson) Fitzp. from a decaying leaf of *Carica*	zoosporangia 17-35 µm diam.; zoospores 3·5-5·0 µm diam.; sub-apical flagellar insertion; oogonia 13-28 µm diam., oogonia and companion cells smooth-walled
Olpidiopsis pythii (E. J. Butler) Karling	various *Pythium* spp. from garden soil	zoosporangia up to 35 µm diam.; zoospores with lateral flagellar insertion; no companion cell; oogonia 19-35 µm diam. with extramural fibrils
Pythiella pythii (Whiffen) M. W. Dick [≡ *Lagenidium pythii* Whiffen]	a digitate sporangial *Pythium* sp. on grass leaf blade in a moistened soil sample	zoosporangia 16-25 µm diam.; zoospores 2·1-2·8 × 3·5-4·1 µm diam.; oogonia 12-23 µm diam., oospores aplerotic, 7-15 µm diam.
Pythiella vernalis Couch	*Pythium dictyosporum* [presumed to be *P. dictyosporum* Racib. *sensu* Sparrow] and *Pythium* sp. [as *Pythium gracile*] parasitic in *Spirogyra areolata* Lagerh. and *Spirogyra* sp.]	zoospores 3·7-4·0 µm diam.; periplasmic oosporogenesis; oogonia 11-19 µm diam.; oospores aplerotic, 9-15 µm diam.
Rozellopsis inflata (E. J. Butler) Karling	*Pythium intermedium*	zoosporangia 50-80 × 90-120 µm long; zoospores biflagellate, 2·0 × 4·0 µm; anisokont with longer posterior flagellum; resting spores not known
Rozella barrettii Karling	*Phytophthora cactorum* (Lebert & Cohn) J. Schröt.	all morphological details lacking; resting spores not known; no illustrations
Rozella cuculus (E. J. Butler) Sparrow	*Pythium intermedium*	zoosporangia 19-24 µm diam.; zoospores 2·0 × 4·0 µm; resting spores 12-18 µm diam., wall smooth, pigmented
Rozella irregularis (E. J. Butler) Sparrow	*Pythium vexans* de Bary (*Phytophthora* ?) from soil with roots of *Ophioglossum*	zoosporangia 23 µm diam.; zoospores 2·0 × 4·0 µm; resting spores 11-15 µm diam., pigmented, with short spines
Rozella laevis Karling	*Pythium* F-group [as '*Pythium gracile* Schenk']	zoosporangia 10-20 × 30-112 µm; zoospores 1·5 × 3·0 µm; resting spores 11-18 µm diam.
Rozella longicollis Karling	a *Pythium* sp. from a soil sample	zoosporangia 20-60 µm diam.; zoospores 1·8 × 3·0 µm; resting spores 12-17 µm diam., punctate/spiny
Dictyomorpha pythiensis (N. Sarkar & Dayal) M. W. Dick [≡ *Dictyomorpha dioica* Mullins var. *pythiensis* N. Sarkar & Dayal]	*Pythium aphanidermatum* (Edson) Fitzp.	zoosporangia 12-20 µm diam.; zoospores posteriorly uniflagellate 2·5-3·8 × 3·0 µm; resting spores reticulate, 15-19 µm diam.
Sorodiscus cokeri Goldie-Sm.	in *Pythium proliferum* de Bary from grassy soil under *Carica*	zoosporangia 5·3-8·8 µm diam.; anisokont whiplash flagella; resting spores 3·1-4·1 µm diam.
Woronina pythii Goldie-Sm.	an unidentified species of *Pythium* from a damp swale soil sample [said to be a 'strain	multiple interconnecting zoosporangia; zoosporangia 4·9-8·8 µm diam.; anisokont whiplash flagella; resting spores 3·7-5·6 µm diam., thick walled

THE *SAPROLEGNIALES*

The *Saprolegniales* is the principal order of the *Saprolegniomycetidae*, which, in this classification and according to molecular biological data, contains myceliar and non-myceliar orders. The latter are discussed in PART IV. Molecular biological studies (Dick *et al.*, 1999; Riethmuller *et al.*, 1999; Petersen & Rosendahl, 2000; and unpublished work) are revealing that the long-established genera, based on zoosporangial form and zoospore encystment patterns, do not have monophyletic origins. The data are not sufficient to reconstruct a new generic classification at present; the classification is outdated but cannot yet be superceded. As with the *Peronosporales* and *Pythiales*, the species are omitted from PART V but are listed in APPENDIX I.

The *Saprolegniaceae*

Monographs of the *Saprolegniaceae sensu lato* have been written at regular intervals, and have been supplemented by monographs of the three largest genera, *Achlya* (Johnson, 1956), *Aphanomyces* (Scott, 1961) and *Saprolegnia* (Seymour, 1970). I list below the citations from these three works; together with later taxonomic additions. In particular, I have drawn attention to some of the more important taxa, together with their bibliographies.

Citations from Seymour (1970), adjusted for spelling of authors' names, dates and index order, for *Saprolegnia*, are as follows:

Agersborg (1933); Aleem, Ruivo & Théodorides (1953); Allison (1950); Apinis (1929, 1935, 1960, 1964); Archer (1867); Atkinson (1897); Bail (1861); Bangham (1933); Beneke (1948*b*); Beneke & Rogers (1962); Beneke & Schmitt (1961); Berkeley (1864); Beverwijk (1948); Bhargava & Srivastava (1966); Blank & Tiffney (1936); Boedjin (1921, 1923); Braun (1851, 1853, 1856); Bretsnyder (1943); Brown (1938); Buell & Weston (1947); Burton (1939); Carlson (1920); Cejp (1932*a*, 1934, 1959*a*, *b*); Chaudhuri & Banerjee (1942); Chaudhuri & Kochhar (1935); Chaudhuri *et al.* (1947); Chiapelli (1933); Chidamboram (1942); Claussen (1908); Clinton (1894); Coker, W. C. (1923, 1927); Coker & Braxton (1926); Coker & Matthews (1937); Collins (1920); Cooke, M.C. (1871); Cooke, W.B. (1954); Cooke & Bartsch (1959, 1960); Cookson (1937); Cornu (1872*b*); Cotner (1930); Couch, J. N. (1941); Couch R. A. (1951); Crooks (1937); Dangeard (1890*b*, 1916); Davis & Lazar (1940); Dayal (1958, 1959, 1960, 1961*a*, *b*, *c*, *d*); Dayal & Tandon (1962, 1963); Dayal & Thakur Ji (1965, 1966); De Bary (1860, 1881*a*, 1883; 1888); De-La-Rue (1869); Dick (1960, 1962, 1963, 1964, 1965, 1966); Dick & Newby (1961); Dissmann (1931); Dop (1905); Dudka (1964); Duncan (1876); Elliott (1968); Emerson & Cantino (1948); Emoto (1923); Fischer (1892*b*); Forbes (1935*a*, *b*); Frank (1881); Frey (1950); Galloway (1891); Gäumann (1918); Gilman (1957); Goldie-Smith (1952, 1956*a*); Goldsmith (1948); Graff (1928); Gruithuisen (1821); Hardy (1910); Hartog (1887); Harvey (1925, 1927*b*, 1928, 1930, 1942, 1952); Häyrén (1903, 1928, 1943, 1945, 1954, 1956); Hine (1878-1879); Hoffman (1949); Höhnk (1933, 1935*a*, *b*, 1939, 1952*a*, *b*, 1953, 1956*a*, *b*, 1958, 1960, 1962); Höhnk & Bock (1954); Holland (1958); Horn (1904); Hoshina & Ookuba (1956); Hughes (1959, 1962); Humphrey (1893); Huxley (1882); Istvánffi (1895); Ito, S. (1936); Ito, T. (1942, 1944); Ivimey Cook & Forbes (1933); Ivimey Cook & Morgan (1934); Johannes (1949, 1955); Johnson (1950*b*, 1956*a*, *b*, 1968); Johnson & Surratt (1955); Johnson & Sparrow (1961); Johnston (1917, 1921); Kanouse (1927*a*, *b*, 1932); Kauffman (1906, 1908, 1916, 1921); Kelman (1947); Klebs (1899); Kobayasi & Ookubo (1952); Koch (1966); Krause (1960); Kützing (1843, 1849); Larsen (1931); Lechmere (1910, 1911*a*, *b*); Leitgeb (1868, 1869); Lennon (1954); Lindstedt (1872); Lounsbury (1927); Lund (1934); Mackel (1928); Manton, Clarke & Greenwood (1951); Massee (1891); Matthews (1935); Maurizio (1894, 1899); Meier & Webster (1954); Meyen (1839); Migula (1903); Milanez & Beneke (1966); Miller (1965*b*); Milovtzova (1935); Minden (1911-1915, 1916); Monsma (1937); Moore (1908-1909); Moreau & Moreau (1935*a*, 1939, 1948); Morgan (1939); Murgoci (1947); Murray (1885); Nagai (1931); Nagai & Takahashi (1962); Nees von Esenbeck (1823); Newby (1948*a*, *b*); Obel (1910*a*, *b*); O'Bier (1960); Ookubo (1954); Ookubo & Kobayasi (1955); Ou (1940); Paterson (1967); Perrott (1960); Petersen (1909, 1910); Pieters (1915*a*, *b*, *c*); Poitras (1955); Povah (1935); Pringsheim (1851, 1858, 1860, 1873); Radais (1898); Ramsbottom (1916*a*); Raper (1937); Reinke (1869); Reinsch (1878); Reischer (1949*a*); Remy (1948); Richter (1937); Rioux & Achard (1956); Roberts, R. E. (1963); Robin (1853); Rodway (1897); Rossy-Valderrama (1955, 1956); Rothert (1888); Saccardo (1888); Saksena & Bhargava (1944); Salvin (1940, 1941); Sawada (1912*a*, 1919); Schkorbatow (1923); Schmitt & Beneke (1962); Schnetzler (1887); Schröter (1885-1889, 1897); Schröter & Schneider (1870); Scott (1961); Scott & O'Bier (1962); Scott, Powell & Seymour (1963); Scott, Seymour & Warren (1963); Shen & Siang (1948); Shibuya (1959); Slifkin (1961);

Smith, W. G. (1878*a*, *b*, 1884*b*); Sparrow (1936*b*, 1952*a*, *b*, 1960); Stirling (1877-1878, 1878-1879); Stoll (1936); Stpiczńska (1963); Suzuki (1960, 1961*a*, *b*, *c*, *d*, 1962); Sydow & Butler (1907); Szaniszlo (1965); Szwanke (1938); Talukdar & Baruah (1952); Tempère (1904); TeStrake (1958, 1959); Thuret (1850); Tiesenhausen (1912); Tiffney (1939*a*, *b*); Tiffney & Wolf (1937); Trelease (1884); Valkanov (1931*a*); Viégas & Teixeira (1943); Vlk (1939); Walentowicz (1885); Walz (1870); Ward (1939*a*); Wemmer (1954); Weston (1919); Willoughby (1962); Wilson (1960); Wolf (1937*a*, *b*, 1944); Wolf & Wolf (1941); Wunder (1938); Yung & Stenton (1964); Ziegler (1948*a*, 1952, 1953, 1958).

Since then, the following binomials have been erected: *S. blelhamensis* (Mil'ko, 1979), *S. irregularis* (Johnson & Seymour, 1975), *S. shikotsuensis* (Hatai, Egusa & Awakura, 1977), *S. subeccentrica* (Mil'ko, 1979), *S. bhargavae* (Khulbe & Verma, 1987) and *S. polymorpha* (Willoughby, 1998).

The most important species of this genus is *Saprolegnia parasitica*, one of the causes of disease in salmonid and other fish. Willoughby (1994) has reviewed the literature for this fungus and other fungi implicated in fish diseases and his citations for *Saprolegnia* are listed below:

Alabi (1971*a*, *b*); Alderman (1982*a*, 1985); Alderman & Polglaze (1986*b*); Amlacher (1965); Bermùdez (1980); Bly & Clem (1991); Bly *et al.* (1992, 1993); Bootsma (1973); Bruno & Stamps (1987); Cohen & Coffey (1986); Coker (1923); Copland & Willoughby (1982); Cross & Willoughby (1989); Domsch, Gams & Anderson (1980); Fijan (1969); Goldie-Smith (1954); Hara & Pillay (1962); Hatai & Egusa (1977); Hatai, Egusa & Awakura (1977); Hatai, Egusa & Nomura (1977); Hatai *et al.* (1977); Hatai & Willoughby (1988); Hatai, Willoughby & Beakes (1990); Hoshina, Sano & Sunayama (1960); Johansson, Svensson & Fridberg (1982); Johnson (1972); Kimura *et al.* (1980); Macmillan (1985); Neish & Hughes (1980); Noga & Dykstra (1986); Nolard-Tintigner (1973, 1974); Pickering & Christie (1980); Pickering & Pottinger (1985); Prowse (1951); Puckridge *et al.* (1989); Richards & Pickering (1978, 1979); Roberts, R. J. (1989); Roberts *et al.* (1970, 1971); Roberts, Willoughby & Chinabut (1993); Schäperclaus (1954); Scott (1961); Scott & Warren (1964); Shaffer, Padgett & Celio (1990); Shanor (1940); Srivastava, R. C. (1980); Srivastava, G. & Srivastava (1975, 1977); Srivastava, R. & Srivastava (1978); Vishniac & Nigrelli (1957); Willoughby (1978, 1994); Willoughby & Copland (1984); Willoughby & Lilley (1992); Willoughby & Roberts (1992*a*, 1992*b*); Wolke (1975); Wood & Willoughby (1986); Wood, Willoughby, & Beakes (1988); Yuasa & Hatai (1995).

The disease of salmonid fish was first reported by fly-fishermen in the nineteenth century in Irish and Scottish river systems, where it caused periodic, but not always annual, damage to game-fish stocks (salmon and trout). Accounts of the disease have accumulated over a long period and the disease has a large bibliography (Willoughby, 1994). *Saprolegnia parasitica* is now a noteworthy disease in fish farms and hatcheries in Europe, North America and south east Asia. Subsequent to the description by Coker (1923), several other related fish parasites Tiffney & Wolf, 1937; Tiffney, 1939; Scott, 1964; Kitancharoen *et al.*, 1995) and fish hosts have been reported (Bly *et al.*, 1992; Scott & Warren, 1964; Söderhäll *et al.*, 1991; Tiffney, 1939*a*, *b*; Tiffney & Wolf, 1937). Of the additional genera reported, the more common are *Achlya* and *Aphanomyces*. Other species of *Saprolegnia* (*S. shikotsuensis*, Hatai, Egusa & Awakura, 1977; *S. polymorpha*, Willoughby, 1998) are also associated with the disease. Although the most common disease syndrome is of massive superficial lesions, as described in the introductory paragraphs, visceral mycoses have also been reported (Davis & Lasar, 1940; Hatai & Egusa, 1977). The geographic range is now world-wide (Willoughby, 1994), with additional reports from India (Lilley & Roberts, 1997) and Australia (Fraser, Callinan & Calder, 1992).

One of the most difficult problems in the pathology of this disease is the determination of those organisms that are causal, and the dismissal of those organisms that are opportunistic wound-site secondary invaders. Many of the associated saprolegniaceous fungi fall into the latter category, or have not been unequivocally shown to be primary pathogens. The greatest confusion has arisen between the disease-causing *Saprolegnia parasitica* and the opportunist saprobe *S. diclina*. The distinctions between these two species have been discussed at the physiological (Beakes & Ford, 1983; Hatai & Hoshiai, 1993; Hatai, Willoughby & Beakes, 1990; Wood, Willoughby & Beakes, 1988), ultrastructural (Beakes, 1983; Dick, 1990; Söderhäll *et al.*, 1991) and molecular biological (Yuasa & Hatai, 1995) levels. The cyst ornamentation of the principal-form zoospore, consisting of tufts of long, flexuous 'boathook' hairs, is distinctive and diagnostic (Beakes, 1983; Söderhäll *et al.*, 1991), although other saprobic *Saprolegnia* species may have intermediate ornamentation (see pp. 46-47) and not all species of saprobic *Saprolegnia* species have yet been examined. Similarly, the problem with molecular methodology will be the ability to screen the causal agent from opportunists when so few species of the genus have been analysed.

Study of the environmental constraints on disease establishment has only just been touched upon (Bly *et al.*, 1993). Much work has still to carried out on the mycotic aspects of the disease when it is epidemic in fish-farm culture (Roberts, Willoughby & Chinabut, 1993). For many years the standard treatment for this disease has been the application of the toxic Malachite Green (see Alderman, 1985), but a more sophisticated approach to fungicide application is now being put forward (Willoughby & Roberts, 1992).
The following citations can be added to those of Willoughby (1994):

Hatai & Hoshiai (1992); Izumikawa *et al.* (1999); Kitancharoen, Yamamoto & Hatai (1997, 1998); Kitancharoen, Yuasa & Hatai (1995); Kitancharoen *et al.* (1997); Min, Hatai & Bai (1994); Yuasa & Hatai (1994, 1995, 1996).

Citations from Johnson (1956*b*), adjusted for spelling of authors' names, dates and index order, for *Achlya* are as follows:

Abe (1928); Agardh (1824); Apinis (1929); Archer (1867); Areschong (1844, 1845); Atkinson (1897); Bail (1861); Beneke (1948*a*, *b*); Beneke & Schmitt (1961); Berkeley (1864); Beverwijk (1948); Blanc (1887); Braun (1851, 1853); Bretsnyder (1943); Brown (1938); Buell & Weston (1947); Büsgen (1882); Butler (1907); Butler & Bisby (1932); Carlson (1929); Carus (1823); Cavara (1897); Cejp (1932*a*, *b*, 1934); Chaudhuri & Kochhar (1935); Chaudhuri & Lotus (1936); Chaze (1925); Cienkowski (1855); Cohn (1854); Coker. (1910, 1912*a*, *b*, 1914*a*, *b*, 1923, 1927); Coker & Braxton (1926); Coker & Couch (1920); Coker & Leitner (1938); Coker & Matthews (1937); Coker & Pemberton (1908); Cooke (1871); Cookson (1937); Cooper (1929*a*); Cornu (1872*b*); Cotner (1930); Couch (1924, 1927, 1939); Crooks (1937); de Bary (1852, 1881*a*, *b*, 1883, 1888); De-La-Rue (1874); Duncan (1876); Earle (1901); Emoto (1923); Fischer (1892*b*); FLORA DANICA (1780); Forbes (1935*a*, *b*); Fückel (1869-1870); Gallaud (1907); Gäumann (1918); Gilman (1945, 1957); Graff (1928); Hamid (1942); Hannover (1842); Hartog (1887, 1888, 1889*b*, 1892, 1895, 1899); Harvey (1925, 1927*b*, 1930, 1942); Harz (1906); Häyrén (1903); Hemmi & Abe (1928); Hildebrand (1867); Hine (1878-1879); Höhnk (1933, 1935*a*); Horn (1904); Humphrey (1893); Indoh (1937); Ito, S. (1936); Ivimey Cook & Forbes (1933); Ivimey Cook & Morgan (1934); Johannes (1949); Johnson (1949, 1950*b*, 1952); Kanouse (1926); Kauffman (1906, 1908, 1916, 1921); Kawaguchi (1942); Klebs (1898, 1899, 1900); Kützing (1843, 1849); Latham (1935); Lindstedt (1872); Lund (1934); Lyngbye (1819); Marshall Ward (1883*a*); Massee (1891); Matthews (1935); Maurizio (1894, 1899); Migula (1903); Milovtzova (1935); Minden (1911-1915, 1916); Monsma (1937); Moore (1908-1909); Moreau & Moreau (1935*a*, *b*); Mücke (1908); Nagai (1931); Nees von Esenbeck (1823); Obel (1910*a*); Patterson (1927*a*); Petersen (1909, 1910); Pieters (1915*a*, *b*); Pringsheim (1851, 1855, 1858, 1860, 1873, 1882); Raciborski (1886); Ramsbottom (1916*a*, *b*); Raper, J.R. (1936, 1937, 1939*a*, *b*, 1940*a*, *b*, 1942, 1950*a*, 1950*b*, 1951); Raper, K.B. (1928); Reinsch (1878); Reischer (1949*a*, *b*, *c*); Remy (1948); Richter (1937); Rogers & Beneke (1962); Rossy-Valderrama (1955); Rothert (1888); Saccardo (1888); Salvin (1940, 1941, 1942); Sawada (1912*b*, 1919); Ščerbak (1910); Schkorbatow (1927); Schlösser (1929); Schroeter (1885-1889); Shanor (1937, 1944); Shanor & Conover (1943); Sherf (1943); Shirai & Hara (1927); Sorokine (1889); Sparrow (1936*b*, 1943, 1950*b*); Stoll (1936); Takahashi (1918); Thuret (1850); Tiesenhausen (1912); Tiffney (1939*b*); Tiffney & Wolf (1937); Trow (1899, 1904); Unger (1843, 1844); Viégas & Teixeira (1943); Ward (1939*b*); Weston (1917, 1919); Willoughby (1962); Wolf (1937*b*, 1938, 1939*a*, *b*, 1941, 1944); Wolf & Wolf (1941); Ziegler (1948*a*, *b*, 1950); Ziegler & Gilpin (1954); Ziegler & Linthicum (1950); Zopf (1890).

Since then, the following binomials have been described: *A. aquatica* (Dayal & Thakur Ji, 1969), *A. benekei* (Furtado, 1965), *A. bonariensis* (Beroqui de Martinez, 1969*b*), *A. brasiliensis* (Milanez, 1965), *A. curvicollis* (Beroqui de Martinez, 1969*a*), *A. echinulata* (Beroqui de Martinez, 1969*b*), *A. heterosexualis* (Barksdale, 1965), *A. formosana* (Chiou & Chang, 1976), *A. kamatii* (Date, 1972), *A. stellata* var. *multispora* (Rai & Misra, 1977), *A. bisexualis* var. *ambisexualis* (Mil'ko, 1983).

Achlya is noteworthy for the work on its heterothallism initated by Raper (1936, 1939*a*, *b*, 1940*a*, *b*, 1942*a*, *b*, 1950*a*, *b*, 1951, 1967) and continued by Barksdale (1960, 1962*b*, 1963*a*, *b*, 1965, 1966; Barksdale & Lasure, 1973; Barksdale *et al.*, 1974) on the pheromone (sex hormone) system. The culmination of this work was the chemical description of the pheromone, antheridiol (Arsenault *et al.*, 1968; Barksdale, Carlile & Machlis, 1965; McMorris, 1978; McMorris & Barksdale, 1967). Further studies have revealed that this hormone is active in cell biochemistry and cell wall physiology in relation to hyphal branching (Barksdale, 1970; Brunt & Silver, 1986; Gow & Gooday, 1987; Horgan, 1981; Kropf, 1986; Kropf *et al.*, 1984; Mullins, 1968; Mullins & Raper, 1967; Musgrave & Nieuwenhuis, 1975; Riehl *et al.*, 1984; Thomas &

McMorris, 1987; Timberlake, 1976). Mullins & Warren (1975) and Sherwood (1971) have added further information on the genetics of heterothallism in the *Saprolegniaceae*.

It is now apparent that *Achlya sensu lato* includes groups of species that are more closely related to other *Saprolegniaceae* and that other genera, including *Thraustotheca*, *Dictyuchus* and *Brevilegnia*, will probably be shown to be congeneric with *Achlya sensu stricto*. *Achlya sensu stricto* contains the type species, *A. prolifera*, which, apart from its achlyoid sporangial dehiscence, is characterized by oospores containing an ooplast and a single coalesced lipid globule (see discussions on the variability and significance of zoospore discharge in various *Achlya* species by Coker, 1923; Johnson, 1956*b*, Padgett & Seymour, 1974). *Achlya*, as well as *Saprolegnia*, has been implicated as a causal agent of fish disease. Willoughby (1994) and Kitancharoen *et al.* (1995) provide references to the relevant literature. The species of *Achlya* involved are usually from the lacustrine, eccentric-oospored group, and therefore likely to be assigned to *Achlya sensu stricto*.

Of the other genera in the *Saprolegniaceae*, *Geolegnia* (distinguished by the formation of multinucleate aplanospores, perhaps produced as a result of deficiencies in the mechanism of cleavage in sporangial protoplasts) should probably be considered with *Brevilegnia*; *Calyptralegnia* (characterized by sporangia with circuscissile papilla deliquescence and opercula) should be considered with sections of *Achlya sensu lato* and *Protoachlya*, *Isoachlya* and *Couchia* with *Saprolegnia sensu lato*. See Muhsin (1994) and Khulbe (1983*b*) for new species in *Calyptralegnia* and *Protoachlya*, respectively. Ritchie & Slovin (1974) have reported on a collection of *Protoachlya polysporus*. Molecular data of Riethmüller, Weiss & Oberwinkler (1999) have suggested that *Scoliolegnia* (Dick, 1969*a*) is indeed an outlier to *Saprolegnia sensu stricto*. *Aplanopsis* is also probably polyphyletic (Vick & Dick, unpublished molecular data); Chou & Chiang (1976) placed the type species *A. terrestris* (Höhnk, 1952) in synonymy with *Pythiopsis* but failed to provide any new generic name or other synonymy for *A. spinosa* (Dick, 1959). The movement of species from one of the traditional genera to another (e.g., Langsam, 1986) on morphological grounds alone cannot be supported. Unfortunately, the molecular data base is not yet sufficiently large to reassign all the species, which were informally segregated by Dick (1973). Until this stage has been reached it is better to retain the traditional genera, while acknowledging that polyphyly is involved.

The *Leptolegniaceae*

This family has recently been erected for three genera previously placed in the *Saprolegniaceae*, namely *Leptolegnia*, *Aphanomyces* and *Plectospira*. Dick (1990*a*) suggested that this group of species represented an early divergence from the *Saprolegniaceae sensu stricto*, and this has now been supported by 18S rDNA data (Dick *et al.*, 1999) and mitochondrial DNA data (Hudspeth *et al.*, 2000) although the 28S rDNA has not been so supportive. The molecular data have been correlated with morphological data by Dick *et al.* (1999) and are summarized in PART VI.

By far the most important genus of the three is *Aphanomyces*. *Aphanomyces* is parasitic in algae, other straminipilous fungi, higher plant roots, teleost fish and crayfish; some other species, including the type species, are saprotrophs. Citations from Scott (1966), adjusted for spelling of authors' names, dates and index order, for *Aphanomyces* are as follows:

Afanasiev (1948); Apinis (1929); Apstein (1910); Arnaudow (1923); Barrett (1912*a*); Bartsch & Wolf (1938); Beneke (1948*b*); Benisch (1940); Beverwijk (1948); Blank & Tiffney (1936); Boning (1932-1933, 1935-1936); Buchholtz (1944); Butler (1907); Carus (1823); Cejp (1932*a*, 1934); Chaudhuri (1942); Chaudhuri & Banerjee (1942); Chaudhur & Kochhar (1935); Chaudhuri & Lotus (1936); Coker (1923, 1927); Coker & Braxton (1926); Coker & Matthews (1937); Cornu (1872*b*); Cotner (1930); Couch (1924, 1926*a*, 1927); Crooks (1937); Cutter (1941); Dangeard (1890*b*); de Bary (1852, 1860, 1881*a*, 1888); de Bary & Woronin (1881); De-La-Rue (1869); Drechsler (1927, 1928*a*, 1929, 1935); Edson (1913, 1915*a*, *b*); Eliason (1928); Fischer

(1892*b*); Forbes (1935*a*, *b*); Fritsch, A. (1895); Fückel, L. (1869-1870); Gicklhorn (1923); Gilman (1957); Haenseler (1926); Hamid (1942); Harvey (1925, 1942); Hildebrand & Koch (1943); Hofer (1906); Höhnk (1935*a*, 1952*c*); Humphrey (1893); Huneycutt (1948, 1952); Ito, S. (1936); Ito, T. (1942); Ivimey Cook & Forbes (1933); Ivimey Cook & Morgan (1934); Johnson (1956*b*); Jones & Drechsler (1925); Kasanowsky (1911); Kauffman (1908, 1916, 1921); Kendrick (1927*a*, *b*); King & Bissonette (1954); Klebs (1899); Kützing (1843); Lacy (1949); Leigy (1850); Leitgeb (1868); Linder (1947); Linford (1927); Lund (1934); McKeen, C.D. (1952); McKeen, W.E. (1949); Massee (1891); Matthews (1935); Minden (1911-1915); Mix (1945); Nagai (1931); Nees von Esenbeck (1823); Nybelin (1931, 1934, 1936); Peters (1906, 1911); Petersen (1909, 1910); Pfitzer (1872); Pieters (1905); Pringsheim (1851, 1858); Prowse (1954*a*); Ramsbottom (1916*a*,*b*); Raper (1937); Reischer (1949*a*); Rennerfelt (1936); Schäperclaus (1927, 1935); Scherffel (1925*a*); Schikora (1903, 1906, 1922); Schneider, C. L. (1958); Shanor & Saslow (1944); Shen & Siang (1948); Skvortzow (1925); Smith, R. L. (1940); Solberg (1925); Sorokine (1876*a*); Sparrow (1930, 1932, 1933*b*, 1936*b*, 1937, 1952*c*, 1952*d*); Sprague, R. (1950); Tiesenhausen (1912); Valkanov (1931*b*, 1931*c*, 1932); Vallin (1933); Warren (1948); Weatherwax (1914); Weston (1919); Whiffen (1938, 1945); de Wildeman (1890*a*); Wille (1899); Wolf (1944); Ziegler (1948*a*).

Since then, the following taxa have been described:

A. euteiches f.s. *phaseoli* (Pfender & Hagedorn, 1972), *A. euteiches* f.s. *pisi* (Pfender & Hagedorn, 1972), *A. keratinophilus* (Seymour & Johnson, 1973*b*), *A. volgensis* (Domashova, 1974), *A. brassicae* (Singh & Pavgi, 1977), *A. pisci* (Srivastava, 1979), *A. piscicida* (Hatai, 1980) *A. frigidophilus* (Kitancharoen & Hatai, 1980), *A. iridis* (Ichitani *et al.*, 1986), *A. invadans* Willoughby, 1995)

The species taxonomy of the causal organism or organisms assigned to the genus *Aphanomyces* is not at all clear. Oospores, which would be diagnostic, have not been described for all binomials, especially those fungi parasitic on fish or crayfish. The fungi are ascribed to the genus because of the production of *Aphanomyces*-like zoosporangia in cultural studies (Shanor & Saslow, 1944; Hatai *et al.*, 1994). The growth rates and temperature optima of pathogenic *Aphanomyces* spp. (including *A. astaci* - see below) are different and the south east Asian fish-pathogenic isolates are thermophilic, growing well at 30°C. Neither isozyme nor molecular biological surveys of numerous isolates have been carried out. Such studies might resolve whether the causal agent is, in fact, a distinct species or more than one species. It has also been suggested (Willoughby, 1994) that the causal agent might be a *Leptolegnia*.

The disease of crayfish caused by *Aphanomyces astaci* was first reported in 1903 (see Schikora, 1906; Scott, 1961) and was apparently the result of the introduction of the american crayfish into european river systems before 1860. At some stage some of the Californian stock is presumed to have been contaminated with its benign parasite, *A. astaci*. The european crayfish lacks immunity to this necrotic parasite and the European and Asia Minor commercial production of crayfish has fallen dramatically and probably irreparably (Alderman, 1986, 1996; Alderman, Holdich & Reeve, 1990). The fungus has an extensive bibliography (see Scott, 1961; Alderman & Polglase, 1986) and considerable work has been published on its physiology (Unestam, 1966*a*, *b*, 1968, 1969*a*, *b*, *c*, 1972, 1981; Unestam & Gleason, 1968; Unestam & Nylund, 1972; Unestam & Svensson, 1971; Unestam & Weiss, 1970), biochemistry (Söderhäll & Ajaxon, 1982; Söderhäll & Cerenius, 1992; including an extensive bibliography; Söderhäll & Smith, 1983, 1986) and pathobiology (Cerenius *et al.*, 1988).

The disease of the european crayfish is noteworthy because of the sudden and total destruction of the population in any given river and lake system, the stream beds being strewn with dead crayfish. There are a few isolated river systems in which the european crayfish is as yet uninfected, but quarantine protocols are liable to be circumvented by the irresponsible introduction of non-native crayfish. Once stock of moderately resistant crayfish are present, the fungal inoculum will remain, so that any reintroduction of the European crayfish stock will immediately become annihilated. The spread of the disease throughout European river systems has been traced from Italy and central Europe into northern Europe and Russia, Scandinavia, Iberia and Asia Minor.

In moderately resistant, north american crayfish, the disease starts as superficial lesions on the exoskeleton, particularly in thinner areas such as the limb and abdomen joints. However, the hyphal infections, resulting from encysted zoospores, are arrested, but not killed, by melanization (Cerenius *et al.*, 1988; Nyhlén & Unestam, 1980; Söderhäll & Ajaxon, 1982). In the susceptible european crayfish this melanin reaction is weak and develops slowly, so that the fungus is able to penetrate the haemocoel, rapidly causing death. The haemocytes of the haemocoel are thought to be responsible for the defense reaction whereby the activation of the crayfish prophenoloxidase enzyme by the fungal infection results in the melanin deposit around, and stasis of, the fungal hyphae.

The range of hosts and their relative susceptibilities have been discussed on a number of occasions (Alderman, 1986, 1996; Alderman, Holdich & Reeve (1990); Lilley & Roberts (1997); Persson, Cerenius & Söderhäll, 1987). In Asia Minor the causal agent of the disease, in at least some cases, appears to be *Saprolegnia parasitica* (Söderhäll *et al.*, 1991); however, the pathobiology appears to be identical.

The first record of a teleost fish disease caused by a straminipilous fungus other than that caused by *Saprolegnia* was in 1944 (Shanor & Saslow, 1944). The disease is apparently due to penetration from the gut (cf. Davis & Lazar, 1940; Hatai & Egusa, 1977), presumably caused by ingestion of the pathogen spores. The infected fish develops an abnormal dorsal hump and the abdominal cavity becomes grossly swollen. This is followed by obvious mycelial development in the dorsal musculature; eruption of hyphae to the outside has not been observed, so the source of the infective unit has not been established. Fish to fish transfer via fungal propagules apparently does not occur (Willoughby, Roberts & Chinabut, 1995); it is possible that zoosporogenesis only occurs after death and initial decay of the host, but such sporulation, which has been sought, has not been found either: the source of infection remains an enigma. More recent studies (Hatai *et al.*, 1977, 1994; Wada, 1994) of the disease syndrome have revealed systemic granulomas.

Epizootic Ulcerative disease Syndrome (EUS) caused by *Aphanomyces* may be due to the same organism (Lilley & Roberts, 1997) and has also been reported (Srivastava, 1980) in estuarine conditions (cf. diseases caused by *Saprolegnia*, reviewed above); there may therefore be some confusion between the identies of the pathogens and the disease symptoms. Enhanced salinity tolerance (Padgett, 1978, 1984; Padgett *et al.*, 1988; TeStrake, 1959) may be a factor in the disease caused by *Aphanomyces* (Shafer, Padgett & Celio, 1990).

Further references to *Aphanomyces* causing disease in fish include:

Alderman (1985); Davis & Lazar (1940); Fraser, Callinan & Calder (1992); Hatai (1980); Hatai *et al.* (1994); Kitancharoen & Hatai (1980, 1997); Macmillan (1985); Neish & Hughes (1980); Roberts (1989); Schäperclaus (1954); Srivastava (1979, 1980); Wada *et al.* (1994); Willoughby (1994, 1995); Yuasa & Hatai (1994, 1995, 1996)

References to *Aphanomyces* and *Plectospira* diseases on dicotyledons (mostly *A. euteiches*) include:

Beute & Lockwood (1967); Bhalla & Mitchell (1970); Bowers & Parke (1993); Burke, Hagedorn & Mitchell (1969, 1970); Burke, Mitchell & Hagedorn (1969); Carley (1970); Delwiche *et al.* (1987); Drechsler (1929); Grau (1991); Hoch & Mitchell (1972); Holub, Grau & Parke (1991); Jones & Drechsler (1925); Kjøller & Rosendahl (1996, 1997, 1998); Larsen, Mansfield-Giese & Bødker (2000); Lewis (1973); Lewis & Papavizas (1971); Llanos & Lockwood (1960); Lockwood (1960); Lockwood & Ballard (1960); Manning & Menzies (1984); Malvick, Grau & Percich (1998); Malvick, & Percich (1998*a*, *b*); Mitchell & Mitchell (1973); Mitchell & Hagedorn (1971); Mitchell & Yang (1966); Munkvold & Carlton (1995); Oyarzun *et al.* (1997); Oyarzun & Van Loon (1989); Papavizas (1966); Papavizas & Ayers (1964, 1974); Papavizas & Davey (1960*a*, *b*); Papavizas & Lewis (1971); Parke & Grau (1992); Pfender (1984); Pfender *et al.* (1984); Pueppke & Van Etten (1974); Rai & Strobel (1966); Rosendahl (1985); Scharen (1960); Scott (1961); Shatla, Yang & Mitchell (1966); Sherwood & Hagedorn (1962); Sundheim (1972).

References to *Aphanomyces* diseases on monocotyledons include:

Ichitani *et al.* (1986) (*Iris*, *Iridaceae*); Ridings & Zettler (1973) (*Echinodorus*, *Alismataceae*)

The vascular-plant host ranges of *Aphanomyces* are very similar to those of the *Peronosporales* (Table III: 3) indicating that there may be a fundamental physiological and biochemical requirement shared by the two sub-classes.

THE *SCLEROSPORALES*

The order *Sclerosporales* contains two families, the *Sclerosporaceae* and the *Verrucalvaceae* (Dick *et al.*, 1984). The *Sclerosporaceae* (the graminicolous downy mildews) was separated from the dicotyledonicolous DMs on morphological and morphogenetic grounds (Table III: 9), of which flagellar ultrastructure may be the most significant criterion. The hosts for the graminicolous DMs are found in the *Poaceae*, and the evolutionarily advanced panicoid grasses in particular. Safeeulla (1976) has provided a general account of the biology and control of the DMs of pearl millet, soghum and finger millet; chapters in Spencer (1981) provide further information on the plant pathology of these fungi.

The full life-history of sexual and asexual phases is not always evident: *Peronosclerospora* is characterized by conidiosporangia which germinate directly; *Pachymetra* is not known to form any asexual structures and *Peronosclerospora philippinensis* is not known to have a sexual phase. *Peronosclerospora globosa* (Kubicek & Kenneth, 1984) does not appear to have been validly published. Further discussion of the taxonomy and phylogeny is in preparation (see Spencer & Dick, 2001). Waterhouse (1964) has provided an earlier summary of diagnoses; see also Jones (1955); Kenneth, Koltin & Wahl (1964); Kubicek & Kenneth (1984); Siradhana *et al.* (1980); Shaw (1978, 1980); Thirumalachar, Shaw & Narasimhan (1953); Thirumalachar & Whitehead (1952); Weston (1921, 1929, 1933); Weston & Uppal (1932).

Discussion of the systematics and evolution of the graminicolous DMs must involve *Pachymetra* and *Verrucalvus* in the *Verrucalvaceae* and reference to pathogens in the *Pythiaceae*. *Pachymetra*, parasitic on roots of sugar cane, and *Verrucalvus*, parasitic on roots of *Pennisetum*, are known only from eastern Australia. *Pachymetra* and *Verrucalvus* are monotypic and can be maintained, with difficulty, in axenic culture (Dick *et al.*, 1984, 1989). *Pythium* species (especially *Pythium aphanidermatum*) may be associated with these genera in the same root system. The genera *Sclerospora* and *Peronosclerospora* are obligate biotrophic parasites, usually found in the aerial parts of the plant. However, like the dicotyledonicolous DMs, infections caused by the DMs of panicoid grasses can also be systemic (Kenneth, 1981). Biphasic culture has been achieved for several genera and species (*Sclerospora graminicola*: Arya & Tiwari, 1967; Bhat & Gowda, 1985; Bhat, Safeeulla & Shaw, 1980; Mauch-Mani, Schwinn & Guggenheim, 1989; Tiwari & Arya, 1967, 1969; Upadhyaya *et al.*, 1992; *Peronosclerospora sorghi*: Gowda & Bhat, 1992; Mauch-Mani, Schwinn & Guggenheim, 1989; Upadhyaya *et al.*, 1992; *Peronosclerospora sacchari*: Chen, Liu & Chao, 1979; and general reviews: Ingram, 1980; Lucas *et al.*, 1991, Lucas, Hayter & Crute, 1995). Axenic culture has been claimed for *Sclerospora graminicola* (Arya & Tiwari, 1969) but this culture is no longer extant and the axenic culture has not been repeated. *Sclerophthora* appears to be an analogue of *Phytophthora* in respect of its necrotrophic parasitic habit. However, good examples of symptomless occurrence in grass and herbaceous roots are found with *Pachymetra* in *Imperata cylindrica* var. *major* (blady grass) in Queensland (R. C. Magarey, pers. comm.). Symptomless association does not imply a 'no yield loss' situation.

The origin of the graminicolous downy mildews

The graminicolous DMs are ***confined*** to grasses. The grass family, *Poaceae*, is very large (>9500 spp. world-wide) and Clayton & Renvoize (1986) suggest that although the origins of the grasses may have been in the Southern Hemisphere (continents derived from Gondwanaland) with the subfamily *Bambusoideae*; the more advanced subfamilies and tribes have evolved in North Africa, South East Asia and Europe. The pasture grasses must have radiated from the tropical/warm temperate climate of South America (Jacobs, Kingston & Jacobs, 1999) to what is now temperate North America into Eurasia in the early Tertiary,

TABLE III: 9. *Ultrastructural, developmental and morphological differences between the downy mildews of grasses and the downy mildews of dicotyledons*

Criterion	DMs of monocotyledons (*Sclerosporales*)	DMs of dicotyledons (*Peronosporaceae*)
Flagellar concertina	short, struts possibly present (pers. comm., L. W. Olson, decd. re: Lange, Olson and Safeeulla, 1984 (*Peronosporales*); cf. Barr & Désaulniers, 1987: *Saprolegniales*)	long, without struts (Barr & Désaulniers, 1987:
Zoosporangial vesicle	precipitative (or homohylic?) if present	plasmamembranic if present
Conidio-sporangiophore persistence	evanescent, not preserved in exsiccatae of hosts	found on exsiccatae or pressed herbarium specimens of hosts
Conidio-sporangiophore basal septation	biconvex septum present	no basal septum
Hyphal diameter	very narrow *ca* 3-5 μm diameter	broad *ca* 10-12 μm diameter
Haustoria	*Sclerosporaceae*: small, spherical	*Peronosporaceae*: large, lobate [*Albuginaceae*: peg-like]
Oogoniogenesis sequence	wall deposition continuous with swelling of initial; followed by septation	swelling of initial; followed by plug formation, then wall thickening
Oosporogenesis	centrifugal (McDonough, 1946*a*, *b*); fundamentally plerotic	centripetal; fundamentally aplerotic
Periplasm	usually absent; exospore wall minimal or non-existent	usually present; contributing to a prominent, often ridged, exospore
Ooplast	translucent, sometimes condensing, not thoroughly described	translucent, not thoroughly described
Oospore wall	usually very thick	relatively thin

moving southwards and eastwards as the climate cooled and habitats were opened up with the burgeoning of the grazing mammals.

The grass subfamily *Panicoideae* contains species with leaf anatomy of both kranz and non-kranz kinds; kranz anatomy is associated with C_4 photosynthesis (see Sage & Monson, 1998). There are two kinds of kranz anatomy:

- **PS** Parenchyma Sheath with chlorenchyma strongly radiating and both sheaths present, starch forming only in the PS
- **MS** Mestome Sheath with chlorenchyma irregular, only the one sheath present, derived from the MS and containing starch

However, almost all of the known hosts of *Sclerosporales* are kranz MS species, irrespective of the tribe and subtribe to which they belong (i.e., tribe *Paniceae*; subtribe *Setariinae*: *Setaria*; subtribe *Cenchrinae*; *Pennisetum*; and all the genera of the subtribes of the tribe *Andropogoneae*). Kranz MS species have evolved with metabolic pathways that have accommodated to the high carbohydrate levels arising from their habitats, which although ranging from marshes, through ruderal habitats to stony deserts, are all habitats with high insolation and possibly unbalanced mineral nutrition. The subfamily produces phytoalexins and is also noted for its flavonoids, which uniquely, are sulphated. Some (e.g., *Cymbopogon* - lemon grass) have essential oils. These habitats are not mesic. The ***water*** regime would appear to be of minor significance. Compare the growing conditions for sugar cane: *Saccharum*: humid conditions, sometimes with seasonal flooding of the soil; cereal grasses, such as *Sorghum* and *Pennisetum* (pearl millet) of arid lands.

The graminicolous DMs, like the dicotyledonicolous DMs, appear to be biochemically cophylogenetically linked with their hosts. Therefore the graminicolous DMs could have evolved when grazing became a selective force for the tropical, panicoid grasses of southern Asia. Thus, while the type of the *Sclerosporales* is *Sclerospora graminicola*, described from *Setaria* from Europe, this is still within the bounds of a biochemical coevolutionary development and with an ancestry located on the South East Asia plate. The occurrence of *Sclerophthora* in grasses of the subfamily *Chloridoideae* is not anomalous because this subfamily is regarded as a monophyletic line which has successfully proliferated throughout the tropics because of its consistent C_4 (kranz PS) metabolism. On the other hand, the occurrence of *Sclerospora* and *Sclerophthora* in the non-kranz genus *Oryza* (which shares a monophyletic clade with bambusoid species) is apparently anomalous on information presently available, although future studies might reveal host-metabolic comparabilities between *Oryza* and the *Panicoideae* because of their marsh/wetland habitats and submerged root systems).

The wetland host habitat (often with periodic, prolonged seasonal droughts) is an obvious, but recent, coevolutionary factor for fungal parasites with zoosporic antecedents. Wetland *Poaceae* could be similar to ruderals in lacking mycorrhiza (but apparently this has not been researched, D. H. Lewis, pers. comm.). The sterol-precursor requirement by a straminipilous fungus can be satisfied from certain hosts which, for various reasons, have accumulations of a variety of secondary metabolites such as essential oils, saponins, and alkaloids. The kind of photosynthetic pathway (C_4, as in sugarcane and sweetcorn), or high insolation, or absence of a mycorrhizal carbon 'sink' could all individually or collectively give rise to this accumulation. Whatever the biochemical interaction accounting for attraction to a particular host and stimulation to germination and colonization by the parasite, there are well-documented examples of parasite-mediated modification of host physiology after establishment. Green ear hyperplasia of pearl millet caused by *Sclerophthora* (Williams, 1984) is one of the clearest examples relating to growth substance induction. The precise mechanisms of the biochemical interference with host growth substance metabolism have not been researched.

The origin of the graminicolous DMs is uncertain: on morphological criteria and molecular biology their antecedents appear to be from the *Saprolegniomycetidae* rather than the *Peronosporomycetidae*, although the molecular data are not yet sufficient. Data from 18S rDNA analyses suggest that *Pachymetra* is associated with the *Leptolegniaceae* (M. A. Spencer & M. W. Dick, unpublished data). Riethmüller *et al.* (1999) place *Pachymetra* with *Aphanomyces* using 28S rDNA. A link between *Verrucalvus* and *Pachymetra* with *Phytophthora* and *Pythium* was not supported by biochemical or mitochondrial DNA restriction patterns (Belkhiri & Dick, 1988; Dick *et al.*, 1989; G. R. Klassen, pers. comm.). Both *Leptolegnia* and *Pythium* are basal clades in each of these sub-classes. The geologic age of origin may be the Quaternary; their centre of origin may have been south-east Asia.

A totally different geographic centre of adaptive radiation (south-east Asia/Oceania/Australasia) has been proposed (Weltzien, 1981) for *Sclerospora* and related genera (Shaw, 1978) and accepted by Dick (1988). The Himalayan uplift (5-23 m.y.B.P.), which drained the Tethys Ocean remnants and resulted in subsequent much colder montane and Siberian steppe climates to the north would also have forced the early tropical grasses and their graminicolous downy mildews southwards into Africa, India, and south-east Asia. This

TABLE III: 10. *The chronology of taxonomic and plant-pathogenic studies of the* Sclerosporales

Date	Taxonomic development
1876	*Protomyces graminicola* [type species of *Sclerospora*] described by Saccardo (1876)
1879	*Sclerospora* Schröt. (type species *S. graminicola*) described by Schröter (1879)
1913	subgenus *Peronosclerospora* S. Ito (type species *P. sacchari*) described by Ito (1913)
1927	*Peronosclerospora* Hara (type species *P. sacchari*) raised to generic rank by Hara (Shirai & Hara, 1927
1953	*Sclerophthora* Thirum. *et al.* (type species *S. macrospora*) was described by Thirumalachar, Shaw & Narasimhanin (1953)
1956	*Dicksonomyces* Thirum. *et al.* (type species *D. sorghi* Thirum. *et al.* was named by Rao, Salam & Thirumalachar (1956). This genus was based on a misidentification (Waterhouse, 1968*c*) and should not be used.
1984	*Verrucalvus* P. Wong & M. W. Dick (type species *V. flavofaciens*) was described by Dick *et al.* (1984).
1984	*Verrucalvaceae* M. W. Dick (type species *V. flavofaciens*) was described by Dick *et al.* (1984).
1984	*Sclerosporales* M. W. Dick (type species *S. graminicola*) was described by Dick *et al.* (1984).
1989	*Pachymetra* B. J. Croft & M. W. Dick (type species *P. chaunorhiza*) was described by Dick *et al.* (1989).

pattern of change could also explain the occurrence of the north-eastern Australian genera *Verrucalvus* (Dick *et al.*, 1984) on the grass *Pennisetum clandestinum* (kikuyu grass, introduced from East Africa) and *Pachymetra* (Dick *et al.*, 1989) on sugarcane, which is thought to have originated in New Guinea.

The absence of any associations with southern temperate grasses makes it unlikely that there was early evolution of these graminicolous fungi from Gondwanaland, which would have required isolated, three-pronged migration and evolution in Africa, India, and Australasia (but not southern South America). A centre of radiation from Africa (Williams, 1984) or India could only be supported if the *Sclerosporales* are considered to be of late Neogene (Miocene - Pliocene) or Quaternary origin. Such an origin is unlikely on morphological evidence from the fungi or biogeography of the grasses.

Present-day geographical distributions are of little value in elucidating evolution. They merely confirm the rapid capabilities or opportunism of these parasites. There has been world-wide spread of DM diseases associated with the crops of corn, millets and sugar cane (*Zea*, *Pennisetum*, *Sorghum*, *Saccharum*), and for the evolution of intraspecific geographic differences in the fungal populations. Distribution by man has resulted in the parasitism of *Zea* by *Peronosclerospora maydis* and of *Sorghum* by *Peronosclerospora sorghi* in India. *Zea* was introduced to South East Asia and is a comparable host to South East Asian *Panicoideae*, having the same kranz MS (C_4) anatomy. *Peronosclerospora sorghi* originated in Africa (Ball & Pike, 1984; Idris & Ball, 1984). Opportunism, inadvertently mediated by prehistoric and modern man, can and has been shown to enable rapid worldwide dissemination and subsequent evolution of diseased communities. In this century accommodations between *Sclerospora graminicola* and agricultural grasses have occurred (Welzien, 1981). Ball & Pike (1984) showed that while fairly stable host-pathogen relationships have developed in pearl millet (*Pennisetum americanum*) since its introduction into India about 3000 years ago, the intercontinental variation in pathogenicity in *Sclerospora graminicola* suggests that further coevolutionary adjustment continues.

In contrast to these recent developments, it should be recalled that cereals were the original crops of agriculture, with at least four independent and long-localized origins: the 'Fertile Crescent', the region including parts of Asia Minor and Mesopotamia (*Hordeum*); South East Asia (*Oryza*); north-east Africa (*Sorghum*) and central America (Zea). Clayton & Renvoize (1986) put the earliest cultivation of any these crops (*Hordeum*) as at a confident date of more than 10,000 years b.p., followed by *Zea* (>7000 years b.p.) and *Sorghum* (>3000 years b.p.).

PART IV

THE LAGENIDIACEOUS FUNGI

AND SIMILAR ORGANISMS

THE LAGENIDIACEOUS FUNGI: INTRODUCTION

Brief accounts of the historical development of generic concepts in the lagenidiaceous fungi and the reasons for the nomenclatural changes made are given below. Literature citations from Sparrow (1960) and Karling (1981*a*) have been updated to include references in PART I and physiological studies. Citations mentioning the genus only, may be found under the type species in PART V. The *Index of Fungi* (CAB IMI) has been cross-referenced up to Volume 6, Part 19 (1999), and the bibliography includes generic citations from the *Information Retrieval Service CAB* and *BIOSIS* files up to February 1994.

The systematics discussion must start with the problems in *Myzocytium* and *Lagenidium*. A reorganization has been made (Dick, 1998) for the parasites of the *Aschelminthes*, the freshwater green algae and the diatoms. A review of the status of *Olpidiopsis* and the marine algal parasites follows. Summaries of the morphology and taxonomy of several small groups of other straminipilous species, with various possible affinities, are given. PART IV concludes with notes on the more distantly related straminipilous and biflagellate groups, the *Labyrinthista*, the *Rozellopsis*-like organisms and the *Plasmodiophoromycetes*. There is also a section dealing with miscellaneous taxa excluded from the straminipilous fungi, but which have sometimes been associated with them.

Many of the organisms discussed below are so incompletely or inadequately described that any researchers must be encouraged to go to the original description and other literature to arrive at a personal assessment of the data available. A repetition of the compiled descriptions already accessible in Sparrow (1960) and Karling (1981*a*) would greatly extend the length of this review without a concomitant increase in objectivity. Rather the reverse, since to provide such descriptions would deter reappraisal of the earlier, and often meagre, observations. Such reappraisal accompanied by new and more detailed studies should improve the genus and species concepts. Features which appear to be diagnostic at this stage are emphasised in PART VI: KEYS. The keys to genera and species summarize much of the information that is in the earlier descriptive texts. I stress that, in the present state of knowledge, my classification can only be an improved, but still interim, treatment.

From the emphasis I give to coevolution, the taxonomic data might be expected to be accompanied by a host index. Host indexes have not been prepared for three reasons. First, there is so little information available on host ranges for individual species. Secondly, in many cases the identification of both fungus and host may be questionable or incomplete. Thirdly, such lists are available in Sparrow (1960) and Karling (1981*a*) and little can be added to them. Host species (as named and with current nomenclature) of the type material are listed under the formal taxonomic entry in PART V. Similarly, geographic data have not been included because none of the organisms is reputed to have a restricted distribution, and no serious ecological data have been compiled.

A full check list of all the fungi of the *Straminipila* is given in APPENDIX 1. Citation references, indexed by specific epithet in the form used by Dick (1990*c*) for *Pythium*, and an alphabetic author index to the formal taxonomic literature are in preparation and will eventually be available elsewhere.

MYZOCYTIUM SENSU LATO AND *LAGENIDIUM SENSU LATO*

The genera *Myzocytium* (Schenk, 1858) and *Lagenidium* (Schenk, 1859*b*) were placed in the *Holochytriaceae* (*Ancylistaceae*) by Fischer (1892), and this family, based on the genus *Ancylistes*, was raised to ordinal status (Schröter, 1897 [1893*a*]; Fitzpatrick, 1930). With the transfer of *Ancylistes* to the *Zygomycotina* (Berdan, 1938), the remaining part of the order, containing flagellate fungi, was left without attribution within a taxonomic hierarchy. A new order, the *Lagenidiales*, was formally established by Karling (1939). This order was typified by the family *Lagenidiaceae*, and the family by the genus *Lagenidium*. Neither *Myzocytium*, which antedates *Lagenidium*, nor *Lagenidium* was clearly characterized by Schenk (1858, 1859*b*).

The history of the generic name *Myzocytium* is linked with *Pythium*. The genus and type species, *Myzocytium proliferum* Schenk, were ***named*** by Schenk (1858). A year later Schenk (1859*b*) ***described*** *Pythium gracile* Schenk, *P. globosum* Schenk and *P. proliferum* Schenk as three new species of Pringsheim's genus *Pythium*, then composed of two species, *P. monospermum* Pringsh. and *P. entophytum* Pringsh. (Pringsheim, 1858). It is unclear whether Schenk intended to reduce *Myzocytium* to synonymy with *Pythium*; Schenk (1859*b*) himself acknowledged that *P. globosum* was near to *P. entophytum* or perhaps only a form of *P. proliferum*. Most authors (e.g. Sparrow, 1960; Karling, 1981*a*) accept that *P. globosum* is a synonym of *M. proliferum*. Waterhouse (1968*b*) indicated that if the importance of thallus form were to be downgraded, then *M. proliferum* could be transferred to *Pythium*, as *P. proliferum* (Schenk) Schenk. It was in his later paper that Schenk (1859*b*) introduced the genus name *Lagenidium*, but without any formally associated specific binomial. Thus, initially, *Myzocytium* was monotypic and *Lagenidium* was a rather vaguely defined concept intermediate between *Myzocytium* and *Pythium*. At this stage Schenk's contributions were cautious and inconclusive, being based solely on asexual material. Sexual reproduction in *Myzocytium proliferum* was first noted by Cornu (1869), and later (Cornu, 1872*b*) he gave a further description of the species under the name *M. globosum* (Schenk) Cornu. Of the two original species of *Pythium*, Zopf (1884) removed *Pythium entophytum* to *Lagenidium sensu* Zopf as *L. entophytum* (Pringsh.) Zopf.

The earliest binomial use of the name *Lagenidium* was by Lindstedt (1872) who made the combination *L. globosum* (Schenk) Lindst. for the second of Schenk's species of *Pythium*, but because the binomial *M. proliferum* has priority for these synonyms, Lindstedt's combination has been rendered inadmissible. Nevertheless, his concept of *Lagenidium* could not be considered to be different from *Myzocytium*. The name *Lagenidium* Schenk was first used for a valid species by Zopf (1878) with his brief (unillustrated) description and naming of *L. rabenhorstii* Zopf.

Ivimey Cook (1935) cited the first binomial of *Lagenidium* as *L. entophytum*, but followed Zopf (1884) (who did not credit Schenk with this combination since he headed his description "*L. entophytum* (Pringsheim)") in listing the species as *L. entophytum* (Pringsheim) Zopf. The full descriptions by Zopf (1884), with accounts of asexual and sexual reproduction, formed the basis for the generic concepts of *Myzocytium* and *Lagenidium*, though 'type species' for these genera were not designated as such until much later.

From the descriptions of *M. proliferum* and *L. rabenhorstii* it is clear that there is a normal oomycetous reproductive system involving the penetration of the oogonium by an antheridial fertilization tube. But Ivimey Cook (1935) did not describe a true fertilization tube for his collection of *L. rabenhorstii*. *Lagenidium rabenhorstii* also has an oospore with "a fairly large nearly centric globule" (Sparrow, 1960: 994) which could be presumed to be an ooplast. The illustrations provided by Zopf (1884: plate XIV, fig.

20 and plate XII, fig. 19 for *Myzocytium* and *Lagenidium* respectively) depict zoospores with lateral flagellar insertion. It is not certain whether there is a flagellar groove, but both species have extra-sporangial zoosporogenesis taking place in a vesicle, which must be presumed to be a homohylic vesicle. The zoospores are thus likely to be of the principal form (Dick, 1973, 1990*a*) and therefore both type species are referable to the *Peronosporomycetidae* (Dick, 1990*a*). However, the presence of a vesicle in *L. rabenhorstii* has been disputed by Atkinson (1909) and Ivimey Cook (1935), while Thompson (1934) described non-vesiculate zoosporogenesis in an unnamed *Myzocytium* parasitic in *Spirogyra* as well as observing vesiculate zoosporogenesis in *M. proliferum*. Also, since the reniform zoospore shape is seen first in the quiescent, newly formed zoospores, which may later assume a pyriform shape when swimming (Ivimey Cook, 1935), some doubt must remain concerning later referrals to either of the two type species.

The generic keys in Fitzpatrick (1930) and Sparrow (1960) stress, for both genera, the vesiculate zoosporangial discharge; and for *Lagenidium*, the branched nature and simple septation of the thallus and the differentiated gametangia; and for *Myzocytium*, the unbranched thallus with thickened septa delimiting inflated segments of the thallus, plus the more or less undifferentiated gametangia. I conclude that these criteria are insufficient to separate *Lagenidium rabenhorstii sensu* Zopf from *Myzocytium*.

Species added to these genera were less clearly characterized, even by Zopf (1884), and the gradually increasing disparities within *Lagenidium*, in particular, stem from that date. More recently, Newell *et al.* (1977) have reiterated Barron (1976*a*), noting that *Myzocytium* is "imprecisely distinguished from *Lagenidium*".

In *Lagenidium sensu lato* thallus form ranges from unicellular and ovoid (*L. netrii*), through toruloid forms with or without septa (*L. entophytum*) and tubular and septate thalli limited to one host cell (*L. rabenhorstii*), to tubular and branched, septate thalli (mycelia) invading several cells (*L. marchalianum*); zoosporogenesis may be extrasporangial in a vesicle (*L. rabenhorstii*) or intrasporangial producing flagellate zoospores (*L. cyclotellae*) or non-flagellate aplanospores (*L. oedogonii*, *L. parthenosporum*); zoospore/zoospore cyst size falls into four broad categories: *ca* 3·5 μm diameter (volume < 25 μm^3) (*L. pythii*); *ca* 5·5 μm diameter (volume *ca* 100 μm^3) (*L. oviparasiticum*); *ca* 9·0 μm diameter (volume *ca* 400 μm^3) (*L. destruens*), or larger (volume > 650 μm^3) (*L. caudatum*). Intrasporangial zoospores may be of the principal form (*L. muenscheri*, *L. pygmaeum* var. *pygmaeoides*) or with sub-apical flagellar insertion (*L. septatum*); zoospores may be diplanetic and dimorphic (*L. enecans*); sexual reproduction may involve differentiated gametangia (*L. rabenhorstii*) or presumed automictic development in undifferentiated parts of the thallus (*L. entophytum*); sexual fusion may be by transfer of a male nucleus via a fertilization tube (*L. rabenhorstii*) or by gametangial copulation through a pore (*L. humanum*), the resulting female gametangium may be unisporous (*L. rabenhorstii*) or polyoosporous (*L. destruens*) and the resulting oospore may be aplerotic (*L. rabenhorstii*) or plerotic (*L. lundiae*); the oospore wall may be smooth (*L. rabenhorstii*) or ornamented (*L. entophytum*). Species have been described from green algal (*L. rabenhorstii*), diatomaceous (*L. brachystomum*), higher plant (*L. muenscheri*), nematode (*L. distylae*), insect (*L. giganteum*) and crustacean (*L. callinectes*) hosts and detritic substrates (e.g. pollen, *L. pygmaeum*; keratin, *L. humanum*) and from freshwater (*L. rabenhorstii*) and marine (*L. callinectes*) environments. Valid binomials have been introduced on the basis of thallus form alone (*L. coenocyticum*); asexual reproduction in the absence of sexual reproduction (*L. netrii*), or sexual characters in the absence of asexual reproduction (*L. reductum*).

While it is accepted that the diagnosis of large and complex genera may require a suite of characters, each of which is inadequate on its own (Leadlay & Heywood, 1990), the breadth of the above range for *Lagenidium sensu lato* is such that this genus could accommodate ***all*** the biflagellate holocarpic fungi! Such a situation is neither practicable nor tenable, and the small amount of ultrastructural and biochemical data that is available reinforces the heterogeneity of *Lagenidium sensu lato*. Narrow generic concepts are helpful in drawing attention to those characters that are available, even if eventually, some amalgamation of small genera may develop from further insights into the evolution of these diminutive parasites.

Many of the species assigned to *Lagenidium* and *Myzocytium* cannot be placed in these genera if the principles discussed in PART I. Of the species excluded from *Lagenidium* and *Myzocytium*, many are correctly placed in the *Peronosporomycetes*, but with various systematic dispositions. It is much less certain whether some of the remainder are, in fact, *Peronosporomycotina*.

At this point it is perhaps helpful to give an alphabetic list (Table IV: 1) of all the binomials which at some time have been assigned to *Lagenidium* (53 binomials) and *Myzocytium* (22 binomials), together with their placement in this work. Nomenclatural changes and synonymy are discussed below and the new taxonomic framework is used thereafter. Pending acceptance of the proposal to conserve *Lagenidium sensu stricto* (see below), *Lagenidium* and *Myzocytium sensu stricto* must remain in the family *Pythiaceae* (order *Pythiales*). Both genera are characterized by the following pythiaceous criteria:

1. only principal-form zoospores are produced
2. sexual reproduction is oogamous with antheridial penetration via a fertilization tube
3. oospores are smooth-walled

Similarity with *Pythium* as opposed to *Phytophthora* is demonstrated by:

1. homohylic vesicular sporangial discharge with zoospore development and maturation within the vesicle
2. absence of periplasm

TABLE IV: 1. Lagenidium *and* Myzocytium *species: alphabetic list of previously used names with synonymy in this text*

Lagenidium americanum	synonym of *Aphanomycopsis entophyta*	
Lagenidium astrum	synonym of *Syzygangia papillosa*	Dick, 1997*b*
Lagenidium brachystomum	synonym of *Ectrogella brachystoma*	
Lagenidium callinectes	synonym of *Salilagenidium callinectes*	
Lagenidium canterae	synonym of *Cystosiphon canterae*	
Lagenidium caudatum	synonym of *Pythium caudatum*	Glockling & Dick, 2000
Lagenidium chaetomorphae	synonym of *Sirolpidium salinum*	
Lagenidium chthamalophilum	synonym of *Salilagenidium chthamalophilum*	
Lagenidium clavatum	synonym of *Aphanomycopsis entophyta*	
Lagenidium closterii	synonym of *Cystosiphon closterii*	
Lagenidium coenocyticum	synonym of *Myzocytium rabenhorstii*	
Lagenidium contortum	synonym of *Aphanomycopsis entophyta*	
Lagenidium culicidum	synonym of *Lagenidium giganteum*	
Lagenidium cyclotellae	synonym of *Ectrogella cyclotellae*	
Lagenidium cylindriforme	synonym of *Aphanomycopsis entophyta*	
Lagenidium destruens	synonym of *Cornumyces destruens*	
Lagenidium distylae	synonym of *Myzocytiopsis distylae*	Dick, 1997*b*
Lagenidium elegans	synonym of *Myzocytiopsis elegans*	Dick, 1997*b*
Lagenidium ellipticum	synonym of *Syzygangia elliptica*	Dick, 1997*b*
Lagenidium enecans	synonym of *Aphanomycopsis bacillariacearum*	
Lagenidium entophytum	synonym of *Aphanomycopsis entophyta*	
Lagenidium entosphaericum	synonym of *Aphanomycopsis entophyta*	

(Table IV: 1, continued)

Lagenidium giganteum	*Lagenidium giganteum*	
Lagenidium globosum	synonym of *Myzocytium proliferum*	
Lagenidium gracile	synonym of *M. rabenhorstii* or *A. entophyta*	
Lagenidium humanum	synonym of *Myzocytiopsis humana*	Dick, 1997*b*
Lagenidium intermedium	synonym of *Pythium podbielkowskii*	
Lagenidium lundiae	synonym of *Pleocystidium lundiae*	
Lagenidium marchalianum	synonym of *Syzygangia marchaliana*	Dick, 1997*b*
Lagenidium microsporum	synonym of *Myzocytiopsis microspora*	Dick, 1997*b*
Lagenidium muenscheri	synonym of *Cornumyces muenscheri*	
Lagenidium myophilum	synonym of *Salilagenidium myophilum*	
Lagenidium netrii	synonym of *Myzocytium netrii*	
Lagenidium nodosum	synonym of *Syzygangia nodosa*	Dick, 1997*b*
Lagenidium obovatum	synonym of *Myzocytium rabenhorstii*	
Lagenidium oedogonii	synonym of *Syzygangia oedogonii*	Dick, 1997*b*
Lagenidium oophilum	synonym of *Myzocytiopsis oophila*	Dick, 1997*b*
	or *M. zoophthora sensu lato*	Glockling & Dick, 1997
Lagenidium oviparasiticum	synonym of *Chlamydomyzium oviparasiticum*	Dick, 1997*b*
Lagenidium papillosum	synonym of *Syzygangia papillosa*	Dick, 1997*b*
Lagenidium parthenosporum	synonym of *Myzocytiopsis parthenospora*	Dick, 1997*b*
Lagenidium podbielkowskii	synonym of *Pythium podbielkowskii*	
Lagenidium pygmaeum	synonym of *Cornumyces pygmaeum*	
Lagenidium pygmaeum var. *pygmaeoides*	excluded from *Lagenidium*; doubtfully assigned as *Pleocystidium pygmaeoides*	
Lagenidium pyriforme	synonym of *Myzocytium rabenhorstii*	
Lagenidium pythii	synonym of *Pythiella pythii*	
Lagenidium rabenhorstii	synonym of *Myzocytium rabenhorstii*	
Lagenidium rabenhorstii sensu Ivimey Cook	synonym of *Syzygangia zygnematicola*	Dick, 1997*b*
Lagenidium reductum	synonym of *Cystosiphon reductum*	
Lagenidium sacculoides	synonym of *Syzygangia sacculoides*	Dick, 1997*b*
Lagenidium septatum	synonym of *Chlamydomyzium septatum*	Dick, 1997*b*
Lagenidium syllae	synonym of *Salilagenidium scyllae*	
Lagenidium syncytiorum	unidentifiable; possible synonym of *Syzygangia marchaliana*?	
Lagenidium thermophilum	synonym of *Salilagenidium thermophilum*	
Lagenidium tortum	synonym of *Syzygangia papillosa*	Dick, 1997*b*
Lagenidium trichophryarum	unidentifiable; excluded from *Lagenidium*	
Lagenidium zopfii	synonym of *Syzygangia dictyospora*	Dick, 1997*b*
Myzocytium anomalum	synonym of *Chlamydomyzium anomalum*	Dick, 1997*b*
Myzocytium anomalum (Dasgupta & John - later homonym!)	synonym of *Myzocytium proliferum*	
Myzocytium entophytum	synonym of *Aphanomycopsis entophytum*	
Myzocytium fijiense	synonym of *Myzocytiopsis fijiensis*	Dick, 1997*b*
	or *M. zoophthora sensu lato*	Glockling & Dick, 1997
Myzocytium globosum	synonym of *Myzocytium proliferum*	
Myzocytium glutinosporum	synonym of *Myzocytiopsis glutinospora*	Dick, 1997*b*
Myzocytium humicola	synonym of *Myzocytiopsis humicola*	Dick, 1997*b*
Myzocytium indicum	synonym of *Myzocytiopsis indica*	Dick, 1997*b*
Myzocytium intermedium	synonym of *Myzocytiopsis intermedia*	Dick, 1997*b*
Myzocytium irregulare	synonym of *Myzocytium megastomum*	
Myzocytium lenticulare	synonym of *Myzocytiopsis lenticularis*	Dick, 1997*b*
Myzocytium lethale	unidentifiable *Myzocytiopsis*	
Myzocytium lineare	unidentifiable *Myzocytium*	
Myzocytium megastomum	*Myzocytium megastomum*	
Myzocytium microsporum	synonym of *Myzocytiopsis microspora*	Dick, 1997*b*
Myzocytium papillatum	synonym of *Myzocytiopsis papillata*	Dick, 1997*b*
Myzocytium polymorphum	synonym of *Myzocytiopsis vermicola*	Dick, 1997*b*
Myzocytium proliferum	*Myzocytium proliferum*	
Myzocytium proliferum var. *vermicola*	synonym of *Myzocytiopsis vermicola*	Dick, 1997*b*
Myzocytium proliferum f. *marinum*	synonym of *Sirolpidium bryopsidis*	
Myzocytium proliferum f. *maximum*	synonym of *Sirolpidium bryopsidis*	

(Table IV: 1, continued)

Myzocytium rotiferum	unidentifiable *Myzocytiopsis*	
Myzocytium subuliforme	rejected as *nomen confusum*	
[*Protascus subuliformis* var. *maupasii*	synonym of *Myzocytiopsis subuliformis*	Dick, 1997*b*
Myzocytium vermicola	synonym of *Myzocytiopsis vermicola*	Dick, 1997*b*
Myzocytium zoophthorum	synonym of *Myzocytiopsis zoophthora*	Dick, 1997*b*
Recently described species		
	Chlamydomyzium internum	Glockling & Dick, 1997
	Chlamydomyzium aplanosporum	Glockling & Dick, 1997
	Chlamydomyzium sphaericum	Glockling & Dick, 2000
	Myzocytiopsis osiris	Glockling & Dick, 2000
	Myzocytiopsis bolata	Glockling & Dick, 2000
New names		
	Myzocytiopsis subuliformis	Dick, 1997*b*
	Syzygangia zygnematicola	Dick, 1997*b*

Further attributions to *Lagenidium* and *Myzocytium*, grouped by host phylum/division

CHLOROPHYTA

Lagenidium sp. (*Euastrum*) Wildeman (1895*b*) unidentifiable
Lagenidium sp. (*Oocystis*) Masters (1971) unidentifiable
Lagenidium sp. (*Oedogonium*) Scherffel (1925*a*: 23) referable to *Syzygangia*
Lagenidium sp. (*Oedogonium*) Karling (1968*a*) referable to *Syzygangia*
Lagenidium sp. (*Chaetomorpha*) Deckenbach (1903) synonym of *Sirolpidium salinum*
Lagenidium sp. (*Aegagropila*) Ookubo (1952) synonym of *Sirolpidium salinum*

RHODOPHYTA

Lagenidium sp. (*Gracilaria*) Johnson (1957*a*) referable to *Pythium*

BACILLARIOPHYTA

Lagenidium sp. (*Pinnularia*) Scherffel (1925) referable to *Syzygangia* or *Aphanomycopsis*
Lagenidium sp. (*Synedra*) Couch (1935*b*) referable to *Ectrogella* or *Aphanomycopsis*
Myzocytium sp. (*Synedra*) Sparrow (1936*b*) referable to *Ectrogella* or *Aphanomycopsis*

DINOMASTIGOTA

Lagenidium sp. (*Ceratium*) Canter & Heaney (1984) possibly referable to *Aphanomycopsis*

ASCHELMINTHES

Lagenidium sp. (*Heterodera*) Kerry & Crump (1980) possibly referable to *Cornumyces* (see p. 189-190)

CRUSTACEA

Lagenidium sp. Cook (1971) (on *Penaeus*) referable to *Salilagenidium*
Lagenidium sp. Lightner & Fontaine (1973) (on *Pennaeus*) referable to *Salilagenidium*
Lagenidium sp. (marine) Lio-Po & Sanvictores (1986) referable to *Salilagenidium*

Lagenidium and *Myzocytium* have been separated by the degree of differentiation of the gametangia and host ranges. On these criteria *Lagenidium sensu stricto* is confined to one species, *L. giganteum*, a parasite of freshwater insect larvae. *Myzocytium sensu stricto* contains *M. proliferum* and *M. megastomum*, with the addition of the vesiculate zygnematicolous species *M. (Lagenidium) rabenhorstii* and *M. (Lagenidium) netrii*.

The four remaining species are parasites of the zygnematalean algae. Several other binomials represent fungi that are doubtfully distinct from one or other of these four species of *Myzocytium*. In addition, there are other species, previously referred to *Lagenidium*, and associated with freshwater algae, which are not deemed to be part of *Myzocytium sensu stricto* because they do not fulfil all of the *Pythium*-related criteria listed above.

Under the rules of the *International Code of Botanical Nomenclature*: Article 10.2, for genera named without an accompanying binomial (example 2), a type must be selected from ***any*** binomial subsequently validly published, *which characterizes the genus* (my italics). *Lagenidium* was not typified until 1959 (Cejp, 1959: 382); I interpret this typification as being "a largely mechanical method of selection" (Article 10.5 (*b*) a "choice that may be superceded"). The concept of the genus adopted by Cejp (1959) was not critical, and differs from that adopted here. The species he selected, *Lagenidium rabenhorstii* Zopf, *sensu lato* (*Lagenidium rabenhorstii sensu* Zopf with *Lagenidium rabenhorstii sensu* Ivimey Cook (≡ *Syzygangia zygnematicola* M. W. Dick) see Dick (1997*b*) and below) could even be regarded as a *nomen confusum*, though I accept the species in the sense of Zopf (1884). The genus name '*Lagenidium*' should be conserved with *Lagenidium giganteum* Couch as its type (holotype) and, at present, only species. A proposal to this effect is being sent to *Taxon*. If this proposal is accepted, the order *Lagenidiales* and family *Lagenidiaceae* remain subsumed in the *Pythiales* and *Pythiaceae* respectively.

Thallus form is too variable to provide a sole morphological systematic guide. The coevolutionary relationships are central to the following discussion: the restricted host ranges must be considered together with the zoosporogenesis and oosporogenesis. Both type species are involved.

Lagenidium species parasitic in insect hosts

Only one species, *Lagenidium giganteum*, falls into this category. There is no dispute that *L. culicidum* is synonymous with *L. giganteum*. This species was originally regarded as a saprotroph or weak parasite (Couch, 1935*b*) and its saprotrophic ability was confirmed by Willoughby (1969). Because it has been available in axenic culture it has received considerable attention. It is biochemically and genetically close to *Pythium sensu lato* that it must be placed in the *Pythiaceae*; its zoospore volume equivalent is *ca* 350 μm^3, which is also comparable with *Pythium*, it differs only in the morphology of the vegetative thallus.

Lagenidium and *Myzocytium* species parasitic in *Zygnematales*

The type species of *Myzocytium* and the species now excluded from *Lagenidium* are parasitoidal parasites (i.e., dependent on biochemical systems of recognition) primarily restricted to various genera within the *Zygnematales* (*Charophyceae*). Records of *L. rabenhorstii* on *Oedogoniales* (*Chlorophyceae*) (Graff, 1928; Karling, 1942*a*) and *Myzocytium proliferum* on *Cladophora* (Constantineanu, 1901) should be critically re-examined in the light of new living material in such hosts, since misidentification may have been involved (see below). The anchor point for this discussion is therefore the restricted host ranges. There appear to be four clusters of binomials associated with the *Zygnematales*:

A Parasitic in vegetative filamentous Zygnemataceae *and* Desmidaceae *and possessing smooth oospores*: *Lagenidium rabenhorstii* was the first of this group to be described, apparently with a wide host range within these algae. The original description of *L. rabenhorstii* indicated that the infection was confined to a single host cell. This was not true for the collection described by Ivimey Cook (1935), and in view of the other deviations from the original diagnosis with respect to the fertilization tube and the absence of a vesicle, it is necessary to consider this collection in relation to taxa placed in the new genus *Syzygangia*, discussed on pp. 185, 197. To obviate any confusion reating to the original description of *L. rabenhorstii*, Dick (1997b) erected the species *Syzygangia zygnematicola* based on the description of Ivimey Cook (1935). Zoosporogenesis was not observed for three species (*L. obovatum*, *L. pyriforme* and *L. coenocytium*) recently described by Dasgupta & John (1988). The oospores of the first two of these species are smooth-walled and of comparable size, the third species is anamorphic; the only differences are minor ones of thallus shape and these three taxa should be regarded as incompletely documented synonyms of *M. rabenhorstii*. Lateral flagellar insertion was reported for *L. netrii*, but this species is apparently anamorphic and the thallus is a single saccate cell, so it could be equally-well placed with either *Myzocytium* or *Syzygangia*. The presence of a well-defined vesicle suggests that the former genus is more appropriate at this stage. The *Lagenidium* species on *Euastrum* (Wildeman, 1895*b*) is unidentifiable since neither zoospores nor oospores were observed.

Lagenidium intermedium differs in the rarity of septa in the more or less filamentous thallus, the apparently intercalary oogonia and its desmid (*Closterium*) host. Without further substantiation of distinct differences in host range, this species, which is incompletely described (no measurements) and rarely reported, cannot be retained. While *L. intermedium* could be regarded as a synonym of *M. rabenhorstii*, it is placed here as a synonym of *Pythium podbielkowskii*, the fungus described by Batko (1973) as *Lagenidium podbielkowskii* on the desmid *Micrasterias*.

Lagenidium podbielkowskii was described from preserved material by Batko (1973) who considered it to be related to *L. closterii*, but it differs from that species in having a smooth oospore 15-21 μm in diameter (*L. closterii* has a smaller ornamented oospore 10-12 μm in diameter) and a tubular filamentous thallus 3·9-5·2 μm in diameter. The description of the fertilization process is confusing and not elucidated by the illustrations (Batko, 1973: fig. 3b), but the oogonia appear to be intercalary, as reported for *L. intermedium*. The 'vesicles' were described as persistent, and figured as spheres at the ends of long exit tubes, but this observation could be an artefact of fixation. Zoospores or their cysts were not described. The filamentous thallus with undifferentiated segments functioning as sporangia, and the smooth, relatively large oospore are characters which render the fungus inseparable from *Pythium*. Batko (1973) commmented on the pyriform swellings under the host cell wall which developed prior to the emergent sporangia, and it may be pertinent that Sparrow (1931*b*) also described pyriform internal appressoria in *Pythium adhaerens*, parasitic in green algae. It seems more probable that *L. podbielkowskii* is a depauperate species of *Pythium* and it is therefore transferred to that genus.

Five of the seven taxa (five binomials, one form and one unnamed taxon) of *Myzocytium* reported on algae (Karling, 1981*a*) are associated with the *Zygnematales*. *Myzocytium proliferum* is most commonly reported as a parasite of both *Zygnemataceae* and *Desmidaceae*, but there are a few early records attributed to this species on *Cladophora* (Walz, 1870; Zopf, 1884; Graff, 1928; Tokunaga, 1934) and one report on *Oedogonium* (Zopf, 1884). Material resembling either of the two species *Myzocytium rabenhorstii* and *M. proliferum* is distinct, but intermediate morphological forms have been described (e.g., *M. netrii*, *M. irregulare*, *M. lineare*, *M. anomalum* - the later homonym of Dasgupta & John, 1988; and possibly *L. gracile* - see below). The diversity of hosts for both species within the two host families (which, if *L. gracile* is included, also widens the host cell-types to include gametangia and zygospores), suggests that when cross-inoculation studies have been made the distinction may not be sustainable, and that a polymorphic biological species, *M. proliferum*, would then include *M. rabenhorstii sensu* Zopf as a synonym. For the present the two species are retained, but the erection of further species differing only in minor characters of the vegetative thallus should be resisted.

Myzocytium megastomum (and its synonym *M. irregulare*) appears to be confined to *Desmidaceae* and Cook (1963) has shown that biological races with different host specificity exist. (See also the description and host range of *M. irregulare* by Batko, 1973.) *Myzocytium megastomum* differs from *M. proliferum* in the expanded (appressorial?) part of the evacuation tube below the host cell wall. However, this morphological feature could be an attribute related to particular hosts. The thallus is more extensive than *M. proliferum*, and is more branched; the thickened septa do not appear to be so prominent, and cleavage has been described as external to the sporangium though not in a vesicle (Karling, 1981*a*: 92) but this may be incorrect (see Rieth, 1954, and H. M. Canter - p. 42); the zoospores contain refractive granules and possess laterally inserted flagella of "about equal length" (Sparrow, 1960). This is a well-illustrated and documented fungus and is retained as a separate species for the present, though its relationship with *M. proliferum* is very close if not conspecific.

B Parasitic in zygotes of filamentous Zygnemataceae:
The first such parasite to be described was *Lagenidium (Pythium) entophytum*, but there does not appear to be any convincing evidence that this species has a sporangial vesicle (Ivimey Cook, 1935; Karling, 1981*a*: 110). Zopf (1884) figures an apparently punctate-walled oospore reminiscent of *Aphanomycopsis*, which Ivimey Cook (1935) and Sparrow (1952*a*, *b*, *c*, 1960) interpret as comprising "low dentations": an SEM study is required to resolve this question. I concur with Karling (1981*a*) who placed the anamorphic taxon *L. americanum* in synonymy with *L. entophytum*. Karling (1981*a*) was also inclined to place *L. gracile* with *L. entophytum*, but this species has a smooth oospore wall, a feature discounted by Karling (1981*a*) but which I suggest may indicate that *L. gracile* has affinity with *M. rabenhorstii*. The oospore of *L. entophytum* is developed within an irregularly-shaped part of the thallus with no clearly defined septal delimitation, and does not appear to be accompanied by antheridia. Dasgupta & John (1988) have described four species, *L. clavatum*, *L. cylindriforme*, *L. contortum* and *L. entosphaericum* with a similarly much branched though mostly non-septate thallus. Zoosporogenesis was not observed in any of their species, but the presence of antheridia was claimed for *L. clavatum*, *L. cylindriforme* and *L. contortum*. The oospores of *L. cylindriforme* and *L. contortum* were described as having a double wall, and a larger oospore diameter was reported for the latter species. Each species was illustrated by a single small drawing. There seems to be no justification for regarding these species as separate entities, and so I place all of them in synonymy with *L. entophytum*, but I acknowledge that if it is ever established that *M. rabenhorstii sensu lato* commonly parasitizes zygotes as well as vegetative cells, then the species with antheridia may be better placed as synonyms of *M. rabenhorstii*.

On the evidence at present available (zoosporogenesis and the lack of discrete gametangial differentiation associated with automictic oospore development), *L. entophytum* has more affinity with the genus *Aphanomycopsis* than with *Myzocytium* and I here transfer it to *Aphanomycopsis* as *A. entophyta* (see p. 233).

C Parasitic in Desmidaceae *or* Zygnemataceae, *and possessing ornamented oospores:*
Lagenidium canterae has vesiculate zoosporogenesis, but it is not certain that the zoospore is of the peronosporomycete principal form (it is reported as almost spherical - compare the description by Ivimey Cook, 1935, for *L. rabenhorstii* (≡ *Syzygangia zygnematicola*), see pp. 20), though it is anisokont with a shorter anterior flagellum. The vegetative thallus is diminutive and non-septate in the desmid host. The oospore is distinctly ornamented and formed by sexual fusion involving a fertilization tube, but it is not stated whether fusion is the result of thallus septation followed by fusion or a double infection. This parasite

may thus deviate from *Myzocytium* on two criteria. *Lagenidium closterii* is less completely described, but is essentially similar to *L. canterae* except that the thallus is more extensive in the larger host, and becomes septate at maturity. The nature of the fertilization tube is not clear for either species, particularly with regard to oogonial penetration. In the absence of any host specificity studies, these taxa could be regarded as conspecific. However, the differences in thallus morphology and host enable the two species to be identified and they are therefore kept as separate entities here. Both species could be related to *Lagenidiopsis* and *Cystosiphon*, but the status of these genera requires further discussion (see p. 185). Alternatively, because of their smaller zoospores, they could be related to *Syzygangia papillosa* (see below and p. 15). Although Canter & Lund (1969) noted that the nature of the vesicle of *L. canterae* differed from that of *M. megastomum*, the vesiculate zoosporogenesis and presence of fertilization tubes suggests a provisional placement in *Cystosiphon*.

D Parasitic in Desmidaceae *or* Zygnemataceae *but with features that cannot be assimilated within the* Pythiaceae *as that family is currently defined:*

Lagenidium sacculoides in *Closterium* has zoospore cleavage initiated within the zoosporangium prior to completion within an extramatrical vesicle. However, the zoospore cysts are exceptionally small (3·5 μm diam.; volume <25 μm^3); the oospore develops ***between*** the gametangia, and is ornamented. Although there are no illustrations of this species, it has no affinity with *Myzocytium* or *Lagenidium sensu stricto*, but does have some resemblances to certain species which have previously been assigned to *Myzocytium* and are here placed in *Myzocytiopsis* and *Syzygangia* (see p. 196).

Lagenidium papillosum in vegetative *Spirogyra* is described as having zoospores with apical flagellar insertion and ornamented oospores, again making affiliation with the *Pythiaceae* very doubtful, although its affiliation with *L. rabenhorstii sensu* Ivimey Cook (*Syzygangia zygnematicola*) could be quite close, since Ivimey Cook (1935) described the zoospores as initially reniform but becoming more or less spherical/pyriform (i.e., polymorphic). *Lagenidium tortum* is inadequately described: this taxon cannot be distinguished from *L. papillosum* and it is probable that *L. astrum* is also conspecific in spite of the exceptionally large oospore measurements reported (Dasgupta & John, 1988). The generic placement of these taxa is presented in Table IV: 1.

Lagenidium lundiae is also exceptional in possessing ovoid (almost pyriform) cysts and an apparently plerotic resting spore. The zoospores are described as grape-seed-like and while unpublished notes (H. M. Canter, pers. comm.) confirm that a walled vesicle was present, supporting photographic data are not available. Karling (1981*a*) infers that Canter & Lund (1969) suggest an affinity with *Myzocytium irregulare*, but this is not justified since Petersen's species was anamorphic and zoospores were not observed. The resting spore was compared with that in a photograph in Pongratz (1966), and an affiliation with those *Olpidiopsis* species here transferred to *Pleocystidium* (species in algae) is preferable (see pp. 45, 51, 215).

Lagenidium species parasitic in freshwater *Chlorophyceae*, *Charales* and higher plants: the status of the genera *Aphanistis*, *Lagenidiopsis* and *Cystosiphon*

There are seven species of *Lagenidium* which come into this host range category: *L. ellipticum*, *L. marchalianum*, *L. muenscheri*, *L. oedogonii*, *L. reductum*, *L. syncytiorum* and *L. zopfii*. All possess features which indicate only the most tenuous relationship with *Pythium*. Where zoosporogenesis has been observed (only in *L. marchalianum*, *L. muenscheri* and *L. oedogonii*), it is described as occurring ***within*** the zoosporangium: zoospore measurements are available for the last two species and indicate a volume equivalent in the region of 60-90 μm^3. The oospores for these three species are smooth-walled. The remaining species have ornamented oospores, except for *Lagenidium syncytiorum* which is anamorphic.

Lagenidium reductum is found in *Chara*, *L. muenscheri* in *Potamogeton* and *L. ellipticum* in moss rhizoids (see also *Olpidiopsis ricciae* du Plessis (Table IV: 7) and *Pleotrachelus wildemanii* Petersen (pp. 253-254). The remaining four species are associated with *Oedogonium*.

References to the occurrence of *L. rabenhorstii* (Graff, 1928; Karling, 1942*a*; ≡ *Syzygangia zygnematicola*) and *M. proliferum* (Constantineanu, 1901) in *Oedogonium* may well be referable to one or other of these taxa, since the whole range of thallus form is encompassed here: *Lagenidium oedogonii* has an ovoid thallus confined to one host cell and at the other extreme *L. marchalianum* has a more extensive and more filamentous thallus passing from cell to cell of the host. The thalli of *L. syncytiorum* and *L. zopfii* also extend through several host cells, but *L. syncytiorum*, being anamorphic, makes any identification or synonymy doubtful, while *L. zopfii* differs from *L. marchalianum* in the irregular diameter of its septate thallus, in possessing ornamented oospores, and in the absence of antheridia. Zoosporogenesis is not known for *L. zopfii*. Although Karling (1981*a*) suggests synonymy between *L. zopfii* and *L. entophytum*, *L. zopfii* is here placed as a synonym of *Syzygangia dictyospora* (see below). Karling (1968*a*) described another asexual *Lagenidium* species (unnamed) on *Oedogonium* sporelings, which he considered distinct because of its barrel-like thallus segments.

Oedogonium is host to a large number of aquatic fungi (Sparrow, 1960, lists 52 binomials): it would be desirable to carry out a comparative survey to establish the individuality of all these species, and to determine those characters of the host that have enabled such diversity to evolve. Reference should also be made to the unnamed *Lagenidium* species mentioned by Scherffel (1925*a*). Table IV: 2 provides a list of these fungi assigned to the uniflagellate and biflagellate groups.

I include here a brief discussion of the reportedly uniflagellate *Aphanistis oedogoniorum* Sorokin* and *A. pellucida* Sorokin (Sorokine, 1883, 1889, 1890). The genus *Aphanistis* was established for parasites of *Oedogonium*. The two species are known only from the original descriptions. The thallus of *Aphanistis* consists of a septate hypha which passes through the cells of the algal filament, forming sporangia only in the oogonial (oospore) cells. The sporangia produced posteriorly uniflagellate zoospores. Two interpretations of the descriptions (Sorokine, 1883, 1889) and drawings (Sorokine, 1883 only) are possible. Either the sporangia and hyphae were separate unrelated entities corresponding to those of *Rhizophydium decipiens* and an unidentifiable hyphomycete; or, the units were part of a single organism, in which case homology with *S. marchaliana* would be possible, notwithstanding the description of zoospore flagellation (the zoospores were not observed for the second species, *A. pellucida*). The absence of resting spores in *Aphanistis* precludes any further attempt at synonymy. Neither species is sufficiently clearly described to be accepted in either the *Chytridiomycetes* or the *Straminipila*.

The observation (Wildeman, 1897*a*) that *S. marchaliana* has zoosporangia with four zoospores echoes that of Raciborski (1891) for *Pythium dictyosporum* Racib. There may be a close relationship between *S. marchaliana* and the record of *P. dictyosporum* by Rieth (1954), but there are problems. *Pythium dictyosporum* was described from *Spirogyra*, and Sparrow (1931*a*) rediscovered the fungus in this host in the U.S.A., but was not able to confirm fully Raciborski's observation on zoospore number. Rieth (1954) found a fungus which he called *P. dictyosporum* in *Oedogonium* but this taxon is not conspecific with Sparrow's fungus, although it has a much closer relationship with *S. marchaliana*. Rieth's fungus differs from *S. marchaliana* in lacking the differentiated gametangia of the latter species, in being heterothallic and

---

* FOOTNOTE: Н. Сорокин used the French form of his name in the publications of 1883 and 1889, and this form is used in the references. In the formal taxonomy the transliterated name (N. Sorokīn) is used, following the *British Standard, No. 2979 (1958)*.

TABLE IV: 2. *Parasites of* Oedogonium

A. Biflagellate and uniflagellate species (under old names) considered in this text:

Lagenidium marchalianum
Lagenidium oedogonii
Lagenidium rabenhorstii
Lagenidium syncytiorum
Lagenidium zopfii
Lagenidium sp. (Karling, 1968*a*)
Myzocytium proliferum
Olpidiopsis decipiens
Olpidiopsis oedogoniorum
Pleotrachelus petersenii
Resticularia oedogonii
Aphanistis oedogoniorum
Aphanistis pellucida
Achlyogeton solatium
Plasmophagus oedogoniorum

B. Uniflagellate species, which are not considered in this text (page numbers in brackets refer to Sparrow, 1960):

Chytridium acuminatum A. Braun (495)
Chytridium aggregatum Karling (518)
Chytridium brevipes A. Braun (521)
Chytridium chaetophilum Scherff. (508)
Chytridium inflatum Sparrow (520)
Chytridium lagenariae Schenk (522)
Chytridium minus C. M. V. D. S. Lacoste & Suringar (533)
Chytridium oblongum A. Braun (537)
Chytridium oedogonii Couch (516)
Chytridium olla A. Braun (493)
Chytridium schenkii (Schenk) Scherff. (514)
Rhizophydium ampullaceum (A. Braun) A. Fisch. (262)
Rhizophydium canterae Sparrow (242)
Rhizophydium chaetiferum (Schaarschm.) A. Fisch. (264)
Rhizophydium constantineanui D. Sacc. (254)
Rhizophydium decipiens (A. Braun) A. Fisch. (301)
Rhizophydium globosum (A. Braun) Rabenh. (237)
Rhizophydium mamillatum (A. Braun) A. Fisch. (260)
Rhizophydium oedogonii P. Richter (305)
Rhizophydium ovatum Couch (291)
Rhizophydium parasiticum S.-C. Shen & W. N. Siang (290)
Rhizophydium setigerum Scherff. (317)
Rhizophydium sphaerocarpum (Zopf) A. Fisch. (273)
Rhizophydium sporoctonum (A. Braun) Berl. & De Toni (305)
Rhizophydium vonmindenii Valkanov (304) [as '*vonmindeni*']
Phlyctochytrium biporosum Couch (334)
Phlyctochytrium chaetiferum Karling (339)
Phlyctochytrium lagenariae (Schenk) Domján (332)
Phlyctochytrium planicorne G. F. Atk. (341)
Phlyctochytrium quadricorne (de Bary) J. Schröt. (343)
Phlyctochytrium unispinum R. A. Paterson (349)
Phlyctidium brevipes (G. F. Atk.) Minden var. *brevipes* (221)
Entophlyctis bulligera (Zopf) A. Fisch. (373)
Entophlyctis confervae-glomeratae (Cienk.) Sparrow (377)
Chladochytrium replicatum Karling (464)
Micromyces oedogonii (Scherff.) Sparrow (203)
Olpidium entophytum (A. Braun) Rabenh. (135)
Olpidium indicum Wall. [*in* W. B. Turner] (156)
Olpidium pusillum (Sorokīn) De Wild. (152)

in having ornamented oospores: Rieth did not observe zoosporogenesis. In neither taxon is there evidence of a fertilization tube. On the other hand, Sparrow's fungus is *Pythium*-like in its zoosporogenesis and the presence of a fertilization tube, but exceptional in the very narrow hyphal diameter (1·8-2·0 μm in the host) and the oospore ornamentation, which Sparrow (1931*a*) described as periplasmic. Unpublished ultrastructural (TEM) data on a South African angiosperm-pathogenic isolate attributable to *P. dictyosporum* (W. Botha, pers. comm.) indicates that the convolutions are part of the oospore wall rather than an exosporial (periplasmic) deposit. These data are also relevant to the discussion on p. 64 and Table III: 5, p. 152 regarding the resurrection of the genus *Cystosiphon*. On the other hand, if more than one taxon has been described under the binomial *P. dictyosporum*, then the observations reported concerning the oospore wall could be interpreted either as a convoluted middle wall layer underlying a thin oosphere membrane such as is found in some *Myzocytiopsis* species, or a gametic membrane such as is found in *Apodachlyella*.

Wildeman (1896*c*) erected the genus *Lagenidiopsis* for the taxon found in *Chara* because of the filamentous nature of the thallus. Karling (1942*a*: 81) placed this genus as a synonym of *Lagenidium* on the grounds that the range of thallus form acceptable in *Lagenidium sensu lato* had been extended since *Lagenidiopsis reducta* was described. Although this species is inadequately described (sporangia and zoosporogenesis unknown) it has an ornamented oospore and a well-developed and conspicuous fertilization tube. The genus *Lagenidiopsis* could equally well be regarded as a synonym of *Cystosiphon* (Roze & Cornu, 1869), in which case the genus *Cystosiphon* would be a suitable repository for *C. pythioides* (the type species, recently rediscovered from damp field soil in Denmark by K. Møller, pers. comm., which Dick (1990*c*) has removed from *Pythium* as a form genus), *L. reducta* (≡ *Lagenidium reductum*) and Sparrow's isolate of *P. dictyosporum*. Two of the species of *Lagenidium* discussed above (*L. canterae* and *L. closterii*) can also be placed in *Cystosiphon*.

The genus *Syzygangia*, typified by *S. marchaliana*, has been erected (Dick, 1997*b*) for the remaining species of *Lagenidium* in green algae and higher plants. These species, which possess poroid copulation resulting in smooth-walled or ornamented oospores and intrasporangial zoosporogenesis, are the phytophagic analogues of *Myzocytiopsis* (see below). They are: *L. papillosum* and *L. sacculoides* in *Zygnematales*; *L. oedogonii*, *L. marchalianum* and Reith's *P. dictyosporum* in *Oedogonium*); *L. ellipticum* in higher plants.

Lagenidium muenscheri differs from the above species in its *Petersenia*-like thallus and larger zoospores. Although it could be placed in *Syzygangia* (and is included in the key to that genus), the combination of characters of this fungus suggests a closer affinity with the saprophyte *Lagenidium pygmaeum* and the parasites of *Saprolegniaceae*, *Lagenidium destruens* and *Petersenia utriculoba* (see pp. 189-190), here included under the genus *Cornumyces*.

Under the formal taxonomy of *Syzygangia*, in the doubtful category, I have included *Achlyopsis entospora* De Wild. because one of the illustrations (Wildeman, 1896*c*: plate IX, fig 3) suggests that the oospore may have a convoluted intermediate membrane. The asexual reproductive system was not found and the species has not been recorded since. Another, perhaps more probable possibility is that this 'species' is the sexual stage of *Plasmoparopsis rigida* De Wild.; if this is the case, then his three new species (all in Wildeman, 1896*c*, and all parasitic in *Chara*) namely *Pythium characearum*, *Achlyopsis entospora* and *Plasmoparopsis rigida* may not only be conspecific, but also congeneric with *Pythiogeton* because of the extremely thick oospore wall (Wildeman, 1896*c*: plates VIII, IX) and relatively narrow hyphae (Wildeman, 1896*c*: plate VIII - XI; see also Voglmayr, Bonner & Dick, 1999).

Lagenidium species parasitic in freshwater cyanobacteria: the status of the genus *Resticularia*

Only one species, *Lagenidium nodosum*, is recorded from cyanobacteria and, apart from the doubts regarding its heterokont status (see p. 32), according to Ingold (1949) it has poroid gametangial fusion, coupled with vesiculate zoosporogenesis. From illustrations in Ingold (1949) the zoospore dimensions are moderately small (4·0 × 7·0 μm; volume equivalent of < 100 μm³); the oospores are smooth-walled, and are figured with two "vesicles", which could be compared with the ooplast and nuclear spot of *Pythium*. *Lagenidium nodosum* resembles *Syzygangia marchaliana* in its ability to pass through several host cells. In spite of the fact that it does not conform with the majority of species placed in *Syzygangia* with respect to zoosporogenesis, in other features Ingold's fungus appears to have affinity with that genus, and it is provisionally transferred to *Syzygangia*.

The genus *Resticularia* Dangeard (1890*b*) was erected for *R. nodosa*, the zoospores of which were described as uniflagellate and the thallus as non-septate. Fritsch (1903) provided a further description which is not compatible with that of Dangeard or of placement in any group of flagellate fungi (Karling, 1942*a*; Ingold, 1949); he also added a second species, *R. boodlei*, which is probably a hyphomycete. The third species of *Resticularia*, *R. oedogonii* (Skvortzow, 1925) is, like most of Skvortzow's taxa, inadequately described, but is tentatively placed in synonymy with *Syzygangia marchaliana*. Because of the confusion with respect to the original diagnosis of *Resticularia*; the doubts regarding the flagellar number and ornamentation, and the extrasporangial cleavage of the zoospores of the type species, neither the genus nor the type species is suitable for retention in preference to *Syzygangia*. A proposal is therefore made here, and to *Taxon*, for the genus *Resticularia* to be *nomen rejiciendum*.

Sparrow (1960: 402, 479) drew attention to the possible similarities between *Coenomyces* (≡ *Deckenbachia*, Yačevskie & Yačevskie, 1931), also parasitic in cyanobacteria (marine and freshwater), and *Aphanistis*. In the present context of doubt concerning the weight that should be attached to flagellar number in these very poorly researched fungi, comparisons with *Syzygangia* would be appropriate.

Lagenidium species parasitic in freshwater diatoms

There are three parasites of diatoms which have been placed in *Lagenidium*. *Lagenidium enecans* was briefly described but not figured by Zopf (1884) and *L. brachystomum* and *L. cyclotellae* were later added by Scherffel (1925*a*). The species are little-known, and both Sparrow (1960) and Karling (1981*a*) relied on Scherffel's descriptions for all three species. From these descriptions it appears that the sporangial contents of *L. enecans* and *L. brachystomum* are ***naked*** protoplasts. From the drawings given by Scherffel (1925*a*: plate 2, figs 66a and 74c) for *L. enecans* and *L. brachystomum*, this protoplast is not spherical as in *Pythium*, but angular. It thus seems likely that the extrasporangial cleavage is different from that of *Pythium*. The descriptions of zoospore morphology are confusing: Sparrow (1960) describes both species as having grape-seed-like zoospores, but Karling (1981*a*) described the former as having egg- or kidney-shaped zoospores and the latter (following Couch, 1935*b*) as kidney-shaped with a ventral groove. Furthermore, Karling (1981*a*: plate 31, fig. 34, and plate 33, fig. 91) figured zoospores from this extrasporangial protoplast as ovoid and called them primary zoospores. In fact they are more or less identical to those of nematophagous *Myzocytium* species which complete their development at the mouth of the exit tube, and those of *Syzygangia zygnematicola* (Ivimey Cook, 1935). The description of sexual reproduction is also unclear, there is no evidence of a fertilization tube and while Sparrow (1960) describes the species as dioecious, the evidence is weak. Sparrow (1960) distinguished the two species only on the length of the sporangial exit tube and the possibly plerotic nature of the oospore in *L. enecans*. If Karling's drawings (Karling, 1981*a*: plate 32, fig. 84; plate 33, fig. 93) are compared, there is in reality very little difference between the supposedly plerotic *L. cyclotellae* and the aplerotic *L. brachystomum*. However, zoospore volume equivalents are different, with that for *L. enecans* being exceptionally large (volume >350 μm^3) compared with those for *L. brachystomum* and *L. cyclotellae* (volume <60 μm^3). Scherffel (1925*a*) distinguished *L. cyclotellae* by thallus shape, but there is probably a continuum of intermediate forms between the species. Scherffel (1925*a*) regarded *L. cyclotellae* as intermediate between *Olpidiopsis* and *Ectrogella*.

Scherffel (1925*a*) based the separation of *L. brachystomum*, *L. cyclotellae*, *Aphanomycopsis bacillacearum* and *Ectrogella* on the kind of sporangial discharge; regarding *Ectrogella* as the type of the family *Ectrogellaceae* containing *Ectrogella* (with *Saprolegnia*-like discharge of very small auxiliary-form zoospores) and *Aphanomycopsis* (with *Aphanomyces*-like discharge). The contrast was made between the saprolegniaceous modes and that of *Lagenidium* species having *Pythium*-like discharge. As has been discussed above, this distinction does not hold for the earlier concept of *Lagenidium sensu lato*. The resemblances are probably all superficial, and when other aspects of the asexual and sexual reproduction are considered, a rearrangement of the generic and family attributions resolves some of the inconsistencies. *Ectrogella* and some of the *Olpidiopsis* species in algae may have affinity with *Syzygangia* (and *Myzocytiopsis*), in which diversity of sporangial discharge is a feature (but see pp. 213-214 - family concept for *Ectrogellaceae*). *Lagenidium brachystomum* and *L. cyclotellae*, the two species with small zoospores and a non-septate thallus, can be transferred to *Ectrogella*. *Lagenidium enecans*, with its much larger zoospores, is more appropriately placed in *Aphanomycopsis* as a synonym of *A. bacillariacearum*.

The three unnamed taxa on diatoms (*Myzocytium* sp. on *Synedra*, Sparrow, 1936*b*; *Lagenidium* sp. on *Synedra*, Couch, 1935*b*; and *Lagenidium* sp. on *Pinnularia*, Scherffel, 1925*a* - the last with oogonia but zoospores not observed) are too little known to be placed with any confidence, but should be considered in relation to *Ectrogella*, *Syzygangia* and *Aphanomycopsis*.

Saprotrophic *Lagenidium* species

The first saprotrophic taxon to be described was *Lagenidium pygmaeum* (Zopf, 1887). Karling (1947*b*) described *L. humanum* and later he described the taxon *L. pygmaeum* var. *pygmaeoides* (Karling, 1981*b*).

The description of *L. humanum* is the most comprehensive of this group: the fungus was isolated from a keratin bait substrate (human skin) in which it formed an extensive thallus. The thallus eventually becomes septate, and the segments disarticulate at maturity as in the lagenidiaceous parasites of nematodes. Karling (1947*b*) stated that the fungus had been obtained in axenic broth culture, but it has not been used for any biochemical study. Zoosporogenesis is variable, usually with partial delimitation occurring within the thallus segment behaving as a sporangium, but zoospores may sometimes be fully formed within the sporangium or differentiated in an external vesicle which is apparently formed ***after*** emergence of the protoplast. The zoospores have lateral flagellar insertion and have a diameter of *ca* 5·5 μm (volume equivalent <100 μm^3). Other segments may function as gametangia, with poroid fusion resulting in a single (rarely two) smooth-walled oospore.

Willoughby (1969) suggested that *Lagenidium humanum* could be synonymous with *L. giganteum*, but the subsequent discovery in the latter of sexual reproduction, and its production in axenic culture (Kerwin & Washino, 1983), demonstrates the great difference between the two species. *Lagenidium humanum* must be excluded from *Lagenidium sensu stricto* on the basis of both zoosporogenesis and oosporogenesis, but has many similarities, including affinity for animal baits (its natural substratum is not known), with *Myzocytiopsis*, with which it was placed by Dick (1997*b*).

The taxonomic position of the remaining saprotrophic taxa is very dubious, and comparisons with other taxa should be made, such as *Aphanomycopsis saprophyticus*, *Ducellieria chodatii*, *Hyphochytrium (Anisolpidium) saprobium* and *Myzocytiopsis microspora sensu* Karling (1968*a*). Morphological details are given in Table IV: 3.

The thallus form of *Lagenidium pygmaeum*, which is known as a saprotroph of pollen grains, is variously contorted, either forming a single sporangium, or becoming septate to form two gametangia. Konno (1972) figured this fungus with an irregular tubular thallus. Zoosporogenesis is initiated intrasporangially. Gametangial fusion is reported as via a fertilization tube between the two adjacent segments of the thallus. A relationship with *Ducellieria* is possible, but the fungus is here placed in *Cornumyces* (see pp. 185, 189-190 and PART V).

Lagenidium pygmaeum var. *pygmaeoides* (Karling, 1981*b*) is not conspecific with *L. pygmaeum* although it has been isolated from a similar substrate. The description of zoospore behaviour is somewhat confused, in that two patterns of zoospore release are described, both involving intrasporangial zoosporogenesis. The description of the resumption of motility from cystospores implies that the cystospore wall is either not formed or is resorbed (i.e., polymorphic?). There is no evidence of dimorphism, and the intrasporangial zoospores have lateral flagellar insertion. Oosporogenesis appears to be plerotic (compare the drawings of apparently parthenogenetic oospores with that of the rare sexually developed oospores - Karling, 1981*d*: fig. 49). The fungus must be excluded from *Lagenidium sensu stricto* but cannot be assigned with confidence to another genus. Its affinities may be with *Pleocystidium*, and this combination is made under the 'doubtful' category.

Lagenidium species parasitic in *Peronosporomycetes*

Two species with straminipilous fungal hosts have been placed in *Lagenidium*, neither of which can be assigned to *Myzocytium* or *Lagenidium* as defined here. The first of these species to be described was *L. pythii* (Whiffen, 1946). *Lagenidium pythii* has a unique combination of characters. The thallus is holocarpic and non-segmented, and is apparently walled from inception. There is no sporangial vesicle, zoosporogenesis is intrasporangial and the zoospores are small (mean diameter *ca* 3·0 μm; volume <25 μm^3) with flagella of equal length. The sexual process originates from a fusion between two separate thalli (cf. *Lagena*, *Ciliatomyces* or *Olpidiopsis*) but the resting spore that develops is aplerotic and the oogonium does not induce external fibrillar or spiny ornamentation. Fertilization is stated to be poroid. The internal structure of the oospore was not described. The fungus cannot remain in *Lagenidium sensu stricto* nor can it be placed in *Olpidiopsis sensu stricto*. The involvement of two separate thalli in sexual reproduction is not characteristic of *Syzygangia* with smooth-walled oospores nor *Myzocytiopsis*. The isokont flagella and aplerotic resting spore also rule out placement with *Gracea*. However, *L. pythii* shares with *Pythiella vernalis* (Couch, 1935*a*) not only a filamentous *Pythium* host, but also the involvement of a companion cell, an aplerotic oospore, and a small zoospore volume. It differs from this species in lacking a diplanetic phase and a fertilization tube. Whiffen (1946) did not comment on the possibility of a periplasmic oosporogenesis as described by Couch (1935*a*). Despite these differences, I consider the two species to be congeneric and transfer *L. pythii* to *Pythiella*.

The only other species of *Pythiella* is *P. besseyi*, which was originally described as an *Ectrogella* because of the several sporangial discharge tubes (Sparrow & Ellison, 1949) but later transferred to *Pythiella* (Sparrow, 1960: 814) because of cytoplasmic appearance. This transfer is not accepted here because *P. besseyi* has no known sexual stage; the thallus tends to be tubular even in the ellipsoidal thallus of the host

TABLE IV: 3. *Lagenidiaceous saprotrophs of pollen grains from freshwater or soil habitats. (Compare zoospore volumes with those of* Olpidium *spp. from similar substrates:* O. luxurians *Zopf (<5μm³),* O. appendiculatum *Karling (15-20 μm³),* O. pendulum *(Tomaschek) A. Fisch. (40 μm³) and* O. longicollum *Uebelm. (>500 μm³), all of which also have resting spores.)*

Species	Zoospores	Thalli	Oospores
Ducellieria chodatii	8·0 × 11·0-14·0 μm (>500 μm³) cysts with long hollow spines	bottle-shaped	plerotic (20-30 μm) heterothallic
Aphanomycopsis punctata	6·0-7·0 × 8·0-9·0 μm (>250 μm³)	coralloid	18-32(41) μm diam.
Aphanomycopsis saprophytica	4·0-5·0 × 6·0-7·0 μm (100 μm³) with gliding movements	coralloid	9-12(22) μm diam.
Cornumyces pygmaeus	5·0 × 8·0 μm (150 μm³)	lobed/allantoid	aplerotic (18-29 μm diam.)
Rozellopsis sp. (Kobayasi *et al.* 1971)	4·0-6·0 × 7·0-8·0 μm (>100 μm³)	no data	no data
Hyphochytrium saprobium	3·5-4·0 × 5·0-6·0 μm (60 μm³)	spherical (8-32 μm diam.)	sexuality absent
Hyphochytrium sp. (Persiel, 1960*b*)	3·0-4·0 × 5·0-8·0 μm (60 μm³)	spherical (8-32 μm diam.)	sexuality absent
Hyphochytrium oceanum	3·3-3·8 × 5·0-5·5 μm (45 μm³)	spherical (30-65μm diam.)	sexuality absent
Pleocystidium pygmaeoides	2·5-3·0 × 4·0-5·0 μm (<25 μm³)	pyriform/deformed (10-24 μm diam.; 1-8 per cell)	plerotic, automictic
Myzocytiopsis microspora	2·0-2·3 × 2·7-3·3 μm (<15 μm³)	(6-8 × 9-11(12) μm) up to 4 segments	oospores 7·8-10·5 μm

(*Pleocystidium parasiticum*); and the conceptual difficulties involved in host location for a holocarpic endoparasite of another holocarpic endoparasite are so enormous that I suspect that autoparasitism may be involved (see also *Olpidiopsis myzocytia*; *Rozella pseudomorpha*). The taxon is placed as a doubtful member of the genus *Pleocystidium*.

Lagenidium destruens (Sparrow, 1950) is not related to *Pythiella pythii*. It has large zoospores (8·0 × 12·0 μm; volume equivalent *ca* 400 μm³) ejected as a mass, with development completed extrasporangially. Oosporogenesis is aplerotic, probably automictic (no antheridia known) and polyoosporous. The fungus could be placed in the *Leptolegniellaceae*, but preferably needs to be considered with several undescribed taxa - the polyoosporous, eccentric-oospored fungus in *Aphanomyces* (Dick, 1970: 453); the polyoosporous parasite of nematodes referred to *Lagenidium* (Kerry & Crump, 1980); the unioosporous, eccentric-oospored parasite of rotifer eggs (Seymour & Johnson, 1973); and the freshwater, non-sexual parasites with lobed,

Petersenia-like thalli, including *Lagenidium muenscheri* (Constantineanu, 1901; Cutter, 1943; Miller, 1962; Sunderam, 1968). It must be excluded from *Myzocytium* or *Lagenidium*. A new genus, *Cornumyces*, *loc. cit.*, is erected for these polyoosporous endoparasites, with a generic diagnosis based on the descriptions for *Lagenidium destruens* and *Petersenia utriculoba*.

The unnamed *Lagenidium* species of Kerry & Crump (1980) has 1-4 eggs per oogonium, but there is insufficient additional information to assign the fungus to either *Myzocytiopsis* or *Cornumyces*, except as a doubtful taxon.

Lagenidium species parasitic in marine *Crustacea*

Two teleomorphic marine fungi associated with *Crustacea* have been placed in *Lagenidium* (*L. callinectes* Couch, *L. chthamalophilum* Johnson). They must be removed from *Lagenidium*, the *Pythiaceae* and the *Pythiales* on the following grounds:

1 mature oospore structure, apparently lacking a central translucent ooplast, and possessing reserve lipids in various aggregations

2 intra-sporangial zoosporogenesis involving a large central vacuole

3 sterol biochemistry

and possibly

4 kinetosome organization and transition zone (with a short series of rings in *Haliphthoros*

However, they share with the *Pythiales* the negative criterion that K-bodies are not known.

The three anamorph species (*L. myophilum* Hatai & Lawhav., *L. syllae* Bian *et al.*, and *L. thermophilum* K. Nakamura *et al.* - Bian *et al.*, 1979; Hatai & Lawhavinit, 1988; Nakamura *et al.*, 1995) and the *Lagenidium* species (unnamed) of Cook (1971), Lightner & Fontaine (1973) and Lio-Po & Sanvictores (1986) are almost certainly congeneric with the teleomorphic species and should also be removed from *Lagenidium*. *Sirolpidium zoophthorum* is also probably related to this group of species, and so the appropriate transfer has been made.

Lagenidium callinectes is a heterokont fungus, but its oospore possesses an incomplete cap of oil droplets reminiscent of the subeccentric configurations of the *Saprolegniaceae* (Dick, 1960). Its sporangial 'vesicle' membrane may be apparent as an artefact. Its zoospores are larger than any other fungus which has been placed in *Lagenidium* except for *L. caudatum* and SEM has revealed lateral flagellar insertion in a groove (Fuller & Jaworski, 1987). The clear figures in Nakamura *et al.* (1995) confirm the relationship between *L. thermophilum* and *L. callinectes*. Crisp, Bland & Bahnweg (1989) discuss the morphological and physiological variation that occurs within *L. callinectes*, and conclude that four subgroups exist. They suggest that *L. syllae* may be conspecific with *L. callinectes*, but do not reduce this species to synonymy. The 'vesicle' of *L. chthamalophilum* is also unlike that of either *Pythium* or *Lagenidium sensu stricto* but may be of similar origin to that of *L. callinectes*, though not persistent after discharge. The oospore reserves are said to be a central or eccentric collection of oil droplets, but may be a small cap of subeccentric droplets since the fungus was described by Johnson (1958) about the same time that he erected the subgenera of *Achlya* based upon fallaceous interpretations of oil droplet distribution (Dick, 1969).

On the basis of oospore structure, kinetosome/transitional zone ultrastructure and sterol metabolism the above teleomorph species have characters which appear to place them closer to the *Saprolegniomycetidae* than to the *Peronosporomycetidae*. However, the marine habit and the absence of K-bodies indicate that the relationship with the freshwater, mostly saprophytic *Saprolegniales* may be less close than is presently presumed from the limited ultrastructural information. (*Aphanomyces astaci* is the only notable freshwater crustacean parasite in a saprolegniomycetideous genus.) Molecular data (unpublished, see Lee & Fuller, 1992) indicated a shorter genetic distance between *L. callinectes* and *L. giganteum* than between *L. callinectes* and other *Saprolegniomycetidae*. However, as discussed above and in Dick (1997*a*; Dick *et al.*, 1999), such a relationship does not necessarily indicate that these marine parasites should be placed in the *Peronosporomycetidae* rather than the *Saprolegniomycetidae*. A new, teleomorph-based genus, family and order are therefore erected for *L. callinectes* and *L. chthamalophilum*, using the genus name *Salilagenidium*, typified by *S. callinectes* because this species has been most thoroughly investigated and characterized. The host ranges of the anamorph species *L. syllae*, *L. myophilum* and *L. thermophilum* need to be investigated before conspecificity can be established.

The relationships between these fungi and *Haliphthoros* are probably close (Crisp *et al.*, 1989), but the latter genus has been placed in its own family, defined by its anamorphic life-history, which may usefully be retained for anamorphic marine saprotrophs.

The fungus described by Atkins (1954*a*) as *Leptolegnia marina* D. Atkins was transferred, with some reservations, to *Leptolegniella* by Dick (1971*b*) on the basis of presumed oospore wall morphology. The removal of the marine species of *Lagenidium* to a new family opens up the possibility that *Leptolegniella marina*, which is also a crustacean parasite, is more appropriately placed in the new genus *Salilagenidium* for which diversity of oospore structure is permissible.

Other fungal genera to be compared with *Salilagenidium*: *Haliphthoros*, *Atkinsiella*, *Halocrusticida* and *Halodaphnea*

Fungi associated with *Urosalpinx* egg cases were first noticed by Vishniac (1955) with the description of *Sirolpidium zoophthorum*. Later, Vishniac (1958) was able to isolate and grow a hyphal fungus axenically and she named this organism *Haliphthoros milfordensis*. The fungus was shown to be capable of infecting eggs and larval stages of *Urosalpinx* and also eggs of *Pinnotheres*. Mycelial and zoosporangial features were described and these, together with the large zoospore size, constituted the family diagnosis for the *Haliphthoraceae*. The presence of monoplanetic zoospores with lateral flagellar insertion comprised the generic diagnosis. Zoosporogenesis was intrasporangial. Endospores were observed but not deemed to be sexual spores. Vishniac (1958) made the genus the type genus of the family *Haliphthoraceae*; but she extended the family diagnosis to include diplanetic fungi, although explicitly excluding sexual reproduction. Vishniac (1958) compared *Haliphthoros* with the fungus described earlier by Atkins (1954*b*) as *Plectospira dubia* D. Atkins, a parasite of *Pinnotheres* and other small *Crustacea*. The latter fungus was renamed as *Atkinsiella dubia* (D. Atkins) Vishniac in honour of Dr Daphne Atkins and included as a second member of the family. Atkins (1954*b*) had noted and drawn stages of the distinctive intrasporangial zoosporogenesis of *A. dubia* and Fuller, Fowles & McLaughlin (1964) were able to provide a much fuller description (see also Sparrow & Gotelli, 1969; Sparrow, 1973*a*). Although there is probably a close relationship between *Haliphthoros* and *Salilagenidium*, based on host range, mycelial appearence and zoospore size ($7{\cdot}2 \times 8{\cdot}5\mu m$ according to Fuller *et al.*, 1964): the inclusion of *Atkinsiella* in the family is more doubtful (compare the latest description, in Nakamura *et al.*, 1995). A second species of *Haliphthoros*, (*H. philippinensis*) was added by Hatai *et al*. (1980) and further species, initially asigned to *Atkinsiella*, have been described. Bian & Egusa (1980), Kitancharoen *et al*. (1994), Nakamura & Hatai (1994, 1995*a*, *b*) and Kitancharoen & Hatai (1995) have described *A. hamanaensis*, *A. awabi*, *A. parasitica*, *A. okinawaensis* and *A. panulirata* respectively from marine crustaceans and molluscs. A parasite of freshwater insect eggs (*A. entomophaga*) was added to the genus by Martin (1977). All of these additional species were transferred to the genus *Halocrusticida* by Nakamura & Hatai (1995*b*). I am in full agreement with the need to separate the marine taxa from *Atkinsiella dubia*, but most unfortunately, these authors selected the freshwater species *A. entomophaga* as the type species of their new genus. Since Dick (1998*a*) placed *A. entomophaga* in the genus *Crypticola*, there was no alternative but to provide a new generic name for *A. hamanaensis*, *A. awabi*, *A. parasitica*, *A. okinawaensis* and *A. panulirata*. *Halocrusticida* was replaced by the name *Halodaphnea* (named after Dr Daphne Atkins), with *H. parasitica* (basionym *A. parasitica* (Nakamura & Hatai) M. W. Dick as the type species.

Zoosporogenesis in *Atkinsiella dubia*, *A. (Crypticola) entomophaga* and *Halodaphnea* may be only superficially similar, in a way analogous to that of *Eurychasma* and *Eurychasmopsis* (Canter & Dick, 1994). In the type species (*A. dubia*) there is a distinctive centripetal protoplasmic contraction followed by an amoeboid stage during which the initials migrate centrifugally along cytoplasmic strands before differentiation is completed (Fuller *et al*., 1964; Nakamura & Hatai, 1995*b*). This stage has not been found in *Halodaphnea*. There is a less close resemblance between the thallus morphology of *A. dubia* and the species of *Halodaphnea*, and despite the description of rhizoids in artificial culture of *A. dubia* (Sparrow & Gotelli, 1969), I regard *A. dubia* and *Halodaphnea* as members of the *Salilagenidiales*. (Rhizoids occur elsewhere in the *Saprolegniomycetidae*, notably in *Leptomitus*.) Ultrastructural studies are required to determine the relationships within the order, particularly in connection with the different zoospore cyst dimensions.

The freshwater parasite of insect eggs, *Crypticola entomophaga*, does not have the distinctive amoeboid stage during zoosporogenesis and has been removed from *Atkinsiella*. It shares with *Crypticola* the characters of freshwater insect hosts and the possession of only a single row of tubular tripartite hairs on the tinsel flagellum (a criterion requiring confirmation), I am therefore unwilling to regard it as congeneric with *Halodaphnea*. The most satisfactory hierarchical arrangement on the available evidence is to place *A. entomophaga* in *Crypticola*, which becomes the type genus of a new family, *incertae sedis* but with possible links to the *Myzocytiopsidales*, the *Crypticolaceae*.

Two remaining fungi, *Sirolpidium zoophthorum* Vishniac (1955*d*, 1958) and *Hyphochytrium peniliae* N. J. Artemczuk & L. M. Zelezinskaya (Artemchzuk & Zelezinskaya, 1969), need to be considered with *Salilagenidium*, *Haliphthoros*, *Atkinsiella* and *Halodaphnea* because of their marine habit and crustacean hosts. Despite the exceptionally small zoospores of *S. zoophthorum* ($2{\cdot}0 \times 5{\cdot}0$ μm, volume equivalent <75 μm^3), I consider that the similarities in habitat and substratum; the culturability; and the morphogenesis involving contraction of parts of the protoplast followed by endothallial wall formation, indicate a greater affinity between *S. zoophthorum* and *Haliphthoros* than between *S. zoophthorum* and *S. bryopsidis* (the type species of *Sirolpidium*). I therefore transfer *S. zoophthorum* to *Haliphthoros* as a doubtful taxon until it is reinvestigated. It is misplaced in *Sirolpidium* as that genus is delimited here. The thallus of *Hyphochytrium peniliae* is mycelial and not septate. The description of this fungus is incomplete and the evidence (particularly with respect to flagellar number) for placing the fungus in *Hyphochytrium* is weak. The sporangiospores are $4{\cdot}6 \times 6{\cdot}9$ μm, which is commensurate with those of *Salilagenidium*, to which genus the fungus is referred under the doubtful category, but a new combination is not made because the name is illegitimate, lacking a Latin description.

Myzocytium and *Lagenidium* species in marine green algae

The marine taxon *Myzocytium proliferum* f. *marinum* Kobasyasi & Ookubo (Kobayasi & Ookubo, 1953) was reported from *Cladophora*. It apparently lacked a fertilization tube. Its disposition is uncertain since its oospore is said to have a large eccentric oil drop, but the illustrations could be interpreted as a subcentric internal vesicle. It is not a form of *M. proliferum* but has affinity with the imperfectly known *Lagenidium* sp. (Ookubo, 1952) on *Aegagropila* and the *Lagenidium* sp. (Deckenbach, 1903) on *Chaetomorpha* (≡ *L. chaetomorphae* Jacz. & P. A. Jacz.). These are the only marine phycophagic *Lagenidium* records, and it is probably significant that they are all on *Cladophorales*.

Reference should be made here to several other poorly known fungi also parasitic on marine *Cladophora*. Most of these will probably prove to be synonymous with taxa transferred to *Sirolpidium*, or regarded as unidentifiable. They are: *Gymnococcus cladophorae* de Bruyne (de Bruyne, 1890), *Achlyogeton salinum* P. A. Dang. on *Cladophora* (Dangeard, 1932), and *Eurychasma* sp. on *Bryopsis* (Kobayas[h]i & Ookubo, 1954*a*). The *Pontisma*-like parasite of *Cladophora* described by Raghukumar (1986, 1987) is apparently distinct from *Sirolpidium bryopsidis* in the extent of its thallus and the disease symptoms it induces.

It is less certain whether the parasite of *Tetraselmis* (marine *Chlorococcales*), named as *Ectrogella marina* (Feldmann & Feldmann, 1956) should be placed with *Sirolpidium bryopsidis*, but zoospore dimensions would support a transfer to *Sirolpidium* and its specific status is retained, though the species is placed in the doubtful category.

Marine *Lagenidium* taxa transferred to *Pythium*

Five species of *Pythium* are reported from marine environments (Dick, 1990*c*); most are exclusively parasitic in red algae (*P. chondricola* A. W. A. M. de Cock, 1986; *P. marinum* F. K. Sparrow, 1934; *P. maritimum* W. Höhnk, 1939; *P. porphyrae* Takahashi & Sasaki *in* Takahashi, Ichitani & Sasaki, 1977; *P. salinum* W. Höhnk, 1953). Some of these are commercially important pathogens (Arasaki, 1947, 1948, 1960*a*, *b*; Arasaki, Akino & Tomiyama, 1968; Fuijita & Zenitani, 1976, 1977*a*, *b*, *c*). The unnamed *Lagenidium* sp. on *Gracilaria* (Johnson, 1957*a*) should be removed from *Lagenidium* and placed near *Pythium porphyrae* from which it differs in its inflated sporangial segments. A binomial cannot be given until the fungus is found again and more fully described.

Lagenidium and *Myzocytium* species parasitic in the *Aschelminthes*

Flagellate fungal parasites of nematodes and rotifers, animals which are grouped together in the super-phylum *Aschelminthes*, are very common but not frequently studied. The first nematophagous species to be assigned to *Myzocytium* was *M. vermicola* (Zopf) Fischer. Since then six species of *Lagenidium* and twelve additional species of *Myzocytium* have been named as parasites of these animals. *Myzocytium vermicola* was described as possessing a fertilization tube (Zopf, 1884), but another interpretation is probable and the sexual morphology of this species is similar to that of the more recently described species of *Myzocytium*.

Lagenidium caudatum is distinct in having a well-defined vesicular sporangial discharge and large zoospores (volume equivalent >650 μm^3). The mycelium is filamentous and extensive. This taxon is not separable from *Pythium sensu stricto* and it is removed to that genus as *Pythium caudatum*. Sparrow (1936*b*) figured *Saprolegnia* and *Aphanomyces* parasitizing nematodes, and the parasitism by *L. caudatum* is comparable. *Pythium caudatum* joins *P. elongatum* and *P. parasiticum* as an anamorph binomial, but pending confirmation of host specificity, it might be better to regard this species as a synonym of *Pythium*-F-Group.

Apart from *Pythium* (*Lagenidium*) *caudatum*, it is probable that all the remaining taxa have zoospores or aplanospores at least partly-formed prior to dehiscence of the sporangium. Descriptions of zoosporogenesis imply either that the vesicle is absent throughout extra-sporangial zoospore delimitation (with flagella apparently developed peripherally to the cleaving mass - cf. *Hyphochytrium*), or that a 'vesicle' only becomes apparent late in the development. The zoospores are also sometimes reported to be amoeboid. Cysts sometimes develop a series of adhesive pads and in other cases the spore, at encystment, is hemispherical and adpressed to the host instead of being spherical.

In the majority of species for which sexual reproduction has been described there is gametangial copulation via a pore. However, in *M. anomalum* and *L. oviparasiticum* thick-walled chlamydospores are described, which may or may not represent a sexual system, and in *L. septatum* thick-walled, plerotic endospores are formed as a result of septation of the thalloidal cell. In this last species small thalloidal cells may form a single thick-walled spore, which would then be indistinguishable from the chlamydospores of *M. anomalum* or *L. oviparasiticum*. In sporogenesis these three taxa span the wide range of development that occurs elsewhere in this group of aschelminth parasites (aplanospores giving rise to zoospores in *M. anomalum*, intrasporangial cleavage resulting in an *Achlya*-like spore ball in *L. oviparasiticum*, intrasporangial zoospores in *L. septatum*). The resting spore formation in *L. septatum* is sufficiently distinct, and perhaps significant, to warrant the erection of a new genus, *Chlamydomyzium*. Studies by Dr S. L. Glockling indicate that chlamydospore formation and ultrastructure in a species related to *M. anomalum* is different from the other nematophagous lagenidiaceous fungi that she has sectioned. Since the description for *M. anomalum* is the most complete, this species, rather than *L. septatum*, is selected as the type species of *Chlamydomyzium*. *Lagenidium oviparasiticum* can be attached to this new genus without problem.

There is neither a single character nor a group of characters that can separate the three remaining nematophagous species placed in *Lagenidium* (*L. distylae*, *L. elegans* and *L. parthenosporum*) from those placed in *Myzocytium*. It has not been established that these species have ***any*** of the diagnostic criteria of the *Peronosporomycetes*, namely:

1. heterokont flagellation within the definition used here
2. lateral flagellar insertion in a groove
3. a fertilization tube
4. a haplomitotic B ploidy cycle

On the contrary, they do possess a number of characters not found in other *Pythiaceae*:

1. intrasporangial zoosporogenesis
2. sub-apical flagellar insertion, without any flagellar groove
3. subspherical or hemispherical zoospore shape
4. plasmogamy effected by the fusion of whole gametangial protoplasts via a pore

Four of the species have an ornamented oospore wall (cf. *Cystosiphon* and *Syzygangia*). The species *Myzocytium rotiferum* Dasgupta & John, though possessing a smooth-walled zygote, is not identifiable because zoosporogenesis was not described.

A new genus, *Myzocytiopsis*, typified by *M. lenticularis* because this species appears to combine many of the morphological attributes exhibited within the genus, has been erected for this group of nematophagous fungi (Dick, 1997*b*). The genus is an analogue of *Syzygangia*, which parasitizes plants, and is used to typify higher taxa. There is a nomenclatural problem in establishing the genus *Myzocytiopsis*. This problem was posed by Barron's literature-based synonymy of *Protascus subuliformis* P. A. Dang. (Dangeard, 1903) with *Myzocytium subuliformis* (P. A. Dang.) G. L. Barron (Barron, 1977*a*). The only other record of Dangeard's fungus is the description by Maupas (1915), which included a teleomorphic stage named as *P. subuliformis* var. *maupasii* Maire (Maire, 1915). Acceptance of the synonymy proposed by Barron (1977*a*) would require the genus *Protascus* to be used if the nematophagous taxa are to be separated from *Myzocytium* and *Lagenidium*. However, Barron (1977*a*) noted that Dangeard (1903) based his description on a non-flagellate anamorph stage. This morphological spore form is common to species of *Myzocytiopsis* and *Gonimochaete*; Barron did not have material of *Myzocytium subuliformis* when the transfer was made to *Myzocytium*. *Protascus* is thus a *nomen confusum*. Barron (1977*a*) also noted and concurred with the concession made by Maupas (1915) and Maire (1915) that the sexual stage found by Maupas could indicate that the 1915 fungus was not unequivocally conspecific with Dangeard's fungus. Maire (1915) stated that if not conspecific the varietal name ***could*** be used (but he did not definitively erect the variety, even though he provided the Latin diagnosis). For purely technical nomenclatural reasons, I reject the synonymy of Barron (1977*a*) based on anamorphic stages, and erected the taxon *Myzocytiopsis subuliformis* based on the sexual stage described by Maupas (1915). The varietal name '*maupasii*' does not have to be used when there is a rank change for the specific epithet. The 'new' use, with different typification, of the specific epithet '*subuliformis*' in *Myzocytiopsis* takes precedence over any subsequent transfer of *Protascus subuliformis* to *Myzocytiopsis*. Thus the use of a prior generic name of confusing etymology is barred until ***indubitable*** material of Dangeard's fungus is rediscovered (which is highly improbable). I am grateful to Dr B. C. Sutton for suggesting this device, and to Prof. V. H. Heywood for further discussion of the problem. The formal proposal to regard *Protascus* as a *nomen confusum* and *nomen rejiciendum* is made here, and will also be made to *Taxon*.

Resumé

The case for the placement of *Lagenidium giganteum* in the *Pythiaceae* is extemely strong. For the other species in *Myzocytium sensu stricto*, which have small zoospores, the present evidence with respect to zoosporogenesis and oosporogenesis indicates an affinity with the *Pythiaceae*, but more ultrastructural studies are needed before their systematic position in the *Pythiaceae* is unequivocal. Most species of *Lagenidium sensu lato* and *Myzocytium sensu lato* do not appear to be closely related to the *Pythiaceae*. If there is reappraisal of either the importance of zoosporogenesis or oosporogenesis, based on ultrastructural, cytological and biochemical studies, then there remains the possibility that *Myzocytium sensu stricto* should be transferred to the *Myzocytiopsidales*. Such a transfer would raise the question of synonymy between *Myzocytium*, based on an anamorph description of vesiculate extrasporangial zoosporogenesis, and *Syzygangia*, based on teleomorphic criteria. In the absence of essential comparative data, the smaller genera proposed here will provide the workable classificatory framework that does not exist at present.

THE NEMATOPHAGOUS 'LAGENIDIACEOUS' FUNGI

Myzocytiopsis, *Chlamydomyzium*, *Gonimochaete* and *Haptoglossa*, with reference to *Syzygangia*

Additional genera, also associated with rotifers and nematodes, must be discussed with *Myzocytiopsis* and *Chlamydomyzium*. *Gonimochaete* and *Haptoglossa* were described by Drechsler (1940, 1946). *Crypticola* (Frances, sweeney & Humber, 1989; Dick, 1998*a*) is a parasite of insect larvae which also has affinity with the *Myzocytiopsidaceae*.

Amongst the species of *Myzocytiopsis* and *Gonimochaete* there is a tendency either for zoospore formation to be replaced by aplanospore production (in *Gonimochaete* and in *M. glutinospora* and *M. subuliformis* zoosporangial zoospores are not known), or for zoospores to have a sequential development from aplanospores (*C. anomalum*). The aplanospore cyst is more or less clavate and secretes an adhesive knob or pad from the narrower end of the spore (Newell *et al.*, 1977; Barron, 1976*c*, 1977*a*). A similarly-shaped aplanospore has been described for *Pseudosphaerita dinobryonis* (Canter, 1968). Sometimes a linear series of knobs may be formed by a single spore. It would be difficult to accommodate in the same genus, on the one hand, species of *Myzocytiopsis* (*M. lenticularis*) with laterally biflagellate zoospores with flagellar insertion in a groove, and on the other hand, other species (*M. humicola*) with sub-apical flagellar attachment, if this variation is confirmed. However, with the lack of critical photographic evidence for most species, and the absence of flagellate cells in some species, new generic diagnostic criteria cannot be constructed. *Myzocytiopsis humicola*, with cysts producing adhesive pads could be transferred to *Gonimochaete*, but only by substantial amendment of the generic concept of the latter genus. Generic assignments must be considered provisional.

Similarly, the sexual stages of *G. lignicola* and *M. glutinospora* differ only in the morphological proportions of the reticulations, and these species are thus also intermediate between *Myzocytiopsis* and *Gonimochaete*. Saikawa & Anazawa (1985) have shown that *Gonimochaete* possesses tubular mitochondrial cristae. *Gonimochaete* is distinguished only by the aggregation of the aplanospore initials into the discharge tube of the sporangium, but the balance of the incidence of other characters probably justifies the retention of the two genera and their placement in the *Myzocytiopsidaceae*.

Life-history studies of the *Myzocytiopsidceae* are incomplete. The germination of the resting spore of species of *Myzocytiopsis* has not been observed (but see *Syzygangia zygnematicola*). The only cytological work on the *Myzocytiopsidales* is that of Dangeard (1906) on *Myzocytium subuliforme sensu* Barron (as *Protascus subuliformis*). It would appear that there is a synchrony of divisions in the vegetative thalli (see Karling, 1981*a*: 103, figs 137, 138), but there is no evidence for the distinctive mitotic configurations of the *Plasmodiophoromycetes*, which are confined to the plasmodial stages of the nuclear cycle (the meiotic spindles do not have persistent nucleoli - see p. 75). Any reduction in the duration of a plasmodial stage would limit the scope for finding such configurations. Differences in mitotic ultrastructure are so extensive within the classes of flagellate fungi that reports, such as that of Dylewski *et al.* (1978), comparing similarities in nuclear division between the plasmodiophorid *Sorosphaera* and the chytrid *Entophlyctis* cannot yet be evaluated for their phylogenetic significance. On the other hand, synchrony of divisions could indicate meiotic prophases in species for which the potential gametangium is multinucleate. Incomplete evidence of meiosis (Glockling, 1994; Dick, 1995) suggests that premeiotic and postmeiotic nuclear abortions precede oospore formation in *Myzocytiopsis*. The presence of DBVs and mitochondria with sharply defined tubular cristae and the zoospore morphology of the type species (Glockling, 1994; Dick & Glockling, 2000; Glockling & Beakes, 2000*b*) provide further evidence that *Myzocytiopsis*, at least, is a member of the *Peronosporomycetes*.

On zoosporogenesis, zoospore shape, amoeboid propensity of the zoospore, refractive granules in the zoospore, flagellar length ratios, gametangial copulation, copulation through a pore and resting spore structure and wall ornamentation, the group *Myzocytiopsis*, *Syzygangia* (parasitic on algae), *Gonimochaete* and possibly *Chlamydomyzium*, is probably homogeneous. There is a tendency for *Syzygangia* to have thalli which are more 'mycelial' in that they become septate but do not disarticulate, whereas thalli of *Myzocytiopsis* have usually become disarticulated into thalloidal cells at maturity, and this, coupled with the very different host ranges and habitats (*Myzocytiopsis* is primarily wet-terrestrial, *Syzygangia* is primarily freshwater aquatic), and the underlying biochemistry which controls the realizable habitat niches, is sufficient justification for establishing the two generic entities, *Myzocytiopsis* and *Syzygangia*, at this stage.

Other biflagellate parasites of the *Aschelminthes*: *Sommerstorffia*, *Synchaetophagus*, *Endosphaerium* and *Hydatinophagus*

Earlier references to nematophagous fungi in general are provided by Dollfus (1946). For completeness, reference should also be made to five other biflagellate parasites of rotifers, *Sommerstorffia*, *Synchaetophagus*, *Endosphaerium* (all monotypic) and *Hydatinophagus* (*H. apsteinii* Valkanov (Valkanov, 1931*a*, *b*, 1932) and *H. americanus* Bartsch & Wolf (Bartsch & Wolf, 1938)). Only *Sommerstorffia* is known from other than the original description, with further reports by Karling (1952), Prowse (1954), Saikawa & Hoshino (1986) and Czeczuga & Próba (1980). However, as noted by Sparrow (1960), there must be considerable doubt about Prowse's delineation of pluriovulate oogonia. Although the tips of the trapping apparatus show some resemblance to the tips of zoopagalean conidia (Drechsler, 1962) and the presence of a holdfast has analogy with some *Trichomycetes*, a zoopagalean affinity is less likely because the evidence for zoosporogenesis is stronger and consistent in detail. Dick (1971*b*) suggested that the fungus could be related to the *Leptolegniellaceae*. The relatively small zoospore cyst size (*ca* 5·5 μm diameter, volume equivalent *ca* 90 μm^3) and the holdfast could be cited against placing *Sommerstorffia* in either the *Saprolegniales* or the *Pythiales*.

The binomial *Synchaetophagus balticus* Apstein (Apstein, 1910) refers to the parasite of *Synchaeta* in the oligohaline Baltic Sea: neither asexual nor sexual reproduction is known. In the light of the range of fungi now known to parasitize rotifers, the organism cannot be referred, even doubtfully, to any fungal class: it must be left *incertae sedis* until rediscovered on *Synchaeta*.

Endosphaerium funiculatum D'Eliscu (D'Eliscu, 1977) is parasitic on rotifers and nematodes in the mantle cavity of a bivalve mollusc. The trapping mechanism of this fungus appears to be adhesive sections of otherwise undifferentiated hyphae. The description of this anamorphic fungus is deficient in many critical details, such as hyphal width and zoosporangial length. There are no illustrations of the zoosporangia, which are described as filamentous, distally septate with "encapsulated zoospores reniform (7-22 μm), laterally biflagellate, but apparently nonmotile". The author places the fungus in the *Pythiaceae* but there seems to be little justification for this, and comparisons with *Ostracoblabe* could also be made. This fungus must also be regarded as *incertae sedis*.

Hydatinophagus apsteinii parasitized *Hydatina* (Bulgarian record) and *H. americanus* parasitized *Monostyla* (Wisconsin, U.S.A. record); Scott (1961) transferred both to *Aphanomyces* (as *A. hydatinae* Valkanov and

A. americanus (Bartsch & Wolf) Scott respectively), but on slender criteria. Zoospore cyst dimensions are not recorded for *H. apsteinii* and this spore form was not even observed for *H. americanus*. Oospore dimensions of *H. americanus* are exceptionally small for an *Aphanomyces*. (Zoospore measurements for the protozoan parasite *Aphanomyces acetinophagus* Bartsch & Wolf (Bartsch & Wolf, 1938) indicate a volume equivalent of *ca* 110 μm^3, which is also exceptionally small for an *Aphanomyces*. The spore-ball of this fungus is comparable to those of *Myzocytiopsis parthenospora* and *Chlamydomyzium oviparasiticum*.) Until rediscovered these taxa are best recorded under their distinctive generic names and left *incertae sedis*, rather than as doubtful members of the genus *Aphanomyces*.

There is nothing to exclude *Synchaetophagus balticus* and *Hydatinophagus americanus* being referable to the *Zoopagales* (see Dick, 1990*b*). Equally, they could be considered, along with *A. acetinophagus* and *A.gordajeverae* Skvortzov (Skvortzow, 1925) in relation to the myzocytiopsidaceous fungi.

Uniflagellate parasites of the *Aschelminthes*: *Catenaria anguillulae* and chytrids

To facilitate comprehensive and comparative studies on nematophagous fungi, reference should also be made to certain species of the *Olpidiaceae* and *Blastocladiales*.

Catenaria anguillulae is the best-known of this group of parasites. Olson & Reichle (1978) have shown that *C. anguillulae* has a brachyallomyces-type life-history (haplomitotic A ploidy cycle) and they conclude that the data of Ichida & Fuller (1968) refer to meiosis not mitosis. On the other hand *C. allomycis* has a cystogenes-type life-history (haplomitotic B ploidy cycle). The *Blastocladiales* is therefore unique in being an order containing all three ploidy cycles (see Dick, 1987).

The morphology of *Catenaria anguillulae* has been well-documented at both light microscope and ultrastructural levels. It is distinct in its catenulate development with intercalary segments between sporangia, and in the production of rhizoids. No confusion should arise between this species and members of the *Myzocytiopsidales*, except, perhaps, with the little-known *M. indica* (Singh, 1973). *Catenaria anguillulae* is also distinct in its host range which includes helminths as well as nematodes. Axenic cultures are available and physiological studies (Nolan, 1970*a*, *b*, *c*) have been carried out. The nature of zoospore adhesion to solid surfaces has been explored (Tunlid *et al.*, 1991), and could provide a basis for comparisons with zoospore attachments by *Myzocytiopsidales*.

A list of chytrids associated with the *Aschelminthes* is given in Table IV: 4. The taxonomy of the non-rhizoidal species is included under *Olpidium* (PART V) for convenience in literature searches.

Achlyogeton, *Septolpidium* and *Bicricium*

Three very poorly known genera constitute the family *Achlyogetonaceae* (Sparrow, 1943) of the *Chytridiales* (Sparrow, 1960). *Achlyogeton* (at present with four species) was described by Schenk (1859*a*); *Bicricium* (at present with three species) by Sorokine (1883), and *Septolpidium* (monotypic) by Sparrow (1933*a*). Many, if not all, of the species will have to be left in the doubtful or unidentifiable categories until material of undoubted conspecificity is found and critically described. I doubt whether such a demonstration of conspecificity will be achieved, given the imperfect nature of the original descriptions. Hence the taxonomic changes proposed below.

TABLE IV: 4. *Chytrids associated with* Aschelminthes *(page numbers in brackets refer to Sparrow, 1960)*

Olpidium allomycetos Karling (143) (occasionally)
Olpidium entophytum (A. Braun) Rabenh. var. *intermedium* Const. (137)
Olpidium granulatum Karling (145)
Olpidium gregarium (Nowak.) J. Schröt. (144)
Olpidium macrosporum (Nowak.) J. Schröt. (155)
Olpidium nematodeae Skvortzov (147)
Olpidium rotiferum Karling (146)
Olpidium sparrowii Dogma (Dogma, 1977)
Olpidium vermicola G. L. Barron & Szijarto (Barron & Szijarto, 1986)
Olpidium zootocum (A. Braun) N. Sorokīn (155)
Endochytrium oophilum Sparrow (573)
Endochytrium operculatum (De Wild.) Karling (569)
Rhizophydium gibbosum (A. Braun) Rabenh. (237)
Rhizophydium vermicola Sparrow (277)
Rhizophydium zoophthorum (P. A. Dang.) A. Fisch. (285)
Phlyctochytrium nematodeae Karling (337)

Karling (1977) added the genus *Myiophagus* to the *Achlyogetonaceae*, but Sparrow (1939*a*, 1960) did not accept that this genus was a flagellate phycomycete. The monotypic genus *Myiophagus* (*M. ucrainicus* (Wize) Sparrow) was based solely on its reticulate resting spore, and is not placeable since it could also be referred to the *Zoopagales*. The zoosporic phase described by Karling (1948, 1977) was not associated with any resting spore.

The septate nature of the thalli in the *Achlyogetonaceae*, the '*Achlya*-like' spore balls formed at the tips of the discharge tubes, and the host ranges (nematodes and green algae) strongly suggest that all species could be related to *Myzocytiopsis* and *Syzygangia*. The species are holocarpic endobionts without rhizoids and anamorphic, with the possible exception of *Bicricium transversum* (Sparrow, 1960: 184, suggests that the resting spore is a mycoparasite). *Bicricium transversum* is figured by Sorokine (1883) as having one of the two cells (a sporangium) with an exit tube to one side of the host cell. The other cell, containing an aplerotic resting spore, also has an exit tube (to the opposite side of the host cell, hence the specific epithet) - a morphological state which has no equivalent except in the *Monoblepharidales*.

The *Achlyogetonaceae* has been placed with the *Olpidiaceae* solely on the very dubious reports of the uniflagellate nature of the zoospore. Sorokine (1883, 1889, 1890) described *Bicricium* and *Achlyogeton rostratum* as uniflagellate. It is quite possible that at that time the smaller of an anisokont pair of flagella may have been missed, particularly if it was very short. The type species of the genus, and therefore of the family, *Achlyogeton entophytum*, was reported to have ***cellulosic*** zoospore cysts, rendering placement in the *Chytridiales* questionable. There are also indications (Karling, 1942*a*: 94; Sparrow, 1960: 183) that flagellar insertion may ***not*** be posterior. The situation vis-à-vis flagellar number and insertion could be the same as for the *Rozellopsidaceae* or *Pleotrachelus*.

The early reports of Sorokine (1883, 1889, 1890), and his erection of the genus *Bicricium*, can be attributed to a failure to appreciate variation in septation. I therefore treat his two species *B. lethale* and *Achlyogeton rostratum* as depauperate, unidentifiable, specimens of *Myzocytiopsis* and *Syzygangia* respectively. Similarly, *A. solatium* is placed as a doubtful synonym of *Syzygangia oedogonii* and *B. naso* is referred as an unidentifiable species of *Syzygangia*, rather than *Myzocytium* (Cejp, 1959) because there is no evidence for sexuality involving a fertilization tube. The description of the parasite of freshwater *Cladophora* (*Achlyogeton entophytum* Schenk, 1859*a*) is not well-characterized, and all references to both *Lagenidium* and *Olpidium* on *Cladophora* will require re-evaluation in the light of new collections. The species is placed as an unidentifiable member of *Syzygangia* because the sexual stage is not known. *Achlyogeton salinum* (Dangeard, 1932) parasitic in marine *Cladophora*, is transferred to *Sirolpidium* as the doubtful species, *Sirolpidium salinum*, to accommodate the *Lagenidium*- or *Pontisma*-like parasites of marine *Cladophorales*.

Septolpidium lineare, in diatoms (Sparrow, 1933*a*), has small uniflagellate zoospores; it is anomalous in the *Chytridiomycetales* and of uncertain affiliation to straminipilous fungi. It cannot be grouped with *Ectrogella*, which has non-septate thalli; the linear, septate thallus and zoospore size are not consistent with an affiliation with *Aphanomycopsis*. For the present the genus is retained, and placed provisionally in the *Myzocytiopsidales*, occupying a position intermediate between *Syzygangia* and *Ectrogella*.

Crypticola

Two parasites of insects could be related to *Myzocytiopsis* and *Gonimochaete* because of their sporangial development: *Crypticola clavulifera* (Frances *et al.*, 1989) and *C. entomophaga* (*Atkinsiella entomophaga* of Martin, 1977). The character of the long sporangial discharge tubes of *Gonimochaete pyriforme* (Barron, 1973: fig. 10) is distinctive, and this feature is also seen in *C. entomophaga*. The lateral development of the sporangium in *C. crypticola* is probably comparable. It is not known whether there are similar mechanisms for sporangial discharge in these three species.

If confirmed, there is a most interesting similarity between the arrangement of flagellar hairs (tripartite hairs in *Crypticola*), which are reported to form a ***single*** row along the flagellar axis for the two insect parasites. This construction might be the same in *Syzygangia nodosa*: either there are three erroneous reports (which is quite possible), or there exists an entirely new group of aquatic fungi. No TEMs of flagella of any of the nematophagous lagenidiaceous fungi have been published, nor have we been successful, and there are no other ultrastructural studies of *Myzocytiopsis*.

Within *Myzocytiopsis*, *Syzygangia*, *Gonimochaete* and *Chlamydomyzium* there are several recurring independent lines of morphological variation: the thallus is not always septate, but when septate the septa are not conspicuously thickened and hyaline, as in the type species of *Myzocytium*. However, the large terminal vacuoles of *Myzocytiopsis osiris* may be homologous with cytoplasmic extremities adjacent to septa in *Myzocytium proliferum*, and with the conspicuous globules of *Crypticola*, in which case the affinities of *Myzocytium sensu stricto* with either the *Pythiales* or the *Myzocytiopsidales* will have to be reconsidered. Segments of septate thalli tend to be inflated, and most species with disarticulating thalli occur in nematodes but not in rotifers. Branching of the thallus is often absent, but can occur, with or without septation.

In contrast to the situation in the *Pythiaceae*, the sporogenetic sequence does not appear to be a conserved character in this group and spore delimitation is frequently completed intrasporangially. Zoospore morphology does not appear to be uniform but differences in descriptions between sub-apical and lateral flagellar insertion and observer interpretation may reflect polymorphy of the zoospore. The occurrence of the principal-form zoospore has not been established. In general the zoospores are ovoid to hemispherical, perhaps with a slight concave curvature of the longitudinal axis, but lacking a deep ventral groove. There is no evidence of dimorphism between the zoospores delimited within the zoosporangium and those derived from aplanospores expelled from the sporangium, or from zoospore cysts in diplanetic species. Zoospore volume equivalents fall into two categories of *ca* 25 μm^3 and *ca* 100 μm^3.

Three kinds of gametangial arrangement, two positions for the zygote and two kinds of zygote morphology are reported in this group of fungi. In most species there is gametangial fusion between two adjacent elements; rarely there may be 'parthenogenetic' development, and rarely there may be union between a branch and another element. The zygote is normally developed in the lumen of the receptive gametangium, but in *Myzocytiopsis subuliformis* (Maupas, 1915: figs 11-13) and *Syzygangia sacculoides* it is sometimes developed in the isthmus between the two gametangia. The spore wall of the zygote is smooth in some species and ornamented in others. The ornamentation usually takes the form of reticulations possibly caused by a collapse of an outer membrane on to the endospore. In some cases the observed refractive disruption compared with smooth-walled spores appears to be partly due to irregularity of an intermediate layer between the outer epispore and the endospore. In both cases there is probably a more or less gradual progression from the apparently smooth to the obviously ornamented state from species to species. The possibility that periplasm is present cannot be ruled out. Occasionally the resting spore is said to be pigmented, yellow-brown, but this may be an artefact of light microscopy.

The pragmatic grounds for separating the extensively studied genus *Pythium* (with smooth-walled oospores) from the little-researched form genus *Cystosiphon* (with ornamented oospores) do not apply in this group. For further discussion of the importance of the generic name *Cystosiphon*.

The protoplasmic arrangement of the mature resting spore is too inadequately documented for generalizations to be made, but the presence of DBVs in TEMs of the new species of *Myzocytiopsis* and *Chlamydomyzium* described here; the prominent vesicles in the chlamydospore of *C. sphaericum*, and the internal cytoplasmic structure of *M. papillata* and *M. lenticularis* (Barron, 1976*b*, *c*) all indicate a straminipilous affinity. Characters, such as those of molecular biology, that would indicate an affinity with either the *Saprolegniomycetidae* or *Peronosporomycetidae* have yet to be presented. The molecular data will need to be assessed against morphological and morphogenetic characters which point to a rather more distant relationship, such as oosporogenesis, zoosporogenesis, doubts concerning the presence of a principal-form zoospore, zoospore polymorphism and flagellar ornamentation.

The possibility that the arrangement of tubular tripartite hairs is anomalous in *Crypticola* also indicates that separation at a supra-generic level is preferable at this stage. A new order is therefore required. Since most of the *Myzocytium*-like species fall into this group, the genus *Myzocytiopsis* will form the stem for family and ordinal names. The order will be referred to hereafter as the *Myzocytiopsidales*, with two families, the *Myzocytiopsidaceae* (Dick, 1997*b*) and *Crypticolaceae* (Dick, 1998*a*). Other families *incertae sedis*, perhaps including the *Ectrogellaceae*, may be provisionally associated with the order.

The order *Myzocytiopsidales* is characterized by plasticity in the control of the precise site of zoosporogenesis, although intrasporangial zoosporogenesis more closely resembling *Phytophthora* (contrast the similarities between *Lagenidium sensu stricto* and *Pythium*) rather than *Saprolegnia* or *Pythium* appears to predominate.

Haptoglossa, its morphology, infection apparatus and relationship to the plasmodiophorids

The genus *Haptoglossa* (type species: *H. heterospora* Drechsler) was first described by Drechsler (1940) as a parasite of numerous species of nematodes obtained from soil litter which had been spread on water-plus-agar plates. He was doubtful about the systematic position of the genus because his species only formed aplanospores from thalli that became transformed into sporangia. More species have now been described from similar habitats: *H. zoospora* Davidson & Barron (Davidson & Barron, 1973); *H. mirabilis* Barron (Barron, 1980); *H. humicola* Barron (Barron, 1981); *H. intermedia* Barron (Barron, 1989); *H. elegans* Barron (Barron, 1990); *H. erumpens* Glockling & Beakes and *H. dickii* Glockling & Beakes (Glockling & Beakes, 2000*a*) and *H. heteromorpha* (Glockling & Beakes, 2000*c*). Barron placed his species in the *Oomycetes*. Material of one species, *H. heterospora*, has been collected from nematodes in marine habitats (Newall, Cefalu & Fell, 1977). Another little-known taxon attacking rotifers, *Pleotrachelus rotatoriorum* Scherffel (Scherffel, 1930), should be attached to *Haptoglossa*. Scherffel made the generic attachment solely on the presence of three exit tubes from an ovoid thallus-sporangium; zoospores were not observed; the taxon is probably not now identifiable.

Haptoglossa is distinct in its combination of biflagellate zoospores or aplanospores; glossoid cells (gun cells); and sporidia (Drechsler, 1940; Barron, 1977*b*, 1980, 1981, 1987, 1989*b*, 1990; Robb & Barron, 1982; Robb & Lee, 1986*a*, *b*; Lee, Vaughan & Durschner-Pelz, 1982; Beakes & Glockling, 1998; Glockling & Beakes, 2000*a*, *b*, *c*, *d*). No species is known to form any sexual or resting spore. There is no evidence, such as straminipilous flagellar ornamentation or strictly tubular cristae in the mitochondria, to justify placement of *Haptoglossa* with other nematophagous biflagellate fungi in the *Myzocytiopsidaceae* or *Leptolegniellaceae* of the *Peronosporomycetes*. TEMs of cells do not show clear profiles of mitochondrial cristae and the cell wall of *Haptoglossa* is appreciably thicker than those of *Myzocytiopsis* (Glockling, 1994). Likewise, there is no evidence for the characteristic kinetids or plate-like mitochondrial cristae typical of the *Chytridiomycetes* (which can be uniflagellate or multiflagellate now that the rumen fungi are placed in this class).

The *Haptoglossa* thallus develops from a sporidium implanted under the cuticle of nematodes or rotifers. The sporidium enlarges to form a spherical or allantoid, walled, non-septate sac which often almost fills the host body cavity (multiple infections result in many smaller thalli). Each thallus becomes transformed into a sporangium with one to three exit tubes. Aplanospores or zoospores are differentiated within the sporangium. The zoospores are elongate, ellipsoid or obpyriform with sub-apical flagellar insertion and ***anisokont*** flagellation (neither flagellum is straminipilous); each flagellum may possess an acronema. The zoospores contain refractive granules which are more obvious than those of *Myzocytiopsis*. In many species the zoospores are heterosporous in size but only one size of zoospore is produced in any one sporangium. Aplanospores may also be of two sizes (Glockling & Beakes, 2000*c*). Observations of swimming zoospores indicate that those of *Haptoglossa* have a much more jerky kind of motion than those of *Myzocytiopsis*, thus a fundamental difference in flagellar ultrastructure would not be unexpected. The heterosporous zoospore production could either be analagous to the heterospory of *Saprolegnia anisospora* (Hallett & Dick, 1976), or the two sizes could represent two phases of the life-history, as with the different sizes of the zoospores produced from the diploid and haploid cysts (primary and secondary zoospores respectively) of *Plasmodiophora brassicae* (Ingram & Tommerup, 1972). After zoospore encystment the cysts of some species may secrete an adhesive pad analagous to those of *Myzocytiopsis* and *Gonimochaete*.

The most distinctive feature of *Haptoglossa* is that the zoospore cyst or aplanospore germinates (without any mitotic division) to produce a uninucleate, tongue-shaped or tricornate, glossoid spore (gun cell) which remains attached to the initial spore shell. The glossoid cell is so named because its shape is determined by a single, elongate extrusome of elaborate ultrastructure. The extrusome develops rapidly and is often bowed between two cornae or coiled back upon itself. The various species of *Haptoglossa* are distinguished using morphological details of the glossoid cell; the sequence of its development from sporogenesis; and presumed host infection strategies.

When the tip of the glossoid cell containing the anterior part of the 'extrusome' is touched by a nematode or rotifer, the major part of the contents of the glossoid cell are ***almost instantaneously*** everted into the host, immediately forming the spherical or ellipsoid, walled sporidium. It is this rapid action that gives the glossoid cell its alternative name - the gun cell.

Structures and functions of the evertible infection apparatus in *Haptoglossa*, *Polymyxa* and *Plasmodiophora* compared with the nematogene/taeniogene pairs of *Polykrikos*

It has been acknowledged that the gun cell of *Haptoglossa* and the simpler infection apparatuses of *Polymyxa* and *Plasmodiophora* should be compared with the extremely elaborate extrusomes of certain dinomastigotes (e.g. *Polykrikos*) and the less well-known, evertible infection tubes of *Endemosarca* and the microsporidians (Aist & Williams, 1971; Robb & Lee, 1985*a*, *b*; and Lee *et al.*, 1992). However, the essential detailed and critical comparisons with respect to morphogenesis, form and function have not been carried out. Comprehensive accounts of the ultrastructure of the glossoid cell have been provided by: Robb & Lee (1986*a*, *b*) for *Haptoglossa mirabilis*, Lee *et al.* (1992) for *H. heterospora* Drechsler; Beakes & Glockling (1998) and Glockling & Beakes (2000*b*) for *H. dickii*. The infection apparatus of *Polymyxa* was described by Keskin & Fuchs (1969), and that of *Plasmodiophora* by Aist & Williams (1971). Claxton *et al.* (1996) provided sufficient details to suggest that *Spongospora nasturtii* (*S. subterranea* f. sp. *nasturtii*) also has a similar infection apparatus. The complexities of the extrusome ultrastructure have resulted in a range of proposed terminologies (see Table IV: 5 and Figure IV: 1).

Beakes & Glockling (1998) have approached the integration of structure and function from the structural observations: I prefer to look at the complexity from a mechanistic, functional standpoint, starting with the observation that the time between release and injection is likely to be measured in milliseconds. There are some fundamental mechanical principles that should be noted:

1 The finer the point (such as the diameter of a needle) at which pressure is applied for penetration, the less the force needed to effect penetration (Hiller, Bruce & Jeronimidis, 1996)
2 The narrower the tube cylinder, the greater the resistance to fluid flow and thus the greater the force (energy) required to move material through the tube. A corollary is the importance of agents that may lower the viscosity. Does protoplasm have a lower surface tension than water?
3 There must be a minimum volume for the nucleus. The greater the deformation of the nucleus to a cylinder of narrow diameter, the longer the nuclear cylinder will become and the greater the force that will be needed to move the protoplast (nucleoplasm and cytoplasm with cytoplasmic organelles) through the tube in a finite time. A corollary will be the slower the time that will be taken to traverse the length of the injection tube with a given (and possibly dissipating or fading) force

Presumably, for this kind of biological material, there will be an optimum set of parameters for: the diameter of the tube; the force necessary to effect entry; the length of the tube; the volume of material to be moved; the time available for transfer. In such a very small spore (*ca* 3-6 μm diameter), ***every*** structure almost certainly has an essential function.

The mechanistic sequences for penetration of plant or animal tissues will have some similarities and some differences. In animal hosts the sporidium is deposited ***inside the haemocoel***, ***with*** a cell wall; in plant tissue the sporidium is deposited ***within the host protoplasm***, ***without*** a cell wall. Many of the differences

TABLE IV: 5. *Suggested homology between terms used in the description of the internal structures of extrusomes of* Haptoglossa, Plasmodiophora, Polymyxa *and* Polykrikos

This account	Other authors: Beakes & Glockling (1998)	Robb & Lee (1986*b*)	Keskin & Fuchs (1969)	Aist & Williams (1971)	Lee *et al.* (1992)	Dodge & Greuet (1987) taeniocyst	Dodge & Greuet (1987) nematocyst
Plasmamembranic/cell wall enation of cyst							
extracellular components							
breech bolster	ring of thickened wall	...	...	...	terminal pore	head	apical meniscus
cytoskeleton coronet	...	...	...	...	...	...	...
tube: outer wall (barrel)	injection tube	bore muzzle	...	...	cap [material]	...	operculum
tube: inner wall (bore)	bore	...	stilett	adhesorium	...		
tube contents (adhesive primer)	fibrous bore-tube lining	fibrous sealing material	fibrillenzylinder	fibrillar adhesive material (ex-rohr cavity)	fibrillar material	tigella	...
apical vesicles and ER	apical beak	apical cone	...	...	...	...	...
intracellular component [PLUG]	bore chamber	bore plug	...	...	cap material	...	plug
	partition layer	...	...	...	...	...	...
Postulated anterior vesicular development							
needle chamber	missile chamber	missile chamber	rohr	rohr	anterior end of injection tube	neck	anterior chamber
needle	missile tip	missile head	stilett	stachel	tapered end of rod	...	stylet and filament
cushion	head	missile head	...	...	button	...	basal plate
skirt	annulus cones	buttresses	rohr	rohr	hollow cone	posterior cnidosome	posterior guide
suppository chamber	missile chamber	injection tube	...	...	vacuole	medulla and cortex	posterior chamber
suppository sporidium wall	missile shaft	missile shaft	stachel	stachel	injection rod	...	...
polymer primer	injection tube	injection tube wall anterior vacuole	...	...	...	...	...
intracellular component [KINETID INVOLVEMENT?]						basal body	basal body
anterior vesicle	apical beak cavity	anterior vesicle	...	...	...	...	...
Postulated posterior vesicular development							
tail housing	tail cylinder	tail housing	schlauch	schlauch	injection tube	...	...
tail	missile tail	tail	...	...	injection rod		
intracellular component							
posterior vacuole	posterior vacuole	posterior vacuole	vacuole	vacuole	vacuole	mucron	'posterior' cap

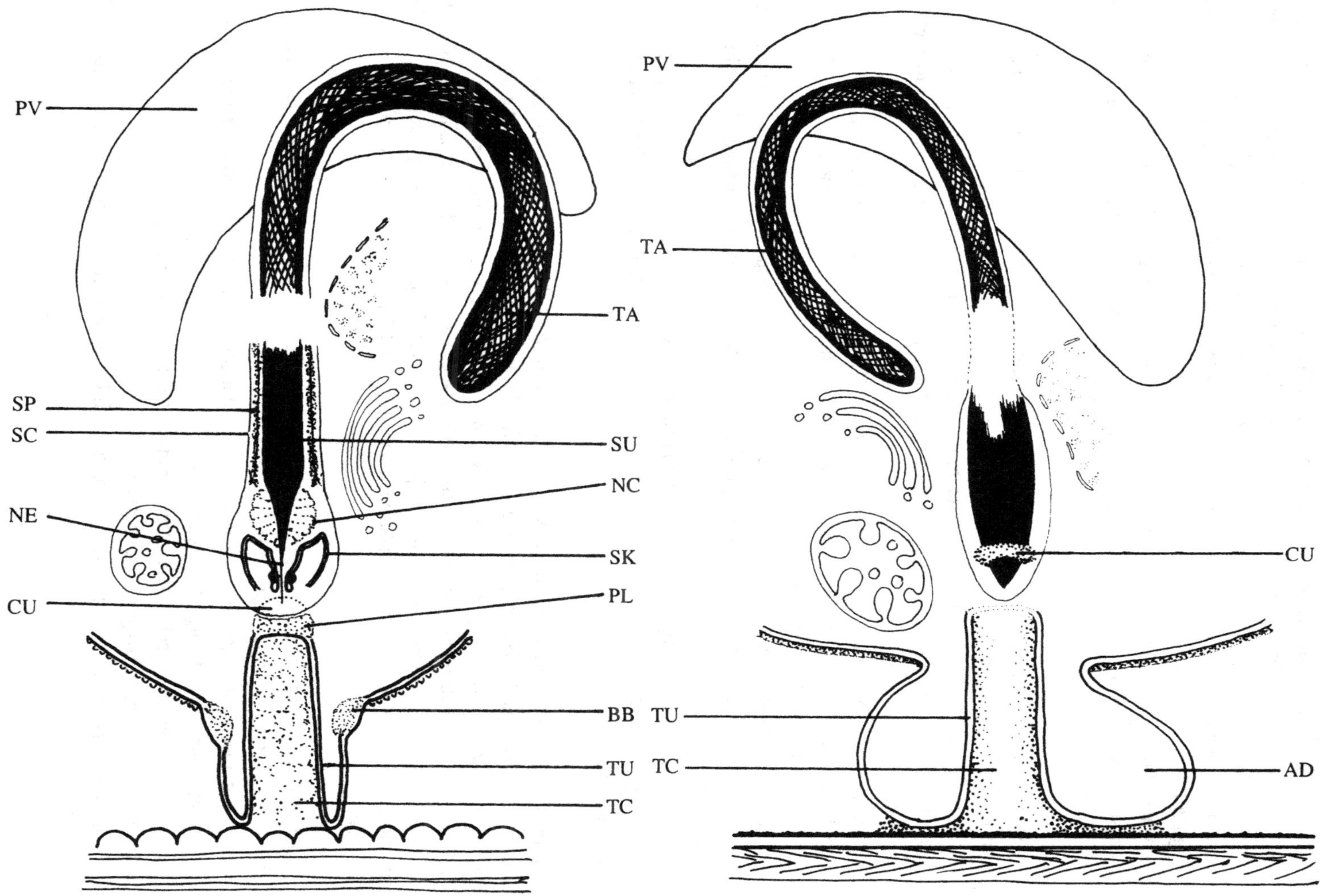

FIGURE IV: 1. *Comparative diagrams of the injection apparatus of* Haptoglossa *(left) and* Plasmodiophora *(right). Refer to Table IV: 5 for these structure names, additional features and their equivalents in other works.* AD *adhesorium;* BB *breach bolster, morphogenetically homologous with the plug;* CU *cushion;* NC *needle chamber;* NE *needle;* PL *plug;* PV *posterior vacuole;* SC *suppository chamber;* SK *skirt;* SP *sporidium wall precursor;* SU *suppository;* TA *tail;* TC *tube contents;* TU *tube.*

between the gun cell of *Haptoglossa* and the infection apparatus of *Plasmodiophora* can be related to the nature of the substrate and the subsequent development of the thallus.

The following steps will be required for attachment to a moving animal and penetration of the ***animal pellicle***, which is highly elastic as well as tough, and injection of the sporidium into the haemocoel.

1 biochemical recognition and ***instant*** attachment: possibly involving liquifaction and polymerization of a 'super-glue' from plug release

2 spring-loaded release of the penetration needle to effect millisecond penetration, followed by rapid dilation of the channel of ingress created by the needle to part fibres of the host pellicle

3 controlled pressure injection (posterior vacuolar pressure, perhaps from a glycerol-containing vacuole), together with a mechanism (cytoskeletal?) for ensuring the passage of ***all*** the vital organelles

4 biochemical protection against reaction to the 'foreign body' nature of the parasitic protoplast (possibly by preformed non-polymerized wall precursor uncovered by the 'inside-out' inversion of the parasite plasmamembrane)

The following steps will be required for penetration of the ***plant cell wall***, which is static, rigid and thick, and injection of the sporidium into the plant protoplast through the ***plasmamembrane***:

1 biochemical recognition and surface attachment: there is evidence of secretions in bore-fluid spread in *Plasmodiophora* possibly involving liquifaction and polymerization of a 'super-glue' from plug release

2 spring-loaded release of the penetration needle to effect millisecond penetration, followed by rapid dilation of the channel of ingress created by the needle **(a)** to part the cellulose fibres of the host wall and **(b)** to penetrate the host plasmamembrane

3 controlled pressure injection (posterior vacuolar pressure, perhaps from a glycerol-containing vacuole), together with a mechanism (cytoskeletal?) for ensuring the passage of ***all*** the vital organelles

4 biochemical protection against reaction to the 'foreign body' nature of the parasitic protoplast (possibly by parasite inversion of the parasite plasmamembrane so that host antigens are ineffective)

The two features critical to penetration are the spring-loaded nature of the penetration 'missile' and the propulsive force providing the 'hypodermic syringe' transfer through the hollow injection tube behind the needle. The former may be provided by the prolonged, often bowed 'tail-piece' to the penetration structure, together with a possible rapid change in molecular architecture affecting the shape of the missile and possibly lubrication of the missile. ***Controlled*** pressure injection is probably provided by posterior vacuolar enlargement and cavitation (note the electron-lucent zone described by all TEM accounts) between the missile and its tail-piece.

There are a number of important morphogenetic developments connected with the act of penetration by this ultrastructurally complex 'extrusome' that are still in question:

1 By what mechanism are the essential cell components (nucleus, centriole, mitochondrion) packaged during the firing of the gun cell?

2 How is the sporidium cell wall material transported and polymerized?

3 Is touch sufficient to trigger the gun cell?

4 From what organellar system is the 'extrusome' initiated and constructed in the glossoid cell?

The ultrastructure of the developing and mature 'extrusome' of *Polykrikos*

In the discussion of the structure and probable function of the various components of the extrusome of *Haptoglossa*, reference can usefully be made to the much more complex nematogene/taeniogene pairs of some dinoflagellates such as *Polykrikos* (see Dodge & Greuet, 1987; Hausmann, 1978; Westfall, Bradbury & Townsend, 1983 for TEMs and explanatory diagrams). The *Polykrikos* cell (*ca* 50 μm long) is many times larger than the cysts of *Haptoglossa* (*ca* 4-8 μm diam.) or the plasmodiophorids (*ca* 2-5 μm diam.) and its organelles could be expected to be larger and more complex.

The dinoflagellate *Polykrikos* has 8 nuclei per cell and 8 kinetids; it produces 8 nematogene/taeniogene pairs. Each nematogene/taeniogene pair is composed, in linear array, of: a nematogene; an intermediary mucoid body; a taeniogene. Associated with each nematogene/taeniogene pair is a kinetid, a dictyosome (Golgi body) and a nucleus. The attached kinetid may be responsible for the polar development of both the nematogene and the taeniogene in *Polykrikos* and this development may be used as a model for endoplasmic morphogenesis. In *Polykrikos* the morphogenesis of the nematocyst from the nematogene and of the taeniocyst from the taeniogene are both initially polarized. The differentiation of the nematogene into the nematocyst is different from that of the taeniogene into the taeniocyst, with differential syntheses occurring simultaneously at the ends proximal or distal to the kinetid. At maturity both the nematocyst and the taeniocyst possess a complex, polar-orientated internal structure. The (presumed) phagotrophic function of the 'captive' thread of *Polykrikos* would of course be different from the everting and penetrating force of the missile and posterior vacuole of *Haptoglossa* but morphogenetic homology may be postulated.

Although Hausmann (1978) treats the nematogene/taeniogene pair as an extrusome, its origin may be fundamentally different from the independently dictyosome-derived extrusomes involved, for example, in zoospore encystment and wall formation. A parallel may be drawn between the 8-nucleate *Polykrikos* with 8 nematogene/taeniogene pairs and the uninucleate gun cell of *Haptoglossa* with its single elaborate extrusome. The involvement of the kinetid in the formation of the nematogene/taeniogene complex of the dinomastigotes suggests that a similar but paired development may occur in *Haptoglossa* and the plasmodiophorids, since the uninucleate zoospore (or aplanospore) of these organisms possesses two kinetids (centrioles). The difference would be that in *Polykrikos* one kinetid is responsible for the initiation of both nematogene and taeniogene, whereas the two kinetids could both be functional in extrusome morphogenesis for *Haptoglossa* and the plasmodiophorids.

The role of the second *Haptoglossa* kinetid has not yet been elucidated: however, if one kinetid is the template for the development of the anterior vesicular complex of needle chamber, suppository chamber and anterior vacuole, it is possible that the second kinetid is the organizer for the tail and the posterior vesicle. The close association of the nucleus and dictyosome with the plasmamembrane of the tail housing is shown by Robb & Lee (1986*a*) and might be equivalent to the intermediate connective (intermediary mucoid body) of the nematocyst/taeniocyst complex of *Polykrikos*.

The ultrastructure of the mature 'extrusome' of *Haptoglossa*

Micrographs in Robb & Barron (1982), Robb & Lee (1986*a*, *b*), Lee *et al*. (1982) and Beakes & Glockling (1998) and interpretative diagrams (Lee *et al*., 1982: fig. 11; Beakes & Glockling, 1998) show the complexity of the mature 'extrusome' apparatus. Figure IV: 1A represents a concensus diagram. The mature gun cell projectile of *Haptoglossa* is composed of three sections, the needle, the suppository and the tail; the first two are anterior to the nucleus and the last is posterior. The needle and suppository occupy a single vesicle, but the development of each part is postulated separately under a series of somewhat arbitrary zones to facilitate discussion rather than to represent morphogenetic sequence or functional deployment, i.e.: .

1 needle chamber
2 suppository or 'bullet' chamber
3 anterior beak, tube ('gun-barrel') and plug
4 'hypodermic syringe' transformation
5 centriole-kinetid involvement
6 tail-piece
7 posterior vesicle

The needle chamber is ultrastructurally the most complex part. Figures IV: 1A (*Haptoglossa*) and IV: 1B (*Plasmodiophora*) enable the recognition of the following: a solid, slightly more electron-opaque central shaft, around which four serial, doughnut-shaped, schizogenous cavities have developed. The three anterior cavities define the anterior **cushion**, the **button** and the skirt-like **washer**. The posterior cavity is an elongate, perhaps lubricating, sleeve. The cushion and button, initially one homogeneous mass, have become separated by a circumscissile split perpendicular to the needle axis, and the cushion likewise becomes separated from the peripheral anterior part of the chamber. The structure termed the skirt has, in the past been termed the **harpoon** or **tether** but instead of forming barbs or backward-pointing rings or ridges, it appears that the anterior attachment to the button shears first, followed by a peripheral concentric rift between the outer layers of the needle chamber so that a ***forward-pointing*** cup with posterior attachment to the button, is delimited. Such a structure resembles a suction plunger.

Contiguous with the needle chamber is the suppository or 'bullet' chamber. This contains an electron-opaque core, the **suppository**, which is considerably wider than, but continuous with, the needle. The needle and the suppository constitute the 'missile'. It thus would have the function of a dilator following penetration by the needle, and the term 'bullet', which suggests a primary penetration organ, is less appropriate. It is possibly significant that the plasmamembrane lining the suppository chasmber is clearly defined and attached to the suppository at the posterior end. It is also apparent that electron-opaque material is present in this chamber between the vesicle plasmamembrane and the electron-opaque core.

Anterior to the needle chamber is a central shaft, the **tube**, **chute**, or **'gun-barrel'**, closed by a **plug**. Unfortunately, the initial ontogeny of the tube is not described in the fungal TEM studies. Lee *et al.* (1992) have pointed out that there is continuity between the outer wall layers of the first-formed cyst and the subsequently formed angular gun cell, but that the outermost wall layer does not cover the dome overlying the apex of the missile. Both groups of authors (Keskin & Fuchs, 1969, and Aist & Williams, 1971, for the plasmodiophorids) (Robb & Barron, 1982; Robb & Lee, 1986*a*, *b*; Lee *et al.*, 1982; Beakes & Glockling, 1998, for *Haptoglossa*) describe a plugged pore in this region at maturity, but do not provide evidence of the way in which plasmamembrane continuity between the cyst plasmamembrane and the tube plasmamembrane is established. Data from Glockling (1994) indicate that the plug is formed between the plasmamembrane and the extrusome membrane at the base of an invagination of the cyst wall, which thus forms the gun barrel. The plug can be presumed to be a differential deposit between the external plasmamembrane and the plasmamembrane of the needle vesicle. In *Haptoglossa* the electron-lucent plug is apparently similar to the electron-lucent swollen breech **bolster** or **torus**. The plug and breech bolster probably develop at the same time and in the same plane. Circumscissile fracture, parallel to the needle axis, of this differentiating electron-lucent deposit would then allow slippage and the growth or extrusion of the tube. The bolster or torus has also been termed a **sphincter** or **buttress** at maturity. The internal and external walls of the chute or gun-barrel are smooth and continuous with the spore cell wall, in contrast to the minutely papillate outermost spore cell wall layer. More recent studies (Beakes & Glockling, 1998) have shown that the length of the gun-barrel, and the resultant **'beak'** of the glossoid spore can develop rapidly, and at a late stage of development, so that an explanation of the papillate outermost wall discontinuity at the apex of the beak is available. The papillae have a more osmiophilic outline and a basal osmiophilic stratum and this layer can be seen to be clearly ruptured, but the papillate layer is not clearly differentiated in the micrographs of Robb & Lee (1985*b*) or Lee *et al.* (1992). The rupture confirms the later extrusive origin of the tube. The plug at the tip of the beak and its torus are thus displaced relative to each other with either the plug or the torus in the more anterior position.

The 'hypodermic syringe' transformation occurs at the base of the 'bullet' or suppository. The region between the suppository base and the tail piece is electron-lucent. As the suppository passes through the channel of ingress it must draw through the vital protoplasmic organelles of the parasite. In so doing, the plasmamembrane of the vesicle will be pulled inside out.

The tail piece structure is much less complex. The most detailed TEMs have been provided by Beakes & Glockling (1998) and show an electron-opaque core and surrounding housing; tangential sections revealing a construction of parallel fibres.

The kinetid connection is not established. However, in order for the discrete (non-membrane-bound), vital protoplasmic organelles (nucleus, centriole, mitochondrion, dictyosome) to be 'packaged' there must be some cytoskeletal involvement. The simplest postulate would be for the microtubules of the cytoskeleton, derived from the microtubule organizing function of the centriole-kinetid, to be employed.

The **posterior vesicle** is morphologically entirely separate but functionally intrinsically inseparable from the mechanism of the infection apparatus. From comparisons with other vesicles for which similar functions have been proposed, it is possible that this vesicle contains glycerol which has the capacity for rapid but not uncontrolled volume increase.

Comparative discussion

In species of *Haptoglossa* (Barron, 1980, 1981) the principal components of the glossoid cell, or gun cell, can be regarded as homologous with the rohr/stachel/schlauch complex of the plasmodiophorids (Robb & Barron, 1982) and these authors discussed the phylogenetic implications. The species of *Haptoglossa* differ from the plasmodiophorids, and from each other, in certain details. *Haptoglossa* shows the greater complexity: in general, the tube (rohr) and missile (stachel) are more elaborate than in the plasmodiophorids, but the anterior vesicular systems may be lacking. Robb & Lee (1986*a*, *b*) regard the complex cone at the tip of the stachel as a harpoon, but Lee *et al.* (1992) suggest that the apparatus may act as a sphincter for controlling the discharge of the needle. Although the ultrastructural details of this region differ amongst the species of *Haptoglossa*, the function and operation of the system are likely to be the same.

In both *Haptoglossa* and the plasmodiophorids the mature encysted spore contains a tubular cavity, the rohr, directed towards the host pellicle or cell wall. The cavity is separated from the environment by a plug. The ontogeny of this cavity is not known: either it develops from a vesicle, or it develops from an invagination of the plasmamembrane. Synthesis of the rohr plug is not accounted for in either case. The diagram in Aist & Williams (1971:fig. 31A) shows the plug as confluent with and not bounded by the cyst wall. The finding by Beakes & Glockling (1998) that the beak of *Haptoglossa* is a late and rapidly developing structure suggests that different morphologies between species can be explained by differential growth of the tube either side of the plug: greater synthesis peripheral to the plug will result in the plug becoming more deep-seatedwithin the tube; more equal synthesis will result in beak and tube elongation with the plug in the apical position.

Synthesis of the plasmodiophorid rod or stachel (needle equivalent) is described as occurring extracellularly within the rohr cavity. The rohr cavity is more likely to be a composite of the tube and the needle chamber of *Haptoglossa*. Internal to the base of the stachel is a narrow continuation of the rohr cavity, the schlauch. The rohr and schlauch plasmamembrane is continuous. The rohr cavity is lined with an extra-plasmamembrane wall precursor and in *Plasmodiophora* eversion and stretching of the rohr wall (the *Haptoglossa* beak equivalent) results in the formation of the adhesorium. The contents of the rohr are also considered to bind the adhesorium to the host. The adhesorium contains dense vesicles and membranes derived from the cyst cytoplasm proximal to the rohr. The development and behaviour of the "stilett" of *Polymyxa* is not clear from the description or micrographs, but appears to be intermediate between *Haptoglossa* and *Plasmodiophora*. Alternatively, if an extrusomal origin for the rohr/stachel is accepted, there must be a mechanism for the complex longitudinal structural differentiation that develops. In *Haptoglossa* the only evidence for eversion comes from the possibly abnormal behaviour of non-functional gun cells; concentration of dense, vesiculate cytoplasm between the barrel and the bore does not occur to anything like the same extent, although a prominent apical vesicle is present beneath the tube.

In the plasmodiophorids further evagination injects the stachel through the root hair wall and its plasmamembrane, and this is followed by the schlauch. As the schlauch everts, its plasmamembrane encloses the nucleus and other organelles. The stachel becomes detached and the uninucleate plasmodium resides ***within*** the host protoplast. The whole process probably depends on the expansion of the prominent posterior vesicle of the spore cyst.

The earliest study (Keskin & Fuchs, 1969), of *Polymyxa* did not show or consider features that are now of essential interest, such as the morphogenesis of the extrusome organelle; an explanation of the eversion procedure, or the relationships of the penetration apparatus with the nucleus or other organelles. However, both Keskin & Fuchs (1969) and Aist & Williams (1971) proposed that the rohr and schlauch of the plasmodiophorids constituted an ***invagination*** of the cyst so that their contents (the stachel and tail) were formed extracellularly, and that there was an extracellular plug to the invagination. The authors' claims were based on the fact that the cell walls of the cyst and the invagination appeared to be confluent and that the inner wall of the rohr evaginated on discharge of the projectile. The hypothesis did not explain how the uninucleate amoeboid parasite was transferred through this wall to the host cytoplasm. Whilst it is possible that the 'extracellular' nature of the rohr (and schlauch?) could be perceived as an invagination equivalent to that of a flagellar vestibule as in *Rozella* (Held, 1981), it then becomes more difficult to account for the subsequent elaborate morphological construction of the stachel outside the cell wall and the development of the plug after invagination. The much later development of the tube of *Haptoglossa* by wall dissolution, plasmamembranic extension and new cell-wall secretion (Robb & Lee, 1985*a*, *b*; Beakes & Glockling, 1998) could also be used as evidence against an extracellular development of the entire projectile housing (the tube together with the rohr-plus-schlauch). In the plasmodiophorids there is evidence of early initiation of the apparatus with very late and rapid final development.

Compared with the plasmodiophorids, the development of the glossoid cell of *Haptoglossa* represents a stage equivalent to the encystment and penetration of the plasmodiophorid zoospore and the act of penetration. The glossoid cell can be formed from either aplanosporic or zoosporic cysts. Its development is rapid and the relatively enormous inclusion is reponsible for the deformed cell shape (Robb & Lee, 1985*ab*) not evident in the plasmodiophorid cysts. Although wall formation around cysts of *Phytophthora* can be very rapid (see references p. 46), it is improbable that the major part of the gun cell inclusion could be formed by invagination of the plasmamembrane, followed by extraplasmamembranic wall secretion, after the emergence of the gun cell protoplast from its precursor cyst. A precursor vesicular origin of the inclusion (thereby an extrusome) is more likely, and the difficulty experienced by Robb & Lee (1985*ab*) in finding very early developmental stages of the inclusion would support this view.

The more elaborate, 'harpoon-like' missile of *Haptoglossa* has a similar ontogeny to that of the stachel of *Plasmodiophora*, its different structure presumably providing a more efficient entry through the tough, elastic animal pellicle, without the need to penetrate rigid cellulose cell walls. The missile is also discarded; it is not seen in association with discharged but unsuccessful gun cells. The intercalary swelling in discharged but unsuccessful gun cells probably represents the junction between the schlauch and rohr equivalents: the undischarged and discharged lengths of the proximal part are comparable. There would be no need for the rohr equivalent to develop into the adhesorium prior to penetration of a plant cell wall by the more streamlined stachel. The tube distal to the intercalary swelling can be equated with the schlauch and its extrusion inside the host animal would place the sporidium deep in the host tissues. Ontogenetic information is needed before a comparison of these structures with the raphids of the raphidophytes and the extrusomes of the cryptophytes and euglenoids can be made, but the closest similarities between elaborate extrusomes appear to be between those of *Haptoglossa* and the plasmodiophorids. In *Haptoglossa* the tube can be curved. This represents a further complication to the mechanism, but does provide an explanation for the construction of the washer or skirt, which would therefore guide the tip of the needle through the central axis of the tube.

The cysts of the plasmodiophorids do not have any thickening of the cyst wall around the orifice of the rohr. In *Haptoglossa* there is a swollen ring of electron-lucent wall material (the torus from the plug initial) which constitutes a turgid, tyre-like breech bolster to the tube of *Haptoglossa*. Differences in functional details of the infection process could therefore be anticipated.

There must be a cytoskeletal coronet or comb of microtubules derived from the kinetosome directed forwards to act as a template for the tube through which the missile will be expelled. Such synthesis provides a role for the anterior vesicle. This cytoskeleton must be established at the time of tube morphogenesis, but subsequent secretion of cell wall material will render the cytoskeleton superfluous in this region and thus it may not persist in mature gun cells.

The kinetosome of flagellates is usually discussed in the context of the transitional plate and flagellum. The transitional plate of many (e.g. peronosporomycetes) but not all (e.g. plasmodiophorids) flagellates is unique in that it is attached to the cell boundary plasmamembrane and thus transects the cytoplasm of the flagellum and the cell. In some straminipilous organisms (*Schizochytrium* - Kazama (1974*a*, 1980; see p. 37) the transitional plate has a complex and cone-like ultrastructure.

The centriole-kinetid hypothesis has the advantage that it is known to initiate polar development within the protoplast. Apart from the polarity of the mature kinetosome, attached to the flagellar axoneme at the distal end and the microtubular organizing centre at proximal end, the formation of the transitional plate occurs at an early stage in the transformation of the centriole into the kinetid, prior to the development of the flagellar axoneme. In *Myzocytiopsis* the transitional plate, developing at one end of the kinetid, does not appear to be attached to a flagellar vesicle (cf. *Phytophthora*). The transitional plate has all the 'attributes' necessary to link: the plasmamembrane; the kinetid, and thorough the kinetid, the cytoskeleton and the nucleus. Such a linkage is essential if the vital contents of an endoparasitic cell are to be packaged and delivered into the cytoplasm or haemocoel of a host. Another requirement of this transfer is that the diameter of the passage through the host cell wall or cuticle is sufficiently wide to enable the passage of centriole, nucleus and mitochondrion. These structures are deformable, but not infinitely so. Any penetration implement must achieve breakthrough and appropriate dilation of the aperture. The minimum diameter could be approximate to that of the centriole. This argument could lead to the astounding speculation that the missile (stachel) is the homologue of the transitional plate. The diameters of the missile, centriole/kinetosome and the flagellum are comparable. Aist & Williams (1971) stated "The stachel, which was pushed aside when the parasite entered the host cell, was also commonly found later embedded in the papilla." - it would be surprising that a mechanism for penetration of the cell wall was not also responsible for breaching the host plasmamembrane. My interpretation would be that the stachel and plasmodiophorid amoeba are both deposited ***within*** the host cytoplasm, and that the rapid formation by the host of the 'papilla' between the host cell wall and the host plasmamembrane by vesicle excretion would also include vesicular excretion of the stachel. Other suggestions could be that the casting of flagella might be a possible analogue of the discarding of the stachel tail and that the semi-conservative renewal and synthesis of the second kinetosome could account for the anisokont condition of the flagellate cell and the differential development of the anterior and posterior parts of the infection apparatus.

The expelling force of the gun cell almost certainly originates from the turgor of the large posterior vesicle as suggested by Keskin & Fuchs (1969), Aist & Williams (1971) and Robb & Lee (1985*ab*). However, the recoil from the gun in *Haptoglossa* must be severe, and it is difficult to reconcile the tube evertion mechanism of spent (non-functional) gun cells in *H. mirabilis* (Barron, 1980; Robb & Barron, 1982) and *H. humicola* (Barron, 1981), with those that have penetrated a host (for which information is not available from any observer). If the bore of the tube everted relative to the barrel of the tube, this would dissipate a considerable part of the thrust of the charge, which comes from within a small cyst only loosely attached to detritic substrata. I suggest that the tip of the barrel probably adheres to the host cuticle to lock the barrel in place, and that evertion of the bore wall (such as occurs with the inner rohr wall in *Plasmodiophora* and *Polymyxa*) does not occur in functional cells of *Haptoglossa*. Drechsler (1940) originally suggested adhesion by the ventral surface of the glossoid tube, prior to injection of the sporidium. The cavity of the bore is occluded by a foamy material, which is apparently more dense against the wall and could be the adhesive primer.

The different properties of an adult nematode cuticle and a root hair cell wall are obvious, hence a less elaborate device would suffice for the latter. In the plasmodiophorids it is not surprising that there may be only a simple tip cushion as in *Plasmodiophora*, or the apparent absence of any tip elaboration as in *Polymyxa*. In *Haptoglossa* the tip of the needle possesses a series of flanges which have been described as the harpoon (Robb & Lee, 1985*a*, *b*) or the sphincter (Lee *et al*., 1992). Robb & Lee (1985*a*, *b*) presume penetration of the entire tip into the host. While a harpoon is more effective at anchoring a missile, it is likely to be less effective than a needle for penetration. The harpoon hypothesis is not compatible with the ultrastructural data since the schizogenous planes are such that a backwardly directed frill would not be produced. Lee *et al*. (1992) suggest a static valve action; there is no need to postulate contractile concentric fibres acting as a sphincter which would act as a brake on the mechanism. Beakes and Glockling (1998) regard the structure as a tether but this would only be necessary if there were a malfunction of the plug unit. I envisage the structure around the needle tip as moving to the apex of the tube where it would enhance the adhesive capacity of extracellular material and provide a guide to the penetration of the cuticle by the needle. The impact of this finely pointed needle, clasped by the washer (or skirt), on to the strong, elastic nematode cuticle would result in penetration, whereas a larger diameter projectile tip might fail to achieve this penetration.

Relationships of *Haptoglossa*

Despite their similar hosts and parasitic habits, any systematic association between *Haptoglossa* and *Myzocytiopsis* is extremely doubtful. Until further detailed comparative ultrastructural studies have been made, an association with plasmodiophorids is equally questionable. The genus should be placed in its own family, the *Haptoglossaceae*, and the family in its own order, the *Haptoglossales*. *Haptoglossa* is an anamorph genus, *incertae sedis*, but the flagellar insertion and non-septate thalli suggest similarities with the *Ectrogellaceae*, a family perhaps also possessing characters of both the the *Myzocytiopsidales* and the *Plasmodiophorales*.

Some comparisons may also be made with *Endemosarca* and the microsporidians, both of which also have eversion mechanisms for the transfer of parasite protoplasm into the host. Microsporidians differ in many fundamental criteria (they have 'bacterial' RNA: 16S + 23S and no 5S; they do not have centrioles or mitochondria (Canning *et al.*, 1992). However, Hirt *et al.* (1997) have suggested that the mitochondrion may have been lost and that the organisms may have fungal affinities. Microsporidians have a rudimentary dictyosome; a chitinous cell wall and a nuclear pore plaque as a spindle organizing centre - the spindle is the only site of microtubules. Microsporidian cells are 2-20 μm diam.; nuclei *ca* 0.5 μm (500 nm) diam.; the polar tube is *ca* 0.1 μm diam., but it dilates to allow passage of ribosomes; the thickness of the plasmamembrane is also distinctive.

The *Endemosarcaceae* (see p. 262) have been tentatively placed close to the plasmodiophorids (Olive, 1975). They are parasites of protozoans. They possess anisokont flagellation and flagellar insertion is sub-apical in a shallow groove. Penetration of the host is by an evertible tube preformed in the infection spore, but many more details of the ultrastructure are required before comparisons can be made.

Ectrogella

Ectrogella was erected by Zopf (1884) for a diatomaceous parasite with an apparently *Saprolegnia*-like zoosporogenesis and a zoosporangium that develops more than one exit tube. Since then three further species have been described from freshwater diatoms, three from marine diatoms and one from a marine green alga (*Chlorodendron*, now *Tetraselmis*).

The type species, *Ectrogella bacillariacearum*, together with *E. gomphonematis* and *E. eunotiae*, are characterized by extremely small zoospores (volume equivalent <20 μm^3) and are said to have posterior refractive granules. On morphological grounds (non-septate thallus, endosporangial zoosporogenesis with multiple exit tubes, small zoospores with posterior refractive granules) there would appear to be considerable resemblance between *Ectrogella* and *Haptoglossa*. Scherffel (1925*a*), working with the type species, described and drew the first-formed zoospores ***and*** those emerging from zoospore cysts as pyriform or ovoid with sub-apical flagellar insertion; therefore there is no evidence of dimorphism. The zoospores emerging from the cysts are shown with different relative flagellar lengths. None of these three species has a sexual or resting stage, and the sexual stage of the marine species *E. perforans* is poorly described. Thus the evidence for placing the genus in the *Saprolegniales* (Sparrow, 1960; Karling, 1981*a*) is tenuous, resting entirely on the intrasporangial cleavage and motility of the zoospores prior to discharge.

The marine *Ectrogella perforans* is straminipilous (Raghu Kumar, 1980*b*), but most of other critical details of the transitional zone have not been determined. Comparisons with *Lagenisma* with respect to life-history, flagellar ultrastructure and molecular biology could be valuable.

Ectrogella is sometimes regarded as being partly epibiotic because of the manner in which the expanding and maturing sporangium forces apart the two frustules of the diatom cell. The distinction between this development and that of a thick-walled base to the exit tube (a 'forcing apparatus') may be host-related, or due to a different emphasis given by the observer. I discount it as a generic taxonomic character.

The restriction to a range of diatomaceous hosts is probably diagnostic for *Ectrogella*. In a coevolutionary perspective *Ectrogella* and the diatoms bridge the freshwater/marine environmental divide. For these reasons I have transferred the diatom parasites, *Lagenidium cyclotellae* (sporangia of which may have more than one exit tube, intrasporangial flagellate motile spores and similar zoosporangial discharge to *Ectrogella*), and *L. brachystomum* (with a sub-frustular thickening to the exit tube and extrasporangial cleavage) to *Ectrogella*, even though Scherffel (1925*a*) figured 'sexual' stages.

Ectrogella monostoma has much larger zoospores (volume equivalent > 150 μm^3), a superficially *Achlya*-like (*Aphanomyces*-like?) aplanospore discharge and a wide diatomaceous host range; it is treated as a synonym of *Aphanomycopsis bacillariacearum*.

Karling (1981*a*) considered that there is considerable scope for synonymy in the genus. I agree: the host range for the type species is such that *E. gomphonematis* and *E. eunotiae* cannot be distinguished from small-thalloid forms of *E. bacillariacearum*. Similarly *E. licmophorae*, from the Adriatic Sea, is probably conspecific with *E. perforans*. *Ectrogella eurychasmoides* appears to differ from *E. perforans* by the emergence of the sporangial exit tube ***through*** the frustule and by the propensity for premature encystment of the zoospores within the sporangium, but it shares the same host genus (*Licmophora*) and is doubtfully distinct. Petersen (1909), Höhnk (1939) and Aleem (1953) contended that their collections of *E. perforans* had only a single posteriorly directed flagellum when swimming, which, according to Sparrow (1960: 809), would indicate that two taxa, *Ectrogella* and *Pleotrachelus*, may be involved in the same host. Such a relationship between these two genera would be problematic; the only diagnostic difference being flagellar number (see discussion of *Pleotrachelus*).

It is possible that the marine taxa *E. perforans* and *E. marina* may be misplaced in *Ectrogella*, though for different reasons. *Ectrogella perforans* has been studied by TEM, but may not be congeneric with the less-well-documented type species. Detailed comparisons of the life-history and critical ultrastructural features vis-à-vis *Lagenisma* would be valuable. *Ectrogella marina* was placed in this genus primarily because of the diplanetic zoospore behaviour, although dimorphism was not established and flagellar number is doubtful; Dangeard (1912) thought that there was only one flagellum, but according to Feldmann & Feldmann (1956) there 'seem to us to be two flagella, laterally inserted'. This species is parasitic in a marine green alga and is more appropriately associated with *Sirolpidium* as a doubtful member of that genus.

The little-known fungus, *Ectrogella lauderiae*, too inadequately described to be other than a doubtful taxon, is parasitic in *Lauderia*, a marine centric diatom related to *Cyclotella* and *Coscinodiscus*, both of which also have 'lagenidiaceous' parasites.

In the absence of any undoubted positive saprolegnialean characters in the type species, the genus is retained as the type of the family *Ectrogellaceae*. *Ectrogella*, as represented by its type species, is regarded as a phytophagic analogue of *Haptoglossa*. The family is doubtfully associated with the *Myzocytiopsidales*: there is the possibility that flagellar ornamentation may be heterogeneous in this order (see discussion above, with reference to the flagellar ornamentation of *Crypticola* compared with the marine *E. perforans*).

The family concept (i.e., genera included in the family) has had a chequered history. Şcherffel (1925*a*) designated the family after his reappraisal of *Ectrogella* (with the inclusion of *Eurychasma*) and placed his other new genus, *Aphanomycopsis*, in the family. Sparrow (1960) included *Ectrogella*, *Eurychasma*, *Eurychasmidium*, *Aphanomycopsis* and *Pythiella* in the family, but later (Sparrow, 1976) excluded *Eurychasma*, *Eurychasmidium* and *Ectrogella eurychasmoides* (without renaming the taxon) while retaining *Pythiella* in the family (albeit as the sole periplasmic genus in the "Saprolegnian galaxy"). Sparrow (1976) and Karling (1981*a*) accepted the placement of *Aphanomycopsis* in the *Leptolegniellaceae* and Karling (1981*a*) regarded the family as monogeneric. The family concept adopted here is essentially that of Karling (1981*a*) except that I do not accept that the family diagnosis has been accurately researched in any respect.

OLPIDIOPSIS AND *PLEOCYSTIDIUM*

The first member of this genus to be recognized and described was *Olpidiopsis saprolegniae* by Braun (1855), who named it as *Chytridium saprolegniae*. Cohn (1854) had seen, described and figured *Peronium aciculare*, which while probably assignable to the genus *Olpidiopsis*, cannot be equated with a particular species because the host was not identified; this genus and species name must therefore be rejected. The genus was established by Cornu (1872*b*) for *O. saprolegniae* and four other parasites of *Saprolegniaceae*. In the following three decades ten species were described from algae, relying on the simple spherical or ellipsoidal thallus form, biflagellate zoospores, and, when present, the sexual union that developed between two contiguous thalli from different zoosporic infections. A feature of the type species is the accretion of extra-mural fibrils (spines) around some thalli, especially thalli functioning as receptive gametangia and becoming resting spores. Fischer (1892*a*) attempted to distinguish between the 'teleomorphic' species (*Olpidiopsis*) and the 'anamorphic' species (*Pseudolpidium*), although he placed forms with echinulate 'resting sporangia' in the latter genus, and both genera were typified by *Olpidiopsis saprolegniae pro parte*. Sparrow (1960) and Karling (1981*a*) included species from three major and distinct groups of hosts: peronosporomycetes; freshwater green algae, and marine red (perhaps rarely chromophyte and green) algae (see Table IV: 6). A few parasites from other hosts have also been assigned to this genus so that it is now difficult to characterize *Olpidiopsis sensu lato* (see Table IV: 7). I have therefore brought *Pleocystidium* C. Fisch (1884*a*) out of synonymy with *Olpidiopsis* for parasites in freshwater algae.

The zoospore cyst is usually spherical and the penetration of the host is by a thin tube, releasing a naked endobiotic uninucleate plasmodium. The empty cyst is persistent outside the host. In the type species several thalli, resulting from multiple infections, are often found in a single host hypha. Infection stimulates considerable hypertrophy of the hypha, with cessation of apical growth. The extent of the infection can affect parasite thallus morphology, which is thus an unreliable taxonomic character: for example, a heavy infection of *O. fusiformis* in a young *Achlya radiosa* colony was seen (unpublished notes) to produce up to ten or more small sub-spherical thalli in some hyphae and single large allantoid thalli in others. In *O. gracilis* Pemberton *et al.* (1990) have described a different pattern of disease development, in which the parasite induces a large diverticulum from the hyphal wall ***opposite*** the site of penetration. Multiple occupancy of this diverticulum by two or more thalli was not explained, but could be due to division in early (plasmodial?) vegetative stages. The plasmodium tends to develop a more or less ovoid to spherical shape and wall formation follows, resulting in a sub-spherical thallus lying free in the lumen of the host. The thallus converts to a single zoosporangium, which in most species produces a single (rarely two) discharge tube(s). Cleavage is intrasporangial. Within the genus *sensu lato* there is variability in zoospore size and shape, flagellar insertion, relative flagellar lengths, and the occurrence of refractive granules in the zoospore.

Heterokont flagellation has been established ***only*** for *Olpidiopsis sensu lato* parasites of the *Peronosporomycetes*, but even species in this host group two different kinds of anisokont flagellation exist. In the species parasitic in *Saprolegnia* (Bortnick *et al.*, 1985) and *Achlya* (Martin & Miller, 1986*b*) the implication is that the anisokont pattern is as in the *Saprolegniales*, with the two flagella sub-equal in length or with the straminipilous flagellum the shorter. However, in *Gracea* (*Olpidiopsis*) *gracilis* the straminipilous flagellum is longer (Dick, 1997*a*, 2000*d*).

The resting spore is normally accompanied by one or more companion cells which function as donor gametangia. The companion cell is always appreciably smaller (volume relationship *ca* 1:3) than the receptor gametangium (oogonium); a feature in contrast to the situation in *Myzocytiopsis*. The gametangia are presumed to arise from multiple zoospore infections which develop into contiguous thalli, rather than cleavage or partition of the single primary thallus. Drawings for algal parasites by Wildeman (1896*b* - *O. fibrillosa*) and Dasgupta & John (1988 - *O. sphaerica*) could indicate thallus fragmentation prior to gametangial formation (cf. *Syzygangia*). The companion cell of *O. appendiculata* is exceptional in being vermiform with a distal dilation (perhaps a non-contiguous thallus with a fertilization hypha - cf. *Eurychasmopsis*?). Fertilization may involve almost total amalgamation of the gametangial protoplasts, and the process is mediated by a very short injection tube or narrow pore (Martin & Miller, 1986*c*), the adjacent protoplasm containing a distinctive array of numerous small vesicles similar to those associated with fertilization in *Phytophthora* and *Achlya*. The resting spore (oospore) is strictly plerotic. Sexual reproduction is therefore apparently distinct from the *Myzocytiopsidales* on several criteria. Barrett (1912) and McLarty (1941*b*) carried out cytological work on *Olpidiopsis* parasites of *Saprolegniaceae*, and Martin & Miller (1986*c*) have shown that the nuclear cycle in *O. varians* is haplomitotic B (diploid mitosis), but the nuclear cycle has not been established for any species in non-saprolegniaceous hosts.

Wildeman (1896*b*) separated the species of *Olpidiopsis* on the basis of the nature of the extra-mural deposit on the resting spore, recognizing four categories: verrucose; spinose; fibrillose and smooth. However, the exosporial investment is often variable within a species, or even within the same infected part of the host. Its taxonomic significance is doubtful unless correlated with other criteria.

TABLE IV: 6. *Tabulations of* Olpidiopsis *species according to hosts*

Olpidiopsis: ***alphabetic list of binomials with hosts in*** **Saprolegniomycetidae**

Olpidiopsis achlyae	*Achlya*
Olpidiopsis aphanomycis	*Aphanomyces*
Olpidiopsis braziliensis	*Achlya*
Olpidiopsis echinata	*Saprolegnia*
Olpidiopsis endogena	*Aphanomyces*
Olpidiopsis fusiformis	*Achlya*
Olpidiopsis fusiformis forma *achlyae*	*Achlya*
Olpidiopsis index	*Achlya*
Olpidiopsis incrassata	*Achlya*
Olpidiopsis irregularis	*Achlya*
Olpidiopsis luxurians	*Aphanomyces*
Olpidiopsis major	*Saprolegnia*
Olpidiopsis milleri	*Saprolegnia*
Olpidiopsis minor	*Achlya*
Olpidiopsis saprolegniae	*Saprolegnia, Achlya*
Olpidiopsis saprolegniae var. *laevis*	*Saprolegnia*
Olpidiopsis saprolegniae var. *indica*	*Saprolegnia*
Olpidiopsis spinosa	*Achlya*
Olpidiopsis varians	*Achlya*
Olpidiopsis verrucosa	*Achlya*
Olpidiopsis vexans	*Saprolegnia*
Olpidiopsis sp. (Johnson, 1972)	*Saprolegnia*
Olpidiopsis sp. (Howard & Johnson, 1969)	*Aphanomyces*
Olpidiopsis sp. (Karling, 1987*b*, 1988)	*Brevilegnia*

(Table IV: 6, continued)

Olpidiopsis: ***alphabetic list of binomials with hosts in*** **Pythium**

Olpidiopsis brevispinosa	*Pythium*
Olpidiopsis curvispinosa	*Pythium*
Olpidiopsis gracilis	*Pythium*
Olpidiopsis gracilis forma *curvispinosa*	*Pythium*
Olpidiopsis indica	*Pythium*
Olpidiopsis pythii	*Pythium*
Olpidiopsis pythii forma *brevispinosa*	*Pythium*

Olpidiopsis: ***alphabetic list of binomials with hosts in other*** **Peronosporomycotina**

Olpidiopsis myzocytia	*Myzocytium proliferum*

Olpidiopsis: ***alphabetic list of binomials with hosts other than*** **Peronosporomycotina**

Olpidiopsis andreei	*Cladophorales*, (also *Ectocarpus*?)
Olpidiopsis antithamnionis	*Antithamnion* (red alga)
Olpidiopsis appendiculata	*Zygnematales*
Olpidiopsis cruciferarum	*Arabis alpina*
Olpidiopsis dangeardii	*Radicilingua* (red alga)
Olpidiopsis deformans	*Draparnaldia* (green alga)
Olpidiopsis decipiens	*Oedogonium*
Olpidiopsis elliptica	*Zygnematales*
Olpidiopsis feldmannii	*Asparagopsis* (*Falkenbergia*), *Trailliella* (red algae)
Olpidiopsis fibrillosa	*Zygnematales*
Olpidiopsis gillii	freshwater diatoms
Olpidiopsis glenodiniana	marine dinoflagellate
Olpidiopsis globosa	*Papulospora* (marine hyphomycete)
Olpidiopsis karlingiae	*Karlingia* (chytrid)
Olpidiopsis longicollis	*Vampyrella* (protist)
Olpidiopsis magnusii	*Ceramium* (red alga)
Olpidiopsis oedogoniarum	*Oedogonium*
Olpidiopsis paradoxa	*Rhizophydium* (chytrid)
Olpidiopsis parasitica	*Zygnematales*
Olpidiopsis ricciae	*Riccia* (bryophyte) rhizoids
Olpidiopsis schenkiana	*Zygnematales*
Olpidiopsis sorokinei	*Tribonema* (golden-green alga)
Olpidiopsis sphaerica	*Zygnematales*
Olpidiopsis sphaeritae	*Sphaerita* (rozellopsid)
Olpidiopsis tumefaciens	*Ceramium* (red alga)
Olpidiopsis ucrainica	arthropod
Olpidiopsis verrucosa var. *radicicola*	*Arabis alpina* (1964).
Olpidiopsis vuilleminii	basidiomycete basidia
Olpidiopsis zopfii	*Zygnematales*
Olpidiopsis sp. (Aleem, 1953)	*Navicula* (diatom)
Olpidiopsis sp. (Johnson, 1957*a*)	*Licmophora* (diatom, see also *Olpidium* in *Melosira*)
Olpidiopsis sp. (Sparrow, 1957)	*Zygnematales* (*Mougeotia*)
Olpidiopsis sp. (Johnson, 1966*d*)	*Navicula* (diatom)
Olpidiopsis sp. (Sparrow, 1969)	*Porphyra* (red alga)

TABLE IV: 7. Olpidiopsis *species: alphabetic list with synonymy in this text*

Olpidiopsis achlyae	see *Olpidiopsis fusiformis sensu lato*
Olpidiopsis andreei	synonym of *Sirolpidium andreei*
Olpidiopsis antithamnionis	synonym of *Pontisma antithamnionis*
Olpidiopsis appendiculata	synonym of *Pleocystidium parasiticum*
Olpidiopsis aphanomycis	*Olpidiopsis aphanomycis*
Olpidiopsis brevispinosa	see *Olpidiopsis pythii sensu lato*
Olpidiopsis braziliensis	see *Olpidiopsis fusiformis sensu lato*
Olpidiopsis cruciferarum	synonym of *Olpidium cruciferarum*
Olpidiopsis curvispinosa	see *Olpidiopsis pythii sensu lato*
Olpidiopsis dangeardii	synonym of *Pontisma dangeardii*
Olpidiopsis deformans	synonym of *Plasmophagus deformans*
Olpidiopsis decipiens	doubtful taxon placed with *Syzygangia*
Olpidiopsis echinata	synonym of *Olpidiopsis saprolegniae*
Olpidiopsis elliptica	synonym of *Pleocystidium parasiticum*
Olpidiopsis endogena	synonym of *Cornumyces endogena*
Olpidiopsis feldmannii	synonym of *Pontisma feldmannii*
Olpidiopsis fibrillosa	synonym of *Pleocystidium parasiticum*
Olpidiopsis fusiformis	see *Olpidiopsis fusiformis sensu lato*
Olpidiopsis fusiformis forma *achlyae*	see *Olpidiopsis fusiformis sensu lato*
Olpidiopsis gillii	synonym of *Ectrogella bacillariacearum*
Olpidiopsis glenodiniana	? synonym of *Sirolpidium glenodinianum*
Olpidiopsis globosa	? synonym of *Sirolpidium globosum*
Olpidiopsis gracilis	synonym of *Gracea gracilis*
Olpidiopsis gracilis forma *curvispinosa*	see *Olpidiopsis pythii sensu lato*
Olpidiopsis index	see *Olpidiopsis saprolegniae sensu lato*
Olpidiopsis incrassata	see *Olpidiopsis fusiformis sensu lato*
Olpidiopsis indica	see *Olpidiopsis saprolegniae sensu lato*
Olpidiopsis irregularis	synonym of *Cornumyces irregularis*
Olpidiopsis karlingiae	synonym of *Cornumyces karlingiae*
Olpidiopsis longicollis	*nomen nudum*, on *Vampyrella*
Olpidiopsis luxurians	see *Olpidiopsis aphanomycis sensu lato*
Olpidiopsis major	see *Olpidiopsis saprolegniae sensu lato*
Olpidiopsis magnusii	synonym of *Pontisma magnusii*
Olpidiopsis milleri	synonym of *Olpidiopsis saprolegniae*
Olpidiopsis minor	see *Olpidiopsis fusiformis sensu lato*
Olpidiopsis myzocytia	synonym of *Myzocytium proliferum*
Olpidiopsis oedogoniarum	synonym of *Syzygangia oedogonii*
Olpidiopsis paradoxa	synonym of *Sirolpidium paradoxum*
Olpidiopsis parasitica	synonym of *Pleocystidium parasiticum*
Olpidiopsis pythii	*Olpidiopsis pythii*
Olpidiopsis pythii forma *brevispinosa*	see *Olpidiopsis pythii sensu lato*
Olpidiopsis ricciae	synonym of *Syzygangia elliptica*
Olpidiopsis saprolegniae	*Olpidiopsis saprolegniae*
Olpidiopsis saprolegniae var. *laevis*	see *Olpidiopsis saprolegniae sensu lato*
Olpidiopsis saprolegniae var. *indica*	see *Olpidiopsis saprolegniae sensu lato*
Olpidiopsis schenkiana	synonym of *Pleocystidium parasiticum*
Olpidiopsis sorokinei	? synonym of *Syzygangia sorokinei*
Olpidiopsis sphaerica	synonym of *Pleocystidium parasiticum*
Olpidiopsis sphaerita	? autoparasitism
Olpidiopsis spinosa	see *Olpidiopsis fusiformis sensu lato*
Olpidiopsis tumefaciens	synonym of *Eurychasmidium tumefaciens*
Olpidiopsis ucrainica	synonym of *Myiophagus ucrainicus*
Olpidiopsis varians	see *Olpidiopsis fusiformis sensu lato*
Olpidiopsis verrucosa	see *Olpidiopsis fusiformis sensu lato*
Olpidiopsis verrucosa	synomym of *Olpidium verrucosa* var. *radicicola*
Olpidiopsis vexans	see *Olpidiopsis saprolegniae sensu lato*
Olpidiopsis vuilleminii	synonym of *Pleotrachelus vuilleminii*
Olpidiopsis zopfii	synonym of *Pleocystidium parasiticum*

(Table IV: 7, continued)

Olpidiopsis sp. (Aleem, 1953)	compare with *Ectrogella perforans*
Olpidiopsis sp. (Johnson, 1957*a*)	synonym of *Ectrogella perforans*
Olpidiopsis sp. (Sparrow, 1957)	synonym of *Pleocystidium parasiticum*
Olpidiopsis sp. (Johnson, 1966*d*)	compare with *Ectrogella perforans*
Olpidiopsis sp. (Sparrow, 1969)	compare with *Pontisma* spp., *Olpidiopsis* sp.
Olpidiopsis sp. (Howard & Johnson, 1969)	compare with *Olpidiopsis aphanomycis sensu lato*
Olpidiopsis sp. (Johnson, 1972)	compare with *Olpidiopsis saprolegniae sensu lato*
Olpidiopsis sp. (Karling, 1987*b*)	compare with *Olpidiopsis fusiformis sensu lato*
Olpidiopsis sp. (Karling, 1988)	compare with *Olpidiopsis fusiformis sensu lato*

Olpidiopsis species parasitic in *Peronosporomycetes*

Host specificity of the oomycete parasites has been studied by Shanor (1940), Whiffen (1942), Johnson (1955*a*), Slifkin (1961), Martin & Miller (1985) and Pemberton *et al.* (1990). For the parasites of *Saprolegniaceae*, *Olpidiopsis saprolegniae* (Shanor, 1940) and "*O. incrassata*" (*O. vexans*?) (Slifkin, 1961) were found to be confined to various species of *Saprolegnia* and *Isoachlya*, while *Achlya* was infected by *O. varians* and *O. fusiformis*. (Martin & Miller (1985) grew *O. varians* in *Aplanopsis terrestris* Höhnk.) The species of *Achlya* infected were from different taxonomic groups (the racemose and eccentric groups of Dick, 1973), but Shanor was unable to establish infection in other *Achlya* species related to the eccentric group (*Achlya sensu stricto*). Nevertheless, there are reports of *O. saprolegniae* and *O. incrassata* on *Achlya* (e.g. Coker, 1923; Cornu, 1872; Sparrow, 1933*c*), *O. varians* from *Isoachlya* (Srivastava, 1964) and *O. fusiformis* from *Saprolegnia* (Sparrow, 1932*b*). Shanor (1940) could not establish infection of *Achlya flagellata* W. C. Coker by *O. fusiformis*, but this species of *Achlya* was infected by *O. achlyae* (McLarty, 1941*a*) and *O. spinosa* Tokunaga (1933). This confusion is still evident as shown by the study of Willoughby & Roberts (1992), who reported a study of *O. incrassata* on and confined to *Saprolegnia*, despite the type host being an *Achlya*.

There is similar confusion regarding the host range limits of *Olpidiopsis* species in *Aphanomyces* and *Pythium*. Various species of *Aphanomyces* have been reported with *Olpidiopsis* parasites but Whiffen (1942) provided evidence of host restriction within the genus. Howard & Johnson (1969) collected *Aphanomyces* with a form of *Olpidiopsis* with distinctive oogonial spines but concluded that a broad species concept was preferable to the erection of a new species. In contast, Whiffen (1942) named two new species of *Pythium* parasites (*O. curvispinosa* and *O. brevispinosa*) on the grounds of host restriction and oogonial spine morphology. More recently Pemberton *et al.* (1990) have shown that one isolate of *Gracea gracilis* was capable of infecting some but not all species of both *Pythium* and *Phytophthora*. Unfortunately, the interaction of host and environmental conditions for infection have not been fully explored on a comparative basis. Within the saprolegnialean hosts there seem to be three broad categories, roughly equivalent to the groups of genera associated with *Saprolegnia sensu lato* (*Saprolegnia* with *Isoachlya*, and racemose *Achlya* species) and *Achlya sensu lato* (*Achlya*, *Dictyuchus* and *Aplanopsis*), and *Aphanomyces*. Suprageneric recognition of these three groups of hosts could be supported with morphological criteria (e.g. Soumati & Dick, 1988). Sparrow (1960) and Karling (1981*a*) both point to possible synonymies in *Olpidiopsis*, and in the present state of our knowledge broad species concepts are preferable when host specificity or extra-mural oogonial ornamentation are the principal variants. Until more thorough morphological and physiological studies have been carried out, I follow Karling (1942*a*) in part, and propose that three of the five species of Cornu (1872) should be retained and regarded *sensu lato*. The other binomials for parasites of the *Saprolegniomycetidae* may possibly eventually receive recognition at a taxonomic rank below that of the species, but at present there is insufficient information to justify the selection of either varietal or form status. For this reason the bionomials are separately listed in PART V under the species *sensu lato*, but are keyed separately in PART VI (GENERAL KEY).

The binomials used for parasites of the *Saprolegniaceae* cannot yet be resolved. If host is considered more important than extramural oogonial deposits, then *O. saprolegniae sensu lato*, *O. fusiformis sensu lato* and *O. aphanomycis sensu lato* should be used, as is done here; if oogonial morphology is considered more important than host range, then *O. saprolegniae sensu lato*, *O. incrassata sensu lato* and *O. index sensu lato* would be the appropriate binomials. For further information, see PART VI (GENERAL KEY).

The selection of the species *Olpidiopsis fusiformis sensu lato* for the parasites of *Achlya* requires explanation. Of the three species described by Cornu (1872), *O. index* was the first, and *O. fusiformis* the last to be described. However, the descriptions and drawings of the first two species are incomplete and I prefer to base the *Achlya* parasites on *O. fusiformis* parasitic in the *Achlya racemosa* group. It is probable that the parasites of the eccentric-egged group of *Achlya* species (*Achlya sensu stricto*) are not conspecific with those of the *Achlya racemosa* group, but the two groups of parasites cannot be clearly distinguished at present. If the distinction can be justified, it would be preferable to use the name *O. spinosa* for parasites of the *Achlya flagellata* (eccentric-egged; *Achlya sensu stricto* - type: *A. prolifera*) group of hosts.

It is not clear to me why Karling (1981*a*) has used the binomial *O. vexans*, a parasite of *Saprolegnia*, in preference to the earlier name *O. incrassata*, which he places as a synonym of *O. vexans*, for parasites of *Achlya*.

The distinctions between the species of *Olpidiopsis* in *Pythiaceae* are more fundamental. The first of the two species of *Olpidiopsis* described by Butler (1907) was *O. pythii*, followed by '*O. gracilis*' (*Gracea gracilis*). The original descriptions of zoosporogenesis and zoospore behaviour by Butler for these two species reveal some differences: the development and morphology for *O. pythii* were equated with those for *O. saprolegniae*, in which a short quiescent phase after emergence was followed by resumption of motility. The zoospores in the latter phase were figured with distinct lateral flagellar insertion. In contrast, the zoospores of *G. gracilis* did not have a quiescent phase, and were figured as pyriform, possibly with sub-apical flagellar attachment - a description consistent with the observations of the isolate studied by Pemberton *et al.* (1990) (Figs 7, 79-81). Neither of these species was regarded as possessing companion cells, but the species described by Whiffen (1942) and Srivastava (1975) have receptive gametangia accompanied by companion cells. The distinctive, predominantly lateral, hyperplasic diverticula of *G. gracilis* can be contrasted with the less characterized terminal and lateral hypertrophy caused by *O. pythii*, *O. curvispinosa*, *O. brevispinosa* and *O. indica*. *Gracea gracilis* is distinct in the relative lengths of its flagella; the kind of host deformation induced; lack of a companion cell, and the greater breadth of its host range. It shares with *Rozellopsis waterhouseae* the possession of a longer anterior flagellum and hosts in *Phytophthora*. It merits a separate genus. The genus *Pseudolpidium*, as defined by Fischer in his key (Fischer, 1892*a*: 16) would have been ideal, but for the fact that the genus was typified by *P. saprolegniae* and is thus a synonym for *Olpidiopsis* (Sparrow, 1960: 927 and references therein). I have therefore proposed the genus *Gracea*, typified by *G. gracilis*, and containing *G. (Rozellopsis) waterhouseae*. The genus may be placed in an intermediate position between the *Olpidiopsidales* and the *Rozellopsidales*.

Olpidiopsis irregularis (Constantineanu, 1901) and *Olpidiopsis endogena* (Karling, 1985) are aplerotic and polyoosporous parasites of *Saprolegnia* and *Achlya*, and *Aphanomyces* respectively. They must be removed from *Olpidiopsis*, but the assignation to *Petersenia* (Sparrow, 1943), which was accepted by Karling (1981*a*), is not followed here. The species are discussed with the other (sometimes polyoosporous) parasites under a new genus, *Cornumyces*.

Olpidiopsis species parasitic in other endobiotic biflagellate fungi

Rieth (1954) described *Olpidiopsis myzocytia* as a parasite of *Myzocytium* in *Mougeotia*. The recognition systems and mechanisms of conjunction for parasitism of holocarpic endoparasites of algae by other obligate holocarpic endoparasites are formidible (see also *Pythiella besseyi* and *Rozella pseudomorpha*) and would be compounded for heterothallic organisms. Homothallic sexual reproduction is described for this species, and this would be contrary to the heterothallic concept for *Olpidiopsis*. However, the morphological development described (with differences in oospore dimensions between host and parasite) would be expected if autoparasitism and depauperate growth were involved. The other developmental data that are available are consistent with this hypothesis (the resting spore appears to be aplerotic and lacks extramural fibrils) and the taxon is rejected until this parasitic association is rediscovered and subjected to TEM studies which reveal critical ultrastructural differences between the *Myzocytium* host and its endobiotic parasite.

Pleocystidium (*Olpidiopsis*) species parasitic in *Zygnematales*

Seven binomials and one unnamed record refer to parasites of the *Zygnematales*. Zoospores have only been observed in *O. schenkiana* and *Pleocystidium parasiticum* Fisch (1884) which Fischer (1892) transferred to *Olpidiopsis* as *O. parasitica*. Differences in opinion as to the number of flagella by Zopf (1884) and Fisch (1884) lead to the erection of the genus *Pseudolpidiopsis* (Minden, 1915) for *Olpidiopsis schenkiana*. Both Sparrow (1960) and Karling (1981*a*) regarded the two species as conspecific, and possessing biflagellate zoospores, placing *O. parasitica* as the synonym, but Schröter (1886*b*) kept them separate because of differences in the numbers of exit tubes and companion cells. Cejp (1959) correctly reversed the synonymy, giving priority to Fischer's combination because Fisch's article appeared in print in fascicle form in 1883, although the bound volume is correctly dated 1884. Scherffel (1925*a*) and Sparrow (1960: 947) describe biflagellate intrasporangial zoospores as emerging and reforming without encystment into "secondary" zoospores (cf. *O. pythii* Butler, 1907; *Petersenia utriculoba* Miller, 1962*a*). At this stage I prefer to assume that the 'polymorphism' is a misinterpretation of a principal-form zoospore with sub-apical flagellar insertion.

The resting spores of these parasites of *Zygnematales* lack extramural spines, and diameters of the resting spore vary from 19-24 μm in *O. parasitica* (Karling, 1981; Sparrow, 1960); 20-25 μm in *O. fibrillosa* (Wildeman, 1896*b*); 16-22 μm in *O. zopfii* (Scherffel, 1925*a*); 15 μm in the unnamed description of Sparrow (1957); 13-15 μm in *O. sphaericis* (Dasgupta & John, 1988) and 9-15 μm in *O. appendiculata* (Dasgupta & John, 1988). The number of companion cells/receptive gametangium is 1-5 in *O. parasitica*, but in the other species the receptive gametangium normally has only a single companion cell. The lack of information on zoospores for these other species, and the presence of only a single sporangial discharge tube will make a comparison with *Olpidium* essential for accurate identification to species, and I therefore append a list of *Olpidium* species (and doubtful species) known to parasitize *Zygnematales*, *Zygnemataceae* (Table IV: 8).

For the present, recognition of a single species parasitic in the filamentous *Zygnematales* is proposed, although I acknowledge the possibility that the other imperfectly known taxa given in synonymy may eventually be more correctly placed in *Syzygangia*.

TABLE IV: 8. Olpidium *and* Micromyces *species known to parasitize* Zygnemataceae *(numbers in parentheses refer to pages in Sparrow, 1960). See PART V.*

Olpidium	*Micromyces*
O. ampullaceum (A. Braun) Rabenh. (262)	*M. cristata* (Scherff.) Sparrow var. *cristata* (201)
O. algarum Sorokin var. *algarum* (132)	*M. cristata* (Scherff.) Sparrow var. *minor* (Scherffel) Sparrow (201)
O. algarum var. *brevirostrum* Sorokīn (132)	*M. echinocystis* Linder (205)
O. endogenum (A. Braun) J. Schröt. (132)	*M. fischeri* (Scherff.) Sparrow (202)
O. entophytum (A. Braun) Rabenh. (135)	*M. grandis* C. E. Mill. (204)
O. hyalothecae Scherff. (137)	*M. laevis* Canter (202)
O. immersum Sorokīn (151)	*M. longispinosus* Couch (198)
O. mesocarpi De Wild. (152)	*M. intermedia* (Canter) Sparrow (200)
O. mougeotia Skvortzov (152)	*M. mesocarpi* De Wild. (195)
O. rostratum De Wild. (152)	*M. minimus* Linder (205)
O. rostriferum Tokun. var. *rostriferum* (134)	*M. mirabilis* (Canter) Sparrow (199)
O. saccatum Sorokīn (138)	*M. oedogonii* (M. Roberts) Sparrow (203)
O. spirogyrae Skvortzov (153)	*M. ovalis* Rieth var. *ovalis* (200)
O. tuba Sorokīn (154)	*M. ovalis* Rieth var. *giganteus* Sparrow & M. E. Barr (200)
O. utriculiforme Scherff. (139)	*M. petersenii* Scherff. (206)
O. zygnemicola Magnus (134)	*M. spirogyrae* Skvortzov (206)
	M. zygnemicola (Cejp) Sparrow (203)
	M. zygogonii P. A. Dang. (195)

The difference in host between the type species of *Olpidiopsis* and *Olpidiopsis parasitica*; the absence of extra-mural oogonial spines, and the large number of companion cells, make it expedient to resurrect *Pleocystidium*, the original genus of Fisch (1884*a*) for this taxon, redefined on its parasitic habit in *Zygnematales*, and the plerotic and smooth resting spore accompanied by one or more companion cells. Petersen (1909) also excluded *Olpidiopsis parasitica* (as *O. schenkiana*) from *Olpidiopsis sensu stricto*.

The fungus described by Canter & Lund (1969) as a desmid parasite with a smooth, plerotic, oospore, and named as a *Lagenidium*, *L. lundiae* (as *L. lundii*), by Karling (1981a) is also tentatively placed in *Pleocystidium*, rather than *Lagenidium sensu stricto* or *Syzygangia*.The fungus originally described as *Ectrogella besseyi* and transferred to *Pythiella besseyi* by Sparrow (1960), is here assigned to *Pleocystidium* because it is mycoparasitic within a zygnemataceous cell, and has more than one discharge tube.

Olpidiopsis species parasitic in other freshwater algae

Other species of *Olpidiopsis* also require reassignment. *Olpidiopsis oedogoniarum* and *O. decipiens* are parasitic in *Oedogoniales*. *Olpidiopsis oedogoniarum* has zoospores which form a hemisphaeric appressorium on the host and an aplerotic oospore. Scherffel (1925*a*) regarded this species as a transition form between *Olpidiopsis* and *Lagenidium* and it is here transferred as a synonym of *Syzygangia oedogonii* because the minor differences in thallus form are not of generic significance.

Olpidiopsis decipiens is claimed to infect only the oogonia of *Oedogonium* and *Bulbochaete*. Dasgupta & John (1988) discuss the possible relationship with *Rhizophydium decipiens* (A. Braun) A. Fisch., but report that the zoospores are biflagellate. Zoosporogenesis is not described and dimensions of the zoospore are not given. The resting spore is not accompanied by a companion cell, and the description is ambiguous with respect to the plerotic or aplerotic nature of the resting spore. The fungus cannot be placed in *Olpidiopsis* on any positive criterion; synonymy with *S. oedogonii* is dubious. It is probably better to place the name under *Syzygangia* as a doubtful taxon, with cross-reference to *Rhizophydium*, notwithstanding the description of biflagellate zoospores.

Olpidiopsis gillii and the unnamed fungus described by Johnson (1957*a*) are parasitic in freshwater and marine diatoms respectively. In the original account Gill (1893) regarded his fungus as an *Ectrogella*, and as this genus is now delimited, a transfer is appropriate with synonymy under *Ectrogella bacillariacearum*. Likewise, Johnson (1957*a*) considered that synonymy with *Ectrogella perforans* was an alternative disposition for his unnamed marine olpidiopsidaceous fungus, and this suggestion is adopted here.

Olpidiopsis deformans, in *Draparnaldia* (Serbinow, 1907), is stated to have a more extended plasmodial phase, with plasmodial divisions, each unit eventually forming a separate sporangium (Karling, 1981: 30). This species has no known resting spore, so its affinities are uncertain, but it must be excluded from *Olpidiopsis* and placed in the *Rozellopsidales* adjacent to *Rozellopsis* (cf. *Rozella coleochaetis*) in *Plasmophagus* (*P. oedogoniorum*) as *Plasmophagus deformans*.

Olpidiopsis sorokinei was described by Wildeman (1890, 1896*a*) as a parasite of the fresh water chromophyte *Tribonema*, and a similar organism has been recorded by Sparrow (1960: 153). Zoospores have never been seen. Sparrow (1960: 153) found resting spores similar to those of *Syzygangia oedogonii*, and I place this fungus as a doubtful member of *Syzygangia*.

Olpidiopsis species parasitic in terrestrial angiosperm roots

Della Torre (1964) described two taxa found in rootlets of *Arabis alpina* L. which he placed in *Olpidiopsis*: one was described as a new species, *O. cruciferarum*, partly based on observations by Peyronel (1924), and the opinions of Ciferri and Mattioli (quoted in Della Torre, 1960); the other as the variety of *O. verrucosa* var. *radicicola*. Although both taxa were described as biflagellate, no evidence has been presented to support this statement. Moreover, the resting spore stages described and illustrated show important differences from *Olpidiopsis sensu stricto*: those from *O. cruciferarum* (17-22 μm diam.) lacked companion cells and had long 'flexuous' spines; those from *O. verrucosa* var. *radicicola* (24-55 μm diam.) had aplerotic resting spores (note that his fig. 5 is merely a magnification of fig. 4 and purports to show cytoplasmic organization but since the photograph is from a lactophenol preparation the distribution described is an artefact). These taxa should be removed from *Olpidiopsis sensu stricto* and compared with *Olpidium* species in roots, *Synchytrium*, and (if proven to be biflagellate) *Lagena*.

Pending a monographic treatment of *Olpidium* species in roots, new combinations for both taxa are, for the present, erected in *Olpidium*, although *O. cruciferarum* may be more closely allied to *Synchytrium*. Note that Lange & Insunza (1977) recognized two species of *Olpidium* in roots but did not discuss the validity of the other taxa.

Olpidiopsis species parasitic in terrestrial fungi

Olpidiopsis vuilleminiae, in a basidiomycete, is not known to be biflagellate and is removed from *Olpidiopsis*. It may be related to the other basidiomycete parasite, *Rozella itersoniliae* (Barr & Bandoni, 1979), which Held (1981) regarded as an *Olpidium* although he did not make the nomenclatural change.

Comparisons between these two species and *Pleotrachelus fulgens* and *P. zopfianus* in *Pilobolus* sugggest that both would be better placed in *Pleotrachelus* as delimited here.

Olpidiopsis globosa (Anastasiou & Churchland, 1968) is a parasite of a saprotrophic marine hyphomycete and is here transferred, with some doubts because of its plerotic oospore, to *Sirolpidium*.

Olpidiopsis species parasitic in dinoflagellates

Olpidiopsis glenodinianum (Dangeard) Johnson is a little-known marine parasite. The only report of a resting spore is that of Johnson (1966*d*); the spore is plerotic, and the oogonium has a few short spines but lacks a companion cell. The zoospores are small (2·0-3·0 × 3·0-4·0 μm) and laterally biflagellate. Its affinities appear to be with the other marine olpidioid forms which are transferred to *Sirolpidium*. Other possible placements could be with *Pleocystidium* because of its plerotic resting spore, but it is marine; or with parasites of protozoa (cf. *Pseudosphaerita*). Other parasites of *Dinomastigota* include the two species of *Aphanomycopsis*.

Olpidiopsis species parasitic in Chytridiomycetes and other aquatic heterotrophic flagellates

Olpidiopsis karlingiae (Karling, 1949), in *Karlingia rosea*, must be removed from *Olpidiopsis sensu stricto* because of the large size of its zoospores (6·0-6·5 × 10·0-10·8 μm; volume equivalent >300 μm^3), absence of companion cells and lack of extramural fibrils on the oogonium as well as the chytridiaceous host substratum. In his description of centrifugal cleavage during zoosporogenesis, there appear to be close similarities with the *Saprolegniomycetidae*. The fungus does not conform well to any genus, but may be closest to the chytridiaceous parasite *Petersenia catenophlyctidis* (Sundaram, 1968) here placed in *Cornumyces*, to which genus *Olpidiopsis karlingiae* is also assigned under the doubtful category.

Olpidiopsis sphaeritae was regarded as a doubtful species by Karling (1949) and Sparrow (1960: 956). In view of the unresolvable confusion in Dangeard's accounts of *Sphaerita* and *Pseudosphaerita*, I think it possible that this form may be a case of autoparasitism. The species is rejected from *Olpidiopsis*.

Olpidiopsis longicollis (Zopf, 1890) is a name applied to an incompletely described parasite of *Vampyrella*. It is a *nomen nudum* and is rejected from *Olpidiopsis*.

Olpidiopsis species parasitic in marine algae

Resting spores are unknown for the five taxa of *Olpidiopsis* reported from marine red algae (*O. antithamnionis* Whittick & South, 1972; *O. dangeardii* Feldmann & Feldmann, 1967; *O. feldmannii* Aleem, 1952; *O. magnusii* Feldmann & Feldmann, 1955, and the unnamed *Olpidiopsis* species of Sparrow, 1969), and zoosporogenesis, where described, is intrasporangial. The zoospores are very small, with a volume equivalent of *ca* 15-25 μm^3, which is comparable with those of other red algal parasites (see Table IV: 9); this comparison is strengthened by their strongly anisokont character. These species are unlikely to be closely related to the type species of *Olpidiopsis*, but constitute a group of posssibly depauperate forms related to *Pontisma* and *Sirolpidium*. The type species of the *Pontismataceae* and *Sirolpidiaceae* have distinctive thallus characters not exhibited by any of the marine *Olpidiopsis* species, but other species of these genera are also known to have simpler thallus morphology. None of the marine species of *Olpidiopsis* is described in sufficient detail to provide a generic type, therefore, rather than erecting a new genus, I have elected to place the parasites of red algae under *Pontisma*, the generic diagnosis of which allows for olpidioid thalli. The parasite of marine green algae (*Olpidiopsis andreei* Lagerh. - Lagerheim, 1900*a*; Karling, 1981*a*: 51) is transferred to *Sirolpidium* (with *Ectrogella marina* (P. A. Dang.) Feldm. & Feldm.-Maz. - Feldmann & Feldmann, 1956), but the fungus on the brown alga *Ectocarpus*, which Sparrow (1936*a*) placed in synonymy with *Olpidiopsis andreei*, is more difficult to place, and is tentatively equated with *Pleotrachelus ectocarpi* Jokl (1916) under the binomial *Anisolpidium joklianum* M. W. Dick (*loc. cit.*) because it possesses a possibly aplerotic resting spore. Unnamed *Olpidiopsis* spp. in diatoms (*Navicula*) (Johnson, 1966*d*; Aleem, 1953) are referred to *Ectrogella*. *Olpidiopsis paradoxa* (Johnson, 1966*e*), a parasite in a marine parasitic chytrid is considered with other marine mycoparasites.

Resumé

There seems to be no doubt from general organelle morphology, including mitochondrial form with tubular cristae (Bortnick *et al.*, 1985; Martin & Miller, 1986*a*, *b*) that *Olpidiopsis saprolegniae* and *O. varians* are straminipilous fungi. The former species is the type of the genus, so there can be no question of the status of this genus in the *Peronosporomycotina*. Details of mature flagellar morphology are still lacking, but kinetosome and rootlet morphology (Bortnick *et al.*, 1985) suggest that the relationship with other *Peronosporomycetes* may be distinct from either the *Saprolegniomycetidae* or *Peronosporomycetidae*. *Olpidiopsis* is the type genus of the *Olpidiopsidaceae*. The family cannot be assigned to any of the subclasses of the *Peronosporomycetes*, but should stand alone at ordinal or even class level, as predicted by Sparrow (1976). The change to ordinal separation, *ordo incertae sedis*, is made here, and at present the family contains only two genera, *Olpidiopsis sensu stricto*, confined to oomycete hosts, and *Pleocystidium*, confined to freshwater conjugate algae.

The order is defined as possessing a holocarpic, non-dividing thallus, sexual reproduction and plerotic oospores. It shares some characters with the *Myzocytiopsidales* (but differs in its plerotic oospores and probably in the ultrastructural details of fertilization and oosporogenesis) and *Saprolegniales* (but differs in details of asexual reproduction). On the other hand, the order shares with the *Rozellopsidales* an initially plasmodial assimilation phase and the stimulation of extra-mural fibril deposition on the resting spores.

TABLE IV: 9. *Alphabetic list of binomials of biflagellate or straminipilous fungi parasitic in marine algae and fungi (synonymous binomials indented)*

RHODOPHYCEAE

Binomial	Host
Chytridium tumefaciens, synonym of *Eurychasmidium tumefaciens*	
Eurychasma sacculus, synonym of *Eurychasmidium sacculus*	
Eurychasma joycei, synonym of *Eurychasmidium joycei*	
Eurychasma sp. (Sparrow, 1969: sp. 2), possible synonym of a *Eurychasmidium* species (parasitic in *Polysiphonia*, *Pterosiphonia*)	
Eurychasmidium dangeardii, synonym of *Pontisma dangeardii*	
Eurychasmidium joycei	*Polysiphonia*, *Pterosiphonia*
Eurychasmidium sacculus	*Halosaccion*, *Rhodymenia*
Eurychasmidium tumefaciens	*Ceramium*
Olpidiopsis antithamnionis, synonym of *Pontisma antithamnionis*	
Olpidiopsis dangeardii, synonym of *Pontisma dangeardii*	
Olpidiopsis feldmannii, synonym of *Pontisma feldmannii*	
Olpidiopsis magnusii, synonym of *Pontisma magnusii*	
Olpidiopsis tumefaciens, synonym of *Eurychasmidium tumefaciens*	
Olpidiopsis sp. (Sparrow, 1969), possible synonym of a species of *Petersenia* or *Pontisma* (parasitic in *Porphyra*)	
Olpidium tumefaciens, synonym of *Eurychasmidium tumefaciens*	
Petersenia lobata	*Callithamnion*, *Ceramium*
Petersenia pollagaster	*Ceramium*
Petersenia palmariae	*Palmaria*
Petersenia sp. (Sparrow, 1936*a*)	*Ceramium*
Petersenia sp. (Aleem, 1950*d*)	*Achrochaetium*
Petersenia sp. (Johnson, 1966*g*)	*Rhizosolenia*
Petersenia sp.(Howard & Johnson, 1968)	*Polysiphonia*
Pleotrachelus inhabilis, synonym of *Pontisma inhabilis*	
Pleotrachelus lobatus, synonym of *Petersenia lobata*	
Pleotrachelus pollagaster, synonym of *Petersenia pollagaster*	
Pleotrachelus tumefaciens, synonym of *Eurychasmidium tumefaciens*	
Pontisma antithamnionis	*Antithamnion*
Pontisma dangeardii	*Radicilingua*
Pontisma feldmannii	*Asparagopsis* (*Falkenbergia*), *Trailliella*
Pontisma inhabilis	*Polysiphonia*
Pontisma lagenidioides	*Ceramium* [*Cladophora* see below]
Pontisma magnusii	*Ceramium*
Pyrrhosorus marinus	*Cystoclonium*
Sirolpidium lagenidioides, synonym of *Pontisma lagenidioides*	

FUCOPHYCEAE AND BACILLARIOPHYCEAE

Binomial	Host
Anisolpidium ectocarpi	*Ectocarpus*
Anisolpidium minutum	*Chorda*
Anisolpidium rosenvingii	*Bachelotia*
Anisolpidium sphacelariarum	*Sphacelaria*, *Cladostephus*
Chytridium sphacelariarum, synonym of *Anisolpidium sphacelariarum*	
Ectrogella dicksonii, synonym of *Eurychasma dicksonii*	
Ectrogella lauderiae	centric diatom
Ectrogella licmophorae, synonym of *Ectrogella*	
Ectrogella perforans	diatom
Eurychasma dicksonii	*Ectocarpales*
Eurychasma lauderiae, synonym of *Ectrogella lauderiae*	
Eurychasma sp. (Sparrow, 1969: sp. 1)	*Feldmannia*, Ectocarpales
Labyrinthula pohlia	*Chorda*
Labyrinthula roscoffiensis	*Taonia*, *Ectocarpus*
Lagenisma coscinodisci	centric diatom
Olpidiopsis andreei sensu Sparrow, synonym of *Sirolpidium ectocarpi*	
Olpidiopsis sp. (Johnson, 1957*a*), synonym of *Ectrogella perforans*	
Olpidium dicksonii, synonym of *Eurychasma dicksonii*	
Olpidium lauderiae, synonym of *Ectrogella lauderiae*	

(Table IV: 9, continued)

Parasite	Host
Olpidium sphacelariarum, synonym of *Anisolpidium sphacelariarum*	
Petersenia andreei sensu Sparrow, synonym of *Sipolpidium ectocarpi*	
Phagomyxa algarum	*Bachelotia*
Pleotrachelus ectocarpi, synonym of *Anisolpidium joklianum*	
Pleotrachelus minutus, synonym of *Anisolpidium minutum*	
Pleotrachelus olpidium, synonym of *Anisolpidium olpidium or A. ectocarpi*	
Pleotrachelus rosenvingii, synonym of *Anisolpidium rosenvingii*	
Pleotrachelus sphacelariarum, synonym of *Anisolpidium sphacelariarum*	
Rhizophydium dicksonii, synonym of *Eurychasma dicksonii*	
Sirolpidium ectocarpi	*Ectocarpus*

CHLOROPHYCEAE

Parasite	Host
Achlyogeton salinum, synonym of *Sirolpidium salinum*	
Bicilium andreei, synonym of *Sirolpidium andreei*	
Ectrogella marina, synonym of *Sirolpidium marinum*	
Eurychasmidium sp. (Konno, 1988), unassignable	
Eurychasma sp. (Kobayashi & Ookubo, 1954)	*Bryopsis*
Labyrinthula coenocystis	*Cladophora*
Labyrinthula chattonii	*Cladophora*
Lagenidium chaetomorphae, synonym of *Sirolpidium salinum*	
Myzocytium proliferum f. *marinum*, synonym of *Sirolpidium bryopsidis*	
Olpidiopsis andreei, synonym of *Sirolpidium andreei*	
Olpidium bryopsidis, synonym of *Sirolpidium bryopsidis*	
Olpidium marinum, synonym of *Sirolpidium marinum*	
Pleotrachelus andreei, synonym of *Sirolpidium andreei*	
Pontisma lagenidioides sensu Kobayashi & Ookubo (1953), synonym of *Sirolpidium salinum*	
Pontisma lagenidioides sensu Raghukumar (1986*a*), synonym of *Sirolpidium salinum*	
Sirolpidium andreei	*Spongomorpha*, *Acrosiphonia*
Sirolpidium bryopsidis	*Bryopsis*, *Cladophora*
Sirolpidium marinum	*Tetraselmis*
Sirolpidium salinum	*Cladophora*

IN OTHER ALGAL HOSTS

Parasite	Host
Sirolpidium zoophthorum, synonym of *Haliphthoros zoophthorum* in clam larvae	
Olpidiopsis paradoxa, synonym of *Sirolpidium paradoxum*, in *Rhizophydium* in *Spongomorpha*	
Pleotrachelus paradoxus, synonym of *Sirolpidium paradoxum*, in *Rhizophydium* in *Spongomorpha*	
Olpidiopsis globosa, synonym of *Sirolpidium globosum*, in *Papulospora*	
Olpidiopsis glenodiniana, synonym of *Sirolpidium glenodinianum*, in *Glenodinium*	
Sirolpidium globosum	*Papulospora* on *Arbutus* debris
Sirolpidium paradoxum	*Rhizophydium* in *Spongomorpha*

IN OTHER HOSTS (NON-MARINE)

Petersenia catenophlyctidis, synonym of *Cornumyces catenophlyctidis* in *Catenophlyctis* (*Chytridiomycetes*)
Petersenia irregulare, synonym of *Cornumyces irregularis*, in *Saprolegnia* and *Achlya*
Petersenia panicicola, doubtful taxon here allied to *Lagena* in grass leaves
Petersenia utriculoba, synonym of *Cornumyces utriculobus*, in *Pythium intermedium*

SIROLPIDIUM, PONTISMA, PETERSENIA, EURYCHASMA AND *EURYCHASMIDIUM*

It is convenient to consider the five genera of marine algal parasites (*Sirolpidium*, *Pontisma*, *Petersenia*, *Eurychasma* and *Eurychasmidium*) together. Freshwater taxa attributed to *Petersenia* are discussed with *Cornumyces* (see Table IV: 9). Comparisons should also be made with *Anisolpidium* and *Phagomyxa*, which are also parasites of the *Ectocarpales*; and to several little-known marine species previously assigned to *Pleotrachelus*. The zoospores of these species of *Pleotrachelus* have not been described, so their correct placement is doubtful, but the thallus walls are presumably of glucan since they were reported to give a positive reaction with the chlor-zinc iodide test (Petersen, 1905), and thus a placement in the *Chytridiomycetes* is unlikely (see also Sparrow, 1960). The marine species of *Olpidiopsis* are included here and all synonyms have been given in Table IV: 7 for easy cross-reference. The marine species of *Olpidium* (listed under *Olpidium* in PART V) should also be compared when new collections are made.

The first of these organisms to be identified and described were *Eurychasmidium tumefaciens* (as a *Chytridium* parasitic on a red alga) by Magnus (1872), and *Eurychasma dicksonii* (as a *Rhizophydium* parasitic on a brown alga) by Wright (1879). Petersen (1905) provided the most comprehensive coverage of these fungi, followed by Sparrow (1934, 1936*a*, 1960, 1969), Johnson & Sparrow (1961) and Karling (1981*a*).

The most recent and complete morphological accounts for individual species related to these five genera are those by Whittick & South (1971, 1972) for the fungus they named as *Olpidiopsis antithamnionis* and Pueschel & Van Der Meer (1985) and Van Der Meer & Pueschel (1985) for *Petersenia palmariae*. This last species is the only one of the group known to be straminipilous. Table IV: 9 provides a list of the taxa discussed in this section, arranged according to host group, together with the synonymy and nomenclatural changes adopted.

Sexual reproduction is unknown for most of the taxa, but even for those in which resting spores have been mentioned, the structures are not clearly described. In most cases the reports refer to thallus segments with thick walls but which are not otherwise differentiated. Three exceptions are fungi transferred here from Myzocytium and *Olpidiopsis* respectively. The parasite of marine *Cladophora* named as *Myzocytium proliferum* forma *marina* (Kobayashi & Ookubo, 1953) was reported as possessing an aplerotic resting spore. The fungus parasitic on *Ectocarpus* and referred to *Petersenia andreei* by Sparrow (1936*a*), and *Olpidiopsis andreei* by Karling (1942*a*) has a smooth-walled, possibly aplerotic (see Sparrow, 1936*a*: 247, fig. 8) resting spore accompanied by a single companion cell. (The placement of this taxon will be anomalous in any existing genus: it does not have a lobed thallus or red algal host (which excludes *Petersenia*); its resting spore may be aplerotic and lacking in extramural fibrils, and it is marine (which excludes *Olpidiopsis sensu stricto*); it is biflagellate (which excludes *Anisolpidium*); it does not cause hypertrophy (which excludes *Eurychasma*), it shows sexuality but has a brown algal host (a combination which excludes *Pontisma* and *Sirolpidium*). In view of the previously accepted synonymy, it is preferable to place the taxon close to *Sirolpidium andreei*, though not synonymous with this species.) The third species is the parasite of a marine conidial fungus, and named as *Olpidiopsis globosa* (Anastasiou & Churchland, 1968).

Until more comparative information becomes available the marine genera, apart from *Sirolpidium*, should be regarded as anamorphic form genera. The delimitation of the genera is thus most conveniently diagnosed first on the basis of host and secondly on thallus form and zoosporogenesis. Sparrow (1934) separated *Petersenia* from *Olpidiopsis* on the grounds of thallus shape, but added "further work may reveal that that they have distinguishing characters of more significant taxonomic worth".

All three major filamentous marine algal classes are known to have biflagellate fungal parasites. There is no conclusive evidence for host specificity, but the greatest diversity of hosts is in the red algae and here there is some evidence for host specificity in relation to *Trailliella* (Aleem, 1952*b*, 1953; Whittick & South, 1971, 1972). However, there are also indications of host restriction where different potential host species are growing together. In the brown algae almost all records relate to host species in the *Ectocarpales* (the exception is *Chorda* in the *Laminariales*), and in the green algae records relate to the *Cladophorales* except for *Tetraselmis* (Dangeard, 1912; Feldmann & Feldmann, 1956) in the *Chlorococcales*. Thus there is a marginal balance in favour of host specificity. On the other hand, Jenneborg (1977) followed Petersen (in Sparrow, 1960: 819) in regarding the red algal parasites *Eurychasma sacculus* and *E. joycei* as synonyms of the brown algal parasite, *E. dicksonii* because there was no detectable discontinuity in sporangial dimensions between these species even though there were distinct host differences (see Table IV: 9). If there is host specificity, then form genera should reflect this phenomenon by reference to the algal class; if not, then thallus shape, zoosporogenesis and hypertrophic symptoms must be used for generic assignation and specific identification, but these are not clearly delimited.

The mature thalli of *Sirolpidium*, *Pontisma*, *Petersenia*, *Eurychasma* and *Eurychasmidium* are endobiotic, usually saccate or lobed, rarely more or less subspherical (olpidioid), and may or may not become septate before conversion into sporangia. The initial development may be plasmodial, as in *Eurychasma dicksonii* (Sparrow, 1934) and *Petersenia palmariae* (Pueschel & Van Der Meer, 1985). The walls of *Petersenia pollagaster* have been shown to give a positive reaction with chlor-zinc iodide (Molina, 1986). The genera *Pontisma*, *Petersenia* and *Sirolpidium*, as represented by their type species, are distinguished by the shape, segmentation and segmental separation of the thallus, but many recorded collections, including those of the type species, have described single, olpidioid, thalli. For this reason I have grouped with these species the marine taxa previously placed in *Olpidiopsis*, and which lack the characters diagnostic for *Olpidiopsis*, such as the extra-mural fibrils on resting spore thalli, or detailed information of zoosporogenesis or zoospore shape and flagellar insertion. Moreover, most of these marine '*Olpidiopsis*' species may have more than one discharge tube, and the same is true for the doubtful marine *Pleotrachelus* species.

Pontisma has an elongate saccate thallus which may become segmented but the segments do not disarticulate before zoosporogenesis. *Petersenia* has a lobed thallus, which does not become septate, and when sporangial, may produce more than one exit tube. The lobes may be short and subspherical, as in *Petersenia pollagaster*, or more tubular as in the type species, *Petersenia lobata* (see figures in Sparrow, 1934). Pueschel & Van Der Meer (1985) suggest that the thallus of *Petersenia* in red algae may penetrate adjacent cells via the intercellular pits of the host cells.

The parasites of *Cladophora* and related genera in the *Cladophorales*, which have been placed in *Sirolpidium* (and also *Achlyogeton*, *Olpidiopsis*, *Lagenidium* and *Eurychasma*!) show a diversity of thallus shapes, but at maturity the zoosporangial thalli are ellipsoidal and, if septation has occurred, the segments have become rounded and more or less disarticulated. In thallus form, but not vesiculate zoosporogenesis or zoospore size, *Sirolpidium* resembles *Myzocytium sensu stricto*. The parasites of *Cladophorales* recorded by Kobayashi & Ookubo (1953, 1954*a*) and Konno (1972) and the recent reports by Raghukumar (1986, 1987) indicate that more than one *Sirolpidium*-like species may be involved in the parasitism of the *Cladophorales*, but it is difficult to reinterpret the earlier descriptions. Until definitive diagnoses, including host ranges, can be formulated, I propose that populations of olpidioid zoosporangia with no indication of thallus division be placed in *Sirolpidium andreei* (species originally placed in *Pleotrachelus*); species with zoosporangia which have been derived from a disarticulated thallus be placed in *S. bryopsidis* (including *Myzocytium proliferum* forma *marinum*), and species with zoosporangia which have been derived from a partially coherent, septate, cylindrical thallus be placed in *S. salinum* (including *Lagenidium chaetomorphae*, *Achlyogeton marinum*, *Pontisma lagenidioides sensu* Kobayashi & Ookubo, 1953; Konno, 1972, and probably C. Raghukumar, 1986*a*). The specific epithet originates from *Achlyogeton salinum* because this is the only valid basionym. The olpidioid parasite of *Ectocarpus* (Sparrow, 1936*a*, Jokl, 1916)) is placed as a doubtful member of *Sirolpidium*, as is the fungus described by Dangeard (1912) as *Olpidium marinum* from a non-cladophoralean green alga, and transferred to *Ectrogella* by Feldmann & Feldmann (1956).

I have adopted an arbitrary approach to the reclassification of the marine biflagellate fungal parasites of parasites of algae, placing these parasites in relation to the photosynthetic host, rather than the immediate fungal host. There is no evidence to justify their retention in the freshwater genus *Olpidiopsis*.

A comparable treatment of the sole marine *Rozella* species (also a mycoparasite of an algal parasite) could also be made, but I prefer to suggest its removal to the *Chytridiomycetes*. Nevertheless, I think it may be helpful, even if artificial, to keep all these imperfectly described mycoparasites of marine algae together.

Infections by *Petersenia*, *Pontisma* and *Sirolpidium* rarely cause hypertrophy. In contrast, *Eurychasma* and *Eurychasmidium* have non-segmented, slightly saccate thalli and cause marked hypertrophy of the host cell. These thalli also produce sporangia with few to many short, broad exit tubes. The zoospores are delimited intrasporangially and the zoospores are motile within the zoosporangium before discharge. The zoospores are small (3·0 × 4·5 µm) or very small (2·0 × 4·0 µm) and ovoid or pyriform. There are discrepancies in the descriptions and drawings of flagellar attachment, even within the same paper (e.g., Sparrow, 1969), and in the relative flagellar lengths, but most drawings indicate a lateral rather than a sub-apical insertion. There is no evidence for dimorphism, and diplanetism is only presumed for the net-sporangium formation in *Eurychasma* (note that the *Eurychasma* sp. described by Kobayas[h]i & Ookubo (1954*a*) produced aplanospores with no known polyplanetism). Sparrow (1934) and Aleem (1950*b*) have indicated that net sporangium formation may not be diagnostic since many sporangia appeared to have effected complete discharge of the first-formed zoospores. Canter & Dick (1994) concluded that net-sporangium formation could occur in taxa which were not closely related. Thus there would be a case for combining the two genera *Eurychasma* and *Eurychasmidium*. However, I prefer to retain the two genera on the basis of host algal group and therefore transfer *Eurychasma sacculus* and *E. joycei* to *Eurychasmidium*, pending resolution of host/parasite specificity.

Zoospore behaviour led Sparrow (1960) to separate *Eurychasma* and *Eurychasmidium* into the *Saprolegniales* and *Pontisma*, *Petersenia* and *Sirolpidium* into the *Lagenidiales*, but I do not believe this to be justified in the absence of evidence for dimorphism in these genera, or given the diversity of zoosporogenesis and zoospore behaviour that has been recorded for various lagenidiaceous fungi. Only *Petersenia palmariae* has been the subject of ultrastructural study (Pueschel & Van Der Meer, 1985; Van Der Meer & Pueschel, 1985), but a complete description of the zoospore morphology was not obtained.

The generic and family distinctions were clearly defined by Petersen (1905, 1909, 1910), who erected the families *Eurychasmataceae* and *Holochytriaceae* (Petersen, 1905), and *Sirolpidiaceae* (*Sirolpidium*) and *Pontismataceae* (*Pontisma*) (Petersen, 1909). Since then, Sparrow (1942), followed by Karling (1942*a*), reverted to the earlier concept of Petersen (1905), but renaming the *Holochytriaceae* (*Sirolpidium* and *Pontisma*) as the *Sirolpidiaceae*. Petersen (1909) listed four families in his *Lagenidiineae*: *Lagenidiaceae* (*Lagenidium*, *Myzocytium*); *Sirolpidiaceae* (*Sirolpidium*); *Pontismataceae* (*Pontisma*), and *Pseudolpidiaceae* (*Pseudolpidium*, *Olpidiopsis*). The families *Sirolpidiaceae* and *Pontismataceae* were designated as families of marine organisms, although Petersen (1909) also erected the genus *Bicilium* for "marine *Olpidium*-shaped and *Pleotrachelus*-shaped forms", which was placed in the *Pseudolpidiaceae*. The *Pontismataceae* is thus the earliest suitable family name for the genera *Pontisma* and *Petersenia*. Until there is a better understanding of the inter-relationships of the *Pontismataceae* and *Sirolpidiaceae*, I prefer to retain the two families as parasites of the red and green algae respectively, especially as it is possible that the *Sirolpidiaceae* is teleomorphic. As far as morphological criteria are concerned, the two families may have closer relationships with the *Myzocytiopsidales* (cf. *Ectrogella* and *Syzygangia*) rather than the

Olpidiopsidales. While flagellar transitional zone details (Van Der Meer & Pueschel, 1985) do not provide information that would assist selection between these two orders, they do indicate a closer resemblance to the *Saprolegniomycetidae* rather than the *Peronosporomycetidae* (cf. zoosporogenesis in *Myzocytiopsis*). *Eurychasma* and *Eurychasmidium* can be placed in Petersen's family *Eurychasmataceae*, but the affinities of this family are much less certain.

THE *LEPTOLEGNIELLACEAE*

The *Leptolegniellaceae* except for *Aphanomycopsis*

The *Leptolegniellaceae* is a family peripheral to the fungi under discussion here and a full morphological and taxonomic review is not given, except for the details in PART VI (KEYS I and IX). The family is characterized by intrasporangial sporogenesis leading to flagellate or aplanosporic asexual spores and oospores with a distinctive wall structure. Most species are associated with animal-derived substrates, especially keratin (skin, scales and hair). The first of these keratinophilic species to be described was *Aphanodictyon papillatum* Huneycutt *ex* Dick (Huneycutt, 1948). The family was established by Dick (1971*b*) for the genera *Leptolegniella, Aphanodictyon* and *Brevilegniella* which had previously been placed in the *Saprolegniaceae*. *Aphanomycopsis* (Scherffel, 1925*a*) was attached to the family with some reservations. This family concept was accepted by Karling (1981*a*).

The family is characterized by thallus diversity ranging from a finely filamentous mycelium with morphologically distinct reproductive systems (*Aphanodictyon*); a similar but more limited intramatrical mycelial system with slightly inflated sporangial axes and intercalary swellings with resting spores (*Brevilegniella*); coralloid thalli behaving as sporangia or with oospores (*Leptolegniella*), or olpidioid saccate thalli subsequently functioning as sporangia or oogonia (*Nematophthora*). Morphological sporangial delimitation is therefore variable, but sporogenesis is intrasporangial and the zoospores are medium-sized (4·0-6·5 × 6·5-11·0 μm; volume equivalent (75)140(240) μm^3). Asexual spore behaviour is diverse and mirrors that found in the *Saprolegniaceae* and *Leptomitaceae*. Axenic culture of *Nematophthora* has been achieved (Graff & Madelin, 1989) and in 1975 I was successful in making an axenic culture of *Aphanodictyon* from a soil sample sent from Brazil, but the culture did not survive more than a few transfers.

The oospores of the species of genera other than *Aphanomycopsis* have a distinctive wall structure, with a thick outer wall, a fluid layer middle layer and a thin innermost membrane. Oospore cytoplasmic structure has not been unequivocally described, but an ooplast is probably present. Oospores are formed in differentiated oogonia accompanied by antheridia (*Aphanodictyon*); intercalary hyphal swellings (*Brevilegniella*), or undifferentiated portions of the thallus (*Leptolegniella*). Kerry & Crump (1980) showed that the nematophagous fungus *Nematophthora gynophila* had similar oospores, but Lopez-Llorca & Duncan (1991: fig. 16) provided SEM data for abnormally small spores ascribed to this species which indicated that the spores had a few scattered minute pores.

TEM and molecular biological studies are lacking, so the systematic position of the family rests solely on light microscopy. In view of the studies on the thallus structure and morphogenesis in *Verrucalvus* (Dick *et al.*, 1984), *Leptomitus* and *Apodachlya* (M. W. Dick, unpublished), and *Eurychasmopsis* (Canter & Dick, 1994) a placement of the family in the *Leptomitales* rather than the *Saprolegniales* now seems preferable. Details of flagellar insertion, flagellar and kinetosome structure, DBVs, etc., are required to confirm this placement.

The inter-relationships of *Aphanomycopsis* and *Cornumyces* with the genera of the *Leptolegniellaceae sensu stricto* are problematic and ultrastructural research is needed to resolve the classification of these genera also. *Aphanomycopsis* is discussed below.

The genus *Cornumyces* is erected primarily for the polyoosporous parasites of aquatic fungi, using *Lagenidium destruens* as the basionym of the type species, but also includes other aquatic parasites which share the following characters:

1 a lobed or irregularly shaped thallus
2 intrasporangial zoosporogenesis producing medium-sized zoospores (flagellar insertion often requiring confirmation)
3 aplerotic oosporogenesis, with or without donor gametangial elements

Links with the *Leptomitaceae* and *Apodachlyellaceae* may be apparent through *Ducellieria* and *Blastulidium*. There are also habitat and morphological similarities between the *Leptolegniellaceae*, *Leptomitaceae* and *Hyphochytriaceae*.

Aphanomycopsis

Aphanomycopsis was erected as a monotypic genus by Scherffel (1925*a*), on the grounds that the zoosporogenesis resembled that of *Aphanomyces* (in comparison with his treatment of *Ectrogella* - a *Saprolegnia* analogue), but in *Aphanomycopsis* there was no mycelial development and the thallus was holocarpic. The aplanospores are preformed in the sporangium and glide out through the exit tube(s) to form a more or less loose spore ball of cysts at the orifice, each cyst then giving rise to a principal-form zoospore. In addition, the type species was described as having one or more resting spores formed in undifferentiated thalli, and without apparent sexual union. The genus is thus potentially polyoosporous.

Scherffel (1925*a*) placed the genus with *Ectrogella* in the *Saprolegniaceae* because he equated the kinds of zoospore production of these holocarpic fungi with that of the better-known genera *Saprolegnia*, *Achlya* and *Aphanomyces*. Tokunaga (1934*a*) proposed that the genus be transferred to the Lagenidiaceae because of its non-mycelial thallus form and holocarpic sexuality. Sparrow (1960) was correct in rejecting this move, pointing out that *Aphanomycopsis* has a typically saprolegniaceous protoplasm and zoosporogenesis, but he was probably incorrect in rejecting the emendment of the genus diagnosis by Karling (1942*a*, *b*) to include the possibility of septation in the thallus, which enabled the fungus studied by Tokunaga (1934*a*) to be included in *Aphanomycopsis*. Sparrow (1960) ignored the heterogeneity of *Lagenidium*, particularly the diversity of zoosporogenesis. Of more significance is the size and shape of the principal-form zoospore, which is typically oomycetous and forms a cyst of approximately 7μm diameter. Dick (1971*b*) tentatively proposed transfer of the genus to the *Leptolegniellaceae* on vegetative morphological criteria, and this move was accepted by Karling (1981*a*). However, Dick (1971*b*) pointed out that the characteristic oospore wall of this family did not seem to be present. Two of the existing species of the genus have a punctate oospore wall, as does *Lagenidium entophytum*, which is here transferred to *Aphanomycopsis*. As at present constituted, the genus has a wide host range comparable to *Aphanomyces*, with diatom and desmid algal parasites, dinoflagellate parasites and keratinophilic and saprophytic taxa from freshwater and soil habitats. The saprophytic taxa should be closely compared with *Hyphochytrium* (see Table IV: 9, p. 202). The zoospore cyst size of the diatom parasite *Ectrogella monostoma* is appreciably larger than that of the type species of *Ectrogella* and this species differs from *Aphanomycopsis bacillariacearum* only in the simpler form of the thallus; it does not have flagellate intrasporangial spores as in the type species of *Ectrogella* and I therefore transfer this species to *Aphanomycopsis*. The genus is retained in the *Leptolegniellaceae* for the present, but a closer affiliation with other genera such as *Cornumyces* is also possible.

Karling (1968*c*) was uncertain as to the generic position of his two new saprophytic (keratinophilic) species, *A. saprophytica* and *A. punctata*, with respect to *Leptolegniella*, but he regarded zoosporogenesis as the more important criterion. These two species lack the distinctive separable endospore membrane typical of the *Leptolegniellaceae sensu stricto*, *A. punctata* having the two layered apparently punctate wall similar to that of the type species. *Aphanomycopsis saprophytica* is less well characterized, and was obtained in axenic culture, but this culture is not, to the best of my knowledge, extant.

The inclusion in this genus of *Aphanomycopsis sexualis* (Martin, 1975*b*), a parasite of midge eggs, is questionable because the presence of antheridia could indicate a closer relationship with *Aphanomyces*, although the cytoplasmic reorganization of mature oospore is not described. The fungus differs from *Crypticola (Atkinsiella) entomophaga* (Martin, 1976), which occurs in similar hosts, in the details of zoosporogenesis and in the size and ornamentation of the zoospore cysts (*A. sexualis* has finely spiny cysts, mean diameter 6·6 μm, while *C. entomophaga* has smooth cysts, mean diameter 10·5 μm).

Further substratum diversity occurs with the two species parasitic in dinoflagellates (*Aphanomycopsis cryptica* and *A. peridiniella*).

Blastulidium and *Blastulidiopsis*

Pérez (1903, 1905) described *Blastulidium paedophthorum* Pérez as a parasite of the embryos of the freshwater crustacean, *Daphnia*. A few years later Chatton (1908) gave more details of this fungus. Nothing further was reported until Sigot (1931) described *Blastulidiopsis chattonii* Sigot from the eggs of *Cyclops*, another freshwater crustacean. The two genera were at that time distinguished on the basis of flagellar number, Chatton (1908) having described *Blastulidium* as uniflagellate, while *Blastulidium* was described as biflagellate. The only other authors to have had first-hand knowledge of this group were Jírovec (1955), Green (1974) and Manier (1976).

Manier (1976) has provided the sole modern substantial light microscope and TEM study. She has shown that *Blastulidium* is biflagellate with subapical flagellar insertion, but did not comment on relative flagellar lengths. The redrawing of the zoospore of *Blastulidiopsis* by Karling (1981*a*: plate 11, fig. 1) is not a true copy of that by Sigot (1931: fig. 3). The latter is of an obpyriform zoospore possessing an anterior refractive globule, and sub-apical flagellar attachment, in a lateral depression, anterior to the nucleus. Sigot (1931) described the flagellar attachment as lateral; Karling (1981*a*) repeated this statement and changed the figure to depict lateral insertion. Sigot (1931) described the flagella as very long (*ca* 15-20 μm) and his drawing suggests an anisokont condition, with the anterior flagellum about two-thirds of the length of the posterior flagellum.

I concur with Manier (1976) that generic separation cannot be maintained on the basis of slight differences in the shape of the thallus between the two genera (*Blastulidium* thalli are initially elongate, becoming septate, the disarticulating segments then assuming a more or less spherical shape with a diameter of *ca* 20μm; *Blastulidiopsis* was described from fixed material as irregularly lobed). Differences in flagellar insertion are spurious. The range of hosts for *Blastulidium* is such that until the parasite of *Cyclops* has been shown to be ***restricted*** to that genus, the two species, each in a monotypic genus, should be regarded as synonyms.

The occurrence of resting spores has not been established. Although Pérez (1903) and Chatton (1908) have mentioned such structures, no data of diagnostic significance have been presented.

The blastula-like development of an apparently single parietal layer of zoospore initials surrounding the less dense medullary cytoplasm is well-figured by Manier (1976: figs 1, 2), but it is probably little different from the centrifugal cleavage around the central vacuole familiar in *Saprolegnia*. In near-mature sporangia the zoospores are closely packed throughout (Manier, 1976: fig. 12). A comparison could also be made with the early developmental stages of *Eurychasmopsis* (Canter *et al.*, 1990, Canter & Dick, 1994).

The zoospores have not been shown to be heterokont. Nevertheless, the probable occurrence of vesicles containing packets of tripartite flagellar hairs in the cytoplasm at a late stage in zoosporogenesis; the presence of "corpuscules dense", which I interpret as dense body vesicles; the tubular mitochondrial cristae, and the pattern of cleavage in the zoosporangium, may be sufficient to place the fungus in the class *Peronosporomycetes*. Furthermore, the dimensions of the zoospores (*ca* 6-8 μm according to Sigot; *ca* 5·5 × 7·5 μm from Manier's illustrations, equivalent to a volume of > 120 μm^3) are within the lower limit for the *Peronosporomycetes*. (Note that there is a discrepancy in magnifications between Manier's light and TEM micrographs.) However, the intrasporangial cleavage and flagellar maturation prior to discharge, coupled with the absence of any discharge vesicle, and the sub-apical flagellar insertion on zoospores within the sporangium, rule out any affiliation with the subclass *Peronosporomycetidae*. The persistent orthogonal arrangement of centrioles (such as occurs in the *Hyphochytriales*) is not consistent with a placement in the *Saprolegniomycetidae*, but she may not have distinguished between centriole rearrangements at different stages of nuclear division and zoospore development. The persistent nuclear membrane is not compatible with a placement in the *Hyphochytriales*.

Any relationship with the marine crustacean parasites is equally uncertain. The genus must remain *incertae sedis* within the *Peronosporomycetes*, though an eventual placement near the *Apodachlyellaceae-Leptolegniellaceae* cannot be dismissed.

Ducellieria

Ducellieria was first described as the non-photosynthetic alga *Coelastrum chodatii* F. Ducell. (Ducellier, 1915). The species was associated with different major algal groups until the genus was erected by Teiling (1957). Teiling (1957) drew attention to the apocytic development with geometric progression of cell number in *Coelastrum*, and the tubular nature of the connections between the cells. In *Ducellieria*, cell number was variable, and Teiling claimed that the intercellular connections were solid. However, the intercellular bridges between adjacent cells in *D. chodatii* have been shown to be constructed from two, thick-walled but hollow spines, one from each cell, in closely parallel arrangement (Hesse, Kusel-Fetzmann & Carniel, 1989).

The transfer of *Coelastrum augustae* var. *armatum* Skuja and *C. tricuspidatum* Borge to *Ducellieria* by Teiling (1957) is not accepted here: the illustrations of *C. augustae* var. *armatum* in Teiling (1957) show that each bridge is a single tubular structure with septa. Hesse *et al.* (1989) did not mention the former taxon, but noted that the latter was known only from the original description.

The assimilative thallus of *Ducellieria* is holocarpic and non-septate with a bottle-shaped thallus, and the type species is characterized by the formation of a pseudo-*Achlya* like spore ball (the 'coelastrum' stage) at the tip of the exit papilla of the zoosporangium (Kusel-Fetzmann & Carniel, 1984; Kusel-Fetzmann & Nouak, 1981; Hesse *et al.*, 1989). The cysts that are formed are clothed with hollow spines and warts which separate the units of the spore-ball, and thus create a fenestrated hollow sphere. Each cyst then germinates by a centripetally directed papilla to produce one large (8·0 × 11·0-14·0 μm) biflagellate, anisokont and heterokont zoospore with lateral flagellar insertion. The zoospores escape from the 'coelastrum' ball through the interstices left by the spiny connections. In this respect there is some resemblance to the centripetal discharge from the net sporangium of *Eurychasmopsis* (Canter *et al.*, 1990, Canter & Dick, 1994). The thallus in attenuated cultures is saccate and lobed, rather like a diminutive *Leptolegniella*. The difference in behaviour between vigorous and attenuated cultures is interesting in that in the latter the 'coelastrum' stage is omitted. The zoospores are preformed in the zoosporangium irrespective of the developmental sequence. There is no evidence of apical or sub-apical flagellar insertion in either case.

Sexual reproduction is apparently plerotic, and the large central structure could be an ooplast with dense bodies (Hesse *et al.*, 1989: fig. 9b). The size of the oospore is such that the dense granulation and Brownian movement characteristic of the saprolegniaceous ooplast should be discernable (cf. discussion in Hesse *et al.*, 1989).

The kinetosome arrangement in the pro-zoosporic 'coelastrum' stage cyst is consistent with an oomycete affiliation, as are the densely packed, tubular mitochondria, but neither the precise arrrangement nor the structure of the tubular tripartite hairs has been established (Hesse *et al.*, 1989).

Despite the conclusions of Hesse *et al.* (1989), placement in a separate family, provisionally assigned to the *Leptomitales* rather than the *Saprolegniales*, is preferred here because of the hollow nature of the cyst spines and the plerotic oospore.

Lagenidium pygmaeum, now transferred to *Cornumyces*, shares with *Ducellieria* the pollen grain habitat (Table IV: 3). Karling (1981*d*) described a variety of *Lagenidium pygmaeum* var. *pygmaeoides*, now transferred (but with some doubts) to *Pleocystidium*, with the following characters:

1 zoospores preformed in the sporangium
2 zoospores discharged in a vesicle
3 zoospores encysting to form angular cysts

and with diplanetism regarded as a possibility, but without dimorphism. However, the zoospores were appreciably smaller than those of *Ducellieria* (2·5-3·0 × 4·0-5·0 μm). Plerotic resting spores, developed without apparent gametic union, were also described. The possibility that *L. pygmaeum* var. *pygmaeoides* is related to (or conspecific with ?) *Ducellieria* cannot be excluded.

THE *HYPHOCHYTRIALES* AND *ANISOLPIDIALES*

The first mention of anteriorly uniflagellate fungi was by Zopf (1884), with his descriptions of *Hyphochytrium* and *Rhizidiomyces*. In the same year Fisch (1884*a*, *b*) described *Reesia* and a species of *Euchytridium* with anteriorly uniflagellate zoospores. These fungi were classified, according to thallus form, with the chytrids. The first suggestions that these fungi may have been misclassified was by Bessey (1942), after Couch (1941) had shown that in *Rhizidiomyces* the single flagellum was straminipilous. Sparrow (1936*a*) reported that the flagellum of *Pleotrachelus sphacelarum* was inserted anteriorly, and later Karling (1943) erected the genus *Anisolpidium*, transferring *P. sphacelarum* to this genus. The ordinal concept based on *Anisolpidium* (*Anisolpidiaceae*, *Anisochytridiales*) and the other families established by Karling (1943) have been accepted by all later reviewers (Sparrow, 1960; Karling, 1977; Fuller, 1990). However, Sparrow (1960) elected to use the ordinal name *Hyphochytriales*, based on the much older family concepts typified by *Hyphochytrium* (Fischer, 1892; Lotsy, 1907). The thallus of the *Hyphochytriales* may be monocentric (*Anisolpidiaceae*), monocentric with one or more rhizoidal systems (*Rhizidiomycetaceae*), or polycentric (*Hyphochytriaceae*).

A total of 35 named and unnamed taxa have been referred to the *Hyphochytriales* (see Table IV: 10). Over 40% of these are known only from the original descriptions, including the type species of *Hyphochytrium*, and the occurrence of several others has been reported only rarely, without information adding to the original description. Most species are known only as anamorphs. Some of these taxa (*Elina*, *Latrostium*, *Cystochytrium*) must be regarded as having doubtful status.

The ***only*** unifying feature of the *Hyphochytriales* is the possession of an anteriorly inserted straminipilous flagellum. In other groups of chromistan organisms this feature, on its own, is not considered sufficient for the erection of a taxon at class level. Compare the flagellation of *Dictyota* (*Fucophyceae*) and *Lithodesmium* (*Bacillariophyceae*), and note that Leadbeater (1989) accepts that various losses of parts of the flagellar apparatus have occurred at different times. The monophyletic status of the *Hyphochytriales* cannot be maintained solely on the light microscope recognition of the straminipilous flagellum. Barr & Allen (1985) have provided considerable detail for the flagellar apparatus of *Rhizidiomyces*, but surprisingly, no TEM study of the flagellum of *Hyphochytrium* has been published. Hallett (1975)* confirmed the straminipilous ornamentation on the single flagellum of a *Hyphochytrium* sp., and also demonstrated the hollow bullate ornamentation of the zoospore cyst (cf. *Leptomitus*).

Zoospore dimensions vary extremely widely from exceptionally small, as in *Hyphochytrium catenoides* (volume equivalent 6 μm^3), up to the very large zoospores of the doubtfully placed *Cystochytrium radicale* (volume equivalent >1000 μm^3), but most of the species of *Hyphochytrium* and *Rhizidiomyces* have zoospores comparable in size to other lagenidiaceous fungi (volume equivalents 30-60 μm^3). The unnamed *Hyphochytrium* studied by Hallett (1975) had zoospores larger than any other described species (>500 μm^3), a fact which has deterred formal generic placement and makes it unwise to speculate whether the distinctive cyst ornamentation is characteristic of the order. The extremely small zoospores of *Anisochytrium* are formed by intrasporangial zoosporogenesis around a central vacuole; extrasporangial completion of zoosporogenesis is more commonly encountered in *Hyphochytrium* and *Rhizidiomyces*, and should be compared with extrasporangial zoosporogenesis as described for a few lagenidiaceous taxa. In *Rhizidiomyces*

---

*FOOTNOTE: The isolate of *Hyphochytrium* used by Hallett (1975) was isolated on snake skin bait from a soil sample collected from disturbed ground on a hilltop near Rio de Janiero, Brazil, and brought back to me by Professor T. M. Harris in May, 1966 (see Table I: 2, p. 24). The same sample also yielded *Aphanodictyon*.

there is a homohylic vesicle (Clay *et al.*, 1991). Karling (1977) has dismissed the genera *Canteromyces* and *Rhizidiomycopsis*, erected by Sparrow (1960) and based on the two kinds of zoosporogenesis pointed out by Canter (1950), because of the variability that he had recorded for various saprotrophic and parasitic species. This may be correct with respect to *Rhizidiomyces* and *Rhizidiomycopsis*, but I consider that the difference in zoosporogenesis, coupled with the different environment and different class of alga parasitized, is sufficient to justify the retention of *Canteromyces*. The habitats of the species of *Rhizidiomyces*/*Rhizidiomycopsis* and of *Hyphochytrium* range between saprotrophic and weakly parasitic to parasitic, but the parasitism appears to be facultative and the biochemical control of parasitism may be rather undeveloped. The major difference between these taxa is the monocentric, rhizoidal development of *Rhizidiomyces* compared with the polycentric, non-rhizoidal development of *Hyphochytrium*. I therefore place all of the terrestrial saprotrophic and parasitic species in one or other of these two genera using this criterion; a few new combinations are hence necessary. The genera *Hyphochytrium*, *Rhizidiomyces* and *Canteromyces* form a coherent group based on zoosporogenesis, zoospore dimensions and anamorphic status. There are several features (chitin synthesis, the combination of rhizoidal and blastic thallus morphogenesis, zoospore cyst morphology) which suggest a common genic complement with the *Leptomitales sensu lato*, but I retain the order *Hyphochytriales* for the *Hyphochytriaceae* and *Rhizidiomycetaceae*.

The distinct status of *Rhizidiomyces* and *Hyphochytrium* with respect to the filamentous *Peronosporomycetes* became apparent with the biochemical approaches of Lovett & Haselby (1971) and LéJohn (1972, 1974). The studies of Fuller and coworkers (Fuller, 1960, 1962, 1966, 1978, 1990; Fuller & Barshad, 1960; Fuller & Jaworski, 1987; Fuller & Reichle, 1965; Barstow *et al.*, 1989; Clay *et al.*, 1991; Olson & Fuller, 1968) have emphasised the distinctiveness of *Rhizidiomyces apophysatus*. The wall biochemistry of these two genera has been studied by Fuller (1960), Fuller & Barshad (1960) and most recently by Clay *et al.* (1991). The walls contain considerable amounts of chitin, but Clay *et al.* (1991) have shown distinct differences in the relative abundance and distribution of the wall components between the two genera. An increasing body of data (Beakes, 1989; Barstow *et al.*, 1989, Clay *et al.*, 1991) indicates an association of *Rhizidiomyces* and *Hyphochytrium* with some sections of the *Peronosporomycetes* rather than treatment as a separate phylum. Although a separate clade, the very small number of well-researched species and their anamorphic status must be taken into consideration when determining hierarchical status. Molecular biological data (see PART I, p. 92; PART II, p. 113-114) suggest that the hyphochytrids are an early-diverging clade of straminipiles more distant than the *Peronosporomycetes*. In hierarchic classifications this may support classification with either class or subclass status. I provide the formal nomenclature for both alternatives. My preference is to place the class *Hyphochytriomycetes* in synonymy with the class *Peronosporomycetes*, but with separation at the subclass level. However, to accommodate alternative views, I have retained the separate class status in the formal taxonomy, formally erecting the subclass *Hyphochytridiomycetidae* on the basis of flagellation and wall composition, so that further heirarchical nomenclature will be unnecessary.

Caution is required when considering the evidence for sexual reproduction: isoplanogametes were described for *Reesia* and *Euchytridium mesocarpi* (Fisch, 1884*a*, *b*), and Johnson (1957*c*) described the fusion of uninucleate endobiotic protoplasts for *Anisolpidium ectocarpii*. Until the fungi described by Fisch are rediscovered, the status of these fungi, and their sexuality must remain doubtful. Johnson (1957*c*) was unable to determine the site of meiosis in *Anisolpidium* so this life-cycle is also doubtful. All other reports of 'resting spores' describe various thick-walled structures, functioning as sporangia, and associated with colonies of sporangial material. It is possible to draw up a continuum of species of *Rhizidiomyces* from those without thick-walled sporangia (e.g., *R. saprophyticus*) to those in which all sporangia are ornamented (e.g., *R. bullatus*). In none of these species is there any suggestion of a teleomorphic life-history.

The description of sexual fusion in *Anisolpidium* bears some resemblance to that for *Lagenisma*, which is also a marine parasite of a chromistan alga, but there is no evidence to support relatedness. The similarities of the marine *Anisolpidium* species to the *Sirolpidiaceae*, *Pontismaceae* and *Eurychasmaceae* in zoospore

TABLE IV: 10. *The habitats of the taxa which have been attached to the* Hyphochytriomycetes *in Sparrow (1960), Karling (1977) and subsequently*

Habitat	Taxon			
Freshwater/terrestrial saprotrophs	*Anisolpidium elongatum*	+		
	Anisolpidium saprobium	+		
	Hyphochytrium catenoides			
	Hyphochytrium oceanum			
	Hyphochytrium sp. (Persiel, 1960)		*	
	Hyphochytrium sp. (Dick, 1990*a*)		*	
	Rhizidiomyces bivellatus			
	Rhizidiomyces bulbosus			
	Rhizidiomyces bullatus			
	Rhizidiomyces coronus			
	Rhizidiomyces hansonii			
	Rhizidiomyces hirsutus			
	Rhizidiomyces sp. (Karling, 1981*b*)		*	
	Rhizidiomycopsis saprophytica			
Parasites, freshwater green algae	*Canteromyces stigeoclonii*			
	Euchytridium mesocarpi		*	
	Hyphochytrium hydrodictyonis		*	
	Reesia cladophorae		*	
	Rhizidiomyces ichneumon		*	
Parasites, aquatic angiosperms	*Cystochytrium radicale*		*	
	Reesia amoeboides		*	
	Reesia lemnae			
Parasites, freshwater *Tribophyceae*	*Latrostium comprimens*			
	Zygorhizidium vaucheriae		*	
Parasites, freshwater *Peronosporomycetes*	*Rhizidiomyces apophysatus*			
	Rhizidiomycopsis japonica			
Parasites, *Chytridiomycetes*	*Rhizidiomyces parasiticus*			
Parasites, *Eumycota*	*Hyphochytrium infestans*		*	
	Rhizidiomycopsis stomatosa		*	
Parasites, marine *Fucophyceae*	*Anisolpidium ectocarpii*			**0**
	Anisolpidium rosenvingii			**0**
	Anisolpidium sphacelarum			**0**
Parasites, marine *Crustacea*	*Hyphochytrium peniliae*		*	**0**
Marine sediment saprotrophs	*Elina marisalba*		*	**0**
	Elina sinorifica		*	**0**

* Taxa known only from the original description.
\+ Subject to nomenclatural change in formal taxonomy.
0 Removed to other orders.

size, zoosporogenesis, marine habitat (but not the chromophyte host) and the difference in flagellar number suggest that these species of *Anisolpidium* should be segregated from the terrestrial and freshwater hyphochytrids at a higher rank, but placed close to the *Sirolpidiaceae*, *Pontismataceae* an *Eurychasmataceae*. I therefore erect the order *Anisolpidiales* for the marine species. The terrestrial species of *Anisolpidium* are not materially different from *Hyphochytrium*. In making these changes, I also implicitly suggest that the *Hyphochytriomycetes*, as previously circumscribed, may be polyphyletic and that loss of the smooth posterior flagellum may have occurred on at least two occasions.

The remaining marine hyphochytrid, *Hyphochytrium peniliae* (Artemchuk & Zelezinskaya, 1969) is poorly described and a *nom. illeg.*, but its affinities are probably with *Salilagenidium*, to which genus it is allied as a doubtful taxon.

Of the other freshwater algal parasites attached to the hyphochytrids (Table IV: 10), there is further diversity in morphology and development. *Reesia* is amoeboid in the early stages of infection, and the suggestion that penetration of the host is by the flagellum could be interpretated as flagellar tip contact. Thus comparison with *Olpidiopsis*, *Pleocystidium*, *Rozella* and *Plasmodiophora* could be made. *Latrostium* is rhizoidal and its resting spore appears to be radially striated (i.e., punctate, cf. *Aphanomycopsis*). *Latrostium* has been reported by Zopf (1894), Wildeman (1895*a*) and Sparrow (1960), but only Zopf found zoospores. The association of *Zygorhizidium vaucheriae* (Rieth, 1967) (and *Olpidiopsis sorokinei* (Wildeman, 1890*c*, 1896*a*)) on the same host raises the question of conspecificity. The genus *Latrostium* must be regarded as of doubtful status and placed *incertae sedis*, but the evidence of the resting spore suggests a link with the endobiotic algal parasites placed in *Aphanomycopsis* rather than any other hyphochytrid in spite of the slender evidence for the presence of a rhizoid. (See also comparisons with *Rhizophydium decipiens* viz-à-viz *Syzygangia*, and consider the possibility of placing the saprophytic *Aphanomycopsis* species closer to *Hyphochytrium*.) Both *Reesia* and *Latrostium* are accorded 'doubtful status', but keyed with *Rhizidiomyces*.

The description of *Cystochytrium* (Ivimey Cook, 1932*a*) is accompanied by clear habit photographs, and cannot be lightly dismissed. However, there are several critical features which require substantiation before the fungus can be assigned without doubt to a higher taxon. As a root parasite it could be considered with *Lagena*, but it has no known sexual system, and its zoospores are much larger. The genus is left *incertae sedis*, but keyed with *Hyphochytrium*.

Elina is discussed with the *Labyrinthista*.

LAGENISMA

The first record of a phycomycetous parasite of centric marine diatoms was a brief mention by Pavillard (1914). Subsequently Parsons (1962), Johnson (1966*a*), Drebes (1966) and Gotelli (1971) referred to similar organisms. Drebes (1968) finally gave the binomial *Lagenisma coscinodisci* Drebes to the parasite, and this description was consolidated by ultrastructural research (Schnepf & Drebes, 1977; Schnepf, Deichgräber & Drebes, 1978*a*, *b*, *c*, *d*, *e*).

Lagenisma coscinodisci is reported to have a number of unusual features, apart from its sexual system discussed on p. 51. The zoospores are almost hemispherical (Schnepf & Drebes, 1977) rather than kidney-shaped, and the flagella of the first-formed zoospores are withdrawn at encystment by direct entry as the kinetosome moves to the centre of the cell. Apparently there is no lateral fusion of the flagellar and cell plasmamembranes, but more suprisingly, no remnant tuft of tripartite flagellar hairs from the straminipilous flagellum is reported outside the plasmamembrane of the encysting spore or on the cyst wall developed subsequently. The cyst wall formed by the first-formed zoospore is spiny, with the spines initially filled with cytoplasm, but the cyst formed after diplanetic behaviour is smooth. The cyst germinates by a fine tube which forces entry between the diatom frustules. The thallus is initially plasmodial (Schnepf *et al.*, 1978*a*) and develops ***between*** the frustule and the diatom plasmamembrane. Plasmamembrane interfaces differ from both the oomycetous haustorial/host interface and the plasmodiophoraceous plasmamembrane complexes. Older thalli are coenocytic, branched-digitate but not septate, and become walled at the time of host plasmamembrane disintegration, prior to zoosporogenesis. The centrioles of the vegetative nuclei are aligned at 180°. Zoospore cleavage is unique, involving elongate cleavage vesicles (Schnepf *et al.*, 1978*c*). Vesicles containing packets of tripartite flagellar hairs are present in the cytoplasm at this stage (Schnepf *et al.*, 1978*c*, *d*).

I am not convinced that the customary attribution (e.g., Karling, 1981*a*) of the fungus described by Johnson (1966*a*) to this taxon is sound. Johnson's fungus differs in three important criteria: the zoospores are smaller, 3·0-5·0 × 4·0-7·0 μm, volume equivalent <50 μm^3 (Drebes, 1966: 5·6-7·5 × 7·1-10·0 μm; Gotelli, 1971: 5·0-7·7 × 7·5-10·0 μm, both with volume equivalents of >190 μm^3, a four-fold difference); the zoosporogenesis takes place outside the zoosporangium (compare Drebes, 1966: fig. 5; Gotelli, 1971: figs 3, 6); and there is some septation in Johnson's fungus. Drebes (1968) has noted the report by Johnson (1966*a*), but Schnepf *et al.* (1978*e*) acknowledge the records of Gotelli (1971) and Grahame (1976) but not those of Parsons (1962) or Johnson (1966*a*) with respect to *Lagenisma*. Johnson's fungus could be more appropriately considered with *Sirolpidium* which may be septate (*Ectrogella* is not septate, but has diatomaceous hosts). On the other hand, the fungus described as "*Petersenia* sp." on *Rhizosolenia* by Johnson (1966*g*) could be referred to *Lagenisma*. Grahame (1976) found *Lagenisma* in the hemicyclical form of *Coscinodiscus*, known as *Hemidiscus hardmaniana* (Grev.) Kuntze (cited as "*Palmeria hardmaniana* [Grev.]" using zoological code practice for the authority).

Lagenisma is a heterokont fungus, but has so many apparently unique features that it must be separated in a monotypic order and family. The order must remain *incertae sedis* within the *Straminipila*. The life-history, which involves thalloid meiosis (Schnepf *et al.*, 1978*b*) and gamete fusion (Schnepf & Drebes, 1977; Schnepf *et al.*, 1978*b*) as opposed to gametangial contact and gametic nucleus transfer, is unlike that of any other biflagellate fungus, though possibly reminiscent of the *Blastocladiales*.

Comparisons with *Ectrogella perforans* could be instructive when more precise information is available for both organisms with respect to life-history studies, including the mechanisms for sexual reproduction, flagellar retraction and zoospore encystment; flagellar and kinetosome ultrastructure; and molecular biology.

LAGENA AND THE *LAGENACEAE*

Lagena

The monotypic genus *Lagena* was originally described by Vanterpool & Ledingham (1930), followed by further observations by Truscott (1933), Macfarlane (1970) and Barr & Désaulniers (1987*b*, 1990*a*). The fullest description is by Barr & Désaulniers (1990*a*). Macfarlane (1970) also cites Sprague (1950) and Antoine & Ricaud (1966) as reporting isolations of *Lagena*. Sparrow (1960) did not give an account of this genus. More recent records in connection with virus transmission are Ijdenberg, Kummert & Lepoivre (1986), Langenberg & Van Der Wal (1986) and Haber, Barr & Platford (1991).

It is difficult to know whether the isolates described by Vanterpool & Ledingham (1930), Truscott (1933), Macfarlane (1970) and Barr & Désaulniers (1990*a*) can all be regarded as conspecific. The concurrent presence of *Pythium* species is a complicating factor. Truscott (1933) commented that the sporangia of his fungus differed from the type in typically being branched. He claimed that infections of the same host cultivar, both from the type locality soil and the Ontario soil, in pots on the same greenhouse bench, could be distinguished from each other. Truscott also implied that multiple oospores in the oogonium were of rare occurrence in his Ontario isolate.

It is clear that *Lagena* is anisokont and heterokont, with the straminipilous flagellum the shorter. The flagellar attachment to the zoospore is subapical with no evidence of a longitudinal groove. Although the kinetosomes are angled at 120°, the kinetosome/rootlet system is distinct in its simplicity from all main types so far described (see pp. 33-35; Figure I: 7e, I: 8d), and cannot be equated with that of *Peronosporomycetidae*, *Saprolegniomycetidae*, *Hyphochytriales*, *Olpidiopsis*, or *Plasmodiophoromycetes*. Zoosporogenesis is probably only superficially 'pythioid' and there is circumstantial evidence for the possibility that zoospores may sometimes be preformed within the zoosporangium. The zoospore is pyriform (6·0-8·0 × 9·0-11·0 μm, equivalent to a volume of *ca* 250 μm^3) with no pronounced amoeboid tendency. The zoospore cyst is spherical, even when attached to the host cell. Penetration is via a penetration peg which induces callus formation followed by host cell wall deposition providing a thickened collar visible with light microscopy. The developing thallus remains attached to the penetration peg and is normally unbranched and tubular, but sometimes coiled. Mitochondria have tubular cristae with constricted attachment to the inner membrane; dense body vesicles are present, their number inversely correlated with the occurrence of large vacuoles; dictyosomes are present in young thalli. Resting spore formation (probably automictic sexual reproduction) is periplasmic, with residual nucleated cytoplasm outside the resting spore wall. The only other lagenidiaceous fungus demonstrated to have nucleated periplasm is *Ciliatomyces* (Foissner, 1987): in *Pythiella* (Couch, 1935*a*) the periplasm was not stated to contain nuclei. The unique method of fertilization by means of a "conjugation tube" (fertilization hypha ?) (Vanterpool & Ledingham, 1930; Truscott, 1933) is probably based on the misinterpretation of multiple infections of the host. The thick endospore wall is not lamellate or in two parts as in the *Leptolegniellaceae*. A large globule, identified as an ooplast by Barr & Desaulniers (1990*a*), is also visible in photographs by Truscott (1933) and Macfarlane (1970). The development of the ooplast has not yet been shown to be derived from coalescence of dense body vesicles as is the case in the *Peronosporomycetes* or *Ciliatomyces*. The globule is often in the position described as eccentric (Dick, 1969, 1971*b*).

The genus name *Lagena* would need to be replaced by *Lagenocystis* Copeland (Copeland, 1956) if the fungus is placed in the Kingdom *Protoctista* because of the earlier use of *Lagena* for a protist (see Jones, 1984, regarding the status of *Lagena* (*Lagenidae*) in the *Foraminifera*).

TABLE IV: 11. *Alphabetic list of fungal parasites of protoctists (excluding protoctistan algae and fungi), with page references to Sparrow (1960) and some other references, see also entries in PART V*

Fungal parasite	Host
Amphicypellus elegans Ingold	*Ceratium* and *Peridinium* (Sparrow, 1960: 546)
Aphanomycopsis cryptica	*Ceratium*
Aphanomycopsis peridiniella	*Peridinium*
Catenaria anguillulae	unidentified protozoa (Sparrow, 1960: 653)
Chytridium lateoperculatum Scherff.	*Vampyrella* (Sparrow, 1960: 495)
Chytridium lecythii (Ingold) Goldie-Sm.	*Lecythium* (Sparrow, 1960: 503)
Ciliatomyces spectabilis	terrestrial *Kahliella*
Ectrogella perforans	*Vorticella* (Sparrow, 1936)
Endemosarca anomala	*Colpoda* (Olive, 1975)
Endemosarca hypsalyxis	*Colpoda* (Olive, 1975)
Endemosarca ubatubensis	*Colpoda* (Olive, 1975)
Endochytrium operculatum (De Wild.) Karling	unidentified protozoa (Sparrow, 1960: 569)
Lagenidium trychophryarum	*Trichophrys* (Karling, 1981*a*: 138)
Myzocytium sp.	*Vorticella* (Karling, 1942*a*: 90; Karling, 1981*a*: 106; Stein, 1851, [1854?], 1859 - not in Sparrow)
Nucleophaga amoebae	*Heterophrys* (Sparrow, 1960: 123)
Olpidiomorpha pseudosporae	*Pseudospora* in *Vaucheria* (Sparrow, 1960: 123)
Olpidiopsis glenodiniana	marine *Glenodinium* (*Pseudolpidium*) [*Olpidium glenodiniana* on *Glenodinium*; *Pseudolpidium glenodiniana* on *Glenodinium*] (Sparrow, 1960: 956)
Olpidiopsis longicollis	*Vampyrella* (Sparrow, 1960: 955)
Olpidium arcellae	*Arcella* (Sparrow, 1960: 154)
Olpidium difflugiae	*Difflugia* (Sparrow, 1960: 154)
Olpidium leptophrydis	*Leptophrys* (Sparrow, 1960: 154)
Olpidium pseudosporearum	*Pseudospora*, *Pseudosporopsis* (Sparrow, 1960: 146)
Olpidium vampyrellae	*Vampyrella* (Sparrow, 1960: 146)
Pseudosphaerita radiata	*Cryptomonas* (Sparrow, 1960: 963; Karling, 1942*a*: 58; Stein, 1878)
Pseudosphaerita euglenae	*Euglena* (Sparrow, 1960: 963; Karling, 1942*a*: 56; Stein, 1878 - not in Sparrow)
Rhizophydium amoebae Karling	*Amoeba* (Sparrow, 1960: 253)
Rhizophydium apiculatum Karling	unidentified protozoa (Sparrow, 1960: 285)
Rhizophydium leptophrydis Scherff.	*Leptophrys* (Sparrow, 1960: 312)
Rhizophydium vampyrellae (P. A. Dang.) Minden	*Vampyrella* (Sparrow, 1960: 252)
Rhizidium vorax (Strasb.) Sparrow	*Chilomonas* (Sparrow, 1960: 413)
Sphaerita dangeardii	*Euglena* (Sparrow, 1960: 125)
Sphaerita endogena	*Amoeba*, *Heterophrys* and *Nuclearia* (Sparrow, 1960: 127)
Sphaerita trachelomonadis	*Trachelomonas* (Sparrow, 1960: 127)
Sphaerita sp.	*Gonyostomum* (Sparrow, 1960: 128)

Lagena can be placed in the *Peronosporomycotina*, but like *Olpidiopis sensu stricto*, differences in flagellar organization and sexual morphology mean that it cannot be assigned to a sub-class within the *Peronosporomycetes*. The new family *Lagenaceae* is erected for the genus.

Petersenia panicicola (Thirumalachar & Lacy, 1951) is an incompletely described fungus found on previously submerged leaves of the aquatic grass *Panicum repens* L. Lobed holocarpic 'resting' thalli with golden-brown walls were found within mesophyll cells. These thalli behaved as zoosporangia releasing laterally biflagellate reniform zoospores (no measurements available). The host and habitat are so different from the marine *Petersenia* species, that, if found again, it should be compared with *Lagena* or *Cornumyces*.

Ciliatomyces

Ciliatomyces is known only from the original report of Foissner & Foissner (1986*a*, *b*) and Foissner (1987). The generic name was changed from *Ciliomyces* to *Ciliatomyces* by Foissner & Foissner (1995). The fungus is an endoparasite of the cysts of the heterotrichous ciliate *Kahliella*. The host was isolated from air-dried "meadow soil" near Salzburg, Austria, so the fungus must be regarded as terrestrial.

The thallus is extensive, tubular (2-6 μm diameter in the host, but up to 15 μm extramatrically), occasionally branched and initially non-septate, rarely extending to 70 μm in length outside the cysts in non-reproductive development. Zoosporangia are formed relatively soon after infection by an expansion of the hyphal tip, followed by protoplasmic migration into the sporangial initial, which is then cut off from the empty thallus by a septum. The zoosporangia are partly extramatrical and possess a papilla with complex ultrastructure. Zoosporogenesis is internal, producing obpyriform zoospores with subapically inserted flagella. Sexual reproduction is by copulation of two contiguous young (pre-mycelial) thalli. Fertilization is effected via a simple pore. Oosporogenesis is periplasmic. Centriole pairs of vegetative nuclei are aligned at 130-180°, but in the zoospores the kinetosomes are at 90°.

The presence in the cytoplasm of mitochondria with tubular cristae and vesicles with packets of tripartite tubular hairs confirm a placement in the straminipilous fungi. The periplasmic oosporogenesis and the possible dense body vesicle origin of the central vesicle of the mature oospore might suggest affiliation with the *Peronosporomycetidae*, *Rhipidiomycetidae* or *Lagena*. However, there is a great difference in hosts, and confirmation of the flagellar transition zone and kinetosome rootlet organization would be desirable. Zoospore shape and flagellar insertion are reminiscent of these features of *Myzocytiopsis* zoospores rather than the typical principal-form zoospores of the *Peronosporomycetes*, as is the poroid fertilization. It should also be noted that the precise morphology and arrangement of the tripartite tubular hairs on the flagellar shaft have not been established.

Comparisons with the anomalous species of *Aphanomyces* parasitic in the suctorian protozoan *Acineta* (Bartsch & Wolf, 1938; Scott, 1961), should be made, and the systematic details are given in PART V under *Ciliatomyces*. Tentative affiliation with *Lagena* is proposed.

It is apparent that there are several reputedly uniflagellate or biflagellate organisms which are non-rhizoidal, holocarpic endobionts in other flagellate fungi or in free-living or parasitic protists. A review of host parasite relations is badly needed. *Pseudosphaerita* and *Eurychasmopsis* are discussed below. The taxonomic position of every one of these parasitic organisms is doubtful. They are grouped together in Table IV: 11, and where different generic associations are possible, these are indicated. Page references to comments in Sparrow (1960) and Karling (1977, 1981*a*) are also given.

Pythiella

Pythiella was first described as a monotypic genus by Couch (1935*a*) for a parasite of *Pythium gracile* and *P. dictyosporum*, both parasitic on *Spirogyra*. The second species of *Pythiella* (Sparrow & Ellison, 1949) was first described as a species of *Ectrogella* parasitic on another *Spirogyra* parasite, *Pleocystidium parasiticum* (as *Olpidiopsis schenkiana*).

The genus is characterized by 'achlyoid' sporangial dehiscence, laterally biflagellate zoospores and sexual reproduction involving a companion cell and periplasmic oosporogenesis. Although the size of the zoospore of *P. vernalis* (<4·0 μm diam.) suggests affinity with *Olpidiopsis sensu stricto*, the flagellar insertion is apparently more oomycetous and the resting spore is not plerotic. Couch (1935*a*) also acknowledged a similarity in vegetative appearence to *Myzocytium*, *Lagenidium* and *Olpidiopsis*. The chlor-zinc iodide staining reaction of the wall of the type species did not accord with the results for *Olpidiopsis schenkiana* (purplish) or *Pythium* (blue).

The type species does not have unequivocal affinity to any of the oomycete orders, or to the *Myzocytiopsidales*, but is best placed with *Ciliatomyces* and *Lagena* in the *Lagenales*.

Pythiella besseyi is much less characterized and no resting spore is known. The thallus is lobate and flagellar insertion is uncertain. The physical problems facing a holocarpic endobiotic parasite of another holocarpic endobiotic parasite in locating and entering its host are enormous and until stronger evidence of the distinctiveness of the parasitic host and its mycoparasite is available, I regard this as autoparasitism and place the fungus in *Pleocystidium* (see *Olpidiopsis myzocytia*; *Olpidiopsis sphaeritae*; *Rozella pseudomorpha*).

Pythiella is here enlarged to accommodate *Lagenidium pythii* as *Pythiella pythii* (Dick, 1998*a*).

THE *ROZELLOPSIDALES*

Rozellopsis (with reference to *Skirgiellopsis, Rozella, Skirgiellia* and *Pleolpidium*)

A fungus referable to *Rozellopsis* was first observed by Nägeli (1846) as a parasite on an unidentified *Saprolegnia*. Berkeley (1864: figs 8, 9) provided clear illustrations of what can now be recognized as *Rozella septigena* in *Saprolegnia dioica*, with his fig. 9 depicting the uniflagellate zoospore. However, the taxonomy starts with the description of four species of *Rozella* by Cornu (1872*b*). Most of the subsequent species have been named by Karling.

Cornu (1872*b*) established *Rozella* as a genus with uniflagellate zoospores, although he also depicted the occurrence of biflagellate zoospores. *Rozella monoblepharidis** was the first-named, and therefore type species of *Rozella*, and this has been the source of the nomenclatural problems. *Rozella monoblepharidis* is the least adequately described of Cornu's species, for he noted that the zoosporangia had already discharged their zoospores when the fungus was observed, so that the zoospores were not described, nor have they been reported subsequently. Sparrow (1938) suggested that *R. rhipidii* would be preferable as the type of the genus, but did not effect any change. Cornu (1872*b*) had originally drawn attention to the anomalous development of *R. septigena*, in which the parasite induced septation of the host thallus. Fischer (1892*a*) found similar parasites in *Saprolegniaceae* with biflagellate zoospores, and although he stressed the importance of flagellar number elsewhere, he complicated the nomenclature by restricting the genus *Rozella* to taxa with septigenous development (and biflagellate zoospores) in contradiction to Cornu's concept, using a new genus, *Pleolpidium*, for those species without septigenous development but with uniflagellate zoospores. *Pleolpidium* is thus a synonym of *Rozella*. Butler (1907) then misapplied Fischer's genus for his biflagellate but non-septigenous *Pleolpidium inflatum*.

More recent workers have replaced this primary emphasis on septigenous development with the criterion of zoospore flagellar number. The species assigned to *Rozella sensu* Fischer were thus left without a valid generic name. Karling (1942*c*) sought to rectify the situation by proposing the generic name *Rozellopsis* for the biflagellate group of species, but failed to provide a Latin diagnosis. This omission was fortuitously remedied by Cejp (1959) in his Latin synopsis for the *Flora ČSR*. Cejp (1959) also proposed the *nomen novum, Rozellopsis fischeri,* for *Rozellopsis septigena*, but this was unnecessary since *Rozella* and *Rozellopsis* are typified by different species (*Rozella monoblepharidis* and *Rozellopsis septigena* respectively), even though confusion is created by the valid binomials *Rozella septigena* Cornu and *Rozellopsis septigena* (Cornu) Karling *ex* Cejp. Finally, Batko (1977) proposed the genera *Skirgiellia* and *Skirgiellopsis* for the septigenous taxa in the uniflagellate and biflagellate series respectively, but again the type species of *Rozellopsis, R. septigena,* as its synonym *R. fischeri*, was used to typify *Skirgiellopsis*, rendering the latter genus a synonym of *Rozellopsis*.

---

*FOOTNOTE: Cornu (1972*b*) appeared to use the long and short versions of the specific epithets *monoblepharidis* and *monoblepharidis-polymorphae* and *rhipidii* and *rhipidii-spinosi* without preference. The selected epithet is therefore determined by 'accepted usage' (*ICBN* as interpreted by authorities at CABI Bioscience). Fischer (1892*a*), Minden (1911*a*), Sparrow (1938) and Karling (1942*c*) used the shorter versions, and their choice, rather than the longer versions used by Sparrow (1960) or Karling (1977, 1981*a*) is followed here, because the shorter version does not imply unsubstantiated species specificity.

Almost all the species placed in *Rozellopsis* and *Rozella* are parasitic in aquatic fungi, a host preference they share with many *Plasmodiophoromycetes*; the few exceptions may be miss-attributed (Held, 1981). Four species have been assigned to *Rozellopsis* (*sensu lato* including *Skirgiellopsis*) and 25 species to *Rozella* (*sensu lato* including *Skirgiellia*). The majority of these 29 species are parasites of either the *Chytridiales* (10 species, of which 4 are anamorphic) (no members of the *Spizellomycetales* are known to be parasitized) or the *Pythiaceae* (7 species, of which 3 are anamorphic). In both of these groups there are no diagnostic morphological features intrinsic to the parasite and which are not related to the host morphology. In the absence of any thorough cross-infection studies or molecular biology, these species can only have a status equivalent to that of *formae speciales*. The species with priority in these two groups are *Rozella polyphagi* and *R. irregularis* respectively.

The rigid interpretation of the uniflagellate or biflagellate condition in these parasites has led to the separation of *Rozella* and *Skirgiellia* into the *Chytridiomycetes* and *Rozellopsis* and *Skirgiellopsis* into the *Peronosporomycetes*. (Batko (1977) placed *Skirgiellia* in the *Achlyogetonaceae* and *Skirgiellopsis* in the *Sirolpidiaceae*.) Neither assignation is entirely satisfactory.

None of the biflagellate species has been shown to be straminipilous. For *Rozellopsis* there are no suggestions of lateral flagellar insertion (compare the descriptions in Butler (1907) for his species *Pleolpidium (Rozellopsis) inflatum* and *Pseudolpidium (Olpidiopsis) pythii*); although the drawings by Prowse (1951) could be interpreted as of bean-shaped zoospores with lateral flagellar insertion, he only describes them as pyriform. If the second centriole of *Rozella* were to develop a functional flagellum then ventral (lateral) flagellation in a groove might be expected for *Rozellopsis*, but this would discount ultrastructural evidence for the flagellar apparatus of *Rozella*. The intrasporangial rotational behaviour of the mature undischarged zoospores of various species of *Rozellopsis* and *Rozella*, noted by many observers, is also apparently identical.

Comparative TEM studies of the zoospores for these species are lacking: only *Rozella allomycis* has been examined in detail (Held, 1975). This zoospore has bilateral symmetry and is unique in possessing a deep tubular flagellar chamber. In living zoospores the flagellum emerges from the chamber on the side (ventral side) of the cell and the second centriole is ventral to, and at an angle of 60° to the functional kinetosome - centriole. It should be noted that this eccentric flagellar arrangement was figured by Cornu (1872*b*) for both *Rozella rhipidii* and *Rozella septigena* and as a result of this bilateral symmetry, these zoospores were described as kidney-shaped. This description should not be confused with the bean-shaped biflagellate zoospores of the *Peronosporomycetes* (see Figure I: 2). Cornu (1872*b*) regarded the biflagellate zoospores of *Rozella septigena* as abnormal due to poor environmental conditions.

The zoospore of *Rozella* is not typically chytridiaceous (Lange & Olson, 1979: 88), although some other chytrids have elongate zoospores; may have more than one lipid globule, and have dispersed ribosomes (see Koch, 1958). The zoospore organization is also distinct in the position and size of the large mitochondrion which displaces the nucleus anteriorly. A striated fibrous root lies between the kinetosome and this mitochondrion. The kinetosome is exceptionally long and the transitional zone is not well-defined, although it possesses struts to the plasmamembrane similar to those of the *Chytridiomycetes*. From light microscopy, *Rozella* and *Rozellopsis* zoospores apparently differ ***only*** in the number of flagella (see also Held, 1981).

It should be recalled that partial flagellar loss is known in heterokont organisms (cf. Leadbeater, 1989, and the discussion under *Hyphochytriales*), and when comparative studies of the kinetosomes are available, it might be possible to confirm or refute the suggestion that flagellar number could be of minor taxonomic importance in this particular group as well.

TEM profiles of encysting zoospores of *Rozella allomycis* are provided by Held (1973). Attachment to the host is initially by the tip of the posterior flagellum in *Rozella* (Held, 1980) and *Rozellopsis* (Prowse, 1951). The exception is *Rozellopsis (Gracea) waterhouseae* which becomes attached by the tip of the longer anterior flagellum (Waterhouse, 1940) (see below). During retraction of the flagellum in *Rozella allomycis* the encysting zoospore assumes a discus-like or hemispherical shape and the flagellum does not appear to be involved in adhesion. However, the position of the ventral flagellar canal may mean that the extrusomes associated with the kinetosomes may still be proximal to the site of adhesion. Similar configurations are reported for other species. After retraction of the flagella, encystment on the host surface is as a sphere, often with a small peg attached to the host; Held (1973) made reference to similarities with *Olpidium*, *Olpidiopsis* and the plasmodiophorids.

Thallus development is initially plasmodial, as in *Olpidiopsis*, but the plasmodial phase persists, as it does in *Olpidium* (Temmink & Campbell, 1968). Butler (1907) accepted the findings of Fischer (1882, 1892*a*) which have now been confirmed ultrastructurally by Held (1980): each uninucleate plasmodial initial develops either into a single or, by vegetative division, into several multinucleate plasmodia which become transformed into sporangia (apocytic coenocytes). In other cases there may be fusions between immature plasmodia (heterothallic coenocytes), resulting from the multiple infections characteristic of these parasites. Butler (1907) suggested that plasmodial fusion occurred in *Pleolpidium (Rozella) irregulare* (a uniflagellate *Rozella*), and Karling (1981*e*) indicated that plasmodial division could occur in this species. It is generally assumed that *Rozellopsis septigena* and *R. simulans*, and *Rozella septigena* and *R. allomycis*, have a protracted plasmodial phase: the occurrence of plasmodial division with compartmentalization of the host would result in the septigenous state. The extent to which the parasite induces compartmentalization of the host is variable (some species occupy host sporangial initials), but the parasite sporangium finally fills any compartment completely. The balance between this hyperplasia and accompanying hypertrophy still requires further data (cf. Pemberton *et al.*, 1990), but the amount of hypertrophy induced is small compared with that induced by *Olpidiopsis*.

Most species of *Rozellopsis* and *Rozella*, with the exception of *Rozella monoblepharidis* and the parasites of *Pythium* (with the possible exception of *Rozella irregularis*), either transform reproductive cells, or cause cessation of vegetative growth with limited hypertrophy. The development of *Rozella cladochytrii* could be compared with that of *R. monoblepharidis* rather than with development in the other chytridiaceous parasites.

The TEM profiles of mitochondria of germinating cysts of *Rozella allomycis* (Held, 1973: fig. 10; 1975) are described and clearly figured as possessing well-developed plate-like or shelf-like cristae, though a flattened, finger-like interpretation is also possible for a few profiles. Those of *Rozella polyphagi* (Powell, 1984) are stated in the legend to be plate-like, like those of the host, but in the text they are noted as morphologically distinct from those of the host (Powell, 1984: fig. 7 and its enlargement, fig. 13; fig. 10). The cristae could be interpreted as flattened, finger-like lobes with constricted attachment at the innner membrane. In both of these species mitochondria are somewhat similar to those of the plasmodiophorids, and to those of *Olpidium brassicae*, which were described as tubular (Temmink & Campbell, 1968). There are undoubtedly discrepancies between the illustrations of cristae and the words used to describe their morphology.

For those taxa of *Rozellopsis* and *Rozella* for which this stage is known, the resting spores appear to be very similar, usually golden or brownish in colour, with exosporial, fine spines. In this respect they resemble the plerotic oogonia of *Olpidiopsis*. However, they all lack companion cells (cf. *Olpidiopsis* versus *Pseudolpidium*), so that if the resting spore is sexually conceived, sexuality is either automictic or dependent on prior plasmodial fusion. Held (1980) refutes the suggestion of Sörgel (1952) that heterothallic union of plasmodia is necessarily involved. Held (1981) also emphasised, as diagnostic, the occurrence of a single resting spore in a host compartment (contrast *Olpidiopsis*).

The similarity in the appearence of host material infected with *Rozella septigena* or *Rozellopsis septigena*, and the suggestion by Cornu (1872*b*), Fischer (1882, 1892*a*) and Minden (1911*a*) that flagellar number is variable, leads to the question as to whether such similarities can really be due to parallel evolution. The diversity of life-histories in non-oomycete biflagellate fungi would make it possible that *Rozella septigena* and *Rozellopsis septigena* could be conspecific, possibly as different phases of the same life-history, or with variable nuclear cycles, and comparisons with *Dictyomorpha* (p. 230) are relevant.

Of the parasites of *Pythium*, the fungus described by Waterhouse (1940), named as *Rozellopsis waterhouseae* Karling (Karling, 1942*c*) and now transferred to *Gracea* (Dick, 1998*a*), was reported as biflagellate with the anterior flagellum the longer. Both species of *Gracea* are parasites of *Pythiaceae* (*Pythium* and *Phytophthora*), *G. gracilis* causing considerable hyperplasia (Pemberton *et al.*, 1990). The resting spores of *Gracea gracilis* and those of *Rozella* and *Rozellopsis* all lack companion cells, but there may be differences in the duration of the plasmodial phase and the extent to which the parasite sporangial wall is developed. The poorly described *Rozella barrettii* is also parasitic in *Phytophthora*. The doubts attached to the classification of the pythiaceous parasites in *Rozellopsis* and *Rozella*; the demonstration of a tinsel flagellum in *G. gracilis*; the relative lengths of the two flagella; the absence of a companion cell and the possibility of plasmodial division in the hyperplasic proliferation are presented as reasons for establishing the new genus, *Gracea*. The biflagellate *Rozellopsis inflata* may either be congeneric with *Gracea*, or part of a biflagellate/uniflagellate pair similar to the other oomycete parasites of *Rozellopsis* and *Rozella*; *Rozella barrettii* could also be placed with these species, but in the doubtful category. If the nature of the sporangial wall membrane relative to the host wall were to be regarded as of low hierarchical significance, as is suggested by placing *Plasmophagus* in this order, then *Gracea* becomes intermediate in some respects between the *Olpidiopsidales* and the *Rozellopsidales*. The removal of possibly straminipilous, as opposed to anisokont, biflagellate fungi from the *Rozellopsidales* enables this order to be placed in an intermediate position with respect to the *Chytridiomycetes*, *Plasmodiophoromycetes* and other plasmodial non-oomycetous biflagellate fungi such as *Endemosarca*.

The remaining species regarded as anomalous by Held (1981) can also be transferred to other genera. *Rozella itersoniliae* was regarded as an *Olpidium* by Held (1981), but I prefer to place it with the other flagellate parasites of the *Eumycota*, in *Pleotrachelus*. Its resting spores are not known.

Similarly, *Rozella coleochaetis* is equally-well placed in *Plasmophagus*, and this was considered as a possibility by Sparrow *et al.* (1965).

Rozella pseudomorpha presents a rather different problem, in that it is difficult to imagine how an endoparasite of the endoparasitic *Lagenidium* or *Myzocytium* in *Spirogyra* could locate its host, unless the infection was by a parasite zoospore in a host zoospore (compare the parasitism of *Dictyomorpha*). The species must be regarded as of doubtful status; possibly having been described as a result of a mixed infection, or autoparasitism. Until rediscovered, the species is better returned to *Olpidium* as a doubtful species, where it was originally placed by Scherffel (1926*b*).

Rozella marina may also represent a case of chytrid mycoparasitism. It is anomalous within *Rozella* both in its marine habit and in possessing posteriorly uniflagellate zoospores. Resting spores, which might confirm its placement in *Rozella*, are not known (Johnson, 1966*f*), and the taxon is placed in the *Chytridiomycetes*.

All the biological evidence (zoospore behaviour; thallus development, including the distinctive modification of the terminus of the parasitized host axes; resting spore structure and coevolutionary aspects of host/parasite associations) points to the relatedness of *Rozellopsis* and *Rozella*, and this was recognized by Held (1981). The only divisive character is the flagellar number, which has yet to be substantiated by comparative ultrastructural examinations. Until evidence is forthcoming on the ultrastructural and molecular biological distinctiveness of *Rozellopsis septigena* and *Rozella septigena*, *Rozellopsis simulans* and *Rozella achlyae*, or *Rozellopsis uliginosa* and *Rozella rhipidii*, I propose that these pairs should eventually be merged. However, it is probably prudent to retain the genus *Rozella* in the interim for the uniflagellate taxa. If the concept of Fischer (1892*a*) followed by Batko (1977) is accepted, the diagnosis of *Rozellopsis* should be emended on the basis of the septigenous behaviour of the plasmodium irrespective of flagellar number, the uniflagellate septigenous species *Rozella allomycis* would either have to be transferred to *Rozellopsis* without a biflagellate analogue or given a new genus, because *Skirgiellia* (typified by *Skirgiellia septigena* (Cornu) A. Batko) would no longer be an acceptable generic alternative.

Both Sparrow (1960) and Karling (1981*a*) consider the possibility that *Rozellopsis* is related to the *Plasmodiophoromycetes*, but prefer to leave the genus doubtfully assigned to the *Peronosporomycetes*. Barr (1980*a*, 1990) regarded *Rozella* as a chytridiomycete, and placed the genus in the *Olpidiaceae*. The *Olpidiaceae* was placed in the *Spizellomycetales*, with slight reservations.

This group of parasites is distinctive, and I think that a more convenient and possibly more natural classification results if both *Rozella* and *Rozellopsis* are transferred to a new family, which should become the type of a new order. The lack of descriptive information for the type species of *Rozella* makes it unwise to use this genus as the type of the higher taxa.

I therefore propose that the genus *Rozellopsis* be used to typify this family and order for the following reasons:

1. the genus has been identified as possessing two flagella, and loss of one flagellum is more probable than flagellar duplication
2. the pattern of zoosporogenesis appears to be identical for the biflagellate and uniflagellate taxa
3. the pattern of parasitism and parasitic effect on the host appear to be identical for the biflagellate and uniflagellate taxa
4. the type of *Rozella* is not known to have zoospores with clearly defined flagellation

Pseudosphaerita and *Sphaerita*

Pseudosphaerita and *Sphaerita* are parasites of protoctists including the euglenids and were first described by Dangeard (1886, 1895). Jahn (1933) provided the most useful summary of the early work, stressing the lack of information on zoospores of species attributed to these genera, and the differences in hosts. Dangeard (1886) originally described *Sphaerita* as a parasite of both rhizopods and euglenids (*Euglena* and *Phacus*). Subsequently (1895) he redescribed *Sphaerita endogena* as a parasite of euglenids and found another euglenid parasite, for which he suggested the name *Pseudosphaerita euglenae*. The thallus of *Pseudosphaerita euglenae* is initially plasmodial and undergoes successive cleavage as is described for *Schizochytrium*, rather than simultaneous cleavage as in *Sphaerita*; this is a method of zoosporogenesis which would exclude *Pseudosphaerita* from the *Peronosporomycetes*. Coincidentally, this distinction between *Sphaerita* and *Pseudosphaerita* was coupled with a difference in flagellar number, the former being uniflagellate and the latter biflagellate. Both *Pseudosphaerita* and *Sphaerita* apparently have anterior or lateral flagellar insertion, but there is no evidence to suggest that the anterior flagellum has tubular tripartite hairs.

Pseudosphaerita has spiny resting spores without companion cells, but resting spores are not known for *Sphaerita*.

Chatton & Brodsky (1909) separated the rhizopod parasite of Dangeard (1886) as *Sphaerita dangeardii*. Species of *Sphaerita* described later related to the kind of cell division, rather than flagellation, and in fact most species of these genera are not known to be flagellate. The parasites of amoebae, mostly found incidentally and named by zoological protistologists, are not described with details which would support fungal diagnoses. These later *Sphaerita* species have similarities to other ill-defined protist parasites under the names *Nucleophaga* and *Morella* (Karling, 1972). The formal taxonomy has been assembled for most of the binomials in *Nucleophaga* or *Morella* because this information has not been given by Karling (1972) or Sparrow (1960); the early binomials were assembled by Jahn (1933). The genera are very poorly understood and reference should be made to accounts in Jahn (1933), Sparrow (1960) and Karling (1972, 1977, 1981*a*). Additional references (Cejp, 1935*b*; Mitchell, 1928; Pérez Reyes & Gomez, 1961; Pérez Reyes, 1963; Puymaly, 1927) summarized by Karling (1981*a*) indicate that *Pseudosphaerita* and *Sphaerita* may be congeneric.

The biflagellate genus *Pseudosphaerita* should be excluded from the *Peronosporomycetes* until the type species has been shown to be straminipilous. The genus may have affinities with the *Rozellopsidales*. In accord with my general thesis that flagellar number, on its own, is not a character that should be given undue weight, I here formally group the species of the two genera under the teleomorph generic name *Pseudosphaerita*, and the anamorph generic name *Sphaerita*. Parallels with *Rozellopsis/Rozella* are close and a placement of both genera in a separate family, the *Pseudosphaeritaceae*, in the *Rozellopsidales* is proposed until the genera are better known.

More recently, Canter (1968) described *Sphaerita dinobryonis*, which has elongate aplanospores (resembling *Gonimochaete*) and spiny resting spores without companion cells. This fungus may therefore have straminipilous affinities (cf. another protist endoparasite, *Eurychasmopsis*).

In PART VI only the parasites of euglenoids are separated. Comparisons with other protistan endoparasites (*Nucleophaga*, *Morella* and microsporidians, etc.) will be essential for correct identifications to be made.

Plasmophagus

Plasmophagus was established by Wildeman (1895*c*) for a parasite of *Oedogonium*. The thallus is apparently plasmodial, filling the host cell, although the sporangial membrane may not be so closely associated with that of the host as is the case in *Rozellopsis*. The zoospores are elongate, and from the figures (Wildeman, 1895*c*: plate VIII, fig. 9) it is possible that the flagellum is sub-posteriorly inserted. Sparrow (1933*b*) reported a similar infection in *Tribonema* (cf. *Olpidiopsis sorokinei*) but was unsure whether his fungus was the same. Neither author found resting spores.

The numerous drawings by Wildeman (1895*c*) indicate that the parasite thallus almost completely fills the host cell, and that rarely there may be two or three thalli tightly packed together. This arrangement is not like that of *Olpidium* or *Olpidiopsis*, and I interpret the drawings and description as representing an algal parasite with features most similar to, but not identical with *Rozellopsis* or *Rozella*. The close-packed thalli may represent a divided plasmodium.

There are two other fungi which should be placed in this genus, *Olpidiopsis deformans* and *Rozella coleochaetis* (Sparrow *et al.*, 1965). Karling (1942*e*: 54) suggested that *Olpidiopsis deformans* may be referable to *Rozellopsis* or *Woronina*, but as an algal parasite with a repeatedly dividing plasmodium, it is better removed from *Olpidiopsis sensu stricto* and placed in *Plasmophagus*. Algal parasitism is anomalous in *Rozellopsis/Rozella* and, in the absence of any strong evidence in favour of placement in either of these genera, transfer of *Rozella coleochaetis* to *Plasmophagus* is also proposed.

Olpidiomorpha

Olpidiomorpha has been treated as a genus of the *Chytridiomycetes* because its zoospores have been described as possessing only a single flagellum. It was reported as a parasite of a protist parasite of freshwater *Vaucheria*. Although *Olpidiomorpha* is known only from the type description of the species *O. pseudosporae* Scherffel (1926*b*), the significant feature of the description is the sub-apically to laterally attached, posteriorly directed flagellum (Scherffel, 1926*b*: plate 28, fig 8; Karling, 1977: 27, figs 22, 23). This flagellum could be inserted in a similar fashion to that of either *Rozella* or the posterior flagellum of the *Plasmodiophorales*. *Olpidiomorpha* is placed here, with *Plasmophagus*, *incertae sedis*, in the order *Rozellopsidales* until details of flagellar insertion are known.

Dictyomorpha

Among the parasites of *Peronosporomycetes*, one of the earliest to be recorded was the fungus now known as *Dictyomorpha* (Pringsheim, 1860; Berkeley, 1864; Cooke, 1871). A full account of the development of this fungus is given by Mullins (1961) and only the salient points are summarized here. The fungus is initially plasmodial and does not cause hypertrophy. The zoospores of the parasite invade hyphal tips or sporangial initials, and subsume the host zoospore initials to form a large number of small parasite sporangia in the host sporangial cavity, one parasite sporangium being equivalent to a host zoospore. The parasite zoospores are uniflagellate and function asexually or fuse as isogametes to form a biflagellate zygote. The sexual system is heteroecious. The zygote invades the host in the same manner as the encysted zoospore, but induces segmentation of the host thallus and develops to give a single aplerotic resting spore in the host segment.

The plasmodial development and uniflagellate zoospore suggest affinity with *Rozella*, but the life-history may be more like that of the *Blastocladiales* or *Plasmodiophorales*. It shares with *Rozellopsis*, *Rozella* and the plasmodiophorid genera *Woronina* and *Octomyxa* a host in the *Saprolegniaceae*, but its resting spore morphology and morphogenesis differ in detail from those of any of these fungi. Mullins (1961) foresaw a possible reduction to congenerity with *Rozella*: placement in the *Rozellopsidales* is justifiable but provisional until ultrastructural information becomes available.

The recent report (Sarkar & Dayal, 1988) of *Dictyomorpha* in *Pythium* differs in two important respects from *D. dioica*: the fungus apparently causes some degree of hypertrophy of the host; resting spores do not appear to be surrounded by individual host-derived walls, and several resting spores are produced in each induced septate section of the hypha. But for its uniflagellate zoospore, comparison with plasmodiophoraceous parasites of *Pythium* would be made. However, the resting cysts are not joined together. The taxon is left as described until more critical details become available, but the change of status made in Dick (1998*a*) is warranted on the basis of host and developmental criteria.

PLEOTRACHELUS

This genus has been treated within the *Chytridiomycetes* because it is described as possessing only a single flagellum. However, the genus should be considered here because of doubts regarding the number and pattern of insertion of the flagella; wall chemistry, and the host ranges. *Pleotrachelus* was erected by Zopf (1884) for a parasite of *Pilobolus*; its thallus was stated to be initially plasmodial, later forming an ellipsoid sporangium with many discharge tubes. Zoosporogenesis was intrasporangial and the zoospores were stated to be posteriorly uniflagellate. A complete list of *Pleotrachelus* binomials is given in Table IV: 12.

Petersen (1905) enlarged the concept of the genus, introducing two named sections of the genus, sections *Integri* and *Lobati* respectively. He included a number of marine parasites which have since been shown to be biflagellate and which are considered under *Pontisma* and *Sirolpidium*. *Pleotrachelus inhabilis* was found in marine red algae and *P. paradoxus* in a chytrid parasite of *Polysiphonia*; both were said to have walls giving a positive chor-zinc iodide reaction (but *Pleotrachelus olpidium*, in brown algae, was reported to give a negative reaction).

Pleotrachelus wildemanii, from moss rhizoids and protonemata, was described as having a single laterally attached, trailing flagellum (Petersen, 1910; Rieth, 1962), a criterion recalling *Olpidiomorpha*. In the most recent report of this taxon, Döbbeler & Itzerott (1983) did not observe the flagellation. On the other hand, the bryophyte host could indicate a relationship with *Syzygangia elliptica*. The taxon may well be an asexual state of this species, perhaps with a very short anterior flagellum which was not detected by these observers. All descriptions of *P. wildemanii* are incomplete, but where comparisons of dimensions are possible, conspecificity with *Olpidiopsis ricciae* and *Syzygangia elliptica* appears highly probable, and is reflected in the synonymy here. Good photographs of the thalli are provided by Rieth (1962).

Sahtiyanci (1962) transferred *Olpidium brassicae* to *Pleotrachelus* because of the large number (1-16) of discharge tubes associated with each sporangium, claiming that this morphology represented the original concept of Woronin (1878) and Dangeard (1886). At the same time, she erected the species *Pleotrachelus bornovanus* and *P. virulentus* for other angiosperm root parasites. The transfer by Sahtiyanci (1962) of *Olpidium brassicae* to *Pleotrachelus* is not accepted. A case should be made (in preparation) for the retypification of *Olpidium* using *Olpidium brassicae* as the type, since this species is of economic importance and therefore likely to be most fully investigated by ultrastructural and molecular biological techniques. Barr (1980*b*) has reduced one of her species to synonymy with *Olpidium brassicae*, and the other has been placed in synonymy with *Olpidium radicale* by Lange & Insunza (1977). The inadequately described *P. radicis* (see Samson (1939) *re Asterocystis radicis* and *Olpidium*, *Pleotrachelus radicis*) could be referred to one or other of these species and Barr (1980*b*) has placed it with *O. brassicae*.

Sparrow (1960) considered that the genus might ultimately be suppressed, with the taxa being assimilated in various uniflagellate and biflagellate genera. Karling (1977) took this step, placing *Pleotrachelus* as a synonym of *Olpidium* because he did not regard the number of discharge tubes as meriting generic separation; but he did not consider the anomalous features of wall biochemistry and flagellar insertion, mentioned above, that were thereby introduced into *Olpidium* by a wholesale synonymy.

Note should be made of the fact that the host of the type species and *P. zopfianus* is *Pilobolus*, a terrestrial coprophilous zygomycete. While I agree with Karling (1977) that the number of sporangial discharge tubes, on its own, is insufficient for the separation of *Pleotrachelus* from *Olpidium*, the distinctive host and habitat of the type species makes it unlikely that this species is congeneric with the type species of *Olpidium* (*O. endogenum* (Braun) Schroeter in freshwater desmids). I prefer to retain the genus *Pleotrachelus* as a form

TABLE IV: 12. Pleotrachelus *species: alphabetic list with synonymy in this text*

Pleotrachelus andreei	synonym of *Sirolpidium andreei*
Pleotrachelus bornovanus	synonym of *Olpidium radicale*
Pleotrachelus brassicae	synonym of *Olpidium brassicae*
Pleotrachelus ectocarpi	synonym of *Anisolpidium joklianum*
Pleotrachelus fulgens	*Pleotrachelus fulgens*
Pleotrachelus inhabilis	synonym of *Pontisma inhabilis*
Pleotrachelus itersoniliae	(synonym: *Rozella itersoniliae*)
Pleotrachelus lobatus	synonym of *Petersenia lobata*
Pleotrachelus minutus	synonym of *Anisolpidium minutum*
Pleotrachelus olpidium	synonym of *Anisolpidium olpidium*
Pleotrachelus paradoxus	synonym of *Sirolpidium paradoxum*
Pleotrachelus petersenii	synonym of *Syzygangia oedogonii*
Pleotrachelus pollagaster	synonym of *Petersenia pollagaster*
Pleotrachelus radicis	synonym of *Olpidium brassicae*
Pleotrachelus rosenvingii	synonym of *Anisolpidium rosenvingii*
Pleotrachelus rotatoriorum	unidentifiable species of *Haptoglossa*
Pleotrachelus sphacelarum	synonym of *Anisolpidium sphacelarum*
Pleotrachelus tumefaciens	synonym of *Eurychasmidium tumefaciens*
Pleotrachelus virulentus	synonym of *Olpidium brassicae*
Pleotrachelus vuilleminii	(synonym: *Olpidiopsis vuilleminii*)
Pleotrachelus wildemanii	synonym of *Syzygangia elliptica*
Pleotrachelus zopfianus	possible synonym of *Pleotrachelus fulgens*

genus for the endoparasites of higher fungi (*P. fulgens* and *P. zopfianus*) and transfer *Rozella itersoniliae* and *Olpidiopsis vuilleminii* to *Pleotrachelus* (the latter transfer made possible by placing *Pleotrachelus* in the *Rozellopsidales*, in which both uniflagellate and biflagellate taxa are included, see discussions under *Rozellopsis* and *Rozella*; *Pseudosphaerita* and *Sphaerita*; *Dictyomorpha*).

Pleotrachelus, in this restricted sense, is placed in the *Rozellopsidales*, and contains three species (*P. fulgens*, *P. itersoniliae* and *P. vuilleminii*).

THE *LABYRINTHISTA* (*LABYRINTHULOMYCETES*)

The summary of the labyrinthulids and thraustochytrids given here is brief because their morphology and taxonomy is peripheral to the main theme of this work. However, the most recent reviews (Olive, 1975; Moss, 1986; Porter, 1990) did not give the formal taxonomy or keys for the group. The taxonomy is confused partly because of the doubts concerning species limits and partly because these groups have been discussed and their morphology interpreted with reference to either zoological or botanical analogues and classifications. The keys included here are based on the reviews cited above, the original descriptions and the partial tabulations, illustrations and keys of Johnson & Sparrow (1961), Pokorny (1967, 1985), Quick (1974*b*), Raghu-kumar (1988*b*) and Ulken (1990).

Labyrinthulids and thraustochytrids are grouped together and separated from other straminipiles because they have a longer anterior flagellum (which has typical straminipilous ornamentation); mitochondria with tubular cristae; produce cell investments of scales, and have ectoplasmic nets associated with sagenogenetosomes. However, not all of the organisms in this group produce zoospores or possess sagenogenetosomes. Organisms identifiable as labyrinthulids were first described by Cienkowski (1867) and classified in a separate family by Haeckel (1868). In the same year Barker (1868) described the genus *Diplophrys*, a genus sometimes affiliated with *Labyrinthula* species. The labyrinthulids remained a little-known group associated with rhizopod *Protozoa* until interest was revived following the epidemic of wasting disease of *Zostera* in the 1930s (Renn, 1934, 1935, 1936*a*, *b*, 1937, 1942; Young, 1938, 1943 and references therein).

Recent molecular biological data (18S rDNA) on the phylogeny of the labyrinthulids (Leipe *et al.*, 1996) should be treated with some caution because the thraustochytrid organism used (*Labyrinthuloides haliotidis*) is aberrant with respect to the arrangement of the straminipilous hairs on the anterior flagellum (see below). The second straminipilous organism used (*Developayella elegans*) is also unusual in the flagellar motion and cell movement.

Sparrow (1936*a*) described the first known thraustochytrid as a member of the *Saprolegniaceae*, and since then numerous other genera and species of thraustochytrids have been reported. Only since the advent of TEM studies has it become apparent that the thraustochytrids were related to the labyrinthulids. While many observations have been made on the labyrinthulids *in situ*, most of the observations on the thraustochytrids have been made from axenic cultures from bait-derived collections and very little is known of their natural substrata or form in these substrata (but cephalopod infections have been reported by Polglase, 1980, and abalone infections have been studied by Bower, 1987*a*, *b*, *c*, *d*; 1989; Bower *et al.*,1989 and Bower *et al.*, 1989). Both the labyrinthulids and thraustochytrids can be cultured easily on specialized agar media (Porter, 1990), but cultural studies have not been able to elucidate the life-histories of these organisms. The life-histories, nuclear cycles, morphogenetic sequences and taxonomy of the labyrinthulids and (to a lesser extent) the thraustochytrids are all inadequately documented.

The part of the life-history that has received adequate comparative attention from light microscope and TEM investigations is the trophic cell. This cell has an investment of dictyosome-derived, 0·5-2·0μm diameter, carbohydrate scales, which is quite distinct from any other straminipilous fungus, although the construction recalls that for the prymnesiophytes and chrysophytes (see Margulis *et al.*, 1990). The principal component of these scales, at least in some species, is L-galactose, an unusual isomer since most organisms produce only D-galactose. The trophic cell is typically uninucleate and contains mitochondria with tubular cristae. The most distinctive organelle of the trophic cell is the sagenogenetosome, which is associated with the ectoplasmic net, but this is not universally present (see and compare with other diagnostic criteria for species of *Althornia*, *Labyrintuloides*, *Diplophrys* and *Corallochytrium*).

The principal differences between the labyrinthulids and the thraustochytrids are the shapes of the trophic cells (spindle-shaped in the labyrinthulids, ovoid/ellipsoid/spherical in the thraustochytrids); the number of layers of scales (single in labyrinthulids), and the extent to which the ectoplasmic net envelops (labyrinthulids) or is polar (thraustochytrids) to the trophic cells. When enveloping, the ectoplasmic net enables the trophic cells to move within the net system (Nakatsuji & Bell, 1980; Nakatsuji *et al.*, 1981), so giving the characteristic gliding motion of the spindle-shaped labyrinthulid cells; however, a few thraustochytrids also show gliding movements, but not ***through*** their net systems.

Compared with the labyrinthulids, the thraustochytrids typically have a much thicker investment of overlapping scales. *Corallochytrium* is regarded as a very doubtful member of the *Labyrinthista* because it lacks these typical scales, has no known zoospore, has no sagenogenetosome (Chamberlain & Moss, 1988); no convincing evidence for mitochondria with tubular cristae was provided by Raghu-kumar (1987). *Corallochytrium* should be compared with *Apicomplexa* and other circumjacent flagellates.

Cell division in the labyrinthulids is initially transverse, becoming oblique to the long axis of the cell with growth, and is regularly followed by cellular separation so that the trophic cells are normally uninucleate, although groups of two or four uninucleate cells can be found following division. In the thraustochytrids nuclear division accompanied by cell cleavage does not usually result in cell separation, and frequently nuclear division is not accompanied by cell cleavage, and this results in an **apocytic phase** or phases (*Schizochytrium* is an exception to this generalization). The apocytic phase is associated with zoospore production, either at an early stage when the **fundament** (basal rudiment) is differentiated from the apocytium, as in the proliferous section of the genus *Thraustochytrium*, which is thus eucarpic with a persistent fundament, or at a later stage when zoospores are differentiated, as in *Ulkenia*.

In the past the behaviour of the trophic cells has caused some terminological confusion because aggregations of cells can be achieved in three ways, and three gross-morphological constructions have been described:

1. cells from different parts of a colony can become appressed together, so that the boundaries between adjacent cells are not apparent, resulting in a 'plasmodium'-like appearence
2. comparable aggregations of cells (in 'plasmodial' or other species) may round up and develop a common envelope of scales (the 'sorus'); the thick outer membrane thus formed distinguishes this aggregation from a 'plasmodium'
3. large groups of cells can become massed by repeated cell division without dispersion, or by immediate encystment of non-dispersed zoospores

Terms which have been used to describe these aggregate constructions are: plasmodia, sori (pseudosori) and sporangia. In all probability there is a continuum between these extremes, and different species may show different extents to which these systems may coexist, even in the same aggregate. In general, plasmodia and sori are found in *Labyrinthula* and sori and sporangia in the Thraustochytriaceae.

In eucarpic species of *Thraustochytrium* the single fundament (basal and persistent rudiment) generates a succession of sporangia. In other species groups of holocarpic thalli, each with a non-persistent **rudiment** and capable of generating new sporangia have been described. These rudiments may develop their own ectoplasmic nets and are thus equivalent to schizogenous cells of the early developmental stages of species of *Schizochytrium* or *Labyrinthuloides*, and in such cases it is not certain whether the thallus should be described as eucarpic. The suggestion that there is a continuum of patterns of morphological development in the thraustochytrids is thus reinforced, but generic delimitation is still well-defined.

The subsequent development from the aggregations may be by direct dispersal as trophic cells; by disintegration of the mass as mixed trophic cells and spherical 'cyst' cells; by direct conversion into biflagellate zoospores or by division (or divisions) of the component cells of the aggregate to give rise to two, eight or more zoospores per cell.

The term **sorus** is used here (cf. Porter, 1990) for all aggregations with a common boundary membrane (including the tetrads of *Schizochytrium*), but I retain the use of **zoosporangium** for any structure in which significant apocytic development leading to zoospore production occurs.

The trophic cell, especially of labyrinthulids, may change shape from the fusiform spindle to a spherical cyst stage, which may subsequently release a new spindle cell.

The flagellated cell is well documented for the *Thraustochytrium* (Kazama, 1972*a*, *b*, 1973, 1974*a*), *Schizochytrium* (Kazama, 1980) and *Labyrinthuloides* (Bower, 1987*a*) but has been less thoroughly studied in the labyrinthulids, where it is part of a sexual cycle (Perkins & Amon, 1969; Moens & Perkins, 1969) (flagellated spores are known for *Labyrinthula vitellina* and *L. algeriensis*). An eye spot is reported to be present in the flagellate cell of the labyrinthulids. Porter (1990) stated that zoospores are known for all species of *Labyrinthuloides* but Quick (1974) did not refer to zoospore morphology. Attention should be drawn to the the TEM study of *Labyrinthuloides haliotidis* (Bower, 1987*a*) in which the tubular tripartite hairs appear to be confined to a 30° arc of the flagellar cross-section, and therefore do not conform to the bilateral array found in other thraustochytrids (Kazama, 1974*a*, 1980).

Sexual reproduction has been shown for one species (Moens & Perkins, 1969) but the life-history has not been explained. Although the flagellate spores (see below) produced are haploid, the ploidy of the trophic cells has not been ascertained.

The labyrinthulids is the smaller group and all species are placed in the genus *Labyrinthula*. The number of species that should be recognized is doubtful because differences in trophic cell size and mass colony colour are said to be variable, and these constitute some of the more important morphological differences. Muehlstein *et al.* (1991) also propose that at least some species are defined on the basis of the habitat: *Labyrinthula zosterae* is confined to parasitism of *Zostera* species; other species have been described from marine algae, but insufficient work has been carried out to confirm host specificity. Conversely, Johnson & Sparrow (1961) considered that some varieties should be accorded specific status, but the varieties in question have not been validly named or described so a *stat. nov.* is not possible and the taxa will require diagnoses and new names. Speciation and generic status in the thraustochytrids is also debatable, especially with respect to *Ulkenia amoeboidea*, *Aplanochytrium kerguelense* and *Labyrinthuloides* (Porter, 1990). The key given here reflects this inconsistency in that not all species of a genus will be keyed together.

The ordinal separation of the small genus *Labyrinthula* from the thraustochytrids can be justified by reference to the shape of the trophic cell; the thickness of the layer of investment scales and the enveloping ectoplasmic net. There are also differences in the details of mitosis between the two groups (Kazama, 1974*b*; Olive, 1975; Perkins, 1970; Porter, 1972). Separation at a higher level has been proposed, but can only be justified when a clearer picture of the sexuality and concommitant role of the flagellated cell has emerged. As far as is known, all thraustochytrids are anamorphic. At present it would appear that the flagellate cell of the thraustochytrid is a mitotically produced and mitotically capable zoospore, while that of the labyrinthulids is a meiotically produced meiospore with unknown subsequent function.

Some other groups of fungi should be compared with the labyrinthulids and thraustochytrids because of similarities in host and habitat. It is also prudent to draw attention to these comparisons because of the incomplete information concerning the morphology and life-histories of both those organisms included and those excluded from the *Labyrinthista*.

In particular, the genus *Elina*, which is customarily listed with the *Hyphochytriales*, is here placed in the *Labyrinthista*. The genus is only known from the original descriptions of the two species by Artem[t]chuk (1972), who placed the genus in the *Rhizidiomycetaceae*. Karling (1979a) suggested, but did not formally place, the genus in synonymy with *Rhizidiomyces*. I agree with Fuller (1990) that the descriptions of zoosporangial wall deliquescence should rule out such a synonymy. Rhizoids are said to be present, but are not figured: they may well be equivalent to the 'rhizoids' of *Thraustochytrium*. A description of zoosporogenesis was not given, nor is there any ultrastructural information.

The anteriorly uniflagellate zoospores of *Elina* were not compared with the biflagellate thraustochytrid zoospores. The latter have a long anterior flagellum and a shorter posterior flagellum, and it is possible that *Elina* may have a vestigial posterior flagellum that was not detected. If flagellar number is accorded a lesser significance, then placement of *Elina* in the *Thraustochytriaceae* is both possible, and made more probable when the marine habitat and zoosporangial wall disintegration are also taken into account. *Elina sinorifica* Artemtchuk has a much thicker (4-5 μm) thallus-sporangial wall, a feature which would also be compatible with thraustochytrids. The genus is therefore transferred to the *Thraustochytriaceae*, and synonymy of both species of *Elina* with species of *Thraustochytrium* may be possible.

The three species of *Anisolpidium sensu stricto* (Karling, 1943; Kny, 1871; Petersen, 1905) also have a single anterior flagellum. They are marine parasites of brown algae here placed in a monotypic family and order (*Anisochytridiales*, provisionally grouped with the *Pontismaceae* here). The remainder of their life-history and morphological characteristics are not well defined and their descriptions should be objectively assessed when reviewing new collections of labyrinthulids and thraustochytrids on marine algae. Similar comparisons may be worthwhile with the marine plasmodiophorids (especially *Plasmodiophora bicaudata* on *Zostera* species and *P. diplantherae* on *Halodule*). The full accounts (Dangeard, 1932; Kobayasi & Ookubo, 1953*b*) of the doubtfully characterized *Labyrinthula chattonii* should be consulted when referring to the other parasites of the marine *Cladophora* species (see further comments under *Phagomyxa* and *Sirolpidium*). Porter (1990) excluded this taxon from *Labyrinthula*, presumably accepting the transfer to *Phagomyxa* by Karling (1944*a*), but I prefer to retain *L. chattonii* in *Labyrinthula* under the doubtful category.

One species of *Diplophrys*, *D. marina* Dykstra & Porter (1984), should be considered with the thraustochytrids, but the placement of the type species is extremely doubtful (the brief original descriptions by Barker, 1867, 1868, do not provide any characters that would now be considered diagnostic, and he did not provide any illustrations). The transfer of a third species of *Diplophrys* to *Sorodiplophrys stercorea* Dykstra & Olive (1975) similarly drew attention to the differences between these terrestrial organisms and the marine labyrinthulids and thraustochytrids. (The taxonomic position of the single freshwater labyrinthulid, *L. cienkowskii*, could also be reconsidered.)

Diplophrys marina has a uninucleate ellipsoidal trophic cell that divides mitotically, a scale investment several layers thick and an ectoplasmic net, but it does not possess a sagenogenetosome, show aggregation of cells, or produce zoospores. The chemical composition of the scales is not known. *Diplophrys marina* is included in the key to the *Labyrinthista* as a doubtfully assigned taxon.

THE *PLASMODIOPHOROMYCETES*

Features of possible significance in any discussion of the interrelationships of the *Plasmodiophoromycetes* with the *Peronosporomycotina* and other straminipilous fungi have been given in *Criteria for Diagnoses*. A detailed summary of the morphology of this class will not given here. Reviews of the class are available in Fitzpatrick (1930), Palm & Burk (1933), Ivimey Cook (1933), Karling (1942*b*), Sparrow (1960 - part only), Karling (1968*e*), Waterhouse (1973), Olive (1975) and Dylewski (1990). My keys have been adapted from those given by Sparrow (1960), Olive (1975) and Karling (1968*e*). However, since Sparrow (1960) only gave the formal taxonomy for the aquatic species, and as Karling (1968*e*) did not give the full synonymy or typification, there are some minor adjustments to be made, and these are supplied here. A few species have economic importance as hypertrophic plant pathogens: *Plasmodiophora brassicae* Woronin (Club Root of crucifers); *Spongospora subterranea* (Wallr.) Ledingham (Powdery Scab of potato); *Spongospora nasturtii* (J. A. Toml.) M. W. Dick *loc. cit.* (Crook Root of watercress). *Polymyxa betae* Keskin infecting sugar beet, and *Polymyxa graminis* Ledingham infecting cereals, but not causing extensive hypertrophic symptoms, should also be included as having economic importance because they are implicated as virus vectors, especially *P. betae* in 'Rhizomania' of beet. Similarly *P. graminis* in cereal roots is an important virus vector (Langenberg & Van Der Wal, 1986). Barr (1981*a*) considered that many records of *Ligniera junci* (Schwartz) Maire & A. Tison may relate to *Polymyxa graminis*. The phytopathological literature on the genetics of resistance, agricultural management, and virology of these five parasites is extensive and I have cited only the 5% of the recent literature which has more direct relevance to the taxonomy and morphology of these species. Abstracting services list about 250 citations for the last seven years, almost all of which is concerned with the pathology and control of *Plasmodiophora brassicae*. The older literature has been comprehensively covered by Karling (1968*e*).

Dylewski (1990) accepts 26 species in 10 genera. (*Pyrrhosorus* and *Phagomyxa* are excluded - see below.) Karling (1968*e*) lists rather more species, and in the formal taxonomy and key I have included several of these doubtful species so that their documentation is available. However, it should be emphasised that many of the species are imperfectly known, and over one third have no known primary or secondary zoospores, thus their inclusion in the class is based on slight circumstantial evidence. The present understanding of the class rests on more or less ***incomplete*** knowledge of less than a dozen species.

Studies on the morphology, cytology, ultrastructure and life-history of the plasmodiophorids are listed under the species in PART V. The morphology of the plasmodia is difficult to circumscribe, except by TEM, and the organelle complement and characteristics have not been established on a comparative basis. The most recent publication on wall structure is that of Yano *et al.* (1994). Knowledge of zoospore ultrastructure is limited to *Plasmodiophora* and *Polymyxa* (Barr & Allen, 1982), and *Spongospora* (Merz, 1992). The elaborate extrusome penetration apparatus of *Plasmodiophora brassicae* (Aist & Williams, 1971) and *Polymyxa betae* (Keskin & Fuchs, 12969) is described above, in relation to *Haptoglossa*. The injected uninucleate protoplast forms a naked coenocytic plasmodium. Aist (1974) and Williams & Yukowa (1967) provide information on the complex nature of the host/parasite plasmamembrane interface. At maturity the plasmodium divides to form groups of sporangiosoral cysts in arrangements characteristic for the different genera. The contents of each component of the sporangiosorus undergoes intrasporangial cleavage to give the biflagellate secondary zoospores. No other plasmodiophorid genus is reported to have this organelle; infection by *Woronina* is effected through a germination tube.

Cystosoral cysts appear golden-brownish in transmitted light and are smooth except in *Plasmodiophora* (spiny), *Ligniera verrucosa* (verrucose) and *Sorodiscus radicicola* (small protrusions). Cysts are spherical except in *Plasmodiophora bicaudata*: this species is exceptional not only in the fusiform shape of the cysts, but also in its marine (as opposed to maritime) habitat in *Zostera*, criteria which deserve reassessment vis-à-vis *Labyrinthula* when *P. bicaudata* is reinvestigated.

References are given for biphasic (callus) culture (Ingram, 1969; Williams *et al.*, 1969; Singh, Sharma & Pandey, 1989; Ikegami, 1992) and the early debate on the possibility that these parasites were vectors of viruses (Kusaba & Toyama, 1970; Toyama & Kusaba (1970), Langenberg & Van Der Wal (1986) under relevant species in PART V (see also Bryngelsson *et al.*, 1988, with respect to DNA movement from host to parasite).

The plasmodiophorids have been the subject of much cytological study, from Horne (1930) to the series of papers by Braselton and co-workers, summarized in Braselton (1992). It is now certain that meiosis occurs immediately prior to cyst formation. It is less certain at what points plasmogamy and karyogamy occur. Tommerup & Ingram (1971) and Ingram & Tommerup (1972) have reported fusion of secondary zoospores to produce a heterokaryotic haplophase cystosoral plasmodium in *Plasmodiophora*. There is no indication that there is a syn-dikaryophase of paired nuclei during the cystosoral plasmodium assimilative phase, nor that the cystosoral plasmodium has to develop from heterothallic origins. At maturity of the cystosoral plasmodium (the transitional stage) karyogamy occurs, immediately followed by meiosis and cyst wall development around each meiotic product. The synchronous meioses of the transitional stage of the cystosoral plasmodium are in some respects comparable with the meioses of an oomycete oogonium, even though the preceding development and ploidy are different. For most species each cystosoral cyst germinates to produce a single primary zoospore, and much recent research has been concerned with the conditions for germination of the resting spores of *Plasmodiophora* (e.g., Einhorn & Bochow, 1990*a*, *b*; Takahashi, 1991; Yano *et al.*, 1991). In species for which several primary zoospores are formed from each cyst there may either be precocious mitosis or the cysts may be syn-gametous. The sporangiosoral stages, which develop from primary zoospore infection, are presumably the culmination of a vegetative gametophytic phase, and interspecific or intergeneric diversity could be expected with respect to the existence or suppression of this phase of the life-history. Ludwig-Muller *et al.* (1999) have investigated a biochemical tool (endogenous glucosinulate content) which could be applied to the taxonomy of these organisms.

The higher taxon nomenclature has had a chequered history, the organisms being initially grouped with the slime moulds. Schröter (1897) listed these organisms (with the exclusion of *Woronina*) under his family *Phytomyxinae*, and Minden (1911*a*) also followed Fischer (1892) in placing *Woronina* with the flagellate fungi as the type genus of the *Woroninaceae*. Maire & Tison (1909) pointed out that with the removal of *Phytomyxa* the family of Schröter should revert to the name used by Zopf (1885), the *Plasmodiophoraceae*. The group was treated as an order, *Plasmodiophorales*, by Stevens (1913) and subsequently Alexopoulos raised the group to class level, first (Alexopoulos, 1952) as the class *Plasmodiophorae* in the phylum *Myxomycophyta*, and secondly (Alexopoulos, 1962), following Cejp (1957), as the class *Plasmodiophoromycetes* in the sub-division *Eumycotina*.

Generic distinctions are based on the aggregates of cysts in cystosori, but there is some overlap (Palm & Burk, 1933) and generic revision may be necessary when the species are more completely described.

Two genera (*Woronina* and *Octomyxa*) are parasitic in the *Peronosporomycetes* and chromophyte algae. A wide range of freshwater aquatic plants are also reported to be hosts of plasmodiophorids (green algae - *Mougeotia*; pteridophytes - *Isoetes*; stoneworts - *Chara*, and angiosperms - *Alismataceae*, *Callitrichaceae*, *Hydrocharitaceae*, *Juncaceae*, *Potamogetonaceae*, *Zannichelliaceae* and other smaller families of the *Najadales*). Terrestrial angiosperm hosts are mainly in the more highly evolved orders of *Caryophyllales* (*Amaranthaceae*, *Chenopodiaceae* - *Beta*), *Brassicales* (*Brassicaceae*), and the tubiflorous orders

Polemoniales (*Solanaceae*), *Scrophulariales* (*Scrophulariaceae* - *Veronica*) and *Asterales* (*Asteraceae*); in this respect the host range of the plasmodiophorids echoes that of the order *Peronosporales* (Dick, 1988).

Host range and host specificity appear to vary widely from species to species, and the status of some plasmodiophorid species may change when more is known of their parasitic abilities. *Woronina polycystis* M. Cornu was first described from *Achlya*, but the fungus studied by Goldie-Smith (1954) was restricted to *Saprolegnia sensu lato* and could not be transferred to *Achlya*. The recent addition of *Woronina leptolegniae* (Karling, 1981*c*) and the host specificity of *Octomyxa* suggest that several plasmodiophorid taxa remain to be described. Tomlinson (1958) has also shown different host specificities between *Spongospora subterranea* and his new varietal taxon *Spongospora subterranea* var. *nasturtii*. In view of this host specificity, coupled with differences in the sporangial stages and the different habitats, I have raised this taxon to specific rank.

The difference in penetration mechanism between *Plasmodiophora* and *Polymyxa*, on the one hand, and that of *Woronina* on the other hand, coupled with the differences in hosts between *Woronina* and *Octomyxa* (*Peronosporomycetes*) and those species with angiosperm hosts, might suggest that supra-generic separation would be appropriate. However, Braselton (1992) has indicated that the chromosomes of *Woronina* and *Plasmodiophora* are similar to each other, but distinct from those of other genera. The latest species of *Woronina* to be described (*W. raii*, Srivastava & Sinha, 1989; *W. charii*, Kiran & Dayal, 1992) are saprotrophs and are rejected from the *Plasmodiophoraceae*; their affinities are not yet determinable. More comparative ultrastructural and developmental studies are required before the inter-relationships both between *Plasmodiophora* and *Woronina*; and with other flagellate fungi are urgently needed. The *Woroninaceae* may need to be removed from synonymy with the *Plasmodiophoraceae*. Molecular biological data (Ward & Adams, 1998: partial 18S rDNA plus ITS regions; Castlebury & Domier, 1998: 18S phylogeny) do not resolve this issue, neither do they resolve the relationships of the *Plasmodiophoromycetes* with other flagellates, although Ward & Adams (1998) suggested distant linkage with the straminipiles. Other molecular studies have focused on the identification of *Plasmodiophora brassicae* in the field using RAPD techniques (Chee *et al.*, 1998; Faggian *et al.*, 1999; Ito *et al* ., 1997, 1999*a*, *b*; Yano *et al.*, 1997). However, the exclusiveness of this procedure has not been sufficiently tested against the occurrence of other soil flagellates, including the straminipilous fungi.

Pyrrhosorus and *Phagomyxa*

These are two monotypic genera parasitic in marine algae, sometimes appended to the *Plasmodiophoromycetes* as doubtful genera. Their habitat makes them widely separated, in coevolutionary terms, from the *Plasmodiophoraceae*, especially if *Plasmodiophora bicaudata* J. Feldmann is also regarded as a doubtful species.

Pyrrhosorus marinus Juel (1901) from the red alga *Cystoclonium*, has laterally biflagellate, isokont zoospores containing a conspicuous orange spot. The presence of a possible eyespot suggests affinity with the labyrinthulids, which also have an eyespot associated with the flagellum. The remaining morphological and life-history characteristics of *Pyrrhosorus* are poorly characterized, but it appears to possess a combination of a plasmodial stage penetrating several adjacent cells, followed by schizogeny to produce uninucleate, spherical or fusiform cells 8 μm in diameter. These enter an aggregation stage to form a sorus, and each cell then divides to form 8 cells which become transformed into small (2·5 × 4·5 μm) zoospores. The latter sequence is very similar to *Labyrinthula* and *Schizochytrium octosporum* and the dimensions of the zoospore are similar. The genus and species are therefore listed in the doubtful category under the

Labyrinthista (*Labyrinthulomycetes*), *Labyrinthulales*. Sparrow (1960) considered that the description may relate to a mixed infection by a labyrinthulid and a protist.

The life-history of *Phagomyxa algarum* (Karling, 1944) from the brown algae *Pylaiella* and *Ectocarpus*, is not completely known. Karling (1968*e*) stated that those elements of the life-history that were reported indicated a similarity with the life-history of *Woronina glomerata*. However, neither cysts nor cystosori were described. The plasmodium was reported to form pseudopodia, possibly phagotrophic within the host cells, and successive division of the plasmodium was described, followed by further nuclear division prior to to the production of sporangia, but these nuclear divisions had normal spindles and the nuclear membrane was persistent. The plasmodium divided into contiguous sporangia, each of which produced one to several anteriorly biflagellate, somewhat amoeboid zoospores, the flagella were reported to be of different lengths, with the anteriorly directed flagellum the shorter. In this last respect comparisons can be made with the the plasmodiophorids, but comparisons with the *Pontismaceae* and the labyrinthulids should not be dismissed until more information becomes available.

Karling (1944*a*) transferred *Labyrinthula chattonii* P. A. Dang. as a second species of *Phagomyxa* on the grounds that Dangeard (1932) had probably described *Labyrinthula chattonii* from a mixture of a *Labyrinthula* and a *Phagomyxa* which had a similar aggregation of sporangiosori. Whether these aggregations could be equated with the sori produced by some species of *Labyrinthula* is debatable (compare the sori of *Labyrinthula roscoffensis* Chadef.). Dangeard's species appears to have some similarities with plasmodiophorids, labyrinthulids or the *Pontismaceae*, but is here retained in *Labyrinthula* in the doubtful category.

The genus *Phagomyxa* must remain *incertae sedis* until rediscovered.

Endemosarcaceae and *Heimerliaceae*

The family *Endomosarcaceae* was established by Erdos & Olive (1971), Erdos (1971, 1972*a*, *b*, 1973) and Olive (1975) for three biflagellate zoosporic parasites of ciliates belonging to the genus *Colpoda*. They regarded the organisms as possibly having affinity to the *Plasmodiophoromycetes*. The zoospores are produced in endobiotic zoosporangia, but they are discharged through an endosporangial tube which is everted at the time of discharge. The flagella do not have any ornamentation; they are anisokont and are inserted sub-apically.

Olive (1975) also refers to the saprotrophic taxa that Arnaud (1948) described under the family name *Heimerliaceae*. Unfortunately these taxa were invalidly published, but I include descriptive details in the identification keys for comparison.

MISCELLANEOUS GENERA

Several taxa which have either been tentatively placed with higher taxa of the *Straminipila* at some time, or have biflagellate stages in their life-histories, are mentioned here so that details of their nomenclature can be recorded.

In addition, the nomenclature of those uniflagellate fungi likely to be found with taxa discussed in detail above (particularly some species of *Catenaria*, *Olpidium* and the aquatic *Synchytriaceae*), is included in this work.

Lagenidicopsis

Lagenidicopsis is monotypic and known only from the original description of an anamorphic state (Artemchuk, 1972). The thallus is small (less than 15 μm in diameter), monocentric and holocarpic; saprotrophic and endobiotic in pine pollen baits. There is no mention of rhizoids. The zoosporangium produces a long discharge tube, and zoosporogenesis takes place outside the zoosporangium in a vesicle. The zoospores are 5·0-7·0 μm long (volume equivalent < 100 μm^3) with lateral flagellar insertion.

This marine saprotroph cannot be placed until better known, but should be considered with other marine taxa with appropriately sized zoospores. It could even be a depauperate holocarpic specimen of *Pythium*. The oligohaline marine habit and extra-sporangial zoospore cleavage are not completely compatible with the suggested synonymy with *Lagenidium pygmaeum* (Karling, 1981*a*).

Other 'oomycete' genera in the 7th and 8th edns of *Ainsworth & Bisby's Dictionary of the Fungi*

There are a few genera, some of which authors had assigned to the oomycetes (e.g., *Plasmoparopsis* De Wild.) but which were omitted from *Ainsworth & Bisby's Dictionary of Fungi, 7th edn*, some which are obsolete (*Trichothrauma* Germ. affin. *Saprolegniales*?) and some which had been unjustifiably affiliated to flagellate fungi in that work (*Sporoclema* - Tiesenhausen, 1912; *Thanatostrea* - Franc & Arvy, 1969; and certain genera of Scherffel, 1925*a*). The nomenclature of both groups of genera (mostly monotypic) is given alphabetically in a separate section to expand the entries in *Ainsworth & Bisby's Dictionary of the Fungi, 8th edn*.

In particular, the monad genera and species erected by Scherffel (1902, 1925*a*, 1926*a*, *b*) in his account of the monad ancestry of the biflagellate fungi (Scherffel, 1925*a*) should be mentioned here. Of these genera *Amoeboaphelidium* Scherff., *Amylophagus* Scherff., *Endospora* Scherff., *Pseudosporopsis* Scherff. and *Spirospora* Scherff. (Scherffel, 1926*b*) were given entries in the *Dictionary of Fungi, 7th edn* but *Aphelidiopsis* Scherff., *Aphelidium* Scherff., *Barbetia* P. A. Dang., *Ectobiella* de Bruyne and *Pseudospora* Cienk. were not listed. Most of these taxa are too little documented to be placed in a modern classification, but the species of *Pseudosporopsis* could be considered with *Ectrogella* and *Haptoglossa* because of the body shape, flagellation and flagellar insertion pertaining to the motile cell.

Nephromyces

Nephromyces is a genus of parasitic or commensal organisms associated with marine ascidian tunicates, and characterized by its biflagellate zoospores. The cells secrete chitin; the mitochondria have tubular cristae. While recent research (Saffo, 1981, 1982; Saffo & Davis, 1982; Saffo & Nelson, 1983; Saffo & Fultz, 1986) is elucidating the biology of one of the three species, it is still too early to suggest a final systematic position for the genus. Relationships could be with the polyflagellate *Spizellomycetales* or the choanomastigotes. The posterior flagellar insertion makes it unlikely that these species will be included in the *Straminipila*.

Dermocystidium: *Hyalochlorella* and *Perkinsus*

These three marine genera no longer need to be considered with the straminipilous fungi. Formerly (Sparrow, 1976), they have been discussed with the labyrinthulids. Poyton (1970*a*, *b*) established that some isolates collected and deposited as unidentified species of *Dermocystidium* were algal in character, and for these the monotypic genus *Hyalochlorella* was erected. Other forms attributed to *Dermocystidium* following Mackin, Owen & Collier (1950) have now been conclusively shown to belong to the protistan phylum *Apicomplexa* (Perkins, 1976*b*) under the name *Perkinsus* (Levine, 1978), following presentation of molecular biological data (Goggin & Barker, 1993; Fong *et al.*, 1993). The nomenclature of these species is provided here for convenience.

Another species of *Dermocystidium*, *Dermocystidium salmonis* H. S. Davis (Davis, 1947), should be regarded as a species of *Ostracoblabe* (see below).

Ostracoblabe

Ostracoblabe has, until now, been a monotypic genus, *O. implexa* (Bournet & Flahault, 1889) being regarded as the causal organism of shell disease in oysters. In culture (Alderman, 1976), the fungus produces coenocytic, non-septate, narrow (2 μm diam.) hyphae, which are exceptionally straight and bear intercalary swellings delimited by septa. The fungus does not possess any of the characters diagnostic for straminipilous fungi. Neither zoospores nor oospores are known, and distinctive ultrastructural features, such as mitochondria with tubular cristae, or DBVs, have not been recorded. Cultures of this species are no longer extant. *Dermocystidium salmonis* (Davis, 1947) should also be regarded as a species of *Ostracoblabe* (D. J. Alderman, pers. comm.), but cultures of this species (originally described from fixed material only) are no longer extant either, (axenic isolation and culture of these taxa are not difficult, according to D. J. Alderman, pers. comm.) so no further information which would lead to taxonomic verification can be obtained; nevertheless, the recombination from *Dermocystidium* to *Ostracoblabe* is formally made here. The genus is excluded from the *Straminipila*.

In view of the report of *Coulterella* in northern marine waters (Höhnk, 1969; Artemchuk 1981: 76) cross-reference should be made to the "fossil to recent" *Cladochytriaceae* described by Zebrowski (1936) from south Australian marine calcareous sands, which included the original description of *Coulterella* as a monotypic genus from the shells of molluscs. *Coulterella petersonii* Zebrowski was also described as possessing straight, very narrow hyphae with interspersed swellings.

Thalassomyces

The genus *Thalassomyces* was erected by Niezabitowski (1913) for two species reported as parasites of deep sea (>90 m) decapods. The fungi were stated to have non-septate filaments terminating in chains of three conidia (*ca* 16 × 20 μm) (see Johnson & Sparrow, 1961). There are very slight resemblances to the fungus described as *Hyphochytrium peniliae* (Artemchuk & Zelezinskaya, 1969) regarded here as a doubtful taxon of the *Salilagenidiales*, but there are no grounds for accepting these two species of *Thalassomyces* in the *Straminipila* at the present time.

Galt & Whisler (1970) described the differentiation of flagellated spores in another species of *Thalassomyces* and concluded that the flagellation was typical of the dinoflagellates.

Twelve species have been assigned to *Thalassomyces* (Boschma, 1959; Collard, 1966; Vader, 1973*b*; Whisler, 1990; *loc. cit.*) but it is not clear whether all should be placed with the dinoflagellates. Some of the species have been described and typified as fungi under the botanical code of nomenclature but others have been loosely described in the zoological literature. Boschma (1959) has made many of the recombinations into this genus but no species was accompanied by a formal *combinatio nova* statement. The genus is excluded from the *Straminipila*, but I have nevertheless provided the formal taxonomy of these species. Keys to genera and species are given by Kane (1964), Collard (1966) and Wing (1975).

Diplophrys and *Sorodiplophrys*

These two genera are sometimes associated with the *Labyrinthista*, but differ from these marine organisms in some important respects. *Diplophrys* was named by Barker (1868) for a freshwater protist. Cienkowsky (1876) described a second species on horse dung, which was made the type of *Sorodiplophrys* when rediscovered by Dykstra & Olive (1975). The third species in this group is *Diplophrys marina*, named by Dykstra & Porter (1984).

The type species *Diplohrys archeri* was very poorly diagnosed by Barker (1867, 1868) and has not been observed since the report of Penard (1902); therefore there is a complete absence of any criteria now considered essential for taxonomic placement. *Sorodiplophrys stercorea* has spindle-shaped cells with ectoplasmic extensions, an investment of scales and shows aggregation of cells to form a sorocarp. However, it is not known to produce zoospores, does not have a sagenogenetosome and the principal constituent of the carbohydrate scales is arabinose (Dykstra, 1976*a*). Mitosis is within a persistent nuclear membrane but the centrioles are composed of nine singlets not triplets (Dykstra, 1976*b*). Mitochondria (Dykstra, 1976*b*: figs 16-18) appear to have inflated finger-like cristae rather than plate-like or strictly tubular cristae.

The third species has thus been placed in a dubious genus and it would be preferable, when its characters are fully known, to be made the type of a new genus. For convenience, *Diplophrys marina* has been considered with the *Labyrinthista*. However, in the absence of fundamental criteria such as zoospores with straminipilous ornamentation, or rigorous ultrastructural comparisons of mitochondria and vacuolar characters (e.g. DBVs); and given the observed differences in detail regarding mitosis, centrioles, wall-scale chemistry, and ectoplasmic net origins, these two genera and three species must remain *incertae sedis*, with respect to both the *Straminipila* and the *Peronosporomycetes*.

PART V

SYSTEMATICS

SYSTEMATICS: INTRODUCTION

E. O. Wilson, in his book *The Diversity of Life* (1992) presented some definitions that stand repetition as a preface to PART V:

"Taxonomy: The science (and art) of the classification of organisms.

Systematics: The scientific study of the diversity of life. Sometimes used synonymously with taxonomy to mean the procedures of pure classification and reconstruction of phylogeny (relationship among species); on other occasions it is used more broadly to cover all aspects of the origins and content of biodiversity."

The format here follows the order of the hierarchical classification as given in Table V: 1, pp. 271-274, rather than the taxonomic discussions in PARTS III and IV. For each category, the type species is given first, followed by the remaining species in chronological order of the earliest basionym and subsequent synonyms. Generic 'synonyms' are sometimes fuller than is required for strict synonym entries and include all partial cross-references to other genera from species synonyms to provide a more complete history of the taxon; they are usually designated [*pro parte typo excluso*]. In the citations the page number refers to the first page of the Latin diagnosis, or in earlier works, any formal diagnosis; in the absence of such diagnoses the page number refers to the first formal naming of the binomial.

Ideally, the type material for each species name should have been inspected. The fulfilment of such a requirement would be both time-consuming and futile because the critical diagnostic features rely upon living material or TEM evidence. Furthermore, Newell *et al.* (1977) note that they have been advised that the type material of Zopf, which would be crucial for the lagenidiaceous fungi, is no longer extant. I have therefore made no attempt to seek the type material, but provide information as to its existence or location. In the absence of such information, I have designated lectotype material for the name, based on published illustrations, thereby augmenting the lectotypifications of Cejp (1959). Typifications of straminipilous species names are transcribed from the original texts where possible. In each case the subheading '**Typification**' refers to the name at the head of the entry. The typifications for a number of non-straminipilous taxa have not been verified.

The hierarchies above Class level (Superclass, Subphylum, Phylum, Subkingdom) within the *Straminipila* (kingdom *Chromista sensu* Cavalier-Smith, 1986) are given in the formal taxonomy, but their listing does not necessarily imply acceptance of all the hierarchical levels for the straminipilous fungi. The use of the word 'phylum' (hence phylogeny) is preferred in this kingdom comprised of both 'animals' and 'plants', rather than the words 'divisio' or 'division' as is normal in the botanical hierarchy.

According to the *International Code of Botanical Nomenclature* (*ICBN*) since 1993, the rules of priority apply to the taxonomic rank of Family and subordinate ranks. The situation for Orders is less clear, although it is usual to apply the rule of priority. Higher taxa, Kingdoms, Phyla and Classes, should be accompanied by a Latin diagnosis and be typified, but the rule of priority does not apply and the selection of a name is by personal preference. The typification of a name based on a genus should use the genus as the type: where other higher taxa have been used, these are indicated in parentheses after the type genus.

The adoption of plant, fungal, protistological or animal higher taxon suffixes is, or could be controversial, since some straminipiles are more akin to one or another of the fungal, algal or animal nutritional life-styles. A pragmatic, perhaps illogical, scheme is preferable for the present, and so I use mixed suffixes for the higher ranks of the peronosporomycetes and the labyrinthulids and thraustochytrids.

With the acceptance of the Kingdom *Straminipila*, which includes certain fungi, algae and monads, problems arise as to the starting dates of the Latin requirement - if the taxon was originally placed as an animal no Latin diagnosis would have been required; if placed as an alga then the crucial date for the requirement of a Latin diagnosis is 1958, but if placed as a fungus then a Latin diagnosis is required on or after 1935. Any organism now newly described within the *Straminipila* could therefore conform either to the published rules of the *Zoological Code* (1985), or the *ICBN*. This is unsatisfactory, particularly with regard to the Latin requirement in the *ICBN* and the formulations of the specific epithets. In particular, problems have arisen when the genus (and perhaps several subsequent species on different occasions) has been described under the *ICBN* while other species have been described under the *Zoological Code*. This is the situation with *Thraustochytrium*, *Labyrinthuloides* and *Thalassomyces*: in the *Index of Fungi* (e.g., *Thalassomyces*: **4**: 291 - *T. nouvelii*; **4**: 319 - *T. fagei*; **4**: 389 - *T. fasciatus*) some of these species are entered with the annotation [*nom. illeg.*] although they may conform to the rules for the *Zoological Code*. This problem will not be resolved by individuals stating an opinion: a joint concensus is ***URGENTLY REQUIRED*** by all interested nomenclaturists within the mycological, phycological and protistological disciplines. As a mycologist and botanist, I have adopted the *ICBN* procedures in this text. I acknowledge that both codes have their advantages and disadvantages: the practice of providing the source date in the citation under the *Zoological Code* could, with advantage, be used by botanists; the practice of indicating a generic re-attribution by parentheses fails to provide an adequate mechanism for tracing the nomenclatural vicissitudes compared with the practice under *ICBN* rules.

Cyrillic script

I have transliterated the Cyrillic letters to Roman using the *British Standard, No. 2979 (1958)* in the REFERENCES AND BIBLIOGRAPHY and APPENDIX 1, except for the optional use of 'č' in preference to 'ch'. It should be recalled that the phonetics relating to 'v' and 'w' and 'j' and 'y' are different between German and English systems, and this accounts for some discrepancies with respect to the names Jaczewski/Yačevskie; Rostovzev/Rostowzew and Voronin/Woronin etc, but the Germanic transliteration 'Woronin' is too well-entrenched to change.

The Russian publication, Yačevskie & Yačevskie (1931), contains several new names at various taxonomic ranks; Sparrow (1960) treated these names as *nomina nuda*, but in most cases these names were adequately diagnosed within the taxonomic treatment, and the 1931 date precludes the Latin requirement. The presentation is therefore equivalent to the Latin version of the nomenclature given by Cejp (1959) which is accepted as validation for names presented between 1935 and 1959 but which lacked Latin diagnoses. I have therefore treated most of these names as having been validly published. The authors cited themselves as "Jacz." (Yačevskie & Yačevskie, 1931), but since it is unclear whether this abbreviation applied to A. A. Yačevskie or to P. A. Yačevskie or to both authors, I include both authors as the authority.

TABLE V: 1. *Hierarchy and sequence of the organisms considered below, complete to family and genus; the type genus is presented first, followed by the other genera in date-order*

Taxon	Family	Genus	page
kingdom ***STRAMINIPILA***			275
[subkingdom ***CHROMOPHYTA***]			
phylum ***HETEROKONTA***			275
class ***LABYRINTHISTA*** (*LABYRINTHULOMYCETES*)			276
order ***LABYRINTHULALES***			
	Labyrinthulaceae	*Labyrinthula*	276
		Pyrrhosorus, genus *incertae sedis* provisionally attached to the *Labyrinthulales*, but sometimes attached to the *Plasmodiophoromycetes*	
order ***THRAUSTOCHYTRIALES***			280
	Thraustochytriaceae	*Thraustochytrium*	280
		Japonochytrium	282
		Schizochytrium	283
		Althornia	283
		Aplanochytrium	283
		Elina	284
		Labyrinthuloides	284
		Ulkenia	285
		Diplophrys [*pro parte*]	285
		Corallochytrium, genus dubiously attached to the *Thraustochytriales*	
		Diplophrys and *Sorodiplophrys*, two genera doubtfully attached to the *Labyrinthista*	
subphylum ***PERONOSPOROMYCOTINA***			288
class ***PERONOSPOROMYCETES***			289
subclass ***PERONOSPOROMYCETIDAE***			290
order ***PERONOSPORALES***			291
	Peronosporaceae	*Peronospora*	291
		Bremia	291
		Plasmopara	291
		Basidiophora	291
		Pseudoperonospora	291
		Paraperonospora	291
		Benua	291
	Albuginaceae	*Albugo*	291
order ***PYTHIALES***			292
	Pythiaceae	*Pythium*	292
		Myzocytium sensu stricto	295
		Lagenidium sensu stricto	297
		Cystosiphon	301
		Phytophthora	302
		Trachsphaera	302
		Diasporangium	302
		Peronophythora	302
		Halophytophthora	302
		Endosphaerium, genus *incertae sedis*	302
	Pythiogetonaceae	*Pythiogeton*	303
		Medusoides	304

STRAMINIPILOUS FUNGI

STRAMINIPILOUS FUNGI

STRAMINIPILA M. W. Dick, **regnum novum**

Etymology: *stramen*, a straw; *pilus*, a hair; with reference to the tubular tripartite hairs.

Diagnosis: plasmamembrana cum squamis straminipilosis; flagella heterokonta, plus minusve anisokonta; flagellum anterius straminipilosum. Aut organismi monophyleticaliter derivati ex organismis vel cum his characteribus quondam natis, vel eos postea nactis.

Holotypus: *Heterokonta* T. Cavalier-Smith [phylum] {*Peronosporomycotina* M. W. Dick [subphylum]}.

Diagnosis: plasmamembrane with straminipilous scales; flagella heterokont, more or less anisokont; anterior flagellum straminipilous. Or, derived monophyletically from organisms that originally possessed, or evolved to possess, these characters.

Typification: *Heterokonta* T. Cavalier-Smith [phylum] {*Peronosporomycotina* M. W. Dick [subphylum]}.

NOTE: Characters which may, in time, also prove to be diagnostic, are: ***particular*** sequences of DNA; ***some*** features of the kinetid; ***certain*** features of mitosis and meiosis; the possession of DBVs (phosphate storage metabolism); DAP lysine synthesis; biochemistry of glucan polymerization.

Coevolutionarily linked endosymbiont characters, diagnostic incorporation dependent upon non-reticulate evolution (i.e., the elimination of cross-incorporation, with or without monophyly, of the endosymbiont organelle): mitochondria with tubular cristae (if mitochondrial phylogeny is monophyletic within the kingdom); and, if photosynthetic, plastids with periplastid membranes and chlorophyll c_1 and/or chlorophyll c_2 (if plastid phylogeny is monophyletic within the kingdom).

CHROMISTA T. Cavalier-Smith, *BioSystems* **14**: 478 (1981) [kingdom].

Diagnosis: "Cellulae eukaryotae. Cristae mitochondriales tubulatae; tubuli teretes vel complanati. Vel reticulum endoplasmiticum plastidae circum plastidam vel mastigonemae tubulatae in cilia minima una."

Typification: *Chromophyta* T. Cavalier-Smith [lectotype] [subkingdom].

Other subkingdom: *Cryptista* [subkingdom and phylum].

CHROMOPHYTA T. Cavalier-Smith, *Advances in Phycological Research*: 338 (1986) [subkingdom].

Typification: *Heterokonta* T. Cavalier-Smith [phylum].

Other phylum: *Haptomonada*.

HETEROKONTA T. Cavalier-Smith, *Advances in Phycological Research*: 338 (1986) [phylum].

non Heterokontae A. Luther, *Bihang till Kongliga Svenska Vetenskaps-Akademiens Handlingar* **24**: 19 (1899) [class].

Typification: *Ochrista* T. Cavalier-Smith [subphylum].

Other subphyla:
Bicoecia
Labyrinthista
Peronosporomycotina (= *Pseudofungi*, *Heterokontimycotina pro parte majore*).

SYSTEMATICS

LABYRINTHISTA T. Cavalier-Smith, *The Chromophyte Algae, Problems and Perspectives*: 385 (1989) [subphylum].
Labyrinthulomycota R. H. Whittaker, *Science, New York* **163**: 155 (1969) [phylum].
Labyrinthulina L. S. Olive, *The Mycetozoans*: 215 (1975) [*nomen nudum*] [subphylum].
Typification: *Labyrinthula* L. Cienkowski (*Labyrinthulea* L. S. Olive *ex* T. Cavalier-Smith [class]).

LABYRINTHISTA T. Cavalier-Smith, *Advances in Phycological Research*: 342 (1986) [superclass].
Typification: *Labyrinthula* L. Cienkowski (*Labyrinthulea* L. S. Olive *ex* T. Cavalier-Smith [class]).

LABYRINTHULOMYCETES [*Dictionary of the Fungi, 7th edn*: 204] (1983) [class].
Labyrinthuleae H. M. Fitzpatrick, *The Lower Fungi: Phycomycetes*: 2 (1930) [*nomen nudum*] [class].
Labyrinthuleae C. J. Alexopoulos, *Introductory Mycology*: 27 (1952) [*nomen nudum*] [class].
Labyrinthulea L. S. Olive *ex* T. Cavalier-Smith, *Advances in Phycological Research*: 342 (1986).
Labyrinthulea L. S. Olive, *The Mycetozoans*: 215 (1975) [*nomen nudum*].
Typification: *Labyrinthulales* E. A. Bessey.

LABYRINTHULIDAE T. Cavalier-Smith, *The Chromophyte Algae, Problems and Perspectives*: 385 (1989) [subclass].
Labyrinthulia N. D. Levine & J. O. Corliss, *Journal of Protozoology, Supplement* **10**: 27 (1963) [subclass].
Typification: *Labyrinthulales* E. A. Bessey.

LABYRINTHULALES E. A. Bessey, *Morphology and Taxonomy of Fungi*: 29 (1950).
Typification: *Labyrinthulaceae* E. Haeckel.

LABYRINTHULACEAE E. [H.] Haeckel, *Jenaische Zeitschrift für Medicin und Naturwissenschaft* **4**: 127 (1868) [as '*Labyrinthuleen*'].
Labyrinthulidae E. [H.] Haeckel ex F. Doflein, *Die Protozoen als Parasiten und Krankheitserreger nach biologischen Gesichtspunkten darsgestellt*: 47 (1901).
Labyrinthulida H. F. Copeland, *A Classification of Lower Organisms*: 203 (1956).
Labyrinthulidae L. S. Olive, *The Mycetozoans*: 215 (1975).
Typification: *Labyrinthula* L. Cienkowski.

LABYRINTHULA L. Cienkowski, *Archiv für Mikroscopische Anatomie und Entwickelungsmechanik* **3**: 275 (1867).
Chlamydomyxa W. Archer, *Quarterly Journal of Microscopical Science* **15**: 107-130 (1875) (**Type species**: *Chlamydomyxa labyrinthuloides* W. Archer - holotype, sole original species).
Pseudoplasmodium H. Molisch, *Science Reports, Tôhoku Imperial University, Sendai, Series 4* **1**: 119 (1926) (**Type species**: *Pseudoplasmodium aurantiacum* H. Molisch - holotype, sole original species).
Labyrinthodictyon A. Valkanov, *Progress in Protozoology, Abstracts of Papers Read at the IIIrd International Congress on Protozoology*: 373 (1969) [as '*Labyrinthodyction*'] (**Type species**: *Labyrinthodictyon magnificum* A. Valkanov - holotype, sole original species).
Labyrinthomyxa O. Duboscq, *Compte rendu des Séances de la Société de Biologie, Paris* **84**: 28 (1921) (**Type species**: *Labyrinthomyxa sauvageaui* O. Duboscq - holotype, sole original species).
Labyrinthorhiza M. Chadefaud, *Comptes Rendus Hebdomadaires des Séances de l'Académie des Sciences, Paris, Série D* **243**: 1795 (1956) (**Type species**: *Labyrinthorhiza cienkowskii* (W. Zopf) M. Chadefaud).
Typification: *Labyrinthula vitellina* L. Cienkowski.

Labyrinthula vitellina L. Cienkowski, *Archiv für Mikroscopische Anatomie und Entwickelungsmechanik* **3**: 275 (1867).
Labyrinthula macrocystis L. Cienkowski var. *vitellina* E. L. Young, *American Journal of Botany* **30**: 592 (1943).
Labyrinthula vitellina L. Cienkowski var. *pacifica* S. W. Watson, (Thesis): 102 (1951) [*nomen nudum*].
? *Chlamydomyxa labyrinthuloides* W. Archer, *Quarterly Journal of Microscopical Science* **15**: 107-130 (1875).
Pseudoplasmodium aurantiacum H. Molisch, *Science Reports, Tôhoku Imperial University, Sendai, Series 4* **1**: 119 (1926).
Typification: type material not designated; lectotype *Archiv für Mikroscopische Anatomie und Entwickelungsmechanik* **3**: plate XV figs 1-3, 5, plate XVI figs 6-9, 10a, 11, 13, plate XVII figs 16, 17, 19 (1867) (iconotype). Type locality: Black Sea, Odessa Harbour, Ukraine. Type habitat: with marine algae.
Identification key: Key XVI; key point 11'.
Literature: Alderman & Polglase (1986), Armiger (1964), Cienkowski (1867, 1876), Hohl (1966), Johnson & Sparrow (1961: 528), Nakatsuji & Bell (1980), Nakatsuji, Nakatsuji & Bell (1981), Molisch (1926), Pokorny (1967), Porter (1969, 1972, 1974, 1986, 1990), Porter & Kirk (1987), Porter & Kochert (1978), Schultze (1867), Sykes & Porter (1973), Valkanov (1929*a*, 1972*a*), Vishniac (1955*b*, *c*), Vishniac & Watson (1953), Watson (1951, 1957), Watson & Ordal (1951), Watson & Raper (1957).

Labyrinthula macrocystis L. Cienkowski, *Archiv für Mikroscopische Anatomie und Entwickelungsmechanik* **3**: 275 (1867).
Labyrinthula macrocystis L. Cienkowski var. *atlantica* H. S. Vishniac & S. W. Watson, *Journal of General Microbiology* **8**: 248 (1953) [*nomen nudum*].
Typification: type material not designated; lectotype *Archiv für Mikroscopische Anatomie und Entwickelungsmechanik* **3**: plate XV fig. 4, plate XVI figs 10b-d, 12, 14, plate XVII figs 15, 16, 18, 20-28 (1867) (iconotype). Type locality: Black Sea, Odessa Harbour, Ukraine. Type habitat: with marine algae.
Identification key: Key XVI; key point 12.
Literature: Aschner (1958), Aschner & Kogan (1959), Cienkowski (1867), Schultze (1867), Hohl (1966), Johnson & Sparrow (1961: 529), Karling (1944), Kobayasi & Ookubo (1953*b*), Pokorny (1967), Porter (1974, 1990), Renn (1934, 1935, 1936*a*, *b*, 1937, 1942), Tutin (1938), Valkanov (1929*a*, 1972*a*), Van der Werff (1938), Watson (1957), Young (1937, 1938, 1943).

Labyrinthula cienkowskii W. Zopf, *Beiträge zur Physiologie und Morphologie niederer Organismen* **2**: 61 (1892).
Labyrinthorhiza cienkowskii (W. Zopf) M. Chadefaud, *Comptes Rendus Hebdomadaires des Séances de l'Académie des Sciences, Paris, Série D* **243**: 1795 (1956).
Typification: type material not designated; lectotype *Beiträge zur Physiologie und Morphologie niederer Organismen* **2**: plate III, figs 4-5 (1892) (iconotype). Type locality: Halle, Germany. Type habitat: parasitic in *Vaucheria terrestris* Lyngbye *emend.* Walz (*Tribophyceae*) [from a seawater aquarium, Kiel - Zopf (1887), according to Schmoller (1966)].
Identification key: Key XVI; key point 5'.
Literature: Chadefaud (1956), Pokorny (1967), Porter (1990), Schmoller (1966), Zopf (1892).

Labyrinthula valkanovii J. S. Karling, *American Journal of Botany* **31**: 50 (1944). [*nom. nov.*]
Labyrinthula zopfii A. Valkanov, *Archiv für Protistenkunde* **67**: 112 (1929) [as'*zopfi*'].
non Labyrinthula zopfii P.-A. Dangeard, *Le Botaniste* **11**: 59 (1910).
Typification: type material not designated; lectotype *Archiv für Protistenkunde* **67**: 114-117, figs 1-7 (1929) (iconotype). Type locality: Black Sea [Bulgarian coast]. Type habitat: from cultures for marine algae.
Identification key: Key XVI; key point 10.
Literature: Dangeard (1932), Jepps (1931), Johnson & Sparrow (1961: 531), Karling (1944), Kobayasi & Ookubo (1953*b*), Pokorny (1967), Porter (1990), Valkanov (1929*a*), Watson (1957), Young (1943).

Labyrinthula algeriensis A. Hollande & M. Enjumet, *Annales des Sciences Naturelles, Zoologie et Biologie Animale, Série 11*, **17**: 357 (1955) [Botanical Code *nom. inval.* as a fungus (post-1935) but not as an alga (pre-1958), Art. 36.1, without Latin diagnosis].
Typification: type material not designated; lectotype *Annales des Sciences Naturelles, Zoologie et Biologie Animale, Série 11* **17**: 357-365, figs 1-10 (1955) (iconotype). Type locality: Matifou, Bay of Algiers, Mediterranean Sea. Type habitat: water sample at 20m depth.
Identification key: Key XVI; key point 4'.
Literature: Amon & Perkins (1968), Hollande & Enjumet (1955), Johnson & Sparrow (1961: 527), Perkins (1972), Pokorny (1967), Porter (1990).

Labyrinthula roscoffensis M. Chadefaud, *Comptes Rendus Hebdomadaires de Séances de l'Académie des Sciences, Paris, Série D* **243**: 1797 (1956) [Zoological Code; Botanical Code: *nom. invalid.*, Art. 36.1, without Latin diagnosis].
Typification: Iconotype *Comptes Rendus Hebdomadaires de Séances de l'Académie des Sciences, Paris, Série D* **243**: 1796, figs 1-5 (1956). Type locality: Roscoff, Brittany, France. Type habitat: parasitic in cells of *Taonia atomaria* J. Agardh (also *Ectocarpus* (*Hincksia*) spp.) (*Fucophyceae*).
Identification key: Key XVI; key point 12'.
Literature: Chadefaud (1956), Johnson & Sparrow (1961: 527), Pokorny (1967), Porter (1990).

Labyrinthula coenocystis H. Schmoller, *Archiv für Mikrobiologie* **36**: 371 (1960).
Typification: type material not designated; lectotype *Archiv für Mikrobiologie* **36**: 367, fig. 1, 368, figs 2-8 (1960) (iconotype). Type locality: Baltic Sea, 1.5km west of Warnmünder Mole, Rostock, Germany. Type habitat: from material of *Cladophora* sp. (*Chlorophyceae*, *Cladophorales*).
Identification key: Key XVI; key point 8'.
Literature: Pokorny (1967), Porter (1990), Schmoller (1960), Schwab-Stey & Schwab (1974*a*, *b*), Stey (1969).

Labyrinthula pohlia (H. Schmoller) M. W. Dick, **comb. nov.**
[(H. Schmoller) Zoological Code];
[Botanical Code: *nom. invalid.*, Art. 36.1, without Latin diagnosis].
Labyrinthomyxa pohlia H. Schmoller, *Archiv für Protistenkunde* **109**: 237 (1966).
Typification: type material not designated; lectotype *Archiv für Protistenkunde* **109**: 232, figs 1-3, 234, figs 4-7, 237, figs 8-10 (1966) (iconotype).Type locality: Rostock, Warnemünde Mole, 2km west, Baltic Sea. Type habitat: parasitic in *Chorda filum* (L.) Stackh. (*Fucophyceae*) from the littoral zone.
Identification key: Key XVI; key point 6'.
Literature: Porter (1990), Schmoller (1966).

Labyrinthula magnifica (A. Valkanov) L. S. Olive, *The Mycetozoans*: 227 (1975).
Labyrinthodictyon magnificum A. Valkanov, *Progress in Protozoology, Abstracts of Papers Read at the IIIrd International Congress on Protozoology*: 373 (1969) [Zoological Code; Botanical Code: *nom. invalid.*, Art. 36.1, without Latin diagnosis].
Typification: type material not designated; lectotype *Archiv für Protistenkunde* **114**: plates 46-50 (1972) (iconotype). Type locality: Black Sea [Bulgarian coast]. Type habitat: extracellular parasite on diatoms from aquaria containing marine algae.
Identification key: Key XVI; key point 9'.
Literature: Olive (1975), Porter (1990), Valkanov (1969, 1972*a*, *b*).

Labyrinthula zosterae D. Porter & L. K. Muehlstein *in* L. K. Muehlstein, D. Porter & F. T. Short, *Mycologia* **83**: 182 (1991).
Typification: type material not designated; lectotype *Mycologia* **83**: 183-186, figs 1-12; 188, figs 13-15 (1991) (iconotype); type culture: #528, Culture Collection, Department of Botany, University of Georgia and #76120 ATCC, U.S.A. Type locality: Westcott Bay, San Juan Island, Washington, U.S.A. Type habitat: from diseased plants of *Zostera* (*Najadales*, *Zosteraceae*).
Identification key: Key XVI; key point 6.
Literature: Muehlstein *et al.* (1991).

DOUBTFUL SPECIES

Labyrinthula chattonii P.-A. Dangeard, *Le Botaniste* **24**: 223 (1932).
Phagomyxa chattonii (P.-A. Dangeard) J. S. Karling, *American Journal of Botany* **31**: 51 (1944) [as '*chattoni*'].
Typification: type material not designated; lectotype *Le Botaniste* **24**: plate XXII, figs 1-11, plate XXIII, figs 1-14 (1932) (iconotype). Type locality: Roscoff, Brittany, France. Type habitat: parasitic in *Cladophora refracta* Aresch. and *C. seriacea* Kütz. or *C. utricosa* Kütz. (as "*C. laetevirens*" [*C. laetevirens* Harv. or *C. laetevirens* (Dillwyn) Kütz. ?) (*Chlorophyceae*, *Cladophorales*).
Identification key: Key XVI; key point 10'.
Literature: Dangeard (1932), Johnson & Sparrow (1961: 531), Karling (1944*a*, 1968*e*: 94, 110), Kobayasi & Ookubo (1953*b*), Pokorny (1967), Porter (1990), Sparrow (1936), Young (1943).

Labyrinthula thaidis B. A. Cox & J. G. Mackin, *Transactions of the American Microscopical Society* **93**: 63 (1974) [as'*thaisi*'].
Typification: Holotype #BP171816, type material stained slides from cultures, National Fungus Collection, USDA, Beltsville, Maryland, U.S.A. Type locality: Galveston, Texas, U.S.A. Type habitat: from gill tissue of *Thais haemastoma floridana* (Conrad) (*Mollusca*, *Gastropoda*).
Identification key: Key XVI; key point 13.
Literature: Cox & Mackin (1974).

Labyrinthula jeremarina L. Rolf *in* B. A. Cox & J. G. Mackin, *Transactions of the American Microscopical Society* **93**: 70 (1974) [*nomen nudum*].
Typification: Type habitat: *Littorina ziczac* (Gmelin) (*Mollusca*, *Gastropoda*).
Identification key: Key XVI; key point 13'.
Literature: Cox & Mackin (1974), Rolf (1967).

DOUBTFUL GENUS (*LABYRINTHULALES*)

PYRRHOSORUS H. O. Juel, *Bihang till Kongliga Svenska Vetenskaps-Akademiens Handlingar* **26** (III, 14): 14 (1901).
Typification: *Pyrrhosorus marinus* H. O. Juel (holotype - sole original species).

Pyrrhosorus marinus H. O. Juel, *Bihang till Kongliga Svenska Vetenskaps-Akademiens Handlingar* **26** (III, 14): 14 (1901).
Typification: type material not designated; lectotype *Bihang till Kongliga Svenska Vetenskaps-Akademiens Handlingar* **26** (III, 14): figs 1-28 (1901) (iconotype). Type locality: Kristineburg, Bohuslän, Sweden. Type habitat: saprotrophic in moribund branches of *Cystoclonium purpureum* (Huds.) Blatt. [as its synonym, *C. purpurascens* Kütz.] (*Rhodophyceae*).
Identification key: Key I; key point 141'.
Literature: Aleem (1953, 1955), Chadefaud (1956), Juel (1901), Karling (1968*e*: 99), Sparrow (1960: 790).

EXCLUDED SPECIES

Chlamydomyxa labyrinthuloides W. Archer, *Quarterly Journal of Microscopical Science* **15**: 107-130 (1875), ? synonym of *Labyrinthula vitellina* L. Cienkowski, *Archiv für Mikroscopische Anatomie und Entwickelungsmechanik* 3: 275 (1867) [the genus is regarded as a synonym of *Labyrinthula*, but the sole species has not apparently been referred to a species of *Labyrinthula*].

Labyrinthodictyon magnificum A. Valkanov, *Progress in Protozoology, Abstracts of Papers Read at the IIIrd International Congress on Protozoology*: 373 (1969) [Zoological Code; Botanical Code: *nom. invalid.*, Art. 36.1, without Latin diagnosis], synonym of *Labyrinthula magnifica* (A. Valkanov) L. S. Olive, *The Mycetozoans*: 227 (1975).

Labyrinthomyxa marina (J. G. Mackin, H. M. Owen & A. Collier) J. G. Mackin & S. M. Ray, *Journal of Invertebrate Pathology* **8**: 545 (1966), synonym of *Perkinsus marinus* (J. G. Mackin, H. M. Owen & A. Collier) N. D. Levine (1978), placed in a monotypic family, order and class of the phylum *Apicomplexa*.

Labyrinthomyxa pohlia H. Schmoller, *Archiv für Protistenkunde* **109**: 237 (1966) [Zoological Code; Botanical Code: *nom. invalid.*, Art. 36.1, without Latin diagnosis], synonym of *Labyrinthula pohlia* (H. Schmoller) M. W. Dick, *loc. cit.* [the species was not reassigned when the genus was placed in synonymy with *Labyrinthula*, or when *Labyrinthomyxa marina* was transferred to *Perkinsus*].

Labyrinthomyxa sauvageaui O. Duboscq, *Compte rendu des Séances de la Société de Biologie, Paris* **84**: 28 (1921), a name based on a mixture of taxa [**Literature**: Dubosq (1921*a*, *b*)].

Labyrinthorhiza cienkowskii (W. Zopf) M. Chadefaud, *Comptes Rendus Hebdomadaires des Séances de l'Académie des Sciences, Paris, Série D* **243**: 1795 (1956), synonym of *Labyrinthula cienkowskii* W. Zopf, *Beiträge zur Physiologie und Morphologie niederer Organismen* **2**: 61 (1892).

Labyrinthorhiza zopfii (P.-A. Dangeard) M. Chadefaud, *Comptes Rendus Hebdomadaires des Séances de l'Académie des Sciences, Paris, Série D* **243**: 1795 (1956), synonym of *Labyrinthula zopfii* P.-A. Dangeard, *Le Botaniste* **11**: 59, plate VI (1910), a protozoan with anastomosing rhizopodia.

Labyrinthula cienkowskii W. Zopf var. *zopfii* E. L. Young, *American Journal of Botany* **30**: 586 (1943), synonym of *Labyrinthula zopfii* P.-A. Dangeard, Dangeard, Le Botaniste **11**: 59 (1910), a protozoan with anastomosing rhizopodia.

Labyrinthula macrocystis L. Cienkowski var. *atlantica* H. S. Vishniac & S. W. Watson, *Journal of General Microbiology* **8**: 248 (1953) [*nomen nudum*], synonym of *Labyrinthula macrocystis* L. Cienkowski, *Archiv für Mikroscopische Anatomie und Entwicklungsmechanik* **3**: 275 (1867). [**Typification**: type material not designated; no information on type locality or type habitat; no illustrations. **Literature**: Vishniac (1955*b*), Vishniac & Watson (1953)].

Labyrinthula macrocystis L. Cienkowski var. *vitellina* E. L. Young, *American Journal of Botany* **30**: 592 (1943), synonym of *Labyrinthula vitellina* L. Cienkowski, *Archiv für Mikroscopische Anatomie und Entwickelungsmechanik* **3**: 275 (1867).

Labyrinthula minuta S. W. Watson & K. B. Raper, *Journal of General Microbiology* **17**: 369 (1957), synonym of *Labyrinthuloides minuta* (S. W. Watson & K. B. Raper) F. O. Perkins, *Mycologia* **66**: 697 (1957).

Labyrinthula minuta S. W. Watson & K. B. Raper var. *atlantica* H. S. Vishniac & S. W. Watson, *Journal of General Microbiology* 8: 248 (1953) [*nomen nudum*], synonym of *Labyrinthuloides minuta* (S. W. Watson & K. B. Raper) F. O. Perkins, *Mycologia* **66**: 697 (1957).

Labyrinthula saliens J. A. Quick, *Transactions of the American Microscopical Society* **93**: 54 (1974), synonym of *Labyrinthuloides saliens* (J. A. Quick) J. A. Quick, *Transactions of the American Microscopical Society* **93**: 345 (1974).

Labyrinthula vitellina L. Cienkowski var. *pacifica* S. W. Watson, (Thesis): 102 (1951) [*nomen nudum*] [**Literature**: Vishniac (1955*b*, *c*), Vishniac & Watson (1953), Watson (1951, 1957), Watson & Ordal (1951)].

Labyrinthula zopfii P.-A. Dangeard, *Le Botaniste* **11**: 59 (1910), a protozoan with anastomosing rhizopodia. [**Synonymy**: *Labyrinthorhiza zopfii* (P.-A. Dangeard) M. Chadefaud, *Comptes Rendus Hebdomadaires des Séances de l'Académie des Sciences, Paris, Série D* **243**: 1795 (1956); *Labyrinthula cienkowskii* W. Zopf var. *zopfii* E. L. Young, *American Journal of Botany* **30**: 592 (1943); *non Labyrinthula zopfii* A. Valkanov, *Archiv für Protistenkunde* **67**: 112 (1929). **Literature**: Chadefaud (1956), Dangeard (1910), Karling (1944), Pokorny (1967), Young (1943)].

Pseudoplasmodium aurantiacum H. Molisch, *Science Reports, Tôhoku Imperial University, Sendai, Series 4* **1**: 119 (1926), synonym of *Labyrinthula vitellina* L. Cienkowski, *Archiv für Mikroscopische Anatomie und Entwickelungsmechanik* 3: 275 (1867).

EXCLUDED GENUS (*LABYRINTHULALES*)

THANATOSTREA A. Franc & L. Arvy, *Comptes Rendus Hebdomadaires des Séances de l'Académie des Sciences, Paris, Série D* **268**: 3189 (1969) (**Type species**: *Thanatostrea polymorpha* A. Franc & L. Arvy).

Thanatostrea polymorpha A. Franc & L. Arvy, *Comptes Rendus Hebdomadaires des Séances de l'Académie des Sciences, Paris, Série D* 268: 3189 (1969) [*nom. invalid.*, account lacks morphological description, Latin diagnosis and illustration; said to resemble *Labyrinthomyxa sauvageaui* O. Duboscq, a name based on a mixture of taxa]. Type locality: Atlantic coast, Portugal. Type habitat: parasitic in *Ostrea edulis* L. (*Mollusca, Bivalvia*)

THRAUSTOCHYTRIDAE T. Cavalier-Smith, ***The Chromophyte Algae, Problems and Perspectives*: 404 (1989) [subclass].**
Typification: *Thraustochytriales* F. K. Sparrow

THRAUSTOCHYTRIALES F. K. Sparrow, ***The Fungi, An Advanced Treatise* IV B: 69 (1973).**
Typification: *Thraustochytriaceae* F. K. Sparrow **(holotype - sole original family).**

THRAUSTOCHYTRIACEAE F. K. Sparrow, ***Aquatic Phycomycetes*: 538 (1943).**
Thraustochytriacea H. F. Copeland, ***The Classification of Lower Organisms*: 82 (1956).**
Thraustochytriidae L. S. Olive, ***The Mycetozoans*: 215 (1975) [*nomen nudum*].**
Typification: *Thraustochytrium* F. K. Sparrow **(holotype - sole original species).**

THRAUSTOCHYTRIUM F. K. Sparrow, ***Biological Bulletin of the Marine Biological Laboratory, Woods Hole* 70: 259 (1936).**
Typification: *Thraustochytrium proliferum* F. K. Sparrow (holotype - sole original species).

Thraustochytrium proliferum F. K. Sparrow, ***Biological Bulletin of the Marine Biological Laboratory, Woods Hole* 70: 259 (1936).**
Typification: type material not designated; lectotype *Biological Bulletin of the Marine Biological Laboratory, Woods Hole* **70**: 249, fig. 2, 257, figs 22-28 (1936) (iconotype). Type locality: Woods Hole, Massachusetts, U.S.A. Type habitat: saprotrophic on decayed material of *Bryopsis plumosa* C. Agardh (*Chlorophyceae*, *Cladophorales*) (also on *Ceramium diaphanum* (Lightfoot) Roth. - *Rhodophyceae*).
Identification key: Key XVI; key point 31.
Literature: Adair & Vishniac (1958), Booth (1969), Gaertner (1964, 1969), Höhnk (1956*b*), Johnson (1957*a*, *b*, 1968), Johnson & Sparrow (1961: 331), Kazama (1972*a*, *b*, 1973, 1974*a*, *b*), Miller & Whitney (1981), Moss (1986*b*), Porter (1990), Porter & Lingle (1992), Sparrow (1936*a*, 1968), Vishniac (1955*a*, *d*).

Thraustochytrium globosum Y. Kobayas[h]i & M. Ookubo, ***Bulletin of the National Science Museum (Tokyo)* 33: 60 (1953).**
Typification: Type in Herbarium, National science Museum, Tokyo. Type locality: Hutomi, Tiba Prefecture, Japan. Type habitat: from marine *Ulothricales* (*Chlorophyceae*).
Identification key: Key XVI; key point 37.
Literature: Artemchuk (1981: 74), Gaertner (1969), Johnson & Sparrow (1961: 332), Kobayashi & Ookubo (1953), Konno (1972), Kuznetsov (1979), Sparrow (1968), Ulken (1964).

Thraustochytrium pachydermum E. Scholz, ***Archiv für Mikrobiologie* 29**: 361 (1958) [*nom. inval.*, Art. 36.1, without Latin diagnosis].
Typification: type material not designated; lectotype *Archiv für Mikrobiologie* **29**: 360, fig. 4a-f (1958) (iconotype). Type locality: Montpellier, France. Type habitat: isolated from water from a salt pan.
Identification key: Key XVI; key point 35.
Literature: Bahnweg & Jäckle (1986), Booth (1969), Gaertner (1966, 1968*a*, *b*, *c*, 1969, 1974*b*), Gaertner & Raghu Kumar* (1980), Johnson & Sparrow (1961: 331), Konno (1969, 1972), Kuznetsov (1979), Scholtz (1958*a*,
b), Sparrow (1968, 1969), Ulken (1969*c*, 1970, 1990).

Thraustochytrium aureum S. Goldstein, ***Archiv für Mikrobiologie* 45: 102 (1963).**
Typification: type material not designated; lectotype *Archiv für Mikrobiologie* **45**: 103, figs 1-2 (1963) (iconotype). Type locality: Woods Hole,
Massachusetts, U.S.A. Type habitat: from sea water baited with *Pseudotsuga* pollen, and Vishniac's medium.
Identification key: Key XVI; key point 26.
Literature: Bahnweg & Jäckle (1986), Booth & Miller (1968*a*), Bremer (1974), Findlay *et al.* (1986), Gaertner (1969), Gaertner & Raghu-Kumar (1980), Goldstein (1963*a*), Goldstein & Belsky (1963), Goldstein, Moriber & Hershenov (1964), Jennings (1986), Paton & Jennings (1989), S. Raghu Kumar (1980), C. Raghukumar, Nagarkar & S. Raghukumar (1992), Sparrow (1968), Ulken (1990), Wethered & Jennings (1985).

Thraustochytrium motivum S. Goldstein, ***American Journal of Botany* 50: 273 (1963).**
Typification: type material not designated; lectotype: *American Journal of Botany* **50**: 272, figs 2-11; 274, fig. 11 (1963) (iconotype). Type locality: Woods Hole, Massachusetts, U.S.A. Type habitat: from sea water baited with *Pseudotsuga* pollen.
Identification key: Key XVI; key point 26'.
Literature: Booth & Miller (1968*a*), Bremer (1974), Gaertner (1969), Goldstein (1963*b*), Perkins (1972), S. Raghu Kumar (1980), Sparrow (1968), Ulken (1990).

---

*FOOTNOTE: See Bibliography for procedure here and in following entries relating to the typography of the name 'Raghu Kumar'.

Thraustochytrium multirudimentale S. Goldstein, *American Journal of Botany* **50**: 273 (1963).
Typification: type material not designated; lectotype: *American Journal of Botany* **50**: 274, figs 12-21 (1963) (iconotype). Type locality: Woods Hole, Massachusetts, U.S.A. Type habitat: from sea water baited with *Pseudotsuga* pollen.
Identification key: Key XVI; key point 25'.
Literature: Bahnweg & Jäckle (1986), Bremer (1974), Chamberlain (1980), Findlay *et al.* (1986), Gaertner (1968*b*, 1969, 1974*b*, 1982), Gaertner & Raghu Kumar (1980), Goldstein (1963*b*), S. Raghu Kumar (1980), Schneider (1968*a*), Sparrow (1968), Ulken (1990).

Thraustochytrium roseum S. Goldstein, *Mycologia* **55**: 801 (1963).
Typification: Slide LP82R, Department of Biology Herbarium, Brooklyn College, Brooklyn, New York, U.S.A. Type locality: Lighthouse Point, Long Island Sound New Haven Connecticut, U.S.A. Type habitat: from sea water plated on modified Vishniac's medium.
Identification key: Key XVI; key point 41.
Literature: Bahnweg & Jäckle (1986), Booth (1969), Booth & Miller (1968*a*), Bremer (1974), Chamberlain (1980), Findlay *et al.* (1986), Gaertner (1966, 1969), Goldstein (1963c), Jennings (1986), Miller & Whitney (1981), Sparrow (1968), Ulken (1981, 1990), Wethered & Jennings (1985).

Thraustochytrium aggregatum A. Ulken, *Veröffentlichungen des Instituts für Meeresforschung in Bremerhaven* **9**: 291 (1965).
Typification: type material not designated; lectotype: *Veröffentlichungen des Instituts für Meeresforschung in Bremerhaven* **9**: plate 2, figs 3-6 (1965) (iconotype). Type locality: Weser Estuary, Germany. Type habitat: from submerged estuarine sediment.
Identification key: Key XVI; key point 47'.
Literature: Bahnweg & Jäckle (1986), Gaertner (1968*a*, *b*, *c*, 1969, 1974*b*, 1982), Gaertner & Raghu-Kumar (1980), Konno (1972), Kuznetsov (1979), Moss (1986*b*), Sparrow (1936*a*, 1968, 1969), Ulken (1964, 1965*a*, 1990).

Thraustochytrium kinnei A. Gaertner, *Helgoländer wissenschaftliche Meeresuntersuchungen* **15**: 186 (1967).
Typification: type material not designated; lectotype *Helgoländer wissenschaftliche Meeresuntersuchungen* **15**: 183-185, figs 1-17 (1967) (iconotype). Type locality: Helgoland, North Sea. Type habitat: from water sample baited with pine pollen.
Identification key: Key XVI; key point 30'.
Literature: Bahnweg & Jäckle (1986), Bremer (1974), Chamberlain (1980), Gaertner (1967, 1968*a*, *b*, *c*, 1969, 1970, 1974*b*, *c*), Gaertner & Raghu Kumar (1980), Harrison & Jones (1974*c*), Moss (1986*b*), Schneider (1968*a*).

Thraustochytrium striatum J. Schneider, *Kieler Meeresforschung* **23**: 16 (1967).
Typification: type material not designated; lectotype *Kieler Meeresforschung* **23**: plate 1, figs 1-9, plate 2, figs 10-14 (1967) (iconotype); culture deposited at the Institut für Meereskunde, Kiel, Germany. Type locality: Western Baltic Sea (Kiel). Type habitat: marine sediment.
Identification key: Key XVI; key point 38.
Literature: Bahnweg & Jäckle (1986), Bremer (1974), Gaertner (1969), Gaertner & Raghu Kumar (1980), Harrison & Jones (1974*b*), Kuznetsov (1979), Moss (1986*b*), Schneider (1967, 1968*b*, 1969, 1971*a*, *b*), Sparrow (1969), Sparrow & Dogma (1973), Ulken (1990).

Thraustochytrium arudimentale N. Ya. Artemchuk, *Veröffentlichungen des Instituts für Meeresforschung in Bremerhaven* **13**: 233 (1972).
Typification: type material not designated, illustration inadequate for lectotypification. Type locality: Kandalashial Bay, Great Salma Strait, White Sea, northern Russia. Type habitat: water baited with pine pollen, lower littoral zone.
Identification key: Key XVI; key point 37'.
Literature: Artem[t]chuk (1972, 1981: 78), Kuznetsov (1979).

Thraustochytrium antarcticum G. Bahnweg & F. K. Sparrow, *American Journal of Botany* **61**: 760 (1974).
Typification: Lectotype designated as *American Journal of Botany* **61**: 756, fig. 3A, B; 760, 761, figs 29-33 (1974) (iconotype). Type locality: south eastern Indian Ocean (Station 11, 'Eltanin' cruise 46, depth 50m). Type habitat: antarctic surface waters baited with pine pollen and dead brine shrimp larvae.
Identification key: Key XVI; key point 29'.
Literature: Bahnweg & Jäckle (1986), Bahnweg & Sparrow (1974), Pugh & Jones (1986).

Thraustochytrium rossii G. Bahnweg & F. K. Sparrow, *American Journal of Botany* **61**: 762 (1974).
Typification: Lectotype designated as *American Journal of Botany* **61**: 756, fig. 1A-G; 757, figs 5-10; 762 (1974) (iconotype). Type locality: south western Pacific Ocean (Station 14, 'Eltanin' cruise 51, depth 100m). Type habitat: antarctic surface waters baited with pine pollen and dead brine shrimp larvae.
Identification key: Key XVI; key point 27'.
Literature: Bahnweg & Jäckle (1986), Bahnweg & Sparrow (1974), Pugh & Jones (1986).

Thraustochytrium kerguelense G. Bahnweg & F. K. Sparrow, *American Journal of Botany* **61**: 762 (1974) [as '*kerguelensis*'].
Typification: Lectotype designated as *American Journal of Botany* **61**: 756, fig. 2A, B; 757, figs 11-14; 762 (1974) (iconotype). Type locality: south eastern Indian Ocean (Station 17, 'Eltanin' cruise 46, depth 90m). Type habitat: from antarctic surface waters baited with pine pollen and dead brine shrimp larvae.
Identification key: Key XVI; key point 28'.
Literature: Bahnweg & Jäckle (1986), Bahnweg & Sparrow (1974), Pugh & Jones (1986).

Thraustochytrium benthicola S. Raghu Kumar, *Transactions of the British mycological Society* 74: 612 (1980).
Typification: type material not designated: lectotype *Transactions of the British mycological Society* 74: 608, figs 1-15; 610, figs 16-18, 613, fig. 46 A-L (1980) (iconotype). Type locality: Fladenground, North Sea. Type habitat: from marine sediments.
Identification key: Key XVI; key point 31.
Literature: Raghu Kumar (1980).

Thraustochytrium indicum D. K. Chakravarty, *Beiheft 63, Nova Hedwigia*: 328, 331 (1979).
Typification: not found.
Identification key: not keyed.
Literature: Chakravarty (1976 - cited as "in print, 1976" in 1979!), 1979)

EXCLUDED SPECIES

Thraustochytrium visurgense A. Ulken, *Veröffentlichungen des Instituts für Meeresforschung in Bremerhaven* 9: 290 (1965), synonym of *Ulkenia visurgensis* (A. Ulken) A. Gaertner, *Veröffentlichungen des Instituts für Meeresforschung in Bremerhaven* 16: 141 (1977).

Thraustochytrium amoeboidum G. Bahnweg & F. K. Sparrow, *American Journal of Botany* 61: 764 (1974), synonym of *Ulkenia amoeboidea* (G. Bahnweg & F. K. Sparrow) A. Gaertner, *Veröffentlichungen des Instituts für Meeresforschung in Bremerhaven* 16: 141 (1977).

JAPONOCHYTRIUM Y. Kobayas[h]i & M. Ookubo, *Bulletin of the National Science Museum (Tokyo)* 33: 60 (1953).
Typification: *Japonochytrium marinum* Y. Kobayas[h]i & M. Ookubo (holotype - sole original species).

Japonochytrium marinum Y. Kobayas[h]i & M. Ookubo, *Bulletin of the National Science Museum (Tokyo)* 33: 60 (1953).
Typification: Type in Herbarium, National Science Museum, Tokyo. Type locality: Kanazawa-hakkei, Yokohama City, Japan. Type habitat: from *Gracilariopsis longissima* (S. G. Gmel.) Steentoft *et al.* [Steentoft, Irvine & Farnham] [as its synonym *Gracilaria confervoides* (L.) Grev.] (*Rhodophyceae*).
Identification key: Key XVI; key point 22'.
Literature: Bahnweg & Jäckle (1986), Bremer (1974), Harrison & Jones (1974*a*), Johnson & Sparrow (1961: 333), Kobayashi & Ookubo (1953), Konno (1972), Miller & Whitney (1981), Sparrow (1960: 832, 1968).

SCHIZOCHYTRIUM S. Goldstein & M. Belsky, *American Journal of Botany* **51**: 72 (1964).
Typification: *Schizochytrium aggregatum* S. Goldstein & M. Belsky (holotype - sole original species).

Schizochytrium aggregatum S. Goldstein & M. Belsky, *American Journal of Botany* **51**: 72 (1964).
Typification: type material not designated; lectotype *American Journal of Botany* **51**: 74-75, figs 1-16 (1964) (iconotype). Type locality: Long Island Sound, Lighthouse Point, New Haven, Connecticut, U.S.A. Type habitat: from sea water baited with *Pseudotsuga* pollen.
Identification key: Key XVI; key point 46.
Literature: Alderman & Polglase (1986), Artemchuk (1981: 75), Bahnweg & Jäckle (1986), Booth (1969), Booth & Miller (1969), Bremer (1974), Darley, Porter & Fuller (1973), Gaertner (1968*b*, 1969, 1974*b*, *c*, 1979, 1982), Goldstein & Belsky (1964), Kazama (1980), Konno (1972), Kuznetsov (1979), Miller & Whitney (1981), Perkins (1972), Porter (1974), S. Raghu-Kumar (1988*a*), Sparrow (1968), Sparrow & Dogma (1973), Ulken (1968*a*, *c*, 1990).

Schizochytrium minutum A. Gaertner, *Veröffentlichungen des Instituts für Meeresforschung in Bremerhaven* **19**: 69 (1972).
Typification: type material not designated; lectotype *Veröffentlichungen des Instituts für Meeresforschung in Bremerhaven* **19**: 63-68, figs 1-5 (1972) (iconotype). Type locality: south coast of Island of Madiera, Atlantic Ocean. Type habitat: from water sample baited with pine pollen.
Identification key: Key XVI; key point 48'.
Literature: Gaertner (1968*c*, 1969, 1974*a*, 1980, 1981), Gaertner & Raghu-Kumar (1980), S. Raghu-Kumar (1988*b*), Ulken (1990).

Schizochytrium octosporum S. Raghu-kumar, *Transactions of the British mycological Society* **90**: 273 (1988).
Typification: Holotype *Transactions of the British mycological Society* **90**: 274, figs 1-13 (1980) (iconotype). Type locality: Rosfjord, Norway (*ex* culture 3283,15). Type habitat: from sea water baited with pine pollen.
Identification key: Key XVI; key point 48.
Literature: Raghu-kumar (1988*a*).

Schizochytrium mangrovei S. Raghu-kumar, *Transactions of the British mycological Society* **90**: 627 (1988).
Typification: Holotype *Transactions of the British mycological Society* **90**: 628, figs 1-19 (1988) (iconotype). Type locality: Orda, Goa, India. Type habitat: from dead leaves of *Rhizophora mucronata* Lam. (*Myrtales*, *Rhizophoraceae*). Cultures (Nos 90, 97 & 98) deposited at the National Institute of Oceanography (NIOCC).
Identification key: Key XVI; key point 45'.
Literature: S. Raghu-kumar (1988*b*).

Schizochytrium limacinum D. Honda & T. Yokochi, *in* Honda *et al.*, *Mycological Research* **102**: 441 (1998).
Typification: Holotype *Mycological Research* **102**: 440, figs 1-4; 441, figs 5-8; 443, figs 31-38; 444, figs 41,45-49 (1998) (iconotype). Type locality: Colonia, Yap Islands, Federated States of Micronesia (9° 32' N, 138° 03' E). Type habitat: from sea water, July 1994. Culture (No. 32693) at IFO.
Identification key: Key XVI; key point 46'.
Literature: Honda *et al.* (1998).

ALTHORNIA E. B. G. Jones & D. J. Alderman, *Nova Hedwigia* **21**: 381 (1971).
Typification: *Althornia crouchii* E. B. G. Jones & D. J. Alderman (holotype - sole original species).

Althornia crouchii E. B. G. Jones & D. J. Alderman, *Nova Hedwigia* **21**: 382 (1971).
Typification: Holotype IMI 148396. Type locality: Althorne Creek, River Crouch, Essex, U.K. Type habitat: isolated from shells of *Ostrea edulis* L. (*Mollusca*, *Bivalvia*).
Identification key: Key XVI; key point 21'.
Literature: Jones & Alderman (1971).

APLANOCHYTRIUM G. Bahnweg & F. K. Sparrow, *Archiv für Mikrobiologie* **81**: 46 (1972).
Typification: *Aplanochytrium kerguelense* G. Bahnweg & F. K. Sparrow (holotype - sole original species).

Aplanochytrium kerguelense G. Bahnweg & F. K. Sparrow, *Archiv für Mikrobiologie* **81**: 46 (1972) [as '*kerguelensis*'].
Typification: Type slides deposited in the Herbarium, University of Michigan, annn Arbor, Michigan, U.S.A. Type locality: Kerguelen Islands, South Indian Ocean (subantarctic). Type habitat: from water sample baited with pine pollen.
Identification key: Key XVI; key point 17'.
Literature: Bahnweg & Jäckle (1986), Bahnweg & Sparrow (1972), Gaertner (1981, 1982), Ulken (1979, 1990), Ulken, Jäckle & Bahnweg (1985).

ELINA N. Ya. Artemchuk, *Veröffentlichungen des Instituts für Meeresforschung in Bremerhaven* **13**: 232 (1972).
Typification: *Elina marisalba* N. J. Artemchuk.

Elina marisalba N. Ya. Artemchuk, *Veröffentlichungen des Instituts für Meeresforschung in Bremerhaven* **13**: 232 (1972).
Typification: type material not designated; lectotype *Veröffentlichungen des Instituts für Meeresforschung in Bremerhaven* **13**: 232 fig. I (1972) (iconotype). Type locality: Kandalakshiel Bay, Great Salma Strait, White Sea, northern Russia. Type habitat: water baited with pine pollen, lower tidal horizon.
Identification key: Key XVI; key point 38'.
Literature: Artem[t]chuk (1972, 1981: 77).

Elina sinorifica N. Ya. Artemchuk, *Veröffentlichungen des Instituts für Meeresforschung in Bremerhaven* **13**: 233 (1972).
Typification: type material not designated; lectotype *Veröffentlichungen des Instituts für Meeresforschung in Bremerhaven* **13**: 233 fig. II (1972) (iconotype). Type locality: Kandalakshiel Bay, Great Salma Strait, White Sea, northern Russia. Type habitat: water baited with pine pollen, sublittoral 15m depth.
Identification key: Key XVI; key point 35'.
Literature: Artem[t]chuk (1972, 1981: 78).

LABYRINTHULOIDES F. O. Perkins, *Archiv für Mikrobiologie* **90**: 3 (1973).
Typification:*Labyrinthuloides yorkensis* F. O. Perkins (holotype - sole original species).

Labyrinthuloides yorkensis F. O. Perkins, *Archiv für Mikrobiologie* **90**: 4 (1973).
Typification: type material not designated; lectotype *Archiv für Mikrobiologie* **90**: 5-15, figs 1-22 (1973) (iconotype); designated type culture: 15-6-2. Type locality: York River, Virginia, U.S.A. Type habitat: isolated from water samples and sediments.
Identification key: Key XVI; key point 18'.
Literature: Bahnweg & Jäckle (1986), Perkins (1972, 1973*a*, *b*), Gaertner (1982), Gaertner & Raghu Kumar (1980), Jacobsen (1983), Moss (1986).

Labyrinthuloides minuta (S. W. Watson & K. B. Raper) F. O. Perkins, *Mycologia* **66**: 697 (1974).
Labyrinthula minuta S. W. Watson & K. B. Raper, *Journal of General Microbiology* **17**: 368(-377) (1957).
Labyrinthula minuta S. W. Watson & K. B. Raper var. *atlantica* H. S. Vishniac & S. W. Watson, *Journal of General Microbiology* **8**: 248 (1953) [*nomen nudum*]. (**Typification**: type material not designated; no information on type locality or type habitat; no illustrations).
Typification: type material not designated; lectotype *Journal of General Microbiology* **17**: plates 1 & 2, figs 1-14 (1957) (iconotype). Type locality: Atlantic Ocean, Woods Hole, Massachusetts, U.S.A. Type habitat: isolated from *Ulva* sp. (*Chlorophyceae*, *Ulvales*).
Identification key: Key XVI; key point 19'.
Literature: Bahnweg & Jäckle (1986), Johnson & Sparrow (1961: 525), Leipe *et al.* (1994), Moss (1986*b*), Perkins (1972, 1974), Pokorny (1967), Porter (1990), S. Raghu-Kumar (1988*a*), Vishniac (1955*a*, *d*, 1957), Vishniac & Watson (1953), Watson & Raper (1957).

Labyrinthuloides saliens (J. A. Quick) J. A. Quick, *Transactions of the American Microscopical Society* **93**: 344 (1974).
Labyrinthula saliens J. A. Quick, *Transactions of the American Microscopical Society* **93**: 54 (1974).
Typification: Holotype slide No. P10-ML-3H89 (1-16-69), Marine Research Laboratory Herbarium, St Petersburg, Florida, U.S.A. Type locality: 30°04'34"N, 84°10'57"W, SW of St Marks Lighthouse, Wakulla Co. Florida, U.S.A. Type habitat: on *Halophila englemannii* Ascherson (*Najadales*, *Cynodoceaceae*).
Identification key: Key XVI; key point 20.
Literature: Quick (1974*a*, *b*, *c*).

Labyrinthuloides schizochytrops J. A. Quick, *Transactions of the American Microscopical Society* **93**: 345 (1974).
Typification: Holotype: slide NO. L9a-RM 10-21-69, Marine Research Laboratory Herbarium, Florida, U.S.A. paratype: National Fungus Collection, USNM, Washington, D.C., U.S.A. Type locality: Bulkhead Point, St George Island, Apalachicola Bay, Franklin County, Florida, U.S.A. (29°40'15''N,84°51'20''W). Type habitat: parasitic on *Halodule wrightii* (Ascherson) Ascherson (syn. *Diplanthera wrightii* Ascherson) (*Najadales*, *Zannichelliaceae*).
Identification key: Key XVI; key point 20'.
Literature: Quick (1974*c*).

Labyrinthuloides haliotidis S. M. Bower, *Canadian Journal of Zoology* **65**: 1997 (1987) [Zoological code; Botanical Code *nom invalid.*, Article 36.1, without Latin diagnosis].
Typification: Type specimens (fixed and stained slides and material in 10% buffered formaldehyde) at National Museum of Natural Science, Ottawa, Canada, Nos NMCP 1986-0893 - 1986-0897. Cultures deposited with at Alfred-Wegener-Insitut für Polar und Meeresforschung, Columbusstrassse, D-2850 Bremerhaven, Germany and School of Biological Sciences, King Henry 1 Street, Portsmouth PO1 2DY, U.K. Type locality: Saanich Inlet, Vancouver Island, British columbia, Canada. Type habitat: parasitic in the muscle and nervous tissue of head and foot of juvenile *Haliotis kamschatkana* Jonas; also occurring in *H. rufescens* Swainson.
Identification key: not keyed; see account of flagellation.
Literature: Bower (1987*a*, *b*, *c*, *d*, 1989), Bower *et al.* (1889, 1989), Leipe *et al.* (1996).

ULKENIA A. Gaertner, *Veröffentlichungen des Instituts für Meeresforschung in Bremerhaven* **16**: 141 (1977).
Typification: *Ulkenia visurgensis* (A. Ulken) A. Gaertner.

Ulkenia visurgensis (A. Ulken) A. Gaertner, *Veröffentlichungen des Instituts für Meeresforschung in Bremerhaven* **16**: 141 (1977).
Thraustochytrium visurgense A. Ulken, *Veröffentlichungen des Instituts für Meeresforschung in Bremerhaven* **9**: 290 (1965).
Typification: type material not designated; lectotype: *Veröffentlichungen des Instituts für Meeresforschung in Bremerhaven* **9**: plate 1, figs a-h; plate 2, figs 1-2 (1965) (iconotype). Type locality: Weser Estuary, Germany. Type habitat: from submerged estuarine sediment.
Identification key: Key XVI; key point 41'.
Literature: Artemchuk (1981: 74), Bahnweg & Jäckle (1986), Booth & Miller (1968*a*), Bremer (1974), Gaertner (1969, 1972, 1974*b*, 1977), Gaertner & Raghu Kumar (1980), Kuznetsov (1979), Moss (1980), Sparrow (1968), Ulken (1964, 1965*a*, 1966, 1990).

Ulkenia amoeboidea (G. Bahnweg & F. K. Sparrow) A. Gaertner, *Veröffentlichungen des Instituts für Meeresforschung in Bremerhaven* **16**: 141 (1977).
Thraustochytrium amoeboidum G. Bahnweg & F. K. Sparrow, *American Journal of Botany* **61**: 764 (1974).
Typification: Lectotype designated as *American Journal of Botany* **61**: 756, fig. 4A-I; 759, figs 15-22; 761, figs 23-28; 762 (1974) (iconotype). Type locality: south western Pacific Ocean (Station 1, 'Eltanin' cruise 51). Type habitat: antarctic surface waters, in brine shrimp cultures.
Identification key: Key XVI; key point 16'.
Literature: Bahnweg & Sparrow (1974), Gaertner (1977), S. Raghu Kumar (1977*a*, 1982*a*, *b*), Ulken (1990).

Ulkenia sarkariana A. Gaertner, *Veröffentlichungen des Instituts für Meeresforschung in Bremerhaven* **16**: 144 (1977).
Typification: type material not designated; lectotype *Veröffentlichungen des Instituts für Meeresforschung in Bremerhaven* **16**: 142-143, figs 1-2 (1977) (iconotype). Type locality: Weser Estuary, near Brinkamahof, close to the river bank (53°35,5'N, 08°31,6'E). Type habitat: saprotrophic in pine pollen baits.
Identification key: Key XVI; key point 39'.
Literature: Bahnweg & Jäckle (1986), Gaertner (1977), Ulken (1990).

Ulkenia profunda A. Gaertner, *Veröffentlichungen des Instituts für Meeresforschung in Bremerhaven* **16**: 146 (1977).
Typification: type material not designated; lectotype *Veröffentlichungen des Instituts für Meeresforschung in Bremerhaven* **16**: 147-148, figs 3-4 (1977) (iconotype). Type locality: Atlantic Ocean, coast of Portugal (37°42N, 09°44,5'W) at 1883m depth. Type habitat: saprotrophic in pine pollen baits.
Identification key: Key XVI; key point 42.
Literature: Bahnweg & Jäckle (1986), Gaertner (1977), Ulken (1990).

Ulkenia radiata A. Gaertner, *Veröffentlichungen des Instituts für Meeresforschung in Bremerhaven* **16**: 150 (1977).
Typification: type material not designated; lectotype *Veröffentlichungen des Instituts für Meeresforschung in Bremerhaven* **16**: 151-152, figs 5-6 (1977) (iconotype). Type locality: North Sea, German Bight, Isle of Heligoland (54°1o,9'N, 07°53E) intertidal pools. Type habitat: saprotrophic in pine pollen baits.
Identification key: Key XVI; key point 42'.
Literature: Gaertner (1977), Ulken (1990).

Ulkenia minuta S. Raghu-Kumar, *Veröffentlichungen des Instituts für Meeresforschung in Bremerhaven* **16**: 160 (1977).
Typification: type material not designated; lectotype *Veröffentlichungen des Instituts für Meeresforschung in Bremerhaven* **16**: 162-164, figs 1a-i, 2a-f, 3a-g (1977) (iconotype). Type culture: Kulturensammlung mariner Pilze Bremerhaven KMPB No. 4996. Type locality: Fladengrund, North Sea (58°35.5'N,00°27.5'E). Type habitat: water baited with pine pollen.
Identification key: Key XVI; key point 43'.
Literature: Gaertner (1982), Gaertner & Raghu Kumar (1980), Goldstein & Belsky (1964), S. Raghu Kumar (1977*a*), Ulken (1990).

TAXA SOMETIMES PLACED WITH THE *LABYRINTHISTA*

DIPLOPHRYS J. Barker, *Quarterly Journal of Microscopical Science, N. S.* **8**: 123 (1868).
Typification: *Diplophrys archeri* J. Barker (holotype - sole original species).

Diplophrys archeri J. Barker, *Quarterly Journal of Microscopical Science, N. S.* **8**: 123 (1868).
Typification: Not possible from the original, unillustrated descriptions.
Identification key: Key I; key point 128'.
Literature: Barker (1867, 1868), Dykstra & Olive (1975), Hertwig & Lesser (1874), Penard (1902), Olive (1902), Olive (1975).

Diplophrys marina M. J. Dykstra & D. Porter, *Mycologia* **76**: 627 (1984).
Typification: Holotype material: coll. D. Porter, L82-12, 10 September 1982; holotype culture deposited with the American Type Culture Collection, Maryland, U.S.A. Type locality: Whidbey Island, Washington, U.S.A. (also from Sapelo Island, Georgia, U.S.A.). Type habitat: associated with marine vegetation, probably as a superficial saprotroph.
Identification key: Key XVI; key point 14'.
Literature: Dykstra & Porter (1984).

EXCLUDED SPECIES

Diplophrys stercorea L. Cienkowski, *Archiv für Mikroscopische Anatomie und Entwickelungsmechanik* **12**: 15-50 (1876), synonym of *Sorodiplophrys stercorea* (L. Cienkowski) L. S. Olive & M. J. Dykstra *in* M. J. Dykstra & L. S. Olive, *Mycologia* **67**: 873 (1975).

SORODIPLOPHRYS L. S. Olive & M. J. Dykstra *in* M. J. Dykstra & L. S. Olive, *Mycologia* **67**: 873 (1975).
Typification: *Sorodiplophrys stercorea* (L. Cienkowski) L. S. Olive & M. J. Dykstra (holotype - sole original species).

Sorodiplophrys stercorea (L. Cienkowski) L. S. Olive & M. J. Dykstra *in* M. J. Dykstra & L. S. Olive, *Mycologia* **67**: 873 (1975).
Diplophrys stercorea L. Cienkowski, *Archiv für Mikroscopische Anatomie und Entwickelungsmechanik* **12**: 15-50 (1876).
Typification: Neotype: NC-72S. Type locality: Russia; neotype locality: North Carolina, U.S.A. Type habitat: from cow and horse dung.
Identification key: Key I; key point 128.
Literature: Cienkowski (1876), Dykstra (1976*a*, *b*), Dykstra & Olive (1975).

DOUBTFUL GENUS OF THE *THRAUSTOCHYTRIALES*

CORALLOCHYTRIUM S. Raghu-kumar, *Botanica Marina (Berlin)* **30**: 86 (1987) [no separate generic diagnosis].
non Coralliochytrium A. Domján, *Folia cryptogamica* **2**: 22 (1936).
Typification: *Corallochytrium limacisporum* S. Raghu-kumar (holotype - sole original species).

Corallochytrium limacisporum S. Raghu-kumar, *Botanica Marina (Berlin)* **30**: 86 (1987).
Typification: The description and fig. 2 p. 85 constitute the type material (iconotype). Cultures are deposited in the NIO Culture Collection under the numbers NIOCC 69-74. Type locality: Agatti Lagoon, Lakshadweep Islands, Arabian Sea. Type habitat: from water sample plated on agar.
Identification key: not keyed.
Literature: Chamberlain & Moss (1988), Johns (1956), Raghu-kumar (1987).

EXCLUDED SPECIES

Coralliochytrium scherffelii A. Domján, *Folia cryptogamica* **2**: 22 (1936), synonym of *Scherffeliomyces scherffelii* (A. Domján) R. M. Johns, *Mycologia* **48**: 437 (1956) (*Chytridiomycetes*).

GENUS SOMETIMES COMPARED WITH THE *THRAUSTOCHYTRIALES*

PHAGOMYXA J. S. Karling, *American Journal of Botany* **31**: 39 (1944).
Typification: *Phagomyxa algarum* J. S. Karling (holotype - sole original species).

Phagomyxa algarum J. S. Karling, *American Journal of Botany* **31**: 39 (1944).
Typification: type material not designated; lectotype *American Journal of Botany* **31**: 40, figs 1-29; 44, figs 30-74 (1944) (iconotype). Type locality: Harker's Island (also at Schackelford and Shark Shoal jetties), Beaufort, North Carolina, U.S.A. Type habitat: parasitic in *Bachelotia antillarum* (Grunow) Gerloff [= *Bachelotia fulvescens* (Bornet) Hamel (Schousb. *ex* Bornet) Kuckuck *ex* Hamel] (basionym: *Pilayella fulvescens* (Schousb.) Bornet)] (*Fucophyceae*) [but see also: Schnepf, 1994, parasite of *Bellerochea malleus* (Brightwell) Heurck (*Bacillariophyceae*)].
Identification key: Key I; key point 142.
Literature: Johnson & Sparrow (1961: 319), Karling (1944*a*), Olive (1975), Porter (1986), Schnepf (1994).

EXCLUDED TAXON

Phagomyxa chattonii (P.-A. Dangeard) J. S. Karling, *American Journal of Botany* **31**: 51 (1944) [as '*chattoni*'], synonym of *Labyrinthula chattonii* P.-A. Dangeard, *Le Botaniste* **24**: 223 (1932).

GENUS OF UNCERTAIN AFFINITY

DEVELOPAYELLA S. M. Tong, *European Journal of Protistology* **31**: 30 (1995) [Zoological Code; Botanical Code *nom invalid.*, Article 36.1, no Latin diagnosis].

Typification: *Developayella elegans* S. M. Tong (holotype - sole original species).

Developayella elegans S. M. Tong, *European Journal of Protistology* **31**: 31 (1995) [Zoological code; Botanical Code *nom invalid.*, Article 36.1, no Latin diagnosis].

Typification: type micrograph *European Journal of Protistology* **31**: 27 fig. 8) (iconotype). Type locality: Southampton Water, U.k (50° 54'N, 01° 25'W) (also collected at 10m 47°N 20°W on 8 July 1989). Type habitat: surface estuarine water, collected 7 August 1992.

Identification Key: not keyed.

Literature: Leipe *et al.* (1996), Tong (1995).

PERONOSPOROMYCOTINA M. W. Dick, **subphylum novum***

Heterokontimycotina M. W. Dick, *Recent Advances in Aquatic Mycology*: 536 (1976) [*pro parte majore*] [subdivision = subphylum].

Pseudofungi T. Cavalier-Smith, *Progress in Phycological Research*: 342 (1986) [subphylum].

non Pseudofungi T. Cavalier-Smith, *The Chromophyte Algae, Problems and Perspectives*: 404 (1989) [subphylum].

Pythiista T. Cavalier-Smith, *Progress in Phycological Research*: 342 (1986) [superclass] (**Typification**: *Oomycetes* G. Winter (1884) [Winter (1879) predates Winter (1884) and is presumably derived from Cohn (1872) and Brefeld (1881)] = (order *Oomycetes* Schroeter 1885, Fischer 1892*a*). **Other class**: *Hypochytrea* T. Cavalier-Smith, *Progress in Phycological Research*: 342 (1986) [*sic* ? *Hyphochytrea*]).

non Pythiista T. Cavalier-Smith, *The Chromophyte Algae, Problems and Perspectives*: 404 (1989) [class].

Fungi parasitici vel saprotrophici, terrestres, in aquis dulcibus vel marinis. **Thallus** eucarpicus vel holocarpicus; intramatricalis vel extramatricalis; filamentosus et ramulosus, cum vel sine haustoria; saccatus (olpidioides) vel coralliformis, interdum monocentricus, cum vel sine rhizoideo. **Reproductio asexualis** per sporangia; sporogenesis intrasporangialis vel extrasporangialis, cum vel sine vesicula; sporangiosporae zoosporicae vel aplanosporicae; zoosporae biflagellatae, raro uniflagellatae; flagellum anterius straminipilosum. **Reproductio sexualis** oogamosa ex contactu gametangiali, vel automictica, raro per gametas isogameticas.

Holotypus: *Peronospora* A. J. K. Corda (*Peronosporomycetes* M. W. Dick).

Fungi parasitic or saprotrophic, terrestrial, freshwater or marine. **Thallus** eucarpic or holocarpic; intramatrical or extramatrical; filamentous and branched, with or without haustoria; saccate (olpidioid) or coralline, sometimes monocentric, with or without rhizoids. **Asexual reproduction** by sporangia; sporogenesis intrasporangial or extrasporangial, with or without a vesicle; sporangiospores zoosporic or aplanosporic; zoospores biflagellate, rarely uniflagellate; anterior flagellum straminipilous. **Sexual reproduction** oogamous as a result of gametangial contact, or automictic, rarely by means of isogamous gametes.

Typification: *Peronospora* A. J. K. Corda (*Peronosporomycetes* M. W. Dick).

Other class of some authors: *Hyphochytriomycetes* F. K. Sparrow *orthogr. emend.* M. W. Dick.

Order ***incertae sedis***: *Lagenismatales* M. W. Dick.

---

*FOOTNOTE: This and the following taxa are above the rank of class. The use of rank between those of class and family did not come into general use until the latter part of the nineteenth entury; it is thus difficult to establish priority of authorities for higher taxa. Dr J. C. David (pers. comm.) has written that Streinz, *Nomenclator Fungorum* (1866), placed the *Peronosporei* at the same (but unstated) hierarchical level as the ascomycetes and basidiomycetes. This could be taken as the basis of the class name. However, Dr David prefers to cite the publications where the ranks of the names are indicated. For this reason he suggests that the author citation for the order *Peronosporales* should be A. Fischer, *Rabenhorst's Kryptogamen-Flora von Deutschland, 2 Aufl., Bd 1, Abt. 4*: 383 (1892) (as '*Peronosporineae*'). The family *Peronosporaceae* (as '*Peronosporei*') should be attributed to A. de Bary, *Annales des Sciences Naturelles, Botanique, Série IV* **20**: 102 (1863) in which he stated that the *Peronosporei* was a "famille naturelle". In his later publications the ranks are vague. The attribution of the *Peronosporaceae* to Warming (*Dictionary of Fungi*, 8th edition, 1995, from *Index Supplement on Family Names* is incorrect).

PERONOSPOROMYCETES M. W. Dick, **classis nova**

Oosporeae F. Cohn, *Hedwigia* **11**: 18 (1872) [*pro parte minore*, order VII of the class *Thallophytae*, see synonymy of *Peronosporomycotina* below].

Oomycetes G. Winter, *Hedwigia* **18**: 5, 9 (1879) [*pro parte majore*] [class].

Oomyceten O. Brefeld, *Botanische Untersuchungen über Schimmelpilze. IV Heft: 11*: 162 (1881) [*pro parte majore*; sections 1, 4 and 5 were *Chytridiaceen*, *Entophothoreen* and *Ustilagineen* respectively] [class].

Oomycetes G. Winter, *Rabenhorst's Kryptogamen-Flora, 2 Aufl., Bd 1, Abt. 1*: 32 (1884)* [*pro parte majore*] [class].

Oomycetes J. Schroeter, *Kryptogamen-Flora von Schlesien. [1885-1889]*: 88 (1885) [order].

Oomycetes H. F. Copeland, *The Classification of Lower Organisms*: 76 (1956) [class].

Oomycetales J. A. von Arx, *Genera of Fungi in Cultivation* 3edn: 33 (1981) [order].

Phykomycophyta A. Pascher, *Beihefte Botanische Centralblatt* **48**: 330 (1931) [*pro parte*].

Biflagellatae F. K. Sparrow, *Aquatic Phycomycetes*: 487 (1943).

Pythiistea T. Cavalier-Smith, *The Chromophyte Algae, Problems and Perspectives*: 385 (1989) [class].

Oomycetidae T. Cavalier-Smith, *The Chromophyte Algae, Problems and Perspectives*: 404 (1989) [subclass].

Fungi cum cyclo nucleari haplomitotico-B. Fungi parasitici vel saprotrophici, in facie assimilativa a pariete circumdati (osmophagici), terrestres, in aquis dulcibus vel marinis. **Thallus** eucarpicus vel holocarpicus; intramatricalis vel extramatricalis; filamentosus et ramulosus, cum vel sine haustoria; raro septatus; saccatus (olpidioides) vel coralliformis, interdum monocentricus; cum vel sine rhizoideo; cum DAP lysina synthese; lanosterolium ex squalina oxide directe formatum; mitochondria cum cristis tubularibus. **Reproductio asexualis** per sporangia;sporogenesis intrasporangialis vel extrasporangialis, cum vel sine vesicula; sporangiosporae zoosporicae vel aplanosporicae; zoosporae biflagellatae et heterokontae, plus minusve anisokontae, flagellum anterius curtius, raro uniflagellatum et insertum in parte anteriore; flagellum anterius straminipilosum, cum pilis tripartitis tubularibus biseriatim; discus transitionalis in plano super planum plasmamembranae, cum pilula centrali; structura kinetides basalis cum sex partibus, inter quas quattuor radices: prima (R1) triplex cum microtubulis costatis versus partem anteriorem; altera (R2) duplex plana versus anteriorem; tertia (R3) taeniata et multifilata versus posteriorem; quarta (R4) cum 3-6 microtubulis versus posteriorem. **Reproductio sexualis** homothallica vel heterothallica, oogamosa ex contactu gametangiali, vel automictica; fissurae gametangiales centripetales vel centrifugales; gametae nudae vel cum pariete, aut non discretae; fecundatio per hypham fecundationis, tubum fecundationis aut copulatio gametangialis per porum.

Holotypus: *Peronospora* A. J. K. Corda (*Peronosporales* A. Fischer).

Fungi with a haplomitotic-B nuclear cycle. Fungi parasitic or saprotrophic, assimilative phase bounded by a cell wall (osmophagic); terrestrial, freshwater or marine. **Thallus** eucarpic or holocarpic; intramatrical or extramatrical; filamentous, with or without haustoria; rarely septate; saccate (olpidioid) or coralline, sometimes monocentric; with or without rhizoids; DAP lysine synthesis; lanosterol formed directly from squaline oxide; mitochondria with tubular cristae. **Asexual reproduction** by sporangia; sporogenesis intrasporangial or extrasporangial, with or without a vesicle; sporangiospores zoosporic or aplanosporic; zoospores biflagellate and heterokont, more or less anisokont, anterior flagellum the shorter, rarely uniflagellate and flagellum anteriorly inserted; anterior flagellum straminipilous, with tripartite tubular hairs in two rows; transitional plate of the flagellum in a plane above that of the plasmamembrane, with a central bead; kinetid base structure with six parts including four roots - R1 triplet with ribbed microtubules anteriorly directed, R2 simple doublet anteriorly directed, R3 band-shaped and multistranded posteriorly directed, R4 3-6 microtubules posteriorly directed. **Sexual reproduction** homothallic or heterothallic, oogamous as a result of gametangial contact, or automictic; gametangial cleavage centripetal or centrifugal; gametes naked or walled, or not discrete; fertilization by means of a fertilization hypha, a fertilization tube, or gametangial copulation through a pore.

Typification: *Peronospora* A. J. K. Corda (*Peronosporales* A. Fischer).

Other orders: *Anisolpidiales* M. W. Dick, *Hyphochytriales* F. K. Sparrow, *Leptomitales* B. B. Kanouse, *Myzocytiopsidales* M. W. Dick, *Olpidiopsidales* M. W. Dick, *Pythiales* M. W. Dick, *Rhipidiales* M. W. Dick, *Salilagenidiales* M. W. Dick, *Saprolegniales* A. Fischer, *Sclerosporales* M. W. Dick.

Subclasses: *Peronosporomycetidae* M. W. Dick, *Saprolegniomycetidae* M. W. Dick, *Rhipidiomycetidae* M. W. Dick, *Hyphochytriomycetidae* M. W. Dick.

* FOOTNOTE: There are alternative and equally correct title pages to *Rabenhorst's Kryptogamen-Flora* (see references): that quoted by Cavalier-Smith (1989) is on the recto (iii); but while the recto is bibliographically preferable, that quoted above, and printed on the preceding verso (ii), is the more usual form for the many generic and specific names first used in that work.

PERONOSPOROMYCETIDAE **M. W. Dick, subclassis nova**
Peronosporomycetidae M. W. Dick *ex* M. W. Dick *in* M. W. Dick, P. T. W. Wong & G. Clark, *Botanical Journal of the Linnean Society* **89**: 192 (1984) [subclass].
Leucosporeae F. Cohn, *Hedwigia* **11**: 18 (1872) [order VII, sectio 1 of the class *Thallophytae* (three families: *Chytridiaceae, Peronosporaceae, Saprolegniaceae) pro parte majore*; sections 2 and 3 comprised the colonial, coenocytic and oogonial green algae and the brown algae respectively].

Fungi cum cyclo nucleari haplomitotico-B. Fungi parasitici (in primis in angiospermatibus) vel saprotrophici, usi $SO_4^=$, fontibus N variabilibus, terrestres, in aquis dulcibus vel marinis. **Thallus** eucarpicus; intramatricalis vel extramatricalis; filamentosus et ramulosus, cum vel sine haustoria; raro septatus; raro saccatus (olpidioides) vel coralliformis; pars fibrillaris parietis β-1,3- and β-1,6-glucan, glucosaminum absens; sterolis synthesis ex mevalonato impossibilis; membrana nuclearis persistens usque ad telophasem. **Reproductio asexualis** per sporangia; sporogenesis intrasporangialis vel extrasporangialis, cum vel sine vesicula; sporangiosporae zoosporicae vel aplanosporicae; zoosporae formae principalis, raro polyplaneticae, heterokontae, plus minusve anisokontae, flagellum anterius curtius; flagellum anterius straminipilosum; zona transitionalis cum helice longa et sine tigillis; sporangiophorae distinctae praesentes vel absentes; K_2-grana non reperta. **Reproductio sexualis** per oogonia; saepe obnoxia exogenis sterolibus antecursoribus; homothallica vel heterothallica; meioses contemporales; oogamosa ex contactu gametangiali; fecundatio per tubum fecundationis, vel automictica, gametae discretae non formatae; oosporogenesis centripetalis; cum vel sine periplasmate nucleato; oosporae persaepe singulae, apleroticae vel quasi pleroticae, nunquam plene pleroticae; oospora regulatim cum pariete laevi, cum ooplasta solida et translucenta, distributio copiae lipidae regulatim per guttulas lipidas minutas; numerus chromosomatis basalis $x = 5$.
Holotypus: *Peronospora* A. J. K. Corda (*Peronosporales* A. Fischer).

Fungi with a haplomitotic-B nuclear cycle. Fungi parasitic (primarily in angiosperms) or saprotrophic; utilizing $SO_4^=$, variable N sources, terrestrial, freshwater or marine. **Thallus** eucarpic; intramatrical or extramatrical; filamentous, with or without haustoria; rarely septate; rarely saccate (olpidioid) or coralline; fibrillar component of wall β-1,3- and β-1,6-glucan, glucosamine absent; sterol synthesis from mevalonate not possible; nuclear membrane persistent to telophase. **Asexual reproduction** by sporangia; sporogenesis intrasporangial or extrasporangial, with or without a vesicle; sporangiospores zoosporic; zoospores of principal form, heterokont, more or less anisokont, anterior flagellum the shorter; anterior flagellum straminipilous; transitional zone with a long helix and without struts; differentiated sporangiophores present or absent; K_2 bodies not known. **Sexual reproduction** homothallic or heterothallic; meioses synchronous; oogamous as a result of gametangial contact, fertilization by means of a fertilization tube; or automictic; discrete gametes not formed; oosporogenesis centripetal, with or without nucleated periplasm; oogonia thick or thin walled; oospores predominantly single, aplerotic or nearly plerotic, never plerotic; oospores normally with a smooth wall, with a translucent solid ooplast, distribution of oil reserves normally as minute droplets; basal chromosome number $x = 5$.
Typification: *Peronospora* A. J. K. Corda (*Peronosporales* A. Fischer).
Other order: *Pythiales* M. W. Dick.

PERONOSPORALES A. Fischer, *Rabenhorst's Kryptogamen-Flora von Deutschland, 2 Aufl., Bd 1, Abt. 4*: 383 (1892) (as '*Peronosporineae*').
Typification: *Peronospora* A. J. K. Corda (*Peronosporaceae* A. de Bary).

PERONOSPORACEAE A. de Bary, *Annales des Sciences Naturelles, Botanique, Série IV* **20**: 102 (1863)*.
Typification: *Peronospora* A. J. K. Corda.

*FOOTNOTE: The generic entries for the *Peronosporaceae* and *Albuginaceae* given on the following page are not followed by species: see PART III for explanation, but note that alphabetic lists of species with authorities and additional information are given in Tables III: 3 and III: 4 (pp. 137-146) and APPENDIX 1. Fitzpatrick (1930) lists two doubtful genera of *Peronosporaceae*: *Chlorospora* C. L. Spegazzini and *Drepanoconis* J. Schröter & P. C. Hennings. The former is not mentioned by Kochman & Majeski (1970) or Constantinescu (1991*a*). The latter is now placed in the *Cryptobasidiaceae*.

PERONOSPORA A. J. K. Corda, *Icones Fungorum Hucusque Cognitorum* **1**: 20 (1837).
Peronospora A. J. K. Corda subgenus *Peronospora*, *Icones Fungorum Hucusque Cognitorum* **1**: 20 (1837).
Peronospora A. J. K. Corda subgenus *Peronoplasmopara* (A. N. Berlese) V. Skalický, *Novitates Botanicae Universitatis Carolinae (Praha)* **6**: 12 (1990) (**Type species**: *P. celtidis* M. B. Waite) [≡ *Pseudoplasmopara pro parte*].
Other infrageneric taxa:
Peronospora A. J. K. Corda Section *Leiothecae* J. Schröter.
Subsection *Parasiticae* A. de Bary.
Subsection *Effusae* A. de Bary.
Peronospora A. J. K. Corda Section *Calothecae* A. de Bary.
Subsection *Verrucosae* A. Fischer.
Subsection *Reticulatae* A. Fischer.
Typification: *Peronospora rumicis* A. J. K. Corda.

BREMIA E. Regel, *Botanische Zeitung, Berlin, 1 Jahrgang* **39**: 665 (1843).
Typification: *Bremia lactucae* E. Regel var. *lactucae* (holotype - sole original species)

BASIDIOPHORA E. Roze & M. Cornu, *Annales des Sciènces Naturelles, Botanique, Série 5* **11**: 89 (1869).
Gilletia P. A. Saccardo & Penzig [see P. A. Saccardo, *Sylloge Fungorum* **12**: 267 (1897)], (1882) (**Type species**: *Gilletia spinulifera* P. A. Saccardo & J. Therry, *Michelia* **2**: 587 (1882).)
Typification: *Basidiophora entospora* E. Roze & M. Cornu (holotype - sole original species).

PLASMOPARA J. Schröter *in* Cohn, *Kryptogamen-Flora von Schlesien* **3**(1): 236 (1886).
Plasmopara J. Schröter subgenus *plasmopara*, *in* Cohn, *Kryptogamen-Flora von Schlesien* **3**(1): 236 (1886).
Plasmopara J. Schröter subgenus *hyalodendron* J. Schröter, *in* Cohn, *Kryptogamen-Flora von Schlesien* **3**(1): 236 (1886) (**Type species**: *P. viticola* (M. J. Berkeley & M. A. Curtis) A. N. Berlese & G. B. De Toni).
Pseudoplasmopara K. Sawada, *Transactions of the Natural History Society of Formosa* **62**: 77-84 (1922).
Rhysotheca G. W. Wilson, *Bulletin of the Torrey Botanical Club* **34**: 387 (1907) (**Type species**: *R. umbelliferarum* (R. Caspary) G. W. Wilson).
Typification: *Plasmopara pygmaea* (F. Unger) J. Schröter forma *pygmaea*.

PSEUDOPERONOSPORA S. I. Rostovzev, *Flora* **92**: 424 (1903).
Peronoplasmopara (A. N. Berlese) G. P. Clinton, *Connecticut Agricultural Experiment Station, Annual Report, 1904*: (1905) (**Type species**: *P. cubensis* (M. J. Berkeley & M. A. Curtis) G. P. Clinton).
Typification: *Pseudoperonospora cubensis* (M. J. Berkeley & M. A. Curtis) S. I. Rostovzev (holotype - sole original species).

BREMIELLA G. W. Wilson, *Mycologia* **6**: 195 (1914).
Typification: *Bremiella megasperma* (A. N. Berlese) G. W. Wilson (holotype - sole original species).

PARAPERONOSPORA O. Constantinescu, *Sydowia* **41**: 84 (1989).
Typification: *Paraperonospora leptosperma* (A. de Bary) O. Constantinescu (type species).

BENUA O. Constantinescu, *Nova Hedwigia* **66**: 258 (1998).
Typification: *Benua kellermanii* (W. T. Swingle *ex* P. A. Saccardo) O. Constantinescu (holotype - sole original species).

ALBUGINACEAE J. Schröter, *Die Natürlichen Pflanzenfamilien* **I (4)**: 108 (1893).
Typification: *Albugo* (C. H. Persoon) H. F. A. de Roussel (holotype - sole original genus).

ALBUGO (C. H. Persoon) H. F. A. de Roussel, *Flore du Calvados* ed. 2: 47 (1806).
Cystopus J. H. Léveillé, *Annales des Sciences Naturelles, Série III* **8**: 371 (1847).
Typification: *Albugo candida* (J. F. Gmelin) G. Kuntze var. *candida* [type species].

PYTHIALES M. W. Dick *ex* M. W. Dick

Pythiales M. W. Dick *in* M. W. Dick, P. T. W. Wong & G. Clark, *Botanical Journal of the Linnean Society* **89**: 193 (1984) [*pro parte*] (**Type family**: *Pythiaceae* J. Schröter).

Ancylistineae J. Schröter, *Die Natürlichen Pflanzenfamilien* **1(1)**: 88 (1893) [*pro parte*].

Ancylistales H. M. Fitzpatrick, *The Lower Fungi: Phycomycetes*: 39 (1930) [*pro parte*] (**Type family**: *Ancylistaceae* J. Schröter).

Lagenidiales J. S. Karling, *American Journal of Botany* **26**: 518 (1939) [*pro parte minore*] (**Type family**: *Lagenidiaceae* J. S. Karling).

Lagenidialea H. F. Copeland, *The Classification of Lower Organisms*: 81 (1956) [*pro parte minore*].

Fungi parasitici in plantis, fungis vel animalibus, aut saprotrophici, terrestres, in aquis dulcibus vel marinis; parasitae in angiospermatibus et fungis plerumque in cultura axenica. **Thallus** filamentosus et ramulosus, mycelialis vel pseudomycelialis; fluxio cytoplasmica raro testata. **Reproductio asexualis** per sporangia vel conidio-sporangia; zoosporogenesis intrasporangialis cum vel sine vesicula plasmamembranica, aut extrasporangialis in vesicula homohylica; zoosporangia terminalia, rarius in serie, postea percurrentia aut proliferata internaliter vel sympodialiter; sporangiophorae distinctae raro; zoosporae heterokontae, formae principalis, grandiores vel grandes [(50)200-1000(2000)μm^3]. **Reproductio sexualis** per oogonia; partim obnoxia exogenis sterolibus antecursoribus; paries oogoniorum gracilis, oospora persaepe singula, aplerotica vel quasi plerotica, nunquam plene plerotica, periplasma absens vel minime praesens et non permanens; paries oosporicus persaepe laevis, ooplastum solidum et translucens, distributio copiae lipidae regulatim per guttulas lipidas minutas.

Holotypus: *Pythium* N. Pringsheim (*Pythiaceae* J. Schröter) (holotype - sole original family).

Fungi parasitic in plants, fungi or animals, or saprotrophic; terrestrial, freshwater or marine; parasites of higher plants and fungi mostly in axenic culture. **Thallus** mycelial or pseudomycelial; little evidence of cytoplasmic streaming. **Asexual reproduction** by sporangia or conidio-sporangia; zoosporogenesis intrasporangial with or without a plasmamembranic vesicle, or extrasporangial in a homohylic vesicle; zoosporangia terminal, less frequently sequential, then percurrent or by internal or sympodial proliferation; sporangiophores rarely differentiated; zoospores biflagellate and of the principal form, medium-large or large (volume equivalent (50)200-1000(2000)μm^3). **Sexual reproduction** by oogonia; partially dependent on exogenous sterol precursors; oogonia thin-walled, oospore usually single, aplerotic or nearly plerotic, never completely plerotic, periplasm absent or minimal and not persistent; oospore wall almost always smooth, ooplast solid and translucent, distribution of oil reserves as minute droplets.

Typification: *Pythium* N. Pringsheim (*Pythiaceae* J. Schröter) (holotype - sole original family).

Other family: *Pythiogetonaceae* M. W. Dick.

PYTHIACEAE J. Schröter, *Die Natürlichen Pflanzenfamilien* **1(1)**: 96, 104 (1893).

Phytophthoraceae G. H. Pethybridge, *Scientific Proceedings of the Royal Dublin Society* **13**: 556 (1913) (**Type genus**: *Phytophthora* de Bary).

Lagenidiaceae J. Schröter, *Die Naturürlichen Pflanzenfamilien* **1(1)**: 89 (1893) [*pro parte*] (**Type genus**: *Lagenidium* W. Zopf).

Lagenidiacea H. F. Copeland, *The Classification of Lower Organisms*: 82 (1956) [*pro parte*] (**Type genus**: *Lagenidium* W. Zopf).

Peronophythoraceae W. H. Ko, H. S. Chang, H. J. Su, C. C. Chen & L. S. Leu, *Mycologia* **70**: 383 (1978).

Typification: *Pythium* N. Pringsheim.

PYTHIUM N. Pringsheim, *Jahrbuch für wissenschaftlichen Botanik* **1**: 304 (1858) [*nom. conserv.*].

Artotrogus C. Montagne *in* M. J. Berkeley, *Gardeners' Chronicle and Agricultural Gazette, 1845 (38)*: 640 (1845) (**Type species**: *Artotrogus hydnosporus* C. Montagne *in* M. J. Berkeley).

Pythium sensu J. Schröter, *Die Natürlichen Pflanzenfamilien* **1(1)**: 104 (1893) (**Type species**: not designated, first-named species *Pythium proliferum* A. de Bary).

Nematosporangium J. Schröter, *Die Natürlichen Pflanzenfamilien* **1(1)**: 104 (1893) (**Type species**: not designated, first-named species *Nematosporangium gracile* (A. Schenk) J. Schröter).

Eupythium J. A. Nieuwland, *American Midland Naturalist* **4**: 384 (1916) [*nom. nov.*] [*nom. superfl.*] (**Type species**: *Eupythium proliferum* (A. de Bary) J. A. Nieuwland).

Rheosporangium H. A. Edson, *Journal of Agricultural Research* **4**: 291 (1915) (**Type species**: *Rheosporangium aphanidermatum* H. A. Edson - holotype, sole original species).

Sphaerosporangium F. K. Sparrow, *Science, New York* **73**: 42 (1931) (**Type species**: none designated; synonym of *Phytophthora* A. de Bary *pro parte*).

Pseudopythium C. P. Sideris & G. E. Paxton, *Phytopathology* **20**: 951-958 (1930) [*nomen nudum*] [F. P. Mehrlich, *Mycologia* **x**: 453] [*nomen nudum*] (**Type species**: *Pseudopythium phytophthoron* C. P. Sideris & G. E. Paxton) [≡ *Phytophthora cinnamomi*].

non Pythium C. G. Nees von Esenbeck *in* C. G. Carus, *Nova Acta Physico-medica Academiae Caesareae Leopoldino-Carolinae Naturae Curiosorum* **11**: 515 (1823) [*nom. rej.*].

non Pythion C. F. P. Martius, *Flora, Jena* **14**: 459 (1831) [*nom. rej.*].

non Pythius C. S. Rafinesque, *Flora Telluriana* **4**: 116 (1836) [*nom. rej.*].

Infrageneric taxa in synonymy with genus:

Aphragmium A. Fischer, *Rabenhorst's Kryptogamen-Flora, 2 Aufl., Bd 1, Abt. 4*: 396 (1882) [subgenus "untergattung"] (**Type species**: not designated; first-named species *Pythium gracile* A. Schenk).

Nematosporangium A. Fischer, *Rabenhorst's Kryptogamen-Flora, 2 Aufl., Bd 1, Abt. 4*: 396 (1882) [subgenus "untergattung"] (**Type species**: not designated; first-named species *Pythium monospermum* N. Pringsheim).

Sphaerosporangium A. Fischer, *Rabenhorst's Kryptogamen-Flora, 2 Aufl., Bd 1, Abt. 4*: 396 (1882) [subgenus "untergattung"] (**Type species**: not designated; first-named species *Pythium proliferum* A. de Bary).

Metasporangium A. Fischer, *Rabenhorst's Kryptogamen-Flora, 2 Aufl., Bd 1, Abt. 4*: 396 (1882) [section of subgenus *Sphaerosporangium*] (**Type species**: not designated; first-named species *Pythium debaryanum* R. Hesse).

Orthosporangium A. Fischer, *Rabenhorst's Kryptogamen-Flora, 2 Aufl., Bd 1, Abt. 4*: 396 (1882) [section of subgenus *Sphaerosporangium*] (**Type species**: not designated; first-named species *Pythium proliferum* A. de Bary).

Artotrogus (C. Montagne) J. Schröter, *Die Natürlichen Pflanzenfamilien* **1(1)**: 105 (1893) [subgenus of *Pythium sensu* J. Schröter] (**Type species**: not designated, first-named species *Pythium hydnosporum* (C. Montagne) J. Schröter).

Eupythium J. Schröter, *Die Natürlichen Pflanzenfamilien* **1(1)**: 104 (1893) [subgenus of *Pythium sensu* J. Schröter] (**Type species**: not designated, first-named species *Pythium proliferum* A. de Bary).

Aphragmium A. Fischer, *Rabenhorst's Kryptogamen-Flora, 2 Aufl., Bd 1, Abt. 4*: 396 (1882) [subgenus of *Nematosporangium* (A. Fischer) J. Schröter] (**Type species**: not designated, first-named species *Nematosporangium gracile* (A. Schenk) J. Schröter).

Eunematosporangium J. Schröter, *Die Natürlichen Pflanzenfamilien* **1(1)**: 104 (1893) [subgenus of *Nematosporangium* (A. Fischer) J. Schröter] (**Type species**: not designated, only species *Nematosporangium monospermum* (N. Pringsheim) J. Schröter).

Metasporangium H. E. Petersen, *Botanisk Tidsskrift* **29**: 395 (1909) [subgenus *nomen nudum*] (**Type species**: not designated, only species *Pythium debaryanum* R. Hesse).

Orthosporangium H. E. Petersen, *Botanisk Tidsskrift* **29**: 394 (1909) [subgenus *nomen nudum*] (**Type species**: not designated, first-named species *Pythium proliferum* A. de Bary).

Polyandra C. P. Sideris, *Mycologia* **23**: 268 (1931) [section of *Nematosporangium sensu* J. Schröter] (**Type species**: not designated; first-named species *N. arrhenomanes* (C. Drechsler) C. P. Sideris).

Bradyspora C. P. Sideris, *Mycologia* **23**: 270 (1931) [subsection of section *Polyandra* of *Nematosporangium sensu* J. Schröter] (**Type species**: not designated; first-named species *Nematosporangium arrhenomanes* (C. Drechsler) C. P. Sideris).

Hemibradyspora C. P. Sideris, *Mycologia* **23**: 270 (1931) [subsection of section *Polyandra* of *Nematosporangium sensu* J. Schröter] (**Type species**: not designated; first-named species *Nematosporangium spaniogamon* C. P. Sideris).

Tachyspora C. P. Sideris, *Mycologia* **23**: 270 (1931) [subsection of section *Polyandra* of *Nematosporangium sensu* J. Schröter] (**Type species**: not designated; first-named species *Nematosporangium rhizophthoron* C. P. Sideris).

Oligandra C. P. Sideris, *Mycologia* **23**: 271 (1931) [section of *Nematosporangium sensu* J. Schröter] (**Type species**: not designated; first-named species *Nematosporangium aphanidermatum* (H. A. Edson) H. M. Fitzpatrick).

Plethrocomba C. P. Sideris, *Mycologia* **23**: 271 (1931) [subsection of section *Oligandra* of *Nematosporangium sensu* J. Schröter] (**Type species**: not designated; first-named species *Nematosporangium aphanidermatum* (H. A. Edson) H. M. Fitzpatrick).

Oligocomba C. P. Sideris, *Mycologia* **23**: 271 (1931) [subsection of section *Oligandra* of *Nematosporangium sensu* J. Schröter] (**Type species**: not designated; first-named species *Nematosporangium indigoferae* (E. J. Butler) C. P. Sideris).

Platyphalla C. P. Sideris, *Mycologia* **24**: 23 (1932) [subgenus of *Pythium sensu* J. Schröter] (**Type species**: not designated; first-named species *Pythium allantocladon* C. P. Sideris).

Aplerospora C. P. Sideris, *Mycologia* **24**: 23 (1932) [section of subgenus *Platyphalla* of *Pythium sensu* J. Schröter] (**Type species**: not designated; first-named species *Pythium allantocladon* C. P. Sideris).

Leiospora C. P. Sideris, *Mycologia* **24**: 23 (1932) [subsection of section *Aplerospora* of *Pythium sensu* J. Schröter] (**Type species**: not designated; first-named species *Pythium allantocladon* C. P. Sideris).

Acanthospora C. P. Sideris, *Mycologia* **24**: 24 (1932) [subsection of section *Aplerospora* of subgenus *Platyphalla* of *Pythium sensu* J. Schröter] (**Type species**: not designated; first-named species *Pythium acanthophoron* C. P. Sideris).

Plerospora C. P. Sideris, *Mycologia* **24**: 25 (1932) [section of subgenus *Platyphalla* of *Pythium sensu* J. Schröter] (**Type species**: "No species have been found belonging to this section"!).

Stenophalla C. P. Sideris, *Mycologia* **24**: 25 (1932) [subgenus of *Pythium sensu* J. Schröter] (**Type species**: not designated; first-named species *Pythium allantocladon* C. P. Sideris).

Aplerospora C. P. Sideris, *Mycologia* **24**: 25 (1932) [section of subgenus *Stenophalla* of *Pythium sensu* J. Schröter] (**Type species**: not designated; first-named species *Pythium debaryanum* R. Hesse).

Leiospora C. P. Sideris, *Mycologia* **24**: 25 (1932) [subsection of section *Aplerospora* of subgenus *Stenophalla* of *Pythium sensu* J. Schröter] (**Type species**: not designated; first-named species *Pythium debaryanum* R. Hesse).

Polymorphospora C. P. Sideris, *Mycologia* **24**: 26 (1932) [subsection of section *Aplerospora* of subgenus *Stenophalla* of *Pythium sensu* J. Schröter] (**Type species**: not designated; first-named species *Pythium irregulare* C. J. Buisman).

Acanthospora C. P. Sideris, *Mycologia* **24**: 26 (1932) [subsection of section *Aplerospora* of subgenus *Stenophalla* of *Pythium sensu* J. Schröter] (**Type species**: not designated; first-named species *Pythium mamillatum* A. Meurs).

Plerospora C. P. Sideris, *Mycologia* **24**: 27 (1931) [section of subgenus *Stenophalla* of *Pythium sensu* J. Schröter] (**Type species**: not designated; first-named species *Pythium diameson* C. P. Sideris).

Excluded synonyms:

Cystosiphon E. Roze & M. Cornu *Annales des Sciences Naturelles, Botanique, Série 5* **11**: 73 (1869) (**Type species**: *Cystosiphon pythioides* E. Roze & M. Cornu) [ornamented oospore].

Lucidium G. Lohde, *Tageblatt der... Versammlung Deutscher Naturforscher und Aertze in Breslau* **47**: 205 (1874) (**Type species**: *Lucidium pythioides* G. Lohde) [no sporangial vesicle].

Typification: *Pythium monospermum* N. Pringsheim, *Jahrbuch für wissenschaftlichen Botanik* **1**: 288 (1858) (lectotypified by G. M. Waterhouse when the generic name was conserved: *Taxon* **17**: 88 (1968); *International Code of Botanical Nomenclature*: 280 (1978).
Typification of synonymous genera: *Artotrogus hydnosporus* C. Montagne *in* M. J. Berkeley; *Nematosporangium monospermum* (N. Pringsheim) J. Schröter; *Rheosporangium aphanidermatum* H. A. Edson; type species for *Sphaerosporangium* not designated.

Pythium podbielkowskii **(A. Batko) M. W. Dick, comb. nov.**
Lagenidium podbielkowskii A. Bakto, *Acta Mycologica, Warszawa* **9**: 107 (1973).
Lagenidium intermedium É. de Wildeman, *Annales de la Société Belge de Microscopie, Mémoires* **19**: 97 (1895).
Typification: Type material deposited in the Herbarium, University of Warsaw, Poland; designated iconotype *Acta Mycologica, Warszawa* **9**: 108, fig. 2; 109, fig. 3 (1973). Type locality: Celestynów, Warsaw, Poland. Type habitat: parasitic in *Micrasterias denticulata* Ralfs [Bréb. *ex* Ralfs] (*Zygnematales, Desmidiaceae*) from peat ditches. **Typification for the synonym**: Lectotype (Cejp, 1959): *Annales de la Société Belge de Microscopie, Mémoires* **19**: plate 4, figs 10-13 (1895) (iconotype). Type locality: Nancy, France.
Identification key: see p. 442.
Literature: Batko (1973), Cejp (1935, 1959), Ivimey Cook (1935), Karling (1942*a*, 1981*a*: 116), Pelizon & Milanez (1979), Sparrow (1960: 997), Wildeman (1895*b*).

Pythium caudatum **(G. L. Barron) M. W. Dick, comb. nov.**
Lagenidium caudatum G. L. Barron, *Antonie van Leeuwenhoek* **42**: 134 (1976).
Typification: Slide(s) OAC 10904, Herbarium OAC. Type locality: University of Guelph, Ontario, Canada. Type habitat: parasitic on nematodes, from soil in a cattle pen.
Identification key: see p. 442.
Literature: Barron (1976*b*, 1977*b*), Jaffee (1985, 1986*a*, *b*), Karling (1981*a*: 134), Newell *et al.* (1977).

UNNAMED SPECIES REFERABLE TO *PYTHIUM*

Lagenidium sp. T. W. Johnson, *Mycologia* **49**: 396 (1957) in *Gracilaria* (*Rhodophyceae*).

MYZOCYTIUM A. Schenk, *Über das Vorkommen contractiler Zellen im Pflanzenreich*: 10 (1858).
Typification: *Myzocytium proliferum* A. Schenk (holotype - sole original species).

Myzocytium proliferum A. Schenk, *Über das Vorkommen contractiler Zellen im Pflanzenreich*: 10 (1858).
Pythium proliferum A. Schenk, *Verhandlungen der Physikalisch-medicinische Gesellschaft in Würzburg* **9**: 27 (1859).
Pythium globosum A. Schenk, *Verhandlungen der Physikalisch-medicinische Gesellschaft in Würzburg* **9**: 27 (1859).
Pythium globosum J. Walz, *Botanische Zeitung, Berlin* **28**: 553 (1870).
Lagenidium globosum (A. Schenk) K. Lindstedt, *Synopsis der Saprolegnieen und Beobachtung über einige Arten*: 54 (1872).
Myzocytium globosum (A. Schenk) M. Cornu, *Annales des Sciences Naturelles, Botanique, Série 5* **15**: 21 (1872).
Myzocytium anomalum S. N. Dasgupta & R. John, *Bulletin of the Botanical Survey of India* **30**: 57 (1988) (later homonym).
Olpidiopsis myzocytia A. Rieth *ex* K. Cejp, *Flora ČSR, Oomycetes I*: 458 (1959); A. Rieth, *Die Kulturpflanze* **2**: 184 (1954) (**Typification**: Lectotype (Cejp, 1959): *Die Kulturpflanze* **2**: 182, fig. 10A-L (1954) (iconotype). Type locality: Hedersleben, Krs Achersleben, Germany).
Typification: Lectotype (Cejp, 1959): *Verhandlungen der Physikalisch-medicinische Gesellschaft in Würzburg* **9**: plate 1, figs 30-41,47 and 42-46 (1859) (iconotype). Type locality: Würzburg, Germany. Type habitat: parasitic in *Mougeotia* sp. (also *Zygnema* sp.) (*Zygnematales, Zygnemataceae*) from the Botanic Garden, Würzburg. **Typification of synonym**: *M. anomalum*: holotype deposited at the HCIO, Division of Mycology and Plant Pathology, IARI, New Delhi-110 012. HCIO No. 37184.
Identification key: Key II; key point 1.
Literature: Arnaud (1952), Bérczi (1940), Cejp (1932*a*, 1959), Chaudhuri (1931), Chaudhuri *et al.* (1947), Cohn (1856), Constantineanu (1901), Cornu (1869, 1872*b*), Czeczuga (1995*a*), Czeczuga & Woronowicz (1991), Das-Gupta & John (1953), Dasgupta & John (1988), Domján (1936), Graff (1928), Johnson & Howard (1968), Karling (1968*a*, 1981*a*: 40, 90), Konno (1972), Lacy (1955), Lindstedt (1872), Lopez & MacCarthy (1985), Martin (1927), Minden (1915), Mundkur (1938), Petersen (1909, 1910), Pystina (1995), Reinsch (1878), Rieth (1954), Schenk (1858, 1859*b*), Scherffel (1902*b*), Schroeter *in* Cohn (1885), Skvortzow (1927), Sparrow (1932, 1933*b*, 1952*c*, 1960: 953, 974), Sparrow & Barr (1955), Thompson (1934), Tokunaga (1934*a*), Valkanov (1931*a*), Voronichin (1920), Walz (1870), Wildeman (1893*b*, 1895*a*, 1896*b*), Wolf (1944), Yačevskie & Yačevskie (1931: 68), Zopf (1884).

Myzocytium rabenhorstii (W. Zopf) M. W. Dick, **comb. nov.**
Lagenidium rabenhorstii W. Zopf, *Verhandlungen des Botanischen Vereins der Provinz Brandenburg* **20**: 79 (1878).
Lagenidium obovatum S. N. Dasgupta & R. John, *Bulletin of the Botanical Survey of India* **30**: 68 (1988).
Lagenidium pyriforme S. N. Dasgupta & R. John, *Bulletin of the Botanical Survey of India* **30**: 69 (1988).
Lagenidium coenocytium S. N. Dasgupta & R. John, *Bulletin of the Botanical Survey of India* **30**: 74 (1988).
? *Lagenidium gracile* W. Zopf, *Nova Acta der Kaiserlichen Leopoldinisch-carolinischen Akademie der Naturforscher* **47**: 158 (1884).
? *Lagenidium clavatum* S. N. Dasgupta & R. John, *Bulletin of the Botanical Survey of India* **30**: 66 (1988).
? *Lagenidium cylindriforme* S. N. Dasgupta & R. John, *Bulletin of the Botanical Survey of India* **30**: 72 (1988).
? *Lagenidium contortum* S. N. Dasgupta & R. John, *Bulletin of the Botanical Survey of India* **30**: 73 (1988).
non Lagenidium rabenhorstii W. Zopf *sensu* W. R. Ivimey Cook, *Archiv für Protistenkunde* **68**: 63 (1935).
Typification: Lectotype (Cejp, 1959): *Nova Acta der Kaiserlichen Leopoldinisch-carolinischen Akademie der Naturforscher* **47**: plate 1, figs 1-28; plate 2, figs 1-9 (1884) (iconotype). Type locality: Berlin, Germany. Type habitat: parasitic in *Spirogyra* spp. (*Zygnematales, Zygnemataceae*). **Typification of synonyms**: *L. obovatum*: holotype deposited at HCIO, Division of Mycology and Plant Pathology, IARI, New Delhi-110 012. HCIO No. 37120; *L. pyriforme*: holotype deposited at HCIO, Division of Mycology and Plant Pathology, IARI, New Delhi-110 012. HCIO No. 37121; *L. coenocytium*: holotype deposited at HCIO, Division of Mycology and Plant Pathology, IARI, New Delhi-110 012. HCIO No. 37125.
Identification key: Key III; key point 2'.
Literature: Atkinson (1909), Bérczi (1940), Cejp (1959), Constantineanu (1901), Czeczuga (1995*a*), Dasgupta & John (1988), Domján (1936), Graff (1928), Ivimey Cook (1932*a*, 1935), Karling (1942*a*, 1968*a*, 1981*a*: 109, 114), Kobayas[h]i *et al.* (1971), Konno (1972), Linder (1947), Minden (1915), Pystina (1995), Scherffel (1914), Schroeter (1886*b*), Shen & Siang (1948), Sparrow (1932, 1943, 1960: 994,999), Wildeman (1896*c*), Yačevskie & Yačevskie (1931: 67), Zopf (1878, 1884, 1890: 571).

Myzocytium megastomum É. de Wildeman, *Annales de la Société Belge de Microscopie, Mémoires* **17**: 53 (1893).
Myzocytium irregulare H. E. Petersen, *Botanisk Tidsskrift* **29**: 403 (1909).
Ancylistes miurii B. W. Skvortzow, *Archiv für Protistenkunde* **51**: 432 (1925).
Typification: Lectotype (Cejp, 1959): *Annales de la Société Belge de Microscopie, Mémoires* **17**: plate 6, figs 6-10; plate 7, figs 19-20 (1893) (iconotype). Type locality: Vallé du Roannay, Ruy, Ardennes, Belgium. Type habitat: parasitic in *Spirotaenia* sp. (*Zygnematales, Mesotaeniaceae*).
Identification key: Key II; key point 1'.
Literature: Berdan (1938), Canter (1947), Canter & Lund (1948, 1969), Cejp (1933*a*, 1935, 1959), Chaudhuri (1931), Dogma (1978), Karling (1942*a*, 1981*a*: 91, 94), Martin (1927), Milanez (1965), Minden (1915), Pelizon & Milanez (1979), Petersen (1909, 1910), Reinsch (1878), Rieth (1954), Scherffel (1914), Schulz (1922), Skvortzow (1925), Sparrow (1943, 1960: 978, 981), Sparrow & Barr (1955), Sparrow, Paterson & Johns (1965), Wildeman (1893*b*, 1895*a*, 1896*b*), Yačevskie & Yačevskie (1931: 68).

Myzocytium netrii (C. E. Miller) M. W. Dick, **comb. nov.**
Lagenidium netrii C. E. Miller, *Journal of the Elisha Mitchell Scientific Society* **81**: 5 (1965).
Typification: type material not designated; lectotype: *Journal of the Elisha Mitchell Scientific Society* **81**: 8, figs 21-35 (1965) (iconotype). Type locality: Virginia, U.S.A. Type habitat: parasitic in *Netrium* sp. (*Zygnematales, Mesotaeniaceae*).
Identification key: Key II; key point 1'.
Literature: Karling (1981*a*: 122), Miller (1965).

SYSTEMATICS

EXCLUDED SPECIES

Myzocytium anomalum G. L. Barron, *Canadian Journal of Microbiology* **22**: 760 (1976), synonym of *Chlamydomyzium anomalum* (G. L. Barron) M. W. Dick, *Mycological Research* **101**: 880 (1997).

Myzocytium anomalum S. N. Dasgupta & R. John, *Bulletin of the Botanical Survey of India* **30**: 57 (1988) [later homonym], synonym of *Myzocytium proliferum* A. Schenk, *Über das Vorkommen contractiler Zellen im Pflanzenreich*: 10 (1858).

Myzocytium entophytum (N. Pringsheim) M. Cornu, *Annales des Sciences Naturelles, Botanique, Série V* **15**: 21 (1872), synonym of *Aphanomycopsis entophyta* (N. Pringsheim) M. W. Dick, *loc. cit.*

Myzocytium fijiense J. S. Karling, *Mycopathologia et Mycologia Applicata* **37**: 168 (1969) [as '*fijiensis*'], synonym of *Myzocytiopsis fijiensis* (J. S. Karling) M. W. Dick, *Mycological Research* **101**: 880 (1997).

Myzocytium globosum (A. Schenk) M. Cornu, *Annales des Sciences Naturelles, Botanique, Série 5* **15**: 21 (1872), synonym of *Myzocytium proliferum* A. Schenk, *Über das Vorkommen contractiler Zellen im Pflanzenreich*: 10 (1858).

Myzocytium glutinosporum G. L. Barron, *Canadian Journal of Microbiology* **22**: 755 (1976), synonym of *Myzocytiopsis glutinospora* (G. L. Barron) M. W. Dick, *Mycological Research* **101**: 880 (1997).

Myzocytium humicola G. L. Barron & J. G. Percy, *Canadian Journal of Botany* **53**: 1306 (1975), synonym of *Myzocytiopsis humicola* (G. L. Barron & J. G. Percy) M. W. Dick, *Mycological Research* **101**: 879 (1997).

Myzocytium indicum U. P. Singh, *Hydrobiologia* **42**: 449 (1973), synonym of *Myzocytiopsis indica* (U. P. Singh) M. W. Dick, *Mycological Research* **101**: 879 (1997).

Myzocytium intermedium G. L. Barron, *Canadian Journal of Botany* **54**: 2 (1976), synonym of *Myzocytiopsis intermedia* (G. L. Barron) M. W. Dick, *Mycological Research* **101**: 880 (1997).

Myzocytium irregulare H. E. Petersen, *Botanisk Tidsskrift* **29**: 403 (1909), synonym of *Myzocytium megastomum* É. de Wildeman, *Annales de la Société Belge de Microscopie, Mémoires* **17**: 53 (1893).

Myzocytium lenticulare G. L. Barron, *Antonie van Leeuwenhoek* **42**: 132 (1976), synonym of *Myzocytiopsis lenticularis* (G. L. Barron) M. W. Dick, *Mycological Research* **101**: 879 (1997).

Myzocytium microsporum (J. S. Karling) F. K. Sparrow, *Aquatic Phycomycetes*: 980 (1960), synonym of *Myzocytiopsis microspora* (J. S. Karling) M. W. Dick, *Mycological Research* **101**: 879 (1997).

Myzocytium papillatum G. L. Barron, *Canadian Journal of Microbiology* **22**: 752 (1976), synonym of *Myzocytiopsis papillata* (G. L. Barron) M. W. Dick, *Mycological Research* **101**: 880 (1997).

Myzocytium polymorpha E. Maupas, *Bulletin de la Société d'Histoire Naturelle de l'Afrique du Nord* **6**: 34-49 (1915) [*nomen nudum*], synonym of *Myzocytiopsis vermicola* (W. Zopf) M. W. Dick, *Mycological Research* **101**: 879 (1997).

Myzocytium proliferum A. Schenk forma *marinum* Y. Kobayas[h]i & M. Ookubo, *Bulletin of the National Science Museum (Tokyo)* **33**: 64 (1953), synonym of *Sirolpidium bryopsidis* (C. de Bruyne) H. E. Petersen, *Oversigt over det Kongelige Danske Videnskabernes Seskabs Forhandlinger. Medlemmers Arbeider, 1905*: 480 (1905).

Myzocytium proliferum A. Schenk forma *maximum* Y. Kobayas[h]i & M. Ookubo *in* J. S. Karling (1981) [*nomen superfluum*], synonym of *Sirolpidium bryopsidis* (C. de Bruyne) H. E. Petersen, *Oversigt over det Kongelige Danske Videnskabernes Seskabs Forhandlinger. Medlemmers Arbeider, 1905*: 480 (1905).

Myzocytium proliferum A. Schenk var. *vermicola* W. Zopf, *Nova Acta der Kaiserlichen Leopoldinisch-carolinischen Akademie der Naturforscher* **47**: (1884) [as '*vermicolum*' - cf. Sparrow, 1960: 979], synonym of *Myzocytiopsis vermicola* (W. Zopf) M. W. Dick, *Mycological Research* **101**: 879 (1997).

Myzocytium subuliforme (P.-A. Dangeard) G. L. Barron, *Canadian Journal of Botany* **55**: 822 (1977), excluded taxon see p. 195.

Myzocytium vermicola (W. Zopf) A. Fischer, *Rabenhorst's Kryptogamen-Flora, 2 Aufl., Bd 1, Abt. 4*: 75 (1892) [as '*vermicolum*' - cf. Sparrow, 1960: 979], synonym of *Myzocytiopsis vermicola* (W. Zopf) M. W. Dick, *Mycological Research* **101**: 879 (1997).

Myzocytium zoophthorum F. K. Sparrow, *Journal of the Linnean Society of London (Botany)* **50**: 462 (1936), synonym of *Myzocytiopsis zoophthora* (F. K. Sparrow) M. W. Dick, *Mycological Research* **101**: 879 (1997).

UNIDENTIFIABLE SPECIES

Myzocytium lethale (N. Sorokin) K. Cejp, *Flora ČSR, Oomycetes I*:413 (1959), unidentifiable taxon referable to *Myzocytiopsis*.

Myzocytium lineare M. Cornu, *Annales des Sciences Naturelles, Botanique, Série 5* **15**: 21 (1872), unidentifiable [**Literature**: Cejp (1933*a*), Cornu (1872*b*, 1877), Karling (1981*a*: 95), Reinsch (1878), Sparrow (1960: 982), Wildeman (1895*a*)].

Myzocytium rotiferum S. N. Dasgupta & R. John, *Bulletin of the Botanical Survey of India* **30**: 58 (1988), unidentifiable taxon referable to *Myzocytiopsis*.

Myzocytium sp. F. K. Sparrow, *Journal of the Linnean Society (Botany)* **50**: (1936), unidentifiable parasite of a *Synedra* sp. (*Bacillariophyceae*).

Myzocytium sp. J. S. Karling, *Simple Holocarpic Biflagellate Phycomycetes*: 90 (1942), unidentifiable parasite of a *Vorticella* sp. (*Protoctista*, *Ciliata*).

LAGENIDIUM A. Schenk, *Verhandlungen der Physikalisch-medicinische Gesellschaft in Würzburg* **9**: 27 (1859) (*nom. conserv.* proposed).
non Lagenidium sensu W. Zopf, *Verhandlungen des Botanischen Vereins der yProvinz Brandenburg* **20**: 79 (1878).
Excluded synonyms:
Resticularia P.-A. Dangeard, *Le Botaniste* **2**: 96 (1890) (**Type species**: *Resticularia nodosa* P.-A. Dangeard).
Lagenidiopsis É. de Wildeman, *Annales de la Société Belge de Microscopie, Mémoires* **20**: 115 (1896) (**Type species**: *Lagenidiopsis reducta* É. de Wildeman).
Typification: *Lagenidium giganteum* J. N. Couch (proposed lectotype - see text ms pp 115).

Lagenidium giganteum J. N. Couch, *Mycologia* **27**: 376 (1935).
Lagenidium culicidum C. J. Umphlett & C. S. Huang, *Association of Southern Biologists, Bulletin* **17**: 68 (1970) [*nomen nudum*].
Typification: type material not designated: lectotype *Mycologia* **27**: 377 figs 1-14 (1935) (iconotype). Type locality Chapel Hill, North Carolina, U.S.A. Type habitat: weakly parasitic in mosquito larvae (also copepods, *Daphnia*).
Identification key: Key III; key point 2.
Literature: Anonymous (1973, 1979, 1985*a*, *b*), Axtell *et al.* (1983), Axtell & Guzman (1987), Balaraman & Hoti (1986), Bao *et al.* (1996), Becnel, Garcia & Johnson (1996), Bell (1979), Bell, Lee & Domnas (1989), Berbee & Kerwin (1988, 1993), Bertke & Aronson (1992*b*), Bhattacharya & Stickel (1994), Brey (1985), Brey & Lebrun (1987), Brey *et al.* (1986, 1988), Brey & Remaudière (1985), Brown & Washino (1977, 1979), Chen (1995), Couch (1935*b*), Couch & Romney (1973), Christensen *et al.* (1977), Cuda *et al.* (1997), Dean & Domnas (1983*a*, *b*), Domanas, Biswas & Gallagher (1994), Domnas, Fagan & Jaronski (1982), Domnas, Giebel & McInnis (1974), Domnas, Jaronski & Hanton (1986), Domnas, Shipley & Hicks (1976), Domnas, Srebro & Hicks (1977), Federici (1981, 1995), Fetter-Lasko & Washino (1977, 1978, 1979, 1983), Fontaine & Schaefer (1978), Fuller & Gisi (1985), Galagali, Balasubramaniane & Balaraman (1984), Gallagher, Warner & Domnas (1994), Garcia & Sweeney (1983), Giebel & Domnas (1973, 1976), Glenn & Chapman (1978), Goettel, Toohey & Pillai (1983), Golkar *et al.* (1993), Guzman & Axtell (1986, 1987*a*, *b*), Haag & Buckingham (1991), He & Su (1995), Holck (1988), Hoti & Balaraman (1990, 1992), Jaronski (1990), Jaronski & Axtell (1982, 1983*a*, *b*, 1984*a*, *b*), Jaronski *et al.* (1983), Karling (1981*a*: 135), Kerwin (1987, 1995), Kerwin, Dritz & Washino (1988, 1990, 1994), Kerwin, Duddles & Washino (1991), Kerwin, Grant & Berbee (1991), Kerwin, Simmons & Washino (1986*a*, *b*), Kerwin & Washino (1983, 1984, 1985, 1986*a*, *b*, *c*, 1987*a*, *b*, 1988*a*, *b*), Kerwin *et al.* (1983*a*, *b*, 1995, 1996), Kramer (1990), Lacey & Lacey (1990), Lacey & Orr (1994), Lacey & Undeen (1986), Laird *et al.* (1990), Lord & Roberts (1985*a*, *b*, 1986, 1987), McCray (1985), McCray, Umphlett & Fay (1973), McCray *et al.* (1973), McInnis (1971), McInnis & Domnas (1971, 1973, 1974), MacKichan, Tuininga & Kerwin (1994), Martin (1984), Mattingly (1972*a*, *b*), Merriam & Axtell (1982*a*, *b*, 1983), Nabel (1939), Nes *et al.* (1986), Nestrud & Anderson (1994), Newell *et al.* (1977), Olson (1985), Orduz & Axtell (1990, 1991), Orduz *et al.* (1992*a*, *b*), Papierok & Brey (1991), Patel, Rueda & Axtell (1990), Patel *et al.* (1991), Petersen *et al.* (1997), Roberts (1974), Roberts & Panter (1985), Rueda, Patel & Axtell (1990, 1991), Siegel & Shadduck (1987), Simmons, Kerwin & Washino (1987), Smith (1973), Sparrow (1960: 1005), Su (1994), Su & Guo (1994), Su, Guzman & Axtell (1986), Su, Jia & Xu (1997), Suh & Axtell (1999), Umphlett (1973), Umphlett & Huang (1970, 1973), Umphlett & McCray (1975), Van Der Gheynst, May & Karagosian (2000), Vatan-Doost *et al.* (1995), Warner, Sovocool & Domnas (1982, 1983*a*, *b*), Warner *et al.* (1981, 1984, 1986), Washino & Fukushima (1978), Washino & Laird (1983), Washino *et al.* (1976, 1977, 1988), Willoughby (1969), Woodring & Kaya (1992), Woodring, Kaya & Kerwin (1995), Work *et al.* (1985), Wright & Easton (1996), Zou & Su (1996).

EXCLUDED SPECIES

Lagenidium americanum G. F. Atkinson, *Botanical Gazette* **48**: 336 (1909), synonym of *Aphanomycopsis entophyta* (N. Pringsheim) M. W. Dick, *loc. cit.*

Lagenidium astrum S. N. Dasgupta & R. John, *Bulletin of the Botanical Survey of India* **30**: 70 (1988), synonym of *Syzygangia papillosa* (G. Cocconi) M. W. Dick, *Mycological Research* **101**: 881 (1997).

Lagenidium brachystomum A. Scherffel, *Archiv für Protistenkunde* **52**: 21 (1925), synonym of *Ectrogella brachystoma* (A. Scherffel) M. W. Dick, *loc. cit.*

Lagenidium callinectes J. N. Couch, *Journal of the Elisha Mitchell Scientific Society* **58**: 1548 (1942), synonym of *Salilagenidium callinectes* (J. N. Couch) M. W. Dick, *loc. cit.*

Lagenidium canterae J. S. Karling, *Predominantly Holocarpic and Eucarpic Simple Biflagellate Phycomycetes*: 122 (1981) [as '*canterii*'], synonym of *Cystosiphon canterae* (J. S. Karling) M. W. Dick, *loc. cit.*

Lagenidium caudatum G. L. Barron, *Antonie van Leeuwenhoek* **42**: 134 (1976), synonym of *Pythium caudatum* (G. L. Barron) M. W. Dick, *loc. cit.*

Lagenidium chaetomorphae A. A. Yačevskie & P. A. Yačevskie, *Opredelitel' Gribov*: 96 (1931), synonym of *Sirolpidium salinum* (P.-A. Dangeard) M. W. Dick *loc. cit.*

Lagenidium chthamalophilum T. W. Johnson, *Biological Bulletin* **114**: 205 (1958), synonym of *Salilagenidium chthamalophilum* (T. W. Johnson) M. W. Dick, *loc. cit.*

Lagenidium clavatum S. N. Dasgupta & R. John, *Bulletin of the Botanical Survey of India* **30**: 66 (1988) [*nom. inval.*, Art. 36.1, without Latin diagnosis], synonym of *Aphanomycopsis entophyta* (N. Pringsheim) M. W. Dick, *loc. cit.*

Lagenidium closterii É. de Wildeman, *Annales de la Société Belge de Microscopie, Mémoires* **17**: 43 (1893), synonym of *Cystosiphon closterii* (É. de Wildeman) M. W. Dick, *loc. cit.*

Lagenidium coenocytium S. N. Dasgupta & R. John, *Bulletin of the Botanical Survey of India* **30**: 74 (1988), synonym of *Myzocytium rabenhorstii* (W. Zopf) M. W. Dick, *loc. cit.*

Lagenidium contortum S. N. Dasgupta & R. John, *Bulletin of the Botanical Survey of India* **30**: 73 (1988), synonym of *Aphanomycopsis entophyta* (N. Pringsheim) M. W. Dick, *loc. cit.*

Lagenidium culicidum C. J. Umphlett & C. S. Huang, *Association of Southern Biologists, Bulletin* **17**: 68 (1970) [*nomen nudum*], synonym of *Lagenidium giganteum* J. N. Couch, *Mycologia* **27**: 376 (1935).

Lagenidium cyclotellae A. Scherffel, *Archiv für Protistenkunde* **52**: 18 (1925), synonym of *Ectrogella cyclotellae* (A. Scherffel) M. W. Dick, *loc. cit.*

Lagenidium cylindriforme S. N. Dasgupta & R. John, *Bulletin of the Botanical Survey of India* **30**: 72 (1988), synonym of *Aphanomycopsis entophyta* (N. Pringsheim) M. W. Dick, *loc. cit.*

Lagenidium destruens F. K. Sparrow, *Journal of the Washington Academy of Sciences* **40**: 54 (1950), synonym of *Cornumyces destruens* (F. K. Sparrow) M. W. Dick, *loc. cit.*

Lagenidium distylae J. S. Karling, *Lloydia* **7**: 330 (1944), synonym of *Myzocytiopsis distylae* (J. S. Karling) M. W. Dick, *Mycological Research* **101**: 879 (1997).

Lagenidium elegans (E. Perroncito) R. Ciferri, *Atti dell'Istituto Botanicao della Universitata e Laboratorio Crittogamico Pavia, Serie 6* **19**: 62 (1961), synonym of *Myzocytiopsis elegans* (E. Perroncito) M. W. Dick, *Mycological Research* **101**: 880 (1997).

Lagenidium ellipticum É. de Wildeman, *Annales de la Société Belge de Microscopie, Mémoires* **17**: 18 (1893), synonym of *Syzygangia elliptica* (É. de Wildeman) M. W. Dick, *Mycological Research* **101**: 881 (1997).

Lagenidium enecans W. Zopf, *Nova Acta der Kaiserlichen Leopoldinisch-carolinischen Akademie der Naturforscher* **47**: 154 (1884), synonym of *Aphanomycopsis bacillariacearum* A. Scherffel, *Archiv für Protistenkunde* **52**: 11 (1925).

Lagenidium entophytum (N. Pringsheim) W. Zopf, *Nova Acta der Kaiserlichen Leopoldinisch-carolinischen Akademie der Naturforscher* **47**: 154 (1884), synonym of *Aphanomycopsis entophyta* (N. Pringsheim) M. W. Dick, *loc. cit.*

Lagenidium entosphaericum S. N. Dasgupta & R. John, *Bulletin of the Botanical Survey of India* **30**: 76 (1988), synonym of *Aphanomycopsis entophyta* (N. Pringsheim) M. W. Dick, *loc. cit.*

Lagenidium globosum (A. Schenk) K. Lindstedt, *Synopsis der Saprolegnieen und Beobachtung über einige Arten*: 54 (1872), synonym of *Myzocytium proliferum* A. Schenk, *Über das Vorkommen contractiler Zellen im Pflanzenreich*: 10 (1858).

Lagenidium gracile W. Zopf, *Nova Acta der Kaiserlichen Leopoldinisch-carolinischen Akademie der Naturforscher* **47**: 158 (1884), probable synonym of *Myzocytium rabenhorstii* (W. Zopf) M. W. Dick, *loc. cit.*

Lagenidium humanum J. S. Karling, *Mycologia* **39**: 225 (1947), synonym of *Myzocytiopsis humana* (J. S. Karling) M. W. Dick, *Mycological Research* **101**: 879 (1997).

Lagenidium intermedium É. de Wildeman, *Annales de la Société Belge de Microscopie, Mémoires* **19**: 97 (1895), synonym of *Pythium podbielkowskii* (A. Batko) M. W. Dick *loc. cit.*

Lagenidium lundiae J. S. Karling, *Predominantly Holocarpic and Eucarpic Simple Biflagellate Phycomycetes*: 123 (1981) [as '*lundii*'], synonym of *Pleocystidium lundiae* (J. S. Karling) M. W. Dick, *loc. cit.*

Lagenidium marchalianum É. de Wildeman, *Annales de la Société Belge de Microscopie, Mémoires* **21**: 10 (1897), synonym of *Syzygangia marchaliana* (É. de Wildeman) M. W. Dick, *Mycological Research* **101**: 881 (1997).

Lagenidium microsporum J. S. Karling, *Lloydia* **7**: 328 (1944), synonym of *Myzocytiopsis microspora* (J. S. Karling) M. W. Dick, *Mycological Research* **101**: 879 (1997).

Lagenidium muenscheri V. M. Cutter, *Mycologia* **35**: 11 (1943), synonym of *Cornumyces muenscheri* (V. M. Cutter) M. W. Dick, *loc. cit.*

Lagenidium myophilum K. Hatai & O.-a. Lawhavinit, *Transactions of the Mycological Society of Japan* **29**: 177 (1988), synonym of *Salilagenidium myophilum* (K. Hatai & O.-a. Lawhavinit) M. W. Dick, *loc. cit.*

Lagenidium netrii C. E. Miller, *Journal of the Elisha Mitchell Scientific Society* **81**: 5 (1965), synonym of *Myzocytium netrii* (C. E. Miller) M. W. Dick, *loc. cit.*

Lagenidium nodosum (P.-A. Dangeard) C. T. Ingold, *Annals of Botany, London, N.S.* **13**: 442 (1949), synonym of *Syzygangia nodosa* (P.-A. Dangeard) M. W. Dick, *Mycological Research* **101**: 881 (1997).

Lagenidium obovatum S. N. Dasgupta & R. John, *Bulletin of the Botanical Survey of India* **30**: 68 (1988), synonym of *Myzocytium rabenhorstii* (W. Zopf) M. W. Dick, *loc. cit.*

Lagenidium oedogonii A. Scherffel, *Hedwigia* **41**: 105 (1902), synonym of *Syzygangia oedogonii* (A. Scherffel) M. W. Dick, *Mycological Research* **101**: 881 (1997).

Lagenidium oophilum F. K. Sparrow, *Mycologia* **31**: 531 (1939), synonym of *Myzocytiopsis oophila* (F. K. Sparrow) M. W. Dick, *Mycological Research* **101**: 880 (1997).

Lagenidium oviparasiticum G. L. Barron, *Canadian Journal of Botany* **67**: 269 (1989), synonym of *Chlamydomyzium oviparasiticum* (G. L. Barron) M. W. Dick, *Mycological Research* **101**: 880 (1997).

Lagenidium papillosum G. Cocconi, *Rendiconti dell'Accademia delle Scienze dell'Istituto di Bologna, Serie V* **4**: 362 (1894), synonym of *Syzygangia papillosa* (G. Cocconi) M. W. Dick, *Mycological Research* **101**: 881 (1997).

Lagenidium parthenosporum J. S. Karling, *Lloydia* **7**: 329 (1944), synonym of *Myzocytiopsis parthenospora* (J. S. Karling) M. W. Dick, *Mycological Research* **101**: 879 (1997).

Lagenidium podbielkowskii A. Bakto, *Acta Mycologica, Warszawa* **9**: 107 (1973), synonym of *Pythium podbielkowskii* (A. Batko) M. W. Dick *loc. cit.*

Lagenidium pygmaeum W. Zopf, *Abhandlungen der Naturforschung Gesellschaft zu Halle* **17**: 96 (1887), synonym of *Cornumyces pygmaeus* (W. Zopf) M. W. Dick, *loc. cit.*

Lagenidium pygmaeum W. Zopf var. *pygmaeoides* J. S. Karling, *Nova Hedwigia* **35**: 112 (1981) [as '*pygmaeodies*'], synonym of *Pleocystidium pygmaeoides* M. W. Dick *loc. cit.*

Lagenidium pyriforme S. N. Dasgupta & R. John, *Bulletin of the Botanical Survey of India* **30**: 69 (1988), synonym of *Myzocytium rabenhorstii* (W. Zopf) M. W. Dick, *loc. cit.*

Lagenidium pythii A. J. Whiffen, *Journal of the Elisha Mitchell Scientific society* **62**: 54 (1946), synonym of *Pythiella pythii* (A. J. Whiffen) M. W. Dick [*loc. cit.*] (1998).

Lagenidium rabenhorstii sensu W. Zopf, *Verhandlungen des Botanischen Vereins der Provinz Brandenburg* **20**: 79 (1878), synonym of *Myzocytium rabenhorstii* (W. Zopf) M. W. Dick, *loc. cit.*

Lagenidium rabenhorstii W. Zopf *sensu* W. R. Ivimey Cook, *Archiv für Protistenkunde* **68**: 63 (1935), synonym of *Syzygangia zygnematicola* M. W. Dick, *Mycological Research* **101**: 881 (1997).

Lagenidium reductum (É. de Wildeman) J. S. Karling, *Simple Holocarpic Biflagellate Phycomycetes*: 80 (1942), synonym of *Cystosiphon reducta* (É. de Wildeman) M. W. Dick, *loc. cit.*

Lagenidium sacculoides I. L. Serbinov, *La Défense des Plantes. Bulletin du Bureau Permanent des Congrès Entomo-Phytopathologiques de Russie* **2(2)**: 85 (1925), synonym of *Syzygangia sacculoides* (I. L. Serbinov) M. W. Dick, *Mycological Research* **101**: 881 (1997).

Lagenidium scyllae B. Z. Bian, K. Hatai, G. L. Po & S. Egusa, *Transactions of the Mycological Society of Japan* **20**: 117 (1979), synonym of *Salilagenidium scyllae* (B. Z. Bian, K. Hatai, G. L. Po & S. Egusa) M. W. Dick, *loc. cit.*

Lagenidium septatum J. S. Karling, *Mycopathologia et Mycologia Applicata* **37**: 162 (1969), synonym of *Chlamydomyzium septatum* (J. S. Karling) M. W. Dick, *Mycological Research* **101**: 880 (1997).

Lagenidium thermophilum K. Nakamura, M. Nakamura, K. Hatai & Zafran, *Mycoscience* **36**: 400 (1995), synonym of *Salilagenidium thermophilum* (K. Nakamura, M. Nakamura, K. Hatai & Zafran) M. W. Dick, *loc. cit.*

Lagenidium tortum S. N. Dasgupta & R. John, *Bulletin of the Botanical Survey of India* **30**: 71 (1988), synonym of *Syzygangia papillosa* (G. Cocconi) M. W. Dick, *Mycological Research* **101**: 881 (1997).

Lagenidium zopfii É. de Wildeman, *Bulletin des Séances de la Société Belge de Microscopie, Mémoires, 1889-1890* **16**: 139 (1890), synonym of *Syzygangia dictyospora* M. W. Dick, *Mycological Research* **101**: 882 (1997).

Lagenidium sp. C. von Deckenbach, *Flora, Jena* **12**: 253-283 (1903) parasitic in *Chaetomorpha* (*Cladophorales*) synonym of *Sirolpidium salinum* (P.-A. Dangeard) M. W. Dick *loc. cit.*

UNIDENTIFIABLE SPECIES

Lagenidium syncytiorum H. Klebahn, *Jahrbücher für wissenschaftliche Botanik* **24**: 265 (1892), possibly a synonym of *Syzygangia marchaliana* (É. de Wildeman) M. W. Dick, *loc. cit.*

Lagenidium trichophryarum R. Gönnert, *Archiv für Protistenkunde* **86**: 149 (1935) [**Typification**: type material not designated, parasitic in *Trichophrys* sp. [as '*Trichophrya*'] (*Ciliophora, Suctoria*). **Literature**: Canter & Dick (1994), Gönnart (1935), Karling (1981*a*: 114)].

Lagenidium sp. É. de Wildeman, *Annales de la Société Belge de Microscopie, Mémoires* **19**: 75 (1895), parasitic in *Euastrum oblongum* (Grev.) Ralfs (*Zygnematales*) unidentifiable.

Lagenidium sp. A. Scherffel, *Archiv für Protistenkunde* **54**: 246 (1926), parasitic in *Oedogonium* aff. *franklinianum* Wittr. (*Oedogoniales*) and referable to *Syzygangia* (*Myzocytiopsidales, Myzocytiopsidaceae*).

Lagenidium sp. A. Scherffel, *Archiv für Protistenkunde* **52**: 23 (1925), parasitic in *Pinnularia* sp. (*Bacillariophyceae*) and referable to *Ectrogella* (*Myzocytiopsidales, Myzocytiopsidaceae*).

Lagenidium sp. J. N. Couch, *Mycologia* **27**: 385 (1935), parasitic in *Synedra* sp. (*Bacillariophyceae*) and referable to *Ectrogella* (*Myzocytiopsidales, Myzocytiopsidaceae*).

Lagenidium sp. M. Ookubo, *Nagaoa* **2**: 125 (1952), parasitic in *Aegarophila* sp. (*Cladophorales*) and referable to *Sirolpidium* (*Myzocytiopsidales, Sirolpidiaceae*).

Lagenidium sp. T. W. Johnson, *Mycologia* **49**: 396 (1957*a*), parasitic in *Gracilaria* sp. (*Rhodophyceae*) and referable to *Pythium* (*Peronosporomycetes, Pythiales*).

Lagenidium sp. J. S. Karling, *Sydowia* **20**: 195 (1968), parasitic in *Oedogonium* sp. (*Oedogoniales*) and referable to *Syzygangia* (*Myzocytiopsidales, Myzocytiopsidaceae*).

Lagenidium sp. H. L. Cook, *FAO Aquaculture Bulletin* **3**: 13 (1971), parasitic in *Penaeus* sp. (*Crustacea*) and referable to *Salilagenidium* (*Peronosporomycetes, Saprolegniomycetidae, Salilagenidiales*).

Lagenidium sp. M. J. Masters, *Canadian Journal of Botany* **49**: 480 [477, fig. 5e-h; plate I, figs 14-16] (1971), parasitic in *Oocystis* sp. (*Chlorophyceae*) unidentifiable.

Lagenidium sp. D. V. Lightner & C. T. Fontaine, *Journal of Invertebrate Pathology* **22**: 94 (1973), parasitic in *Pennaeus* sp. (*Crustacea*) and referable to *Salilagenidium* (*Peronosporomycetes, Saprolegniomycetidae, Salilagenidiales*).

Lagenidium sp. B. R. Kerry & D. H. Crump, *Transactions of the British mycological Society* 74: 123 (1980), parasitic in *Heterodera* sp. (*Aschelminthes*) and referable to *Cornumyces* (*Saprolegniomycetidae*).

Lagenidium sp. H. M. Canter & S. I. Heaney, *New Phytologist* 97: 605 (1984), parasitic in *Ceratium* sp. (*Dinomastigota*) and referable to *Myzocytiopsis* (*Myzocytiopsidales*, *Myzocytiopsidaceae*).

Lagenidium sp. G. D. Lio-Po & M. E. G. Sanvictores, *Aquaculture* 51: 161-168 (1986), parasitic in *Penaeus* sp. (*Crustacea*) and referable to *Salilagenidium* (*Peronosporomycetes*, *Saprolegniomycetidae*, *Salilagenidiales*).

CYSTOSIPHON E. Roze & M. Cornu, *Annales des Sciences Naturelles, Botanique, Série 5* **11**: 73 (1869).
Pythium N. Pringsheim, *Jahrbuch für Wissenschaftlichen Botanik* **1**: 304 (1858) [*pro parte minore*] (**Type species**: *Pythium monospermum* N. Pringsheim).
Lagenidium W. Zopf, *Verhandlungen des Botanischen Vereins der Provinz Brandenburg* **20**: 79 (1878) [*pro parte minore*] (**Type species**: *Lagenidium rabenhorstii* W. Zopf - holotype, sole original species).
Lagenidiopsis É. de Wildeman, *Annales de la Société Belge de Microscopie, Mémoires* **20**: 115 (1896) (**Type species**: *Lagenidiopsis reducta* É. de Wildeman - holotype, sole original species).
Typification: *Cystosiphon pythioides* E. Roze & M. Cornu (holotype - sole original species).

Cystosiphon pythioides E. Roze & M. Cornu, *Annales des Sciences Naturelles, Botanique, Série 5* **11**: 73 (1869).
Pythium cystosiphon (E. Roze & M. Cornu) K. Lindstedt, *Synopsis der Saprolegniaceen und Beobachtungen über einige Arten*: 50 (1872).
Pythium pythioides (E. Roze & M. Cornu) J. Ramsbottom, *Transactions of the British Mycological Society* **5**: 315 (1916).
Typification: type material not designated; lectotype: *Annales des Sciences Naturelles, Botanique, Série 5* **11**: plate 3, figs 1-20 (1869) (iconotype). Type locality: Romorantin, France. Type habitat: parasitic in *Wolfia arrhiza* (L.) Wimm. (*Lemnaceae*).
Identification key: Key III; key point 2.
Literature: Dick (1990*c*), Lindstedt (1872), Plaats-Niterink (1981), Ramsbottom (1916*a*, *b*), Roze & Cornu (1869), Waterhouse (1967, 1968).

Cystosiphon dictyosporum (M. Raciborski) M. W. Dick, **comb. nov.**
Pythium dictyosporum M. Raciborski, *Anzieger der Akademie der Wissenschaften in Krakau* **8**: 283 (1891).
Nematosporangium dictyosporum (M. Raciborski) J. Schröter, *Die natürlichen Pflanzenfamilien* **1(1)**: 104 (1897).
Nematosporangium dictyosporum (M. Raciborski) A. A. Yačevskie & P. A. Yačevski, *Opredelitel' Gribov*: 96 (1931).
non Pythium dictyosporum M. Raciborski, *Anzieger der Akademie der Wissenschaften in Krakau* **8**: 283 (1891) *sensu* A. Rieth, *Die Kulturpflanze* **2**: 176 (1954).
Typification: type material not designated; lectotype: *Mycologia* **23**: plate 20, figs A-V (1931) (iconotype). Type locality: Kostrze, Tyniec, Poland. Type habitat: parasitic in *Spirogyra nitida* Kütz. (*Zygnematales*, *Zygnemataceae*) in a roadside pond.
Identification key: Key III; key point 2'.
Literature: Dick (1990*c*), Matthews (1931), Middleton (1943), Plaats-Niterink (1981), Sparrow (1931*a*, 1933), Waterhouse (1967, 1968), Wildeman (1895*c*), Yačevskie & Yačevskie (1931: 96).

Cystosiphon closterii (É. de Wildeman) M. W. Dick, **comb. nov.**
Lagenidium closterii É. de Wildeman, *Annales de la Société Belge de Microscopie, Mémoires* **17**: 43 (1893).
Typification: Lectotype (Cejp, 1959): *Annales de la Société Belge de Microscopie, Mémoires* **17**: plate 6, figs 1-5 (1893) (iconotype). Type locality: La Flange a Stoumont u Liège, Belgium. Type habitat: parasitic in *Closterium* sp. (*Zygnematales*, *Desmidiaceae*).
Identification key: Key III; key point 4.
Literature: Cejp (1933*a*, 1959), Couch (1935*b*), Ivimey Cook (1935), Karling (1942*a*, 1949*a*, 1981*a*: 115), Petersen (1909, 1910), Sparrow (1960: 1001), Sparrow & Barr (1955), Wildeman (1893*b*).

Cystosiphon reducta (É. de Wildeman) M. W. Dick, **comb. nov.**
Lagenidiopsis reducta É. de Wildeman, *Annales de la Société Belge de Microscopie, Mémoires* **20**: 115 (1896).
Lagenidium reductum (É. de Wildeman) J. S. Karling, *Simple Holocarpic Biflagellate Phycomycetes*: 80 (1942).
Typification: Lectotype (Cejp, 1959): *Annales de la Société Belge de Microscopie, Mémoires* **20**: plates 6, 7 (1896) (iconotype). Type locality: Marsais de Pinchat, Geneva, Switzerland. Type habitat: parasitic in oogonia of *Chara* and *Characeae* (*Chlorophyceae*, *Charales*).
Identification key: Key III; key point 3'.
Literature: Cejp (1959), Karling (1942*a*, 1981*a*: 116), Sparrow (1960: 1011), Wildeman (1896*c*).

Cystosiphon canterae (J. S. Karling) M. W. Dick, **comb. nov.**
Lagenidium canterae J. S. Karling, *Predominantly Holocarpic and Eucarpic Simple Biflagellate Phycomycetes*: 122 (1981) [as '*canterii*'].
Biflagellate species (1) H. M. Canter & J. W. G. Lund, *Österreichische Botanische Zeitschrift* **116**: 354 (1969).
Typification: Iconotype: *Österreichische Botanische Zeitschrift* **116**: 356, fig. 2 a-o (1969) (redrawn as iconotypus in *Predominantly Holocarpic and Eucarpic Simple Biflagellate Phycomycetes*: 121, plate 33 figs 111-118 (1981); additional illustrations [photographs, excluded by Karling from the type material] *Österreichische Botanische Zeitschrift* **116**: 362, fig. 5 d-e (1969). Type locality: Lake Windermere, Cumbria, U.K. Type habitat: parasitic *Staurastrum* spp. and *Cosmarium* sp. (*Zygnematales*, *Desmidiaceae*).
Identification key: Key III; key point 4'.
Literature: Canter & Lund (1969), Karling (1981*a*: 122).

SYSTEMATICS

NOTE: the species of *Phytophthora*, *Trachysphaera*, *Peronophythora* and *Halophytophthora* are not listed, see PART III for explanation; lists are given in APPENDIX 1.

PHYTOPHTHORA A. de Bary, *Journal of the Royal Agricultural Society of England, Second Series* **12**: 240 (1876).
Typification: *Phytophthora infestans* (J. F. C. Montagne) A. de Bary var. ***infestans*** **[type species].**

TRACHYSPHAERA R. J. Tabor & R. H. Bunting, *Annals of Botany, London* **37**: 156 (1923).
Typification: *Trachysphaera fructigena* R. J. Tabor & R. H. Bunting **[holotype].**

DIASPORANGIUM W. Höhnk, *Beihefte zum Botanischen Zentralblatt (Abteilung A)* **55**, 97 (1936).
Type species: *Diasporangium jonesianum* W. Höhnk (holotype - sole original species).

Diasporangium jonesianum W. Höhnk, *Beihefte zum Botanischen Zentralblatt (Abteilung A)* **55**, 97 (1936).
Typification: type material not designated; lectotype: *Beihefte zum Botanischen Zentralblatt (Abteilung A)* **55**, 94, fig. III, 1-9 (1936) (iconotype [the geographic origin of the material drawn is not specified]). Type locality: Shorewood Hills, Madison, Wisc. USA (also Nordenham, Germany). Type habitat: from [damp?] soil samples, associated with *Pythium* and *Dictyuchus*.
Literature: Höhnk (1936, 1962), Sparrow (1960), Voglmayr *et al.* (1999).

PERONOPHYTHORA C. C. Chen *ex* W. H. Ko, H. S. Chang, H. J. Su, C. C. Chen & L. S. Leu, *Mycologia* **70**: 381 (1978).
Typification: *Peronophythora litchi* C. C. Chen *ex* W. H. Ko, H. S. Chang, H. J. Su, C. C. Chen & L. S. Leu **[holotype].**

HALOPHYTOPHTHORA H. H. Ho & S. C. Jong, *Mycotaxon* **36**: 380 (1990).
Typification: *Halophytophthora vesicula* (C. J. Anastasiou & L. M. Churchland) H. H. Ho & S. C. Jong.

GENUS *INCERTAE SEDIS*

ENDOSPHAERIUM P. N. D'Eliscu, *Journal of Invertebrate Pathology* **30**: 420 (1977).
Typification: *Endosphaerium funiculatum* P. N. D'Eliscu (holotype - sole original species).

Endosphaerium funiculatum P. N. D'Eliscu, *Journal of Invertebrate Pathology* **30**: 420 (1977).
Typification: Holotype slide No. 71819 B.P.I., U.S. National Collection, Agricultural Research Centre, Maryland, U.S.A. (slide X-9-71a of author). Type locality: oxbow pond, Noon Ranch, Arivaca, Arizona, U.S.A. Type habitat: in the mantle cavity of the small freshwater bivalve *Musculium transversum* (Say) (*Mollusca, Bivalvia*).
Identification key: Key I; key point 22'.
Literature: D'Eliscu (1977).

PYTHIOGETONACEAE M. W. Dick, *in* Voglmayr *et al.*, *Mycological Research* **103**: 603 (1999).

Mycelium of narrow hyphae, 1·0-3·5(7·0)μm; saprotrophic in submerged plant debris; probably with facultative or obligate fermentative metabolism; sporangia spherical or bursiform (sometimes not known), often unequally dilated from an intercalary sub-terminal hyphal segment; zoosporogenesis extrasporangial in a detachable homohylic vesicle, formed at the end of an elongate and cylindrically dilated exit papilla; zoospores large (zoospore cysts *ca* 15μm diam) of principal form with lateral flagellar insertion, polyplanetic; initiation and development of gametangia almost simultaneous; oogonia spherical; uniovulate; oospores plerotic with a thick wall.
Typification: *Pythiogeton* Minden

PYTHIOGETON M. von Minden, *Mykologische Untersuchungen und Berichte von Dr Richard Falck* **1**, 241 (1916) [1923].
? *Achlyopsis* É. de Wildeman, *Annales de la Société Belge de Microscopie, Mémoires* **20**, 123 (1896) [**Type species (holotype)**: *Achlyopsis entospora* É. de Wildeman].
? *Plasmoparopsis* É. de Wildeman, *Annales de la Société Belge de Microscopie, Mémoires* **20**, 127 (1896) [**Type species (holotype)**: *Plasmoparopsis rigida* É. de Wildeman].
Type species: *Pythiogeton utriforme* M. von Minden

Pythiogeton utriforme M. von Minden, *Mykologische Untersuchungen und Berichte von Dr Richard Falck* **1**, 242 (1916) [1923] **[type species]**.
Typification: type material not designated; lectotype: *Mykologische Untersuchungen und Berichte von Dr Richard Falck* **1**, plate VI, figs 56-65 (1916) (iconotype). Type locality: Hamburg, Germany (first found at Wrocław [Breslau], Poland). Type habitat: on submerged plant debris.
Literature: Cornu (1872), Czeczuga (1995*b*), Drechsler (1932), Gaertner (1954), Lund (1934), Minden (1916), Newell (1976), Sparrow (1936*b*, 1952*c*, 1960), Thaxter (1895), Voglmayr *et al.* (1999).

Pythiogeton transversum M. von Minden, *Mykologische Untersuchungen und Berichte von Dr Richard Falck* **1**, 242 (1916) [1923].
Typification: type material not designated; lectotype: *Mykologische Untersuchungen und Berichte von Dr Richard Falck* **1**, plate VII, figs 66-72 (1916) (iconotype). Type locality: Hamburg, Germany. Type habitat: on submerged plant debris.
Literature: Drechsler (1932), Minden (1916), Sparrow (1960), Voglmayr *et al.* (1999).

Pythiogeton ramosum M. von Minden, *Mykologische Untersuchungen und Berichte von Dr Richard Falck* **1**, 243 (1916) [1923].
? Pythiogeton sterilis A. Hamid, *Proceedings of the Indian Academy of Science* **15**, 212 (1942).
Typification: type material not designated; lectotype: *Mykologische Untersuchungen und Berichte von Dr Richard Falck* **1**, plate VIII, figs 74-75 (1916) (iconotype). Type locality: Hamburg, Germany. Type habitat: on roots of beetroot (*Beta vulgaris* Tournefort).
Literature: Chaudhuri *et al.* (1947), Hamid (1942), Ito & Nagai (1931), Minden (1916), Shen & Siang (1948), Sparrow (1932, 1960), Voglmayr *et al.* (1999), Watanabe (1974), Wolf (1944).

Pythiogeton autossytum C. Drechsler, *Journal of the Washington Academy of Science* **22**, 448 (1932).
Typification: type material not designated; lectotype: *Journal of the Washington Academy of Science* **22**, 425, fig. 1; 427, fig 2; 429, fig 3; 430, fig. 4; 433, fig. 5 (1932) (iconotype). Type locality: Port Clinton, Ohio, USA. Type habitat: from a community including *Pythium* spp. associated with foot rot of *Typha latifolia* L. (Oct 1931).
Literature: Drechsler (1932), Hsieh & Chang (1976), Sparrow (1960), Voglmayr *et al.* (1999).

Pythiogeton dichotomum Y. Tokunaga, *in* Ito & Tokunaga, *Transactions of the Sapporo Natural History Society (Japan)* **14**, 12 (1935).
Typification: type material not designated; no illustrations. Type locality: Kuroishi, Mutso Province. Type habitat: on decaying plants of rice, *Oryza sativa* L. (collector J. Kimura, May 1930).
Literature: Ito & Tokunaga (1935), Sparrow (1960), Voglmayr *et al.* (1999).

Pythiogeton uniforme A. Lund, *Det Kongelige Danske Videnskabernes Selskab Skrifter, naturvidens kabelig og mathematisk Afdeling, 9 Raekke* **6**, 54 (1934).
Typification: type material not designated; lectotype: *Det Kongelige Danske Videnskabernes Selskab Skrifter, naturvidens kabelig og mathematisk Afdeling, 9 Raekke* **6**, 55, Fig. 28, a-c (1934) (iconotype). Type locality: Jonstup Vang, north east Sealand, Denmark. Type habitat: from sand in a small ditch with running water, pH 7·7, in *Fagus* wood (18 Oct 1931).
Literature; Czeczuga (1994, 1995*a*, *c*), Hsieh & Chang (1976), Lund (1934), Sparrow (1933*b*, 1960), Voglmayr *et al.* (1999), Waterhouse (1942).

Pythiogeton nigrescens A. Batko, *Acta Mycologica, Warszawa* **7**, 241 (1971 [1972]).
Typification: Iconotype: *Acta Mycologica, Warszawa* **7**, 243, figs 1-11; 247, figs 12-23 (1971 [1972]). Type locality: River Vistula near the Miedzeszyn dam, Warsaw, Poland. Type habitat: on submerged detritus (*Sorbus intermedia* Pers. fruit baits) from an old river bed, after *ca* 30 days incubation (Nov 1967).
Literature: Batko (1971*a*), Czeczuga (1994, 1995*a*, *b*, *c*), Czeczuga, Chomutowska & Woronowicz (1990), Voglmayr *et al.* (1999).

Pythiogeton zeae H. J. Jee, H. H. Ho & W. D. Cho, *Mycologia* **92**: 523 (2000).
Typification: NIAST: CPy-8, ex-type culture: ATCC. Type locality: Suwon, Korea. Type habitat: corn (*Zea*) plants with root and basal stalk rot (Jun 1997).
Literature: Jee, Ho & Cho (2000).

SYSTEMATICS

MEDUSOIDES H. Voglmayr, *in* Voglmayr *et al.*, ***Mycological Research*** **103**: 603 (1999).
Typification: *Medusoides argyrocodium* H. Voglmayr [holotype - sole original species]
Etym.: αργυρος silvery, κοδιυμ fleece; from the shiny, silvery appearance in reflected light, due to the fleece-like investment of appendages, and the ability of this investment to trap air.

Medusoides argyrocodium H. Voglmayr, *in* Voglmayr *et al.*, ***Mycological Research*** **103**: 603 (1999).
Typification: USA, Carolina Borealis, Comitatus Macon, Highlands, Lacus Ravenel prope Highlands Biological Station, ad folia submersa diversarum arborum (praecipue *Rhododendron* sp., *Quercus* sp.), conservatur in Herbario WU (WU 17310, holotype).
Literature: Voglmayr *et al.* (1998).

EXCLUDED GENUS

DIASPORANGIUM W. Höhnk, *Beihefte zum Botanischen Zentralblatt (Abteilung A)* **55**, 97 (1936).
Type species: *Diasporangium jonesianum* Höhnk (holotype - sole original species).

Diasporangium jonesianum W. Höhnk, *Beihefte zum Botanischen Zentralblatt (Abteilung A)* **55**, 97 (1936).

OTHER, REJECTED, TAXA COMPARABLE WITH *PYTHIOGETON*

Achlyopsis entospora É. de Wildeman, *Annales de la Société Belge de Microscopie, Mémoires* **20**, 123 (1896).
Typification: type material not designated; lectotype: *Annales de la Société Belge de Microscopie, Mémoires* **20**, plate IX, figs 1-11 (1896) (iconotype). Type locality: Marais de Pinchat, Geneva, Switzerland. Type habitat: inside oogonia of *Chara* (Jul, Aug 1894).
Literature: Voglmayr *et al.* (1999), Wildeman (1896*c*).

Plasmoparopsis rigida É. de Wildeman, *Annales de la Société Belge de Microscopie, Mémoires* **20**, 127 (1896).
Typification: type material not designated; lectotype: *Annales de la Société Belge de Microscopie, Mémoires* **20**, plate X, figs 1-10; plate XI, figs 1-9 (1896) (iconotype). Type locality: Marais de Pinchat, Geneva, Switzerland. Type habitat: oogonia of *Chara* (Jul, Aug 1894).
Literature: Voglmayr *et al.* (1999), Wildeman (1896*c*).

Pythium diacarpum E. J. Butler, *Memoirs of the Department of Agriculture, India, Botanical Series* **1**, 80 (1907).
? *Pythium imperfectum* Cornu, *Annales des Sciences Naturelles, Botanique, Série V* **15**, 13 (1872).
Typification: type material not designated; lectotype: *Memoirs of the Department of Agriculture, India, Botanical Series* **1**, plate II, figs 8-17 (1907) (iconotype). Type locality: Freiburg, Germany. Type habitat: saprotrophic on submerged plant debris.
Literature: Butler (1907), Dick (1990*b*), Drechsler (1932), Plaats-Niterink (1981), Voglmayr *et al.* (1999), Waterhouse (1968*b*).

Pythium gibbosum É. de Wildeman, *Annales de la Société Belge de Microscopie, Mémoires* **20**, 119 (1896).
Typification: type material not designated; lectotype: *Annales de la Société Belge de Microscopie, Mémoires* **20**, plate XII, figs 11-15 (1896) (iconotype). Type locality: Marais de Pinchat, Geneva, Switzerland. Type habitat: inside oogonia of *Chara* (Jul, Aug 1894).
Literature: Dick (1990*b*), Drechsler (1932), Plaats-Niterink (1981), Voglmayr *et al.* (1999), Waterhouse (1968*b*), Wildeman (1896*c*).

RHIPIDIOMYCETIDAE M. W. Dick *ex* M. W. Dick, **subclassis nova.**

Rhipidiomycetidae M. W. Dick *in* M. W. Dick, B. J. Croft, R. C. Magarey, A. W. A. M. de Cock & G. Clark, *Botanical Journal of the Linnean Society* **99**: 110 (1989) [*nomen nudum*] (**Type order**: *Rhipidiales* M. W. Dick).

Leucosporeae F. Cohn, *Hedwigia* **11**: 18 (1872) [order VII, sectio 1 of the class *Thallophytae* (three families: *Chytridiaceae, Peronosporaceae, Saprolegniaceae*) [*pro parte typo excluso*]; sections 2 and 3 comprised the colonial, coenocytic and oogonial green algae and the brown algae respectively].

Fungi cum cyclo nucleari haplomitotico-B. Fungi saprotrophici, facultative aut obligate fermentativi; plerique egentes organico N; saprophyta haud facile culturabilia; in aquis dulcibus vel aquis stagnantibus. **Thallus** eucarpicus, monocentricus; saccatus vel pseudomycelialis, non septatus, cum rhizoideo; pars fibrillaris parietis β-1,3- and β-1,6-glucan; glucosaminum absens; ultrastructura mitotica non reperta. **Reproductio asexualis** per zoosporangia; zoosporogenesis intrasporangialis, plerumque cum vesicula plasmamembranica; zoosporangia terminalia: sporangiophorae absentes; zoosporae formae principalis, heterokontae, plus minusve anisokontae, flagellum anterius curtius; flagellum anterius straminipilosum; zona transitionalis cum helice curta et cum tigillis; K_2-grana non reperta. **Reproductio sexualis** per oogonia; utrum obnoxia exogenis sterolibus antecursoribus non repertum; homothallica vel heterothallica; meioses non repertae; oogamosa ex contactu gametangiali, fecundatio per tubum fecundationis, vel automictica, gametae discretae non formatae; oosporogenesis centripetalis; cum periplasmate persistenti, interdum permanenti; oosporae vulgo apleroticae, saepius singulae, oogonia cum pariete plus minusve crasso; paries oosporae cum exospora plus minusve rugosa, structura ooplasti non reperta, distributio copiae lipidae regulatim per guttulas lipidas minutas; numerus chromosomatis basalis non repertus.
Holotypus: *Rhipidium* M. Cornu (*Rhipidiales* M. W. Dick).

Fungi with a haplomitotic-B nuclear cycle. Fungi saprotrophic, facultatively or obligately fermentative; many with a requirement for organic N; saprophytes not readily culturable; in fresh or stagnant water. **Thallus** eucarpic, monocentric; saccate or pseudomycelial, non septate, with rhizoids; fibrillar component of wall β-1,3- and β-1,6-glucan, glucosamine absent; mitotic ultrastructure not known. **Asexual reproduction** by zoosporangia; zoosporogenesis intrasporangial, sometimes with a plasmamembranic vesicle; zoosporangia terminal; sporangiophores absent; zoospores of the principal form, heterokont, more or less anisokont, anterior flagellum the shorter; anterior flagellum straminipilous; transitional zone with a short helix and with struts; K_2-bodies not known. **Sexual reproduction** by oogonia; not known to be dependent on exogenous sterol precursors; homothallic or heterothallic; pattern of meioses not known; oogamous resulting from gametangial contact, fertilization by means of a fertilization tube, discrete male gametes not formed; oogonia more or less thick-walled, oosporogenesis centripetal, periplasm persistent, sometimes permanent; oospores usually single, aplerotic; oospore wall with a more or less rugose exospore, ooplast structure not known, distribution of oil reserves normally as minute droplets; basal chromosome number not reported.
Typification: *Rhipidium* M. Cornu (*Rhipidiales* M. W. Dick, holotype - sole original order).

RHIPIDIALES M. W. Dick *ex* M. W. Dick, **ordo novum**.

Rhipidiales M. W. Dick *ex* M. W. Dick *in* M. W. Dick, P. T. W. Wong & G. Clark, *Botanical Journal of the Linnean Society* **89**: 193 (1984) (**Type family**: *Rhipidiaceae* F. K. Sparrow).

Leptomiteae K. Lindstedt, *Synopsis der Saprolegniaceen und Beobachtungen über einige Arten*: 35 (1872) [*pro parte typo excluso*] [subfamily of the *Saprolegniaceae*].

Leptomitales B. B. Kanouse, *American Journal of Botany* **14**: 295 (1927) [*pro parte typo excluso*].

Saprolegnina H. F. Copeland, *The Classification of Lower Organisms*: 77 (1956) [*pro parte typo excluso*].

Fungi saprc ophici; in aquis dulcibus vel aquis stagnantibus; facultative aut obligate fermentativi; plerique egentes organico N; saprophyta haud facile culturabilia. **Thallus** eucarpicus, monocentricus; saccatus vel pseudomycelialis, cum rhizoideo. **Reproductio asexualis** per zoosporangia; zoosporogenesis intrasporangialis plerumque cum vesicula plasmamembranica; zoosporangia terminalia; sporangiophorae absentes; zoosporae heterokontae, grandes ((135)600-1200μm^3). **Reproductio sexualis** per oogonia, utrum obnoxia exogenis sterolibus antecursoribus non repertum; paries oogonialis plus minusve crassus, periplasma persistens, interdum permanens; oosporae plerumque singulares, apleroticae; paries oosporae cum exospora plus minusve rugosa, structura ooplasti non reperta, distributio copiae lipidae regulatim per guttulas lipidas minutas, utrum cogantur in guttulas maiores non repertum.
Holotypus: *Rhipidium* M. Cornu (*Rhipidiaceae* F. K. Sparrow *ex* K. Cejp, holotypus).

Fungi, saprotrophic, in fresh or stagnant water; facultatively or obligately fermentative; many with a requirement for organic N; saprophytes not readily culturable. **Thallus** eucarpic, monocentric; saccate or pseudomycelial, with rhizoids. **Asexual reproduction** by zoosporangia; zoosporogenesis intrasporangial sometimes with a plasmamembranic vesicle; zoosporangia terminal; sporangiophores absent; zoospores heterokont, large (volume equivalent (135)600-1200μm^3). **Sexual reproduction** by oogonia, not known to be dependent on exogenous sterol precursors; oogonia more or less thick-walled, oospores usually single, aplerotic, periplasm persistent, sometimes permanent; oospore wall with a more or less rugose exospore, ooplast structure not known, distribution of oil reserves as minute droplets, aggregation into larger droplets not established.
Typification: *Rhipidium* M. Cornu (*Rhipidiaceae* F. K. Sparrow *ex* K. Cejp, holotype - sole original order).

SYSTEMATICS

RHIPIDIACEAE F. K. Sparrow *ex* K. Cejp, *Flora ČSR, B Řada Mykologicko-Lichenologická, Svazek 2, Oomycetes I*: 449 (1959); *Aquatic Phycomycetes*: 572 (1943).
Mindeniellaceae K. Cejp, *Flora ČSR, B Řada Mykologicko-Lichenologická, Svazek 2, Oomycetes I*: 301 & 448 (1959) (**Type genus**: *Mindeniella* B. B. Kanouse - holotype, sole original genus).
Typification: *Rhipidium* M. Cornu.

RHIPIDIUM M. Cornu, *Bulletin de la Société Botanique de France* **18**: 58 (1871).
non Rhipidium sensu J. J. Bernhardi *in* H. A. Schrader, *Journal für die Botanik* **2**: 127 (1800/1).
non Rhipidium W. R. Wallroth, *Flora Cryptogamica Germaniae* **2**: 742 (1833).
? *Hyphomycetarum* P. F. Reinsch, *Contributiones ad Algologiam et Fungologiam. Volume I* : plate 14 (1875) (**Type species**: not designated - *Naegelia* "species II" P. F. Reinsch, *Jahrbücher für wissenschaftliche Botanik* **11**: 298 (1877).
Typification: *Rhipidium interruptum* M. Cornu.

Rhipidium interruptum M. Cornu, *Bulletin de la Société Botanique de France* **18**: 58 (1871).
Rhipidium continuum M. Cornu, *Annales des Sciences Naturelles, Botanique, Série V* **15**: 15 (1872).
Rhipidium europaeum M. von Minden, *Kryptogamenflora der Mark Brandenburg* **5(4)**: 597 (1912).
Rhipidium europaeum M. von Minden, *Kryptogamenflora der Mark Brandenburg* **5(4)**: 597 (1912) var. *interruptum* M. von Minden, *Mykologische Untersuchungen und Berichte* **14**: 342 (1916).
Rhipidium europaeum M. von Minden, *Kryptogamenflora der Mark Brandenburg* **5(4)**: 597 (1912) forma *attenuatum* B. B. Kanouse, *American Journal of Botany* **14**: 342 (1927).
Rhipidium europaeum M. von Minden, *Kryptogamenflora der Mark Brandenburg* **5(4)**: 597 (1912) var. *compactum* E. J. Forbes, *Transactions of the British mycological Society* **19**: 235 (1935).
Rhipidium interruptum M. Cornu, *Bulletin de la Société Botanique de France* **18**: 58 (1871) forma *attenuatum* (B. B. Kanouse) W. C. Coker *In* W. C. Coker & V. D. Matthews, *North American Flora* **2(1)**: 67 (1937).
? *Rhipidium compactum* V. D. Matthews, *Journal of the Elisha Mitchell Scientific Society* **52**: 292 (1936) [*nom. invalid.*, Art 36.1, without Latin diagnosis] [Lectotype (Cejp, 1959) *Transactions of the British Mycological Society* **19**: plate 10, fig. 11 (1935) (iconotype)].
? *Naegelia* "species II" P. F. Reinsch, *Jahrbücher für wissenschaftliche Botanik* **11**: 298 (1877).
Typification: Lectotype (Cejp, 1959) *Mykologische Untersuchungen und Berichte* **2**: plate 2, figs 9-20 (1916) (iconotye). Type locality: France (iconotype Hamburg, Germany). Type habitat: on twigs.
Identification key: Key XIII; key point 4'.
Literature: Behrens (1931), Boedjin (1923), Cejp (1959: 318), Coker & Matthews (1937), Cornu (1871, 1872*b*), Crooks (1937), Dick (1973*b*), Fischer (1892*b*: 376), Forbes (1935*a*, *b*), Höhnk (1935), Indoh (1953), Kanouse (1926, 1927*a*, *b*), Laibach (1927), Lund (1934), Minden (1912, 1916), Petersen (1909, 1910), Pystina (1995), Sparrow (1936*b*, 1960: 895, 898), Valkanov (1931*a*), Waterhouse (1942).

Rhipidium thaxteri M. von Minden, *Kryptogamenflora der Mark Brandenburg* **5(4)**: 600 (1912).
Typification: Iconotype (Cejp, 1959): *Mykologische Untersuchungen und Berichte* **2**: plate 3, figs 22-24 (1916) (iconotype). Type locality: Hamburg, Germany (Walterdorf Island). Type habitat: submerged fruits.
Identification key: Key XIII; key point 3'.
Literature: Behrens (1931), Cejp (1959: 313), Forbes (1935*ab*), Höhnk (1935), Indoh (1953), Kanouse (1927*a*, *b*), Laibach (1927), Lund (1934), Milanez & Trufem (1984); Minden (1912, 1916), Petersen (1909, 1910), Shen & Siang (1948), Sparrow (1936*b*, 1960: 898), Valkanov (1931*a*).

Rhipidium americanum R. Thaxter, *Botanical Gazette* **21**: 327 (1896).
Typification: Lectotype (Cejp, 1959) *Botanical Gazette* **21**: plate 21, figs 1-15 (1896) (iconotype). Type locality: Cambridge, Massachusetts and Kittery Point, Maine, U.S.A. Type habitat: "on various decaying vegetable substances in ponds and ditches".
Identification key: Key XIII; key point 4.
Literature: Apinis (1929), Beverwijk (1948), Cejp (1932*b*, 1959: 315), Coker & Matthews (1937), Crooks (1937), Emerson (1950), Forbes (1935*a*, *b*), Indoh (1953), Kanouse (1927*a*, *b*), Lund (1934), Minden (1912, 1916), Pystina (1995), Sparrow (1932, 1933*b*, 1936*b*, 1960: 899), Sparrow & Barr (1955), Thaxter (1896*b*), Waterhouse (1942).

Rhipidium parthenosporum B. B. Kanouse, *American Journal of Botany* **14**: 344 (1927).
Typification: type material not designated; lectotype *American Journal of Botany* **14**: plate 48, figs 34-37 (1927) (iconotype). Type locality: Ann Arbor, Michigan, U.S.A. Type habitat: on decaying apple, submerged at lake shore.
Identification key: Key XIII; key point 14.
Literature: Cejp (1959: 321), Coker & Matthews (1937), Kanouse (1927*b*), Lund (1934), Sparrow (1936*b*, 1960: 900).

DOUBTFUL TAXON

Rhipidium compactum V. D. Matthews, *Journal of the Elisha Mitchell Scientific Society* **52**: 292 (1936) [*nom. invalid.*, Art 36.1, without Latin diagnosis] [on *Vaccinium* and *Amelanchier* fruits, Mountain Lake, Giles County Virginia, U.S.A.].

EXCLUDED SPECIES

Rhipidium continuum M. Cornu, *Annales des Sciences Naturelles, Botanique, Série V* **15**: 15 (1872), synonym of *Rhipidium interruptum* M. Cornu, *Annales des Sciences Naturelles, Botanique, Série V* **15**: 15 (1872).

Rhipidium elongatum M. Cornu, *Annales des Sciences Naturelles, Botanique, Série V* **15**: 15 (1872), synonym of *Sapromyces elongatus* (M. Cornu) W. C. Coker *in* W. C. Coker & V. D. Matthews, *North American Flora* **2(1)**: 62 (1937).

Rhipidium europaeum M. von Minden, *Kryptogamenflora der Mark Brandenburg* **5(4)**: 597 (1912) var. *interruptum* M. von Minden, *Mykologische Untersuchungen und Berichte* **14**: 342 (1916), synonym of *Rhipidium interruptum* M. Cornu, *Annales des Sciences Naturelles, Botanique, Série V* **15**: 15 (1872).

Rhipidium europaeum M. von Minden, *Kryptogamenflora der Mark Brandenburg* **5(4)**: 597 (1912) forma *attenuatum* B. B. Kanouse, *American Journal of Botany* **14**: 342 (1927), synonym of *Rhipidium interruptum* M. Cornu, *Annales des Sciences Naturelles, Botanique, Série V* **15**: 15 (1872).

Rhipidium europaeum M. von Minden, *Kryptogamenflora der Mark Brandenburg* **5(4)**: 597 (1912) var. *compactum* E. J. Forbes, *Transactions of the British mycological Society* **19**: 235 (1935) synonym of *Rhipidium interruptum* M. Cornu, *Annales des Sciences Naturelles, Botanique, Série V* **15**: 15 (1872).

Rhipidium interruptum M. Cornu, *Annales des Sciences Naturelles, Botanique, Série V* **15**: 15 (1872) forma *attenuatum* B. B. Kanouse, *American Journal of Botany* **14**: 342 (1927) synonym of *Rhipidium interruptum* M. Cornu, *Annales des Sciences Naturelles, Botanique, Série V* **15**: 15 (1872).

Rhipidium interruptum M. Cornu, *Annales des Sciences Naturelles, Botanique, Série V* **15**: 15 (1872) var. *compactum* E. J. Forbes, *Transactions of the British mycological Society* **19**: 235 (1935) synonym of *Rhipidium interruptum* M. Cornu, *Annales des Sciences Naturelles, Botanique, Série V* **15**: 15 (1872).

Rhipidium spinosum M. Cornu, *Annales des Sciences Naturelles, Botanique, Série V* **15**: 15 (1872), synonym of *Araiospora spinosa* (M. Cornu) R. Thaxter, *Botanical Gazette* **21**: 326 (1896).

? *Naegelia* "species II" P. F. Reinsch, *Jahrbücher für Wissenschaftliche Botanik* **11**: 298 (1877), synonym of *Rhipidium interruptum* M. Cornu, *Annales des Sciences Naturelles, Botanique, Série V* **15**: 15 (1872).

SAPROMYCES K. Fritsch, *Österreichische Botanische Zeitschrift* **43**: 420 (1893).
Rhipidium M. Cornu, *Annales des Sciences Naturelles, Botanique, Série V* **15**: 15 (1872) [*pro parte*].
Naegelia P. F. Reinsch, *Jahrbücher für wissenschaftliche Botanik* **11**: 298 (1877).
non Naegelia sensu [G.] L. Rabenhorst, *Deutschlands Kryptogamenflora* **I**: 85 (1844) (syn: *Schinzia penicillata* Nägeli).
non Naegelia sensu J. Lindley, *Edwards Botanical Register* **31**: 40 (1845) (syn: *Cotoneaster denticulata* H.B.K., belonging to the *Rosaceae*).
non Naegelia sensu A. Moritzi, *Systema Verzeichniss der von H. Zollinger in den Jahren 1842-1844 auf Java gesammelten Pflanzen*: 20 (1845-1846) (*Naegelia dubia* Zoll. & Mor., belonging to the *Rhamnaceae*).
non Naegelia sensu E. A. von Regel, *Flora, Jena, 1848*: 249 (1848) (syn: *Smithiantha* O. Kuntze, belonging to the *Gesneriaceae*).
Naegeliella J. Schröter, *Die Natürlichen Pflanzenfamilien* **1(1)**: 103 (1893) [*pro parte typo excluso*].
non Naegeliella sensu C. F. J. E. Correns, *Berichte der Deutschen Botanischen Gesellschaft* **10**: 636 (1892 [1893]) (a genus of the *Chrysophyceae*).
? *Hyphomycetarum* P. F. Reinsch, *Contributiones ad Algologiam et Fungologiam. Volume I* : plate 14 (1875) (**Type species**: not designated - *Naegelia* "species II" P. F. Reinsch, *Jahrbücher für wissenschaftliche Botanik* **11**: 298 (1877).
Typification: *Sapromyces elongatus* (M. Cornu) R. Thaxter.

Sapromyces elongatus (M. Cornu) W. C. Coker *in* W. C. Coker & V. D. Matthews, *North American Flora* **2(1)**: 62 (1937).
Rhipidium elongatum M. Cornu, *Annales des Sciences Naturelles, Botanique, Série V* **15**: 15 (1872).
Naegelia "species I" P. F. Reinsch, *Jahrbücher für wissenschaftliche Botanik* **11**: 298 (1877).
Naegeliella reinschii J. Schröter, *Die Natürlichen Pflanzenfamilien* **1(1)**: 103 (1893).
Sapromyces reinschii (J. Schröter) K. Fritsch, *Österreichische Botanische Zeitschrift* **43**: 421 (1893).
? *Naegelia* "species II" P. F. Reinsch, *Jahrbücher für wissenschaftliche Botanik* **11**: 298 (1877).
Typification: Lectotype (Cejp, 1959) *Mykologische Untersuchungen und Berichte* **2**: plate 7, fig. 73 (1916) (iconotype). Type locality: Sologne, France (iconotype: Lüneburg Heath). Type habitat: on *Viscum* stems and algae (also on cones and twigs of *Pinus* sp.).
Identification key: Key XIII; key point 9.
Literature: Apinis (1929), Bishop (1940), Cejp (1932*c*, 1936, 1959: 306), Coker (1923), Coker & Matthews (1937), Cornu (1872*b*), Crooks (1937), Fischer (1892*b*: 376), Fritsch (1892, 1893), Gleason (1968*b*), Gleason & Stuart (1970), Golueke (1957), Graff (1928), Gwynne-Vaughan & Barnes (1937), Indoh (1953), Kanouse (1927*b*), Kevorkian (1935), Lee & Fuller (1992), Lund (1934), Matthews (1935), Milanez & Trufem (1981), Minden (1912, 1916), Moore (1908-9), Pau & Aronson (1970), Petersen (1910), Pystina (1995), Reinsch (1875, 1878), Richter (1937), Sparrow (1932, 1936*b*, 1960: 884), Tiesenhausen (1912), Thaxter (1894, 1896*b*), Vaziri-Tehrani & Dick (1980*a*, *b*, *c*).

Sapromyces androgynus R. Thaxter, *Botanical Gazette* **21**: 329 (1896).
Typification: Lectotype (Cejp, 1959) *Botanical Gazette* **21**: plate 22, figs 16-19 (1896) (iconotype). Type locality: Cambridge, Massachusetts, U.S.A. Type habitat: "on submerged sticks in ponds and ditches".
Identification key: Key XIII; key point 9'.
Literature: Cejp (1936, 1959: 310), Coker & Matthews (1937), Emerson (1950), Fuller & Jaworski (1987), Gleason & Stuart (1970), Gotelli & Hanson (1987), Kanouse (1927*b*), Kevorkian (1935), Milanez (1968), Milanez & Trufem (1984), Minden (1912), Pystina (1995), Sparrow (1932, 1960: 886), Thaxter (1896*b*), Weston (1938).

Sapromyces indicus M. O. P. Iyengar, K. Ramakrishnan & C. V. Subramanian, *Journal of the Indian Botanical Society* **34**: 144 (1955).
Typification: Type material Herbarium MUBL No. 1219 (coll. M. O. P. Iyengar, 02:03:1952). Type locality: Kambakkam Hills, 100 km north of Madras, India. Type habitat: on unspecified leaves from a stream.
Identification key: Key XIII; key point 14'.
Literature: Iyengar, Ramakrishnan & Subramanian (1955), Sparrow (1960: 886).

EXCLUDED SPECIES

Sapromyces dubius K. Fritsch, *Österreichische Botanische Zeitschrift* **43**: 421 (1893), a name for *Naegelia* ("eine andere verwandte Form") P. F. Reinsch, *Jahrbücher für wissenschaftliche Botanik* **11**: 290, plate XV, figs 7-11 (1877).

? *Hyphomycetarum* P. F. Reinsch, *Contributiones ad Algologiam et Fungologiam. Volume I* : plate 14 (1875), synonym of *Sapromyces* K. Fritsch, *Österreichische Botanische Zeitschrift* **43**: 420 (1893).

Naegeliella reinschii J. Schröter, *Die Natürlichen Pflanzenfamilien* **1(1)**: 103 (1893), synonym of *Sapromyces elongatus* (M. Cornu) W. C. Coker *in* W. C. Coker & V. D. Matthews, *North American Flora* **2(1)**: 62 (1937).

Sapromyces reinschii (J. Schröter) K. Fritsch, *Österreichische Botanische Zeitschrift* **43**: 420 (1893), synonym of *Sapromyces elongatus* (M. Cornu) W. C. Coker *in* W. C. Coker & V. D. Matthews, *North American Flora* **2(1)**: 62 (1937).

? *Naegelia* "species II" P. F. Reinsch, *Jahrbücher für Wissenschaftliche Botanik* **11**: 298 (1877), synonym of *Sapromyces elongatus* (M. Cornu) W. C. Coker *in* W. C. Coker & V. D. Matthews, *North American Flora* **2(1)**: 62 (1937).

ARAIOSPORA (M. Cornu) R. Thaxter, *Botanical Gazette* **21**: 326 (1896).
Rhipidium M. Cornu, *Bulletin de la Société Botanique de France* **18**: 58 (1871) [*pro parte*].
Typification: *Araiospora spinosa* (M. Cornu) R. Thaxter.

Araiospora spinosa (M. Cornu) R. Thaxter, *Botanical Gazette* **21**: 326 (1896).
Rhipidium spinosum M. Cornu, *Bulletin de la Société Botanique de France* **18**: 59 (1871).
Typification: Lectotype (Cejp, 1959) *Mykologische Untersuchungen und Berichte* **2**: plate 1, figs 1-8 (1916) (iconotype). Type locality: France (iconotype: Waltersdorf Island, Hamburg, Germany). Type habitat: on decaying oak and alder twigs and *Nymphaea* rhizomes.
Identification key: Key XIII; key point 12.
Literature: Bertke & Aronson (1985), Cejp (1959: 323), Cornu (1871), Fischer (1892*b*: 377), Gleason (1968*b*), Gleason & Stuart (1970), Kanouse (1927*b*), Minden (1902, 1912, 1916), Pystina (1995), Sparrow (1960: 889), Thaxter (1896*b*).

Araiospora pulchra R. Thaxter, *Botanical Gazette* **21**: 328 (1896).
Typification: Lectotype (Cejp, 1959) *Botanical Gazette* **21**: plate 23, figs 20-25 (1896) (iconotype). Type locality: Cambridge, Massachusetts, U.S.A. Type habitat: on decaying twigs.
Identification key: Key XIII; key point 11'.
Literature: Cejp (1959: 325), Coker & Matthews (1937), Kanouse (1927*b*), King (1903), Minden (1912), Pystina (1995), Sparrow (1932, 1960: 891), Thaxter (1896*b*).

Araiospora coronata D. Linder, *Mycologia* **18**: 176 (1926).
Typification: Type material deposited in the Farlow Herbarium, Boston, Massachussetts, U.S.A., and in the private herbarium of D. Linder (#258) (*Mycologia* **18**: plate 21, figs 1-14 (1926), iconotype). Type locality: Georgetown, Guiana, South America. Type habitat: on submerged fruits of *Spondias lutea* L. (hog plum) (*Anacardiaceae*) (also on bignoniaceous fruits, Bartica, Essequibo River).
Identification key: Key XIII; key point 13'.
Literature: Linder (1926), Sparrow (1960: 893).

Araiospora streptandra A. G. Kevorkian, *Mycologia* **26**: 152 (1934).
var. ***streptandra***.
Typification: Type material deposited in the Farlow Herbarium, Boston, Massachussetts, U.S.A. Type locality: Kingston, Rhode Island, U.S.A. Type habitat: on submerged twigs of *Prunus* sp. and *Salix* sp.
Identification key: Key XIII; key point 12'.
Literature: Coker & Matthews (1937), Kevorkian (1934), Sparrow (1960: 892), Webster (1943).

Araiospora streptandra A. G. Kevorkian, *Mycologia* **26**: 152 (1934) var. ***echinulospora*** L. Shanor & L. S. Olive, *Mycologia* **34**: 540 (1942).
Typification: type material not designated; lectotype *Mycologia* **34**: 538, figs 1-11 (1942) (iconotype). Type locality: Harbison Lake, Highlands, North Carolina, U.S.A. Type habitat: on submerged twigs of *Betula lenta* L. (cherry birch).
Identification key: Key XIII; key point 12'.
Literature: Kevorkian (1934), Shanor & Olive (1942), Sparrow (1960: 892).

MINDENIELLA B. B. Kanouse, *American Journal of Botany* **14**: 301 (1927).
Typification: *Mindeniella spinospora* B. B. Kanouse.

Mindeniella spinospora B. B. Kanouse, *American Journal of Botany* **14**: 301 (1927).
Typification: Lectotype (Cejp, 1959) *American Journal of Botany* **14**: plate XXXIV, figs 17-26 (1927) (iconotype). Type locality: Ann Arbor, Michigan, U.S.A. Type habitat: saprotrophic on submerged fruits of *Crataegus* sp. and apple in stagnant water at a lake shore.
Identification key: Key XIII; key point 7.
Literature: Bertke & Aronson (1985), Cejp (1959: 325), Coker & Matthews (1937), Gleason (1968*b*), Höhnk (1938), Kanouse (1927*a*), Pystina (1995), Sparrow (1960: 855, 901), Sparrow & Barr (1955), Sparrow & Cutter (1941).

Mindeniella asymmetrica T. W. Johnson, *American Journal of Botany* **38**: 74 (1951).
Typification: Type material: preserved specimens deposited in the Herbarium, University of Michigan, Ann Arbor, Michigan, U.S.A. Type locality: Huron River, Ann Arbor, Michigan, U.S.A. Type habitat: saprotrophic on submerged apples.
Identification key: Key XIII; key point 7'.
Literature: Johnson (1951*a*), Sparrow (1960: 903).

AQUALINDERELLA R. Emerson & W. H. Weston, *American Journal of Botany* **54**: 715 (1967).
Typification: *Aqualinderella fermentans* R. Emerson & W. H. Weston.

Aqualinderella fermentans R. Emerson & W. H. Weston, *American Journal of Botany* **54**: 715 (1967).
Typification: Type material: culture CR78. Type locality: swamp ditches and lagoons on the coastal plains, Margarita near Zent, Province of Limón, Costa Rica (also Louisiana, Florida & southern Illinois, U.S.A. and in equatorial Africa - Monrovia, Liberia). Type habitat: on submerged whole fruits of cashew (*Anacardium occidentale* L.) (*Anacardiaceae*) suspended in stagnant water, but also known from many other kinds of fruit in anaerobic waters.
Identification key: Key XIII; key point 2'.
Literature: Emerson & Held (1969), Emerson & Weston (1967), Held (1970), Held & Emerson (1970), Held *et al.* (1969), Stuehling & Ziegler (1958).

NELLYMYCES A. Batko, *Acta Mycologica, Warszawa* **7**: 256 (1971).
Typification: *Nellymyces megaceros* A. Batko.

Nellymyces megaceros A. Batko, *Acta Mycologica, Warszawa* **7**: 256 (1971).
Typification: Iconotype *Acta Mycologica, Warszawa* **7**: 258, fig. 1a (1971), "paratypes: *Acta Mycologica, Warszawa* **7**: 258, figs 1b, c; 2a-e; 3a-f; 4a-i (1971) and preserved material in the Herbarium, Institute of Botany, University of Warsaw, Poland". Type locality: Kampinos National Park, Warsaw, Poland. Type habitat: on submerged twigs and branches of *Alnus glutinosa* Gaertn. (*Betulaceae*).
Identification key: Key XIII; key point 6'.
Literature: Batko (1971*b*, 1977).

SAPROLEGNIOMYCETIDAE M. W. Dick *ex* M. W. Dick, **subclassis nova**
Saprolegniomycetidae M. W. Dick *in* M. W. Dick, P. T. W. Wong & G. Clark, *Botanical Journal of the Linnean Society* **89**: 193 (1984) [no Latin diagnosis] (**Type order**: *Saprolegniales* A. Fischer).
Leucosporeae F. Cohn, *Hedwigia* **11**: 18 (1872) [order VII, sectio 1 of the class *Thallophytae* (three families: *Chytridiaceae, Peronosporaceae, Saprolegniaceae*) *pro parte majore*, sections 2 and 3 comprised the colonial, coenocytic and oogonial green algae and the brown algae respectively].

Fungi cum cyclo nucleari haplomitotico-B. Fungi parasitici in plantis, fungis aut animalibus, vel saprotrophici, plerique egentes organico S; parasiti in angiospermatibus obligati vel facultativi; saprophyta non semper culturabilia; terrestres, in aquis dulcibus vel marinis. **Thallus** eucarpicus vel holocarpicus; mycelialis, plerumque cum hyphis latissimis vel perangustis, coralliformis vel pseudomycelialis (blasticus), raro saccatus, raro septatus, raro cum rhizoideo; fluxio cytoplasmica manifesta in hyphis latis vel pseudomycelia blasticis; pars fibrillaris parietis β-1,3- and β-1,6-glucan, glucosaminum praesens; sterolis synthesis ex mevalonato possibilis; membrana nuclearis persistens usque ad telophasem. **Reproductio asexualis** per sporangia; zoosporogenesis intrasporangialis sine vesicula plasmamembranica aut extrasporangialis cum vesicula precipitativa (?); zoosporangia terminalia, rarius in serie, postea percurrentia aut proliferata internaliter vel sympodialiter; sporangiophorae absentes; zoosporae heterokontae, plus minusve anisokontae, flagellum anterius curtius; flagellum anterius straminipilosum; zona transitionalis cum helice curta et cum tigillis, zoosporae alias dimorphicae, alias polyplaneticae, alias cum dimorphicae tum polyplaneticae; K_2-grana praesentia. **Reproductio sexualis** per oogonia; utrum obnoxia exogenis sterolibus antecursoribus non repertum; homothallica vel heterothallica; meioses contemporales; oogamosa ex contactu gametangiali; fecundatio per tubum fecundationis, raro per gametogenesim intra gametangia et per hypham fecundationis, vel automictica; oogonia cum pariete plus minusve crasso, oosporogenesis centrifugalis, sine periplasmate nucleato, oosporae singulae vel gregatim, oosporae apleroticae vel pleroticae; paries oosporae paene semper laevis, ooplastum granulatum et fluidum, distributio copiae lipidae regulatim per guttulas lipidas minutas aut coactas in unam vel plures guttulas maiores; numerus chromosomatis basalis $x = 3$.
Holotypus: *Saprolegnia* C. G. D. Nees von Esenbeck (*Saprolegniales* A. Fischer).

Fungi with a haplomitotic-B nuclear cycle. Fungi parasitic in plants, fungi or animals, or saprotrophic, many with a requirement for organic S; parasites of angiosperms obligate or facultative; saprophytes not always culturable; terrestrial, freshwater or marine. **Thallus** eucarpic or holocarpic; mycelial, sometimes with exceptionally wide or exceptionally narrow hyphae, coralloid or pseudomycelial (blastic), rarely saccate, rarely septate, rarely with rhizoids; prominent cytoplasmic streaming in wide hyphae or blastic pseudomycelia; fibrillar component of wall β-1,3- and β-1,6-glucan, glucosamine present; sterol synthesis from mevalonate possible; nuclear membrane persistent to telophase. **Asexual reproduction** by sporangia; zoosporogenesis intrasporangial without a plasmamembranic vesicle or extrasporangial in a precipitative (?) vesicle; zoosporangia terminal, less frequently sequential, then percurrent or by internal or sympodial proliferation; sporangiophores absent; zoospores heterokont, more or less anisokont, anterior flagellum the shorter; anterior flagellum straminipilous; transitional zone with a short helix and with struts, zoospores sometimes dimorphic, sometimes polyplanetic, and sometimes dimorphic and polyplanetic; K_2 bodies present. **Sexual reproduction** by oogonia; not known to be dependent on exogenous sterol precursors; homothallic or heterothallic; meioses synchronous; oogamous resulting from gametangial contact, fertilization by means of a fertilization tube, rarely by means of gametogenesis within the gametangia and the production of a fertilization hypha, or automictic; oogonia more or less thick-walled, oosporogenesis centrifugal, periplasm absent or minimal and not persistent; oospores single or numerous, aplerotic or plerotic, oospore wall almost always smooth, ooplast granular and fluid, distribution of oil reserves as minute droplets or aggregated into one or a few large droplets; basal chromosome number $x = 3$.
Typification: *Saprolegnia* C. G. D. Nees von Esenbeck (*Saprolegniales* A. Fischer).
Other orders: *Leptomitales* B. B. Kanouse, *Salilagenidiales* M. W. Dick, *Sclerosporales* M. W. Dick.

SAPROLEGNIALES A. Fischer, *Rabenhorst's Kryptogamen-Flora, 2 Aufl., Bd 1, Abt. 4*: 310 (1892).
Saprolegniineae J. Schröter, *Die Natürlichen Pflanzenfamilien* **1(1)**: 93 (1893) [sub-order].
Saprolegniineae D. H. Campbell, *A University Textbook of Botany*: 153 (1902).
Saprolegnina H. F. Copeland, *The Classification of Lower Organisms*: 77 (1956) [*pro parte*].
Eremospermeae F. T. Kützing, *Phycologia Generalis*: 146 (1843) [*pro parte*].
Oosporeae F. Cohn, *Hedwigia* **11**: 18 (1872) [*pro parte typo excluso*].
Oomycetes J. Schröter, *Die Natürlichen Pflanzenfamilien* **1(1)**: 93 (1893) [*pro parte typo excluso*].
Siphonomycetae C. E. Bessey, *University of Nebraska Studies* **7**: 286 (1907).
Typification: *Saprolegnia* C. G. D. Nees von Esenbeck (*Saprolegniaceae* N. Pringsheim, holotype - sole original family). Other family: *Leptolegniaceae* M. W. Dick.

SAPROLEGNIACEAE N. Pringsheim, *Jahrbuch für wissenschaftlichen Botanik* **1**: 284 (1858).
Saprolegnieae F. T. Kützing, *Phycologia Generalis*: 157 (1843).
Saprolegniea H. F. Copeland, *The Classification of Lower Organisms*: 77 (1956).
Typification: *Saprolegnia* C. G. D. Nees von Esenbeck.

NOTE: the species of *Saprolegnia*, *Achlya*, *Dictyuchus*, *Pythiopsis*, *Aplanes*, *Thraustotheca*, *Isoachlya*, *Protoachlya*, *Geolegnia*, *Brevilegnia*, *Calyptralegnia*, *Aplanopsis*, *Scoliolegnia* and *Couchia* are not listed, see PART III for explanation; lists are given in APPENDIX 1.

SAPROLEGNIA C. G. D. Nees von Esenbeck *in* Carus, *Nova Acta Physico-Medica Academiae Caesareae Leopoldino-carolinae Naturae Curiosorum* **11**: 513 (1823) ***sensu lato***.
Typification: *Saprolegnia ferax* (F. von P. Gruithuisen) F. T. Kützing.

ACHLYA C. G. D. Nees von Esenbeck *in* Carus, *Nova Acta Physico-Medica Academiae Caesareae Leopoldino-carolinae Naturae Curiosorum* **11**: 514 (1823).
Typification: *Achlya prolifera* C. G. D. Nees von Esenbeck.

DICTYUCHUS H. Leitgeb, *Botanische Zeitung, Berlin, 26 Jahrgang* **31**: 503 (1868).
Typification: *Dictyuchus monosporus* H. Leitgeb.

PYTHIOPSIS A. de Bary, *Botanische Zeitung, Berlin, 46 Jahrgang* **38**: 609 (1888).
Typification: *Pythiopsis cymosa* A. de Bary.

APLANES A. de Bary, *Botanische Zeitung, Berlin, 46 Jahrgang* **39**: 613 (1888).
Typification: *Aplanes braunii* (P. F. Reinsch) A. de Bary (≡ *Aplanes androgynus* (W. Archer) J. E. Humphrey).

THRAUSTOTHECA J. E. Humphrey, *Transactions of the American Philosophical Society, New Series* **17**: 131 (1893).
Typification: *Thraustotheca clavata* (A. de Bary *ex* M. Büsgen) J. E. Humphrey.

ISOACHLYA C. H. Kauffman, *American Journal of Botany* **8**: 231 (1921).
Typification: *Isoachlya toruloides* C. H. Kauffman & W. C. Coker.

PROTOACHLYA W. C. Coker, *Saprolegniaceae*: 90 (1923).
Typification: *Protoachlya paradoxa* W. C. Coker.

SOMMERSTORFFIA N. Arnaudow, *Flora, Jena* **116**: 109 (1923).
Typification: *Sommerstorffia spinosa* N. Arnaudow (holotype - sole original species).

Sommerstorffia spinosa N. Arnaudow, *Flora, Jena* **116**: 109 (1923).
Typification: type material not designated; lectotype *Flora, Jena* **116**: 109-113, figs 1-5 (1923) (iconotype). Type locality: Sofia, Bulgaria. Type habitat: freshwater sample, epiphytic on *Cladophora* in July (*Chlorophyceae*, *Cladophorales*).
Identification key: Key I; key point 21'.
Literature: Arnaudow (1923*a*, *b*), Czeczuga (1993), Czeczuga & Próba (1980), Czeczuga, Próba & Brzozowska (1984), Karling (1952), Miller (1965), Prowse (1954), Pystina (1995), Saikawa & Hoshino (1986), Sparrow (1929), Valkanov (1931*b*, 1932).

GEOLEGNIA W. C. Coker & J. V. Harvey *in* Harvey, *Journal of the Elisha Mitchell Scientific Society* **41**: 154 (1925).
Typification: *Geolegnia inflata* W. C. Coker & J. V. Harvey.

BREVILEGNIA W. C. Coker & J. N. Couch *in* Coker, *Journal of the Elisha Mitchell Scientific Society* **42**: 212 (1927).
Typification: *Brevilegnia subclavata* J. N. Couch **[type species]**

CALYPTRALEGNIA W. C. Coker, *Journal of the Elisha Mitchell Scientific Society* **42**: 219 (1927).
Typification: *Calyptralegnia achlyoides* (W. C. Coker & J. N. Couch) W. C. Coker **[holotype]**.

SYSTEMATICS

HYDATINOPHAGUS A. Valkanov [A. V'lkanov'], *Godishnik na Sofiiskiya Universitet (Jahrbuch der Sofianer Universität)* **27**: 228 (1931).
Typification: *Hydatinophagus apsteinii* A. Valkanov (holotype - sole original species).

Hydatinophagus apsteinii A. Valkanov [A. V'lkanov'], *Godishnik na Sofiiskiya Universitet (Jahrbuch der Sofiana Universität)* **27**: 231 (1931).
Aphanomyces hydatinae A. Valkanov, *Archiv für Protistenkunde* **74**: 17 (1931).
Typification: type material not designated; lectotype *Godishnik na Sofiiskiya Universitet (Jahrbuch der Sofiana Universität)* **27**: figs 1-11 (1931) (iconotype). Type locality: Sofia, Bulgaria. Type habitat: parasitic in *Epiphanes senta* (Müller) [as its synonym, *Hydatina senta* Ehrenb.] (*Aschelminthes, Rotifera*).
Identification key: Key I; key point 23.
Literature: Hudson & Gosse (1889), Konsuloff (1908), Scott (1961), Valkanov (1931*b*, *c*, 1932).

Hydatinophagus americanus A. F. Bartsch & F. T. Wolf, *American Journal of Botany* **25**: 394 (1938).
Aphanomyces americanus (A. F. Bartsch & F. T. Wolf) W. W. Scott, *Virginia Agricultural Experiment station, Technical Bulletin* **151**: 36 (1961).
Typification: type material not designated; lectotype *American Journal of Botany* **25**: 393, figs 10-11 (1938) (iconotype). Type locality: Lake Wingra, Madison, Wisconsin, U.S.A. Type habitat: parasitic in *Monostyla* sp. (*Aschelminthes, Rotifera*).
Identification key: Key I; key point 23'.
Literature: Bartsch & Wolf (1938), Scott (1961).

APLANOPSIS W. Höhnk, *Veröffentlichungen des Instituts für Meeresforschung in Bremerhaven* **1**: 127 (1952).
Typification: *Aplanopsis terrestris* W. Höhnk (holotype - sole original species).

SCOLIOLEGNIA M. W. Dick, *Botanical Journal of the Linnean Society* **62**: 257 (1969).
Typification: *Scoliolegnia asterophora* (A. de Bary) M. W. Dick.

COUCHIA W. W. Martin, *Mycologia* **73**: 1143 (1981).
Typification: *Couchia circumplexa* W. W. Martin (holotype - sole original species).

DOUBTFUL OR REJECTED GENERA

JARAIA B. Němec, *Rospavy České Akademie Císaře Františka Josefa Pro Vedy, Třída II* **21** Čislo 8: 1 (1912).
Typification: *Jaraia salicis* B. Němec (holotype - sole original species).

HAMIDIA M. Chaudhuri, *Proceedings of the Indian Academy of Sciences* **15**: 227 (1942).
Typification: *Hamidia indica* M. Chaudhuri (holotype - sole original species).

CLADOLEGNIA H. Johannes, *Feddes Repertorium* **58**: 211 (1955).
Typification: *Cladolegnia intermedia* (W. C. Coker) H. Johannes.

LEPTOLEGNIACEAE M. W. Dick, *in* Dick *et al.*, *Mycological Research* **103**: 1124 (1999).
Saprolegniaceae N. Pringsheim, *Jahrbuch für wissenschaftlichen Botanik* **1**: 284 (1858) [*pro parte typo excluso*].
Typification: *Leptolegnia* de Bary.
Other genera: *Aphanomyces* de Bary, *Plectospira* Drechsler.

NOTE: the species of *Leptolegnia*, *Aphanomyces* and *Plectospira* are not listed, see PART III for explanation; lists are given in APPENDIX 1.

LEPTOLEGNIA A. de Bary, *Botanische Zeitung, Berlin, 46 Jahrgang* **38**: 609 (1888).
Typification: *Leptolegnia caudata* A. de Bary (holotype - sole original species).

APHANOMYCES A. de Bary, *Jahrbücher für Wissenschaftliche Botanik* **2**: 178 (1860).
Typification: *Aphanomyces stellatus* A. de Bary **[typus *non designatus*!! species]**.
Aphanomyces A. de Bary, subgenus *Aphanomyces* (**Type species**: *Aphanomyces stellatus* A. de Bary)
[Subgenera in Scott (1961) were not typified.]

PLECTOSPIRA C. Drechsler, *Journal of Agricultural Research* **34**: 294 (1927).
Typification: *Plectospira myriandra* C. Drechsler **[holotype - sole original species]**.

SCLEROSPORALES M. W. Dick *in* Dick, Wong & Clark, *Botanical Journal of the Linnean Society* **89**: 173 (1984).

NOTE: the species of *Sclerospora*, *Peronosclerospora*, *Sclerophthora*, *Pachymetra* and *Verrucalvus* are not listed, see PART III for explanation; lists are given in APPENDIX 1.

SCLEROSPORACEAE M. W. Dick *in* Dick, Wong & Clark, *Botanical Journal of the Linnean Society* **89**: 174 (1984).

SCLEROSPORA J. Schröter, *Hedwigia* **18**: 86 (1879).
Typification: *Sclerospora graminicola* (P. A. Saccardo) J. Schröter.

PERONOSCLEROSPORA (S. Ito) K. Hara *in* Shirai & Hara, *A List of Japanes Fungi Hitherto Known*: (1927).
Typification: *Peronoscleropora sacchari* (T. Miyake) K. Hara.

VERRUCALVACEAE M. W. Dick *in* Dick, Wong & Clark, *Botanical Journal of the Linnean Society* **89**: 174 (1984)

VERRUCALVUS P. Wong & M. W. Dick *in* Dick, Wong & Clark, *Botanical Journal of the Linnean Society* **89**: 174 (1984).
Typification: *Verrucalvus flavofaciens* P. Wong & M. W. Dick (holotype - sole original species).

SCLEROPHTHORA M. J. Thirumalachar, C. G. Shaw & M. J. S. Narasimhan, *Bulletin of the Torrey Botanical Club* **80**: 304 (1953).
Typification: *Sclerophthora macrospora* (P. A. Saccardo) M. J. Thirumalachar, C. G. Shaw & M. J. S. Narasimhan.

PACHYMETRA B. J. Croft & M. W. Dick *in* Dick, Croft, Magarey, de Cock & Clark, *Botanical Journal of the Linnean Society* **99**: 100 (1989).
Typification: *Pachymetra chaunorhiza* B. J. Croft & M. W. Dick **[holotype - sole original species]**.

REJECTED GENUS

DICKSONOMYCES M.J. Thirumalachar, P.N. Rao & M.A. Salam *in* Rao, Salam & Thirumalachar, *Mycologia* **48**: 863 (1956).
Typification: *Dicksonomyces sorghi* M.J. Thirumalachar, P.N. Rao & M.A. Salam (holotype - sole original species).

SALILAGENIDIALES M. W. Dick, **ordo novum**
Lagenidiales J. S. Karling, *American Journal of Botany* **26**: 518 (1939) [*pro parte typo excluso*].
Eurychasmales F. K. Sparrow *in* E. B. G. Jones, *Recent Advances on Aquatic Mycology*: 220 (1976) [*pro parte minore, nomen nudum*].

Fungi parasitici in Crustaceis et Molluscis marinis, aut raro saprobi marini; plerumque culturabilis. **Thallus** holocarpicus vel eucarpicus, coralliformis vel pseudomycelialis, raro saccatus, interdum septatus, interdum cum rhizoideis. **Reproductio asexualis** per segmenta thalloidea sporangialia; zoosporogenesis intrasporangialis sine vesicula plasmamembranica, vel extrasporangialis cum vesicula precipitativa (?); zoosporae heterokontae et anisokontae, grandiores ((20)150-900μm^3). **Reproductio sexualis** per oogonia vel absens, plerumque partim obnoxia exogenis sterolibus antecursoribus; oogonia cum pariete plus minusve crasso, oosporae plerumque singulae, apleroticae, periplasma absens vel minime praesens et non permanens; paries oosporae paene semper laevis, structura ooplasti ignota, distributio copiae lipidae per guttulas lipidas minutas aut coactas in unam vel plures guttulas maiores.
Holotypus: *Salilagenidium* M. W. Dick (*Salilagenidiaceae* M. W. Dick).

Fungi parasitic in marine Crustacea or Mollusca, or rarely marine saprobes; sometimes culturable. **Thallus** holocarpic or eucarpic, coralloid or pseudomycelial, rarely saccate, sometimes septate, sometimes with rhizoids. **Asexual reproduction** by sporangial thalloid segments; zoosporogenesis intrasporangial without a plasmamembranic vesicle or extrasporangial in a precipitative (?) vesicle; zoospores heterokont and isokont, medium-large (volume equivalent (20)150-900μm^3). **Sexual reproduction** by oogonia or absent, sometimes partially dependent on exogenous sterol precursors; oogonia more or less thick-walled, oospores usually single, aplerotic, periplasm absent or minimal and not persistent; oospore wall almost always smooth, ooplast details lacking, distribution of oil reserves as minute droplets or aggregated into one or a few large droplets.
Typification: *Salilagenidium* M. W. Dick (*Salilagenidiaceae* M. W. Dick).
Other family: *Haliphthoraceae* H. S. Vishniac.

SALILAGENIDIACEAE M. W. Dick, **familia nova**

Fungi parasitici in Crustaceis et Molluscis marinis; plerumque culturabiles. **Thallus** coralliformis vel pseudomycelialis, plerumque septatus, raro cum rhizoideo. **Reproductio asexualis** per segmenta thalloidea sporangialia; zoosporogenesis extrasporangialis in vesicula precipitativa (?); zoosporae heterokontae et isokontae, grandiores (300-900μm^3). **Reproductio sexualis** per oogonia et antheridia, raro absens, interdum partim obnoxia exogenis sterolibus antecursoribus; oogonia cum pariete plus minusve crasso, oosporae plerumque singulae, apleroticae, periplasma absens vel minime praesens et non permanens; paries oosporae paene semper laevis, structura ooplasti non reperta, distributio copiae lipidae per guttulas lipidas minutas aut coactas in unam vel plures guttulas maiores.
Holotypus: *Salilagenidium* M. W. Dick.

Fungi parasitic in marine Crustacea and Mollusca; sometimes culturable. **Thallus** coralloid or pseudomycelial, sometimes septate, rarely with rhizoids. **Asexual reproduction** by sporangial thalloid segments; zoosporogenesis extrasporangial in a precipitative (?) vesicle; zoospores heterokont and isokont, medium-large (volume equivalent 300-900μm^3). **Sexual reproduction** by oogonia and antheridia, rarely absent, sometimes partially dependent on exogenous sterol precursors; oogonia more or less thick-walled, oospores usually single, aplerotic, periplasm absent or minimal and not persistent; oospore wall almost always smooth, ooplast details lacking, distribution of oil reserves as minute droplets or aggregated into one or a few large droplets.
Typification: *Salilagenidium* M. W. Dick.

SALILAGENIDIUM M. W. Dick, **genus novum**
ETYMOLOGY: *Lagenidium*-like fungi from saline habitats.

Fungi parasitici in Crustaceis et Molluscis marinis; plerumque culturabiles. Thallus coralliformis vel pseudomycelialis, plerumque septatus. **Reproductio asexualis** per segmenta thalloidea sporangialia; zoosporogenesis extrasporangialis in vesicula precipitativa (?); zoosporae heterokontae et isokontae. **Reproductio sexualis** per oogonia; plerumque partim obnoxia exogenis sterolibus antecursoribus; oogonia cum pariete plus minusve crasso, oosporae plerumque singulae, apleroticae, periplasma absens vel minime praesens et non permanens; paries oosporae paene semper laevis, structura ooplasti non reperta, distributio copiae lipidae per guttulas lipidas minutas aut coactas in unam vel plures guttulas maiores.
Holotypus: *Salilagenidium callinectes* (J. N. Couch) M. W. Dick.

Fungi parasitic in marine Crustacea and Mollusca; sometimes culturable. **Thallus** coralloid or pseudomycelial, sometimes septate. **Asexual reproduction** by sporangial thalloid segments; zoosporogenesis extrasporangial in a precipitative (?) vesicle; zoospores heterokont and isokont. **Sexual reproduction** by oogonia; sometimes partially dependent on exogenous sterol precursors; oogonia more or less thick-walled, oospores usually single, aplerotic, periplasm absent or minimal and not persistent; oospore wall almost always smooth, ooplast details lacking, distribution of oil reserves as minute droplets or aggregated into one or a few large droplets.
Typification: *Salilagenidium callinectes* (J. N. Couch) M. W. Dick.

Salilagenidium callinectes (J. N. Couch) M. W. Dick, **comb. nov.**
Lagenidium callinectes J. N. Couch, *Journal of the Elisha Mitchell Scientific Society* **58**: 158 (1942).
Typification: A neotype has been designated by Crisp *et al.* (1989) as the preserved specimen of isolate L1 of Bland & Amerson (1973*a*) and deposited at BPI. A neotype culture has been designated by Crisp *et al.* (1989) as ATCC 24973. Type locality: East Lynnhaven, Virginia, U.S.A. Type habitat: parasitic in eggs and newly hatched young of the blue crab *Callinectes sapidus* Rathbun (*Crustacea*).
Identification key: Key XI; key point 14.
Literature: Alderman (1986), Alderman & Polglase (1986), Amerson & Bland (1973), Armstrong, Buchanan & Caldewell (1976), Bahnweg & Bland (1980), Bahnweg & Gotelli (1980), Berg & Patterson (1983, 1986), Bertke & Aronson (1992*a*), Bian *et al.* (1979), Bland & Amerson (1973*a*, *b*), Cassels & Krebs (1983), Chien & Lin (1985), Colorni (1985), Cook (1971), Couch (1942), Crisp & Bland (1980, 1989, 1990), Dahle (1982), Domnas, Biswas & Gallagher (1994), Domnas & Warner (1991), Fisher, Nilson & Shleser (1983), Fuller (1996), Fuller, Fowles & McLaughlin (1964), Ganaros (1957), Gil-Turnes & Fenical (1992), Gil-Turnes, Hay & Fenical (1989), Gotelli (1974*a*, *b*), Hatai (1986), Johnson & Bonner (1960), Johnson & Sparrow (1961: 345), Karling (1981*a*: 135), Lightner (1981), Lightner & Fontaine (1973), Lio-Po, Lavilla & Trillo-Llobrera (1978), Lio-Po *et al.* (1982), Nakamura & Hatai (1995), Nes (1987), Nes *et al.* (1986), Newell *et al.* (1977), Nilson, Fisher & Shleser (1976), Polglase, Alderman & Richards (1986), Porter (1986), Redfield & Vincent (1979), Rogers-Talbert, (1948), Ruch & Bland (1979), Sandoz, Rogers & Newcombe (1944), Shields & Kuris (1987), Sindermann (1977), Sparrow (1960: 1005), Warner, Sovocool & Domnas (1982, 1983*a*, *b*), Villegas & Ti (1978), Warner *et al.* (1981).

Salilagenidium chthamalophilum (T. W. Johnson) M. W. Dick, **comb. nov.**
Lagenidium chthamalophilum T. W. Johnson, *Biological Bulletin of the Marine Biological Laboratory, Woods Hole* **114**: 205 (1958).
Typification: type material not designated; lectotype: *Biological Bulletin of the Marine Biological Laboratory, Woods Hole* **114**: 206, figs 1-9; 208, figs 10-27 (1958) (iconotype). Type locality: Beaufort Inlet, North Carolina, U.S.A. Type habitat: parasitic in "*Chthamalus fragilis denticulata*" (*Chthamalus fragilis* Darwin, *Crustacea*).
Identification key: Key X; key point 14'.
Literature: Alderman & Polglase (1986), Bertke & Aronson (1992*a*), Fuller, Fowles & McLaughlin (1964), Johnson (1958, 1960), Johnson & Sparrow (1961: 344), Karling (1981*a*: 136), Newell *et al.* (1977).

Salilagenidium myophilum (K. Hatai & O.-a. Lawhavinit) M. W. Dick, **comb. nov.**
Lagenidium myophilum K. Hatai & O.-a. Lawhavinit, *Transactions of the Mycological Society of Japan* **29**: 177 (1988).
Typification: Holotype NJM 8601 in the herbarium, Department of Fish Diseases, Nippon Veterinary and Zootechnical College, Tokyo. A culture is maintained in the same department. Type locality: Ishikawa Prefecture, Japan. Type habitat: from the abdominal muscles and swimmerets of the adult northern shrimp *Pandalus borealis* Krøyer (*Crustacea*) in culture at the Japan Seafarming Association.
Identification key: Key X; key point 12.
Literature: Hatai & Lawhavinit (1988), Nakamura *et al.* (1994).

Salilagenidium scyllae (B. Z. Bian, K. Hatai, G. L. Po & S. Egusa) M. W. Dick, **comb. nov.**
Lagenidium scyllae B. Z. Bian, K. Hatai, G. L. Po & S. Egusa, *Transactions of the Mycological Society of Japan* **20**: 117 (1979).
Typification: Holotype SANK 15078 deposited in the herbarium of the Central Research Laboratories, Sankyo Co. Ltd, Tokyo, Japan. A subculture is maintained in the same laboratory. Type locality: Tigbauan, Iloito, Philippines. Type habitat: parasitic in eggs and larvae of the mangrove crab *Scylla serrata* (Forsskål) (*Crustacea*).
Identification key: Key X; key point 11'.
Literature: Alderman & Polglase (1986), Bian *et al.* (1979), Karling (1981*a*: 246).

Salilagenidium marinum (D. Atkins) M. W. Dick, **comb. nov.**
Leptolegnia marina D. Atkins, *Journal of the Marine Biological Association of the United Kingdom* **33**: 622 (1954).
Leptolegniella marina (D. Atkins) M. W. Dick, *Transactions of the British Mycological Society* **57**: 422 (1971).
Typification: Type material has been deposited in the British Museum (Natural History). Type locality: Camel Estuary, St Issey Cliff, Padstow, Cornwall, U.K. Type habitat: parasitic in eggs of the crab *Mytilus edulis* L. (*Crustacea*).
Identification key: Key X; key point 13'.
Literature: Atkins (1929, 1954*a*), Dick (1971*b*), Johnson & Pinschmidt (1963).

Salilagenidium thermophilum (K. Nakamura, M. Nakamura, K. Hatai & Zafran) M. W. Dick, **comb. nov.**
Lagenidium thermophilum K. Nakamura, M. Nakamura, K. Hatai & Zafran, *Mycoscience* **36**: 400 (1995).
Typification: Holotype NJM 9338, colony exsiccata, and subcultures. Type locality: Bali, Indonesia. Type habitat: from eggs of the mangrove crab *Scylla serrata* (Forsskål) (*Crustacea*).
Identification key: Key X; key point 11'.
Literature: Nakamura *et al.* (1995).

UNIDENTIFIED SPECIES REFERABLE TO *SALILAGENIDIUM*

Lagenidium sp. H. L. Cook, *FAO Aquaculture Bulletin* 3: 13 (1971), parasitic on the brown shrimp *Penaeus aztecus* Ives from Freeport, Texas, U.S.A.

Lagenidium sp. D. V. Lightner & C. T. Fontaine, *Journal of Invertebrate Pathology* 22: 94 (1973), parasitic on the white shrimp *Penaeus setiferus* (L.), U.S.A., pure cultures lodged with Dr A. J. Domnas.

Lagenidium sp. G. D. Lio-Po & M. E. G. Sanvictores, *Aquaculture* 51: 161 (1986), parasitic on a shrimp *Penaeus* sp., Philippines.

DOUBTFUL SPECIES

'***Hyphochytrium peniliae***' N. Ya. Artemchuk & L. M. Zelezinskaya, *Mikologiya i Fitopatologiya* 3: 357 (1969) [*nom. inval*. Art. 36.1, without Latin diagnosis].

Typification: Type material preserved in the Institute of Oceanography, Academy of Sciences, Moscow, Russia, illustrations inadequate for lectotypification. Type locality: north west Black Sea. Type habitat; parasitic in planktonic crayfish, *Penilia avirostris* (Dana) (*Crustacea*).

Identification key: Key X; key point 12'.

Literature: Alderman & Polglase (1986), Artemchuk (1981: 55), Artemchuk & Zelezinskaya (1969), Karling (1977: 398), Porter (1986).

HALIPHTHORACEAE H. S. Vishniac, *Mycologia* **50**: 75 (1958).
Atkinsiellaceae F. K. Sparrow *in* E. B. G. Jones, *Recent Advances on Aquatic Mycology*: 215, 220 (1976) [*nomen nudum*].
Typification: *Haliphthoros* H. S. Vishniac.

HALIPHTHOROS H. S. Vishniac, *Mycologia* **50**: 75 (1958).
Typification: *Haliphthoros milfordensis* H. S. Vishniac (holotype - sole original species).

Haliphthoros milfordensis H. S. Vishniac, *Mycologia* **50**: 75 (1958).
Typification: The description and illustrations in Vishniac (1958) were designated as constituting the type material (iconotype). Type locality: Milford, Connecticut, U.S.A. Type habitat: parasitic in the egg cases of the oyster drill *Urosalpinx cinerea* (Say) (*Mollusca*).
Identification key: Key X; key point 9.
Literature: Abrahams & Brown (1977), Alderman & Polglase (1986), Bahnweg (1980), Bahnweg & Bland (1980), Crisp & Bland (19891990), Fischer, Nilson & Schleser (1975), Hatai (1982), Hatai, Rhoobunjongde & Wada (1992), Izumikawa *et al.* (1999), Kaji *et al.* (1991), Nakamura & Hatai (1995), Nes (1987), Porter (1986), Tharp & Bland (1977), Vishniac (1958).

Haliphthoros philippinensis K. Hatai, B. Z. Bian, M. C. L. Baticados & S. Egusa, *Transactions of the Mycological Society of Japan* **21**: 48 (1980).
Typification: Holotype SANK 15178 in the Herbarium, Central Research Laboratories, Sankyo Co. Ltd., Tokyo. Culture IMI 241639, International Mycological Institute, Kew, Surrey, U. K. Type locality: Tigbauan, Iloilo, Philippines. Type habitat: parasitic in larvae of the tiger prawn *Penaeus monodon* Fabricius (*Crustacea*).
Identification key: Key X; key point 9'.
Literature: Alderman & Polglase (1986), Hatai *et al.* (1980), Lio-Po & Sanvictores (1986).

Haliphthoros hiroshimaensis K. Hatai *et al.*, *nomen nudum* (Marine Mycology Symposium, Portsmouth, 1995).

DOUBTFUL SPECIES REFERABLE TO *HALIPHTHOROS*

Haliphthoros zoophthorum (H. S. Vishniac) M. W. Dick, **comb. nov.**
Sirolpidium zoophthorum H. S. Vishniac, *Mycologia* **47**: 641 (1955).
Typification: type material not designated; lectotype *Mycologia* **47**: 634, figs 1-4, 639, figs 5-8 (1955). Type locality: Milford, Connecticut, U.S.A. Type habitat: parasitic in larvae of *Venus* sp. (also *Crassostrea* sp.) clams (*Mollusca*).
Identification key: Key X; key point 8'.
Literature: Alderman & Polglase (1986), Davis *et al.* (1954), Davis & Loosanoff (1954), Ganaros (1957), Johnson & Sparrow (1961: 349), Karling (1981*a*: 58), Sparrow (1960: 968), Vishniac (1955*d*).

ATKINSIELLA H. S. Vishniac, *Mycologia* **50**; 75 (1958).
Typification: *Atkinsiella dubia* (D. Atkins) H. S. Vishniac (holotype - sole original species).

Atkinsiella dubia (D. Atkins) H. S. Vishniac, *Mycologia* **50**: 75 (1958).
Plectospira dubia D. Atkins, *Journal of the Marine Biological Association of the United Kingdom* **33**: 731 (1954*b*).
Typification: "Deposited in the British Museum (Natural History)". Type locality: Plymouth, Devon, U.K. Type habitat: parasitic in the eggs of the pea crab *Pinnotheres pisum* (L.) (*Crustacea*).
Identification key: Key X; key point 2.
Literature: Alderman (1976), Alderman & Polglase (1986), Aronson & Fuller (1969), Atkins (1954*b*), Fuller, Fowles & McLaughlin (1964), Nes (1987), Schatz (1984), Sparrow (1973*a*), Sparrow & Gotelli (1969), Vishniac (1958).

EXCLUDED SPECIES

Atkinsiella hamanaensis B. Z. Bian & S. Egusa, *Journal of Fish Diseases* **3**: 379 (1980), synonym of *Halodaphnea hamanaensis* (B. Z. Bian & S. Egusa) M. W. Dick, *loc. cit.* (≡ *Halocrusticida hamanaensis* (B. Z. Bian & S. Egusa) K. Nakamura & K. Hatai, *Mycoscience* **36**: 437 (1995).

Atkinsiella awabi N. Kitancharoen, K. Nakamura, S. Wada & K. Hatai, *Mycoscience* **35**: 267 (1994), synonym of *Halodaphnea awabi* (N. Kitancharoen, K. Nakamura, S. Wada & K. Hatai) M. W. Dick, *loc. cit.* (≡ *Halocrusticida awabi* (N. Kitancharoen, K. Nakamura, S. Wada & K. Hatai) K. Nakamura & K. Hatai, *Mycoscience* **36**: 437 (1995).

Atkinsiella parasitica K. Nakamura & K. Hatai, *Mycoscience* **35**: 387 (1994), synonym of *Halodaphnea parasitica* (K. Nakamura & K. Hatai) M. W. Dick, *loc. cit.* (≡ *Halocrusticida parasitica* (K. Nakamura & K. Hatai) K. Nakamura & K. Hatai, *Mycoscience* **36**: 437 (1995).

Atkinsiella okinawaensis K. Nakamura & K. Hatai, *Mycoscience* **36**: 89 (1995), synonym of *Halodaphnea okinawaensis* (K. Nakamura & K. Hatai) M. W. Dick, *loc. cit.* (≡ *Halocrusticida okinawaensis* (K. Nakamura & K. Hatai) K. Nakamura & K. Hatai, *Mycoscience* **36**: 437 (1995).

Atkinsiella panulirata N. Kitancharoen & K. Hatai, *Mycoscience* **36**: 100 (1995), synonym of *Halodaphnea panulirata* (N. Kitancharoen & K. Hatai) M. W. Dick, *loc. cit.* (≡ *Halocrusticida panulirata* (N. Kitancharoen & K. Hatai) K. Nakamura & K. Hatai, *Mycoscience* **36**: 437 (1995).

REJECTED SPECIES

Atkinsiella entomophaga W. W. Martin, *American Journal of Botany* **64**: 767 (1977), synonym of *Crypticola entomophaga* (W. W. Martin) M. W. Dick *loc. cit.*

HALODAPHNEA M. W. Dick, *Mycological Research* **102**: 1065 (1998).
Atkinsiella H. S. Vishniac, *Mycologia* **50**: 75 (1958) [*pro parte typo excluso*] (**Type species** *Atkinsiella dubia* (D. Atkins) H. S. Vishniac).
Halocrusticida K. Nakamura & K. Hatai, *Mycoscience* **36**: 437 (1995) [*pro parte typo excluso*] (**Type species** *Halocrusticida entomophaga* (W. W. Martin) K. Nakamura & K. Hatai).
Typification: *Halodaphnea parasitica* (K. Nakamura & K. Hatai) M. W. Dick.

Halodaphnea parasitica (K. Nakamura & K. Hatai) M. W. Dick, *Mycological Research* **102**: 1065 (1998).
Atkinsiella parasitica K. Nakamura & K. Hatai, *Mycoscience* **35**: 387 (1994).
Halocrusticida parasitica (K. Nakamura & K. Hatai) K. Nakamura & K. Hatai, *Mycoscience* **36**: 437 (1995).
Typification: Holotype *colonia exsiccata* in NJM 9231, type culture at Nippon Veterinary and Animal Science University, Tokyo, Japan. Type locality: Chiba Prefecture, Japan. Type habitat: isolated from a marine rotifer, *Brachionus plicatilis* Müller (*Aschelminthes, Rotifera*).
Identification key: Key X; key point 6'.
Literature: Comps, Menu & Moreau (1993), Dick (1998), Nakamura & Hatai (1994, 1995*a*, *b*).

Halodaphnea hamanaensis (B. Z. Bian & S. Egusa) M. W. Dick, *Mycological Research* **102**: 1065 (1998).
Atkinsiella hamanaensis B. Z. Bian & S. Egusa, *Journal of Fish Diseases* **3**: 379 (1980).
Halocrusticida hamanaensis (B. Z. Bian & S. Egusa) K. Nakamura & K. Hatai, *Mycoscience* **36**: 437 (1995).
Typification: Holotype IMI 241640; a subculture from the type maintained at the Laboratory of Aquaculture Biology, Department of Fisheries, Faculty of Agriculture, University of Tokyo, Tokyo, Japan. Type locality: Hamanako, Schizuoka, Schizuoka Prefecture, Japan. Type habitat: parasitic in ova of *Scylla serrata* (Forsskål) (*Crustacea*).
Identification key: Key X; key point 5.
Literature: Bian & Egusa (1980), Dick (1998), Nakamura & Hatai (1995*b*), Porter (1986).

Halodaphnea awabi (N. Kitancharoen, K. Nakamura, S. Wada & K. Hatai) M. W. Dick, *Mycological Research* **102**: 1065 (1998).
Atkinsiella awabi N. Kitancharoen, K. Nakamura, S. Wada & K. Hatai, *Mycoscience* **35**: 267 (1994).
Halocrusticida awabi (N. Kitancharoen, K. Nakamura, S. Wada & K. Hatai) K. Nakamura & K. Hatai, *Mycoscience* **36**: 437 (1995).
Typification: Holotype *colonia exsiccata* in NJM 9354, type culture at Nippon Veterinary and Animal Science University, Tokyo, Japan. Type locality: Yamaguchi Prefecture, Japan. Type habitat: isolated from stocked abalone, *Haliotis sieboldii* Reeve (*Mollusca*).
Identification key: Key X; key point 3'.
Literature: Dick (1998), Nakamura & Hatai (1995*b*), Kitancharoen *et al.* (1994).

Halodaphnea okinawaensis (K. Nakamura & K. Hatai) M. W. Dick, *Mycological Research* **102**: 1065 (1998).
Atkinsiella okinawaensis K. Nakamura & K. Hatai, *Mycoscience* **36**: 89 (1995).
Halocrusticida okinawaensis (K. Nakamura & K. Hatai) K. Nakamura & K. Hatai, *Mycoscience* **36**: 437 (1995).
Typification: Holotype *colonia exsiccata* in NJM 9435, type culture at Nippon Veterinary and Animal Science University, Tokyo, Japan. Type locality: Okinawa Prefecture, Japan. Type habitat: isolated from a marine cab, *Portunus pelagicus* L. (*Crustacea, Decapoda*).
Identification key: Key X; key point 6.
Literature: Dick (1998), Nakamura & Hatai (1995*a*, *b*).

Halodaphnea panulirata (N. Kitancharoen & K. Hatai) M. W. Dick, *Mycological Research* **102**: 1065 (1998).
Atkinsiella panulirata N. Kitancharoen & K. Hatai, *Mycoscience* **36**: 100 (1995).
Halocrusticida panulirata (N. Kitancharoen & K. Hatai) K. Nakamura & K. Hatai, *Mycoscience* **36**: 437 (1995).
Typification: Holotype *colonia exsiccata* in NJM 9439, type culture at Nippon Veterinary and Animal Science University, Tokyo, Japan. Type locality: Minami Izu Station, Japan Sea Farming Association, Shizuoka Prefecture, Japan. Type habitat: isolated from philozoma of spiny lobster, *Panulirus japonicus* (von Siebold) (*Crustacea, Decapoda*).
Identification key: Key X; key point 5'.
Literature: Dick (1998), Izumikawa *et al.* (1999), Kitancharoen & Hatai (1995).

REJECTED TAXA

HALOCRUSTICIDA K. Nakamura & K. Hatai, *Mycoscience* **36**: 437 (1995) [*pro parte typo excluso*] (**Type species** *Halocrusticida entomophaga* (W. W. Martin) K. Nakamura & K. Hatai), type species transferred to *Crypticola*, *loc. cit.*

Halocrusticida entomophaga (W. W. Martin) K. Nakamura & K. Hatai, *Mycoscience* **36**: 437 (1995), synonym of *Crypticola entomophaga* (W. W. Martin) M. W. Dick, *Mycological Research* **102**: 1064 (1998).

LEPTOMITALES B. B. Kanouse, *American Journal of Botany* **14**: 295 (1927).
Leptomiteae K. Lindstedt, *Synopsis der Saprolegniaceen und Beobachtungen über einige Arten*: 35 (1872) [*pro parte*] [subfamily of the *Saprolegniaceae*].
Saprolegnina H. F. Copeland, *The Classification of Lower Organisms*: 77 (1956) [*pro parte*].
Typification: *Leptomitus* C. A. Agardh (*Leptomitaceae* J. Schröter).

LEPTOMITACEAE J. Schröter, *Die Natürlichen Pflanzenfamilien* **1(1)**: 96, 101 (1893).
Leptomiteae F. T. Kützing, *Phycologia Generalis*: 150 (1843).
Leptomiteae K. Lindstedt, *Synopsis der Saprolegniaceen und Beobachtungen über einige Arten*: 35 (1872) [*pro parte*] [subfamily of the *Saprolegniaceae*].
Leptomiteae J. E. Humphrey, *Transactions of the American Philosophical Society, New series* **17**: 153 (1893).
Leptomitea H. F. Copeland, *The Classification of Lower Organisms*: 79 (1956).
Typification: *Leptomitus* C. A. Agardh.
Other families: *Apodachlyellaceae* M. W. Dick, *Leptolegniellaceae* M. W. Dick, *Ducellieriaceae* M. W. Dick.

LEPTOMITUS (A. W. Roth) C. A. Agardh, *Systema Algarum*: 47 (1824).
Conferva A. W. Roth, *Catalecta Botanica* **2**: 216 (1800) [*pro parte typo excluso*].
Saprolegnia C. G. D. Nees von Esenbeck *in* C. G. Carus, *Nova Acta Physico-Medica Academiae Caesareae Leopoldino Carolinae Naturae Curiosorum* **11**: 513 (1823) [*pro parte typo excluso*].
Apodya M. Cornu, *Annales des Sciences Naturelles, Botanique, Série V* **15**: 14 (1872) (**Type species**: *Apodya lactea* (A. W. Roth) M. Cornu) [*pro parte*].
Typification: *Leptomitus lacteus* (A. W. Roth) C. A. Agardh, (1824).

Leptomitus lacteus (A. W. Roth) C. A. Agardh, *Systema Algarum*: 47 (1824).
Conferva lactea A. W. Roth, *Catalecta Botanica* **2**: 216 (1800): p.217: "*Observ*. Huic elegantissimae Algae respectu articulorum fimill-[copy unclear] est *Conferva vu*-[copy unclear]-*lis*. Flora Dan. Tab. 771. fig. 1. quoad ramos vero multo. Iaxiores minithe virgatos ramulosque laterales brevissimos spinaeformes, colorem denique et loicum natalem fatis diversa".
**Conferva lactea* A. W. Roth, [? *Beyträge zur Botanik* (1782)].
**Conferva pusilla* A. W. Roth, [? *Beyträge zu Botanik* (1782)].
Saprolegnia lactea (A. W. Roth) C. A. Agardh *in* A. Braun, *Betrachtung über die Erscheinung der Verjüngung in der Natur, insbesondere in der Lebens- und Bildungsgesichte der Pflanze, Ed.* 2: 287 (1851) [as '*Saprolegnia lactea* C. A. Agardh'].
Saprolegnia lactea (A. W. Roth) N. Pringsheim, *Jahrbücher für wissenschaftliche Botanik* **2**: 228 (1860) [as '*Saprolegnia lactea* (C. A. Agardh) N. Pringsheim'].
Apodya lactea (A. W. Roth) M. Cornu, *Annales des Sciences Naturelles, Botanique, Série V* **15**: 14 (1872) [as '*Leptomitus lacteus* Ag.'].
Saprolegnia libertiae (J. B. G. G. M. Bory) F. T. Kützing, *Species Algarum*: 160 (1849).
Leptomitus libertiae (J. B. G. G. M. Bory) C. A. Agardh, *Systema Algarum*: 49 (1824).
Saprolegnia dichotoma J. N. Suhr *in* Brentel, *Flora Germanica, exsicc. 206* [*in* Minden (1912)].
Saprolegnia corcagiensis M. M. Hartog, *Quarterly Journal of Microscopical Science, Series II* **27**: 429 (1887).
Typification: Lectotype (Cejp, 1959) *Jahrbücher für wissenschaftliche Botanik* **2**: plate 23, figs 6-10; plate 25, figs 1-6 (1960) (iconotype); exsiccata: Roth *in* Rabenhorst *Algen Sachsens* No. 587, No. 144. Type locality: Germany. Type habitat: freshwater. Roth's description (1800) and Dillwyn's illustration (1809) constitute the earliest binomial characterization of any straminipilous fungus.
Identification key: Key XII; key point 1'.
Literature: Agardh (1824), Apinis (1929), Aronson & Lin (1978), Braun (1851), Büsgen (1882), Cejp (1959: 289), Cohn (1854), Coker (1923), Coker & Matthews (1937), Cornu (1872*b*), Czeczuga (1995*a*, *b*, *c*), Dangeard (1890-91), Dick (1971*c*, 1973*b*, 1990*a*), Dillwyn (1809), Dorries & Haase (1930), Fischer (1892*b*: 370), Fuller & Jaworski (1987), Gleason (1968*a*, *b*), Guilliermond (1922), Hartog (1886-7), Humphrey (1893), Kato *et al.* (1990), Kevorkian (1935), Kolkwitz (1901, 1903), Kützing (1843, 1849), Lee & Fuller (1992), Minden (1912), Pringsheim (1860, 1883), Pystina (1995), Radais (1898), Ramsbottom (1916), Roth (1800), Saccardo, Berlese & de Toni (1888), Schade (1940), Schade & Thimann (1940), Schroeter (1885), Sparrow (1960: 873), Tiegs & Dorries (1926), Trommsdorf (1918), Tulloch (1934), Vaziri-Tehrani & Dick (1980*a*, *b*, *c*), Zopf (1890: 373).

---

*FOOTNOTES: The citation for *Conferva lactea* "Roth, *Beiträge zu Botanik* (1789)" (Minden, 1912) was probably an editorial error. Dillwyn (1809: legend to plate 79) listed "*C. lactea*. Roth. Cat. Bot. I. p. 216. III. p. 292. *C. pusilla*. Fl. Germ. III. pars. i. p.524", but this is probably a mistaken interpretation of the use of the Italic font in Roth (1800):

"Conferva *pusilla* filamentis ... Roth Flora Germ. Tom. 3. Pars 1. n.37. ... Filamenta *minora*..." (Roth, 1800).

Ramsbottom (1916*c*: 346) reported that he could not trace either of these epithets in either of Roth's "*Beytrage*" publications (Roth, 1782, 1783; 1802).

It is just possible that Roth (1800) was recognizing a distinction in size between *Leptomitus* ('*Conferva lactea*') and *Apodachlya* ('*Conferva pusilla*'), but since the filament with the epithet '*pusilla*' cannot be equated with either *A. brachynema* or *A. pyrifera*, this epithet must lapse as a *nomen confusum*.

EXCLUDED SPECIES

Conferva lactea A. W. Roth, *Beyträge zu Botanik* (1782, 1783), synonym of *Leptomitus lacteus* (A. W. Roth) C. A. Agardh, *Systema Algarum*: 47 (1824).

Conferva lactea A. W. Roth, *Catalecta Botanica* 2: 216 (1800), synonym of *Leptomitus lacteus* (A. W. Roth) C. A. Agardh, *Systema Algarum*: 47 (1824).

Apodya brachynema (F. Hildebrand) M. Cornu, *Annales des Sciences Naturelles, Botanique, Série V* 15: 14 (1872), synonym of *Apodachlya brachynema* (F. Hildebrand) N. Pringsheim, *Berichte der Deutschen Botanischen Gesellschaft* 1: 289 (1883).

Apodya lactea M. Cornu, *Annales des Sciences Naturelles, Botanique, Série V* 15: 14 (1872), synonym of *Leptomitus lacteus* (A. W. Roth) C. A. Agardh, *Systema Algarum*: 47 (1824).

Leptomitus libertiae (J. B. G. G. M. Bory) C. A. Agardh, *Systema Algarum*: 49 (1824) [*Exsiccatae*, Libert, *Plantes Crypt. Ard.* (1797)], synonym of *Leptomitus lacteus* (A. W. Roth) C. A. Agardh, *Systema Algarum*: 47 (1824).

Leptomitus brachynema F. Hildebrand, *Jahrbücher für Wissenschaftliche Botanik* 6: 261 (1867), synonym of *Apodachlya brachynema* (F. Hildebrand) N. Pringsheim, *Berichte der Deutschen Botanischen Gesellschaft* 1: 289 (1883).

Leptomitus pyriferus (W. Zopf) W. Zopf, *in* A. Schenk, *Handbuch der Botanik, vol. 4*: 569 (1890), synonym of *Apodachlya pyrifera* W. Zopf, *Nova Acta der Kaiserlichen Leopoldinisch-Carolinischen Deutschen Akademie der Naturforscher* 52: 362 (367) (1888), quoted as "*Leptomitus piriferus* Zopf, *in* A. Schenk, *Handbuch der Botanik, vol. 4*: 299 (1890)" by Sparrow (1960: 879).

A large number of remaining binomials are unidentifiable, but need to be listed so that use of the specific epithets can be avoided in future:
"...and many other early records of sterile submerged hyphae." (Humphrey, 1893); Cejp (1959: 326) noted most of the following, but without the citation to the authority. Kützing (1843) refers to plates in "*Erdmann's Journal*" but without author, date or volume number, these I have not traced, but give as quoted. A few others perhaps exist as exsiccatae, but have not been traced.

Leptomitus achantiformis B. Biasoletto, *Di Alcune Alghe Microscopiche, Saggio del Dr. B. Biasoletto*: 55, plate XXVII (1832).

Leptomitus ammoniatus F. T. Kützing, *Phycologia generalis*: 155 (1843); F. T. Kützing, [*Erdmann's*] *Journal für Praktische Chemie*: plate II, fig. XVI.

Leptomitus aureo-stannicus F. T. Kützing, *Phycologia generalis*: 154 (1843); F. T. Kützing, [*Erdmann's*] *Journal für Praktische Chemie*: plate III, fig. XXIV.

Leptomitus brevis C. A. Agardh, *Systema Algarum*: 48 (1824).

Leptomitus ceratophylli C. A. Agardh, *Systema Algarum*: 48 (1824).

Leptomitus cerevisiae J. E. Duby [Exsiccatae ?, perhaps see Duby in Candolle (1828)].

Leptomitus divergens C. A. Agardh, *Systema Algarum*: 48 (1824).

Leptomitus doriae V. Cesati [*Exsiccatae, Rabenhorst, Algen Sachsens*, 575, perhaps see Cesati (1840)].

Leptomitus elaeophilus F. T. Kützing, *Phycologia generalis*: 155 (1843); F. T. Kützing, [*Erdmann's*] *Journal für Praktische Chemie*: plate II, fig. XV.

Leptomitus glutini-farinulae J. E. Duby, [Exsiccatae ?, perhaps see Duby in Candolle (1828)].

Leptomitus incompositus C. A. Agardh [Given as *Systema Algarum* (1824), but the name does not occur in this work].

Leptomitus juniperinus B. Biasoletto, *Di Alcune Alghe Microscopiche, Saggio del Dr. B. Biasoletto*: 36, plate XIII (1832).

Leptomitus lanugo C. A. Agardh, *Systema Algarum*: 47 [as p. 49 between pp. 46 & 48] (1824).

Leptomitus lavendulae B. Biasoletto, *Di Alcune Alghe Microscopiche, Saggio del Dr. B. Biasoletto*: 53, plate XXV (1832).

Leptomitus malti-cerevisiae J. E. Duby [Exsiccatae ?, perhaps see Duby in Candolle (1828)].

Leptomitus malti-juniperini J. E. Duby [Exsiccatae ?, perhaps see Duby in Candolle (1828)].

Leptomitus minutissimus C. A. Agardh, *Systema Algarum*: 47 [as p. 49 between pp. 46 & 48] (1824).

Leptomitus nanus C. A. Agardh, *Systema Algarum*: 49 (1824).

Leptomitus naphae F. T. Kützing, *Phycologia generalis*: 155 (1843).

Leptomitus phosphoratus F. T. Kützing, *Phycologia generalis*: 155 (1843); F. T. Kützing, [*Erdmann's*] *Journal für Praktische Chemie*: plate II, fig. XVIIIb.

Leptomitus pinnatus B. Biasoletto, *Di Alcune Alghe Microscopiche, Saggio del Dr. B. Biasoletto*: 42, plate XXIII (1832) [Kützing, *Phycologia generalis*: 156 (1843)].

Leptomitus plantaginis B. Biasoletto, *Di Alcune Alghe Microscopiche, Saggio del Dr. B. Biasoletto*: 40, plate XVI (1832) [Kützing, *Phycologia generalis*: 155 (1843)].

Leptomitus polychrous B. Biasoletto, *Di Alcune Alghe Microscopiche, Saggio del Dr. B. Biasoletto*: 41, (1832) [Kützing, *Phycologia generalis*: 156 (1843)].

Leptomitus pulegii B. Biasoletto, *Di Alcune Alghe Microscopiche, Saggio del Dr. B. Biasoletto*: 41, plate XV (1832).

Leptomitus rubescens L. A. de Brébisson, [*Exsiccatae, Algues de la France*, 306. At Falaise ?].

Leptomitus rubi idaei B. Biasoletto, *Di Alcune Alghe Microscopiche, Saggio del Dr. B. Biasoletto*: 52, plate XXIV (1832) [Kützing, *Phycologia generalis*: 155 (1843)].

Leptomitus saccharicola F. T. Kützing, *Phycologia generalis*: 154 (1843).

Leptomitus salviae B. Biasoletto, *Di Alcune Alghe Microscopiche, Saggio del Dr. B. Biasoletto*: 52, plate XXIII (1832).

Leptomitus spinosus B. Biasoletto, *Di Alcune Alghe Microscopiche, Saggio del Dr. B. Biasoletto*: 54, plate XXVI (1832).

Leptomitus tiliae B. Biasoletto, *Di Alcune Alghe Microscopiche, Saggio del Dr. B. Biasoletto*: 38, plate XIV (1832).

Leptomitus tuberosus F. T. Kützing, *Phycologia generalis*: 155 (1843).

Leptomitus valerianae F. T. Kützing, [Mentioned by Cejp (1959) but not found in Kützing (1843), perhaps Kützing (1849)?].

Leptomitus violaceus F. T. Kützing, *Phycologia generalis*: 154 (1843); F. T. Kützing, [*Erdmann's*] *Journal für Praktische Chemie*: plate III, fig. XXII.

Leptomitus vitreus C. A. Agardh, *Systema Algarum*: 48 (1824) [as '*vitraeus*' in Cejp (1959)].

APODACHLYA N. Pringsheim, *Berichte der Deutschen Botanischen Gesellschaft* **1**: 289 (1883).
Leptomitus (A. W. Roth) C. A. Agardh, *Systema Algarum*: 47 (1824) (**Type species**: *Leptomitus lacteus* (A. W. Roth) C. A. Agardh) [*pro parte typo excluso*].
Apodya M. Cornu, *Annales des Sciences Naturelles. Botanique, Série V* **15**: 15 (1872) (**Type species**: *Apodya lactea* M. Cornu) [*pro parte typo excluso*].
Typification: *Apodachlya brachynema* (F. Hildebrand) N. Pringsheim.

Apodachlya brachynema (F. Hildebrand) N. Pringsheim, *Berichte der Deutschen Botanischen Gesellschaft* **1**: 289 (1883).
Leptomitus brachynema F. Hildebrand, *Jahrbücher für wissenschaftliche Botanik* **6**: 261 (1867).
Apodya brachynema (F. Hildebrand) M. Cornu, *Annales des Sciences Naturelles, Botanique, Série V* **15**: 14 (1872).
Apodachlya brachynema (F. Hildebrand) N. Pringsheim, *Berichte der Deutschen Botanischen Gesellschaft* **1**: 289 (1883) var. *major* M. Tiesenhausen, *Archiv für Hydrobiologie und Planktonkunde* **7**: 296 (1912) [Lectotype (Cejp, 1959) *Archiv für Hydrobiologie und Planktonkunde* **7**: fig. 20 (1912) (iconotype)].
Apodachlya brachynema (F. Hildebrand) N. Pringsheim, *Berichte der Deutschen Botanischen Gesellschaft* **1**: 289 (1883) forma *major* F. K. Sparrow, *Aquatic Phycomycetes, 2nd revised edn*: 877 (1960).
Apodachlya punctata M. von Minden, *Kryptogamenflora der Mark Brandenburg* **5(4)**: 586 (1912) [Lectotype (Cejp, 1959) *Kryptogamenflora der Mark Brandenburg* **5(4)**: fig. 15*b-d* (1912) (iconotype)].
Apodachlya seriata A. Lund, *Det Kongelige Danske Videnskabernes Selskab Skrifter, naturvidens kabelig og mathematisk Afdeling, 9 Raekke* **6**: 34 (1934) [Lectotype (Cejp, 1959) *Det Kongelige Danske Videnskabernes Selskab Skrifter, naturvidens kabelig og mathematisk Afdeling, 9 Raekke* **6**: fig. 14 (1934) (iconotype)].
Typification: Lectotype (Cejp, 1959) *Jahrbücher für wissenschaftliche Botanik* **6**: plate 16, figs 12-23 (1867) (iconotype). Type locality: Roisdorf, Germany. Type habitat: on twigs.
Identification key: Key XII; key point 3'.
Literature: Apinis (1929), Barbier (1950), Cejp (1959: 293, 294, 300), Coker (1923), Coker & Leitner (1938), Coker & Matthews (1937), Dick (1969, 1973*b*, 1990*a*), Fischer (1892*b*: 374), Gilpin (1954), Gleason (1968*a, b*), Gleason & Stuart (1970), Gleason *et al.* (1970), Harvey (1930), Hildebrand (1867), Höhnk (1935, 1938), Huizar & Aronson (1985, 1986), Humphrey (1893), Indoh (1953), Jacobs (1982), Kanouse (1927), Kato *et al.* (1990), Kevorkian (1935), Lee & Aronson (1975), Lin & Aronson (1970), Lund (1934), Pringsheim (1883), Pystina (1995), Shen & Siang (1948), Sietsma, Eveleigh & Haskins (1969), Sparrow (1932, 1933*b*, 1960: 876, 877, 880), Tiesenhausen (1912), Thaxter (1896*b*), Vaziri-Tehrani & Dick (1980*a, b, c*), Wolf (1944).

Apodachlya pyrifera W. Zopf, *Nova Acta der Kaiserlichen Leopoldinisch-Carolinischen Deutschen Akademie der Naturforscher* **52**: 362 (1888).
Leptomitus pyriferus (W. Zopf) W. Zopf, *in* A. Schenk, *Handbuch der Botanik, vol. 4*: 569 (1890).
Apodachlya pyrifera W. Zopf, *Nova Acta der Kaiserlichen Leopoldinisch-Carolinischen Deutschen Akademie der Naturforscher* **52**: 362 (1888) var. *macrosporangia* M. Tiesenhausen, *Archiv für Hydrobiologie und Planktonkunde* **7**: 295 (1912) [Lectotype (Cejp, 1959) *Archiv für Hydrobiologie und Planktonkunde* **7**: fig. 19*a-c* (1912) (iconotype)].
Apodachlya pyrifera W. Zopf, *Nova Acta der Kaiserlichen Leopoldinisch-Carolinischen Deutschen Akademie der Naturforscher* **52**: 362 (1888) forma *macrosporangia* F. K. Sparrow, *Aquatic Phycomycetes, 2nd revised edn*: 879 (1960).
Typification: Lectotype (Cejp, 1959) *Nova Acta der Kaiserlichen Leopoldinisch-Carolinischen Deutschen Akademie der Naturforscher* **52**: plate 21, figs 1-21 (1888) (iconotype). Type locality: Berlin, Germany. Type habitat: from sewage water, with dead *Chara* (also on *Picea* sp. needles - Tiesenhausen, 1912).
Identification key: Key XII; key point 3.
Literature: Cejp (1932*b*, 1959: 297), Czeczuga (1994), Dick (1969, 1973*b*, 1990*a*), Fischer (1892*b*: 374), Fuller & Jaworski (1987), Höhnk (1935), Humphrey (1893), Jacobs (1982), Kanouse (1927*b*), Karling (1968*c*), Klassen, McNabb & Dick (1987), Lund (1934), Minden (1912), Petersen (1909, 1910), Pystina (1995), Randolph & Powell (1992), Richter (1937), Shen & Siang (1948), Sparrow (1932, 1933*b*, 1960: 879), Tiesenhausen (1912), Thaxter (1896*b*), Vaziri-Tehrani & Dick (1980*a, b, c*), Wolf (1939), Zopf (1890).

Apodachlya minima W. C. Coker & J. Leitner *ex* K. Cejp, *Flora ČSR, B Řada Mykologicko-Lichenologická, Svazek 2, Oomycetes I*: 447 (1959).

Coker, W. C. & J. Leitner, *Journal of the Elisha Mitchell Scientific Society* **54**: 313 (1938).

Typification: type material not designated; lectotype *Journal of the Elisha Mitchell Scientific Society* **54**: plate 39, figs 1-19 (1938) and *Transactions of the British Mycological Society* **38**: 416, fig. 1a-j (1955) (iconotype). Type locality: Chapel Hill, North Carolina, U.S.A. Type habitat: stream water baited with hemp seed.

Identification key: Key XII; key point 5.

Literature: Cejp (1959), Coker & Leitner (1938), Dick (1966, 1973*b*, 1990*a*), Gleason (1968*b*), Gleason & Stuart (1970), Gleason *et al*. (1970), Jacobs (1982), Johnson (1955*c*), Sparrow (1960: 878), Vaziri-Tehrani & Dick (1980*a*, *b*, *c*).

EXCLUDED SPECIES

Apodya brachynema (F. Hildebrand) M. Cornu, *Annales des Sciences Naturelles, Botanique, Série V* **15**: 14 (1872), synonym of *Apodachlya brachynema* (F. Hildebrand) N. Pringsheim, *Berichte der Deutschen Botanischen Gesellschaft* **1**: 289 (1883).

Apodachlya brachynema (F. Hildebrand) N. Pringsheim, *Berichte der Deutschen Botanischen Gesellschaft* **1**: 289 (1883) forma *major* F. K. Sparrow, *Aquatic Phycomycetes, 2nd revised edn*: 877 (1960), synonym of *Apodachlya brachynema* (F. Hildebrand) N. Pringsheim, *Berichte der Deutschen Botanischen Gesellschaft* **1**: 289 (1883).

Apodachlya brachynema (F. Hildebrand) N. Pringsheim, *Berichte der Deutschen Botanischen Gesellschaft* **1**: 289 (1883) var. *major* M. Tiesenhausen, *Archiv für Hydrobiologie und Planktonkunde* **7**: 296 (1912), synonym of *Apodachlya brachynema* (F. Hildebrand) N. Pringsheim, *Berichte der Deutschen Botanischen Gesellschaft* **1**: 289 (1883).

Apodachlya pyrifera W. Zopf, *Nova Acta der Kaiserlichen Leopoldinisch-Carolinischen Deutschen Akademie der Naturforscher* **52**: 362 (1888) forma *macrosporangia* F. K. Sparrow, *Aquatic Phycomycetes, 2nd revised edn*: 879 (1960), synonym of *Apodachlya pyrifera* W. Zopf, *Nova Acta der Kaiserlichen Leopoldinisch-Carolinischen Deutschen Akademie der Naturforscher* **52**: 362 (1888).

Apodachlya pyrifera W. Zopf, *Nova Acta der Kaiserlichen Leopoldinisch-Carolinischen Deutschen Akademie der Naturforscher* **52**: 362 (1888) var. *macrosporangia* M. Tiesenhausen, *Archiv für Hydrobiologie und Planktonkunde* **7**: 295 (1912), synonym of *Apodachlya pyrifera* W. Zopf, *Nova Acta der Kaiserlichen Leopoldinisch-Carolinischen Deutschen Akademie der Naturforscher* **52**: 362 (1888).

Apodachlya punctata M. von Minden, *Kryptogamenflora der Mark Brandenburg* **5(4)**: 586 (1912), synonym of *Apodachlya brachynema* (F. Hildebrand) N. Pringsheim, *Berichte der Deutschen Botanischen Gesellschaft* **1**: 289 (1883).

Apodachlya seriata A. Lund, *Det Kongelige Danske Videnskabernes Selskab Skrifter, naturvidens kabelig og mathematisk Afdeling, 9 Raekke* **6**: 34 (1934), synonym of *Apodachlya brachynema* (F. Hildebrand) N. Pringsheim, *Berichte der Deutschen Botanischen Gesellschaft* **1**: 289 (1883).

PLEROGONE M. W. Dick, *Botanical Journal of the Linnean Society* **93**: 228 (1986).
Typification: *Plerogone helodes* M. W. Dick.

Plerogone helodes M. W. Dick, *Botanical Journal of the Linnean Society* **93**: 228 (1986).
Apodachlyella completa (J. E. Humphrey) H. Indoh *sensu* Dick, *Journal of the Linnean Society (Botany)* **59**: 57-60 (1964).
Typification: Holotype: *Journal of the Linnean Society (Botany)* **59**: plate 1, figs 1-12; plate 2, figs 14-19; plate 3, figs 22-30 (1964) (iconotype); culture (ex-type) APCC 5004a (IMI 308001/2); lectotype: slides deposited at the University of Reading, prepared from culture APC 5004a, 1986. Type locality: Blelham Bog North (Grid. Ref. NY366007), Ambleside, Cumbria, U.K. Type habitat: mud/water sample from a *Myrica-Erica-Narthecium-Sphagnum* community.
Identification key: Key XII; key point 5'.
Literature*: Bertke & Aronson (1980), Dick (1964, 1966, 1969, 1986), Fuller & Jaworski (1987), Gleason & Stuart (1970), Win-Tin & Dick (1975).

APODACHLYELLACEAE M. W. Dick, *Botanical Journal of the Linnean Society* 93: 226 (1986).
Typification: *Apodachlyella* H. Indoh (holotype - sole original genus).

APODACHLYELLA H. Indoh, *Science Reports of the Tokyo Bunrika Daigaku, Section B* **4**: 47 (1939).
Typification: *Apodachlyella completa* (J. E. Humphrey) H. Indoh (holotype - sole original species).

Apodachlyella completa (J. E. Humphrey) H. Indoh, *Science Reports of the Tokyo Bunrika Daigaku, Section B* **4**: 47 (1939).
Apodachlya completa J. E. Humphrey, *Transactions of the American Philosophical Society, New Series* **17**: 137 (1893).
non Apodachlyella completa (J. E. Humphrey) H. Indoh, *Science Reports of the Tokyo Bunrika Daigaku, Section B* **4**: 47 (1939) *sensu* Dick, *Journal of the Linnean Society (Botany)* **59**: 57-60 (1964) [**syn**: *Plerogone helodes* M. W. Dick, *Botanical Journal of the Linnean Society* **93**: 228 (1986)].
Typification: Type material deposited in the Tokyo Bunrika Daigaku (iconotype: *Science Reports of the Tokyo Bunrika Daigaku, Section B* **4**: plate VII, figs 1-11 (1939)). Type locality: Otsuka, Tokyo, Japan. Type habitat: from fresh water and soil, saprotrophic on baits of flies and hemp seeds.
Identification key: Key XII; key point 6.
Literature*: Dick (1986), Humphrey (1893), Indoh (1939), Loncore, Brooks & Homola (1987).

EURYCHASMOPSIS H. M. Canter & M. W. Dick, *Mycological Research* **98**: 106 (1994).
Typification: *Eurychasmopsis multisecunda* H. M. Canter (holotype - sole original species).

Eurychasmopsis multisecunda H. M. Canter *in* H. M. Canter & M. W. Dick, *Mycological Research* **98**: 107 (1994).
Typification: Iconotype *Mycological Research* **98**: figs 16-73, 108, 110, 112-114 (1994). Type locality: in Lake Windermere, Cumbria, U.K. Type habitat: parasitic in *Podophrya* sp. (*Protozoa, Suctoria*).
Identification key: Key I; key point 19'; Key XII; key point 6'.
Literature: Canter & Dick (1994), Canter, Heaney & Lund (1990), Czeczuga (1995*b*).

---

*FOOTNOTE: Most of the non-taxonomic literature on *Apodachlyella completa* (Humphrey) Indoh relates to the culture of *Plerogone helodes* M. W. Dick; this culture was widely distributed in the U.S.A.

SYSTEMATICS

LEPTOLEGNIELLACEAE M. W. Dick, *Transactions of the British Mycological Society* **57**: 421 (1971).
Ectrogellaceae A. Scherffel, *Archiv für Protistenkunde* **52**: 6 (1925) [*pro parte minore typo excluso*].
Ectrogellacea H. F. Copeland, *The Classification of Lower Organisms*: 77 (1956) [*pro parte minore typo excluso*].
Typification: *Leptolegniella* M. B. Huneycutt.

LEPTOLEGNIELLA M. B. Huneycutt, *Journal of the Elisha Mitchell Scientific Society* **68**: 110 (1952).
Typification: *Leptolegniella keratinophila* M. B. Huneycutt (holotype - sole original species).

Leptolegniella keratinophila M. B. Huneycutt, *Journal of the Elisha Mitchell Scientific Society* **68**: 110 (1952) [as '*keratinophilum*'].
Typification: type material not designated; lectotype *Journal of the Elisha Mitchell Scientific Society* **68**: plate 15, figs 1-8 (1952) (iconotype). Type locality: Wilson County, Tennessee, U.S.A. Type habitat: saprotrophic on keratinized materials from a sheep pasture.
Identification key: Key IX; key point 8.
Literature: Czeczuga (1995*a*, *b*, *c*), Czeczuga & Muszyńska (1994), Dick (1971*b*), Huneycutt (1952), Johnson (1968), Karling (1968*a*, *c*, *f*, 1976*a*, *b*, 1981*b*, *d*, *e*, 1984, 1987*b*, 1988), Scott, Seymour & Warren (1963), Sparrow (1957), Sparrow, Paterson & Johns (1965).

Leptolegniella exoospora W. D. Kane, *Mycologia* **58**: 909 (1966) [as '*exoosporus*'].
Typification: Slide in Mycology Slide Collection No. 102K, Herbarium, University of North Carolina, Chapel Hill, N. C., U.S.A. Type locality: St Joseph County, Indiana, U.S.A. Type habitat: from a soil sample.
Identification key: Key IX; key point 6.
Literature: Kane (1966), Karling (1968*a*, *c*, 1987*b*), Sparrow (1975).

Leptolegniella exogena J. S. Karling, *Nova Hedwigia* **45**: 433 (1987).
Typification: Iconotype *Nova Hedwigia* **45**: 439, figs 1-17 (1987). Type locality: Thailand. Type habitat: from a soil sample (in this paper the soil samples (2, 12, 16, 17) and localities (1-8) cannot be cross-referenced from the information published).
Identification key: Key IX; key point 6'.
Literature: Karling (1987, 1988).

Leptolegniella angularis J. S. Karling, *Nova Hedwigia* **45**: 434 (1987) [as '*angulare*'].
Typification: Iconotype *Nova Hedwigia* **45**: 441, figs 35-45 (1987). Type locality: Thailand. Type habitat: from a soil sample (in this paper the soil sample (23) and localities (1-8) cannot be cross-referenced from the information published).
Identification key: Key IX; key point 7'.
Literature: Karling (1987, 1988).

Leptolegniella abnormis J. S. Karling, *Nova Hedwigia* **46**: 243 (1988).
Typification: Iconotype *Nova Hedwigia* **46**: 248, figs 35-48 (1988). Type locality: Australia (soil samples and their localities are listed for other records, but not for this species).
Identification key: Key IX; key point 8'.
Literature: Karling (1988).

EXCLUDED SPECIES

Leptolegniella piligena M. Ookubo & Y. Kobayasi, *Nagaoa* 5: 4 (1955), synonym of ***Leptolegnia piligena*** (M. Ookubo & Y. Kobayasi) J. S. Karling, ***Mycologia*** **60**: 279 (1968) [**Typification**: type material not designated; lectotype *Nagaoa* 5: 5,fig.3A-G (1955) (iconotype). Type locality: Ninomiya, Tiba Prefecture, Japan. Type habitat: rice field soil baited with human hair. **Literature**: Czeczuga & Muszyńska (1994), Karling (1968*d*), Ookubo & Kobayasi (1955).]

Leptolegniella marina (D. Atkins) M. W. Dick, *Transactions of the British Mycological Society* **57**: 422 (1971), synonym of ***Salilagenidium marinum*** (D. Atkins) M. W. Dick *loc. cit.*

APHANODICTYON M. B. Huneycutt *ex* M. W. Dick, *Transactions of the British Mycological Society* **57**: 422 (1971); *Journal of the Elisha Mitchell Scientific Society* **64**: 279 (1948).
Typification: *Aphanodictyon papillatum* M. B. Huneycutt *ex* M. W. Dick (holotype - sole original species).

Aphanodictyon papillatum M. B. Huneycutt *ex* M. W. Dick, *Transactions of the British Mycological Society* **57**: 422 (1971); *Journal of the Elisha Mitchell Scientific Society* **64**: 279 (1948).
Typification: type material not designated; lectotype *Journal of the Elisha Mitchell Scientific Society* **64**: plate 35, figs 1-13; plate 36, figs 14-23 (1948) (iconotype). Type locality: Chapel Hill, North Carolina, U.S.A. Type habitat: saprotrophic on keratinized materials from pasture.
Identification key: Key IX; key point 2.
Literature: Czeczuga (1995*c*), Czeczuga & Muszyńska (1994), Czeczuga & Woronowicz (1991), Dick (1971*b*), Gaertner (1954), Huneycutt (1948), Karling (1984, 1988), Sparrow (1950, 1952*c*).

BREVILEGNIELLA M. W. Dick, *Papers of the Michigan Academy of Science, Arts, and Letters* **46**: 195 (1961).
Typification: *Brevilegniella keratinophila* M. W. Dick (holotype - sole original species).

Brevilegniella keratinophila M. W. Dick, *Papers of the Michigan Academy of Science, Arts, and Letters* **46**: 195 (1961).
Typification: Type specimens deposited at the University of Michigan Herbarium (catalogue number not known), isotype slides with M. W. Dick, University of Reading. Type locality: University of Michigan Campus, Ann Arbor, Michigan, U.S.A. Type habitat: soil sample flooded and baited with human hair.
Identification key: Key IX; key point 2'.
Literature: Dick (1961), Sparrow (1957), Sparrow & Dogma (1973), Sparrow, Paterson & Johns (1965).

NEMATOPHTHORA B. R. Kerry & D. H. Crump, *Transactions of the British Mycological Society* **74**: 120 (1980).
Typification: *Nematophthora gynophila* B. R. Kerry & D. H. Crump (holotype -sole original species).

Nematophthora gynophila B. R. Kerry & D. H. Crump, *Transactions of the British Mycological Society* **74**: 120 (1980).
Typification: Holotype IMI 234655, isotype slides with M. W. Dick, University of Reading and D. J. Hooper, Rothamsted Experimental Station. Type locality: Butt Close Field, Woburn, Bedfordshire, U.K. Type habitat: parasitic on females of *Heterodera avenae* Wollenweber (*Aschelminthes, Nematoda*).
Identification key: Key I; key point 20'.
Literature: Boag & Lopez-Llorca (1989), Crump (1991), Crump & Moore (1990), Crump, Sayre & Young (1983), Graff & Madelin (1989), Kerry (1974, 1975, 1980, 1981), Kerry & Crump (1980), Kerry, Crump & Mullen (1980, 1982), Lopez-Llorca & Duncan (1991), Nordbring-Hertz (1988).

APHANOMYCOPSIS A. Scherffel, *Archiv für Protistenkunde* **52**: 11 (1925).
? *Cymbanche* E. Pfitzer, *Sizungsberichte der niederrheinischen Gesellschaft Natur- und Heilkunde, Bonn, 1869*: 221 (**Type species**: *Cymbanche fockei* Pfitzer - holotype sole original species) [see *Ectrogella*].
Typification: *Aphanomycopsis bacillariacearum* A. Scherffel (holotype - sole original species).

Aphanomycopsis bacillariacearum A. Scherffel, *Archiv für Protistenkunde* **52**: 14 (1925).
Lagenidium enecans W. Zopf, *Nova Acta der Kaiserlichen Leopoldinisch-carolinischen Akademie der Naturforscher* **47**: 154 (1884). [**Lectotypification** (Cejp, 1959); *Archiv für Protistenkunde* **52**: plate 2, figs 60-69 (1925) (iconotype)].
Ectrogella monostoma A. Scherffel, *Archiv für Protistenkunde* **52**: 8 (1925) [**Lectotypification** (Cejp, 1959); *Archiv für Protistenkunde* **52**: plate 1, figs 10-19 (1925) (iconotype)].
Typification: Lectotype (Cejp, 1959): *Archiv für Protistenkunde* **52**: plate 1, figs 31-35 (1925) (iconotype). Type locality: Igló, Slovenia. Type habitat: parasitic in *Pinnularia viridis* Kütz. (*Bacillariophyceae*) and other diatoms (*Epithemia, Cymbella, Nitzschia*).
Identification key: Key I; key point 44'.
Literature: Canter (1949), Cejp (1959), Dick (1971*b*), Friedmann (1952), Johnson (1968), Kadlubrowska (1970), Karling (1968*c*, 1981*a*: 114, 156), Scherffel (1902*a*, 1925), Skvortzow (1925), Sparrow (1933*c*, 1936*a*, 1960: 825, 989), Sparrow & Barr (1955), Tokunaga (1934*a*), Wildeman (1893*a*, *b*), Yačevskie & Yačevskie (1931: 67), Zopf (1894).

Aphanomycopsis entophyta (N. Pringsheim) M. W. Dick, **comb. nov.**
Pythium entophytum N. Pringsheim, *Jahrbuch für wissenschaftlichen Botanik* **1**: 289 (1858).
Myzocytium entophytum (N. Pringsheim) M. Cornu, *Annales des Sciences Naturelles, Botanique, Série V* **15**: 21 (1872).
Lagenidium entophytum (N. Pringsheim) W. Zopf, *Nova Acta der Kaiserlichen Leopoldinisch-carolinischen Akademie der Naturforscher* **47**: 154 (1884).
Lagenidium gracile W. Zopf, *Nova Acta der Kaiserlichen Leopoldinisch-carolinischen Akademie der Naturforscher* **47**: 158 (1884).
Lagenidium americanum G. F. Atkinson, *Botanical Gazette* **48**: 336 (1909).
Lagenidium clavatum S. N. Dasgupta & R. John, *Bulletin of the Botanical Survey of India* **30**: 66 (1988) [*nom. inval.*, Art. 36.1, without Latin diagnosis].
Lagenidium cylindriforme S. N. Dasgupta & R. John, *Bulletin of the Botanical Survey of India* **30**: 72 (1988).
Lagenidium contortum S. N. Dasgupta & R. John, *Bulletin of the Botanical Survey of India* **30**: 73 (1988).
Lagenidium entosphaericum S. N. Dasgupta & R. John, *Bulletin of the Botanical Survey of India* **30**: 76 (1988).
Typification: Lectotype (Cejp, 1959): *Jahrbuch für wissenschaftlichen Botanik* **1**: plate 21, fig 1 (1858) (iconotype). Type locality: Germany. Type habitat: parasitic in *Spirogyra* sp. (*Zygnematales, Zygnemataceae*). **Typification of synonyms**: *L. clavatum*: holotype deposited at HCIO, Division of Mycology and Plant Pathology, IARI, New Delhi-110 012. HCIO No. 37119; *L. cylindriforme*: Holotype deposited at HCIO, Division of Mycology and Plant Pathology, IARI, New Delhi-110 012. HCIO No. 37124; *L. contortum*: Holotype deposited at HCIO, Division of Mycology and Plant Pathology, IARI, New Delhi-110 012. HCIO No. 37126; *L. entosphaericum*: Holotype deposited at HCIO, Division of Mycology and Plant Pathology, IARI, New Delhi-110 012. HCIO No. 37128.
Identification key: Key I; key point 46.
Literature: Atkinson (1909), Aleem (1952*a*), Atkinson (1909*a*), Cejp (1959), Cornu (1872*b*), Dasgupta & John (1988), Ivimey Cook (1935), Karling (1981*a*: 110, 114), Konno (1972), Lacy (1949), Minden (1915), Pringsheim (1858), Pystina (1995), Scherffel (1904), Schulz (1923), Shen & Siang (1948), Sparrow (1952*c*, 1960: 998, 999), Voronichin (1920), Wildeman (1895*b*, 1896*a*,*b*), Wolf (1944), Yačevskie & Yačevskie (1931: 67), Zopf (1884).

Aphanomycopsis desmidiella H. M. Canter, *Transactions of the British Mycological Society* **32**: 166 (1949).
Typification: type material not designated: lectotype: *Transactions of the British Mycological Society* **32**: 165, fig. 3 (1949) (iconotype). Type locality: Batemanfold, Lancashire, U.K. Type habitat: parasitic in *Netrium digitus* (Ehrenb.) Itzigsch & Rothe (*Zygnematales, Mesotaeniaceae*).
Identification key: Key I; key point 46'.
Literature: Canter (1949), Cejp (1959), Dick (1971*b*), Friedmann (1952), Karling (1981*a*: 158), Pystina (1995), Sparrow (1960: 827).

Aphanomycopsis saprophytica J. S. Karling, *Mycologia* **60**: 273 (1968).
Typification: Slides of type specimen 140 in Herbarium of Purdue University, Lafayette, Indiana, U.S.A. Type locality: Pitcairn Island, Oceania. Type habitat: saprotrophic in bleached *Zea* leaves and *Pinus* pollen grains from soil sample P13.
Identification keys: Key I; key point 26', Key XI; key point 4'.
Literature: Dick (1971*b*), Karling (1968*c*, *f*, 1981*a*: 158, 1984, 1987*b*).

Aphanomycopsis punctata J. S. Karling, *Mycologia* **60**: 273 (1968),
Typification: Slides of type specimen 141 in Herbarium of Purdue University, Lafayette, Indiana, U.S.A. Type locality: Rarotonga, Cook Islands, Oceania. Type habitat: saprotrophic in snake skin from soil sample R11.
Identification key: Key IX; key point 4.
Literature: Dick (1971*b*), Karling (1968*c*, *f*, 1969, 1970, 1981*a*: 160), Persiel (1960*a*, *b*).

Aphanomycopsis sexualis W. W. Martin, *Mycologia* **67**: 923 (1975).
Typification: Holotype slide in the Mycology Herbarium, Department of Botany, University of North Carolina, Chapel Hill, North Carolina, U.S.A. Type locality: Stagg Creek, Hanover County, Virginia, U.S.A. Type habitat: parasitic in eggs of *Chironomidae* (*Insecta*).
Identification key: Key I; key point 12'.
Literature: Karling (1981*a*: 162), Martin (1975*b*, 1991).

Aphanomycopsis cryptica H. M. Canter *in* H. M. Canter & S. I. Heaney, *New Phytologist* **97**: 602 (1984).
Typification: Slide IMI 284024. Type locality: Esthwaite Water, Cumbria, U.K. Type habitat: parasitic in *Ceratium hirundinella* (O. F. Müll.) Bergh. (*Dinomastigota*, *Gonyaulacales*).
Identification key: Key I; key point 42.
Literature: Canter & Heaney (1984).

Aphanomycopsis peridiniella A. Boltovskoy & A. M. Arambarri *in* A. Boltovskoy, *Limnobios* **2**: 643 (1984).
Typification: Type: LPS 42677. Type locality: Turbera Barco de Piedra, Tierra del Fuego, Argentina. Type habitat: parasitic in cysts of *Peridinium willei* Huitf.-Kaas (*Dinomastigota*).
Identification key: Key I; key point 42'.
Literature: Boltovskoy (1984).

UNIDENTIFIABLE SPECIES

CYMBANCHE E. Pfitzer, *Sizungsberichte der niederrheinischen Gesellschaft Natur- und Heilkunde, Bonn, 1869*: 221 (**Type species**: *Cymbanche fockei* E. Pfitzer - holotype, sole original species).

Cymbanche fockei E. Pfitzer, *Sizungsberichte der niederrheinischen Gesellschaft Natur- und Heilkunde, Bonn, 1869*: 221, parasitic in diatoms, see *Ectrogella*.

GENUS *INCERTAE SEDIS*

CORNUMYCES M. W. Dick, **genus novum**
ETYMOLOGY: Named after Maxim Cornu.

Fungi parasitici in fungis flagellatis aut saprotrophici; terrestres vel in aquis dulcibus. **Thallus** holocarpicus, coralliformis, raro allantoides; non septatus, unusquisque thallus se transformans vel in sporangium vel in gametangium. **Reproductio asexualis** per zoosporangia; zoosporogenesis intrasporangialis; zoosporae grandiores (5·0 × 7·5 μm; (40)175 μm^3). **Reproductio sexualis** per oogonia, gametangia donantia parum cognita; homothallica vel heterothallica, automictica; oosporae singulae vel plures per oogonium, aplerotica, structura ooplasticae non reperta, distributio copiae lipidae guttulas lipidas minutas aut coactas in unam guttulam maiorem.
Holotypus: *Cornumyces destruens* (F. K. Sparrow) M. W. Dick.

Fungi parasitic in flagellate fungi or saprotrophic; terrestrial or freshwater. **Thallus** holocarpic; coralloid, rarely allantoid, non septate; transformed into a sporangium or an oogonium. **Asexual reproduction** by zoosporangia, zoosporogenesis intrasporangial; zoospores medium-sized (*ca* 5·0 × 7·5 μm) (volume equivalent (40)175 μm^3). **Sexual reproduction** by oogonia, donor gametangia not well known; homothallic, automictic; oospores one to several, aplerotic, ooplast structure not reported, distribution of lipid reserve material as minute droplets or aggregated into a single large droplet.
Typification: *Cornumyces destruens* (F. K. Sparrow) M. W. Dick.

Cornumyces destruens (F. K. Sparrow) M. W. Dick, **comb. nov.**
Lagenidium destruens F. K. Sparrow, *Journal of the Washington Academy of Sciences* **40**: 54 (1950).
Typification: type material not designated: lectotype *Journal of the Washington Academy of Sciences* **40**: 53, figs 15-24 (1950) (iconotype). Type locality: Las Villas Province, Cuba. Type habitat: parasitic in hyphae and reproductive organs of *Achlya* sp. (*Peronosporomycetes, Saprolegniaceae*).
Identification key: Key I; key point 36'.
Literature: Karling (1981*a*: 128), Sparrow (1950, 1952*c*, 1960: 991).

Cornumyces pygmaeus (W. Zopf) M. W. Dick, **comb. nov.**
Lagenidium pygmaeum W. Zopf, *Abhandlungen der Naturforschung Gesellschaft zu Halle* **17**: 96 (1887).
Typification: Lectotype (Cejp, 1959): *Abhandlungen der Naturforschung Gesellschaft zu Halle* **17**: plate 1, figs 29-31, plate 2 figs 1-12 (1887) (iconotype). Type locality: Saal, Germany. Type habitat: saprotrophic in pollen.
Identification key: Key I; key point 27'.
Literature: Booth (1971*a*, *b*), Cejp (1959), Dogma (1975), Gaertner (1954), Gaertner & Sparrow (1966), Harder & Gallwitz-Uebelmesser (1959), Howard & Johnson (1969), Karling (1941*b*, *c*, 1949, 1968*a*, 1969, 1981*a*: 126, 1981*d*, 1987*b*), Kobayas[h]i & Konno (1970), Kobayas[h]i *et al.* (1971), Konno (1972), Maurizio (1895), Miller (1961), Petersen (1909, 1910), Pongratz (1966), Pystina (1995), Schulz (1923), Sparrow (1943, 1952*c*, 1960: 993), Sparrow & Dogma (1973), Sparrow & Johnson (1959), Sparrow & Koch (1959), Voronichin (1920), Wildeman (1895*a*, 1896*a*), Yačevskie & Yačevskie (1931: 66), Zopf (1887).

Cornumyces irregularis (J. C. Constantineanu) M. W. Dick, **comb. nov.**
Olpidiopsis irregularis J. C. Constantineanu, *Revue Générale de Botanique* **13**: 373 (1901).
Petersenia irregularis (J. C. Constantineanu) F. K. Sparrow, *Aquatic Phycomycetes* p. 634 (1943) [no resting spores].
Typification: Lectotype (Cejp, 1959): *Revue Générale de Botanique* **13**: fig. 76 A-K (1901) (iconotype). Type locality: Roumania. Type habitat: parasitic in *Saprolegnia* sp. and *Achlya* sp. (*Peronosporomycetes, Saprolegniaceae*).
Identification key: Key I; key point 37'.
Literature: Cejp (1959), Constantineanu (1901), Harder & Gallwitz-Uebelmesser (1959), Johnson (1977), Karling (1942*a*, 1981*a*: 61), Milanez (1969), Milanez & Val (1969), Sparrow (1934, 1943, 1960: 961).

Cornumyces muenscheri (V. M. Cutter) M. W. Dick, **comb. nov.**
Lagenidium muenscheri V. M. Cutter, *Mycologia* **35**: 11 (1943).
Typification: Type #1001 (collector: W. C. Muenscher) in the herbarium of V. M. Cutter, at Cornell University, Ithaca, New York, U.S.A. Type locality: Oquaga Lake, Broome County, New York, U.S.A. Type habitat: parasitic in root hairs of *Potamogeton spirillus* Tuckerman (*Potamogetonaceae*).
Identification keys: Key I; key point 31, Key VIII; key point 2'.
Literature: Cutter (1943), Karling (1981*a*: 127), Sparrow (1960: 984).

Cornumyces utriculobus (C. E. Miller) M. W. Dick, **comb. nov.**
Petersenia utriculoba C. E. Miller, *Mycologia* **54**: 422 (1962).
Typification: type material not designated; lectotype *Mycologia* **54**: 424, figs 1-11, 426, figs 12-27 (1962) (iconotype). Type locality: House Creek, Copperas Cove, Coryell County, Texas, U.S.A. Type habitat: parasitic in vegetative hyphae of *Pythium intermedium* de Bary (*Peronosporomycetes*, *Pythiaceae*) from dry oat field soil.
Identification key: Key I; key point 38'.
Literature: Karling (1981*a*: 61), Miller (1962*a*).

SYSTEMATICS

Cornumyces catenophlyctidis (B. M. Sundaram) M. W. Dick, **comb. nov.**
Petersenia catenophlyctidis B. M. Sundaram, *Journal of the Indian Botanical Society* **47**: 144 (1968).
Typification: Type: Herb. MUBL No. 2004 (slides). Type locality: Poonemalle, Madras, India. Type habitat: parasitic in *Catenophlyctis variabilis* Karling (*Chytridiomycetes*) on snake-skin bait in a moistened sample of rice-field soil.
Identification key: Key I; key point 34.
Literature: Karling (1981*a*: 62), Sundaram (1968).

Cornumyces endogenus (J. S. Karling) M. W. Dick, **comb. nov.**
Olpidiopsis endogena J. S. Karling, *Nova Hedwigia* **43**: 7 (1985).
Typification: Iconotype *Nova Hedwigia* **43**: 10, figs 1-18 (1985). Type locality: Ross Biological Reserve, Purdue University, Indiana, U.S.A. Type habitat: parasitic in *Aphanomyces laevis* de Bary (*Peronosporomycetes, Saprolegniales*) from a soil sample.
Identification key: Key I; key point 37.
Literature: Karling (1985).

DOUBTFUL SPECIES

Cornumyces karlingiae (J. S. Karling) M. W. Dick, **comb. nov.**
Olpidiopsis karlingiae J. S. Karling, *Mycologia* **41**: 271 (1949).
Typification: type material not designated; lectotype *Mycologia* **41**: 272, figs 1-17 (1949) (iconotype). Type locality: Frederick, Maryland, U.S.A. Type habitat: parasitic in *Karlingia rosea* (de Bary & Woronin) A. E. Johanson on onion skin bait, water sample from a stagnant pool near a pig pen, April, 1948.
Identification key: Key I; key point 34'.
Literature: Dogma (1975), Karling (1949, 1981*a*: 38). Sparrow (1960: 946), Sparrow & Dogma (1973).

UNNAMED TAXA WITH POSSIBLE AFFINITY TO *CORNUMYCES*

Dick (1970: 453): a polyoosporic fungus with eccentric oospores parasitic in *Aphanomyces* (*Cornumyces endogenus*?).

Seymour & Johnson (1973): a unioosporous fungus with eccentric oospores parasitic in rotifer eggs.

Kerry & Crump (1980): a polyoosporous fungus parasitic in nematode eggs.

DUCELLIERIACEAE M. W. Dick, **familia nova**

Fungi saprotrophici, in aquis dulcibus, in granis pollinis. **Thallus** holocarpicus, coralliformis, allantoides vel pseudomycelialis; unusquisque thallus se transformans vel in sporangium vel in gametangium. **Reproductio asexualis** per sporangia; zoosporogenesis intrasporangialis cum vesicula precipitativa (?), liberans planontes in cystas se transformantes, quae inter se connexae per cornua tubularia sphaeram vacuam configurant; zoosporae heterokontae, formae principalis, flagellis in parte laterali insertis, flagellum anterius curtius; zoosporae grandiores (180-400 μm^3). **Reproductio sexualis** per oogonia; heterothallica, fecundatio non reperta; oosporae pleroticae, distributio copiae ooplasticae et lipidae non reperta.
Holotypus: *Ducellieria* E. Teiling.

Fungi saprotrophic, freshwater, in pollen grains. Thallus holocarpic, coralloid, allantoid or pseudomycelial, each thallus becoming transformed into a sporangium or gametangium. **Asexual reproduction** by sporangia; zoosporogenesis intrasporangial, releasing planonts which form cysts linked to form a hollow sphere, cysts interlinked by tubular horns; zoospores heterokont, of the principal form with lateral flagellar insertion, anterior flagellum the shorter; zoospores medium-large (volume equivalent 180-400 μm^3). **Sexual reproduction** by oogonia; heterothallic, mode of fertilization not established; oospores plerotic, distribution of ooplast and oil reserve materials not established.
Typification: *Ducellieria* E. Teiling (holotype - sole original genus).

DUCELLIERIA E. Teiling, *Svensk Botanisk Tidskrift* **51**: 209 (1957).
Typification: *Ducellieria chodatii* (F. Ducellier) E. Teiling.

Ducellieria chodatii (F. Ducellier) E. Teiling, *Svensk Botanisk Tidskrift* **51**: 209 (1957) [as '*chodati*'].
Coelastrum chodatii F. Ducellier, *Bulletin de la Société Botanique de Genève, 2 Serié* **7**: 73 (1915) [as '*chodati*'].
Typification: type material not designated: lectotype *Plant Systematics and Evolution* **165**: 3,fig.2; 4, fig.3; 5, fig. 5; 7, fig. 6; 9, fig.8; 10, fig. 9 (1989) (iconotype). Type locality: lakes of Lens-Montana, Valais, Switzerland. Type habitat: in stagnant water with various green algae (alt. 1500m).
Identification key: Key I; key point 27.
Literature: Bourelly (1968), Couté (1984), Ducellier (1915), Ettl (1978), Komárek & Fott (1983), Hesse, Kusel-Fetzmann & Carniel (1989), Kusel-Fetzmann & Carniel (1984), Kusel-Fetzmann & Nouak (1981), Teiling (1957).

REJECTED SPECIES

The following organisms are algae and may require new binomials if the basionym is unsuitable:

Ducellieria chodatii (F. Ducellier) E. Teiling var. *armata* (H. Skuja) E. Teiling, *Svensk Botanisk Tidskrift* **51**: 211 (1957), a synonym of *Coelastrum augustae* H. Skuja var. *armatum* H. Skuja, *Acta Horti Universitatis Latviensis* **7**: 55 (1934).

Ducellieria tricuspidata (O. Borge) E. Teiling, *Svensk Botanisk Tidskrift* **51**: 211 (1957), a synonym of *Coelastrum tricuspidatum* O. Borge, *Arkiv för Botanik* **28**: 21 (1936).

ANAMORPHIC GENUS *INCERTAE SEDIS*

BLASTULIDIUM C. Pérez, *Compte rendu des Séances de la Société de Biologie, Paris* **55**: 716 (1903).
Blastulidiopsis A. Sigot, *Comptes Rendus de la Société de Biologie de Strasbourg* **108**: 37 (1931) (**Type species**: *Blastulidiopsis chattonii* A. Sigot - holotype sole original species).
Typification: *Blastulidium paedophthorum* C. Pérez (holotype - sole original species).

Blastulidium paedophthorum C. Pérez, *Compte rendu des Séances de la Société de Biologie* **55**: 716 (1903)
Blastulidiopsis chattonii A. Sigot, *Comptes Rendus de la Société de Biologie de Strasbourg* **108**: 37 (1931).
Typification: type material not designated: lectotype *Protistologica* **12**: 227, figs 1-5; 229, figs 6-8; 230, fig. 9; 231, fig. 10; 232-235, figs 11-17 (1976) (iconotype). Type locality: Lake Gradignan, Bordeaux, France. Type habitat: parasitic in embryos of *Daphnia obtusa* Kurz. (*Crustacea*). **Typification of synonym**: *B. chattonii*; lectotype (Cejp, 1959): *Comptes Rendus de la Société de Biologie de Strasbourg* **108**: figs 1-3 (1931) (iconotype).
Identification key: Key I; key point 13'.
Literature: Chatton (1908), Green (1974), Jirovec (1955), Karling (1981*a*: 49-51), Manier (1976), Pérez (1903, 1905), Sigot (1931), Sparrow (1960: 181-182, Yačevskie & Yačevskie (1931: 29).

HYPHOCHYTRIOMYCETES F. K. Sparrow *emend.* M. W. Dick, *Zoosporic Plant Pathogens*: 285 (1983) [class].
Hyphochytridiomycetes F. K. Sparrow, *Mycologia* **50**: 811 (1958) [class].
Hyphochytridiomycota R. H. Whittaker, *Science, New York* **163**: 154 (1969) [phylum].
Typification: *Hyphochytrium* W. Zopf (*Hyphochytriales* F. K. Sparrow, holotype - sole original order).

HYPHOCHYTRIOMYCETIDAE M. W. Dick, **subclassis et status novae**
Hypochytridae T. Cavalier-Smith, *The Chromophyte Algae Problems and Perspectives*: 385 (1989) [*pro parte*] (typification not given).

Cyclus nuclearis ignotus. Fungi saprotrophici vel parasitici; terrestres vel in aquis dulcibus. **Thallus** holocarpicus, sphericus, saccatus vel coralliformis, non septatus, cum vel sine rhizoideis; pars fibrillaris parietis chitine formata; membrana nuclearis partim amissa in mitose. **Reproductio asexualis** per sporangia; cum pariete laevi aut crasso; zoosporogenesis extrasporangialis, cum vel sine vesicula homohylica; zoosporae uniflagellatae, flagellum insertum in parte anteriore, straminipilosum; zona transitionalis cum helice longa et sine tigillis; K_2-grana non reperta. **Reproductio sexualis** absens.
Holotypus: *Hyphochytrium* W. Zopf (*Hyphochytriales* F. K. Sparrow).

Nuclear cycle not known. Fungi saprotrophic or parasitic, terrestrial or freshwater. **Thallus** spherical, saccate or coralloid, non septate, with or without rhizoids; fibrillar component of cell walls with chitin; nuclear membrane partly lost during mitosis. **Asexual reproduction** by thick or thin-walled sporangia; zoosporogenesis extrasporangial, with or without a homohylic vesicle; zoospores with a single, anteriorly inserted, straminipilous flagellum; transitional zone with a short helix and with struts; K_2 bodies not known. **Sexual reproduction** absent.
Typification:*Hyphochytrium* W. Zopf (*Hyphochytriales* F. K. Sparrow, holotype - sole original order).

HYPHOCHYTRIALES F. K. Sparrow, *Aquatic Phycomycetes, Second Revised Edition*: 743 (1960) [*sensu stricto*].
Hyphochytriales E. A. Bessey, *Morphology and Taxonomy of Fungi*: 69 (1950) [*pro parte*] (**Type family**: *Hyphochytriaceae* A. Fischer *emend.* J. S. Karling).
Hyphochytrialea H. F. Copeland, *A Classification of Lower Organisms*: 69 (1956) [*pro parte*] (**Type family**: *Hyphochytriaceae* A. Fischer *emend.* J. S. Karling).
Hyphochytriales F. K. Sparrow, *Aquatic Phycomycetes, 2nd Revised Edition*: 743 (1960) [*pro parte*] (**Type family**: *Hyphochytriaceae* A. Fischer *emend.* J. S. Karling).
Anisochytridiales J. S. Karling, *American Journal of Botany* **30**: 641 (1943) [*pro parte*] (**Type family**: no type family designated, first-mentioned family: *Anisolpidiaceae* J. S. Karling).
Typification: *Hyphochytrium* W. Zopf (*Hyphochytriaceae* A. Fischer *emend.* J. S. Karling).

HYPHOCHYTRIACEAE A. Fischer, *Rabenhorst's Kryptogamen-Flora, Aufl., Bd 1, Abt. 4*: 131 (1892) *emend.* J. S. Karling, *American Journal of Botany* **26**: 518 (1939).
Hyphochytridiaceae J. P. Lotsy, *Vorträge über botanische Stammengeschichte. I.* (1907) [the original use of the syllable 'id'] (**Type genus**: *Hyphochytrium* W. Zopf).
Hyphochytriacea H. F. Copeland, *A Classification of Lower Organisms*: 69 (1956) (**Type genus**: *Hyphochytrium* W. Zopf).
Typification: *Hyphochytrium* W. Zopf.

HYPHOCHYTRIUM W. Zopf, *Nova Acta der Kaiserlichen Leopoldinisch-carolinischen Akademie der Naturforscher* **47**: 187 (1884).
Hyphophagus (W. Zopf) M. von Minden, *Kryptogamenflora der Mark Brandenburg und angrenzender Gebiete* **5**: 420 (1911) (**Type species**: *Hyphophagus infestans* (W. Zopf) M. von Minden).
Typification: *Hyphochytrium infestans* W. Zopf (holotype - sole original species).

Hyphochytrium infestans W. Zopf, *Nova Acta der Kaiserlichen Leopoldinisch-carolinischen Akademie der Naturforscher* **47**: 187 (1884).
Hyphophagus infestans (W. Zopf) M. von Minden, *Kryptogamenflora der Mark Brandenburg und angrenzender Gebiete* **5**: 420 (1911).
Typification: type material not designated: lectotype *Nova Acta der Kaiserlichen Leopoldinisch-carolinischen Akademie der Naturforscher* **47**: plate 7, figs 13-20 (1884) (iconotype). Type locality: Berlin, Germany. Type habitat: in the fruiting body of ?*Helotium* sp. (*Ascomycetes*).
Identification key: Key XII; key point 9.
Literature: Karling (1977: 398), Minden (1911*b*), Schröter (1893), Sparrow (1960: 761), Zopf (1884).

Hyphochytrium hydrodictyi A. Valkanov, *Archiv für Protistenkunde* **67**: 122 (1929) [as '*hydrodictii*'].
Typification: type material not designated; lectotype *Archiv für Protistenkunde* **67**: 123-127, figs 1-11 (1929) (iconotype). Type locality: environs of Sofia, Bulgaria. Type habitat: in *Hydrodictyon reticulatum* (L.) Lagerh. (*Chlorophyceae*), collected summer, 1928.
Identification key: Key XII; key point 9'.
Literature: Karling (1977: 398), Sparrow (1960: 763), Valkanov (1929).

Hyphochytrium catenoides J. S. Karling, *American Journal of Botany* **26**: 513 (1939).
Typification: type material not designated; lectotype *American Journal of Botany* **26**: 514, figs 1-18 (1939) (iconotype). Type locality: New York City, N.Y., U.S.A. Type habitat: saprotrophic on *Zea* leaf baits with *Physoderma* resting spore (*Chytridiomycetes*) inoculum.
Identification key: Key XII; key point 8'.
Literature: Ayers & Lumsden (1977), Bean *et al.* (1972), Cooney *et al.* (1985), Czeczuga (1995*a*), Czeczuga & Muszyńska (1994), Czeczuga & Woronowicz (1991), Ellis-Evans (1985), Fuller & Jaworski (1987), Harder & Persiel (1962), Karling (1939, 1964*a*, 1967*a*, 1968*b*, 1977: 400, 1981*b*, 1987*b*), Lange & Olson (1979), Siang (1949), Sparrow (1960: 762), Van de Peer *et al.* (1996), Van der Auwera *et al.* (1995), Vogel (1961, 1964), Warner *et al.* (1983), Weete *et al.* (1989), Willoubý (1971), Wynn & Epton (1979).

Hyphochytrium oceanum J. S. Karling, *Sydowia* **20**: 139 (1967).
Typification: type material not designated: lectotype *Sydowia* **20**: plate 26, figs 11-24; plate 27, figs 25-32; plate 28, fig. 33 (1967) (iconotype). Type locality not identified from the various localities in Oceania from which the fungus was recovered. Type habitat: saprotrophic in grass leaf baits floated on soil samples P16, P17.
Identification key: Key XII; key point 7.
Literature: Karling (1967*a*, 1968*b*, 1977: 400, 1981*b*, 1984).

Hyphochytrium saprobium (J. S. Karling) M. W. Dick, **comb. nov.**
Anisolpidium saprobium J. S. Karling, *Journal of the Elisha Mitchell Scientific Society* **84**: 166 (1968).
Typification: type material not designated; lectotype *Journal of the Elisha Mitchell Scientific Society* **84**: 168, figs 1-18 (1968) (iconotype). Type locality: Pitcairn Island, Oceania. Type habitat: saprotrophic in cleared pollen grains of *Pinus* (samples P16, P17).
Identification key: Key XII; key point 3.
Literature: El-Nagdy & Abdel-Hafez (1990), Karling (1968*b*, 1977: 386, 1981*b*), Lange & Olson (1979), Sparrow & Dogma (1973).

Hyphochytrium elongatum (J. S. Karling) M. W. Dick, **comb. nov.**
Anisolpidium elongatum J. S. Karling, *Nova Hedwigia* **28**: 217 (1976).
Typification: type material not designated; lectotype *Nova Hedwigia* **28**: 227, figs 32-44 (1976) (iconotype). Type locality: Swan Lake, Sun City Center, Ruskin, Florida, U.S.A. Type habitat: saprotrophic on snake skin, human corneum and grass leaves in watered soil.
Identification key: Key XII; key point 3'.
Literature: Karling (1976*b*, 1981*b*, 1988).

UNNAMED SPECIES

Hyphochytrium sp. I. Persiel, *Archiv für Mikrobiologie* **36**: 301 (1960)
Literature: Persiel (1960*a*, *b*), Persiel *et al.* (1966).

Hyphochytrium sp. I. C. Hallett (1975); *loc. cit.*
Literature: Hallett (1975).

EXCLUDED SPECIES

HYPHOPHAGUS (W. Zopf) M. von Minden, *Kryptogamenflora der Mark Brandenburg und angrenzender Gebiete* 5: 420 (1911) (**Type species**: *Hyphophagus infestans* (W. Zopf) M. von Minden), a superfluous name for *Hyphochytrium*.

Hyphophagus infestans (W. Zopf) M. von Minden, *Kryptogamenflora der Mark Brandenburg und angrenzender Gebiete* 5: 420 (1911), synonym of *Hyphochytrium infestans* W. Zopf, *Nova Acta der Kaiserlichen Leopoldinisch-carolinischen Akademie der Naturforscher* 47: 187 (1884).

Hyphochytrium peniliae N. Ya. Artemchuk & L. M. Zelezinskaya, *Mikologiya i Fitopatologiya* 3: 357 (1969), referable to *Salilagenidium* M. W. Dick, *loc. cit.*

CANTERIOMYCES F. K. Sparrow, *Aquatic Phycomycetes, 2nd revised edition*: 750 (1960).
Typification: *Canteriomyces stigeoclonii* (É. de Wildeman) F. K. Sparrow (holotype - sole original species).

Canteriomyces stigeoclonii (É. de Wildeman) F. K. Sparrow, *Aquatic Phycomycetes, 2nd revised edition*: 750 (1960).
Olpidium stigeoclonii É. de Wildeman, *Mémoires de l'Herbier Bossier* **15**: 3 (1900).
Anisolpidium stigeoclonii (É. de Wildeman) H. M. Canter, *Transactions of the British Mycological Society* **33**: 343 (1950).
Typification: type material not designated; lectotype *Transactions of the British Mycological Society* **33**: 336-341, figs 1-6, plates 24-26 (1950) (iconotype). Type locality: Belgium. Type habitat: parasitic in *Stigeoclonium* sp. (*Chlorophyceae, Chaetophorales*).
Identification key: Key XII; key point 4.
Literature: Canter (1950), Dangeard (1886*a*, 1890*c*), Karling (1941*c*), Minden (1911*a*), Sparrow (1960: 750), Wildeman (1900, 1931).

DOUBTFUL GENUS

CYSTOCHYTRIUM W. R. Ivimey Cook, *Transactions of the British Mycological Society* **16**: 251 (1932).
Typification: *Cystochytrium radicale* W. R. Ivimey Cook (holotype - sole original species).

Cystochytrium radicale W. R. Ivimey Cook, *Transactions of the British Mycological Society* **16**: 251 (1932).
Typification: type material not designated; lectotype *Transactions of the British Mycological Society* **16**: 248-249, figs 1-19, plate 10, figs 1-6 (1932) (iconotype). Type locality: Sevenoaks, Kent, U.K. Type habitat: parasitic in roots of the aquatic angiosperm *Veronica beccabunga* L. (*Scrophulariaceae*).
Identification keys: Key I; key point 31', Key XIII; key point 4'.
Literature: Ivimey Cook (1929, 1932*a*), Karling (1977; 404).

RHIZIDIOMYCETACEAE J. S. Karling, *American Journal of Botany* **30**: 643 (1943).
Typification: *Rhizidiomyces* W. Zopf (holotype - sole original genus).

RHIZIDIOMYCES W. Zopf, *Nova Acta der Kaiserlichen Leopoldinisch-carolinischen Akademie der Naturforscher* **47**: 188 (1884).
Rhizidiomycopsis F. K. Sparrow, *Aquatic Phycomycetes, 2nd Revised Edition*: 757 (1960) (**Type species**: *Rhizidiomycopsis japonica* (Y. Kobayashi & M. Ookubo) F. K. Sparrow - holotype, sole original species).
Typification: *Rhizidiomyces apophysatus* W. Zopf (holotype - sole original species).

Rhizidiomyces apophysatus W. Zopf, *Nova Acta der Kaiserlichen Leopoldinisch-carolinischen Akademie der Naturforscher* **47**: 188 (1884).
Typification: type material not designated: lectotype *Nova Acta der Kaiserlichen Leopoldinisch-carolinischen Akademie der Naturforscher* **47**: plate 9, figs 1-70 (1884) (iconotype). Type locality: Berlin, Germany. Type habitat: epiparasitic on oogonia of *Saprolegniaceae* (*Saprolegnia ferax* (Gruith.) Thur., *S. asterophora* de Bary and *A. polyandra* de Bary) (*Peronosporomycetes*).
Identification key: Key XIII; key point 11.
Literature: Barstow *et al.* (1989), Bean *et al.* (1972), Bérczi (1940), Cejp (1934, 1959), Chaudhuri & Kochlar (1935), Clay *et al.* (1991), Coker (1923), Couch (1939*a*), Czeczuga (1995*a*, *c*), Czeczuga & Muszyńska (1994), Dawe & Kuhn (1983*a*, *b*), Forbes (1935*b*), Fuller (1960, 1962, 1966, 1978, 1990), Fuller & Barshad (1960), Fuller & Jaworski (1987), Fuller & Reichle (1965), Gaertner (1954), Johnson (1956), Karling (1943, 1944*b*, *d*, 1964*a*, 1977: 388, 1967*a*, 1981*b*, *d*), Konno (1972), Lange & Olson (1979), LéJohn (1972, 1974), Lovell & Haselby (1971), Olson & Fuller (1968), Schröter (1892), Sneh, Humble & Lockwood (1977), Sparrow (1932, 1960: 754), Sparrow & Dogma (1973), Tokunaga (1934*b*), Vogel (1964), Warner *et al.* (1983), Weete *et al.* (1989), Yačevskie & Yačevskie (1931: 94), Zopf (1884).

Rhizidiomyces ichneumon C. Gobi, *Scriptis Botanicis Horti Universitatis Impericolis Petropolitanae* **15**: 234 [in Russian], 258 [in German] (1900).
Typification: type material not designated; lectotype *Scriptis Botanicis Horti Universitatis Impericolis Petropolitanae* **15**: plate VI, figs 1-28, plate VII, figs 29-39 (1900) (iconotype). Type locality: either Russia (Sparrow, 1960) or Finland (cf. *Pythium tenue* C. Gobi - Waterhouse, 1968). Type habitat: epiparasitic in *Chloromonas globulosa* (M. Perty) Gobi (*Chlorophyceae*, *Volvocales*) from water from a peat moor.
Identification key: Key XIII; key point 13'.
Literature: Gobi (1900*a*, *b*), Karling (1977: 388), Sparrow (1960: 756), Yačevskie & Yačevskie (1931: 95).

Rhizidiomyces bivellatus K. Nabel, *Archiv für Mikrobiologie* **10**: 537 (1939) [no Latin diagnosis].
Typification: type material not designated; lectotype *Archiv für Mikrobiologie* **10**: 520-534, figs 1-7 (1939) (iconotype). Type locality: probably Central America (Haiti, Venezuela, Mexico), perhaps Yugoslavia. Type habitat: from soil samples flooded and baited with sterilized *Drosophila* flies.
Identification key: Key XIII; key point 4.
Literature: Czeczuga (1995*c*), Nabel (1939), Karling (1943, 1944*b*, *d*, 1964*a*, 1967*a*, 1968*b*, 1977: 390, 1981*b*, *d*, 1984, 1987*a*, *b*), Sparrow (1960: 754).

Rhizidiomyces hansonii J. S. Karling, *American Journal of Botany* **31**: 396 (1944).
Typification: type material not designated; lectotype *American Journal of Botany* **31**: 394, figs 35-64 (1944) (iconotype). Type locality: San Carlos, Matto Grosso, Brazil. Type habitat: saprotrophic on *Zea* leaf baits in soil with vegetable debris.
Identification key: Key XIII; key point 3'.
Literature: Karling (1944*b*, 1964*a*, 1967*a*, 1968*b*, 1977: 392, 1987*b*), Sparrow (1960: 755).

Rhizidiomyces hirsutus J. S. Karling, *Bulletin of the Torrey Botanical Club* **72**: 47 (1945).
Typification: type material not designated; lectotype *Bulletin of the Torrey Botanical Club* **72**: 49, figs 1-19 (1945) (iconotype). Type locality: Seringal Monte Alegre, Acre Territory, Brazil. Type habitat: saprotrophic in moist leaf mould.
Identification key: Key XIII; key point 6'.
Literature: Karling (1945, 1964*a*, 1967*a*, 1968*b*, 1976*a*, *b*, 1977: 392, 1981*b*, *d*, *e*, 1984, 1987*a*, *b*, 1988), Sparrow (1960: 756).

Rhizidiomyces japonicus Y. Kobayashi & M. Ôkubo *in* Y. Ogura, *Scientific Researches of the Ozegahara Moor, 1954*: 573 (1954).
Rhizidiomycopsis japonica (Y. Kobayashi & M. Ookubo) F. K. Sparrow, *Aquatic Phycomycetes, 2nd revised edition*: 757 (1960).
Typification: Type in Herbarium of the National Science Museum, Tokyo, Japan. Type locality: Ozegahara Moor, Nikko National Park, Japan. Type habitat: parasitic on oogonia of *Achlya* sp. (as *Aplanes* sp.) (*Peronosporomycetes*, *Saprolegniaceae*).
Identification key: Key XIII; key point 11'.
Literature: Fuller (1990), Karling (1977), Kobayashi & Ookubo (1954), Konno (1972), Sparrow (1960: 757).

Rhizidiomyces parasiticus J. S. Karling, *Sydowia* **17**: 194 (1964).
Typification: Type slide No. 120 PU (Herbarium Purdue University). Iso-iconotype *Sydowia* **17**: 194, figs 1-6 (1964). Type locality: Mandapam Camp, Madras State, India. Type habitat: parasitic on sporangia of *Rhizophlyctis* sp. (*Chytridiomycetes*) from brackish soil, also from non-brackish soil.
Identification key: Key XIII; key point 12'.
Literature: Karling (1964*a*, 1977: 392).

Rhizidiomyces saprophyticus (J. S. Karling) J. S. Karling, *Chytridiomycetarum Iconographia*: 394 (1977).
Rhizidiomycopsis saprophytica J. S. Karling, *Sydowia* **20**: 137 (1967).
Typification: type material not designated: lectotype *Sydowia* **20**: plate 26, figs 1-10 (1967) (iconotype). Type locality; Waipoua River, Waipoua Kauri Forest, Auckland Province, New Zealand. Type habitat: saprotrophic on bleached *Zea* leaves in soil sample.
Identification key: Key XIII; key point 5'.
Literature: Karling (1967*a*, 1968, 1977).

Rhizidiomyces bulbosus J. S. Karling, *Journal of the Elisha Mitchell Scientific Society* **84**: 167 (1968).
Typification: type material not designated; lectotype *Journal of the Elisha Mitchell Scientific Society* **84**: 170, figs 19-42 (1968) (iconotype). Type locality: Niue, Oceania. Type habitat: soil sample N12, saprotrophic on *Zea* leaves, snake skin and pollen grains (also from Pitcairn and Cook Islands).
Identification key: Key XIII; key point 8.
Literature: Karling (1968*a*, *b*, 1977: 394, 1981*b*)

Rhizidiomyces coronus J. S. Karling, *Journal of the Elisha Mitchell Scientific Society* **84**: 173 (1968).
Typification: type material not designated; lectotype *Journal of the Elisha Mitchell Scientific Society* **84**: 170, figs 43-45, 172, figs 46,47 (1968) (iconotype). Type locality: Aitutaki, Cook Islands, Oceania. Type habitat: soil sample A11, saprotrophic on *Zea* leaves and pollen grains.
Identification key: Key XIII; key point 7'.
Literature: Karling (1968*a*, *b*, 1977: 392, 1981*b*)

Rhizidiomyces bullatus (F. K. Sparrow) J. S. Karling, *Nova Hedwigia* **34**: 672 (1981).
Rhizidiomycopsis bullatus F. K. Sparrow; *Mycologia* **69**: 665 (1977).
Typification: Type specimen in University of Michigan Herbarium (Sparrow Collection); iconotype *Mycologia* **69**: 662 (1977). Type locality; Emmet County, Michigan, U.S.A. Type habitat: in *Pinus* and *Liquidambar* pollen bait in *Sphagnum* bog sample.
Identification key: Key XIII; key point 4'.
Literature: Karling (1981*b*, *e*, 1984, 1987*a*, *b*, 1988), Sparrow (1977*a*).

Rhizidiomyces stomatosus (F. K. Sparrow) M. W. Dick, **comb. nov.**
Rhizidiomycopsis stomatosa F. K. Sparrow, *Mycologia* **69**: 1057 (1977).
Typification: Type specimen in University of Michigan Herbarium (Sparrow Collection); iconotype *Mycologia* **69**: 1054, figs 1-5 (1977). Type locality; Gainsville, Florida, U.S.A. Type habitat: on azygospores of *Gigaspora margaritae* W. N. Becker & I. R. Hall (*Zygomycotina*, *Endogonales*).
Identification key: Key XIII; key point 12.
Literature: Fuller (1990), Schenck & Nicolson (1977), Sparrow (1977*b*).

UNNAMED SPECIES

Rhizidiomyces sp. J. S. Karling, *Nova Hedwigia* **34**: 674 (1981).
Literature: Karling (1981*b*).

EXCLUDED SPECIES

RHIZIDIOMYCOPSIS F. K. Sparrow, *Aquatic Phycomycetes, 2nd revised edition*: 757 (1960), synonym of *Rhizidiomyces* W. Zopf, *Nova Acta der Kaiserlichen Leopoldinisch-carolinischen Akademie der Naturforscher* **47**: 188 (1884).

Rhizidiomycopsis japonica (Y. Kobayashi & M. Ookubo) F. K. Sparrow, *Aquatic Phycomycetes, 2nd revised edition*: 757 (1960), synonym of *Rhizidiomyces japonicus* Y. Kobayashi & M. Ôkubo *in* Y. Ogura, *Scientific Researches of the Ozegahara Moor, 1954*: 573 (1954).

Rhizidiomyces dangeardii J. L. Serbinow, *Scriptis Botanicis Horti Universitatis Impericolis Petropolitanae* **24**: 162 (1907), possible synonym of *Saccomyces endogenus* (L. Nowakowski) F. K. Sparrow, *Aquatic Phycomycetes*: 246 (1943), zoospores not known. **Literature**: Karling (1977), Minden (1911*a*), Sparrow (1960: 352), Wildeman (1895*b*), Yačevskie & Yačevskie (1931: 42).

Rhizidiomyces spirogyrae É. de Wildeman, *Annales de la Société Belge de Microscopie, Mémoires* **19**: 111 (1895), possible synonym of *Phlyctochytrium spirogyrae* É. de Wildeman, zoospores not known. **Literature**: Minden (1911*a*), Sparrow (1960: 352), Wildeman (1895*b*).

DOUBTFUL GENERA AND SPECIES

LATROSTIUM W. Zopf, *Beiträge zur Physiologie Morphologie Niederer Organismen* **4**: 65 (1894).
Typification: *Latrostium comprimens* W. Zopf (holotype - sole original species).

Latrostium comprimens W. Zopf, *Beiträge zur Physiologie Morphologie Niederer Organismen* **4**: 65 (1894).
Typification: type material not designated; lectotype *Beiträge zur Physiologie Morphologie Niederer Organismen* **4**: plate III, figs 6-19 (1894) (iconotype). Type locality: Halle, Germany. Type habitat: parasitic in the oogonia of *Vaucheria* sp. (*Tribophyceae*).
Identification key: Key XIII; key point 14.
Literature: Sparrow (1960: 758), Karling (1977: 398), Minden (1911*a*), Wildeman (1895*a*), Zopf (1894).

REESSIA C. Fisch, *Sitzungsberichte der Physicalische-medicinischen Societät zu Erlangen* **16**: 32 (1884).
Typification: *Reesia amoeboides* C. Fisch (holotype - sole original species).

Reessia amoeboides C. Fisch, *Sitzungsberichte der Physicalische-medicinischen Societät zu Erlangen* **16**: 32 (1884).
Typification: type material not designated; lectotype *Sitzungsberichte der Physicalische-medicinischen Societät zu Erlangen* **16**: figs 1-6 (1884) (iconotype). Type locality: Erlangen, Germany. Type habitat: parasitic in *Lemna minor* L. (*Lemnaceae*).
Literature: Fisch (1884*a*), Karling (1977: 28), Minden (1911*a*), Schröter (1892), Schwartz & Ivimey Cook (1928), Sparrow (1960: 765).

Reessia lemnae (C. Fisch) J. S. Karling, *American Journal of botany* **30**: 642 (1943).
Chytridium lemnae C. Fisch, *Sitzungsberichte der Physicalische-medicinischen Societät zu Erlangen* **16**: 43 (1884).
Olpidium lemnae (C. Fisch) J. Schroeter *in* F. Cohn, *Kryptogamen-Flora von Schlesien* **3**: 181 (1885).
Typification: type material not designated; lectotype *Sitzungsberichte der Physicalische-medicinischen Societät zu Erlangen* **16**: figs 7-9 (1884) (iconotype). Type locality: Erlangen, Germany. Type habitat: parasitic in *Lemna minor* L. (*Lemnaceae*).
Literature: Fisch (1884*a*), Karling (1943, 1977: 28), Schroeter (1886), Schröter (1892), Sparrow (1960: 765), Wagner (1969), Wildeman (1896*a*).

EXCLUDED SPECIES

CATENARIOPSIS J. N. Couch, *American Journal of Botany* **28**: 707 (1941) [*nomen nudum*], see Karling (1943), Sparrow (1960: 766).

Reesia cladophorae C. Fisch, *Sitzungsberichte der Physicalische-medicinischen Societät zu Erlangen* **16**: 101 (1884), a synonym of *Olpidium entophytum* A. Braun, *Abhandlungen Berlin Akademie, 1856*: 589 (1856).
Literature: Fisch (1884*b*), Karling (1977: 28), Minden (1911*a*), Sparrow (1960: 764).

'*Euchytridium mesocarpi*' C. Fisch, *Sitzungsberichte der Physicalische-medicinischen Societät zu Erlangen* **16**: 101 (1884).
Chytridium mesocarpi (C. Fisch) A. Fischer, *Rabenhorst's Kryptogamen-Flora, 2 Aufl., Bd 1, Abt. 4*: 126 (1892).
[***CHYTRIDIUM*** A. Braun, *Betrachtungen über die Erscheinung der Verjüngung in der Natur,...*: 198 (1851) (**Type species**: *Chytridium olla* A. Braun); *Chytridium* subgenus *Euchytridium* A. Braun, *Physicalische Abhandlungen der Königlichen Akademie der Wissenschaften zu Berlin, 1855*: 74 (1856) (**type species**: *Chytridium olla* A. Braun); *Euchytridium* N. Sorokin, *Archives Botaniques du Nord de la France* **2**: 21 (1883) (**Type species**: *Euchytridium olla* (A. Braun) N. Sorokin.]
Identification key: Key XIV; key point 14'.
Literature: Fisch (1884*b*), Fischer (1892), Karling (1977: 392), Sparrow (1960: 766).

EXCLUDED MARINE SPECIES

Elina marisalba N. Ya. Artemchuk, *Veröffentlichungen des Instituts für Meeresforschung in Bremerhaven* **13**: 232 (1972), is transferred to the *Labyrinthista*.

Elina sinorifica N. Ya. Artemchuk, *Veröffentlichungen des Instituts für Meeresforschung in Bremerhaven* **13**: 233 (1972), is transferred to the *Labyrinthista*.

Olpidium maritimum W. Höhnk & A. A. Aleem, *Veröffentlichungen der Instituts für Meeresforschung in Bremerhaven* **2**: 227 (1953) [as "*O. maritium* Höhnk & Allem" in Kuznetsov (1979)], a synonym of *Olpidium pendulum* W. Zopf, *Chytridiales* (Sparrow & Johnson, 1961).

Zygorhizidium vaucheriae A. Rieth, *Biologisches Zentralblatt* **86**: 444 (1967) [*nom. invalid.*, Art. 36.1, without Latin diagnosis] is not a hyphochytrid (Karling, 1977: 404) but was correctly placed in the *Chytridiomycetes*.
Typification: type material not designated; lectotype *Biologisches Zentralblatt* **86**: 436, fig. 1a-n, 437, plate 1a-f, 438, plate 2g-k, 439, fig. 2a-l (1967) (iconotype). Type locality: Gatersleben, Germany. Type habitat: parasitic on *Vaucheria debaryana* Woronin (also *V. woroniniana* Heering). (*Tribophyceae*).
Literature: Karling (1977: 404), Rieth (1967).

FAMILY *INCERTAE SEDIS* WITHIN THE *PERONOSPOROMYCETES*

LAGENACEAE M. W. Dick, **familia nova**

Fungi parasitici in angiospermatibus, fungi flagellati aut Protoctista; terrestres et in aquis dulcibus. **Thallus** holocarpicus, allantoideus, non septatus; unusquisque thallus se transformans vel in zoosporangium vel in gametangium. **Reproductio asexualis** per zoosporangia; zoosporogenesis intrasporangialis; zoosporae grandiores (>250 μm^3), heterokontae et anisokontae; flagellis in parte subapicali laterali insertis, flagellum anterius curtius; zona transitionalis sine helice curta et sine tigillis, discus transitionalis in plano super planum plasmamembranae, sine pilula centrali. **Reproductio sexualis** per gametangia; heterothallica vel automictica; oosporogenesis periplasmica, oosporae singulae vel gregatim, periplasma nucleatum persistens; oosporae *ca* 15 μm diam., paries oosporae crassus. **Holotypus**: *Lagena* T. C. Vanterpool & G. A. Ledingham.

Fungi parasitic in angiosperms, flagellate fungi or Protoctista, terrestrial and freshwater. **Thallus** holocarpic, allantoid, non septate, becoming transformed into a sporangium or a gametangium. **Asexual reproduction** by zoosporangia; zoosporogenesis intrasporangial; zoospores large 7·0 × 11·0 μm (volume equivalent >250 μm^3), heterokont and anisokont; flagellar insertion subapical and lateral, anterior flagellum the shorter; transitional zone without helix or struts, transitional plate above the plane of the plasmamembrane, lacking a central bead. **Sexual reproduction** by gametangia; heterothallic or automictic; oosporogenesis periplasmic, with one to several oospores, nucleated periplasm persistent; oospores *ca* 15 μm diam., oospore wall thick.
Typification: *Lagena* T. C. Vanterpool & G. A. Ledingham (holotype - sole original genus).

LAGENA T. C. Vanterpool & G. A. Ledingham, *Canadian journal of Research* **2**: 192 (1930).
Lagenocystis H. F. Copeland, *The Classification of Lower Organisms*: 82 (1956) [*nom. nova*].
Typification: *Lagena radicicola* T. C. Vanterpool & G. A. Ledingham (holotype - sole original species).

Lagena radicicola T. C. Vanterpool & G. A. Ledingham, *Canadian Journal of Research* **2**: 192 (1930).
Lagenocystis radicicola (T. C. Vanterpool & G. A. Ledingham) H. F. Copeland, *The Classification of Lower Organisms*: 82 (1956).
Typification: type material not designated; lectotype *Canadian journal of Research* **2**: 175, fig. 2, plate I, figs 1-10, 180, fig 5, 182, fig. 6, plate II, figs 1-9, 185, fig. 7 (1930) (iconotype). Type locality: Regina, Saskatchewan, Canada. Type habitat: parasitic in roots of *Triticum aestivum* L. (also *Triticum durum* Desf., *Hordeum vulgare* L., *Secale cereale* L. *Zea mays* L.) (*Poaceae*).
Identification key: Key I; key point 30.
Literature: Antoine & Ricaud (1966), Barr (1990), Barr & Désaulniers (1987*b*, 1990*a*), Barr & Slykhius (1976), Copeland (1956), Haber, Barr & Platford (1991), Ijdenburg *et al.* (1986), Karling (1981*a*: 142), Langenburg & Van Der Wal (1986), Macfarlane (1970), Nolt *et al.* (1981), Slykhius & Barr (1978), Sprague (1950), Tomlinson, Zitter & Clay (1985), Truscott (1933), Vanterpool & Ledingham (1930), Zhang & Langenberg (1988).

DOUBTFUL SPECIES ASSIGNED HERE:

'***Petersenia panicicola***' M. J. Thirumalachar & R. C. Lacy, *Mycopathologia et Mycologia Applicata* **6**: 75 (1951).
Typification: Type in Herbarium of Cryptophytes, India and Orient, New Delhi, India; Isotypes at IMI Egham, Surrey, U.K. and USDA, Beltsville, Maryland, U.S.A. Type habitat: in previously partially submerged leaves of *Panicum repens* L. (*Poaceae*). Leaf spot of *Panicum*.
Identification key: Key I; key point 30'.
Literature: Karling (1981*a*: 62), Thirumalachar & Lacy (1951).

EXCLUDED SPECIES

LAGENOCYSTIS H. F. Copeland, *The Classification of Lower Organisms*: 82 (1956), synonym of *Lagena* T. C. Vanterpool & G. A. Ledingham, *Canadian Journal of Research* **2**: 192 (1930).

Lagenocystis radicicola (T. C. Vanterpool & G. A. Ledingham) H. F. Copeland, *The Classification of Lower Organisms*: 82 (1956), synonym of *Lagena radicicola* T. C. Vanterpool & G. A. Ledingham, *Canadian Journal of Research* **2**: 192 (1930).

PYTHIELLA J. N. Couch, *Mycologia* **27**: 160 (1935).
Lagenidium W. Zopf, *Verhandlungen des Botanischen Vereins der Provinz Brandenburg* **20**: 79 (1878) [*pro parte minore*] (**Type species**: *Lagenidium rabenhorstii* W. Zopf - holotype, sole original species).
Typification: *Pythiella vernalis* J. N. Couch (holotype - sole original species)

Pythiella vernalis J. N. Couch, *Mycologia* **27**: 160 (1935)
Typification: Preserved material on slides, University of North Carolina Herbarium. Type locality: Chapel Hill [?], North Carolina, U.S.A. Type habitat: parasitic in *Pythium dictyosporum* [presumed to be *P. dictyosporum* Racib. *sensu* Sparrow] and *Pythium* sp. [as *Pythium gracile*] (*Peronosporomycetes*, *Pythiales*) parasitic in *Spirogyra areolata* Lagerh. and *Spirogyra* sp. (*Zygnematales*, *Zygnemataceae*).
Identification key: Key I; key point 39.
Literature: Couch (1935*a*), Karling (1949, 1981*a*: 52), Pystina (1995), Sparrow (1960: 812), Sparrow & Ellison (1949).

Pythiella pythii (A. J. Whiffen) M. W. Dick, **comb. nov.**
Lagenidium pythii A. J. Whiffen, *Journal of the Elisha Mitchell Scientific Society* **62**: 54 (1946).
Typification: type material not designated; lectotype *Journal of the Elisha Mitchell Scientific Society* **62**: plate 7, figs 6, 7, 9, 10-12, 13, 14 (1946) (iconotype). Type locality: Soledad (Cienfuegos), Cuba (coll. S. B. Salvin). Type habitat: parasitic in a digitate sporangial *Pythium* sp. (*Peronosporomycetes*, *Pythiales*) on grass leaf bait in a moistened soil sample.
Identification key: Key I; key point 39'.
Literature: Johnson (1968), Karling (1981*a*: 127, 1987*b*), Pystina (1995), Sparrow (1960: 991), Whiffen (1946).

EXCLUDED SPECIES

Pythiella besseyi (F. K. Sparrow & B. Ellison), F. K. Sparrow, *Aquatic Phycomycetes, 2nd revised edition*: 814 (1960), synonym of *Pleocystidium besseyi* (F. K. Sparrow & B. Ellison) M. W. Dick, *loc. cit.*

CILIATOMYCES I. Foissner & W. Foissner, *ΦYton (Horn, Austria)* **35**: 115 (1995).
Ciliatomyces I. Foissner & W. Foissner, *Zeitschrift für Protistenkunde* **72**: 52 (1986).
Typification: *Ciliomyces spectabilis* I. Foissner & W. Foissner (holotype - sole original species).

Ciliatomyces spectabilis I. Foissner & W. Foissner, *ΦYton (Horn, Austria)* **35**: 115 (1995).
Ciliatomyces spectabilis I. Foissner & W. Foissner, *Zeitschrift für Protistenkunde* **72**: 53 (1986).
Typification: Embedded material has been deposited in the Upper Austria Museum, Linz, Austria. Type locality: Schaming, Eugendorf, Salzburg, Austria. Type habitat: parasitic in cysts of *Kahliella simplex* (Horvath) (*Protozoa*, *Ciliata*) from air-dried meadow soil, after re-moistening.
Identification key: Key I; key point 19.
Literature: Foissner (1987), Foissner & Foissner (1986*a*, *b*, 1995).

SPECIES FOR COMPARISON

Aphanomyces acinetophagus A. F. Bartsch & F. T. Wolf, *American Journal of Botany* **25**: 394 (1938).
Typification: type material not designated; lectotype *American Journal of Botany* **25**: 393, figs 1-9 (1938) (iconotype). Type locality: Yahara River, Madison, Wisconsin, U.S.A. Type habitat: parasitic in *Acineta flava* Claparède & Lachmann (*Protozoa*, *Suctoria*).
Literature: Canter & Dick (1993), Bartsch & Wolf (1938), Scott (1961).

ORDER *INCERTAE SEDIS*

OLPIDIOPSIDALES M. W. Dick, **ordo novum**

Olpidiopsidales M. W. Dick *in* K. A. Pirozynski & D. L. Hawksworth, *Coevolution of Fungi with Plants and Animals*: 37 (1988) [*nomen nudum*].

Lagenidiales J. S. Karling, *American Journal of Botany* **26**: 518 (1939) [*pro parte typo excluso*] (**Type family**: *Lagenidiaceae* J. S. Karling).

Fungi in aquis dulcibus et terris udis, endoparasitici obligate in fungis flagellatis aut algis viridibus. **Thallus** holocarpicus, in facie assimilativa ad initium sine pariete cellulae (osmophagicus), olpidioideis, subsphaericus vel allantoideis, non septatus, sine rhizoideo; thalli saepe gregatim; unusquisque thallus se transformans in structuram reproductionis capacem. **Reproductio asexualis** per zoosporangia; zoosporogenesis intrasporangialis, zoosporae heterokontae, formae principalis, flagellis in parte subapicali insertis; zoosporae pusillae (20-30 μm^3). **Reproductio sexualis** heterothallica, thalli magni fungentes ut oogonia, unumquidque oogonium cum uno vel pluribus thallis parvis contiguis se transformantibus in antheridia (cellulas sociales); oosporae pleroticae, distributio copiae penariae nondum pro certo cognita.

Holotypus: *Olpidium* M. Cornu (*Olpidiopsidaceae* F. K. Sparrow *ex* K. Cejp).

Freshwater and wet-terrestrial fungi, obligately endoparasitic in flagellate fungi or green algae. **Thallus** holocarpic, initially plasmodial (osmophagic), olpidioid, subspherical or allantoid, non septate, without rhizoids; often many thalli together; each thallus becoming transformed into a reproductive structure. **Asexual reproduction** by zoosporangia; zoosporogenesis intrasporangial, zoospores heterokont, of the principal form with subapical flagellar insertion; zoospores small (volume equivalent 20-30 μm^3). **Sexual reproduction** heterothallic, larger thalli functioning as oogonia, each with one or more smaller contiguous thalli functioning as antheridia (companion cells); oospores plerotic, distribution of reserve materials not established.

Typification: *Olpidium* M. Cornu (*Olpidiopsidaceae* F. K. Sparrow *ex* K. Cejp) (holotype - sole original family).

OLPIDIOPSIDACEAE F. K. Sparrow *ex* K. Cejp, *Flora ČSR, Oomycetes I*: 349 & 455 (1959); *Mycologia* **34**: 113 (1942).

Pseudolpidiaceae H. E. Petersen, *Botanisk Tidsskrift* **29**: 357 (1909) (**Type genus**: *Pseudolpidium* A. Fischer, **other genera**: *Olpidiopsis*, *Bicilium*).

Olpidiopsidacea H. F. Copeland, *The Classification of Lower Organisms*: 81 (1956) (**Type genus**: *Olpidiopsis* M. Cornu).

Typification: *Olpidiopsis* M. Cornu (holotype - sole original genus).

OLPIDIOPSIS M. Cornu, *Annales des Sciences Naturelles, Botanique, Série V* **15**: 114, 127 (1872) [*sensu stricto*].

Chytridium A. Braun, *Betrachtungen über die Erscheinung der Verjüngung in der Natur*...: 198 (1851) [*pro parte minore typo excluso*] (**Type species**: *Chytridium olla* A. Braun).

Peronium F. Cohn, *Verhandlungen der Kaiserlichen Leopoldinish-carolinisch Akadmie der Naturforscher* **24**: 159 (1854) [*typo excluso*] (**Type species**: *Peronium aciculare* F. Cohn, ['in *Saprolegniaceae*', specific synonymy not possible).

Diplophysa J. Schroeter *in* F. Cohn, *Kryptogamen-Flora von Schlesien [1885-1889]* **3(1)**: 195 (1886), subgenus 1 *Euolpidiopsis* C. Fisch [as 'E. Fischer'] (**Type species**: *Diplophysa saprolegniae* (M. Cornu) J. Schröter, **syn**. *Olpidiopsis saprolegniae* M. Cornu *pro parte*).

Olpidiopsis (M. Cornu *pro parte*) A. Fischer, *Rabenhorst's Kryptogamen-Flora, 2 Aufl., Bd 1, Abt. 4*: 37 (1892) (**Type species**: *Olpidiopsis saprolegniae* M. Cornu *pro parte*).

Pseudolpidium A. Fischer, *Rabenhorst's Kryptogamen-Flora, 2 Aufl., Bd 1, Abt. 4*: 33 (1892) (**Type species**: *Pseudolpidium saprolegniae* (M. Cornu *pro parte*) A. Fischer).

Pleotrachelus W. Zopf secto *integri* H. E. Petersen, *Oversigt over det Kongelige Danske Videnskabernes Selskabs Forhandlinger. Medlemmers Arbeiter, 1905*: 448 (1905) (**Type species**: *Pleotrachelus andreei* G. Lagerheim).

Pseudolpidiopsis M. von Minden [*pro parte typo excluso*], *Kryptogamenflora der Mark Brandenburg* **5**: 255 (1915) (**Type species**: *Pseudolpidiopsis schenkiana* (W. Zopf) M. von Minden).

Excluded synonyms:

Pleocystidium C. Fisch, *Sitzungsberichte der Physicalish-Medicinischen Societät zu Erlangen* **16**: 66 (1884) (**Type species**: *Pleocystidium parasiticum* C. Fisch) [see *Pleocystidium*].

Bicilium H. E. Petersen *Botanisk Tidsskrift* **29**: 357 (1909). (**Type species**: *Bicilium andreei* ([N.] G. Lagerheim) H. E. Petersen) [see *Eurychasma*].

Pseudolpidiopsis M. von Minden [*pro parte*], *Kryptogamenflora der Mark Brandenburg* **5**: 255 (1911) (**Type species**: *Pseudolpidiopsis schenkiana* (W. Zopf) M. von Minden) [see *Pleocystidium*].

Pseudolpidiella K. Cejp, *Flora ČSR, Oomycetes I*: 460 (1959) (**Type species**: *Pseudolpidiella glenodiniana* (P.-A. Dangeard) K. Cejp) [see *Sirolpidium*].

Typification: *Olpidiopsis saprolegniae* (A. Braun) M. Cornu.

Olpidiopsis saprolegniae (A. Braun) M. Cornu, *Annales des Sciences Naturelles, Botanique, Série V* **15**: 145 (1872) [*sensu lato*].
Chytridium saprolegniae A. Braun, *Bericht über die zur Bekanntmachung geeigneten verhandlungen der Königlichen Preussischen Akademie der Wissenschaften zu Berlin, 1855*: 384 (1855).
Olpidium saprolegniae (A. Braun) L. Rabenhorst, *Flora Europaea Algarum* **3**: 283 (1868).
Diplophysa saprolegniae (M. Cornu) J. Schroeter, *in* F. [J.] Cohn, *Kryptogamen-Flora von Schlesien [1885-1889]* **3(1)**: 195 (1886).
Pseudolpidium saprolegniae A. Fischer, *Rabenhorst's Kryptogamen-Flora, 2 Aufl., Bd 1, Abt. 4*: 35 (1892).
Olpidiopsis milleri A. A. Yačevskie & P. A. Yačevskie, *Opredelitel' Gribov*: 62 (1931).
Olpidiopsis echinata H. E. Petersen, *Botanisk Tidsskrift* **29**: 405 (1909).
non Olpidiopsis irregularis J. C. Constantineanu, *Revue Générale de Botanique* **13**: 373 (1901).
Typification: Lectotype (Cejp, 1959): *Bericht über die zur Bekanntmachung geeigneten verhandlungen der Königlichen Preussischen Akademie der Wissenschaften zu Berlin, 1855*: plate 5, fig. 23 (1855) (iconotype). Type locality: Germany. Type habitat: parasitic in *Saprolegnia* sp. (*Peronosporomycetes*, *Saprolegniaceae*).
Identification key: Key I; key point 71, 74'.
Literature: see under subsidiary taxa.

Olpidiopsis saprolegniae (A. Braun) M. Cornu, *Annales des Sciences Naturelles, Botanique, Série V* **15**: 145 (1872).
var. ***saprolegniae***.
Typification: Lectotype (Cejp, 1959): *Bericht über die zur Bekanntmachung geeigneten verhandlungen der Königlichen Preussischen Akademie der Wissenschaften zu Berlin, 1855*: plate 5, fig. 23 (1855) (iconotype). Type locality: Germany. Type habitat: parasitic in *Saprolegnia* sp. (*Peronosporomycetes*, *Saprolegniaceae*).
Identification key: Key I; key point 74'.
Literature: Arnaud (1952), Barrett (1912), Beneke & Rogers (1962), Beverwijk (1948), Bortnick, Powell & Bangert (1985), Braun (1855, 1856), Butler (1907), Cejp (1934, 1959), Chaudhuri *et al.* (1947), Cienkowski (1855), Coker (1923), Constantineanu (1901), Cornu (1872*b*), Couch (1941), Czeczuga (1994, 1995*a*, *b*, *c*), Czeczuga, Chomutowska & Woronowicz (1990), Dangeard (1890-91*b*), Davis (1914), Diehl (1935), Domján (1936), Fischer (1892), Forbes (1935*b*), Gilman & Archer (1929), Graff (1928), Harder & Gallwitz-Uebelmesser (1959), Harvey (1942), Johnson (1956, 1968, 1972, 1977), Karling (1966, 1968*a*, 1981*a*: 10), Konno (1972), Lopez & MacCarthy (1985), Maneval (1927), Miller (in Yacevskie 1931), Minden (1911*a*), Nägeli (1846), Petersen (1909, 1910), Powell, Lehnen & Bortnick (1985), Pringsheim (1860), Rogers, Milanez & Beneke (1970), Sawada (1912), Schmitt & Beneke (1962), Schroeter *in* Cohn (1886), Schröter (1892), Schwarze (1922), Shanor (1940), Shen & Siang (1948), Sorokine (1883, 1889, 1890), Sparrow (1932, 1933*b*, 1960: 930), Srivastava (1966), Srivastava & Bhargava (1963), Tokunaga (1933*a*), Valkanov (1931*a*), Wolf & Wolf (1941), Yačevskie & Yačevskie (1931: 61, 62).

Olpidiopsis saprolegniae (A. Braun) M. Cornu var. ***laevis*** W. C. Coker, *The Saprolegniaceae, with notes on other Water Molds*: 185 (1923) [as '*levis*'].
Typification: Lectotype (Cejp, 1959): *The Saprolegniaceae, with notes on other Water Molds*: plate 62, figs 1-6 (1923) (iconotype). Type locality: North Carolina, U.S.A. Type habitat: parasitic in *Saprolegnia ferax* (Gruith.) Thur. (*Peronosporomycetes*, *Saprolegniaceae*).
Identification key: Key I; key point 75.
Literature: Coker (1923), Johnson (1956, 1972, 1977), Karling (1981*a*: 14), Sparrow (1960: 933).

Olpidiopsis saprolegniae (A. Braun) M. Cornu var. ***indica*** R. Dayal & J. Thakur, *Sydowia* **22**: 280 (1968 [1969]).
Olpidiopsis saprolegniae (A. Braun) M. Cornu var. *indica* J. Thakur, *Indian Phytopathology* **20**: 386 (1967) [*nomen nudum*].
Typification: type material not designated; lectotype *Sydowia* **22**: 280, figs 6-11 (1968) (iconotype); type culture RD21. Type locality: Varanasi, Uttar Pradesh, India. Type habitat: parasitic in *Saprolegnia* sp. (*Peronosporomycetes*, *Saprolegniaceae*) from channel water near Medical College, January 1966.
Identification key: Key I; key point 75'.
Literature: Dayal & Thakur Ji (1968), Karling (1981*a*: 14), Thakur Ji (1967).

SPECIES SOMETIMES SEPARATED FROM *OLPIDIOPSIS SAPROLEGNIAE*

Olpidiopsis vexans J. T. Barrett, ***Annals of Botany, London* 26**: 231 (1912).

Olpidiopsis saprolegniae A. Fischer, *Rabenhorst's Kryptogamen-Flora, 2 Aufl., Bd 1, Abt. 4*: 38 (1892).

Olpidiopsis incrassata M. Cornu, *Annales des Sciences Naturelles, Botanique, Série V* **15**: 146 (1872) [of some authors].

Pseudolpidium incrassata (M. Cornu) A. Fischer, *Rabenhorst's Kryptogamen-Flora, 2 Aufl., Bd 1, Abt. 4*: 37 (1892) [of some authors].

Olpidiopsis major A. Maurizio, *Jahresbericht der Naturforschenden Gesellschaft Graubündens* **38**: 15 (1895).

Olpidiopsis verrucosa T. W. Johnson, *Journal of the Elisha Mitchell Scientific Society* **71**: 60 (1955).

? *Olpidiopsis saprolegniae* (A. Braun) M. Cornu var. *indica* R. Dayal & J. Thakur, *Sydowia* **22**: 280 (1968 [1969]).

Typification: type material not designated; lectotype *Annals of Botany, London* **26**: plates XXIII, figs 15, 21; XXIV, figs 26, 31-39; XXV, figs 40-42, 44, 47-71; XXVI, fig. 75 (1912) (iconotype). Type locality: Ithaca, New York, U.S.A. Type habitat: parasitic in *Saprolegnia ferax* (Gruith.) Thur. (*Peronosporomycetes, Saprolegniaceae*).

Identification key: Key I; key point 75'.

Literature: Barrett (1912), Cejp (1959), Cornu (1872*b*), Fischer (1892), Graff (1928), Howard *et al.* (1969), Johnson (1955*a*, 1972, 1977), Karling (1942*a*, 1981*a*: 14), Konno (1972), Maurizio (1895), Minden (1915), Petersen (1909, 1910), Shanor (1940), Slifkin (1961, 1962, 1963, 1968), Sorokine (1883, 1889, 1890), Sparrow (1933*c*, 1960: 940), Srivastava & Srivastava (1977).

Olpidiopsis fusiformis M. Cornu, ***Annales des Sciences Naturelles, Botanique, Série V* 15**: 147 (1872) [*sensu lato*].

Pseudolpidium fusiforme A. Fischer, *Rabenhorst's Kryptogamen-Flora, 2 Aufl., Bd 1, Abt. 4*: 35 (1892).

Olpidiopsis minor A. Fischer, *Rabenhorst's Kryptogamen-Flora, 2 Aufl., Bd 1, Abt. 4*: 39 (1892).

Pseudolpidium stellatum K. Sawada *ex* Y. Tokunaga, *Transactions of the Sapporo Natural History Society (Japan)* **13**: 23 (1933); *Special Report. Formosa Agricultural Experiment Station* 3: 70 (1912) [as '*steldlatum*'], lectotype *Special Report. Formosa Agricultural Experiment Station* 3: plate VIII figs 11-16 (1912) (iconotype). Type locality: Japan. Type habitat: parasitic in *Achlya prolifera* Nees (*Peronosporomycetes, Saprolegniaceae*).

Typification: Lectotype (Cejp, 1959): *Annales des Sciences Naturelles, Botanique, Série V* **15**: plate 3, fig. 11; plate 4, figs 1-4 (1872) (iconotype). Type locality: Romorantin, Loir-et-Cher, France. Type habitat: parasitic in *Achlya racemosa* Hildebr. (*Peronosporomycetes, Saprolegniaceae*).

Identification key: Key I; key points 83', 85'.

Literature: Butler (1907), Cejp (1934, 1959), Chaudhuri *et al.* (1947), Cornu (1872*b*), Fischer (1892), Johnson (1956, 1972, 1977), Karling (1966, 1981*a*: 18), Konno (1972), Matthews (1935), Milanez (1967), Milanez & Beneke (1968), Minden (1915), Petersen (1909, 1910), Reinsch (1878), Rogers Milanez & Beneke (1970), Sawada (1912), Shanor (1939*b*, 1940), Shen & Siang (1948), Sorokine (1883, 1889, 1890), Sparrow (1932*b*, 1936*a*, 1960: 934), Srivastava (1966), Tokunaga (1933*a*), Yačevskie & Yačevskie (1931: 61, 62).

Olpidiopsis incrassata M. Cornu, ***Annales des Sciences Naturelles, Botanique, Série 5* 15**: 146 (1872).

Pseudolpidium incrassatum (M. Cornu) A. Fischer, *Rabenhorst's Kryptogamen-Flora, 2 Aufl., Bd 1, Abt. 4*: 37 (1892).

Olpidiopsis major A. Maurizio, *Jahresbericht der Naturforschenden Gesellschaft Graubündens* **38**: 15 (1895).

Typification: Lectotype (Cejp, 1959): *Annales des Sciences Naturelles, Botanique, Série V* **15**: plate 4, fig. 12 (1872) (iconotype). Type locality: Villeherviers, Loir-et-Cher, France. Type habitat: parasitic in *Achlya racemosa* Hildebr. (*Peronosporomycetes, Saprolegniaceae*) [*A. leucosperma* Cornu, the first-recorded host, is unidentifiable].

Identification key: Key I; key point 84'.

Literature: Barrett (1912), Cejp (1959), Cornu (1872*b*), Fischer (1892), Karling (1981*a*: 14), Maurizio (1895), Sparrow (1960: 938, 940), Willoughby & Roberts (1992), Yačevskie & Yačevskie (1931: 61).

Olpidiopsis index M. Cornu, ***Annales des Sciences Naturelles, Botanique, Série V* 15**: 145 (1872).

Typification: type material not designated; status of species doubtful. Type locality: Sologne, France. Type habitat: parasitic in an unidentified *Achlya* sp. (*Peronosporomycetes, Saprolegniaceae*).

Identification key: Key I; key point 85.

Literature: Cornu (1872*b*), Karling (1981*a*: 19), Sorokine (1883, 1889, 1890), Sparrow (1960: 936), Yačevskie & Yačevskie (1931: 62).

Olpidiopsis spinosa Y. Tokunaga, *Transactions of the Sapporo Natural History Society (Japan)* **13**: 23 (1933).
Typification: type material not designated; lectotype *Transactions of the Sapporo Natural History Society (Japan)* **13**: plate 2, figs10,11 (1933) (iconotype). Type locality: Kotoni, Ishikari Province, Hokkaido, Japan. Type habitat: parasitic in *Achlya flagellata* Coker (*Peronosporomycetes*, *Saprolegniaceae*).
Identification key: Key I; key points 83, 86, 88.
Literature: Karling (1981*a*: 20), Konno (1972), Sparrow (1960: 937), Tokunaga (1933*a*).

Olpidiopsis varians L. Shanor, *Journal of the Elisha Mitchell Scientific Society* **55**: 171 (1939).
Typification: type material not designated; lectotype *Journal of the Elisha Mitchell Scientific Society* **55**: 169, fig. 1A-E, plate 24, figs 1-13 (1939) (iconotype). Type locality: Burgaw, North Carolina, U.S.A. Type habitat: parasitic in *Achlya flagellata* Coker (*Peronosporomycetes*, *Saprolegniaceae*).
Identification key: Key I; key point 88'.
Literature: Beneke & Schmidt (1961), Held (1972*b*), Johnson (1972, 1977), Karling (1981*a*: 19), Martin & Miller (1982, 1985, 1986*a*, *b*, *c*), Rogers, Milanez & Beneke (1970), Shanor (1939*a*, *b*, 1940), Sparrow (1960: 937), Srivastava (1964).

Olpidiopsis achlyae D. A. McLarty, *Bulletin of the Torrey Botanical Club* **68**: 62 (1941).
Olpidiopsis fusiformis M. Cornu forma *achlyae* (D. A. McLarty) K. Cejp, *Flora ČSR, Oomycetes I*: 458 (1959) [not typified by Cejp, 1959].
Typification: type material not designated; lectotype *Bulletin of the Torrey Botanical Club* **68**: 52, figs 1-11, 55, figs 12-26 (1941) (iconotype). Type locality: London, Ontario, Canada. Type habitat: parasitic in *Achlya flagellata* Coker (*Peronosporomycetes*, *Saprolegniaceae*).
Identification key: Key I; key point 89.
Literature: Das-Gupta & John (1953), Dayal & Thakur Ji (1968), Johnson (1972, 1977), Karling (1949, 1966, 1968*a*, 1981*a*: 20), Manoharachary (1985), McLarty (1941*a*, *b*), Rogers, Milanez & Beneke (1970), Sparrow (1960: 935) Thakur Ji (1967).

Olpidiopsis verrucosa T. W. Johnson, *Journal of the Elisha Mitchell Scientific Society* **71**: 60 (1955).
non Olpidiopsis verrucosa Johnson var. *radicicola* B. Della Torre, *Rivista di Patologia Vegetale, Padova, Serie 3* **4**: 91 (1964).
Typification: type material not designated; lectotype *Journal of the Elisha Mitchell Scientific Society* **71**: 61, figs 1-24 (1955) (iconotype). Type locality: Bagley tower, Tallahatchie Experimental Forest, Lafayette County, Mississippi, U.S.A. Type habitat: parasitic in *Achlya glomerata* Coker (*Peronosporomycetes*, *Saprolegniaceae*), from a soil sample.
Identification key: Key I; key point 89'.
Literature: Cejp (1959), Johnson (1955*a*, 1956), Karling (1981*a*: 21), Sparrow (1960: 954).

Olpidiopsis braziliensis F. K. Sparrow, *Aquatic Phycomycetes, 2nd Revised Edition*: 933 (1960).
Pseudolpidiopsis achlyae A. P. Viégas & A. R. Teixeira, *Bragantia* **3**: 225 (1943).
Typification: Type in Laboratório de Botânica, Instituto Agronômico, Sao Paulo. Type locality: Campinas, Province Sao Paulo, Brazil. Type habitat; parasitic in hyphae and sporangia of *Achlya* sp. (*Peronosporomycetes*, *Saprolegniaceae*).
Identification key: Key I; key point 86'.
Literature: Karling (1981*a*: 21), Sparrow (1960: 933), Viégas & Teixeira (1943).

Olpidiopsis aphanomycis M. Cornu, *Annales des Sciences Naturelles, Botanique, Série V* **15**: 148 (1872) [*sensu lato*].
Pseudolpidium aphanomycis (M. Cornu) A. Fischer, *Rabenhorst's Kryptogamen-Flora, 2 Aufl., Bd 1, Abt. 4*: 37 (1892).
Olpidiopsis luxurians J. T. Barrett, *Annals of Botany, London* **26**: 231 (1912).
Typification: Lectotype (Cejp, 1959): *Annales des Sciences Naturelles, Botanique, Série V* **15**: plate 4, figs 5-11 (1872) (iconotype). Type locality: Paris, France. Type habitat: parasitic in *Aphanomyces* sp. (*Peronosporomycetes*, *Saprolegniales*).
Identification key: Key I; key points 71', 73'.
Literature: Barrett (1912), Butler (1907), Cejp (1959), Chaudhuri *et al.* (1947), Cornu (1872*b*), Czeczuga (1994), Dangeard (1890-91*b*), Fischer (1892), Howard & Johnson (1969), Johnson (1977), Karling (1942*a*, 1949, 1966, 1968*a*, 1976*b*, 1981*a*: 21, 1981*b*), Kiran & Dayal (1983), Miller (1961, 1965), Minden (1915), Petersen (1909, 1910), Scott *et al.* (1963), Shanor (1939*b*), Shen & Siang (1948), Sparrow (1960: 942), Sparrow, Paterson & Johns (1965), Whiffen (1942*b*).

SPECIES SOMETIMES SEPARATED FROM *OLPIDIOPSIS APHANOMYCIS*

Olpidiopsis luxurians J. T. Barrett, *Annals of Botany, London* **26**: 231 (1912).
Typification: type material not designated; lectotype *Annals of Botany, London* **26**: plates XXIII, figs 1, 3-7, 10-14, 17, 18, 21B-23; XXIV, figs 24, 25, 28-30; XXV, figs 43, 45, 46; XXVI, figs 74, 76-90 (1912) (iconotype). Type locality: Ithaca, New York, U.S.A. Type habitat: parasitic in *Aphanomyces laevis* de Bary (*Peronosporomycetes*, *Saprolegniales*).
Identification key: Key I; key point 73.
Literature: Barrett (1912), Karling (1942*a*, 1968*d*, 1981*a*: 21), Konno (1972), Sparrow (1952*c*, 1960: 941).

Olpidiopsis pythii (E. J. Butler) J. S. Karling, *The Simple Holocarpic Biflagellate Phycomycetes*: 47 (1942) [*sensu lato*].
Pseudolpidium pythii E. J. Butler, *Memoirs of the Department of Agriculture, India, Botanical Series* **1**: 129 (1907).
Typification: Lectotype (Cejp, 1959): *Memoirs of the Department of Agriculture, India, Botanical Series* **1**: plate 7, figs 9-16 (1907) (iconotype). Type locality: Antibes, France. Type habitat: parasitic in various *Pythium* spp. (*Peronosporomycetes*, *Pythiaceae*) from garden soil.
Identification key: Key I; key point 61'.
Literature: Butler (1907), Cejp (1959), Chaudhuri *et al.* (1947), Dangeard (1890-91*b*), Dogma (1975), Johnson (1977), Karling (1942, 1949*a*, 1966, 1968*a*, 1969, 1981*a*: 24), Konno (1972), Miller (1962), Minden (1915), Remy (1948), Sparrow (1936*a*, 1957, 1960: 944), Tokunaga (1933*a*), Yačevskie & Yačevskie (1931: 61).

SPECIES SOMETIMES SEPARATED FROM *OLPIDIOPSIS PYTHII SENSU STRICTO*

Olpidiopsis pythii (E. J. Butler) J. S. Karling forma ***brevispinosa*** (A. J. Whiffen) K. Cejp, *Flora ČSR, Oomycetes I*: 371 (1959).
Olpidiopsis brevispinosa A. J. Whiffen, *American Journal of Botany* **29**: 610 (1942).
Typification: Lectotype Cejp (1959): *American Journal of Botany* **29**: 608, figs 2, 22, 27 (1942) (iconotype) [for further details see under *Olpidiopsis brevispinosa*].
Identification key: Key I;, key point 63.
Literature: Cejp (1959), Karling (1981*a*: 25), Whiffen (1942).

SPECIES SOMETIMES SEPARATED FROM *OLPIDIOPSIS PYTHII*

Olpidiopsis brevispinosa A. J. Whiffen, *American Journal of Botany* **29**: 610 (1942).
Olpidiopsis pythii (E. J. Butler) J. S. Karling forma *brevispinosa* (A. J. Whiffen) K. Cejp, *Flora ČSR, Oomycetes I*: 371 (1959).
Typification: Lectotype (Cejp, 1959): *American Journal of Botany* **29**: 608, fig. 2, 609 figs 22, 27 (1942) (iconotype). Type locality: Delhi, Louisiana, U.S.A. (coll. P. Couch). Type habitat: parasitic in a *Pythium* sp. (*Peronosporomycetes*, *Pythiaceae*).
Identification key: Key I; key point 63.
Literature: Cejp (1959), Karling (1968*a*, 1979*b*, 1981*a*: 25), Sparrow (1960: 945), Whiffen (1942).

TAXON SOMETIMES NOT ACCORDED SPECIFIC RANK

Olpidiopsis curvispinosa A. J. Whiffen, *American Journal of Botany* **29**: 610 (1942).
Olpidiopsis gracilis (E. J. Butler) J. S. Karling forma *curvispinosa* (A. J. Whiffen) K. Cejp, *Flora ČSR, Oomycetes I*: 370 (1959).
Typification: type material not designated; forma incompletely lectotypified by Cejp (1959), lectotype *American Journal of Botany* **29**: 608, figs 1,5,6, 609, fig. 21 (1942) (iconotype). Type locality: Chapel Hill, North Carolina, U.S.A. Type habitat: parasitic in a *Pythium* sp. (*Peronosporomycetes*, *Pythiaceae*).
Identification key: Key I; key point 63'.
Literature: Cejp (1959), Chaudhuri *et al.* (1947), Karling (1981*a*: 25, 1981*e*, 1988), Sparrow (1957, 1960: 945), Whiffen (1942).

Olpidiopsis indica G. C. Srivastava, *Current Science, Bangalore, India* **44**: 642 (1975).
Typification: Type slide(s): Indian Type Culture Collection, IARA, New Delhi, India; cultures deposited in the Mycological Laboratory, Department of Botany, St Andrew's College, Gorakhpur, India. Type locality: Gorakhpur, Uttar Pradesh, India. Type habitat: parasitic in *Pythium aphanidermatum* (Edson) Fitzp. (*Peronosporomycetes*, *Pythiaceae*), from a decaying leaf of *Carica*.
Identification key: Key I; key point 62'.
Literature: Karling (1981*a*: 25), Pystina (1995), Srivastava (1975).

EXCLUDED SPECIES

Olpidiopsis andreei ([N.] G. Lagerheim) J. S. Karling, *The Simple Holocarpic Biflagellate Phycomycetes*: 51 (1942) [*pro parte*, *sensu* Lagerheim and Petersen], synonym of *Sirolpidium andreei* ([N.] G. Lagerheim) M. W. Dick, *loc. cit.*

Olpidiopsis andreei ([N.] G. Lagerheim) J. S. Karling, *The Simple Holocarpic Biflagellate Phycomycetes*: 51 (1942) [*pro parte*, *sensu* Sparrow], synonym of *Sirolpidium ectocarpi* (M. Jokl) M. W. Dick, *loc. cit.*

Olpidiopsis antithamnionis A. Whittick & G. R. South, *Archiv für Mikrobiologie* **82**: 354 (1972), synonym of *Pontisma antithamnionis* (A. Whittick & G. R. South) M. W. Dick, *loc. cit.*

Olpidiopsis appendiculata É. de Wildeman, *La Notarisia* **10**: 34 (1895), synonym of *Pleocystidium parasiticum* C. Fisch, *Sitzungsberichte der Physicalisch-Medicinischen Societät zu Erlangen* **16**: 67 (1884).

Olpidiopsis cruciferarum B. Della Torre, *Rivista di Patologia Vegetale, Padova, Serie 3* **4**: 83 (1964), synonym of *Olpidium cruciferarum* (B. Della Torre) M. W. Dick *loc. cit.*

Olpidiopsis dangeardii J. Feldmann & G. Feldmann, *Le Botaniste* **50**: 183 (1967), synonym of *Pontisma dangeardii* (J. Feldmann & G. Feldmann) M. W. Dick, *loc. cit.*

Olpidiopsis decipiens S. N. Dasgupta & R. John, *Bulletin of the Botanical Survey of India* **30**: 59 (1988), synonym of an unidentifiable *Syzygangia* sp. (see ms pp. 155, 162, 423).

Olpidiopsis deformans (I. L. Serbinov*) J. S. Karling, *Predominantly Holocarpic and Eucarpic Simple Biflagellate Phycomycetes*: 30 (1981), synonym of *Plasmophagus deformans* (I. L. Serbinov) M. W. Dick, *loc. cit.*

Olpidiopsis echinata H. E. Petersen, *Botanisk Tidsskrift* **29**: 405 (1909), synonym of *Olpidiopsis saprolegniae* (A. Braun) M. Cornu, *Annales des Sciences Naturelles, Botanique, Série 5* **15**: 145 (1872).

Olpidiopsis elliptica (J. Schroeter) A. Fischer, *Rabenhorst's Kryptogamen-Flora, 2 Aufl., Bd 1, Abt. 4*: 41 (1892), synonym of *Pleocystidium parasiticum* C. Fisch, *Sitzungsberichte der Physicalisch- Medicinischen Societät zu Erlangen* **16**: 67 (1884).

Olpidiopsis endogena J. S. Karling, *Nova Hedwigia* **43**: 7 (1985), synonym of *Cornumyces endogenus* (J. S. Karling) M. W. Dick, *loc. cit.*

Olpidiopsis feldmannii A. A. Aleem, *Comptes Rendus Hebdomadaires de Séances de l'Académie des Sciences, Paris, Série D* **235**: 1250 (1952) [as '*feldmanni*'], synonym of *Pontisma feldmannii* (A. A. Aleem) M. W. Dick, *loc. cit.*

Olpidiopsis fibrillosa É. de Wildeman, *La Notarisia* **10**: 34 (1895), synonym of *Pleocystidium parasiticum* C. Fisch, *Sitzungsberichte der Physicalisch- Medicinischen Societät zu Erlangen* **16**: 67 (1884).

Olpidiopsis fusiformis M. Cornu forma *achlyae* (D. A. McLarty) K. Cejp, *Flora ČSR, Oomycetes I*: 458 (1959), synonym of *Olpidiopsis achlyae* D. A. McLarty, *Bulletin of the Torrey Botanical Club* **68**: 62 (1941).

Olpidiopsis fusiformis M. Cornu var. *oedogoniarum* N. Sorokin, *Archives Botaniques du Nord de la France* **2**: 30 (1883), synonym of *Syzygangia oedogonii* (A. Scherffel) M. W. Dick, *loc. cit.*

Olpidiopsis gillii (É. de Wildeman) I. Friedmann, *Österreichische Botanische Zeitschrift* **99**: 200 (1952), synonym of *Ectrogella bacillariacearum* W. Zopf, *Nova Acta der Kaiserlichen Leopoldinisch-carolinischen Akademie der Naturforscher* **47**: 175 (1884).

Olpidiopsis glenodiniana (P.-A. Dangeard) T. W. Johnson, *Nova Hedwigia* **10**: 586 (1966), synonym of *Sirolpidium glenodinianum* (P.-A. Dangeard) M. W. Dick, *loc. cit.*

Olpidiopsis globosa C. J. Anastasiou & L. M. Churchland, *Syesis* **1**: 81 (1968), synonym of *Sirolpidium globosum* (C. J. Anastasiou & L. M. Churchland) M. W. Dick, *loc. cit.*

Olpidiopsis gracilis (E. J. Butler) J. S. Karling, *The Simple Holocarpic Biflagellate Phycomycetes*: 47 (1942), synonym of *Gracea gracilis* (E. J. Butler) M. W. Dick, *loc. cit.*

Olpidiopsis irregularis J. C. Constantineanu, *Revue Générale de Botanique* **13**: 373 (1901), synonym of *Cornumyces irregularis* (J. C. Constantineanu) M. W. Dick, *loc. cit.*

Olpidiopsis karlingiae J. S. Karling, *Mycologia* **41**: 271 (1949), synonym of *Cornumyces karlingiae* (J. S. Karling) M. W. Dick, *loc. cit.*

Olpidiopsis longicollis W. Zopf, *in* A. Schenk, *Handbuch de Botanik* **4**: 508 (1890) [*nomen nudum*, **Literature**: Karling (1981*a*: 40), Schenk (1890), Sparrow (1960: 955)].

---

*FOOTNOTE: The authority is given here following the *British Standard, No. 2979 (1958)* English transliteration of И. Л. Сербинов. In the *References* the name is spelt as given, either I. L. Serbinov or J. L. Serbinow.

Olpidiopsis magnusii J. Feldmann & G. Feldmann, *Revue de Mycologie* **20**: 243 (1955), synonym of *Pontisma magnusii* (J. Feldmann & G. Feldmann) M. W. Dick, *loc. cit.*

Olpidiopsis major A. Maurizio, *Jahresbericht der Naturforschenden Gesellschaft Graubündens* **38**: 15 (1895), synonym of *Olpidiopsis incrassata* M. Cornu, *Annales des Sciences Naturelles, Botanique, Série V* **15**: 146 (1872).

Olpidiopsis milleri A. A. Yačevskie & P. A. Yačevskie, *Opredelitel' Gribov*: 61, 62 (1931), basionym *Olpidiopsis* sp. Miller [basionym not identified] [**Typification**: on *Saprolegnia monoica* N. Pringsheim (*Peronosporomycetes, Saprolegniaceae*)], synonym of *Olpidiopsis saprolegniae* (A. Braun) M. Cornu, *Annales des Sciences Naturelles, Botanique, Série V* **15**: 145 (1872).

Olpidiopsis minor A. Fischer, *Rabenhorst's Kryptogamen-Flora, 2 Aufl., Bd 1, Abt. 4*: 39 (1892), synonym of *Olpidiopsis fusiformis* M. Cornu, *Annales des Sciences Naturelles, Botanique, Série V* **15**: 147 (1872).

Olpidiopsis myžocytia A. Rieth *ex* K. Cejp, *Flora ČSR, Oomycetes I*: 458 (1959); A. Rieth, *Die Kulturpflanze* **2**: 184 (1954), synonym of *Myzocytium proliferum* A. Schenk, *Über das Vorkommen contractiler Zellen im Pflanzenreich*: 10 (1858).

Olpidiopsis paradoxa (H. E. Petersen) T. W. Johnson, *American Journal of Botany* **53**: 949 (1966), synonym of *Sirolpidium paradoxum* (H. E. Petersen) M. W. Dick, *loc. cit.*

Olpidiopsis parasitica (C. Fisch) A. Fischer, *Rabenhorst's Kryptogamen-Flora, 2 Aufl., Bd 1, Abt. 4*: 40 (1892), synonym of *Pleocystidium parasiticum* C. Fisch, *Sitzungsberichte der Physicalisch- Medicinischen Societät zu Erlangen* **16**: 67 (1884).

Olpidiopsis oedogoniarum (N. Sorokin) A. Scherffel, *Archiv für Protistenkunde* **52**: 103?8 (1925), synonym of *Syzygangia oedogonii* (A. Scherffel) M. W. Dick, *loc. cit.*

Olpidiopsis ricciae S. J. du Plessis, *Annals of Botany, London* **47**: 761 (1933), synonym of *Syzygangia elliptica* (É. de Wildeman) M. W. Dick, *loc. cit.*

Olpidiopsis schenkiana W. Zopf, *Nova Acta der Kaiserlichen Leopoldinisch-carolinischen Akademie der Naturforscher* **47**: 168 (1884), synonym of *Pleocystidium parasiticum* C. Fisch, *Sitzungsberichte der Physicalisch- Medicinischen Societät zu Erlangen* **16**: 67 (1884).

Olpidiopsis sorokinei É. de Wildeman, *Annales de la Société Belge de Microscopie, Mémoires* **14**: 22 (1890), possible synonym of *Syzygangia oedogonii*, also synonym of *Olpidium sorokinei* (É. de Wildeman) É. de Wildeman, *Bulletin de la Société Royale de Botanique de Belge, Mémoires* **35**: 16 (1896).

Olpidiopsis sphaerica S. N. Dasgupta & R. John, *Bulletin of the Botanical Survey of India* **30**: 62 (1988) [as '*sphaericis*'], synonym of *Pleocystidium parasiticum* C. Fisch, *Sitzungsberichte der Physicalisch- Medicinischen Societät zu Erlangen* **16**: 67 (1884).

Olpidiopsis sphaeritae (P.-A. Dangeard) J. Schröter, *Die natürlichen Pflanzenfamilien* **1(1)**: 69 (1897), unidentifiable, probably autoparasitism in *Pseudosphaerita* or *Sphaerita*.

Olpidiopsis tumefaciens (P. Magnus) K. Cejp, *Flora ČSR, Oomycetes I* p. 377 (1959), synonym of *Eurychasmidium tumefaciens* (P. Magnus) F. K. Sparrow, *Biological Bulletin of the Marine Biological Laboratory, Woods Hole* **70**: 241 (1936).

Olpidiopsis ucrainica C. Wize, *Bulletin International de l'Académie des Sciences de Cracovie, Classe des Sciences Mathématiques et Naturelles*] 1904: **10**: 715 (1904), synonym of *Myiophagus ucrainicus* (C. Wize) F. K. Sparrow, *Mycologia* **31**: 443 (1939).

Olpidiopsis verrucosa T. W. Johnson, *Journal of the Elisha Mitchell Scientific Society* **71**: 60 (1955) var. *radicicola* B. Della Torre, *Revista di Patologia Vegetale, Padova, Serie 3* **4**: 91 (1964), synonym of *Olpidium verrucosum* (B. Della Torre) M. W. Dick *loc. cit.*

Olpidiopsis vuilleminii G. Arnaud, *Bulletin trimestriel de la Société Mycologique de France* 68: 183 (1952) [*nom. inval.*, Art. 36.1, without Latin diagnosis], synonym of *Pleotrachelus vuilleminii* (G. Arnaud) M. W. Dick, *loc. cit.*

Olpidiopsis zopfii É. de Wildeman, *La Notarisia* **10**: 34 (1895), synonym of *Pleocystidium parasiticum* C. Fisch, *Sitzungsberichte der Physicalisch- Medicinischen Societät zu Erlangen* **16**: 67 (1884).

UNNAMED SPECIES OF *OLPIDIOPSIS*

Olpidiopsis sp. T. W. Johnson, *Mycologia* **49**: 397 (1957), synonym of *Ectrogella perforans* H. E. Petersen, *Oversigt over det Kongelige Danske Videnskabernes Selskabs Forhandlinger. Medlemmers Arbeiter, 1905*: 466 (1905).
Olpidiopsis sp. (Johnson, 1972; Karling, 1981*a*: 16) in *Saprolegnia*, referable to *Olpidiopsis*.
Olpidiopsis sp. (Karling, 1987, 1988) in *Brevilegnia*, referable to *Olpidiopsis*.
Olpidiopsis sp. (Howard & Johnson, 1969; Karling (1981*a*: 24) in *Aphanomyces* sp., referable to *Olpidiopsis*.
Olpidiopsis sp. (Sparrow, 1957; Karling, 1981*a*: 36), referable to *Syzygangia* in *Mougeotia*.
Olpidiopsis sp. (Sparrow, 1969; Karling, 1981*a*: 38), referable to *Petersenia* or *Pontisma* in *Porphyra*.
Olpidiopsis sp. (Johnson, 1966*d*), referable to *Ectrogella*.
Olpidiopsis sp. (Aleem, 1953), referable to *Ectrogella*.

EXCLUDED GENERA

DIPLOPHYSA J. Schroeter, *Kryptogamen-Flora von Schlesien [1885-1889]* 3(1): 195 (1886), synonym of *Olpidiopsis* M. Cornu, *Annales des Sciences Naturelles, Botanique, Série V* **15**: 114, 127 (1872) [*sensu lato*], synonym of *Pleocystidium* C. Fisch.

PSEUDOLPIDIUM A. Fischer, *Rabenhorst's Kryptogamen-Flora, 2 Aufl., Bd 1, Abt. 4*: 33 (1892), synonym of *Olpidiopsis* M. Cornu, *Annales des Sciences Naturelles, Botanique, Série V* **15**: 114, 127 (1872) [*pro parte*].

PLEOCYSTIDIUM C. Fisch, *Sitzungsberichte der Physicalisch-Medicinischen Societät zu Erlangen* **16**: 60 (1884).

Diplophysa J. Schroeter, *Kryptogamen-Flora von Schlesien [1885-1889]* **3(1)**: 195 (1886) (**Type species**: *Diplophysa saprolegniae* (M. Cornu) J. Schröter, **syn.** *Olpidiopsis saprolegniae* M. Cornu *pro parte*); subgenus 2 *Pleocystidium* C. Fisch [as 'E. Fischer'], *Sitzungsberichte der Physicalisch-Medicinischen Societät zu Erlangen* **16**: 67 (1884) (**Type species**: *Diplophysa schenkiana* (C. Fisch) J. Schröter).

Olpidiopsis (M. Cornu *pro parte*) A. Fischer, *Rabenhorst's Kryptogamen-Flora, 2 Aufl., Bd 1, Abt. 4*: 37 (1892) (**Type species**: *Olpidiopsis saprolegniae* M. Cornu *pro parte*).

Pseudolpidiopsis M. von Minden, *Kryptogamenflora der Mark Brandenburg* 5: 255 (1911) (**Type species**: *Pseudolpidiopsis schenkiana* (W. Zopf) M. von Minden).

Pleolpidiopsis M. von Minden *in* A. A. Yačevskie & P. A. Yačevskie, *Opredelitel' Gribov*: 16 (1931) [a repeated misprint for *Pseudolpidiopsis* M. von Minden, *Kryptogamenflora der Mark Brandenburg* 5: 255 (1911)].

Typification: *Pleocystidium parasiticum* C. Fisch.

Pleocystidium parasiticum C. Fisch, *Sitzungsberichte der Physicalisch- Medicinischen Societät zu Erlangen* **16**: 60 (1884).

Olpidiopsis schenkiana W. Zopf, *Nova Acta der Kaiserlichen Leopoldinisch-carolinischen Akademie der Naturforscher* **47**: 168 (1884).

Olpidiopsis parasitica (C. Fisch) A. Fischer, *Rabenhorst's Kryptogamen-Flora, 2 Aufl., Bd 1, Abt. 4*: 40 (1892).

Diplophysa schenkiana (W. Zopf) J. Schröter, *Die natürlichen Pflanzenfamilien* **1(1)**: 85 (1893).

Pseudolpidiopsis schenkiana (W. Zopf) M. von Minden, *Kryptogamenflora der Mark Brandenburg* **5**: 257 (1911).

Pseudolpidiopsis parasitica (C. Fisch) M. von Minden, *Kryptogamenflora der Mark Brandenburg* **5**: 258 (1911).

Olpidiopsis elliptica (J. Schroeter) A. Fischer, *Rabenhorst's Kryptogamen-Flora, 2 Aufl., Bd 1, Abt. 4*: 41 (1892).

Diplophysa elliptica J. Schroeter, *Kryptogamen-Flora von Schlesien [1885-1889]* **3(1)**: 196 (1886).

Pseudolpidiopsis elliptica (J. Schroeter) M. von Minden, *Kryptogamenflora der Mark Brandenburg* **5(2)**: 260 (1911).

Pleolpidiopsis schenkiana (C. Fisch) M. von Minden *in* A. A. Yačevskie & P. A. Yačevskie, *Opredelitel' Gribov*: 16 (1931) [a repeated misprint for *Pseudolpidiopsis schenkiana* (C. Fisch) M. von Minden, *Kryptogamenflora der Mark Brandenburg* 5: 255 (1911)].

Olpidiopsis zopfii É. de Wildeman, *La Notarisia* **10**: 34 (1895) (**typification**: lectotype (Cejp, 1959): *Annales de la Société Belge de Microscopie, Mémoires* **20**: plate 1, figs 1-3, 5-7 (1896) (iconotype); type locality: Houffalize, Luxembourg)

Pseudolpidiopsis zopfii (É. de Wildeman) M. von Minden, *Kryptogamenflora der Mark Brandenburg* **5**: 259 (1911).

Olpidiopsis fibrillosa É. de Wildeman, *La Notarisia* **10**: 34 (1895) (**typification**: lectotype (Cejp, 1959): *Annales de la Société Belge de Microscopie, Mémoires* **20**: plate 13, fig. 18; plate 14, fig. 19 (1896*b*) (iconotype); type locality: Houffalize, Luxembourg).

Pseudolpidiopsis fibrillosa (É. de Wildeman) M. von Minden, *Kryptogamenflora der Mark Brandenburg* **5**: 259 (1911).

Olpidiopsis appendiculata É. de Wildeman, *La Notarisia* **10**: 34 (1895) (**typification**: Lectotype (Cejp, 1959): *Annales de la Société Belge de Microscopie, Mémoires* **20**: plate 1, figs 4, 8-12 (1896*b*) (iconotype); type locality: Beloeil, Belgium).

Pseudolpidiopsis appendiculata (É. de Wildeman) M. von Minden, *Kryptogamenflora der Mark Brandenburg* **5**: 259 (1911).

Olpidiopsis sphaericis S. N. Dasgupta & R. John, *Bulletin of the Botanical Survey of India* **30**: 62 (1988) (**typification**: holotype deposited at HCIO, Division of Mycology and Plant Pathology, IARI, New Delhi-110 012. HCIO No. 37181).

Typification: Lectotype (Cejp, 1959): *Sitzungsberichte der Physicalisch-Medicinischen Societät zu Erlangen* **16**: figs 24-39 (1884) (iconotype). Type locality: Berlin, Germany. Type habitat: parasitic in *Spirogyra* sp. (*Zygnematales, Zygnemataceae*).

Identification key: Key I; key point 98.

Literature: Bérczi (1940), Butler (1907), Cejp (1959), Chaudhuri *et al.* (1947), Constantineanu (1901), Dasgupta & John (1988), Domján (1936), Fisch (1884a), Fischer (1892), Johnson (1972), Karling (1942*a*, 1966, 1981*a*: 28, 35), Konno (1972), Krenner (1935), Minden (1911*a*), Pystina (1995), Scherffel (1925), Schroeter (1885, 1897), Shen & Siang (1948), Sparrow (1957, 1960: 947, 950, 951, 954), Sparrow & Ellison (1949), Sparrow & Koch (1959), Tokunaga (1933*b*), Wildeman (1890*c*, 1895*d*, 1896*b*), Yačevskie & Yačevskie (1931: 16), Zopf (1884, 1890: 555).

Pleocystidium lundiae **(J. S. Karling) M. W. Dick, comb. nov.**

Lagenidium lundiae J. S. Karling, *Predominantly Holocarpic and Eucarpic Simple Biflagellate Phycomycetes*: 123 (1981) [as '*lundii*', named after Mrs J. W. G. Lund (Dr Hilda Canter)].

Biflagellatre species (2) H. M. Canter & J. W. G. Lund, *Österreichische Botanische Zeitschrift* **116**: 355 (1969).

Typification: Iconotype *Österreichische Botanische Zeitschrift* **116**: 358, fig. 3 a-l (1969); redrawn as the iconotypus in *Predominantly Holocarpic and Eucarpic Simple Biflagellate Phycomycetes*: 125, plate 34, figs 119-126 (1981); excluded illustrations [photographs] in *Österreichische Botanische Zeitschrift* **116**: 364, fig.5 f-g (1969). Preserved material may be available from Dr H. M. Canter (FBA) or Dr C. Soeder. Type locality: Titisee, Germany. Type habitat: *Staurastrum* sp. cf. *gracile* (*Zygnematales*, *Desmidiaceae*) [also in *Cosmarium* sp., *Staurodesmus* sp. and *Xanthidium* sp.].

Identification key: Key I; key point 97'.

Literature: Canter & Lund (1969), Karling (1981*a*: 123), Pongratz (1966), Skuja (1956).

Pleocystidium pygmaeoides **(J. S. Karling) M. W. Dick, comb. et stat. nov.**

Lagenidium pygmaeum var. *pygmaeoides* J. S. Karling, *Nova Hedwigia* **35**: 112 (1981) [as '*pygmaeodies*'].

Typification: Iconotype *Nova Hedwigia* **35**: 110-111, figs 20-54 (1981). Type locality: Kelaniya, Sri Lanka. Type habitat: parasitic [saprotrophic?] in pine pollen grains and germ tubes.

Identification key: Key I; key point 126'.

Literature: Karling (1981*d*).

UNIDENTIFIED SPECIES

Olpidiopsis sp. in *Mougeotia* (Sparrow, 1957; Karling 1981*a*: 36).

EXCLUDED SPECIES

PSEUDOLPIDIOPSIS M. von Minden, *Kryptogamenflora der Mark Brandenburg* **5**: 255 (1911), synonym of *Olpidiopsis* M. Cornu, *Annales des Sciences Naturelles, Botanique, Série V* **15**: 114, 127 (1872) [*pro parte*].

Pleolpidiopsis ramosa (J. L. Serbinov) A. A. Yačevskie & P. A. Yačevskie, *Opredelitel' Gribov*: 16 (1931) (**syn**: *Olpidium ramosum* J. L. Serbinov, *Leningradskoe oshchestvo estestvoispytateleĭ Trudy* **30**: 255 (1899), description based on a mixture of unidentifiable taxa (Sparrow, 1960: 157).

GENUS *INCERTAE SEDIS* INTERMEDIATE BETWEEN THE *OLPIDIOPSIDALES* AND THE *ROZELLOPSIDALES*

GRACEA M. W. Dick, *Mycological Research* **101**: 882 (1997).

Olpidiopsis M. Cornu, *Annales des Sciences Naturelles, Botanique, Série V* **15**: 114, 127 (1872) [*pro parte typo excluso*] (**Type species**: *Olpidiopsis saprolegniae* (A. Braun) M. Cornu).

Pseudolpidium A. Fischer, *Rabenhorst's Kryptogamen-Flora, 2 Aufl., Bd 1, Abt. 4*: 33 (1892) [*pro parte typo excluso*] (**Type species**: *Pseudolpidium saprolegniae* (A. Braun) A. Fischer).

Typification: *Gracea gracilis* (E. J. Butler) M. W. Dick.

Gracea gracilis (E. J. Butler) M. W. Dick, *Mycological Research* **101**: 882 (1997).

Pseudolpidium gracile E. J. Butler, *Memoirs of the Department of Agriculture, India, Botanical Series* **1**: 131 (1907).

Olpidiopsis gracilis (E. J. Butler) J. S. Karling, *The Simple Holocarpic Biflagellate Phycomycetes*: 47 (1942) [as '*gracile*'].

Typification: Lectotype (Cejp, 1959): *Memoirs of the Department of Agriculture, India, Botanical Series* **1**: plate 7, figs 1-8 (1907) (iconotype). Type locality: Antibes, France. Type habitat: parasitic in *Pythium intermedium* de Bary (*Peronosporomycetes, Pythiaceae*) from garden soil.

Identification key: Key I; key point 64.

Literature: Butler (1907), Dick (1997*a*, *b*), Gaertner (1954), Harder & Gallwitz-Uebelmesser (1959), Karling (1942*a*, 1968*a*, 1981*a*: 24, 1987, 1988), Milanez (1970), Miller (1962*b*), Pemberton *et al.* (1991), Persiel (1960*a*), Remy (1948), Sparrow (1960: 943), Sparrow & Dogma (1973), Whiffen (1942).

Gracea waterhouseae (J. S. Karling) M. W. Dick, *Mycological Research* **101**: 882 (1997).

Rozellopsis waterhouseae J. S. Karling, *Mycologia* **34**: 206 (1942) [as '*waterhousei*', named after Miss (Dr) Grace Waterhouse; this combination is one of the few in which the binomial echoes the given and surnames of an authority].

Typification: Lectotype (Cejp, 1959): *Simple Holocarpic Biflagellate Phycomycetes*: 15, plate 4, figs 5-14 (1942) (iconotype). Type locality: Hogsmill River, Surrey, U.K. Type habitat: parasitic in *Phytophthora cryptogea* Pethybr. & Laff. and *P. megasperma* Drechsler (*Peronosporomycetes, Pythiales*).

Identification key: Key I; key point 64'.

Literature: Dick (1997*b*), Held (1981), Karling (1942*a*, *c*, 1981*a*: 43), Sparrow (1960: 926), Waterhouse (1940, 1942).

DOUBTFUL TAXON

Zoospores figured by Seaman, Larson & Walker (1961).

ORDER *INCERTAE SEDIS*

MYZOCYTIOPSIDALES **M. W. Dick, ordo novum**

Myzocytiopsidales M. W. Dick *in* K. A. Pirozynski & D. L. Hawksworth, *Coevolution of Fungi with Plants and Animals*: 37 (1988) [*nomen nudum*].

Lagenidiales J. S. Karling, *American Journal of Botany* **26**: 518 (1939) [*pro parte typo excluso*] (**Type family**: *Lagenidiaceae* J. Scröter).

Lagenidialea H. F. Copeland, *The Classification of Lower Organisms* : 81 (1956) [*pro parte typo excluso*] (**Type family**: *Lagenidiacea* H. F. Copeland).

Fungi terrestres et in aquis dulcibus, raro marinis; endoparasitici obligati (non culturabiles) in *Aschelminthibus*, *Arthropodibus* aut algis. **Thallus** denique holocarpicus, pseudomycelialis, interdum inflatus, raro mycelialis, coralliformis; septatus, sine rhizoideo; segmenta thalloidea se transformantia in structuras reproductionis capaces; mitochondria cum cristis tubularibus; DBVs praesentes. **Reproductio asexualis** per sporangia; zoosporogenesis intrasporangialis, interdum sporangiosporas inchoatas emittens; sporangiosporae zoosporicae vel aplanosporicae; zoosporae heterokontae non repertae, formae principalis (?), flagellis in parte lateraliter insertis; zoosporae pusillae vel parviores (30-150μm^3). **Reproductio sexualis** per oogonia; homothallica, per segmenta thalloidea contigua et plus minusve aequalia, fungentia pro gametangiis donantibus et recipientibus, vel automictica; oosporae apleroticae; parietes oosporae laeves vel ornamentati; distributio copiae penariae nondum pro certo cognita.
Holotypus: *Myzocytium* M. W. Dick (*Myzocytiopsidaceae* M. W. Dick).

Terrestrial and freshwater fungi, rarely marine; obligate endoparasites of *Aschelminthes*, *Arthropoda* or algae. **Thallus** holocarpic over a period of time, pseudomycelial, sometimes inflated, rarely mycelial, coralloid; septate, without rhizoids; thalloidal segments transformed into reproductive structures; mitochondria with tubular cristae; DBVs present. **Asexual reproduction** by sporangia; zoosporogenesis intrasporangial, sometimes discharging incompletely developed sporangiospores; sporangiospores zoosporic or aplanosporic; zoospores not proven to be heterokont, of the principal form (?), with lateral flagellar insertion; zoospores small or medium small (volune equivalent 30-150μm^3). **Sexual reproduction** homothallic and oogamous, contiguous and more or less equal thalloid segments functioning as donor and receptor gametangia, or automictic; oospores aplerotic; oospore walls smooth or ornamented; distribution of reserve materials not established.
Typification: *Myzocytium* M. W. Dick (*Myzocytiopsidaceae* M. W. Dick).
Other family: *Crypticolaceae* M. W. Dick.

MYZOCYTIOPSIDACEAE M. W. Dick, *Mycological Research* **101**: 878 (1997).

Lagenidiaceae J Schröter, *Die Natürlichen Pflanzenfamilien* **1(1)**: 89 (1893) [*pro parte typo excluso*] (**Type genus**: *Lagenidium* W. Zopf).

Lagenidiacea H. F. Copeland, *The Classification of Lower Organisms*: 82 (1956) [*pro parte typo excluso*] (**Type genus**: *Lagenidium* W. Zopf).

? *Achlyogetonaceae* F. K. Sparrow, *Mycologia* **34**: 114 (1942) (**Type genus**: *Achlyogeton* A. Schenk).

? *Achlyogetonacea* H. F. Copeland, *The Classification of Lower Organisms*: 117 (1956) [*pro parte*] (**Type genus**: *Achlyogeton* A. Schenk).

Typification: *Myzocytiopsis* M. W. Dick.

MYZOCYTIOPSIS M. W. Dick, *Mycological Research* **101**: 878 (1997).

Bicricium N. Sorokin, *Archives Botaniques du Nord de la France* **2**: 35 (1883) [*pro parte typo excluso*] (**Type species**: *B. lethale* N. Sorokin).

Myzocytium A. Schenk, *Über das Vorkommen contractiler Zellen im Pflanzenreich*: 10 (1858) [*pro parte typo excluso*] (**Type species**: *M. proliferum* A. Schenk).

Lagenidium W. Zopf, *Verhandlungen des Botanischen Vereins der Provinz Brandenberg* **20**: 79 (1878) [*pro parte typo excluso*] (**Type species**: *L. rabenhorstii* W. Zopf).

? *Achlyogeton* A. Schenk, *Botanische Zeitung, Berlin* **17**: 398 (1859) [*pro parte typo excluso*] (**Type species**: *A. entophytum* A. Schenk).

Typification: *Myzocytiopsis lenticularis* (G. L. Barron) M. W. Dick.

Myzocytiopsis lenticularis (G. L. Barron) M. W. Dick, *Mycological Research* **101**: 879 (1997).
Myzocytium lenticulare G. L. Barron, *Antonie van Leeuwenhoek* **42**: 132 (1976).
Typification: Slide(s) OAC 10903, Herbarium OAC. Type locality: University of Guelph, Ontario, Canada. Type habitat: parasitic on nematodes from soil in a cattle pen.
Identification key: Key IV; key point 4'.
Literature: Barron (1976*b*, 1977*b*), Dick (1995, 1997*b*), Karling (1981*a*: 98), Newell *et al.* (1977).

Myzocytiopsis vermicola (W. Zopf) M. W. Dick, *Mycological Research* **101**: 879 (1997).

Myzocytium vermicola (W. Zopf) A. Fischer, *Rabenhorst's Kryptogamen-Flora, 2 Aufl., Bd 1, Abt. 4*: 71 (1892) [as '*vermicolum*'].

Myzocytium proliferum A. Schenk var. *vermicola* W. Zopf, *Nova Acta der Kaiserlichen Leopoldinisch-carolinischen Akademie der Naturforscher* **47**: 167 (1884) [as '*vermicolum*'].

? *Myzocytium polymorpha* E. Maupas, *Bulletin de la Société d'Histoire Naturelle de l'Afrique du Nord* **6**: 34-49 (1915) [*nomen nudum*].

Typification: type material not designated; lectotype: *Nova Acta der Kaiserlichen Leopoldinisch-carolinischen Akademie der Naturforscher* **47**: plate XIV, figs 35-37 (1884) (iconotype). Type locality: Berlin, Germany. Type habitat: parasitic in *Anguillula* sp. (*Aschelminthes, Nematoda*) with *Harposporium anguillulae* Lohde.

Identification key: Key IV; key point 4.

Literature: Barron (1977*b*), Dangeard (1903*b*), Davidson & Barron (1973), Dick (1997*b*), Fischer (1892), Karling (1942*a*, 1981*a*: 100, 102), Maupas (1915), Newell *et al.* (1977), Porter (1986), Pystina (1995), Sparrow (1960: 979), Yu (1977), Valkanov (1931*a*), Yačevskie & Yačevskie (1931: 68), Zopf (1884).

Myzocytiopsis subuliformis M. W. Dick, *Mycological Research* **101**: 879 (1997).

Protascus subuliformis E. Maupas var. *maupasii* R. Maire, *Bulletin de la Société d'Histoire Naturelle de l'Afrique du Nord* **6**: 50 (1915) [teleomorph].

Protascus subuliformis E. Maupas, *Bulletin de la Société d'Histoire Naturelle de l'Afrique du Nord* **6**: 34 (1915) [*pro parte* - teleomorph, but not anamorph].

non Protascus subuliformis P.-A. Dangeard, *Comptes Rendus Hebdomadaires des Séances de l'Académie des Sciences, Paris, Série D* **136**: 627 (1903) [anamorph].

non Myzocytium subuliforme (P.-A. Dangeard) G. L. Barron, *Canadian Journal of Botany* **55**: 822 (1977) [anamorph].

Typification: type material of *P. subuliforme* var. *maupasii* not designated, lectotype *Bulletin de la Société d'Histoire Naturelle de l'Afrique du Nord* **6**: 35 Fig 1; 39 Figs 2-5; 41 Figs 6-7; 43 Figs 8-9; 45 Figs 10-13 (1915) (iconotype). Type locality: unknown (from nematodes collected from wrapping and packing materials around a parcel sent to Algeria from Paris, France). Type habitat: parasitic in *Rhabditis teres* Schn. and *Rhabditis giardii* E. Maupas (*Aschelminthes, Nematoda*).

Identification key: Key IV; key point 11'.

Literature: Barron (1977*a*, *b*), Dangeard (1903*a*, 1906), Dick (1997*b*), Fitzpatrick (1930), Karling (1981*a*: 105), Maire (1915), Maupas (1915), Ramsbottom (1915).

Myzocytiopsis zoophthora (F. K. Sparrow) M. W. Dick, *Mycological Research* **101**: 879 (1997).

Myzocytium zoophthorum F. K. Sparrow, *Journal of the Linnean Society of London (Botany)* **50**: 462 (1936).

? *Myzocytiopsis oophila* (F. K. Sparrow) M. W. Dick, *loc. cit.* [≡ *Lagenidium oophilum* F. K. Sparrow, *Mycologia* **31**: 531 (1939)].

? *Myzocytiopsis fijiensis* (J. S. Karling) M. W. Dick, *loc. cit.* [≡ *Myzocytium fijiense* J. S. Karling, *Mycopathologia et Mycologia Applicata* **37**: 168 (1969)].

Typification: Lectotype (Cejp, 1959): plate 19, figs 1-14 (1936) (iconotype). Type locality: Cambridge, U.K. Type habitat: parasitic in rotifers and rotifer eggs.

Identification key: Key IV; key point 13.

Literature: Dick (1997*b*), Glockling & Dick (1997), Karling (1981*a*: 95), Konno (1972), Newell *et al.* (1977), Sparrow (1936*b*, 1952*c*, 1960: 979), Sparrow, Paterson & Johns (1965).

Myzocytiopsis microspora (J. S. Karling) M. W. Dick, *Mycological Research* **101**: 879 (1997).

Lagenidium microsporum J. S. Karling, *Lloydia* **7**: 328 (1944).

Myzocytium microsporum (J. S. Karling) F. K. Sparrow, *Aquatic Phycomycetes*: 980 (1960).

Typification: type material not designated; lectotype *Lloydia* **7**: 332, plate 1, figs 1-34 (1944) (iconotype). Type locality: Flores Nabuco near Manaus, Amazonas, Brazil. Type habitat: parasitic in bodies of *Distyla* sp. (*Aschelminthes, Rotifera*) from a soil sample.

Identification key: Key IV; key point 13'.

Literature: Czeczuga (1995*a*), Czeczuga, Chomutowska & Woronowicz (1990), Dick (1997*b*), Karling (1944*c*, 1968*a*, 1981*a*: 96), Sparrow (1960: 980).

Myzocytiopsis parthenospora (J. S. Karling) M. W. Dick, *Mycological Research* **101**: 879 (1997).

Lagenidium parthenosporum J. S. Karling, *Lloydia* **7**: 329 (1944).

Typification: type material not designated; lectotype *Lloydia* **7**: 336, plate 2, figs 35-68 (1944) (iconotype). Type locality: Campo Grande, Matto Grosso, Brazil. Type habitat: parasitic in eggs and bodies of *Distyla* sp. and *Philodina* sp. (*Aschelminthes, Rotifera*), eggs of *Chaetonotus larus* O. Müller (*Gasterotricha*) and bodies of *Heterodera* sp. (*Aschelminthes, Nematoda*) from a soil sample.

Identification key: Key IV; key point 6.

Literature: Dick (1997*b*), Karling (1944*c*, 1981*a*: 132), Newell *et al.* (1977), Sparrow (1960: 996).

Myzocytiopsis distylae (J. S. Karling) M. W. Dick, *Mycological Research* **101**: 879 (1997).
Lagenidium distylae J. S. Karling, *Lloydia* **7**: 330 (1944).
Typification: type material not designated; lectotype *Lloydia* **7**: 340, plate 3, figs 69-107 (1944) (iconotype). Type locality: Igarape Raimundo at Manaus, Amazonas, Brazil. Type habitat: parasitic in eggs of *Distyla* sp. (*Aschelminthes, Rotifera*) from a water sample.
Identification key: Key IV; key point 10.
Literature: Dick (1997*b*), Glockling & Dick (1997), Karling (1944*c*, 1981*a*: 130), Newell *et al.* (1977), Sparrow (1960: 996).

Myzocytiopsis humana (J. S. Karling) M. W. Dick, *Mycological Research* **101**: 879 (1997).
Lagenidium humanum J. S. Karling, *Mycologia* **39**: 225 (1947).
Typification: type material not designated; lectotype *Mycologia* **39**: 226 figs 1-39 (1947) (iconotype). Type locality: Briarcliff, New York, U.S.A. Type habitat: saprotrophic from a cemetery soil baited with dead human skin in water.
Identification key: Key IV; key point 10'.
Literature: Booth (1971*a*), Czeczuga & Muszyńska (1994), Dick (1997*b*), Gaertner (1954), Karling (1947, 1949, 1976*b*, 1981*a*: 140, 1981*b*, 1987*b*, 1988), Milanez & Beneke (1968), Sparrow (1957, 1960: 1005).

Myzocytiopsis indica (U. P. Singh) M. W. Dick, *Mycological Research* **101**: 879 (1997).
Myzocytium indicum U. P. Singh, *Hydrobiologia* **42**: 449 (1973).
Typification: MSP No. 339, isotype slide deposited at IMI. Type locality: Varanasi, Uttar Pradesh, India. Type habitat: in rotifers, October 1964.
Identification key: Key IV; key point 15.
Literature: Dick (1997*b*), Karling (1981*a*: 105), Newell *et al.* (1977), Singh (1973).

Myzocytiopsis humicola (G. L. Barron & J. G. Percy) M. W. Dick, *Mycological Research* **101**: 879 (1997).
Myzocytium humicola G. L. Barron & J. G. Percy, *Canadian Journal of Botany* **53**: 1306 (1975).
Typification: Slides OAC 10901, Herbarium OAC. Type locality: University of Guelph, Ontario, Canada. Type habitat: from soil in a cattle pen.
Identification key: Key IV; key point 18'.
Literature: Barron (1977*b*), Barron & Percy (1975), Dick (1997*b*), Karling (1981*a*: 102), Newell *et al.* (1977), Prasad & Dayal (1986), Saikawa & Anazawa (1985), Saxena & Mukerji (1991).

Myzocytiopsis papillata (G. L. Barron) M. W. Dick, *Mycological Research* **101**: 880 (1997).
Myzocytium papillatum G. L. Barron, *Canadian Journal of Microbiology* **22**: 752 (1976).
Typification: Slide(s) OAC 10904, Herbarium OAC. Type locality: Aberfoyle, Ontario, Canada. Type habitat: parasitic on *Rhabditis terricola* Dujardin (*Aschelminthes, Nematoda*) from barnyard soil.
Identification key: Key IV; key point 18.
Literature: Barron (1976*c*, 1977*b*), Dick (1997*b*), Karling (1981*a*: 104), Newell *et al.* (1977), Prasad & Dayal (1985), Saxena & Mukerji (1991).

Myzocytiopsis glutinospora (G. L. Barron) M. W. Dick, *Mycological Research* **101**: 880 (1997).
Myzocytium glutinosporum G. L. Barron, *Canadian Journal of Microbiology* **22**: 755 (1976).
Typification: Slide(s) OAC 10905, Herbarium OAC. Type locality: Aberfoyle, Ontario, Canada. Type habitat: parasitic on *Rhabditis terricola* Dujardin (*Aschelminthes, Nematoda*) from barnyard soil.
Identification key: Key IV; key point 2'.
Literature: Barron (1976*c*, 1977*b*), Dick (1997*b*), Karling (1981*a*: 104), Newell *et al.* (1977).

Myzocytiopsis intermedia (G. L. Barron) M. W. Dick, *Mycological Research* **101**: 880 (1997).
Myzocytium intermedium G. L. Barron, *Canadian Journal of Botany* **54**: 2 (1976).
Typification: Slide(s) OAC 10902, Herbarium OAC. Type locality: Bruce Peninsula, Ontario, Canada. Type habitat: parasitic on *Rhabditis* sp. (*Aschelminthes, Nematoda*) in a soil sample from a birch-maple-poplar wood.
Identification key: Key IV; key point 12'.
Literature: Barron (1976*a*, 1977*b*), Dick (1997*b*), Karling (1981*a*: 104), Newell *et al.* (1977).

Myzocytiopsis osiris S. L. Glockling *in* Dick & Glockling, *Botanical Journal of the Linnean Society* **[in press*]**.
Typification: IMI 366219 (Glockling, 1994; sample 32: 14 June 1991, fungus found 11 Sept 1991), from rabbit dung, beechwoods, Berkshire. Iconotype: *Botanical Journal of the Linnean Society* **[in press]**.
Identification key: Key IV; key point 5'.
Literature: Dick (1995), Dick & Glockling (1999), Glockling (1994), Glockling & Dick (2000).

Myzocytiopsis bolata S. L. Glockling, *in* Dick & Glockling, *Botanical Journal of the Linnean Society* **[in press*]**.
Typification: IMI 366221 (Glockling, 1994; sample 65: 7 July 1992, fungus found 11 July 1992), red deer dung, Wellington Country Park, Hampshire. Iconotype: *Botanical Journal of the Linnean Society* **102 [in press]**.
Identification key: Key IV; key point 17'.
Literature: Dick (1995), Dick & Glockling (1999), Glockling (1994), Glockling & Dick (2000).
Identification key: Key IV; key point 6'.
Literature: Dick & Glockling (1999), Glockling (1994), Glockling & Dick (1998).

DOUBTFUL SPECIES

Myzocytiopsis elegans (E. Perroncito) M. W. Dick, *Mycological Research* **101**: 880 (1997).
Lagenidium elegans (E. Perroncito) R. Ciferri, *Atti dell'Istituto Botanicao della Universitata e Laboratorio Crittogamico Pavia Serie 6* **19**: 62 (1961).
Chytridium elegans E. Perroncito, *Centralblatt für Bacteriologie, Parasitenkunde, Infektionskrankheit* **4**: 295 (1888).
Woronina elegans (E. Perroncito) A. Fischer, *Rabenhorst's Kryptogamenflora, 2. Aufl., Bd 1, Abt. 4*: 66 (1892).
Typification: type material not verified. Type locality: Vinardio, Italy. Type habitat: parasitic in *Philodina roseola* Ehrenb. (*Aschelminthes*, "*Rotatoriaceae*").
Identification key: Key IV; key point 15'.
Literature: Ciferri (1962), Dick (1997*b*), Fischer (1892), Karling (1942*a*, 1968*e*: 94, 1981a: 134), Minden (1911*a*), Perroncito (1888), Sparrow (1960: 781), Yačevskie & Yačevskie (1931: 64).

Myzocytiopsis oophila (F. K. Sparrow) M. W. Dick, *Mycological Research* **101**: 880 (1997).
Lagenidium oophilum F. K. Sparrow, *Mycologia* **31**: 531 (1939).
Typification: type material not designated; lectotype *Mycologia* **31**: 531, figs 1-15 (1939) (iconotype). Type locality: Huron River, Ann Arbor, Michigan, U.S.A. Type habitat: parasitic in eggs and embryos of rotifers from aquatic debris.
Identification key: Key IV; key point 17.
Literature: Dick (1997*b*), Glockling & Dick (1997), Karling (1981*a*: 128), Newell *et al.* (1977), Sparrow (1939*b*, 1960: 992).

---

*FOOTNOTES:

Myzocytiopsis osiris Glockling, **sp. nov.**: **Thallus** holocarpicus, extensus, ramosus; compositus in primis hypharum latarum, diametro uniformi, 5-6μm diam.; deinde septatus factus, generans segmenta longa sporangialia, aut curta oogonialia, gradatim dissociantia, extremitatibus rotundatis. **Reproductio asexualis** per sporangia; sporogenesis intrasporangialis; segmenta sporangialia 45-145μm × 9-16μm; tubus evacuationis 8-12μm longus, 3·0-4·0μm diam., curtus, cylindricalis, raro bini per sporangium, extrudens per integumentum hospitiale; sporangiosporae aplanosporicales, sphaericae 3·0-3·5 diam. **Reproductio sexualis** automictica, gametangia donantia (antheridia) absentia; segmenta gametangialia curta, 22-26μm longa serie longa, unumquidque in tertia parte centrali inflatius crescens cum vesiculis terminalibus manifeste dissociantibus; oosporae primo sphaericae, laeves, deinde se transformantes in superficium porcatum, cum 5-7 porcis, interdum cum porca coronali; 12-18μm diam.; cum pariete pluristrato.
Holotypus: IMI 366219 [Glockling, 1994; sample 32 (collected 14 June 1991, fungus found 19/20 June 1991), rabbit dung, beechwoods, Berkshire, UK].

Myzocytiopsis bolata Glockling, **sp. nov.**: **Thallus** holocarpicus, cylindricus, diametro uniformis, 20-60μm longus × 8(28)μm diam., raro lobatus et dilatatus. **Reproductio asexualis** per zoosporangia; zoosporogenesis intrasporangialis; segmenta zoosporangialia 20-60 × 8(28)μm; tubus evacuationis 10-14μm longus × 5·0μm diam., cylindricalis, extrudens per integumentum hospitiale; zoosporae biflagellatae, se transformantes in superficie hospitis in cystas sphaericas 5·0μm diam., integumentum nematodiale penetrantes per hypham perangustam germinantem 1·0μm diam. et usque ad 7·0-8·0μm longam. **Reproductio sexualis** automictica, gametangia donantia (antheridia) absentia; oogonia abundantia; oosporae apleroticae, pusillae, sphaericae, 12μm diam.
Holotypus: IMI 366221 [Glockling, 1994; sample 65 (collected 7 July 1992, fungus found 11 July 1992), red deer dung, Wellington Country Park, Hampshire, UK).

Myzocytiopsis fijiensis (J. S. Karling) M. W. Dick, *Mycological Research* **101**: 880 (1997).
Myzocytium fijiense J. S. Karling, *Mycopathologia et Mycologia Applicata* **37**: 168 (1969) [as *'fijiensis'*].
Typification: Type slide No. 146, Herbarium, Purdue University (PUL). Type locality: Viti Levu, Fiji. Type habitat: parasitic in eggs of *Distyla* sp. (*Aschelminthes, Rotifera*) in soil sample VL6.
Identification key: Key IV; key point 9'.
Literature: Dick (1997*b*), Glockling & Dick (1997), Karling (1969, 1981*a*: 95), Newell *et al.* (1977).

UNIDENTIFIABLE TAXA REFERABLE TO THE GENUS *MYZOCYTIOPSIS*, BUT NOT TO SPECIES AND THEREFORE EXCLUDED

Myzocytium lethale (N. Sorokin) K. Cejp, *Flora ČSR, Oomycetes I*:413 (1959).
Bicricium lethale N. Sorokin, *Archives Botaniques du Nord de la France* **2**: 35 (1883). **Typification**: type material not designated; Cejp (1959) used the figures for *Myzocytium proliferum* var. *vermicola* (Zopf, 1884) to typify this name. **Literature**: Cejp (1959), Karling (1977: 59), Sorokine (1883, 1889, 1890), Sparrow (1960: 188, 979), Yačevskie & Yačevskie (1931: 69).

Achlyogeton rostratum N. Sorokin, *Archives Botaniques du Nord de la France* **2**: 36 (1883). **Typification**: type material not designated. **Literature**: Karling (1977: 59), Sorokine (1883, 1889, 1890), Sparrow (1960: 185).

Myzocytium rotiferum S. N. Dasgupta & R. John, *Bulletin of the Botanical Survey of India* **30**: 58 (1988). **Typification**: Holotype deposited at HCIO, Division of Mycology and Plant Pathology, IARI, New Delhi-110 012. HCIO No. 37183. **Literature**: Dasgupta & John (1988).

Myzocytium polymorphum E. Maupas, *Bulletin de la Société d'Histoire Naturelle de l'Afrique du Nord* **6**: 34-49 (1915) [*nomen nudum*]. **Literature**: Karling (1981*a*: 102), Maupas (1915).

UNNAMED SPECIES POSSIBLY REFERABLE TO *MYZOCYTIOPSIS*

Lagenidium sp. (Canter & Heaney, 1984), parasitic in *Ceratium* sp. (*Dinomastigota*), zoosporogenesis intrasporangial, zoospores small (*ca* 3μm), oogonium with donor thallus of comparable size, oospore with ornamented wall layer.

Olpidium sp. (Sparrow, Paterson & Johns, 1965), zoospores not seen, no companion cell, sporangia and resting spores present.

EXCLUDED TAXA

BICRICIUM N. Sorokin, *Archives Botaniques du Nord de la France* **2**: 35 (1883), generic synonym of *Myzocytiopsis* M. W. Dick, *loc. cit.*, type species not determinable - see previous section.

Myzocytium subuliforme (P.-A. Dangeard) G. L. Barron, *Canadian Journal of Botany* **55**: 822 (1977), anamorphic *nomen confusum*.

PROTASCUS P. A. Dangeard, *Comptes Rendus Hebdomadaires des Séances de l'Académie des Sciences, Paris, Série D* **136**: 627 (1903) (**Type species**: *P. subuliformis* P. A. Dangeard, anamorphic *nomen confusum*) [*nomen rejiciendus* proposed].

Protascus subuliformis P.-A. Dangeard, *Comptes Rendus Hebdomadaires des Séances de l'Académie des Sciences, Paris, Série D* **136**: 627 (1903), synonym of *Myzocytium subuliforme* (P.-A. Dangeard) G. L. Barron, anamorphic *nomen confusum [nomen rejiciendus* proposed].

Protascus colorans P. C. van der Wolk, *Mycologische Centralblatt* **3**: 153 (1913), genus illegitimate after synonymy of the type species with *Myzocytiopsis subuliformis loc. cit.*, synonym of *Wolkia colorans* (P. C. van der Wolk) J. Ramsbottom *Transactions of the British Mycological Society* **5**: 144 (1915*a*) [as '*decolorans*'], a probable member of the *Endomycetaceae* (*Ascomycetes*).

GONIMOCHAETE C. Drechsler, *Bulletin of the Torrey Botanical Club* **73**: 14 (1946).
Typification: *G. horridula* C. Drechsler (holotype - sole original species).

Gonimochaete ***horridula*** C. Drechsler, *Bulletin of the Torrey Botanical Club* **73**: 15 (1946).
Typification: type material not designated; lectotype *Bulletin of the Torrey Botanical Club* **73**: 2, figs 128, 5, figs 29-36 (1946) (iconotype). Type locality: Georgetown, Delaware, U.S.A. Type habitat: parasitic in *Acrobeloides* sp. (affin. *A. buetschlii* (De Man) Thorne) (*Aschelminthes, Nematoda*) on decaying leaves of *Acer rubrum* L.
Identification key: Key V; key point 1'.
Literature: Barron (1977*b*), Drechsler (1946), Duddington (1955, 1958), Guima & Cooke (1972), Karling (1981*a*: 152), McCullough (1977), Miura (1973), Shilova (1985).

Gonimochaete ***pyriformis*** G. L. Barron, *Canadian Journal of Botany* **51**: 2451 (1973) [as '*pyriforme*'].
Typification: type material not designated; lectotype *Canadian Journal of Botany* **51**: plates I and II facing pp. 2452 and 2453 (1973) (iconotype). Type locality: University of Guelph, Ontario, Canada (iconotype). Type habitat: parasitic in *Diploscapter* spp. (*Aschelminthes, Nematoda*) from soil in a tropical greenhouse.
Identification key: Key V; key point 3'.
Literature: Barron (1973), Karling (1981*a*: 152), Saikawa & Anazawa (1985).

Gonimochaete ***latitubus*** S. Y. Newell, R. Cefalu & J. W. Fell, *Bulletin of Marine Science* **27**: 189 (1977).
Typification: Type slide(s) and photographs deposited in Herbarium NY. Type locality: Florida, U.S.A., marine littoral. Type habitat: parasitic in *Rhabditis marina* Bastian (*Aschelminthes, Nematoda*).
Identification key: Key V; key point 2'.
Literature: Karling (1981*a*: 154), Newell, Cefalu & Fell (1977).

Gonimochaete ***lignicola*** G. L. Barron, *Mycologia* **77**:17 (1985).
Typification: Slides OAC 10840, Herbarium OAC. Type locality: Eramosa River, Guelph, Ontario, Canada. Type habitat: parasitic in nematodes from river bank soil.
Identification key: Key V; key point 3.
Literature: Barron (1985).

CHLAMYDOMYZIUM M. W. Dick, *Mycological Research* **101**: 880 (1997).
Myzocytium A. Schenk, *Über das Vorkommen contractiler Zellen im Pflanzenreich*: 10 (1858) [*pro parte typo excluso*] (**Type species**: *M. proliferum* A. Schenk).
Lagenidium W. Zopf, *Verhandlungen des Botanischen Vereins der Provinz Brandenberg* **20**: 79 (1878) [*pro parte typo excluso*] (**Type species**: *L. rabenhorstii* W. Zopf).
Typification: *Chlamydomyzium anomalum* (G. L. Barron) M. W. Dick.

Chlamydomyzium anomalum (G. L. Barron) M. W. Dick, *Mycological Research* **101**: 880 (1997).
Myzocytium anomalum G. L. Barron, *Canadian Journal of Microbiology* **22**: 760 (1976).
Typification: Slide(s) OAC 10906, Herbarium OAC. **Type locality**: Aberfoyle, Ontario, Canada. **Type habitat**: parasitic on an unidentified nematode from barnyard soil.
Identification key: Key VI; key point 3'.
Literature: Barron (1976*c*, 1977*b*), Dick (1997*b*), Karling (1981*a*: 98), Newell *et al.* (1977).

Chlamydomyzium septatum (J. S. Karling) M. W. Dick, *Mycological Research* **101**: 880 (1997).
Lagenidium septatum J. S. Karling, *Mycopathologia et Mycologia Applicata* **37**: 162 (1969).
Typification: Type specimen slide no. 145 Herbarium Purdue University (PUL). **Type locality**: Nuie Island (also Pitcairn Is.), Pacific Ocean. **Type habitat**: parasitic in rotifer eggs and sporangia of *Catenaria anguillulae* Sorokīn on snake skin, from soil sample NI 1.
Identification key: Key VI; key point 1'.
Literature: Dick (1997*b*), Karling (1969, 1981*a*: 132), Newell *et al.* (1977).

Chlamydomyzium oviparasiticum (G. L. Barron) M. W. Dick, *Mycological Research* **101**: 880 (1997).
Lagenidium oviparasiticum G. L. Barron, *Canadian Journal of Botany* **67**: 269 (1989).
Typification: "Diagrams in text constitute the type material" *Canadian Journal of Botany* **67**: 269 (1989) (iconotype). **Type locality**: Aberfoyle, Ontario, Canada. **Type habitat**: parasitic in the eggs of *Adineta* spp. (*Aschelminthes*, *Rotifera*) from farmyard soil.
Identification key: Key VI; key point 3.
Literature: Barron (1989*a*), Dick (1997*b*).

Chlamydomyzium internum S. L. Glockling, *Mycological Research* **101**: 885 (1997).
Typification: Slide(s) IMI 370558. **Type locality**: Tsukuba, Ibaraki Prefecture, Japan. Parasitic in *Distylae* rotifers recovered from organic debris with pine needles.
Identification key: Key VI; key point 3'.
Literature: Glockling & Dick (1997).

Chlamydomyzium aplanosporum S. L. Glockling, *Mycological Research* **101**: 885 (1997).
Typification: Slide(s) IMI 370999. **Type locality**: Tsukuba, Ibaraki Prefecture, Japan. Parasitic in *Distylae* rotifers recovered from rotting straw.
Identification key: Key VI; key point 3'.
Literature: Glockling & Dick (1997).

Chlamydomyzium sphaericum S. L. Glockling, *in* Dick & Glockling, *Botanical Journal of the Linnean Society* [**in press**[*]].
Typification: IMI 366223 (Glockling, 1994; sample 65 (7 July 1992, fungus found 11 July 1992), red deer dung, Wellington Country Park, Hampshire; present in 6 other collections). Iconotype: *Botanical Journal of the Linnean Society* [**in press**].
Identification key: Key VI; key point 2'.
Literature: Dick & Glockling (1999), Glockling (1994), Glockling & Dick (2000).

---

[*]FOOTNOTE:

Chlamydomyzium sphaericum Glockling, **sp. nov.**: **Thallus** holocarpicus, extensus, ramosus; compositus in primis hypharum latarum, diametro uniformi, 5-6μm diam.; septatus factus, se transformans in segmenta dilata et lobata, pro sporangiis fungentia, et pro chlamydosporangiis fungentia, extremitatibus rotundatis, gradatim dissociantia. **Reproductio asexualis** per sporangia; sporogenesis intrasporangialis; segmenta sporangialia 45-145μm longa × 9-16μm diam.; tubus evacuationis 7-12μm longus, cum labro pandente et crassiore viso 4·5μm diam., extrudens per integumentum hospitiale; zoosporae 5-6 per sporangium, biflagellatae 5·5-6·0μm diam., liberatae ut fascis compactus et lente movens; motus paulatim crescens ex actione flagellari; motus saltans (pulicis modo). **Reproductio perdurans** per chlamydosporas (oosporas automicticas?), gametangia ignota; celeriter formatas per transformationam segmentorum thallorum plus minusve separatorum (segmenta 10-25 × 10-15μm), saepe post disintegrationem integumenti hospitialis; chlamydosporae 16-18μm diam.; paries gracilis; multistratus, leviter undulatus factus.
Holotypus: IMI 366223 [Glockling, 1994; sample 62 (collected 27 Apr 1992, fungus found 29 May 1992), rabbit dung, mixed deciduous woodland, Hampshire, UK].

SYZYGANGIA M. W. Dick, *Mycological Research* **101**: 880 (1997).

Lagenidium W. Zopf, *Verhandlungen des Botanischen Vereins der Provinz Brandenberg* **20**: 79 (1878) [*pro parte typo excluso*] (**Type species**: *L. rabenhorstii* W. Zopf).

Bicricium N. Sorokin, *Archives Botaniques du Nord de la France* **2**: 35 (1883) [*pro parte typo excluso*] (**Type species**: *B. lethale* N. Sorokin).

Pythium N. Pringsheim, *Jahrbuch für wissenschaftlichen Botanik* **1**: 304 (1858) [*pro parte minore typo excluso*] (**Type species**: *Pythium monospermum* N. Pringsheim).

Pleotrachelus W. Zopf, *Nova Acta der Kaiserlichen Leopoldinisch-carolinischen Akademie der Naturforscher* **47**: 173 (1884) [*pro parte typo excluso*] (**Type species**: *Pleotrachelus fulgens* W. Zopf).

Olpidiopsis M. Cornu, *Annales des Sciences Naturelles, Botanique, Série 5* **15**: 114 (1872) [*pro parte typo excluso*] (**Type species**: *Olpidiopsis saprolegniae* (A. Braun) M. Cornu).

Olpidium (A. Braun) L. Rabenhorst, *Flora Europea Algarum* **3**: 283?8 (1868) [*pro parte minore typo excluso*] (**Type species**: *Olpidium endogenum* (A. Braun) J. Schroeter).

? *Achlyogeton* A. Schenk, *Botanische Zeitung, Berlin* **17**: 398 (1859) [*typo excluso*] (**Type species**: *A. entophytum* A. Schenk).

? *Aphanistis* N. Sorokin, *Archives Botaniques du Nord de la France* **2**: 33 (1883) [*typo excluso*] (**Type species**: *Aphanistis oedogoniarum* N. Sorokin).

? *Resticularia* P.-A. Dangeard, *Le Botaniste* **2**: 96 (1890) [*typo excluso*] (**Type species**: *Resticularia nodosa* P.-A. Dangeard) [*nomen rejiciendus* proposed].

? *Achlyopsis* É. de Wildeman, *Annales de la Société Belge de Microscopie, Mémoires* **20**: 123 (1896) [*typo excluso*] (**Type species**: *Achlyopsis entospora* É. de Wildeman).

Type species: *Syzygangia marchaliana* (É. de Wildeman) M. W. Dick.

Syzygangia marchaliana (É. de Wildeman) M. W. Dick, *Mycological Research* **101**: 881 (1997).

Lagenidium marchalianum É. de Wildeman, *Annales de la Société Belge de Microscopie, Mémoires* **21**: 10 (1897).

? *Aphanistis oedogoniorum* N. Sorokin, *Archives Botaniques du Nord de la France* **2**: 33 (1883).

? *Aphanistis pellucida* N. Sorokin, *Archives Botaniques du Nord de la France* **2**: 33 (1883).

? *Resticularia oedogonii* B. W. Skvortzow, *Archiv für Protistenkunde* **51**: 432 (1925) [*nomen rejiciendum* proposed].

Typification: Lectotype (Cejp, 1959): *Annales de la Société Belge de Microscopie, Mémoires* **21**: plate 1, figs 1-9 (1897) (iconotype). Type locality Belgium. Type habitat: parasitic in *Oedogonium* sp. (*Chlorophyceae, Oedogoniales*).

Identification key: Key VII; key point 6.

Literature: Cejp (1959), Couch (1935*b*), Czeczuga (1994, 1995*a*), Dick (1997*b*), Ivimey Cook (1935), Karling (1981*a*: 118), Sparrow (1943, 1960: 1003), Wildeman (1897*a*), Yačevskie & Yačevskie (1931: 47).

Syzygangia nodosa (P.-A. Dangeard) M. W. Dick, *Mycological Research* **101**: 881 (1997).

? *Resticularia nodosa* P.-A. Dangeard, *Le botaniste* **2**: 99 (1890) [*nomen rejiciendum* proposed].

Lagenidium nodosum (P.-A. Dangeard) C. T. Ingold, *Annals of Botany, London, N.S.* **13**: 442 (1949).

Typification: type material not designated; lectotype *Annals of Botany, London, N.S.* **13**: 437: fig. 1; 438-9: figs 2-3; 441: fig. 4 (1949) (iconotype). Type localities: type locality: *R. nodosa*: Caen, Normandy, France; iconotype locality: *L. nodosum*: Sevenoaks, Kent, U.K.. Type habitats: *R. nodosa*: parasitic in *Lyngbya aestuarii* Liebm. (*Cyanobacteria*) "dans les fossés que traversent le champ de causes de Caen"; *L. nodosum*: parasitic in *Lyngbya* sp.

Identification key: Key VII; key point 3'.

Literature: Dangeard (1890*a*), Dick (1997*b*), Ingold (1949), Karling (1981*a*: 119), Sparrow (1932, 1960:1000), Yačevskie & Yačevskie (1931: 47).

Syzygangia elliptica (É. de Wildeman) M. W. Dick, *Mycological Research* **101**: 881 (1997).

Lagenidium ellipticum É. de Wildeman, *Annales de la Société Belge de Microscopie, Mémoires* **17**: 8 (1893).

Pleotrachelus wildemanii H. E. Petersen, *Botanisk Tidsskrift* **29**: 422 (1909) [as '*wildemani*'].

Olpidiopsis ricciae S. J. du Plessis, *Annals of Botany, London* **47**: 761 (1933).

Typification: Lectotype (Cejp, 1959): *Annales de la Société Belge de Microscopie, Mémoires* **17**: plate 1, figs 1-11 (1893) (iconotype). Type locality: Belgium. Type habitat: rhizoids of mosses.

Identification key: Key VII; key point 2.

Literature: Dick (1997*b*), Döbbeler & Itzerott (1983), Karling (1981*a*: 41, 127), Plessis (1933), Sparrow (1960: 66, 159, 160, 905, 1008), Wildeman (1893*a*).

Syzygangia papillosa (G. Cocconi) M. W. Dick, *Mycological Research* **101**: 881 (1997).
Lagenidium papillosum G. Cocconi, *Rendiconti dell'Accademia delle Scienze dell'Istituto di Bologna. Serie V* **4**: 362 (1894).
Lagenidium astrum S. N. Dasgupta & R. John, *Bulletin of the Botanical Survey of India* **30**: 70 (1988).
Lagenidium tortum S. N. Dasgupta & R. John, *Bulletin of the Botanical Survey of India* **30**: 71 (1988).
Typification: type material not designated; lectotype *Rendiconti dell'Accademia delle Scienze dell'Institute di Bologna. Serie V* **4**: 370, figs 1-3 (1894) (iconotype). Type locality: Bologna, Italy. Type habitat: parasitic in *Spirogyra* sp. (*Zygnematales, Zygnemataceae*).
Typification of synonyms: *L. tortum*: holotype deposited at HCIO, Division of Mycology and Plant Pathology, IARI, New Delhi-110 012. HCIO No. 37123; *L. astrum*: holotype deposited at HCIO, Division of Mycology and Plant Pathology, IARI, New Delhi-110 012. HCIO No. 37122.
Identification key: Key VII; key point 7.
Literature: Cocconi (1894), Dick (1997*b*), Ivimey Cook (1935), Karling (1981*a*: 119), Minden (1915), Pystina (1995), Saccardo (1898), Sparrow (1960: 996), Voronichin (1920), Yačevskie & Yačevskie (1931: 67).

Syzygangia oedogonii (A. Scherffel) M. W. Dick, *Mycological Research* **101**: 881 (1997).
Lagenidium oedogonii A. Scherffel, *Hedwigia* **41**: 105 (1902).
Olpidiopsis fusiformis M. Cornu var. *oedogoniarum* N. Sorokin, *Archives Botaniques du Nord de la France* **2**: 27 (1883).
Olpidium oedogoniarum (N. Sorokin) É. de Wildeman, *Annales de la Société Belge de Microscopie, Mémoires* **18**: 154 (1894).
Olpidiopsis oedogoniarum (N. Sorokin) A. Scherffel, *Archiv für Protistenkunde* **52**: 103 (1925).
Pleotrachelus petersenii A. Lund, *Botanisk Tidsskrift* **41**: 241 (1930).
? *Olpidiopsis sorokinei* É. de Wildeman, *Annales de la Société Belge de Microscopie, Mémoires* **14**: 22 (1890).
? *Achlyogeton solatium* M. Cornu, *Bulletin de la Société Botanique de France* **17**: 297 (1870).
Typification: Lectotype (Cejp, 1959): *Archiv für Protistenkunde* **52**: plate 5, figs 209-219 (1925) (iconotype). Type locality: Spišská, Nová Ves, Slovenia. Type habitat: parasitic in *Oedogonium* sp. (*Chlorophyceae, Oedogoniales*). **Typification of synonym**: *O. fusiformis* var. *oedogoniarum* lectotype (Cejp, 1959): *Archives Botaniques du Nord de la France* **2**: fig. 31 (1883) (iconotype).
Identification key: Key VII; key point 5'.
Literature: Bérczi (1940), Cornu (1870), Couch (1935*b*), Dick (1997*b*), Domján (1936), Karling (1981*a*: 28, 118), Pystina (1995), Rieth (1954), Scherffel (1902, 1925), Sorokine (1883, 1889, 1890), Sparrow (1933*c*, 1943, 1960: 156, 948, 990), Wildeman (1894).

Syzygangia sacculoides (I. L. Serbinov) M. W. Dick, *Mycological Research* **101**: 881 (1997).
Lagenidium sacculoides I. L. Serbinov, *La Défense des Plantes. Bulletin du Bureau Permanent des Congrès Entomo-Phytopathologiques de Russie* **2(2)**: 85 (1925).
Typification: Type material not designated (no illustrations in Serbinov (1925), one small illustration of two host cells in Yačevskie & Yačevskie (1931: 67, fig. 105), nothing noted in Cejp, 1959). Type locality: Menzelinsk, Ufimsk, Russia. Type habitat: parasitic in *Closterium ralfsii* Bréb. var. *hybridum* Rabenh. (*Zygnematales, Desmidaceae*).
Identification key: Key VII; key point 7'.
Literature: Cejp (1959), Dick (1997*b*), Jaczewski (1931), Karling (1942*a*: 122), Serbinov (1925), Sparrow (1960: 1008), Yačevskie & Yačevskie (1931: 67).

Syzygangia zygnematicola M. W. Dick, *Mycological Research* **101**: 881 (1997).
Lagenidium rabenhorstii W. Zopf *sensu* W. R. Ivimey Cook, *Archiv für Protistenkunde* **86**: 63 (1935).
Typification: *Archiv für Protistenkunde* **86**: plate 1, figs 1-14; plate 4, figs 1-8 (1935) (iconotype). Type locality: Marshfield, Glamorgan, Wales. Type habitat: parasitic in *Spyrogyra* sp. (*Charophyceae, Zygnematales, Zygnemataceae*).
Identification key: Key VII; key point 6'.
Literature: Atkinson (1909), Bérczi (1940), Cejp (1959), Constantineanu (1901), Dasgupta & John (1988), Dick (1997*b*), Domján (1936), Graff (1928), Ivimey Cook (1932*a*, 1935), Karling (1942*a*, 1968*a*, 1981*a*: 109, 114), Konno (1972), Linder (1947), Minden (1915), Pystina (1995), Scherffel (1914), Schroeter *in* Cohn (1885), Shen & Siang (1948), Sparrow (1932*b*, 1943, 1960: 994,999), Thompson (1934), Wildeman (1896*c*), Yačevskie & Yačevskie (1931: 67), Zopf (1878, 1884).

Syzygangia dictyospora M. W. Dick, *Mycological Research* **101**: 882 (1997).
Pythium dictyosporum M. Raciborski, *Anzieger der Akademie der Wissenschaften in Krakau* **8**: 283 (1891) *sensu* A. Rieth, *Die Kulturpflanze* **2**: 176 (1954).
? *Lagenidium zopfii* É. de Wildeman, *Bulletin des Séances de la Société Belge de Microscopie, Mémoires, 1889-1890* **16**: 139 (1890).
non Pythium dictyosporum M. Raciborski, *Anzieger der Akademie der Wissenschaften in Krakau* **8**: 283 (1891) *sensu* F. K. Sparrow, *Mycologia* **23**: 191-203 (1931).
Typification: lectotype *Die Kulturpflanze* **2**: 177, fig. 8 A-G (1954) (iconotype). Type locality: Tübingen, Germany. Type habitat: parasitic in *Oedogonium intermedium* Wittr. (*Chlorophyceae, Oedogoniales*) from the Botanic Garden, Tübingen, Germany. [Typification by Cejp (1959) of *L. zopfii* by drawings of *L. syncytiorum* is not accepted.]
Identification key: Key VII; key point 6'.
Literature: Dick (1997*b*), Karling (1981*a*: 114), Raciborski (1891), Rieth (1954), Sparrow (1960: 1009), Wildeman (1890*a*).

SPECIES DOUBTFULLY REFERABLE TO EITHER *SYZYGANGIA* OR *PYTHIOGETON*

ACHLYOPSIS É. de Wildeman, *Annales de la Société Belge de Microscopie, Mémoires* **20**: 123 (1896) (**Type species**: *Achlyopsis entospora* É. de Wildeman).

Achlyopsis entospora É. de Wildeman, *Annales de la Société Belge de Microscopie, Mémoires* **20**: 123 (1896).
Typification: type material not designated, lectotype *Annales de la Société Belge de Microscopie, Mémoires* **20**: plate IX, figs 1-11 (1896) (iconotype). Type locality: Marais de Pinchat, Geneva, Switzerland (July and August, 1894). Type habitat: parasitic in the oogonia of *Chara* species (*Charophyceae*, *Charales*).
Literature: de Wildeman (1896*c*).

UNIDENTIFIABLE TAXA REFERABLE TO THE GENUS *SYZYGANGIA*, BUT NOT TO SPECIES AND THEREFORE EXCLUDED

ACHLYOGETON A. Schenk, *Botanische Zeitung, Berlin, 17 Jahrgang* **46**: 398 (1859) (**Type species**: *A. entophytum* A. Schenk).

Achlyogeton entophytum A. Schenk, *Botanische Zeitung, Berlin, 17 Jahrgang* **46**: 398 (1859). **Typification**: type material not designated; parasitic in *Cladophora* sp. **Literature**: Karling (1942*a*: 94), Schenk (1859*a*), Sparrow (1960: 184).

Bicricium transversum N. Sorokin, *Archives Botaniques du Nord de la France* **2**: 35 (1883). **Typification**: type material not designated; parasitic in *Cladophora* sp. **Literature**: Yačevskie & Yačevskie (1931: 48, 69).

Bicricium naso N. Sorokin, *Archives Botaniques du Nord de la France* **2**: 35 (1883). **Typification**: type material not designated. **Literature**: Fischer (1892), Karling (1942*a*, 1977: 59, 1981*a*: 102), Minden (1915), Scherffel (1926*a*), Sorokine (1883, 1889, 1890), Sparrow (1960: 188), Yačevskie & Yačevskie (1931: 69).

Olpidiopsis sorokinei (É. de Wildeman) É. de Wildeman, *Annales de la Société Belge de Microscopie, Mémoires* **14**: 22 (1890) (**Syn**: *Olpidium sorokinei* É. de Wildeman, *Bulletin de la Société Royale de Botanique de Belge, Mémoires* **35**: 16 (1896)). **Typification**: not verified. Type locality: not verified. Type habitat: parasitic in *Tribonema* sp. (*Tribophyceae*). **Literature**: Karling (1981*a*: 35), Sparrow (1960: 955), Wildeman (1890*c*, 1896*a*).

Lagenidium syncytiorum H. Klebahn, *Jahrbücher für wissenschaftliche Botanik* **24**: 265 (1892). **Typification**: Lectotype (Cejp, 1959): *Jahrbücher für wissenschaftliche Botanik* **24**: plate 3, figs 22-24 (1892) (iconotype). **Literature**: Klebahn (1892), Karling (1981*a*: 115), Sparrow (1960: 1009), Wildeman (1895*c*).

Olpidiopsis decipiens S. N. Dasgupta & R. John, *Bulletin of the Botanical Survey of India* **30**: 59 (1988). **Typification**: Holotype deposited at HCIO, Division of Mycology and Plant Pathology, IARI, New Delhi-110 012. HCIO No. 37179. Type locality: Lucknow, Utter Pradesh, India. Type habitat: parasitic only in oogonia of *Oedogonium* sp. (*Chlorophyceae*, *Oedogoniales*) from a tank in the gardens of the Botany Department, Lucknow University, India. **Literature**: Dasgupta & John (1988).

Lagenidium sp. J. S. Karling, *Sydowia* **20**: 195 (1968). **Literature**: Karling (1968*a*).

EXCLUDED GENERA

APHANISTIS N. Sorokin, *Archives Botaniques du Nord de la France* **2**: 33 (1883), indeterminable but possible synonym of *Syzygangia* M. W. Dick, *loc. cit.*, type species very doubtfully assigned to *S. marchaliana* (É. de Wildeman) M. W. Dick.

RESTICULARIA P.-A. Dangeard, *Le Botaniste* **2**: 96 (1890), probable synonym of *Syzygangia* M. W. Dick, *loc. cit.*

Resticularia boodlei F. E. Fritsch, *Annals of Botany, London* **17**: 661 (1903): type material not designated, probably a hyphomycete (Sparrow, 1960: 1001; Yačevskie & Yačevskie, 1931: 47).

GENUS *INCERTAE SEDIS*

SEPTOLPIDIUM F. K. Sparrow, *Transactions of the British Mycological Society* **18**: 215 (1933).
Typification: *Septolpidium lineare* F. K. Sparrow (holotype - sole original species).

Septolpidium lineare F. K. Sparrow, *Transactions of the British Mycological Society* **18**: 215 (1933)
Typification: type material not designated; lectotype *Journal of the Linnean Society of London (Botany)* **50**: 429, fig. 1 a-f (1936) (iconotype). Type locality: Chapman's Garden, Emmanuel College, Cambridge, U.K. Type habitat: parasitic in *Synedra* sp. (*Bacillariophyceae*).
Identification key: Key I; key point 43'.
Literature: Czeczuga (1995*a*), Karling (1977: 60), Paterson (1958*a*), Sparrow (1933*a*, 1936*b*, 1960: 187).

CRYPTICOLACEAE M. W. Dick, *Mycological Research* **102**: 1064 (1998).
Typification: *Crypticola* R. A. Humber, S. P. Frances & A. W. Sweeney (holotype - sole original genus).

CRYPTICOLA R. A. Humber, S. P. Frances & A. W. Sweeney *in* S. P. Frances, A. W. Sweeney & R. A. Humber, *Journal of Invertebrate Pathology* **54**: 107 (1989).
Atkinsiella H. S. Vishniac, *Mycologia* **50**: 75 (1958) [*pro parte typo excluso*] (**Type species**: *A. dubia* (D. Atkins) H. S. Vishniac).
Halocrusticida K. Nakamura & K. Hatai, *Mycoscience* **36**: 437 (1995) [*pro parte*] (**Type species**: *Halocrusticida entomophaga* (W. W. Martin) K. Nakamura & K. Hatai).
Typification: *Crypticola clavulifera* R. A. Humber, S. P. Frances & A. W. Sweeney.

Crypticola clavulifera R. A. Humber, S. P. Frances & A. W. Sweeney *in* S. P. Frances, A. W. Sweeney & R. A. Humber, *Journal of Invertebrate Pathology* **54**: 107 (1989).
Typification: Holotype slide(s) and dried culture ARSEF 1940 in the Agricultural Research Service Collection of Entomopathogenic Fungi, Ithaca, New York, U.S.A. A culture of the ex-type collection is deposited under the same number at the same institute. Type locality: Millaa Millaa Falls, Queensland, Australia. Type habitat: parasitic in larvae of *Aedes kochi* (Dönitz) (*Diptera, Culicidae*) also in larvae of *Forcipomyia* (subgenus *Phytohelea*) *marksae* Tokunaga (*Diptera, Ceratopogonidae*) in the leaf axils of *Colocasia macrorhiza* Schott (*Araceae*).
Identification key: Key I; key point 14.
Literature: Dick (1998), Frances (1991), Frances *et al.* (1989).

Crypticola entomophaga (W. W. Martin) M. W. Dick, *Mycological Research* **102**: 1064 (1998).
Atkinsiella entomophaga W. W. Martin, *American Journal of Botany* **64**: 767 (1977).
Halocrusticida entomophaga (W. W. Martin) K. Nakamura & K. Hatai, *Mycoscience* **36**: 437 (1995).
Typification: Holotype slide and isotype slides deposited in the private herbarium of W. W. Martin. Type locality: 5km southeast of Hanover, Virginia, U.S.A. Type habitat: parasitic in larvae of *Diptera* (*Chironominae* and *Tanypodinae*) also in caddis flies.
Identification key: Key I; key point 14'.
Literature: Dick (1998), Martin (1977, 1981*b*, 1984, 1991).

FAMILY *INCERTAE SEDIS*

SIROLPIDIACEAE H. E. Petersen, *Botanisk Tidsskrift* **29**: 357 (1909) [typified by *Sirolpidium* but without other diagnosis].
Sirolpidiacea H. F. Copeland, *The Classification of Lower Organisms*: 81 (1956) [*pro parte*] (**Type genus**: *Sirolpidium* H. E. Petersen).
Typification: *Sirolpidium* H. E. Petersen (holotype - sole original genus).

SIROLPIDIUM H. E. Petersen, *Oversigt over det Kongelige Danske Videnskabernes Seskabs Forhandlinger. Medlemmers Arbeider, 1905*: 478 (1905).
Myzocytium A. Schenk, *Über das Vorkommen contractiler Zellen im Pflanzenreich*: 10 (1858) [*pro parte minore typo excluso*] (**Type species**: *Myzocytium proliferum* A. Schenk).
Achlyogeton A. Schenk, *Botanische Zeitung, Berlin* **17**: 398 (1859) [*pro parte typo excluso*] (**Type species**: *A. entophytum* A. Schenk).
Olpidiopsis M. Cornu, *Annales des Sciences Naturelles, Botanique, Série 5* **15**: 114, 127 (1872) [*pro parte minore typo excluso*] (**Type species**: *Olpidiopsis saprolegniae* (A. Braun) M. Cornu).
Pleotrachelus W. Zopf, *Nova Acta der Kaiserlichen Leopoldinisch-carolinischen Akademie der Naturforscher* **47**: 173 (1884) sect. *integri* H. E. Petersen, *Oversigt over det Kongelige Danske Videnskabernes Selskabs Forhandlinger. Medlemmers Arbeider, 1905*: 444 (1905) [*pro parte typo excluso*] (**Type species**: *Pleotrachelus fulgens* W. Zopf - holotype, sole original species; sect. *integri*: *Pleotrachelus andreei* [N.] G. Lagerheim - first-named species).
Bicilium H. E. Petersen, *Botanisk Tidsskrift* **29**: 357 (1909) (**Type species**: *Bicilium andreei* (G. Lagerheim) H. E. Petersen - holotype, sole original species).
Pseudolpidiella K. Cejp, *Flora ČSR, Oomycetes I*: 460 (1959) [*pro parte typo excluso*] (**Type species**: *Pseudolpidiella glenodiniana* (P.-A. Dangeard) K. Cejp).
Typification: *Sirolpidium bryopsidis* (C. de Bruyne) H. E. Petersen (holotype - sole original species).

Sirolpidium bryopsidis (C. de Bruyne) H. E. Petersen, *Oversigt over det Kongelige Danske Videnskabernes Seskabs Forhandlinger. Medlemmers Arbeider, 1905*: 480 (1905).
Olpidium bryopsidis C. de Bruyne, *Archives de Biologie* **10**: 85 (1890).
Myzocytium proliferum A. Schenk forma *marinum* Y. Kobayas[h]i & M. Ookubo, *Bulletin of the National Science Museum (Tokyo)* **33**: 64 (1953) (**Typification**: type material not designated; type locality: Wada-ura, Tiba, Japan; type habitat: parasitic in *Cladophora japonica* Yam.).
Pontisma lagenidioides H. E. Petersen, *Oversigt over det Kongelige Danske Videnskabernes Selskabs Forhandlinger. Medlemmers Arbeider, 1905*: 482 (1905) [*sensu* Kobayas[h]i & Ookubo (1953)].
Myzocytium proliferum A. Schenk forma *maximum* Y. Kobayas[h]i & M. Ookubo *in* J. S. Karling (1981) [*nomen superfluum*].
Typification: type material not designated; lectotype *Archives de Biologie* **10**: plate V, figs 1-15 (1890) (iconotype). Type locality: Bay of Naples, Italy. Type habitat: parasitic in *Bryopsis plumosa* C. Agardh (*Chlorophyceae*, *Cladophorales*).
Identification key: Key XI; key point 21.
Literature: Aleem (1950*d*, 1952*a*, 1953), de Bruyne (1890), Deckenbach (1903), Höhnk (1956), Jaczewski (1931), Johnson (1957*a*, 1968), Johnson & Sparrow (1961: 348), Karling (1968*a*, 1981*a*: 57, 91), Kobayashi & Ookubo (1953), Konno (1972), Ookubo (1952), Petersen (1905), Pystina (1995), Sparrow (1934, 1936*a*, 1960: 968, 976, 1010).

Sirolpidium andreei ([N.] G. Lagerheim) M. W. Dick, **comb. nov.**
Pleotrachelus andreei [N.] G. Lagerheim, *YMER: Tidskrift utgifven af Svenska Sällskapet för Antropologi och Geografi; Nittonde Årgången 1899*: 436 (1900) [also in N. G. Lagerheim, *Meddelanden från Stockholms Högskola Botaniska Institut* **3**: 12 (1900)].
Bicilium andreei ([N.] G. Lagerheim) H. E. Petersen, *Botanisk Tidsskrift* **29**: 357 (1909).
Olpidiopsis andreei ([N.] G. Lagerheim) J. S. Karling, *The Simple Holocarpic Biflagellate Phycomycetes*: 51 (1942) [*pro parte*].
non Olpidiopsis andreei ([N.] G. Lagerheim) J. S. Karling, *The Simple Holocarpic Biflagellate Phycomycetes*: 51 (1942) [*sensu* Sparrow].
non Petersenia andreei ([N.] G. Lagerheim) F. K. Sparrow, *Biological Bulletin of the Marine Biological Laboratory, Woods Hole* **70**: 245 (1936).
Typification: type material not designated; lectotype *YMER: Tidskrift utgifven af Svenska Sällskapet för Antropologi och Geografi; Nittonde Årgången 1899*: figure p. 436 (1900) (iconotype) [*Meddelanden från Stockholms Högskola Botaniska Institut* **3**: 12 (1900) is identical]. Type locality: beach of King Charles Land, Arctic Ocean, 11 Sept. 1899. Type habitat: parasitic in *Spongomorpha* sp. (also *Acrosiphonia* sp.) (*Chlorophyceae*, *Cladophorales*).
Identification key: Key XI; key point 21'.
Literature: Aleem (1950*d*, 1953), Cejp (1959: 376), Johnson (1972), Jokl (1916), Johnson & Howard (1968), Johnson & Sparrow (1961: 342), Karling (1942*a*, 1981*a*: 29), Lagerheim (1900*a*, *b*), Petersen (1905), Porter (1986), Pystina (1995), Sparrow (1934, 1936*b*, 1960: 951).

Sirolpidium salinum (P.-A. Dangeard) M. W. Dick, **comb. nov.**
Achlyogeton salinum P.-A. Dangeard, *Le Botaniste* **24**: 240 (1932).
Lagenidium chaetomorphae A. A. Yačevskie & P. A. Yačevskie, *Opredelitel' Gribov*: 67 (1931) [? *nomen nudum*].
? *Lagenidium* sp. [*Coenomyces consuens* C. von Deckenbach ?], *Flora, Jena* **12**: 253-283 (1903).
Typification: type material not designated; lectotype *Le Botaniste* **24**: plate XXIV, figs 1-3 (1932) (iconotype). Type locality: Roscoff, Brittany, France. Type habitat: parasitic in marine *Cladophora seriacea* Kütz. or *Cladophora utriculosa* Kütz. (as "*Cladophora laetevirens*" [*C. laetevirens* Harv. or *C. laetevirens* (Dillwyn) Kütz. ?]) (also *Cladophora fracta* (O. F. Müll.) Kütz. var. *flavescens* (Roth) Collins (as *Cladophora flavescens* Harv.) (*Chlorophyceae, Cladophorales*).
Identification key: Key XI; key point 18'.
Literature: Dangeard (1932), Deckenback (1903), Karling (1981*a*: 126), Kobayas[h]i & Ookubo (1954*a*), Sparrow (1960: 1010), Yačevskie & Yačevskie (1931: 67).

Sirolpidium paradoxum (H. E. Petersen) M. W. Dick, **comb. nov.**
Pleotrachelus paradoxus H. E. Petersen, *Oversigt over det Kongelige Danske Videnskabernes Selskabs Forhandlinger. Medlemmers Arbeiter, 1905*: 459 (1905).
Typification: type material not designated; lectotype *Oversigt over det Kongelige Danske Videnskabernes Selskabs Forhandlinger. Medlemmers Arbeiter, 1905*: 457, fig. IV, 7-8 (1905) (iconotype). Type locality: Christianssund, Norway. Type habitat: parasitic in *Rhizophydium discinctum* H. E. Petersen (*Chytridiomycetes*) on *Spongomorpha vernalis* Kjellm. (*Chlorophyceae, Cladophorales*) (coll. F. Börgesen).
Identification key: Key XI; key point 22.
Literature: Johnson (1966*e*), Karling (1981*a*: 245), Petersen (1905), Sparrow (1960: 118, 163).

DOUBTFUL TAXA

Sirolpidium glenodinianum (P.-A. Dangeard) M. W. Dick, **comb. nov.**
Olpidium glenodinianum P.-A. Dangeard, *Journal de Botanique* **2**: 131 (1888).
Pseudolpidium glenodinianum (P.-A. Dangeard) A. Fischer, *Rabenhorst's Kryptogamen-Flora, 2 Aufl., Bd 1, Abt. 4*: 35 (1892).
Pseudolpidiella glenodiniana (P.-A. Dangeard) K. Cejp, *Flora ČSR, Oomycetes I:* p. 460 (1959).
Olpidiopsis glenodiniana T. W. Johnson, *Nova Hedwigia* **10**: 586 (1966).
Typification: Selection of the lectotype from the two available will need to be made when the species is rediscovered: (Cejp, 1959): *Journal de Botanique* **2**: plate 5, figs 6-10 (1888) (iconotype). Type locality: France; (Cejp, 1959): *Scripta Botanica Horti Universitatis Impericolis Petropolitani* **24**: plate 1, figs 1-12; plate 4, figs 16-28 (1907) (iconotype). Type locality: parasitic in *Glenodinium* sp. (*Dinomastigota*), Jelagina, Russia.
Identification key: not keyed.
Literature: Cejp (1959), Dangeard (1888), Fischer (1892), Johnson (1966*d*), Karling (1981*a*: 30, 34), Serbinov (1907), Sparrow (1960: 956).

Sirolpidium marinum (P.-A. Dangeard) M. W. Dick, **comb. nov.**
Olpidium marinum P.-A. Dangeard, *Le Botaniste, Série 12*: 16 (1912).
Ectrogella marina (P.-A. Dangeard) J. Feldmann & G. Feldmann, *Revue de Mycologie* **20**: 238 (1956).
Typification: type material not designated; lectotype *Revue de Mycologie* **20**: 239 fig. 2 (1956) (iconotype). Type locality: Roscoff, Brittany, France. Type habitat: parasitic in *Tetraselmis subsalsum* Davis (as *Chlorodendron subsalsum* (Davis) Senn) (marine *Chlorophyceae, Chlorococcales*).
Identification key: Key XI; key point 20'.
Literature: Dangeard (1912), Feldmann & Feldmann (1956), Johnson & Sparrow (1961: 352), Karling (1981*a*: 75), Sparrow (1960: 156, 811).

Sirolpidium globosum (C. J. Anastasiou & L. M. Churchland) M. W. Dick, **comb. nov.**
Olpidiopsis globosa C. J. Anastasiou & L. M. Churchland, *Syesis* **1**: 81 (1968).
Typification: type material not designated; lectotype *Syesis* **1**: 82, figs 1-13; 84, figs 14-19 (1968) (iconotype). Type locality: Horseshoe Bay, Vancouver, British Columbia, Canada. Type habitat: parasitic in *Papulospora halima* Anastasiou (conidial hyphomyçete fungus) saprotrophic on submerged *Arbutus* leaves and twigs (*Ericaceae*).
Identification key: Key XI; key point 22'.
Literature: Anastasiou & Churchland (1968), Karling (1981*a*: 245), Porter (1986).

EXCLUDED SPECIES

Sirolpidium lagenidioides (H. E. Petersen) J. S. Karling, *Simple Holocarpic Biflagellate Phycomycetes*: 66 (1942), synonym of ***Pontisma lagenidioides*** H. E. Petersen, *Oversigt over det Kongelige Danske Videnskabernes Seskabs Forhandlinger. Medlemmers Arbeider, 1905*: 482 (1905).

Sirolpidium zoophthorum H. S. Vishniac, *Mycologia* **47**: 641 (1955), synonym of ***Haliphthoros zoophthorum*** (H. S. Vishniac) M. W. Dick, *loc. cit.*

FAMILY *INCERTAE SEDIS*

PONTISMATACEAE H. E. Petersen, *Botanisk Tidsskrift* **29**: 357 (1909) [as '*Pontismaceae*', typified by *Pontisma* but without other diagnosis].

Sirolpidiaceae H. E. Petersen, *Oversigt over det Kongelige Danske Videnskabernes Seskabs Forhandlinger. Medlemmers Arbeider, 1905*: 478 (1905) [*pro parte typo excluso, sensu* F. K. Sparrow, *Mycologia* **34**: 113 (1942) and J. S. Karling, *Simple Holocarpic Biflagellate Phycomycetes*: 63 (1942)] (**Type genus**: *Sirolpidium* H. E. Petersen).

Sirolpidiacea H. F. Copeland, *The Classification of Lower Organisms*: 81 (1956) [*pro parte typo excluso*] (**Type genus**: *Sirolpidium* H. E. Petersen).

Typification: *Pontisma* H. E. Petersen (holotype - sole original genus).

PONTISMA H. E. Petersen, *Oversigt over det Kongelige Danske Videnskabernes Selskabs Forhandlinger. Medlemmers Arbeider, 1905*: 482 (1905).

Olpidiopsis M. Cornu, *Annales des Sciences Naturelles, Botanique, Série V* **15**: 114, 127 (1872) [*pro parte typo excluso*] (**Type species**: *Olpidiopsis saprolegniae* (A. Braun) M. Cornu).

Pleotrachelus W. Zopf, *Nova Acta der Kaiserlichen Leopoldinisch-carolinischen Akademie der Naturforscher* **47**: 173 (1884) sect. *integri* H. E. Petersen, *Oversigt over det Kongelige Danske Videnskabernes Selskabs Forhandlinger. Medlemmers Arbeider, 1905*: 444 (1905) [*pro parte typo excluso*] (**Type species**: *Pleotrachelus fulgens* W. Zopf - holotype, sole original species; sect. *integri*: *Pleotrachelus andreei* [N.] G. Lagerheim - first-named species).

Sirolpidium H. E. Petersen, *Oversigt over det Kongelige Danske Videnskabernes Seskabs Forhandlinger. Medlemmers Arbeider, 1905*: 478 (1905) [*pro parte, sensu* Karling] (**Type species**: *Sirolpidium bryopsidis* (C. de Bruyne) H. E. Petersen).

Typification: *Pontisma lagenidioides* H. E. Petersen (holotype - sole original species).

Pontisma lagenidioides H. E. Petersen, *Oversigt over det Kongelige Danske Videnskabernes Selskabs Forhandlinger. Medlemmers Arbeider, 1905*: 482 (1905).

Sirolpidium lagenidioides (H. E. Petersen) J. S. Karling, *Simple Holocarpic Biflagellate Phycomycetes*: 66 (1942).

Typification: type material not designated; lectotype *Oversigt over det Kongelige Danske Videnskabernes Selskabs Forhandlinger. Medlemmers Arbeider, 1905*: 483, fig 10 (1905) (iconotype). Type locality: Frederikshavn, Denmark. Type habitat: parasitic in *Ceramium rubrum* C. Agardh (*Rhodophyceae*).

Identification key: Key XI; key point 10'.

Literature: Aleem (1950*c*, *d*, 1953), Feldmann (1954), Feldmann & Feldmann (1955), Höhnk (1939, 1956), Johnson (1968), Karling (1942*a*, 1981*a*: 57), Kobayashi & Ookubo (1953), Konno (1972), Petersen (1905), Pystina (1995), Sparrow (1934, 1936*a*, 1960: 971).

Pontisma inhabilis (H. E. Petersen) M. W. Dick, **comb. nov.**

Pleotrachelus inhabilis H. E. Petersen, *Oversigt over det Kongelige Danske Videnskabernes Selskabs Forhandlinger. Medlemmers Arbeiter, 1905*: 456 (1905).

Typification: type material not designated; lectotype *Oversigt over det Kongelige Danske Videnskabernes Selskabs Forhandlinger. Medlemmers Arbeiter, 1905*: 457, fig IV, 1-6 (1905) (iconotype). Type locality: Busserev, near Frederikshavn, Denmark. Type habitat: parasitic in *Polysiphonia fibrillosa* (Dillwyn) Sprengel (as its synonym, *Polysiphonia violacea* (Roth) Sprengel) (*Rhodophyceae*).

Identification key: Key XI; key point 14'.

Literature: Johnson & Howard (1968), Karling (1977: 14), Petersen (1905), Sparrow (1960: 118, 161).

Pontisma feldmannii (A. A. Aleem) M. W. Dick, **comb. nov.**

Olpidiopsis feldmannii A. A. Aleem, *Comptes Rendus Hebdomadaires de Séances de l'Académie des Sciences, Paris, Série D* **235**: 1250 (1952) [as '*feldmanni*' - the author did not state whether the intention was to name the fungus in honour of G. Feldmann ('*feldmanniae*') or J. Feldmann ('*feldmannii*')] [Botanical Code, algae, no Latin diagnosis required].

Typification: type material not designated; lectotype *Comptes Rendus Hebdomadaires de Séances de l'Académie des Sciences, Paris, Série D* **235**: 1251, figs a-e (1952) (iconotype). Type locality: Mediterranean Sea, Banyules-sur-Mer, France. Type habitat: parasitic in *Asparagopsis armata* Harv. (as its synonym, *Falkenburgia rufolanosa* (Harv.) F. Schmitz (*Rhodophyceae*), at 25-30m depth.

Identification key: Key XI; key point 14'.

Literature: Aleem (1952, 1953), Dixon (1963), Johnson (1972), Johnson & Howard (1968), Johnson & Sparrow (1961: 343), Karling (1981*a*: 32), Sparrow (1960: 952, 1969).

Pontisma magnusii (J. Feldmann & G. Feldmann) M. W. Dick, **comb. nov.**

Olpidiopsis magnusii J. Feldmann & G. Feldmann, *Revue de Mycologie* **20**: 243 (1955).

Typification: type material not designated: lectotype *Revue de Mycologie* **20**: 244, fig 4 (1956) (iconotype). Type locality: Roscoff, Brittany, France. Type habitat: parasitic in *Ceramium flabelligerum* J. G. Agardh (*Rhodophyceae*).

Identification key: Key XI; key point 17.

Literature: Feldmann & Feldmann (1955), Johnson & Sparrow (1961: 343), Karling (1981*a*: 36), Sparrow (1960: 953).

Pontisma dangeardii (J. Feldmann & G. Feldmann) M. W. Dick, **comb. nov.**

Olpidiopsis dangeardii J. Feldmann & G. Feldmann, *Le Botaniste* **50**: 193 (1967).

Typification: type material not designated; lectotype *Le Botaniste* **50**: 199, plate II, figs A-C, plate III, figs A-I, plate IV, figs A-G (1967) (iconotype). Type locality: Mediterranean Sea, Roche Toreilles, Argelès-sur-Mer, Banyules-sur-Mer, France. Type habitat: parasitic in *Radicilingua reptans* (Hol.) Papenfuss (*Rhodophyceae*, *Delesseriaceae*) at 30m depth.

Identification key: Key XI; key point 16'.

Literature: Feldmann & Feldmann (1967), Karling (1981a: 35).

Pontisma antithamnionis (A. Whittick & G. R. South) M. W. Dick, **comb. nov.**

Olpidiopsis antithamnionis A. Whittick & G. R. South, *Archiv für Mikrobiologie* **82**: 354 (1972).

Typification: Type NFLD No. 6114. Type locality: Boone Point, Bay Bulls, Newfoundland, Canada. Type habitat: parasitic in *Antithamnion floccosum* (O. F. Müll.) Kleen (*Rhodophyceae*).

Identification key: Key XI; key point 17'.

Literature: Karling (1981*a*: 34), Porter (1986), Whittick & South (1971, 1972).

EXCLUDED SPECIES REFERABLE TO THE *CHYTRIDIOMYCETES*

Rozella marina (F. K. Sparrow) F. K. Sparrow, *Mycologia* **30**: 377 (1938).

Pleolpidium marinum F. K. Sparrow, *Biological Bulletin of the Marine Biological Laboratory, Woods Hole* **70**: 256 (1936).

Rozella maximum J. S. Karling, *American Journal of Botany* **29**: 24 (1942) [*nomen nudum*].

Typification: type material not designated; lectotype *Biological Bulletin of the Marine Biological Laboratory, Woods Hole* **70**: 257 figs 32, 33 (1936) (iconotype). Type locality: Woods Hole, Massachusetts, U.S.A. Type habitat: parasitic in *Chytridium polsiphoniae* Cohn parasitic on *Polysiphonia fibrillosa* (Dillwyn) Sprengel (*Rhodophyceae*).

Identification key: Key XII; key point 13'.

Literature: Held (1981), Johnson (1966*f*), Johnson & Sparrow (1961: 305), Karling (1942*a*, 1977: 20), Sparrow (1936*a*, 1938, 1960: 176, 181).

PETERSENIA F. K. Sparrow, *Dansk Botanisk Arkiv* **8**: 13 (1934).

Pleotrachelus W. Zopf, *Nova Acta der Kaiserlichen Leopoldinisch-carolinischen Akademie der Naturforscher* **47**: 173 (1884) sect. *lobati* H. E. Petersen, *Oversigt over det Kongelige Danske Videnskabernes Selskabs Forhandlinger. Medlemmers Arbeider, 1905*: 444 (1905) [*pro parte typo excluso*] (**Type species**: *Pleotrachelus fulgens* W. Zopf - holotype, sole original species; sect. *integri*: *Pleotrachelus andreei* [N.] G. Lagerheim - first-named species).

Typification: *Petersenia lobata* (H. E. Petersen) F. K. Sparrow.

Petersenia lobata (H. E. Petersen) F. K. Sparrow, *Dansk Botanisk Arkiv* **8**: 13 (1934).

Pleotrachelus lobatus H. E. Petersen, *Oversigt over det Kongelige Danske Videnskabernes Selskabs Forhandlinger. Medlemmers Arbeiter, 1905*: 460 (1905).

Typification: type material not designated; lectotype *Oversigt over det Kongelige Danske Videnskabernes Selskabs Forhandlinger. Medlemmers Arbeiter, 1905*: 460, fig. V, 1-7 (1905) (iconotype). Type locality: Banc de Tönneberg, N. E. of Laesø, Denmark. Type habitat: parasitic in *Callithamnion corymbosum* Lyngb. (*Rhodophyceae*).

Identification key: Key XI; key point 8.

Literature: Aleem (1950*d*, 1953), Cejp (1959), Feldmann (1954), Feldmann & Feldmann (1940), Johnson (1957*a*, 1966*d*), Johnson & Howard (1968), Johnson & Sparrow (1961: 340), Karling (1981*a*: 60), Kuznetsov (1979), Petersen (1905), Pystina (1995), Sparrow (1934, 1936*a*, 1960: 958), Vishniac (1956), Wright (1879).

Petersenia pollagaster (H. E. Petersen) F. K. Sparrow, *Dansk Botanisk Arkiv* **8**: 15 (1934).

Pleotrachelus pollagaster H. E. Petersen, *Oversigt over det Kongelige Danske Videnskabernes Selskabs Forhandlinger. Medlemmers Arbeiter, 1905*: 462 (1905).

Typification: type material not designated; lectotype *Oversigt over det Kongelige Danske Videnskabernes Selskabs Forhandlinger. Medlemmers Arbeiter, 1905*: 463, fig. VI, 1-5 (1905) (iconotype). Type locality: Frederikshavn, Denmark. Type habitat: parasitic in *Ceramium rubrum* C. Agardh (*Rhodophyceae*).

Identification key: Key XI; key point 8'.

Literature: Craigie (1984), Johnson & Howard (1968), Johnson & Sparrow (1961: 341), Karling (1981*a*: 61), Molina (1986), Molina, Hughes & Craigie (1988), Petersen (1905), Porter (1986), Sparrow (1934, 1960: 959).

Petersenia palmariae J. P. Van Der Meer & C. M. Pueschel, *Canadian Journal of Botany* **63**: 407 (1985).

Typification: type material not designated; lectotype *Canadian Journal of Botany* **63**: 406, figs 1-12; 410, figs 1-3; 412-415, figs 4-15; 417, figs 16-17 (1985) (iconotype). Type locality: Whiffen Spit, Sooke, Vancouver Island, B.C., Canada. Type habitat: parasitic in fronds of *Palmaria mollis* (Setchell & Gardner) Van Der Meer & Bird (*Rhodophyceae*).

Identification key: Key XI; key point 7'.

Literature: Porter (1986), Pueschel & Van Der Meer (1985), Van Der Meer & Pueschel (1985).

UNIDENTIFIED SPECIES

Petersenia sp. in *Achrochaetium* (Aleem, 1950).

Petersenia sp. in *Polysiphonia* (Howard & Johnson, 1968).

EXCLUDED SPECIES

Petersenia andreei ([N.] G. Lagerheim) F. K. Sparrow, *Biological Bulletin of the Marine Biological Laboratory, Woods Hole* **70**: 245 (1936) [**synonymy**: *Olpidiopsis andreei* ([N.] G. Lagerheim) J. S. Karling, *The Simple Holocarpic Biflagellate Phycomycetes*: 51 (1942) [*pro parte*]; *non Sirolpidium andreei* ([N.] G. Lagerheim) M. W. Dick, *loc. cit.*; *non Pleotrachelus andreei* [N.] G. Lagerheim, *YMER: Tidskrift utgifven af Svenska Sällskapet för Antropologi och Geografi; Nittonde Årgången 1899*: 436 (1900); *non Bicilium andreei* (N. G. Lagerheim) H. E. Petersen, *Botanisk Tidsskrift* **29**: 357 (1909)], synonym of *Anisolpidium joklianum* M. W. Dick, *loc. cit.*

Petersenia catenophlyctidis B. M. Sundaram, *Journal of the Indian Botanical Society* **47**: 144 (1968), synonym of *Cornumyces catenophlyctidis* (B. M. Sundaram) M. W. Dick, *loc. cit.*

Petersenia irregularis (J. C. Constantineanu) F. K. Sparrow, *Aquatic Phycomycetes* p. 634 (1943), synonym of *Cornumyces irregularis* (J. C. Constantineau) M. W. Dick, *loc. cit.*

Petersenia panicicola M. J. Thirumalachar & R. C. Lacy, *Mycopathologia et Mycologia Applicata* **6**: 75 (1951), possibly related to *Lagena* [see p. 336].

Petersenia utriculoba C. E. Miller, *Mycologia* **54**: 422 (1962), synonym of *Cornumyces utriculobus* (C. E. Miller) M. W. Dick, *loc. cit.*

Petersenia sp. in *Rhizosolenia* (*Bacillariophyceae*) (Johnson, 1966).

Petersenia sp. in *Laminaria* (Schatz, 1984).

FAMILY *INCERTAE SEDIS*

EURYCHASMATACEAE H. E. Petersen, *Oversigt over det Kongelige Danske Videnskabernes Selskabs Forhandlinger. Medlemmers Arbeider, 1905*: 445 (1905) [as '*Eurychasmaceae*', typified by *Eurychasma* but without other diagnosis].

Eurychasmatales F. K. Sparrow *in* E. B. G. Jones, *Recent Advances in Aquatic Mycology*: 220 (1976) [as '*Eurychasmales*', order *pro parte*, families *Eurychasmataceae* and *Atkinsiellaceae*] [*nomen nudum*].

Ectrogellacea H. F. Copeland, *The Classification of Lower Organisms*: 81 (1956) [*pro parte typo excluso*] (**Type genus**: *Ectrogella* A. Scherffel).

Typification: *Eurychasma* H. E. Petersen (holotype - sole original genus).

EURYCHASMA P. Magnus, *Hedwigia* **44**: 348 (1905).

Typification: *Eurychasma dicksonii* (E. P. Wright) P. Magnus.

Eurychasma dicksonii (E. P. Wright) P. Magnus, *Hedwigia* **44**: 348 (1905).

Rhizophydium dicksonii E. P. Wright, *Transactions of the Royal Irish Academy, Dublin (Science)* **26**: 374 (1879).

Olpidium dicksonii (E. P. Wright) N. Wille var. *striariae* N. Wille, *Videnskabsselskabets Skrifter. I. Mathematisk-naturvidenskabelig Klasse. 1899. No. 3. Christiana*: 2 (1899).

Ectrogella dicksonii (E. P. Wright) A. Scherffel, *Archiv für Protistenkunde* **52**: 4 (1925).

Typification: type material not designated; lectotype *Dansk Botanisk Arkiv* **8**: plate 1, figs A-S (1934) (iconotype). Type locality: Ireland. Type habitat: parasitic in *Hincksia granulosa* (J. E. Sm.) P. C. Silva (as *Ectocarpus granulosus* (Engler) C. Agardh (*Fucophyceae*).

Identification key: Key XI; key point 3'.

Literature: Adams & Pethybridge (1910), Aleem (1950*a*, *b*, *d*, 1953), Aysel & Uçar (1990), Dangeard (1934), Feldmann (1954), Hariot (1889), Hauck (1878), Jenneborg (1977), Johnson (1909), Johnson & Sparrow (1961: 324), Karling (1981*a*: 77), Konno & Tanaka (1988), Löwenthal (1905), Magnus (1905), Pedersén (1983), Petersen (1905), Porter (1986), Pystina (1995), Rattray (1887), Scherffel (1925*a*), Sparrow (1934, 1960: 815), Weete *et al.* (1981, 1982), Wille (1899), Wright (1879).

UNIDENTIFIABLE SPECIES

Eurychasma sp. in *Polysiphonia* sp. (Sparrow, 1969: 141).

EXCLUDED SPECIES

Eurychasma lauderiae (H. H. Gran) H. E. Petersen, *Oversigt over det Kongelige Danske Videnskabernes Selskabs Forhandlinger. Medlemmers Arbeiter, 1905*: (1905), synonym of *Ectrogella lauderiae* (H. H. Gran) A. Scherffel, *Archiv für Protistenkunde* 52: 3 (1925).

Eurychasma sp. in *Bryopsis* sp. (Kobayashi & Ookubo, 1954).

EURYCHASMIDIUM F. K. Sparrow, *Biological Bulletin of the Marine Biological Laboratory, Woods Hole* **70**: 241 (1936).
Typification: *Eurychasmidium tumefaciens* (P. Magnus) F. K. Sparrow (holotype - sole original species).

Eurychasmidium tumefaciens (P. Magnus) F. K. Sparrow, *Biological Bulletin of the Marine Biological Laboratory, Woods Hole* **70**: 241 (1936).
Chytridium tumefaciens P. Magnus, *Sitzungs-Bericht der Gesellschaft Naturforschender Freunde zu Berlin, 1872*: 87 (1872).
Olpidium tumefaciens (P. Magnus) E. P. Wright, *Transactions of the Royal Irish Academy, Dublin (Science)* **26**: 360 (1879).
Pleotrachelus tumefaciens (P. Magnus) H. E. Petersen, *Oversigt over det Kongelige Danske Videnskabernes Selskabs Forhandlinger. Medlemmers Arbeiter, 1905*: 456 (1905).
Olpidiopsis tumefaciens (P. Magnus) K. Cejp, *Flora ČSR, Oomycetes I* p. 377 (1959).
Typification: type material not designated; lectotype *Biological Bulletin of the Marine Biological Laboratory, Woods Hole* **70**: 246: figs 14-21, plate I fig 1 (1936) (iconotype). Type locality: off-shore waters, Italy. Type habitat: parasitic in *Ceramium spiniferum* (*Rhodophyceae*).
Identification key: Key XI; key point 5.
Literature: Cejp (1959), Czeczuga (1995*b*), Johnson & Sparrow (1961: 327), Karling (1981*a*: 81), Magnus (1872, 1875), Petersen (1905), Porter (1986), Sparrow (1936*a*, 1960: 822), Wildeman (1900), Wright (1879).

Eurychasmidium sacculum (H. E. Petersen) M. W. Dick, **comb. nov.**
Eurychasma sacculus H. E. Petersen, *Oversigt over det Kongelige Danske Videnskabernes Selskabs Forhandlinger. Medlemmers Arbeiter, 1905*: 477 (1905).
Typification: type material not designated; lectotype *Oversigt over det Kongelige Danske Videnskabernes Selskabs Forhandlinger. Medlemmers Arbeiter, 1905*: 473, fig. 8:5,8,9 (1905) (iconotype). Type locality: Umanarsuak, Greenland, east coast. Type habitat: parasitic in *Halosaccion ramentaceum* (*Rhodophyceae*, *Palmariales*).
Identification key: Key XI; key point 4'.
Literature: Höhnk (1956), Jenneborg (1977), Johnson & Sparrow (1961: 353), Karling (1981*a*: 80), Petersen (1905), Sparrow (1960: 819).

Eurychasmidium joycei (F. K. Sparrow) M. W. Dick, **comb. nov.**
Eurychasma joycei F. K. Sparrow, *Archiv für Mikrobiologie* **66**: 144 (1969).
Typification: type material not designated; lectotype (iconotype). Type locality: San Juan Islands, Washington, U.S.A. Type habitat: parasitic in *Polysiphonia* sp. (*Rhodophyceae*, *Ceramiales*).
Identification key: Key XI; key point 5'.
Literature: Karling (1981*a*: 80), Sparrow (1969).

ORDER *INCERTAE SEDIS*

ANISOLPIDIALES M. W. Dick, **ordo novum**
Anisochytridiales J. S. Karling, *American Journal of Botany* **30**: 641 (1943) [*pro parte* - order not subject to rules of priority] (**Type family**: *Anisolpidiaceae* J. S. Karling *ex* M. W. Dick).
Hyphochytriales E. A. Bessey, *Morphology and Taxonomy of Fungi*: 69 (1950) [*pro parte typo excluso*] (**Type family**: *Hyphochytriaceae* A. Fischer *emend.* J. S. Karling).
Hyphochytrialea H. F. Copeland, *The Classification of Lower Organisms*: 69 (1956) [*pro parte typo excluso*] (**Type family**: *Hyphochytriaceae* A. Fischer *emend.* J. S. Karling).
Hyphochytriales F. K. Sparrow, *Aquatic Phycomycetes, 2nd Revised Edition*: 743 (1960) [*pro parte typo excluso*] (**Type family**: *Hyphochytriaceae* A. Fischer *emend.* J. S. Karling).
non Hyphochytriales F. K. Sparrow, *Aquatic Phycomycetes, 2nd Revised Edition*: 743 (1960) [*sensu lato*] (**Type family**: *Hyphochytriaceae* A. Fischer *emend.* J. S. Karling).

Fungi in aquis marinis, endoparasitici obligati in algis marinis. **Thallus** holocarpicus, olpidioides vel saccatus, parietes cellulae glucanis egentes, unusquique se transformans in structuram reproductonis capacem. **Reproductio asexualis** per zoosporangia, zoosporogenesis intrasporangialis, zooosporae pyriformes, uniflagellatae; flagellum insertum in parte anteriore, flagellum straminipilosum (?). **Sporae perdurantae** cum parietibus aliquot crassioribus.
Holotypus: *Anisolpidium* J. S. Karling (*Anisolpidiaceae* J. S. Karling *ex* M. W. Dick).

Marine fungi, obligate endoparasites of marine algae (*Fucophyceae*). **Thallus** holocarpic, olpidioid or saccate, cell walls lacking glucans [chlor-zinc iodide negative], transformed into reproductive structures. **Asexual reproduction** by zoospores; zoosporogenesis intrasporangial; zoospores pyriform, uniflagellate; flagellum insertion anterior, flagellum straminipilous (?). **Resting spores** with slightly thicker walls.
Typification: *Anisolpidium* J. S. Karling (*Anisolpidiaceae* J. S. Karling *ex* M. W. Dick) (holotype - sole original family).

ANISOLPIDIACEAE J. S. Karling *ex* M. W. Dick, *loc. cit.*
Anisolpidiaceae J. S. Karling, *American Journal of Botany* **30**: 641 (1943) [*nom. inval.* Art. 36.1, without Latin diagnosis].
Typification: *Anisolpidium* J. S. Karling (holotype - sole original genus).

Fungi in aquis marinis, endoparasitici obligati in algis marinis. **Thallus** holocarpicus, olpidioides vel saccatus, parietes cellulae glucanis egentes, unusquique se transformans in structuram reproductonis capacem. **Reproductio asexualis** per zoosporangia, zoosporogenesis intrasporangialis, zooosporae pyriformes, uniflagellatae; flagellum insertum in parte anteriore, flagellum straminipilosum (?). **Sporae perdurantae** cum parietibus aliquot crassioribus.
Holotypus: *Anisolpidium* J. S. Karling.

Marine fungi, obligate endoparasites of marine algae (*Fucophyceae*). **Thallus** holocarpic, olpidioid or saccate, cell walls lacking glucans [chlor-zinc iodide negative], transformed into reproductive structures. **Asexual reproduction** by zoospores; zoosporogenesis intrasporangial; zoospores pyriform, uniflagellate; flagellum insertion anterior, flagellum straminipilous (?). **Resting spores** with slightly thicker walls.
Typification: *Anisolpidium* J. S. Karling (holotype - sole original family).

ANISOLPIDIUM J. S. Karling, *American Journal of Botany* **30**: 637 (1943).
Pleotrachelus W. Zopf, *Nova Acta der Kaiserlichen Leopoldinisch-carolinischen Akademie der Naturforscher* **47**: 173 (1884) [*pro parte typo excluso*] (**Type species**: *Pleotrachelus fulgens* W. Zopf - holotype, sole original species).
Olpidiopsis M. Cornu, *Annales des Sciences Naturelles, Botanique, Série V* **15**: 114, 127 (1872) [*pro parte minore typo excluso*; *sensu* F. K. Sparrow, *Aquatic Phycomycetes*: 629 (1943)] (**Type species**: *Olpidiopsis saprolegniae* (A. Braun) M. Cornu).
Typification: *Anisolpidium ectocarpi* J. S. Karling.

Anisolpidium ectocarpi J. S. Karling, *American Journal of Botany* **30**: 637 (1943).
Typification: type material not designated; lectotype *American Journal of Botany* **30**: 638, figs 1-217 (1943) (iconotype). Type locality: Beaufort, North Carolina, U.S.A. Type habitat: parasitic in the brown alga *Hincksia mitchellae* (Harv.) P. C. Silva (as its synonym, *Ectocarpus mitchellae* Harv.) (*Fucophyceae*).
Identification key: Key XI; key point 23'.
Literature: Johnson (1957*c*), Karling (1943, 1977: 384), Sparrow (1960: 748).

Anisolpidium rosenvingii (H. E. Petersen) J. S. Karling, *American Journal of Botany* **30**: 641 (1943).
Pleotrachelus rosenvingii H. E. Petersen, *Oversigt over det Kongelige Danske Videnskabernes Selskabs Forhandlinger. Medlemmers Arbeiter, 1905*: 453 (1905).
Typification: type material not designated; lectotype *Oversigt over det Kongelige Danske Videnskabernes Selskabs Forhandlinger. Medlemmers Arbeiter, 1905*: 455, fig. III: 2-4 (1905) (iconotype). Type locality: Julianehaab, east coast of Greenland. Type habitat: parasitic in *Pilayella littoralis* (L.) Kjellm. (*Fucophyceae*).
Identification key: Key XI; key point 24.
Literature: Aleem (1953), Karling (1943, 1977: 384), Petersen (1905), Sparrow (1934*c*, 1960: 749).

Anisolpidium sphacelarum (L. Kny) J. S. Karling, *American Journal of botany* **30**: 641 (1943) [as '*sphacellarum*'].
Chytridium sphacelarum L. Kny, *Sitzungs-Berichte der Gesellschaft naturforschender Freunde zu Berlin, 1871*: 97 (1871) [as '*sphacellarum*'].
Olpidium sphacelarum (L. Kny) L. Kny, *Hedwigia* **112**: 87 (1872) [as '*sphacellarum*'].
Olpidium sphacelarum (L. Kny) A. Fischer, *Rabenhorst's Kryptogamen-Flora, 2 Aufl., Bd 1, Abt. 4*: 26 (1892) [as '*sphacellarum*'].
Pleotrachelus sphacelarum (L. Kny) H. E. Petersen, *Oversigt Kongelige Danske Videnskabernes Selskabs Forhandl Medlemmers Arbeiter 1905*: 452 (1905) [as '*sphacellarum*'].
Olpidiopsis sphacelarum (L. Kny) F. K. Sparrow, *Aquatic Phycomycetes*: 629 (1943) [as '*sphacellarum*'].
Typification: type material not designated; lectotype (iconotype). Type locality: Cotes septentrionale, France. Type habitat: parasitic in *Cladostephus spongiosus* J. G. Agardh (*Fucophyceae*).
Identification key: Key XI; key point 24'.
Literature: Aleem (1953), Farlow & Seymour (1888), Feldmann (1954), Fischer (1892), Johnson (1909), Karling (1943, 1977: 384), Kny (1871, 1872), Magnus (1872*a*, *b*, 1873, 1875), Petersen (1905), Pringsheim (1855), Seymour (1929), Sparrow (1936*a*, 1943, 1960: 747).

DOUBTFUL SPECIES

Anisolpidium joklianum M. W. Dick, **nom. nov.**

Pleotrachelus ectocarpi M. Jokl, *Österreichische Botanische Zeitschrift* **66**: 267 (1916).

Petersenia andreei ([N.] G. Lagerheim) F. K. Sparrow, *Biological Bulletin of the Marine Biological Laboratory, Woods Hole* **70**: 245 (1936).

Olpidiopsis andreei ([N.] G. Lagerheim) J. S. Karling, *The Simple Holocarpic Biflagellate Phycomycetes*: 51 (1942) [*sensu* Sparrow].
non Anisolpidium ectocarpi J. S. Karling, *American Journal of Botany* **30**: 637 (1943).
Typification: type material not designated; lectotype *Biological Bulletin of the Marine Biological Laboratory, Woods Hole* **70**: 244, figs 1-13 (1936) (iconotype). Type locality: Gulf of Trieste, Adriatic Sea (Mediterranean). Type habitat: parasitic in *Hincksia granulosa* (J. E. Sm.) P. C. Silva (as *Ectocarpus granulosus* (Engler) C. Agardh (*Fucophyceae*).
Identification key: Key XI; addendum.
Literature: Jokl (1916), Karling (1981*a*: 29), Sparrow (1934, 1936*a*, 1960: 951)

Anisolpidium minutum (H. E. Petersen) M. W. Dick, **comb. nov.**
Pleotrachelus minutus H. E. Petersen, *Oversigt over det Kongelige Danske Videnskabernes Selskabs Forhandlinger. Medlemmers Arbeiter, 1905*: 451 (1905).
Typification: type material not designated; lectotype *Oversigt over det Kongelige Danske Videnskabernes Selskabs Forhandlinger. Medlemmers Arbeiter, 1905*: 452, fig II, 1-4 (1905) (iconotype). Type locality: Isles of Hirholmene, Denmark. Type habitat: parasitic in *Chorda filum* (L.) Stackh. (*Fucophyceae*).
Identification key: Key XI; addendum.
Literature: Petersen (1905), Sparrow (1960: 162).

Anisolpidium olpidium (H. E. Petersen) M. W. Dick, **comb. nov.**
Pleotrachelus olpidium H. E. Petersen, *Oversigt over det Kongelige Danske Videnskabernes Selskabs Forhandlinger. Medlemmers Arbeiter, 1905*: 455 (1905).
Typification: type material not designated; lectotype *Oversigt over det Kongelige Danske Videnskabernes Selskabs Forhandlinger. Medlemmers Arbeiter, 1905*: 455, fig III, 7-9 (1905) (iconotype). Type locality: Faroes Islands, Denmark. Type habitat: parasitic in *Ectocarpus siliculosus* (Dillwyn) Lyngbye (as its synonym, *Ectocarpus confervoides* (Roth) Le Jol.) (*Fucophyceae*).
Identification key: Key XI; addendum.
Literature: Petersen (1905), Sparrow (1960: 163), El-Nagdy & Abdel-Hafez (1990).

EXCLUDED SPECIES

Anisopidium saprobium J. S. Karling, *Journal of the Elisha Mitchell Scientific Society* **84**: 111 (1968), synonym of *Hyphochytrium saprobium* (J. S. Karling) M. W. Dick *loc. cit.*

Anisolpidium elongatum J. S. Karling, *Nova Hedwigia* **28**: 217 (1976), synonym of *Hyphochytrium elongatum* (J. S. Karling) M. W. Dick *loc. cit.*

Anisolpidium stigeoclonii (É. de Wildeman) H. M. Canter, *Transactions of the British Mycological Society* **33**: 343 (1950), synonym of *Canteriomyces stigeoclonii* (É. de Wildeman) F. K. Sparrow, *Aquatic Phycomycetes, 2nd revised edition*: 750 (1960).

ORDER *INCERTAE SEDIS* PROVISIONALLY ATTACHED TO THE *PERONOSPOROMYCOTINA*

LAGENISMATALES M. W. Dick, **ordo novum**

Fungi parasitici in Bacillariophyceis marinis. **Thallus** holocarpicus, inter frustulum hospitale et plasmamembranam hospitalem, in facie assimilativa sine pariete cellulae; coralliformis, allantoideis vel pseudomycelialis; cum pariete in maturitate, unusquisque thallus se transformans vel in zoosporangium vel in gametangium. **Reproductio asexualis** per zoosporangia; zoosporogenesis intrasporangialis; zoosporae non dimorphicae; plus minusve hemisphaericae, flagellis in parte laterali insertis, heterokontae et isokontae; discus transitionalis in plano super planum plasmamembranae; zoosporae parviores ($< 150\ \mu m^3$). **Reproductio sexualis** oogamosa ex contactu cystarum gametalium; gametae praesentes, flagellatae; isogameticae; fecundatio per hyphas fecundationis inter duas cystas gameticas; spora perdurans in cysta recipienti formata, aplerotica, plus minusve cylindricalis.
Holotypus: *Lagenisma* G. Drebes (*Lagenismataceae* M. W. Dick).

Fungi parasitic in marine Bacillariophyceae. **Thallus** holocarpic, between the host frustule and host plasmamembrane, lacking a cell wall during the assimilative phase; coralloid, allantoid or pseudomycelial; walled at maturity, becoming transformed into a zoosporangium or a gametangium. **Asexual reproduction** by zoosporangia; zoosporogenesis intrasporangial; zoospores not dimorphic; more or less hemispherical, flagellar insertion lateral; heterokont and isokont; transitional plate above the plane of the plasmamembrane; zoospores medium-small (volume equivalent $< 150\ \mu m^3$). **Sexual reproduction** oogamous by means of contact between gametic cysts; gametes present, flagellate, isogametic; fertilization by a fertilization hypha between two gametic cysts; resting spore formed in receptor cyst, aplerotic, more or less cylindrical.
Typification: *Lagenisma* G. Drebes (*Lagenismataceae* M. W. Dick, holotype - sole original family).

LAGENISMATACEAE M. W. Dick, **familia nova**

Fungi parasitici in Bacillariophyceis marinis. Thallus holocarpicus, inter frustulum hospitale et plasmamembranam hospitalem, in facie assimilativa sine pariete cellulae; coralliformis, allantoideis vel pseudomycelialis, non septatus; cum pariete in maturitate, unusquisque thallus se transformans vel in zoosporangium vel in gametangium. **Reproductio asexualis** per zoosporangia; zoosporogenesis intrasporangialis; zoosporae diplaneticae sed non dimorphicae; plus minusve hemisphaericae; cystae primariae cum spinis cavis; flagellis in parte laterali insertis, heterokontae et isokontae; discus transitionalis in plano super planum plasmamembranae; zoosporae parviores ($<150\mu m^3$). **Reproductio sexualis** oogamosa ex contactu cystarum gametalium; gametae praesentes, flagellatae isogameticae; fecundatio per hyphas fecundationis inter duas cystas gameticas; spora perdurans in cysta recipienti forma.
ta, aplerotica, plus minusve cylindricalis.
Holotypus: *Lagenisma* G. Drebes.

Fungi parasitic in marine Bacillariophyceae. **Thallus** holocarpic, between the host frustule and host plasmamembrane, lacking a cell wall, pseudomycelial, coralloid or allantoid, non septate, walled at maturity, becoming transformed into a zoosporangium or a gametangium. **Asexual reproduction** by zoosporangia; zoosporogenesis intrasporangial; zoospores diplanetic but not dimorphic; more or less hemispherical; primary cysts with hollow spines; flagellar insertion lateral; heterokont and isokont; transitional plate above the plane of the plasmamembrane; zoospores medium-small (volume equivalent $<150\mu m^3$). **Sexual reproduction** oogamous by means of contact between gametic cysts; gametes present, flagellate, isogametic; fertilization by a fertilization hypha between two gametic cysts; resting spore formed in receptor cyst, aplerotic, more or less cylindrical.
Typification: *Lagenisma* G. Drebes (holotype - sole original genus).

LAGENISMA G. Drebes, *Veröffentlichungen des Instituts für Meeresforschung in Bremerhaven, Sonderband* **3**: 68 (1968).
Typification: *Lagenisma coscinodisci* G. Drebes (holotype - sole original species).

Lagenisma coscinodisci G. Drebes, *Veröffentlichungen des Instituts für Meeresforschung in Bremerhaven, Sonderband* **3**: 68 (1968).
Typification: type material not designated; lectotype *Helgoländer wissenschaft Meeresuntersuchungen* **13**:428-433, figs 1-6; **29**:293,295-298, figs 1-5; *Archives of Microbiology* **116**: 143-144, figs 9-23; *Protoplasma* **94**: 273, figs 32-34 (iconotype). Type locality: North Sea near Heligoland. Type habitat: parasitic in *Coscinodiscus concinnus* W. Sm. (also *Coscinodiscus granii* Gough) (*Bacillariophyceae*).
Identification key: Key I; key point 142'.
Literature: Chakravarty (1969, 1974), Drebes (1966, 1968, 1969), Drebes & Schnepf (1979), Gotelli (1971), Grahame (1976), Johnson (1966*a*, *g*), Karling (1981*a*: 143), Lopez & MacCarthy (1985), Parsons (1962), Pavillard (1914), Porter (1986), Schnepf, Deichgräber & Drebes (1978*a*, *b*, *c*, *d*, *e*), Schnepf & Drebes (1977), Schnepf & Heinzmann (1980), Wetsteyn & Peperzak (1991).

SYSTEMATICS

DOUBTFUL GENERA *INCERTAE SEDIS* IN THE *STRAMINIPILA*

SYNCHAETOPHAGUS C. Apstein, *Wissenschaftliche Meeresuntersuchungen der Kommission zur Wissenschaftlichen Untersuchungen der Deutschen Meere, Abteilung Kiel* **12**: 166 (1910) [no separate generic diagnosis].
Typification: *Synchaetophagus balticus* C. Apstein (holotype - sole original species).

Synchaetophagus balticus C. Apstein, *Wissenschaftliche Meeresuntersuchungen der Kommission zur Wissenschaftlichen Untersuchungen der Deutschen Meere, Abteilung Kiel* **12**: 166 (1910).
Typification: type material not designated; lectotype *Wissenschaftliche Meeresuntersuchungen der Kommission zur Wissenschaftlichen Untersuchungen der Deutschen Meere, Abteilung Kiel* **12**: 163, figs 1-2; 164, figs 3-8 (1910) (iconotype). Type locality: Baltic Sea, near Kiel, Germany. Type habitat: parasitic in *Synchaeta monopus* Plate (*Aschelminthes, Rotifera*).
Literature: Apstein (1910).

LAGENIDICOPSIS N. Ya. Artemchuk*, *Veröffentlichungen des Instituts für Meeresforschung in Bremerhaven* **13**: 235 (1972).
Typification: *Lagenidicopsis arctica* N. Ya. Artemchuk (holotype - sole original species).

Lagenidicopsis arctica N. Ya. Artemchuk, *Veröffentlichungen des Instituts für Meeresforschung in Bremerhaven* **13**: 235 (1972).
Typification: type material not designated; drawings are not adequate to be designated as the lectotype. Type locality: Kandalashial Bay, Great Salma Strait, White Sea, northern Russia. Type habitat: saprotrophic in pine pollen baits.
Literature: Artem[t]chuk (1972, 1981: 78), Karling (1981*a*: 148), Kuznetsov (1979), Pystina (1995).

---

*FOOTNOTE:
The authority is given here following the *British Standard, No. 2979 (1958)* English transliteration of Н. Я. Артемчук. In the German text of the original article the name is spelt N. J. Artemtchuk. See also *Appendix 2* for the different spelling in 'recommended form'.

ORDER *INCERTAE SEDIS*

ROZELLOPSIDALES M. W. Dick, **ordo novum**

Lagenidiales J. S. Karling, *American Journal of Botany* **26**: 518 (1939) [*pro parte typo excluso*] (**Type family**: *Lagenidiaceae* J. S. Karling).

Olpidiacea H. F. Copeland, *The Classification of Lower Organisms*: 118 (1956) [*pro parte typo excluso*] (**Type family**: *Monolpidiaceae* A. Fischer).

Olpidiopsidaceae F. K. Sparrow *ex* K. Cejp, *Flora ČSR, Oomycetes I*: 349 & 455 (1959); *Mycologia* **34**: 113 (1942) [*pro parte typo excluso*].

Olpidiaceae J. Schröter, *Die natürlichen Pflanzenfamilien* **1(1)**: 67 (1892) [*pro parte typo excluso*].

Woroninales A. A. Yačevskie & P. A. Yačevskie, *Opredelitel' Gribov*: 60 (1931) [*pro parte typo excluso*] (**Type family**: *Woroninaceae* M. von Minden).

Fungi in aquis dulcibus, endoparasitici obligati in fungis flagellatis, fungis eumycotis, algis aut protoctistis. **Thallus** holocarpicus, plasmodialis post contagionem et in facie assimilativa; plasmodium nonnumquam divisionis vegetativae, fortasse et anastomosis capax; thallus se transformans aut in zoosporangium pleroticum, raro quasi pleroticum, aut in sporam perdurantem apleroticam. **Reproductio asexualis** per sporangia, sporogenesis intrasporangialis, zoosporas vel aplanosporas ferens; zoosporae motiles intra zoosporangium; zoosporae pusillae (volumine pari ad (3)10-20(75) μm^3), pyriformes, saepe cum granulis posterioribus refractivis, biflagellatae aut uniflagellatae; zoosporae uniflagellatae cum flagello in posteriorem partem verso, sed ventre posteriore inserto; zoosporae biflagellatae anisokontae, nondum straminipilosae demonstratae. **Reproductio sexualis** nondum pro certo cognita, meiosis non reperta; sporae perdurantes plerumque cum fibrillis extramuralibus ab hospite derivatis; sine cellula sociali; distributio copiae penariae nondum pro certo cognita.

Holotypus: *Rozellopsis* J. S. Karling *ex* K. Cejp (*Rozellopsidaceae* M. W. Dick).

Freshwater fungi, obligate endoparasites of flagellate fungi, eumycote fungi, algae or protoctists. **Thallus** holocarpic, plasmodial following infection and during the assimilative phase; plasmodium sometimes capable of vegetative division, possibly of anastomosis; becoming transformed either into a sporangium or sporangia usually filling the host cell cavity, or into a single resting spore which does not fill the host cell cavity. **Asexual reproduction** by sporangia; sporogenesis intrasporangial producing zoospores or aplanospores; zoospores motile within the zoosporangium; zoospores small (volume equivalent (3)10-20(75) μm^3), pyriform, often with posterior refractive granules, with one or two flagella; uniflagellate zoospores with a posteriorly directed, but posterio-ventrally inserted flagellum; biflagellate zoospores anisokont, not proven to be heterokont or straminipilous. **Sexual reproduction** not established, meiosis not known; resting spores usually with host-derived extramural fibrils; not accompanied by a companion cell; distribution of reserve materials not established.

Typification: *Rozellopsis* J. S. Karling *ex* K. Cejp (*Rozellopsidaceae* M. W. Dick).

Other family: *Pseudosphaeritaceae* M. W. Dick.

ROZELLOPSIDACEAE M. W. Dick, **familia nova**

Olpidiaceae J. Schröter, *Die natürlichen Pflanzenfamilien* **1(1)**: 67 (1892) [*pro parte typo excluso*] (**Type genus**: *Olpidium* (A. Braun) L. Rabenhorst).

Olpidiopsidaceae F. K. Sparrow *ex* K. Cejp, *Flora ČSR, Oomycetes I*: 349 & 455 (1959); *Mycologia* **34**: 113 (1942) [*pro parte typo excluso*] (**Type genus**: *Olpidiopsis* M. Cornu).

Woroninaceae M. von Minden, *Kryptogamenflora Mark Brandenburg* **5**: 227 (1911) (**Type genus**: *Woronina* M. Cornu) [*pro parte typo excluso*].

Fungi endoparasitici in *Peronosporomycetes* vel *Chytridiomycetes*. **Thallus** holocarpicus, plasmodialis post contagionem et in facie assimilativa; plasmodium thallorum divisibiles, fortasse et anastomosis; thallus se transformans in zoosporangium pleroticum, raro quasi pleroticum, autim sporam perdurantem pro aplerotica fungentem. **Reproductio asexualis** per zoosporangia; zoosporogenesis intrasporangialis; zoosporae motiles intra zoosporangium et intra zoosporangio circumvolutae ante emissionem; zoosporae pusillae vel parviores (volumine pari ad (3)10-20(75) μm^3), pyriformes, saepe cum granulis posterioribus refractivis, uniflagellatae aut biflagellatae; uniflagellatae zoosporae cum flagello in posteriorem partem verso, sed ventre posteriore inserto;, zoosporae biflagellatae anisokontae, non heterokontae demonstratae, pyriformes (35-75 μm^3). **Reproductio sexualis** nondum pro certo cognita, meiosis non reperta; sporae perdurantes cum fibrillis extramuralibus ab hospite derivatis; sine cellula sociali; distributio copiae penariae nondum pro certo cognita.

Holotypus: *Rozellopsis* J. S. Karling *ex* K. Cejp.

Fungi obligate endoparasites in *Peronosporomycetes* or *Chytridiomycetes*. **Thallus** holocarpic; plasmodial following infection and during the assimilative phase; plasmodium sometimes capable of vegetative division, possibly of anastomosis; transformed either into a sporangium occupying the entire host cell cavity without contraction or division into discrete walled segments or sporangia, or into a single resting spore which does not fill the host cell cavity. **Asexual reproduction** by zoospores; zoosporogenesis intrasporangial; zoospores motile and swirling within the zoosporangium before discharge; zoospores small (volume equivalent (3)10-20(75) μm^3),

pyriform, often with posterior refractive granules, with one or two flagella; uniflagellate zoospores with a posteriorly directed, but posterio-ventrally inserted flagellum; biflagellate zoospores anisokont, not proven to be heterokont (35-75 μm^3). **Sexual reproduction** not established, meiosis not known; resting spores single in a host segment; usually with host-derived extramural fibrils; not accompanied by a companion cell, formed either non-sexually or by an automictic process.
Typification: *Rozellopsis* J. S. Karling *ex* K. Cejp.

ROZELLOPSIS J. S. Karling *ex* K. Cejp, *Flora ČSR, B Řada Mykologicko-Lichenologická, Svazek 2, Oomycetes I*: 454 (1959).
Rozellopsis J. S. Karling, *Mycologia* **34**: 205 (1942).
Rozella M. Cornu, *Annales des Sciences Naturelles, Botanique, Série V* **15**: 114, 148 (1872) [*pro parte typo excluso*] (**Type species**: *Rozella monoblepharidis* M. Cornu).
Pleolpidium A. Fischer, *Rabenhorst's Kryptogamen-Flora, 2 Aufl., Bd 1, Abt. 4*: 43 (1892) [*pro parte minore typo excluso*] (**Type species**: *Pleolpidium monoblepharidis* (M. Cornu) A. Fischer).
Rozella M. Cornu *sensu* M. von Minden, *Kryptogamenflora Mark Brandenburg* **5**: 270 (1911) (**Type species**: *Rozella septigena* M. Cornu).
Skirgiellopsis A. Batko, *Acta Mycologica, Warszawa* **13**: 322 (1977) (**Type species**: *Skirgiellopsis fischeri* (K. Cejp) A. Batko).
Typification: *Rozellopsis septigena* J. S. Karling *ex* K. Cejp.

Rozellopsis septigena J. S. Karling ex K. Cejp, *Flora ČSR, B Řada Mykologicko-Lichenologická, Svazek 2, Oomycetes I*: 455 (1959).
Rozellopsis septigena J. S. Karling, *Mycologia* **34**: 206 (1942).
Rozella septigena M. Cornu, *Annales des Sciences Naturelles, Botanique, Série V* **15**: 163 (1872) [*pro parte*].
Rozella septigena M. Cornu, *sensu* A. Fischer, *Jahrbücher für wissenschaftliche Botanik* **13**: 221 (1882).
Rozellopsis fischeri K. Cejp, *Flora ČSR, B Řada Mykologicko-Lichenologicá, Svazek 2, Oomycetes I*: 455 (1959).
Skirgiellopsis fischeri (K. Cejp) A. Batko, *Acta Mycologica, Warszawa* **13**: 323 (1977).
Typification: Lectotype for *R. fischeri* (Cejp, 1959): *Jahrbücher für wissenschaftliche Botanik* **13**: plate 14, fig. 19, plate 15 (1882) (iconotype). Lectotype locality: Germany. Type habitat: parasitic in *Saprolegnia* spp. (*Peronosporomycetes, Saprolegniales*).
Identification key: Key I; key point 71'.
Literature: Cornu (1872*b*), Dick (1970), Fischer (1882, 1892), Karling (1942*a*, *c*, 1981*a*: 43), Minden (1911*a*), Prowse (1951), Pystina (1995), Sparrow (1938, 1952*d*, 1960: 924), Willoughby & Roberts (1992).

Rozellopsis simulans (A. Fischer) J. S. Karling, *American Journal of Botany* **29**: 34 (1942).
Rozella simulans A. Fischer, *Jahrbücher für Wissenschaftliche Botanik* **13**: 365 (1882).
Skirgiellopsis simulans (A. Fischer) A. Batko, *Acta Mycologica, Warszawa* **13**: 323 (1977).
Typification: Lectotype (Cejp, 1959): *Transactions of the Sapporo Natural History Society (Japan)* **13**: plate 2, figs 12-14 (1933) (iconotype). Lectotype locality: Iwashiro, Japan. Type habitat: parasitic in *Achlya* spp. (*Peronosporomycetes, Saprolegniaceae*).
Identification key: Key I; key point 78.
Literature: Fischer (1892), Karling (1942*c*, *d*, 1981*a*: 43), Konno (1972), Maurizio (1895), Minden (1911*a*), Sparrow (1938, 1960; 925), Tokunaga (1933*a*).

Rozellopsis inflata (E. J. Butler) J. S. Karling, *American Journal of Botany* **29**: 34 (1942).
Pleolpidium inflatum E. J. Butler, *Memoirs of the Department of Agriculture, India, Botanical Series* **1**: 126 (1907).
Typification: Lectotype (Cejp, 1959): iconotype *Memoirs of the Department of Agriculture, India, Botanical Series* **1**: plate 7, figs 17-21 (1907). Type locality: Antibes, France. Type habitat: parasitic in *Pythium intermedium* de Bary (*Peronosporomycetes, Pythiales*).
Identification key: Key I; key point 55'.
Literature: Butler (1907), Czeczuga (1994, 1995*b*), Czeczuga & Woronowicz (1991), Held (1981), Karling (1942*c*, *d*, 1966, 1968*d*, 1969, 1981*a*: 43, 1987), Prowse (1954), Pystina (1995), Remy (1948), Sparrow (1938, 1960: 925).

Rozellopsis uliginosa A. Batko, *Acta Mycologica, Warszawa* **13**: 317 (1977).
Typification: Iconotype: *Acta Mycologica, Warszawa* **13**: 318, fig. 5a-c (1977). Type locality: Kampinos National Park, Poland. Type habitat: parasitic in *Nellymyces megaceros* A. Batko (*Peronosporomycetes, Rhipidiales*).
Identification key: Key I; key point 52.
Literature: Batko (1977), Held (1981), Karling (1981*a*: 245).

DOUBTFUL SPECIES

Rozellopsis sp. (Kobayas[h]i *et al.*, 1971: 7 fig 2, J; 18) [see Table 7, ms p. 125 - Lagenidiaceous saprotrophs of pollen grains from freshwater or soil habitats].

EXCLUDED SPECIES

Rozellopsis waterhouseae J. S. Karling, *Mycologia* **34**: 206 (1942), synonym of *Gracea waterhouseae* (J. S. Karling) M. W. Dick, *loc. cit.*

EXCLUDED GENUS (IN SYNONYMY)

SKIRGIELLOPSIS A. Batko, *Acta Mycologica, Warszawa* **13**: 322 (1977) (**Type species**: *Skirgiellopsis fischeri* (K. Cejp) A. Batko).

ROZELLA M. Cornu, *Annales des Sciences Naturelles, Botanique, Série V* **15**: 114, 148 (1872).
Rozea M. Cornu, *Bulletin de la Société Botanique de France* **19**: 71 (1872) [later homonym].
Rozia M. Cornu, *Annales des Sciences Naturelles, Botanique, Série V* **15**: 114, (1872) [misspelling of *Rozea* M. Cornu].
Pleolpidium A. Fischer, *Rabenhorst's Kryptogamen-Flora, 2 Aufl., Bd 1, Abt. 4*: 43 (1892) (**Type species**: *Pleolpidium monoblepharidis* (M. Cornu) A. Fischer).
Skirgiellia A. Batko, *Acta Mycologica, Warszawa* **13**: 321 (1977) (**Type species**: *Skirgiellia septigena* (M. Cornu) A. Batko).
Typification: *Rozella monoblepharidis* M. Cornu.

Rozella monoblepharidis M. Cornu, *Annales des Sciences Naturelles, Botanique, Série V* **15**: 150 (1872) [also as '*monoblepharidis-polymorphae*'].
Pleolpidium monoblepharidis (M. Cornu) A. Fischer, *Rabenhorst's Kryptogamen-Flora, 2 Aufl., Bd 1, Abt. 4*: 44 (1892).
Typification: type material not designated; lectotype *Annales des Sciences Naturelles, Botanique, Série V* **15**: plate 4, figs 13-18 (1872) (iconotype). Type locality: Solonge, France. Type habitat: parasitic in *Monoblepharis polymorpha* Cornu (*Chytridiomycetes, Monoblepharidales*).
Identification key: Key I; key point 113'.
Literature: Cornu (1872*a*, *b*), Fischer (1892), Karling (1942*c*), Laibach (1927), Minden (1911*a*), Sparrow (1938, 1960: 168), Sparrow & Barr (1955).

Rozella rhipidii M. Cornu, *Annales des Sciences Naturelles, Botanique, Série V* **15**: 153 (1872) [also as '*rhipidii-spinosi*'].
Pleolpidium rhipidii (M. Cornu) A. Fischer, *Rabenhorst's Kryptogamen-Flora, 2 Aufl., Bd 1, Abt. 4*: 44 (1892).
Pleolpidium araiosporae (M. Cornu) M. von Minden, *Kryptogamenflora Mark Brandenburg* **5**: 252 (1911) [no basionym, a *nom. nov.* erected because *R. spinosum* is a synonym for *Araiospora spinosa*].
Typification: type material not designated; lectotype *Annales des Sciences Naturelles, Botanique, Série V* **15**: plate 5, figs 1-9 (1872) (iconotype). Type locality: Solonge, France. Type habitat: parasitic in *Araiospora spinosa* (Cornu) Thaxt. (*Peronosporomycetes, Rhipidiales*).
Identification key: Key I; key point 52'.
Literature: Cornu (1872*b*), Fischer (1892), Held (1981), Karling (1942*c*), Minden (1911*a*), Sparrow (1938, 1960: 168).

Rozella apodachlyae M. Cornu, *Annales des Sciences Naturelles, Botanique, Série V* **15**: 161 (1872) [as '*apodyae*'].
Pleolpidium apodachlyae (M. Cornu) A. Fischer, *Rabenhorst's Kryptogamen-Flora, 2 Aufl., Bd 1, Abt. 4*: 45 (1892) [as '*apodyae*'].
Typification: type material not designated; lectotype *Annales des Sciences Naturelles, Botanique, Série V* **15**: plate 5, figs 10-14 (1872) (iconotype). Type locality: Solonge, France. Type habitat: parasitic in *Apodachlya brachynema* Pringsh. (*Peronosporomycetes, Leptomitales*).
Identification key: Key I; key point 67'.
Literature: Cornu (1872*b*), Fischer (1892), Held (1981), Karling (1942*c*), Sparrow (1938, 1960: 169).

Rozella septigena M. Cornu, *Annales des Sciences Naturelles, Botanique, Série V* **15**: 163 (1872).
non Rozella septigena M. Cornu *sensu* A. Fischer, *Jahrbucher für wissenschaftliche Botanik* **13**: 321 (1882).
Skirgiellia septigena (M. Cornu) A. Batko, *Acta Mycologia, Warszawa* **13**: 322 (1977).
Typification: type material not designated; lectotype *Annales des Sciences Naturelles, Botanique, Série V* **15**: plate 6, figs 1-14 (1872) (iconotype). Type locality: lake at Chaville, Solonge, France. Type habitat: parasitic in *Achlya racemosa* Hildebr. (*Peronosporomycetes, Saprolegniales*).
Identification key: Key I; key point 78'.
Literature: Cejp (1934), Cornu (1872*b*), Dangeard (1890*b*), Held (1981), Howard & Johnson (1969), Karling (1942*c*), Pringsheim (1860), Sorokine (1883), Sparrow (1938, 1960: 176), Yačevskie & Yačevskie (1931: 63).

Rozella irregularis (E. J. Butler) F. K. Sparrow, *Mycologia* **30**: 377 (1938).
Pleolpidium irregulare E. J. Butler, *Memoirs of the Department of Agriculture, India, Botanical Series* **1**: 123 (1907).
Typification: type material not designated; lectotype *Memoirs of the Department of Agriculture, India, Botanical Series* **1**: plate VIII, figs 1-12 (1907) (iconotype). Type locality: Kew, U.K. Type habitat: parasitic in *Pythium vexans* de Bary (*Peronosporomycetes, Pythiales*) from soil with roots of *Ophioglossum* (*Pteridophyta*).
Identification key: Key I; key point 57.
Literature: Butler (1907), Gaertner (1954), Held (1981), Karling (1942*b*: 198, *c*, 1981*e*, 1987*b*), Shen & Siang (1948), Sparrow (1938, 1960: 171).

Rozella cuculus (E. J. Butler) F. K. Sparrow, *Mycologia* **30**: 377 (1938).
Chytridium simulans P.-A. Dangeard, *Le Botaniste* **5**: 21 (1896).
Pleolpidium cuculus E. J. Butler, *Memoirs of the Department of Agriculture, India, Botanical Series* **1**: 125 (1907).
Pleolpidium tuberculorum P. Vuillemin, *Bulletin des Séances de la Société des Sciences (et de la Réunion Biologique) de Nancy* **10**: 30-45 (1909).
Typification: type material not designated; lectotype *Memoirs of the Department of Agriculture, India, Botanical Series* **1**: plate VII, figs 22-25 (1907) (iconotype). Type locality: Cork, Ireland (also Antibes, France). Type habitat: parasitic in *Pythium intermedium* de Bary (*Peronosporomycetes, Pythiales*).
Identification key: Key I; key point 58'.
Literature: Butler (1907), Dangeard (1896), Held (1981), Karling (1942*b*: 198, *c*, 1965, 1968*d*, 1965), Sparrow (1938, 1960: 170), Tokunaga (1933*b*), Vuillemin (1909), Yačevskie & Yačevskie (1931: 15).

Rozella blastocladiae (M. von Minden) F. K. Sparrow, *Mycologia* **30**: 377 (1938).
Pleolpidium blastocladiae M. von Minden, *Kryptogamenflora Mark Brandenburg* **5**: 253 (1911).
Typification: type material not designated; lectotype 424,fig. 26 c-d (1909) (iconotype). Type locality: Hamburg, Germany. Type habitat: parasitic in *Blastocladia pringsheimii* Reinsch (*Chytridiomycetes, Blastocladiales*).
Identification key: Key I; key point 114'.
Literature: Held (1981), Laibach (1927), Karling (1942*c*), Minden (1911*a*, 1916), Petersen (1909, 1910), Sparrow (1938, 1960: 169), Thaxter (1896), Waterhouse (1942).

Rozella polyphagi (F. K. Sparrow) F. K. Sparrow, *Mycologia* **30**: 377 (1938).
Pleolpidium polyphagi F. K. Sparrow, *Transactions of the British mycological Society* **18**: 215 (1933).
Typification: type material not designated; lectotype *Journal of the Linnean Society (Botany)*: plate 14, figs 19-20 (1936) (iconotype). Type locality: Cambridge, U.K. Type habitat: parasitic in prosporangia of *Polyphagus euglenae* Nowak. (*Chytridiomycetes, Chytridiales*) from water in a ditch behind the Leys School, Cambridge, U.K.
Identification key: Key I; key point 122'.
Literature: Held (1981), Karling (1942*c*), Powell (1984), Sparrow (1933*a*, 1936*b*, 1938, 1960: 173).

Rozella allomycis F. K. Foust, *Journal of the Elisha Mitchell Scientific Society* **53**: 198 (1937).
Skirgiellia allomycis (M. Cornu) A. Batko, *Acta Mycologica* **13**: 322 (1977).
Rozella allomycetes K. Nabel, *Archiv für Mikrobiologie* **10**: 527 (1939) [*nomen nudum*].
Typification: type material not designated; lectotype *Journal of the Elisha Mitchell Scientific Society* **53**: plates 22,23 (1937) (iconotype). Type locality: Battle Grove, Coker Arboretum, Chapel Hill, North Carolina, U.S.A. Type habitat: parasitic in *Allomyces* sp. (*Chytridiomycetes, Blastocladiales*) from soil.
Identification key: Key I; key point 114.
Literature: Céspedes & Castillo (198?), Foust (1937), Gaertner (1954), Held (1972*a*, *b*, 1973*a*, *b*, 1974, 1975, 1980, 1981), Karling (1942*c*, 1948, 1964*c*, 1976*a*, 1977: 20, 1987), Kobayas[h]i & Konno (1970), Lange & Olson (1979), Milanez (1984), Nabel (1939), Perry (1967), Shen & Siang (1948), Sörgel (1952), Sparrow (1938, 1943, 1960: 178, 180), Wolf (1941).

Rozella cladochytrii J. S. Karling, *Torreya* **41**: 105 (1941).
Typification: type material not designated; original description lacked illustrations. Type locality: State Park Lake, Bastrop, Texas, U.S.A. Type habitat: parasitic in *Nowakowskiella profusum* Karling (also in *N. elegans* (Nowak.) J. Schröt., *N. ramosum* E. J. Butler, *Cladochytrium replicatum* Karling, *Cladochytrium crassum* A. B. Hillegas, *Cladochytrium hyalinum* H. B. Berdan) (*Chytridiomycetes, Chytridiales*).
Identification key: Key I; key point 119'.
Literature: Held (1981), Karling (1941*b*, 1942*c*, *d*, 1944*d*, 1964*c*, 1965, 1977: 20), Sparrow (1960: 174).

Rozella endochytrii J. S. Karling, *Torreya* **41**: 106 (1941).
Typification: type material not designated; original description lacked illustrations. Type locality: Karling estate, Austin, Texas, U.S.A. Type habitat: parasitic in *Endochytrium operculatum* (De Wild.) Karling (*Chytridiomycetes, Chytridiales*).
Identification key: Key I; key point 121'.

Literature: Held (1981), Karling (1941*b*, 1942*c*, *d*, 1944*d*), Sparrow (1960: 173).

Rozella rhizophlyctidis J. S. Karling, *American Journal of Botany* **29**: 32 (1942) [as '*rhizophlictii*'].
Typification: type material not designated; lectotype *American Journal of Botany* **29**: 30, figs 36-47 (1942) (iconotype). Type locality: Austin, Texas, U.S.A. Type habitat: parasitic in *Rhizophlyctis petersenii* Sparrow (*Chytridiomycetes, Chytridiales*).
Identification key: Key I; key point 120.
Literature: Held (1981), Karling (1942*d*), 1944*d*, 1964*c*, 1965), Sparrow (1960: 174), Ward (1939).

Rozella laevis J. S. Karling, *Mycologia* **34**: 201 (1942).
Typification: type material not designated, no illustrations. Type locality: New Kent County, Virginia, U.S.A. Type habitat: parasitic in *Pythium* F-group [as "*Pythium gracile* Schenk"] (*Peronosporomycetes, Pythiales*).
Identification key: Key I; key point 58.
Literature: Held (1981), Karling (1942*c*, 1944*d*, 1964*c*, 1965, 1988), Sparrow (1960: 171), Sparrow, Paterson & Johns (1965).

Rozella achlyae L. Shanor, *Journal of the Elisha Mitchell Scientific Society* **58**: 100 (1942).
Skirgiellia achlyae (M. Cornu) A. Batko, *Acta Mycologica, Warszawa* **13**: 322 (1977).
Typification: type material not designated; lectotype *Journal of the Elisha Mitchell Scientific Society* **58**: plate 17, figs 1-7 (1942). Type locality: Harbison Lake, Highlands, North Carolina, U.S.A. Type habitat: parasitic in *Achlya flagellata* Coker (*Peronosporomycetes, Saprolegniales*).
Identification key: Key I; key point 77'.
Literature: Held (1981), Johnson (1955*b*), Karling (1942*b*, 1944*d*, 1948), Shanor (1942), Sparrow (1960: 177).

Rozella rhizophydii J. S. Karling, *Mycologia* **36**: 645 (1944).
Typification: type material not designated; lectotype *Mycologia* **36**: 645, figs 20-28 (1944) (iconotype). Type locality: Manaus, Amazonas, Brazil. Type habitat: parasitic in *Rhizophydium globosum* (A. Braun) Rabenh. (*Chytridiomycetes, Chytridiales*).
Identification key: Key I; key point 122.
Literature: Held (1981), Karling (1941*b*, 1942*a*, *c*, 1944*d*, 1948, 1968*d*), Miller (1965), Sparrow (1960: 174).

Rozella chytriomycetis J. S. Karling, *Mycologia* **38**: 107 (1946) [as '*chytriomycii*'].
Typification: type material not designated; lectotype *Mycologia* **38**: 104, figs 9-19 (1946) (iconotype). Type locality: Candlewood Lake, Connecticutt, U.S.A. Type habitat: parasitic in *Chytriomyces hyalinus* Karling (*Chytridiomycetes, Chytridiales*).
Identification key: Key I; key point 120'.
Literature: Held (1981), Karling (1946*a*, 1947*a*, 1965, 1977: 20), Sparrow (1960: 172).

Rozella canterae F. K. Sparrow, *Aquatic Phycomycetes, 2nd Revised Edition*: 172 (1960).
Typification: type material not designated; lectotype *Transactions of the British mycological Society* **33**: 357, fig. 3g-n; plate 29, figs 4-5 (1950) (iconotype). Type locality: Cumbria, U.K. Type habitat: parasitic in sporangia of *Chytridium oedogonii* Couch (*Chytridiomycetes, Chytridiales*).
Identification key: Key I; key point 118'.
Literature: Canter (1950), Held (1981), Karling (1946*a*, 1965), Sparrow (1960: 172).

Rozella longicollis J. S. Karling, *Sydowia* **19**: 218 (1965).
Typification: type material not designated; lectotype *Sydowia* **19**: plate XLVI, figs21-29 (1965) (iconotype). Type locality: Tosswill Farm, Hawkes Bay Province, New Zealand. Type habitat: parasitic in a *Pythium* sp. (*Peronosporomycetes, Pythiales*) from a soil sample.
Identification key: Key I; key point 57'.
Literature: Held (1981), Karling (1965, 1976*a*).

Rozella parva H. M. Canter, *Botanical Journal of the Linnean Society* **62**: 275 (1969).
Typification: Holotype IMI 131673. Type locality: Blelham Tarn, Cumbria, U.K. Type habitat: parasitic in *Zygorhizidium affluens* Canter (*Chytridiomycetes, Chytridiales*).
Identification key: Key I; key point 117'.
Literature: Canter (1969, 1979).

Rozella diplophlyctidis J. S. Karling, *Nova Hedwigia* **45**: 529 (1987).
Typification: Iconotype *Nova Hedwigia* **45**: 533, figs 1-6 (1987). Type locality: Ross Biological Reserve, Purdue University, Indiana, U.S.A. Type habitat: parasitic in *Diplophlyctis intestina* J. Schröt. (*Chytridiomycetes, Chytridiales*) growing on internodes of *Nitella flexilis* (L.) C. Agardh (*Chlorophyceae, Charales*).
Identification key: Key I; key point 118.
Literature: Karling (1987*b*).

DOUBTFUL SPECIES

Rozella barrettii J. S. Karling, *Mycologia* **34**: 202 (1942).
Pleolpidium sp. J. T. Barrett, *Phytopathology* **24**: 1138 (1934).
Typification: type material not designated, no illustrations. Type locality: California, U.S.A. Type habitat: parasitic in *Phytophthora cactorum* (Lebert & Cohn) J. Schröt. (*Peronosporomycetes, Pythiales*).
Identification key: Key I; key point 54'.
Literature: Barrett (1934), Karling (1942*c*), Sparrow (1960: 180).

EXCLUDED SPECIES

Rozella allomycis K. Nabel, *Archiv für Mikrobiologie* **10**: 527 (1939) [as '*allomycetes*'] [*nomen nudum*], synonym of *R. allomycis* F. K. Foust, *Journal of the Elisha Mitchell Scientific Society* **53**: 198 (1937).

Rozella coleochaetis F. K. Sparrow, R. A. Paterson & R. M. Johns, *Papers of the Michigan Academy of Science, Arts, and Letters* **50**: 118 (1965), synonym of *Plasmophagus coleochaetis* (F. K. Sparrow, R. A. Paterson & R. M. Johns) M. W. Dick, *loc. cit.*

Rozella itersoniliae D. J. S. Barr & R. Bandoni, *Mycologia* **71**: 1261 (1979), synonym of *Pleotrachelus itersoniliae* (D. J. S. Barr & R. Bandoni) M. W. Dick, *loc. cit.*

Rozella marina (F. K. Sparrow) F. K. Sparrow, *Mycologia* **30**: 377 (1938) (**Syn**: *Pleolpidium marinum* F. K. Sparrow, *Biological Bulletin of the Marine Biological Laboratory, Woods Hole* **70**: 256 (1936)), a member of the *Chytridiomycetes*.

Rozella maxima J. S. Karling, *American Journal of Botany* **29**: 24 (1942) [as '*maximum*', a typographical error for *R. marina* (F. K. Sparrow) F. K. Sparrow].

Rozella pseudomorpha (A. Scherffel) F. K. Sparrow, *Aquatic Phycomycetes*: 124 (1943) (**Syn**: *Olpidium pseudomorphum* A. Scherffel, *Archiv für Protistenkunde* **54**: 510 (1926). **Literature**: Held (1981), Scherffel (1926*b*), Sparrow (1960: 175), excluded as an improbable mycoparasitic phenomenon. [See p. 249 and also p. 245 re- *Pythiella besseyi*; 221 re- *Olpidiopsis myzocytia*; 224 re- *Olpidiopsis sphaeritae*; returned to *Olpidium* until rediscovered.]

Rozella sp. (Dogma, 1978), see *R. pseudomorpha* (A. Scherffel) F. K. Sparrow.

Rozella sp. (Karling, 1964*c*), parasitic in *Phlyctorhiza variabilis* Karling.

Rozella sp. (Karling, 1976*a*), see *R. longicollis* J. S. Karling.

Rozella sp. (Kobayas[h]i & Konno, 1970: fig. 2, L,M), see *R. allomycis* Foust [said to differ from *R. allomycis* in shape and density of spines and colour of resting spores].

Rozella sp. (Schulz, 1922), see *Plasmophagus coleochaetis* (F. K. Sparrow, R. A. Paterson & R. M. Johns) M. W. Dick, *loc. cit.*

EXCLUDED GENERA (IN SYNONYMY)

ROZEA M. Cornu, *Bulletin de la Société Botanique de France* **19**: 71 (1872) [homonym accepted by Cornu (1872*b*) deferring to Bescherelle (Cornu (1872*a*: 71 and 76); *non Rozea* Bescherelle, *Mémoires de la Société des Sciences Naturelles de Cherbourg* **6**: 241 (1872) (**Type species**: *Rozea andrieuxii* (C. Müller) Besch., *Musci, Brachytheciaceae*].

PLEOLPIDIUM A. Fischer, *Rabenhorst's Kryptogamen-Flora, 2 Aufl., Bd 1, Abt. 4*: 43 (1892) (**Type species**: *Pleolpidium monoblepharidis* (M. Cornu) A. Fischer).

SKIRGIELLIA A. Batko, *Acta Mycologia, Warszawa* **13**: 321 (1977) (**Type species**: *Skirgiellia septigena* (M. Cornu) A. Batko).

DICTYOMORPHA J. T. Mullins, *American Journal of Botany* **48**: 378 (1961).
Pringsheimiella J. N. Couch, *Journal of the Elisha Mitchell Scientific Society* **55**: 409 (1939) (**Type species**: *Pringsheimiella dioica* J. N. Couch).
Typification: *Dictyomorpha dioica* (J. N. Couch) J. T. Mullins (holotype - sole original species).

Dictyomorpha dioica (J. N. Couch) J. T. Mullins, *American Journal of Botany* **48**: 378 (1961).
Pringsheimiella dioica J. N. Couch, *Journal of the Elisha Mitchell Scientific Society* **55**: 410 (1939).
Typification: Slides prepared by J. N. Couch, Nos. 169, 218 in the Herbarium of the University of North Carolina constitute the type material; "cotype" [*sic*] slides prepared by J. T. Mullins, Nos. 141, 181. Type locality: Chapel Hill, North Carolina, U.S.A. Type habitat: parasitic in *Achlya* (*A. flagellata* Coker, also in *Thraustotheca clavata* (de Bary) Humphrey (Mullins, 1961) (*Peronosporomycetes*, *Saprolegniales*).
Identification key: Key I; key point 82'.
Literature: Berkeley (1864), Cooke (1871), Cornu (1872), Couch (1939), Czeczuga (1995*a*), Fischer (1892), Hidalgo (1962), Karling (1977: 20), Mullins (1960*a*, *b*, 1961, 1965*a*, *b*), Pringsheim (1860), Sparrow (1960: 157).

Dictyomorpha pythiensis (N. Sarkar & R. Dayal) M. W. Dick, **stat. nov.**
Dictyomorpha dioica (J. N. Couch) J. T. Mullins var. *pythiensis* N. Sarkar & R. Dayal, *Proceedings of the National Academy of Sciences, India, Section B (Biological Sciences)* **58**: 404 (1988).
Typification: Slide No. 2, deposited in the Herbarium of Division of Mycology and Plant Pathology, New Delhi, HC10 33831. Type locality: Kurukshtra Pond, Varanasi, U. P., India. Type habitat: parasitic in *Pythium aphanidermatum* (Edson) Fitzp. (*Peronosporomycetes*, *Pythiales*).
Identification key: Key I; key point 65'.
Literature: Sarkar & Dayal (1988).

PSEUDOSPHAERITACEAE M. W. Dick, **familia nova**

Olpidiaceae J. Schröter, *Die natürlichen Pflanzenfamilien* **1**(1): 67 (1892) [*pro parte typo excluso*] (**Type genus**: *Olpidium* (A. Braun) L. Rabenhorst).

Olpidiacea H. F. Copeland, *The Classification of Lower Organisms*: 118 (1956) [*pro parte typo excluso*].

Olpidiopsidaceae F. K. Sparrow *ex* K. Cejp, *Flora ČSR, Oomycetes I*: 349 & 455 (1959); *Mycologia* **34**: 113 (1942) [*pro parte typo excluso*] (**Type genus**: *Olpidiopsis* M. Cornu).

Endoparasiti obligati in protoctistis (rhizopodibus, euglenidibus, cryptomonadibus), aut fungis eumycotis, aut algis viridibus. **Thallus** holocarpicus; plasmodialis post contagionem et nonnumquam in facie assimilativa; plasmodium nonnumquam divisionis vegetativae capax; se transformans aut in sporangium parietatum, aut in plura, non omnino cavitatem cellulae hospitialis repleta; aut in sporam singulam perdurantem, non omnino cavitatem cellulae hospitialis repletam. **Reproductio asexualis** per sporangia; sporogenesis intrasporangialis aut per divisionem liberam nuclei et postea per fissionem, aut per fissionem binariam successivam, zoosporas aut aplanosporas generantem; zoosporae parvae (volumine pari ad 1·5-3·0(35·0) μm^3), ellipsoideae, uniflagellatae aut biflagellatae; zoosporae biflagellatae anisokontae, flagello anteriore curtiore, non heterokontae demonstratae; zoosporae uniflagellatae cum flagellis variatim insertis; aplanosporae elongatae et arcuatae. **Reproductio sexualis** non demonstrata, meiosis non reperta; sporae perdurantes cum fibrillis extramuralibus ab hospite derivatis; sine cellula sociali; distributio copiae penariae nondum pro certo cognita.
Holotypus: *Pseudosphaerita* P.-A. Dangeard.

Obligate endoparasites of protoctists (rhizopods, euglenids, cryptomonads), eumycote fungi or green algae. **Thallus** holocarpic; plasmodial following infection and sometimes during the assimilative phase; plasmodium sometimes capable of vegetative division; transformed either into a walled sporangium or sporangia, not completely filling the host cell cavity; or into a single resting spore, not completely filling the host cell cavity. **Asexual reproduction** by sporangia; sporogenesis intrasporangial either by free nuclear division followed by cleavage, or successive binary cleavage producing zoospores or aplanospores; zoospores small (volume equivalent 1·5-3·0(35·0) μm^3), ellipsoid, with one or two flagella; biflagellate zoospores anisokont, with a shorter anterior flagellum, not proven to be heterokont; uniflagellate zoospores with variable flagellar insertion; aplanospores elongate and arcuate. **Sexual reproduction** not established, meiosis not known; resting spores with host-derived extramural fibrils, not accompanied by a companion cell; reserve globules not known.
Typification: *Pseudosphaerita* P.-A. Dangeard.
Other genera: *Plasmophagus* É. de Wildeman, *Pleotrachelus* W. Zopf.

PSEUDOSPHAERITA P.-A. Dangeard, *Le Botaniste* **4**: 243 (1895).
Typification: *Pseudosphaerita euglenae* P.-A. Dangeard.

Pseudosphaerita euglenae P.-A. Dangeard, *Le Botaniste* **4**: 243 (1895).
Pseudosphaerita sphaeritae P.-A. Dangeard *in* A. A. Yačevskie & P. A. Yačevskie, *Opredelitel' Gribov*: 61 (1931) [misprint of specific epithet].
Typification: Lectotype (Cejp, 1959): *Le Botaniste* **4**: fig. 9 (1895) (iconotype). Type locality: France. Type habitat: parasitic in *Euglena* sp. (*Protoctista, Sarcomastigophora, Euglenida*).
Identification key: Key I; key point 102'.
Literature: Cejp (1959), Dangeard (1895, 1933), Doflein (1907), Karling (1941*a*, 1942*a*, 1981*a*: 46), Mitchell (1928), Pérez Reyes (1963), Puymaly (1927), Sparrow (1960: 963), Stein (1878), Yačevskie & Yačevskie (1931: 61).

Pseudosphaerita radiata (P.-A. Dangeard) J. S. Karling, *Simple Holocarpic Biflagellate Phycomycetes*: 36 (1942).
Sphaerita radiata P.-A. Dangeard, *Le Botaniste* **2**: 54 (1890).
Typification: Lectotype (Cejp, 1959): *Le Botaniste* **2**: plate 2, fig. 20 (1890) (iconotype). Type locality: France. Type habitat: parasitic in *Cryptomonas ovata* Ehrenb. (*Cryptophyceae, Cryptomona*).
Identification key: Key I; key point 107.
Literature: Cejp (1959), Dangeard (1890*a*), Karling (1942*a*, 1981*a*: 48), Sparrow (1960: 963), Stein (1851, 1854, 1859).

Pseudosphaerita dinobryonis (H. M. Canter) M. W. Dick **comb. nov.**
Sphaerita dinobryonis H. M. Canter, *Journal of the Elisha Mitchell Scientific Society* **84**: 60 (1968) [as '*dynobryoni*'].
Typification: Type IMI 124294, Egham, Surrey, U.K. Type locality: Grasmere, Cumbria, U.K. Type habitat: parasitic in *Dinobryon sertularia* Ehrenb. (*Chrysophyceae, Ochromonadales*) (also from Rydal Water) (also parasitic in *Dinobryon divergens* Imhof, from Esthwaite Watere, Loughrigg Tarn and Windermere, Cumbria, U.K.).
Identification key: Key I; key point 99'.
Literature: Canter (1968).

DOUBTFUL SPECIES

Pseudosphaerita drylii R. Pérez-Reyes, M. Madrazo-Garibay & E. López Ochoterena, ***Revista Latino-americana de Microbiologia (Mexico)*** **27**: 90 (1985) [as '*dryli*', named after Professor Stanislaw Dryl] [*nom. inval.*, Art. 36.1, without Latin diagnosis].
Typification: type material not designated; mature sporangia not found, zoospores "biflagellate". Type locality: Iztapalapa, Mexico. Type habitat: parasitic in *Phacus acuminatus* Stokes (*Protoctista, Sarcomastigophora, Euglenida*).
Identification key: Key I; key point 102.
Literature: Pérez-Reyes, Madrazo-Garibay & López Ochoterena (1985).

SPHAERITA P.-A. Dangeard, ***Bulletin de la Société de Botanique, France*** **33**: 241 (1886).
Typification: *Sphaerita endogena* P.-A. Dangeard.

Sphaerita endogena P.-A. Dangeard, ***Annales des Sciences Naturelles, Botanique, Série VII*** **4**: 277 (1886) [*pro parte*].
Typification: type material not designated; [*Sphaerita endogena* - Dangeard 1889 p. 46 plate II figs 11-19 (14,19 - 1 flagellum) III 1-2,7-8 (2 flagella) Type locality: Caen, France. Type habitat: parasitic in *Euglena sanguinea* Ehrenb. (*Protoctista, Sarcomastigophora, Euglenida*).
Identification key: Key I; key point 108'.
Literature: Chatton & Brodsky (1909), Dangeard (1886*a*, 1889, 1933), Jahn (1933), Karling (1941*a*, 1965, 1977: 24), Pérez Reyes (1963), Sparrow (1960: 126, 127), Yačevskie & Yačevskie (1931: 61).

Sphaerita dangeardii E. Chatton & A. Brodsky, ***Archiv für Protistenkunde*** **17**: 8 (1909).
Sphaerita endogena P.-A. Dangeard, ***Annales des Sciences Naturelles, Botanique, Série VII*** **4**: 277 (1886) [*pro parte*].
Typification: not determinable.
Identification key: Key I; key point 104.
Literature: Chatton & Brodsky (1909), Constantineau (1901), Dangeard (1886*a*, *b*, 1889, 1933), Golubeva (1982), Jahn (1933), Karling (1965), Mitchell (1928), Pérez Reyes (1963), Serbinow (1907), Sparrow (1960: 126), Valkanov (1931*a*).

Sphaerita minor A. Da Cunha & J. Muniz, ***Brazil-Medico*** **36**: 386 (1923).
Typification: not determinable. Type habitat: parasitic in *Trichomonas* sp. (*Parabasilia*).
Identification key: Key I; key point 108'.
Literature: Da Cunha & Muniz (1923), Jahn (1933).

Sphaerita nucleophaga O. Mattes, ***Archiv für Protistenkunde*** **47**: 413-430 (1924).
Typification: not determinable. Type habitat: parasitic in *Thecamoeba sphaeronucleolus* (Greeff) [as "*Amoeba sphaeronucleolus*", also in "*Amoeba terricola*")] (*Protozoa, Rhizopoda*).
Identification key: Key I; key point 108'.
Literature: Jahn (1933), Karling (1972, 1977: 24), Mattes (1924), Sparrow (1960: 126, 127).

Sphaerita amoebae O. Mattes, ***Archiv für Protistenkunde*** **47**: 413-430 (1924).
Typification: not determinable. Type habitat: parasitic in *Thecamoeba sphaeronucleolus* (Greeff) [as "*Amoeba sphaeronucleolus*"] (*Protozoa, Rhizopoda*).
Identification key: Key I; key point 108'.
Literature: Jahn (1933), Mattes (1924).

Sphaerita plasmophaga O. Mattes, ***Archiv für Protistenkunde*** **47**: 413-430 (1924).
Typification: not determinable. Type habitat: parasitic in *Thecamoeba sphaeronucleolus* (Greeff) [as "*Amoeba sphaeronucleolus*"] (*Protozoa, Rhizopoda*).
Identification key: Key I; key point 108'.
Literature: Jahn (1933), Mattes (1924).

Sphaerita normeti A. Lwoff, ***Bulletin de la Société de Pathologie Exotique*** **18**: 18 (1925).
Typification: not determinable. Type habitat: parasitic in *Entamoeba histolytica* Schaedinn [as "*Entamoeba coli*"] (*Protozoa, Rhizopoda*).
Identification key: Key I; key point 108.
Literature: Jahn (1933), Mattes (1924).

Sphaerita entamoebae E. R. Becker, *Biological Bulletin of the Marine Biological Laboratory, Woods Hole* **50**: 447 (1926).
Morella entamoebae (E. R. Becker) R. Pérez Reyes, *Revista de la Sociedad Mexicana de Historia Natural* **24**: 5 (1963) [*nomen nudum*].
Typification: not determinable. Type habitat: parasitic in *Entamoeba citelli* E. R. Becker (*Protozoa, Rhizopoda*) from the ground squirrel *Citellus*.
Identification key: Key I; key point 108.
Literature: Becker (1926), Jahn (1933), Karling (1972, 1977: 24), Pérez Reyes (1963), Sparrow (1960: 126, 127).

Sphaerita trachelomonadis B. W. Skvortzow, *Archiv für Protistenkunde* **57**: 205 (1927).
Typification: type material not designated; lectotype *Archiv für Protistenkunde* **57**: 205, fig. 2 (1927) (iconotype). Type locality: Harbin, North Manchuria, China. Type habitat: parasitic in *Trachelomonas teres* Maskell var. *glabra* Skvortzov (1927) and *T. swirenkoi* Skvortzov (*Protoctista, Sarcomastigophora, Euglenida*).
Identification key: Key I; key point 104'.
Literature: Jahn (1933), Skvortzow (1927), Sparrow (1960: 127), Yačevskie & Yačevskie (1931: 61).

Sphaerita phaci T. L. Jahn, *Archiv für Protistenkunde* **79**: 350 (1933).
Pseudosphaerita phaci T. L. Jahn, *Archiv für Protistenkunde* **79**: 350 (1933) [as quoted in Pérez-Reyes *et al.* (1985)].
Typification: type material not designated; lectotype *Archiv für Protistenkunde* **79**: plate 16, figs 1-9 (1933) (iconotype). Type locality: pond in van Courtlandt Park, New York City, New York, U.S.A. Type habitat: parasitic in *Phacus pleuronectes* (O. S. Müller) Dujardin (*Protoctista, Sarcomastigophora, Euglenida*).
Identification key: Key I; key point 103'.
Literature: Jahn (1933), Pérez-Reyes, Madrazo-Garibay & López Ochoterena (1985).

UNIDENTIFIED SPECIES

Sphaerita sp. (Karling, 1941*a*) parasitic in *Euglena* (*Euglenida*).

Sphaerita sp. (Fott, 1952) parasitic in *Gonyostomium* (*Chloromonadineae*).

GENERA *INCERTAE SEDIS* IN THE *ROZELLOPSIDALES*

PLASMOPHAGUS É. de Wildeman, *Annales de la Société Belge de Microscopie* **19**: 223 (1895).
Typification: *Plasmophagus oedogoniorum* de Wildeman (holotype - sole original species).

Plasmophagus oedogoniorum É. de Wildeman, *Annales de la Société Belge de Microscopie* **19**: 224 (1895).
Typification: type material not designated; lectotype *Annales de la Société Belge de Microscopie* **19**: plate VIII, figs 1-8, plate IX, figs 1-9 (1895) (iconotype). Type locality: Botanic Garden ponds, Nancy, France. Type habitat: parasitic in *Oedogonium* sp. (*Chlorophyceae, Oedogoniales*).
Identification key: Key I; key point 94'.
Literature: Karling (1977: 22), Schultz-Danzig (1923), Sparrow (1933*b*, 1960: 165), Wildeman (1895*c*).

Plasmophagus deformans (I. L. Serbinov) M. W. Dick, **comb. nov.**
Pseudolpidium deformans I. L. Serbinov, *Scripta Botanica Horti Universitatis Impericolis Petropolitani* **24**: 154 (1907).
Olpidiopsis deformans (I. L. Serbinov) J. S. Karling, *Predominantly Holocarpic and Eucarpic Simple Biflagellate Phycomycetes*: 30 (1981).
Typification:type material not designated; lectotype *Predominantly Holocarpic and Eucarpic Simple Biflagellate Phycomycetes*: 31, figs 175-184 (1981) (iconotype). Type locality: Russia. Type habitat: parasitic in cells of *Draparnaldia glomerata* (Vaucher) C. Agardh (*Chlorophyceae, Chaetophorales*).
Identification key: Key I; key point 94.
Literature: Karling (1981*a*: 30), Pystina (1995), Serbinow (1907), Sparrow (1960: 956), Sparrow, Paterson & Johns (1965), Yačevskie & Yačevskie (1931: 60).

Plasmophagus coleochaetis (F. K. Sparrow, R. A. Paterson & R. M. Johns) M. W. Dick, **comb. nov.**
Rozella coleochaetis F. K. Sparrow, R. A. Paterson & R. M. Johns, *Papers of the Michigan Academy of Science, Arts, and Letters* **50**: 118 (1965).
Typification: type material not designated; lectotype *Papers of the Michigan Academy of Science, Arts, and Letters* **50**: 116, figs A-E (1965) (iconotype). Type locality: Smith's Bog, Cheboygan County, Michigan, U.S.A. Type habitat: parasitic in *Coleochaete* sp. (*Charophyceae, Coleochaetales*).
Identification key: Key I; key point 96'.
Literature: Held (1981), Schulz (1922), Sparrow, Paterson & Johns (1965).

UNIDENTIFIED SPECIES REFERABLE TO *PLASMOPHAGUS*

Rozella sp. (Schulz, 1922).

PLEOTRACHELUS W. Zopf, *Nova Acta der Kaiserlichen Leopoldinisch-carolinischen Akademie der Naturforscher* **47**: 173 (1884).
Olpidiopsis M. Cornu, *Annales des Sciences Naturelles, Botanique, Série V* **15**: 114, 127 (1872) [*pro parte typo excluso*] (**Type species**: *Olpidiopsis saprolegniae* M. Cornu).
Rozella M. Cornu, *Annales des Sciences Naturelles, Botanique, Série V* **15**: 114, 148 (1872) [*pro parte typo excluso*] (**Type species**: *Rozella monoblepharidis* M. Cornu).
Typification: *Pleotrachelus fulgens* W. Zopf (holotype - sole original species).

Pleotrachelus fulgens W. Zopf, *Nova Acta der Kaiserlichen Leopoldinisch-carolinischen Akademie der Naturforscher* **47**: 173 (1884).
Olpidium fulgens (W. Zopf) J. S. Karling, *Chytridiomycetarum Iconographia*: 18 (1977).
Typification: type material not designated; lectotype *Nova Acta der Kaiserlichen Leopoldinisch-carolinischen Akademie der Naturforscher* **47**: plate 5, figs 25-36 (1884) (iconotype). Type locality: Germany. Type habitat: parasitic in *Pilobolus crystallinus* (Tode) Van Tieghem (*Zygomycotina*).
Identification key: Key I; key point 124.
Literature: Karling (1977: 18), Sparrow (1960: 159, 247), Yačevskie & Yačevskie (1931: 20), Zopf (1884, 1892).

Pleotrachelus zopfianus F. Morini, *Memorie della Reale Accademia delle Scienze dell'Istituto di Bologna. Seria VI* **10**: 301 (1913).
Typification: type material not verified.
Identification key: Key I; key point 124'.
Literature: Morini (1913), Sparrow (1960: 159, 160, 247)

Pleotrachelus vuilleminii (G. Arnaud) M. W. Dick, **comb. nov.**
Olpidiopsis vuilleminii G. Arnaud, *Bulletin trimestriel de la Société Mycologique de France* **68**: 182 (1952) [*nom. inval.*, Art. 36.1, without Latin diagnosis] [as '*vuilleminiae*'].
Typification: type material not designated; lectotype *Bulletin trimestriel de la Société Mycologique de France* **68**: 183, fig. 1c (1952) (iconotype). Type locality: Le Pin, Gard, France. Type habitat: parasitic in *Corticium comedens* Fr. (*Basidiomycetes*) on a fallen *Quercus* branch (*Fagaceae*).
Identification key: Key I; key point 125'.
Literature: Arnaud (1952), Karling (1981*a*: 40), Sparrow (1960: 954).

Pleotrachelus itersoniliae (D. J. S. Barr & R. Bandoni) M. W. Dick, **comb. nov.**
Rozella itersoniliae D. J. S. Barr & R. Bandoni, *Mycologia* 71: 1261 (1979).
Typification: Type *Mycologia* **71**: 1262, figs 1-4 (1979) (photographs DAOM 172110) (iconotype). Type locality: British Columbia, Canada. Type habitat: parasitic in *Itersonilia perplexans* Derx (*Basidiomycetes*, anamorphic genus) on a *Taraxacum* leaf (*Asteraceae*).
Identification key: Key I; key point 125.
Literature: Barr & Bandoni (1979), Held (1981).

SPECIES EXCLUDED FROM *PLEOTRACHELUS*

Pleotrachelus andreei [N.] G. Lagerheim, *YMER: Tidskrift utgifven af Svenska Sällskapet för Antropololgi och Geografi; Nittonde Årgången 1899*: 436 (1900), synonym of *Sirolpidium andreei* ([N.] G. Lagerheim) M. W. Dick, *loc. cit.*

Pleotrachelus bornovanus S. Sahtiyanci, *Archiv für Mikrobiologie* **41**: 216 (1962), synonym of *Olpidium radicale* E. J. Schwartz & W. R. Ivimey Cook, *Transactions of the British mycological Society* **13**: 216 (1928).

Pleotrachelus brassicae (M. Woronin) S. Sahtiyanci, *Archiv für Mikrobiologie* **41**: 198 (1962), synonym of *Olpidium brassicae* (M. Woronin) P.-A. Dangeard, *Annales des Sciences Naturelles, Botanique* **7**: 241-341 (1886).

Pleotrachelus ectocarpi M. Jokl, *Österreichische Botanische Zeitschrift* **66**: 267 (1916), synonym of *Sirolpidium ectocarpi* (M. Jokl) M. W. Dick, *loc. cit.*

Pleotrachelus inhabilis H. E. Petersen, *Oversigt over det Kongelige Danske Videnskabernes Selskabs Forhandlinger. Medlemmers Arbeiter, 1905*: 456 (1905), synonym of *Pontisma inhabilis* (H. E. Petersen) M. W. Dick, *loc. cit.*

Pleotrachelus lobatus H. E. Petersen, *Oversigt over det Kongelige Danske Videnskabernes Selskabs Forhandlinger. Medlemmers Arbeiter, 1905*: 460 (1905), synonym of *Petersenia lobata* (H. E. Petersen) F. K. Sparrow, *Dansk Botanisk Arkiv* **8**: 13 (1934).

Pleotrachelus minutus H. E. Petersen, *Oversigt over det Kongelige Danske Videnskabernes Selskabs Forhandlinger. Medlemmers Arbeiter, 1905*: 451 (1905), synonym of *Anisolpidium minutum* (H. E. Petersen) M. W. Dick, *loc. cit.*

Pleotrachelus olpidium H. E. Petersen, *Oversigt over det Kongelige Danske Videnskabernes Selskabs Forhandlinger. Medlemmers Arbeiter, 1905*: 455 (1905), synonym of *Anisolpidium olpidium* (H. E. Petersen) M. W. Dick, *loc. cit.*

Pleotrachelus paradoxus H. E. Petersen, *Oversigt over det Kongelige Danske Videnskabernes Selskabs Forhandlinger. Medlemmers Arbeiter, 1905*: 459 (1905), synonym of *Sirolpidium paradoxum* (H. E. Petersen) M. W. Dick, *loc. cit.*

Pleotrachelus petersenii A. Lund, *Botanisk Tidsskrift* **41**: 241 (1930), synonym of *Syzygangia oedogonii* (A. Scherffel) M. W. Dick, *loc. cit.*

Pleotrachelus pollagaster H. E. Petersen, *Oversigt over det Kongelige Danske Videnskabernes Selskabs Forhandlinger. Medlemmers Arbeiter, 1905*: 462 (1905), synonym of *Petersenia pollagaster* (H. E. Petersen) F. K. Sparrow, *Dansk Botanisk Arkiv* **8**: 15 (1934).

Pleotrachelus radicis É. de Wildeman, 17: 23 (1893), synonym of *Olpidium brassicae* (M. Woronin) P.-A. Dangeard, *Annales des Sciences Naturelles, Botanique* **7**: 241-341 (1886).

Pleotrachelus rosenvingii H. E. Petersen, *Oversigt over det Kongelige Danske Videnskabernes Selskabs Forhandlinger. Medlemmers Arbeiter, 1905*: 453 (1905), synonym of *Anisolpidium rosenvingii* (H. E. Petersen) J. S. Karling, *American Journal of Botany* **30**: 641 (1943).

Pleotrachelus rotatoriorum A. Scherffel, *Abstracts of Communications, V International Botanical Congress, Cambridge, 1930*: 222 (1930), an unidentifiable member of the genus *Haptoglossa*.

Pleotrachelus sphacelarum (L. Kny) H. E. Petersen, *Oversigt over det Kongelige Danske Videnskabernes Selskabs Forhandlinger. Medlemmers Arbeiter, 1905*: 453 (1905) (syn: *Chytridium sphacelarum* L. Kny, *Sitzungs-Berichte der Gesellschaft naturforschender Freunde zu Berlin, 1871*: 97 (1871) [as '*sphacellarum*']), synonym of *Anisolpidium sphacelarum* (L. Kny) J. S. Karling, *American Journal of Botany* **30**: 641 (1943).

Pleotrachelus tumefaciens (P. Magnus) E. Petersen, *Oversigt over det Kongelige Danske Videnskabernes Selskabs Forhandlinger. Medlemmers Arbeiter, 1905*: 456 (1905) (syn: *Chytridium tumefaciens* P. Magnus, *Sitzungs-Berichte der Gesellschaft naturforschender Freunde zu Berlin, 1871*: 97 (1871)), synonym of *Eurychasmidium tumefaciens* (P. Magnus) F. K. Sparrow, *Biological Bulletin of the Woods Hole Marine Biological Laboratory* **70**: 241 (1936).

Pleotrachelus virulentus S. Sahtiyanci, *Archiv für Mikrobiologie* **41**: 206 (1962), synonym of *Olpidium brassicae* (M. Woronin) P.-A. Dangeard, *Annales des Sciences Naturelles, Botanique* **7**: 327 (1886).

Pleotrachelus wildemanii H. E. Petersen, *Botanisk Tidsskrift* **29**: 422 (1909) [as '*wildemani*'], synonym of *Syzygangia elliptica* (É. de Wildeman) M. W. Dick, *loc. cit.*

OLPIDIOMORPHA A. Scherffel, *Archiv für Protistenkunde* **54**: 515 (1926).
Typification: *Olpidiomorpha pseudosporae* A. Scherffel (holotype - sole original species).

Olpidiomorpha pseudosporae A. Scherffel, *Archiv für Protistenkunde* **54**: 515 (1926).
Typification: type material not designated; lectotype *Archiv für Protistenkunde* **54**: plate 28, figs 7-8 (1926) (iconotype). Type locality: Igló, Hungary. Type habitat: parasitic within the zoocyst of *Pseudospora leptoderma* Scherff. (*Protoctista*) parasitic in *Vaucheria* sp. (*Tribophyceae*).
Identification key: Key I; key point 107'.
Literature: Karling (1977: 26), Scherffel (1926*b*), Sparrow (1960: 123).

PLASMODIOPHOROMYCETES K. Cejp, *Houby I*: 48 (1957) [see Sparrow 1973*b*: 61], C. J. Alexopoulos, *Introductory Mycology, 2nd edn*: 36 (1962).
Plasmodiophoreae C. J. Alexopoulos, *Introductory Mycology*: 65 (1952).
Plasmodiophoreae C. J. Alexopoulos, *Introductory Mycology, 2nd edn*: 86 (1962).
Phytomyxea T. Cavalier-Smith, *Journal of Eukaryotic Microbiology* **40**: 609-615 (1993) [superphylum *Opalomyxa*, phylum *Opalozoa*]
Typification: *Plasmodiophora* M. Woronin (*Plasmodiophorales* F. L. Stevens).

PLASMODIOPHORALES F. L. Stevens, *The Fungi Which Cause Plant Disease*: 5 (1913).
Plasmodiophorales A. A. Jaczewski & P. A. Jaczewski, *Opredelitel' Gribov*: 9 (1931).
Woroninales A. A. Jaczewski & P. A. Jaczewski, *Opredelitel' Gribov*: 60 (1931) [*pro parte*].
Plasmodiophorida T. Cavalier-Smith, *Journal of Eukaryotic Microbiology* **40**: 609-615 (1993).
Phagomyxida T. Cavalier-Smith, *Journal of Eukaryotic Microbiology* **40**: 609-615 (1993).
Typification: *Plasmodiophora* M. Woronin (*Plasmodiophoraceae* W. Zopf).

PLASMODIOPHORACEAE W. Zopf, *Encyklopaedie der Naturwissenschaften* **3(2)**: 129 (1885).
Woroninaceae M. von Minden, *Kryptogamenflora Mark Brandenburg* **5**: 227 (1911) (**Type genus**: *Woronina* M. Cornu) [*pro parte*].
Reessiaceae C. Gobi, [? see Yačevskie & Yačevskie (1931: 62), **Type genus**: *Reessia* C. Fisch {not in Gobi (1900), *Scriptis Botanicis Horti Universitatis Impericolis Petropolitanae* **15**: 191-226 nor Gobi (1900), *Scriptis Botanicis Horti Universitatis Impericolis Petropolitanae* **15**: 227-272}].
Typification: *Plasmodiophora* M. Woronin.

PLASMODIOPHORA M. Woronin, *Trudy Imperatorskago Sankt-Peterburgskago Obshchestva Estestvoispytatelei. St. Peterburg* **8**: 169 [name], 184 [description] (1877).
Schinzia C. W. Nägeli, *Linnea* **16**: 281 (1842) (**Type species**: *Schinzia cellulicola* C. W. Nägeli [*Ustilaginales*]) *sensu* M. Woronin [see *Schinzia alni* M. Woronin], *Mémoires de l'Académie Impériale des Sciences de St.-Pétersbourg, VIIe Série* (*Zapiski Imperatorskoĭ Akademii nauk*) **10**: 6 (1866) [*pro parte typo excluso; non Schinzia* A. W. Dennstedt, *Schlüssel zum Hortus Indicus Malabaricus* (1818): 24 [*Schinzia inconspicua*] - generic names introduced in this work are *nomina nuda*].
Ostenfeldiella C. Ferdinandsen & Ö. Winge, *Annals of Botany, London* **28**: 648 (1914) (**Type species**: *Ostenfeldiella diplantherae* C. Ferdinandsen & Ö. Winge).
Typification: *Plasmodiophora brassicae* M. Woronin.

Plasmodiophora brassicae M. Woronin, *Trudy Imperatorskago Sankt-Peterburgskago Obshchestva Estestvoispytatelei. S. Peterburg* **8**: 169 [name], 184 [description] (1877).
Typification: type material not designated; lectotype *Trudy Imperatorskago Sankt-Peterburgskago Obshchestva Estestvoispytatelei. S. Peterburg* **8**: plates I-VI (1877) (iconotype). Type locality: St Petersburg, Russia. Type habitat: parasitic in *Brassica* cultivars (*Brassicales, Brassicaeae*).
Identification key: Key XVIII; key point 1'.
Selected Literature: Aist (1974), Aist & Williams (1971), Arnold *et al.* (1996), Braselton (1982), Brunchorst (1887*b*), Bryngelsson *et al.* (1988), Buczacki (1977, 1983), Buczacki & Clay (1984), Buczacki & Moxham (1980, 1983), Castlebury, Maddox & Glawe (1994), Colhoun (1958), Dehkuijzen (1979), Dixon & Robinson (1986), Dobson & Gabrielson (1983), Einhorn & Bochow (1990*a*, *b*), Fraser & Buczacki (1983), Gabrielson & Robak (1988), Garber & Aist (1979*a*, *b*), Ikegami (1992), Ingram & Tommerup (1972), Ivimey Cook (1928*b*, *c*), Ivimey Cook & Schwartz (1930), Jones (1928), Jones & Ingram (1982), Karling (1967*b*, 1968*e*: 35), Keen, Reddy & Williams (1969), Köle (1955), Köle & Gielink (1961, 1962, 1963), Macfarlane (1970*a*), Maire & Tison (1909), Mithen & Magrath (1992), Nabel (1939), Narisawa *et al.* (1996), Olive (1975), Scott (1985), Seaman, Larson & Walker (1961), Singh, Sharma & Pandey (1989), Suzuki *et al.* (1992), Takahashi (1991), Takahashi & Yamaguchi (1988), Thornton, Jarvis & Cooke (1991), Tinggal & Webster (1981), Tommerup & Ingram (1971), Viégas & Teixeira (1943), Webster & Dixon (1991*a*, *b*), Williams, Aist & Aist (1971), Williams & McNabola (1967, 1970), Williams, Reddy & Strandburg (1969), Williams & Yukawa (1967), Woronin (1877, 1878), Yačevskie & Yačevskie (1931: 9), Yano *et al.* (1991).

Plasmodiophora diplantherae (C. Ferdinandsen & Ö. Winge) W. R. Ivimey-Cook, *Hong Kong Naturalist (Supplement No. 1)*: 34 (1932).

Ostenfeldiella diplantherae C. Ferdinandsen & Ö. Winge, *Annals of Botany* **28**: 648 (1914).

Typification: type material not designated; lectotype *Annals of Botany* **28**: 644-647, figs 1-4 (1914) (iconotype). Type locality: St Croix, West Indies. Type habitat: parasitic in *Halodule wrightii* (Ascherson) Ascherson (syn.: *Diplanthera wrightii* Ascherson) (*Najadales, Cynodoceaceae*).

Identification key: Key XVIII; key point 12'.

Literature: Braselton & Short (1985), Ferdinandsen & Winge (1914), Ivimey-Cook (1932*b*, 1933), Johnson & Sparrow (1961: 319), Karling (1968*e*: 42), Sparrow (1960: 783).

Plasmodiophora bicaudata J. Feldmann, *Bulletin de la Société d'Histoire Naturelle de l'Afrique du Nord* **31**: 173 (1941).

Typification: type material not designated; lectotype *Bulletin de la Société d'Histoire Naturelle de l'Afrique du Nord* **31**: 172, fig. 1 (1941) and *Revue génerale de Botanique* **63**: plates XX-XXIV (1956) (iconotype). Type locality: Atlantic seaboard, Mauritania, Africa. Type habitat: parasitic in *Zostera nana* Roth. (*Najadales, Zosteraceae*).

Identification key: Key XVIII; key point 1'.

Literature: G. Feldmann (1956), J. Feldmann (1941, 1953), Johnson & Sparrow (1961: 318), Karling (1968*e*: 43), Sparrow (1960: 784).

Plasmodiophora maritima G. Feldmann, *Revue Générale de Botanique* **63**: 16 (1958).

Typification: Type material in the collection of G. Feldmann. Type locality: Penzé Estuary, Roscoff, Brittany, France. Type habitat: parasitic in *Triglochin maritimum* L. (*Juncaginaceae*).

Identification key: Key XVIII; key point 1'.

Literature: Feldmann (1958), Johnson & Sparrow (1961: 318), Karling (1968*e*: 43), Maire & Tison (1911*a*), Molliard (1909).

DOUBTFUL SPECIES

Plasmodiophora halophilae C. Ferdinandsen & Ö. Winge, *Centralblatt für Bakteriologie, Parasitenkunde und Infektionskrankheit, Abt. 2* **37**: 167 (1913).

Typification: type material not verified. Type locality: Noesa Kembangan Island, Indonesia. Type habitat: parasitic in petioles of *Halophila ovalis* (R. Br.) Hook. f. (*Hydrocharitaceae*) causing pea-like galls.

Identification key: Key XVIII; key point 12'.

Literature: Feldmann (1936), Ferdinandsen & Winge (1913), Johnson & Sparrow (1961: 318), Karling (1968*e*: 42), Sparrow (1960: 784).

Plasmodiophora ficirepentis A. Andreucci, *Archivio Botanico per la Systematica, Fitogeographica e Genetica* **2**: 26 (1926).

Typification: type material not verified. Type locality: Italy. Type habitat: parasitic in roots of *Ficus pumila* L. (syn.: *Ficus repens* auct. non Willd.) (*Moraceae*).

Identification key: Key XVIII; key point 1'.

Literature: Andreucci (1926), Ivimey Cook (1933), Karling (1968*e*: 42).

EXCLUDED SPECIES

Plasmodiophora alni (M. Woronin) H. Möller, *Berichte der Deutschen Botanische Gesellschaft* **3**: 102 (1885) (synonyms: *Schinzia alni* M. Woronin, *Mémoires de l'Académie Impériale des Sciences de St.-Pétersbourg, VIIe Série (Zapiski Imperatorskoï Akademii nauk)* **10**: 6 (1866); *Frankia subtilis* J. Brunchorst, *Untersuchungen aus den Botanischen Institut, Tübingen* **2**: 174 (1886); *Frankiella alni* R. Maire & A. Tison, *Annales Mycologici* **7**: 243 (1909)). **Literature**: Brunchorst (1886), Gardner (1965), Karling (1968*e*: 43), Maire & Tison (1909), Moeller (1885, 1890), Woronin (1866, 1867). [An actinomycete according to Gardner (1965).]

Plasmodiophora californicae P. Viala & C. Sauvageau, *Comptes Rendus Hebdomadaires des Séances de l'Academie des Sciences, Paris* **115**: 68 (1892) [as '*californica*']. **Literature**: Karling (1968*e*: 45), Viala & Sauvageau (1892*b*, *c*). [Not a separate organism - host tyloses etc.]

Plasmodiophora elaeagni J. Schroeter, *Kryptogamen-Flora von Schlesien [1885-1889]* **3(1)**: 134 (1885) [as '*Plasmidiophora eleagni*']. **Literature**: Gardner (1965), Karling (1968*e*: 44), Maire & Tison (1909), Schroeter (1885), Schröter (1897). [An actinomycete according to Gardner (1965).]

Plasmodiophora humuli T. W. Kirk, *New Zealand Department of Agriculture. The Fourteenth Report of the Department of Agriculture, 1906* **14**: 402 (1906) [*nomen nudum*] [the name also occurs under the habit photograph on the 5th of 6 unnumbered plates between pages 364 and 365]. **Literature**: Karling (1968*e*: 45), Kirk (1906). [Disease caused by *Pseudomonas*.]

Plasmodiophora lewisii P. M. Jones, *Ohio Journal of Science* **30**: 296 (1930). [Excluded, bacteriophagic, see Jones (1930), Karling (1968*e*: 46).]

Plasmodiophora orchidis G. Massee, *Annals of Botany, London* **9**: 170 (1895). **Literature**: Karling (1968*e*: 45), Massee (1895). [Not a separate organism - host tyloses etc.]

Plasmodiophora solani W. Brehmer & J. Bärner, *Arbeiten aus der Biologischen Reichsanstalt für Land- und Forstwirtschaft* **18**: 52 (1930). **Literature**: Brehmer & Bärner (1930), Karling (1968*e*: 46). [Unidentifiable.]

Plasmodiophora tabaci P. M. Jones, *Botanical Gazette* **81**: 446 (1926) [no formal description; 446, fig. 1; 455, fig. 2; plates XXXIV-XXXVII, figs 3-75; origin U.S.A., possibly Chicago, Illinois] (synonym: *Plasmodiophora tabaci* P. M. Jones, *Phytopathology* **16**: 67 (1926) [*nomen nudum*]). **Literature**: Jones (1926*a*, *b*), Karling (1968*e*: 45). [More doubtfully excluded, but possibly a protozoan/myxomycete?]

Plasmodiophora theae H. M. Fitzpatrick, *The Lower Fungi: Phycomycetes*: 59 (1930) [*nomen nudum*], a synonym for the excluded taxon *Sorosphaera theae* N. N. Speschnew, *Die Pilzparasiten des Teestrauches*: 22 (1907).

Plasmodiophora tomati G. Abbey, *Journal of Horticulture, Series 3* **30**: 360 (1895). **Literature**: Abbey (1895), Karling (1968*e*: 45). [Not a parasitic organism.]

Plasmodiophora vascularum J. Matz, *Journal of the Puerto Rico Agricultural Experiment Station* **4**: 45 (1920). **Literature**: Matz (1920), Karling (1968*e*: 45). [**Typification**: Type locality: Mayagüez, Río Piedras, Puerto Rico. Type habitat: in the vascular system of *Saccharum officinarum* L. Symptoms caused by protozoans, see also synonyms *Ligniera vascularum* (J. Matz) M. T. Cook, *Journal of the Department of Agriculture, Puerto Rico* **13**: 19 (1929) and *Sorosphaera vascularum* (J. Matz) M. T. Cook, *Journal of the Department of Agriculture, Puerto Rico* **21**: 81 (1937).]

Plasmodiophora vitis P. Viala & C. Sauvageau, *Comptes Rendus Hebdomadaires des Séances de l'Academie des Sciences, Paris* **114**: 1559 (1892). **Literature**: Karling (1968*e*: 44), Viala & Sauvageau (1892*a*, *c*). [Not a separate organism - host tyloses etc.]

WORONINA M. Cornu, *Annales des Sciences Naturelles, Botanique, Série V* **15**: 114, 176 (1872).
Chytridium A. Braun, *Betrachtungen über die Erscheinung der Verjüngung in der Natur, insbesondere in der Lebens- und Bildungsgesischte der Pflanze*: 198 (1851) [*pro parte minore typo excluso*] (**Type species**: *Chytridium olla* A. Braun).
Typification: *Woronina polycystis* M. Cornu (holotype - sole original species).

Woronina polycystis M. Cornu, *Annales des Sciences Naturelles, Botanique, Série V* **15**: 176 (1872).
Woronina polycystis M. Cornu var. *scalariformis* H. E. Petersen, *Botanisk Tidsskrift* **29**: 426 (1909).
Typification: Lectotype (Cejp, 1959): *Annales des Sciences Naturelles, Botanique, Série V* **15**: plate 7, figs 1-19 (1872) (iconotype). Type locality: Chaville, Versailles, France. Type habitat: parasitic in *Saprolegnia* sp. (*Peronosporomycetes, Saprolegniaceae*).
Identification key: Key XVIII; key point 1', Key I, 69', 80'.
Literature: Dangeard (1890*b*), Fischer (1882), Forbes (1935), Goldie-Smith (1954), Ivimey Cook & Nicholson (1933), Karling (1942*a*, 1967*b*, 1968*e*: 88, 1981*b*), Maurizio (1895), Minden (1911*a*), Petersen (1909, 1910), Shen & Siang (1948), Sorokine (1883), Sparrow (1933*b*, 1960: 776), Sparrow, Paterson & Johns (1965), Willoughby & Roberts (1992), Yačevskie & Yačevskie (1931: 64), Zopf (1894).

Woronina glomerata (M. Cornu) A. Fischer, *Rabenhorst's Kryptogamen-Flora, 2 Aufl., Bd 1, Abt. 4*: 67 (1892).
Chytridium glomeratum M. Cornu, *Annales des Sciences Naturelles, Botanique, Série V* **15**: 187 (1872).
Typification: Lectotype (Cejp, 1959): *Annales des Sciences Naturelles, Botanique, Série V* **15**: plate 7, figs 20-22 (1872) (iconotype). Type locality: Chateauneuf-sur-Loire, France. Type habitat: parasitic in *Vaucheria* sp. (*Tribophyceae*).
Identification key: Key XVIII; key point 1', Key I, 92'.
Literature: Cornu (1872), Fischer (1892), Karling (1942*a*, 1968*e*: 92), Minden (1911*a*), Sparrow, 1960: 779), Tokunaga (1933*a*), Yačevskie & Yačevskie (1931: 64), Zopf (1894).

Woronina pythii E. K. Goldie-Smith, *Journal of the Elisha Mitchell Scientific Society* **72**: 348 (1956).
Typification: type material not designated: lectotype *Journal of the Elisha Michell Scientific Society* **72**: 351-352, figs 1-35 (1956) (iconotype). Type locality: Fort Lauderdale, Florida, U.S.A. Type habitat: parasitic in an unidentified species of *Pythium* [said to be a "strain of *Pythium proliferum* Schenk" which would have been a *Myzocytium* species!] (*Peronosporomycetes, Pythiales*) from a damp swale soil sample.
Identification key: Key XVIII; key point 1', Key I, 66.
Literature: Braselton & Dylewski (1986), Dylewski & Miller(1983*a*, *b*), Dylewski, Miller & Braselton (1978), Goldie-Smith (1956), Karling (1964*b*, 1967*b*, 1968*e*: 90, 1969, 1981*b*, 1984, 1987*a*, *b*), Miller (1962*b*), Miller & Dylewski (1980, 1983*a*, *b*, 1984*a*, *b*), Persiel (1960*a*), Sparrow (1960: 781).

Woronina leptolegniae J. S. Karling, *Nova Hedwigia* **35**: 20 (1981).
Typification: type material not designated: lectotype *Nova Hedwigia* **35**: 21, figs 1-24 (1981) (iconotype). Type locality: Isla Orchedia, Rio Carrao, Venezuela. Type habitat: parasitic in the hyphae of *Leptolegnia caudata* de Bary (*Peronosporomycetes*, *Saprolegniales*), from a soil sample.
Identification key: Key XVIII; key point 1'.
Literature: Karling (1981*b*).

DOUBTFUL SPECIES

Woronina aggregata W. Zopf, *Beiträge zur Physiologie und Morphologie Niederer Organismen* **4**: 60 (1894).
Typification: type material not designated; lectotype *Beiträge zur Physiologie und Morphologie Niederer Organismen* **4**: 60, figs (1894) (iconotype). Type locality: Berlin, Germany. Type habitat: parasitic in *Mougeotia* sp. (*Zygnematales*, *Zygnemataceae*).
Identification key: Key XVIII; key point 1', Key I, 98'.
Literature: Cornu (1872), Fischer (1892), Lopez & MacCarthy (1985), Karling (1942*a*, 1968*e*: 92), Minden (1911*a*), Sparrow (1960: 780), Tokunaga (1933*a*), Valkanov (1931*a*), Voronichin (1920), Yačevskie & P. A. Yačevskie (1931: 64), Zopf (1894).

Woronina sorokiniana A. A. Jaczewski & P. A. Jaczewski, *Opredelitel' Gribov*: 64 (1931) [*nomen nudum*].
Typification: parasitic in *Achlya racemosa* Hildebr. (*Peronosporomycetes*, *Saprolegniaceae*).
Literature: Yačevskie & P. A. Yačevskie (1931: 64).

Woronina asterina Y. Tokunaga, *Transactions of the Sapporo Natural History Society (Japan)* **13**: 26 (1933).
Typification: type material not designated; *Transactions of the Sapporo Natural History Society (Japan)* **13**: plate 2, figs 15-16 (1933) lectotype (iconotype). Type locality: Assabu Hiyama, Oshima Province, Japan. Type habitat: parasitic in *Achlya americana* Coker (*Peronosporomycetes*, *Saprolegniaceae*).
Literature: Karling (1942*a*, 1968*e*: 94), Sparrow (1960: 779), Tokunaga (1933*a*).

EXCLUDED SPECIES

Woronina elegans (E. Perroncito) A. Fischer, *Rabenhorst's Kryptogamen-Flora, 2 Aufl., Bd 1, Abt. 4*: 66 (1892), a synonym for the doubtful taxon *Myzocytiopsis elegans* (E. Perroncito) M. W. Dick *loc. cit.*

Woronina raii O. N. Srivastava & R. K. Sinha, *Acta Botanica Indica* **17**: 259 (1989). **Typification**: *Acta Botanica Indica* **17**: 259, figs 1-4 (1989) (iconotype). Type locality: Gudri bazar road, Lohardaga, Ranchi, India. Type habitat: saprotrophic on *Cedrus* wood in a moderately polluted perennial man-made kachha pond. [Unidentifiable: identification stated to be confirmed by Professors B. Rai and C. T. Ingold; the iconotype is inadequate for any identification, and its saprotrophic nature makes it unlikely that the fungus belongs to this genus.]

Woronina charae Kiran & Dayal, *Indian Phytopathology* **45**: 176 (1992).
Typification: Slides and preserved material (HC10 30508) Benares Hindu University, Varanasi, India. Type locality: Varanasi, India. Type habitat: from decomposed *Chara* branches, November 1990.
Identification key: not keyed: possibly a saprotroph.
Literature: Kiran & Dayal (1992).

SPONGOSPORA J. Brunchorst, *Bergens Museums Årsberetning, 1886*: 225 (1887).

Sorosporium F. Rudolphi, *Linnaea, ein Journal für Botanik* **4**: 117 (1829) [*pro parte minore typo excluso*] (*Ustilaginales*. **Type species**: *Sorosporium saponariae* F. Rudolphi - holotype, sole original species).

Protomyces F. Unger, *Die Exantheme der Pflanzen...*: (1833) [*pro parte minore typo excluso*] (*Taphrinales*. **Type species**: not designated).

Erysibe R. W. Wallroth, *Linnaea, ein Journal für Botanik* **16**: 332 (1842) [*pro parte minore typo excluso*] (*Uredinales*. **Type species**: not designated).

Rhizosporium L. Rabenhorst, *Archiv der Pharmazie und Berichte der Deutschen pharmazentischen Gesellschaft* **83**: 300 (1843) [*pro parte minore typo excluso*] (*Hyphomycetes*. **Type species**: not designated).

Tuburcinia E. M. Fries [*nom rej.* ≡ *Urocystis* L. Rabenhorst *ex* K. W. E. L. Fuckel] (as: *Tubercinia scabies* M. J. Berkeley, *Journal of the Horticultural Society of London* **1**: 33 (1846)).

Clathrosorus C. Ferdinandson & Ö. Winge, *Annals of Botany, London* **34**: 468 (1920) (**Type species**: *Clathrosorus campanulae* C. Ferdinandson & Ö. Winge).

Typification: *Spongospora subterranea* (R. W. Wallroth) G. A. Lagerheim.

Spongospora subterranea (R. W. Wallroth) [N.] G. Lagerheim, *Journal of Mycology* **7**: 103 (1891).

Erysibe subterranea R. W. Wallroth, *Linnaea* **16**: 332 (1842).

Protomyces tuberum-solani C. F. P. Martius, *Die Kartoffeln-Epidemie der letzten Jahr oder die Stockfäule und Ründe der Kartoffeln, Geschildert und in ihren ursachlichen Verhältnissen erörtert*: 28 (1842).

Rhizosporium solani L. Rabenhorst, *Archiv der Pharmazie und Berichte der Deutschen pharmazentischen Gesellschaft* **83**: 300 (1843).

Tuburcinia scabies M. J. Berkeley, *Journal of the Horticultural Society of London* **1**: 33 (1846).

Sorosporium scabies A. A. Fischer von Waldheim, *Apèrcu Systématique des Ustilaginées*: 51 (1877).

Spongospora solani J. Brunchorst, *Bergens Museums Årsberetning, 1886*: 225 (1887).

Spongospora scabies G. Massee, *Journal of the Board of Agriculture and Fisheries (England)* **15**: 592 (1908).

Spongospora subterranea radicicola C. Blattny, *Sbornik Vyzhumn ch ústaru zemědělsk ch RČS* **137**: 21 (1935).

Spongospora subterranea tubericola C. Blattny, *Sbornik Vyzhumn ch ústaru zemědělsk ch RČS* **137**: 21 (1935).

Typification: type material not designated but probably extant at the University of Strasburg, exsiccatae collection of R. W. Wallroth [see *Mycologia* **70**: 216 (1978)]; lectotype *Linnaea* **16**: 332 (1842) (iconotype). Type locality: ? Type habitat: parasitic in potato (*Solanales*, *Solanaceae*).

Identification key: Key XVIII; key point 1'.

Literature: Berkeley (1846), Blattny (1935), Braselton (1992), Brunchorst (1887*a*), Clay & Walsh (1990), Diriwächter, Herter & Gindrat (1979), Diriwächter & Parbery (1991), Flett (1983), Harrison *et al.* (1993), Heim (1960), Karling (1967*b*, 1968*e*: 71), Kole & Gielink (1963), Kunkel (1915), Lagerheim (1891), Lahert & Kavanagh (1985), Martius (1842*a*, *b*), Massee (1908), Merz (1990, 1992), Olive (1975), Osborn (1911), Rabenhorst (1843), Tomlinson (1958), Viégas & Teixeira (1943), Waldheim (1877), Wallroth (1842*a*, *b*) Yačevskie & Yačevskie (1931: 13).

Spongospora nasturtii (J. A. Tomlinson) M. W. Dick, **stat. nov.**

Spongospora subterranea (R. W. Wallroth) [N.] G. Lagerheim var. *nasturtii* J. A. Tomlinson, *Transactions of the British mycological Society* **31**: 498 (1958).

Typification: Type specimen: IMI 74293, Egham, Surrey, U.K. Type locality: Bere Regis, Dorset, U.K. Type habitat: parasitic in the roots of *Nasturtium officinale* R. Br. (*Brassicales*, *Brassicaceae*).

Identification key: Key XVIII; key point 12'.

Literature: Claxton *et al.* (1996), Clay & Walsh (1990, 1997), Karling (1968*e*: 72), Tomlinson (1958), Walsh & Phelps (1991).

DOUBTFUL SPECIES

Spongospora campanulae (C. Ferdinandson & Ö. Winge) W. R. Ivimey Cook, *Archiv für Protistenkunde* **80**: 215 (1933).

Clathrosorus campanulae C. Ferdinandson & Ö. Winge, *Annals of Botany, London* **34**: 468 (1920).

Typification: Type locality: Denmark. Type habitat: parasitic in roots of *Campanula rupunculoides* L. (*Asteridae*, *Campanulales*, *Campanulaceae*).

Literature: Ferdinandson & Winge (1920), Ivimey Cook (1933), Karling (1968*e*: 72), Yačevskie & Yačevskie (1931).

Spongospora cotulae J. T. Barrett *in* J. S. Karling, *The Plasmodiophorales*: 74 (1968).

Typification: Type locality: San Francisco, California, U.S.A. Type habitat: parasitic in roots of *Cotula australis* (Sieber *ex* Spreng.) Hook. f. (*Asteridae*, *Asteraceae*).

Literature: Karling (1968*e*: 74).

EXCLUDED SPECIES

Spongospora urticae (N. A. Naumov) A. A. Yačevskie & P. A. Yačevskie, *Opredelitel' Gribov*: 14 (1931): presumably the basionym attributed to "Naoumov" is *Sorosphaera urticae* N. A. Na[o]umov[a], but it is also possible that this is a later *comb. nov.* for *Synchytrium urticae* N. Sorokin, [*Contrib 3rd Mtg Russian Sci Kiev Trans* 39-42] (1873). The status of this and the even later homonym *Synchytrium urticae* M. T. Cook, *Mycologia* 44: 827 (1952) is discussed, with Sorokin's illustration, by Karling (1953*b*).

Spongospora verrucosa (R. Maire & A. Tison) A. A. Yačevskie & P. A. Yačevskie, *Opredelitel' Gribov*: 14 (1931), synonym of *Ligniera verrucosa* R. Maire & A. Tison, *Comptes Rendus Hebdomadaires des Séances de l'Académie des Sciences, Paris, Série D* 152: 206 (1911).

TETRAMYXA K. Goebel, *Flora, Jena* **67**: 517 (1884).
Molliardia R. Maire & A. Tison, *Annales Mycologici* **9**: 238 (1911) (**Type species**: *Molliardia triglochinis* (M. Molliard) R. Maire & A. Tison).
Thecaphora W. A. Setchell, *Mycologia* **16**: 243 (1924) (**Type species**: *Thecaphora rupppiae* W. A. Setchell).
Typification: *Tetramyxa parasitica* K. Goebel.

Tetramyxa parasitica K. Goebel, *Flora, Jena* **67**: 517 (1884).
Thecaphora ruppiae W. A. Setchell, *Mycologia* **16**: 243 (1924).
Typification: type material not verified; parasitic in roots of *Zannichellia*, *Potamogeton* and *Ruppia* spp. (*Najadales*).
Identification key: Key XVIII; key point 1'.
Literature: Goebel (1884), Johnson & Sparrow (1961: 317), Karling (1942*b*: 37, 1967*b*, 1968*e*: 47), Luther (1950), Maire & Tison (1909), Setchell (1924), Sparrow (1960: 785), Yačevskie & Yačevskie (1931: 10).

Tetramyxa marina Y. Lipkin & A. Avidor, *Nova Hedwigia* **72**: 809 (1974).
Typification: Type material PHYT No. 2261, Phycological Laboratory, Department of Botany, Tel Aviv University, Tel Aviv, Israel. Type habitat: parasitic in the petioles of *Halophila stipulacea* (Forssk.) Aschers. Type locality: Shorat el Manqata, Gulf of Elat, The Red Sea, coll. 20 December 1969.
Identification key: not keyed.
Literature: Lipkin & Avidor (1974).

DOUBTFUL SPECIES

Tetramyxa rhizophaga D. Lihnell, *Symbolae Botanicae Upsalienses* **5**: 1-11 (1942).
Typification: type material not verified. Type locality: Sweden. Type habitat: parasitic in roots of *Juniperus communis* L. (*Gymnospermae*, *Cupressaceae*).
Literature: Karling (1968*e*: 48), Lihnell (1942).

Tetramyxa triglochinis M. Molliard, *Bulletin trimestriel de la Société de Botanique, France* **56**: 24 (1909).
Molliardia triglochinis (M. Molliard) R. Maire & A. Tison *Annales Mycologici* **9**: 238 (1911).
Typification: type material not verified. Type locality: France. Type habitat: parasitic in the stems, ovaries and stamens of *Triglochin maritimum* L. (*Juncaginaceae*).
Literature: Karling (1968*e*: 48), Molliard (1909), Yačevskie & Yačevskie (1931: 11).

Tetramyxa elaeagni Y. Yendo & K. Takase, *Bulletin of Sericulture and Silk Industry, Uyeda, Japan* **4**: 5 (1932).
Typification: type material not verified. Type locality: Japan. Type habitat: parasitic in roots of *Elaeagnus multiflora* Thunb. (*Elaeagnaceae*), causing galls.
Literature: Karling (19687*e*: 50), Yendo & Takase (1932).

SOROSPHAERA J. Schroeter, *Kryptogamen-Flora von Schlesien* **3(1)**: 135 (1886).

Tuburcinia E. M. Fries [*nom rej.* ≡ *Urocystis* L. Rabenhorst *ex* K. W. E. L. Fuckel] (as: *Tubercinia veronicae* J. Schroeter, *Beiträge zur Biologie der Pflanzen* **2**: 249-283 (1877)).

Sorosporium F. Rudolphi (1829) [*pro parte typo excluso*] (*Ustilaginales*) (**Type species**: *Sorosporium saponariae* F. Rudolphi).

Typification: *Sorosphaera veronicae* (J. Schroeter) J. Schroeter (holotype - sole original species).

Sorosphaera veronicae (J. Schroeter) J. Schroeter, *Kryptogamen-Flora von Schlesien [1885-1889]* **3(1)**: 135 (1886).

Tuburcinia veronicae J. Schroeter, *Beiträge zur Biologie der Pflanzen* **2**: 249-283 (1877).

Sorosporium veronicae G. Winter, *Die Pilze Deutschlands, Oesterreichs und der Schweitz*: 103 (1884).

Typification: type material not verified. Type habitat: parasitic in roots of *Veronica beccabunga* L. (*Lamiales*, *Scrophulariaceae*).

Identification key: Key XVIII; key point 12'.

Literature: Braselton & Miller (1973, 1975, 1978), Braselton, Miller & Pechak (1975), Dylewski, Braselton & Miller (1978), Harris, Braselton & Miller (1980), Ivimey Cook (1928*c*), Karling (1967*b*, 1968*e*: 55), Maire & Tison (1909), Miller (1958*a*, *b*), Olive (1975), Schroeter (1877, 1885), Sparrow (1960: 790), Talley, Miller & Braselton (1978), Webb (1935), Winter (1884), Yačevskie & Yačevskie (1931: 13).

Sorosphaera radicalis W. R. Ivimey Cook & E. J. Schwartz, *Annals of Botany, London* **43**: 86 (1929) [as '*radicale*'].

Typification: type material not verified. Type habitat: parasitic in aquatic *Poaceae* (*Poa*, *Molinia*, *Catabrosa*).

Identification key: Key XVIII; key point 1'.

Literature: Ivimey Cook (1928*c*, 1933), Ivimey Cook & Schwartz (1929), Karling (1968*e*: 58).

EXCLUDED SPECIES

Sorosphaera theae N. N. Speschnew, *Die Pilzparasiten des Teestrauches*: 22 (1907) [Probably a physiological disease, see Karling (1968*e*: 60).].

Sorosphaera vascularum (J. Matz) M. T. Cook, *Journal of the Department of Agriculture, Puerto Rico* **21**: 81 (1937), see under the excluded taxon *Plasmodiophora vascularum* J. Matz.

Sorosphaera urticae N. A. Naumov [in Yačevskie & Yačevskie (1931: 14); see note under *Spongospora urticae* ms p. 565. I have been unable to locate this citation (see Karling, 1968*e*)].

LIGNIERA R. Maire & A. Tison, *Comptes Rendus Hebdomadaires des Séances de l'Académie des Sciences, Paris, Série D* **152**: 206 (1911).

Rhizomyxa A. Borzí, Rhizomyxa, *Nuovo Ficomicete*: 6 (1884) [*pro parte typo excluso*] (**Type species**: *Rhizomyxa hypogeae* A. Borzi).

Sorosphaera J. Schroeter, *Kryptogamen-Flora von Schlesien [1885-1889]* **3(1)**: 135 (1886) [*pro parte typo excluso*] (**Type species**: *Sorosphaera veronicae* (J. Schröter) J. Schröter (holotype - sole original species).

Sorolpidium B. Němec, *Berichte der Deutschen Botanischen Gesellschaft* **29**: 48 (1911) (**Type species**: *Sorolpidium betae* B. Němec - holotype, sole original species).

Anisomyxa B. Němec, *Rospavy České Akademie, Pro Vědy Slovestnost a Uměni, Třída II, Praha*, **21** Čislo **35**: 1 (1913) (**Type species**: *Anisomyxa plantaginis* B. Němec - holotype, sole original species).

Spongospora (R. Maire & A. Tison) A. A. Yačevskie & P. A. Yačevskie, *Opredelitel' Gribov*: 14 (1931) [*pro parte*] (**Type species**: *Spongospora subterranea* (R. W. Wallroth) [N.] G. Lagerheim).

Typification: *Ligniera verrucosa* R. Maire & A. Tison.

Ligniera verrucosa R. Maire & A. Tison, *Comptes Rendus Hebdomadaires des Séances de l'Académie des Sciences, Paris, Série D* **152**: 206 (1911).

Spongospora verrucosa (R. Maire & A. Tison) A. A. Yačevskie & P. A. Yačevskie, *Opredelitel' Gribov*: 14 (1931).

Typification: type material not verified.

Identification key: Key XVIII; key point 12'.

Literature: Braselton (1989*a*), Fron & Gaillat (1925), Ivimey Cook (1928*c*), Karling (1968*e*: 77), Maire & Tison (1911*b*), Miller (1959*a*, *b*).

Ligniera junci (E. [J.] Schwartz) R. Maire & A. Tison, *Comptes Rendus Hebdomadaires des Séances de l'Académie des Sciences, Paris, Série D* **152**: 206 (1911).
Sorosphaera junci E. [J.] Schwartz, *Annals of Botany, London* **24**: 512 (1910).
Sorosphaera graminis E. [J.] Schwartz, *Annals of Botany, London* **25**: 791 (1911).
Ligniera radicalis R. Maire & A. Tison, *Comptes Rendus Hebdomadaires des Séances de l'Académie des Sciences, Paris, Série D* **152**: 206 (1911).
Ligniera graminis (E. [J.] Schwartz) Ö. Winge, *Arkiv för Botanik* **12**: 15 (1912).
Ligniera bellidis E. [J.] Schwartz, *Annals of Botany, London* **28**: 232 (1914).
Ligniera menthae E. [J.] Schwartz, *Annals of Botany, London* **28**: 232 (1914).
Ligniera alismatis E. [J.] Schwartz, *Annals of Botany, London* **28**: 233 (1914).
Typification: type material not verified. Type habitat: *Juncus* spp. (*Juncaceae*) (**Typification of synonyms**: *L. bellidis* parasitic in *Chrysanthemum leucanthemum* L., Fawke common, Sevenoaks, Kent, U.K.; *L. menthae* parasitic in *Mentha pulegium* L., Chislehurst, Kent, U.K.; *L. alismatis* parasitic in *Alisma plantago-aquatica* L., Dunton Green, Kent, U.K. [no illustrations]).
Identification key: Key XVIII; key point 1'.
Literature: Barr (1981*a*), Ivimey Cook (1927, 1928*a, c*), Karling (1968*e*: 80), Maire & Tison (1911*b*), Olive (1975), Schwartz (1910, 1911, 1914), Sparrow (1960: 785), Winge (1912), Yačevskie & Yačevskie (1931: 13).

Ligniera isoetes B. T. Palm, *Svensk Botanisk Tidskrift* **12**: 228 (1918).
Typification: type material not verified. Type habitat: *Isoetes lacustris* L. (*Pteridophyta, Isoetaceae*).
Identification key: Key XVIII; key point 12'.
Literature: Karling (1968*e*: 81), Palm (1918), Sparrow (1960: 785).

Ligniera pilorum G. Fron & D. F. Gaillat, *Bulletin trimestriel de la Société Mycologique de France* **41**: 390 (1925).
Typification: type material not designated; lectotype *Bulletin trimestriel de la Société Mycologique de France* **41**: plate X, figs 1-8 (1925) (iconotype). Type locality: Saint Germain, Paris, France. Type habitat: parasitic in *Poa annua* L. (*Poaceae*).
Identification key: Key XVIII; key point 1'.
Literature: Haber, Barr & Platford (1991), Fron & Gaillat (1925), Ijdenberg, Kummert & Lepoivre (1986), Karling (1967*b*, 1968*e*: 80), Maire & Tison (1911*ab*).

Ligniera betae (B. Němec) J. S. Karling, *The Plasmodiophorales*: 81 (1968).
Sorolpidium betae B. Němec, *Rospavy České Akademie, Pro Vědy Slovestnost a Umĕni, Třída II, Praha*, **20** Čislo **12**: 1 (1911).
Typification: type material not verified.
Identification key: Key XVIII; key point 12'.
Literature: Karling (1968*e*: 81), Němec (1911*b*).

DOUBTFUL SPECIES

Ligniera hypogeae (A. Borzí) J. S. Karling, *The Plasmodiophorales*: 84 (1968).
Rhizomyxa hypogeae A. Borzí, Rhizomyxa, *Nuova Ficomicete*: 6 (1884) [*pro parte*].
Typification: type material not verified.
Literature: Borzi (1884), Karling (1968*e*: 84).

Ligniera plantaginis (B. Němec) J. S. Karling, *The Plasmodiophorales*: 85 (1968).
Anisomyxa plantaginis B. Němec, *Rospavy České Akademie, Pro Vědy Slovestnost a Umĕni, Třída II, Praha*, **21** Čislo **35**: 1 (1913).
Typification: type material not verified.
Literature: Karling (1968*e*: 85), Němec (1913).

EXCLUDED SPECIES

Ligniera vascularum (J. Matz) M. T. Cook, *Journal of the Department of Agriculture, Puerto Rico* **13**: 19 (1929), see under the excluded species *Plasmodiophora vascularum* J. Matz (see p. 387).

SORODISCUS [N.] G. Lagerheim & Ø. Winge *in* Ø. Winge, *Arkiv för Botanik* **12**: 23 (1912).
Typification: *Sorodiscus callitrichis* [N.] G. Lagerheim & Ø. Winge (holotype - sole original species).

Sorodiscus callitrichis [N.] G. Lagerheim & Ø. Winge *in* Ø. Winge *Arkiv för Botanik* **12**: 23 (1912).
Typification: type material not verified. Type habitat *Callitriche* spp. (*Asteridiae*, *Callitrichales*).
Identification key: Key XVIII; key point 1'.
Literature: Karling (1968*e*: 64), Winge (1912), Yačevskie & Yačevskie (1931).

Sorodiscus radicicola W. R. Ivimey Cook, *Annales Mycologici* **29**: 313-324 (1931) [as '*radicicolus*'].
Typification: type material not verified. Type locality: Pretoria, South Africa. Type habitat: parasitic in roots of *Cleome gynandra* L. (syn.: *Gynandropsis gynandra* DC.) (*Brassicales*, *Brassicaceae* [*Capparaceae*]).
Identification key: Key XVIII; key point 12'.
Literature: Ivimey Cook (1931), Karling (1968*e*: 66).

Sorodiscus karlingii W. R. Ivimey Cook, *Archiv für Protistenkunde* **80**: 207 (1933).
Typification: type material not verified. Type locality: U.S.A. Type habitat: parasitic in *Chara contraria* Kütz. [A. Braun *ex* Kütz.] (*Chlorophyta*, *Charophyceae*).
Identification key: Key XVIII; key point 1'.
Literature: Ivimey Cook (1933), Karling (1928, 1968*e*: 66), Sparrow, 1960: 788).

Sorodiscus cokeri E. K. Goldie-Smith, *Journal of the Elisha Mitchell Scientific Society* **67**: 108 (1951).
Typification: type material not designated; lectotype *Journal of the Elisha Mitchell Scientific Society* **67**: plates 1 & 2 (1951) (iconotype). Type locality: Fort Lauderdale, Florida, U.S.A. Type habitat: parasitic in *Pythium proliferum* de Bary (*Peronosporomycetes*, *Pythiaceae*) from grassy lot under *Carica papaya* L. (*Cucurbitales*, *Caricaceae*).
Identification key: Key XVIII; key point 12'.
Literature: Gaertner (1954), Goldie-Smith (1951, 1956), Karling (1968*e*: 66, 1984, 1987), Sparrow (1960: 789), Sparrow & Dogma (1973).

EXCLUDED SPECIES

Sorodiscus heterantherae C. C. Wernham, *Mycologia* **27**: 272 (1935) [as '*heterantheri*'], synonym of *Membranosorus heterantherae* C. H. Ostenfeld & H. E. Petersen, *Zeitschrift für Botanik* **23**: 17 (1930).

MEMBRANOSORUS C. H. Ostenfeld & H. E. Petersen, *Zeitschrift für Botanik* **23**: 17 (1930).
Typification: *Membranosorus heterantherae* C. H. Ostenfeld & H. E. Petersen (holotype - sole original species).

Membranosorus heterantherae C. H. Ostenfeld & H. E. Petersen, *Zeitschrift für Botanik* **23**: 17 (1930).
Sorodiscus heterantherae C. C. Wernham, *Mycologia* **27**: 272 (1935) [as '*heterantheri*'].
Typification: Type locality: Ontario, Canada. Type habitat: parasitic in *Heteranthera dubia* (Jacq.) MacMill. (*Liliales*, *Pontaderiaceae*).
Identification key: Key XVIII; key point 1'.
Literature: Braselton (1989*b*), Karling (1968*e*: 67), Wernham (1935).

POLYMYXA G. A. Ledingham, *Canadian Journal of Research, Section C* **17**: 50 (1939).
Typification: *Polymyxa graminis* G. A. Ledingham (holotype - sole original species).

Polymyxa graminis G. A. Ledingham, *Canadian Journal of Research, Section C* **17**: 50 (1939).
Typification: Type specimens deposited in the Herbarium of the Botany Department, University of Toronto. Type locality: Ontario, Canada. Type habitat: parasitic in roots of *Triticum aestivum* L., *T. durum* Desf., *Hordeum vulgare* L. and *Secale cereale* L. (*Poaceae*).
Identification key: Key XVI; key point 12'.
Literature: Barr & Allan (1982), Barr & Slykhuis (1976), Braselton (1984), Chen *et al.* (1990, 1991), Haber, Barr & Platford (1991), Ijdenberg, Kummert & Lepoivre (1986), Karling (1967*b*, 1968*e*), Langenberg & Van Der Wal (1986), Littlefield, *et al.* (1998), Maraite, Goffart & Bastin (1988), Nolt *et al.* (1981), Pruett & Waller (1989), Slykhius & Barr (1978), Toyama & Kusaba (1970), Yili, Wanhe & Jianping (1990), Ward & Adams (1998).

Polymyxa betae B. Keskin, *Archiv für Mikrobiologie* **49**: 370 (1964).
Typification: type material not designated; lectotype *Archiv für Mikrobiologie* **49**: 351, figs 2-3, 354, fig. 4, 357, fig. 5, 362-3, figs 6-7 (1964) (iconotype). Type locality: Regierungsbezirk Hildesheim, Einbeck, Germany. Type habitat: parasitic without hypertrophy in *Beta vulgaris* Tournefort (*Caryophyllales*, *Chenopodiaceae*) and other members of this family.
Identification key: Key XVI; key point 12'.
Literature: Anonymous (1991), Barr (1981*b*), Barr & Asher (1992, 1996), Blunt & Asher (1989), Blunt, Asher & Gilligan (1991, 1992), Braselton (1983), Buttner (1992), Ciafardini & Marotta (1988, 1989*a*, *b*), Ivanovic (1988*a*, *b*), Karling (1968*e*: 96), Keskin (1971), Keskin & Fuchs (1969), Kastirr (1991), Ledingham (1939), Mutasa *et al.* (1996), Olive (1975), Payne & Asher (1990), Vurbanov & Kroumov (1989), Ward & Adams (1998).

OCTOMYXA J. N. Couch, J. Leitner & A. Whiffen, *Journal of the Elisha Mitchell Scientific Society* **55**: 400 (1939).
Typification: *Octomyxa achlyae* J. N. Couch, J. Leitner & A. Whiffen (holotype - sole original species).

Octomyxa achlyae J. N. Couch, J. Leitner & A. Whiffen, *Journal of the Elisha Mitchell Scientific Society* **55**: 400 (1939).
Typification: type material not designated; lectotype *Journal of the Elisha Mitchell Scientific Society* **55**: plates 47, 48 (1939) (iconotype). Type locality: Hollow Rock, Durham Co., North Carolina, U.S.A. Type habitat: parasitic in *Achlya glomerata* Coker (*Peronosporomycetes*, *Saprolegniaceae*).
Identification key: Key XVI; key point 12'.
Literature: Couch, Leitner & Whiffen (1939), Karling (1968*e*: 51), Olive (1975), Sparrow (1960: 782), Whiffen (1939*a*, *b*).

Octomyxa brevilegniae W. R. Pendergrass, *Journal of the Elisha Mitchell Scientific Society* **64**: 133 (1948).
Typification: type material not designated; lectotype *Journal of the Elisha Mitchell Scientific Society* **64**: plate 15, figs 1-7 (1948) (iconotype). Type locality: Florence, South Carolina, U.S.A. Type habitat: parasitic in *Brevilegnia linearis* Coker & Braxton (*Peronosporomycetes*, *Saprolegniales*), from a soil sample.
Identification key: Key XVI; key point 12'.
Literature: Goldie-Smith (1951, 1954, 1956), Karling (1968*e*: 52), Pendergrass (1948, 1950), Sherwood (1968), Sparrow (1960: 783).

EXCLUDED GENERA AND SPECIES

TREMATOPHLYCTIS N. Patouillard, *Bulletin trimestriel de la Société Mycologique de France* **34**: 86 (1918). (**Type species**: *Trematophlyctis leptodesmiae* N. Patouillard - holotype, sole original species).

Trematophlyctis leptodesmiae N. Patouillard, *Bulletin trimestriel de la Société Mycologique de France* **34**: 86 (1918). **Literature**: Karling (1968*e*: 99), Patouillard (1918). [Possibly a species of the Synchytriaceae.]

CYSTOSPORA J. A. Elliott, *Delaware Agriculture Experiment Station Bulletin* **114**: 15 (1916). (**Type species**: *Cystospora batatas* J. A. Elliott - holotype, sole original species).

Cystospora batatas J. A. Elliott, *Delaware Agriculture Experiment Station Bulletin* **114**: 15 (1916) [as '*batata*']. **Literature**: Elliott (1916), Karling (1968*e*: 100). [Possibly a physiological symptom.]

PELTOMYCES L. Léger, *Comptes Rendus Hebdomadaires des Séances de l'Académie des Sciences, Paris, Série D* **149**: 239 (1909). (**Type species**: *Peltomyces hyalinus* L. Léger).

Peltomyces hyalinus L. Léger, *Comptes Rendus Hebdomadaires des Séances de l'Académie des Sciences, Paris, Série D* **149**: 239 (1909). **Literature**: Karling (1968*e*: 101), Léger (1909). [Unclassifiable organism parasitic in the malpighian tubes of insects.]

Peltomyces blatella L. Léger, *Comptes Rendus Hebdomadaires des Séances de l'Académie des Sciences, Paris, Série D* **149**: 239 (1909). **Literature**: Karling (1968*e*: 101), Léger (1909). [Unclassifiable organism parasitic in the malpighian tubes of insects.]

Peltomyces forficulae L. Léger, *Comptes Rendus Hebdomadaires des Séances de l'Académie des Sciences, Paris, Série D* **149**: 239 (1909). **Literature**: Karling (1968*e*: 101), Léger (1909). [Unclassifiable organism parasitic in the malpighian tubes of insects.]

SPOROMYXA L. Léger, *Archiv für Protistenkunde* **12**: 111 (1908). (**Type species**: *Sporomyxa scauri* L. Léger - holotype, sole original species).

Sporomyxa scauri L. Léger, *Archiv für Protistenkunde* **12**: 111 (1908). **Literature**: Karling (1968*e*: 101), Léger (1908). [Unclassifiable organism parasitic in insects.]

Sporomyxa tenebriones P. Rietschel, *Archiv für Protistenkunde* **86**: 349-358 (1936). **Literature**: Karling (1968*e*: 101), Rietschel (1936). [Unclassifiable organism parasitic in insects.]

GENUS OF PROBABLE PLASMODIOPHOROMYCETE AFFINITY SOMETIMES COMPARED WITH THE *THRAUSTOCHYTRIALES*

PHAGOMYXA J. S. Karling, *American Journal of Botany* **31**: 39 (1944).
Typification: *Phagomyxa algarum* J. S. Karling (holotype - sole original species).

Phagomyxa algarum J. S. Karling, *American Journal of Botany* **31**: 39 (1944).
Typification: type material not designated; lectotype *American Journal of Botany* **31**: 40, figs 1-29; 44, figs 30-74 (1944) (iconotype). Type locality: Harker's Island (also at Schackelford and Shark Shoal jetties), Beaufort, North Carolina, U.S.A. Type habitat: parasitic in *Bachelotia antillarum* (Grunow) Gerloff [≡ *Bachelotia fulvescens* (Bornet) Hamel [(Schousb. *ex* Bornet) Kuckuck *ex* Hamel] (basionym: *Pilayella fulvescens* (Schousb.) Bornet)] (*Fucophyceae*) [but see also: Schnepf, 1994, parasite of *Bellerochea malleus* (Brightwell) Heurck (*Bacillariophyceae*)].
Identification key: Key I; key point 142.
Literature: Johnson & Sparrow (1961: 319), Karling (1944*a*), Olive (1975), Porter (1986), Schnepf (1994).

EXCLUDED SPECIES

Phagomyxa chattonii (P.-A. Dangeard) J. S. Karling, *American Journal of Botany* **31**: 51 (1944) [as '*chattoni*'], synonym of *Labyrinthula chattonii* P.-A. Dangeard, *Le Botaniste* **24**: 223 (1932).

PYRROSORUS H. O. Juel, *Bihang till Kongliga Svenska Vetenskaps-Akademiens Handlingar* **26** (III, 14): 14 (1901).
Typification: *Pyrrhosorus marinus* H. O. Juel (holotype - sole original species).

Pyrrhosorus marinus H. O. Juel, *Bihang Kongliga Svenska Vetenskaps-Akademiens Handlingar* **26** (III, 14): 14 (1901).
Typification: type material not designated; lectotype *Bihang Kongliga Svenska Vetenskaps-Akademiens Handlingar* **26** (III, 14): figs 1-11 (1901) (iconotype). Type locality: Sweden. Type habitat: parasitic in *Cystoclonium purpureum* (Huds.) Blatt. [syn.: *C. purpurescens* Kütz.] (*Rhodophyceae*).
Identification key: Key XVI; key point 12'.
Literature: Aleem (1953, 1955), Juel (1901), Karling (1968*e*: 99), Sparrow (1960: 790).

HAPTOGLOSSALES **M. W. Dick, ordo novo.**

Fungi in animalibus parasitantes, vel in protoctistis photosyntheticis aut non photosyntheticis incolantes. Thallus non-septatus, in statu maturitate in sporangium mutans. Sporangium uno vel plures tubis exitibus nonulli, dispersia sporarum dissolutione sine tubis exitibus rara. Sporangia cum aplanosporis aut zoosporis. Zoosporae biflagellatae, anisokontae sed non heterokontae, alter vel ambo flagellarum acronemate praeditae; insertio flagellarum subapicalis, striata vel estriata; interdum zoosporae magnitudinum duarum sunt. Cystides sporarum extrusomata elaborata, quae hospites penetrantes, formantes. Sexualitas et historia vitae ignota.
Typus: *Haptoglossa* C. Drechsler [holotypus].

Parasitic in animals or endoparasitic in non-photosynthetic or photosynthetic protoctists. Thallus non-septate, converting into a sporangium at maturity. Sporangium usually with one to several exit tubes, rarely disintegrating without exit tubes to release spores. Sporangia with aplanospores or zoospores. Zoospores biflagellate, anisokont but not heterokont, one or both flagella with an acronema; flagellar insertion sub-apical, with or without a groove; zoospores sometimes of two sizes. Spore cysts often developing elaborate extrusomes for penetration of hosts. Sexuality and life-history not known.
Typification: *Haptoglossa* C. Drechsler [holotype].

HAPTOGLOSSACEAE **M. W. Dick, familia nova.**

Fungi in invertebratis (praecipue Aschelminthibus) parasitantes. Thallus non-septatus, in statu maturitate in sporangium mutans. Sporangium uno vel plures tubis exitibus nonulli, dispersio sporarum dissolutione sine tubis exitibus rara. Sporangia cum aplanosporis aut zoosporis. Zoosporae biflagellatae, anisokontae sed non heterokontae, alter vel ambo flagellarum acronemate praeditae; insertio flagellarum subapicalis, striata vel estriata; interdum zoosporae magnitudinum duarum sunt. Sexualitas et historia vitae ignota.
Typus: *Haptoglossa* C. Drechsler [holotyus].

Parasitic in invertebrates (mosly *Aschelminthes*). Thallus non-septate, converting into a sporangium at maturity. Sporangium usually with one to several exit tubes, rarely disintegrating without exit tubes to release spores. Sporangia with aplanospores or zoospores. Zoospores biflagellate, anisokont but not heterokont, one or both flagella with an acronema; flagellar insertion sub-apical without a groove; zoospores frequently of two sizes. Sexuality and life-history not known.
Typification: *Haptoglossa* C. Drechsler [holotype].

HAPTOGLOSSA C. Drechsler, *Journal of the Washington Academy of Sciences* **30**: 246 (1940).
Pleotrachelus W. Zopf, *Nova Acta der Kaiserlichen Leopoldinisch-carolinischen Akademie der Naturforscher* **47**: 173 (1884) [*pro parte minore typo excluso*] (**Type species**: *Pleotrachelus fulgens* W. Zopf).
? *Pseudosporopsis* A. Scherffel, *Archiv für Protistenkunde* **52**: 70 (1925) [*pro parte typo excluso*].
Typification: *H. heterospora* C. Drechsler (holotype - sole original species)

Haptoglossa heterospora C. Drechsler, *Journal of the Washington Academy of Sciences* **30**: 246 (1940).
Typification: type material not designated; lectotype: *Journal of the Washington Academy of Sciences* **30**: 244, fig. 1 A-Q (1940) (iconotype). Type locality not identified. Type habitat: parasitic in many species of nematodes found in agar-water-soil litter cultures.
Identification key: Key XVII; key point 2.
Literature: Barron (1977*b*), Campos & Silva (1990), Davidson & Barron (1973), Drechsler (1940), Duddington (1954, 1955, 1958), Duerschner (1983), Esser (1976), Estey & Olthrof (1965), Giuma & Cooke (1972), Gray, Wyborn & Smith (1982), Karling (1981*a*: 149), Landmann (1986), Lee, Vaughan & Durschner-Pelz (1992), McCullough (1977), Miura (1970), Newell, Cefalu & Fell (1977), Norton (1962), Porter (1986), Roccuzzo & Cianio (1991), Shephard (1956), Silva & Campos (1991*a*, *b*, 1992).

Haptoglossa zoospora J. G. N. Davidson & G. L. Barron, *Canadian Journal of Botany* **51**: 1322 (1973).
Typification: Slide(s) 10870, Herbarium OAC. Type locality: not stated, presumably Ontario, Canada. Type habitat: parasitic on nematodes from farmyard soil.
Identification key: Key XVII; key point 6'.
Literature: Barron (1977*b*), Davidson & Barron (1973), Duerschner (1983), Karling (1981*a*: 149), Landmann (1987).

Haptoglossa mirabilis G. L. Barron, *Mycologia* **72**: 1187 (1980).
Typification: Slide(s) OAC 10799, Herbarium OAC. Type locality: University of Guelph, Ontario, Canada. Type habitat: parasitic on *Adineta* spp. (*Aschelminthes*, *Rotifera*) from soil in a tropical greenhouse.
Identification key: Key XVII; key point 7.
Literature: Barron (1980, 1987), Karling (1981*a*: 246), Robb & Barron (1982), Robb & Lee (1986*a*, *b*).

Haptoglossa humicola G. L. Barron, *Canadian Journal of Botany* **59**: 1449 (1981).
Typification: Slide(s) OAC 10820, Herbarium OAC. Type locality: University of Guelph, Ontario, Canada. Type habitat: parasitic on bdelloid rotifers from soil in a tropical greenhouse.
Identification key: Key XVII; key point 7'.
Literature: Barron (1981).

Haptoglossa intermedia G. L. Barron, *Canadian Journal of Botany* **67**: 1645 (1989).
Typification: "Diagrams in the text constitute the type material" *Canadian Journal of Botany* **67**: 1646, figs 1-7, 1647, figs 8-15 (1989) (iconotype). Type locality: Guelph, Ontario. Canada. Type habitat: parasitic in bdelloid rotifers and nematodes from woodland soil.
Identification key: Key XVII; key point 5'.
Literature: Barron (1989*b*), Saikawa, Oyama & Yamaguchi (1991).

Haptoglossa elegans G. L. Barron, *Canadian Journal of Botany* **68**: 435 (1990).
Typification: Preserved material in Herbarium OAC (10846). Type locality: Waitakere, North Island, New Zealand. Type habitat: parasitic in bdelloid rotifers from debris under a tree fern.
Identification key: Key XVII; key point 2'.
Literature: Barron (1990), Saikawa, Oyama & Yamaguchi (1991).

Haptoglossa erumpens Glockling, *Mycological Research* **104**: 100 (2000).
Typification: (IMI 380405). Type locality: Northumberland, U.K. Type habitat: isolated from nematodes in cattle manure.
Identification key: Key XVII; key point 3'.
Literature: Glockling & Beakes (2000*a*, *b*).

Haptoglossa dickii Glockling, *Mycological Research* **104**: 104 (2000).
Typification: (IMI 379744). Type locality: Northumberland, U.K. Type habitat: isolated from nematodes in cattle manure.
Identification key: Key XVII; key point 6.
Literature: Glockling & Beakes (2000*a*, *b*).

Haptoglossa heteromorpha Glockling, *Mycologia* **92**: 750 (2000).
Typification: (IMI 380136). Type locality: Leazes Moor (Ord. Serv.: NZ 240 650), Northumberland, U.K., collected 8 Nov 1998. Type habitat: from *Rhabditis* nematodes isolated from cow dung.
Identification key: not keyed.
Literature: Glockling & Beakes (2000*c*, *d*).

UNIDENTIFIABLE SPECIES POSSIBLY REFERABLE TO THE GENUS *HAPTOGLOSSA*, BUT NOT TO SPECIES AND THEREFORE EXCLUDED

Pseudosporopsis rotatoriorum A. Scherffel, *Archiv für Protistenkunde* **52**: 82 (1925). **Typification**: type material not designated. **Literature**: Scherffel (1925*a*).

Pleotrachelus rotatoriorum A. Scherffel, *Abstracts of Communications, V International Botanical Congress, Cambridge, 1930*: 222 (1930). **Typification**: type material not designated. **Literature**: Scherffel (1930, 1931), Sparrow (1960: 164).

FAMILY *INCERTAE SEDIS*

ECTROGELLACEAE A. Scherffel, ***Archiv für Protistenkunde* 52**: 6 (1925) [*sensu stricto*].

Ectrogellaceae A. Scherffel, ***Archiv für Protistenkunde* 52**: 6 (1925) [*pro parte*] (**Type genus**: *Ectrogella* A. Scherffel, including *Eurychasma* P. Magnus).

Ectrogellacea H. F. Copeland, ***The Classification of Lower Organisms***: 81 (1956) [*pro parte*] (**Type genus**: *Ectrogella* A. Scherffel).

? *Eurychasmatales* F. K. Sparrow *in* E. B. G. Jones, ***Recent Advances in Aquatic Mycology***: 220 (1976) [*pro parte typo excluso*] [as '*Eurychasmales*'] [*nomen nudum*, cf. *Eurychasmataceae* and *Atkinsiellaceae*].

Typification: *Ectrogella* A. Scherffel (holotype - sole original genus).

ECTROGELLA W. Zopf, ***Nova Acta der Kaiserlichen Leopoldinisch-carolinischen Akademie der Naturforscher* 47**: 175 (1884).

Lagenidium W. Zopf, ***Verhandlungen des Botanischen Vereins der Provinz Brandenberg* 20**: 79 (1878) [*pro parte typo excluso*] (**Type species**: *L. rabenhorstii* W. Zopf).

Olpidiopsis M. Cornu, ***Annales des Sciences Naturelles, Botanique, Série 5* 15**: 114 -?145 (1872) [*pro parte typo excluso*] (**Type species**: *Olpidiopsis saprolegniae* M. Cornu).

Olpidium (A. Braun) L. Rabenhorst, ***Flora Europea Algarum* 3**: 283?8 (1868) [*pro parte minore typo excluso*] (**Type species**: *Olpidium endogenum* (A. Braun) J. Schroeter).

? *Cymbanche* E. Pfitzer, ***Sitzungberichte der niederrheinischen Gesellschaft Natur- und Heilkunde, Bonn, 1869***: 221 (1869) (**Type species**: *Cymbanche fockei* E. Pfitzer - holotype, sole original species).

Excluded synonym:

Pseudolpidiella K. Cejp, ***Flora ČSR, Oomycetes I***: 460 (1959) [*pro parte typo excluso*] (**Type species**: *Pseudolpidiella glenodiniana* (P.-A. Dangeard) K. Cejp).

Typification: *Ectrogella bacillariacearum* W. Zopf (holotype - sole original species).

Ectrogella bacillariacearum W. Zopf, *Nova Acta der Kaiserlichen Leopoldinisch-carolinischen Akademie der Naturforscher* **47**: 175 (1884).

Ectrogella gomphonematis A. Scherffel, *Archiv für Protistenkunde* **52**: 9 (1925).

Ectrogella eunotiae I. Friedmann, *Österreichische Botanische Zeitschrift* **99**: 200 (1952).

Olpidium gillii É. de Wildeman, *Annales de la Société Belge de Microscopie, Mémoires* **20**: 42 (1896).

Olpidiopsis gillii (É. de Wildeman) I. Friedmann, *Österreichische Botanische Zeitschrift* **99**: 200 (1952).

Pseudolpidiella gillii (É. de Wildeman) K. Cejp, *Flora ČSR, Oomycetes I* p. 460 (1959).

Olpidium sp. C. H. Gill, *Journal of Microscopical Science, London, 1893* plate 1, figs 1-8 (1893).

Typification: Lectotype (Cejp, 1959): *Nova Acta der Kaiserlichen Leopoldinisch-carolinischen Akademie der Naturforscher* **47**: plate 16, figs 1-24 (1884) (iconotype). Type locality: Berlin, Germany. Type habitat: parasitic in *Synedra ulna* (Nitzsch) Ehrenb. (also in *S. lunularis* Ehrenb., *Gomphonema* sp. *Pinnularia* sp. *Meridion* sp. and *Nitzschia* sp.) (*Bacillariophyceae*). **Typification of synonyms**: *Pseudolpidiella gillii*: lectotype (Cejp, 1959): *Journal of Microscopical Science* plate 1 figs 1-8 (1893) (iconotype); *Ectrogella eunotiae*: lectotype (Cejp, 1959): *Österreichische Botanische Zeitschrift* **99**: fig. 4 (1952).

Identification key: Key VIII; key point 3'.

Literature: Atkinson (1909*a*), Domjan (1936), Friedmann (1952), Geitler (1949), Gill (1893), Heurck (1899), Karling (1942*a*, 1981*a*: 34, 69, 72, 75), Paterson (1958*b*), Pystina (1995), Scherffel (1902, 1925), Sparrow (1933*c*, 1960: 805, 807, 810, 946), Wildeman (1890*c*, 1894, 1895a, 1896*b*), Wolf (1944), Zopf (1884).

Ectrogella perforans H. E. Petersen, *Oversigt over det Kongelige Danske Videnskabernes Selskabs Forhandlinger. Medlemmers Arbeiter, 1905*: 466 (1905).

Ectrogella licmophorae A. Scherffel, *Archiv für Protistenkunde* **52**: 10 (1925).

Olpidiopsis sp. T. W. Johnson, *Mycologia* **49**: 397 (1957).

Typification: type material not designated; lectotype *Oversigt over det Kongelige Danske Videnskabernes Selskabs Forhandlinger. Medlemmers Arbeiter, 1905*: 467, fig VII, 1-5 (1905) (iconotype). Type locality: Frederickshavn, Denmark. Type habitat: parasitic in *Licmophora lyngbyei* (Kütz.) Grun. (also *Synedra ulna* (Nitzsch) Ehrenb.) (*Bacillariophyceae*).

Identification key: Key VIII; key point 4.

Literature: Aleem (1950*c*, *d*, 1953), Höhnk (1939), Johnson (1957*a*, 1968), Johnson & Sparrow (1961: 325, 326), Karling (1942*a*, 1981*a*: 36, 70), Kuznetsov (1979), Petersen (1905), Porter (1986), Pystina (1995), C. Raghu Kumar (1978, 1980*a*, *b*, 1983), Scherffel (1925), Sparrow (1934, 1936*b*, 1960: 809, 1969).

Ectrogella brachystoma (A. Scherffel) M. W. Dick, **comb. nov.**
Lagenidium brachystomum A. Scherffel, *Archiv für Protistenkunde* **52**: 21 (1925).
Typification: Lectotype (Cejp, 1959): *Archiv für Protistenkunde* **52**: plate 2, figs 70-85 (1925) (iconotype). Type locality: Balaton, Mad'arsko, Hungary. Type habitat: parasitic in *Synedra ulna* (Nitzsch) Ehrenb. (also *Cymbella* sp., *Gomphonema* sp. and *Nitzschia* sp.) (*Bacillariophyceae*).
Identification key: Key VIII; key point 1'.
Literature: Couch (1935*b*), Ivimey Cook (1935), Karling (1981*a*: 119), Scherffel (1925), Sparrow (1960: 988).

Ectrogella cyclotellae (A. Scherffel) M. W. Dick, **comb. nov.**
Lagenidium cyclotellae A. Scherffel, *Archiv für Protistenkunde* **52**: 18 (1925).
Typification: *Archiv für Protistenkunde* **52**: plate 2, figs 49-59 (1925). Type locality: Spišská Nová Ves, Slovenia. Type habitat: parasitic in *Cyclotella kuetzingiana* Chauvin (*Bacillariophyceae*).
Identification key: Key VIII; key point 3.
Literature: Cejp (1959), Ivimey Cook (1935), Karling (1981*a*: 119), Scherffel (1925), Sparrow (1960: 986).

Ectrogella eurychasmoides J. Feldmann & G. Feldmann, *Revue de Mycologie* **20**: 238 (1956).
Typification: type material not designated; lectotype *Revue de Mycologie* **20**: 235, fig. 1 (1956) (iconotype). Type locality: Roscoff, France. Type habitat: parasitic on epiphytic *Licmophora lyngbyei* (Kütz.) Grun. on *Pilayella* sp. (*Fucophyceae*).
Identification key: Key VIII; key point 4'.
Literature: Dangeard (1912), Feldmann & Feldmann (1956), Johnson & Sparrow (1961: 326), Karling (1981*a*: 75), Sparrow (1960: 811, 1976).

DOUBTFUL SPECIES

Ectrogella lauderiae (H. H. Gran) A. Scherffel, *Archiv für Protistenkunde* **52**: 4 (1925).
Olpidium lauderiae H. H. Gran, *Nytt Magasin for Naturvidenskapene* **38**: 123 (1900).
Eurychasma lauderiae (H. H. Gran) H. E. Petersen, *Oversigt over det Kongelige Danske Videnskabernes Selskabs Forhandlinger. Medlemmers Arbeiter, 1905*: 469 (1905).
Typification: type material not designated; lectotype *Nytt Magasin for Naturvidenskapene* **38**: plate IX, figs 8-9 (1900) (iconotype). Type locality: Norway. Type habitat: parasitic in *Lauderia borealis* Gran (*Bacillariophyceae*).
Literature: Aleem (1953), Gran (1900), Petersen (1905), Scherffel (1925), Sparrow (1960: 149).

UNNAMED SPECIES POSSIBLY REFERABLE TO *ECTROGELLA*

Myzocytium sp. on *Synedra* (Sparrow, 1936*b*).

Lagenidium sp. on *Synedra* (Couch, 1935*b*).

Lagenidium sp. on *Pinnularia* (Scherffel, 1925*a*).

Olpidiopsis sp. on *Navicula* (Aleem, 1953).

Olpidiopsis sp. on *Navicula* (Johnson, 1966*d*).

Petersenia sp. on *Rhizosolenia* (Johnson, 1966*g*).

REJECTED SPECIES

Ectrogella besseyi, F. K. Sparrow & B. Ellison, *Mycologia* 41: 33 (1949), synonym of *Pleocystidium besseyi* (F. K. Sparrow & B. Ellison) M. W. Dick, *loc. cit.*

Ectrogella eunotiae I. Friedmann, *Österreichische Botanische Zeitschrift* 99: 200 (1952), synonym of *Ectrogella bacillariacearum* W. Zopf, *Nova Acta der Kaiserlichen Leopoldinisch-carolinischen Akademie der Naturforscher* 47: 175 (1884).

Ectrogella gomphonematis A. Scherffel, *Archiv für Protistenkunde* 52: 9 (1925), synonym of *Ectrogella bacillariacearum* W. Zopf, *Nova Acta der Kaiserlichen Leopoldinisch-carolinischen Akademie der Naturforscher* 47: 175 (1884).

Ectrogella licmophorae A. Scherffel, *Archiv für Protistenkunde* 52: 10 (1925), synonym of *Ectrogella perforans* H. E. Petersen, *Oversigt over det Kongelige Danske Videnskabernes Selskabs Forhandlinger. Medlemmers Arbeiter, 1905*: 466 (1905).

Ectrogella monostoma A. Scherffel, *Archiv für Protistenkunde* 52: 8 (1925) [**Lectotypification** (Cejp, 1959); *Archiv für Protistenkunde* 52: plate 1, figs 10-19 (1925) (iconotype)], synonym of *Aphanomycopsis bacillariacearum* A. Scherffel, *Archiv für Protistenkunde* 52: 14 (1925).

Ectrogella marina (P.-A. Dangeard) J. Feldmann & G. Feldmann, *Revue de Mycologie* 20: 238 (1956), synonym of *Sirolpidium marinum* (P.-A. Dangeard) M. W. Dick, *loc. cit.*

REJECTED GENUS

CYMBANCHE E. Pfitzer, *Sitzungberichte der niederrheinischen Gesellschaft Natur- und Heilkunde, Bonn, 1869*: 221 (1869) (Type species: *Cymbanche fockei* E. Pfitzer - holotype, sole original species).

Cymbanche fockei E. Pfitzer, *Sitzungberichte der niederrheinischen Gesellschaft Natur- und Heilkunde, Bonn, 1869*: 221 (1869), an unidentifiable parasite of diatoms, see also *Aphanomycopsis*.

ENDEMOSARCACEAE L. S. Olive & G. W. Erdos *in* G. W. Erdos & L. S. Olive, *Mycologia* **63**: 877 (1971).
Endemosarcidae L. S. Olive & G. W. Erdos *in* G. W. Erdos & L. S. Olive, *Mycologia* **63**: 877 (1971).
Typification: *Endemosarca* L. S. Olive & G. W. Erdos (holotype - sole original genus).

ENDEMOSARCA L. S. Olive & G. W. Erdos *in* G. W. Erdos & L. S. Olive, *Mycologia* **63**: 878 (1971).
Typification: *Endemosarca hypsalyxis* G. W. Erdos & L. S. Olive.

Endemosarca hypsalyxis G. W. Erdos & L. S. Olive, *Mycologia* **63**: 878 (1971).
Typification: L. S. Olive Hi-64-16. Type locality: Jocassee, South Carolina, U.S.A. (13 Aug 1964) (also from Brazil, Ecuador, Hawaii, Fiji and Malaysia). Type habitat: parasitic in *Colpoda* spp. (*Protozoa, Ciliata*) from fallen flowers of *Althaea* sp. (*Malvaceae*).
Identification key: Key I; key point 110.
Literature: Erdos & Olive (1971), Olive (1975).

Endemosarca ubatubensis G. W. Erdos & L. S. Olive, *Mycologia* **63**: 878 (1971).
Typification: L. S. Olive Br-67-148. Type locality: Ubatuba National Forest, Sao Paulo State, Brazil (Jan 1967). Type habitat: parasitic in *Colpoda* spp. (*Protozoa, Ciliata*) from fallen inflorescences of *Cecropia* sp. (*Moraceae*).
Identification key: Key I; key point 110'.
Literature: Erdos & Olive (1971), Olive (1975).

Endemosarca anomala G. W. Erdos, *Mycologia* **65**: 229 (1973).
Typification: L. S. Olive Sey-71-C-2. Type locality: Cousin Island, Seychelles Islands (17 Jun 1971). Type habitat: parasitic in *Colpoda* spp. (*Protozoa, Ciliata*) from old fruits of *Annona muricata* L. (*Annonaceae*).
Identification key: Key I; key point 109'.
Literature: Erdos (1973), Olive (1975).

HEIMERLIACEAE G. Arnaud, *Le Botaniste* **34**: 42 (1949) [Botanical Code *nom. invalid.*, Art. 36.1, without Latin diagnosis].
Typification: *Heimerlia* F. von Höhnel.

HEIMERLIA F. von Höhnel, *Annales Mycologici* **1**: 391 (1903).
Typification: *Heimerlia hyalina* F. von Höhnel (holotype - sole original species).

Heimerlia hyalina F. von Höhnel, *Annales Mycologici* **1**: 391 (1903).
Typification: type material not designated, no illustrations. Type locality: Wienerwald, Pressbaum, Austria. Type habitat: on rotting wood of *Quercus* and *Carpinus*.
Identification key: Key I; key point 134.
Literature: Arnaud (1949*a*, *b*), Höhnel (1903), Olive (1975).

RHABDOCYSTIS G. Arnaud, *Le Botaniste* **34**: 42 (1949) [Botanical Code *nom. invalid.*, Art. 36.1, without Latin diagnosis].
Typification: *Rhabdocystis peniophorae* G. Arnaud.

Rhabdocystis peniophorae G. Arnaud, *Le Botaniste* **34**: 42 (1949) [Botanical Code *nom. invalid.*, Art. 36.1, without Latin diagnosis].
Typification: type material not designated.
Identification key: Key I; key point 132'.
Literature: Arnaud (1949*a*, *b*), Olive (1975).

Rhabdocystis columnaris G. Arnaud, *Le Botaniste* **34**: 42 (1949) [Botanical Code *nom. invalid.*, Art. 36.1, without Latin diagnosis] [as '*columnare*'].
Typification: type material not designated.
Identification key: Key I; key point 132.
Literature: Arnaud (1949*a*, *b*), Olive (1975).

PYGMOMYCES G. Arnaud, *Le Botaniste* **34**: 43 (1949) [Botanical Code *nom. invalid.*, Art. 36.1, without Latin diagnosis].
Typification: *Pygmomyces microsporus* G. Arnaud (holotype - sole original species).

Pygmomyces microsporus G. Arnaud, *Le Botaniste* **34**: 43 (1949) [Botanical Code *nom. invalid.*, Art. 36.1, without Latin diagnosis].
Typification: type material not designated.
Identification key: Key I; key point 131'.
Literature: Arnaud (1949*a*,*b*), Olive (1975).

HELIOMYCOPSIS G. Arnaud, *Le Botaniste* **34**: 44 (1949) [Botanical Code *nom. invalid.*, Art. 36.1, without Latin diagnosis].
Typification: *Heliomycopsis fusisporus* G. Arnaud (holotype - sole original species).

Heliomycopsis fusispora G. Arnaud, *Le Botaniste* **34**: 44 (1949) [Botanical Code *nom. invalid.*, Art. 36.1, without Latin diagnosis] ['*fusisporus*'].
Typification: type material not designated.
Identification key: Key I; key point 134'.
Literature: Arnaud (1949*a*, *b*), Olive (1975).

CALOSPEIRA G. Arnaud, *Le Botaniste* **34**: 45 (1949) [Botanical Code *nom. invalid.*, Art. 36.1, without Latin diagnosis].
Typification: *Calospeira elegans* G. Arnaud (holotype - sole original species).

Calospeira elegans G. Arnaud, *Le Botaniste* **34**: 45 (1949) [Botanical Code *nom. invalid.*, Art. 36.1, without Latin diagnosis].
Typification: type material not designated.
Identification key: Key I; key point 133'.
Literature: Arnaud (1949*a*, *b*).

ENIGMA G. Arnaud, *Le Botaniste* **34**: 49 (1949) [Botanical Code *nom. invalid.*, Art. 36.1, without Latin diagnosis].
Typification: *Enigma vorticelloides* G. Arnaud (holotype - sole original species).

Enigma vorticelloides G. Arnaud, *Le Botaniste* **34**: 49 (1949) [Botanical Code *nom. invalid.*, Art. 36.1, without Latin diagnosis].
Typification: type material not designated.
Identification key: Key I; key point 129'.I
Literature: Arnaud (1949*a*, *b*), Olive (1975).

GENUS *INCERTAE SEDIS* SOMETIMES PLACED IN THE *PROTISTA* OR NEAR THE *LABYRINTHISTA*

DIPLOPHRYS J. Barker, *Quarterly Journal of Microscopical Science, N. S.* **8**: 123 (1868).
Typification: *Diplophrys archeri* J. Barker (holotype - sole original species).

Diplophrys archeri J. Barker, *Quarterly Journal of Microscopical Science, N. S.* **8**: 123 (1868).
Typification: Not possible from the original, unillustrated descriptions.
Identification key: Key I; key point 128'.
Literature: Barker (1867, 1868), Dykstra & Olive (1975), Hertwig & Lesser (1874), Penard (1902), Olive (1902), Olive (1975).

Diplophrys marina M. J. Dykstra & D. Porter, *Mycologia* **76**: 627 (1984).
Typification: Holotype material: coll. D. Porter, L82-12, 10 September 1982; holotype culture deposited with the American Type Culture Collection, Maryland, U.S.A. Type locality: Whidbey Island, Washington, U.S.A. (also from Sapelo Island, Georgia, U.S.A.). Type habitat: associated with marine vegetation, probably as a superficial saprotroph.
Identification key: Key XV; key point 14'.
Literature: Dykstra & Porter (1984).

EXCLUDED SPECIES

Diplophrys stercorea L. Cienkowski, *Archiv für Mikroscopische Anatomie und Entwickelungsmechanik* **12**: 15-50 (1876), synonym of *Sorodiplophrys stercorea* (L. Cienkowski) L. S. Olive & M. J. Dykstra *in* M. J. Dykstra & L. S. Olive, *Mycologia* **67**: 873 (1975).

SORODIPLOPHRYS L. S. Olive & M. J. Dykstra *in* M. J. Dykstra & L. S. Olive, *Mycologia* **67**: 873 (1975).
Typification: *Sorodiplophrys stercorea* (L. Cienkowski) L. S. Olive & M. J. Dykstra (holotype - sole original species).

Sorodiplophrys stercorea (L. Cienkowski) L. S. Olive & M. J. Dykstra *in* M. J. Dykstra & L. S. Olive, *Mycologia* **67**: 873 (1975).
Diplophrys stercorea L. Cienkowski, *Archiv für Mikroscopische Anatomie und Entwickelungsmechanik* **12**: 15-50 (1876).
Typification: Neotype: NC-72S. Type locality: Russia; neotype locality: North Carolina, U.S.A. Type habitat: from cow and horse dung.
Identification key: Key I; key point 128.
Literature: Cienkowski (1876), Dykstra (1976*a*, *b*), Dykstra & Olive (1975).

GENUS OF UNCERTAIN AFFINITY

DEVELOPAYELLA S. M. Tong, *European Journal of Protistology* **31**: 30 (1995) [Zoological Code; Botanical Code *nom invalid.*, Article 36.1, no Latin diagnosis].

Typification: *Developayella elegans* S. M. Tong (holotype - sole original species).

Developayella elegans S. M. Tong, *European Journal of Protistology* **31**: 31 (1995) [Zoological code; Botanical Code *nom invalid.*, Article 36.1, no Latin diagnosis].

Typification: type micrograph *European Journal of Protistology* **31**: 27 fig. 8) (iconotype). Type locality: Southampton Water, U.k (50° 54'N, 01° 25'W) (also collected at 10m 47°N 20°W on 8 July 1989). Type habitat: surface estuarine water, collected 7 August 1992.

Literature: Leipe *et al.* (1996), Tong (1995).

ORGANISMS NOW EXCLUDED FROM THE KINGDOM *STRAMINIPILA*: KINGDOMS *ANIMALIA* AND *PROTOCTISTA*

Taxonomy of Perkinsus *and* Dermocystidium*: protoctists of the* Apiclompexa

PERKINSUS N. D. Levine, *Journal of Parasitology* **64**: 549 (1978).
Labyrinthomyxa J. G. Mackin & S. M. Ray, *Journal of Invertebrate Pathology* **8**: 545 (1966).
Dermocystidium C. Pérez, *Compte-rendu des Séances de la Société de Biologie* **64**: 738 (1908) [*pro parte typo excluso*].
Typification: *Perkinsus marinus* J. G. Mackin, H. M. Owen & A. Collier) N. D. Levine (holotype - sole original species).

Perkinsus marinus (J. G. Mackin, H. M. Owen & A. Collier) N. D. Levine, *Journal of Parasitology* **64**: 549 (1978).
Labyrinthomyxa marina (J. G. Mackin, H. M. Owen & A. Collier) J. G. Mackin & S. M. Ray, *Journal of Invertebrate Pathology* **8**: 545 (1966).
Dermocystidium marinum J. G. Mackin, H. M. Owen & A. Collier, *Science* **111**: 328 (1950) [Zoological Code; Botanical Code *nom. invalid.*, Art. 36.1, without Latin diagnosis].
Typification: Type: "slides in collections of authors". Type locality: Sugar House Bend, Barataria Bay, Jefferson Parish, Louisiana, U.S.A. Type habitat: estuarine conditions, parasitic in *Crassostrea virginica* (Gmelin) (*Mollusca, Bivalvia*).
Selected literature: Bahnweg & Jäckle (1986), Chu & Greene (1989), Chu, La Payre & Burreson (1993), Fong *et al.* (1993), Gauthier & Vasta (1993), Goggin & Barker (1993), Goggin, Sewell & Lester (1990), Hatai (1986), Kleinschuster & Swink (1993), Levine (1978), Mackin (1962), Mackin & Boswell (1956), Mackin, Owen & Collier (1950), Mackin & Ray (1966), Pérez (1908), Perkins (1968, 1969, 1976*b*), Sprague (1954), White *et al.* (1989).

Perkinsus olsenii R. J. G. Lester & G. H. G. Davis, *Journal of Invertebrate Pathology* **37**: 184 (1981) [as '*olseni*'] [Zoological Code; Botanical Code *nom. invalid.*, Art. 36.1, without Latin diagnosis].
Typification: Type: USNM Catalogue No. 30478; British Museum Accession No. 1980:11:4:1-3; South Australian Museum No. E 1594. Type locality: near Port Lincoln, South Australia, 35°56'S 135°59'W . Type habitat: [parasitic (?)] in *Haliotis ruber* Leach (*Mollusca, Prosobranchia*).
Literature: Goggin & Barker (1993), Goggin, Sewell & Lester (1990), Lester (1986), Lester & Davis (1981).

Perkinsus atlanticus C. Azevedo, *Journal of Parasitology* **75**: 633 (1989) [Zoological Code; Botanical Code *nom. invalid.*, Art. 36.1, without Latin diagnosis].
Typification: Slides (Accession No. 41601) with paratypes deposited in the International Protozoan Type Slide Collection, Smithsonian Institution, Washington, DC, U.S.A. (also in collection of C. Azevedo). Type locality: Atlantic Coast, estuarine region near Faro, Portugal. Type habitat: [parasitic (?)] in the clam *Ruditapes decussatus* L. (*Mollusca, Bivalvia*).
Literature: Azevedo (1989, 1990), Azevedo, Corral & Cachola (1990), Goggin & Barker (1993), Goggin, Sewell & Lester (1990), Rodriguez, Godoy & Navas (1994).

Perkinsus karlssonii S. E. McGladdery, R. J. Cawthorn & B. C. Bradford, *Diseases of Aquatic Organisms* **10**: 129 (1991) [as '*karlssoni*'] [Zoological Code; Botanical Code *nom. invalid.*, Art. 36.1, without Latin diagnosis].
Typification: Slides with hapantotypes deposited in Canadian Museum of Nature, Ottawa, Ontario, Canada, Nos. CMN 1990-0026 to 0028 and 0031. Type locality: Tradacadie Bay, Prince Edward Island, Gulf of St Lawrence (also Atlantic Nova scotia), Canada. Type habitat: [parasitic (?)] in the bay scallop *Argopecten irradians* (Lamarck) (*Mollusca, Bivalvia*).
Literature: Goggin & Barker (1993), Goggin, Sewell & Lester (1990), Whyte, Cawthorn & Bradford (1991), Whyte, Cawthorn & McGladdery (1994), Whyte *et al.* (1993).

DERMOCYSTIDIUM C. Pérez, *Compte-rendu des Séances de la Société de Biologie* **64**: 738 (1908).
Dermocystis C. Pérez, *Compte-rendu des Séances de la Société de Biologie* **63**: 445 (1907) [later homonym].
Typification: *Dermocystidium pusula* (C. Pérez) C. Pérez.

Dermocystidium pusula (C. Pérez) C. Pérez, *Compte-rendu des Séances de la Société de Biologie* **64**: 738 (1908).
Dermocystis pusula C. Pérez, *Compte-rendu des Séances de la Société de Biologie* **63**: 445 (1907).
Typification: type material not verified; the organism is not in the kingdom *Straminipila*.
Literature: Beauchamp (1914), Dunkerly (1914), Jírovec (1939), Mackin, Owen & Collier (1950), Pérez (1907, 1908, 1913).

Dermocystidium branchialis L. Léger, *Comptes Rendus Hebdomadaires des Séances de l'Académie des Sciences, Paris, Série D* **158**: 807 (1914).
Typification: type material not verified; the organism is not in the kingdom *Straminipila*.
Literature: Jírovec (1939), Léger (1914), Mackin, Owen & Collier (1950),

Dermocystidium ranae E. Guyénot & A. Naville, *Revue Suisse de Zoologie, 1921-1922* **29**: 144 (1922).
Typification: type material not verified; the organism is not in the kingdom *Straminipila*.
Literature: Guyénot & Naville (1922), Jírovec (1939), Mackin, Owen & Collier (1950).

Dermocystidium vejdovskyi O. Jírovec, *Archiv für Protistenkunde* **92**: 145 (1939) [Zoological Code; Botanical Code *nom. invalid.*, Art. 36.1, without Latin diagnosis].
Typification: type material not designated; lectotype *Archiv für Protistenkunde* **92**: 139, fig. 1; 140, fig. 2; 141, fig. 3; 143, fig. 4; 145, fig. 5 (1939) (iconotype). Type locality: Teich Malá Kuš, Lnáře-Teiche, Böhmen, Čzechoslovakia. Type habitat: parasitic in Pike (*Esox lucius* L.) (*Pisces, Esocidae*). The organism is not in the kingdom *Straminipila*.
Literature: Jírovec (1939), Mackin, Owen & Collier (1950),

Dermocystidium daphniae O. Jírovec, *Archiv für Protistenkunde* **92**: 145 (1939) [Zoological Code; Botanical Code *nom. invalid.*, Art. 36.1, without Latin diagnosis].
Dermocystidium daphniae (W. Rühberg) [the Zoological Code citation, but Rühberg, *Archiv für Protistenkunde* **80**: 72-100 (1933) does not give his 'yeast' either a generic name or a specific epithet].
Typification: type material not designated, illustrations insufficiently detailed for iconotypification. Type locality: Rostock, Germany. Type habitat: parasitic in *Daphnia magna* Straus (*Crustacea, Cladocera*). The organism is not in the kingdom *Straminipila*.
Literature: Jírovec (1936), Mackin, Owen & Collier (1950), Rühberg (1933).

Dermocystidium cochliopodii A. Valkanov, *Nova Hedwigia* **12**: 393 (1966).
Typification: type material not designated; lectotype *Nova Hedwigia* **12**: 390-392, figs 1-4 (1966) (iconotype). Type locality: Black Sea. Type habitat: parasitic in a rhizopod, *Cochliopodus* sp. (*Protoctista, Testacea*). The organism is not in the kingdom *Straminipila*.
Literature: Valkanov (1966).

EXCLUDED SPECIES

Dermocystidium salmonis H. S. Davis, *United States Department of the Interior. Fish and Wildlife Service, Fishery Bulletin, 41* 51: 25 (1947), synonym of *Ostracoblabe salmonis* (H. S. Davis) M. W. Dick, *loc. cit.*

Genus of the Chlorophyceae *sometimes placed with* Dermocystidium: Hyalochlorella

HYALOCHLORELLA R. O. Poyton, *Journal of General Microbiology* **62**: 185 (1970) [no separate generic diagnosis].
Typification: *Hyalochlorella marina* R. O. Poyton (holotype - sole original species).

Hyalochlorella marina R. O. Poyton, *Journal of General Microbiology* **62**: 185 (1970).
Dermocystidium sp. *sensu* Vishniac ATCC 16323.
Dermocystidium sp. *sensu* Goldstein & Moriber ATCC 16324 (*Archiv für Mikrobiologie* 53: 1-11, 1966).
Typification: type material not designated; lectotype *Journal of General Microbiology* **62**: plate 1, figs 1-14, plate 2, figs 15-21 (1970) (iconotype); type culture strain 66-6A deposited in the Indiana University Algal Culture Collection (paratype culture 69-1A (provenance: Moss Beach, California, U.S.A.) also deposited in the same culture collection). Type locality: Newport, Rhode Island, U.S.A. Type habitat: from filaments of *Polysiphonia* sp. (*Rhodophyceae*).
Literature: Alderman (1974), Bremer (1976), Fuller & Poyton (1964), Goldstein & Moriber (1966), Jones & Harrison (1976), Perkins (1976), Poyton (1970*a*, *b*), Sparrow (1976).

Non-flagellate organisms sometimes confused with **Dermocystidium**

OSTRACOBLABE E. Bornet & C. Flahault, ***Bulletin de la Société Botanique de France, Série 2* 11**: **CXLVII-CLXXVI (Congrès de Botanique Tenu à Paris, Août 1889) (1889).**
Typification: *Ostracoblabe implexa* E. Bornet & C. Flahault (holotype - sole original species).

Ostracoblabe implexa E. Bornet & C. Flahault, *Bulletin de la Société Botanique de France, Série 2* **11**: CXLVII-CLXXVI (Congrès de Botanique Tenu à Paris, Août 1889) (1889).
Verrucaria consequens W. Nylander, *Flora oder allegemeine botanische Zeitung herausgegeben der königlichen bayerischen botanischen Gesellschaft in Regensburg* [later *Flora, Jena*]: 357 (1864) *sensu* E. Bornet, *Journal de Botanique* **5**: 399 (1891).
Typification: type material not verified; the organism is not in the kingdom *Straminipila*.
Literature: Alderman (1971, 1976), Alderman & Jones (1971), Alderman & Polglase (1986), Bornet (1891), Bornet & Flahault (1889), Dollfus (1921), Durve & Bal (1960), Galtsoff (1942), Jones & Harrison (1976), Karling (1981*a*: 208), Korringa (1951, 1952), Nylander (1864), Ramsbottom (1916*b*), Sprague (1972).

DOUBTFUL SPECIES

Ostracoblabe salmonis (H. S. Davis) M. W. Dick, **comb. nov.**
Dermocystidium salmonis H. S. Davis, *United States Department of the Interior. Fish and Wildlife Service, Fishery Bulletin, 41* **51**: 25 (1947).
Typification: type material not designated; lectotype *United States Department of the Interior. Fish and Wildlife Service, Fishery Bulletin, 41* **51**: plate 6, figs 59-66; plate 14, figs 210-213 (1947) (iconotype). Type locality: Sacramento River, Balls Ferry, California, U.S.A. Type habitat: parasitic in gills of Chinook Salmon (*Oncorhynchus tshawytscha* (Walbaum)) (*Pisces, Salmonidae*). The organism is not in the kingdom *Straminipila*.
Literature: Allen *et al.* (1968), Davis (1947), Olson, Dungan & Holt (1991), Mackin, Owen & Collier (1950).

Biflagellate parasites possibly related to the dinoflagellates and originally placed in the fungi: **Thalassomyces**

ELLOBIOPSIDA, protoctists *incertae sedis*

THALASSOMYCES E. L. Niezabitowski, *Kosmos, Lwów* **38**: 1568 (1913).
Ellobiopsis M. Caullery, *Bulletin Scientifique de la France et de la Belgique, Paris* **44**: 202 (1910) (**Type species**: *Ellobiopsis chattonii* M. Caullery [as '*chattoni*']) [*pro parte*].
Staphylocystis H. Coutière, *Bulletin Scientifique de la France et de la Belgique, Paris* **45**: 189 (1911) (**Type species**: *Staphylocystis racemosus* (H. Coutière) H. Coutière, **basionym**: *Ellobiopsis racemosus* H. Coutière) [later homonym].
Amallocystis L. Fage, *Archives de Zoologie Expérimentale et Générale* **78**: 146 (1936) (**Type species**: *Amallocystis fasciatus* L. Fage).
Typification: *Thalassomyces spiczakovii* E. L. Niezabitowski.

Thalassomyces spiczakovii E. L. Niezabitowski, *Kosmos, Lwów* **38**: 1568 (1913).
Typification: type material not verified. Type locality: Villefranche-sur-Mer, Mediterranean Sea. Type habitat: parasitic on *Pasiphaea sivado* (Risso) (*Crustacea, Decapoda*) at a depth of 90-1000m.
Literature: Collard (1966), Galt & Whisler (1970), Johnson & Sparrow (1961: 364), Niezabitowski (1913), Vader (1973*a*, *b*), Whisler (1990).

Thalassomyces racemosus (H. Coutière) P. Bergan, *Universitet I Bergen Årbok, 1952, Naturvitenskapelig Rekke* **14**: 11 (1953).

Ellobiopsis racemosus H. Coutière, *Comptes Rendus Hebdomadaires des Séances de l'Académie des Sciences, Paris, Série D* **152**: 411 (1911).

Staphylocystis racemosus (H. Coutière) H. Coutière, *Bulletin Scientifique de la France et de la Belgique, Paris* **45**: 189 (1911).

Thalassomyces batei E. L. Niezabitowski, *Kosmos, Lwów* **38**: 1569 (1913) (type locality: Pacific Ocean, Fiji Islands; type habitat: parasitic on *Pasiphaea cristata* Bate (*Crustacea, Decapoda*) at a depth of 576m).

non Staphylocystis racemosus sensu R. MacDonald, *Journal of the Marine Biological Association of the United Kingdom* **14**: 780 (1927) [Boschma (1949)].

Typification: type material not verified. Type habitat: parasitic on *Acanthephyra sexspinosa* Kemp (also on *Hymenodora glacialis* Buchholz, *Pasiphaea cristata* Bate and *P. tarda* Krøyer) (*Crustacea, Decapoda*).

Literature: Bate (1888), Bergan (1953), Boschma (1959), Collard (1966), Coutière (1911*a*, *b*), Galt & Whisler (1970), Johnson & Sparrow (1961: 364), Macdonald (1927), Vader (1973*a*, *b*).

Thalassomyces fasciatus (L. Fage) H. Boschma, *Atlantide Report* **5**: 167 (1959).

Amallocystis fasciatus L. Fage, *Archives de Zoologie Expérimentale et Générale* **78**: 151 (1936).

Typification: Type locality: Dana Expeditions, station 3593, off the Fiji Islands. Type habitat: parasitic on *Gnathophausia zoea* Willemoes-Suhm (*Crustacea, Mysidacea*).

Literature: Boschma (1959), Collard (1966), Fage (1936), Vader (1973*b*).

Thalassomyces capillosus (L. Fage) H. Boschma, *Atlantide Report* **5**: 166 (1959).

Amallocystis capillosus L. Fage, *Travaux de la Station Zoologique de Wimereaux, Paris* **13**: 287 (1938).

Typification: type material not verified. Type locality: Skagerak between trhe Baltic and North Seas (also Stait of Magellan - Boschma, 1949). Type habitat: parasitic on *Pasiphaea tarda* Krøyer (and other *Pasiphaea* spp.) (*Crustacea, Decapoda*).

Literature: Bergan (1953, 1961), Boschma (1949, 1959), Collard (1966), Dahl (1951), Fage (1938), Vader (1973*a*, *b*), Whisler (1990), Wing (1975).

Thalassomyces fagei (H. Boschma) H. Boschma, *Atlantide Report* **5**: 168 (1959).

Amallocystis fagei H. Boschma, *Proceedings. Koninklijke nederlandse Akademie van Wetenschappen, Amsterdam* **51**: 448 (1948).

Amallocystis sp. *sensu* H. Einarsson, *Dana-Report, No. 27*: 1-191 (1945).

Staphylocystis racemosus sensu R. MacDonald, *Journal of the Marine Biological Association of the United Kingdom* **14**: 780 (1927) [Boschma (1949)].

Ellobiopsis fagei (H. Boschma) R. Hovasse, *Comptes Rendus Hebdomadaires des Séances de l'Académie des Sciences, Paris, Série D* **233**: 980 (1951).

Typification: type material not designated. Type locality: Antarctic. Type habitat: parasitic on *Euphausia vallentini* Stebbing [**syn.** *Clausocalanus arcuicornis* (Dana)] (*Crustacea, Euphausiacea*).

Literature: Boschma (1948, 1949, 1959), Einarsson (1945), Galt & Whisler (1970), Hovasse (1951), Macdonald (1927), Mauchline (1966), Vader (1973*b*), Whisler (1990), Wing (1975).

Thalassomyces umbellatus (H. Boschma) H. Boschma, *Atlantide Report*: **5**: 162 (1959).

Amallocystis umbellatus H. Boschma, *Discovery Reports*: **25**: 294 (1949).

Typification: Type specimen: *Discovery Reports*: **25**: figs 6-9a, 10; plate LXLI, figs 1, 2 (1949). Type locality: Discovery Expedition, Station 692 (02°02'N 30°08'W), Atlantic Ocean, tropical region, east of Brazil (also station 81, south Atlantic, off Capetown, parasitic on *Haplophorus novae-zeelandiae* DeMan). Type habitat: parasitic on *Oplophorus grimaldii* (Coutière) (as *Haplophorus grimaldii* Coutière) (*Crustacea, Decapoda*) at a depth of up to 350m.

Literature: Boschma (1949, 1959), Collard (1966), Galt & Whisler (1970), Vader (1973*b*), Whisler (1990).

Thalassomyces boschmai (H. Nouvel) H. Boschma, *Atlantide Report* **5**: 171 (1959).

Amallocystis boschmai H. Nouvel, *Vie et Milieu* **4**: 57 (1954).

Typification: type material not verified. Type locality: Cape Palos, south east coast of Spain. Type habitat: parasitic on *Leptomysis gracilis* (Sars) (*Crustacea, Mysidacea*).

Literature: Collard (1966), Galt & Whisler (1970), Nouvel (1954), Nouvel & Hoenigman (1955), Vader (1973*b*), Wing (1975).

Thalassomyces nouvelii (J. Hoenigman) H. Boschma, *Atlantide Report* **5**: 172 (1959) [as '*nouveli*'].

Amallocystis nouvelii J. Hoenigman, *Bioloski Vestnik, Ljubljana* **3**: 111 (1954) [as '*nouveli*'].

Typification: type material not verified. Type locality: Adriatic Sea. Type habitat: parasitic on *Anchialina agilis* (Sars) (*Crustacea, Mysidacea*).

Literature: Hoenigman (1954, 1958), Vader (1973*b*), Wing (1975).

Thalassomyces niezabitowskii (J. Hoenigman) W. Vader, *Sarsia* **52**: 178 (1973).
Amallocystis niezabitowskii J. Hoenigman, *Rapport et procès-verbaux des Réunions. Commission Internationale pour l'Exploration scientifique de la Mer Méditerranée* **15**: 341 (1960) [Zoological Code; Botanical Code *nom. invalid.*, Art. 36.1, without Latin diagnosis].
Typification: type material not designated; no illustrations. Type locality: Jabuka, Adriatic Sea. Type habitat: parasitic on *Leptomysis megalops* Zimmer (*Crustacea, Mysidacea*).
Literature: Hoenigman (1960), Vader (1973*a*, *b*), Wing (1975).

Thalassomyces marsupii J. E. Kane, *New Zealand Journal of Science* **7**: 292 (1964).
Typification: Holotype 1962.6.7.1. Natural History Museum (London); 5 paratypes: paratype 1: NZOI 10 (New Zealand Oceanographic Institute, Wellington, New Zealand), paratypes 2-5: 1962.6.7.3-6. Natural History Museum (London). Type locality: Pacific Ocean, Discovery Station 940, 38°24.8'S, 173°41'N. Type habitat: parasitic on *Parathemisto gaudichaudii* (Guér.) (also on other genera of amphipods) (*Crustacea, Amphipoda*).
Literature: Galt & Whisler (1970), Kane (1964), Vader (1973*b*), Vader & Kane (1968), Whisler (1990), Wing (1975).

Thalassomyces californiensis S. Collard, *Proceedings. Koninklijke nederlandse Akademie van Wetenschappen, Amsterdam* **69**: 41 (1966).
Typification: Type BPC 71729 National Fungus Collections, U.S. National Herbarium, Beltsville, Maryland, U.S.A. Type locality: Santa Barbara Channel, 34°11' N 119°56' W, Pacific Ocean. Type habitat: parasitic on *Pasiphaea emarginata* Rathbun (*Crustacea, Decapoda*) at a depth of 360-516m.
Literature: Collard (1966), Galt & Whisler (1970), Vader (1973*a*, *b*), Whisler (1990), Wing (1975).

Thalassomyces albatrossi B.L. Wing, *Fishery Bulletin* **73**: 175 (1975).
Thallassomyces fasciatus sensu S. Collard, *Proceedings. Koninklijke nederlandse Akademie van Wetenschappen, Amsterdam* **69**: 41 (1966).
Thalassomyces sp. W. Vader, *Sarsia* **52**: 178 (1973).
? *Amallocystis fasciatus sensu* W. M. Tattersall, *United States National Museum, Bulletin* **201**: 1-292 (1951).
Typification: Type (USNM 24366 (holotype) and USNM 243367 (paratype) slides) from USNM 82439; deposited in Division of Crustacea, Smithsonian Institution, U.S. National Museum, Washington, D.C., U.S.A. Type locality: Albatross Station 4867, lat. 36°31'N, long. 129°46' E; depth 150 fathoms, USNM 82439. Type habitat: parasitic on *Stilomysis major* Tattersall (*Crustacea, Mysidacea*).
Literature: Vader (1973*b*), Whisler (1990), Wing (1975).

A total of 13 species of *Thalassomyces* have been named. Other ellobiopsid genera are *Ellobiopsis* (*E. chattonii* Caullery, *E. elongata* Steuer), *Parallobiopsis* (*P. coutieri* Collin), *Rhizellobiopsis* (*R. eupraxiae* (Zachs) Hovasse) and *Ellobiocystis* (*E. caridarum* (Coutière) Coutière, *E. villosus* Coutière, *E. tuberosus* Coutière, *E. filicollis* Coutière, *E. catenatus* Coutière, *E. tenuis* Coutière, *E. mysidarum* Coutière) (Boschma, 1959; Caullery, 1910; Collin, 1913; Coutière, 1911*b*; Hoenigman, 1958; Hovasse, 1926; Steuer, 1932; Whisler, 1990; Wing, 1975).

Miscellaneous genera which have at some time been associated with higher taxa now placed in the Straminipila

NEPHROMYCES A. Giard, *Comptes Rendus Hebdomadaires des Séances de l'Académie des Sciences, Paris* **106**: 1180 (1888).
non Nephromyces M. Plehn, *Zeitschrift für Fischerei* **18**: 53 (1916) (**Type species**: *Nephromyces piscium* M. Plehn [as'*Nephromyces piscinus*']) [see Alderman (1976: 252)].
Typification: *Nephromyces molgularum* A. Giard.

Nephromyces molgularum A. Giard, *Comptes Rendus Hebdomadaires des Séances de l'Académie des Sciences, Paris* **106**: 1180 (1888).
Typification: type material not verified. Type habitat: parasitic in *Moluga manhattensis* (De Kay) (tunicate worm).
Literature: Alderman (1976), Giard (1888), Saffo (1981, 1982), Saffo & Davis (1982), Saffo & Fultz (1986), Saffo & Nelson (1983).

Nephromyces sorokinii A. Giard, *Comptes Rendus Hebdomadaires des Séances de l'Académie des Sciences, Paris* **106**: 1180 (1888) [as '*sorokini*'].
Typification: type material not verified. Type habitat: parasitic in *Molgula complanata* Alder & Hancock [≡ *Lithonephrya eugyranda* L.-D.] (tunicate worm).
Literature: Alderman (1976), Giard (1888).

Nephromyces roscovitanus A. Giard, *Comptes Rendus Hebdomadaires des Séances de l'Académie des Sciences, Paris* **106**: 1182 (1888).
Typification: type material not verified. Type habitat: parasitic in *Molgula occulta* Kupffer [≡ *Anurella roscovitana* L.-D.] (tunicate worm).
Literature: Alderman (1976), Giard (1888).

MYIOPHAGUS R. Thaxter *in* F. K. Sparrow, *Mycologia* **31**: 443 (1939) [as '*Myrophagus*'].
Typification: *Myiophagus ucrainicus* (C. Wize) F. K. Sparrow (holotype - sole original species).

Myiophagus ucrainicus (C. Wize) F. K. Sparrow, *Mycologia* **31**: 443 (1939).
Olpidiopsis ucrainica C. Wize, *Bulletin International de l'Académie des Sciences de Crocovie, Classes des Sciences Mathématiques et Naturelles* **10**: 715 (1904).
Typification: type material not verified.
Literature: Karling (1948, 1981*a*: 38), Sparrow (1939*a*, 1960: 183), Wize (1904), Yačevskie & Yačevskie (1931: 62).

DECKENBACHIA A. A. Yačevskie & P. A. Yačevskie, *Opredelitel' Gribov*: 31 (1931).
Typification: *Deckenbachia consuens* (C. von Deckenbach) A. A. Yačevskie & P. A. Yačevskie (holotype - sole original species).

Deckenbachia consuens (C. von Deckenbach) A. A. Yačevskie & P. A. Yačevskie, *Opredelitel' Gribov*: 31 (1931).
Coenomyces consuens C. von Deckenbach, *Flora, Marburg* **92**: 253 (1903).
Typification: type material not verified. Type habitat: parasitic in *Calothrix parasitica* (Chauvin) Thur. [as "*Calathria parasitica*"] (*Cyanobacteria*).
Literature: Deckenbach (1903), Yačevskie & Yačevskie (1931).

NUCLEOPHAGA P.-A. Dangeard, *Le Botaniste* **4**: 214 (1895).
Typification: *Nucleophaga amoebae* P.-A. Dangeard (holotype - sole original species).

Nucleophaga amoebae P.-A. Dangeard, *Le Botaniste* **4**: 214 (1895).
Typification: type material not verified. Type locality: France. Type habitat: parasitic in the hypertrophied nucleus of *Thecamoeba verrucosa* (Gläser) [as its synonym, *Amoeba verrucosa* Gläser] (*Protozoa*).
Literature: Dangeard (1895), Sparrow (1960: 123).

Nucleophaga hypertrophica G. Epstein, *Archives de la Société russe de Protistologie* **1**: 76 (1922).
Nucleophaga intestinalis S. L. Brug, *Mededelingen van den Dienst der Volksgezondheit in Nederlandisch-Indië* **4**: 520 (1926).
Typification: type material not verified.
Literature: Brug (1926), Brumpt & Lavier (1935), Doflein (1907), Epstein (1922), Karling (1972, 1977: 32), Kirby (1941), Sassuchin (1934).

Nucleophaga peranemae A. Hollande & H. H. de Balsac, *Archives de Zoologie Expérimentale et Générale* **82**: 38 (1942) [Zoological Code; Botanical Code *nom. invalid.*, Art. 36.1, without Latin diagnosis].
Typification: type material not verified. Type habitat: parasitic in *Peranema trichophorum* (Ehrenb.) Stein (*Protoctista, Sarcomastigophora, Euglenida*).
Literature: Doflein (1907), Hollande & de Balsac (1942), Karling (1972), Sassuchin (1934).

Other species of *Nucleophaga* are described in the protozoological literature.

MORELLA R. Pérez Reyes, *Revista de la Sociedad Mexicana de Historia Natural (Mexico)* **24**: 5 (1963) [*nomen nudum*; also a later homonym, *non Morella* Lour. (1790) (*Myricaceae*)].

Typification: *Morella endamoebae* (E. R. Becker) R. Perez-Reyes [*nomen nudum*] (holotype - sole original species).

Morella endamoebae R. Pérez Reyes, *Revista de la Sociedad Mexicana de Historia Natural (Mexico)* **24**: 5 (1963) [*nomen nudum*] (holotype - sole original species).

Sphaerita endamoebae E. R. Becker, *Biological Bulletin of the Marine Biological Laboratory, Woods Hole* **50**: 447 (1926) [*nomen nudum*].

Typification: see under *Sphaerita entamoebae*.

Literature: Becker (1926), Karling (1972), Kirby (1941), Pérez Reyes (1963).

Taxa here excluded from the Straminipila, *but mentioned in* the "Dictionary of the Fungi, 7th edn" *as having affinity to taxa now placed in the* Straminipila

AMOEBOAPHELIDIUM A. Scherffel, *Archiv für Protistenkunde* **52**: 52 (1925) ["? Mastigomycotina - *Dictionary of the Fungi, 7th edn*"].

Typification: *Amoeboaphelidium achnanthidis* A. Scherffel (holotype - sole original species).

Amoeboaphelidium achnanthidis A. Scherffel, *Archiv für Protistenkunde* **52**: 52 (1925) [a non-flagellate monad parasitic in diatoms].

Typification: iconotypification not possible; no illustrations. Type locality: Igló, Hungary. Type habitat: parasitic in *Achnanthes lanceolata* Bréb. (*Bacillariophyceae*).

Literature: Scherffel (1925*a*).

AMYLOPHAGUS A. Scherffel, *Archiv für Protistenkunde* **52**: 87 (1925) ["? Mastigomycotina - *Dictionary of the Fungi, 7th edn*"].

Typification: *Amylophagus algarum* A. Scherffel (holotype - sole original species).

Amylophagus algarum A. Scherffel, *Archiv für Protistenkunde* **52**: 87 (1925) [an anisokont biflagellate monad parasitic in algae (*Spirogyra* and *Oedogonium*)].

Typification: type material not designated; lectotype *Archiv für Protistenkunde* **52**: plate 4, figs 169-187 (1925) (iconotype). Type locality: Hungary. Type habitat: parasitic in *Spirogyra* sp. (*Zygnematales*, *Zygnemataceae*).

Literature: Scherffel (1925*a*).

APHELIDIUM W. Zopf, *Die Pilzethiere oder Schleimpilze ... mit ... Schnitten*: 127 (1885) [genus cited without affiliation - *Dictionary of the Fungi, 7th edn*].

Typification: *Aphelidium deformans* W. Zopf.

Aphelidium deformans W. Zopf, *Die Pilzethiere oder Schleimpilze ... mit ... Schnitten*: 127 (1885).

Typification: type material not designated; lectotype *Die Pilzethiere oder Schleimpilze ... mit ... Schnitten*: 127, fig. 48, I-IV (1885) (iconotype). Type locality: Halle, Germany. Type habitat: parasitic in a *Coleochaete* sp. (*C. soluta* Pringsh. or *C. irregularis* Pringsh.) (*Charophyceae*, *Coleochaetales*).

Literature: Scherffel 1925*a*), Zopf (1885*a* [1887]).

Aphelidium melosirae A. Scherffel, *Hedwigia* **41**: 106 (1902) [a uniflagellate monad parasitic in *Melosira varians* Agardh (*Bacillariophyceae*)].

Typification: type material not designated; lectotype *Archiv für Protistenkunde* **52**: plate 2, figs 87-90; plate 3, figs 91-101 (1925) (iconotype). Type locality: Igló, Hungary. Type habitat: parasitic in *Melosira varians* C. Agardh (*Bacillariophyceae*).

Literature: Scherffel (1902*a*, 1925*a*).

Aphelidium tribonemae A. Scherffel, *Archiv für Protistenkunde* **52**: 44 (1925) [a uniflagellate monad parasitic in *Tribonema*].
Typification: type material not designated; lectotype *Archiv für Protistenkunde* **52**: plate 3, figs 102-112 (1925) (iconotype). Type locality: Hungary. Type habitat: parasitic in *Tribonema* sp. (*Tribophyceae*).
Literature: Scherffel (1925*a*).

Aphelidium chaetophorae A. Scherffel, *Archiv für Protistenkunde* **52**: 47 (1925) [a uniflagellate monad parasitic in *Chaetophora*].
Typification: type material not designated; lectotype *Archiv für Protistenkunde* **52**: plate 3, figs 113-122 (1925) (iconotype). Type locality: Hungary. Type habitat: parasitic in *Chaetophora* sp. (*Chlorophyceae, Chaetophorales*).
Literature: Scherffel (1925*a*).

APHELIDIOPSIS A. Scherffel, *Archiv für Protistenkunde* **52**: 44 (1925) [entry omitted from *Dictionary of the Fungi, 7th edn*].
Typification: *Aphelidiopsis epithemiae* A. Scherffel (holotype - sole original species).

Aphelidiopsis epithemiae A. Scherffel, *Archiv für Protistenkunde* **52**: 54 (1925) [a biflagellate monad parasitic in diatoms].
Typification: type material not designated; lectotype *Archiv für Protistenkunde* **52**: plate 3, figs 123-130 (1925) (iconotype). Type locality: Igló, Hungary. Type habitat: parasitic in *Epithemia turgida* (Ehrenb.) Kütz. (*Bacillariophyceae*).
Literature: Scherffel (1925*a*).

ENDOSPORA A. Scherffel, *Archiv für Protistenkunde* **52**: 89 (1925).
Typification: *Endospora ovalis* A. Scherffel (holotype - sole original species).

Endospora ovalis A. Scherffel, *Archiv für Protistenkunde* **52**: 89 (1925) [non-flagellate monad parasitic in *Amylophagus* (see above) in an *Oedogonium*].
Typification: type material not designated; lectotype *Archiv für Protistenkunde* **52**: plate 3, figs 188-190 (1925) (iconotype). Type locality: Hungary. Type habitat: parasitic in *Amylophagus algarum* Scherff. parasitic in an *Oedogonium* sp. (*Chlorophyceae, Oedogoniales*).
Literature: Scherffel (1925*a*).

PLASMOPAROPSIS É. de Wildeman, *Annales de la Société de Botanique de Belge de Microscopie, Mémoires* **20**: 127 (1896) [entry omitted from *Dictionary of the Fungi, 7th edn*, but probably a straminipilous fungus].
Typification: *Plasmoparopsis rigida* É. de Wildeman (holotype - sole original species).

Plasmoparopsis rigida É. de Wildeman, *Annales de la Société de Botanique de Belge de Microscopie, Mémoires* **20**: 128 (1896) [probably an anamorphic species of *Pythiogeton*, but see also under *Syzygangia* (? synonym: *Achlyopsis entospora* É. de Wildeman); possible teleomorph *Pythium characearum* É. de Wildeman].
Typification: type material not designated; lectotype *Annales de la Société de Botanique de Belge de Microscopie, Mémoires* **20**: plate X, figs 1-10; plate XI, figs 5-8 (1896) (iconotype). Type locality: Marais de Pinchat, Geneva, Switzerland. Type habitat: parasitic in the oogonia of *Chara* species (*Chlorophyceae, Charophyta*).
Literature: de Wildeman (1896*c*).

PSEUDOSPORA L. Cienkowski, *Bulletin de la Classe Physico-Mathématique de l'Académie Impériale des Sciences de St.-Pétersbourg (Izvêstîya Imperatorskoî akademîî nauk, St Petersburg)* **17**: 88 (1859), L. Cienkowski, *Mélanges Biologiques* **3**: 8 (1858) [entry omitted from *Dictionary of the Fungi, 7th edn*, but see *Pseudosporopsis* below].
Typification: *Pseudospora parasitica* L. Cienkowski (holotype - sole original species).

Pseudospora parasitica L. Cienkowski, *Bulletin de la Classe Physico-Mathématique de l'Académie Impériale des Sciences de St.-Pétersbourg (Izvêstiya Imperatorskoî akademiî nauk, St.-Petersburg)* **17**: 88 (1859), L. Cienkowski, *Mélanges Biologiques* **3**: 8 (1858) [see above].

Typification: type material not designated; lectotype *Archiv für Protistenkunde* **52**: plate 3, figs 152-154; plate 4, figs 155-157 (1925) (iconotype). Lectotype locality: Igló, Hungary. Lectotype habitat: parasitic in *Spirogyra* sp. (*Zygnematales, Zygnemataceae*).
Literature: Cienkowski (1858, 1859, 1865), Scherffel (1925*a*), Zopf (1885*a*, [1887]).

Pseudospora aculeata W. Zopf, *Die Pilzethiere oder Schleimpilze ... mit ... Schnitten*: 117 (1885).
Typification: type material not designated; lectotype *Die Pilzethiere oder Schleimpilze ... mit ... Schnitten*: 127, fig. 44, I-IV (1885) (iconotype). Type locality: Halle, Germany. Type habitat: parasitic in an *Oedogonium* sp. (*Chlorophyceae, Chlorophyta, Oedogoniales*).
Literature: Zopf (1885*a* [1887]).

Pseudospora maligna W. Zopf, *Die Pilzethiere oder Schleimpilze ... mit ... Schnitten*: 120 (1885).
Typification: iconotypification not possible; no illustrations. Type locality: Halle, Germany. Type habitat: parasitic in an *Hypnum* sp. (*Musci*).
Literature: Zopf (1885*a* [1887]).

Pseudospora benedenii C. de Bruyne, *Archives de Biologie* **10**: 17 (1890) [as '*benedeni*'].
Typification: type material not designated; lectotype *Archives de Biologie* **10**: plate III, figs 1-14 (1890) (iconotype). Type locality: Bay of Naples, Italy. Type habitat: parasitic in *Cladophora gracilis* Kütz. (*Chlorophyceae, Cladophorales*).
Literature: de Bruyne (1890).

Pseudospora edax C. de Bruyne, *Archives de Biologie* **10**: 54 (1890).
Typification: type material not designated; lectotype *Archives de Biologie* **10**: plate III, figs 23-26 (1890) (iconotype). Type locality: Bay of Naples, Italy. Type habitat: parasitic in *Cladophora gracilis* Kütz. (*Chlorophyceae, Cladophorales*).
Literature: de Bruyne (1890).

Pseudospora leptoderma A. Scherffel, *Archiv für Protistenkunde* **52**: 44 (1925).
Typification: type material not designated; lectotype *Archiv für Protistenkunde* **52**: plate 3, figs 131-140 (1925) (iconotype). Type locality: Hungary. Type habitat: parasitic in *Vaucheria* sp. and *Zygnema* sp. (*Tribophyceae* and *Zygnematales* respectively).
Literature: Scherffel (1925*a*).

Pseudospora myzocytioides A. Scherffel, *Archiv für Protistenkunde* **52**: 44 (1925).
Typification: type material not designated; lectotype *Archiv für Protistenkunde* **52**: plate 3, figs 141-143 (1925) (iconotype). Type locality: Csorbaër-Sees in the high Tátra, Hungary. Type habitat: parasitic in *Zygogonium* sp. (*Zygnematales, Zygnemataceae*).
Literature: Scherffel (1925*a*).

EXCLUDED SPECIES

Pseudospora bacillariacearum W. Zopf, *Die Pilzethiere oder Schleimpilze ... mit ... Schnitten*: 120 (1885), a synonym of *Pseudosporopsis bacillariacearum* (W. Zopf) A. Scherffel, *Archiv für Protistenkunde* 52: 72 (1925) [an anisokont biflagellate monad parasitic in diatoms].

PSEUDOSPOROPSIS A. Scherffel, *Archiv für Protistenkunde* **52**: 70 (1925) ["? Mastigomycotina - *Dictionary of the Fungi, 7th edn*"].
Typification: *Pseudosporopsis bacillariacearum* (W. Zopf) A. Scherffel.

Pseudosporopsis bacillariacearum (W. Zopf) A. Scherffel, *Archiv für Protistenkunde* **52**: 72 (1925) [an anisokont biflagellate monad parasitic in diatoms].
Pseudospora bacillariacearum W. Zopf, *Die Pilzethiere oder Schleimpilze ... mit ... Schnitten*: 120 (1885).
Typification: type material not designated; lectotype *Archiv für Protistenkunde* **52**: plate 3, figs 144-151 (1925) (iconotype). Lectotype locality: Hungary. Lectotype habitat: parasitic in various genera of *Pennales* (*Bacillariophyceae*).
Literature: Scherffel (1925*a*), Zopf (1885*a* [1887]).

Pseudosporopsis rotatoriorum A. Scherffel, *Archiv für Protistenkunde* **52**: 82 (1925), an anisokont biflagellate monad parasitic in nematodes [see also under *Haptoglossa*].
Typification: type material not designated; lectotype *Archiv für Protistenkunde* **52**: plate 4, figs 165-168 (1925) (iconotype). Type locality: Hungary. Type habitat: parasitic in nematodes (*Aschelminthes, Nematoda*).

RHIZOBLEPHARIS P.-A. Dangeard, *Le Botaniste* **7**: 86 (1900) [a septate mycelium parasitizing several amoebae; probably referable to the *Zoopagales*, here excluded from the *Straminipila*].
Typification: *Rhizoblepharis amoebae* P.-A. Dangeard (holotype - sole original species).

Rhizoblepharis amoebae P.-A. Dangeard, *Le Botaniste* **7**: 86 (1900).
Typification: type material not designated; lectotype *Le Botaniste* **7**: plate III, figs 30-31 (facing p. 82) (1900) (iconotype). Type locality: France. Type habitat: parasitic in *Amoeba binucleata* Grüb. (*Protoctista*, *Rhizopoda*).

SPIROSPORA A. Scherffel, *Archiv für Protistenkunde* **54**: 522 (1926).
Typification: *Spirospora paradoxa* A. Scherffel (holotype - sole original species).

Spirospora paradoxa A. Scherffel, *Archiv für Protistenkunde* **54**: 522 (1926).
Typification: type material not designated; lectotype *Archiv für Protistenkunde* **54**: plate 28, figs 21-25 (1926) (iconotype). Type locality: Igló, Hungary. Type habitat: parasitic in cysts of *Vampyrella* sp. (possibly *Vampyrella spirogyrae* Cienk.) (*Protoctista*).
Literature: Scherffel (1925*b*).

SPOROCLEMA M. Tiesenhausen, *Archiv für Hydrobiologie und Planktonkunde* **7**: 302 (1912) [inexplicably assigned to the *Oomycetes* in *Dictionary of the Fungi, 7th edn*: 363].
Typification: *Sporoclema piriforme* M. Tiesenhausen (holotype - sole original species).

Sporoclema piriforme M. Tiesenhausen, *Archiv für Hydrobiologie und Planktonkunde* **7**: 302 (1912), the author placed the fungus in the *Mucorales* with *Sepedonium natans* Tiesenh., but it may be a hyphomycete.

TRICHOTHRAUMA E. Germain, *Compte rendu des Séances de la Société de Biologie, Paris* **2**: 158 (1850) *incertae sedis*.
Typification: *Trichothrauma dermale* E. Germain (holotype - sole original species).

Trichothrauma dermale E. Germain, *Compte rendu des Séances de la Société de Biologie, Paris* **2**: 158 (1850), *incertae sedis* [parasitic on the skin of eels, possibly saprolegniaceous (hyphae of irregular diameter, no reproductive organs observed].

TAXONOMY OF UNIFLAGELLATE FUNGI OF RELEVANCE TO THIS WORK

Catenaria (Blastocladiales)

CATENARIA N. Sorokin, *Annales des Sciences Naturelles, Botanique, Série VI* **4**: 67 (1876).
Typification: *Catenaria anguillulae* N. Sorokin (holotype - sole origibnal species).

Catenaria anguillulae N. Sorokin, *Annales des Sciences Naturelles, Botanique, Série VI* **4**: 67 (1876).
Typification: type material not designated: lectotype *Mycologia* **37**: 175, figs 43-48; 179, figs 49-60, 181, figs 61-78 (1945) (iconotype). Lectotype locality: Angleton, Texas, U.S.A. (May 1944, coll. J. W. Olsen). Lectotype habitat: parasitic in sheep liver fluke eggs.
Literature: Barron (1977*b*), Boosalis & Mankau (1965), Buckley & Clapham (1929), Butler (1928), Butler & Buckley (1927), Butler & Humphries (1932), Carter (1858), Chong & Barr (1974), Constantineanu (1901), Couch (1945), Czeczuga (1995*a*), Dangeard (1884-85*b*, 1886*a*), Esser & Ridings (1973), Fuller & Jaworski (1987), Guima & Cooke (1972), Jaffee (1985, 1986*a*, *b*), Jansson & Thiman (1992), Karling (1934, 1946*b*, 1967*b*, 1968*f*, 1976*b*, 1981*b*, *e*, 1984, 1987*a*, *b*), Landmann (1986), Nolan (1966, 1970*a*, *b*, *c*), Olson & Reichle (1978), Olson *et al.* (1978), Prasad & Dayal (1986), Raghu Kumar (1977*b*), Roy (1982), Samota *et al.* (1988), Sayre & Keeley (1969), Seurat (1920), Singh (1989), Singh & Gupta (1986), Singh *et al.*, (1993, 1996), Sorokine (1876, 1883), Sparrow (1932, 1952*b*, *d*, 1957, 1960: 653), Sparrow, Paterson & Johns (1965), Stirling & Platzer (1978), Ulken (1977, 1990), Villot (1874), Yačevskie & Yačevskie (1931: 49).

Catenaria allomycis J. N. Couch, *Mycologia* **37**: 171 (1945).
Typification: type material not designated: lectotype *Mycologia* **37**: 167, figs 1-11; 169, figs 12-31, 172, figs 32-41 (1945) (iconotype). Type locality: Texas, U.S.A. (2 Jan 1942). Type habitat: parasitic in *Allomyces anomalus* R. Emers. (*Blastocladiales*) from soil.
Literature: Couch (1945), Sparrow (1960: 655).

Catenaria sphaerocarpa J. S. Karling, *American Journal of Botany* **25**: 328 (1938).
Typification: type material not designated; lectotype *American Journal of Botany* **25**: 329, fig 1; 330, figs 2-20 (1938) (iconotype). Type locality: New York City, U.S.A. Type habitat: saprotrophic in dead and cooked cells of *Hydrodictyon reticulatum* (L.) Lagerh. (also in other *Chlorophyceae*, *Charophyceae* and root tips of *Allium*, *Elodea* and *Zea*).
Literature: Couch (1945), Czeczuga (1995*a*), Karling (1938), Sparrow (1943, 1952*b*, 1960: 657).

Catenaria vermicola W. Birchfield, *Mycopathologia* **13**: 332 (1960).
Typification: Type specimen No. 71634, U.S. Department of Agriculture National Fungus Collections; paratypes in New York Botanic Garden and Royal Botanic Gardens, Kew, U.K. Type locality: Shreveport, Louisiana, U.S.A. Type habitat: parasitic in *Xiphinema chambersi* Thorne (*Aschelminthes*, *Nematoda*) from St Augustine grass.
Literature: Birchfield (1960), Tribe (1977*a*).

Catenaria verrucosa J. S. Karling, *Sydowia* **20**: 144 (1967).
Typification: type material not designated; lectotype *Sydowia* **20**: plate 29, figs 1,2 (1967) (iconotype). Type locality: New Zealand. Type habitat: saprotrophic in snake skin in soil sample AKC.
Literature: Czeczuga (1995*a*), Karling (1967*b*).

Catenaria indica U. P. Singh & M. S. Pavgi, *Mycologia* **62**: 589 (1970).
Typification: "a type slide is being deposited at IMI". Type locality: Varanasi, Uttar Pradesh, India. Type habitat: on snake skin (keratin) bait, 16 Nov. 1965, U. P. Singh No. MSP 363.
Literature: Singh & Pavgi (1970).

Catenaria auxiliaris (J. Kühn) H. T. Tribe, *Transactions of the British mycological Society* **69**: 370 (1977).
Tarychium auxiliarum J. Kühn, *Zeitschrift des Vereins für die Rübenzucker-Industrie des Deutschen Reich*: 457 (1877) [*pro parte*].
Tarichium auxiliare J. Kühn, *Berichte aus em physiologischen Laboratorium und der Versuchanstalt des landwirthschaftelichen Instituts der Universität Halle* **3**: 136 (1881) [*pro parte*].
Typification: "Slides of *Catenaria auxiliaris* have been deposited at...[IMI, Egham, U.K]...(IMI 212206) and...[CBS, Baarn, Netherlands]. These function as neotypes since it is virtually certain that no material was preserved". Iconotype: *Transactions of the British Mycological Society* **69**: 371, figs 1-6; 372, figs 7-14 (1977). Type locality: Germany. Type habitat: parasitic in *Heterodera schachtii* Schmidt (*Aschelminthes*, *Nematoda*).
Literature: Bursnall & Tribe (1974), Doberski & Tribe (1978), Kerry & Crump (1977), Korab (1929), Kühn (1877, 1881), MacLeod & Müller-Kögler (1970), Rozsypal (1934), Tribe (1977*a*, *b*, *c*).

Catenaria species on midge eggs

Catenaria spinosa W. W. Martin, *Mycologia* **68**: 264 (1975).
Typification: Holotype slide in the mycology herbarium, Botany Department, University of North Carolina, Chapel Hill, U.S.A., isotype slides in the herbarium of W. W. Martin. Type locality: creek near Hanover, Virginia, U.S.A. Type habitat: parasitic in eggs of unidentified species of *Chironomus* (*Diptera*).
Literature: Martin (1975*a*).

Catenaria uncinata W. W. Martin, *Mycologia* **70**: 461 (1978).
Typification: Holotype slide and isotype slides in the herbarium of W. W. Martin. Type locality: lake near Hanover, Virginia, U.S.A. Type habitat: parasitic in the eggs of *Glyptotendipes lobiferus* Say. (*Diptera*).
Literature: Martin (1978).

Catenaria ramosa W. W. Martin, *Mycologia* **70**: 463 (1978).
Typification: Holotype slide and isotype slides in the herbarium of W. W. Martin. Type locality: streams near Hanover, Virginia, U.S.A. Type habitat: parasitic in the eggs of *Dicrotendipes modestus* Say. (*Diptera*).
Literature: Martin (1978).

EXCLUDED SPECIES

Catenaria pygmaea I. L. Serbinov, *Scriptis Botanicis Horti Universitatis Impericolis Petropolitanae* **24**: 161 (1907), synonym of *Entophlyctis pygmaea* (I. L. Serbinov) F. K. Sparrow, *Aquatic Phycomycetes*: 256 (1943); parasitic in *Mougeotia*, Sparrow (1960: 376), Yačevskie & Yačevskie (1931: 49).

Olpidium (Chytridiales)

In this section I list all binomials in *Olpidium* irrespective of synonymy, or the validity of the species. For those species dealt with more fully elsewhere in this text, only the basionym to a binomial in *Olpidium* is given and reference is made to the binomial accepted here. These entries are indented.

OLPIDIUM (A. Braun) L. Rabenhorst, *Flora Europaea Algarum* **3**: 288 (1868).

Chytridium subgenus *Olpidium* A. Braun, *Physicalische Abhandlungen der königlichen Akademie der Wissenschaften zu Berlin, 1855*: 75 (1856) (**Type species**: not named; the first-named species was *Chytridium apiculatum* A. Braun, now a synonym of *Entophlyctis apiculata* (A. Braun) L. Rabenhorst; the second species listed was *Chytridium endogenum* A. Braun; the third and last species was *Chytridium saprolegniae* A. Braun, now a synonym of *Olpidiopsis saprolegniae* (A. Braun) M. Cornu, the type species of *Olpidiopsis*).

Olpidium (A. Braun) L. Rabenhorst *sensu* J. Schroeter, *Kryptogamen-Flora von Schlesien [1885-1889]* **3(1)**: 180 (1886) (**Type species**: first-named species *Olpidium endogenum* (A. Braun) J. Schroeter) [Sparrow (1960: 128-130) accepts this concept as definitive].

Cyphidium P. Magnus, *Wissenschaft Meeresuntersuchungen der Kommission zur wissenschaftlichen Untersuchung der deutschen Meere, Abteilung Kiel* **2-3**: 77 (1875) (**Type species**: no binomial established, but see *Olpidium plumulae*).

Olpidiella [N.] G. Lagerheim, *Journal de Botanique* **2**: 438 (1888) (**Type species**: *Olpidiella uredinis* [N.] G. Lagerheim).

Endolpidium É. de Wildeman, *Annales de la Société de Botanique de Belge de Microscopie, Mémoires* **18**: 153 (1894) (**Type species**: *Endolpidium hormisciae* É. de Wildeman [genus, but not type species, placed as a synonym by Sparrow, 1960]).

Typification: *Olpidium endogenum* (A. Braun) J. Schroeter.

Olpidium *species parasitic in* Zygnematales (Charophyceae) *(15 binomials and 1 variety)*

Olpidium endogenum (A. Braun) J. Schroeter, *Kryptogamen-Flora von Schlesien. [1885-1889]* **3(1)**: 180 (1886).

Chytridium endogenum A. Braun, *Bericht über die zur Bekanntmachung geeigneten verhandlungen der Königlichen Preussischen Akademie der Wissenschaften zu Berlin, 1855*: 384 (1855).

Chytridium intestinum A. Braun, *Bericht über die zur Bekanntmachung geeigneten verhandlungen der Königlichen Preussischen Akademie der Wissenschaften zu Berlin, 1855*: 589 (1856).

Olpidium intestinum (A. Braun) L. Rabenhorst, *Flora Europaea Algarum* **3**: 283 (1868).

Olpidiella endogena (A. Braun) [N.] G. Lagerheim, *Journal de Botanique* **2**: 438 (1888).

Typification: type material not verified. Type locality: Leipzig, Germany. Type habitat: parasitic in *Closterium lunula* (O. F. Müll.) Nitzsch (*Zygnematales, Desmidaceae*).

Literature: Braun (1855, 1856), Fischer (1892*a*), Kiran & Dayal (1992), Konno (1972), Lagerheim (1888), Rabenhorst (1868), Schroeter (1885), Sparrow (1960: 132), Wildeman (1896*a*).

Olpidium ampullaceum (A. Braun) L. Rabenhorst, *Flora Europaea Algarum* **3**: 282 (1868).

Chytridium ampullaceum A. Braun, *Bericht über die zur Bekanntmachung geeigneten verhandlungen der Königlichen Preussischen Akademie der Wissenschaften zu Berlin, 1855*: 384 (1855).

Sphaerostylidium ampullaceum A. Braun, *Physicalische Abhandlungen der Königlichen Akademie der Wissenschaften zu Berlin, 1855*: 66 (1856).

Sphaerostylidium ampullaceum (A. Braun) N. Sorokin, *Archives Botaniques du Nord de la France* **2**: 20 (1883).

Rhizophydium ampullaceum (A. Braun) A. Fischer, *Rabenhorst's Kryptogamen-Flora, 2 Aufl., Bd 1, Abt. 4*: 101 (1892).

Typification: type material not verified: a species of *Rhizophydium.*

Literature: Braun (1855, 1856*a*), Fischer (1892*a*), Gaertner (1954), Graff (1928), Rabenhorst (1868), Sorokin (1883), Sparrow (1960: 262).

Olpidium entophytum (A. Braun) L. Rabenhorst, *Flora Europaea Algarum* **3**: 283 (1868).

Chytridium entophytum A. Braun, *Bericht über die zur Bekanntmachung geeigneten verhandlungen der Königlichen Preussischen Akademie der Wissenschaften zu Berlin, 1855*: 589 (1856).

Reesia cladophorae C. Fisch, *Sitzungsberichte der Physicalische-medicinischen Societät zu Erlangen* **16**: 101 (1884).

Typification: type material not verified. Type locality: Germany. Type habitat: parasitic in (marine?) *Vaucheria* spp. (*Tribophyceae*) [see Wildeman (1896*a*) for the broader species concept and synonymy].

Literature: Braun (1856*b*), Fischer (1892*a*), Konno (1972), Rabenhorst (1868), Sparrow (1960: 135), Wildeman (1896*a*).

Olpidium saccatum N. Sorokin, *Archives Botaniques du Nord de la France* **2**: 28 (1883).
Typification: type material not verified. Type locality: Tashkent, Asiatic Russia. Type habitat: parasitic in *Cosmarium botrytis* Menegh. (*Zygnematales, Desmidaceae*).
Literature: Kobayas[h]i & Ookubo (1954*b*), Konno (1972), Linder (1947), Scherffel (1926*a*), Sorokine (1883, 1889), Sparrow (1960: 138), Sparrow & Barr (1955), Wildeman (1893*b*, 1896*a*).

Olpidium algarum N. Sorokin, *Archives Botaniques du Nord de la France* **2**: 28 (1883).
var. ***algarum***.
Olpidium algarum N. Sorokin var. *longirostrum* N. Sorokin, *Archives Botaniques du Nord de la France* **2**: 28 (1883) [the first of two varieties described].
Typification: type material not verified. Type locality: Russia. Type habitat: parasitic in "Confervacées" [- a *Tribonema*-like filament according to Sparrow, 1960: 150] (*Tribophyceae*).
Literature: Fischer (1892), Minden (1915), Sorokine (1883, 1889), Sparrow (1960: 150).

Olpidium algarum N. Sorokin, *Archives Botaniques du Nord de la France* **2**: 28 (1883) var. ***brevirostrum*** N. Sorokin, *Archives Botaniques du Nord de la France* **2**: 28 (1883).
Typification: type material not verified. Type localities: Kazan and Tashkent, Asiatic Russia. Type habitat: parasitic in *Sphaerozosma vertebratum* (Bréb.) Ralfs (*Zygnematales, Desmidaceae*).
Literature: Fischer (1892), Minden (1915), Sorokine (1883, 1889), Sparrow (1960: 151), Wildeman (1893*b*).

Olpidium immersum N. Sorokin, *Archives Botaniques du Nord de la France* **2**: 29 (1883).
Typification: type material not verified. Type locality: Tashkent, Asiatic Russia. Type habitat: parasitic in *Staurastrum* spp. and other desmids (*Zygnematales, Desmidaceae*).
Literature: Sorokine (1883, 1889), Scherffel (1926*a*), Sparrow (1960: 151), Wildeman (1893*b*, 1895*a*, 1896*a*).

Olpidium tuba N. Sorokin, *Archives Botaniques du Nord de la France* **2**: 31 (1883).
Typification: type material not verified. Type locality: Kazan and Tashkent, Asiatic Russia. Type habitat: parasitic in "Confervacées" (*Zygnematales, Desmidaceae*).
Literature: Sorokine (1883, 1889), Sparrow (1960: 154).

Olpidium zygnemicola P. Magnus, *Verhandlungen der Botanischer Vereins der Provinz Brandenburg* **26**: 79 (1885).
Typification: type material not verified. Type locality: Germany. Type habitat: parasitic in *Zygnema* sp. (*Zygnematales, Zygnemataceae*).
Literature: Fischer (1892*a*), Magnus (1885), Sparrow (1960: 134), Wildeman (1896*a*).

Olpidium rostratum É. de Wildeman, *La Notisaria* **10**: 35 (1895).
Typification: type material not designated; lectotype *Annales de la Société de Botanique de Belge de Microscopie, Mémoires* **20**: 40, fig. 1 (1896) (iconotype). Type locality: between Rustoën and Mälkevold, Norway. Type habitat: parasitic in *Closterium* sp. (*Zygnematales, Desmidaceae*).
Literature: Sparrow (1960: 152), Wildeman (1895*d*, 1896*b*).

Olpidium mesocarpi É. de Wildeman, *Annales de la Société de Botanique de Belge de Microscopie, Mémoires* **20**: 45 (1896).
Typification: type material not designated; lectotype *Annales de la Société de Botanique de Belge de Microscopie, Mémoires* **20**: plate I, figs 13-16 (1896) (iconotype). Type locality: Beloeil, Belgium. Type habitat: parasitic in *Mougeotia* sp. [às "*Mesocarpus* sp."] (*Zygnematales, Zygnemataceae*).
Literature: Sparrow (1960: 152), Wildeman (1896*b*).

Olpidium spirogyrae B. W. Skvortzow, *Archiv für Protistenkunde* **51**: 429 (1925).
Typification: type material not verified. Type locality: Manchuria. Type habitat: parasitic in *Spirogyra inflata* (Vaucher) Kütz. (*Zygnematales, Zygnemataceae*).
Literature: Skvortzow (1925), Sparrow (1960: 153).

Olpidium mougeotia B. W. Skvortzow, *Archiv für Protistenkunde* **51**: 430 (1925).
Typification: type material not verified. Type locality: Manchuria. Type habitat: parasitic in *Mougeotia scalaris* Hassall (*Zygnematales, Zygnemataceae*).
Literature: Skvortzow (1925), Sparrow (1960: 152).

Olpidium hyalothecae A. Scherffel, *Archiv für Protistenkunde* **54**: 200 (1926).

Typification: type material not verified. Type locality: Hungary. Type habitat: parasitic in *Hyalotheca dissiliens* (J. E. Sm.) Bréb. (*Zygnematales, Desmidaceae*).

Literature: Canter (1949), Scherffel (1926*ab*), Sparrow (1960: 137), Sparrow & Barr (1955).

Olpidium utriculiforme A. Scherffel, *Archiv für Protistenkunde* **54**: 213 (1926).

Typification: type material not verified. Type locality: Hungary. Type habitat: parasitic in *Cosmarium botrytis* (Menegh.) Ralfs (*Zygnematales, Desmidaceae*).

Literature: Canter (1949), Linder (1947), Scherffel (1926*ab*), Sparrow (1960: 139).

Olpidium rostriferum Y. Tokunaga, *Transactions of the Sapporo Natural History Society (Japan)* **13**: 80 (1933) var. ***rostriferum***.

non Olpidium rostriferum Y. Tokunaga, *Transactions of the Sapporo Natural History Society (Japan)* **13**: 80 (1933) var. *indicum* U. Kiran & R. Dayal, *Proceedings of the National Academy of Sciences, India* **62**: 296 (1992) [*nom. invalid* Art. 37.4] [as var. '*indica*'].

Typification: type material not verified. Type locality: Japan. Type habitat: parasitic in *Spirogyra juergensii* Kütz. [as '*jurgensii*'] (*Zygnematales, Zygnemataceae*).

Literature: Konno (1972), Tokunaga (1933*b*), Sparrow (1960: 134).

Olpidium *species parasitic in other freshwater algae (9 binomials)*

Olpidium apiculatum (A. Braun) L. Rabenhorst, *Flora Europaea Algarum* **3**: 283 (1868).

Chytridium apiculatum A. Braun, *Bericht über die zur Bekanntmachung geeigneten verhandlungen der Königlichen Preussischen Akademie der Wissenschaften zu Berlin, 1855*: 383 (1855).

Rhizidium apiculatum (A. Braun) W. Zopf, *Nova Acta der Kaiserlichen Leopoldinisch-carolinischen Akademie der Naturforscher* **47**: 207 (1884).

Entophlyctis apiculata (A. Braun) A. Fischer, *Rabenhorst's Kryptogamen-Flora, 2 Aufl., Bd 1, Abt. 4*: 117 (1892).

Typification: type material not verified. Type locality: Germany. Type habitat: parasitic in *Gloeococcus* (*Chlamydomonas*) spp. (*Chlorophyceae, Volvocales*).

Literature: a species of *Entophlyctis* see Sparrow (1960: 372).

Olpidium indicum N. Wallich *in* W. B. Turner, *Kongliga Svenska Vetenskaps-Akademiens Handlingar* **25** (**5**): 164 (1892).

Typification: W. B. Turner, *Kongliga Svenska Vetenskaps-Akademiens Handlingar* **25** (5): plate XXI, fig. 8 (1892), ectoparasitic, a choanomastigote, type material not verified.

Literature: Sparrow (1960: 156), Turner (1892) [excluded: a choanomastigote].

Olpidium pusillum (N. Sorokin) É. de Wildeman, *Bulletin de la Société Royale de Botanique de Belgique, Bruxelles* **35**: 16 (1896).

Chytridium pusillum N. Sorokin, *Archives Botaniques du Nord de la France* **2**: 22 (1883).

Typification: type material not verified. Type locality: Kazan, Asiatic Russia. Type habitat: parasitic in *Oedogonium* sp. (*Chlorophyceae, Oedogoniales*).

Literature: Sorokine (1883, 1889), Sparrow (1960: 152), Wildeman (1896*a*).

> ***Olpidium oedogoniarum*** (N. Sorokin) É. de Wildeman, *Annales de la Société Belge de Microscopie, Mémoires* **18**: 154 (1894).
>
> *Olpidiopsis fusiformis* M. Cornu var. *oedogoniarum* N. Sorokin, *Archives Botaniques du Nord de la France* **2**: 27 (1883).
>
> For full synonymy see under *Syzygangia oedogonii* (A. Scherffel) M. W. Dick *loc. cit.*
>
> **Typification**: see under *Syzygangia oedogonii*.
>
> **Literature**: see under *Syzygangia oedogonii*.

Olpidium sorokinei (É. de Wildeman) É. de Wildeman, *Bulletin de la Société Royale de Botanique de Belgique, Bruxelles* **35**: 16 (1896).

Olpidiopsis sorokinei É. de Wildeman, *Annales de la Société de Belge de Microscopie, Mémoires* **14**: 22 (1890).

Typification: type material not verified. Type locality: Lacken, Belgium. Type habitat: parasitic in *Tribonema bombycinum* (C. Agardh) Derb. & Sol. (*Tribophyceae*).

Literature: Sparrow (1960: 153), Wildeman (1890*b*, 1896*a*).

Olpidium gillii É. de Wildeman, *Annales de la Société Belge de Microscopie, Mémoires* **20**: 42 (1896).
For full synonymy see under *Ectrogella bacillariacearum* W. Zopf.
Typification: see under *Ectrogella bacillariacearum*.
Literature: see under *Ectrogella bacillariacearum*.

Olpidium euglenae P.-A. Dangeard, *Le Botaniste* **4**: 248 (1894-5).
Typification: type material not verified. Type locality: Poitiers, France. Type habitat: parasitic in *Euglena* sp. (*Protoctista, Sarcomastigophora, Euglenida*).
Literature: Dangeard (1894-5), Sparrow (1960: 139), Wildeman (1896*a*).

Olpidium hantzschiae B. W. Skvortzow, *Archiv für Protistenkunde* **51**: 430 (1925).
Typification: type material not verified. Type locality: Manchuria. Type habitat: parasitic in *Hantzschia amphioxys* (Ehrenb.) Grun. (*Bacillariophyceae*).
Literature: Skvortzow (1925), Sparrow (1960: 139).

Olpidium sphaeropleae Y. Tokunaga, *Transactions of the Sapporo Natural History Society (Japan)* **13**: 81 (1933).
Typification: type material not verified. Type locality: Japan. Type habitat: parasitic in "*Sphaeroplea annua*" [= *S. annulina* (Roth) C. Agardh?] (*Ulothricales, Ulothricaceae*).
Literature: Konno (1972), Tokunaga (1933*b*), Sparrow (1960: 137).

Olpidium pseudoeuglenae S. N. Dasgupta & R. John, *Bulletin of the Botanical Survey of India* **30**: 5 (1988).
Typification: Holotype deposited at HCIO, Division of Mycology and Plant Pathology, IARI, New Delhi-110 012. HCIO No. 37090. Type locality: Lucknow, India. Type habitat: parasitic in *Euglena* sp. (*Protoctista, Sarcomastigophora, Euglenida*).
Literature: Dasgupta & John (1988).

Olpidium *species parasitic in marine algae - compare with marine straminipilous fungi (12 binomials)*

Olpidium entophytum (A. Braun) L. Rabenhorst, *Flora Europaea Algarum* **3**: 283 (1868).
Chytridium entophytum A. Braun, *Bericht über die zur Bekanntmachung geeigneten verhandlungen der Königlichen Preussischen Akademie der Wissenschaften zu Berlin, 1855*: 589 (1856).
Typification: type material not verified. Type locality: Germany. Type habitat: parasitic in (marine?) *Vaucheria* spp. (*Tribophyceae*).
Literature: Braun (1856*b*), Fischer (1892*a*), Konno (1972), Rabenhorst (1868), Sparrow (1960: 135), Wildeman (1896*a*).

Olpidium plumulae (F. Cohn) A. Fischer, *Rabenhorst's Kryptogamen-Flora, 2 Aufl., Bd 1, Abt. 4*: 27 (1892).
Chytridium plumulae F. Cohn, *Hedwigia* **4**: 169 (1865).
Chytridium antithamnii F. Cohn, *Archiv für Mikroscopische Anatomie und Entwicklungsmechanik* **3**: 59 (1867).
Phlyctidium plumulae (F. Cohn) L. Rabenhorst, *Flora Europaea Algarum* **3**: 279 (1868).
Cyphidium plumulae (F. Cohn) A. Fischer, *Rabenhorst's Kryptogamen-Flora, 2 Aufl., Bd 1, Abt. 4*: 27 (1892) [*nomen nudum*, as *Cyphidium plumulae* P. Magnus, *Wissenschaft Meeresuntersuchungen der Kommission zur wissenschaftlichen Untersuchung der deutschen Meere, Abteilung Kiel* **2-3**: 77 (1875).
Typification: type material not verified. Type locality: shore of Heligoland, North Sea. Type habitat: parasitic in *Antithamnion plumula* Thur. (*Rhodophyceae*).
Literature: Cohn (1865, 1867), Fischer (1892), Magnus (1875), Rabenhorst (1868), Sparrow (1960: 156), Wildeman (1896*a*).

Olpidium entosphaericum (F. Cohn) A. Fischer, *Rabenhorst's Kryptogamen-Flora, 2 Aufl., Bd 1, Abt. 4*: 27 (1892).
Chytridium entosphaericum F. Cohn, *Hedwigia* **4**: 170 (1865).
Typification: type material not verified. Type locality: Germany. Type habitat: parasitic in marine algae *Bangia fuscopurpurea* (Dillw.) Lyngb. (*Rhodophyceae*) (and "*Hormidium penicilliformis*" [≡ *Urospora penicilliformis* (Roth) Aresch.] (*Cladophorales*) [compare with *Olpidium aggregatum*]).
Literature: Cohn (1865), Fischer (1892*a*), Schultze (1867), Sparrow (1960: 148), Wildeman (1896*a*).

Olpidium sphacelarum (L. Kny) L. Kny, *Hedwigia* **112**: 87 (1872) [as '*sphacellarum*'].
Chytridium sphacelarum L. Kny, *Sitzungs-Berichte der Gesellschaft naturforschender Freunde zu Berlin, 1871*: 97 (1871) [as '*sphacellarum*'].
For full synonymy see under *Anisolpidium sphacelarum* (L. Kny) J. S. Karling [as '*sphacellarum*'].
Typification: see under *Anisolpidium sphacelarum*.
Literature: see under see under *Anisolpidium sphacelarum*.

Olpidium tumefaciens (P. Magnus) E. P. Wright, *Transactions of the Royal Irish Academy, Dublin (Science)* **26**: 360 (1879).

Chytridium tumefaciens P. Magnus, *Sitzungs-Bericht der Gesellschaft Naturforschender Freunde zu Berlin, 1872*: 87 (1872).

For full synonymy see under *Eurychasmidium tumefaciens* (P. Magnus) F. K. Sparrow.

Typification: see under *Eurychasmidium tumefaciens*.

Literature: see under *Eurychasmidium tumefaciens*.

Olpidium dicksonii (E. P. Wright) N. Wille, *Videnskabsselskabets Skrifter. I. Math.-naturv. Klasse. 1899. No. 3. Christiana*: 2 (1899) var. ***dicksonii*** [by default, see *Olpidium dicksonii* (E. P. Wright) N. Wille var. *striariae*].

Rhizophydium dicksonii E. P. Wright, *Transactions of the Royal Irish Academy, Dublin (Science)* **26**: 374 (1879).

For full synonymy see under *Eurychasma dicksonii* (E. P. Wright) P. Magnus.

Typification: see under *Eurychasma dicksonii*.

Literature: see under *Eurychasma dicksonii*.

Olpidium bryopsidis C. de Bruyne, *Archives de Biologie* **10**: 85 (1890).

For full synonymy see under *Sirolpidium bryopsidis* (C. de Bruyne) H. E. Petersen.

Typification: *see under Sirolpidium bryopsidis*.

Literature: see under *Sirolpidium bryopsidis*.

Olpidium lacerans C. de Bruyne, *Archives de Biologie* **10**: 104 (1890) [*nomen nudum*].

Typification: type material not designated; see *Archives de Biologie* **10**: plate V, figs 28-31 (1890). Type locality: Bay of Naples, Italy. Type habitat: parasitic in an unnamed marine [chlorophycean] alga.

Literature: de Bruyne (1890), Sparrow (1960: 156) [excluded: the zoospores contain chlorophyll residues, the figures are of a monad].

Olpidium aggregatum P.-A. Dangeard, *Le Botaniste* **2**: 237 (1890-91).

? *Endolpidium hormisciae* É. de Wildeman, *Annales de la Société de Botanique de Belge de Microscopie, Mémoires* **18**: 153 (1894).

Typification: type material not verified. Type locality: France. Type habitat: parasitic in marine *Cladophora* sp. [as "*Urospora* sp." (= *Hormiscia zonata* ?) marine or brackishwater *Cladophorales*] (*Cladophorales*).

Literature: Dangeard (1890-91), Fischer (1892*a*), Sparrow (1960: 148), Wildeman (1896*a*).

Olpidium lauderiae H. H. Gran, *Nytt Magasin for Naturvidenskapene* **38**: 123 (1900).

For full synonymy see under *Ectrogella lauderiae* (H. H. Gran) A. Scherffel.

Typification: see under *Ectrogella lauderiae*.

Literature: see under *Ectrogella lauderiae*.

Olpidium laguncula H. E. Petersen, *Oversigt over det Kongelige Danske Videnskabernes Selskabs Forhandlinger. Medlemmers Arbeiter, 1905*: 466 (1905).

Typification: type material not verified. Type locality: Denmark. Type habitat: parasitic in *Dumontia contorta* (S. G. Gmel.) Rupr. [as its synonym *D. filiformis* Grev. (≡ *D. incrassata* J. V. Lamour.)] (*Rhodophyceae*).

Literature: Aleem (1953), Petersen (1905), Sparrow (1960: 149).

Olpidium marinum P.-A. Dangeard, *Le Botaniste, Série 12*: xvi (1912).

For full synonymy see under *Sirolpidium marinum* (P.-A. Dangeard) M. W. Dick *loc. cit*.

Typification: see under *Sirolpidium marinum*.

Literature: see under *Sirolpidium marinum*.

Olpidium *species parasitic (rarely saprobic) in the roots and leaves of higher plants (15 binomials and 1 variety)*

Olpidium simulans A. de Bary & M. Woronin, *Berichte der Naturforschenden Gesellschaft zu Freiburg* **3**: 50 (1865) [also *Annales des Sciences Naturelles, Botanique, Série V* **3**: 239-269 (1865)].
Chytridium simulans A. de Bary & M. Woronin, *Berichte der Naturforschenden Gesellschaft zu Freiburg* **3**: 50 (1865).
Typification: type material not designated; lectotype *Berichte der Naturforschenden Gesellschaft zu Freiburg* **3**: plate I, figs 11-16 (1865) (iconotype). Type locality: Germany. Type habitat: parasitic in leaves of *Taraxacum officinale* Wigg. (*Asteridae*, *Asteraceae*).
Literature: de Bary & Woronin (1865*a*, *b*), Fischer (1892*a*), Rabenhorst (1868), Schwartz & Ivimey Cook (1928).

Olpidium brassicae (M. Woronin) P.-A. Dangeard, *Annales des Sciences Naturelles, Botanique, Série VII* **4**: 327 (1886).
Chytridium brassicae M. Woronin, *Jahrbücher für Wissenschaftliche Botanik* **11**: 557 (1878) [*pro parte*,?].
Plasmodiophora brassicae M. Woronin, *Jahrbücher für Wissenschaftliche Botanik* **11**: 557 (1878) [*pro parte*,?].
Olpidium brassicae (M. Woronin) A. Fischer, *Rabenhorst's Kryptogamen-Flora, 2 Aufl., Bd 1, Abt. 4*: 28 (1892).
Pleotrachelus brassicae (M. Woronin) S. Sahtiyanci, *Archiv für Mikrobiologie* **41**: 198 (1962).
Asterocystis radicis É. de Wildeman, *Annales de la Société de Botanique de Belge de Microscopie, Mémoires* **17**: 21 (1893).
Pleotrachelus radicis É. de Wildeman, *Annales de la Société de Botanique de Belge de Microscopie, Mémoires* **17**: 23 (1893).
Olpidiaster radicis (É. de Wildeman) A. Pascher, *Beihefte Botanische Centralblatt* **35**: 578 (1917).
Olpidium borzii É. de Wildeman, *Annales de la Société de Botanique de Belge de Microscopie, Mémoires* **17**: 19 (1893).
Olpidium radicicola (É. de Wildeman) É. de Wildeman, *Annales de la Société de Botanique de Belge de Microscopie, Mémoires* **20**: 43 (1893) [as '*radicicolum*'].
Pleotrachelus virulentus S. Sahtiyanci, *Archiv für Mikrobiologie* **41**: 198 (1962).
Typification: type material not designated; lectotype *Fungi Canadenses No. 176*: figs 1-5 (1980) (iconotype); original illustrations: *Jahrbücher für Wissenschaftliche Botanik* **11**: plate XXXI, figs 12-18 (1878). Type locality: St Petersburg, Russia; lectotype locality: Bradford Marsh, Ontario, Canada. Type habitat: parasitic in *Brassica* roots (*Brassicales*, *Brassicaceae*); lectotype habitat: parasitic on cultivated *Daucus carota* L. subsp. *sativus* (Hoffm.) Arcangeli (*Apiales*, *Apiaceae*).
Recent and taxonomic literature: Barr (1980*b*), Barr & Slykhuis (1976), Bartlett (1928), Dangeard (1886*a*), de Bary & Woronin (1865*a*, *b*), Fischer (1892*a*), Haber, Barr & Platford (1991), Hiruki (1994), Ijdenberg, Kummert & Lepoivre (1986), Kassanis & Macfarlane (1964), Němec (1912), Sahtiyanci (1962), Sparrow (1960: 163), Teakle & Nienhaus (1974), Temmink & Campbell (1968), Temmink *et al.* (1970), Tomlinson, Zitter & Clay (1985), Wildeman (1893*a*, 1896*a*), Woronin (1878), Yačevskie & Yačevskie (1931).

> ***Olpidium lemnae*** C. Fisch, *Sitzungsberichte der Physicalische-medicinischen Societät zu Erlangen* **16**: 43 (1884).
> For full synonymy see under *Reesia lemnae* (C. Fisch) J. S. Karling, *American Journal of Botany* **30**: 642 (1943).
> **Typification**: see under *Reesia lemnae*.
> **Literature**: see under *Reesia lemnae*.

Olpidium trifolii (G. Passerini) J. Schroeter, *Kryptogamen-Flora von Schlesien [1885-1889]* **3(1)**: 180 (1886).
Synchytrium trifolii G. Passerini, *Giornale Botanico Italiano* **13**: 274 (1881).
[A *Synchytrium*, additional synonymy in Karling (1964*d*)].
Typification: type material not verified. Type locality: Italy. Type habitat: parasitic in *Trifolium pratense* L. (*Fabales*, *Fabaceae*).
Literature: Karling (1964*d*), Kusano (1929), Passerini (1881), Schroeter (1886*b*), Schwartz & Ivimey Cook (1928) [a doubtful species of *Physoderma* from a mixed fungal infection?].

Olpidium borzii É. de Wildeman, *Annales de la Société de Botanique de Belge de Microscopie, Mémoires* **17**: 19 (1893).
Typification: type material not designated. Type locality: Belgium. Type habitat: parasitic in root epidermal cells of *Brassica oleracea* L. and *Capsella bursa-pastoris* (L.) Medic. (*Brassicales*, *Brassicaceae*).
Literature: Barr (1980*b*), Wildeman (1893*a*) [synonym of *Olpidium brassicae* (M. Woronin) P.-A. Dangeard].

Olpidium radicicola (É. de Wildeman) É. de Wildeman, *Annales de la Société de Botanique de Belge de Microscopie, Mémoires* **20**: 43 (1896) [as '*radicicolum*'].
Pleotrachelus radicis É. de Wildeman, *Annales de la Société de Botanique de Belge de Microscopie, Mémoires* **17**: 23 (1893).
Typification: type material not designated. Type locality: Belgium. Type habitat: parasitic in roots of *Thalaspi arvense* L. (*Brassicales*, *Brassicaceae*).
Literature: Barr (1980*b*), de Wildeman (1896*b*, 1893*a*) [synonym of *Olpidium brassicae* (M. Woronin) P.-A. Dangeard].

Olpidium salicorniae B. Němec, *Rospavy České Akademie, Praha, Věd a Um cl 2* **20**: 2 (1911).
Typification: type material not verified. Type locality: Czech Republic (?). Type habitat: parasitic in *Salicornia stricta* Dum. [as '*S. herbacea*'] (*Caryophyllales, Chenopodiaceae*).
Literature: Němec (1911*c*), Schwartz & Ivimey Cook (1928).

Olpidium viciae A. Kusano, *Journal of the College of Agriculture, Imperial University of Tokyo* **4**: 183 (1912).
Typification: type material not verified. Type locality: Komaba, Tokyo, Japan. Type habitat: parasitic in *Vicia unijuga* Al. Br. (*Fabales, Fabaceae*).
Literature: Kusano (1912), Schwartz & Ivimey Cook (1928).

Olpidium radicale E. J. Schwartz & W. R. Ivimey Cook, *Transactions of the British Mycological Society* **13**: 216 (1928).
Olpidium cucurbitacearum D. J. S. Barr & H. F. Dias *in* D. J. S. Barr, *Canadian Journal of Botany* **46**: 1087 (1968).
Pleotrachelus bornovanus S. Sahtiyanci, *Archiv für Mikrobiologie* **41**: 216 (1962).
Typification: type material not designated; lectotype *Transactions of the British Mycological Society* **13**: plates 13-15 (1928). Type locality: Dunton Green, Kent, U.K. Type habitat: parasitic in roots of *Veronica beccabunga* L. (*Lamiales, Scrophulariaceae*), also experimentally induced into *Poa annua* L. and *Lepidium sativum* L.).
Literature: Barr (1968), Lange & Insunza (1977), Lange & Olson (1978), Ivimey Cook & Collins (1935), McLean *et al.* (1994), Sahtiyanci (1962), Schwartz & Ivimey Cook (1928), Stobbs *et al.* (1982).

Olpidium agrostis K. Sampson, *Transactions of the British Mycological Society* **17**: 192 (1932).
Typification: type material not designated; lectotype *Transactions of the British Mycological Society* **17**: 183, fig. 1; 185; figs 2-4; 186, fig. 5; plate X, figs 1-14; plate XI, figs 15-24; plate XII, figs 25-31 (1932) (iconotype). Type locality: Bingley, Yorkshire, U.K. Type habitat: parasitic in root hairs of *Agrostis stolonifer* L. (*Poaceae*).
Literature: Ivimey Cook & Collins (1935), Sampson (1932).

Olpidium majus W. R. Ivimey Cook & W. B. Collins, *Annales Mycologici* **33**: 76 (1935).
Typification: Type material deposited in the Department of Botany, University of Wales at Cardiff, Gwent, U.K. (also in the herbarium of W. R. Ivimey Cook). Type locality: Llandaff, Gwent, U.K. Type habitat: parasitic in *Cucumis sativus* L. (*Cucurbitales, Cucurbitaceae*) with a *Pythium* species.
Literature: Barr (1968), Ivimey Cook & Collins (1935).

Olpidium cucurbitacearum D. J. S. Barr & H. F. Dias *in* D. J. S. Barr, *Canadian Journal of Botany* **46**: 1087 (1968).
Typification: Iconotype *Canadian Journal of Botany* **46**: plate 1, figs 1-12; plate 2, figs 13-22 (1968) from living material isolate L5-B7-5; dried infected roots deposited at DAOM No 117739. Type locality: Harrow, Ontario, Canada. Type habitat: parasitic in roots of *Cucumis sativus* L. (also in *C. melo* L., *C. pepo* L., *C. maxima* L. and *Citrullus vulgaris* Schrad.) (*Cucurbitales, Cucurbitaceae*).
Literature: Barr (1968), Barr & Hadland-Hartman (1977), Lange & Insunza (1977).

Olpidium cruciferarum (B. Della Torre) M. W. Dick, **comb. nov**.
Olpidiopsis cruciferarum B. Della Torre, *Revista di Patologia vegetale, Padova, Serie 3* **4**: 83 (1964).
Typification: Iconotype *Rivista di Patologia vegetale, Padova, Serie 3* **4**: 81, fig. 1 (1964). Type locality: Italy. Type habitat: parasitic in rootlets of *Arabis alpina* L. (*Brassicales, Brassicaceae*).
Literature: Della Torre (1964), Peyronel (1924).

Olpidium verrucosum (B. Della Torre) M. W. Dick, **comb. et stat. nov.**
Olpidiopsis verrucosa Johnson var. *radicicola* B. Della Torre, *Revista di Patologia vegetale, Padova, Serie 3* **4**: 91 (1964)
Typification: Iconotype *Rivista di Patologia vegetale, Padova, Serie 3* **4**: 85, fig. 2; 85, fig. 3; 87, fig. 4; 88, fig. 5 (1964). Type locality: Italy. Type habitat: parasitic in rootlets of *Arabis alpina* L. (*Brassicales, Brassicaceae*).
Literature: Della Torre (1964).

Olpidium indicum U. Kiran & R. Dayal, *Proceedings of the National Academy of Sciences, India* **62**: 296 (1992) [*nom. invalid* Art. 37.4].
non Olpidium indianum J. S. Karling, *Sydowia* **17**: 303 (1964) [as '*indicum*', author's handwritten correction on reprint and later entry in printed text].
non Olpidium indicum N. Wallich *in* W. B. Turner, *Kongliga Svenska Vetenskaps-Akademiens Handlingar* **25** (**5**): 164 (1892).
Typification: Type material (slide No. 42) deposited at HCIO: 30575, New Delhi, and in the Department of Mycology and Plant Pathology, Institute of Agricultural Sciences, Banaras Hindu University, Varanasi, India (iconotype: *Proceedings of the National Academy of Sciences, India* **62**: plate 1, fig. 2 (1992)). Type locality: Ramkatora, Varanasi, Uttar Pradesh, India. Type habitat: "parasitic on root hairs of fallen angiospermic plants ... in pond water".
Literature: Kiran & Dayal (1992).

Olpidium rostriferum Y. Tokunaga, *Transactions of the Sapporo Natural History Society (Japan)* **13**: 80 (1933) var. ***indicum*** U. Kiran & R. Dayal, *Proceedings of the National Academy of Sciences, India* **62**: 296 (1992) [*nom. invalid* Art. 37.4] [as var. '*indica*'].
non Olpidium rostriferum Y. Tokunaga, *Transactions of the Sapporo Natural History Society (Japan)* **13**: 80 (1933) var. *rostriferum*.
Typification: Type material (slide No. 43) deposited at HCIO: 30572, New Delhi, and in the Department of Mycology and Plant Pathology, Institute of Agricultural Sciences, Banaras Hindu University, Varanasi, India (iconotype: *Proceedings of the National Academy of Sciences, India* **62**: plate 1, fig. 3 (1992)). Type locality: Ramkatora, Varanasi, Uttar Pradesh, India. Type habitat: saprobic "on angiospermic leaves ... in pond water".
Literature: Kiran & Dayal (1992).

Olpidium *species parasitic in fungi (7 binomials)*

Olpidium saprolegniae (A. Braun) L. Rabenhorst, *Flora Europaea Algarum* **3**: 283 (1868) [synonym not given in Sparrow (1960)].
Chytridium saprolegniae A. Braun, *Bericht über die zur Bekanntmachung geeigneten verhandlungen der Königlichen Preussischen Akademie der Wissenschaften zu Berlin, 1855*: 384 (1855).
For later synonyms see under *Olpidiopsis saprolegniae* (A. Braun) M. Cornu.
Typification: see under *Olpidiopsis saprolegniae*.
Literature: see under *Olpidiopsis saprolegniae*.

Olpidium uredinis ([N.] G. Lagerheim) A. Fischer, *Rabenhorst's Kryptogamen-Flora, 2 Aufl., Bd 1, Abt. 4*: 30 (1892).
Olpidiella uredinis [N.] G. Lagerheim, *Journal de Botanique* **2**: 438 (1888).
Typification: type material not designated; lectotype *Journal de Botanique* **2**: plate X, figs 1-15 (1888) (iconotype). Type locality: Friburg-Brisgau, Germany. Type habitat: parasitic in uredospores of *Uredo airae* G. Lagerheim [binomial established in a later paper], *Puccinia violae* (Schumach.) DC and *P. coronata* Corda (as *P. rhamni* (Pers.) Wettst. (*Uredinales*).
Literature: Fischer (1892), Lagerheim (1888), Sparrow (1960: 142), Wildeman (1896*a*).

Olpidium borzianum F. Morini *in* A. N. Berlese & J. B. DeToni *in* P. A. Saccardo, *Sylloge Fungorum...* VII 1- 312 (1888?)
non Olpidium borzii É. de Wildeman, *Annales de la Société de Botanique de Belge de Microscopie, Mémoires* **17**: 19 (1893).
Typification: type material not verified. Type locality: Bologna, Italy. Type habitat: parasitic in *Saprolegnia* sp. (*Saprolegniaceae*).
Literature: Fischer (1892: 31), Minden (1911*a*), not in Sparrow (1960), Wildeman (1896*a*).

Olpidium pseudomorphum A. Scherffel, *Archiv für Protistenkunde* **54**: 510 (1926).
For full synonymy see under *Rozella pseudomorpha* (A. Scherffel) F. K. Sparrow, *Aquatic Phycomycetes*: 124 (1943).
Typification: see under *Rozella pseudomorpha* [excluded species of *Rozella*].
Literature: see under *Rozella pseudomorpha*.

Olpidium allomycetos J. S. Karling, *American Journal of Botany* **35**: 508 (1948).
Typification: type material not designated; lectotype *American Journal of Botany* **35**: 5, figs 1-32 (1948) (iconotype). Type locality: Baton Rouge, Louisiana, U.S.A. Type habitat: parasitic in *Allomyces anomalus* R. Emerson (and by cross infection in *A. arbuscula* E. J. Butler, *A.* × *javanicus* Kniep, *A. cystogenes* R. Emers., *A. moniliformis* Coker & Braxton, *Rhizophlyctis rosea* (de Bary & Woronin) A. Fisch. and rotifer eggs).
Literature: Karling (1948), Sparrow (1952*a*, 1960: 143).

Olpidium rhizophlyctidis F. K. Sparrow, *Mycologia* **40**: 449 (1948).
Typification: type material not designated; lectotype *Mycologia* **40**: 448, figs 7, 8; 451, figs 12, 14, 18, 19 (1948) (iconotype). Type locality: Marshall Islands, Pacific Ocean. Type habitat: parasitic in *Rhizophlyctis* spp. (*Chytridiomycetes*).
Literature: Sparrow (1952*a*, 1960: 144).

Olpidium indianum J. S. Karling, *Sydowia* **17**: 303 (1964) [as '*indicum*', author's handwritten correction "*indianum*" on reprint, and see later correctly spelt entry in printed text].
non Olpidium indicum N. Wallich *in* W. B. Turner, *Kongliga Svenska Vetenskaps-Akademiens Handlingar* **25 (5)**: 164 (1892).
non Olpidium indicum U. Kiran & R. Dayal, *Proceedings of the National Academy of Sciences, India* **62**: 296 (1992) [*nom. invalid* Art. 37.4].
Typification: Type slide PU No. 103 [Purdue University, Indiana, U.S.A.]. Type locality: Rhamnad District, Madras State, India. Type habitat: parasitic in thalli of *Phlyctorhiza variabilis* Karling and *Rhizophlyctis fuscis* Karling (*Chytridiomycetes*).
Literature: Karling (1964*c*).

Olpidium *species in pollen (6 binomials)*

Olpidium pendulum W. Zopf *in* A. Schenk, *Handbuch der Botanik, vol. 4*: 555 (1890).
Typification: type material not designated; lectotype *Handbuch der Botanik* (ed. A. Schenk), vol. 4: 556, fig. 66, I-V (1890) (iconotype). Type locality: Germany. Type habitat: in pine pollen.
Literature: Fischer (1892), Gaertner (1954*b*), Graff (1928), Karling (1981*b*), Konno (1972), Sparrow (1952, 1957, 1960: 140), Voronichin (1920), Wildeman (1896*a*), Zopf (1890).

Olpidium luxurians (A. Tomaschek) A. Fischer, *Rabenhorst's Kryptogamen-Flora, 2 Aufl., Bd 1, Abt. 4*: 29 (1892). (1890).
Chytridium luxurians A. Tomaschek, *Sitsungsbericht der Akademie der Wissenschaften, Mathematische-naturwissenschaftliche Klasse,* Wien **78**: 204 (1878-9).
Chytridium diplochytrium A. Tomaschek, *Sitsungsbericht der Akademie der Wissenschaften, Mathematische-naturwissenschaftliche Klasse, Wien* **78**: 204 (1878-9).
Chytridium pollinis-typhae forma *latifoliae* A. Tomaschek, *Sitsungsbericht der Akademie der Wissenschaften, Mathematische-naturwissenschaftliche Klasse, Wien* **78**: 203 (1878-9).
Olpidium diplochytrium (A. Tomaschek) J. Schroeter *in* F. Cohn, *Kryptogamen-Flora von Schlesien* **3(1)**: 181 (1885).
Diplochytrium sp. A. Tomaschek, *Sitsungsbericht der Akademie der Wissenschaften, Mathematische-naturwissenschaftliche Klasse, Wien* **78**: 198 (1878-1879).
Olpidiella diplochytrium [N.] G. Lagerheim, *Journal de Botanique* **2**: 439 (1888).
Typification: type material not verified. Type locality: Germany. Type habitat: in pine pollen (and pollen of other gymnosperms and angiosperms).
Literature: Fischer (1892), Gaertner (1954*b*), Graff (1928), Karling (1981*b*), Sparrow (1952, 1957, 1960: 142), Tomaschek (1878-1879), Voronichin (1920), Wildeman (1896*a*), Zopf (1890).

Olpidium maritimum W. Höhnk & A. A. Aleem, *Veröffentlich des Instituts für Meeresforschung in Bremerhaven* **2**: 227 (1953).
Typification: type material *Veröffentlich des Instituts für Meeresforschung in Bremerhaven* **2**: 227, figs 1-7 (1953) (iconotype). Type locality: Germany. Type habitat: saprobic in pine pollen baitings of samples of brackish soil.
Literature: Höhnk & Aleem (1953), Sparrow (1960: 141).

Olpidium longicollum E.-R. Uebelmesser, *Archiv für Mikrobiologie* **25**: 309 (1956).
Olpidium sp. R. Harder & E. Uebelmesser, *Archiv für Mikrobiologie* **22**: 89 (1955).
Typification: type material not designated: lectotype *Archiv für Mikrobiologie* **25**: 309, fig. 1 (1956) (iconotype). Type locality: North Sea (Wangerooge). Type habitat: saprobic in pine pollen.
Literature: Harder & Uebelmesser (1955), Sparrow, Paterson & Johns (1965), Uebelmesser (1956).

Olpidium appendiculatum J. S. Karling, *Sydowia* **19**: 214 (1965).
Typification: type material not designated; lectotype *Sydowia* **19**: plate XLVI, figs 1-13 (1965) (iconotype). Type locality: New Zealand. Type habitat: saprobic in pollen of *Pinus* and *Phyllocladus* added with water to sand and moss sample.
Literature: Karling (1965).

Olpidium ramosum I. L. Serbinov, *Trudy. St. Petersburg. Obschestvo estestvoispytatelei* **30**: 255 (1899).
Typification: Type material not determined. Type locality: Russia. Type habitat: saprobic in *Pinus* pollen.
Literature: Serbinow (1899), Sparrow (1960: 157).

Olpidium *species parasitic in rotifers & nematodes (12 binomials and 1 variety)*

Olpidium entophytum (A. Braun) L. Rabenhorst, *Flora Europaea Algarum* **3**: 283 (1868) var. *intermedium* J. C. Constantineanu, *Revue Générale de Botanique* **13**: 371 (1901).
Typification: type material not designated; lectotype *Lloydia* **9**: 5, figs 1-18 (1946) (iconotype). Type locality: San Carlos, Matto Grosso, Brazil. Type habitat: parasitic in rotifer eggs (*Aschelminthes, Rotifera*).
Literature: Constantineanu (1901), Karling (1946), Sparrow (1960: 137).

Olpidium zootoctum (A. Braun) N. Sorokin, *Archives Botaniques du Nord de la France* **2**: 31 (1883).
Chytridium zootoctum A. Braun, *Bericht über die zur Bekanntmachung geeigneten verhandlungen der Königlichen Preussischen Akademie der Wissenschaften zu Berlin, 1855*: 591 (1856).
Typification: type material not verified. Type locality: Germany. Type habitat: parasitic in *Anguillula* sp. (*Aschelminthes, Nematoda*).
Literature: Sorokine (1883, 1889), Sparrow (1960: 154), Wildeman (1896*a*).

Olpidium gregarium (L. Nowakowski) J. Schroeter, *Kryptogamen-Flora von Schlesien* **3(1)**: 182 (1885).
Chytridium gregarium L. Nowakowski, *Beiträge zur Biologie der Pflanzen* **2**: 77 (1876).
Typification: type material not verified. Type locality: Wrocław (Breslau), Poland. Type habitat: parasitic in rotifer eggs (*Aschelminthes, Rotifera*).
Literature: Bérczi (1940), Butler (1907), Fischer (1892), Karling (1941?), Konno (1972), Nowakowski (1876), Petersen (1909, 1910), Schroeter (1885), Sparrow (1936*b*, 1957, 1960: 144), Sparrow & Barr (1955), Tokunaga (1933*b*), Wildeman (1896*a*).

Olpidium macrosporum (L. Nowakowski) J. Schroeter, *Kryptogamen-Flora von Schlesien* **3(1)**: 182 (1885).
Chytridium macrosporum L. Nowakowski, *Beiträge zur Biologie der Pflanzen* **2**: 79 (1876).
Typification: type material not verified. Type locality: Wrocław (Breslau), Poland. Type habitat: parasitic in rotifer eggs (*Aschelminthes, Rotifera*).
Literature: Fischer (1892), Nowakowski (1876), Schroeter (1885), Sparrow (1960: 155), Wildeman (1896*a*).

Olpidium nematodeae B. W. Skvortzow, *Archiv für Protistenkunde* **57**: 204 (1927).
Typification: type material not designated. Type locality: Manchuria. Type habitat: parasitic in nematodes (*Aschelminthes, Nematoda*).
Literature: Korab (1929), Rozsypal (1934), Skvortzow (1927), Sparrow (1960: 147), Tribe (1977*a*, *c*).

Olpidium granulatum J. S. Karling, *Lloydia* **9**: 3 (1946).
Typification: type material not designated; lectotype *Lloydia* **9**: 5, figs 1-18 (1946) (iconotype). Type locality: San Carlos, Matto Grosso, Brazil. Type habitat: parasitic in rotifer eggs (*Aschelminthes, Rotifera*).
Literature: Karling (1946), Kiran & Dayal (1992), Sparrow (1960: 145).

Olpidium rotiferum J. S. Karling, *Lloydia* **9**: 6 (1946).
Typification: type material not designated; lectotype *Lloydia* **9**: 5, figs 19-34 (1946) (iconotype). Type locality: Brazil. Type habitat: parasitic in adult rotifers and rotifer eggs (*Aschelminthes, Rotifera*).
Literature: Karling (1946*b*), Sparrow (1960: 146).

Olpidium sparrowii I. J. Dogma, *Kalikasan, Philippine Journal of Biology* **6**: 10 (1977).
Typification: Iconotype *Kalikasan, Philippine Journal of Biology* **6**: 15, figs 1-16 (1977); semipermanent mounts nos. 806-808 in author's collection. Type locality: Putiao, Sorsogon Province, Luzon, Philippine Islands (coll. M. B. Castillo, 13 Jun 1973). Type habitat: on eggs of *Lecane* species (*Aschelminthes, Rotifera*) from cultivated soil.
Literature: Dogma (1977).

Olpidium vermicola G. L. Barron & E. Szijarto, *Mycologia* **78**: 972 (1986).
Typification: Type slides in Herbarium OAC (10841). Type locality: Lake Manitowabing, Ontario, Canada. Type habitat: endoparasitic in nematode eggs (*Aschelminthes, Nematoda*) from rotting wood.
Literature: Barron & Szijtarto (1986).

Olpidium longum S. N. Dasgupta & R. John, *Bulletin of the Botanical Survey of India* **30**: 4 (1988).
Typification: Holotype deposited at HCIO, Division of Mycology and Plant Pathology, IARI, New Delhi-110 012. HCIO No. 37088. Type locality: Lucknow, India. Type habitat: "parasitic in the eggs of certain [aquatic] animalcules".
Literature: Dasgupta & John (1988).

Olpidium incognitum S. N. Dasgupta & R. John, *Bulletin of the Botanical Survey of India* **30**: 5 (1988).
Typification: Holotype deposited at HCIO, Division of Mycology and Plant Pathology, IARI, New Delhi-110 012. HCIO No. 37089. Type locality: Lucknow, India. Type habitat: "parasitic in the egg of a rotifer" (*Aschelminthes, Rotifera*).
Literature: Dasgupta & John (1988).

Olpidium poreferum U. Kiran & R. Dayal, *Proceedings of the National Academy of Sciences, India* **62**: 295 (1992) [*nom. invalid* Art. 37.4].
Typification: Type material (slide No. 41) deposited at HCIO: 30576, New Delhi, and in the Department of Mycology and Plant Pathology, Institute of Agricultural Sciences, Banaras Hindu University, Varanasi, India (iconotype: *Proceedings of the National Academy of Sciences, India* **62**: plate 1, fig. 1 (1992)). Type locality: Agriculture Farm pond, Varanasi, Uttar Pradesh, India. Type habitat: parasitic on rotifers (*Aschelminthes, Rotifera*).
Literature: Kiran & Dayal (1992).

Olpidium paradoxum S. L. Glockling, *Mycological Research* **101**: 207 (1997).
Typification: Type material: slides deposited at IMI 273919. Type locality: Forestry & Forest Products Institute, Ibaraki, Japan. Type habitat: from pond water.
Literature: Dick & Glockling (1998); Glockling (1997).

See also *Olpidium allomycetos* J. S. Karling.

Olpidium *species parasitic in protoctists (8 binomials)*

Olpidium arcellae N. Sorokin, *Archives Botaniques du Nord de la France* **2**: 32 (1883).
Typification: type material not designated, lectotype *Archives Botaniques du Nord de la France* **2**: 32, fig. 40 (1883) (iconotype). Type locality: Tashkent, Asiatic Russia. Type habitat: saprotrophic on *Arcella* sp. (*Protoctista*).
Literature: Sorokine (1883, 1889), Sparrow (1960: 154), Wildeman (1896*a*).

Olpidium vampyrellae A. Scherffel, *Archiv für Protistenkunde* **54**: 168 (1926).
Typification: type material not designated; lectotype *Archiv für Protistenkunde* **54**: plate 9, figs 2-5 (1926) (iconotype). Type locality: Igló, Hungary. Type habitat: parasitic in zoocysts of *Vampyrella* spp. (*Protoctista*).
Literature: Scherffel (1926*a*), Sparrow (1960: 146).

Olpidium pseudosporeanum A. Scherffel, *Archiv für Protistenkunde* **54**: 170 (1926).
Typification: type material not designated; lectotype *Archiv für Protistenkunde* **54**: plate 9, figs 6-7 (1926) (iconotype). Type locality: Hungary. Type habitat: parasitic in zoocysts of *Pseudosporopsis bacillariacearum* Scherffel and *Pseudospora parasitica* Cienkowski (*Protoctista*).
Literature: Scherffel (1926*a*), Sparrow (1960: 146).

Olpidium difflugiae A. Scherffel, *Archiv für Protistenkunde* **54**: 168 (1926).
Typification: type material not designated; lectotype *Archiv für Protistenkunde* **54**: plate 9, fig 1 (1926). Type locality: Hungary. Type habitat: parasitic in *Difflugia* sp. (*Protoctista*).
Literature: Scherffel (1926*a*), Sparrow (1960: 154).

Olpidium leptophrydis A. Scherffel, *Archiv für Protistenkunde* **54**: 171 (1926).
Typification: type material not designated; lectotype *Archiv für Protistenkunde* **54**:plate fig 171 (1926) (iconotype). Type locality: Hungary. Type habitat: parasitic in zoocysts of *Leptophrys vorax* (Cienk.) (*Protoctista*).
Literature: Scherffel (1926*a*), Sparrow (1960: 154).

Olpidium ellipticum (P.-A. Dangeard) P. A. Saccardo & G. B. Traverso, *Sylloge Fungorum* **20**: 217 (1911) [excluded: a monad which ingests solid particles].
Minutularia elliptica P.-A. Dangeard, *Le Botaniste* **2**: 241 (1890-91).
Typification: type material not verified.
Literature: Dangeard (1890-91), Saccardo & Traverso (1911), Sparrow (1960: 156).

Other chytrids on rotifer eggs

Phlyctochytrium nematodae J. S. Karling, *Lloydia* **9**: 7 (1946).
Typification: type material not designated; lectotype *Lloydia* **9**: 9, figs 35-38 (1946) (iconotype). Type locality: San Carlos, Matto Grosso, Brazil. Type habitat: parasitic in adults, eggs and cysts of nematodes.
Literature: Karling (1946), Sparrow (1960: 337).

Rhizophydium zoophthorum (P.-A. Dangeard) A. Fischer, *Rabenhorst's Kryptogamen-Flora, 2 Aufl., Bd 1, Abt. 4*: 94 (1892).
Chytridium zoophthorum P.-A. Dangeard, *Bulletin de la Société Botanique de France* **34**: xxii (1887), also *Le Botaniste* **1**: plate 3, figs 10, 21 (1889).
Typification: type material not designated; lectotype *Le Botaniste* **1**: plate 3, figs 10, 21 (1889) (iconotype). Type locality: France. Type habitat: in cultures of rotifers (adults and eggs).
Literature: Dangeard (1887), Fischer (1892*a*), Karling (1946), Minden (1911*a*), Sparrow (1960: 285), Willoughby (1959).

Rhizophydium gibbosum (W. Zopf) A. Fischer, *Rabenhorst's Kryptogamen-Flora, 2 Aufl., Bd 1, Abt. 4*: 102 (1892).
Rhizophyton gibbosum W. Zopf, *Nova Acta der Kaiserlichen Leopoldinisch-Carolinischen Deutschen Akademie der Naturforscher* **52**: 344 (1888).
Typification: plate 20, figs 8-20 (1888) (iconotype). Type locality: Germany. Type habitat: parasitic on rotifer eggs (and parasitic on various algae and nematodes according to other authors).
Literature: Karling (1946), Rieth (1954), Scherffel (1904), Sparrow (1960: 299), Zopf (1888).

Endochytrium operculatum (É. de Wildeman) J. S. Karling, *American Journal of Botany* **24**: 353 (1937).
Rhizophlyctis operculata É. de Wildeman, *Annales de la Société de Botanique de Belge de Microscopie, Mémoires* **19**: 108 (1895).
? *Entophlyctis maxima* P.-A. Dangeard, *Le Botaniste* **24**: 242 (1932).
Typification: type material not designated; lectotype *American Journal of Botany* **24**: 354, figs 1-38; 357, figs 39-53 (1937) (iconotype). Type locality: France Belgium. Type habitat: saprotrophic in tissues of higher plants (also in various algae, protists and rotifer and nematode eggs according to later reports).
Literature: Dangeard (1932), Karling (1937, 1946), Sparrow (1960: 569), Wildeman (1895).

Aquatic Synchytriaceae *parasitic in algae*

Karling (1967*a*) stated that the genus *Synchytrium*, with over 200 species, was the largest genus of the *Chytridiomycetes* and type family of the *Synchytriaceae*. Since Karling (1964*d*) provided a monograph of this genus, it is not appropriate to provide the formal taxonomy for all these species here. However, I give below the formal taxonomy of all species published after 1964. The generic boundaries of the parasites of algae are still in dispute: Sparrow (1960) placed the genus *Micromycopsis* in synonymy with *Micromyces*, Karling (1964*d*) placed the genus *Micromyces* as part of a subgenus of *Synchytrium*, but Subramanian (1974) retained the separate genera, and made *Micromycopsis* the type of a new family, the *Micromycopsidaceae*. For convenience, I follow the generic concepts of Sparrow (1960), but give the formal taxonomy of the genus and subgenera of Karling since this was not given in full in Karling (1964*d*).

SYNCHYTRIUM A. de Bary & M. Woronin, *Berichte der Naturforschenden Gesellschaft zu Freiburg* **3**: 46 (1865).
Micromyces P.-A. Dangeard, *Le Botaniste* **1**: 55 (1889) (**Type species**: *Micromyces zygogonii* P.-A. Dangeard).
Pycnochytrium (A. de Bary) J. Schröter, *Die Natürlichen Pflanzenfamilien* **1(1)**: 73 (1893) (**Type species**: *Pycnochytrium succisae* (A. de Bary & M. Woronin) J. Schröter).
Pycnochytrium (A. de Bary) J. Schröter, *Die Natürlichen Pflanzenfamilien* **1(1)**: 73 (1893) subgenus *Mesochytrium* J. Schröter, *Die Natürlichen Pflanzenfamilien* **1(1)**: 73 (1893) (**Type species**: *Pycnochytrium succisae* (A. de Bary & M. Woronin) J. Schröter).
Pycnochytrium (A. de Bary) J. Schröter, *Die Natürlichen Pflanzenfamilien* **1(1)**: 73 (1893) subgenus *Chrysochytrium* J. Schröter, *Die Natürlichen Pflanzenfamilien* **1(1)**: 74 (1893) (**Type species**: *Pycnochytrium aureum* (J. Schröter) J. Schröter).
Pycnochytrium (A. de Bary) J. Schröter, *Die Natürlichen Pflanzenfamilien* **1(1)**: 73 (1893) subgenus *Leucochytrium* J. Schröter, *Die Natürlichen Pflanzenfamilien* **1(1)**: 74 (1893) (**Type species**: *Pycnochytrium anemones* (A. De Candolle) J. Schröter).
Chrysophlyctis K. Schilberszky, *Berichte der Deutschen Botanischen Gesellschaft* **14**: 37 (1896) (**Type species**: *Chrysophlyctis endobiotica* K. Schilberszky - holotype, sole original species).
Woroninella M. Raciborski, *Zeitschrift für Pflanzenkrankheiten und Pflanzenschutz* **8**: 195 (1898) (**Type species**: *Woroninella psophocarpi* M. Raciborski - holotype, sole original species).
Miyabella S. Ito & Y. Homma, *Botanical Magazine (Tokyo)* **40**: 110-113 (1926) (**Type species**: *Miyabella aecidioides* (Peck) S. Ito & Y. Homma).
Typification: *Synchytrium taraxaci* A. de Bary & M. Woronin.

SUBGENERA

Synchytrium A. de Bary & M. Woronin, *Berichte der Naturforschenden Gesellschaft zu Freiburg* **3**: 46 (1865) subgenus ***Synchytrium***.
Eusynchytrium A. de Bary, *Vergleichende Morphologie und Biologie der Pilze, Mycetozoan und Bacterien*: 180 (1884) [p. 167 in English translation].
Typification: *Synchytrium taraxaci* A. de Bary & M. Woronin.

Synchytrium A. de Bary & M. Woronin, *Berichte der Naturforschenden Gesellschaft zu Freiburg* **3**: 46 (1865) subgenus ***Microsynchytrium*** J. S. Karling, *Mycologia* **50**: 374 (1958).
Micromyces P.-A. Dangeard, *Le Botaniste* **1**: 55 (1889) (**Type species**: *Micromyces zygogonii* P.-A. Dangeard - holotype, sole original species).
Typification: not designated, first-named species *Synchytrium zygogonii* (P.-A. Dangeard) J. S. Karling.

Synchytrium A. de Bary & M. Woronin, *Berichte der Naturforschenden Gesellschaft zu Freiburg* **3**: 46 (1865) subgenus ***Mesochytrium*** J. Schröter, *Die Natürlichen Pflanzenfamilien* **1(1)**: 73 (1893).
Pycnochytrium (A. de Bary) J. Schröter, *Die Natürlichen Pflanzenfamilien* **1(1)**: 73 (1893) subgenus *Mesochytrium* J. Schröter, *Die Natürlichen Pflanzenfamilien* **1(1)**: 73 (1893) (**Type species**: *Synchytrium succisae* (A. de Bary & M. Woronin) J. Schröter).
Chrysophlyctis K. Schilberszky, *Beriche der Deutschen Botanischen Gesellschaft* **14**: 37 (1896) (**Type species**: *Chrysophlyctis endobiotica* K. Schilberszky - holotype, sole original species).
Typification: *Synchytrium succisae* A. de Bary & M. Woronin.

Synchytrium A. de Bary & M. Woronin, *Berichte der Naturforschenden Gesellschaft zu Freiburg* **3**: 46 (1863) subgenus ***Exosynchytrium*** J. S. Karling, *Mycologia* **50**: 375 (1958).
Typification: *Synchytrium callirroeae* J. S. Karling.

Synchytrium A. de Bary & M. Woronin, *Berichte der Naturforschenden Gesellschaft zu Freiburg* **3**: 46 (1863) subgenus ***Pycnochytrium*** A. de Bary, *Vergleichende Morphologie und Biologie der Pilze, Mycetozoan und Bacterien*: 180 (1884) [p. 167 in English translation].

Pycnochytrium (A. de Bary) J. Schröter, *Die Natürlichen Pflanzenfamilien* **1(1)**: 73 (1893) subgenus *Chrysochytrium* J. Schröter, *Die Natürlichen Pflanzenfamilien* **1(1)**: 74 (1893) (**Type species**: *Pycnochytrium aureum* (J. Schröter) J. Schröter).
Pycnochytrium (A. de Bary) J. Schröter, *Die Natürlichen Pflanzenfamilien* **1(1)**: 73 (1893) subgenus *Leucochytrium* J. Schröter, *Die Natürlichen Pflanzenfamilien* **1(1)**: 74 (1893) (**Type species**: *Pycnochytrium anemones* (A. De Candolle) J. Schröter).
Typification: *Synchytrium aureum* (J. Schröter) J. Schröter.

Synchytrium A. de Bary & M. Woronin, *Berichte der Naturforschenden Gesellschaft zu Freiburg* **3**: 46 (1863) subgenus ***Woroninella*** (M. Raciborski) J. S. Karling, *Synchytrium*: 324 (1964).
Woroninella M. Raciborski, *Zeitschrift für Pflanzenkrankheiten und Pflanzenschutz* **8**: 195 (1898) (**Type species**: *Woroninella psophocarpi* M. Raciborski - holotype, sole original species).
Miyabella S. Ito & Y. Homma, *Botanical Magazine (Tokyo)* **40**: 110-113 (1926) (**Type species**: *Miyabella aecidioides* (Peck) S. Ito & Y. Homma).
Typification: type species not designated, priority of generic synonymy would indicate *Synchytrium psophocarpi* M. Raciborski.

Synchytrium - Micromyces synonymy

Synchytrium longispinosus (J. N. Couch) J. S. Karling, *Mycologia* **45**: 278 (1953).
Micromyces longispinosus J. N. Couch, *Mycologia* **29**: 592 (1937).
Micromyces zygogonii P.-A. Dangeard *sensu* J. N. Couch, *Journal of the Elisha Mitchell Scientific Society* **46**: 231 (1931).
Olpidiopsis sp. Y. Kobayas[h]i & M. Ookubo, *Reports of the Ozegahara General Scientific Survey Commission 1954*: 571 (1954) [Sparrow, 1960: 199].
Typification: type material not designated; lectotype *Mycologia* **29**: 593 figs 9-14 (1937) (iconotype). Type locality: Cold Spring Harbor, Long Island, New York, U.S.A. Type habitat: parasitic in *Spirogyra* sp. (*Zygnematales*, *Zygnemataceae*).
Literature: Couch (1931), Karling (1953*a*, 1964*d*), Kobayas[h]i & Ookubo (1954*a*?), Linder (1947), Sparrow (1960: 198), Sparrow, Paterson & Johns (1965).

Synchytrium ovale (A. Rieth) J. S. Karling, *Mycologia* **45**: 278 (1953).
Micromyces ovalis A. Rieth, *Österreichische Botanische Zeitschrift* **97**: 516 (1950).
var. ***ovale***.
Typification: type material not verified. Type locality: Tübingen, Germany. Type habitat: parasitic in *Mougeotia* sp. (*Zygnematales*, *Zygnemataceae*).
Literature: Karling (1953*a*, 1964*d*), Miller (1965), Rieth (1950), Sparrow (1960: 200).

Synchytrium ovale (Rieth) J. S. Karling, *Mycologia* **45**: 278 (1953) var. ***giganteum*** (F. K. Sparrow & M. E. Barr) J. S. Karling, *Synchytrium*: 124 (1964).
Micromyces ovalis A. Rieth, *Österreichische Botanische Zeitschrift* **97**: 516 (1950) var. *giganteus* F. K. Sparrow & M. E. Barr, *Mycologia* **47**: 546 (1955).
Typification: type material not designated; lectotype *Mycologia* **47**: 548, fig 21 (1955) (iconotype). Type locality: 'Smith's Bog', Cheboygan County, Michigan, U.S.A. Type habitat: parasitic in *Zygnema* sp. (*Zygnematales*, *Zygnemataceae*).
Literature: Karling (1964*d*), Rieth (1950), Sparrow (1960: 200), Sparrow & Barr (1955).

Synchytrium grande (C. E. Miller) J. S. Karling, *Synchytrium*: 124 (1964).
Micromyces grandis C. E. Miller, *Journal of the Elisha Mitchell Scientific Society* **71**: 247 (1955).
Typification: type material not designated; lectotype *Journal of the Elisha Mitchell Scientific Society* **71**: 248, figs 1-9; 250, figs 10-29 (1955) (iconotype). Type locality: University Lake, Orange County, North Carolina, U.S.A. (11 Nov 1953). Type habitat: parasitic in *Spirogyra* sp. (*Zygnematales*, *Zygnemataceae*).
Literature: Karling (1964*d*), Miller (1955), Rieth (1956), Sparrow (1960: 204).

Recent Synchytrium species not in Karling (1964*d*)

Synchytrium cylistae S. D. Patil & T. S. Mahabalé, *Journal of the University of Poona* **26**: 69 (1964).
Typification: type material deposited in the Herbarium, Poona University, Poona, India. Type locality: Khandala, District of Poona, India. Type habitat: parasitic in *Paracalyx scariosus* (Roxb.) Ali [as its synonym *Cylista scariosa* Roxb.] (*Fabales*, *Fabaceae*, *Papilionoideae*).
Literature: Patil & Mahabalé (1964).

Synchytrium balsaminae S. D. Patil & T. S. Mahabalé, *Journal of the University of Poona* **26**: 70 (1964) [as '*balsimini*'].
Typification: type material deposited in the Herbarium, Poona University, Poona, India. Type locality: Sinhgad, District of Poona, India. Type habitat: parasitic in *Impatiens dalzellii* Hook. f. & Thoms. (*Geraniales, Balsaminaceae*).
Literature: Patil & Mahabalé (1964).

Synchytrium desmodii S. D. Patil & T. S. Mahabalé, *Journal of the University of Poona* **26**: 72 (1964) [*nom. superfl.*, probably a synonym of *Synchytrium desmodiae* Munasinghe on other species parasitic on *Desmodium*].
Typification: type material deposited in the Herbarium, Poona University, Poona, India. Type locality: Mangale, District of Sangli, India. Type habitat: parasitic in *Desmodium diffusum* DC (*Fabales, Fabaceae*).
Literature: Patil & Mahabalé (1964).

Synchytrium haplanthi S. D. Patil & T. S. Mahabalé, *Journal of the University of Poona* **26**: 73 (1964).
Typification: type material deposited in the Herbarium, Poona University, Poona, India. Type locality: Khandala, District of Poona, India. Type habitat: parasitic in *Haplanthodes verticillaris* (Roxb.) R. B. Majumdar [as its synonym, *Haplanthus verticillaris* (Roxb.) Nees] (*Lamiales, Acanthaceae*).
Literature: Patil & Mahabalé (1964).

Synchytrium pogostemonis S. D. Patil & T. S. Mahabalé, *Journal of the University of Poona* **26**: 74 (1964).
Typification: type material deposited in the Herbarium, Poona University, Poona, India. Type locality: Khandala, District of Poona, India. Type habitat: parasitic in *Pogostemon parviflorus* Benth. (*Lamiales, Labiatae*).
Literature: Patil & Mahabalé (1964).

Synchytrium senecionis S. D. Patil & T. S. Mahabalé, *Journal of the University of Poona* **26**: 75 (1964).
Typification: type material deposited in the Herbarium, Poona University, Poona, India. Type locality: Khandala, District of Poona, India. Type habitat: parasitic in *Senecio grahamii* Hook. f. (*Asterideae, Asteraceae*).
Literature: Patil & Mahabalé (1964).

Synchytrium hemigraphidis S. D. Patil & T. S. Mahabalé, *Journal of the University of Poona* **26**: 76 (1964).
Typification: type material deposited in the Herbarium, Poona University, Poona, India. Type locality: Bhaja, District of Poona, India. Type habitat: parasitic in *Hemigraphis latebrosa* Nees (*Lamiales, Acanthaceae*).
Literature: Patil & Mahabalé (1964).

Synchytrium tragiae S. D. Patil & T. S. Mahabalé, *Journal of the University of Poona* **26**: 77 (1964).
Typification: type material deposited in the Herbarium, Poona University, Poona, India. Type locality: Khandala, District of Poona, India. Type habitat: parasitic in *Tragia involucrata* L. (*Malpighiales, Euphorbiaceae*).
Literature: Patil & Mahabalé (1964).

Synchytrium kumaonense S. C. Gupta & B. C. L. Sah, *Indian Phytopathology* **16**: 378 (1964) [as '*kumaonensis*'].
Typification: type material deposited at the Herb. Crypt. Indiae Orient., I.A.R.I., New Delhi, India, and also at IMI, Egham, Surrey, U.K. Type locality: Naini Tal, Kumaon, India. Type habitat: parasitic in *Dipsacus inermis* Wall. (*Dipsacales, Dipsacaceae*).
Literature: Gupta & Sah (1964).

Synchytrium hypochoeridis J. S. Karling, *Sydowia* **20**: 51 (1967).
Typification: Type specimen in PDD, Mount Albert, Auckland, New Zealand, isotype in PU (Illinois, U.S.A.). Type locality: Sharp's Bush, Auckland, New Zealand (also collected in Australia). Type habitat: parasitic in *Hypochoeris radicata* L. (*Asterideae, Asteraceae*).
Literature: Karling (1967).

Synchytrium leontodontis J. S. Karling, *Sydowia* **20**: 55 (1967).
Typification: Type specimen no. 24308, Herbarium, Plant Disease Division, DSIR, Auckland, New Zealand. Type locality: Upper Hutt, Wellington, New Zealand. Type habitat: parasitic in *Leontodon leysseri* (Wallr.) G. Beck (*Asterideae, Asteraceae*).
Literature: Karling (1967).

Synchytrium limosellae J. S. Karling, *Sydowia* **20**: 56 (1967).
Typification: Type specimen no. 25212, Herbarium, Plant Disease Division, DSIR, Auckland, New Zealand. Type habitat: parasitic in *Limosella lineata* Gluck (*Lamiales, Scrophulariaceae*).
Literature: Karling (1967).

Synchytrium malvacearum S. D. Bharadwaj & R. Dwivedi, *Proceedings of the 56th Indian Scientific Congress* **3**: 306 (1969) [*nom. invalid.*, Art. 36.1, without Latin diagnosis].
Typification: type material not designated; no illustrations. Type locality: Varanasi, Uttar Pradesh, India. Type habitat: parasitic in *Abutilon indicum* Sweet (*Malvales, Malvaceae*).
Literature: Bharadwaj & Dwivedi (1969).

Synchytrium smithiae M. S. Patil, *Botanique, Nagpur, India* **9**: 203 (1978).
Typification: Type locality: India. Type habitat: parasitic in *Smithia geminiflora* Roth. [as its synonym, *S. conferta* Sm.], *S. setulosa* Dalz. and *Crotalaria latibrosa* L. [as its synonym, *Heylandia latibrosa* (L.) A. R. DC.] (all *Fabales, Fabaceae*).
Literature: Patil (1978).

Synchytrium urgineae M. S. Patil, *Botanique, Nagpur, India* **9**: 204 (1978) [as '*urginiae*'].
Typification: Type locality: India. Type habitat: parasitic in *Cyanotis tuberosa* Schult. f. var. *adscendens* C. B. Clarke [as "*Cynotis tyserosa* var. *ascendense*"] (*Commelinales, Commelinaceae*) and *Urginea indica* Kunth (*Liliales, Hyacinthaceae*).
Literature: Patil (1978).

Synchytrium asteracanthae M. S. Patil, *Indian Phytopathology* **40**: 156 (1987) [1988].
Typification: Type locality: India. Type habitat: parasitic in *Hygrophila auriculata* (Schumach.) Heine [as its synonym: *Asteracantha longifolia* Nees] (*Lamiales, Acanthaceae*).
Literature: Patil (1987).

Synchytrium tephrosiae M. S. Patil, *Indian Phytopathology* **40**: 158 (1987) [1988].
Typification: Type locality: India. Type habitat: parasitic in *Tephrosia apollinea* Link (*Fabales, Fabaceae*).
Literature: Patil (1987).

MICROMYCES P.-A. Dangeard, *Le Botaniste* **1**: 55 (1889).
Micromycopsis A. Scherffel, *Archiv für Protistenkunde* **54**: 202 (1926) (**Type species**: *Micromycopsis cristata* A. Scherffel).
Synchytrium subgenus *Microsynchytrium* J. S. Karling, *Mycologia* **45**: 278 (1953) [*nom. inval.* Art. 36.1, without Latin diagnosis] (**Type species**: not designated; first-named species *Synchytrium zygogonii* (P.-A. Dangeard) J. S. Karling).
Typification: *Micromyces zygogonii* P.-A. Dangeard (holotype - sole original species).

Micromyces zygogonii P.-A. Dangeard, *Le Botaniste* **1**: 52 (1889).
Synchytrium zygogonii (P.-A. Dangeard) J. S. Karling, *Mycologia* **45**: 278 (1953).
non Micromyces zygogonii P.-A. Dangeard *sensu* J. N. Couch, *Journal of the Elisha Mitchell Scientific Society* **46**: 231 (1931).
Typification: type material not designated; lectotype *Le Botaniste* **1**: plate 2, figs 1-10 (1889) (iconotype). Type locality: Mézidon (also Moult-Argences, Mutrécy), Caen, Normandy, France. Type habitat: parasitic in *Zygnema pectinatum* C. Agardh (*Zygnematales, Zygnemataceae*).
Literature: Canter (1949*a*), Couch (1931, 1937), Dangeard (1889), Johnson & Heidt (1937), Howard (1968), Karling (1953*a*), Linder (1947), Miller (1965), Sparrow (1943, 1952, 1960: 195), Yačevskie & Yačevskie (1931).

Micromyces mesocarpi É. de Wildeman, *Mémoires de l'Herbier Boissier, 1900* **3**: 1 (1900).
Typification: type material not verified [see Karling (1953*a*) and Sparrow, 1960)].
Literature: Wildeman (1900*a*), Karling (1953*a*), Linder (1947), Sparrow (1960: 195).

Micromyces cristata (A. Scherffel) F. K. Sparrow, *Aquatic Phycomycetes, 2nd revised edn*: 201 (1960).
var. ***cristata***.
Micromycopsis cristata A. Scherffel, *Archiv für Protistenkunde* **54**: 202 (1926) var. *cristata*.
Typification: type material not designated; lectotype *Archiv für Protistenkunde* **54**: plate 10, figs 69-76 (1926) (iconotype). Type locality: Ungarn, Slovakia. Type habitat: parasitic in *Hyalotheca dubia* Kütz. (also *Hyalotheca dissiliens* (J. E. Sm.) Bréb.) (*Zygnematales, Desmidaceae*).
Literature: Linder (1947), Rieth (1956), Scherffel (1926*a*), Sparrow (1960: 201).

Micromyces cristata (A. Scherffel) F. K. Sparrow, *Aquatic Phycomycetes, 2nd revised edn* (1960) var. ***minor*** (A. Scherffel) F. K. Sparrow, *Aquatic Phycomycetes, 2nd revised edn*: 201 (1960).
Micromycopsis cristata A. Scherffel var. *minor* F. K. Sparrow, *Mycologia* **24**: 273 (1932).
Typification:type material not designated; lectotype *Mycologia* **24**: 271, fig. 1d-e (1932) (iconotype). Type locality: Cold Spring Harbor, Long Island, New York, U.S.A. Type habitat: parasitic in *Spirogyra* sp. (*Zygnematales, Zygnemataceae*).
Literature: Linder (1947), Rieth (1956), Scherffel (1926*a*), Sparrow (1932, 1960: 201).

Micromyces fischeri (A. Scherffel) F. K. Sparrow, *Aquatic Phycomycetes, 2nd revised edn*: 202 (1960).
Micromycopsis fischeri A. Scherffel, *Archiv für Protistenkunde* **54**: 208 (1926) [as '*fischerii*'].
Typification: type material not designated; lectotype *Archiv für Protistenkunde* **54**: plate 10, fig. 77 (1926) (iconotype). Type locality: Csorbaer-See-Gegend, Hungary. Type habitat: parasitic in *Zygogonium* sp. (*Zygnematales, Zygnemataceae*).
Literature: Canter (1949*a*), Linder (1947), Scherffel (1926*a*), Sparrow (1960: 202).

Micromyces petersenii A. Scherffel, *Archiv für Protistenkunde* **54**: 209 (1926).
Synchytrium petersenii (A. Scherffel) J. S. Karling, *Mycologia* **45**: 278 (1953).
Typification: type material not designated; lectotype *Archiv für Protistenkunde* **54**: plate 10, figs 78-79 (1926) (iconotype). Type locality: Hungary. Type habitat: parasitic in *Mougeotia* sp. (*Zygnematales, Zygnemataceae*).
Literature: Canter (1949*a*), Couch (1931), Karling (1953*a*), Scherffel (1926*a*), Sparrow (1960: 206).

Micromyces spirogyrae B. W. Skvortzow, *Archiv für Protistenkunde* **51**: 428-433 (1925); *Archiv für Protistenkunde* **57**: 206 (1927).
Synchytrium spirogyrae (B. W. Skvortzow) J. S. Karling, *Mycologia* **45**: 278 (1953).
Typification: type material not designated; lectotype *Archiv für Protistenkunde* **57**: 206, fig. 9 (1927) (iconotype). Type locality: Harbin, north Manchuria, China. Type habitat: parasitic in *Spirogyra* sp. (*Zygnematales, Zygnemataceae*).
Literature: Karling (1953*a*), Linder (1947), Skvortzow (1927), Sparrow (1960: 206), Yačevskie & Yačevskie (1931).

Micromyces zygnemicola (K. Cejp) F. K. Sparrow, *Aquatic Phycomycetes 2nd revised edn*: 203 (1960) [as '*zygnaemicola*'].
Micromycopsis zygnemicola K. Cejp, *Rozpravy České Akademie, Praha, Věd a Um cl 2* **42**: 2 (1932) [as '*zygnaemicola*'].
Typification: type material not designated; lectotype *Rozpravy České Akademie, Praha, Věd a Um cl 2* **42**: plate 1, figs 5-10; plate 2, figs 3-4 (1932). Type locality: Germany. Type habitat: parasitic in *Zygnema stellinum* (Vaucher) C. Agardh [as '*Zygnema stellata*'] (*Zygnematales, Zygnemataceae*).
Literature: Cejp (1933), Sparrow (1960: 203).

Micromyces longispinosus J. N. Couch, *Mycologia* **29**: 592 (1937).
Micromyces zygogonii P.-A. Dangeard *sensu* J. N. Couch, *Journal of the Elisha Mitchell Scientific Society* **46**: 231 (1931).
Synchytrium longispinosus (J. N. Couch) J. S. Karling, *Mycologia* **45**: 278 (1953).
Olpidiopsis sp. Y. Kobayas[h]i & M. Ookubo, *Reports of the Ozegahara Gen Sci Surv Comm 1954*: 571 (1954) [Sparrow, 1960: 199].
Typification: *Mycologia* **29**: 592? figs 9-14 (1937). Type locality: Cold Spring Harbor, Long Island, New York, U.S.A. Type habitat: parasitic in *Spirogyra* sp. (*Zygnematales, Zygnemataceae*).
Literature: Couch (1931), Karling (1953*a*), Kobayashi & Ookubo (1954*b*), Linder (1947), Sparrow (1960: 198), Sparrow, Paterson & Johns (1965).

Micromyces minimus D. H. Linder, *Bulletin No. 97, Biological Series No. 26 National Museum of Canada*: 239 (1947).
Typification: Type slide No 2708. Type locality: South Baffin, Lake Harbour, coll. Polunin, 28 July 1936. Type habitat: parasitic in *Zygnema* sp. (*Zygnematales, Zygnemataceae*).
Literature: Linder (1947), Sparrow (1960: 205).

Micromyces echinocystis D. H. Linder, *Bulletin No. 97, Biological Series No 26. National Museum of Canada*: 240 (1947).
Typification: Type slide No 2709. Type locality: North Quebec, Wakeham Bay, coll. Polunin, 29 July 1936. Type habitat: parasitic in *Zygnema* sp. and *Mougeotia* sp. (*Zygnematales, Zygnemataceae*).
Literature: Linder (1947), Sparrow (1960: 205).

Micromyces wheldenii D. H. Linder, *Bulletin No. 97, Biological Series No 26. National Museum of Canada*: 240 (1947).
Typification: Type slide No 2716. Type locality: North Quebec, Wakeham Bay, coll. Polunin, 29 July 1936. Type habitat: parasitic in *Zygnema* sp. (*Zygnematales, Zygnemataceae*).
Literature: Linder (1947), Sparrow (1960: 206).

Micromyces laevis H. M. Canter, *Transactions of the British Mycological Society* **32**: 88 (1949).
Synchytrium laevis (H. M. Canter) J. S. Karling, *Mycologia* **45**: 278 (1953).
Typification: type material not designated; lectotype *Transactions of the British Mycological Society* **32**: plate X, figs 7-8 (1949) (iconotype). Type locality: Clay Pond, Wray Castle, Windermere, Cumbria, UK. Type habitat: parasitic in *Mougeotia* sp. (*Zygnematales*, *Zygnemataceae*).
Literature: Canter (1949*a*), Karling (1953*a*), Rieth (1956), Sparrow (1960: 202).

Micromyces intermedia (H. M. Canter) F. K. Sparrow, *Aquatic Phycomycetes, 2nd revised edn*: 200 (1960).
Micromycopsis intermedia H. M. Canter, *Transactions of the British Mycological Society* **32**: 79 (1949).
Typification: type material not designated; lectotype *Transactions of the British Mycological Society* **32**: 78, fig. 5; plate X, figs 9-10 (1949) (iconotype). Type locality: Blelham Bog, Cumbria, UK. Type habitat: parasitic in various species of *Zygnema* (*Zygnematales*, *Zygnemataceae*).
Literature: Canter (1949*a*), Sparrow (1960: 200).

Micromyces mirabilis (H. M. Canter) F. K. Sparrow, *Aquatic Phycomycetes, 2nd revised edn*: 199 (1960).
Micromycopsis mirabilis H. M. Canter, *Transactions of the British Mycological Society* **32**: 82 (1949).
Typification: type material not designated; lectotype *Transactions of the British Mycological Society* **32**: 80, fig. 6; 81, fig. 7; plate XI (1949) (iconotype). Type locality: Sphagnum pool, Blelham Tarn, Cumbria, UK. Type habitat: parasitic in *Closterium lunula* (O. F. Müll.) Nitzsch, *C. dianae* Ehrenb., *C. costatum* Corda and *C. kuetzingii* Bréb. (*Zygnematales*, *Desmidaceae*).
Literature: Canter (1949*a*), Sparrow (1960: 199), Sparrow, Paterson & Johns (1965).

Micromyces ovalis A. Rieth, *Österreichische Botanische Zeitschrift* **97**: 516 (1950) var. ***ovalis***.
Synchytrium ovale (A. Rieth) J. S. Karling, *Mycologia* **45**: 278 (1953).
Typification: type material not verified. Type locality: Tübingen, Germany. Type habitat: parasitic in *Mougeotia* sp. (*Zygnematales*, *Zygnemataceae*).
Literature: Karling (1953*a*), Miller (1965), Rieth (1950), Sparrow (1960: 200).

Micromyces ovalis A. Rieth, *Österreichische Botanische Zeitschrift* **97**: 516 (1950) var. ***giganteus*** F. K. Sparrow & M. E. Barr, *Mycologia* **47**: 546 (1955).
Synchytrium ovale (A. Rieth) J. S. Karling, *Mycologia* **45**: 278 (1953) var. ***giganteum*** (F. K. Sparrow & M. E. Barr) J. S. Karling, *Synchytrium*: 124 (1964).
Typification: type material not designated; lectotype *Mycologia* **47**: 548, fig 21 (1955) (iconotype). Type locality: 'Smith's Bog', Cheboygan County, Michigan, U.S.A. Type habitat: parasitic in *Zygnema* sp. (*Zygnematales*, *Zygnemataceae*).
Literature: Sparrow (1960: 200), Sparrow & Barr (1955).

Micromyces grandis C. E. Miller, *Journal of the Elisha Mitchell Scientific Society* **71**: 247 (1955).
Synchytrium grande (C. E. Miller) J. S. Karling, *Synchytrium*: 124 (1964).
Typification: type material not designated; lectotype *Journal of the Elisha Mitchell Scientific Society* **71**: 248, figs 1-9; 250, figs 10-29 (1955) (iconotype). Type locality: University Lake, Orange County, North Carolina, U.S.A. (11 Nov 1953). Type habitat: parasitic in *Spirogyra* sp. (*Zygnematales*, *Zygnemataceae*).
Literature: Karling (1964*d*), Miller (1955), Rieth (1956), Sparrow (1960: 204).

Micromyces oedogonii (M. Roberts) F. K. Sparrow, *Aquatic Phycomycetes 2nd revised edn*: 203 (1960).
Micromycopsis oedogonii M. Roberts, *Transactions of the British Mycological Society* **36**: 320 (1953).
Typification: holotype IMI 50464, material deposited at IMI, Egham, Surrey, U.K. Type locality: Bolder Mere, Wisley, Surrey, U.K. Type habitat: parasitic in *Oedogonium* sp. (*Chlorophyceae*, *Oedogoniales*).
Literature: Roberts (1953), Sparrow (1960: 203).

FAMILY IN SYNONYMY

MICROMYCOPSIDACEAE C. V. Subramanian, *Current Science, India* **43**: 723 (1974) (**Type genus:** *Micromycopsis* A. Scherffel).

GENUS IN SYNONYMY

MICROMYCOPSIS A. Scherffel, *Archiv für Protistenkunde* **54**: 202 (1926) (**Type species:** *Micromycopsis cristata* A. Scherffel).

Micromycopsis cristata A. Scherffel, *Archiv für Protistenkunde* **54**: 202 (1926) var. *cristata*, synonym of *Micromyces cristata* (A. Scherffel) F. K. Sparrow var. *cristata*, *Aquatic Phycomycetes, 2nd revised edn*: 201 (1960) var. *cristata*.

Micromycopsis cristata A. Scherffel var. *minor* F. K. Sparrow, *Mycologia* **24**: 273 (1932), synonym of *Micromyces cristata* (A. Scherffel) F. K. Sparrow var. *minor* (A. Scherffel) F. K. Sparrow, *Aquatic Phycomycetes, 2nd revised edn*: 201 (1960).

Micromycopsis fischeri A. Scherffel, *Archiv für Protistenkunde* **54**: 208 (1926) [as '*fischerii*'], synonym of *Micromyces fischeri* (A. Scherffel) F. K. Sparrow, *Aquatic Phycomycetes, 2nd revised edn*: 202 (1960).

Micromycopsis zygnemicola K. Cejp, *Rozpravy České Akademie, Praha, Věd a Um cl 2* **42**: 2 (1932), synonym of *Micromyces zygnemicola* (K. Cejp) F. K. Sparrow, *Aquatic Phycomycetes 2nd revised edn*: 203 (1960).

Micromycopsis intermedia H. M. Canter, *Transactions of the British Mycological Society* **32**: 79 (1949), synonym of *Micromyces intermedia* (H. M. Canter) F. K. Sparrow, *Aquatic Phycomycetes, 2nd revised edn*: 200 (1960).

Micromycopsis mirabilis H. M. Canter, *Transactions of the British Mycological Society* **32**: 82 (1949), synonym of *Micromyces mirabilis* (H. M. Canter) F. K Sparrow, *Aquatic Phycomycetes, 2nd revised edn*: 20?2 (1960).

Micromycopsis oedogonii M. Roberts, *Transactions of the British Mycological Society* **36**: 320 (1953), synonym of *Micromyces oedogonii* (M. Roberts) F. K. Sparrow, *Aquatic Phycomycetes 2nd revised edn*: 203 (1960).

ENDODESMIDIUM H. M. Canter, *Transactions of the British Mycological Society* **32**: 69-94 (1949).
Typification: *Endodesmidium formosum* H. M. Canter (holotype - sole original species).

Endodesmidium formosum H. M. Canter, *Transactions of the British Mycological Society* **32**: 72 (1949).
Typification: type material not designated; lectotype *Transactions of the British Mycological Society* **32**: 70, fig. 1; 72, fig. 2; plates VII, VIII (1949) (iconotype). Type locality: Rusland Bog, Lancashire, UK. Type habitat: parasitic in *Netrium oblongum* (de Bary) Lütkemüller, *Cylindrocystis crassa* de Bary and *C. brebissonii* (Ralfs) de Bary [Menegh. *ex* Ralfs] (*Zygnematales*, *Desmidaceae*).
Literature: Canter (1949*a*), Sparrow (1960: 191).

PART VI

KEYS

SYNOPTIC (NON-DICHOTOMOUS) KEY TO THE CLASSES, ORDERS AND FAMILIES OF STRAMINIPILOUS AND OTHER HETEROTROPHS

NOTE: The following key has been edited and augmented from the dichotomous key I wrote for Hawksworth *et al.* (1995), incorporating information from Cantino (1955), Cantino & Turian (1959) and Dick *et al.* (1984, 1999).

A' of 2 Thallus or assimilative plasmodium bounded by a fibrillar cell wall, at least at maturity; or if naked then lacking naked protoplasmic extensions; cell wall components, where known, predominantly ß-1,3- and ß-1,6-glucans, but sometimes with chitin, (*Peronosporomycetes* and *Hyphochytriomycetidae*); freshwater or marine [straminipilous-derived condition assumed (unjustifiably?) in some taxa]. **B**

B' of 3 Motile cells (or zoospore cysts or aplanospores) small or very small ($<60\mu m^3$ volume), with one flagellum, or two anisokont flagella, but neither flagellum known to be straminipilous, with sub-apical flagellar insertion; obligate endobiotic parasites; thalli walled, olpidioid or allantoid, rarely septate. **S**

B'' of 3 Motile cells small or very small, with one flagellum, or two anisokont flagella, but neither flagellum known to be straminipilous; thalli naked during the assimilative phase (excluded from the *Peronosporomycetes sensu lato*) freshwater [but see also **B'** (*Anisolpidiaceae*)]. **(Rozellopsidales, Plasmodiophorales) X**

STRAMINIPILOUS ORGANISMS (FUNGI AND HETEROTROPHS)

B''' of 3 Motile cells large or small, with one straminipilous flagellum or two isokont or anisokont flagella, one of which is straminipilous; flagellar insertion diverse; thalli normally walled during the assimilative phase. **(Peronosporomycotina) D**

D' of 2 Zoospores biflagellate; flagellar attachment sub-apical or lateral; sexuality present or not known; life-history haplomitotic B; freshwater, terrestrial in soil or on roots or stems, or marine. **(Peronosporomycetes *sensu stricto*, excluding Hyphochytriomycetes or Peronosporomycetes *sensu lato*, but excluding Hyphochytriomycetidae) E**

E' of 2 Thallus walled, at least towards the end of the assimilative phase; thallus, if endoparasitic, within the host plasmamembrane; gametes, if discrete, non-motile. **(Peronosporomycetes *sensu stricto*) G**

STRAMINIPILOUS FUNGI

D'' of 2 Zoospores anteriorly uniflagellate; flagellar attachment apical; sexuality not known; life-history (ploidy) not known; freshwater or terrestrial. **(Hyphochytriomycetes, Hyphochytriomycetidae, Hyphochytriales) F**

F' of 2 Thallus intramatrical, monocentric or polycentric, with proliferating branches, rhizoidal system absent; sporangia with thin walls.
Genera: *Canteromyces, Cystochytrium, Hyphochytrium* **Hyphochytriaceae**

F'' of 2 Thallus epibiotic, normally monocentric, rarely apophysate, with rhizoids; sporangia with thin or thick walls, the latter sometimes bullately ornamented.
Genera: *Latrostium, Reessia, Rhizidiomyces* **Rhizidiomycetaceae**

PERONOSPOROMYCETES *SENSU STRICTO*

NOTES: The definitive selection of the diagnostic criteria for the three sub-classes and orders *incertae sedis* requires much more comparative study of physiological, biochemical and molecular characteristics.

The subclasses may be of very different taxonomic breadths, with the *Peronosporomycetidae* and *Rhipidiomycetidae* representing derived developments. It is therefore to be expected that, like any other ***single*** taxonomic character, molecular biological data of a single criterion (e.g. the small subunit of the rDNA) will show greater diversity within the *Saprolegniomycetidae*, compared with its diversity within the *Peronosporomycetidae*; some components of the former could then be ***expected*** to clade closer to the latter than to other components of the former (cf. *Salilagenidiales*). The extent of the biodiversity, species richness and capacity for, and rates of, sexual ***and*** non-sexual genetic change (e.g. ephemeral anamorphs such as the hyphochytrids) must be taken into account when assessing molecular biological cladograms.

Rejustified left: G - R

G'of 6 Thallus filamentous, mycelial with branched hyphae, occasionally septate (rarely of very narrow diameter (*Pythiogeton*, *Verrucalvus*), or allantoid to coralloid with or without septa (*Lagenidium*, *Myzocytium*) but then with extrasporangial homohylic vesicular discharge); zoosporogenesis intrasporangial or extrasporangial in a homohylic vesicle; oogonia and antheridia usually differentiated; oogonia sometimes pluriovulate; oosporogenesis with or without periplasm; saprobic or parasitic. . **(Peronosporomycetidae, Saprolegniomycetidae *pro parte*) H**

H' of 2 Thallus mycelial and intercellular with haustoria; zoosporogenesis, when present, by internal cleavage, otherwise asexual reproduction by deciduous conidiosporangia; conidiosporangiophores well-differentiated, persistent. Oosporogenesis centripetal with persistent periplasm; oogonia thin-walled, oospore single, aplerotic with a well-defined exospore wall layer derived from persistent periplasm (and therefore well-preserved in herbarium material). Possibly all are dependent on exogenous sources of sterols. Obligate parasites. Mostly stem and leaf parasites of *Dicotyledones*. **(Peronosporales) I**

I'of 2 Thallus mycelial and intercellular with large, lobate haustoria. Asexual reproduction by deciduous conidiosporangia or conidia borne on conidiosporangiophores; conidiosporangia pedicellate; conidiosporangiophores dichotomously branched, monopodially branched, or unbranched and clavate; conidiogenesis simultaneous; zoosporogenesis, when present, internal within a plasmamembranic membrane, release by operculate or poroid discharge. The dicotyledonous downy mildews.
Genera: *Basidiophora*, *Benua*, *Bremia*, *Bremiella*, *Paraperonospora*, *Peronospora*, *Plasmopara*, *Pseudoperonospora*. **Peronosporaceae**

I'of 2 Thallus mycelial and intercellular with small, spherical or peg-like haustoria. Asexual reproduction by deciduous sporangia, conidiosporangia or conidia borne on unbranched conidiophores. Conidiogenesis sequential and percurrent. Zoosporogenesis internal with papillate discharge. The white blister rusts.
Genus: *Albugo*. **Albuginaceae**

H''of 2 Oosporogenesis centrifugal or centripetal with insignificant periplasm; conidio-sporangiophores rarely differentiated. **J**

J' of 4 Thallus mycelial with little evidence of cytoplasmic streaming, or monocentric and pseudomycelial; hyphae of uniform diameter (6-10(15) μm diam.);; zoosporangium formation terminal, less frequently sequential, then percurrent or by internal or sympodial proliferation; sporangia rarely absent; sporangiophores rarely differentiated; zoosporogenesis either by internal cleavage without vesicular discharge or with a plasmamembranic vesicle or by external cleavage in a homohylic vesicle; antheridial gametogenesis following oogonial gametogenesis; oogonial periplasm minimal and not persistent; oogonia modally uniovulate, oogonia (with very few exceptions) thin-walled; oospores never strictly plerotic; oospores with a hyaline ooplast, lipid coalescence minimal. Able to utilize $SO_4^{=}$; variable ability to utilize different inorganic N sources; aerobic metabolism. Evidence of partial dependence on exogenous sterol precursors. Parasites or saprotrophs; parasites mostly in axenic culture. Some members parasitic on fungi and some on animals. Freshwater or marine. Basal chromosome number $x = 5$.
Genera: *Cystosiphon*, *Diasporangium*, *Endosphaerium*, *Halophytophthora*, *Lagenidium sensu strictissimo*, *Myzocytium sensu strictissimo*, *Peronophythora*, *Phytophthora*, *Pythium*, *Trachysphaera*.
Peronosporomycetidae *pro parte*: Pythiales, Pythiaceae

J'' of 4 Thallus mycelial, with or without inflated sinuses, perhaps rhizoidal; hyphae very slender (usually <5 μm diam.); zoosporogenesis by external cleavage in a detached homohylic vesicle, or absent; oogonial and antheridial morphogenesis nearly simultaneous; oogonia thick-walled, uniovulate; oospore plerotic with a very thick endospore wall resorbed on germination. Saprotrophic, probably with anaerobic metabolism.
Genera: *Medusoides, Pythiogeton.* **Peronosporomycetidae** *pro parte*: **Pythiales, Pythiogetonaceae**

J''' of 4 Hyphae often stout with diameter increasing with age (up to 150μm diam.), most *ca* 20μm diam.; zoosporogenesis intrasporangial and centrifugal; oosporogenesis centrifugal; antheridial gametogenesis following oogonial gametogenesis, or absent; oogonia often pluriovulate, oospores usually aplerotic; oospore with a fluid and granular ooplast, lipid coalescence variable (limited or to a single eccentric globule); unable to utilize $SO_4^=$ or NO_3^-; basal chromosome number $x = 3$. **K**

K' of 2 Mycelium filamentous, often of wide diameter (*ca* 20-40 μm), usually with considerable later increase (to *ca* 150 μm), especially basally, due to intussusception of wall material, zoosporangia cylindrical, naviculate or almost spherical, renewal internal, cymose or basipetal, zoosporogenesis completed within the zoosporangium, oogonia commonly with simple pits, smooth or papillate, frequently pluriovulate, oospore with a fluid ooplast with many granular inclusions in Brownian motion.
Genera: *Achlya, Aplanopsis, Brevilegnia, Calyptralegnia, Dictyuchus, Geolegnia, Isoachlya, Saprolegnia, Scoliolegnia, Thraustotheca.*
Saprolegniomycetidae *pro parte*: **Saprolegniales, Saprolegniaceae** *sensu stricto*

K'' of 2 Mycelium filamentous, slightly sinuous, diameter *ca* 15-25 μm, with little or no basal intussusception, zoosporangia undifferentiated or scarcely differentiated from hyphal elements, sometimes branched, renewal, if any usually basipetal, zoosporogenesis frequently incomplete at the moment of discharge, oogonia lacking pits, smooth or papillate, usually uniovulate, oospore with a semi-solid ooplast with few granular inclusions.
Genera: *Aphanomyces, Leptolegnia, Plectospira.*
Saprolegniomycetidae *pro parte*: **Saprolegniales, Leptolegniaceae**

J'''' of 4 All species are known only as parasites of Poaceae. Mycelium of very narrow (<5 μm diameter) hyphae, with granular cytoplasm, cytoplasmic streaming visible where wide enough; zoosporogenesis intrasporangial or absent; discharge vesicles not formed; oogonia very thick-walled, often verrucate, with a single, often plerotic oospore; periplasm minimal or absent; ooplast homogeneous; lipid coalescence limited. Two families. Parasites of *Poaceae.* **(Saprolegniomycetidae** *pro parte*: **Sclerosporales) L**

L' of 2 Parasitic, not culturable. Mycelium with peglike or digitate haustoria; zoosporangia or conidia borne on strongly inflated conidiosporangiophores; conidiosporangiophores pseudo-dichotomously branched; conidiosporangiophores evanescent (and therefore seldom well-preserved in herbarium material) [graminicolous downy mildews]; zoosporangium formation sequential or more or less simultaneous on inflated sporangiophores, maturation more or less simultaneous. Zoospore release, if known, by operculate discharge. Graminicolous downy mildews.
Genera: *Peronosclerospora, Sclerospora.* **Sclerosporaceae**

L'' of 2 Parasitic but culturable. Mycelium without haustoria; asexual phase sometimes absent; sporophores poorly differentiated; sporangium formation sequential, either by internal or sympodial renewal. Zoospore release, if known, by papillate discharge. Graminicolous root parasites.
Genera: *Pachymetra, Sclerophthora, Verrucalvus.* **Verrucalvaceae**

G'' of 6 Thallus, blastic, coralloid, allantoid or olpidioid (rarely pseudomycelial); zoosporogenesis usually intrasporangial but if extrasporangial then without a homohylic vesicle; zoospores (or zoospore cysts) medium-small (>60μm³ volume) or large (>300μm³ volume); gametangia frequently not morphologically differentiated; oosporogenesis without nucleated periplasm. **(Saprolegniomycetidae** *pro parte*: **Leptomitales** (?), **Salilagenidiales) M**

M' of 2 Freshwater or terrestrial families (more rarely including genera with some marine species); thallus blastic, allantoid or coralloid to pseudomycelial; zoospores medium-large or large (>175μm³ volume). **N**

N' of 2 Thallus blastic, rarely olpidioid; sexual reproduction homothallic, automictic or heterothallic; donor gametangia (when present) smaller than the oogonium. **O**

O' of 2 Thallus blastic, composed of allantoid segments connected by simple constrictions, occasionally with a few rhizoids; 'cellulin' granules (chitin positive) usually present in constricted thalli; zoospore cysts sometimes with short hollow protrusions; donor and receptor gametangia differentiated, gametangia terminal or intercalary, apparently delimited by plugs of wall material at the subtending constrictions; separable gametic membranes not distinguished; fertilization by means of a short fertilization tube; oospores single and plerotic (oogonia rarely pluriovulate - *Plerogone*); oospores with a nearly hyaline ooplast and diverse patterns of lipid coalescence; rarely anamorphic (*Leptomitus*); often found on corticated twigs at lake margins; partial inablity to utilize NH_4^+; metabolism strongly oxidative; basal chromosome number probably $x = 4$].
Genera: *Apodachlya*, *Leptomitus*, *Plerogone*. **Leptomitaceae**

O'' of 2 Thallus blastic or olpidioid; separable gametic membranes present; fertilization by means of a fertilization hypha; oospores aplerotic; olpidioid thalli heterothallic [sometimes parasitic in aquatic protoctists].
Genera: *Apodachlyella*, *Eurychasmopsis*. **Apodachlyellaceae**

N'' of 2 Thallus tubular with swellings or coralloid (rarely of unconstricted hyphae of very narrow ($<5\mu m$) diameter; gametangia little differentiated. **P**

P' of 2 Homothallic, often automictic; gametangial zones usually little or not differentiated, often without delimiting septa, intercalary, as lateral diverticula, or rarely differentiated and terminal (*Aphanodictyon*); pluriovulate; oospore wall layers separated by fluid zones; zoosporogenesis variable; oogonia often pluriovulate; oospores aplerotic; keratinophilic saprophytes or parasites of algae (*Aphanomycopsis*) or nematodes (*Nematophagus*).
Genera: *Aphanodictyon*, *Aphanomycopsis*, *Brevilegniella*, *Leptolegniella*, *Nematophthora*. **Leptolegniellaceae**

STRAMINIPILOUS FUNGI (*INCERTAE SEDIS*)

P'' of 2 Heterothallic (?); zoospores emerging to form a fenestrated hollow sphere of cysts, connected by hollow protrusions from each cyst; oogonia uniovulate; oospore plerotic (?); saprophytic in pollen floating on lakes [monotypic].
Genus: *Ducellieria*. **Ducellieriaceae**

M'' of 2 Marine families; thallus pseudomyceliar or coralloid, cytoplasm often with prominent granulation; zoosporogenesis intrasporangial or with a precipitative vesicle; oospore with a multilayered wall and granular ooplast, lipid droplets condensed to varing degrees, zoospores with lateral flagellar insertion at least in the second phase in polyplanetic species; parasitic in *Crustacea* or saprotrophic on marine sustrata. **(Salilagenidiales) Q**

Q' of 2 Thallus more or less tubular, rarely septate; zoospores medium-large (>275μm^3 volume); teleomorphic; oospores with granular ooplasts and lipid coalescence [most species parasitic in *Crustacea*].
Genera: *Halodaphnea*, *Salilagenidium*. **Salilagenidiaceae**

Q'' of 2 Thallus tubular or saccate, sometimes septate; zoospores medium-small (<275μm^3 volume), not proven to be straminipilous; anamorphic; facultatively parasitic or saprotrophic.
Genus: *Haliphthoros*. **Haliphthoraceae**

G''' of 6 Thallus polar with an inflated basal cell and rhizoids; thalloid segments separated by short narrow thick-walled isthmuses; zoosporogenesis intrasporangial; zoosporangia with a single exit tube and sometimes an evanescent protoplasmamembranic vesicle; zooospores of the principal form; gametangia differentiated into oogonia and antheridia; oosporogenesis periplasmic with nucleated periplasm; oogonia uniovulate; saprotrophic; facultatively or obligately fermentative; requirement for organic N.
Genera: *Aqualinderella*, *Araiospora*, *Mindeniella*, *Nellymyces*, *Rhipidium*, *Sapromyces*
Rhipidiomycetidae, Rhipidiales, Rhipidiaceae

G'''' of 6 Thalli more or less spherical, olpidioid, rarely tubular, never septate; zoosporangia with a single exit tube; zoospores with sub-apical flagellar insertion; transitional zone with rootlets of the principal zoospore form; teleomorphic, sexual reproduction presumed to be heterothallic with donor gametangial thalli of very different volume from the receptor thalli (*ca* 10%); fertilization via an insipient fertilization tube,; oogonia uniovulate; oospores plerotic; often inducing extramural fibrillar deposits on oogonial, companion cell and sporangial thalli [freshwater parasites of *Peronosporomycetes* and *Chlorophyceae*].
Genera: *Gracea*, *Olpidiopsis*, *Pleocystidium*. **Olpidiopsidales** (order *incertae sedis*), **Olpidiopsidaceae**

G''''' of 6 Thallus allantoid, non-polar and without rhizoids; zoosporogenesis intrasporangial or with a plasmamembranic vesicle; zoospores pyriform with subapical flagellar insertion; transitional zone lacking a central bead; gametangia not differentiated, oosporogenesis with nucleated periplasm; presumed to be automictic; endoparasitic, endobiotic [monotypic].
Genus: *Lagena*. **Lagenaceae** (family *incertae sedis*)

G'''''' of 6 Thallus initially tubular, sometimes branched, becoming septate and often disarticulating; zoosporangia normally with a single exit tube; zoospores medium-small (<150μm^3 volume); zoospore ultrastructure not known; sexual reproduction homothallic or automictic with gametangia of more or less equal volume; fertilization by copulation through a pore; oogonia possibly periplasmic; normally uniovulate; oospore ooplast and lipid reserve distribution not established. **Myzocytiopsidales** (order *incertae sedis*) **R**

R' of 2 Straminipilous flagella with tripartite tubular hairs presumed to be arranged in two rows; parasites in nematodes and rotifers, or algae.
Genera: *Gonimochaete*, *Myzocytiopsis*, *Syzygangia*. **Myzocytiopsidaceae**

R'' of 2 Straminipilous flagella with tripartite tubular hairs arranged in a single row [this feature requires critical reevaluation]; parasitic in *Insecta*.
Genus: *Crypticola*. **Crypticolaceae**

PERONOSPOROMYCOTINA (INCERTAE SEDIS)

E'' of 2 Thallus plasmodial, unwalled throughout the assimilative phase; thallus between the host plasmamembrane and frustule or cell wall of the host; gametes motile [marine].
Genus: *Lagenisma*. **Lagenismatales** (order *incertae sedis*), **Lagenismataceae**

DOUBTFULLY STRAMINIPILOUS HETEROTROPHS

S' of 2 Parasitic in marine algae, or in fungi parasitic in marine algae, or in freshwater; zoospores very small ($<25\mu m^3$ volume); mostly anamorphic [mary diagnostric details not recorded]. **T**

T' of 2 Zoospores biflagellate; flagellar insertion not well-defined or precisely documented. **U**

U' of 2 Rarely causing hypertrophy of the host cell; zoosporogenesis intrasporangial but zoospores not known to encyst prematurely to form a 'net' sporangium. **V**

V' of 2 Parasitic in *Rhodophyceae* [thallus lobed and non-septate (*Petersenia*) or tubular and more or less septate (*Pontisma*); anamorphic].
Genera: *Petersenia, Pontisma.* **Pontismataceae** (family *incertae sedis*)

V'' of 2 Primarily parasitic in *Chlorophyceae*, rarely *Fucophyceae* or marine fungal parasites associated with these algae [thallus often septate, segments not disarticulating; teleomorphic].
Genera: *Sirolpidium.* **Sirolpidiaceae** (family *incertae sedis*)

U'' of 2 Causing marked hypertrophy of host cell; zoosporogenesis often with a 'net sporangium' (intrasporangial encystment of the first planetic stage) [anamorphic; parasitic in *Fucophyceae* or *Rhodophyceae*].
Genera: *Eurychasma, Eurychasmidium.* **Eurychasmataceae** (family *incertae sedis*)

T'' of 2 Zoospores reputedly uniflagellate; flagellar insertion not well-defined [reputedly teleomorphic; resting spores reported with or without companion cells; parasitic in *Fucophyceae*].
Genus: *Anisolpidium.* **Anisolpidiales, Anisolpidiaceae** (family *incertae sedis*)

NOT PROVEN TO BE, AND PERHAPS UNLIKELY TO BE, STRAMINIPILOUS HETEROTROPHS

S'' of 2 Parasitic in terrestrial or marine *Aschelminthes*, or in marine or freshwater *Bacillariophyceae*; thallus never septate; zoosporangia frequently with more than one exit tube; anamorphic; extramural fibrillar deposits not known. **W**

W' of 2 Parasitic in marine and freshwater *Bacillariophyceae* (sexuality reported for *Ectrogella perforans*, a species which may not be congeneric with the type species - cf. *Lagenismataceae*).
Genus: *Ectrogella*, (*Septolpidium*?). **Ectrogellaceae** (family *incertae sedis*)

W'' of 2 Parasitic in marine and freshwater *Aschelminthes* [sporangiospores transforming into glossoid (gun cell) spores - cf. *Plasmodiophoraceae*].
Genus: *Haptoglossa.* **Haptoglossaceae** (family *incertae sedis*)

NOT LIKELY TO BE STRAMINIPILOUS HETEROTROPHS

X' of 2 Plasmodium filling the host or host segment completely [transformed into a single sporangium; alternatively developing into a single aplerotic resting spore with extramural fibrillar ornamentation; plasmodial mitosis not described; meiosis not reported]. **(Rozellopsidales) Y**

Y' of 2 Parasitic in fungi (especially *Peronosporomycetes* and *Chytridiomycetes*) and algae; plasmodium (and asexual sporangium) filling or almost filling the host compartment, plasmodium sometimes possibly schizogenous; zoospores uniflagellate or biflagellate; resting spores formed without any known union; often with fibrillar or spiny ornamentation; aplerotic in the host compartment].
Genera: *Rozella, Rozellopsis.* **Rozellopsidaceae**

Y'' of 2 Parasitic in protozoa; plasmodium not filling host cell, sometimes schizogenous; zoospores uniflagellate or biflagellate [many diagnostic details imprecisely recorded].
Genera: *Pseudosphaerita, Sphaerita.* **Pseudophaeritaceae**

X'' of 2 Plasmodium not filling the host or host segment completely. **(Plasmodiophorales) Z**

Z' of 2 Flagellar insertion superficial; transitional plate at the same level as the cell plasmamembrane; plasmodium transformed into a sporangium complex or sorus, with discharge tubes developing prior to discharge; plasmodial mitosis with a central persistent nucleolus (cruciform mitosis); meiosis recorded; parasitic in angiosperm roots or *Peronosporomycetidae.*
Genera: *Octomyxa, Plasmodiophora, Sorodiscus, Sorosphaera, Tetramyxa, Woronina.* **Plasmodiophoraceae**

Z'' of 2 Flagellar insertion in a groove; transitional plate not studied; plasmodium transformed into a single sporangium with an evertible discharge tube; most details of mitosis not known, but nucleolus degenerating during division and nuclear membrane persistent; meiosis not known; parasitic in protozoa.
Genus: *Endemosarca.* **Endemosarcaceae**

LABYRINTHISTA

A'' of 2 Thallus or thallus cells bounded by a wall composed of carbohydrate (L-galactose) scales, absorbtive system usually a naked ectoplasmic net emenating from a sagenogenetosome; marine [straminipilous with the anterior flagellum the longer, but not fungi]. **(Labyrinthista) C**

C' of 2 Trophic cells normally uninucleate, bounded by one or two layers of scales; trophic cells embedded within the ectoplasmic net.
Genus: *Labyrinthula.* **Labyrinthulales, Labyrinthulaceae**

C'' of 2 Trophic cells apocytic (becoming multinucleate), bounded by several layers of scales; trophic cells superficial to the ectoplasmic net.
Genera: *Althornia, Japonochytrium, Schizochytrium, Thraustochytrium, Ulkenia.*
Thraustochytriales, Thraustochytriaceae

KEYS TO GENERA AND SPECIES

NOTES:

Dichotomous keys do not provide an ideal means of identification, but have the advantage of familiarity. For some genera it is possible to prepare Venn Diagram keys which may help to clarify relationships. The dichotomous keys to the genera given below are unlikely to be foolproof because the genera are defined on suites of characters, some of which are shared with other genera, and the data available may be incomplete. There are two reasons for this incompleteness. First, the original descriptions may lack information, or provide information without the precision now required. Secondly, either the specimens that are available to be keyed out may not possess all the attributes needed by the key criteria, or the taxon itself may be constitutionally deficient with respect to certain criteria. Family placement of species of *Cystosiphon* and *Myzocytium sensu stricto* remains questionable.

I have arranged this artificial identification system as a series of keys in which both habitat and morphology are important. Sometimes the only reliable character is provided by the organism acting as host. The first key is to genera and species of monotypic genera or genera with only one or two species. The keys following are to species in the larger or more important lagenidiaceous genera; or to species in orders, families or genera where it is desirable to keep the species concerned in a group. The treatment of 'doubtful taxa' varies from group to group, some are included in these keys, but for others the information available is insufficient for a practicable key. Key I provides the only comprehensive account of the straminipilous and plasmodiophoromycetous parasites of the *Peronosporomycetidae* and *Saprolegniomycetidae*.

These keys contain condensed species descriptions which replace the long species descriptions as given in Sparrow (1960) and Karling (1981a). The full descriptions can be obtained by back-tracing the key points from the species. The additional information for a species, or a group of species for which there is no appropriate alternative in the dichotomy, is given in square brackets, together with other information that may be of value. However, deficiencies in the original descriptions may result in apparent inconsistencies in the style in which this information is provided.

Keys to genera and species of the better-known families of the *Peronosporomycotina* are not given in this work, for reasons discussed in *Part III*, above. The families with mycelia composed of hyphae are omitted from these keys (namely: *Albuginaceae*, *Leptolegniaceae*, *Peronosporaceae*, *Pythiaceae*, *Pythiogetonaceae*, *Saprolegniaceae*, *Sclerosporaceae* and *Verrucalvaceae*). Key point 2, Key I, leads, without any dichotomous entries (see preceding synoptic key), to keys of the three better-known families with blastic or monocentric thalli (*Apodachlyellaceae*, *Leptomitaceae*, *Rhipidiaceae*).

The *Pythiaceae* is a large family and only a few of the taxa discussed in *Part IV* need to be mentioned; two of these species are placed in *Pythium*, but it would not be appropriate to copy the whole of the keys to *Pythium* (Dick, 1990*c*) for these two species. They would key out as follows: *Pythium caudatum* at key point 107' (Dick, 1990*c*: 45) with *Pythium* F-group (Revision of Van der Plaats-Niterink's Key); *Pythium podbielkowskii* at key point 13' (Dick, 1990*c*: 14) with *Pythium dissotocum*, from which it is easily distinguished by habitat, oogonium shape, oospore dimensions and antheridial characters.

KEY I: GENERAL KEY TO FAMILIES, GENERA, AND SPECIES IN GENERA WITH VERY FEW SPECIES

1 Freshwater or terrestrial fungi; saprotrophic or parasitic 2

1' Marine fungi; saprotrophic or parasitic 135

2(1) Thallus walled throughout the assimilative phase from commencement of cyst germination (cyst often remaining attached to the young thallus) or rarely initially plasmodial but rapidly becoming walled (*Eurychasmopsis* - see key point 19'); thallus more or less mycelial, olpidioid or allantoid or coralloid or monocentric and rhizoidal or blastic; septate or non-septate, thalloidal segments disarticulating or not (*Pythiales, Saprolegniales, Leptomitales, Rhipidiales, Hyphochytriales, Myzocytiopsidales*) 3

[Note: the myceliar *Pythiales* and *Saprolegniales* are not keyed here.]

(*Leptomitaceae* and *Apodachlyellaceae*; **KEY XII**)

(*Rhipidiaceae*; **KEY XIII**)

2'(1) Thallus remaining plasmodial throughout the assimilative phase, or initially plasmodial and gradually becoming walled, if becoming walled then not septate (but may induce host septation); thallus more or less olpidioid, allantoid or coralloid; assimilative coenocytes occasionally dividing; not blastic, not rhizoidal (*Olpidiopsidales, Rozellopsidales, Plasmodiophorales*) 47

3(2) Zoosporogenesis extrasporangial in a homohylic vesicle, releasing biflagellate or uniflagellate straminipilous zoospores; plant or animal parasites (sometimes including aquatic insects or aschelminths) or saprotrophs (*Pythiales, Hyphochytriales*) 4

3'(2) Sporogenesis intrasporangial or only partly extrasporangial (except for *Lagena* - see key point 30), releasing motile (except for *Aphanomycopsis* - see key points 25', 26', 42, 42', 44', 46, 46') biflagellate (except for *Septolpidium* - see key point 43') zoospores of various morphology (possibly with a quiescent phase at mouth of exit tube), or aplanospores; plant or animal parasites (principally aschelminths) or saprotrophs (various orders) 9

4(3) Zoospores biflagellate and of the principal form, volume equivalent (<50)70-400(>500) μm^3; sexual reproduction usually present, fertilization by means of a fertilization tube (*Pythiales*) 5

4'(3) Zoospores anteriorly uniflagellate; zoospore volume equivalent 30-60 μm^3; sexuality not known (*Hyphochytriales*) 8

5(4) Oospores with smooth walls 6

5'(4) Oospores with ornamented walls (*Cystosiphon*) **KEY III**

6(5) Thallus holocarpic or eucarpic, filamentous but rarely branched, septate and sometimes inflated, or pseudomycelial and septate, especially when parasitic in animals; when parasitic in plants, usually confined to a single host cell; rarely olpidioid; gametangia (oogonia and antheridia) more or less similar in shape, frequently developing from contiguous intercalary or intercalary and terminal segments 7

6'(5) Thallus eucarpic, mycelial, extensive with hyphae of uniform diameter, branched even when parasitic in a unicell or nematode, septa absent or infrequent; gametangia (oogonia and antheridia) well-defined and morphologically distinct, oogonia terminal, lateral or intercalary, with or without antheridia; homothallic with monoclinous or diclinous antheridia, or heterothallic (*Pythium* - **not keyed here**)

7(6) Thallus normally unbranched, septate with more or less thickened septal walls, thallus segments becoming inflated and partially disarticulating, rarely olpidioid; gametangia not morphologically differentiated [parasitic in *Zygnematales*] (*Myzocytium sensu stricto*) **KEY II**

7'(6) Thallus unbranched or sparsely branched, septate with unthickened septal walls, thallus segments somewhat inflated, cohering throughout; gametangia with some differentiation between oogonia and antheridia [parasitic in *Insecta*] (*Lagenidium sensu stricto*) **KEY II**

8(4') Thallus intramatrical, monocentric or polycentric, with proliferating branches, rhizoidal system absent; sporangia thin-walled (*Hyphochytrium, Canteriomyces* and *Cystochytrium*) **KEY XIV**

8'(4') Thallus epibiotic, normally monocentric, with one (rarely more) rhizoidal system, sometimes with an apophysis; sporangia with thin or thick walls, the latter sometimes bullately ornamented (*Rhizidiomyces, Latrostium* and *Reesia*) **KEY XV**

9(3') Thallus entirely endobiotic (parasites), or intramatrical except for reproductive hyphae or extramatrical (saprotrophs); usually holocarpic 10

9'(3') Thallus mostly epibiotic (parasites), except for assimilative hyphae; initially eucarpic 21

10(9) Parasitic in aschelminths, insects or other invertebrates [sporangia producing aplanospores or biflagellate zoospores; sexual or resting spore reproduction present or absent] 11

10'(9) Parasitic in plant cells (green algae, diatoms, aquatic fungi and angiosperms), or saprotrophic in or on plant or animal substrata [sporangia producing biflagellate (rarely uniflagellate) zoospores; sexual or resting spore reproduction present or absent] 24

11(10) Parasites of aquatic arthropods (*Diptera* eggs; *Daphnia* etc.) 12

11'(10) Parasites of aschelminths or ciliate protozoa [sporangiospores biflagellate zoospores (but not proven to be straminipilous), or aplanospores] 15

12(11) Sexual reproduction not known(*Blastulidium* and *Crypticola*) 13

12'(11) Sexual reproduction present [parasitic in eggs of *Chironomidae*; thallus holocarpic, richly branched, non-septate except for delimitation of gametangia; becoming transformed into a sporangium, 88-120 × 206-280 μm; sporogenesis intrasporangial, initials gliding out and encysting at orifices of exit tubes; cysts (6·0)6·6(7·3) μm diam.; zoospores 4·2-6·3 × 5·4-8·3 μm (volume equivalent *ca* 150 μm^3); gametangia developed from parts of the sporangial thallus, with or without sepation, oogonia irregular, pyriform or dumbbell-shaped (44-72 × 82-240 μm), antheridia slender and septate, fertilization by means of a fertilization tube; oospores 44-72 × 52-170 μm, oospore wall unlayered, 2·2-4·3 μm thick] [compare with *Cornumyces*] **Aphanomycopsis sexualis**

13(12) Parasitic in insects [sporangiospores biflagellate zoospores, but reported to have an anterior flagellum with a single row of hairs (uniseriate?, straminipilous?)] (*Crypticola*) 14

13'(12) Parasitic in freshwater *Crustacea* (*Daphnia* embryos) [thalli developing as ovoid or elongate chains of cells, 10-12 × 18-25(100) μm, separated by septa, thalloid segments functioning as zoosporangia; sporangiospores delimited around a large central vacuole; zoospores 3·0-5·0 × 7·0-7·5 μm (volume equivalent < 100 μm^3), anisokont (?), flagella long (> 15 μm), with sub-apical flagellar insertion] **Blastulidium paedophthorum** (monotypic genus)

14(13) Parasitic in nematoceran *Diptera* in wet terrestrial tropical environments; thallus pseudomycelial with hyphae 8-12 μm diam., becoming septate with thalloid segments 50-60 μm long, unilaterally inflated or branched, forming sporangia; sporangia globose-ovoid, 25-42 μm diam., zoospores few (6-15 per sporangium); zoospores monoplanetic, reniform, laterally biflagellate, 5·0-6·0 × 8·0-10·0 μm (volume equivalent *ca* < 175 μm^3) **Crypticola clavulifera** (type species)

14'(13) Parasitic in eggs of midges and caddis flies in temperate waters; thallus holocarpic, saccate, becoming transformed into a zoosporangium 20-319 × 18-118 μm, with many zoospores; zoospores diplanetic (possibly dimorphic), first-formed zoospore with subapical flagellar attachment, (6·0)6·9(8·5) × (8·5)11·6(160) μm, posterior flagellum the longer; cysts (9·0)10·5(13·5) μm diam., second-formed zoospore (5·5)7·0(9·0) × (7·0)9·5(12·5) μm (volume equivalent *ca* > 225 μm^3), with lateral flagellar insertion **Crypticola entomophaga**

15(11') Thallus initially tubular, sometimes branched, becoming septate, eventually holocarpic; thalloid segments disarticulating, rarely olpidioid and non-septate, each segment becoming transformed into a sporangium or a gametangium; gametangial copulation by a pore [oospores sometimes with complex wall layers]; parasitic in adult nematodes and rotifers, occasionally in their eggs, rarely in other invertebrates 16

15'(11') Thallus always olpidioid or more or less coralloid, non-septate or septate only at the delimitation of reproductive structures, holocarpic; becoming transformed into a sporangium or oogonium; method of gametangial union by a fertilization hypha or not described; parasitic in ciliate protozoa or in nematode eggs, sometimes in adult nematodes and rotifers 18

16(15) Thallus initially tubular, sometimes branched, becoming septate, thalloid segments more or less disarticulating, segments becoming transformed into sporangia or gametangia; sporangia with short exit tubes of more or less even diameter with hyaline protoplasm; sporangiospores either zoospores or spherical aplanospores 17

16'(15) Thallus tubular, dividing by successive bipartitioning, segments becoming transformed into sporangia, rarely into gametangia; sporangia with broad, long, often tapering exit tubes containing most of the sporogenous protoplasm; zoospores not known, aplanospores elongate (*Gonimochaete*) **KEY V**

17(16) Most species in aschelminth hosts, but some in other invertebrates or on animal substrata; thalloid segments becoming transformed into sporangia or gametangia; oogonia with or without antheridial gametangia, oospores aplerotic with smooth or ornamented walls (*Myzocytiopsis*) **KEY IV**

17'(16) Species normally confined to aschelminth hosts (rarely in tardigrades); thalloidal segments transformed into sporangia or chlamydospores; gametangia and oospores not known (*Chlamydomyzium*) **KEY VI**

18(15') Parasitic in aschelminths [thallus allantoid or filamentous coralloid, becoming transformed into a zoosporangium or oogonium (donor gametangia absent); sexual reproduction present or absent] 20

18'(15') Parasitic in ciliate protozoa [see also key points 42, 42', 99] 19

19(18') Parasitic in *Kahniella* cysts from soil; thallus with assimilative phase more or less spherical, producing coralloid hyphae (intramatrical 'sporangiophores') 2-6 μm diam. giving rise to zoosporangia [zoosporangia pyriform, 22-30 × 30-44 μm, separated from the hyphae by a septum; zoosporogenesis intrasporangial without a central vacuole; zoospores 3·0 × 6·0 μm, tapered posteriorly, flagellar insertion subapical; zoospore cysts 4·0 μm diam.; sexual reproduction heterothallic between two young adjacent thalli, oosporogenesis periplasmic, oospore 10-15 μm diam., with a smooth or slightly irregular two-layered wall up to 2 μm thick, and a central reserve globule 3·5-7·0 μm in diam.] **Ciliatomyces spectabilis** (monotypic genus)

19'(18') Parasitic in *Podophrya* on *Nassula* from freshwater; thallus initially plasmodial, quickly becoming walled, spherical or allantoid, (8)10-38(80)μm diam.; transformed into a sporangium or a gametangium [zoosporangia with intrasporangial sporogenesis; zoospores encysting peripherally within the zoosporangium; zoospores ellipsoidal, diplanetic, dimorphic (?), 3·0 × 5·0 μm; heterothallic, gametangia not contiguous; oogonia ovoid (15)20(44) μm diam., with 1(2) oospore(s); antheridia allantoid, 6-10 × 18-52 μm, with up to 8 antheridial cells with fertilization hyphae; oospores (16)18(22) μm diam., with a thick two-layered wall and two large refractive globules (ooplast and lipid reserve)] [see also KEY XII] **Eurychasmopsis multisecunda** (monotypic genus)

20(18) Parasitic in adult nematodes and rotifers; thallus allantoid [thallus becoming transformed into a zoosporangium; zoosporès with sub-apical flagellar insertion; sexual reproduction not known] *(Haptoglossa)* **KEY XVI**

20'(18) Parasitic in nematode eggs [also in axenic culture]; thallus filamentous with coralloid hyphae [hyphae (8)12(15) μm diam., becoming septate before oosporogenesis; extramatrical hyphae narrow (6 μm diam.) functioning as zoosporangial exit tubes; zoospores laterally biflagellate, 7·0-11·0 μm, occasionally diplanetic; sexual reproduction in lateral diverticula 35 μm long, without antheridia; oospores single 20μm diam., with a very thick (4 μm) double wall with a thin separable endospore membrane and a reserve body 9 μm diam.] **Nematophthora gynophila** (monotypic genus)

21(9') Mycelium extensive, lacking specialized hyphal branches with glutinous tips; not known to be anchored to substrata by hyphal secretions **22**

21'(9') Mycelium of limited extent, with specialized hyphal branches with tapered glutinous tips; anchored to the substratum (often a *Cladophora* branch) [thallus non-septate, 5·6-10·4 μm diam.; parasitizing rotifers (*Monostyla, Distyla* and *Colurus*); zoospore initials emerge and encyst; cysts 6.8-10.2 μm diam.; zoospores of principal form, 6·0-6·7 × 9·0-10·4 μm; oogonia crenulate, with a single oospore **Sommerstorffia spinosa** (monotypic genus)

22(21) Parasitic on rotifers *(Hydatinophagus)* **23**

22'(21) Parasitic on the mantle cavity of freshwater bivalves (*Musculium*) [hyphae forming a rope-like strand becoming anchored between host gills; hyphae 200-950 μm long; sporangia not described, but separated from hyphae by septa; zoosporogenesis not described; zoospores biflagellate but not known to be motile, 7·0-22·0 μm long] [doubtful taxon, *incertae sedis*, known only from original description] **Endosphaerium funiculatum** (monotypic genus)

23(22) Parasitizing rotifers (*Hydatina*); antheridia present [no measurements available, known only from original description] **Hydatinophagus apsteinii** (type species)

23'(22) Parasitizing rotifers (*Monostylus*); antheridia not observed [intramatrical mycelium extensive, branched, hyphae 2-4 μm diam.; asexual reproduction not observed; oogonia spherical, 8·5-11·2 μm diam., oospores aplerotic, with a single oospore 7-9 μm diam., known only from original description] **Hydatinophagus americanus**

24(10') Saprotrophs **25**

24'(10') Parasites **28**

25(24) Saprotrophic (parasitic?) in plant debris (usually isolated using pollen grain bait) [see also key point 126', *Pleocystidium pygmaeoides*] **26**

25'(24) Saprotrophic in animal substrata (usually isolated using keratin substrata, e.g. snake skin or hair) (majority of the *Leptolegniellaceae*, see also key point 34' and Keys IV: *Myzocytiopsis* and XIV: *Hyphochytrium*) **KEY IX**

26(25) Thallus lobed or olpidioid; heterothallic or homothallic with donor gametangia; oospores large (> 18 μm diam.) **27**

26'(25) Thallus pseudomycelial or coralloid; apparently automictic; oospores small (< 15 μm diam.) [saprotrophic in *Zea* leaves and pine pollen; thallus of intramatrical hyphae of irregular diameter (5-40 μm diam.), much branched, occasionally septate; thalloid segments becoming transformed into sporangia (15-200 × 6-8 μm), or more or less inflated vesicles (<40 μm diam.) functioning as gametangia; sporogenesis intrasporangial, initials gliding out and encysting, cysts 5·0-8·0 μm diam.; zoospores (4·0-5·0 × 6·0-7·0 μm), with lateral flagellar insertion, diplanetic (?); oospores spherical or subspherical, 9-12(22) μm diam., with hyaline wall 3·0-5·0 μm thick] **Aphanomycopsis saprophytica**

27(26) Saprotrophic in pollen grains; thallus bottle-shaped, transformed into a zoosporangium, zoospores 8·0 × 11·0-14·0 μm, with lateral flagellar insertion; zoospore cysts 6·0-8·0 μm diam., with hollow prolongations, the spore ball forming a hollow fenestrated sphere; resting spore formed after multiple colonization of a pollen grain (heterothallic ?); plerotic ?; oospore 20-30 μm diam., with a large central globule (ooplast?) **Ducellieria chodatii** (monotypic genus)

27'(26) Saprotrophic in pollen grains; thallus with irregular lobes or allantoid, transformed into a zoosporangium; zoospores 5·0 × 8·0 μm, with lateral flagellar insertion; gametangia formed by septation of the thallus, rarely heterothallic, donor gametangium forming a fertilization tube; oospore aplerotic, spherical, 18-29 μm diam., with large globule [and/or double wall?] **Cornumyces pygmaeus**

28(24') Freshwater or terrestrial; parasitic in higher plants [see also *Jaraia salicis* (Němec, 1931)] **29**

28'(24') Freshwater; parasitic in other aquatic organisms (fungi, algae, protozoa) **32**

29(28) Terrestrial; parasitic in *Poaceae* **30**

29'(28) Freshwater; parasitic in roots of *Veronica* or *Potamogeton* **31**

30(29) Parasitic in roots of *Poaceae*; thallus broadly tubular, rarely branched, remaining attached to the penetration tube by a collar, 10-12 × 25 μm up to 7-16 × 200 μm, non-septate, transformed into a sporangium or an oogonium [zoosporogenesis apparently extrasporangial in a vesicle; zoospores 7·0 × 11·0 μm; sexual reproduction probably automictic, perhaps rarely heterothallic; oosporogenesis periplasmic, with up to 4 oospores per oogonium; oospores ellipsoid/cylindrical (7)12(14) × 18(26) μm with walls (1·0)1·5-2·0 μm thick, with an ooplast and partially fused lipid globules] **Lagena radicicola** (monotypic genus)

30'(29) Parasitic in leaves of an aquatic/littoral grass, *Panicum repens* after inundation, not causing hypertrophy; thallus lobed, initially plasmodial, transformed into a yellow/brown resting sporangium, 10-15 × 17-27 μm, with walls 1·7 μm thick; germinating by a short exit tube to release biflagellate, reniform zoospores [dimensions not given; zoosporogenesis apparently intrasporangial; sexual reproduction not known; known only from original description; not a *Petersenia*] '*Petersenia panicicola*'

31(29') Parasitic in root hairs of *Potamogeton*, causing hypertrophy; thallus lobed 12-90 μm; zoospores fully formed in the zoosporangium, 3·0-5·0 × 6·0-11·0 μm (volume equivalent <300 μm³) with numerous minute refringent globules, isokont with lateral flagellar insertion; oogonium lobed, without donor gametangium, 40-110 μm, oospore smooth, *ca* 24 μm diam., with an eccentric globule [see also KEY VII] **Cornumyces muenscheri**

31'(29') Parasitic in roots of *Veronica* (*Scrophulariaceae*); thallus more or less spherical, transformed into a zoosporangium [zoosporangial dimensions not given; zoospores 12·0-15·0 μm diam. (volume equivalent >1000 μm³); sexual reproduction not described] [see also KEY XIV] **Cystochytrium radicale** (monotypic genus)

32(28') Parasitic in aquatic fungi (*Peronosporomycetes, Chytridiomycetes*) [see also key point 49' for other parasites of *Peronosporomycetes*] 33

32'(28') Parasitic in diatoms, green algae or protozoa (dinoflagellates) 40

33(32) Parasitic in *Peronosporomycetes*, causing hypertrophy 35

33'(32) Parasitic in *Chytridiomycetes*, not causing hypertrophy 34

34(33') Parasitic in *Catenophlyctis* saprotrophic on snake skin bait; thallus irregularly lobed, 43·0-137·6 × 77·4-156·8 μm, becoming transformed into a zoosporangium; zoosporogenesis intrasporangial; zoospores bean-shaped with lateral flagellar insertion, 3·0 × 5·0-7·0 μm; sexual reproduction not known [compare with keratinophilic fungi, Keys X, XIII and keypoint 25'] **Cornumyces catenophlyctidis**

34'(33') Parasitic in *Karlingia rosea* on onion skin bait; thalli spherical, ovoid or allantoid, (8)50-60(120) μm diam., becoming transformed into a zoosporangium or oogonium; zoosporogenesis intrasporangial (centrifugal); zoospores reniform with lateral flagellar insertion, 6·0-6·5 × 10·0-10·8 μm; resting spores (8)13 × 19(23) μm, wall hyaline, smooth, 1·8-2·2 μm thick **Cornumyces karlingiae**

35(33) Parasitic in *Peronosporomycetidae* [*Pythium*] 38

35'(33) Parasitic in *Saprolegniomycetidae* 36

36(35') Parasitic in hyphae of *Saprolegnia* or *Aphanomyces* 37

36'(35') Parasitic in hyphae and reproductive organs of *Achlya*, sometimes causing host septation; thallus of short finger-like branches or lobes, non-septate, holocarpic; zoosporogenesis partly extrasporangial; zoospores 8·0 × 12·0 μm; resting spores formed in series in branches of the thallus, rectangular or ellipsoidal, 8-12 × 10-15 μm, thick-walled **Cornumyces destruens** (type species)

37(36) Parasitic in hyphae of *Aphanomyces*; thallus allantoid or angular, (10)15-42 × 20-54(90) μm; zoosporogenesis intrasporangial; zoosporangia 1-30 in a hypertrophic swelling, usually with 12 exit tube up to 60-180μm long; zoospores 4·5 × 6·5 μm, cysts 4·0-5·0 μm diam.; oogonia of same size and shape as zoosporangia with 1-12 oospores; antheridia (heterothallic thalli?) 1-4 to an oogonium, 7-18μm diam.; oospores spherical, 8-20 μm diam. or ellipsoid, 6-10 × 15-18 μm, wall hyaline, smooth, 2-3 μm thick [known only from original description] **Cornumyces endogenus**

37'(36) Parasitic in hyphae of *Saprolegnia*; thallus allantoid or lobed; zoosporogenesis intrasporangial; zoospores 4·5-5·0 μm diam.; sexual reproduction not known [known only from original description?] **Cornumyces irregularis**

38(35) Thallus olpidioid; heterothallic; oosporogenesis periplasmic; oogonia uniovulate; oospores aplerotic [oospores smooth-walled] 39

38'(35) Thallus lobed, rarely olpidioid; automictic (?); oosporogenesis not described; oogonia pluriovulate [parasitic in *Pythium intermedium*; thallus of irregular lobes and vesicles (initially lacking a wall?), up to 87 μm diam., or olpidioid 10-12 μm diam., becoming transformed into a zoosporangium or an oogonium; zoosporogenesis intrasporangial, zoospores motile, then quiescent at and after discharge, gradually assuming motility, 4·0 × 5·0-6·0 μm, flagellar insertion lateral; oogonium with 1-20 oospores, 12-17 μm diam.] **Cornumyces utriculobus**

39(38) Parasitic in filamentous-sporangiate *Pythium* species parasitic in *Spirogyra* from water, causing hypertrophy with 1-4 thalli per swelling; thallus spherical, 10-30 μm diam., holocarpic, becoming transformed into a sporangium or a gametangium; zoospores 3·7-4·0 μm diam., diplanetic; heterothallic between two contiguous thalli, fertilization by means of a fertilization tube; oogonium 11·0-18·5 μm diam.; oospores 9-15 μm diam. **Pythiella vernalis** (type species)

39'(38) Parasitic in digitate-sporangiate *Pythium* species from soil; thallus spherical or pyriform, 16-25 μm diam., zoosporogenesis intrasporangial; zoospores 2·1-2·8 × 3·5-4·2 μm; isokont; oogonium 12-23 μm diam., oospore 7-15 μm diam. **Pythiella pythii**

40(32') Parasitic in green algae; thallus more or less mycelial, less frequently coralloid, rarely olpidioid, frequently septate 45

40'(32') Parasitic in chromophyte algae or dinoflagellates; thallus coralloid or olpidioid, rarely septate 41

41(40') Parasitic in diatoms 43

41'(40') Parasitic in dinoflagellates [thallus endobiotic, holocarpic, lobed, non-septate, transformed into either a single zoosporangium or a single oogonium; oospores single, aplerotic, more or less spherical, wall smooth] [see also *Lagenidium* sp. (Canter & Heaney, 1984)] 42

42(41') Freshwater (lakes); parasitic in *Ceratium* [thallus up to 240 μm long, 4-12 μm diam., zoosporangium with 1(2) short exit tubes 3-7 μm long, 4 μm diam., monoplanetic and monomorphic, zoospores with sub-apical flagellar insertion; cysts 6·0 μm diam.; oogonia with 1 to several oospores, oospore 9-18 μm diam. or very irregularly lobed conforming to the thallus shape, wall smooth, contents with clusters of peripheral globules] **Aphanomycopsis cryptica**

42'(41') Freshwater (peat bogs); parasitic in *Peridinium* [with up to 9 thalli per host cell; thallus 24-53 μm diam., with exit tubes 10-120 μm long, 2·5-5·5 μm diam.; oogonia with single oospores, oospore 15-35 μm diam., with a thick wall] **Aphanomycopsis peridiniella**

43(41) Zoospores biflagellate; thallus non-septate, becoming transformed into a zoosporangium or oogonium, or rarely septate with intermixed sporangia and oogonia 44

43'(41) Zoospores uniflagellate; thallus becoming septate [thallus cylindrical 5-12 μm diam., 75-130 μm long, thalloid segments 8-37 μm long, transformed into zoosporangia; zoospores spherical, 3·0-4·0 μm diam. with 1-4 refringent granules; sexual reproduction not known; parasitic in *Synedra*] **Septolpidium lineare** (monotypic genus)

44(43) Thallus cylindrical, ovoid to vermiform, rarely lobed, non-septate, becoming transformed into a sporangium (or gametangium?); zoosporangia often with more than one (up to 10) exit tube(s); zoospores biflagellate, small (volume equivalent $<60\ \mu m^3$); sexual reproduction doubtful [marine sp?: (always dioecious?), oospore plerotic or aplerotic, smooth-walled] (*Ectrogella*) **KEY VIII**

44'(43) Thallus pseudomycelial or coralloid, 6-10 × 150(240) μm; exit tube up to 75 × 9 μm; sporangiospores encysting at the mouth of the long exit tubes; cysts (8)10-12 μm diam.; zoospores laterally biflagellate; oospores 1 to several per thallus, spherical (14)20(24) μm diam., oospore wall thick **Aphanomycopsis bacillariacearum** (type species)

45(40) Thallus coralloid, usually confined to a single host cell 46

45'(40) Thallus mycelial or pseudomycelial, often passing through several host cells, rarely olpidioid (*Syzygangia*) **KEY VII**

46(45) Parasitic in zygotes of *Zygnematales*; thallus coralloid and lobed, becoming septate, thalloid segments transformed into zoosporangia or oogonia; zoospores 4·0-6·0 × 5·0-7·0 μm, with lateral flagellar insertion; oospores single, aplerotic, donor gametangia absent, 13·0-16·5 μm diam., with a thick dentate/striate outer wall **Aphanomycopsis entophyta**

46'(45) Parasitic in desmids (*Netrium*) in peat bog pools; thallus endobiotic, holocarpic, lobed, non-septate, transformed into a sporangium or an oogonium; zoosporangium 5-13 μm diam., with 1(2) exit tubes, 10-185 × 5 μm diam., sporangiospores diplanetic, encysting as a cluster of 10-60 cysts at the mouth of the exit tube, cysts 6·0-8·0 μm diam., zoospores with lateral flagellar insertion; oogonium pluriovulate oospores 9-20 μm diam., with a two-layered (outer radially striate?) wall **Aphanomycopsis desmidiella**

47(2') Parasitic in higher plants, or terrestrial saprotrophs (*Plasmodiophorales* and possibly related taxa) **KEY XVIII**

47'(2') Parasitic in algae, protozoa or fungi (category includes certain *Plasmodiophorales*), or aquatic saprotrophs (on pollen grains) 48

48(47') Parasitic in fungi (*Mycota, Chytridiomycetes* and *Peronosporomycetes*) [57 taxa] 49

48'(47') Parasitic in algae or protozoa, or saprotrophic (parasitic?) in pollen grains [26 taxa] 90

49(48) Parasitic in *Mycota* [*Pleotrachelus*, 4 taxa] or *Chytridiomycetes* [*Rozella*, 12 taxa] 111

49'(48) Parasitic in *Peronosporomycetes* [41 taxa] 50

50(49') Parasitic in *Saprolegniomycetidae* [24 taxa] 67

50'(49') Parasitic in *Peronosporomycetidae* [15 taxa] or *Rhipidiomycetidae* [2 taxa] 51

51(50') Parasitic in *Pythiales* [confined to *Pythium* and *Phytophthora* as far as is known; but see *Rozella* sp. (Dogma, 1978), parasitic in *Myzocytium*, with zoosporangia 21-30 × 12-22 μm; zoospores 2·0 × 4·0 μm; resting spores not known] 53

51'(50') Parasitic in *Rhipidiales* 52

52(51') Zoospores biflagellate [parasitic thallus completely filling the sporangium of *Nellymyces megaceros*, host sporangia distinguishable by the absence of a papilla (poroid dehiscence), zoosporangia 50-135 × 30-54 μm; zoospores smaller than those of the host, but no dimensions given, anisokont with lateral flagellar insertion; resting spores not known; known only from the original description] **Rozellopsis uliginosa**

52'(51') Zoospores uniflagellate [parasitic thallus completely filling the smooth hypertrophied sporangium of *Araiospora spinosa*, host sporangia distinguishable by the absence of a papilla (poroid dehiscence); zoospores elongate, smaller than those of the host, but no dimensions given; resting spores minutely spinulose*, single and aplerotic in spiny sporangia of host, no dimensions given] **Rozella rhipidii**

53(51) Zoosporangium completely filling (except possibly in the case of *Rozella laevis*, keypoint 58) the host thallus segment; zoosporangium wall, if present, not distinguishable from the host wall 54

53'(51) Zoosporangium not filling the host thallus segment or hypertrophic swelling; zoosporangium wall clearly distinct from host wall 59

54(53) Parasitic in *Pythium* 55

54'(53) Parasitic in *Phytophthora* (*P. cactorum*) [zoosporangial measurements lacking; zoospores not described; resting spores not known] **Rozella barrettii** [doubtful taxon]

---

* FOOTNOTE: The reliability of the characters of extra-mural fibrils in the *Olpidiopsidaceae* and the *Rozellopsidaceae* is doubtful: presence or absence of these fibrils; their hair-like or spine-like appearence, and their length all need to be re-evaluated as taxonomic characters. The host ranges of most of the species in these families are also more or less unresearched, but such data that are available indicate that infection of a limited range of species ***within*** a host genus is the norm. Other morphological criteria, on their own, are not sufficient to construct a key, and I have therefore had to use extra-mural fibril and host range information in this key.

55(54) Resting spores known 56

55'(54) Resting spores not reported [parasitic in *Pythium intermedium*, causing much hypertrophy; zoosporangia not in chains, up to 85 μm long; zoospores biflagellate, narrow anteriorly, 2·0-2·5 × 4·0 μm] **Rozellopsis inflata**

56(55) Resting spore wall smooth 58

56'(55) Resting spore wall with short spines 57

57(56') Resting spores numerous in, but not filling hyphal swellings [parasitic in hypertrophic swellings of hyphae of *Pythium vexans* (?); zoosporangia 23 μm diam.; zoospores 2·0 × 4·0 μm, obclavate with narrow anterior end and a prominent posterior globule; resting spores 11-15(18) μm diam., with short regular spines, wall pigmented, moderately thick] **Rozella irregularis**

57'(56') Resting spores single [parasitic in *Pythium*; zoosporangia 20-60 μm diam., with long exit tubes; zoospores 1·5-2·0 × 2·5-3·0 μm; resting spores single, 12-17 μm diam., wall punctate to spiny, rarely smooth [see also *Rozella* sp. (Karling, 1976*a*), parasitic in *Pythium proliferum*; zoospores 1·6-1·9 × 2·6-3·2 μm; resting spores not known] **Rozella longicollis**

58(56) Resting spore wall hyaline [parasitic in *Pythium gracile* (*sic*) causing marked hypertrophy, sporangia solitary, spherical or clavate, 10-20 × 30-112 μm, with 1-3 exit papillae, zoospores obclavate, 1·5-1·8 × 2·9-3·3 μm, occasionally with more than one flagellum; resting spores 11-18 μm diam., wall 1·5-2·0 μm thick] **Rozella laevis**

58'(56) Resting spore wall pigmented [parasitic in sporangia of *Pythium intermedium* and hyphae of *P. monospermum*, sporangium 19·2-24·0 μm diam.; zoospores 2·0 × 4·0 μm obclavate; resting spores 12-18 μm diam., single and aplerotic in host swelling, with a smooth pigmented wall] **Rozella cuculus**

59(53') Biflagellate; flagella heterokont, anisokont or sub-isokont; anterior flagellum straminipilous 60

59'(53') Uniflagellate or if biflagellate then flagella not heterokont; flagella markedly anisokont; anterior flagellum not straminipilous 65

60(59) Thalli in hyperplasic lateral swellings in the *Pythium* or *Phytophthora* host, or hypertrophic swellings in *Phytophthora*; anterior flagellum longer than the posterior flagellum; oogonium without companion cell; oogonium wall smooth 64

60'(59) Thalli in terminal or intercalary hypertrophic swellings of *Pythium* host; anterior flagellum shorter than posterior flagellum; oogonium usually with a companion cell; oogonium wall usually with extramural fibrils or spines 61

61(60') Oogonia with companion cells; oogonia with or without extramural fibrils 62

61'(60') Oogonia without companion cells; oogonia with extramural fibrils [thalli single or numerous up to 50μm diam.; oogonia 19·2-30·0 μm diam., oogonium wall covered with fine extramural fibrils or spines] **Olpidiopsis pythii**

62(61) Thalli (oogonia and sporangia) usually covered with extramural fibrils 63

62'(61) Thalli (oogonia and sporangia) smooth [parasitic in *Pythium aphanidermatum*; thalli 17-28 μm diam.; zoospores 3·5-5·0 μm long; oogonium with 1-3 companion cells, oogonium 13-28 μm diam., oogonium wall 1·7-3·4 μm thick] **Olpidiopsis indica**

63(62) Thalli 9-130 μm diam.; zoospores 2·0-3·5 × 3·0-4·5 μm; oogonia with 1(2) companion cells; oogonia 10-45 μm diam., oogonium wall 0·7-1·3 μm thick **Olpidiopsis brevispinosa**

63'(62) Thalli 12-68 μm diam.; zoospore dimensions not known; oogonia with 1 companion cell, oogonia 17-24 μm diam., oogonium wall thick **Olpidiopsis curvispinosa**

64(60) Parasitic in *Pythium intermedium* and other species of *Pythium* and *Phytophthora* causing lateral hyperplasia; oogonia smooth walled (*contra* Karling, 1981*a*) [thalli 1-40 in a swelling, 4-52 μm diam.; zoospores 3·0-5·0 μm diam.; resting spores 12-27 μm diam.] **Gracea gracilis** (type species)

64'(60) Parasitic in *Phytophthora cryptogea* and *P. megasperma*; resting spores not known [thalli up to 74 μm diam.; zoospores 5·0-8·0 μm long] **Gracea waterhouseae**

65(59') Resting spores smooth-walled; causing hypertrophic hyphal swellings to a greater or lesser extent [parasitic in a wide range, but not all, species of *Pythium sensu lato*] (see also KEY XVIII *Plasmodiophorales*) 66

65'(59') Resting spores finely reticulate (spiny?); causing minimal hypertrophy [parasitic in oogonia and hyphae of *Pythium aphanidermatum*, inducing catenulate septation of hyphae; zoosporangia up to 8 per oogonium, 12-20 μm diam., zoospores elongate 2·5-3·8 × 3·0 μm; resting spores up to 8 per hyphal compartment, separate, spherical, 14-19 μm diam., with a large eccentric lipid (?) body, wall two-layered, outer wall finely reticulate] **Dictyomorpha pythiensis**

66(65) Resting spores loosely clustered in a cystosorus; inducing spherical, intercalary, often catenulate hypertrophic swellings [zoosporangia small, clustered (glomerulate?), 4·3-9·0 μm diam.; resting spore 3·7-5·6 μm diam., wall thick] [see also KEY XVIII, key point 26'] **Woronina pythii**

66'(65) Resting spores closely pressed together to form oval discs one spore thick, discs frequently paired; inducing variable degrees of hypertrophy [zoosporangia 5-9 μm diam.; resting spore 3·0-4·0 μm diam., wall thick] [see also KEY XVIII, key point 12'] **Sorodiscus cokeri**

67(50) Parasitic in *Saprolegniales* 68

67'(50) Parasitic in *Leptomitales* [parasitic thallus completely filling the sporangium of *Apodachlya brachynema*, host sporangia distinguishable by the absence of a papilla (poroid dehiscence); zoospores elongate, smaller than those of the host, but no dimensions given; resting spores minutely spinulose, single and aplerotic in sporangia or oogonia of host, no dimensions given] **Rozella apodachlyae**

68(67) Parasitic in *Achlya sensu lato*, including *Dictyuchus*, *Brevilegnia* and *Geolegnia* [*Olpidiopsis*, *Rozellopsidales* or *Plasmodiophorales* **76**

68'(67) Parasitic in *Saprolegnia*, *Leptolegnia* or *Aphanomyces* [*Olpidiopsis*, *Rozellopsidales* or *Plasmodiophorales*] **69**

69(68') Zoosporangium completely filling the host thallus segment; zoosporangium; or thalli olpidioid; resting spores formed from oogonia and companion cells (*Rozellopsis* or *Olpidiopsis*) **71**

69'(68') Thalli forming glomeruloid sporangiosori, zoosporangium not filling the host thallus segment or hypertrophic swelling] (*Plasmodiophorales*) **70**

70(69') Parasitic in various species of *Saprolegnia* and *Isoachlya*; primary plasmodium transformed into a sporangiosorus; zoosporangia 8-33 μm diam., interconnected; resting spores several - > 100 in a cystosorus, 4·0-8·6 μm diam., walls smooth, pigmented golden-brown [may not be conspecific with *Woronina polycystis* in *Achlya* - see host-range studies, Goldie-Smith (1954)] [see also KEY XVIII, key point 27] **Woronina polycystis**

70'(69') Parasitic in *Leptolegnia caudata*; primary plasmodium lunate or hemispherical, 10-130 μm diam., transformed into a sporangiosorus; zoosporangia 4- > 100 in a sporangiosorus, 4·5-8·4 μm diam., interconnected, sometimes linear; resting spores angular, 4-6 μm diam., walls smooth, hyaline [see also KEY XVIII, key point 27'] **Woronina leptolegniae**

71(69) Assimilative thallus not filling host hyphae, not inducing host septation **72**

71'(69) Assimilative thallus completely filling compartments of host hyphae, host septation parasite induced [parasitic in *Saprolegnia*; zoospores 3·0-4·0 × 5·0-8·0 μm, anterior flagellum 5 μm, posterior flagellum 10 μm; resting spores spinulose, single and aplerotic in host compartment, 20 μm diam. **Rozellopsis septigena** (type species)

72(71) Parasitic in *Saprolegnia* [usually in terminal, occasionally intercalary hypertrophic swellings] (*Olpidiopsis saprolegniae sensu lato*) **74**

72'(71) Parasitic in *Aphanomyces* (*Olpidiopsis aphanomycis sensu lato*) **73**

73(72') Companion cells present [parasitic in *Aphanomyces laevis*, sporangia one or several, with 1-3 long exit tubes; oogonia spherical, with a thick endospore wall and a thin exospore wall with extramural conical spines up to 2·5 μm long; companion cells 10-25 μm diam.; oospores 25-50 μm diam.] [placed in synonymy with *O. aphanomycis* by Karling (1981*a*)] **Olpidiopsis luxurians**

73'(72') Companion cells not known or absent [parasitic in unidentified species of *Aphanomyces* (*Aphanomyces laevis* according to some authors); sporangia spherical, solitary or in groups of 2-3, with a single discharge tube; oogonia not known] [zoosporangia 34-94 μm diam.; oogonia ellipsoid, 24-37 × 28-43 μm, wall thick, two-layered, endospore wall 2·3 μm thick, epispore with broad-based extramural fibril spines 2-4μm long; companion cells absent; parasitic in *A. cladogamus* but not *A. laevis* according to Whiffen (1942)] **Olpidiopsis aphanomycis**

74(72) Oogonia smooth or with tuberculate extramural deposits **75**

74'(72) Oogonia with extramural fibrils [thalli spherical or ellipsoid, 3-36 × 45-50(150) μm; zoospores 2·0-4·0 μm long; oogonium spherical, 47-107 μm diam., with extramural fibrils 5-10 μm long; with 1-2 companion cells, 18-28 μm diam., with thin smooth walls (attributions to host taxa in other genera require reassessment)] **Olpidiopsis saprolegniae** var. **saprolegniae** (type species)

75(74) Thalli and oogonia spherical or ellipsoid, walls thin, smooth; with 1-2 companion cells [no measurements available] **Olpidiopsis saprolegniae** var. **laevis**

75'(74) Thalli 15-137 × 18-203 μm, with 1-3 exit tubes; zoospores 2·0-5·0 μm long; oogonia 29-74 μm diam., with a tuberculate deposit; 1-3 companion cells 18-37 μm diam. **Olpidiopsis saprolegniae** var. **indica**

[*Olpidiopsis vexans* Barrett (1912) can scarcely be distinguished from this taxon except in the larger maximum number of zoosporangial exit tubes (up to 6) and the possibly thinner oogonial/oospore wall, but see discussions in Sparrow (1960) and Karling (1981*a*) and pp. 219-220, 340]

76(68) Zoosporangium not filling the host thallus segment or hypertrophic swelling **79**

76'(68) Zoosporangium completely filling the host thallus segment, zoosporangium wall, if present, not distinguishable from the host wall **77**

77(76') Parasitic in centric or subcentric *Achlya* species **78**

77'(76') Parasitic in eccentric *Achlya* species [parasitic in *Achlya flagellata* and *Dictyuchus* species; zoosporangia variable in size depending on host hypha, 72-112 × 30-72 μm; zoospores 2·0-3·0 × 3·0-4·0 μm; resting spores 1- several in a host segment, spiny, (12·6)15·8-17·3(23·7) μm excluding spines, spines 1·6-2·3 μm long; resting spore germination to give posteriorly uniflagellate zoospores] **Rozella achlyae**

78(77') Zoospores biflagellate [parasitic in centric or subcentric *Achlya* species (also *A. flagellata* ?), plasmodia inducing septation of host, with up to 15 contiguous sporangial segments; zoosporangia 25-90 × 60-250 μm; zoospores 2·4 × 6·0 μm; resting spore spiny, 15 μm diam.] **Rozellopsis simulans**

78'(77') Zoospores uniflagellate [parasitic in *Achlya racemosa*; zoosporangia in chains 20-77 × 6-29 μm; zoospores 3·0-4·0 × 5·0-8·0 μm; resting spores spiny, 16-20 μm diam.] **Rozella septigena**

79(76) Zoosporangia single or many, not aggregated to form a sporangiosorus **82**

79'(76) Zoosporangia aggregated or conglomerate, forming a sporangiosorus (*Plasmodiophorales*) **80**

80(79') Forming cystosori with few (< 8) cysts 81

80'(79') Forming cystosori with numerous cysts [parasitic in various species of *Achlya* (including centric and eccentric species); primary plasmodium transformed into a sporangiosorus; zoosporangia 8-33 μm diam., interconnected; resting spores several - > 100 in a cystosorus, 4·0-8·6 μm diam., walls smooth, pigmented golden-brown] [may not be conspecific with *Woronina polycystis* in *Achlya* - host-range studies, Goldie-Smith, 1954; see also the doubtful species *W. asterina* in *A. americana* and *W. sorokiniana* in *A. racemosa*] [see also KEY XVIII, key point 27] **Woronina polycystis**

81(80) Parasitic in *Achlya* [mean cyst diameter < 3·0 μm (2·4-3·2 μm diam.); mean sporangial diam. > 10·0 μm (6·0-16·0 μm diam.); zoospore dimensions not known] [see also KEY XVIII, key point 8] **Octomyxa achlyae**

81'(80) Parasitic in *Brevilegnia linearis* [mean cyst diameter > 3·5 μm (2·9-4·9 μm diam.); mean sporangial diam. < 8·0 μm (5·1-7·6 μm diam.)] [see also KEY XVIII, key point 8'] **Octomyxa brevilegniae**

82(79) Zoospores sub-anteriorly biflagellate (*Olpidiopsis*) 82

82'(79) Zoospores posteriorly uniflagellate [parasitic in *Achlya flagellata*, but causing little or no hypertrophy; thalli small, crowded, transformed into zoosporangia, 15-20 μm diam.; zoospores minute, 1·5-3·5 × 3·0 μm; resting spores formed following the fusion of isoplanogametes which infect the host in the same manner as the zoospores, but which induce host wall formation, resting spores aplerotic within this host cell, one or more per host cell, 8-11 μm diam., with a finely reticulate outer wall] **Dictyomorpha dioica**

83(82) Parasitic in eccentric *Achlya* species (or unidentified *Achlya* species) (*Olpidiopsis spinosa sensu lato*) 86

83'(82) Parasitic in centric or subcentric *Achlya* species (*Olpidiopsis fusiformis sensu lato*) 84

84(83') Oogonia and sometimes companion cells with a spiny extramural deposit 84

84'(83') Oogonium with a irregular or tuberculate extramural deposit, companion cells always smooth walled [parasitic in *A. racemosa* and other racemose-group *Achlya* species; much confused with *O. vexans* in the literature, dimensions probably similar to those of other species in *Achlya*, see keypoints 88'-89'] **Olpidiopsis incrassata**

85(84) Oogonium with a fine fibrillar extramural deposit; companion cell with a similar but less dense investment [dimensions probably similar to those of other species in *Achlya*, see keypoints 88'-89'] **Olpidiopsis index**

85'(84) Oogonium with a coarsely spiny extramural deposit [dimensions probably similar to those of other species in *Achlya*, see keypoints 88'-89'] **Olpidiopsis fusiformis**

86(83) Oogonium with a spiny, fibrillar or verrucose extramural deposit (*Olpidiopsis spinosa sensu lato*) 87

86'(83) Oogonium with a smooth wall (lacking any extramural deposit?) [sporangium 16-60 × 20-70 μm 1-2 exit tubes up to 28 μm long; oogonium 40-44 μm diam. wall 2 μm thick, smooth hyaline companion cell single 20-30 μm diam.] **Olpidiopsis braziliensis**

87(86) At least some oogonia with an irregular or verrucose extramural deposit 89

87'(86) Oogonia, and often companion cells and zoosporangia with a spiny or fibrillar extramural deposit 88

88(87') Oogonium and companion cell spiny, oogonium spherical, 51-73 μm diam.; parasitic in *A. flagellata* **Olpidiopsis spinosa**

88'(87') Oogonium coarsely spiny; parasitic in *A. flagellata*; zoosporangia smooth or spiny, (40)80(140) × (60)200(350) μm; zoospores (2·3)2·8(3·0) × (3·8)4·2(4·6) μm; oogonia (26)52-61(83) μm; companion cells 1-2, (17)26(30) μm diam., with scattered spines **Olpidiopsis varians**

89(87) Oogonia and zoosporangia with a variable extramural deposit, smooth, fibrillar, spiny or verrucose; parasitic in *A. flagellata*; zoospores (2·3)3·1(5·7) × (2·9)4·2(4·3) μm; oogonia 41 × 50(122) μm; companion cells 0-3, 15-25 × 12-30 μm **Olpidiopsis achlyae**

89'(87) Oogonium with an irregular or rugulose extramural deposit; parasitic in *A. glomerata*; zoosporangia 28-40 × 83-103 μm; zoospores 2·4-3·0 × 3·6-4·8 μm; oogonia (22)42-50 × 49-56(67) μm; companion cells 1-5, 11-28 μm diam., smooth-walled; oospore germinating to give zoospores **Olpidiopsis verrucosa**

90(48') Parasitic in algae or protozoa 91

90'(48') Saprotrophic (or parasitic? in pollen grains) 126

91(90) Parasitic in protozoa 99

91'(90) Parasitic in algae 92

92(91') Parasitic in *Chlorophyta* or *Charophyta* 93

92'(91') Parasitic in *Tribophyta* (*Vaucheria*) [mean cyst diameter > 12·0μm; secondary zoospore 2·0-2·6μm diam.] [see also KEY XVIII, key point 25'; note the parasite of a protozoan parasite of *Vaucheria*, *Olpidiomorpha pseudosporae* (key point 107')] **Woronina glomerata**

93(92) Parasitic in *Charophyta* (*Zygnematales*, *Coleochaetales*, *Charales*) 95

93'(92) Parasitic in *Chlorophyta* (*Chaetophorales*, *Oedogoniales*) 94

94(93') Parasitic in *Draparnaldia* [thalli 1-11 per host cell; zoosporangia 14·5-27·0 × 30·5-47·5 μm; zoospores 3·2-4·8 μm diam., amoeboid, with 2 laterally inserted flagella; resting spores not known] **Plasmophagus deformans**

94'(93') Parasitic in *Oedogonium* [zoosporangia 8-11 × 20-23 μm; zoospores 2·0 × 3·0 μm; resting spores not known] **Plasmophagus oedogoniorum** (type species)

95(93) Parasitic in *Zygnematales* or *Coleochaetales* **96**

95'(93) Parasitic in *Charales* (*Chara*) [forming cystosori, cysts 4·0-9·0 μm diam.] [see also KEY XVIII, key point 14'] **Sorodiscus karlingii**

96(95) Parasitic in *Zygnematales* **97**

96'(95) Parasitic in *Coleochaete* [thallus spherical to clavate, (15)33-40 × 40-42(55) μm diam.; zoospores fusiform, 2·0-2·5 × 6·0 7·0 μm; resting spore not known] **Plasmophagus coleochaetis**

97(96) Parasitic in *Zygnemataceae* **98**

97'(96) Parasitic in *Desmidaceae* [thalli lobed in conformity with the desmid semicells, transformed into a zoosporangium (discharge tube 6-8 μm long, 3-6 μm diam.) or an oogonium; zoospores (4·0)5·0 × 6·0 μm; oogonia 9·0-11·5 × 28·0 -34·0 μm; companion cells not known] **Pleocystidium lundiae**

98(97) Parasitic in *Spirogyra* and *Mougeotia* [zoosporangia ellipsoid, 17 × 32-41 μm, usually with a single discharge tube; zoospores 7·0 × 8·0 μm, biflagellate; oogonia smooth-walled, 19-24 μm diam.; companion cells 1-4, 10·0-16·6 μm diam.] **Pleocystidium parasiticum** (type species)

98'(97) Parasitic in *Mougeotia* [forming a round cluster of 10-20 cysts; zoospores discharged through exit tubes] [doubtful species; see also KEY XVIII, key point 25'] **Woronina aggregata**

99(91) Sporangiospores zoosporic or aplanosporic, spherical or pyriform **100**

99'(91) Sporangiospores aplanosporic, arcuate with a knob at one end [parasitic in *Dinobryon*; sporangia spherical, 6·0-8·5 μm diam., aplanospores 0·7 × 3·5-4·0 μm; dehiscence not known; resting spore more or less spherical, (4·3)4·0(4·7) × (6·5)7·0(8·0) μm, with a large hyaline globule, wall thick, smooth or minutely hirsute] [compare with *Gonimochaete*, Key V] **Pseudosphaerita dinobryonis**

100(99) Parasitic in amoebae or non-euglenoid flagellates [resting spores not known] **105**

100'(99) Parasitic in euglenoids **101**

101(100') Plasmodium apocytic, developing from a uninucleate stage to a multinucleate stage **103**

101'(100') Plasmodium said to be schizogenous, dividing after every nuclear division [zoospores said to be biflagellate] (*Pseudosphaerita*) **102**

102(101') Parasitic in *Phacus*; zoosporangia 16-20 μm diam.; zoospores 1·5 μm diam.; resting spores not known **Pseudosphaerita drylii**

102'(101') Parasitic in *Euglena*; sporangia 1-several in a host cell, spherical or irregular; zoospores pyriform, 2·5-3·0 × 6·0 μm; resting spores minutely spiny **Pseudosphaerita euglenae** (type species)

103(101) Resting spores spiny **104**

103'(101) Resting spores not known [parasitic in *Phacus*; uninucleate stage elongate, small (2 × 8-9 μm) and with terminal vacuoles; zoospores not known] **Sphaerita phaci**

104(103') Parasitic in *Euglena*; zoosporangium spherical; zoospores spherical or ovoid, 1·5-2·0 μm diam,; resting spore 8 μm diam. **Sphaerita dangeardii**

104'(103') Parasitic in *Trachelomonas*; sporangia (18)20(29) μm diam.; resting spores present [doubtful species, no other diagnostic details] **Sphaerita trachelomonadis**

105(100) Parasitic in amoebae; zoospores not known, of if known then flagellation not known **108**

105'(100) Parasitic in non-euglenoid flagellates; resting spores not known **106**

106(105') Parasitic in cryptomonads or parasites of *Vaucheria* **107**

106'(105') Parasitic in flagellate protozoa (*Colpoda*) from terrestrial plant detritus [with biflagellate zoospores] (*Endemosarca*) **109**

107(106') Parasitic in cryptomonads [sporangia 1-3 in a host cell, released by disintegration of host cell; zoospores elongate [dimensions not known] **Pseudosphaerita radiata**

107'(106') Parasitic in *Pseudospora* zoocysts parasitic in *Vaucheria*, zoosporangium ovoid, 8-14 μm diam., with a single exit tube; zoosporogenesis intrasporangial; zoospores obpyriform, 2·0 × 3·0-4·0 μm, uniflagellate with a single trailing sub-apically inserted flagellum [known only from original description] **Olpidiomorpha pseudosporae** (monotypic genus)

108(105) Parasitic in gut amoebae [not diagnostically distinct] **Sphaerita normeti**
Sphaerita entamoebae

108'(105) Parasitic in amoebae [not diagnostically distinct] **Sphaerita endogena** (type species)
Sphaerita minor
Sphaerita amoebae
Sphaerita plasmophaga
Sphaerita nucleophaga

109(106') Zoosporangial discharge tube present **110**

109'(106') Zoosporangial discharge tube absent [zoosporangium (12)16(23) μm diam.; zoospores (1·5)1·9(2·4) × (4·2)4·9(5·4) μm; encysted spores (2·2)2·9(3·8) × (3·3)4·3(5·0) μm] **Endemosarca anomala**

110(109) Zoosporangial discharge tube slender, (2·0)3·6(5·0) × (42)58(74) μm [zooporangium (15)19(23) μm diam.; zoospores (1·9)2·9(4·0) × (4·7)5·6(6·6) μm; encysted spores (2·0)2·1(3·0) × (4·0)4·8(5·0) μm] **Endemosarca hypsalyxis**

110'(109) Zoosporangial discharge tube broad, (10·0)11·6(17·5) × (15)20(30) μm [zoosporangium (25)32(45) μm diam.; zoospores (1·7)2·7(3·8) × (4·7)5·7(7·0) μm; encysted spores (2·5)2·9(3·0) × (5·0)5·6(6·0) μm] **Endemosarca ubatubensis**

111(49) Parasitic in *Mycota*; zoospores uniflagellate **123**

111'(49) Parasitic in *Chytridiomycetes* **112**

112(111') Parasitic in *Chytridiales* (or *Spizellomycetales*?) **115**

112'(111') Parasitic in *Monoblepharidales* or *Blastocladiales* **113**

113(112') Parasitic in *Blastocladiales* **114**

113'(112') Parasitic in *Monoblepharidales* [parasitic in *Monoblepharis polymorpha*; zoosporangial measurements lacking; zoospores not described; resting spore measurements lacking] **Rozella monoblepharidis** (type species)

114(113) Parasitic in *Allomyces* [parasitic in *Allomyces*; zoosporangia 20-250 × 12-90 μm; zoospores 2·0-3·0 × 3·5-4·0 μm; resting spores spiny 12-20 μm diam.] [*Rozella* sp. (Kobashi & Konno, 1970) with zoosporangia 21-35 × 14-16 μm; zoospore details not recorded; resting spores 12-18 μm diam. will also key out here] **Rozella allomycis**

114'(113) Parasitic in *Blastocladia* [parasitic in *Blastocladia pringsheimii*; zoosporangial measurements lacking; zoospores not observed; resting spore measurements lacking] **Rozella blastocladiae**

115(112) Resting spores absent **121**

115'(112) Resting spores present **116**

116(115') Parasitic in saprotrophic chytrids [*Cladochytrium* may occasionally be parasitic in green algae] **119**

116'(115') Parasitic in parasitic chytrids **117**

117(116) Parasitic in chytrids parasitic on *Chlorophyta* or *Charophyta* **118**

117'(116) Parasitic in chytrids parasitic on *Bacillariophyta* [parasitic in *Zygorhizidium affluens* on *Asterionella*; zoosporangia 4·0-9·5 μm diam., with one exit papilla; zoospores, with a broader anterior end containing 2-5 refractive granules, 1·8 × 3·5 μm; resting spores with an outer rugose wall, 5-11 μm diam.] **Rozella parva**

118(117) Parasitic in *Diplophlyctis intestina* on *Nitella*; zoosporangia 18-28 μm diam. with 1-4 discharge tubes; zoospores pyriform, 1·8 × 2·5 μm; resting spores echinulate, 16-22 μm diam. **Rozella diplophlyctidis**

118'(117) Parasitic in *Chytridium oedogonii* on *Oedogonium*, not causing hypertrophy; zoosporangia 29-39 × 13-15 μm; zoospores 1·5 × 3·0 μm; resting spores 13-20 × 8-13 μm with reticulations and spines **Rozella canterae**

119(116) Host thallus monocentric [parasite developing in host sporangium derived from host zoospore cyst] **120**

119'(116) Host thallus polycentric [parasitic in several species of *Nowakowskiella* and *Cladochytrium*, causing slight to marked hypertrophy of the rhizomycelium; zoosporangia 10-40 × 10-35 μm; zoospores 3·0-3·5 × 1·8-2·0 μm; resting spores 8-22 μm diam., smooth-walled or spiny] **Rozella cladochytrii**

120(119) Parasitic in *Rhizophlyctis petersenii*; zoospores 2·5-3·0 × 1·5-2·0 μm; resting spores 14-18 μm diam., spiny [zoosporangium dimensions not given] **Rozella rhizophlyctidis**

120'(119) Parasitic in *Chytriomyces hyalinus*, not causing hypertrophy; zoosporangia 40 × 10 μm; zoospores 1·5 × 3·0 μm; resting spores single, almost plerotic, 7-20 μm diam., strongly pigmented, usually echinulate **Rozella chytriomycetis**

121(115) Parasitic in parasitic chytrid sporangia **122**

121'(115) Parasitic in sporangia of saprotrophic chytrids [parasitic in *Endochytrium operculatum*, not causing hypertrophy; zoosporangia 15-500 μm; zoospores 1·5 × 3·4-4·0 μm] **Rozella endochytrii**

122(121) Parasitic in *Rhizophydium globosum*, parasitic on green algae; zoosporangia 15-30 μm diam, with 1-3 exit papillae; zoospores with a small posterior globule 2·0-2·5 × 3·0-4·0 μm **Rozella rhizophydii**

122'(121) Parasitic in *Polyphagus euglenae*, parasitic on euglenoids; zoosporangia 20-48 μm; zoospores 1·5-2·0 × 2·0-3·0 μm **Rozella polyphagi**

123(111) Parasitic in *Zygomycotina* **124**

123'(111) Parasitic in *Basidiomycotina* **125**

124(123) Parasitic in *Pilobolus crystallinus* and *P. kleinii*; zoosporangia 1-80 in hypertrophied parts of host, 6-250 μm diam., with colourless or pigmented cellulosic wall and 10-30 exit tubes; zoospores 2·2-3·0 μm diam., obovoid with posteriorly inserted flagellum; resting spore not known **Pleotrachelus fulgens** (type species)

124'(123) Parasitic in trophocysts of *Pilobolus pirottianus*; sporangium 41-47 μm diam., pale yellowish orange, with 1-2 exit tubes; zoospores 3·0-4·0 μm long, ovoid; resting spore spherical, thick-walled [doubtfully distinct from the type species according to Sparrow (1960: 160)] **Pleotrachelus zopfianus**

125(123') Parasitic in *Itersonilia perplexans* (*Basidiomycetes*); sporangia lacking a definite wall; zoospores globose 2·0-2·5 μm diam., resting spores not known [known only from the original collection] **Pleotrachelus itersoniliae**

125'(123') Parasitic in basidia of *Corticium comedens* (*Basidiomycetes*); sporangia ovoid [no other details recorded] **Pleotrachelus vuilleminii**

126(90') Fungi saprotrophic in terrestrial detritus (dung, litter) or freshwater (*Diplophrys*) 127

126'(90') Fungi saprotrophic (parasitic ?) in germinating pollen grain tubes [thallus spherical, pyriform or deformed 15-40 μm diam.; zoospores 2·5-3·0 × 4·0-5·0 μm; cysts subspherical or angular, 3·5-4·5 μm diam.; resting spores 1-8 in a host cell, apparently asexual and plerotic, 10-24 μm diam. with a thick, hyaline, smooth wall] **Pleocystidium pygmaeoides**

127(126) Sorocarps [sporocarps] on pedicels (*Heimerliaceae*) 129

127'(126) Sorocarps sessile [or sorocytes single] 128

128(127') Sorocytes ellipsoid, 2·4-4·8 × 4·8-9·6 μm, uninucleate, enlarging to vegetative cells 4·8-9·6 × 7·2-16·8 μm with fine branched ectoplasmic extensions; sorocytes aggregating in single file or abreast to form sorocarps [sorocarps on cow and horse dung, golden yellow, 100-430 μm diam.] **Sorodiplophrys stercorea**

128'(127') Sorocytes [cells] broadly ellipsoid [no measurements] with two, not always diametrically opposed, groups of filiform pseudopodia, points of origin sometimes replaced by small depressions [cell with a subcentric amber-coloured globule] **Diplophrys archeri**

129(127) Pedicel simple 130

129'(127) Pedicel with a collarette [fructification 1000 μm long, lacking a columella but with a conical collarette 100 μm diam., sporangia 65μm diam., spores ellipsoid 6·0-7·0 μm long] **Enigma vorticelloides** *nom. illeg.* (monotypic genus)

130(129) Spores spherical, <2·5 μm diam. 131

130'(129) Spores ellipsoidal, >3·0 μm long 133

131(130) Columella conical (rarely absent) 132

131'(130) Columella cylindrical, lobed [fructification 265 μm long, sporangium with a short cylindrical lobed columella, sporangia 40-65μm diam., spores ovoid 1·8-2·0μm long] **Pygmomyces microsporus** *nom. illeg.* (monotypic genus)

132(131) Sporangium with a conical columella [fructification 250 μm long, sporangia 50 μm diam., spores spherical 2·0-2·2 μm diam.] **Rhabdocystis columnaris** *nom. illeg.*

132'(131) Sporangium lacking a columella [fructification 330 μm long, sporangia 50 μm diam., spores spherical 1·25 μm diam.] **Rhabdocystis peniophorae** *nom. illeg.*

133(130') Pedicel straight 134

133'(130') Pedicel twisted below sporangium [fructification 730-1200μm long, twisted below the sporangium, sporangium with a columella, sporangia 100-150μm diam., spores ellipsoid 3·0-3·5μm long] **Calospeira elegans** *nom. illeg.* (monotypic genus)

134(133) Columella extending through and beyond the sporangium [fructification 260-500μm long, sporangia 66μm diam., spores ellipsoidal 4·0-5·0μm diam.] **Heimerlia hyalina** (monotypic genus)

134'(133) Sporangium lacking a persistent columella [fructification 300μm long, sporangia 75μm diam., spores ellipsoid 2·5-3·0μm long with gelatinous tips] **Heliomycopsis fusispora** *nom. illeg.* (monotypic genus)

135(1') Thallus initially plasmodial or remaining plasmodial throughout the assimilative phase; thallus more or less olpidioid or allantoid or coralloid [assimilative coenocytes occasionally dividing; zoospores anisokont, not known to be straminipilous] 140

135'(1') Thallus with an investment of scales or a wall throughout the assimilative phase; thallus monocentric or more or less mycelial, olpidioid or allantoid or coralloid [septate or non-septate, thalloidal segments disarticulating or not](*Labyrinthista*, *Pythiales*. *Anisolpidiales*, *Myzocytiopsidales*) 136

136(135) Thallus walled, more or less mycelial, allantoid or olpidioid; zoospores heterokont (straminipilous) and more or less anisokont, anterior flagellum the shorter; or biflagellate, but with flagellar details not yet determined 137

136'(35) Thallus with an investment of scales, usually with sagenogenetosomes and ectoplasmic nets; zoospores heterokont (straminipilous) and anisokont, anterior flagellum the longer; or aplanosporic (*Labyrinthista*) **KEY XVI**

137(136) Parasitic in algae (rarely parasitic in parasitic fungi of algae) 139

137'(136) Parasitic in marine animals or saprotrophic 138

138(137') Parasitic in marine *Crustacea* (rarely *Rotifera*) or saprotrophic (Salilagenidiales) **KEY X**

138'(137') Parasitic in nematodes (Keys IV, V) **16**

[return to key points 16-17', or go directly to Keys IV, V]

139(137) Parasitic in thalloid or filamentous red, brown or green algae **KEY XI**
139'(137) Parasitic in diatoms (*Ectrogella*) **KEY VIII**

140(135) Parasitic in algae 141
140'(135) Parasitic in vascular plants (*Plasmodiophorales*) **KEY XVIII**

141(140) Parasitic in chromophyte algae (*Fucophyceae, Bacillariophyceae*) 142
141'(140) Parasitic in *Cystoclonium* (*Rhodophyta*) [plasmodium penetrating several adjacent cells, followed by schizogeny to produce fusiform cells, aggregating into a sorus, each cell then producing 8 zoospores, 2·5 × 4·5 μm, with eyespot] **Pyrrhosorus marinus** (monotypic genus)

142(141) Parasitic in *Bachelotia* (*Pilayella*) and *Hincksia* and *Ectocarpus*; plasmodium presumed to be inside the host plasmamembrane adjacent to plastids and nucleus, partly or completely filling the host cell; oblong, 15-25 × 7-130 μm, schizogenous producing 1-5 plasmodia in a host cell, each part becoming transformed into sporangiosori, 5-50 × 10-125 μm, each of which produces 1-several elongate-pyriform biflagellate zoospores, 2·0-2·5 × 3·5-4·0 μm; flagella sub-apically inserted **Phagomyxa algarum** (monotypic genus)
142'(141) Parasitic in diatoms; plasmodial phase situated between the frustule and the plasmamembrane, tubular and branched, siphons (6)10-20(36) μm diam., rhizoids not known, septa not known; becoming walled and transformed into a zoosporangium; zoospores 5·0-6·2 × 6·4-8·0(10·0) μm, biflagellate, flagella laterally inserted [diplanetic, first- formed cyst spiny; sexual reproduction by means of copulation between encysted, previously motile gametes] **Lagenisma coscinodisci** (monotypic genus)

KEY II: *MYZOCYTIUM* AND *LAGENIDIUM*

1 Parasitic in *Zygnematales* 2
1' Parasitic in aquatic insects (especially mosquitoes) [Thallus branched, septate, segments more or less swollen but not disarticulating; in insects; zoospores 9·5 μm long (volume equivalent *ca* 350 μm³); oospores 18-30 μm diam. **Lagenidium giganteum** (proposed new type species)

2(1) Thallus tubular, rarely branched, becoming septate, confined to a single host cell; sexual reproduction known 3
2'(1) Thallus of a single ovoid holocarpic cell [thallus transformed into a sporangium; zoospore 7·0 μm long (volume equivalent 70 μm³); sexual reproduction not known; in *Netrium* (*Mesotaeniaceae*, saccoderm desmids)] **Lagenidium netrii**

3(2) Thallus subspherical or ellipsoid, segments becoming inflated with thickened refractive-hyaline septa (Myzocytium *sensu stricto*) 4
3'(2) Thallus segments elongate, scarcely inflated and septa not noticably thickened or refractive-hyaline, occasionally branched [parasitic in *Spirogyra*; zoospores 8·5 μm long (volume equivalent 170 μm³); oospores 10-15 μm diam. **Myzocytium (Lagenidium) rabenhorstii**

4(3) Thallus rapidly developing thickened hyaline septa; sporangia 15-20 × 13-16 μm; sporangial exit tube of uniform diameter; zoospores 10·0 μm long (volume equivalent 180 μm³); oospores 8-25 μm diam. **Myzocytium proliferum** (type species)
4'(3) Thallus, when young, with segments sometimes separated by refractive plugs or septa; sporangia 19-32 × 13-15 μm; sporangial exit tube with an inflated appressorium below the host wall; zoospores 6·5 μm long (volume equivalent 70 μm³); oospores 13-15 μm diam. **Myzocytium megastomum**

KEY III: *CYSTOSIPHON*

1 Thallus passing through several host cells [thallus more or less filamentous, septate] 2
1' Thallus confined to one host cell 3

2(1) Sporangia more or less spherical; in *Lemnaceae* [known only from original description; zoospore cyst *ca* 9.0 μm diam. (volume equivalent >350 μm³); oospore *ca* 18 μm diam.] **Cystosiphon pythioides** (type species)
2'(1) Sporangia more or less indifferentiated intercalary thallus segments; in *Zygnemataceae* [zoospore volume equivalent <250 μm³; flagella resorbed; oospore *ca* 17·5 μm diam. (Sparrow), *ca* 14·5 μm diam. (de Wildeman)] **Cystosiphon dictyosporum**

3(1') In vegetative cells of desmids [see also *Syzygangia papillosa*, the morphology of the fertilization process is critical] 4
3'(1') In oogonia of *Chara* [oospore *ca* 16 μm diam.] **Cystosiphon reducta**

4(3) Thallus extensive, more or less mycelial, septate at maturity; in *Closterium*; zoospore volume equivalent <50 μm³; oospore *ca* 11 μm diam. **Cystosiphon closterii**
4'(3) Thallus saccate, lobed, non-septate; in *Staurastrum*, *Cosmarium*; zoospore volume equivalent >100 μm³; oospore *ca* 14 μm diam. **Cystosiphon canterae**

KEY IV: *MYZOCYTIOPSIS*

1 Oospores distinctly ornamented with a reticulate or echinulate oospore membrane (not necessarily the outermost spore membrane) 2
1' Oospores smooth-walled (or merely the inner surface of the endospore slightly irregular) 5

2(1) Thalloidal segments broadly cylindrical (L:B ratio > 1:2); zoospores or aplanospores produced 3
2'(1) Thalloidal segments sub-spherical (14-30 × 18-38 μm); zoospores produced, but motile phase very brief, encysting shortly after emergence or within the zoosporangium forming aplanospores [zoospores 8·0-12·0 μm long, pyriform with lateral flagellar insertion; aplanospores elongate, secreting adhesive pads; oospores 15-22 μm diam.] **Myzocytiopsis glutinospora**

3(2) Sporangiospore cysts ovoid or hemispherical [intrasporangial spores flagellate or becoming flagellate in a precipitative vesicle]; antheridial gametangia produced 4
3'(2) Sporangiospore cysts spherical; antheridial gametangia not formed. [Oospore 8-16 μm diam. or 5-8 × 12-20 μm (volume equivalent <1000 μm³), wall deeply wrinkled [thallus of up to 5 segments, 8-12 × 20-90 μm; intrasporangial aplanospores (4·5-6·0 μm diam.) forming an *Achlya*-like spore-ball from which polyplanetic zoospores are produced; zoospores 4·0-5·0 × 5·5-6·5 μm, volume equivalent >60 μm³] **Myzocytiopsis parthenospora**

4(3) Sporangial thalloidal segments smaller (4-6 × 5-37 μm) than oogonial thalloidal segments (15-20 × 21-28 μm); zoospores fully formed in the sporangium [oospores 17-22 μm diam.] **Myzocytiopsis vermicola**
4'(3) Sporangial and oogonial thalloidal segments of comparable size (8-16 × 15-50 μm); zoosporogenesis either within the sporangium or development completed at the exit tube orifice after a quiescent phase [zoospores 2·0-3·0 × 4·0-5·0 × 8·0-12·0 μm, reniform with lateral flagellar insertion; forming hemispherical cysts on the pellicle of the host; oospores 14-22 μm diam.] **Myzocytiopsis lenticularis** (type species)

5(1') Intrasporangial spores not motile within the sporangium; aplanosporic, or behaving in an *Achlya*-like manner, or completing development at the mouth of the exit tube 6
5'(1') Intrasporangial spores flagellate, motile within and swimming away from the sporangium 13

6(5) Sporogenesis intrasporangial (spore initials largely defined before emergence) 9
6'(5) Sporogenesis extrasporangial (sporogenous protoplasm extruded with only rudimentary initial delimitation) 7

7(6') Thallus extensive, becoming septate 8

7'(6') Thallus reduced, of 1-4 ovoid thalloidal segments 14-20 × 30-36 μm [zoospores reniform, 4·0-5·0 × 6·5-7·0 μm; oospores 10-16 μm diam.; parasitic in rotifer eggs] **Myzocytiopsis fijiensis**

8(6) Thallus tubular, seldom branched, of 2-8 segments 6-9 × 30-40 μm diam.; oospores 12-18 μm diam.; in rotifer eggs [zoospores 6·0 × 8·0 μm; oogonia spindle-shaped, antheridia elongate] **Myzocytiopsis distylae**

8'(6) Thallus extensive, richly branched, segments inflated but not disarticulating, 4-20 μm wide; sporangial segments 15-200 μm long; oospores 10-18 μm diam.; saprobic in keratin substrata [zoospores 3·5-4·5 × 7·5-8·0 μm, reniform] **Myzocytiopsis humana**

9(6) Intrasporangial spores developing into zoospores 11

9'(6) Intrasporangial spores strictly aplanosporic 10

10(9') Aplanospores clavate, 3·5 × 17·0-20·5 μm [adhesive pads presumed to be present]; copulating gametangia laterally orientated (from non-adjacent thalloidal segments?) [thallus becoming segmented with up to 9 thalloidal segments; oospores 15-30 μm diam.] **Myzocytiopsis subuliformis**

10'(9') Aplanospores spherical, 3·0-3·5 μm diam.; gametangia automictic, oogonial segments 22-28 μm, inflated only in the central zone, leaving distinctive vesiculate terminal parts which eventually collapse to form a ridged envelope around the oospore [thallus tubular, extensive and branched; sporangial segments 9-16 × 45-145 μm, intrasporangial aplanospores forming a loose aggregation on discharge; oospores 12-18 μm diam.] **Myzocytiopsis osiris**

11(9) Planonts becoming flagellate after planont emergence but without a quiescent immobile phase; in rotifers 12

11'(9) Planonts completing maturity after a quiescent phase at the mouth of the exit tube; cysts spherical, producing medium-large zoospores (5·0-6·0 × 8·0-10·0 μm, volume equivalent *ca* 140 μm³) with lateral flagellar insertion; in nematodes [oogonial and antheridial segments adjacent; oospores 10-17 μm diam.] **Myzocytiopsis intermedia**

12(11) Zoospores medium-large (6·0-7·0 × 10·0-11·0 μm, volume equivalent *ca* 140 μm³); mean oospore diameter >13 μm (12-15 μm diam.) [thalloidal segments 5-17 μm wide, length variable] **Myzocytiopsis zoophthora**

12'(11) Zoospores small, bean-shaped (2·5 × 3·5 μm, volume equivalent *ca* 20 μm³); mean oospore diam. <12 μm (7-14 μm diam. or 7-9 × 10-11 μm) [thallus usually non-septate, 5-15 × 6-25 μm; cf. *Haptoglossa*] **Myzocytiopsis microspora**

13(5') Zoospore size medium small or medium-large (volume equivalent >60 μm³) 15

13'(5') Zoospore size small (volume equivalent <30 μm³) 14

14(13') Sporangia with short exit tubes; mean oospore diameter >17 μm (6-27 μm diam.); antheridial gametangia not known [thalloidal segments 5-6(10), 11-23 × 11-27 μm; zoospores 2·2-4·4 × 3·3-5·5 μm; a doubtful species because the sporangial thalli are separated by isthmuses, cf. *Catenaria*] **Myzocytiopsis indica**

14'(13') Sporangia producing very long exit tubes; oospores not known [very early (1888) description, cf. *Gonimochaete*] **Myzocytiopsis elegans**

15(13) Thallus tubular, becoming septate, thalloidal cells disarticulating; oospores formed; primarily in nematodes 17

15'(13) Thallus cylindrical, lobed or ellipsoidal, non-septate, ellipsoidal or lobed, usually non-septate 16

16(15') Thallus ellipsoid or lobed, 12-25 × 20-40 μm, non-septate; zoospores 6·0 ×8·0 μm with lateral flagellar insertion; sexuality not known; primarily in rotifer eggs and embryos, also in nematode eggs **Myzocytiopsis oophila**

16'(15') Thallus cylindrical or lobed, 8(28) ×20-60 μm; sporangiospore cysts 5·0 μm diam.; antheridial gametangia absent; oospores *ca* 12 μm diam.; in adult nematodes **Myzocytiopsis bolata**

17(15) Thalloidal segments limoniform, 10-22 × 20-40 μm; zoospores 8·0-10·0 μm long, forming hemispherical cysts on the pellicle of the host (cf. *M. lenticularis*); oogonia spindle-shaped; oospores 12-22 μm diam. **Myzocytiopsis papillata**

17'(15) Thalloidal cells subspherical, 15-30 × 20-40 μm; zoospores 4·5-5·0 × 5·0-11·0 μm, pyriform with sub-apical flagellar insertion, forming clavate cysts 5·0-6·0 × 10·0-17·0 μm, with adhesive pads at the narrower end (cf. *M. glutinospora*); oospores 12-20 μm diam. **Myzocytiopsis humicola**

KEY V: *GONIMOCHAETE*

1 Thallus filamentous, becoming septate, thallus segments subsequently swelling or becoming disarticulated 2

1' Thallus undergoing successive bipartitioning accompanied by disarticulation to form ovoid prosporangia, 5-10 × 6-33μm [exit tubes 10-100μm long 2·8-5·0μm wide; aplanospores small (3·0-4·2 × 4·0-9·5μm, volume equivalent *ca* 50μm³, with adhesive pads 1·5-2·5μm diam.); secondary cysts producing biflagellate zoospores] **Gonimochaete horridula** (type species)

2(1) Thallus filamentous, septate, disarticulating, each segment becoming ovoid; exit tube broadly cylindrical, at least proximal to the prosporangium, resulting in a flask-shaped prosporangium; freshwater or soil; aplanospores large (volume equivalent >200μm³) 3

2'(1) Thallus filamentous, not disarticulating; segments 7-28 × 21-468μm, forming very broad, mammiform exit tubes 14-75μm long, 5-44μm wide, on the still-recognisable thallus segments; marine; aplanospores small, 3·0-4·0 × 5·0-7·0μm (volume equivalent <50μm³) **Gonimochaete latitubus**

3(2) Aplanospore formation confined to an expanded distal part of the exit tube; sporangium separated from the prosporangial bulb by a septum [thalloidal segments 9-30 ×15-60μm; exit tubes 100-136μm long, 5·5-12·0μm wide; aplanospores 4·0-5·0 × 12·0-16·0μm; antheridial segments smaller and narrower than oogonial segments; oospores 17-22μm diam., oospore wall ornamented] **Gonimochaete lignicola**

3'(2) Aplanospore formation occupying the whole of the exit tube; no well-defined septum delimiting the sporangium from the prosporangial bulb [thalloidal segments 25-45μm diam. or 10-20 × 25-45μm; exit tubes up to 110μm long, 6-8μm wide; aplanospores 4·0-6·0 × 10·0-18·0μm, with adhesive knobs 2·0-2·5μm diam.] **Gonimochaete pyriformis**

KEY VI: *CHLAMYDOMYZIUM*

1 Thallus becoming septate and disarticulating; thalloidal segment forming a single chlamydospore 4

1' Thallus non-septate 2

2(1') Sporangial thallus forming aplanospores only 3

2'(1') Sporangial thallus forming zoospores [thalloidal segment 12-18 × 22-38μm, becoming a sporangium or forming several chlamydospores by internal partition, 10-14 × 17-20μm [intrasporangial zoospores 4·0-5·0 × 6·0-7·2μm, reniform with sub-apical flagellar insertion] **Chlamydomyzium septatum**

3(2) Aplanospores large 3·0-4·0 × 5·0-7·0μm, ovoid; thallus broadly cylindrical 15-30 × 25-55μm; chlamydospores 13-18μm diam. **Chlamydomyzium internum**

3'(2) Aplanospores small 2·2-2·8 × 3·8-5·0μm, ovoid to reniform; thallus cylindrical 12-16 × 30-45μm; chlamydospores 13-17μm diam. **Chlamydomyzium aplanosporum**

4(1) Intrasporangial production of aplanospores which germinate to produce zoospores [chlamydospores thick-walled] 5

4'(1) Intrasporangial production of zoospores [zoosporangial exit tubes with a flared rim]; chlamydospores 16-18μm diam. thin-walled, formed singly in thalloidal segments [thallus extensive, branched, thalloidal segments 5-15 × 10-30μm; zoospores 5·5-6·0 long; in nematodes] **Chlamydomyzium sphaericum**

5(4) Aplanospores (cysts) 5·0-5·5(7·0)μm diam., aggregating at the mouth of the exit tube, germinating to produce medium-sized zoospores, grape-seed-like, 4·5-5·5(6·0) × 8·0-8·5μm (volume equivalent *ca* 70μm³); chlamydospores developed after multiple infections, 14-17μm diam.; in rotifer eggs [thalloidal segments 2 or more, cylindrical, 17-20 × 15-25μm] **Chlamydomyzium oviparasiticum**

5'(4) Aplanospores 4·0-6·0 × 8·0-10·0μm, dispersing freely, germinating to produce large zoospores (8·0-10·0μm long, volume equivalent >110μm³); chlamydospores 12-18μm diam.; in nematodes [thalloidal segments spindle-shaped; sporangia 10-15 × 25-55μm] **Chlamydomyzium anomalum** (type species)

KEY VII: *SYZYGANGIA*

1 Parasitic in moss rhizoids or aquatic angiosperms [donor gametangia (antheridia) not known] 2
1' Parasitic in freshwater green algae or cyanobacteria 3

2(1) Parasitic in moss rhizoids [zoosporogenesis not known; oospore ornamented, ellipsoid] **Syzygangia elliptica**
2'(1) Parasitic in *Potamogeton* [zoosporogenesis intrasporangial; zoospore diam. *ca* 8·0 μm (volume equivalent >250 μm³); oospore smooth, *ca* 24 μm diam.] **[Cornumyces muenscheri]**

3(1') Parasitic in green algae 4
3'(1') Parasitic in cyanobacteria [thallus passing through several host cells; zoospore volume equivalent <100 μm³; oospore *ca* 15 μm diam.] **Syzygangia nodosa**

4(3) Parasitic in *Oedogoniales* 5
4'(3) Parasitic in *Zygnematales* 7

5(4) Vegetative thallus more or less mycelial, septate, passing through several cells; oospore *ca* 11μm diam. 6
5'(4) Vegetative thallus ovoid or slightly toruloid, non-septate, confined to a single host cell; oospore *ca* 13 μm diam. **Syzygangia oedogonii**

6(5) Oospore smooth-walled, gametangia differentiated [zoosporogenesis intrasporangial; zoospore dimensions not known] **Syzygangia marchaliana** (type species)
6'(5) Oospore wall ornamented, gametangia undifferentiated intercalary segments of two different adjacent thalli [zoosporogenesis not known] **Syzygangia dictyospora**

7(4') Zoospore cysts >4·5 μm diam. (volume equivalent >50 μm³); oospore formed within one gametangium [see also *Cystosiphon*, the morphology of the fertilization process is critical] 8
7'(4') Zoospore cysts <4·0 μm diam. (volume equivalent <35 μm³); oospore formed between the two gametangia, oospore wall ornamented **Syzygangia sacculoides**

8(7) Oospore wall ornamented **Syzygangia papillosa**
8'(7) Oospore wall smooth **Syzygangia zygnematicola**

KEY VIII: *ECTROGELLA*

NOTE THAT THIS GENUS MAY BE POLYPHYLETIC, WITH SOME SPECIES POSSIBLY HAVING AFFINITY TO THE PLASMODIOPHOROMYCETES

1 Sporangiogenesis complete within the sporangium prior to discharge [zoosporogenesis intrasporangial and zoospores motile within the zoosporangium] 2
1' Sporogenesis initiated within the sporangium, completed at mouth of the exit tube(s), exit tubes 1-2; [thallus ovoid or tubular, non-septate; dioecious; oospore plerotic, 11-24 x 6-10 μm] **Ectrogella brachystoma**

2(1) Marine species [primarily in *Licmophora*] 4
2'(1) Freshwater species 3

3(2') Thallus ovoid or irregularly saccate; transformed into a zoosporangium with a short (3 × 3 μm) exit tube [zoospores 5-6 in a zoosporangium, 3·5 × 6·0μm, flagellar insertion sub-apical; dioecious, oospore plerotic, *ca* 10 μm diam.] **Ectrogella cyclotellae**
3'(2') Thallus ovoid or vermiform; transformed into a zoosporangium up to 200 μm long; with 1-10 short, thick-walled exit tubes [zoospores pyriform, 2·0 × 4·0 μm flagellar insertion lateral; resting spore not known] **Ectrogella bacillariacearum** (type species)

4(2) Sporangia with 1-10 exit tubes [thallus 20-23 × 26-40 μm, transformed into a zoosporangium; zoospores pyriform, 2·0 × 3·0 μm, flagellar attachment anterior (?); resting spores 14-19 μm diam., with smooth, hyaline thick wall (2·5-3·0 μm)] **Ectrogella perforans**
4'(2) Sporangia with 1-2 exit tubes [thallus 35-40 μm long; zoospores diplanetic, cysts 2·0-2·5 μm diam.; resting spores 16-18 μm diam., with a smooth lamellate wall 3·0-3·5 μm thick] **Ectrogella eurychasmoides**

KEY IX: KERATINOPHILIC *LEPTOLEGNIELLACEAE*

1 Thallus broad, coralloid [saprotrophic, usually on keratinized substrata] 3
1' Thallus of a fine mycelium, extensive, intra- or extramatrical 2

2(1') Oogonia terminal, extramatrical; antheridia present [saprotrophic on keratinic substrata (shavings of horse hoof); hyphae slender (1)3-5(6) μm diam., much branched; sporangia terminal, subtending hypha 2 μm diam., spherical or subspherical/clavate, (7)12-18(26) × (12)33(70) μm; sporangiospores encysting within the sporangium, zoospores of the principal form emerging as in *Dictyuchus*, (4)6(12) × (6)9(15-24) μm; oogonia terminal or intercalary 14-35 × 15-39(59) μm, variably papillate, pluriovulate with 1-7 oospores; antheridia monoclinous or diclinous; oospores 12·5-20·0 × 14·0-24·0 μm, oospore wall 1-5 μm thick but not uniform, reserve globule 6·8-9·3 μm diam.] **Aphanodictyon papillatum** (monotypic genus)

2'(1') Oogonia intercalary, intramatrical, antheridia not known [saprotrophic in hair baits, terrestrial; mycelium mostly intramatrical, extramatrical hyphae 6 μm diam., usually unbranched, 115-215(450) μm long, becoming transformed into sporangia; zoospore initials discharged as aplanospores or encysting within the sporangium (cf. *Brevilegnia*), in 1-2 rows, 6·0-7·0 μm diam., gametangia (oogonia) intramatrical, delimited by septa, (9)13-14(19) μm diam., oospores 1-2(7), (6)7-9(10) μm diam.] **Brevilegniella keratinophila** (monotypic genus)

3(1) Sporangiospores expelled from sporangium, encysting in a spore ball, zoospores emerging from these cysts [apparently automictic] 4
3'(1) Zoospores motile within the zoosporangium, usually discharging as in *Leptolegnia caudata*, but sometimes encysting within the sporangium 5

4(3) Oospores large (>20 μm diam.), 18-32(41) μm diam., with a two-layered punctate wall 4·0-5·5 μm thick [intramatrical on snake skin bait; hyphae irregular up to 15 μm diam., branched, occasionally septate; thalloid segments becoming transformed into sporangia (5-8 × 20-180 μm), or more or less inflated vesicles (<50 μm diam.) functioning as oogonia; sporogenesis intrasporangial, initials gliding out and encysting, cysts 6·0-8·0 μm diam.; zoospores (6·0-7·0 × 8·0-9·0 μm), with lateral flagellar insertion; oogonia with 1-7 oospores] **Aphanomycopsis punctata**

4'(3) Oospores small (<15 μm diam.), spherical or subspherical, 9-12(22) μm diam., with hyaline wall 3·0-5·0 μm thick [thallus pseudomycelial or coralloid; saprotrophic in *Zea* leaves and pine pollen; thallus of intramatrical hyphae of irregular diameter (5-40 μm diam.), much branched, occasionally septate; thalloid segments becoming transformed into sporangia (15-200 × 6-8 μm), or more or less inflated vesicles (<40 μm diam.) functioning as gametangia; sporogenesis intrasporangial, initials gliding out and encysting, cysts 5·0-8·0 μm diam.; zoospores (4·0-5·0 × 6·0-7·0 μm with lateral flagellar insertion, diplanetic (?)] **Aphanomycopsis saprophytica**

5(3') Oospores formed in undifferentiated sections of the thallus 7
5'(3') Oospores formed in lateral outgrowths from the thallus 6

6(5') Oogonia with 1-14 aplerotic oospores; oospores 12-20μm diam.; in hair baits; mycelium mostly intramatrical, hyphae 7-10μm [zoosporangia more or less the same diameter as hyphae, sometimes branched, major part extramatrical, extending up to 450μm from substratum; zoosporogenesis intrasporangial; zoospores 6·5 × 11·0μm, diplanetic, dimorphic; oogonia extramatrical, iregularly clavate or lobed; antheridia usually absent, but occasionally possible monoclinous or epigynous antheridial hyphae present] **Leptolegniella exoospora**

6'(5') Oogonia plerotic with 1 oospore; oospores 15-25 μm diam.; mycelium mostly extramatrical, occasionally septate and constricted at intervals, hyphae 8-20 μm diam. [terrestrial; zoosporangia more or less the same diameter as the hyphae; zoosporogenesis intrasporangial, emergent, forming a loose globular mass or encysting in chains within the zoosporangia; zoospores 5·0 × 8·0 μm; oogonia lateral, spherical; antheridia not observed] **Leptolegniella exogena**

7(5) Oospores small (<12 μm diam.) 8
7'(5) Oospores large (>15 μm diam.) [saprotrophic in snake skin baits; mycelium intramatrical and extramatrical, hyphae 10-30 μm diam.; zoosporangia undifferentiated, emergent from the substrate, 8-15 μm diam.; zoospores 5·0-6·5 μm diam., emergent or encysting within the zoosporangium; oogonia developing from undefined hyphal sections; oospore number variable per section, shape irregular, conforming to the restricting hyphal contours, 15-28 μm diam., wall 2·0-2·8 μm thick] **Leptolegniella angularis**

8(7) Zoospores present, diplanetic [on hair, skin and horn shavings; mycelium mostly intramatrical, hyphae 5-20(25) μm diam., rarely septate, rhizoids present, slender, tapered with blunt apices; zoosporangia formed from undiffentiated parts of the thallus, branched, with exit tubes 5·0-13·5(18·5) μm diam., and up to 395μm long; zoospore cysts 5·4-7·2(13·0) μm diam.; undifferentiated parts of the thallus functioning as gametangia (oogonia); oospores 9-12(23) μm diam., sometimes irregular, oospore wall 0·5-2·0 μm thick, reserve globule 3·5-4·0(5·8) μm diam.] **Leptolegniella keratinophila** (type species)

8'(7) Zoospores not known, sporangiospores encysting within sporangia [mycelium variable, hyphae 5-70 μm diam., undefined sectionstransformed into sporangia and/or oogonia; gametangia not morphologically differentiated, antheridia not known; oospores 8-12 μm diam.] **Leptolegniella abnormis**

KEY X: *SALILAGENIDIACEAE* AND *HALIPHTHORACEAE*

1 Thallus more or less mycelial, branched, mean hyphal diam. <24 μm, usually <15 μm, septate or sparingly septate; non-rhizoidal, culturable; eucarpic or holocarpic with time; sexual reproduction present or absent 7

1' Thallus more or less inflated, lobed, branches with mean diam. >25 μm; septa rare or absent; rarely rhizoidal (in culture), culturable; holocarpic with time (doubtfully eucarpic); sexual reproduction absent 2

2(1') Thallus irregularly and broadly tubular, rhizoids absent; sporangiogenesis not reported to have a centripetal contraction phase; zoospores with lateral flagellar insertion; mean diameter of zoospore cysts <6 μm (volume equivalent <120 μm³), or if greater then parasitic in molluscs [these species may be closer to *Haliphthoros*, but resemble *A. dubia* in their habit and the long tapering exit tubes] (*Halodaphnea*) 3

2'(1') Thallus inflated, septa normally absent; sporangiogenesis intrasporangial, often with a late, marked centripetal contraction, initials amoeboid at first on fine cytoplasmic strands; mean diameter of zoospore cysts >7 μm (volume equivalent >180 μm³) [parasitic in eggs of crabs *Pinnotheres* and other crustaceans (*Crangon, Gonoplax, Leander, Macropodia, Paguristes, Portunus & Typton*); rhizoids cut off by septa occasionally present in culture but not observed in natural substrata; thallus lobes stout, 27-50 μm, tips swollen, up to 100 μm diam., zoosporangia 50-400 × 10-30 μm, with one or more exit tubes, distal part (up to 50μm) hyaline; occasionally proliferous; zoospores pyriform or slipper-shaped (with lateral flagellation?), diplanetic (but not dimorphic?), 10-12 μm long (4·0-6·0 × 6·0-8·2 μm; (4·0)4·8(6·0) × 8·7(10·0) μm), first-formed cysts 7·0-8·0(9·0) μm (7·4 μm) diam., second-formed cysts 6·0-7·0 μm (6·8 μm) diam., germ tube 1·7μm diam., gemmae present] [this genus may not belong in the Haliphthoraceae] **Atkinsiella dubia** (monotypic genus in this text)

3(2) Zoospore cysts *ca* 5·0 μm diam. (volume equivalent <75 μm³); colonies more or less saccate with 1-3 zoosporangial exit tube(s); optimum temperarture for growth >24°C [parasitic in various *Crustacea*] 4

3'(2) Zoospore cysts 8·0 μm diam. (volume equivalent >250 μm³); colonies filamentous with 1 (rarely 2) exit tube(s) produced from each zoosporangium; optimum temperature for growth 20°C(5-25°C) [parasitic in *Haliotis sieboldii* (abalone); hyphae stout, irregular 16-41(140) μm diam., branched, non-septate, becoming septate to delimit zoosporangia; zoospores pyriform 4·0-8·0 × 7·0-12·0 μm, diplanetic, isokont; zoospore cysts germinating by means of a fine filament 62-295 μm long before broadening to form the thallus] **Halodaphnea awabi**

4(3) Zoosporangial exit tubes 1-3, always unbranched 6

4'(3) Zoosporangial exit tubes normally single, sometimes branched 5

5(4') Colony pigmentated, grey to light brown; optimum temperature for growth 30-32°C; [parasitic in eggs of *Scylla*; hyphae (12)26(40) μm diam., zoosporangia 42-1150 × 5-15 μm; zoospores pyriform or slipper-shaped, diplanetic but not dimorphic, (3·8)4·5(5·0) × (5·0)6·3(10·0) μm, cysts (4·5)5·0(7·5) μm diam.] **Halodaphnea hamanaensis**

5'(4') Colony hyaline; optimum temperature for growth 25°C(15-30°C) [parasitic in *Panulirus japonicus*; holocarpic, hyphae stout, branched and septate, 10-22(64) μm diam., subthalli transformed into zoosporangia or gemmae; zoospores pyriform or reniform 4·0-5·0 × 7·0-10·0 μm, diplanetic; zoospore cysts 5·0-7·0 μm diam., germinating by means of a fine filament 14-253 μm long before broadening to form the thallus] **Halodaphnea panulirata**

6(4) Zoosporangia with 1-several broad discharge tubes, (6)8-9(10) × (40)200-300(510) μm containing several ranks of zoospores [parasitic in *Portunus pelagicus*; holocarpic, hyphae stout, becoming septate with age, 10-38 μm diam., subthalli transformed into zoosporangia or thick-walled gemmae, 22-190 μm diam.; zoosporogenesis intrasporangial;; zoospores pyriform or subglobose (4·0)4·7(6·5) × (5·0)6·3(8·0) μm, diplanetic; zoospore cysts (4·0)5·2(7·0) μm diam., germinating by means of a fine filament 5-190 μm long before broadening to form the thallus; optimum temperature for growth 25°C (20-30°C)] **Halodaphnea okinawaensis**

6'(4) Zoosporangia with 1-2 infrequently-branched discharge tubes [discharge tubes straight, wavy or coiled, usually with a cone-like base, 6-14 × 20-780 μm, [parasitic in *Brachionus plicatilis*; usually holocarpic (eucarpic with age or in sub-optimal growth temperatures), hyphae stout, saccate, becoming septate with age, 15-50 μm diam., subthalli transformed into sporangia or developing into thick-walled gemmae, 40-200 μm diam.; zoosporogenesis intrasporangial; zoospores pyriform (4·0)4·6(5·6) × (4·8)6·0(7·4) μm, monoplanetic, isokont; zoospore cysts (4.8)5.5(6.0) μm diam., germinating by means of a fine filament 8-250μm long before broadening to form the thallus; optimum temperature for growth 25°C(20-30°C)] **Halodaphnea parasitica** (type species)

The description of **H. hiroshimaensis** has not yet been published.

7(1) Thallus septate but not usually with endothallial contraction and new wall formation; zoospore size large (volume equivalent >275 μm³); sexual reproduction present or absent (*Salilagenidium*) 10

7'(1) Thallus elements rounding up to form new endothallial walls; zoospore size variable (volume equivalent <250 μm³); sexual reproduction absent (*Haliphthoros*) 8

8(7') Parasitic in crustaceans or molluscs; zoospore size medium-large (volume equivalent > 100 μm³); resting bodies not known 9

8'(7') Parasitic in larvae of molluscs (*Venus*); zoospore size small (volume equivalent < 75 μm³); 'resting bodies' 40 × 45 μm up to 80 × 90 μm formed as lateral diverticula (oogonia?) [thallus initially filamentous, 10-15 × 82 μm, becoming both inflated and with tapered extremities, resembling rhizoids; intrathallial walled bodies (31-33 × 46-56 μm); zoosporangia intramatrical, 5 × 15-142 μm; zoospores 2·0 × 5·0 μm, anisokont, with a shorter straminipilous flagellum] **Haliphthoros zoophthorum**

9(8) Parasitic in eggs of molluscs (*Urosalpinx*, *Haliotis*), and crustaceans (*Penaeus*); sporangia with short exit tubes (7·0)8·0(14·0) μm long; zoospores monoplanetic [holocarpic; hyphae (7·0)14·0-18·8(40·0) μm; zoospores pyriform, subspherical or elongate, (5·0)6·8-7·2(10·6) × (6·7)8·5(12·2) μm, monoplanetic, cysts (6·8)7·8-8·5(8·6-20·0) μm, very slender germ tube 0·5(1·8-2·2) μm diam.] **Haliphthoros milfordensis** (type species)

9'(8) Parasitic in larvae of prawns (*Penaeus*); sporangia with long exit tubes 620 × (7·50)-(12·5) μm; zoospores polyplanetic, dimorphic(?) [hyphae stout, branched, irregular non-septate (10·0)21·0(37·5) μm, intrathallial bodies (190 × 100 μm) not disarticulating, remaining connected like beads; first-formed zoospores pyriform, subspherical or elongate, with lateral flagellar insertion, 5·0-7·5 × 7·5-12·5 μm, - second-formed zoospores slipper-shaped and slightly shorter (10·0μm), cysts 5·0-7·5(12·5) μm, very slender germ tube] **Haliphthoros philippinensis**

10(7) Sexual reproduction not known 11

10'(7) Sexual reproduction known 13

11(10) Zoospores of medium size (volume equivalent < 400 μm³) 12

11'(10) Zoospores of large size (volume equivalent > 500 μm³) [parasitic in eggs and larvae of crabs (*Scylla*)] **Salilagenidium scyllae**
Salilagenidium thermophilum

These two species are scarcely separable from the published descriptions:

Salilagenidium scyllae: hyphae thick, irregular, branched, 7·5-17·0(40·0) μm diam., sparingly septate - non segmented, thalloid elements becoming sporangia; zoosporangial discharge tubes short or long, 37-500 × 4-10 μm, apex dilating to form a deliquescent vesicle; cytoplasm not filling vesicle prior to cleavage, vesicle not persistent, zoospores reniform, pyriform, ovoid or oblong, (7·0)10·0(15·0) × (8·0)12·5(17·5) μm, released by deliquescence of vesicle or singly through a pore on the vesicle, monoplanetic; cysts (7·5)10·0(15·0) μm diam., cyst wall 1·5 μm thick]

Salilagenidium thermophilum: hyphae thick, irregular, branched, 8·0-24·0(40·0) μm diam., non-septate, becoming sporangia; zoosporangial discharge tubes 34-440 × 6-14 μm, vesiclular membrane not apparent; cytoplasic cleavage completed after discharge, 'vesicle' 36-80 μm diam.; zoospores pyriform to subglobose, laterally biflagellate, 8·0-14·0 (mean 10·3) × 10·0-16·0 (mean 13·3) μm; monoplanetic; cysts 6·0-16·04 μm diam.; thermotolerant (15)30-40(45)°C

12(11) Parasitic in muscles and swimmerets of crustaceans (shrimps) (*Pandalus*), holocarpic, culturable; hyphae wide, irregular, branched, 7·0-10·0 μm diam., partial cleavage within the sporangium, discharge tube 86-240 × 7-10 μm, vesicle formed, zoosporogenic protoplasm not filling vesicle, zoospores released by rupture of vesicle, vesicle persistent, 9·6 × 12·9 μm globose, reniform, pyriform or elongate, cysts 5·5-12·0 μm **Salilagenidium myophilum**

12'(11) Parasitic in stomach of crayfish (*Penilia*); mycelium non-septate, of uniform diameter, 4·2-5·2(7·0) μm, with homogeneous protoplasm; zoosporangia spherical or ellipsoid, on short side branches, smooth walled; zoosporogenesis intrasporangial, released through a pore over a period of 2 min, swimming away 30 min later; zoospores 30-50 in a zoosporangium, of irregular shape, with an anteriorly inserted flagellum [known only from original locality, see original description] '*Hyphochytrium peniliae*' *nom. illeg.*

13(10') Zoosporogenesis partly extrasporangial in a vesicle; antheridia absent; culturable 14

13'(10') Zoosporogenesis intrasporangial; zoospore initials in a single row distally; antheridia present, hypogynous [parasitic in eggs of crabs (*Mytilus*) and possibly lamellibranchs *Barnea* and *Cardium*; hyphae 7·5-20·0(40·0) μm diam.; zoosporangia undifferentiated from hyphae, occasionally proliferous; zoospores pyriform, 8·0-14·0 μm long, cysts 6·0-11·0 μm diam., diplanetic; oogonia rare, oospores single, nearly plerotic, 17·5-30·0(37·0) μm diam., oospore wall two layered, 7·5 μm thick] . **Salilagenidium marinum**

14(13) Parasitic in eggs and larvae of crabs (*Callinectes*, *Limulu*), and barnacles (*Chelonibia*); hyphae sparingly septate, extramatrical hyphae (5)8-14(50)μm diam., 1030 μm long, aseptate; zoosporangium with vesicle persistent after discharge; tip of exit tube gelatinizing and contents flowing into the thick gelatinous envelope, never filling it, vesicle persistent, zoospores 9·3 × 12·5 μm; cysts 8·0-10·0(11·3) μm diam., monoplanetic, germ tube 2·5 μm diam., intercalary, automictic, resting cells (18)25(36) μm diam., wall 3 μm thick, subeccentric oil reserve **Salilagenidium callinectes** (type species)

14'(13) Parasitic in barnacles (*Chthamalus*); hyphae stout, irregular, branched, highly vacuolate 10-18(39) μm diam., becoming segmented, segments behaving as zoosporangia, zoosporogenesis within a vesicle formed from the swelling of the exit tube apex, cytoplasm entering the vesicle only after the latter is completely developed, cytoplasm not filling the vesicle, vesicle disappearing immediately after discharge, zoospores reniform, 6·8-8·5 × 8·5-10·2 μm, oogonia intercalary or terminal, 19-47 μm diam., antheridia absent, oospores 1(2), aplerotic (16)21-25(27) μm diam. **Salilagenidium chthamalophilum**

KEY XI: *PONTISMATACEAE, SIROLPIDIACEAE, EURYCHASMATACEAE* AND *ANISOLPIDIACEAE*

1 Zoospores biflagellate 2
1' Zoospores anteriorly uniflagellate, or flagellation not known and parasitic in brown algae (*Anisolpidiaceae*) 23

2(1) Parasitism causing little or no hypertrophy of host cells; intrasporangial zoospores discharged to the environment directly; thallus lobed, tubular, segmented with or without a tendency to disarticulation or more or less spherical 6
2'(1) Parasitism causing marked hypertrophy of host cells; intrasporangial zoospores with a tendency to encyst within the zoosporangia, their cyst walls forming a reticulate pattern against the zoosporangium wall after discharge (a net sporangium); thallus saccate or more or less spherical, non-septate [anamorphic] (*Eurychasmataceae*) 3

3(2') Parasitic in red algae (*Eurychasmidium*) 4
3'(2') Parasitic in brown algae (*Ectocarpales*: *Ectocarpus, Hincksia, Pilayella, Stictyosiphon* and *Striaria*); thallus saccate, 25-80 × 20-40 μm; zoosporangia with 1-2 discharge tubes; zoospores 3·0 × 5·0 μm; resting spores not known [see also '*Eurychasma*' sp. (Sparrow, 1969: 139; zoospores 4·0-5·0 μm diam.); and compare marine '*Pleotrachelus*' species below] **Eurychasma dicksonii** (monotypic genus in this text)

4(3) Parasitic in *Ceramiales*; in superficial cells 5
4'(3) Parasitic in *Rhodymeniales* and *Palmariales*; primarily in non-superficial cells [parasitic in *Halosaccion* and *Rhodymenia*, primarily in non-superficial cell layers; thallus saccate or irregular, 48-184 μm; zoospores not known; known only from original description] **Eurychasmidium sacculus**

5(4) Parasitic in *Ceramium rubrum* epiphytic on *Fucus*, primarily in nodal cells; thallus 100-110 μm; zoosporangia with many discharge tubes; zoospores 3·0 × 5·0 μm **Eurychasmidium tumefaciens** (type species)
5'(4) Parasitic in *Polysiphonia* and *Pterosiphonia*, primarily in apical cells; thallus 50-60 × 15-20 μm; zoosporangia with 1-4 discharge tubes; zoospores encysting within the zoosporangium, 2·0 × 4·0 μm [known only from original description, see also '*Eurychasma*' sp. (Sparrow, 1969: 141) with 'net'sporangia on the same host genera] **Eurychasmidium joycei**

6(2) Thallus tubular and septate, or spherical 'olpidioid', but not lobed 9
6'(2) Thallus lobed, non-septate; in red algae (*Petersenia*) 7

7(6') Parasitic in *Ceramiales* 8
7'(6') Parasitic in *Palmariales* [parasitic in *Palmaria*; thallus penetrating several host cells (24)77(135) × (21)10(17) μm; zoosporangium with a single discharge tube; zoospores formed within the zoosporangium, 2·5-3·0 × 4·0-4·5 μm **Petersenia palmariae**

8(7) Thallus form and dimensions variable, simple or lobed, up to 30-192 × 15-30 μm; zoosporangia with 1-3 discharge tubes; parasitic in *Callithamnion corymbosum* and *Spermothamnion turnerii* [zoospores motile within the zoosporangium, but not filling the zoosporangium, 3·0 × 4·5 μm] **Petersenia lobata** (type species)
8'(7) Thallus tightly lobed, 35-43 × 104-112 μm, with 1 or more discharge tubes; parasitic in *Ceramium rubrum* [zoospores biflagellate, dimensions not known] **Petersenia pollagaster**

9(6) Parasitism primarily confined to green algae (*Cladophorales*), but also including parasites of brown algae and fungi on substrates other than red algae (*Sirolpidium*) 18
9'(6) Parasitism confined to red algae (mostly *Ceramiales*) or fungal parasites of red algae (*Pontisma*) 10

10(9') Thalli more or less spherical-cylindrical, possibly dividing and disarticulating when several to a host cell, each segment behaving as a zoosporangium 11
10'(9') Thalli tubular, branched, becoming septate and segmented but not disarticulating, each segment behaving as a zoosporangium [parasitic in *Ceramium*; thallus 20-40(200) × 11-30 μm; zoosporangia with one or more discharge tubes; zoospores motile within zoosporangium, 2·5-3·0 × 4·5-7·0 μm; resting spores (not known in red algal material), 27 μm diam.] **Pontisma lagenidioides** (type species)

11(10) Zoospores very small (< 3·5 μm in greatest dimension) 12
11'(10) Zoospores larger (> 4·0 μm in length) 15

12(11) Thallus large (> 45 μm³) 13
12'(11) Thallus small (< 30 μm³) [parasitic on *Achrochaetium* (*Nemalionales*) on *Codium*; thallus tubular (septate?); zoosporangia 15-25 × 10-12 μm; zoospores 2·0 × 3·0 μm] **Pontisma** sp. (?*Petersenia* sp., Aleem, 1950)

13(12) Parasitic in red algae 14
13'(12) Mycoparasitic in a chytrid [parasitic in *Chytridium* parasitic on *Polysiphonia*; thallus 30-45 μm; zoosporangium with 1-3 discharge tubes; resting spores not known] '*Rozella marina*'

14(13) Parasitic in *Asparagopsis* (*Falkenbergia*) epiphytic on *Cystoseira* (also in *Trailliella*); 1-3 thalli per host cell, subspherical 15(20)-30(50) μm with 1(2-3) discharge tubes; zoospores 1·5-2·0 × 2·0-2·5 μm **Pontisma feldmannii**

14'(13) Parasitic in *Polysiphonia fibrillosa* and *P. elongata*; one or more thalli per host cell [chlor-zinc iodide test on thallus wall positive]; thallus subspherical or slightly lobed, 20-68 μm diam. (39-60 × 30-40 μm), with 1-3 discharge tubes; zoospores 2·0-3·0 μm diam., flagellation not known **Pontisma inhabilis**

15(11') Thallus large (>45 μm³) 16

15'(11') Thallus small (<30 μm³) [parasitic in *Porphyra*; thallus 17 × 12-13 μm; zoospores 2·5 × 5·0 μm] '*Olpidiopsis*' sp. (Sparrow, 1969)

16(15) Parasitic is *Ceramiaceae*; zoospores medium-large (*ca* 5·0 μm long) 17

16'(15) Parasitic in *Delesseriaceae*; zoospores small (<4·0 μm long) [parasitic in *Radicilingua*; thallus 25-40 μm; zoosporangia with 1-3 discharge tubes; zoospores 3·0-4·0 μm diam.] **Pontisma dangeardii**

17(16) Parasitic in *Ceramium*; up to 5 thalli per host cell; thallus 45 μm diam.; zoospores 5·0 μm long **Pontisma magnusii**

17'(16) Parasitic in *Antithamnion* (could not be induced to infect *Trailliella*) 1-3(8) thalli per host cell; thallus (5)20-40 μm diam., or ellipsoidal, 20-50(80) × 40-80(100) μm; exit tubes 1(2-3), 5-10(20) × 5-12(20) μm long; zoospores motile within zosporangium, zoosporangium, 3·0 × 5·0 μm **Pontisma antithamnionis**

18(9) Thallus olpidioid, or disarticulating into olpidioid zoosporangial units 19

18'(9) Thallus septate, not disarticulating; parasitic in *Cladophora*, *Aegagropila* and *Chaetomorpha* [doubtful species, zoospore flagellation to be established] **Sirolpidium salinum**

19(18) Parasitic in *Cladophorales* or *Chlorococcales* 20

19'(18) Parasitic in fungi or *Ectocarpales* 22

20(19) Parasitic in *Cladophorales* 21

20'(19) Parasitic in *Chlorococcales* [parasitic in *Tetraselmis*; thallus 8-12 × 7-8 μm; zoosporangia with 1 discharge tube; zoospores 3·0-4·0 μm] **Sirolpidium marinum**

21(20) Parasitic in *Bryopsis* and *Cladophora*; thallus becoming septate, segments disarticulating, chlor-zinc iodide test on thallus wall weakly positive; thalli segments transformed into zoosporangia (40-90)60 × 15(10-20) μm, with long discharge tubes; zoospores 2·0-3·0 × 4·0-6·0 μm; resting spores not known (spherical, 27 μm diam.) [includes *Pontisma lagenidioides sensu* Kobayashi & Ookubo, 1953; *Myzocytium proliferum* forma *marinum* Kobayashi & Ookubo, 1953] **Sirolpidium bryopsidis** (type species)

[See also *Eurychasma* sp. (Kobayashi & Ookubo, 1953), parasitic in *Bryopsis*; thallus 25-50 μm; sporangia with 5-10 discharge tubes; aplanospores aggregating near sporangium orifice; resting spore 50 μm diam.]

21'(20) Parasitic in *Spongomorpha* and *Acrosiphonia* spp. (Petersen, 1905); thalli spherical, 9-39(80) μm diam., with 1-3 discharge tubes; zoospores ellipsoidal, 3·0-4·0 smooth, 12-23 μm; resting spores ? **Sirolpidium andreei**

22(19) Parasitic in *Rhizophydium discinctum* parasitic on *Spongomorpha vernalis*; thalli subspherical or irregular, walls chlor-zinc iodide positive, dimensions not provided, with 1-2 discharge tubes; zoospores not observed; resting spores not known **Sirolpidium paradoxum**

22'(19) Parasitic in *Papulospora* on submerged leaves of *Arbutus*; causing hypertrophy of hypha; thalli 1-7 per cell; thallus 17-91 × 15-89 μm; zoosporangia with 1-3 discharge tubes; resting spores plerotic, 17-73 μm diam., with or without companion cells 8-25 μm diam.; zoospores 3·5 μm diam. **Sirolpidium globosum**

23(1') Sporangia large, 17-70 μm diam.; exit tubes >6 μm diam. 24

23'(1') Sporangia small, (8)15 × 20(35) μm; exit tubes 5 μm diam. [sporangia solitary or numerous in a host cell; resting spores without companion cells, 10-14 μm diam.; chlor-zinc iodide test on thallus wall negative; parasitic in *Ectocarpus* spp.] **Anisolpidium ectocarpi** (type species)

24(23) Parasitic in *Pilayella littoralis*; variable numbers of thalli per host cell; thalli subspherical, 20-68 × 24-41 μm, with 1-3 discharge tubes; zoospores 2·0-3·0 μm; resting spores not known [from Petersen, 1905] **Anisolpidium rosenvingii**

24'(23) Parasitic in *Sphacelaria*, *Cladostephus* and *Chaetopteris*; variable numbers of thalli per host cell; thalli subspherical, 17-52 μm diam., with 1-3 discharge tubes; zoospores not known; resting spores not known [from Petersen, 1905] **Anisolpidium sphacelarum**

ADDENDUM TO ANISOLPIDIUM

Parasitic in *Ectocarpus siliculosus*; one or more thalli per host cell; chlor-zinc iodide test on thallus wall positive; thallus subspherical, 9-30μm diam., with 1-3 discharge tubes; zoospores not known; resting spores single, aplerotic, 9μm diam. **Anisolpidium olpidium**
'*Pleotrachelus olpidium*'

Parasitic in *Chorda filum*; one or more thalli per host cell; thalli spherical or cylindrical, 16-24 × 12-16μm, with 1-3 discharge tubes; zoospores not known; resting spores not known **Anisolpidium minutum**
'*Pleotrachelus minutus*'

Parasitic in *Ectocarpus siliculosus*; thalli 3-40μm diam; single discharge tube 6-64μm long, 3·5-10·0μm wide at base, 3·0-6·3μm at tip; prolonged period of intrasporangial movement; resting spore variable in size, 12-17 or 20-30μm diam. with (rarely without) companion cell, companion cell smaller than resting spore, 5μm diam., or up to 12μm diam. **Anisolpidium joklianum**
'*Pleotrachelus ectocarpi*' (including *Petersenia andreei sensu* Sparrow)

KEY XII: *LEPTOMITACEAE* AND *APODACHLYELLACEAE*

1 Loment ('hyphal' segment) dimensions very variable, but means not exceeding 150 × 16 μm on established mycelia; rhizoids not known; germination blastic; cellulin granules small, 1-several per loment; oogonia usually present; zoosporangia present or not known **2**

1' Loment ('hyphal' segment) dimensions very variable, but means *ca* 200 × 20 μm on established mycelia; rhizoids sometimes present; germination semi-blastic; cellulin granules large, usually one to a loment; zoosporangia 400 × 8-16(48) μm; zoospore cysts 10-11 μm diam.; not known to form oogonia or antheridia **Leptomitus lacteus** (type species)

2(1) Zoosporangia rare or not known [thallus holocarpic in one species] **4**

2'(1) Zoosporangia abundant [thallus eucarpic; oogonia borne on short lateral branches of 2-3 short segments; oospore wall thin] **3**

3(2') Mean loment size 66-184 × 8·5 μm, with up to *ca* 30 nuclei per loment* [other sources: up to 430 × 4-10 μm]; zoosporangia (12)24 × 12-20 μm; zoospore cysts 11 μm diam.; oogonia 35(63) × 14-36 μm or 27-32 μm diam.; oospores 25-33 μm diam.; oospore lipids coalesce to a single drop **Apodachlya pyrifera**

3'(2') Mean loment size 39-99 × 16·0 μm, with up to *ca* 13 nuclei per loment [other sources: (15)20-185(537) × (4)6-13(23) μm]; zoosporangia 23-76(97) × (15)23-44 μm; zoospore cysts 8.5-10 μm diam.; oogonia 23-29(30) μm diam.; oospores slightly smaller; oospore lipids form peripheral layers inside the endospore **Apodachlya brachynema** (type species)

[*Apodachlya punctata* and *Apodachlya seriata* are subsumed in *Apodachlya brachynema*, see Jacobs (1982)]

4(2) Antheridium not applied to the oogonium surface; fertilization hyphae from cells within the antheridium; gametes with an additional investing membrane [saprotrophic or parasitic] (*Apodachlyellaceae*) **6**

4'(2) Antheridium applied to the oogonium surface; fertilization tube from antheridium; gametes without any additional investing membranes [saprotrophic] **5**

5(4) Oogonia uniovulate, oogonia and oospores 12-16 μm diam.; zoosporangia rare, formed in fascicles on side branches; mean loment dimensions (47)100-170(430) × 3·4-8·5(20) μm (larger on cornmeal agar)]; zoosporangia 23-76(97) × (15)23-44 μm; zoospore cysts 8.5-10 μm diam.; oospore lipids coalesce to form one or a few droplets; oospore wall thin **Apodachlya minima**

5'(4) Oogonia pluriovulate, (22)36-52(95) μm diam.; zoosporangia not known; loments irregular with long and short loments interspersed; antheridium monoclinous, closely laterally applied over a broad arc of the oogonium; antheridial stalk 2-4 μm diam., equivalent to an uninflated segment; fertilization tube to oospheres; oospore lipids coalesce to form a single droplet [segments < 100 × (7)9(14) μm, more or less irregular, with 4-5 cellulin granules to each segnent; oogonia deformed-pyriform due to the tightly packed spherical oospores; oospores (1)4-5(16) to an oogonium; oospores (16)23-25(31) μm; oospore wall thick (*ca* 4 μm)] **Plerogone helodes** (type species) 2

6(1) Eucarpic; homothallic; saprotrophic; oogonia pluriovulate, terminal on side branches or intercalary; zoosporangia not known [mean loment dimension range 130-200 × 9-12 μm; oogonia, when spherical, 23-48 μm diam., with 2-12 oospores, when intercalary, with many more oospores; oospores 16-24 μm diam.; oospore lipids coalesce to form a single droplet; oospore wall thick; antheridia with 4-10 antheridial cells, 4-5 μm diam.] **Apodachlyella completa** (type species)

6'(1) Holocarpic; heterothallic; parasitic; oogonia uniovulate; zoosporangia rare [parasitic in *Podophrya* on *Nassula* from freshwater; thallus initially plasmodial, quickly becoming walled, spherical or allantoid, (8)10-38(80)μm diam.; transformed into a sporangium or a gametangium [zoosporangia with intrasporangial sporogenesis; zoospores encysting peripherally within the zoosporangium; zoospores ellipsoidal, diplanetic, dimorphic (?), 3·0 × 5·0 μm; heterothallic, gametangia not contiguous; oogonia ovoid (15)20(44) μm diam., with 1(2) oospore(s); antheridia allantoid, 6-10 × 18-52 μm, with up to 8 antheridial cells with fertilization hyphae; oospores (16)18(22) μm diam., with a thick two-layered wall and two large refractive globules (ooplast and lipid reserve)] [see also KEY I, key point 19'] **Eurychasmopsis multisecunda** (type species)

---

*FOOTNOTE: Some segment dimensions and nuclear numbers (and cellulin granule numbers per loment) for *Apodachlya* quoted in this key are taken from unpublished morphometric data obtained by Miss Carol Coley (BSc dissertation, University of Reading, 1977).

KEY XIII: *RHIPIDIACEAE*

1 Basal cell large, saccate and lobed; pseudomycelial segments absent or not well developed, or with sympodial branching without constrictions; spiny sporangia not known 2

1' Basal cell cylindrical-truncate, clavate or inconspicuous, with a clearly defined long axis, unbranched or rarely with one or two lobes; pseudomycelial segments extensive, whorled or umbellate with elongate constrictions; if pseudomycelial segments absent, then with smooth and spiny sporangia 5

2(1) Pseudomycelial segments present, 7-14 μm in diam.; oogonia pedicellate; antheridia present [zooospore volume 850-1050μm³] 3

2'(1) Pseudomycelial segments absent; oogonia sessile; antheridia absent [basal cell 100-1000(1500) × 25-50 μm, with long cylindrical-clavate branches 2000 × 150 μm, often developing on one side of primary lobes, all without any elongate constrictions; wall very thick, (2)5-8(10) μm thick; sporangia (66)110(142) × (40)54(68) μm; zoospores large (16)19(21) × (11)13(16) μm; oogonia (54)66(82) × (54)62(70) μm with fine bristles; oospores (36)48(56) × (33)45(56) μm with exospore of broad shallow ridges] **Aqualinderella fermentans** (type species)

3(2) Oogonia borne on subtending pseudomycelial segments or directly on the basal cell 4

3'(2) Oogonia and antheridia both borne on long irregular pseudomycelial stalks [sporangia 50-89 × 20-38 μm; oogonia 57-62 × 47-57 μm; oospores 33 × 43 μm] **Rhipidium thaxteri**

4(3) Antheridial origin closely monoclinous [sporangia 30-90 × 20-46 μm; oogonia 33 × 55 μm; oospores 28 × 46 μm] **Rhipidium americanum**

4'(3) Antheridial origin variable (generally very variable morphology) [sporangia 40-78 × 18-47 μm; oogonia 40 × 60 μm; oospores 27 × 50 μm, exospore wall up to 17 μm thick] **Rhipidium interruptum** (type species)

5(1') Pseudomycelial segments present with intermediate elongate constrictions; sexuality well established, with oogonia and antheridia 8

5'(1') Pseudomycelial segments absent; sexuality uncertain [resting sporangia with parthenospores have been reported] 6

6(5') Basal cell small, clavate or rarely two-lobed (mean < 800 μm long); smooth and spiny sporangia comparable in size; spiny sporangia with spines scattered over distal half of the sporangia 7

6'(5') Basal cell large, cylindrical (750-1600 × 60-110 μm); spiny sporangia larger than smooth sporangia (spiny sporangia (110)153(175) × (65)75(90) μm, smooth sporangia (50)100(135) × (30)42(54) μm); spines in two periapical whorls; 'oogonia' 45-62 × 31-53 μm, 'parthenospores' 24-44 μm diam. [N.B. the golden colour of these parthenospores with their fibrillar coat (Batko, 1971) and later observations (Batko, 1977) suggest that these 'parthenospores' could be the resting spores of *Rozellopsis uliginosa*] **Nellymyces megaceros** (type species)

7(6) Zoospores of medium size (volume *ca* 600 μm³); basal cell 200-850 × 100-200 μm; sporangia 70-250 × 35-75 μm, sometimes with spines 14-30 μm long; resting spores doubtfully identified **Mindeniella spinosa** (type species)

7'(6) Zoospores of small size (volume *ca* 200 μm³); basal cell angled, distal limb 175-515 μm, proximal limb 181-290 μm, breadth 91-143 μm; sporangia (47)60-70(88) × (32)40-50(65) μm; resting 'sporangia' (38)56-70(87) μm diam. **Mindeniella asymmetrica**

8(5) Basal cell truncate; spiny sporangia often present 10

8'(5) Basal cell slender, inconspicuous; spiny sporangia not known 9

9(8') Antheridia diclinous; basal cell 300-1200 × 15-30 μm; pseudomycelial segments up to 620 μm long, 8-25 μm diam.; sporangia in whorls of 2-6 at apices of pseudomycelial segments, 35-200 × 10-30 μm; oogonia 32-63 × 26-40 μm; oospores 20-34 μm diam. **Sapromyces elongatus** (type species)

9'(8') Antheridia monoclinous; basal cell 115-247 × 16-23 μm; pseudomycelial segments up to 180 μm long, 8-16 μm diam.; sporangia single or in small clusters at apices of pseudomycelial segments, 49-109 × 16-30 μm; oogonia 35-50 × 27-30 μm; oospores 20-26 μm diam. **Sapromyces androgynus**

10(8) Oogonia not known, or if present, then lacking any periplasmic cell-like involucre to the oospore 13

10'(8) Oogonia with alveolar periplasm forming a cell-like involucre to the oospore 11

11(10') Antheridia diclinous; basal cell squat (*ca* 800 × 160 μm) 12

11'(10') Antheridia monoclinous; basal cell elongate (*ca* 800-1500 × 25-50 μm) [pseudomycelial segments 275-2750 μm long; smooth sporangia 120-175 × 30-35 μm; spiny sporangia more spherical, 60-70 × 45-48 μm, spines 10-35 μm long; oogonia 30-60 μm; oospores 35-45 μm diam.] **Araiospora pulchra** (type species)

12(11) Smooth sporangia 90-150 × 45-60 μm; spiny sporangia 100-150 × 40-80 μm, spines 60-70 μm long [oogonial and oospore dimensions not given] **Araiospora spinosa**

12'(11) Smooth sporangia 79-111 × 29-49 μm; spiny sporangia 60-78 × 46-63 μm, spines 15-30 μm long [oogonia (52)60-64(68) μm diam.; oospore (39)44-46 μm diam. *A. streptandra* var. *echinulosphaera* differs in its smaller and more spherical spiny sporangia (29-49 μm diam., spines 3-20 μm long) and more toruloid antheridia] **Araiospora streptandra** var. **streptandra**

13(10) Spiny sporangia absent 14
13'(10) Spiny sporangia present, with a small peri-apical ring of 4-6 short spines 7-9 μm long [smooth sporangia 63-85 × 11-16 μm; spiny sporangia 63-130 × 12-26 μm; sexual reproduction not known; basal cell 826-846 × 45-48 μm] **Araiospora coronata**

14(13) Basal cell large, 800-1000 × 25-30 μm; sporangia (21)80(144) × (14)23(30) μm; antheridia absent, 'oogonia' 52-54 μm, plerotic?] **Rhipidium parthenosporum**
14'(13) Basal cell small, (42)56(80) × (26)42(63) μm; sporangia (21)80(144) × (14)23(30) μm; antheridia diclinous; oogonial dimensions not given; oospore 28-32 μm **Sapromyces indicus**

KEY XIV: *HYPHOCHYTRIUM*, *CANTERIOMYCES* AND *CYSTOCHYTRIUM*

1 Primarily polycentric 5
1' Primarily monocentric 2

2(1') Saprotrophic 3
2'(1') Parasitic in green plants [thallus not completely filling the host cell] 4

3(2') Saprotrophic in pollen, 1-5 thalli per grain; thallus spherical, 8-32 μm diam.; exit tube 4-7 × 20-60 μm; zoospores 3·5-4·0 × 5·0-6·0 μm (volume equivalent *ca* 60 μm³) [cf. *Cornumyces pygmaeus*, *Pleocystidium pygmaeoides* and *Ducellieria chodatii*] **Hyphochytrium saprobium**
3'(2') Saprotrophic in keratin or grass leaf baits; thallus elongate, rarely septate, 40-400 μm long; zoospores 2·8-4·0 × 5·0-6·5 μm (volume equivalent 30 μm³) **Hyphochytrium elongatum**

4(2') Parasitic in green algae (*Stigeoclonium* and *Draparnaldia*) [sporangia (9)13-33(46) μm diam.; zoospores 2·5 × 5·0 μm (volume equivalent *ca* 16 μm³)] **Canteriomyces stigeoclonii** (monotypic genus)
4'(2') Parasitic in roots of an aquatic angiosperm (*Veronica beccabunga*) [sporangial dimensions not given; zoospores 12·0-15·0 μm diam. (volume equivalent > 1000 μm³)] **Cystochytrium radicale** (monotypic genus)

5(1) Zoosporogenesis completed outside the sporangium; saprotrophic 6
5'(1) Zoosporogenesis completed within the sporangium; parasitic [known only from original descriptions] 9

6(5) Thallus extensive, pseudomycelial with intercalary sporangia 8
6'(5) Thallus of restricted extent, with few septa, sporangia frequently terminal 7

7(6') Saprotrophic in grass leaf baits [sporangia 30-65 μm diam.; zoospores 3·3-3·8 × 5·0-5·5 μm (volume equivalent *ca* 45 μm³)] **Hyphochytrium oceanum**
7'(6') Saprotrophic in *Corylus* pollen [sporangia (8)12-18(23) μm diam.; zoospores 3·0-4·0 × 5·0-8·0 μm (volume equivalent *ca* 60 μm³] **Hyphochytrium** sp. (Persiel, 1960*b*)

8(6) Zoospores > 10 μm diam. (volume equivalent > 500 μm³) [sporangia dimensions not recorded] **Hyphochytrium** sp. (Dick, *loc. cit.*)
8'(6) Zoospores 1·5 × 3·5 μm (volume equivalent *ca* 6 μm³) [sporangia 10-35 μm diam.] **Hyphochytrium catenoides**

9(5') Parasitic in a terrestrial ascocarp [measurements not available] **Hyphochytrium infestans** (type species)
9'(5') Parasitic in a green alga (*Hydrodictyon*) [measurements not available] **Hyphochytrium hydrodictyonis**

KEY XV: *RHIZIDIOMYCES*, INCLUDING THE DOUBTFUL TAXA *LATROSTIUM* AND '*EUCHYTRIDIUM*', BUT NOT *REESIA*

1 Primarily parasitic in fungi or freshwater angiosperms or algae; rhizoidal system apophysate [except *R. parasiticus* and *Reesia*] 9

1' Primarily saprotrophic on a variety of baits (snake skin, *Zea* leaves, pollen grains); rhizoidal system without an apophysis 2

2(1') Exit tubes 1-3, short (length < two-thirds diameter of sporangium); thick-walled, or resting sporangia not known 5

2'(1') Exit tube 1 rarely more, long (length > diameter of sporangium); sporangia or resting sporangia with thick irregular walls 3

3(2') Sporangia sometimes with two or more exit tubes; resting sporangium wall with more or less discrete plaques or bullate knobs 4

3'(2') Sporangia with a single exit tube; resting sporangium wall irregularly thickened [sporangia (8)10-60 μm diam.; zoospores 4·0-4·5 × 5·5-6·5 μm (volume equivalent *ca* 60 μm³)] **Rhizidiomyces hansonii**

4(3) Zoosporangia smooth-walled, resting zoosporangia thick-walled; tendency to polycentric development [sporangia 80-100 μm diam.; zoospores 5·5-6·0 × 7·0-8·0 μm (volume equivalent *ca* 130 μm³)] **Rhizidiomyces bivellatus**

4'(3) All sporangia studded with bullate thickenings; never polycentric [sporangia 12-50(102) μm diam.; zoospores 3·0 × 6·0-7·0 μm (volume equivalent *ca* 30 μm³)] **Rhizidiomyces bullatus**

5(2) Zoosporogenesis completed outside the sporangium 6

5'(2) Zoosporogenesis intrasporangial [sporangia 16-36 μm diam.; zoospores 3·0-5·0 × 4·0-5·5 μm (volume equivalent *ca* 60 μm³)] **Rhizidiomyces saprophyticus**

6(5) Sporangia smooth; never polycentric 7

6'(5) Sporangia with many long filaments; occasionally polycentric [sporangia (6)8-60(180) μm diam.; zoospores 3·0-4·0 × 6·0-8·0 μm (volume equivalent *ca* 45 μm³)] **Rhizidiomyces hirsutus**

7(6) Sporangia with a smooth wall 8

7'(6) Sporangia with a halo of mucilage [sporangia 20-63 μm diam; exit tubes *ca* 3; zoospores 3·0-3·6 × 5·0-5·5 μm diam; (volume equivalent *ca* 45 μm³)] **Rhizidiomyces coronus**

8(7) Rhizoid single [sporangia 15-62 μm diam; exit tubes 1-3(4); zoospores 3·0-3·8 × 4·0-4·5 μm (volume equivalent *ca* 40 μm³)] **Rhizidiomyces bulbosus**

8'(7) Rhizoids 2-5 [sporangia 5-70 μm diam.; zoospores 4·0-4·2 × 4·8-5·0 μm (volume equivalent *ca* 45 μm³)] **Rhizidiomyces sp.** (Karling, 1981*b*)

9(1) Parasitic on aquatic angiosperms or algae [very rarely reported] 13

9'(1) Parasitic on fungi 10

10(9') Epiparasitic on oogonia of *Peronosporomycetes* [exit tube short] 11

10'(9') Parasitic on *Zygomycetes* or *Chytridiomycetes* [known only from original descriptions] 12

11(10) Zoosporogenesis competed extrasporangially; with septum delimiting sporangium from the apophysis; exit tube short; sporangia spherical or ovoid, 18-45 μm diam.; zoospores 3·7 × 6·0 μm (volume equivalent *ca* 60 μm³)] **Rhizidiomyces apophysatus** (type species)

11'(10) Zoosporogenesis completed intrasporangially; lacking septum between sporangium and apophysis; exit tube very short, often merely an aperture; sporangia spherical, 10-15 μm diam., zoospores 3·0 × 6·0 μm (volume equivalent *ca* 45 μm³)] **Rhizidiomyces japonicus**

12(10') Terrestrial, on zygotes of *Endogonales* (*Gigaspora*) [sporangia 20-150 μm diam.; zoospores 6·0-7·0 × 10·0-12·5 μm (volume equivalent 250 μm³)] **Rhizidiomyces stomatosus**

12'(10') Terrestrial/aquatic, on sporangia of *Chytridiales* (*Rhizidiomyces*) [sporangia (12)20-40 μm diam.; zoospores 3·5-4·2 × 5·7-7·0 μm (volume equivalent 50 μm³)] **Rhizidiomyces parasiticus**

13(9) Sporangia non-apophysate 14

13'(9) Sporangia apophysate [epibiotic on *Chlamydomonas*; sporangia 9-16 μm diam.; zoospores 3·0 × 6·0 μm (volume equivalent *ca* 45 μm³] [known only from original description] **Rhizidiomyces ichneumon**

14(13) Sporangial dehiscence non-operculate, epibiotic on *Vaucheria* oospores within the oogonia; sporangial dimensions not given; zoospores 2·5-3·0 μm diam. (volume equivalent 25 μm³)] [cf. *Rhizophydium decipiens* on *Oedogonium* oospores] **Laterostium comprimens** (doubtful monotypic genus)

14'(13) Sporangial dehiscence operculate [epibiotic on *Mougeotia*; neither measurements nor drawings reported; known only from original description] '*Euchytridium mesocarpi*'

KEY XVI: *LABYRINTHISTA*

1 Trophic cells completely surrounded or embedded within the common sheet- or tube-like anastomosing ectoplasmic net; trophic cells predominantly spindle-shaped; cells normally uninucleate; cell investment of one or two layers of overlapping scales; flagellate cells (meiospores), when produced, with an eyespot *(Labyrinthulales)* 2

1' Trophic cells superficial to the radiating, branching rod-like strands of the ectoplasmic net, ectoplasmic nets of adjacent cells not normally confluent; trophic cells predominantly spherical; cells initially uninucleate, usually becoming apocytic, either directly to form a multinucleate thallus, or after an intervening period of mitotic cellular division; cell investment of many layers of overlapping scales; flagellate cells (zoospores) without an eyespot *(Thraustochytriales)* 14

2(1) Trophic cells isolated from plant debris 3

2'(1) Trophic cells isolated from marine animal material [doubtful taxa] 13

3(2) Trophic cells either forming cell aggregates without a common outer membrane, or lacking any aggregate stage 4

3'(2) Trophic cells forming cell or cyst aggregates (sori) with a common outer membrane 7

4(3) Sporocytes not known to form zoospores 5

4'(3) Sporocytes forming 4 zoospores **Labyrinthula algeriensis**

5(4) Trophic cells not forming aggregates; marine 6

5'(4) Trophic cells forming naked aggregates; freshwater, associated with *Vaucheria terrestris* [trophic cells 11-20 μm; plane of cell division transverse; aggregates of 1-12(100) sporocytes; sporocyte walls thin; seldom reported, cf. *Diplophrys archeri*] **Labyrinthula cienkowskii**

6(5) Trophic cells 3-5 × 15-20 μm; confined to *Zostera marina* (marine angiosperm) **Labyrinthula zosterae**

6'(5) Trophic cells 4-4 × 5-9 μm; confined to *Chorda filum* (brown alga) [some trophic cells becoming single cysts 6 μm diam., germinating to give an amoeboid stage, 6 × 10-20 μm] **Labyrinthula pohlia**

7(3') Sporocyte aggregates (sori) with a thick wall 8

7'(3') Sporocyte aggregates (sori) with a thin wall 11

8(7) Colour of colony in the mass hyaline 9

8'(7) Colour of colony in the mass light green [trophic cells 3-6 × 10-15 μm; each sporocyte of the sorus developing into a trophic cell] **Labyrinthula coenocystis**

9(8) Sporocyte aggregates small, often confined to a cluster of several sporocytes within a host cell; associated with various marine algae (and marine angiosperms?) 10

9'(8) Sporocyte aggregates very large (up to 1000μm diam.) with a very thick (20-30 μm) common envelope; associated with diatoms [trophic cells 15-18 μm long; sporocytes large (6-8 × 18-25 μm), with a thick envelope] **Labyrinthula magnifica**

10(9) In marine brown algae [trophic cells <8 μm long; sporocyte walls thin; sporocyte development unknown] **Labyrinthula valkanovii**

10'(9) In marine *Cladophora* [trophic cell dimensions not known; each sporocyte producing 4 trophic cells; doubtful taxon] **Labyrinthula chattonii**

11(7') Trophic cells large (>20 μm long); sporocytes producing (1-)4 trophic cells, flagellate cells not known; colour of colony in the mass hyaline or pale yellow 12

11'(7') Trophic cells small (<18 μm long); sporocytes forming 8 haploid flagellate cells; colour of colony in the mass yellow/red [trophic cells 4-5 × 15-17 μm] **Labyrinthula vitellina** (type species)

12(11) Colour of colony in the mass pale yellow; trophic cells 4-5 × 18-25 μm; sporocytes with thick walls; associated with various hosts **Labyrinthula macrocystis**

12'(11) Colour of colony in the mass hyaline; trophic cells 15-30 μm long; sporocytes with thin walls; confined to brown algae **Labyrinthula roscoffensis**

13(2') Trophic cells isolated from *Thais* [possibly synonymous with *Labyrinthoides schizochytrops*] **Labyrinthula thaidisi**

13'(2') Trophic cells isolated from *Littorina* **Labyrinthula jeremarina** [*nomen nudum*]

14(1') Forming sporangia by cellular division or apocytic development, or with aggregates of sporocytes (sori) with a common wall 15

14'(1') Not forming sporangia or aggregates of sporocytes [zoospores not known, sagenogenetosomes not known; cells ovoid, 3·7-5·9 × 5·1-8·5 μm, with prominent refractive eccentric granules; investment of a few layers of overlapping scales] **Diplophrys marina**

15(14) Young trophic cells dividing to give motile secondary trophic cells dispersing on the radiating ectoplasmic network 16

15'(14) Young trophic cells apocytic or cellularly mitotic but non-dispersing (static) with a radiating ectoplasmic network 21

16(15) Zoospores not known, or not forming a conspicuous part of the life-history 17

16'(15) Zoospores abundantly produced [zoosporangia 20-25 μm diam.; wall persistent with an apical pore; zoosporangioplasm expelled and cleaving into *ca* 64 cells 3·5-5·0 μm diam., which encyst and then behave as 'zoosporangia', each releasing a zoospore 2·5-3·5 × 5·0-6·5 μm] **Ulkenia amoeboidea**

17(16) Reproduction by trophic cells either from sporangia, or from cell aggregates or sori; zoospores produced rarely in all known species 18

17'(16) Reproduction by aplanospores from sporangia; sporangia single, not forming aggregates, plasmodia or sori; zoospores not known [main wall scale carbohydrate: fucose; sporangia 8-20μm diam.; aplanospores 1·8-6·5μm diam.] **Aplanochytrium kerguelense** (monotypic genus)

18(17) Apocytic phase extended, sporangia developing by progressive synchronized enlargement, nuclear division and cleavage 19

18'(17) Apocytic phase very short, sporangiferous sori developing by enlargement of sporocytes followed by more or less simultaneous nuclear division and cleavage [trophic cells spherical or subspherical, 3-6μm diam.; sori 15-18μm diam.; plasmodia rare] **Labyrinthuloides yorkensis** (type species)

19(18) Sori present; on aquatic angiosperms 20

19'(18) Sori absent; in green algae [trophic cells subspherical or elliptical, 6-10μm diam.; sori 10-20μm diam.; plasmodia sometimes dominant] **Labyrinthuloides minuta**

20(19) Trophic cells ovoid, tapered posteriorly, 5-8μm diam.; sori 10-40μm diam.; on *Halophila* **Labyrinthuloides saliens**

20'(19) Trophic cells spherical, 4-7μm diam.; sori 10-90μm diam.; on *Halodule* [cf. *Plasmodiophora diplantherae* on the same host] **Labyrinthuloides schizochytrops**

21(15') Sagenogenetosomes present; associated with algal debris or plant (pollen) baits 22

21'(15') Sagenogenetosomes absent; associated with shells of *Ostrea* **Althornia crouchii** (monotypic genus)

22(21) Subsoral apophysis absent; most frequently collected on pollen bait or agar plated sea water 23

22'(21) Subsoral apophysis present; associated with red algae [on *Gracilaria*; sorus a zoosporangium; zoosporangium 16-30 × 10-20μm, thick-walled, wall persistent except for an apical pore; apophysis 10-13 × 5-10 μm; zoospores 4·0-5·0 μm diam., motile within the sporangium] **Japonochytrium marinum** (monotypic genus)

23(22) Trophic cell or thallus eucarpic with a persistent (apocytic) fundament giving successive zoosporangial production 24

23'(22) Trophic cell or thallus monocarpic and non-proliferous 32

24(23) Zoospores fully motile at time of release from zoosporangium 25

24'(23) Zoospospores with a quiescent phase after release from zoosporangium before swimming away 27

25 (24) Fundament (rudiment) single, with an ectoplasmic net (eucarpic) 26

25'(24) 2(3-4) rudiments each with an ectoplasmic net (each rudiment monocarpic) [zoosporangium 11-28 × 14-37 μm; wall persistent, zoospores released through ruptures; zoospores 2·0-3·5 × 4·5-7·5 μm] **Thraustochytrium multirudimentale**

26(25) Zoosporangium small (8-17 μm diam.); wall thin, opening distally [colonies golden in mass; zoospores 2·0-4·0 × 3·0-6·0 μm] **Thraustochytrium aureum**

26'(25) Zoosporangium large (7-28 × 15-35 μm); wall cracking open [zoospores 2·0-4·0 × 3·5-6·0 μm] **Thraustochytrium motivum**

27(24') Zoosporangia <30 μm diam.; fundament single, rarely <10 rudiments [zoosporangium thin-walled] 28

27'(24') Zoosporangia >35 μm diam.; rudiments numerous [zoosporangia 40-50 μm diam.; 6-50 rudiments which become independent thalli; wall completely dispersing; zoospores forming flagella after a quiescent state; zoospores 2·5-3·5 × 4·8-5·5 μm] **Thraustochytrium rossii**

28(27) Fundament single 29

28'(27) 3-10 successively forming rudiments [zoosporangia 20-30 μm diam.; zoospores discharged as flagellated bodies in a clump, almost instantly swimming away; zoospores 2·5-3·0 × 5·0-7·0 μm] **Thraustochytrium kerguelense**

29(28) Zoosporangial wall persisting as partial remnants, especially near fundament 30

29'(28) Zoosporangial wall completely disintegrating at maturity [zoosporangia 18-25μm diam.; zoospores discharged as flagellated bodies in a clump, almost instantly swimming away; zoospores 3·0 × 5·0 μm] **Thraustochytrium antarcticum**

30(29) Cleavage planes random and simultaneous 31

30'(29) Cleavage planes primarily radial, central vacuole expanding to separate zoospores [zoosporangium 14-17 × 17-19μm; rupturing in one or more places; zoospores 2·5-3·0 × 4·5-4·8 μm; cf. *T. striatum*] **Thraustochytrium kinnei**

31(30) Zoospores flagellate before discharge [zoosporangial diameter very variable (8-30μm); zoospores 3·0-3·5 × 3·5-6·3μm] **Thraustochytrium benthicola**

31'(30) Zoospores becoming flagellate after discharge [zoosporangia 10-20 × 15-29μm; zoospores 4·0μm diam.] **Thraustochytrium proliferum** (type species)

32(23') Multinucleate zoosporangium developing directly from a primary trophic cell 33

32'(23') Multinucleate zoosporangia developing indirectly from a primary trophic cell after a period of mitotic cellular division in the sorus (zoosporangia thence developing, and sometimes termed secondary zoosporangia) 43

33(32) Multinucleate zoosporangial contents cleaved within the zoosporangium before release as quiescent or motile zoospores 34
33'(32) Multinucleate zoosporangial contents released before cleavage as an amoeboid (undulate) protoplast 39

34(33) Zoosporangial wall thick (>4 μm) [zoospores with a quiescent phase before swimming away] 35
34'(33) Zoosporangial wall thin (<2 μm) 36

35(34) Zoospores with two flagella; zoosporangia 15-30 μm diam. [wall thick (3-5 μm); protoplast with large lipid globules; zoospores 5·0-7·0 μm long] **Thraustochytrium pachydermum**
35'(34) Zoospores reputed to have a single anterior flagellum; zoosporangia 8-20 μm diam. **Elina sinorifica**

36(34') Zoosporangial wall partially persistent 37
36'(34') Zoosporangial wall completely dispersing at the time of zoospore release 38

37(36) Basal part of zoosporangial wall persistent [zoosporangia 5-10 × 5-20 μm; zoospores 2·5 × 3·0-4·0 μm; on marine *Ulothricales*] **Thraustochytrium globosum**
37'(36) Zoosporangium wall disintegrating at the apex only; zoosporangia 15 μm diam. [zoospores with a quiescent phase before swimming away] **Thraustochytrium arudimentale**

38(36') Zoospores with two flagella; sporangia 14·0-26·0 × 14·6-22·0 μm; zoospores 2·0-3·7 × 3·4-6·7 μm [discharge and cleavage by means of an enlarging central vacuole with radiating cleavage planes, cf. *T. kinnei*] **Thraustochytrium striatum**
38'(36') Zoospores reputed to have a single anterior flagellum; zoosporangia 20-30 μm diam.; wall thin; zoospores 4·0-5·0 × 6·0-7·0 μm **Elina marisalba** (type species)

39(33') Zoosporangium wall dispersing at time of protoplasmic release 40
39'(33') Zoosporangium wall partially persistent after protoplasmic release [zoospores 2·9 × 4·2 μm] **Ulkenia sarkariana**

40(39) Zoosporangium wall thick (>2·0 μm) 41
40'(39) Zoosporangium wall thin (<1·0 μm) 42

41(40) Colour of colony in the mass pink; zoosporoplasm released in a partially cleaved state [zoosporangium 14-46 μm diam.; zoosporangium wall very thick (2-6 μm); zoospores 2·5-4·0 × 4·5-7·5 μm] **Thraustochytrium roseum**
41'(40) Colour of colony in the mass hyaline; zoosporoplasm released in an amoeboid state [zoosporangium 15-20 μm diam.; zoosporangium wall thick (2 μm); zoospores 2·5-3·5 × 4·0-5·5 μm] **Ulkenia visurgensis** (type species)

42(38') Zoosporangia 16-22 μm diam.; contents discharged in an amoeboid state, dividing to give 1-5 rudiments and numerous zoospores; zoospores 2·9-4·0 × 3·5-4·8 μm **Ulkenia profunda**
42'(38') Zoosporangia 8-18 μm diam.; contents discharged in a star-like mass, dividing by centripetal cleavage to give zoospores; zoospores 3·0-4·0 × 5·0-8·0 μm **Ulkenia radiata**

43(39) Secondary assimilative cells developing ectoplasmic nets and behaving as secondary assimilative cells before developing into zoosporangia 44
43'(39) Secondary assimilative cells developing directly into secondary zoosporangia [zoosporangia 6-10μm diam.; giving 4-20 secondary zoosporangia; secondary zoosporangia (sporocytes) producing 4-12 zoospores; zoospores with a quiescent phase] **Ulkenia minuta**

44(43) Individual mature cells >6 μm diam. 45
44'(43) Individual mature cells <5 μm diam. 46

45(44) Cell masses (sori) of many sporocytes; trophic cells 6-15 μm diam.; each sporocyte undergoing mitosis to produce 2-8 zoospores 46
45'(44) Cell masses (sori) of few sporocytes (<12); trophic cells 6-9 μm diam.; each sporocyte becoming transformed into a single zoospore; [zoospore 2·6-3·3 × 4·1-4·8 μm] **Schizochytrium mangrovei**

46(45) Zoospores 2·5-5·0 × 4·0-7·5 μm; trophic cells 6-12 μm diam. [cell masses (sori) of many sporocytes, 30-140 μm diam.; zoosporangial tetrads 13-29μm diam.; each zoosporangium producing 32-64 zoospores; colony orange in mass] **Schizochytrium aggregatum** (type species)
46'(45) Zoospores 5·0-7·0 × 6·0-8·5 μm; trophic cells spherical, 7-15μm diam. [zoosporangia 12-24 μm diam. producing 16-64 zoospores; trophic cells forming limaciform binucleate amoeboid cells 12·0-20·0 × 5·0-8·0 μm with fine pseudopodia; rounded amoeboid cell forming 8 smaller zoospores, 3·5-5·0 × 4·5-6·0 μm] **Schizochytrium limacinum**

47(44') Large cell masses (irregular sori) of many sporocytes 48
47'(44') Small compact sori 10-15 μm diam. releasing aggregates of small sporocytes [sporocytes producing <10 zoospores; zoospores 2·5 3·5 × 4·0-5·5 μm] **Thraustochytrium aggregatum**

48(47) Cell masses (sori) of many sporocytes (*ca* 100); each sporocyte producing 8 zoospores [trophic cells 4-5μm diam.; zoospores 1·5-2·5 × 3·5-5·0 μm] **Schizochytrium octosporum**
48'(47) Cell masses (sori) composed of numerous minute (<8 μm diam.) tetrads or octads of sporocytes; each sporocyte giving rise to 2 zoospores [zoospores 1·7 × 2·5 × 5·0 μm] **Schizochytrium minutum**

KEY XVII: *HAPTOGLOSSA*

1 Zoospores formed within the zoosporangium; heterosporous, rarely homosporous but always homosporous in any one zoosporangium [zoospore cyst volume eqivalent within the range 50-120 µm^3] 3

1' Aplanospores formed within the sporangium; heterosporous, but homosporous in any one sporangium 2

2(1) Aplanospores expelled and drifting away from sporangium discharge tube; glossoid cell unequally three-lobed; aplanospore volume equivalents <75 and >300 µm^3 [thalli 7-350 × 4-18 µm, each becoming transformed into a single sporangium; exit tubes short and broad, 1-4 µm long × 4-7 µm broad; number of exit tubes 1-6; aplanospores heterosporous, either 10·0 × 8·0 µm or 6·0 × 4·5 µm; glossoid spores triangular-cushion-shaped, 5·0-8·0 × 4·0-7·0 × 5·0-6·0 µm (or 5·0-5·0 µm diam. with a glossoid lobe 3·0-8·0 µm long); sporidia (tertiary spores) 6·0-9·0 × 2·8-4·0 µm (dimensions not given in original description)] **Haptoglossa heterospora** (type species)

2'(1) Aplanospores collecting in a spherical ball at the tip of the sporangium discharge tube; glossoid cell elongate awl or Y-shaped; aplanospore volume equivalents <30 and >100 µm^3 [thalli 60-95 × 20-40 µm, each becoming transformed into a single sporangium; normally one exit tube; aplanospores heterosporus, 5·0-6·5 µm diam. or 3·3-4·0 µm diam.; glossoid spores 13·5-17·2 × 1·4-1·8 µm; sporidia (tertiary spores) 4·0-5·0 × 2·3-3·3 µm; not known to infect nematodes] **Haptoglossa elegans**

3(1') Sporangia with an evacuation tube; heterosporous 4

3'(1') Sporangia without an evacuation tube; homosporous [thalli 50-200 × 5-15 µm, each becoming transformed into a single sporangium; exit tubes non-existent; aplanospores spherical 3·5-4·5 µm diam.; glossoid spores broadly arcuate, tapered to both ends 6-9 µm long (10-15 µm along curved outer edge) × 2-3 µm; sporidia not observed] **Haptoglossa erumpens**

4(3) Glossoid cell trapezoid or elongate 7

4'(3) Glossoid cell cone-shaped or more or less triangular 5

5(3') Flagella strongly anisokont (ratio 1:2 or 3); sporidia ovoid 6

5'(3') Flagella more or less isokont; sporidia fusiform [thalli 50-80 × 25-45 µm, each becoming transformed into a single sporangium; exit tubes 10-20(30) × 10 µm; number of exit tubes 1(4); primary zoospores 7·5-10·0 × 3·0-4·0 µm; encysted spores spherical, 4·0 µm diam.; glossoid development cylindrical-triangulate, 8·5-11·5 × 3·5 µm; sporidia (tertiary spores): dimensions not given] **Haptoglossa intermedia**

6(5) Evacuation tubes short, unbranched; angle of flagellar kinetosomes 110°; flagellar insertion sub-apical [thalli cylindrical, occasionally lobed, 25-300 × 10-20 µm, each becoming transformed into a single sporangium; exit tubrs 5-10 × 4 µm; number of exit tubes 1-2(3); primary zoospores 5 × 2 µm, anisokont, 5 µm and 10 µm respectively; encysted spores 2·8-4·0 µm diam.; glossoid spores broadly rounded, curved and tapered to a reflexed tip 7-8 × 4 µm; sporidia broadly ovoid, 3·25 × 2·75 µm] **Haptoglossa dickii**

6'(5) Evacuation tubes long and sometimes branched; angle of flagellar kinetosomes 140°; flagellar insertion said to be apical [thalli 20-450 × 10-60 µm, each becoming transformed into a single sporangium; exit tubes 20-50 × 5-15 µm; number of exit tubes 1-2; primary zoospores 6·5-9·0 × 4·0-6·5 µm; encysted spores broadly conical, 3·0-4·0 × 5·0-12·5 µm; glossoid development cylindrical; sporidia (tertiary spores) 4·5-5·0 × 2·0-3·0 µm] **Haptoglossa zoospora**

7(4) Glossoid cell trapezoid, with elongate apical lobe and 'nose-cone'; thallus cylindrical [thalli 90 × 50 µm, each becoming transformed into a single sporangium; exit tubes 10-20(30) × 10 µm; number of exit tubes 1(4); primary zoospores 7·0-10·0(12·0) × 3·5-4·5 µm, diplanetic; ; encysted spores spherical; glossoid development skittle-shaped, 15-18 × 3·2-4·2 µm; sporidia (tertiary spores): dimensions not given] **Haptoglossa mirabilis**

7'(4) Glossoid cell spindle-shaped with elongate posterior lobe; thallus asymmetric with somewhat tapered ends [thalli often asymmetrical, 40-62 × 18-26 µm, each becoming transformed into a single sporangium; 1-several short exit tubes; primary zoospores 8 µm long or 4-5 µm diam.; glossoid spores (8)10-12(14) × 3·0-3·5 µm; sporidia ("smaller than *H. mirabilis*") 3·0-4·0 × 2·5 µm] **Haptoglossa humicola**

KEY XVIII: *PLASMODIOPHOROMYCETES*

1 Cysts in small or large cystosori [cysts separable in *Woronina*; cysts not known for two species of *Ligniera*] 6

1' Cysts separate, free or loose in clusters [causing hypertrophy and hyperplasia] (*Plasmodiophora*) 2

2(1') Cysts smooth [neither primary nor secondary zoospores known] 4

2'(1') Cysts spiny 3

3(2') Cysts spherical (1·6-4·3μm diam., mean 3·9μm diam.), covered in spines; primary zoospores 2·5-3·5μm diam.; sporangia 6·0-6·5μm diam.; secondary zoospores 1·9-3·1μm diam.; parasitic in roots of numerous herbs, mostly *Brassicaceae* (*Capparales*);causing 'Club Root' of crucifers **Plasmodiophora brassicae** (type species)

3'(2') Cysts ovoid or fusiform (3·0-7·0 μm diam.), with a bristle at one or both ends; sporangia unknown; neither primary nor secondary zoospores known; parasitic in *Zostera* (a marine zoospores known; parasitic in monocotyledonous angiosperm) **Plasmodiophora bicaudata**

4(2) Cysts >4·0 μm diam. [neither sporangia nor primary or secondary zoospores known] 5

4'(2) Cysts <2·0 μm diam. [parasitic in *Ficus* roots; a doubtful species, reported from Italy] **Plasmodiophora ficirepentis**

5(4) Cysts 7·0-8·0 μm diam.; parasitic in *Triglochin* stems (a salt marsh plant) [see also the doubtful species *Tetramyxa triglochinis*] **Plasmodiophora maritima**

5'(4) Cysts 4·0-5·0 μm diam.; parasitic in stems of the aquatic monocotyledon *Halodule* (*Cynodoceaceae*) **Plasmodiophora diplantherae***

[* The doubtful species, *P. halophilae*, in the aquatic monocotyledon *Halophila ovalis* (*Hydrocharitaceae*) from Indonesia, would also key out here.]

6(1) Cystosori of large numbers of cysts 10

6'(1) Cystosori of 2, 4 or 8 cysts 7

7(6') Cysts in diads, triads or tetrads, germinating to give amoeboid (non-flagellate?) planonts; neither sporangia nor primary or secondary zoospores known; parasitic in vascular plants (*Tetramyxa*) 9

7'(6') Cysts in octets, germinating to give zoospores; sporangia present;parasitic in *Saprolegniaceae* (*Peronosporomycetes*) (*Octomyxa*) 8

8(7) Mean cyst diameter <3·0μm (2·4-3·2 μm diam.); mean sporangial diam. >10·0 μm (6·0-16·0 μm diam.); zoospore dimensions not known; parasitic in *Achlya* **Octomyxa achlyae** (type species)

8'(7) Mean cyst diameter >3·5μm (2·9-4·9 μm diam.); mean sporangial diam. <8·0 μm (5·1-7·6 μm diam.); parasitic in *Brevilegnia* and *Geolegnia* **Octomyxa brevilegniaee**

9(7') Parasitic in aquatic angiosperms; cysts polyhedral, 3·5-7·0 μm diam. **Tetramyxa parasitica** (type species)

9'(7') Parasitic in terrestrial gymnosperm roots; cysts spherical, ellipsoidal and angular, 4·0-6·0 μm diam. [known only from original description] **Tetramyxa rhizophagus**

10(6) Cystosori composed of cysts forming a hollow sphere or irregular spongy aggregates 15

10'(6) Cystosori composed of cysts forming a single or double layer of closely packed cysts, forming sheets or discs 11

11(10') Cystosori composed of a double layer of tightly packed cysts (*Sorodiscus*) 12

11'(10') Cystosori composed of a single layer of cysts, usually lining the host cell wall [one host cell may contain more than one cystosorus; cysts usually in multiples of 4; cysts 3·5-5·0 μm diam.; neither sporangia nor primary or secondary zoospores known] **Membranosorus heteranthera**e (monotypic genus)

12(11') Parasitic in green plants: sporangia unknown 13

12'(11') Parasitic in *Pythium*; sporangia present, 5·3-8·8 μm diam. [cysts 3·1 - 4·1 μm diam.; each cyst germinating to give a single primary zoospore] **Sorodiscus cokeri**

13(12) Mean cyst diameter 5·0 μm; parasitic in aquatic green plants 14

13'(12) Mean cyst diameter <4·0 μm; parasitic in *Gynandropsis* (terrestrial, *Capparidaceae*) [cysts 3·8-4·2 × 3·2-3·6 μm, outer wall layer with blunt protrusions; primary zoospore 2·5-3·5 μm diam.] **Sorodiscus radicicola**

14(13) Parasitic in *Chara* (stonewort); cysts 4·0-9·0 μm diam. **Sorodiscus karlingii**

14'(13) Parasitic in *Callitriche* (aquatic angiosperm); cysts 4·0-5·0 × 6·0-7·0 μm diam., with an apical collar **Sorodiscus callitrichis** (type species)

15(10) Cystosori in hollow spheres; causing hypertrophy of host 28

15'(10) Cystosori in masses of indefinite shape and size; not usually causing hypertrophy of host 16

16(15') Sporangia large, in elongate or lobed sporangiosori, producing distinct exit tubes 23

16'(15') Sporangia small, in loose sporangiosori (*Ligniera*) 17

17(16') Cysts not known; mean sporangial diameter < 14·0 μm [doubtful species] 22

17'(16') Cysts known; mean sporangial diameter either < 10·0 μm or > 15·0 μm 18

18(16') Sporangia small, 10·0 μm diam. 19

18'(16') Sporangia 15·0-20·0 μm diam. [cysts 4·0-7·0 μm diam., producing a single primary zoospore, dimensions not known; secondary zoospores 3·5 × 4·5 μm; parasitic in various monocotyledonous and dicotyledonous herbs] **Ligniera junci**

19(18) Cysts large, >4·0 μm diam., smooth; sporangia, 6·0 μm diam., or not known 20

19'(18) Cysts small, 3·5 μm diam., walls verrucose; sporangia 4·0-9·0 μm diam., interconnected [zoospore dimensions not known; parasitic in various monocotyledonous and dicotyledonous herbs] **Ligniera verrucosa** (type species)

20(19) Parasitic in terrestrial angiosperms 21

20'(19) Parasitic in *Isoetes* (aquatic pteridophyte) [cysts 5·0 × 6·0-8·0 μm; neither sporangia nor primary or secondary zoospores known; known only from original description] **Ligniera isoetes**

21(20) Parasitic in roots of various herbs [cysts 4·2-4·6 × 5·0-5·2 μm, each producing more than one primary zoospore, zoospore dimensions not known] **Ligniera betae**

21'(20) Parasitic in root hairs of the grass *Poa* [cysts and sporangia 4·0-6·0 μm diam.; secondary zoospores 3·0 × 3·8 μm; known only from original description] **Ligniera pilorum**

22(17) Parasitic in *Plantago lanceolata*; winter sporangia 10·5-15·0 μm diam., spring sporangia 4·5-6·0 μm, producing 4 secondary zoospores, 1·5 × 1·8 μm [known only from Czechoslovakia] **Ligniera plantaginis**

22'(17) Parasitic in various monocotyledonous and dicotyledonous herbs; sporangia 5·0-6·0 μm diam., producing up to 25 zoospores, zoospore dimensions not known **Ligniera hypogeae**

23(16) Zoosporangia single or in sporangiosori, zoosporangia spherical or composed of clusters of interconnecting spherical lobes; primary and secondary zoospores small, <3·5 μm diam.; not known to have elaborate penetration extrusomes (*Woronina*) 25

23'(16) Zoosporangia single or in sporangiosori, zoosporangia spherical or irregular, but not with interconnecting lobes; primary and secondary zoospores larger >4·0 μm diam.; infection by means of an elaborate extrusome (*Polymyxa*) 24

24(23') Cysts of cystosori contiguous and fused; sporangiosorus of several lobed sporangia; parasitic in grasses [cysts 4·0-7·0 μm diam.; producing a single primary zoospore, pyrifirm or slightly reniform, 4·0-5·0 μm diam.] **Polymyxa graminis** (type species)

24'(23') Cysts of cystosori loosely held together by exospore material; sporangia not united into a sporangiosorus; parasitic in *Chenopodiaceae* and *Amaranthaceae* [cysts 4·2-5·0 μm diam., producing a single primary zoospore, pyriform or slightly reniform, 4·5-5·0μm diam.] **Polymyxa betae**

25(23) Mean cyst diameter < 10·0 μm; parasitic in *Peronosporomycetes* 26

25'(25') Mean cyst diameter > 12·0 μm; parasitic in *Vaucheria* [secondary zoospore 2·0-2·6 μm diam.] **Woronina glomerata**

26(25) Parasitic in *Saprolegniaceae*; sporangia 8·0-33·0 μm diam. 27

26'(25) Parasitic in *Pythium*; sporangia 4·3-9·0 μm diam. [cysts 3·5-5·6 μm diam., producing a single primary zoospore] **Woronona pythii**

27(26) Sporangiosorus 30-60 × 100-476 μm; cysts 4·0-8·6 μm diam. [cysts producing 1 or more primary zoospores [NOTE: the parasites in *Saprolegnia* (type host) and *Achlya* may not be conspecific (see host range studies, Goldie-Smith, 1954] **Woronina polycystis** (type species)

27'(26) Sporangiosorus 12-120 μm [sporangia sometimes linear]; cysts 4·0-6·0 μm diam. [secondary zoospores 2·5-4·0 μm diam. (zoospore cysts 3·0-3·5 μm diam.); parasitic in *Leptolegnia caudata*] **Woronina leptolegniae**

28(15) Cystosori composed of spongy balls (*Spongospora*) 30

28'(15) Cystosori composed of hollow spheres (*Sorosphaera*) 29

29(28') Mean cyst diameter >5·0 μm (4·0-5·0 × 8·0-9·0 μm diam.); primary zoospore diam. >4·0 μm (4·0-4·8 μm diam.); secondary zoospores 5·6-7·0 μm diam.; parasitic in *Veronica* (*Scrophulariaceae*) **Sorosphaera veronicae** (type species)

29'(28') Mean cyst diameter <4·0 μm (3·0-4·0 μm diam.); primary zoospore diam. <3·0 μm (2·0-3·0 μm diam.); secondary zoospores 3·2-4·2 μm diam.; parasitic in *Poaceae* [known only from two collections] **Sorosphaera radicalis**

30(28) Cysts smooth-walled 31

30'(28 Cysts slightly verrucose [cystosori 25-50 μm diam.; cysts 4·0-5·5 μm diam.; parasitic in *Campanula* (*Campanulaceae*); neither sporangia nor primary or secondary zoospores known; known only from origainal collection] **Spongospora campanulae**

31(30) Parasitic in *Brassicaceae* or *Asteraceae*; sporangia producing 4 or 8-10 secondary zoospores 32

31'(30) Parasitic in *Solanaceae* (rarely other hosts); sporangia producing variable numbers of secondary zoospores [cystosori 19-85 μm diam.; cysts 3·5-4·5 μm diam.; sporangia single or numerous, loosely united into sporangiosori; secondary zoospores 2·5-4·6 μm diam.; causing 'Powdery Scab' of potatoes] **Spongospora subterranea** (type species)

32(31) **Parasitic in *Cotula* (*Asteraceae*) [cystosori and cyst dimensions not known; primary zoospores 3·5-4·0 μm; sporangia 6·3 μm diam., loosely united into sporangiosori, producing 8 - 10 secondary zoospores; known only from original collection]** **Spongospora cotulae**

32'(31) **Parasitic in *Nasturtium* (*Brassicaceae*) [cystosori 12-82 μm diam.; cysts 3·3-4·4 μm diam.; sporangiosori large, sporangia numerous, 6·0-12·0 μm diam., each producing 4 secondary zoospores; secondary zoospores 2·4-6·0 μm diam.; causing 'Crook Root' of watercress]** **Spongospora nasturtii**

APPENDICES AND GLOSSARY

APPENDIX 1

LISTS OF BINOMIALS IN CURRENT USE

STRAMINIPILOUS FUNGI, STRAMINIPILOUS HETEROTROPHS, POSSIBLY STRAMINIPILOUS HETEROTROPHS, AND OTHERS

NOTES:

The following lists of genera are listed in alphabetic order under families; the species are listed in alphabetic order under each genus, with the type species identified.

For uniformity; it is desirable that the recommendations of Brummitt & Powell (1992) or Kirk & Ansell (1992) for authority abbreviations are followed, but it should be noted that some authors do not fully accept these recommendations. Unfortunately, the recent monographs on *Peronospora* (Kochman & Majewski, 1970; Constantinescu, 1991*a*); earlier entries in the *Index of Fungi*; the check-lists for *Pythium* (Dick, 1990*b*), and the key to *Phytophthora* (Stamps *et al.*, 1990), did not use these forms of author abbreviations consistently. Erwin & Ribeiro (1996) adopted the potentially misleading style of citing the authors of the publication, rather than the authority for the binomial, after the masthead binomial.

Note that the spelling and use of accents in 'authority' notation may differ from the forms used in published articles, for example: Itô for Ito; Plaäts-Nit. for Plaats-Niterink; Skvortzov for Skvortzow; Sorokīn for Sorokin/Sorokine). Korf (1996) has been followed in that '*ex*' combinations have been shortened and '*et al.*' is used for authorities of three or more names (though these details are given in square brackets after the entry). For more recent taxa (and for most binomials in *Plasmopara*) the page reference to the *Index of Fungi* is given, with part numbers for the incomplete volume 6.

Many intra-specific taxa (e.g., Skalický, 1964; Skidmore and Ingram, 1985) are without justification in the sense that they have not been named in accordance with the *ICBN* rules; but because they may have some value for plant pathologists where they are supported by herbarium specimens (Constantinescu, 1991*a*), they have been included. Authors of non-taxonomic articles should note that the use of the binomial alone, where varietal names have been published, automatically includes all varieties and this may not be the intention! For example, the use of *Pythium ultimum* alone will include both *P. ultimum* var. *ultimum* and *P. ultimum* var. *sporangiiferum*. It should be remembered that while subspecies, varieties and forms (*formae*) have formal taxonomic status, with recommendations governing the use of one or other of these intraspecific categories, *formae speciales* are not governed by *ICBN* rules and recommendations. The list for *Plasmopara*, but not that for *Peronospora*, includes all of these *formae* and *formae speciales* names for the purposes of retrieval, but without taxonomic reappraisal.

Check-lists for the principal genera of the dicotyledonicolous downy mildews are not readily available (but see below) although taxonomic reviews of some of the genera of the Peronosporales can be found in Barreto & Dick (1991), Biga (1955), Constantinescu (1979, 1989, 1991*b*, 1992, 1996*a*, *b*, 1998), de Bary (1863), Gäumann (1923), Gustavsson (1959*a*, *b*), Kochman & Majewski (1970), Săvulescu (1962), Săvulescu & Săvulescu (1952), Shaw (1978, 1981), Skalický (1966) and Waterhouse & Brothers (1981). Waterhouse (1964) provided an earlier summary of the graminicolous downy mildews. Since then, Dick *et al.* (1984, 1989) and Dick (1988, 1990*a*, 1995) have proposed a reclassification of these fungi into the Sclerosporales. Săvulescu (1966) and Hall (1996) have reviewed species concepts in the downy mildews. For brevity, excluded taxa and accepted synonyms are not given here, except where they have been used in the CABI database. The treatment of the *Peronosporaceae* here is not consistent from genus to genus because of the disparate numbers of species in genera, particularly *Peronospora* and *Plasmopara*; more detailed information on the treatment adopted is given under the genus name. The most complete, annotated, strictly alphabetic list of *Peronospora* names is in Constantinescu (1991*a*). There is no recent compilation for names in *Plasmopara*. *Albugo* has not been reviewed since the account by Biga (1955). The most recent molecular phylogeny of the angiosperms (Angiosperm Phylogeny Group (APG), 1998) has been used for host entries.

PERONOSPOROMYCETIDAE: *PERONOSPORALES*: *PERONOSPORACEAE*

BASIDIOPHORA Roze & Cornu

Basidiophora entospora Roze & Cornu **[type species]** [Saccardo: *Sylloge Fungorum* **6**: 239]
Basidiophora kellermanii (Sacc.) G. W. Wilson [= *Benua kellermanii*] [Swingle *ex* Sacc.]
Basidiophora montana R. W. Barreto [*Index of Fungi* **6(4)**: 196]

BENUA Constantinescu

Benua kellermanii (Sacc.) Constant. **[type species; monotypic]** [Swingle *ex* Sacc.] [*Index of Fungi* **6(17)**: 900]

BREMIA Regel

NOTE: Some authorities would regard *Bremia sensu stricto* as monotypic, but see *B. graminicola*, known only from the original description. All other taxa have hosts in *Asteraceae*. Infra-specific taxa may be of dubious standing. Synonymy has not been evaluated.

Bremia betae H. C. Bai & X. Y. Cheng [*nom. invalid.*] [*Index of Fungi* **6(8)**: 422]
Bremia centaurea Syd.
Bremia cicerbitae C. J. Li & Z. Q. Yuan [*Index of Fungi* **6(20)**: 1085]
Bremia cirsii (Uljan.) J. F. Tao & Y. N. Yu [Jacz. *ex* Uljan.] [*Index of Fungi* **6(5)**: 260]
Bremia elliptica Sawada
Bremia gemminata (Unger) Kochman & T. Majewski [*Index of Fungi* **4**: 4]
Bremia graminicola Naumov
Bremia graminicola Naumov var. *graminicola*
bremia graminicola Naumov var. *indica* M. K. Patel [*Index of Fungi* **1**: 277]
Bremia lactucae Regel var. *lactucae* **[type species]**
Bremia lactucae Regel var. *arctii* Uljan. [*Index of Fungi* **4**: 4]
Bremia lactucae Regel var. *cardui* Uljan. [*Index of Fungi* **4**: 4]
Bremia lactucae Regel var. *hedypnoidis* Uljan. [*Index of Fungi* **4**: 4]
Bremia lactucae Regel var. *pterothecae* Uljan. [*Index of Fungi* **4**: 4]
Bremia lactucae Regel var. *willemetiae* Uljan. [*Index of Fungi* **4**: 4]
Bremia lactucae Regel var. *xeranthemi* Uljan. [*Index of Fungi* **4**: 4]
Bremia lactucae Regel forma *mulgedii* C. B. Benua [*Index of Fungi* **4**: 303]
Bremia lactucae Regel forma *taraxaci* L. Ling & F. L. Tai [*Index of Fungi* **1**: 226]
Bremia lactucae Regel forma *taraxaci* C. B. Benua [later homonym] [*Index of Fungi* **4**: 303]
Bremia lactucae Regel forma specialis *carthami* M. O. Milovtzova
Bremia lactucae Regel forma specialis *centaureae* D. I. Skidmore & D. S. Ingram [*Index of Fungi* **5**: 470]
Bremia lactucae Regel forma specialis *chinensis* L. Ling & M. C. Tai
Bremia lactucae Regel forma specialis *cirsii* D. I. Skidmore & D. S. Ingram [*Index of Fungi* **5**: 470]
Bremia lactucae Regel forma specialis *crepidis* D. I. Skidmore & D. S. Ingram [*Index of Fungi* **5**: 470]
Bremia lactucae Regel forma specialis *hieracii* D. I. Skidmore & D. S. Ingram [*Index of Fungi* **5**: 470]
Bremia lactucae Regel forma specialis *lapsanae* D. I. Skidmore & D. S. Ingram [*Index of Fungi* **5**: 470]
Bremia lactucae Regel forma specialis *leontodi* D. I. Skidmore & D. S. Ingram [*Index of Fungi* **5**: 470]
Bremia lactucae Regel forma specialis *picridis* D. I. Skidmore & D. S. Ingram [*Index of Fungi* **5**: 470]
Bremia lactucae Regel forma specialis *ovata* L. Ling & M. C. Tai
Bremia lactucae Regel forma specialis *senecionis* D. I. Skidmore & D. S. Ingram [*Index of Fungi* **5**: 470]
Bremia lactucae Regel forma specialis *sonchi* D. I. Skidmore & D. S. Ingram [*Index of Fungi* **5**: 470]
Bremia lactucae Regel forma specialis *sonchicola* L. Ling & M. C. Tai
Bremia lactucae Regel forma specialis *taraxaci* L. Ling & M. C. Tai
Bremia lactucae Regel forma specialis *taraxaci* D. I. Skidmore & D. S. Ingram [*Index of Fungi* **5**: 470]
Bremia lagoceridis Y. N. Yu & J. F. Tao [*Index of Fungi* **6(5)**: 260]
Bremia lapsanae Syd. [as '*lampsanae*']
Bremia leibnitziae J. F. Tao & Y. Qin [*Index of Fungi* **5**: 280]
Bremia moehringiae T. R. Liu & C. K. Pai [*Index of Fungi* **5**: 421]
Bremia microspora Sawada

Bremia ovata Sawada
Bremia picridis S. Ito & Tokun.
Bremia picridis-hieracioidis Savinceva [*Index of Fungi* **4**: 303]
Bremia saussurea Sawada
Bremia sonchi Sawada
Bremia stellata (Desm.) Kochman & T. Majewski [*Index of Fungi* **4**: 4]
Bremia taraxaci S. Ito & Tokun.
Bremia tulasnei (Hoffm.) Syd.

BREMIELLA G. W. Wilson

NOTE: See Constantinescu (1979; 1991*b*).

Bremiella baudysii (Skalický) Constant. & Negrean [*Index of Fungi* **2**: 197; **4**: 589]
Bremiella megasperma (Berl.) G. W. Wilson **[type species]**
  Bremiella oenanthes J. F. Tao & Y. Qin [= *B. baudysii*] [*Index of Fungi* **5**: 421]
Bremiella sphaerosperma Constant. [*Index of Fungi* **6(4)**: 196]

PARAPERONOSPORA Constant.

NOTE: All hosts in Asteracae. See Constantinescu (1989; 1996*a*).

Paraperonospora apiculata Constant. [*Index of Fungi* **6 (14)**: 768]
Paraperonospora artemisiae-annuae (L. Ling & M. C. Tai) Constant. [*Index of Fungi* **1**: 168; **3**: 415; **5**: 1065]
Paraperonospora artemisiae-biennis (Gäum.) Constant. [Saccardo: *Sylloge Fungorum* **26**: 31;*Index of Fungi* **3**: 415; **5**: 1065]
Paraperonospora chrysanthemi-coronarii (Sawada) Constant. [Petrak's Lists **6**; *Index of Fungi* **5**: 190]
Paraperonospora leptosperma (de Bary) Constant. **[type species]** [*Index of Fungi* **3**: 415; **5**: 1065; **6(16)**: 773]
Paraperonospora minor (Săvul. & Rayss) Constant. [*Index of Fungi* **5**: 1065]
Paraperonospora multiformis (J. F. Tao & Y. Qin) Constant. [*Index of Fungi* **5**: 1065]
Paraperonospora sulphurea (Gäum.) Constant. [*Index of Fungi* **3**: 415; **5**: 1065]
Paraperonospora tanaceti (Gäum.) Constant. [*Index of Fungi* **3**: 415; **5**: 1065]

PERONOSPORA Corda

Peronospora rumicis Corda **[type species]**

NOTES:

Perhaps *ca* 75 species, 800 binomials and infra-specific taxa proposed; for the full list of binomials see Constantinescu (1991*a*) and *Index of Fungi* thereafter. *Index of Fungi* references are mostly excluded because full information is available in Constantinescu (1991*a*). Dr Constantinescu writes (*in litt.*, 10 Jul 96): "I do not have a check-list of taxonomically correct or accepted *Peronospora* species. Such a list can only be made after a taxonomic study of the fungi involved. There is no such study covering the whole genus *Peronospora*".

In this long list, species are entered alphabetically, with their host order (and family if different from the type family). For host-relationships with other downy mildews, see tables in PART III.

Names for *Peronospora* as used in Kochman & Majewski (1970) (i.e., including *Pseudoperonospora*: are referenced by [page number]-[entry number] to descriptions);
valid [bold-face] binomials for *Peronospora sensu stricto*, as moderated by Constantinescu (1991*a*) are justified left; updated from *Index of Fungi* to Vol 6 (19);
indented entries are those generally considered conspecific, either ≡ (nomenclatural synonym) or = (taxonomic synonym) [with relevant binomial with priority];
synonyms used in the CABI database have been included;
F is used to indicate information available in Farr *et al.* (1989);
valid, infra-specific taxa listed in Constantinescu (1991*a*) and Farr *et al.* (1989) are omitted.

  Peronospora achilleae Săvul. & Vánky [= *Paraperonospora leptosperma*]

[*Index of Fungi* **2**: 515] [host(s) in Asterales]
Peronospora achlydis S. Ito & Tokun. [host(s) in Ranunculales (Berberidaceae)]
Peronospora aconiti Ý. N. Yu [*Index of Fungi* **4**: 632] [host(s) in Ranunculales]
Peronospora actinostemmatis (Sawada) Skalický [= *Pseudoperonospora cubensis*]
[*Index of Fungi* **3**: 413] [host(s) in Cucurbitales]
Peronospora aestivalis Syd. forma *aestivalis* [201-134] [host(s) in Fabales]
Peronospora aethionomatis Simonyan [*nom. invalid.*] [*Index of Fungi* **4**: 417] [host(s) in Brassicales]
Peronospora affinis Rossmann [140-58] [host(s) in Ranunculales (Papaveraceae)]
Peronospora agrestis Gaüm. [260-210] [host(s) in Lamiales (Scrophulariaceae)]
Peronospora agrimoniae Syd. [= *P. sparsa*] [183-111] [host(s) in Rosales]
Peronospora agrorum Gaüm. [224-162] [host(s) in Ericales (Primulaceae)]
Peronospora agrostemmatis Gaüm. [102-11] [host(s) in Caryophyllales]
Peronospora akatsukae S. Ito & Muray. [host(s) in Lamiales (Plantaginaceae)]
Peronospora alchemillae G. H. Otth [=*P. sparsa*] [184-112] [host(s) in Rosales]
Peronospora alliariae-wasabi Gaüm. [133-49] [host(s) in Brassicales]
Peronospora alliorum Fuckel [= *P. destructor*]
[Saccardo: *Sylloge Fungorum* **7**: 257] [host(s) in Asparagales (Alliaceae)]
Peronospora alkannae Vienn.-Bourg. [host(s) in Boraginaceae]
Peronospora alpestris Gaüm. [133-49] [host(s) in Malvales (Cistaceae)]
Peronospora alpicola Gaüm. [133-49] [host(s) in Ranunculales]
Peronospora alpina Johans. [≡ *Plasmopara alpina*]
[Saccardo: *Sylloge Fungorum* **9**: 343 [host(s) in Ranunculales]
Peronospora alsinearum Casp. [118-32; **F**] [host(s) in Caryophyllales]
Peronospora alta Fuckel [262-213; **F**] [host(s) in Lamiales (Plantaginaceae)]
Peronospora alyssi-calycini Gaüm. [145-53] [host(s) in Brassicales]
Peronospora alyssi-incani Gaüm. [host(s) in Brassicales]
Peronospora alyssi-maritimi Kochman [host(s) in Brassicales]
Peronospora amaranthi Gaüm. [130-44] [host(s) in Caryophyllales (Amaranthaceae)]
Peronospora americana Gaüm. [**F**] [host(s) in Caryophyllales (Polygonaceae)]
Peronospora amethysteae Lebedeva [in Jacz. & Jacz.] [*Index of Fungi* **5**: 841] [host(s) in Lamiales]
Peronospora amethysteae Jacz. & P. A. Jacz. [later entry]
[*Index of Fungi* **6(4)**: 218] [host(s) in Lamiales]
Peronospora anagallidis J. Schröt. [Saccardo: *Sylloge Fungorum* **7**: 248; **F**] [host(s) in Ericales (Primulaceae)]
Peronospora anchusae Ziling [host(s) in Boraginaceae]
Peronospora andicola Speg. [host(s) in Lamiales (Scrophulariaceae)]
Peronospora andina Speg. [host(s) in Malpighiales (Euphorbiaceae)]
Peronospora androsaces Niessl [223-161] [host(s) in Ericales (Primulaceae)]
Peronospora anemonones Tramier [*nom. invalid.*] [host(s) in Ranunculales]
Peronospora anthemidis Gäum. [= *Paraperonospora leptosperma*]
[Saccardo: *Sylloge Fungorum* **26**: 30] [host(s) in Asterales]
Peronospora antirrhini J. Schröt. [246-194; **F**] [host(s) in Lamiales (Scrophulariaceae)]
Peronospora aparines (de Bary) Gaüm. [≡ *P. calotheca* var. *aparines*]
[228-169] [host(s) in Gentianales (Rubiaceae)]
Peronospora apiospora G. Poirault [host(s) in Ranunculales]
Peronospora aquatica Gaüm. [257-207] [host(s) in Lamiales (Scrophulariaceae)]
Peronospora arabidis-alpinae Gaüm. [147-65] [host(s) in Brassicales]
Peronospora arabidis-glabrae Gaüm. [host(s) in Brassicales]
Peronospora arabidis-hirsutae Gaüm. [148-67] [host(s) in Brassicales]
Peronospora arabidis-oxyphyllae Gaüm. [host(s) in Brassicales]
Peronospora arabidis-strictae Jacz. & P. A. Jacz. [host(s) in Brassicales]
Peronospora arabidis-turritae Gaüm. [host(s) in Brassicales]
Peronospora arabidopsidis Gaüm. [149-68] [host(s) in Brassicales]
Peronospora arborescens (Berk.) Casp. [143-61; **F**] [host(s) in Ranunculales (Papaveraceae)]
Peronospora arenariae (Berk.) Tul. var. *arenariae* [112-25] [host(s) in Caryophyllales]
Peronospora arenariae (Berk.) Tul. var. *macrospora* Farl. [*Index of Fungi* **6(3)**: 155] host(s) in Caryophyllales]
Peronospora argemones Gaüm. [142-60] [host(s) in Ranunculales (Papaveraceae)]
Peronospora arnebiae Golovin [host(s) in Boraginaceae]
Peronospora artemisiae-annuae L. Ling & M. C. Tai [≡ *Paraperonospora artemisiae-annuae*]
[host(s) in Asterales]
Peronospora artemisiae-biennis Gäum. [≡ *Paraperonospora artemisiae-biennis*] [host(s) in Asterales]

Peronospora arthurii Farl. [222-159; **F**] [host(s) in Myrtales (Onagraceae)]
Peronospora arvensis Gaüm. [259-209] [host(s) in Lamiales (Scrophulariaceae)]
Peronospora asperuginis J. Schröt. [231-172] [host(s) in Boraginaceae]
Peronospora astragali Syd. [host(s) in Fabales]
Peronospora astragali-purpurei Mayor & Vienn.-Bourg. [host(s) in Fabales]
Peronospora astragalina Syd. [191-121] [host(s) in Fabales]
Peronospora atlantica Gaüm. [host(s) in Caryophyllales]
Peronospora atriplicis-hastatae Săvul. & Rayss [host(s) in Caryophyllales (Chenopodiaceae/Amaranthaceae)]
Peronospora atriplicis-hortensis Săvul. & Rayss [host(s) in Caryophyllales (Chenopodiaceae/Amaranthaceae)]
Peronospora atriplicis-tataricae Oescu & Rădul. [host(s) in Caryophyllales (Chenopodiaceae/Amaranthaceae)]
Peronospora atriplicis-halimi Săvul. & Rayss [host(s) in Caryophyllales (Chenopodiaceae/Amaranthaceae)]
Peronospora aubrietae Mayor [host(s) in Brassicales]
Peronospora australis Speg. [≡ *Plasmopara australis*]
[Saccardo: *Sylloge Fungorum* **7**: 260-261] [host(s) in Cucurbitales]
Peronospora axyridis C. Benois [*Index f Fungi* **4**: 315] [host(s) in Caryophyllales (Chenopodiaceae/Amaranthaceae)]
Peronospora barbareae Gaüm. [151-70] [host(s) in Brassicales]
Peronospora beccarii Pass. [= *P. conglomerata*] [*Index of Fungi* **6(3)**: 155] [host(s) in Geraniales]
Peronospora berteroae Gaüm. [152-71] [host(s) in Brassicales]
Peronospora betae J. G. Kühn [= *P. farinosa*] [host(s) in Caryophyllales (Chenopodiaceae/Amaranthaceae)]
Peronospora biscutellae Gaüm. [152-72] [host(s) in Brassicales]
Peronospora bocconiae Syd. [host(s) in Ranunculales (Papaveraceae)]
Peronospora boehmeriae G. Y. Yin & Z. S. Yang [*Index of Fungi* **6(11)**: 621] [host(s) in Rosales (Urticaceae)]
Peronospora bohemica Gaüm. [host(s) in Caryophyllales (Chenopodiaceae/Amaranthaceae)]
Peronospora boni-henrici Gaüm. [125-39] [host(s) in Caryophyllales (Chenopodiaceae/Amaranthaceae)]
Peronospora borealis Gaüm. [host(s) in Gentiamales (Rubiaceae)]
Peronospora borreriae Lagerh. [≡ *Plasmopara borreriae*]
[Saccardo: *Sylloge Fungorum* **11**: 243] [host(s) in Gentianales (Rubiaceae)]
Peronospora bothriospermi Sawada [host(s) in Boraginaceae]
Peronospora brachycomes Enkina [= *Paraperonospora leptosperma*]
[*Index of Fungi* **3**: 561] [host(s) in Asterales]
Peronospora brassicae Gaüm. [≡ *P. parasitica brassicae*] [153-73] [host(s) in Brassicales]
Peronospora buhrii Săvul. & Vánky [= *Paraperonospora leptosperma*]
[*Index of Fungi* **2**: 515] [host(s) in Asterales]
Peronospora bulbocapni Beck [138-55] [host(s) in Ranunculales (Papaveraceae)]
Peronospora buniadis Gaüm. [155-74] [host(s) in Brassicales]
Peronospora cactorum Lebert & Cohn [as 'E. Cohn & Lebert'] [≡ *Phytophthora cactorum*]
[Saccardo: *Sylloge Fungorum* **7**: 238] [host(s) in Caryophyllales (Cactaceae)]
Peronospora cakiles Savile [host(s) in Brassicales]
Peronospora calaminthae Fuckel [238-182] [host(s) in Lamiales]
Peronospora calepinae Gaüm. [host(s) in Brassicales]
Peronospora calindriniae Speg. [host(s) in Caryophyllales (Portulacaceae)]
Peronospora calotheca de Bary var. *calotheca* [227-168] [host(s) in Gentianales (Rubiaceae)]
Peronospora calotheca de Bary var. *molluginis* (Fuckel) de Bary
[*Index of Fungi* **6(3)**: 155] [host(s) in Caryophyllales (Molluginaceae)]
Peronospora calotheca de Bary var. *sherardiae* (Fuckel) de Bary [*Index of Fungi* **6(3)**: 155] [host(s) in Gentianales (Rubiaceae)]
Peronospora calotheca de Bary var. *galii-vaillantii* de Bary [*Index of Fungi* **6(3)**: 155] [host(s) in Gentianales (Rubiaceae)]
Peronospora camelinae Gaüm. [156-75] [host(s) in Brassicales]
Peronospora campanulae G. Nicholas & Aggéry [*nom. invalid.*] [host(s) in Asterales (Campanulaceae)]
Peronospora campestris Gaüm. [103-12] [host(s) in Caryophyllales]
Peronospora candida Fuckel [223-160; **F**] [host(s) in Ericales (Primulaceae)]
Peronospora canadensis Gaüm. [host(s) in Lamiales (Scrophulariaceae)]
Peronospora canescens C. Benois [*Index of Fungi* **4**: 315] [host(s) in Lamiales (Plantaginaceae)]
Peronospora cannabina G. H. Otth [≡ *Pseudoperonospora cannabina*]
[93-1] [host(s) in Rosales (Urticaceae)]
Peronospora canscorina Thite & M. S. Patil [*Index of Fungi* **4**: 384] [host(s) in Gentianales]
Peronospora capparidis Sawada [*nom. invalid.*] [host(s) in Brassicales (Brassicaceae/Capparaceae)]
Peronospora capsici J. F. Tao & T. J. Li [*Index of Fungi* **5**: 90] [host(s) in Solanales]
Peronospora cardamines-laciniatae A. Gaüm. [host(s) in Brassicales]
Peronospora cardaminopsis A. Gustavsson [160-79] [host(s) in Brassicales]
Peronospora cardariae-repentis N. P. Golovina [host(s) in Brassicales]

Peronospora carnicola Gäum. [*nom. invalid.*] [host(s) in Gentianales]
Peronospora carpoceratis Byzova [host(s) in Brassicales]
Peronospora celsiae Syd. & P. Syd. [host(s) in Lamiales (Scrophulariaceae)]
Peronospora celtidis Waite [= *Pseudoperonospora celtidis*] [Saccardo: *Sylloge Fungorum* **11**: 243] [host(s) in Rosales (Ulmaceae)]
Peronospora centranthi Massenot [*nom. invalid.*] [host(s) in Dipsacales (Valerianaceae)]
Peronospora cephalariae Vincens [host(s) in Dipsacales]
Peronospora cephalariae-laevigatae Săvul. & Rayss [host(s) in Dipsacales]
Peronospora cerastii-anomali Săvul. & Rayss [host(s) in Caryophyllales]
Peronospora cerastii-brachypetali Săvul. & Rayss [107-17] [host(s) in Caryophyllales]
Peronospora cerastii-glandulosi S. Ito & Tokun. [host(s) in Caryophyllales]
Peronospora ceratocarpi Kalymb. [*nom. invalid.*] [*Index of Fungi* **3**: 193] [host(s) in Caryophyllales (Chenopodiaceae/Amaranthaceae)]
Peronospora cerinthes Uljan. [as '*cerinthe*'] [*Index of Fungi* **6(4)**: 218] [232-173] [host(s) in Boraginaceae]
Peronospora chamaesphaci Kalymb. [host(s) in Lamiales]
Peronospora chamaesycis G. W. Wilson [= *P. euphorbiae*] [host(s) in Malpighiales (Euphorbiaceae)]
Peronospora chartomatis Golvin & Kalymb. [host(s) in Brassicales]
Peronospora cheiranthi Gaüm. [161-80] [host(s) in Brassicales]
Peronospora chelidonii Jacz. & P. A. Jacz. [T. Miyabe ex Jacz & P. A. Jacz.] [*Index of Fungi* **6(4)**: 218] [host(s) in Ranunculales (Papaveraceae)]
Peronospora chenopodii Schltdl. [123-37] [host(s) in Caryophyllales (Chenopodiaceae/Amaranthaceae)]
Peronospora chenopodii-ambrosioides Golinia [*nom. invalid.*] [host(s) in Caryophyllales (Chenopodiaceae/Amaranthaceae)]
Peronospora chenopodii-ficifolii Sawada [host(s) in Caryophyllales (Chenopodiaceae/Amaranthaceae)]
Peronospora chenopodii-glauci Gaüm. [125-38] [host(s) in Caryophyllales (Chenopodiaceae/Amaranthaceae)]
Peronospora chenopodii-opulifolii Săvul. & Rayss [host(s) in Caryophyllales (Chenopodiaceae/Amaranthaceae)]
Peronospora chenopodii-polyspermi Gaüm. [126-40] [host(s) in Caryophyllales (Chenopodiaceae/Amaranthaceae)]
Peronospora chenopodii-rubri Gaüm. [host(s) in Caryophyllales (Chenopodiaceae/Amaranthaceae)]
Peronospora chenopodii-urbici Săvul. & Rayss [host(s) in Caryophyllales (Chenopodiaceae/Amaranthaceae)]
Peronospora chenopodii-vulvariae Săvul. & Rayss [host(s) in Caryophyllales (Chenopodiaceae/Amaranthaceae)]
Peronospora chlorae de Bary forma *chlorae* [host(s) in Gentianales]
Peronospora chlorae de Bary forma *cicendii* Molliard [*Index of Fungi* **6(3)**: 155] [host(s) in Gentianales]
Peronospora chorisporae Gäum. [host(s) in Brassicales]
Peronospora chrysosplenii Fuckel [181-108] [host(s) in Saxifragales]
Peronospora cilicia Bremer & Gäum. [host(s) in Fabales]
Peronospora claytoniae Farl. [F] [host(s) in Caryophyllales (Portulacaceae)]
Peronospora clinopodii Terui [host(s) in Lamiales]
Peronospora cochleariae Gaüm. [149-69; **F**] [host(s) in Brassicales]
Peronospora conglomerata Fuckel [**F**] [host(s) in Geraniales]
Peronospora conii Tul. [*nom. invalid.*; ? = *Plasmopara umbelliferarum sensu lato*] [Saccardo: *Sylloge Fungorum* **7**: 241] [host(s) in Apiales]
Peronospora consolidae Jacz. & P. A. Jacz. [Lagerh. ex Jacz & P . A. Jacz.] [131-45] [host(s) in Ranunculales]
Peronospora constantineanui Săvul. & Rayss [host(s) in Caryophyllales (Plumbaginaceae)]
Peronospora convolvuli J. C. Lindq. [host(s) in Solanales (Convolvulaceae)]
Peronospora coringiae Gäum. [host(s) in Brassicales]
Peronospora corollae Tranzschel [267-218] [host(s) in Asterales (Campanulaceae)]
Peronospora coronillae Gaüm. [193-123] [host(s) in Fabales]
Peronospora coronillae-minimae Vienn.-Bourg. [*nom. invalid.*] [host(s) in Fabales]
Peronospora coronillicola C. Camara & Oliviera [host(s) in Fabales]
Peronospora coronopi Gaüm. [162-82] [host(s) in Brassicales]
Peronospora coronopi-procumbentis Vienn.-Bourg. [*nom. invalid.*] [host(s) in Brassicales]
Peronospora cortusae Gaüm. & S. Blumer [224-163] [host(s) in Ericales (Primulaceae)]
Peronospora corydalis de Bary [139-56; **F**] [host(s) in Ranunculales (Papaveraceae)]
Peronospora corydalis-intermediae Gaüm. [140-57] [host(s) in Ranunculales (Papaveraceae)]
Peronospora crambes Jacz. & P. A. Jacz. [host(s) in Brassicales]
Peronospora crispula Fuckel [180-107] [host(s) in Brassicales (Resedaceae)]
Peronospora cristata Tranzschel [host(s) in Ranunculales (Papaveraceae)]
Peronospora crossostephii Sawada [= *Paraperonospora multiformis*] [*Index of Fungi* **2**: 395] [host(s) in Asterales]
Peronospora crucianellae Maire [host(s) in Gentianales (Rubiaceae)]

Peronospora cryptosporae Annal. [host(s) in Brassicales]
Peronospora cucubali S. Ito & Tokun. [host(s) in Caryophyllales]
Peronospora cucumeris (Sawada) Skalický [= *Pseudoperonospora cubensis*] [*Index of Fungi* **3**: 413] [host(s) in Cucurbitales]
Peronospora cynoglossi Swingle [Burrill *ex* Swingle] [233-174; **F**] [host(s) in Boraginaceae]
Peronospora cyparissiae de Bary [218-154] [host(s) in Malpighiales (Euphorbiaceae)]
Peronospora cyperi Ideta [*non P. cyperi* Miyabe & Ideta ≡ *Phytophthora cyperi*] [**F**] [host(s) in Poales (Cyperaceae)]
Peronospora cytisi Rostr. [195-126] [host(s) in Fabales]
Peronospora danica Gäum. [host(s) in Asterales]
Peronospora daturae Hulea [host(s) in Caryophyllales (Chenopodiaceae/Amaranthaceae)]
Peronospora davisii C. G. Shaw [**F**] [host(s) in Lamiales]
Peronospora debaryi E. S. Salmon & Ware [97-4; **F**] [host(s) in Rosales (Urticaceae)]
Peronospora densa Rabenh. [≡ *Plasmopara densa*] [Saccardo: *Sylloge Fungorum* **7**: 243] [host(s) in Lamiales (Scrophulariaceae)]
Peronospora dentariae Rabenh. [159-78] [host(s) in Brassicales]
Peronospora dentariae-macrophyllae Gaüm. [158-77] [host(s) in Brassicales]
Peronospora desertorum Jacz. & P. A. Jacz. [host(s) in Brassicales]
Peronospora desmodii S. Ito & Tokun. [host(s) in Fabales]
Peronospora destructor (Berk.) Casp. [270-222; **F**] [host(s) in Asparagales (Alliaceae)]
Peronospora devastatrix Casp. [= *Phytophthora infestans*] [Saccardo: *Sylloge Fungorum* **7**: 237] [host(s) in Solanales]
Peronospora dianthi de Bary [113-26] [host(s) in Caryophyllales]
Peronospora dianthicola Barthelet [*nom. invalid.*] [107-18; **F**] [host(s) in Caryophyllales]
Peronospora dicentrae Syd. [host(s) in Ranunculales (Papaveraceae)]
Peronospora digitalis Gaüm. [247-195] [host(s) in Lamiales (Scrophulariaceae)]
Peronospora dimorphothecae Săvul. & Vánky [= *Paraperonospora leptosperma*] [*Index of Fungi* **2**: 515] [host(s) in Asterales]
Peronospora dipeltae Jacz. & P. A. Jacz. [host(s) in Fabales]
Peronospora diplachnis Milovtz. [? = *Sclerospora graminicola*] [host(s) in Poales (Poaceae)]
Peronospora diplotaxidis Gaüm. [163-84] [host(s) in Brassicales]
Peronospora dipsaci de Bary [Tul. *ex* de Bary] [264-215; **F**] [host(s) in Dipsacales]
Peronospora diptychocarpi Kalymb. [host(s) in Brassicales]
Peronospora dorycnii Uljan. [host(s) in Fabales]
Peronospora drabae Gaüm. var. *drabae* [164-85] [host(s) in Brassicales]
Peronospora drabae-majusculae Lindtner [host(s) in Brassicales]
Peronospora dracocephali C. J. Li & Zhen Y. Zhao [*Index of Fungi* **6(19)**: 1048] [host(s) in Lamiales]
Peronospora dubia A. Berl. [host(s) in Solanales]
Peronospora ducometii Siemaszko & Jank. [98-6] [host(s) in Caryophyllales (Polygonaceae)]
Peronospora echinopsili Tunkina [host(s) in Caryophyllales (Chenopodiaceae/Amaranthaceae)]
Peronospora echii (Krieger) Jacz. & P. A. Jacz. [= *P. myosotidis*] [host(s) in Boraginaceae]
Peronospora echinospermi (Swingle) Swingle [= *P. cynoglossi*] [233-175] [host(s) in Boraginaceae]
Peronospora effusa (Grev.) Tul. var. *effusa* [129-43] [host(s) in Caryophyllales (Chenopodiaceae/Amaranthaceae)]
Peronospora effusa (Grev.) Tul. var. *polygoni* Thüm. [*Index of Fungi* **6(3)**: 155]
Peronospora effusa-ciconia Becc. [host(s) in Geraniales]
Peronospora eigii Rayss [host(s) in Brassicales]
Peronospora elsholtziae T. R. Liu & C. K. Pai [*Index of Fungi* **5**: 445] host(s) in Lamiales]
Peronospora embergeri Mayor & Vienn.-Bourg. [host(s) in Malpighiales (Euphorbiaceae)]
Peronospora epilobii G. H. Otth [≡ *Plasmopara epilobii*] [Saccardo: *Sylloge Fungorum* **11**: xxiii] [host(s) in Myrtales (Onagraceae)]
Peronospora epilobii Rabenh. [≡ *Plasmopara epilobii*] [Saccardo: *Sylloge Fungorum* **7**: 243] [host(s) in Myrtales (Onagraceae)]
Peronospora epiphylla (Pers.) Pat. & Lagerh. [= *P. farinosa*] [*Index of Fungi* **6(3)**: 155] [host(s) in Caryophyllales (Chenopodiaceae/Amaranthaceae)]
Peronospora eranthidis-hyemalis (Pass.) A. Fisch. [as '*eranthidis*'] [= *P. myosuri*] [Saccardo: *Sylloge Fungorum* **17**: 521] [host(s) in Ranunculales]
Peronospora erini Vienn.-Bourg. [*nom. invalid.*] [host(s) in Lamiales (Scrophulariaceae)]
Peronospora erinicola Durrieu [host(s) in Asterales (Campanulaceae)]
Peronospora eriogoni H. Solheim & Gilb. [*Index of Fungi* **4**: 221; **F**] [host(s) in Caryophyllales (Polygonaceae)]
Peronospora eritrichii S. Ito & Tokun. [host(s) in Boraginaceae]
Peronospora erodii Fuckel [215-151; **F**] [host(s) in Geraniales]

Peronospora erophilae Gaüm. [165-86] [host(s) in Brassicales]
Peronospora erucastri Gaüm. [166-87] [host(s) in Brassicales]
Peronospora ervi A. Gustavsson [= *P. viciae sensu lato*] [214-150] [host(s) in Fabales]
Peronospora erysimi Gaüm. [166-88] [host(s) in Brassicales]
Peronospora erythraeae Gaüm. [J. G. Kühn *ex* Gaüm.] [226-165] [host(s) in Gentianales]
Peronospora esperaussensis Mayor [host(s) in Fabales]
Peronospora esulae Gäum. [host(s) in Malpighiales (Euphorbiaceae)]
Peronospora euclidii Săvul. & Rayss [167-89] [host(s) in Brassicales]
Peronospora euphorbiae Fuckel [219-155; F] [host(s) in Malpighiales (Euphorbiaceae)]
Peronospora euphorbii-glyptospermae Gäum. [host(s) in Malpighiales (Euphorbiaceae)]
Peronospora euphorbii-thymifoliae Sawada [host(s) in Malpighiales (Euphorbiaceae)]
Peronospora eurotiae Kalymb. [host(s) in Caryophyllales (Chenopodiaceae/Amaranthaceae)]
Peronospora fabae Jacz. & Sergeeva [213-148] [host(s) in Fabales]
Peronospora fagi R. Hartig [≡ *Phytophthora fagi* (? = *P. quercina*)]
[Saccardo: *Sylloge Fungorum* 7: 238] [host(s) in Fagales]
Peronospora fagopyri Jacz. & P. A. Jacz. [*Index of Fungi* 6(4): 218] [host(s) in Caryophyllales (Polygonaceae)]
Peronospora farinosa (Fr.) Fr. *sensu lato* [(Fr. : Fr.) Fr.] [F] [host(s) in Caryophyllales (Chenopodiaceae/Amaranthaceae)]
Peronospora farinosa (Fr.) Fr. *sensu stricto* [(Fr. : Fr.) Fr.] [host(s) in Caryophyllales (Chenopodiaceae/Amaranthaceae)]
Peronospora favargeri Mayor & Vienn.-Bourg. [host(s) in Malpighiales (Euphorbiaceae)]
Peronospora ficariae de Bary [Tul. *ex* de Bary] [137-54] [host(s) in Ranunculales]
Peronospora fintelmannii Casp. [= *Phytophthora infestans*]
[Saccardo: *Sylloge Fungorum* 15: 241] [host(s) in Solanales]
Peronospora flava Gaüm. [249-198; F] [host(s) in Lamiales (Scrophulariaceae)]
Peronospora floerkeae Kellerm. [F] [host(s) in Brassicales (Limnanthaceae)]
Peronospora fontana A. Gustavsson [*Index of Fungi* 5: 773] [host(s) in Caryophyllales]
Peronospora fragariae Roze & Cornu [= *P. sparsa*] [F] [host(s) in Rosales]
Peronospora fritzii J. Schröt. [host(s) in Solanales (Convolvulaceae)]
Peronospora fugitai S. Ito & Tokun. [host(s) in Asparagales (Alliaceae)]
Peronospora fulva Syd. [198-130] [host(s) in Fabales]
Peronospora gaeumanniana Jaap [host(s) in Brassicales]
Peronospora gaeumannii Mundk. [= *P. indica* Gaum.] [host(s) in Ranunculales (Papaveraceae)]
Peronospora galegae Săvul. & Rayss [193-124] [host(s) in Fabales]
Peronospora galeopsidis Lobik [*Index of Fungi* 6(4): 218] [238-183] [host(s) in Lamiales]
Peronospora galii Fuckel [= *P. calotheca*] [229-170] [host(s) in Gentianales (Rubiaceae)]
Peronospora galii-anglici Uljan. [host(s) in Gentianales (Rubiaceae)]
Peronospora galii-pedemontani Săvul. & Rayss [host(s) in Gentianales (Rubiaceae)]
Peronospora galii-rubioides Săvul. & Rayss [host(s) in Gentianales (Rubiaceae)]
Peronospora galii-trifidii S. Ito & Tokun. [host(s) in Gentianales (Rubiaceae)]
Peronospora galii-veri Gäum. [host(s) in Gentianales (Rubiaceae)]
Peronospora galligena S. Blumer [146-64] [host(s) in Brassicales]
Peronospora ganglioniformis Berk.) de Bary [as '*gangliformis*'] [= *Bremia lactucae*]
[Saccardo: *Sylloge Fungorum* 7: 244] [host(s) in Asterales]
Peronospora ganglioniformis (Berk.) Tul. [as 'Tul.'] [= *Bremia lactucae*]
[Saccardo: *Sylloge Fungorum* 7: 244] [host(s) in Asterales]
Peronospora gei Syd. [=*P. sparsa*] [184-113] [host(s) in Rosales]
Peronospora gentianiae Rostr. [*Index of Fungi* 6(3): 155] [226-166] [host(s) in Gentianales]
Peronospora geranii Peck [≡ *Plasmopara geranii*]
[Saccardo: *Sylloge Fungorum* 7: 242] [host(s) in Geraniales]
Peronospora gigantea Gaüm. [135-51] [host(s) in Ranunculales]
Peronospora giliae Ellis & Everh. [= *P. phlogina*] [host(s) in Ericales (Polemoniaceae)]
Peronospora glacialis (A. Blytt) Gaüm. [=*P. ficariae*] [135-50] [host(s) in Ranunculales]
Peronospora glaucii Lobik [*Index of Fungi* 6(4): 218] [141-59] [host(s) in Ranunculales (Papaveraceae)]
Peronospora glechomae Oescu & Rădul. [host(s) in Lamiales]
Peronospora glechomatis (Willi Krieg.) T. Majewski [≡ *P. lamii* var. *glechomae*]
[239-184] [host(s) in Lamiales]
Peronospora goldbachiae Jacz. & P. A. Jacz. [host(s) in Brassicales]
Peronospora golovinii Kalymb. [host(s) in Brassicales]
Peronospora gonolobii Lagerh. [≡ *Plasmopara gonolobi*]
[Saccardo: *Sylloge Fungorum* 11: 243] [host(s) in Gentianales (Apocynaceae/Asclepiadiaceae)]
Peronospora graminicola (Sacc.) Sacc. [≡ *Sclerospora graminicola*]

[Saccardo: *Sylloge Fungorum* **7**: 238] [host(s) in Poales (Poaceae)]
Peronospora gregoriae S. Blumer [host(s) in Ericales (Primulaceae)]
Peronospora grisea (Unger) Unger [257-208; **F**] [host(s) in Lamiales (Scrophulariaceae)]
Peronospora gypsophilae Jacz. & P. A. Jacz. [108-19] [host(s) in Caryophyllales]
Peronospora halstedii Farl. [≡ *Plasmopara halstedii*] [Saccardo: *Sylloge Fungorum* **7**: 242] [host(s) in Asterales]
Peronospora hariotii Gaüm. [246-193] [host(s) in Lamiales (Buddlejaceae/Loganaceae, *p.p.*)]
Peronospora hedeomatis Kellerm. & Swingle [**F**] [host(s) in Lamiales]
Peronospora helianthi Rostr. [Saccardo: *Sylloge Fungorum* **26**: 39] [host(s) in Asterales]
Peronospora helichrysi Togashi & Egami [= *Paraperonospora sulphurea*] [Petrak's Lists **7**] [host(s) in Asterales]
Peronospora heliophilae E. Müll. & Poelt [host(s) in Brassicales]
Peronospora hellebori-purpurascentis Săvul. & Rayss [host(s) in Ranunculales]
Peronospora helvetica Gaüm. [host(s) in Caryophyllales]
Peronospora hepaticae de Bary [= *Plasmopara pygmaea*] [Saccardo: *Sylloge Fungorum* **7**: 253] [host(s) in Ranunculales]
Peronospora herniariae de Bary [108-20] [host(s) in Caryophyllales (Caryophyllaceae/Illecebraceae)
Peronospora hesperidis Gaüm. [168-90] [host(s) in Brassicales]
Peronospora hiemalis Gaüm. [137-53; **F**] [host(s) in Ranunculales]
Peronospora hiratsukae S. Ito & Tokun. [host(s) in Gentianales (Rubiaceae)]
Peronospora holostei de Bary [Casp. *ex* de Bary] [109-21] [host(s) in Caryophyllales]
Peronospora hommae S. Ito & Tokun. [host(s) in Gentianales (Rubiaceae)]
Peronospora honckenyae (Syd. & P. Syd.) Syd. [= *P. alsinearum* var. *honckeniae*] [110-22] [host(s) in Caryophyllales]
Peronospora hornungiae A. Gustavsson [host(s) in Brassicales]
Peronospora humuli (Miyabe & M. Takah.) Skalický [≡ *Pseudoperonospora humuli*] [94-2] [host(s) in Rosales (Urticaceae)]
Peronospora hydrophylli Waite [**F**] [host(s) in Boraginaceae/Hydrophyllaceae]
Peronospora hymenolobi Annal. [host(s) in Brassicales]
Peronospora hyoscyami de Bary [**F**] [host(s) in Solanales]
Peronospora hylomeconis S. Ito & Tokun. [host(s) in Ranunculales (Papaveraceae)]
Peronospora hypecoi Bremer [*nom. illegit.*] [Bremer in Petrak] [*Index of Fungi* **2**: 150] [host(s) in Ranunculales (Papaveraceae)]
Peronospora hypecoi Jacz. & P. A. Jacz. [as '*hypecoumis*'] [host(s) in Ranunculales (Papaveraceae)]
Peronospora hypericifoliae Sinha & Mathur [host(s) in Malpighiales (Euphorbiaceae)]
Peronospora ibarakii S. Ito & Muray. [=*P. sparsa*] [host(s) in Rosales]
Peronospora iberidis Gaüm. [Gaüm. *ex* Gaüm.] [169-91] [host(s) in Brassicales]
Peronospora ibrahimovii T. M. Achundov [host(s) in Lamiales]
Peronospora illinoensis Farl. [= *Plasmopara obducens*] [Saccardo: *Sylloge Fungorum* **7**: 261] [host(s) in Rosales (Urticaceae)]
Peronospora illyrica Gaüm. [host(s) in Ranunculales]
Peronospora impatientis Ellis & Everh. [= *Plasmopara obducens*] [Saccardo: *Sylloge Fungorum* **9**: 243] [host(s) in Ericales (Balsaminaceae)]
Peronospora infestans (Mont.) de Bary [as 'de Bary'] [≡ *Phytophthora infestans*] [Saccardo: *Sylloge Fungorum* **7**: 237] [host(s) in Solanales]
Peronospora infestans (Mont.) Casp. [as 'Casp.'] [≡ *Phytophthora infestans*] [Saccardo: *Sylloge Fungorum* **1**: 224] [host(s) in Solanales]
Peronospora insubrica Gaüm. [host(s) in Gentianales (Rubiaceae)]
Peronospora indica Syd. & P. Syd. [host(s) in Lamiales (Scrophulariaceae)]
Peronospora iolotanica Kolosch. [host(s) in Caryophyllales (Chenopodiaceae/Amaranthaceae)]
Peronospora iranica Petr. & Esfand. [host(s) in Brassicales]
Peronospora isatidis Gaüm. var. *isatidis* [169-92] [host(s) in Brassicales]
Peronospora iwatensis S. Ito & Muray. [host(s) in Ranunculales]
Peronospora jaapiana Magnus [100-9] [host(s) in Caryophyllales (Polygonaceae)]
Peronospora jacksonii C. G. Shaw [252-202; **F**] [host(s) in Lamiales (Scrophulariaceae)]
Peronospora jaczewskii Săvul. & Rayss [= *P. gypsophilae*] [host(s) in Caryophyllales]
Peronospora jordanovii R. Krusch. [host(s) in Brassicales]
Peronospora karelii Bremer & Gaüm. [host(s) in Dipsacales]
Peronospora kellermanii Fuckel [as '*kellermannii*'] [≡ *Benua kellermanii*] [Saccardo: *Sylloge Fungorum* **7**: 263] [host(s) in Asterales]

Peronospora knautiae J. Schröt. var. *knautiae* [Fuckel *ex* J. Schröt.] [265-216] [host(s) in Dipsacales]
Peronospora knautiae var. *longispora* Vaner & E. G. Dimitrova [*nom. invalid.*] [*Index of Fungi* 6(8): 442] [host(s) in Dipsacales]
Peronospora kochiae Gaüm. [host(s) in Caryophyllales (Chenopodiaceae/Amaranthaceae)]
Peronospora kochiae-prostratae Sandu & Iacob [= *P. kochiae-scopariae*] [*Index of Fungi* 4: 315] [host(s) in Caryophyllales (Chenopodiaceae/Amaranthaceae)]
Peronospora kochiae-scopariae Kochman & T. Majewski [127-41] [host(s) in Caryophyllales (Chenopodiaceae/Amaranthaceae)]
Peronospora lagerheimii Gaüm. [192-122] [host(s) in Fabales]
Peronospora lallemantiae Golovin & Kalymb. [host(s) in Lamiales]
Peronospora lamii A. Braun [240-185; **F**] [host(s) in Lamiales]
Peronospora lamii var. *glechomae* Willi Krieg. [as '*glechomatis*'] [≡ *P. glechomae*] [host(s) in Lamiales] [Saccardo's Omissions: 48] [host(s) in Lamiales]
Peronospora lapponica Lagerh. [Saccardo: *Sylloge Fungorum* 9: 344; 248-196] [host(s) in Lamiales (Scrophulariaceae)]
Peronospora lanceolatae Gapon. [*nom. invalid.*] [*Index of Fungi* 4: 352] [host(s) in Lamiales (Plantaginaceae)]
Peronospora lathyri-aphacae Săvul. & Rayss [host(s) in Fabales]
Peronospora lathyri-hirsuti Săvul. & Rayss [host(s) in Fabales]
Peronospora lathyri-humilis C. Benois [*Index of Fungi* 4: 315] [host(s) in Fabales]
Peronospora lathyri-maritimi Jermal. [host(s) in Fabales]
Peronospora lathyri-palustris Gaüm. var. *palustris* [**F**] [host(s) in Fabales]
Peronospora lathyri-pisiformis M. I. Nikol. [host(s) in Fabales]
Peronospora lathyri-rosei Osipian [host(s) in Fabales]
Peronospora lathyri-verni A. Gustavsson [197-128] [host(s) in Fabales]
Peronospora lathyri-versicolaris Săvul. & Rayss [host(s) in Fabales]
Peronospora lathyrina Vienn.-Bourg. [host(s) in Fabales]
Peronospora lentis Gaüm. [199-131] [host(s) in Fabales]
Peronospora leonuri T. R. Liu & C. K. Pai [*Index of Fungi* 5: 445] [host(s) in Lamiales]
Peronospora lepigoni Fuckel [117-31] [host(s) in Caryophyllales]
Peronospora leptalei Kolosch. [host(s) in Brassicales]
Peronospora lepidii (McAlpine) G. W. Wilson [≡ *Peronospora parasitica* var. *lepidii*] [170-93] [host(s) in Brassicales]
Peronospora lepidii-perfoliati Săvul. & Rayss [171-94] [host(s) in Brassicales]
Peronospora lepidii-sativi Gaüm. [host(s) in Brassicales]
Peronospora lepidii-virginici Gaüm. [host(s) in Brassicales]
Peronospora leptoclada Sacc. [221-157] [host(s) in Malvales (Cistaceae)]
Peronospora leptosperma de Bary [≡ *Paraperonospora leptosperma*] [*Index of Fungi* 5: 445; 773] [**F**] [host(s) in Asterales]
Peronospora leptopyri C. K. Pai [host(s) in Ranunculales]
Peronospora lilii Stenina [very dubious taxon] [host(s) in Liliales]
Peronospora limonii Simonyan [host(s) in Caryophyllales (Plumbaginaceae)]
Peronospora linariae Fuckel [250-199; **F**] [host(s) in Lamiales (Scrophulariaceae)]
Peronospora linariae-genistifoliae Săvul. & Rayss [249-197] [host(s) in Lamiales (Scrophulariaceae)]
Peronospora lini J. Schröt. [217-153; **F**] [host(s) in Malpighiales (Linaceae)
Peronospora lithospermi Gaüm. [234-176] [host(s) in Boraginaceae]
Peronospora litoralis Gaüm. [120-34] [host(s) in Caryophyllales (Chenopodiaceae/Amaranthaceae)]
Peronospora litwinowiae Kalymb. [host(s) in Brassicales]
Peronospora lobulariae Ubrizsy & Vörös [171-95] [host(s) in Brassicales]
Peronospora lophanthi Farl. [**F**] [host(s) in Lamiales]
Peronospora lotorum Syd. [200-133; **F**] [host(s) in Fabales]
Peronospora luffae (Sawada) Skalický [= *Pseudoperonospora cubensis*] [*Index of Fungi* 3: 413] [host(s) in Cucurbitales]
Peronospora lunulariae Gaüm. [172-96] [host(s) in Brassicales]
Peronospora lychnitis Gaüm. [110-23] [host(s) in Caryophyllales]
Peronospora lycii L. Ling & M. C. Tai [host(s) in Solanales]
Peronospora macrocarpa Corda [= *Plasmopara pygmaea*] [Saccardo: *Sylloge Fungorum* 7: 240] [host(s) in Ranunculales]
Peronospora macrocarpa Corda *sensu* Rabenh. [= *Plasmopara umbelliferarum*] [Saccardo: *Sylloge Fungorum* 7: 240] [host(s) in Apiales]
Peronospora macrospora (Unger) Unger [= *Plasmopara umbelliferarum*] [Saccardo: *Sylloge Fungorum* 7: 241] [host(s) in Apiales]
Peronospora malcolmiae Lobik [*Index of Fungi* 6(4): 218] [host(s) in Brassicales]

Peronospora malyi Lindtner [host(s) in Brassicales]
Peronospora manshurica (Naumov) Syd. [≡ *Peronospora trifoliorum* var. *manshurica*] [194-125; **F**] [host(s) in Fabales]
Peronospora matthiolae Gaüm. [173-97] [host(s) in Brassicales]
Peronospora maublancii Săvul. & Rayss [host(s) in Brassicales]
    Peronospora maydis Racib. [≡ *Peronosclerospora maydis*]
    [Saccardo: *Sylloge Fungorum* **14**: 460] [host(s) in Poales (Poaceae)]
Peronospora mayorii Gaüm. var. *mayorii* [214-149; **F**] [host(s) in Fabales]
Peronospora meconopsidis Mayor [replaces *Peronospora mayorii* Gaüm.] [214-149; **F**] [host(s) in Ranunculales (Papaveraceae)]
Peronospora media Gaüm. [host(s) in Caryophyllales]
Peronospora medicaginis-minimae Gapon. [host(s) in Fabales]
Peronospora medicaginis-orbicularis Rayss [host(s) in Fabales]
Peronospora medicaginis-tianschanicae Gapon. [host(s) in Fabales]
    Peronospora megasperma Berl. [≡ *Bremiella megasperma*]
    [Saccardo: *Sylloge Fungorum* **14**: 458] [host(s) in Malpighiales (Violaceae)]
Peronospora melampyri (Bucholz) Davis [251-200] [host(s) in Lamiales (Scrophulariaceae)]
Peronospora melampyri-cristati Săvul. & Rayss [host(s) in Lamiales (Scrophulariaceae)]
Peronospora melandrii Gaüm. [111-24] [host(s) in Caryophyllales]
Peronospora melandrii-noctiflori Săvul. & Rayss [host(s) in Caryophyllales]
Peronospora meliloti Syd. [203-136] [host(s) in Fabales]
Peronospora melissiti Byzova & Dejeva [host(s) in Fabales]
Peronospora menioci Kolosch. [host(s) in Brassicales]
Peronospora menthae X. Y. Cheng & H. C. Bai [*nom. invalid.*]
    [*Index of Fungi* **6(8)**: 442; **6(14)**: 604] [host(s) in Lamiales]
Peronospora mesembryanthemi Verwoerd [host(s) in Caryophyllales (Aizoaceae)]
Peronospora microulae Y. R. Meng & G. Y. Yin [*Index of Fungi* **5**: 1066] [host(s) in Boraginaceae]
Peronospora minima G. W. Wilson [182-109] [host(s) in Saxifragales]
    Peronospora minor (Casp.) Gaüm. [= *P. effusa* var. *minor*]
    [120-35] [host(s) in Caryophyllales (Chenopodiaceae/Amaranthaceae)]
    Peronospora minor (Casp.) Gäum. [≡ *P. farinosa*]
    [host(s) in Caryophyllales (Chenopodiaceae/Amaranthaceae)]
    Peronospora minuta Vienn.-Bourg. [*nom. invalid.*] [host(s) in Brassicales]
Peronospora mirabilis Jacz. & P. A. Jacz. [host(s) in Ericales (Primulaceae)]
Peronospora moenchiae C. Camara & Oliviera [host(s) in Caryophyllales]
    Peronospora momordicae (Sawada) Skalický [= *Pseudoperonospora cubensis*]
    [*Index of Fungi* **3**: 413] [host(s) in Cucurbitales]
Peronospora monolepidis Gaüm. [host(s) in Caryophyllales (Chenopodiaceae/Amaranthaceae)]
Peronospora moreaui Rayss [host(s) in Fabales]
Peronospora muralis Gaüm. [host(s) in Caryophyllales (Chenopodiaceae/Amaranthaceae)]
Peronospora myagri Mayor [host(s) in Brassicales]
Peronospora myosuri Fuckel [133-48] [host(s) in Ranunculales]
Peronospora myosotidis de Bary var. *myosotidis* [235-177; **F**] [host(s) in Boraginaceae]
Peronospora narbonensis Gaüm. [**F**] [host(s) in Fabales]
Peronospora nasturtii-aquatici Gaüm. [174-98] [host(s) in Brassicales]
Peronospora nasturtii-montani Gaüm. [host(s) in Brassicales]
Peronospora nasturtii-palustris S. Ito & Tokun. [host(s) in Brassicales]
Peronospora nemophilae C. G. Shaw [**F**] [host(s) in Boraginaceae/Hydrophyllaceae]
Peronospora nesliae Gaüm. [174-99] [host(s) in Brassicales]
Peronospora nicotianae Speg. [host(s) in Solanales]
Peronospora niessliana A. Berl. [144-62] [host(s) in Brassicales]
Peronospora nitens Oescu & Rădul. [host(s) in Caryophyllales (Chenopodiaceae/Amaranthaceae)]
Peronospora noneae Jacz. & Sergeeva [as '*nonneae*'] [236-178] [host(s) in Boraginaceae]
Peronospora norvegica Gaüm. [as '*norwegica*'] [host(s) in Brassicales]
    Peronospora obducens J. Schröt. [≡ *Plasmopara obducens*]
    [Saccardo: *Sylloge Fungorum* **7**: 242] [host(s) in Ericales (Balsaminaceae)]
Peronospora oerteliana J. G. Kühn [225-164] [host(s) in Ericales (Primulaceae)]
Peronospora obionis-verruciferae Săvul. & Rayss [host(s) in Caryophyllales (Chenopodiaceae/Amaranthaceae)]
Peronospora obovata Bonord. [115-29; **F**] [host(s) in Caryophyllales]
    Peronospora ochracea Ces. [*nom. invalid.*] [147-66] [host(s) in Brassicales]
    Peronospora ochroleuca Ces. [= *Peronospora ochracea* Ces.] [*nom. invalid.*][147-66] [host(s) in Brassicales]
Peronospora omphalodis Gaüm. [236-179] [host(s) in Boraginaceae]

Peronospora ononidis G. W. Wilson [204-138] [host(s) in Fabales]
Peronospora ornithopi Gaüm. [205-139] [host(s) in Fabales]
Peronospora orobi Gaüm. [198-129] [host(s) in Fabales]
Peronospora orontii J. Schröt. [*nom. invalid.*] [host(s) in Lamiales (Scrophulariaceae)]
Peronospora oxalidis Verwoerd & du Plessis [host(s) in Oxalidales]
Peronospora oxybaphi Ellis & Kellerm. [F] [host(s) in Caryophyllales (Nyctaginaceae)]
Peronospora oxytropidis Gaüm. [206-140] [host(s) in Fabales]
Peronospora pachyphragmatis-macrophylli Savinceva [*Index of Fungi* **4**: 315] [host(s) in Brassicales]
Peronospora palustris Gaüm. [host(s) in Lamiales (Scrophulariaceae)]
Peronospora panacis Bunkina [*nomen nudum*] [≡ *Plasmopara panacis* Bondartsev & Bunkina] [*Index of Fungi* **3**: 56] [host(s) in Apiales (Araliaceae)]
Peronospora papaveris Tul. [*nom. invalid.*] [Saccardo: *Sylloge Fungorum* **7**: 251] [host(s) in Ranunculales (Papaveraceae)]
Peronospora papaveris-pilosi Vienn.-Bourg. [*nom. invalid.*] [host(s) in Ranunculales (Papaveraceae)]
Peronospora parasitica (Fr.) Fr. var. *parasitica* [(Pers. *ex* Fr.) Fr.] [156-76] [host(s) in Brassicales]
Peronospora parasitica (Fr.) Fr. var. *niessleana* Berl. [*Index of Fungi* **6(3)**: 155]
Peronospora parietariae Vanev & E. G. Dimitrova [*Index of Fungi* **5**: 841] [host(s) in Rosales (Urticaceae)]
Peronospora parva Gaüm. [119-33] [host(s) in Caryophyllales]
Peronospora parvula Jacz. & P. A. Jacz. [≡ *Plasmopara parvula*] [W. G. Schneid. *ex* Jacz. & P. A. Jacz.][132-47] [host(s) in Ranunculales]
Peronospora patriniae Kalymb. [host(s) in Dipsacales (Valerianaceae)]
Peronospora paula A. Gustavsson [106-16] [host(s) in Caryophyllales]
Peronospora pedicularis Palm [host(s) in Lamiales (Scrophulariaceae)]
Peronospora pennsylvanica Gaüm. [F] [host(s) in Ranunculales]
Peronospora perillae Miyabe in S. Ito & Tokun. [host(s) in Lamiales]
Peronospora phlogina Dietel & Holw. [F] [host(s) in Ericales (Polemoniaceae)]
Peronospora phyteumatis Fuckel [268-219] [host(s) in Asterales (Campanulaceae)]
Peronospora plantaginis Underw. [F] [host(s) in Lamiales (Plantaginaceae)]
Peronospora plantaginis Burrill [264-21] [host(s) in Lamiales (Plantaginaceae)]
Peronospora pocutica T. Majewski [253-203] [host(s) in Lamiales (Scrophulariaceae)]
Peronospora podagrariae G. H. Otth [≡ *Plasmopara podagrariae*] [Saccardo: *Sylloge Fungorum* **11**: xxiii] [host(s) in Apiales]
Peronospora polycarpi Mayor & Vienn.-Bourg. [as '*polycarponis*'] [host(s) in Caryophyllales]
Peronospora polygoni (Thüm.) A. Fisch. [≡ *P. effusa* var. *polygoni*] [99-7] [host(s) in Caryophyllales (Polygonaceae)]
Peronospora polygoni-convolvuli A. Gustavsson [99-8] [host(s) in Caryophyllales (Polygonaceae)]
Peronospora portoricensis (Lamkey) Skalický [≡ *Plasmopara portoricensis*] [*Index of Fungi* **3**: 413] [host(s) in Sapindales (Meliaceae)]
Peronospora pospelovii Gaponenko [probably an *Aspergillus*] [*Index of Fungi* **4**: 353] [host(s) in Asterales]
Peronospora potentillae de Bary var. *potentillae* [185-114; F] [host(s) in Rosales]
Peronospora potentillae de Bary var. *gei* Trail [*Index of Fungi* **6(3)**: 155]
Peronospora potentillae-americanae Gäum. [= *P. sparsa* (*P. potentillae-anserinae*)] [Petrak's Lists **3**: [542]] [host(s) in Rosales]
Peronospora potentillae-anserinae Gaüm. [= *P. sparsa*] [188-117] [host(s) in Rosales]
Peronospora potentillae-reptantis Gaüm. [=*P. sparsa*] [187-116] [host(s) in Rosales]
Peronospora potentillae-sterilis Gaüm. [=*P. sparsa*] [186-115] [host(s) in Rosales]
Peronospora pseudostellariae G. Y. Yin & Z. S. Yang [*Index of Fungi* **6(11)**: 621] [host(s) in Caryophyllales]
Peronospora pulveracea Fuckel [131-46] [host(s) in Ranunculales]
Peronospora pygmaea Unger [≡ *Plasmopara pygmaea*] [Saccardo: *Sylloge Fungorum* **7**: 240] [host(s) in Ranunculales]
Peronospora phacae Gäum. [host(s) in Fabales]
Peronospora pisi Syd. [≡ *P. viciae*] [206-141; F] [host(s) in Fabales]
Peronospora pratensis Syd. [host(s) in Fabales]
Peronospora pulmonariae Gaüm. [? = *P. myosotidis* f. *pulmonariae*] [237-180] [host(s) in Boraginaceae]
Peronospora radii de Bary [269-221] [host(s) in Asterales]
Peronospora ranunculi Gaüm. [136-52; F] [host(s) in Ranunculales]
Peronospora ranunculi-carpatici Săvul. & Rayss [host(s) in Ranunculales]
Peronospora ranunculi-flabellati Vienn.-Bourg. [*nom. invalid.*] [host(s) in Ranunculales]
Peronospora ranunculi-oxyspermi Jacz. & Sergeeva [host(s) in Ranunculales]
Peronospora ranunculi-peduncularis Roiv. [host(s) in Ranunculales]
Peronospora ranunculi-sardoi Săvul. & Rayss [host(s) in Ranunculales]

Peronospora ranunculi-steveni Săvul. & Rayss [host(s) in Ranunculales]
Peronospora rapistri Jacz. & Sergeeva [as '*rapistrae*'] [host(s) in Brassicales]
Peronospora rhaetica Gaüm. [176-102] [host(s) in Brassicales]
Peronospora ribicola J. Schröt. [≡ *Plasmopara ribicola*] [Saccardo: *Sylloge Fungorum* 7: 243] [host(s) in Saxifragales (Grossulariaceae)]
Peronospora rocheliae Kalymb. [host(s) in Boraginaceae]
Peronospora roemeriae Zaprom. [host(s) in Ranunculales (Papaveraceae)]
Peronospora romanica Săvul. & Rayss [≡ *P. aestivalis* f. *medicaginis-falcatae*] [=*P. trifoliorum sensu lato*] [202-135] [host(s) in Fabales]
Peronospora rossica Gaüm. [host(s) in Lamiales]
Peronospora rorippae-islandicae Gaüm. [176-101] [host(s) in Brassicales]
Peronospora rosae-gallicae Săvul. & Rayss [≡ *P. sparsa*] [host(s) in Rosales]
Peronospora rubi Rabenh. [=*P. sparsa*] [190-119; **F**] [host(s) in Rosales]
Peronospora rubiae Gäum. [host(s) in Gentianales (Rubiaceae)]
Peronospora ruegeriae Gaüm. [204-137] [host(s) in Fabales]
Peronospora rugosa Jacz. & P. A. Jacz. [host(s) in Boraginaceae]
Peronospora rumicis Corda **[type species]** [101-10] [host(s) in Caryophyllales (Polygonaceae)]
Peronospora rumicis-rosei Rayss [host(s) in Caryophyllales (Polygonaceae)]
Peronospora sakamotoi S. Ito & Tokun. [host(s) in Gentianales (Rubiaceae)]
Peronospora salicorniae Jenkina [host(s) in Caryophyllales (Chenopodiaceae/Amaranthaceae)]
Peronospora sanguisorbae Gaüm. [=*P. sparsa*] [191-120] [host(s) in Rosales]
Peronospora satarensis M. S. Patil [host(s) in Lamiales (Scrophulariaceae)]
Peronospora saturejae-hortensis Osipian [host(s) in Lamiales]
Peronospora savulescui Rayss [host(s) in Fabales]
Peronospora savulescui Sandu [*nom. invalid.*] [host(s) in Brassicales]
Peronospora saxatilis Gaüm. [host(s) in Lamiales (Scrophulariaceae)]
Peronospora saxifragae Bubák [182-110; **F**] [host(s) in Saxifragales]
Peronospora schachtii Fuckel [≡ *P. farinosa*] [122-36] [host(s) in Caryophyllales (Chenopodiaceae/Amaranthaceae)]
Peronospora schleidenii Unger [= *P. destructor*] [host(s) in Asparagales (Alliaceae)]
Peronospora scleranthi Rabenh. [Rabenh. ex J. Schröt. in Cohn] [113-27] [host(s) in Caryophyllales (Caryophyllaceae/Illecebraceae)
Peronospora scutellariae Bejlin [242-188] [host(s) in Lamiales]
Peronospora sempervivi Schenk. [= *Phytophthora cactorum*] [Saccardo: *Sylloge Fungorum* 7: 238] [host(s) in Saxifragales (Crassulaceae)]
Peronospora senecionis Fuckel [= *P. parasitica*] [host(s) in Brassicales]
Peronospora senneniana Gonz. Frag. & Sacc. [195-127] [host(s) in Fabales]
Peronospora sepium Gaüm. [≡ *P. viciae*] [211-146] [host(s) in Fabales]
Peronospora septentrionalis Gaüm. [104-13] [host(s) in Caryophyllales]
Peronospora setariae Pass. [= *Sclerospora graminicola*] [Saccardo: *Sylloge Fungorum* 7: 238] [host(s) in Poales (Poaceae)]
Peronospora seymourii Burrill [**F**] [host(s) in Gentianales (Rubiaceae)]
Peronospora sherardiae Fuckel [230-171] [host(s) in Gentianales (Rubiaceae)]
Peronospora sicyicola Peck [= *Plasmopara australis*] [Saccardo: *Sylloge Fungorum* 7: 260] [host(s) in Cucurbitales]
Peronospora sideritidis Byzova [host(s) in Lamiales]
Peronospora silenes G. W. Wilson [≡ *P. arenariae* var. *macrospora*] [*Index of Fungi* **6**(3): 155] [114-28; **F**] [host(s) in Caryophyllales]
Peronospora silvatica Gaüm. [host(s) in Gentianales (Rubiaceae)]
Peronospora silvestris Gaüm. [host(s) in Lamiales (Scrophulariaceae)]
Peronospora simplex Peck [= *Basidiophora entospora*] Sacc. *Syllog. Fung.*7: 239] [host(s) in Asterales]
Peronospora sinensis D. Z. Tang [*Index of Fungi* **5**: 495] [host(s) in Caryophyllales (Polygonaceae)]
Peronospora simonianii Osipian [host(s) in Brassicales]
Peronospora sisymbrii-intermedii Gaüm. [host(s) in Brassicales]
Peronospora sisymbrii-loeselii Gaüm. [host(s) in Brassicales]
Peronospora sisymbrii-officinalis Gaüm. [177-103] [host(s) in Brassicales]
Peronospora sisymbrii-orientalis Gaüm. [host(s) in Brassicales]
Peronospora sisymbrii-sophiae Gaüm. var. *sisymbrii-sophiae* [163-83] [host(s) in Brassicales]
Peronospora sojae Lehman & F. A. Wolf [≡ *P. manschurica*] [host(s) in Fabales]
Peronospora solenanthi Byzova [host(s) in Boraginaceae]

Peronospora sonchi Gapon. [a doubtful taxon - ? *Bremia*, see Constantinescu, 1991*a*] [*Index of Fungi* **2**: 442] [host(s) in Asterales]
Peronospora sophiae-pinnatae Gaüm. [host(s) in Brassicales]
Peronospora sordida Berk. & Broome var. *sordida* [253-204; F] [host(s) in Lamiales (Scrophulariaceae)]
Peronospora sordida Berk. & Broome var. *odontitis-serotinae* C. Massal. [*Index of Fungi* **6(3)**: 155]
Peronospora sparsa Berk. *sensu lato* [188-118; F] [host(s) in Rosales]
Peronospora sparsa Berk. *sensu stricto* [host(s) in Rosales]
Peronospora speculariae Gaüm. [268-220] [host(s) in Asterales (Campanulaceae)]
Peronospora spinaciae Laubert [≡ *P. farinosa*] [host(s) in Caryophyllales (Chenopodiaceae/Amaranthaceae)]
Peronospora stachydis Syd. [243-189] [host(s) in Lamiales]
Peronospora statices Lobik [host(s) in Caryophyllales (Plumbaginaceae)]
Peronospora stellariae de Bary [*nom. invalid.*] [host(s) in Caryophyllales]
Peronospora stellariae-aquaticae Sawada [host(s) in Caryophyllales]
Peronospora stellariae-radiantis S. Ito & Tokun. [host(s) in Caryophyllales]
Peronospora stellariae-uliginosae Sawada [host(s) in Caryophyllales]
Peronospora stigmaticola Raunk. [241-186; F] [host(s) in Lamiales]
Peronospora streptolomatis Golovin & Kalymb. [as '*streptolomatae*'] [host(s) in Brassicales]
Peronospora sulphurea Gäum. [≡ *Paraperonospora sulphurea*] Sacc. *Syllog. Fung.***26**: 55] [host(s) in Asterales]
Peronospora swinglei Ellis & Kellerm. [241-187; F] [host(s) in Lamiales]
Peronospora symphyti Gaüm. [237-181] [host(s) in Boraginaceae]
Peronospora syreniae Jenkina [*nom. invalid.*] [host(s) in Brassicales]
Peronospora syreniae-cuspidatae Oescu & Rădul. [*Index of Fungi* **6(3)**: 155] [host(s) in Brassicales]
Peronospora tabacina D. B. Adam [≡ *P. hyoscyami*] [245-192; F] [host(s) in Solanales]
Peronospora takahashii S. Ito [host(s) in Brassicales]
Peronospora tanaceti Gäum. [≡ *Paraperonospora tanaceti*] [Saccardo: *Sylloge Fungorum* **26**: 56] [host(s) in Asterales]
Peronospora tatarica Săvul. & Rayss [host(s) in Caryophyllales (Chenopodiaceae/Amaranthaceae)]
Peronospora taurica Jacz. & P. A. Jacz. [host(s) in Brassicales]
Peronospora tauscheriae Kalymb. [host(s) in Brassicales]
Peronospora teesdaliae Gaüm. [as '*teesdalae*'] [178-104] [host(s) in Brassicales]
Peronospora teloxydis Jacz. & P. A. Jacz. [host(s) in Caryophyllales (Chenopodiaceae/Amaranthaceae)]
Peronospora teucrii Y. S. Paul & L. N. Bhardwaj [later homonym] [*Index of Fungi* **5**: 897] [243-190] [host(s) in Lamiales]
Peronospora teucrii Gaüm. [*Petrak's Lists* **3**; *Index of Fungi* **5**: 982] [243-190] [host(s) in Lamiales]
Peronospora tetragonolobi Gaüm. [200-132] [host(s) in Fabales]
Peronospora tetragonolobi-palestini Rayss [host(s) in Fabales]
Peronospora thesii Lagerh. [97-5] [host(s) in Santales]
Peronospora thlaspeos-alpestris Gaüm. [179-106] [host(s) in Brassicales]
Peronospora thlaspeos-arvensis Gaüm. [179-105] [host(s) in Brassicales]
Peronospora thlaspeos-perfoliati Gaüm. [host(s) in Brassicales]
Peronospora thymi Syd. [*nom. invalid.*] [host(s) in Lamiales]
Peronospora thyrocarpi L. Ling & M. C. Tai [as '*thyrocarpii*'] [host(s) in Boraginaceae]
Peronospora tomentosa Fuckel [= *P alsinearum*] [105-15] [host(s) in Caryophyllales]
Peronospora tornensis Gaüm. [host(s) in Caryophyllales]
Peronospora torulariae Domashova [*nom. invalid.*] [*Index of Fungi* **3**: 333] [host(s) in Brassicales]
Peronospora tozziae S. Blumer [254-253] [host(s) in Lamiales (Scrophulariaceae)]
Peronospora tranzscheliana Bakhtin [252-201] [host(s) in Lamiales (Scrophulariaceae)]
Peronospora tribulina Pass. [host(s) in Zygophyllaceae]
Peronospora trichotoma Massee [unidentifiable, but cross-reference to *Phytophthora colocasiae*] [Saccardo: *Sylloge Fungorum* **9**: 344] [host(s) in Alismatales (Araceae)]
Peronospora trifolii-alpestris Gäum. [host(s) in Fabales]
Peronospora trifolii-arvensis Syd. [210-145] [host(s) in Fabales]
Peronospora trifolii-cherleri Rayss [*nom. invalid.*] [*Index of Fungi* **3**: 120] [host(s) in Fabales]
Peronospora trifolii-clypeati Rayss [*nom. invalid.*] [*Index of Fungi* **3**: 120] [host(s) in Fabales]
Peronospora trifolii-formosi Rayss [host(s) in Fabales]
Peronospora trifolii-hybridi Gaüm. [209-144] [host(s) in Fabales]
Peronospora trifolii-minoris Gaüm. [host(s) in Fabales]
Peronospora trifolii-pratensis A. Gustavsson [host(s) in Fabales]
Peronospora trifolii-pilularis Rayss [*nom. invalid.*] [*Index of Fungi* **3**: 120] [host(s) in Fabales]
Peronospora trifolii-purpurei Rayss [*nom. invalid.*] [*Index of Fungi* **3**: 120] [host(s) in Fabales]

Peronospora trifolii-repentis Syd. [host(s) in Fabales]
Peronospora trifoliorum de Bary *sensu lato* [208-143; **F**] [host(s) in Fabales]
Peronospora trifoliorum de Bary *sensu stricto* [host(s) in Fabales]
Peronospora trifurcata Unger [= *Phytophthora infestans*]
[Saccardo: *Sylloge Fungorum* **7**: 237] [host(s) in Solanales]
Peronospora trigonellae Gäum. [host(s) in Fabales]
Peronospora trigonotidis S.Ito & Tokun. [host(s) in Boraginaceae]
Peronospora trivialis Gäum. [host(s) in Caryophyllales]
Peronospora turritidis Gäum. [host(s) in Brassicales]
Peronospora uljanishchevii Tunkina [host(s) in Boraginaceae]
Peronospora umbelliferarum var. *berkeleyi* Cas. [as '*berkelei*'] [≡ *Plassmopara umbelliferarum*]
[*Index of Fungi* **6(4)**: 218]
Peronospora uralensis Jacz. & P. A. Jacz. [host(s) in Caryophyllales]
Peronospora ursiniae Săvul. & Vánky [= *Paraperonospora leptosperma*]
[*Index of Fungi* **2**: 515] [host(s) in Asterales]
Peronospora urticae (Lib.) Casp. [≡ *Pseudoperonospora urticae*] [96-3] [host(s) in Rosales (Urticaceae)]
Peronospora ussuriensis Jacz. & P. A. Jacz. [new name for *P. effusa* var. *manshurica* Naumov]
[host(s) in Caryophyllales (Chenopodiaceae/Amaranthaceae)]
Peronospora valerianae Trail [261-211; **F**] [host(s) in Dipsacales (Valerianaceae)]
Peronospora valerianellae Fuckel [261-212] [host(s) in Dipsacales (Valerianaceae)]
Peronospora valesiaca Gäum. [host(s) in Malpighiales (Euphorbiaceae)]
Peronospora variabilis Gäum. [host(s) in Caryophyllales (Chenopodiaceae/Amaranthaceae)]
Peronospora verbasci Gaüm. [256-206] [host(s) in Lamiales (Scrophulariaceae)]
Peronospora verna Gaüm. [host(s) in Lamiales (Scrophulariaceae)]
Peronospora vernalis Gaüm. [116-30] [host(s) in Caryophyllales]
Peronospora veronicae-cymbalariae Rayss [host(s) in Lamiales (Scrophulariaceae)]
Peronospora vexans Gäum. [*nom. invalid.*] [host(s) in Caryophyllales]
Peronospora viciae (Berk.) Casp. *sensu lato* [212-147; **F**] [host(s) in Fabales]
Peronospora viciae (Berk.) Casp. *sensu stricto* [host(s) in Fabales]
Peronospora viciae forma specialis *pisi* (Syd.) Boerema & Verh. [*Index of Fungi* **5**: 304] [host(s) in Fabales]
Peronospora viciae-sativae Gaüm. [≡ *P. viciae*] [host(s) in Fabales]
Peronospora viciae-venosae Jacz. & P. A. Jacz. [host(s) in Fabales]
Peronospora viciicola Campbell [as '*vicicola*'] [*nom. invalid.*] [**F**] [host(s) in Fabales]
Peronospora viennotii Mayor [as '*viennoti*'] [207-142] [host(s) in Fabales]
Peronospora vincae J. Schröt. [227-167] [host(s) in Gentianales (Apocynaceae)]
Peronospora violacea Berk. [variously cited as 'Berk. *ex* Cooke' or Berk. in Cooke (preferred)] [266-217] [host(s) in Dipsacales]
Peronospora violae J. Schröt. [de Bary *ex* J. Schröt.] [220-156] [host(s) in Malpighiales (Violaceae)]
Peronospora vistulensis Wróbl. 128-42] [host(s) in Caryophyllales (Chenopodiaceae/Amaranthaceae)]
Peronospora viticola (Berk. & M. A. Curtis) de Bary [≡ *Plasmopara viticola*]
[Saccardo: *Sylloge Fungorum* **3**: 111] [host(s) in Vitaceae]
Peronospora vvedenskyi Golovin [host(s) in Brassicales]
Peronospora whippleae Ellis & Everh. [**F**] [host(s) in Saxifragales]
Peronospora yamadana Togashi [host(s) in Ranunculales]
Peronospora ziziphorae Byzova [host(s) in Lamiales]

Peronospora names in the CABI data-base, but excluded from *Peronospora*

Peronospora barcinonae Ferrán [as 'Ferraris', = *P. ferrani* Gimeno]
[name used in connection with the agent for cholera] [Saccardo: *Sylloge Fungorum* **15**: 240]
Peronospora dilachnis [see *P. diplachnis* on Poaceae] [Petrak's Lists **8**]
Peronospora exigua Wm. G. Sm. [a hyphomycete, = *Ovularia sphaeroidea*[Saccardo: *Sylloge Fungorum* **11**: 598]
Peronospora gossypina Averna-Saccá [a hyphomycete, *Olpitrichum tenellum*] [Petrak's Lists **1**: [157/233]]
Peronospora interstitialis Berk. & Broome [a hyphomycete, *Ramularia primulae*] [Saccardo: *Sylloge Fungorum* **7**: 259]
Peronospora lutea Carmona y Valle [a bacterium] [Saccardo: *Sylloge Fungorum* **8**: 1054]
Peronospora muscorum Sorokīn [very doubtful taxon, on moss] [Saccardo: *Sylloge Fungorum* **9**: 343]
Peronospora nivea (Unger) Unger [as 'Unger']
[excluded (*nomen confusum*): see Costantinescu, 1991*a*] [Saccardo: *Sylloge Fungorum* **7**: 244]
Peronospora nivea de Bary [not listed in Constantinescu, 1991*a*] [Saccardo: *Sylloge Fungorum* **7**: 240]
Peronospora obliqua Cooke [a hyphomycete, *Ovularia obliqua*] [Saccardo: *Sylloge Fungorum* **4**: 145]
Peronospora rufibasis Berk. & Broome [a hyphomycete, *Ramularia rufibasis*] [Saccardo: *Sylloge Fungorum* **7**: 261]

PLASMOPARA J. Schröt.

NOTE:

All names from *Sylloge Fungorum, Petrak's Lists* and *Index of Fungi* are included here, except for a few recombinations now listed under other genera. Infra-specific taxa may be of dubious standing. Synonymy has not been evaluated. Hosts of thirteen species (in as many families) may be found under *Peronospora* synonyms; families are given below when not the type of the order; for other symbols, see note under *Peronospora*.

Plasmopara acalyphae (G. W. Wilson) G. W. Wilson [Saccardo: *Sylloge Fungorum* **24**: 64; **F**] [host(s) in Malpighiales (Euphorbiaceae)]

Plasmopara achilleae (Săvul. & L. Vánky) Skalický [≡ *Paraperonospora leptosperma*] [*Index of Fungi* **3**: 415; 70-11]

Plasmopara achyranthis J. F. Tao & Y. Qin [*Index of Fungi* **5**: 306] [host(s) in Caryophyllales (Amaranthaceae)]

Plasmopara aegopodii (Casp.) Trotter [= *P. umbelliferarum*] [Saccardo: *Sylloge Fungorum* **24**: 65]

Plasmopara affinis Novot. forma *affinis* [*Index of Fungi* **3**: 255] [host(s) in Asterales]

Plasmopara affinis Novot. forma *silphii* Novot. [*Index of Fungi* **3**: 255]

Plasmopara alpina (C. J. Johanson) A. Blytt [host(s) in Ranunculales]

Plasmopara ammi Constant. [*Index of Fungi* **3**: 492] [host(s) in Apiales]

Plasmopara amurensis [no authority; no reference, but in the CABI database; ? ≡ *P. viticola* var. *amurensis*]

Plasmopara anemones-dichotomae Benua [*Index of Fungi* **4**: 316] [host(s) in Ranunculales]

Plasmopara anemones-nemorosae Săvul. & O. Săvul. [*Index of Fungi* **3**: 364] [host(s) in Ranunculales]

Plasmopara anemones-ranunculoidis Săvul. & O. Săvul. [*Index of Fungi* **3**: 364] [host(s) in Ranunculales]

Plasmopara anethi Jermal. [*Index of Fungi* **3**: 415] [host(s) in Apiales]

Plasmopara angelicae (Casp.) Trotter [Saccardo: *Sylloge Fungorum* **24**: 65] [host(s) in Apiales]

Plasmopara angustiterminalis Novot. forma *angustiterminalis* [*Index of Fungi* **3**: 255; 76-17] [host(s) in Asterales]

Plasmopara angustiterminalis Novot. forma *ambrosiae* Novot. [*Index of Fungi* **3**: 255]

Plasmopara angustiterminalis Novot. forma *bidentis* Novot. [*Index of Fungi* **3**: 255]

Plasmopara anthemidis (Gäum.) Skalický [≡ *Paraperonospora leptosperma*] [*Index of Fungi* **3**: 415; 72-13]

Plasmopara apii Săvul. & O. Săvul. [*Index of Fungi* **3**: 364] [host(s) in Apiales]

Plasmopara archangelicae Gapon. [*Index of Fungi* **4**: 354] [host(s) in Apiales]

Plasmopara artemisiae-annuae (L. Ling & M. C. Tai) Skalický [≡ *Paraperonospora artemisiae-annuae*]

Plasmopara artemisiae-biennis (Gäum.) Skalický [≡ *Paraperonospora artemisiae-biennis*]

Plasmopara asterea Novot. forma *asterea* [*Index of Fungi* **3**: 255] [host(s) in Asterales]

Plasmopara asterea Novot. forma *callistephi* Novot. [= *P. callistephi*] [*Index of Fungi* **3**: 255]

Plasmopara asterea Novot. forma *galatellae* Novot. [= *P. galatellae*] [*Index of Fungi* **3**: 255]

Plasmopara asterea Novot. forma *heteropappi* Novot. [= *P. heteropappi*] [*Index of Fungi* **3**: 255]

Plasmopara asystasiae Vienn.-Bourg. [*Index of Fungi* **2**: 249] [host(s) in Lamiales (Acanthaceae)]

Plasmopara australis (Speg.) Swingle [Saccardo: *Sylloge Fungorum* **9**: 342; **12**: 575; **20**: 422; **F**] [host(s) in Cucurbitales]

Plasmopara baicalensis Jacz. & P. A. Jacz. [host(s) in Ranunculales]

Plasmopara baudysii Skalický [≡ *Bremiella baudysii*] [*Index of Fungi* **2**: 197]

Plasmopara bidentis (Novot.) Novot. [host(s) in Asterales]

Plasmopara borreriae (Lagerh.) Constant. [*Index of Fungi* **6(4)**: 221] [host(s) in Gentianales (Rubiaceae)]

Plasmopara buhrii (Săvul. & L. Vánky) Skalický [≡ *Paraperonospora leptosperma*] [*Index of Fungi* **3**: 415]

Plasmopara calaminthae S. H. Ou [*Index of Fungi* **1**: 45] [host(s) in Lamiales]

Plasmopara callistephi (Novot.) Novot. [host(s) in Asterales]

Plasmopara carlottae Savile [*Index of Fungi* **3**: 334] [host(s) in Apiales]

Plasmopara carthami Negru [*Index of Fungi* **3**: 469] [host(s) in Asterales]

Plasmopara caucalis Săvul. & O. Săvul. [*Index of Fungi* **3**: 364] [host(s) in Apiales]

Plasmopara cenolophii Jermal. [*Index of Fungi* **3**: 415] [host(s) in Asterales]

Plasmopara centaureae-mollis T. Majewski [*Index of Fungi* **3**: 515; 73-14] [host(s) in Asterales]

Plasmopara cephalophora Davis [Saccardo: *Sylloge Fungorum* **24**: 64; **F**] [host(s) in Lamiales]

Plasmopara cercidis C. G. Shaw [*Index of Fungi* **2**: 61; **F**] [host(s) in Fabales]

Plasmopara chaerophyllii (Casp.) Trotter [Saccardo: *Sylloge Fungorum* **24**: 65] [host(s) in Apiales]

Plasmopara chinensis Gorlenko [*Index of Fungi* **3**: 534] [host(s) in Vitaceae]

Plasmopara chrysanthemi-coronarii Sawada [≡ *Paraperonospora chrysanthemi-coronarii*] [Petrak's Lists]

Plasmopara cimicifugae S. Ito & Tokun. [host(s) in Ranunculales]

Plasmopara cissi Vienn.-Bourg. [*Index of Fungi* **2**: 249] [host(s) in Vitaceae]

Plasmopara conii (Casp.) Trotter [≡ *P. umbelliferarum sensu lato*] [Saccardo: *Sylloge Fungorum* **24**: 65]

Plasmopara conii (Wart.) Cif. & C. Camera [later homonym] [*Index of Fungi* **3**: 224]

Plasmopara cruseae C. G. Shaw & Safeeulla [= *P. borreriae*] [*Index of Fungi* **3**: 195]

Plasmopara crustosa (Fr.) Jørst. [*nomen confusum*, Constantinescu (1992); excluded] [*Index of Fungi* **3**: 224; **F**]
Plasmopara cryptotaeniae J. F. Tao & Y. Qin [*Index of Fungi* **5**: 606] [host(s) in Apiales]
Plasmopara cubensis (Berk. & M. A. Curtis) Humphrey var. *cubensis* [≡ *Pseudoperonospora cubensis*] [Saccardo: *Sylloge Fungorum* **17**: 520; **20**: 422, 1288]
[The two varieties listed below do not appear either to have been transferred or reduced to synonymy.]
Plasmopara cubensis (Berk. & M. A. Curtis) Humphrey var. *atra* Zimm. [Saccardo: *Sylloge Fungorum* **17**: 520; **20**: 422]
Plasmopara cubensis (Berk. & M. A. Curtis) Humphrey var. *tweriensis* Rostovzev [Saccardo: *Sylloge Fungorum* **17**: 520; **20**: 422]
Plasmopara curta (Berk.) Skalický subspecies *curta* [≡ *Plasmopara pygmaea*] [*Index of Fungi* **2**: 173]
Plasmopara curta (Berk.) Skalický subspecies *orientalis* Skalický [*Index of Fungi* **2**: 173]
Plasmopara curta (Berk.) Skalický var. *fusca* (Peck) Skalický [*Index of Fungi* **2**: 173]
Plasmopara curta (Berk.) Skalický forma *curta* [*Index of Fungi* **2**: 173]
Plasmopara curta (Berk.) Skalický forma *hellebori* Săvul. & Rayss [*Index of Fungi* **2**: 173]
Plasmopara dahurici Benua [as '*dahurica*'] [*Index of Fungi* **4**: 316] [hos(s) in Apiales]
Plasmopara dauci Săvul. & O. Săvul. [*Index of Fungi* **3**: 364] [host(s) in Apiales]
Plasmopara delphinii (Gapon.) Novot. [host(s) in Ranunculales]
Plasmopara densa (Rabenh.) J. Schröt. [Saccardo: *Sylloge Fungorum* **7**: 243; **12**: 575; **20**: 422] [67-8] [host(s) in Lamiales]
Plasmopara elatostemmatis (Togashi & F. Onuma) S. Ito & Tokun. [as 'elatostematis'] [host(s) in Rosales (Urticaceae)]
Plasmopara elsholtziae J. F. Tao & Y. Qin [cf. *Pseudoperonospora elsholtziae*]] [*Index of Fungi* **5**: 306] [host(s) in Lamiales]
Plasmopara entospora (Roze & Cornu) J. Schröt. [≡ *Basidiophora entospora*] [Saccardo: *Sylloge Fungorum* **7**: 239; **12**: 575; **20**: 423, 1288]
Plasmopara epilobii (G. H. Otth) Sacc. & P. Syd. [as '(G. Otth) Schröt.'] [63-6; **F**] [host(s) in Myrtales (Onagraceae)]
Plasmopara galatellae (Novot.) Novot. [host(s) in Asterales]
Plasmopara galinsogae L. Campb. [host(s) in Asterales]
Plasmopara geranii (Peck) Berl. & De Toni [**F**] [host(s) in Geraniales]
Plasmopara geranii-pratensis Săvul. & O. Săvul. [*Index of Fungi* **3**: 364] [host(s) in Geraniales]
Plasmopara geranii-sylvatici Săvul. & O. Săvul. [*Index of Fungi* **3**: 364] [[host(s) in Geraniales]
Plasmopara gnaphalii Novot. [*Index of Fungi* **3**: 255] [host(s) in Asterales]
Plasmopara gonolobi (Lagerh.) Swingle [**F**] [host(s) in Gentianales (Apocynaceae)]
Plasmopara halstedii (Farl.) Berl. & De Toni [**F**] [host(s) in Asterales - *Verbena*!]
Plasmopara harae S. Ito & Muray. [*Index of Fungi* **2**: 41] [host(s) in Cornales (Hydrangeaceae)]
Plasmopara helianthi Novot. forma *helianthi* [*Index of Fungi* **3**: 364; 75-16] [host(s) in Asterales]
Plasmopara helianthi Novot. forma *patens* Novot. [*Index of Fungi* **3**: 364]
Plasmopara helianthi Novot. forma *perennis* Novot. [*Index of Fungi* **3**: 364]
Plasmopara helichrysi (S. Ito & Tokun.) J. F. Tao & Y. Qin [≡ *Paraperonospora sulphurea*] [Togashi & Egami *ex* S. Ito & Tokun.] [*Index of Fungi* **5**: 776]
Plasmopara heliocarpi Lagerh. [host(s) in Malvales (Grewioideae) (formerly Tiliaceae)]
Plasmopara hellebori-purpurascentis Săvul. & O. Săvul. [*Index of Fungi* **3**: 364] [host(s) in Ranunculales]
Plasmopara hepaticae Casp. [=*P. pygmaea*]
Plasmopara hepaticae (Casp.) C. G. Shaw [= *P. hepaticae*] [*Index of Fungi* **1**: 267]
Plasmopara heteropappi (Novot.) Novot. [host(s) in Asterales]
Plasmopara humuli Miyabe & Takah. [≡ *Pseudoperonospora humuli*] [Saccardo: *Sylloge Fungorum* **21**: 861]
Plasmopara illinoensis (Farl.) Davis [**F**] [host(s) in Rosales (Urticaceae)]
Plasmopara impatientis (Ellis & Everh.) Berl. [= *P. obducens*]
Plasmopara isopyri Skalický [*Index of Fungi* **2**: 173] [host(s) in Ranunculales]
Plasmopara isopyri-thalictroidis (Săvul. & Rayss) Săvul. & O. Săvul. [*Index of Fungi* **3**: 364] [host(s) in Ranunculales]
Plasmopara justiciae (Sawada) Skalický [host(s) in Lamiales (Acanthacerae)]
Plasmopara kellermanii (Ellis & Halst.) Swingle [as '*kellermannii*'] [Saccardo: *Sylloge Fungorum* **9**: 342]
Plasmopara lactucae-radicis Stangh. & Gilb. [*Index of Fungi* **5**: 842] [host(s) in Asterales]
Plasmopara laserpitii (Wartenw.) Săvul. & Rayss [host(s) in Apiales]
Plasmopara latifolii Savile [*Index of Fungi* **3**: 195; **F**] [host(s) in Myrtales (Onagraceae)]
Plasmopara leptosperma (de Bary) Skalický [≡ *Paraperonospora leptosperma*][*Index of Fungi* **3**: 415; 71-12]
Plasmopara megasperma (Berl.) Berl. [≡ *Bremiella megasperma*]
Plasmopara megasperma Săvul. [later homonym] [*Index of Fungi* **1**: 231; **3**: 255]
Plasmopara mei-foeniculi Săvul. & O. Săvul. [*Index of Fungi* **3**: 364] [host(s) in Apiales]
Plasmopara melampyri Bucholtz [≡ *Peronospora melampyri*]
Plasmopara mikaniae Vienn.-Bourg. [*Index of Fungi* **2**: 249] [host(s) in Asterales]
Plasmopara miyakeana S. Ito & Tokun. [host(s) in Rosales (Urticaceae)]

Plasmopara myosotidis C. G. Shaw [*Index of Fungi* 2: 61; F] [host(s) in Boraginaceae]
Plasmopara nakanoi S. Ito & Muray. [*Index of Fungi* 2: 41] [host(s) in Ranunculales (Papaveraceae)]
Plasmopara nivea (Unger) J. Schröt. [*nomen confusum*, see *P. umbelliferum sensu lato*]
Plasmopara obducens (J. Schröt.) J. Schröt. [60-4; F] [host(s) in Ericales (Balsiminaceae)]
Plasmopara oenanthes J. F. Tao & Y. Qin [*Index of Fungi* 5: 306] [host(s) in Apiales]
Plasmopara oplismeni Vienn.-Bourg. [*Index of Fungi* 2: 517] [host(s) in Poales (Poaceae, Paniceae)]
Plasmopara palmae L. Campb. [host(s) in Asterales]
Plasmopara panacis Bondartsev & Bunkina [*Index of Fungi* 3: 56] [host(s) in Apiales (Araliaceae)]
Plasmopara parvula (Jacz. & P. A. Jacz.) Skalický [as '(Schneider) Skalický'] [≡ *Peronospora parvula*] [W. G. Schneid. *ex* Jacz. & P. A. Jacz.]
Plasmopara pastinacae Săvul. & O. Săvul. [*Index of Fungi* 3: 364] [host(s) in Apiales]
Plasmopara paulowniae C. C. Chen [*Index of Fungi* 4: 251] [host(s) in Lamiales (Scrophulariaceae)]
Plasmopara penniseti R. G. Kenneth & J. Kranz [*Index of Fungi* 4: 223] [host(s) in Poales (Poaceae, Paniceae)]
Plasmopara petasitidis S. Ito & Tokun. [host(s) in Asterales]
Plasmopara petroselini Săvul. & O. Săvul. [*Index of Fungi* 3: 364] [host(s) in Apiales]
Plasmopara peucedani Nannf. [host(s) in Apiales]
Plasmopara phrymae S. Ito & Hara [*Index of Fungi* 2: 41] [host(s) in Lamiales ([Phrymaceae] Verbenaceae]
Plasmopara pileae (Gäum.) Jacz. & P. A. Jacz. [Petrak's Lists: Supplement] [host(s) in Rosales (Urticaceae)]
Plasmopara pileae S. Ito & Tokun. [later homonym] [*Index of Fungi* 4: 286]
Plasmopara pimpinellae Săvul. & O. Săvul. var. *pimpinellae* [*Index of Fungi* 3: 364] [host(s) in Apiales]
Plasmopara pimpinellae Săvul. & O. Săvul. var. *maioris* B. Wrońska [*Index of Fungi* 5: 690]
Plasmopara plantaginicola T. R. Liu & C. K. Pai [*Index of Fungi* 5: 448] [host(s) in Lamiales (Plantaginaceae)]
Plasmopara plectranthi L. Ling & M. C. Tai [*Index of Fungi* 1: 169] [host(s) in Lamiales]
Plasmopara plectranthi A. D. Sharma & Munjal [later homonym] [*Index of Fungi* 4: 635]
Plasmopara podagrariae (G. H. Otth) Nannf. [*Index of Fungi* 2: 26] [host(s) in Apiales]
Plasmopara portoricensis (Lamkey) Waterh. [*Index of Fungi* 5: 92] [host(s) in Sapindales (Meliaceae)]
Plasmopara pusilla (de Bary) J. Schröt. [59-2] [host(s) in Geraniales]
Plasmopara pygmaea (Unger) J. Schröt. forma *pygmaea* **[type species]** [57-1; F] [host(s) in Ranunculales]
Plasmopara pygmaea (Unger) J. Schröt. forma *anemones* Gapon. [*Index of Fungi* 4: 354]
Plasmopara pygmaea (Unger) J. Schröt. forma *delphinii* Gapon. [*Index of Fungi* 4: 354]
Plasmopara pygmaea (Unger) J. Schröt. forma *hellebori* T. Săvul. & Rayss [*Index of Fungi* 3: 534]
Plasmopara pygmaea (Unger) J. Schröt. forma *isopyri-thalictroidis* [Petrak's Lists, Supplement]
Plasmopara pyrethri Dudka & Burdjuk. [= *Paraperonospora sulphurea*] [*Index of Fungi* 4: 635]
Plasmopara ribicola Davis [J. Schröt. *ex* Davis] [60-3; F] [host(s) in Saxifragales (Grossulariaceae)]
Plasmopara sambucinae Nelen [*Index of Fungi* 3: 415] [host(s) in Dipsacales (Caprifoliaceae)]
Plasmopara sanguisorbae C. J. Li *et al.* [C. J. Li, Z. Q. Yuan & Zhen Y. Zhao] [*Index of Fungi* 6(12): 676] [host(s) in Rosales]
Plasmopara saniculae Săvul. & O. Săvul. [*Index of Fungi* 3: 364] [host(s) in Apiales]
Plasmopara satarensis P. B. Chavan & U. V. Kulkarni [*Index of Fungi* 4: 418] [host(s) in Malvales (Tiliaceae)]
Plasmopara satureiae F. L. Tai & C. T. Wei [host(s) in Lamiales]
Plasmopara saussureae Novot. [*Index of Fungi* 3: 255] [host(s) in Asterales]
Plasmopara savulescui Novot. [*Index of Fungi* 3: 255; 75-15] [host(s) in Asterales]
Plasmopara selini B. Wrońska [*Index of Fungi* 5: 690] [host(s) in Apiales]
Plasmopara sigesbeckiae (Lagerh.) J. F. Tao [*nom. invalid.*] [= *Peronospora*] [*Index of Fungi* 5: 776]
Plasmopara sii Gapon. [*Index of Fungi* 4: 354] [host(s) in Apiales]
Plasmopara silai Săvul. & O. Săvul. [*Index of Fungi* 3: 364] [host(s) in Apiales]
Plasmopara skvortovii Miura [host(s) in Malvales]
Plasmopara smyrnii Săvul. & M. Bechet [*Index of Fungi* 4: 72] [host(s) in Apiales]
Plasmopara solidaginis Novot. [*Index of Fungi* 3: 255; 78-19] [host(s) in Asterales]
Plasmopara sordida [no authority, no reference in CABI database]
Plasmopara sphaerosperma Săvul. [*Index of Fungi* 1: 231; 77-18] [host(s) in Asterales]
Plasmopara spilanthicola Syd. [host(s) in Asterales]
Plasmopara sulphurea (Gäum.) Skalický [≡ *Paraperonospora sulphurea*] [*Index of Fungi* 3; 415; 70-10]
Plasmopara tanaceti (Gäum.) Skalický [≡ *Paraperonospora tanaceti*] [*Index of Fungi* 3: 415; 69-9]
Plasmopara triumfettae A. D. Sharma & Munjal [*Index of Fungi* 4: 635] [host(s) in Malvales (Grewioideae) (formerly Tiliaceae)]
Plasmopara umbelliferanum (Casp.) Wartenw. var. *umbelliferarum* [64-7] [J. Schröt. *ex* Wartenw.] [host(s) in Apiales]
[This species may include the doubtful *P. crustosa* and *P. nivea* (Constantinescu, 1992)]
Plasmopara umbelliferarum (Casp.) J. Schröt. var. *hacquetiae* Skalický [*Index of Fungi* 2: 197]
Plasmopara ursiniae (Săvul. & L. Vánky) Skalický [= *Paraperonospora leptosperma*][*Index of Fungi* 3: 415]
Plasmopara venezuelana Chardón [host(s) in Brassicales]

Plasmopara vernoniae-chinensis Sawada [host(s) in Asterales]
Plasmopara viburni Peck [F] [host(s) in Dipsacales (Caprifoliaceae)]
Plasmopara vincetoxici Ellis & Everh. [host(s) in Gentianales (Apocynaceae)]
Plasmopara viticola (Berk. & M. A. Curtis) Berl. & De Toni var. *viticola* [62-5; **F**] [host(s) in Vitaceae]
Plasmopara viticola (Berk. & M. A. Curtis) Berl. & De Toni var. *americana* N. P. Golovina [*Index of Fungi* **2**: 398]
Plasmopara viticola (Berk. & M. A. Curtis) Berl. & De Toni var. *amurensis* N. P. Golovina [*Index of Fungi* **2**: 398]
Plasmopara viticola (Berk. & M. A. Curtis) Berl. & De Toni var. *parthica* N. P. Golovina [*Index of Fungi* **2**: 398]
Plasmopara viticola (Berk. & M. A. Curtis) Berl. & De Toni forma *aestivalis-labruscae* Săvul. [*Index of Fungi* **3**: 364]
Plasmopara viticola (Berk. & M. A. Curtis) Berl. & De Toni forma *sylvestris* Săvul. [*Index of Fungi* **3**: 364]
Plasmopara viticola (Berk. & M. A. Curtis) Berl. & De Toni forma *viniferae-ampelopsidis* Săvul. [*Index of Fungi* **3**: 364]
Plasmopara wartenweileri Skalický [*Index of Fungi* **2**: 173] [host(s) in Ranunculales]
Plasmopara wildemaniana Henn. var. *wildemaniana*
[Saccardo: *Sylloge Fungorum* **21**: 861; **24**: 64] [host(s) in Lamiales (Acanthaceae)]
Plasmopara wildemaniana Henn. var. *macrospora* Sawada [Saccardo: *Sylloge Fungorum* **24**: 64]
Plasmopara yunnanensis J. F. Tao & Y. Qin [*Index of Fungi* **5**: 776] [host(s) in Asterales]

PSEUDOPERONOSPORA Rostovzev

NOTE:

Hosts principally in different families of the Rosales (Cannabaceae, Celtidaceae, Ulmaceae, Urticaceae).

Pseudoperonospora aethiomenatis (Simonyan) Waterh. [= *Peronospora parasitica*] [*Index of Fungi* **5**: 93]
Pseudoperonospora elsholtziae D. Z. Tang [cf. *Plasmopara elsholtziae*] [*Index of Fungi* **5**: 403] [host(s) in Lamiales]
Pseudoperonospora cannabina (G. H. Otth) Curzi
Pseudoperonospora cassiae Waterh. [*Index of Fungi* **5**: 93]
Pseudoperonospora celtidis (Waite) G. W. Wilson
Pseudoperonospora cubensis (Berk. & M. A. Curtis) Rostovzev **[type species]**
Pseudoperonospora humuli (Miyabe & M. Takah.) G. W. Wilson
Pseudoperonospora urticae (Berk.) E. S. Salmon & Ware [Lib. *ex* Berk.]

PERONOSPOROMYCETIDAE: *PERONOSPORALES*: *ALBUGINACEAE*

ALBUGO (Pers.) Roussel

Albugo achyranthis (Henn.) Miyabe [host(s) in Caryophyllales (Amaranthaceae)]
Albugo aechmantherae Z.-y. Zhang & Y.-x. Wang [*Index of Fungi* **5**: 377] [in Lamiales (Scrophulariaceae)]
Albugo amaranthi (Schwein.) Kuntze [host(s) in Caryophyllales (Amaranthaceae)]
Albugo austro-africana Syd. & P. Syd. [host(s) in Caryophyllales (Aizoaceae)]
Albugo bliti (Biv.) Kuntze [host(s) in Caryophyllales (Amaranthaceae)]
Albugo candida (J. F. Gmel.) Kuntze var. *candida* **[type species]** [(J. F. Gmel.: Pers.)] [host(s) in Brassicales]
Albugo candida (J. F. Gmel.) Kuntze var. *macrospora* Togashi [(J. F. Gmel.: Pers.)] [host(s) in Brassicales]
Albugo capparidis (de Bary) Cif. [host(s) in Brassicales]
Albugo caryophyllacearum (Wallr.) Cif. & Biga [host(s) in Caryophyllales]
Albugo centaurii (Hansf.) Cif. & Biga [host(s) in Gentianales]
Albugo chardiniae Bremer & Petr. [host(s) in Asterales]
Albugo chardonii W. Weston [host(s) in Brassicales]
Albugo cynoglossi (Unamuno) Cif. & Biga [host(s) in Boraginaceae]
Albugo eomeconis Z. Y. Zhang & Ying X. Wang [*Index of Fungi* **5**: 73] [host(s) in Ranunculales (Papaveraceae)]
Albugo eurotiae Tranzschel [host(s) in Caryophyllales (Amaranthaceae)]
Albugo evansii Syd. [host(s) in Lamiales (Scrophulariaceae)]
Albugo evolvulae (Damle) Cif. & Biga [host(s) in Lamiales]
Albugo gomphrenae (Speg.) Cif. & Biga [host(s) in Caryophyllales (Amaranthaceae)]

Albugo hyoscyami Z. Y. Zhang *et al.*
[Z. Y. Zhang, Ying X. Wang & Z. S. Fu] [*Index of Fungi* 5: 571] [host(s) in Solanales (Convolvulaceae)]
Albugo ipomoeae-aquaticae Sawada [host(s) in Solanales (Convolvulaceae)]
Albugo ipomoeae-hardwikii Sawada [host(s) in Solanales (Convolvulaceae)]
Albugo ipomoeae-panduratae (Schwein.) Swingle var. *ipomoeae-panduratae* [host(s) in Solanales (Convolvulaceae)]
Albugo ipomoeae-panduratae (Schwein.) Swingle var. *tillaceae* Cif. & Biga [host(s) in Solanales (Convolvulaceae)]
Albugo ipomoeae-pes-caprae Cif. [host(s) in Solanales (Convolvulaceae)]
Albugo keeneri Solheim & Gilb. [*Index of Fungi* 4: 531] [host(s) in Ranunculales (Papaveraceae)]
Albugo lepidii A. N. S. Rao [*Index of Fungi* 5: 41] [host(s) in Brassicales]
Albugo mangenotii Mayor & Vienn.-Bourg. [= *A. molluginis*]
[*Index of Fungi* 2: 51] [host(s) in Caryophyllales]
Albugo mauginii (Parisi) Cif. & Biga [host(s) in Fabales]
Albugo minor (Speg.) Cif. [host(s) in Solanales (Convolvulaceae)]
Albugo molluginis S. Ito [also cited as S. Ito & Tokun. (*in* Ito & Tokunaga)] [host(s) in Caryophyllales (Molluginaceae)]
Albugo mysorensis (Thirum. & Safeeulla) Vasudeva [*nom. invalid.*]
[*Index of Fungi* 3: 455] [host(s) in Caryophyllales (Molluginaceae)]
Albugo occidentalis G. W. Wilson [host(s) in Caryophyllales (Amaranthaceae)]
Albugo pes-tigridis Gharse [as '(Verma) Gharse'] [*Index of Fungi* 3: 269] [host(s) in Solanales (Convolvulaceae)]
Albugo pileae J. F. Tao & Y. Qin [*Index of Fungi* 5: 277] [host(s) in Rosales (Urticaceae)]
Albugo platensis (Speg.) Swingle [host(s) in Caryophyllales (Nyctaginaceae)]
Albugo polygoni Z. D. Jiang & P. K. Chi [*Index of Fungi* 6(10): 533] [host(s) in Caryophyllales (Polygonaceae)]
Albugo portulacae (DC.) Kuntze [host(s) in Caryophyllales (Amaranthaceae)]
Albugo portulacearum (Schltdl.) Kochman & T. Majewski [*Index of Fungi* 4: 2] [host(s) in Caryophyllales (Portulacaceae)]
Albugo quadrata (Kalchbr. & Cooke) Kuntze [later homonym] [host(s) in Lamiales (Acanthaceae)]
Albugo quadrata (Wallr.) S. D. Baker [? = *Albugo bliti*] [*Index of Fungi* 2: 207] [host(s) in Caryophyllales]
Albugo resedae (Jacz.) Cif. & Biga [host(s) in Brassicales (Resedaceae)]
Albugo solivae J. Schröt. [host(s) in Asterales]
Albugo solivarum (Speg.) Herter [*Index of Fungi* 4: 301] [host(s) in Asterales]
Albugo spinulosus (de Bary) Herter [*Index of Fungi* 4: 301] [host(s) in Asterales]
Albugo swertiae (Berl. & Kom.) G. W. Wilson [host(s) in Gentianales]
Albugo tilleae (Lagerh.) Cif. & Biga [host(s) in Saxifragales (Crassulaceae)]
Albugo tragopogonis (Pers.) Gray var. *tragopogonis* [host(s) in Asterales]
Albugo tragopogonis (Pers.) Gray var. *cirsii* Cif. & Biga [host(s) in Asterales]
Albugo tragopogonis (Pers.) Gray var. *inulae* Cif. & Biga [host(s) in Asterales]
Albugo tragopogonis (Pers.) Gray var. *pyrethri* Cif. & Biga [host(s) in Asterales]
Albugo tragopogonis (Pers.) Gray var. *xeranthemi-annui* Săvul. & Rayss [host(s) in Asterales]
Albugo trianthemae G. W. Wilson [host(s) in Caryophyllales (Aizoaceae)]
Albugo tropica Lagerh. [host(s) in Piperales]

PERONOSPOROMYCETIDAE: *PYTHIALES*: *PYTHIACEAE*

NOTE:

When there has been a fully comprehensive molecular biological account of all the genera listed here, together with a full sampling of *all* the different sections of *Pythium sensu lato* and *Phytophthora sensu lato*, new monophyletic genera will have to be created and rearranged among the orders *Pythiales* and *Peronosporales*.

CYSTOSIPHON Roze & Cornu

NOTE: = *Pythium sensu lato*; this genus, with amended diagnosis, could provide the earliest valid generic name for species with spherical zoosporangia. As presently constituted, the filamentous-sporangiate *P. dictyosporum* is included here.

Cystosiphon canterae (Karling) M. W. Dick [*loc. cit.*]
Cystosiphon closterii (De Wild.) M. W. Dick [*loc. cit.*]
Cystosiphon dictyosporum (Racib.) M. W. Dick [*loc. cit.*]
Cystosiphon pythioides Roze & Cornu [type species]
Cystosiphon reducta (de Wild.) M. W. Dick [*loc. cit.*]

APPENDIX 1: BINOMIALS: ALBUGINACEAE

DIASPORANGIUM Höhnk

Diasporangium jonesianum Höhnk **[type species; monotypic]**

ENDOSPHAERIUM D'Eliscu [doubtful genus]

Endosphaerium funiculatum D'Eliscu **[type species; monotypic]**

HALOPHYTOPHTHORA H. H. Ho & S. C. Jong

Halophytophthora avicenniae (Gerr.-Corn. & J. A. Simpson) H. H. Ho & S. C. Jong [*Index of Fungi* **5**: 358; **6(1)**: 14]
Halophytophthora bahamensis (Fell & Master) H. H. Ho & S. C. Jong [*Index of Fungi* **4**: 417; **6(1)**: 14]
Halophytophthora batemanensis (Gerr.-Corn. & J. A. Simpson) H. H. Ho & S. C. Jong [*Index of Fungi* **5**: 358; **6(1)**: 14]
Halophytophthora epistomium (Fell & Master) H. H. Ho & S. C. Jong [*Index of Fungi* **4**: 417; **6(1)**: 14]
Halophytophthora exoprolifera H. H. Ho *et al.* [H. H. Ho, A. Nakagiri & S. Y. Newell] [*Index of Fungi* **6(5)**: 269]
Halophytophthora kandeliae H. H. Ho *et al.*[H. H. Ho, S. Y. Hsieh & H. S. Chang] [*Index of Fungi* **6(2)**: 76; **6(4)**: 208]
Halophytophthora masteri A. Nakagiri & S. Y. Newell [*Index of Fungi* **6(10)**: 547]
Halophytophthora mycoparasitica (Fell & Master) H. H. Ho & S. C. Jong [*Index of Fungi* **4**: 417; **6(1)**: 14]
Halophytophthora operculata (Pegg & Alcorn) H. H. Ho & S. C. Jong [*Index of Fungi* **5**: 210; **6(1)**: 14]
Halophytophthora polymorphica (Gerr.-Corn. & J. A. Simpson) H. H. Ho & S. C. Jong [*Index of Fungi* **5**: 358; **6(1)**: 14]
Halophytophthora spinosa (Fell & Master) H. H. Ho & S. C. Jong var. *spinosa* [*Index of Fungi* **4**: 417; **6(1)**: 14]
Halophytophthora spinosa (Fell & Master) H. H. Ho & S. C. Jong var. *lobata* (Fell & Master) H. H. Ho & S. C. Jong [*Index of Fungi* **4**: 417; **6(1)**: 14]
Halophytophthora tartarea A. Nakagiri & S. Y. Newell [*Index of Fungi* **6(10)**: 547]
Halophytophthora vesicula (Anastasiou & Churchl.) H. H. Ho & S. C. Jong **[type species]** [*Index of Fungi* **3**: 534; **6(1)**: 14]

LAGENIDIUM Zopf

NOTE: Retypification of the genus by *L. giganteum* is recommended: the genus would then be monotypic. The existing 'type' species being transferred to *Myzocytium*.

Lagenidium giganteum Couch **[proposed type species; monotypic]**

MYZOCYTIUM Schenk

Myzocytium megastomum De Wild.
Myzocytium proliferum Schenk **[type species]**
Myzocytium netrii (C. E. Mill.) M. W. Dick [*loc. cit.*]
Myzocytium rabenhorstii Zopf [type species of *Lagenidium*]

PERONOPHYTHORA W. H. Ko *et al.* [C. C. Chen *ex* W. H. Ko, H. S. Chang, H. J. Su, C. C. Chen & L. S. Leu]

NOTE: Probably referable to *Phytophthora sensu lato*.

Peronophythora litchi W. H. Ko *et al.* **[type species; monotypic]**
[C. C. Chen *ex* W. H. Ko, H. S. Chang, H. J. Su, C. C. Chen & L. S. Leu] [*Index of Fungi* **4**: 541]

STRAMINIPILOUS FUNGI

PHYTOPHTHORA de Bary

NOTES:

Species groups in Stamps *et al.* (1990) justified right.

Erwin & Ribeiro (1996) cite *publications*, not authorities, after the binomial.

Phytophthora nicotianae and *P. parasitica* re: Erwin & Ribeiro (1996: 391). There is no separate and distinct Code of Nomenclature governing the nomenclature of straminipilous organisms (Chromista) other than the Zoological Code (for heterotrophs without a cell wall) and the Botanical Code. The convention has been to use the Botanical Code for both the photosynthetic organisms and the fungal organisms, but taxa (especially Labyrinthista) have been described under either code. Unilateral action by workers in *Phytophthora* would create, within this kingdom, a chaotic nomenclatural situation, which must be avoided. Therefore the use of *P. nicotianae* should prevail over *P. parasitica* unless or until *P. parasitica* is formally conserved. The extensive use, over many years, of both these names inevitably weakens a submission in support of *P. parasitica*, but a workshop of interested parties could achieve consensus.

Phytophthora arecae Rosenbaum **[II]**
Phytophthora boehmeriae Sawada **[II]**
Phytophthora botryosa Chee **[II]**
Phytophthora cactorum (Lebert & Cohn) J. Schröt. **[I]**
Phytophthora cajani K. S. Amin *et al.* [K. S. Amin, Baldev & F. J. Williams] **[VI]** [*Index of Fungi* **4**: 542]
Phytophthora castanae Katsura & S. Uchida **[II]** [*Index of Fungi* **4**: 514]
Phytophthora castanae (L. Mangin) I. MacFarl. [*nom. invalid.*] [*Index of Fungi* **4**: 635]
Phytophthora cinchonae Sawada [*Index of Fungi* **4**: 71]
Phytophthora cambivora (Petri) Buisman **[VI]**
Phytophthora capsici Leonian **[II]**
Phytophthora cinnamomi Rands var. *cinnamomi* **[VI]**
Phytophthora cinnamomi Rands var. *parvispora* Kröber & R. Marwitz [*Index of Fungi* **6(6)**: 347]
Phytophthora citricola Sawada **[III]**
Phytophthora citrophthora (R.E.Sm. & E.H.Sm.) Leonian **[II]**
Phytophthora clandestina P. A. Taylor *et al.* [P. A. Taylor, Pascoe & F. C. Greenh.] **[I]** [*Index of Fungi* **5**: 447]
Phytophthora colocasiae Racib. **[IV]**
Phytophthora cryptogea Pethybr. & Laff. **[VI]**
Phytophthora cryptogea Pethybr. & Laff. forma specialis *begoniae* Kröber [*Index of Fungi* **5**: 210]
Phytophthora cyperi (Ideta) S. Ito **[III]**
Phytophthora cyperi-bulbosi Seethal. & K. Ramakr. **[III]**
Phytophthora drechsleri Tucker var. *drechsleri* **[VI]**
Phytophthora drechsleri Tucker var. *cajani* M. Pal *et al.* [M. Pal, Grewal & A. K. Sarbhoy] [*Index of Fungi* **4**: 71]
Phytophthora eriugena Clancy & Kavanagh [*nom. invalid.*] **[IV]** [*Index of Fungi* **4**: 635]
Phytophthora erythroseptica Pethybr. var. *erythroseptica* **[VI]**
Phytophthora erythroseptica Pethybr. var. *drechsleri* (Tucker) Sarej. [*Index of Fungi* **4**: 71]
Phytophthora erythroseptica Pethybr. var. *pisi* Bywater & Hickman
Phytophthora fagi R. Hartig [≡ *Peronospora fagi*, ? = *Phytophthora quercina*]
Phytophthora formosana Sawada [*Index of Fungi* **4**: 41]
Phytophthora fragariae Hickman var. *fragariae* **[V]**
Phytophthora fragariae Hickman var. *oryzobladis* J. S. Wang & J. Y. Lu [*Index of Fungi* **4**: 542]
Phytophthora fragariae Hickman var. *rubi* W. F. Wilcox & Deacon [*Index of Fungi* **6(6)**: 347]
Phytophthora gonapodyides (H. E. Petersen) Buisman **[VI]**
Phytophthora heveae A. W. Thomps. **[II]**
Phytophthora hibernalis Carne **[IV]**
Phytophthora humicola W. H. Ko & Ann **[V]** [*Index of Fungi* **5**: 447]
Phytophthora idaei D. M. Kennedy **[I]** [*Index of Fungi* **6(10)**: 560]
Phytophthora ilicis Buddenh. & Roy A. Young **[IV]**
Phytophthora imperfecta Sarej. var. *imperfecta* [*Index of Fungi* **4**: 72]
Phytophthora imperfecta Sarej. var. *citrophthora* (R. E. Sm & E. H. Sm.) Sarej. [*Index of Fungi* **4**: 72, 176]
Phytophthora imperfecta Sarej. var. *nicotianae* (Breda de Haan) Sarej. [*Index of Fungi* **4**: 72]
Phytophthora infestans (Mont.) de Bary var. *infestans* **[type species]** **[IV]**
Phytophthora infestans (Mont.) de Bary var. *phaseoli* (Thaxt.) Leonian [*Index of Fungi* **4**: 72]
Phytophthora infestans (Mont.) de Bary f. sp. *mirabilis* E. M. Möller & De Cock [*Index of Fungi* **6(8)**: 445]

Phytophthora inflata Caros. & Tucker [III]
Phytophthora insolita Ann & W. H. Ko [V] [*Index of Fungi* 5: 91]
Phytophthora iranica Ershad [I] [*Index of Fungi* 4: 72]
Phytophthora italica S. O. Cacciola *et al.* [S. O. Cacciola, G. Magnano di San Lio & A. Belisario] [I] [*Index of Fungi* 6(14): 770]
Phytophthora japonica Waterh. [VI] [*Index of Fungi* 4: 316]
  Phytophthora katsurae W. H. Ko & H. S. Chang [≡ *P. castanae*] [II] [*Index of Fungi* 4: 635; 6(16): 868]
Phytophthora lateralis Tucker & Milbrath [II]
Phytophthora leersiae H. H. Ho & H. S. Chang [Sawada *ex* H. H. Ho & H. S. Chang] [*Index of Fungi* 6(4): 221]
Phytophthora lepironiae Sawada [III]
Phytophthora lycopersici Sawada [*Index of Fungi* 4: 41]
Phytophthora macrochlamydospora J. A. G. Irwin [III/IV] [*Index of Fungi* 6(4): 221]
Phytophthora meadii McRae [II]
Phytophthora medicaginis E. M. Hansen & D. P. Maxwell [V] [*Index of Fungi* 6(3): 159]
Phytophthora megakarya Brasier & M. J. Griffin [II] [*Index of Fungi* 4: 571]
Phytophthora megasperma Drechsler var. *megasperma* [V]
  Phytophthora megasperma Drechsler var. *sojae* (Kaufm. & Gerd.) A. A. Hildebr. [*P. sojae* Kaufm. & Gerd.]
Phytophthora megasperma Drechsler forma specialis *glycines* T.-L. Kuan & Erwin [*nom. illeg.*] [*Index of Fungi* 5: 171]
Phytophthora megasperma Drechsler forma specialis *medicaginis* T.-L. Kuan & Erwin [*nom. illeg.*] [*Index of Fungi* 5: 171]
Phytophthora megasperma Drechsler forma specialis *trifolii* R. G. Pratt [*nom. illeg.*] [*Index of Fungi* 5: 134]
Phytophthora melonis Katsura [VI] [*Index of Fungi* 4: 514]
Phytophthora mexicana H. H. Hotson & Hartge [II]
Phytophthora mirabilis Galindo & H. R. Hohl [IV] [*Index of Fungi* 5: 606]
Phytophthora multivesiculata Ilieva *et al.* [Ilieva, Manin't Veld, W. Veenb.-Rijksa & R. Pieters] [*Index of Fungi* 6(18): 992]
Phytophthora nicotianae Breda de Haan var. *nicotianae* [II]
Phytophthora nicotianae Breda de Haan var. *parasitica* (Dastur) Waterh.
  Phytophthora oryzae (Brizi) K. Hara [≡ *P. japonica*] [*Index of Fungi* 4: 72]
  Phytophthora oryzae (S. Ito & Nagai) Waterh. [later homonym] [*Index of Fungi* 4: 176]
Phytophthora palmivora (E. J. Butler) E. J. Butler var. *palmivora* [II]
Phytophthora palmivora (E. J. Butler) E. J. Butler var. *heterocystica* Babacauh [*Index of Fungi* 5: 358, 899]
  Phytophthora parasitica Dastur var. *parasitica* [II]
  Phytophthora parasitica Dastur var. *capsici* (Leonian) Sarej. [*Index of Fungi* 4: 72]
  Phytophthora parasitica Dastur var. *colocasiae* (Racib.) Sarej. [*Index of Fungi* 4: 72]
Phytophthora phaseoli Thaxt. [IV]
Phytophthora polymorpha [*Index of Fungi* 5: 358]
Phytophthora porri Foister [III]
Phytophthora primulae J. A. Toml. [III]
Phytophthora pseudotsugae Hamm & E. M. Hansen [I] [*Index of Fungi* 5: 358]
Phytophthora quercina T. Jung [I]
Phytophthora quininea Crand. [V]
  Phytophthora oryzae (Brizi) K. Hara [] [*Index of Fungi* 4: 72]
  Phytophthora oryzae (Ito & Nagai) [later homonym] [*Index of Fungi* 4: 176]
Phytophthora richardiae Buisman [VI]
Phytophthora ricini Sawada [*Index of Fungi* 4: 41]
Phytophthora sinensis Y. N. Yu & W. Y. Zhuang [VI] [*Index of Fungi* 5: 171]
Phytophthora sojae Kaufm. & Gerd. [= *P. megasperma* Drechsler var. *sojae* (Kaufm. & Gerd.) A. A. Hildebr.] [V]
Phytophthora syringae (Kleb.) Kleb. [III]
Phytophthora tentacula Kröber & R. Marwitz [I] [*Index of Fungi* 6(6): 347]
Phytophthora trifolii E. M. Hansen & D. P. Maxwell [V] [*Index of Fungi* 6(3): 159]
Phytophthora undulata (H. E. Petersen) M. W. Dick [VI] [*Index of Fungi* 5: 1067]
Phytophthora verrucosa Alcock & Foister [V]
Phytophthora vignae Purss [VI]
Phytophthora vignae Purss forma specialis *adzukicola* S. Tsuchiya *et al.*
[S. Tsuchiya, M. Yanagawa & A. Ogoshi] [*Index of Fungi* 5: 606]
Phytophthora vignae Purss forma specialis *medicaginis* S. Tsuchiya *et al.*
[S. Tsuchiya, M. Yanagawa & A. Ogoshi] [*Index of Fungi* 5: 606]

PYTHIUM **Pringsh. [*Pythium* sensu lato]**

NOTE: For a complete list of binomials and full citations see Dick (1990*c*); citations 1907-1948 are given in Table III: ; information in Plaats-Niterink (1981) is incomplete.

Pythium acanthicum Drechsler
Pythium acanthophoron Sideris
Pythium acrogynum Y. N. Yu [*Index of Fungi* 4: 252]
Pythium adhaerens Sparrow
Pythium amasculinum Y. N. Yu [*Index of Fungi* 4: 252]
Pythium anandrum Drechsler
Pythium angustatum Sparrow
Pythium aphanidermatum (Edson) Fitzp.
Pythium apleroticum Tokun.
Pythium aquatile Höhnk
Pythium aristosporum Vanterp.
Pythium arrhenomanes Drechsler [*Index of Fungi* 6(14): 772]
Pythium australe M. W. Dick [*nomen nudum*] [*Index of Fungi* 6(3): 162]
Pythium betae M. Takah. [*Index of Fungi* 4: 183]
Pythium boreale R. L. Duan [*Index of Fungi* 5: 451]
Pythium buismaniae Plaäts-Nit. [*Index of Fungi* 5: 137]
Pythium capillosum B. Paul var. *capillosum* [*Index of Fungi* 5: 695]
Pythium capillosum B. Paul var. *helicoides* B. Paul [*Index of Fungi* 5: 779]
Pythium catenulatum V. D. Matthews
Pythium caudatum (G. L. Barron) M. W. Dick [*Index of Fungi* 4: 411]
Pythium chamaehyphon Sideris
Pythium chondricola De Cock [*Index of Fungi* 5: 499]
Pythium coloratum Vaartaja
Pythium conidiophorum Jokl
Pythium connatum Y. N. Yu [*Index of Fungi* 4: 252]
Pythium contiguanum B. Paul [≡ *P. drechsleri* B. Paul]
Pythium cucumerinum Bakhariev [*Index of Fungi* 5: 404]
Pythium cryptogynum B. Paul [*Index of Fungi* 6(2): 94]
Pythium cylindrosporum B. Paul [*Index of Fungi* 6(5): 288]
Pythium debaryanum R. Hesse
Pythium deliense Meurs
Pythium diclinum Tokun.
Pythium dictyosporum Racib.
Pythium dimorphum F. F. Hendrix & W. A. Campb. [*Index of Fungi* 4: 106]
Pythium dissimile Vaartaja
Pythium dissotocum Drechsler
Pythium drechsleri B. Paul [later homonym - see *Pythium contiguanum*] [*Index of Fungi* 5: 903; 990]
Pythium drechsleri S. Rajgopalan & K. Ramakr. [*Index of Fungi* 4: 252]
Pythium echinulatum V. D. Matthews
Pythium erinaceus G. I. Robertson [as '*erinaceum*'] [= *Pythium echinulatum* ?] [*Index of Fungi* 4: 636]
Pythium flevoense Plaäts-Nit. [*Index of Fungi* 4: 224]
Pythium fluminum D. Park var. *fluminum* [*Index of Fungi* 4: 515]
Pythium fluminum D. Park var. *flavum* D. Park [*Index of Fungi* 4: 515]
Pythium folliculosum B. Paul [*Index of Fungi* 6(4): 225]
Pythium graminicola Subraman.
Pythium grandisporangium Fell & Master [*Index of Fungi* 4: 419]
Pythium helicandrum Drechsler
Pythium helicoides Drechsler
Pythium helicum T. Itô
Pythium hemmianum M. Takah.
Pythium heterothallicum W. A. Campb. & F. F. Hendrix
Pythium hydnosporum (Mont.) J. Schröt.
Pythium hypoandrum Y. N. Yu & Y. L. Wang [*Index of Fungi* 5: 779]
Pythium hypogynum Middleton
Pythium indigoferae E. J. Butler

Pythium inflatum V. D. Matthews
Pythium insidiosum De Cock *et al.* [De Cock, L. Mend., A. A. Padhye, Ajello & Kaufman] [*Index of Fungi* **5**: 695]
Pythium intermedium de Bary
Pythium irregulare Buisman
Pythium iwayamae S. Ito
Pythium kunmingense Y. N. Yu [*Index of Fungi* **4**: 252]
Pythium lobatum S. Rajagopalan & K. Ramakr. [*Index of Fungi* **4**: 252]
Pythium lucens Ali-Shtayeh [*Index of Fungi* **5**: 499]
Pythium lutarium Ali-Shtayeh [*Index of Fungi* **5**: 499]
Pythium macrosporum Vaartaja & Plaäts-Nit. [*Index of Fungi* **5**: 137]
Pythium mamillatum Meurs
Pythium marinum Sparrow
Pythium maritimum Höhnk
Pythium marsipium Drechsler
Pythium mastophorum Drechsler
Pythium megalacanthum de Bary
Pythium middletonii Sparrow
Pythium middletonii S. Rajagopalan & K. Ramakr. [later homonym] [*Index of Fungi* **4**: 252, 386]
Pythium minor Ali-Shtayeh [*Index of Fungi* **5**: 499]
Pythium monospermum Pringsh. **[type species]**
Pythium multisporum Poitras
Pythium mycoparasiticum Deacon *et al.* [Deacon, S. A. K. Laing & L. A. Berry] [*Index of Fungi* **6(3)**: 162; **6(5)**: 228]
Pythium myriotylum Drechsler
Pythium nagae S. Ito & Tokun.
Pythium nodosum B. Paul *et al.* [B. Paul, D. Galland, T. Bhatnagar & H. Dulieu] [*Index of Fungi* **6(18)**: 994]
Pythium nunn Lifsh. *et al.* [Lifsh., Stangh. & R. Baker] [*Index of Fungi* **5**: 362]
Pythium oedochilum Drechsler
Pythium okanoganense P. E. Lipps [*Index of Fungi* **5**: 93]
Pythium oligandrum Drechsler
Pythium opalinum M. W. Dick (*nomen nudum*) [*Index of Fungi* **6(3)**: 162]
Pythium ornamentatum B. Paul [*Index of Fungi* **5**: 779]
Pythium orthogonon C. Ahrens [*Index of Fungi* **4**: 73]
Pythium ostracodes Drechsler
Pythium pachycaule Ali-Shtayeh var. *pachycaule* [*Index of Fungi* **5**: 499]
Pythium pachycaule Ali-Shtayeh var. *ramificatum* B. Paul [*Index of Fungi* **6(6)**: 351]
Pythium paddicum Hirane
Pythium palingenes Drechsler
Pythium papillatum V. D. Matthews
Pythium parasiticum S. Rajagopalan & K. Ramakr. [*Index of Fungi* **4**: 252]
Pythium paroecandrum Drechsler
Pythium parvum Ali-Shtayeh [*Index of Fungi* **5**: 499]
Pythium periilum Drechsler
Pythium periplocum Drechsler
Pythium perniciosum Serbinow
Pythium perplexum H. Kouyeas & Theoh. [*Index of Fungi* **4**: 636]
Pythium pleroticum T. Itô
Pythium plurisporum G. Abad *et al.* [= *Pythium minor* ?] [G. Abad, H. D. Shew, L. F. Grand & L. T. Lucas] [*Index of Fungi* **6(13)**: 729]
Pythium podbielkowskii (A. Batko) M. W. Dick [*Index of Fungi* **4**: 215]
Pythium polycarpum B. Paul [*Index of Fungi* **5**: 555]
Pythium polymastum Drechsler
Pythium polypapillatum T. Itô
Pythium polytylum Drechsler
Pythium porphyrae M. Takah. & M. Sasaki [*Index of Fungi* **4**: 515]
Pythium prolatum F. F. Hendrix & W. A. Campb.
Pythium pulchrum Minden
Pythium pyrilobum Vaartaja
Pythium radiosum B. Paul [*Index of Fungi* **6(5)**: 288]
Pythium ramificatum B. Paul [*Index of Fungi* **5**: 695]
Pythium rostratum E. J. Butler

Pythium salinum Höhnk
Pythium salpingophorum Drechsler
Pythium scleroteichum Drechsler
Pythium sinense Y. N. Yu [*Index of Fungi* 4: 252]
Pythium spinosum Sawada
Pythium splendens Hans Braun
Pythium sulcatum R. G. Pratt & J. E. Mitch. [*Index of Fungi* 4: 224]
Pythium sylvaticum W. A. Campb. & F. F. Hendrix
Pythium tardicrescens Vanterp.
Pythium tenue Gobi
Pythium toruloides B. Paul [*Index of Fungi* 5: 555]
Pythium torulosum Coker & P. Patt.
Pythium tracheiphilum A. Matta
Pythium uladhum D. Park [*Index of Fungi* 4: 515]
Pythium ultimum Trow var. *ultimum*
Pythium ultimum Trow var. *sporangiiferum* Drechsler
Pythium uncinulatum Plaäts-Nit. & I. Blok [*Index of Fungi* 4: 574]
Pythium vanterpoolii V. Kouyeas & H. Kouyeas
Pythium vexans de Bary var. *vexans*
Pythium vexans de Bary var. *minutum* G. S. Mer & Khulbe [*Index of Fungi* 5: 310]
Pythium violae Chesters & Hickman
Pythium volutum Vanterp. & Truscott
Pythium zingiberis M. Takah.

TRACHYSPHAERA Tabor & Bunting

NOTE: Probably referable to *Phytophthora sensu lato*.

Trachysphaera fructigena Tabor & Bunting **[type species; monotypic]**

PERONOSPOROMYCETIDAE: *PYTHIALES*: *PYTHIOGETONACEAE*

MEDUSOIDES Voglmayr

Medusoides argyrocodium Voglmayr **[type species; monotypic]**

PYTHIOGETON Minden

Pythiogeton autossytum Drechsler
Pythiogeton dichotomum Tokun.
Pythiogeton nigrescens A. Batko
Pythiogeton ramosum Minden
Pythiogeton transversum Minden
Pythiogeton uniforme A. Lund
Pythiogeton utriforme Minden **[type species]**
Pythiogeton zeae H. J. Jee *et al.*

[H. J. Jee, H. H. Ho & W. D. Cho]

RHIPIDIOMYCETIDAE: *RHIPIDIALES*: *RHIPIDIACEAE*

AQUALINDERELLA R. Emers. & W. Weston

Aqualinderella fermentans R. Emers. & W. Weston **[type species; monotypic]** [*Index of Fungi* 3: 445]

ARAIOSPORA (Cornu) Thaxt.

Araiospora coronata Linder
Araiospora pulchra Thaxt.
Araiospora spinosa (Cornu) Thaxt. **[type species]**
Araiospora streptandra Kevorkian var. *streptandra*
Araiospora streptandra Kevorkian var. *echinulospora* Shanor & L. S. Olive

MINDENIELLA Kanouse

Mindeniella asymmetrica T. W. Johnson
Mindeniella spinospora Kanouse **[type species]**

NELLYMYCES A. Batko

Nellymyces megaceros A. Batko **[type species; monotypic]** [*Index of Fungi* 4: 136]

RHIPIDIUM Cornu

Rhipidium americanum Thaxt.
Rhipidium interruptum Cornu **[type species]**
Rhipidium parthenosporum Kanouse
Rhipidium thaxteri Minden

SAPROMYCES Fritsch

Sapromyces androgynus Thaxt.
Sapromyces elongatus (Cornu) Coker **[type species]**
Sapromyces indicus M. O. P. Iyengar *et al.* [M. O. P. Iyengar, K. Ramakr. & Subram.]

SAPROLEGNIOMYCETIDAE: *SAPROLEGNIALES*: *SAPROLEGNIACEAE*

NOTE: *Saprolegniaceae sensu lato* has now been divided into two families, *Saprolegniaceae sensu stricto* and *Leptolegniaceae* (*Leptolegnia*, *Aphanomyces* and *Plectospira*) on morphological and molecular biological criteria (Dick *et al.*, 1999). *Sensu lato* concepts for the genera *Achlya* and *Saprolegnia* are polyphyletic: the species of the *Achlya*-like and *Saprolegnia*-like genera will require both supra-generic and sub-generic separation. These decisions must await publication of the molecular biological data becoming available from various laboratories, but for provisional suggestions see Dick (1973*a*).

For taxonomic reviews of the genera of the *Saprolegniaceae*, see Cejp (1959), Dick (1973*a*, 1990*a*), Johnson (1956*b*), Scott (1961), Seymour (1970) and Sparrow (1960).

SAPROLEGNIACEAE sensu stricto

ACHLYA Nees *sensu lato*

NOTE: Polyphyletic, some species will need to be moved to other genera, species in other genera may need to be brought in to *Achlya sensu stricto*.

Achlya abortiva Coker & Braxton
Achlya ambisexualis J. R. Raper
Achlya americana Humphrey
Achlya apiculata de Bary
Achlya aquatica Dayal & J. Thakur [*Index of Fungi* **3**: 545]
Achlya benekei J. S. Furtado [*Index of Fungi* **3**: 399]
Achlya bisexualis Coker & Couch var. *bisexualis*
Achlya bisexualis Coker & Couch var. *ambisexualis* (J. R. Raper) A. A. Mil'ko [*Index of Fungi* **5**: 277]
Achlya bispora (Couch) Langsam [*Index of Fungi* **5**: 517]
Achlya bonariensis Beroqui [*Index of Fungi* **3**: 523]
Achlya braziliensis A. I. Milanez [*Index of Fungi* **3**: 399]
Achlya cambrica (Trow) T. W. Johnson
Achlya caroliniana Coker
Achlya colorata Pringsh.
Achlya conspicua Coker
Achlya crenulata Ziegler
Achlya curvicollis Beroqui [*Index of Fungi* **3**: 545]
Achlya debaryana Humphrey
Achlya dictyoidea Prasher [*nom invalid.*] [*Index of Fungi* **6(18)**: 971]
Achlya diffusa T. W. Johnson [J. V. Harv. *ex* T. W. Johnson]
Achlya dubia Coker
Achlya echinulata Beroqui [*Index of Fungi* **3**: 523]
Achlya flagellata Coker
Achlya formosana T.-S. Chiou & H. S. Chang [*Index of Fungi* **4**: 467]
Achlya glomerata Coker
Achlya heterosexualis Whiffen [*Index of Fungi* **3**: 347]
Achlya hypogyna Coker & J. J. Pemberton
Achlya inflata Coker
Achlya intricata Beneke
Achlya kamatii Date [as '*kamatti*'] [*Index of Fungi* **4**: 203]
Achlya klebsiana Pieters
Achlya lobata Ziegler & Gilpin
Achlya megasperma Humphrey
Achlya mucronata Ziegler
Achlya oblongata de Bary
Achlya oligacantha de Bary
Achlya orion Coker & Couch

Achlya oviparvula A. L. Rogers & Beneke
Achlya papillosa Humphrey
Achlya polyandra Hildebr.
Achlya prolifera Nees **[type species]**
Achlya proliferoides Coker
Achlya pseudoradiosa A. L. Rogers & Beneke
Achlya racemosa Hildebr.
Achlya radiosa Maurizio
Achlya recurva Cornu
Achlya recurva Cornu *sensu* Latham
Achlya rodrigueziana F. T. Wolf
Achlya sparrowii Reischer
Achlya spinosa de Bary
Achlya spiracaulis T. W. Johnson
Achlya stellata de Bary var. *stellata*
Achlya stellata de Bary var. *multispora* T.-S. Chiou & H. S. Chang [*Index of Fungi* **4**: 467]
Achlya subterranea Coker & Braxton
Achlya treleaseana (Humphrey) Kauffman
Achlya turfosa Johannes

APLANES Humphrey

Aplanes androgynus (W. Archer) Humphrey **[type species]**

NOTE: An *Achlya*, see Dick (1973*a*), but the combination has yet to be made, pending molecular data for the various sections of the genus *Achlya*.

APLANOPSIS Höhnk

Aplanopsis spinosa M. W. Dick
Aplanopsis terrestris Höhnk **[type species]** [see entry under *Pythiopsis*]

BREVILEGNIA Coker & Couch

Brevilegnia bispora Couch
Brevilegnia crassa Rossy-Vald.
Brevilegnia diclina J. V. Harv.
Brevilegnia globosa Ziegler
Brevilegnia indica S. K. Prabhuji & S. K. Sinha [*nom. illeg.*] [*Index of Fungi* **6(11)**: 598]
Brevilegnia irregularis Rossy-Vald.
Brevilegnia linearis Coker & Braxton
Brevilegnia longicaulis T. W. Johnson
Brevilegnia megasperma J. V. Harv.
Brevilegnia minutandra Höhnk
Brevilegnia montana (Coker & Braxton) T. W.Johnson [≡ *B. unisperma* var. *montana*] [*Index of Fungi* **4**: 469]
Brevilegnia parvispora Höhnk
Brevilegnia subclavata Couch **[type species]**
Brevilegnia unisperma (Coker & Braxton) Coker
Brevilegnia variabilis Indoh [≡ *Dictyuchus variabilis*] [*Index of Fungi* **4**: 536]

CALYPTRALEGNIA Coker

Calyptralegnia achlyoides (Coker & Couch) Coker **[type species]**
Calyptralegnia basraensis T. M. Muhsin
Calyptralegnia ripariensis Höhnk

COUCHIA W. W. Martin

Couchia circumplexa W. W. Martin **[type species, monotypic genus]** [*Index of Fungi* 5: 157]

NOTE: Probably an *Isoachlya*, but the combination has yet to be made, and should be held over, pending availability of molecular data for the various sections of the family.

DICTYUCHUS Leitg.

Dictyuchus achlyoides Coker & Alexander
Dictyuchus lucknowensis J. N. Rai & J. K. Misra [*Index of Fungi* 4: 536]
Dictyuchus magnusii Lindst.
Dictyuchus missouriensis Couch
Dictyuchus missouriensis Couch var. *moruzii* N.Toma [*Index of Fungi* 4: 32]
Dictyuchus monosporus Leitg. **[type species]**
Dictyuchus pisci Khulbe & Sati [*Index of Fungi* 5: 338]
Dictyuchus pseudoachlyoides Beneke
Dictyuchus pseudodictyon Coker & Braxton
Dictyuchus variabilis (Indoh) T.-S. Chiou & H. S. Chang [≡ *Brevilegnia variabilis*] [*Index of Fungi* 4: 536]

GEOLEGNIA Coker & J. V. Harv.

Geolegnia inflata Coker & J. V. Harv. **[type species]**
Geolegnia intermedia Höhnk
Geolegnia septisporangia Coker & J. V. Harv.

HYDATINOPHAGUS Valkanov

Hydatinophagus americanus A. F. Bartsch & F. T. Wolf
Hydatinophagus apsteinii Valkanov **[type species]**

ISOACHLYA Kauffman

NOTE: *Isoachlya* includes *Cladolegnia*; probably a few species of *Saprolegnia*, and *Couchia*]

Isoachlya eccentrica Coker
Isoachlya itoana Nagai
Isoachlya luxurians (Bhargava & G. C. Srivast.) R. L. Seym.
Isoachlya monilifera (de Bary) Kauffman
Saprolegnia richteri R. L. Seym.
[W. Richter *ex* R. L. Seym.] [An *Isoachlya*, see Dick (1973*a*), but the combination has yet to be made.]
Isoachlya subterranea Dissmann
Isoachlya torulosa de Bary
Isoachlya toruloides Kauffman & Coker **[type species]**
Isoachlya unispora Coker & Couch

PROTOACHLYA Coker

Protoachlya hypogyna Shanor & Conover
Protoachlya oryzae Khulbe [*Index of Fungi* 5: 260]
Protoachlya paradoxa Coker **[type species]**
Protoachlya polysporus (Lindst.) Apinis

PYTHIOPSIS de Bary

Pythiopsis cymosa de Bary **[type species]**
Pythiopsis humphreyana Coker
Pythiopsis papillata M. Ôkubo & Kobayasi
Pythiopsis subcentrica T.-S. Chiou & H. S. Chang [*Index of Fungi* **4**: 483]
Pythiopsis terrestris (Höhnk) Chiou & Chang [as 'Chiou & Chang', ≡ *Aplanopsis terrestris*] [*Index of Fungi* **4**: 483]
NOTE: *Comb. nov.* is valid; but, if the transfer is justified, the authors failed to provide a new generic name for the remaining species of *Aplanopsis*.

SAPROLEGNIA Nees *sensu lato* **[type genus]**

NOTE: Polyphyletic. In the *sensu lato* concept (Seymour, 1970) the genera *Cladolegnia, Isoachlya* and *Scoliolegnia* are included. Molecular data indicates that *Scoliolegnia* should not be included, and *Saprolegnia anisospora* may also need to be removed. Further revisions may be expected.

Saprolegnia anisospora de Bary
Saprolegnia australis R. F. Elliott [*Index of Fungi* **3**: 495]
Saprolegnia bhargavae Khulbe & B. L. Verma [as '*bhargavii*'] [*Index of Fungi* **5**: 614]
Saprolegnia blelhamensis (M. W. Dick) Milko [≡ *Scoliolegnia blelhamensis*] [*Index of Fungi* **4**: 637]
Saprolegnia chandigarensis Prasher [*nom invalid.*] [*Index of Fungi* **6(18)**: 999]
Saprolegnia diclina Humphrey
Saprolegnia exigua Murgoci [*Index of Fungi* **4**: 184]
Saprolegnia ferax (Gruith.) Thur. **[type species]**
Saprolegnia furcata Maurizio
Saprolegnia glomerata (Tiesenh.) A. Lund
Saprolegnia hypogyna (Pringsh.) de Bary
Saprolegnia irregularis T. W. Johnson & R. L. Seym. [*Index of Fungi* **4**: 387]
Saprolegnia lativica Apinis
Saprolegnia litoralis Coker
Saprolegnia megasperma Coker
Saprolegnia parasitica Coker
Saprolegnia polymorpha Willoughby [*Index of Fungi* **6(18)**: 999]
Saprolegnia shikotsuensis Hatai *et al.* [Hatai, Egusa & Awakura] [*Index of Fungi* **4**: 575]
Saprolegnia subeccentrica (M. W. Dick) Milko [≡ *Scoliolegnia subeccentrica*] [*Index of Fungi* **4**: 637]
Saprolegnia terrestris R. L. Seym. [Cookson *ex* R. L. Seym.]
Saprolegnia turfosa (Minden) Gäum.
Saprolegnia uliginosa Johannes

SCOLIOLEGNIA M. W. Dick

Scoliolegnia asterophora (de Bary) M. W. Dick **[type species]**
Scoliolegnia blelhamensis M. W. Dick [*Index of Fungi* **4**: 637]
Scoliolegnia centrica Khulbe [*Index of Fungi* **5**: 364]
Scoliolegnia depauperata M. W. Dick [*nomen nudum* - a spiny form of *Pythiopsis cymosa*]
Scoliolegnia subeccentrica M. W. Dick [*Index of Fungi* **4**: 637]

SOMMERSTORFFIA Arnautov

Sommerstorffia spinosa Arnautov **[type species, monotypic genus]**

THRAUSTOTHECA Humphrey

Thraustotheca clavata (de Bary) Humphrey **[type species]**
Thraustotheca irregularis Coker & M. W. Ward
Thraustotheca primoachlya Coker & Couch

SAPROLEGNIOMYCETIDAE: *SAPROLEGNIALES*: *LEPTOLEGNIACEAE*

APHANOMYCES de Bary

Aphanomyces acinetophagus A. F. Bartsh & F. T. Wolf
Aphanomyces americanus (A. F. Bartsch & F. T. Wolf) W. W. Scott
Aphanomyces amphigynus Cutter
Aphanomyces apophysii Lacy
Aphanomyces astaci Schikora
Aphanomyces bosminae W. W. Scott
Aphanomyces brassicae S. L. Singh & M. S. Pavgi [*Index of Fungi* 4: 498]
Aphanomyces camptostylus Drechsler
Aphanomyces cladogamus Drechsler
Aphanomyces cochlioides Drechsler
Aphanomyces daphniae Prowse
Aphanomyces euteiches Drechsler
Aphanomyces euteiches Drechsler forma specialis *phaseoli* W. F. Pfender & D. J. Hagedorn [*Index of Fungi* 5: 152]
Aphanomyces euteiches Drechsler forma specialis *pisi* W. F. Pfender & D. J. Hagedorn [*Index of Fungi* 5: 152]
Aphanomyces exoparasiticus Coker & Couch
Aphanomyces frigidophilus N. Kitancharoen & Hatai [*Index of Fungi* 6(16): 898]
Aphanomyces helicoides Minden
Aphanomyces hydatinae Valkanov
Aphanomyces invadans Willoughby *et al.* [as '*invaderis*'] [Willoughby, R. J. Roberts & Chinabut] [*Index of Fungi* 6(13): 706]
Aphanomyces iridis Ichit. & Tak. Kodama [*Index of Fungi* 5: 574]
Aphanomyces irregulare W. W. Scott
Aphanomyces laevis de Bary
Aphanomyces keratinophilus (M. Ôkubo & Kobayasi) R. L. Seym. & T. W. Johnson [*Index of Fungi* 4: 269]
Aphanomyces norvegicus Wille
Aphanomyces ovidestruens Gickelh.
Aphanomyces parasiticus Coker
Aphanomyces patersonii W. W. Scott
Aphanomyces phycophilus de Bary
Aphanomyces pisci R. C. Srivastava [*Index of Fungi* 4: 588]
Aphanomyces piscicida Hatai [*Index of Fungi* 6(15): 796]
Aphanomyces raphani J. B. Kendr.
Aphanomyces scaber de Bary
Aphanomyces sparrowii Cutter
Aphanomyces stellatus de Bary [type species]
Aphanomyces volgensis Domashova [*Index of Fungi* 4: 302]

LEPTOLEGNIA de Bary [type genus]

Leptolegnia baltica Höhnk & Vallin
  Leptolegnia bandoniensis A. P. Swan [A *Saprolegnia*, but not listed in Seymour (1971)]
Leptolegnia caudata de Bary [type species]
Leptolegnia chapmanii R. L. Seym. [*Index of Fungi* 5: 394]
Leptolegnia eccentrica Coker & V. D. Matthews
  Leptolegnia pontica Artemczuk
Leptolegnia subterranea Coker & J. V. Harv.

PLECTOSPIRA Drechsler

Plectospira agama Drechsler
Plectospira gemmifera Drechsler
Plectospira myriandra Drechsler [type species]
Plectospira plagiocaula Drechsler

SAPROLEGNIOMYCETIDAE: *SCLEROSPORALES*: *SCLEROSPORACEAE*

NOTE: For taxonomic reviews of the genera of the *Sclerosporales*, see Dick *et al.* (1984, 1989) and Waterhouse (1964).

SCLEROSPORA J. Schröt.

Sclerospora butleri W. Weston
Sclerospora graminicola (Sacc.) J. Schröt. **[type species]**
Sclerospora iseilmatis Thirum. & Naras.
Sclerospora northii W. Weston
Sclerospora secalina Naumov

PERONOSCLEROSPORA (S. Ito) Hara

NOTE: Hara *in* Shirai, *not* Shirai & Hara (Shaw & Waterhouse, 1980); *not* C. G. Shaw (as in Hawksworth *et al.* (1995), see also type species.

Peronosclerospora dichanthiicola (Thirum. & Naras.) C. G. Shaw [*Index of Fungi* **4**: 541, 570; **5**: 444]
Peronosclerospora globosa Kubicek & R. G. Kenneth [*nom. illeg.*] [*Index of Fungi* **5**: 399]
Peronosclerospora heteropogonis Siradhana *et al.* [Siradhana, Dange, Rathore & S. D. Singh] [*Index of Fungi* **5**: 20]
Peronosclerospora maydis (Racib.) C. G. Shaw [*Index of Fungi* **4**: 541; **5**: 444]
Peronosclerospora miscanthi (T. Miyake) C. G. Shaw [*Index of Fungi* **4**: 541; **5**: 444]
Peronosclerospora noblei (W. Weston) C. G. Shaw [*Index of Fungi* **5**: 20]
Peronosclerospora philippinensis (W. Weston) C. G. Shaw [*Index of Fungi* **4**: 541; **5**: 444]
Peronosclerospora sacchari (T. Miyake) Hara **[type species]** [*Index of Fungi* **4**: 541, 570; **5**: 445]
Peronosclerospora sorghi (W. Weston & Uppal) C. G. Shaw [*Index of Fungi* **4**: 541]
Peronosclerospora spontanea (W. Weston) C. G. Shaw [*Index of Fungi* **4**: 541; **5**: 445]
Peronosclerospora westonii (M. C. Sriniv. *et al.*) C. G. Shaw [M. C. Sriniv., Naras. & Thirum.] [*Index of Fungi* **4**: 541; **5**: 445]
Peronosclerospora zeae Yao [*nom. illeg.*]

SAPROLEGNIOMYCETIDAE: *SCLEROSPORALES*: *VERRUCALVACEAE*

PACHYMETRA B. J. Croft & M. W. Dick

Pachymetra chaunorhiza B. J. Croft & M. W. Dick **[type species; monotypic genus]** [*Index of Fungi* **5**: 897]

SCLEROPHTHORA Thirum. *et al.* [Thirum., C. G. Shaw & Naras.]

Sclerophthora cryophila W. Jones
Sclerophthora farlowii (Griffiths) R. G. Kenneth
Sclerophthora lolii R. G. Kenneth [*Index of Fungi* **3**: 285]
Sclerophthora macrospora (Sacc.) Thirum. *et al.* **[type species]** [Thirum., C. G. Shaw & Naras.]
Sclerophthora rayssiae R. G. Kenneth *et al.* var. *rayssiae* [R. G. Kenneth, Koltin & Wahl] [*Index of Fungi* **3**: 285]
Sclerophthora rayssiae R. G. Kenneth *et al.* var. *zeae* Payak & Naras. [R. G. Kenneth, Koltin & Wahl] [*Index of Fungi* **3**: 447]

VERRUCALVUS P. Wong & M. W. Dick

Verrucalvus flavofaciens P. Wong & M. W. Dick **[type species; monotypic genus]** [*Index of Fungi* **5**: 458]

SAPROLEGNIOMYCETIDAE: *LEPTOMITALES*: *LEPTOMITACEAE*

APODACHLYA Pringsh.

Apodachlya brachynema (Hildebr.) Pringsh. **[type species]**
Apodachlya minima Cejp [Coker & J. Leitn. *ex* Cejp]
Apodachlya pyrifera Zopf

LEPTOMITUS (Roth) C. Agardh

Leptomitus lacteus (Roth) C. Agardh **[type species, monotypic genus]**

PLEROGONE M. W. Dick

Plerogone helodes M. W. Dick **[type species, monotypic genus]** [*Index of Fungi* **5**: 552]

SAPROLEGNIOMYCETIDAE: *LEPTOMITALES*: *APODACHLYELLACEAE*

APODACHLYELLA Indoh

Apodachlyella completa (Humphrey) Indoh **[type species, monotypic genus]**

EURYCHASMOPSIS Canter & M. W. Dick

Eurychasmopsis multisecunda Canter **[type species, monotypic genus]** [*Imdex of fungi* **6(8)**: 430]

SAPROLEGNIOMYCETIDAE: *LEPTOMITALES*: *LEPTOLEGNIELLACEAE*

APHANODICTYON M. W. Dick [Huneycutt *ex* M. W. Dick]

Aphanodictyon papillatum M. W. Dick **[type species, monotypic]** [Huneycutt *ex* M. W. Dick] [*Index of Fungi* **4**: 91]

BREVILEGNIELLA M. W. Dick

Brevilegniella keratinophila M. W. Dick **[type species, monotypic]**

LEPTOLEGNIELLA Huneycutt

Leptolegniella abnormis Karling [*Index of Fungi* **5**: 764]
Leptolegniella angularis Karling [*Index of Fungi* **5**: 764]
Leptolegniella exogena Karling [*Index of Fungi* **5**: 764]
Leptolegniella exoospora W. D. Kane
Leptolegniella keratinophila Huneycutt **[type species]**

NEMATOPHTHORA Kerry & D. H. Crump

Nematophthora gynophila Kerry & D. H. Crump **[type species, monotypic genus]** [*Index of Fungi* 5: 19]

APHANOMYCOPSIS Scherff.

Aphanomycopsis bacillariacearum Scherff. **[type species]**
Aphanomycopsis cryptica Canter [*Index of Fungi* 5: 419]
Aphanomycopsis desmidiella Canter
Aphanomycopsis entophyta (Pringsh.) M. W. Dick
Aphanomycopsis peridiniella Boltovskoy & A. M. Arambarri [*Index of Fungi* 5: 519]
Aphanomycopsis punctata Karling
Aphanomycopsis saprophytica Karling
Aphanomycopsis sexualis W. W. Martin [*Index of Fungi* 4: 370]

CORNUMYCES M. W. Dick

Cornumyces catenophlyctidis (B. M. Sundaram) M. W. Dick [*loc. cit.*]
Cornumyces destruens (Sparrow) M. W. Dick **[type species]** [*loc. cit.*]
Cornumyces endogenus (Karling) M. W. Dick [*loc. cit.*]
Cornumyces irregularis (Const.) M. W. Dick [*loc. cit.*]
Cornumyces karlingiae (Karling) M. W. Dick [*loc. cit.*]
Cornumyces muenscheri (Cutter) M. W. Dick [*loc. cit.*]
Cornumyces pygmaeus (Zopf) M. W. Dick [*loc. cit.*]
Cornumyces utriculobus (C. E. Mill.) M. W. Dick [*loc. cit.*]

SAPROLEGNIOMYCETIDAE: *LEPTOMITALES*: *DUCELLIERIACEAE*

DUCELLIERIA Teiling

Ducellieria chodatii (F. Ducell.) Teiling **[type species]**

BLASTULIDIUM Pérez

Blastulidium paedophthorum Pérez **[type species, monotypic genus]**

ORDER *INCERTAE SEDIS*: *SALILAGENIDIALES*: *SALILAGENIDIACEAE*

SALILAGENIDIUM M. W. Dick

Salilagenidium callinectes (Couch) M. W. Dick **[type species]**
Salilagenidium chthamalophilum (T. W. Johnson) M. W. Dick
Salilagenidium marinum (D. Atkins) M. W. Dick
Salilagenidium myophilum (Hatai & Lawhav.) M. W. Dick
Salilagenidium scyllae (Bian *et al.*) M. W. Dick [Bian, Hatai, Po & Egusa] [*Index of Fungi* **4**: 629;...]
Salilagenidium thermophilum (K. Nakamura *et al.*) M. W. Dick [(K. Nakamura, M. Nakamura, Hatai & Zafran)] [*Index of Fungi* **6(13)**: 721]

ORDER *INCERTAE SEDIS*: *SALILAGENIDIALES*: *HALIPHTHORACEAE*

HALIPHTHOROS Vishniac

Haliphthoros hiroshimaensis (Hatai, in preparation)
Haliphthoros milfordensis Vishniac **[type species]** [*Index of Fungi* **5**: 14]
Haliphthoros philippinensis Hatai *et al.* [Hatai, Bian, M. C. Batic. & Egusa] [*Index of Fungi* **5**: 14]
Haliphthoros zoophthorum (Vishniac) M. W. Dick [*loc. cit.*]

HALODAPHNEA M. W. Dick

Halodaphnea awabi (N. Kitancharoen *et al.*) M. W. Dick [(N. Kitancharoen, K. Nakamura, S. Wada & Hatai)] [*Index of Fungi* **6(10)**: 535; **6(13)**: 717; **6(18)**: 980]
Halodaphnea hamanaensis (Bian & Egusa) M. W. Dick [*Index of Fungi* **5**: 279; **6(13)**: 717; **6(18)**: 980]
Halodaphnea okinawaensis (K. Nakamura & Hatai) M. W. Dick [*Index of Fungi* **6(10)**: 597; **6(13)**: 717; **6(18)**: 980]
Halodaphnea panulirata (N. Kitancharoen & Hatai) M. W. Dick [*Index of Fungi* **6(10)**: 597; **6(13)**: 717; **6(18)**: 980]
Halodaphnea parasitica (K. Nakamura & Hatai) M. W. Dick **[type species]** [*Index of Fungi* **6(10)**: 597; **6(13)**: 717; **6(18)**: 980]

ATKINSIELLA Vishniac

Atkinsiella dubia (D. Atkins) Vishniac **[type species, monotypic genus]**

FAMILY *INCERTAE SEDIS*: *LAGENACEAE*

LAGENA Vanterp. & Ledingham

Lagena radicicola Vanterp. & Ledingham **[type species, monotypic genus]**

PYTHIELLA Couch

Pythiella pythii (Whiffen) M. W. Dick [*loc. cit.*]
Pythiella vernalis Couch **[type species]**

CILIATOMYCES I. Foissner & W. Foissner

Ciliatomyces spectabilis I. Foissner & W. Foissner **[type species, monotypic genus]** [*Index of Fungi* **5**: 581; **6(17)**: 903]

ORDER *INCERTAE SEDIS*: *OLPIDIOPSIDALES*: *OLPIDIOPSIDACEAE*

OLPIDIOPSIS Cornu

Olpidiopsis achlyae D. A. McLarty
Olpidiopsis aphanomycis Cornu
Olpidiopsis braziliensis Sparrow
Olpidiopsis brevispinosa Whiffen
Olpidiopsis curvispinosa Whiffen
Olpidiopsis fusiformis Cornu
Olpidiopsis incrassata Cornu
Olpidiopsis index Cornu
Olpidiopsis indica G. C. Srivast. [*Index of Fungi* **4**: 384]
Olpidiopsis luxurians Barrett
Olpidiopsis pythii (E. J. Butler) Karling
Olpidiopsis pythii (E. J. Butler) Karling forma *brevispinosa* (Whiffen) Cejp
Olpidiopsis saprolegniae (A. Braun) Cornu var. *saprolegniae* **[type species]**
Olpidiopsis saprolegniae (A. Braun) Cornu var. *laevis* Coker
Olpidiopsis saprolegniae (A. Braun) Cornu var. *indica* Dayal & J. Thakur
Olpidiopsis spinosa Tokun.
Olpidiopsis varians Shanor
Olpidiopsis verrucosa T. W. Johnson
Olpidiopsis vexans Barrett

PLEOCYSTIDIUM C. Fisch

Pleocystidium lundiae (Karling) M. W. Dick [*Index of Fungi* **5**: 203] [*loc. cit.*]
Pleocystidium parasiticum C. Fisch **[type species]**
Pleocystidium pygmaeoides (Karling) M. W. Dick [*Index of Fungi* **5**: 164] [*loc. cit.*]

GRACEA M. W. Dick [genus *incertae sedis*]

Gracea gracilis (E. J. Butler) M. W. Dick **[type species]** [*Index of Fungi* **6(16)**: 853]
Gracea waterhouseae (Karling) M. W. Dick [*Index of Fungi* **6(16)**: 853]

ORDER *INCERTAE SEDIS*: *MYZOCYTIOPSIDALES*: *MYZOCYTIOPSIDACEAE*

CHLAMYDOMYZIUM M. W. Dick

Chlamydomyzium anomalum (G. L. Barron) M. W. Dick **[type species]** [*Index of Fungi* **4**: 414; **6(16)**: 849]
Chlamydomyzium aplanosporum Glockling [*Index of Fungi*: **6(16)**: 849]
Chlamydomyzium internum Glockling [*Index of Fungi*: **6(16)**: 849]
Chlamydomyzium oviparasiticum (G. L. Barron) M. W. Dick [*Index of Fungi* **6(16)**: 849]
Chlamydomyzium septatum (Karling) M. W. Dick [*Index of Fungi* **6(16)**: 849]
Chlamydomyzium sphaericum Glockling

GONIMOCHAETE Drechsler

Gonimochaete horridula Drechsler **[type species]**
Gonimochaete latitubus S. Y. Newell *et al.* [S. Y. Newell, Cefalu & Fell]
Gonimochaete lignicola G. L. Barron
Gonimochaete pyriformis G. L. Barron

MYZOCYTIOPSIS M. W. Dick

Myzocytiopsis bolata Glockling
Myzocytiopsis distylae (Karling) M. W. Dick [*Index of Fungi* **6(16)**: 682]
Myzocytiopsis elegans (Perronc.) M. W. Dick [*Index of Fungi* **6(16)**: 682]
Myzocytiopsis fijiensis (Karling) M. W. Dick [*Index of Fungi* **6(16)**: 682]
Myzocytiopsis glutinospora (G. L. Barron) M. W. Dick [*Index of Fungi* **4**: 414; **6(16)**: 682]
Myzocytiopsis humana (Karling) M. W. Dick [*Index of Fungi* **6(16)**: 682]
Myzocytiopsis humicola (G. L. Barron & Percy) M. W. Dick [*Index of Fungi* **4**: 383; **6(16)**: 682]
Myzocytiopsis indica (U. P. Singh) M. W. Dick [*Index of Fungi* **4**: 313; **6(16)**: 682]
Myzocytiopsis intermedia (G. L. Barron) M. W. Dick [*Index of Fungi* **4**: 414; **6(16)**: 682]
Myzocytiopsis lenticularis (G. L. Barron) M. W. Dick **[type species]** [*Index of Fungi* **4**: 414; **6(16)**: 682]
Myzocytiopsis microspora (Karling) M. W. Dick [*Index of Fungi* **6(16)**: 682]
Myzocytiopsis oophila (Sparrow) M. W. Dick [*Index of Fungi* **6(16)**: 682]
Myzocytiopsis osiris Glockling
Myzocytiopsis papillata (G. L. Barron) M. W. Dick [*Index of Fungi* **4**: 414; **6(16)**: 682]
Myzocytiopsis parthenospora (Karling) M. W. Dick [*Index of Fungi* **6(16)**: 682]
Myzocytiopsis subuliformis M. W. Dick [*Index of Fungi*: **4**: 480; **6(16)**: 682]
Myzocytiopsis vermicola (Zopf) M. W. Dick [*Index of Fungi* **6(16)**: 683]
Myzocytiopsis zoophthora (Sparrow) M. W. Dick [*Index of Fungi* **6(16)**: 683]

SYZYGANGIA M. W. Dick

Syzygangia dictyospora M. W. Dick [*Index of Fungi* **6(16)**: 872]
Syzygangia elliptica (De Wild.) M. W. Dick [*Index of Fungi* **6(16)**: 873]
Syzygangia marchaliana (De Wild.) M. W. Dick **[type species]** [*Index of Fungi* **6(16)**: 873]
Syzygangia nodosa (P. A. Dang.) M. W. Dick [*Index of Fungi* **6(16)**: 873]
Syzygangia oedogonii (Scherff.) M. W. Dick [*Index of Fungi* **6(16)**: 873]
Syzygangia papillosa (Cocc.) M. W. Dick [*Index of Fungi* **6(16)**: 873]
Syzygangia sacculoides (Serbinow) M. W. Dick [*Index of Fungi* **6(16)**: 873]
Syzygangia zygnematicola M. W. Dick [*Index of Fungi* **6(16)**: 873]

MYZOCYTIOPSIDALES: *CRYPTICOLACEAE*

CRYPTICOLA Humber *et al.* [Humber, Frances & A. W. Sweeney]

Crypticola clavulifera Humber *et al.* **[type species]** [Humber, Frances & A. W. Sweeney] [*Index of Fungi* **6(1)**: 11]
Crypticola entomophaga (W. W. Martin) M. W. Dick [*Index of Fungi* **6(18)**: 977]

FAMILY *INCERTAE SEDIS*: *SIROLPIDIACEAE*

SIROLPIDIUM H. E. Petersen

Sirolpidium andreei (Lagerh.) M. W. Dick
Sirolpidium bryopsidis (de Bruyne) H. E. Petersen **[type species]**
Sirolpidium glenodinianum (P. A. Dang.) M. W. Dick
Sirolpidium globosum (Anastasiou & Churchl.) M. W. Dick
Sirolpidium marinum (P. A. Dang.) M. W. Dick
Sirolpidium paradoxum (H. E. Petersen) M. W. Dick
Sirolpidium salinum (P. A. Dang.) M. W. Dick

FAMILY *INCERTAE SEDIS*: *PONTISMATACEAE*

PONTISMA H. E. Petersen

Pontisma antithamnionis (Whittick & South) M. W. Dick [*Index of Fungi* 4: 220]
Pontisma dangeardii (Feldm. & Feldm.-Maz.) M. W. Dick
Pontisma feldmannii (Aleem) M. W. Dick
Pontisma inhabilis (H. E. Petersen) M. W. Dick
Pontisma lagenidioides H. E. Petersen **[type species]**
Pontisma magnusii (Feldm. & Feldm.-Maz.) M. W. Dick

PETERSENIA Sparrow

Petersenia lobata (H. E. Petersen) Sparrow **[type species]**
Petersenia palmariae Vanderm. & Pueschel [*Index of Fungi* 5: 445]
Petersenia panicicola Thirum. & R. C. Lacy
Petersenia pollagaster (H. E. Petersen) Sparrow

FAMILY *INCERTAE SEDIS*: *EURYCHASMATACEAE*

EURYCHASMA Magnus

Eurychasma dicksonii (E. P. Wright) Magnus **[type species, monotypic genus]**

EURYCHASMIDIUM Sparrow

Eurychasmidium joycei (Sparrow) M. W. Dick
Eurychasmidium sacculum (H. E. Petersen) M. W. Dick
Eurychasmidium tumefaciens (Magnus) Sparrow **[type species]**

ORDER *INCERTAE SEDIS*: *ANISOLPIDIALES*: *ANISOLPIDIACEAE*

ANISOLPIDIUM Karling

Anisolpidium ectocarpi Karling **[type species]**
Anisolpidium joklianum M. W. Dick
Anisolpidium minutum (H. E. Petersen) M. W. Dick
Anisolpidium olpidium (H. E. Petersen) M. W. Dick
Anisolpidium rosenvingii (H. E. Petersen) Karling
Anisolpidium sphacelarum (Kny) Karling

PHAGOMYXA Karling

Phagomyxa algarum Karling **[type species]**

PERONOSPOROMYCETES: *LAGENISMATALES*: *LAGENISMATACEAE*

LAGENISMA Drebes

Lagenisma coscinodisci Drebes **[type species, monotypic genus]**

HYPHOCHYTRIOMYCETES: *HYPHOCHYTRIOMYCETIDAE*: *HYPHOCHYTRIALES*: *HYPHOCHYTRIACEAE*

HYPHOCHYTRIUM Zopf

Hyphochytrium catenoides Karling
Hyphochytrium elongatum (Karling) M. W. Dick [*Index of Fungi* **4**: 467]
Hyphochytrium hydrodictyi Valkanov
Hyphochytrium infestans Zopf **[type species]**
Hyphochytrium oceanum Karling
Hyphochytrium peniliae N. J. Artemczuk & L. M. Zelezinskaya
Hyphochytrium saprobium (Karling) M. W. Dick

CANTEROMYCES Sparrow

Canteriomyces stigeoclonii (De Wild.) Sparrow **[type species, monotypic genus]**

CYSTOCHYTRIUM Ivimey Cook

Cystochytrium radicale Ivimey Cook **[type species, monotypic genus]**

HYPHOCHYTRIALES: *RHIZIDIOMYCETACEAE*

RHIZIDIOMYCES Zopf

Rhizidiomyces apophysatus Zopf **[type species]**
Rhizidiomyces bivellatus Nabel
Rhizidiomyces bulbosus Karling
Rhizidiomyces bullatus (Sparrow) Karling
Rhizidiomyces coronus Karling
Rhizidiomyces hansonii Karling
Rhizidiomyces hirsutus Karling
Rhizidiomyces ichneumon Gobi
Rhizidiomyces japonicus Kobayasi & M. Ôkubo
Rhizidiomyces parasiticus Karling
Rhizidiomyces saprophyticus (Karling) Karling
Rhizidiomyces stomatosus (Sparrow) M. W. Dick

LATROSTIUM Zopf

Latrostium comprimens Zopf **[type species, monotypic genus]**

LABYRINTHISTA: *LABYRINTHULALES*: *LABYRINTHULACEAE*

LABYRINTHULA Cienk.

Labyrinthula algeriensis A. Hollande & Enjumet [*Index of Fungi* **4**: 381]
Labyrinthula chattonii P. A. Dang.
Labyrinthula cienkowskii Zopf
Labyrinthula coenocystis Schmoller [*Index of Fungi* **4**: 381]
Labyrinthula jeremarina L. Rolf
Labyrinthula macrocystis Cienk.
Labyrinthula magnifica (Valkanov) L. S. Olive [*Index of fungi* **4**: 381]
Labyrinthula pohlia (Schmoller) M. W. Dick
Labyrinthula roscoffensis Chadef.
Labyrinthula thaidis B. A. Cox & Mackin [*Index of Fungi* **4**: 381]
Labyrinthula valkanovii Karling
Labyrinthula vitellina Cienk. **[type species]**
Labyrinthula zosterae D. Porter & Muehlst. [*Index of fungi* **6(3)**: 148]

LABYRINTHISTA: *THRAUSTOCHYTRIALES*: *THRAUSTOCHYTRIACEAE*

ALTHORNIA E. B. G. Jones & Alderman

Althornia crouchii E. B. G. Jones & Alderman **[type species, monotypic genus]** [*Index of fungi* **4**: 156]

APLANOCHYTRIUM Bahnweg & Sparrow

Aplanochytrium kerguelense Bahnweg & Sparrow **[type species, monotypic genus]**

ELINA N. J. Artemczuk

Elina marisalba N. J. Artemczuk **[type species]**
Elina sinorifica N. J. Artemczuk

JAPONOCHYTRIUM Kobayasi & M. Ôkubo

Japonochytrium marinum Kobayasi & M. Ôkubo **[type species, monotypic genus]**

LABYRINTHULOIDES F. O. Perkins

Labyrinthuloides haliotidis S. M. Bower [*Index of Fungi* **6(15)**: 807]
Labyrinthuloides minuta (S. W. Watson & Raper) F. O. Perkins [*Index of Fungi* **4**: 311]
Labyrinthuloides saliens (J. A. Quick) J. A. Quick
Labyrinthuloides schizochytrops J. A. Quick
Labyrinthuloides yorkensis F. O. Perkins **[type species]**

SCHIZOCHYTRIUM S. Goldst. & Belsky

Schizochytrium aggregatum S. Goldst. & Belsky **[type species]**
Schizochytrium limacinum D. Honda & Yokochi [*Index of fungi* **6(17)**: 935]
Schizochytrium mangrovei Raghuk. [*Index of Fungi* **5**: 844]
Schizochytrium minutum A. Gaertn.
Schizochytrium octosporum Raghuk. [*Index of Fungi* **5**: 781]

THRAUSTOCHYTRIUM Sparrow

Thraustochytrium aggregatum Ulken
Thraustochytrium antarcticum Bahnweg & Sparrow [*Index of Fungi* **4**: 319]
Thraustochytrium arudimentale N. J. Artemczuk
Thraustochytrium aureum S. Goldst.
Thraustochytrium benthicola Raghuk. [*Index of Fungi* **5**: 27]
Thraustochytrium globosum Kobayasi & M. Ôkubo
Thraustochytrium indicum Chakravaty [*Index of Fungi* **4**: 639]
Thraustochytrium kerguelense Bahnweg & Sparrow [*Index of Fungi* **4**: 319]
Thraustochytrium kinnei A. Gaertn.
Thraustochytrium motivum S. Goldst.
Thraustochytrium multirudimentale S. Goldst.
Thraustochytrium pachydermum E. Scholz
Thraustochytrium proliferum Sparrow **[type species]**
Thraustochytrium roseum S. Goldst.
Thraustochytrium rossii Bahnweg & Sparrow [*Index of Fungi* **4**: 319]
Thraustochytrium striatum J. Schneider

ULKENIA A. Gaertn.

Ulkenia amoeboidea (Bahnweg & Sparrow) A. Gaertn. [*Index of Fungi* **4**: 319; 519]
Ulkenia minuta Raghuk. [*Index of Fungi* **4**: 519]
Ulkenia profunda A. Gaertn. [*Index of Fungi* **4**: 519]
Ulkenia radiata A. Gaertn. [*Index of Fungi* **4**: 519]
Ulkenia sarkariana A. Gaertn. [*Index of Fungi* **4**: 519]
Ulkenia visurgensis (Ulken) A. Gaertn. **[type species]** [*Index of Fungi* **4**: 519]

DIPLOPHRYS J. Barker

{*Diplophrys archeri* J. Barker **[type species]**}
Diplophrys marina Dykstra & D. Porter

DEVELOPAYELLA S. M. Tong

Developayella elegans S. M. Tong

PLASMODIOPHOROMYCETES: *PLASMODIOPHORALES*: *PLASMODIOPHORACEAE*

LIGNIERA Maire & A. Tison

Ligniera betae (Němec) Karling
Ligniera hypogeae (Borzi) Karling
Ligniera isoetes Palm
Ligniera junci (Schwartz) Maire & A. Tison
Ligniera pilorum Fron & D. F. Gaillat
Ligniera plantaginis (Němec) Karling
Ligniera verrucosa Maire & A. Tison **[type species]**

MEMBRANOSORUS Ostenf. & H. E. Petersen

Membranosorus heterantherae Ostenf. & H. E. Petersen **[type species, monotypic genus]**

OCTOMYXA Couch *et al.* [Couch, J. Leitn. & Whiffen]

Octomyxa brevilegniae Pend.
Octomyxa achlyae Couch *et al.* **[type species]** [Couch, J. Leitn. & Whiffen]

PLASMODIOPHORA Woronin

Plasmodiophora bicaudata Feldm.
Plasmodiophora brassicae Woronin **[type species]**
Plasmodiophora diplantherae (Ferd. & Winge) Ivimey-Cook
Plasmodiophora ficirepentis Andreucci
Plasmodiophora halophilae Ferd. & Winge
Plasmodiophora maritima Feldm.-Maz.

POLYMYXA Ledingham

Polymyxa betae Keskin
Polymyxa betae Keskin forma specialis *portulacae* H. Abi & T. Ui [*Index of Fungi* 5: 691]
Polymyxa graminis Ledingham **[type species]**

SORODISCUS Lagerh. & Winge

Sorodiscus callitrichis Lagerh. & Winge **[type species]**
Sorodiscus cokeri Goldie-Sm.
Sorodiscus karlingii Ivimey Cook
Sorodiscus radicicola Ivimey Cook

SOROSPHAERA J. Schröt.

Sorosphaera radicalis Ivimey Cook & Schwartz
Sorosphaera veronicae (J. Schröt.) J. Schröt. **[type species]**

SPONGOSPORA Brunch.

Spongospora cotulae Barrett
Spongospora campanulae (Ferd. & Winge) Ivimey Cook
Spongospora nasturtii (J. A. Toml.) M. W. Dick
Spongospora subterranea (Wallr.) Lagerh. **[type species]**

TETRAMYXA K. I. Goebel

Tetramyxa elaeagni Y. Yendo & K. Takase
Tetramyxa marina Lipkin & Avidor [*Index of Fungi* **4**: 319]
Tetramyxa parasitica K. I. Goebel **[type species]**
Tetramyxa rhizophaga Lihnell
Tetramyxa triglochinis Molliard

WORONINA Cornu

Woronina aggregata Zopf
Woronina asterina Tokun.
Woronina charae Kiran & Dayal [*Index of Fungi* **6(8)**: 456]
Woronina glomerata (Cornu) A. Fisch.
Woronina leptolegniae Karling
Woronina polycystis Cornu **[type species]**
Woronina pythii Goldie-Sm.
Woronina sorokiniana Jacz. & P. A. Jacz.

PYRRHOSORUS Juel

Pyrrhosorus marinus Juel **[type species, monotypic genus]**

PLASMODIOPHOROMYCETES: HAPTOGLOSSALES: HAPTOGLOSSACEAE

HAPTOGLOSSA Drechsler

Haptoglossa elegans G. L. Barron [*Index of Fungi* **6(1)**: 14]
Haptoglossa heterospora Drechsler **[type species]**
Haptoglossa humicola G. L. Barron
Haptoglossa intermedia G. L. Barron [*Index of Fungi* **5**: 963]
Haptoglossa mirabilis G. L. Barron [*Index of Fungi* **5**: 84]
Haptoglossa northumbrica S. L. Glockling [in press, in Glockling & Beakes]
Haptoglossa polymorpha S. L. Glockling [in press, in Glockling & Beakes]
Haptoglossa zoospora J. G. N. Davidson & G. L. Barron

FAMILY *INCERTAE SEDIS*: *ECTROGELLACEAE*

ECTROGELLA Zopf

Ectrogella bacillariacearum Zopf **[type species]**
Ectrogella brachystoma (Scherff.) M. W. Dick
Ectrogella cyclotellae (Scherff.) M. W. Dick
Ectrogella eurychasmoides Feldm. & Feldm.-Maz.
Ectrogella lauderiae (Gran) Scherff.
Ectrogella perforans H. E. Petersen

SEPTOLPIDIUM Sparrow

Septolpidium lineare Sparrow **[type species, monotypic genus]**

FAMILY *INCERTAE SEDIS ENDEMOSARCACEAE*

ENDEMOSARCA L. S. Olive & Erdos

Endemosarca anomaia Erdos
Endemosarca hypsalyxis Erdos & L. S. Olive **[type species]**
Endemosarca ubatubensis Erdos & L. S. Olive

ORDER *INCERTAE SEDIS*: *ROZELLOPSIDALES*: *ROZELLOPSIDACEAE*

PSEUDOSPHAERITA P. A. Dang.

Pseudosphaerita dinobryonis (Canter) M. W. Dick
Pseudosphaerita drylii Pérez Reyes, Madr.-Gar. & E. L. Ocht.
Pseudosphaerita euglenae P. A. Dang. **[type species]**
Pseudosphaerita radiata (P. A. Dang.) Karling

ROZELLA Cornu

Rozella achlyae Shanor
Rozella allomycis F. K. Faust
Rozella apodachlyae Cornu
Rozella barrettii Karling
Rozella blastocladiae (Minden) Sparrow
Rozella canterae Sparrow
Rozella chytriomycetis Karling
Rozella cladochytrii Karling
Rozella cuculus (E. J. Butler) Sparrow
Rozella diplophlyctidis Karling
Rozella endochytrii Karling
Rozella irregularis (E. J. Butler) Sparrow
Rozella laevis Karling
Rozella longicollis Karling
Rozella monoblepharidis Cornu **[type species]**

Rozella parva Canter
Rozella polyphagi (Sparrow) Sparrow
Rozella rhipidii Cornu
Rozella rhizophlyctidis Karling
Rozella rhizophydii Karling
Rozella septigena Cornu

ROZELLOPSIS Cejp [Karling ex Cejp]

Rozellopsis inflata (E. J. Butler) Karling
Rozellopsis septigena Cejp **[type species]** [Karling ex Cejp]
Rozellopsis simulans (A. Fisch.) Karling
Rozellopsis uliginosa A. Batko [*Index of Fungi* **4**: 544]

SPHAERITA P. A. Dang.

Sphaerita amoebae O. Mattes
Sphaerita dangeardii Chatton & A. Brodskii
Sphaerita endogena P. A. Dang. **[type species]**
Sphaerita entamoebae E. R. Becker
Sphaerita minor A. M. Cunha & Muniz
Sphaerita normeti A. Lwoff
Sphaerita nucleophaga O. Mattes
Sphaerita phaci T. L. Jahn
Sphaerita plasmophaga O. Mattes
Sphaerita trachelomonadis Skvortzov

PLASMOPHAGUS De Wild.

Plasmophagus coleochaetis (Sparrow *et al.*) M. W. Dick [(Sparrow, R. A. Paterson & Johns)]
Plasmophagus deformans (Serbinow) M. W. Dick
Plasmophagus oedogoniorum De Wild. **[type species]**

DICTYOMORPHA Mullins

Dictyomorpha dioica (Couch) Mullins **[type species]**
Dictyomorpha pythiensis (N. Sarkar & Dayal) M. W. Dick [*Index of Fungi* **6(3)**: 140; *loc. cit.*]

PLEOTRACHELUS Zopf

Pleotrachelus fulgens Zopf **[type species]**
Pleotrachelus itersoniliae (D. J. S. Barr & Bandoni) M. W. Dick
Pleotrachelus vuilleminii (G. Arnaud) M. W. Dick
Pleotrachelus zopfianus Morini

LAGENIDICOPSIS N. J. Artemczuk

Lagenidicopsis arctica N. J. Artemczuk **[type species, monotypic genus]**

PLASMOPAROPSIS De Wild.

Plasmoparopsis rigida De Wild. **[type species, monotypic genus]**

SYNCHAETOPHAGUS **Apstein**

Synchaetophagus balticus Apstein **[type species, monotypic genus]**

SORODIPLOPHRYS **Cienk.**

Sorodiplophrys stercorea (Cienk.) L. S. Olive & Dykstra

SPIROSPORA **Scherff.**

Spirospora paradoxa Scherff.

FAMILY *INCERTAE SEDIS*: *HEIMERLIACEAE*

CALOSPEIRA **G. Arnaud**

Calospeira elegans G. Arnaud

ENIGMA **G. Arnaud**

Enigma vorticelloides G. Arnaud

HEIMERLIA **Höhn.**

Heimerlia hyalina Höhn.

HELIOMYCOPSIS **G. Arnaud**

Heliomycopsis fusispora G. Arnaud

PYGMOMYCES **G. Arnaud**

Pygmomyces microsporus G. Arnaud

RHABDOCYSTIS **G. Arnaud**

Rhabdocystis columnaris G. Arnaud
Rhabdocystis peniophorae G. Arnaud

CHYTRIDIOMYCETES

MICROMYCES P. A. Dang.

Micromyces cristata (Scherff.) Sparrow var. ***cristata***
Micromyces cristata (Scherff.) Sparrow var. ***minor*** (Scherff.) Sparrow
Micromyces echinocystis Linder
Micromyces fischeri (Scherff.) Sparrow
Micromyces grandis C. E. Miller
Micromyces intermedia (Canter) Sparrow
Micromyces laevis Canter
Micromyces longispinosus Couch
Micromyces mesocarpi De Wild.
Micromyces minimus Linder
Micromyces mirabilis (Canter) Sparrow
Micromyces oedogonii (M. Roberts) Sparrow
Micromyces ovalis Rieth var. ***giganteus*** Sparrow & M. E. Barr
Micromyces ovalis Rieth var. ***ovalis***
Micromyces petersenii Scherff.
Micromyces spirogyrae Skvortzov
Micromyces wheldenii Linder
Micromyces zygnemicola (Cejp) Sparrow
Micromyces zygogonii P. A. Dang.

ENDOCHYTRIUM Sparrow

Endochytrium digitatum Karling
Endochytrium operculatum (De Wild.) Karling
Endochytrium pseudodistomum (Scherff.) Karling
Endochytrium ramosum Sparrow

ENDODESMIUM Canter

Endodesmidium formosum Canter

REESIA C. Fisch

Reessia amoeboides C. Fisch
Reessia lemnae (C. Fisch) Karling

FOSSIL FUNGI

[See caveat in Dick (1988).]

PALAEOPHTHORA **Singhai**

Palaeophthora mahgaonensis Singhai

PERONOSPORITES **W. G. Sm.**

Peronosporites W. G. Sm. [Carboniferous; Miocene]

PERONOSPOROIDES **E. W. Berry**

Peronosporoides E. W. Berry [Oligocene]

PROPYTHIUM **M. K. Elias**

Propythium carbonarium M. K. Elias [*Index of Fungi* 5: 260]

APPENDIX 2

RECOMMENDED FORM FOR TAXONOMIC AUTHORITIES

NOTES:

Some workers prefer not to use abbreviated forms for authorities: there are advantages and disadvantages to the application of the following entries. The principal merit for using the ***recommended abbreviated form*** of the authority name is that only one author can possibly be referred to, but this only applies within botanical circles. As soon as zoological authorities may be involved the system breaks down because there is no procedure for uniting authorities under the two codes (the *Zoological Code* and the *ICBN*). In the past, authorities have been abbreviated in a haphazard manner (often without forename initials) and when this has happened confusion between different authorities can arise, usually because authors are unaware of the possible confusion.

This list is compiled and augmented from Brummitt & Powell (1992) and Kirk & Ansell (1992) for authorities of names considered in this work. A few entries are given for mycologists who have made contributions to our knowledge of the straminipilous fungi; and for those authorities for other plant names, the authorities for which may be confused with those for straminipilous fungus names. References to mycologists in the *Dictionary of Fungi, 8th edition* are cited as 'Hawksw. *et al.* (1995)'; abbreviations for obituaries are as follows: *TBMS - Transactions of the British Mycological Society*; *Mycol. Res. - Mycological Research*; *Mycologia* - unabbreviated.

For historic reasons, the transliterations of Cyrillic characters in Brummitt & Powell (1992) are neither consistent nor always in accord with those of the *British Standard, No. 2979 (1958)*, as recommended for use by *Hart's Rules..38th Edition.* (Oxford University Press) or *Copy Editing* (Cambridge University Press). The *British Standard* transliterations given in the *References* are repeated here.

Information on authorities is often incomplete: it would be helpful if readers could send me information to improve the record on disk.

Symbols

[*] Authorities given in Kirk & Ansell (1992) but which are not listed in Brummitt & Powell (1992)

[ ] Authorities given in Brummitt & Powell (1992) but which are not listed in Kirk & Ansell (1992)

[...] discrepancies between the two lists are indicated within the abbreviation; the first given, or fuller version is used here.

Authorities for whom no standard form has yet been published are usually given in full in APPENDIX 1, even though they may have been presented in a 'recommended abbreviated form' in the *Index of Fungi*. A few such omissions are of early authors and for these cases the organism concerned is listed in the third column of the table. Likewise, names which have been presented in the zoological literature will not have a standard form in these reference works, but their authorities could be a source of confusion and so the genus or species is also listed in the third column. Authorities known only for binomials in *Peronospora* are also annotated in the third column.

Name and initials (as in publication)	*Standard abbreviation (Brummitt & Powell, 1992)*	*Biographical details: full names and Obituary dates where known (male or female - ♂, ♀) (further inf., reference)*
G. Abad	------	Gloria Z. Abad (see *Pythium plurisporium*)
G. Abbey	------	G. Abbey (see *Plasmodiophora tomati*)
T. M. Achundov	**T. M. Achundov**	T. M. Achundov (pub. 1956 - *Peronospora ibrahimovii*)
D. B. Adam	**D. B. Adam**	David Bonar Adam (pub. 1924 - *Peronospora tabacina* ≡ *P. hyoscyami*)
C. Ahrens	------	C. Ahrens (see *Pythium orthogonon*)
C. A. Agardh	**C. Agardh**	Carl Adolph Agardh (1785-1859)

J. G. Agardh	**J. [G.] Agardh**		Jakob Georg Agardh (1813-1901)
B. Aggéry	**Aggéry**		Anna Berthe Emma Aggéry (pub. 1892 - *Peronoplasmopara densa*)
L. Ajello	**Ajello**		Libero Ajello (1916-)
C. R. Alcock	**Alcock**		Charles Raymond Alcock (1921-)
J. L. Alcorn	**Alcorn**		John Leonard Alcorn (1937-)
D. J. Alderman	**Alderman**		David James Alderman (1943-)
A. A. Aleem	**Aleem**		Anwar Abdel Aleem
A. Alexander	**Alexander**		Arthur Alexander
C. J. Alexopoulos	**Alexop.**		Constantine John Alexopoulos (1907-1986) *Mycologia* **79**: 163-165 (1987); *TBMS* **90**: 153-158 (1988)
M. S. Ali-Shtayeh	**Ali-Shtayeh**		Mohammed Saleem Ali-Shtayeh (1948-)
K. S. Amin	**K. S. Amin**		K. S. Amin
C. J. Anastasiou	**Anastasiou**		C. J. Anastasiou
A. Andreucci	**Andreucci**		A. Andreucci
P. J. Ann	**Ann**		Pao-Jen Ann
S. A. Annalijev	**Annal.**		S. A. Annalijev (see *Peronospora cryptosporae*)
A. Apinis	**Apinis**		Arveds Eduards Apinis (1907-1979)
C. Apstein	**Apstein**		Carl Apstein (1862-1950)
A. M. Arambarri	**Aramb.**		Angélica M. Arambarri (1945-)
W. Archer	**W. Archer !**	[ ]	William Archer (1827-1897) (see *Achlya cornuta*, *Saprolegnia androgynus*)
W. A. Archer	**W. Archer**		William Andrew Archer (1894-1973)
G. Arnaud	**G. Arnaud**		Gabriel Arnaud (1882-1957)
N. Arnaudow	**Arnautov**		Nikola (Nicholas) N. Arnaudow [Н. АРНАУДОВЪ, N. Arnaudov'] (1887-1961)
N. Ya. Artem[t]chuk	**N. J. Artemczuk**		N. Ya. Artem[t]chuk [Н. Я. АРТЕМЧУК, N. Ya. Artemchuk]
L. Arvy	**Arvy**		Lucie Arvy
J. A. von Arx	**Arx**		Joseph Adolf von Arx (1922-1988) Hawksw. *et al.* (1995)
D. Atkins	**D. Atkins**		Daphne Atkins (1896-1961)
G. F. Atkinson	**G. F. Atk.**		George Francis Atkinson (1854-1918) Hawksw. *et al.* (1995)
A. Avidor	**Avidor**		A. Avidor (see *Tetramyxa marina*)
C. Azevedo Menezes	**Menezes**		Carlos Azevedo de Menezes (1863-1928)
C. Azevedo	------		C. Azevedo (see *Perkinsus atlanticus*)
P. C. Azevedo	**P. C. Azevedo**		P. C. Azevedo
K. D. Babacauh	**Babacauh**		K. D. Babacauh
V. S. Bachtin	**Bakhtin**		V. S. Bachtin [Bakhtin] (pub. 1925 - *Peronospora tranzcheliana*)
G. Bahnweg	**Bahnweg**		Günther Bahnweg
H. C. Bai [H.-c. Bai]	**H. C. Bai**		H. C. Bai (see *Peronospora menthae*)
R. Baker	------		Ralph Baker (see *Pythium nunn*)
S. D. Baker	**S. D. Baker**		Shirley D. Baker (see *Albugo quadrata*)
D. Bakhariev	**Bakhariev**		D. Bakhariev
M. S. Balakrishnan	**M. S. Balakr.**		M. S. Balakrishnan
B. Baldev	**Baldev**		B. Baldev
H. H. de Balsac	------		H. Heim de Balsac (see *Nucleophaga peranemae*)
R. Bandoni	**Bandoni**		Robert J. Bandoni
J. Barker	**J. Barker**	[ ]	J. Barker (1830-1907; see *Diplophrys archeri*)
J. Bärner	------		J. Bärner (see *Plasmodiophora solani*)
D. J. S. Barr	**D. J. S. Barr**		Donald J. S. Barr (1937-)
M. E. Barr	**M. E. Barr**		Margaret. E. Barr
R. W. Barreto	**R. W. Barreto**		Robert Weingart Barreto (1955-)
J. T. Barrett	**Barrett**		James Theophilus Barrett (1876-)
G. L. Barron	**G. L. Barron**		George L. Barron
J. Barthelet	**Barthelet**		Jean Jules Barthelet (pub. 1933 - *Peronospora dianthicola*)
A. F. Bartsch	**A. F. Bartsch**		Alfred Frank Bartsch (1913-)
A. de Bary	**de Bary**		Heinrich Anton de Bary (1831-1888) *Mycologia* **70**: 222-252 (1978)
M. C. L. Baticados	**[M. C.] Batic.**		Ma. Cecilia L. Baticados
A. Batko	------		Andrzej Batko (see *Nellymyces megaceros*, *Pythiogeton nigrescens*)
S. A. A. B. Batko	**Batko**		Stanislaw Albert Anthony Benedict Batko (1904-)
O. Beccari	**Becc.**		Odoardo Beccari (1843-1920; see *Peronospora effusa ciconia* - trinomial)
I. Bechet	**I. Bechet**		Ion Bechet
M. Bechet	**M. Bechet**		Maria Bechet (1928-)
G. Beck	**Beck**		Gunther von Mannagetta und Lerchenau Beck (1856-1931; see *Per. bulbocapni*)
E. R. Becker	**E. R. Becker**	[ ]	Elery Ronald Becker (1896-1962; see *Sphaerita entamoebae*)
W. N. Becker	**W. N. Becker**		W. N. Becker
J. Bejlin [I. T. Beilin]	**Bejlin**		J. Bejlin (I. T. Beilin) (pub. 1924 - *Peronospora scutellariae*)
A. Belisario	**A. Belisario**		A. Belisario (see *Phytophthora italica*)
M. Belsky	**Belsky**		M. Belsky
E. S. Beneke	**Beneke**		Everett S. Beneke
C. Benois			C. Benois (see *Peronospora canescens*; and var. *jacutica* of various *Peronospora* species)
C. B. Benua	------		Charles Benois Benua (1885-1943)
K. A. Benua	**Benua**	[ ]	K. A. Benua
P. Bergan	------		Per Bergan (see *Thalassomyces racemosus*)
M. J. Berkeley	**Berk.**		Miles Joseph Berkeley (1803-1889) Hawksw. *et al.* (1995)

APPENDIX 2: TAXONOMIC AUTHORITY ABBREVIATIONS

A. Berlese	**A. Berl.**		Antonio Berlese (1863-1927)	
A. N. Berlese	**Berl.**		Augusto Napoleone Berlese (1864-1903)	Hawksw. *et al.* (1995)
M. E. Beroqui de Martínez	**Beroqui**		Martha E. Beroqui de Martínez	
E. W. Berry	**E. W. Berry**	[ ]	E. W. Berry [fossil fungi]	
L. A. Berry	**L. A. Berry**		L. A. Berry	
C. E. Bessey	**Bessey**		Charles Edwin Bessey (1845-1915)	
E. A. Bessey	**E. A. Bessey**		Ernst Athearn Bessey (1877-1957)	*Mycologia* **50**: 1-5 (1958)
A. Van Beverwijk	**Beverw.**		Agathe L. Van Beverwijk (1907-1963)	*Mycologia* **56**: 641-644 (1964)
S. D. Bharadwaj	**Bharadwaj**		S. D. Bharadwaj (pub. 1977 - *Peronospora*)	
L. N. Bhardwaj	**L. N. Bhardwaj**		L. N. Bhardwaj (pub. 1987 - *Peronospora teucrii*)	
K. S. Bhargava	**Bhargava**		K. S. Bhargava	
B. Z. Bian	**Bian**		Bo Zhong Bian	
B. Biasoletto	**B. Biasol.**	[ ]	Bartolomeo Biasoletto (1793-1859; see *Leptomitus* spp.)	
M. L. B. Biga	**Biga**		M. L. Bestagno Biga	
W. Birchfield	**Birchf.**		Wray Birchfield	
H. Bishop	------		Harlow Bishop [data on *Sapromyces*]	
E. M. Blackwell	------		Elizabeth Marianne Blackwell (1889-1973) [data on *Phytophthora*]	*TBMS* **61**: 611 (19)
I. Blok	**I. Blok**		Ida Blok	
S. Blumer	**S. Blumer**		Samuel von Blumer (pub. 1926 - *Peronospora cortusae*; *P. galigena*; *P. tozziae*)	
A. G. Blytt	**A. Blytt**		Axel Gudbrand Blytt (1843-1898)	
M. N. Blytt	**Blytt**		Mathias Numsen Blytt (1789-1862)	
A. Bivona-Bernardi	**Biv.**		Antonius de Bivona-Bernardi (1774-1837)	
C. Blattny	------		C. Blattny (see *Spongospora solani radicicis*)	
G. H. Boerema	**Boerema**		Gerhard H. Boerema (pub. 1962 - *Peronospora parasitica* f. sp. *galligena*)	
A. Boltovskoy	**Boltovskoy**		Andrés Boltovskoy	
A. S. Bondartsev	**Bondartsev**		Appollinaris Semenovich Bondartsev [Bondarzew]	
H. F. Bonorden	**Bonord.**		Hermann Friedrich Bonorden (1801-1884; see *Peronospora obovata*)	
O. Borge	------		O. Borge (see '*Ducellieria tricuspidata*' [an alga])	
E. Bornet	**Bornet**		Jean-Baptiste Édouard Bornet (1828-1911)	
J. B. G. G. M. Bory	**Bory**		Jean Baptiste Georges Geneviève Marcellin Bory (1778-1846)	
A. Borzí	**Borzí**		Antonio Borzí (1852-1921; see *Rhizomyxa hypogeae*)	
H. Boschma	**Boschma**	[ ]	Hildebrand Boschma (1893-1976; see *Thalassomyces* spp.)	
S. M. Bower	------		Susan M. Bower (see *Labyrinthuloides haliotodis*)	
B. C. Bradford	------		B. C. Bradford (see *Perkinsus karlssonii*)	
T. Brandt	**T. Brandt**		T. Brandt	
C. M. Brasier	**Brasier**		Clive Michael Brasier (1942-)	
A. Braun	**A. Braun**		Alexander Carl Heinrich Braun (1805-1877)	
H. Braun	**Hans Braun**		Hans Braun (1895-)	
H. H. Braxton	**Braxton**		Herman Harrison Braxton	
L. A. de Brébisson	**Bréb.**		Louis Alphonse de Brébisson (1798-1872)	
J. van Breda de Haan	**Breda de Haan**		Jacob van Breda de Haan (1866-1917)	
O. Brefeld	**Bref.**		Julius Oscar Brefeld (1839-1925)	Hawksw. *et al.* (1995)
W. Brehmer	**Brehmer**	[ ]	Wilhelm Georg Baptist Alexander von Brehmer (1883- ; see *Plasmodiophora solani*)	
H. Bremer	**Bremer**		Hans Bremer	
U. Brizi	**Brizi**		Ugo Brizi (1868-1949)	
A. Brodsky	**Brodskii**	[ ]	Abram Lvovič Brodskii (1883- ; see *Sphaerita dangeardii*)	
C. E. Broome	**Broome**		Christopher Edmund Broome (1812-1886; see *Peronospora sordida*)	
S. L. Brug	**Brug**	[ ]	S. L. Brug (see *Nucleophaga hypertrophica*)	
J. Brunchorst	**Brunch.**		Jörgen Brunchorst (1862-1917)	
F. Bubák	**Bubák**		František Bubák (1865-1925; see *Peronospora saxifragae*)	
F. V. Bucholz	**Bucholz**		Feodor (Fedor) Vladimirovic Bucholz (1872-1924)	
I. W. Buddenhagen	**Buddenh.**		Ivan W. Buddenhagen	
C. J. Buisman	**Buisman**		Christine Johanna Buisman (1900-1936)	
I. A. Bunkina	**Bunkina**		I. A. Bunkina	
R. H. Bunting	**Bunting**		R. H. Bunting	
L. Burdyukova	**Burdjuk.**		Ludmila I. Burdyukova	
T. J. Burrill	**Burrill**		Thomas Jonathan Burrill (1839-1916; see *Peronospora cynoglossi*; *P. plantaginis*)	
E. J. Butler	**E. J. Butler**		Edwin John Butler (1874-1943)	Hawksw. *et al.* (1995)
W. J. Byford	**Byford**		W. J. Byford (pub. 1963 - *Peronospora farinosa* formae)	
J. Bywater	**Bywater**		Joan Bywater	
Z. M. Byzova	**Byzova**		Z. M. Byzova (pub. 1960 see *Peronospora carpoceratis*, *P. sideritidis*)	
S. O. Cacciola	**S. O. Cacciola**		S. O. Cacciola (see *Phytophthora italica*)	
E. De S. Da Câmara	**Sousa Da Câmara**		Manuel Emmanuele De Sousa Da Câmara (1871-1955)	
C. Camera	------		C. Camera (see *Plasmopara conii*, *P. coronillicola*)	
F. Cámera	**Cámera**		Fernando Cámera (1906-1985)	
D. H. Campbell	**Campb.**		Douglas Houghton Campbell (1859-1953)	
L. Campbell	**L. Campb.**		Leo Campbell (1894-)	
W. A. Campbell	**W. A. Campb.**		William Andrew Campbell (1906-)	
A. de Candolle	**A. DC.**		Alphonse Louis Pierre Pyramus de Candolle (1806-1893)	
A. de Candolle	**C. DC.**		Anne Casimir Pyramus de Candolle (1836-1918)	
A. de Candolle	**DC.**		Augustin Pyramus de Candolle (1778-1841)	

A. de Candolle	**Aug. DC.**		Richard Émil Augustin de Candolle (1868-1920)	
H. M. Canter	**Canter**		Hilda Mabel Canter (1922-)	
W. M. Carne	**Carne**		Walter Mervyn Carne (1885-1952)	
N. E. Caroselli	**Caros.**		Nestor E. Caroselli	
R. Caspary	**Casp.**		Johann Xaver Robert Caspary (1818-1887)	
M. Caullery	**Caullery**		Maurice Caullery (1868-1952)	
T. Cavalier-Smith	**Caval.-Sm.**	[ ]	Tom Cavalier-Smith (see higher taxa)	
R. J. Cawthorn	------		R. J. Cawthorn (see *Perkinsus karlssonii*)	
K. Cejp	**Cejp**		Karel Cejp (1900-1979)	
R. Cefalu	**Cefalu**		Rose C. Cefalu (1938-)	
V. Cesati	**Ces.**		Vincento de Cesati (1806-1883)	
M. Chadefaud	**Chadef.**		Marius Chadefaud (1900-)	
H.-S. Chang	**H. S. Chang**		Ho-Shii Chang (1938-)	
C. E. Chardón	**Chardón**		Carlos Eugenio Chardón [Palacios] (1897-1965)	
E. Chatton	**Chatton**		Édouard Chatton (1883-1947)	
P. B. Chavan	**P. B. Chavan**		P. B. Chavan	
K. H. Chee	**Chee**		K. H. Chee	
C. C. Chen	**C. C. Chen**		Chi Chang Chen (1903-)	
G. M. Cheney	**G. M. Cheney**		G. M. Cheney	
X. Y. Cheng	**X. Y. Cheng**		X. Y. Cheng (pub. 1985 - *Peronospora menthae*)	
C. G. C. Chesters	**Chesters**		Charles Geddes Coull Chesters (1904-1993)	*Mycol. Res.* **97**: 1023-1024 (1993)
P. K. Chi	**P. K. Chi**		Pei Kun Chi	
S. Chinabut	**S. Chinabut**		Supranee Chinabut (1947- ; see *Aphanomyces invadans*)	
T.-S. Chiou	**T.-S. Chiou**	[ ]?	Tin-Son Chiou (see *Pythiopsis subcentrica*, *P. terrestris*)	
L. M. Churchland	**Churchl.**		L. M. Churchland	
L. Cienkowski	**Cienk.**		Leo de Cienkowski (1822-1887)	
R. Cifferi	**Cif.**		Raffaele Cifferi (1897-1964)	*Mycologia* **57**: 198-201 (1965)
K. J. Clany	**Clancy**		K. J. Clancy	
G. P. Clinton	**G. P. Clinton**		George Perkins Clinton (1867-1937; see *Peronoplasmopara cubensis*)	
G. Cocconi	**Cocc.**		Girolamo Cocconi (1822-1904)	
F. Cohn	**Cohn**		Ferdinand Julius Cohn (1828-1898)	
W. C. Coker	**Coker**		William Chambers Coker (1872-1953)	*Mycologia* **46**: 372-383 (1954)
L. C. Coleman	**L. C. Coleman**		Leslie Charles Coleman (1887-)	
S. Collard	**Collard**		Sneed B. Collard	
A. Collier	**Collier**	[ ]	Albert Walker Collier (1910- ; see *Perkinsus marinus*)	
W. B. Collins	**W. B. Collins**		W. B. Collins	
J. T. Conover	**Conover**		John T. Conover (1918-)	
J. C. Constantineanu	**Const.**		Ion (Joan, Ioan) C. Constantineanu (1860-1931)	
O. Constantinescu	**Constant.**		Ovidiu Constantinescu (1933-)	
M. T. Cook	**M. T. Cook**		Melville Thurston Cook (1869-1952)	
M. C. Cooke	**Cooke**		Mordecai Cubitt Cooke (1824-1914)	Hawksw. *et al.* (1995)
I. Cookson	**Cookson**		Isabel Clifton Cookson (1893-1973)	
H. F. Copeland	**H. F. Copel.**		Herbert Faulkner Copeland (1902-1968)	
A. C. I. Corda	**Corda**		August Karl Joseph Corda (1809-1872)	
J. O. Corliss	**Corliss**	[ ]	John O. Corliss (1922- ; see higher taxa)	
M. Cornu	**Cornu**		Marie Maxime Cornu (1843-1901)	
C. F. J. E. Correns	**Correns**		Carl Franz Joseph Erich Correns (1864-1933)	
J. N. Couch	**Couch**		John Nathaniel Couch (1896-1986)	*Mycologia* **81**: 181-189 (1989)
H. Coutière	**Coutière**		H. Coutière	
B. A. Cox	**B. A. Cox**		B. A. Cox	
B. S. Crandall	**Crand.**		Bowen Sinclair Crandall (1909-)	
B. J. Croft	**B. J. Croft**		Barry J. Croft	
D. H. Crump	**D. H. Crump**		D. H. Crump	
A. da Cuhna	**A. M. Cuhna**	[ ]	Aristides Marques da Cuhna (see *Sphaerita minor*)	
M. A. Curtis	**M. A. Curtis**		Moses Ashley Curtis (1808-1872)	
M. Curzi	**Curzi**		Mario Curzi (1898-1944)	
V. M. Cutter	**Cutter**		Victor Macomber Cutter (1917-1962)	*Mycologia* **54**: 457-459 (1962)
K. Damle	**Damle**		Kamal Damle	
V. P. Damle	**V. P. Damle**	[ ]	Vasudeo Purushottam Damle (see *Albugo evolvulus*)	
S. R. S. Dange	**Dange**		S. R. S. Dange	
P.-A. Dangeard	**P. A. Dang.**		Pierre-Augustin Clement Dangeard (1862-1947)	Hawksw. *et al.* (1995)
S. N. Dasgupta	**S. N. Dasgupta**		S. N. Dasgupta	
J. F. Dastur	**Dastur**		Jehangir Fardunji Dastur (1886-1952)	
K. G. Date	**Date**		K. G. Date (see *Achlya kamatii*)	
J. G. N. Davidson	**J. G. N. Davidson**		John G. N. Davidson	
G. H. G. Davis	------		G. H. G. Davis (see *Perkinsus olsenii*)	
H. S. Davis	**H. S. Davis**	[ ]	Herbert Spencer Davis (1875- ; see *Ostracoblabe salmonis*)	
J. J. Davis	**Davis**		John Jefferson Davis (1852-1937)	
W. H. Davis	**W. H. Davis**		William Harold Davis (1876-1948)	
R. Dayal	**Dayal**		Ram Dayal (1931-)	
J. W. Deacon	**Deacon**		James W. Deacon	
C. de Bruyne	**de Bruyne**		C. de Bruyne	

APPENDIX 2: TAXONOMIC AUTHORITY ABBREVIATIONS

A. W. A. M. de Cock	**De Cock**		Arthur W. A. M. de Cock (1950-)	
C. von Deckenbach	**Deckenb.**	[ ]	Const. von Deckenbach (see *Deckenbachia consuens*)	
K. N. von Deckenbach	**K. N. Deckenb.**		K. N. von Deckenbach	
N. G. Dejeva	**Dejeva**		N. G. Dejeva (pub. 1967 - *Peronospora melissiti*)	
E. De La Rue	------		Eugen De La Rue (see *Pythium fimbriatum*)	
P. N. D'Eliscu	**D'Eliscu**		Peter N. D'Eliscu	
B. Della Torre	**Della Torre**		B. Della Torre	
K. B. Deshpande	**K. B. Deshp.**		K. B. Deshpande	
K. S. Deshpande	**K. S. Deshp.**		K. S. Deshpande	
J. B. H. J. Desmazières	**Desm.**		Jean Baptiste Henri Joseph Desmazières (1786-1862)	
G. B. De Toni	**De Toni**		Giovanni Batista De Toni (1864-1924)	
H. F. Dias	------		H. F. Dias (see *Olpidium cucurbitacearum*)	
M. W. Dick	**M. W. Dick**		Michael William Dick (1934-)	
P. Dietel	**Dietel**		Paul Dietel (1860-1947; see *Peronospora phlogina*)	
L. W. Dillwyn	**Dillwyn**		Lewis Weston Dillwyn (1778-1855)	
E. G. Dimitrova	**E. G. Dimitrova**		Evtimija G. Dimitrova (pub. 1980 - *Peronospora parietariae*)	
E. Dissmann	**Dissmann**		E. Dissmann	
F. Doflein	**Doflein**	[ ]	Franz Doflein (1873-1925; see *Labyrinthulidae*)	
I. J. Dogma	**Dogma**		Irineo J. Dogma	
A. A. Domashova	**Domashova**		A. A. Domashova (pub. 1957 - see *Aphanomyces volgaensis*)	
A. Domján	**Domján**		Anna Domján	
G. Drebes	**Drebes**		Gerhard Drebes	
C. Drechsler	**Drechsler**		Charles Frank Drechsler (1892-1986)	*Mycologia* **79**: 345-352 (1987)
R. L. Duan	**R. L. Duan**	[ ]	R. L. Duan (see *Pythium boreale*)	
O. Duboscq	**Duboscq**		Octave Joseph Duboscq (1868-)	
J. E. Duby	**Duby**		Jean Étienne Duby (1798-1885)	
F. Ducellier	**F. Ducell.**	[ ]	Francois Ducellier (see *Ducellieria chodatii*)	
I. Dudka	**Dudka**		Irena A. Dudka (1934-)	
S. J. du Plessis	**du Plessis**		S. J du Plessis (pub. 1933 - *Peronospora oxalidis*)	
G. Durriieu	**Durriieu**		Guy Durriieu (1931- ; see *Peronospora erinicola*)	
R. Dwivedi	------		R. Dwivedi (see *Synchytrium malvacearum*)	
R. S. Dwivedi	**R. S. Dwivedi**		R. S. Dwivedi	
M. J. Dykstra	**Dykstra**		Michael J. Dykstra	
A. Dżanuzakov	**Dzhanuz.**		A. Dżanuzakov [Dżhanuzokov] (pub. 1964 - *Peronospora brassicae* formae)	
H. A. Edson	**Edson**		Howard Austin Edson (1875-1960)	
T. Egami	**Egami**		T. Egami	
S. Egusa	**Egusa**		Syuzo Egusa	
C. Ehrenberg	**Ehrenb.**		Christian Gottfried Ehrenberg (1795-1876)	
P. F. Elenev	**Elenev**		P. F. Elenev (pub. 1923 - *Peronospora fagopyri* = *Peronospora ducometii*)	
M. K. Elias	------		M. K. Elias [fossil genus *Propythium*]	
J. A. Elliott	**J. A. Elliott**		John Asbury Elliott (1887-1923)	
R. F. Elliott	**R. F. Elliott**		Ruth F. Elliott	
J. B. Ellis	**Ellis**		Job Bicknell Ellis (1829-1905)	
B. Ellison	**Ellison**		B. R. Ellison	
R. Emerson	**R. Emers.**		Ralph Emerson (1912-1979)	*Mycologia* **72**: 857-867 (1980)
M. Enjumet	**Enjumet**	[ ]	Monique Enjumet (see *Labyrinthula algeriensis*)	
H. Epstein	------		H. Epstein (see *Nucleophaga hypertrophica*)	
G. W. Erdos	**Erdos**		Gregory W. Erdos	
D. Ershad	**Ershad**		Djafar Ershad (1936-)	
D. C. Erwin	**Erwin**		Donald C. Erwin	
E. Esfandiari	**Esfand.**		Esfandiar von Esfandiari (1909- ; see *Peronospora iranica*)	
B. M. Everhart	**Everh.**		Benjamin Matlack Everhart (1818-1904)	
L. Fage	**Fage**	[ ]	Louis Fage (1883-1964; see *Thalassomyces* spp)	
W. G. Farlow	**Farl.**		William Gilson Farlow	
F. K. Faust	**F. K. Faust**		Frances K. Faust	
G. Feldmann	**Feldm.-Maz.**		Geneviève Feldmann-Mazoyer	
J. Feldmann	**Feldm.**		Jean Feldmann (1905-1978)	
J. W. Fell	**Fell**		Jack W. Fell	
C. Ferdinansen	**Ferd.**		Carl Christian Frederic Ferdinansen (1879-1944)	
C. Fisch	**C. Fisch**		Carl Fisch (1879-)	
A. Fischer	**A. Fisch.**		Alfred Fischer (1858-1913)	
A. A. Fischer von Waldhem	**A. A. Fisch. Waldh.**		Alexandr Alexandrovich Fischer von Waldheim (1839-1920)	
H. M. Fitzpatrick	**Fitzp.**		Harry Morton Fitzpatrick (1886-1949)	*Mycologia* **43**: 249-266 (1951)
C. Flahault	**Flahault**		Charles Henri Marie Flahault (1852-1935)	
I. Foissner	**I. Foissner**		Ilse Foissner	
W. Foissner	**W. Foissner**		Wilhelm Foissner	
C. E. Foister	**Foister**		Charles Edward Foister	
E. J. Forbes	**E. J. Forbes**		Evelyn J. Forbes	
B. Fott	**Fott**	[ ]	Bohuslav Fott (1908-1976) [algae]	
A. Franc	**A. Franc**		A. Franc	
S. P. Frances	**Frances**		Stephen P. Frances	
I. Friedmann	**Friedmann**		Emerich Imre Friedmann (1921-)	

E. M. Fries	**Fr.**		Elias Magnus Fries (1794-1878; see *Peronospora parasitica*)
Fries	**Fr. f.**		Fries (see *Peronospora farinosa*)
F. E. Fritsch	**F. E. Fritsch**	[ ]	Felix Eugen Fritsch (1864-1934; see *Resticularia boodlei*)
K. Fritsch	**Fritsch**		Karl Fritsch (1832-1893)
G. Fron	**Fron**		Georges Fron (1870-1957)
Z.-s. Fu	**Z. S. Fu**	[*]	Z.-s. Fu
L. Fuckel	**Fuckel**		Karl Wilhelm Gottlieb Leopold Fuckel (1821- 1876; see *Peronospora* spp.)
J. S. Furtado	**J. S. Furtado**		João Salvador Furtado
A. Gaertner	**A. Gaertn.**		Alwin Gaertner
D. F. Gaillat	------		D. F. Gaillat (see *Ligniera pilorum*)
J. Galindo[-A.]	**Galindo**		Jorge Galindo-A.
N. I. Gaponenko	------ [Gapon.]		N. I. Gaponenko (see *Plasmopara archangelicae*)
E. A. Gaümann	**Gaüm.**		Ernest [Ernst] Albert Gaümann (1893-1963)
			[Hawksw. *et al.* (1995); *TBMS* **47**: 459 (19); *Mycologia* **57**: 1-5 (1965)]
J. W. Gerdemann	**Gerd.**		James W. Gerdemann (1921-)
E. Germain	**Germ.**		Jacques Nicolas Ernest Germain de Saint-Pierre (1815-1882)
L. Gerrettson-Cornell	**Gerr.-Corn.**		L. Gerrettson-Cornell
A. Giard	**Giard**		Alfred-Matheiu Giard (1846-1908)
J. Gicklhorn	**Gickelh. [Gicklh.]**		Joseph Gicklhorn (1891-1957)
R. L. Gilbertson	**Gilb.**		Robert Lee Gilbertson (1925-)
R. H. Gilpin	**Gilpin**		R. H. Gilpin
S. L. Glockling	**Glockling**	[*]	Sally Lorna Glockling (1960- ; see *Myzocytiopsis* spp. and *Haptoglossa* spp.)
J. F. Gmelin	**J. F. Gmel.**		Johann Friedrich Gmelin (1748-1804)
C. Gobi	**Gobi**		Christophos [Khristofor] Jacoblewitsch Yakovlevich Gobi (1847-1919)
			[Хр. ГОБИ, Khr. Gobi]
K. Goebel	**K. I. Goebel**	[ ]	Karl Immanuel Eberhard Goebel (1855-1932; see *Tetramyxa parasitica*)
E. K. Goldie-Smith	**Goldie-Sm.**		Elizabeth Kathleen Goldie-Smith (1922-1995)
			[Disney, H. (1999); *FBA News, No.* **7**: 5]
S. Goldstein	**S. Goldst.**		Solomon Goldstein (1929-)
A. Golenia	**Golenia**		A. Golenia
			(pub. 1962 - *Peronospora chenopodii-ambrosioidis* = *P. chenopodii-polyspermae*)
P. N. Golvin	**Golvin**		P. N. Golvin
N. P. Golovina	**N. P. Golovina**		N. P. Golovina
W. P. Golovina	**W. P. Golovina**		W. P. Golovina
O. G. Golubeva	**Golubeva**		Olga G. Golubeva [О. Г. ГОЛУБЕВА, O. G. Golubeva]
R. Gönnert	**Gönnert**	[ ]	R. Gönnert (see *Lagenidium trichophryarum*)
R. Gonzáles Fragoso	**Gonz. Frag.**		Romualdo Gonzáles Fragoso (1862-1928; see *Peronospora senneniana*)
M. V. Gorlenko	**Gorlenko**	[ ]	Michail Vladimirovich Gorlenko (1908- ; see *Plasmopara chinensis*)
H. H. Gran	**Gran**		Haaken Hasberg Gran (1870-1955)
L. F. Grand	**Grand**		Larry F. Grand
F. A. Grant	**F. A. Grant**		Freeman Augustus Grant (1898-)
S. F. Gray	**Gray**		Samuel Frederick Gray (1766-1828)
F. C. Greenhalgh	------ [F. C. Greenh.]		F. C. Greenhalgh (see *Phytophthora clandestina*)
G. N. Greenhalgh	**Greenh.**		Geoffry N. Greenhalgh
R. K. Greville	**Grev.**		Robert Kaye Greville (1794-1866; see *Botrytis effusa* = *Peronospora effusa*)
J. S. Grewal	**Grewal**		J. S. Grewal
M. J. Griffin	**M. J. Griffin**		Michael James Griffin
D. Griffiths	**Griffiths**		David Griffiths (1867-1935)
F. Gruithuisen	**Gruith.**		Franz von Paula Gruithuisen (1774-1852)
S. C. Gupta	**S. C. Gupta**		Satish Chandra Gupta (1937-)
A. Gustavsson	**A. Gustavsson**		Arne Gustavsson (pub. 1959 -
			Peronospora cardaminopsis; *P. ervi*; *P. lathyri-verni*; *P. polygoni-convolvuli*)
E. Guyénot	------		E. Guyénot (see *Dermocystidium ranae*)
J. Haan	------		J. Haan (see *Pythium destruens*)
E. H. Haeckel	**Haeckel**	[ ]	Ernst Heinrich Philipp August Haeckel (1834-1919; see higher taxa)
D. J. Hagedorn	------		Donald J. Hagedorn (see *Aphanomyces euteiches* formae)
B. D. Halsted	**Halst.**		Byron David Halsted (1852-1918)
C. J. J. van Hall	**C. J. J. Hall**		Constant Johann Jakob van Hall (1875-)
I. C. Hallett [sp]	**I. C. Hallett**		Ian Charles Hallett (1951-)
P. Hamm	**Hamm**		Phil B. Hamm
E. M. Hansen	**E. M. Hansen**		Everett M. Hansen (1946-)
E. S. Hansen	**E. S. Hansen**		Eric Steen Hansen (1943-)
C. G. Hansford	**Hansf.**		Clifford Gerald Hansford (1900-1966)
L. M. Hartge	**Hartge**		Lena Medora Hartge
K. Hara	**Hara**		Kanesuke Hara (1885-1962)
M. M. Hartog	------		Marcus Manuel Hartog (1851-1924; see *Saprolegnia corcagiensis*)
J. V. Harvey	**J. V. Harv.**		James Vernon Harvey (1901-)
R. H. Haskins	**Haskins**		Reginald Hinton Haskins
K. Hatai	**Hatai**		Kishio Hatai
R. Heim	**R. Heim**		Roger [Jean] Heim (1900-1979)
			Hawksw. *et al.* (1995); *Mycologia* **72**: 1063-1064 (1980); *TBMS* **76**: 177-179 (1981)
F. F. Hendrix	**F. F. Hendrix**		Floyd Fuller Hendrix

APPENDIX 2: TAXONOMIC AUTHORITY ABBREVIATIONS

P. Hennings	**Henn.**		Paul Christoph Hennings (1841-1908)	
W. G. F. Herter	**Herter**		Wilhelm [Guillermo] Gustav[o] Franz [Francis] Herter (1884-1958)	
R. Hesse	**R. Hesse**		Rudolph [Rudolf] Hesse (1844-1912)	
C. J. Hickman	**Hickman**		Clarence James Hickman (1914-1980)	*TBMS* **79**: 187-188 (1982)
F. Hildebrand	**Hildebr.**		Friedrich Hermann Gustav Hildebrand (1835-1915)	
A. A. Hildebrand	**A. A. Hildebr.**		Alexander Anderson Hildebrand (1896-)	
S. Hirane	**Hirane**		Seiichi Hirane	
H. H. Ho	**H. H. Ho**		Hon Hing Ho	
J. Hoenigman	------		Janez Hoenigman (see *Thalassomyces nouvelii*)	
G. R. Hoerner	**Hoerner**		Godfrey Richard Hoerner (1893- ; see *Pseudoperonospora cannabina* [superfluous synonym])	
G. F. Hoffmann	**Hoffm.**		George Franz Hoffmann (1761-1826) (see *Bremia*)	
H. Hoffmann	**H. Hoffm.**		Heinrich Karl [Carl] Hermann Hoffmann (1819-1891)	
H. R. Hohl	**H. R. Hohl**		Hans Rudolph Hohl	
F. von Höhnel	**Höhn.**		Franz Xavier Rudolf von Höhnel (1852-1920)	Hawksw. *et al.* (1995)
W. Höhnk	**Höhnk**		Johann Willy Georg von Höhnk (1904-)	
A. Hollande	**A. Hollande**	[ ]	André Hollande [algae[see *Labyrinthula algeriensis* and *Nucleophaga peranemae*)	
A. Hollande	**Hollande**		A.-Ch. Hollande [fungi]	
E. W. D. Holway	**Holw.**		Edward Willet Dorland Holway (1853-1923; see *Peronospora phlogina*)	
Y. Homma	**Homma**		Yasu Homma (1892-)	
D. Honda	**D. Honda**		Daiski Honda (see *Schizochytrium limacinum*)	
L. J. Hoogkamer	------		L. J. Hoogkamer (see *Pythium destruens*)	
H. H. Hotson	**H. H. Hotson**		Hugh Howison Hotson (1916-)	
R. Hovasse	**Hovasse**		Raymond Hovasse (1895-)	
J. Hruby	**Hruby**		Johan Hruby (1882-1964; see *Peronospora alyssi-arduini* = *P. galligena*)	
S. Y. Hsieh	**S. Y. Hsieh**		S. Y. Hsieh	
C. C. Huang	**C. C. Huang**	[ ]	Cheng Chiu Huang	
C. S. Huang	------		C. S. Huang (see *Lagenidium culicium*)	
A. Hulea	**Hulea**		Ana Hulea (1915- ; see *Peronospora daturae* = *P. chenopodii*)	
R. A. Humber	**Humber**		Richard A. Humber (1947-)	
J. E. Humphrey	**Humphrey**		James Ellis Humphrey (1861-1897)	
T. Humphrey	------		T. Humphrey (see *Pythium afertile*)	
M. B. Huneycutt	**Huneycutt**		Maeburn Bruce Huneycutt	
T. Ichitani	**Ichit.**		Takio Ichitani	
A. Ideta	**Ideta**		Arata Ideta (1870-)	
H. Indoh	**Indoh**		Hiroharu Indoh	
C. T. Ingold	**Ingold**		Cecil Terence Ingold (1905-)	
D. S. Ingram	------		David S. Ingram (see *Bremia lactucae* formae)	
J. A. G. Irwin	**J. A. G. Irwin**	[*]	J. A. G. Irwin (see *Phytophthora macrochlamydospora*)	
S. Ito	**S. Ito**		Seiya Ito (1883-1962)	
T. Ito	**T. Itô**		Tokutarô Itô (1868-1941)	
W. R. Ivimey Cook	**Ivimey Cook**		Walter Robert Ivimey Cook (1902-1952)	
M. O. P. Iyengar	**M. O. P. Iyengar**		Mandeyam Osuri Parthasarathy Iyengar (1886-1963)	
O. Jaap	**Jaap**		Otto Jaap (1864-1922; see *Peronospora gaeumanniana*)	
A. A. Jaczewski	**Jacz.**		Arthur Louis Arthurovič Jaczewski (1863-1932; see A. A. Yačeveskie)	
P. A. Jaczewski	**P. A. Jacz.**		P. A. Jaczewski (see P. A. Yačeveskie)	
T. L. Jahn	**T. L. Jahn**	[ ]	Theodore Louis Jahn (1905-1979; see *Sphaerita phaci*)	
A. C. Jain	**A. C. Jain**		A. C. Jain	
K. Jankowska	**Jank.**		Krystyna Jankowska (pub. 1929 - *Peronospora ducometii*)	
T. V. Jenkina	**Jenkina**		T. V. Jenkina (pub. 1966 - *Peronospora salicorniae*)	
S. I. Jermalaviczjute	**Jermal.**		S. I. Jermalaviczjute	
Z. D. Jiang	**Z. D. Jiang**		Zi De Jiang	
O. Jírovec	**Jírovec**		Otto Jírovec	
H. Johannes	**Johannes**		Heinrich Johannes	
A. E. Johanson	**A. E. Johanson**		Alfhild E. Johanson (see *Karlingia rosea*)	
C. J. Johanson	**Johanson**		Carl Johan Johanson (1858-1888)	
R. John	**R. John**		Rachel John	
R. M. Johns	**Johns**		Robert Marvin Johns (1928-1963)	*Mycologia* **56**: 309 (1964)
T. W. Johnson	**T. W. Johnson**		Terry Walter Johnson (1923-)	
M. Jokl	**Jokl**		Milla Jokl	
E. B. G. Jones	**E. B. G. Jones**		Evan Benjamin Gareth Jones (1937-)	
F. R. Jones	**F. R. Jones**		Fred Reuel Jones (1884-)	
P. M. Jones	**P. M. Jones**		Philip Malory Jones (1892-)	
W. Jones	**W. Jones**		Walter Jones	
S. C. Jong	**S. C. Jong**		Shung-Chang Jong (1936-)	
I. Jørstad	**Jørst.**		Ivar Jørstad (1887-1967)	*Mycologia* **63**: 697-700 (1971)
H. O. Juel	**Juel**		Hans Oscar Juel (1863-1931)	
T. Jung	T. Jung		T. Jung (see *Phytophthora quercina*)	
K. Kalchbrenner	**Kalchbr.**		Károly [Karl] Kalchbrenner (1807-1886)	
B. K. Kalymbetov	**Kalymb.**		B. K. Kalymbetov (pub. 1952 - *Peronospora chamaesphaci*)	
J. E. Kane	------		Jasmine E. Kane (see *Thalassomyces marsupii*)	
W. D. Kane	------		Wilma D. Kane (see *Leptolegniella exoospora*)	

B. B. Kanouse	**Kanouse**	Bessie Bernice Kanouse (1889-1969)	
J. S. Karling	**Karling**	John Sidney Karling (1898-)	
E. I. Karpova-Benua	**Karp.-Benua**	E. I. Karpova-Benua (pub. 1973 - *Peronospora* spp.)	
K. Katsura	**Katsura**	Kiichi Katsura	
C. H. Kauffman	**Kauffman**	Calvin Henry Kauffman (1869-1931)	*Mycologia* **24**: 265-267 (1932)
L. Kaufman	**Kaufman**	L. Kaufman	
M. J. Kaufmann	**Kaufm.**	M. J. Kaufmann (see *Phytophthora sojae*)	
J. A. Kavanagh	**Kavanagh**	James Aloysius Kavanagh (1937-)	
Y. Kawase	**Kawase**	Y. Kawase	
K. Keissler	**Keissl.**	Karl [Carl] von Keissler (1872-1965; see *Peronospora farinosa* = *P. chenopodii*)	
W. A. Kellermann	**Kellerm.**	William Ashbrook Kellermann (1850-1908; see *Peronospora swinglei*)	
J. B. Kendrick	**J. B. Kendr.**	James Blair Kendrick (1893-)	
W. B. Kendrick	**W. B. Kendr.**	William Bryce Kendrick	
D. M. Kennedy	------	Diana M. Kennedy (see *Phytophthora idaei*)	
R. G. Kenneth	**R. G. Kenneth**	Robert G. Kenneth	
B. R. Kerry	**Kerry**	Brian R. Kerry	
B. Keskin	**Keskin**	B. Keskin	
A. G. Kevorkian	**Kevorkian**	Arthur George Kevorkian (1905-)	
R. D. Khulbe	**Khulbe**	R. D. Khulbe	
U. Kiran	**Kiran**	Usha Kiran	
T. W. Kirk	------	T. W. Kirk (see *Plasmodiophora humuli*)	
N. Kitancharoen	------	Nilubol Kitancharoen (see *Halodaphnea awabi*)	
H. Klebahn	**Kleb.**	Heinrich Klebahn (1859-1942)	
H. Kniep	**Kniep**	Karl Johannes Hans Kniep (1881-1930)	
L. Kny	**Kny**	Carl Ignaz Leopold Kny (1841-1916)	
W.-H. Ko	**W. H. Ko**	Wen-Hsiung Ko	
Y. Kobayas[[h]i	**Kobayasi**	Yosio Kobayas[h]i	
J. Kochman	**Kochman**	Józef Kochman	
T. Kodama	**K. Kodama**	K. Kodama (pub. 1962)	
T. Kodama	**Tak. Kodama**	Takashi Kodama (pub. 1986)	
L. M. Koloschina	**L. M. Kolos[c]hina**	L. M. Kolos[c]hina (see *Peronospora iolotanica* = *P. minor*)	
Y. Koltin	**Koltin**	Y. Koltin	
V. L. Komarov	**Kom.**	Vladimir Leont[j]evich Komarov (1869-1945)	
K. Konno	**K. Konno**	Kaoru Konno	
H. Kouyeas	**H. Kouyeas**	Hebe Kouyeas	
V. Kouyeas	**V. Kouyeas**	V. Kouyeas	
C. A. Kranz	**Kranz**	Cajetan Anton Kranz (1839-1886)	
J. Kranz	**J. Kranz**	Jürgen von Kranz	
W. Krieger	**Willi Krieg.**	Willi Krieger (1886-1954; see *Peronospora lamii* var. *glechomatis* = *P. glechomatis*)	
H. Kröber	**Kröber**	Heinz Kröber (see *Phytophthora tentacula*)	
R. Kruschewa	**R. Krusch.**	R. Kruschewa (pub. 1959 - *Peronospora jordanovii*)	
T.-L. Kuan	------	Ta-Li Kuan (see *Phytophthora megasperma* formae)	
Q. B. Kubicek	**Kubicek**	Q. B. Kubicek	
J. Kühn	**J. G. Kühn**	Julius Gotthelf Kühn (1825-1910)	Hawksw. *et al.* (1995)
R. D. Kulbe	**R. D. Kulbe**	R. D. Kulbe (see *Saprolegnia bhargavae*)	
G. S. Kulkarni	**Kulk.**	Gopal Subrao Kulkarni (1884-)	
U. V. Kulkarni	**U. V. Kulk.**	U. V. Kulkarni	
C. E. O. Kuntze	**Kuntze**	Carl (Karl) Ernst (Eduard) Otto Kuntze (1843-1907)	
J. Kunze	**J. Kunze**	Johannes Kunze (?-1881)	
G. Kunze	**Kunze**	Gustav Kunze (1793-1851)	
A. Kusano	------	A. Kusano (see *Olpidium viciae*)	
S. Kusano	**S. Kusano**	Shunsuke Kusano (1874-1962)	
F. T. Kützing	**Kütz.**	Friedrich Traugott Kützing (1807-1893)	
R. C. Lacy	**Lacy**	R. C. Lacy	
C. M. V. D. S. Lacoste	------	C. M. V. D. S. Lacoste (see *Chytridium minus*)	
H. A. Lafferty	**Laff.**	Henry Aloysius Lafferty (1891-1954)	
[N.] G. Lagerheim	**Lagerh.**	Nils Gustaf von de Lagerheim (1860-1926)	
F. Laibach	**Laib.**	Fredrich Laibach (1885-)	
S. A. K. Laing	**S. A. K. Laing**	S. A. K. Laing	
E. M. R. Lamkey	**Lamkey**	Ernest Michael Rudolph Lamkey (1890-)	
D. M. Langsam	**Langsam**	Deborah M. Langsam	
D. H. Latham	**Latham**	Dennis Harold Latham	
R. Laubert	**Laubert**	Karl Richard Laubert (1870- ; see *Peronospora spinaciae* = *P. effusa*)	
O.-a. Lawhavinit	**Lawhav.**	Ong-ard Lawhavinit	
L. A. Lebedeva	**Lebedeva**	Lydia Alexsandrovna Lebedeva (see *Peronospora* spp.)	
H. Lebert	**Lebert**	Hermann Lebert (1813-1878)	
M. Lechtova-Trinka	**Lecht.-Trnka**	M. Lechtova-Trinka	
G. A. Ledingham	**Ledingham**	George Aleck Ledingham (1903-1962)	*Mycologia* **55**: 365-370 (1963)
L. Léger	**L. Léger**	L. Léger	
S. G. Lehman	**Lehman**	Samuel George Lehman (1887- ; see *Peronospora sojae* = *P. manshurica*)	
H. Leitgeb	**Leitg.**	Hubert Leitgeb (1835-1888)	
J. Leitner	**J. Leitn.**	Jane Leitner	

Name	Abbreviation		Details	Reference
L. H. Leonian	**Leonian**		Leon Hatchig Leonian (1888-1945)	
R. J. G. Lester	------		R. J. G. Lester (see *Perkinsus olsenii*)	
L. S. Leu	------		Lii S. Leu (see *Peronophythora litchi*)	
J. H. Léveillé	**Lév.**		Joseph-Henri Léveillé (1796-1870; see *Cystopus candida*; *Botrytis violacea*)	
N. D. Levine	**Levine**	[ ]	Norman Dion Levine (1912- ; see *Perkinsus marinus*)	
T. I. Li	**T. I. Li**		T. I. Li (see *Peronospora capsici*)	
M.-A. Libert	**Lib.**		Marie-Anne Libert (1782-1865)	
R. Lifshitz	**Lifsh.**		R. Lifshitz	
D. Lihnell	**Lihnell**		D. Lihnell	
F. A. Lilienfeld	------		F. A. Lilienfeld (see *Pythium haplomitri*)	
D. Linder	**Linder**		David Hunt Linder (1889-1946)	*Mycologia* **39**: 113-144 (1947)
J. Lindley	**Lindl.**		John Lindley (1799-1865)	
J. C. Lindquist	**J. C. Lindq.**		Juan Carlos Lindquist (1899- ; see *Peronospora convolvuli*)	
K. Lindstedt	**Lindst.**		Karl Lindstedt (1846-?)	
V. Lindtner	**Lindtner**		Vojtech H. Lindtner (1904-1965; see *Peronospora drabae-majusculae* = *P. erophilae*)	
L. Ling	**L. Ling**		Lee Ling (1911-)	
H. F. Link	**Link**		Johann Heinrich Friedrich Link (1767-1851; see *Caeoma candidum* ≡ *Albugo candida*)	
Y. Lipkin	**Lipkin**		Yaakov Lipkin (see *Tetramyxa marina*)	
P. E. Lipps	**P. E. Lipps**		Patrick E. Lipps	
T.-r. Liu	**T. R. Liu**		T.-r. Liu (pub. 1985 - *Peronospora elsholtziae*)	
A. I. Lobik	**Lobik**		Alexis Iulianovich Lobik (pub. 1928 - *Peronospora galeopsidis*; *Peronospora glauci*)	
G. Lohde	**Lohde**	[ ]	Georg Lohde (see *Lucidium pythioides*)	
E. López Ochoterena	**E. L. Ocht.**		Eucario López Ochoterena	
J. P. Lotsy	**Lotsy**		Johannes Paulus Lotsy (1867-1931)	
J.-y. Lu	------		Jai-yun Lu (see *Phytophthora fragariae* var. *oryzyobladis*)	
L. T. Lucas	**L. T. Lucas**		Leon T. Lucas	
A. Lund	**A. Lund**		Aage Lund	
S. Lundell	**S. Lundell**		Seth Lundell (1892-1966: see *Peronospora* sp.)	
A. Lwoff	**A. Lwoff**	[ ]	André Lwoff (1902- ; see *Sphaerita normeti*)	
D. McAlpine	**McAlpine**		Daniel McAlpine (1849-1932; see *Peronospora lepidii*; *P. parasitica* var. *lepidii*)	
R. MacDonald	------		R. MacDonald (see *Thalassomyces fagei*)	
I. Macfarlane [MacFarlane]	**I. MacFarl.**		Ian Macfarlane	
S. E. McGladdery	------		S. E. McGladdery (see *Perkinsus karlssonii*)	
J. G. Mackin	**Mackin**		J. G. Mackin	
D. A. McLarty	------		D. A. McLarty (see *Olpidiopsis achlyae*)	
H. R. McLarty	**McLarty**		H. R. McLarty	
W. McRae	**McRae**		William McRae (1878-1952)	
M. Madrazo-Garibay	**Madr.-Gar.**		Marcela Madrazo-Garibay	
G. Magnano di San Lio	**G. Magnano di San Lio**		G. Magnano di San Lio (see *Phytophthora italica*)	
P. Magnus	**Magnus**		Paul Wilhelm Magnus (1844-1914)	
T. S. Mahabalé	**Mahab.**		Tryambak Shankur Mahabalé (1909-1983)	
R. Maire	**Maire**		René Charles Joseph Ernest Maire (1878-1949)	Hawksw. *et al.* (1995)
T. Majewski	**T. Majewski**		Tomasz Majewski (1940-)	
L. A. Mangin	**L. Mangin**		Louis Alexandre Mangin (1852-1937)	
G. W. Martin	**G. W. Martin**		George Willard Martin (1886-1971)	*Mycologia* **65**:(5) (1973)
W. W. Martin	**W. W. Martin**		W. Wallace Martin	
C. F. P. Martius	**Mart.**		Carl Friedrich Phil. Sigm. Martius (1829-1899)	
C. F. P. Martius	**C. Mart.**		Carl Friedrich Philipp von Martius (1794-1868)	
R. Marwitz	------		R. Marwitz (see *Phytophthora tentacula*)	
C. Massalongo	**C. Massal**		Caro Benigno Massalongo (1852-1928; see *Peronospora* sp.)	
G. Massee	**Massee**		George Edward Massee (1850-1917)	Hawksw. *et al.* (1995)
Massenot	------		Massenot (see *Peronospora centranthi*)	
I. M. Master	**Master**		Isadore M. Master (1945-1995)	
M. J. Masters	**M. J. Mast.**		Margaret J. Masters (see *Lagenidium* sp.)	
S. B. Mathur	**S. B. Mathur**		S. B. Mathur (see *Peronospora hypericifoliae*)	
A. Matta	**A. Matta**		Alberto Matta	
O. Mattes	------		O. Mattes (see *Sphaerita nucleophaga*)	
V. D. Matthews	**V. D. Matthews**		Velma Dare Matthews (1904-1958)	
J. Matz	**Matz**		Julius Matz (1886-)	
E. Maupas	------		E. Maupas (see *Myzocytiopsis (Protascus) subuliformis*)	
A. Maurizio	**Maurizio**		Adam M. Maurizio (1862-1941)	
D. P. Maxwell	------		D. P. Maxwell (see *Phytophthora medicaginis*)	
E. Mayor	**Mayor**		Eugéne Mayor (1877-1976)	
L. de Mendoza	**L. Mend.**		L. de Mendoza	
Y. R. Meng	**Y. R. Meng**		You Ru Meng (pub. 1989 - *Peronospora microulae*)	
G. S. Mer	**G. S. Mer**		G. S. Mer	
A. Meurs	**Meurs**		Abraham Meurs (1904-)	
J. T. Middleton	**Middleton**		John Tylor Middleton (1912-)	
A. I. Milanez	**A. I. Milanez**		Adauto Ivo Milanez (1937-)	
J. A. Milbrath	**Milbrath**		J. A. Milbrath	
A. A. Mil'ko	**Milko**		A. A. Mil'ko [А. А. МИЛЪКО, A. A. Mil'ko]	

C. E. Miller	**C. E. Mill.**		Charles Edward Miller (1925-1984)	*Mycologia* **77**: 341-344 (1985)
M. O. Milovtzova	**Milovtz.**	[*]	M. O. Milovtzova [М. О. МІЛОВЦОВА, M. O. Milovtzova]	
M. von Minden	**Minden**		Max D. von Minden (1871-)	
J. K. Misra	**J. K. Misra**		J. K. Misra	
J. E. Mitchell	**J. E. Mitch.**		John E. Mitchell	
K. Miura	**K. Miura**		Koichiro Miura	
M. Miura	**Miura**		Michiya Miura	
K. Miyabe	**Miyabe**		Kingo Miyabe (1860-1951)	
T. Miyabe	**T. Miyabe**		Tsutome Miyabe (1880-1921)	
T. Miyake	**T. Miyake**		Tsutomu Miyake (1880-) [same person as above ?]	
H. Moeller	------		Henry Moeller (see *Plasmodiophora alni*)	
H. Molisch	**Molisch**	[*]	Hans Molisch (1856-1937; see *Pseudoplasmodium aurantiacum*)	
E. M. Möller	------		E. M. Möller (see *Phytophthora infestans* forma *mirabilis*)	
M. Molliard	**Molliard**		Marin Molliard (1866-1944)	
C. Montagne	**Mont.**		Jean Pierre Francois Camille Montagne (1784-1866)	Hawksw. *et al.* (1995)
F. Morini	**Morini**		Fausto Morini (1858-1927)	
A. Moritzi	**Moritzi**		Alexandre Moritzi (1807-1850)	
L. K. Muelstein	**Muelst.**	[*]	Lisa K. Muelstein	
T. M. Muhsin	------		Tawfik Muhsin (see *Calyptralegnia basraensis*)	
E. Müller	**E. Müll.**		Emil von Müller (1920- ; see *Peronospora heliophilae*)	
J. T. Mullins	**Mullins**		J. Thomas Mullins	
H. L. Munasinghe	**Munas.**		H. L. Munasinghe	
B. B. Mundkur	**Mundk.**		Bhalchendra Bhavanishankar Mundkur (1896-1952; see *Peronospora gaeumannii* = *P. indica* Gaum.)	
J. Muniz	**Muniz**		Julio Muniz	
R. L. Munjal	**Munjal**		R. L. Munjal	
D. Murayama	**Muray.**	[*]	D. Murayama	
A. A. Murgoci	------		A. Antoniu Murgoci (pub. 1947 - *Saprolegnia exigua*)	
K. Nabel	**Nabel**		Kurt Nabel	
M. Nagai	**Nagai**		Masaji Nagai (1905-1966)	
A. Nakagiri	**Nakagiri**		Akira Nakagiri (1957-)	
K. Nakamura	------		Kazuyo Nakamura (see *Halodaphnea parasitica*)	
M. Nakamura	------		Miho Nakamura (see *Salilagenidium thermophilum*)	
JA. Nannfeldt	**Nannf.**		JohnAxel (Johan) Frithiof Nannfeldt (1904-1985)	*TBMS* **87**:347-352 (1986)
M. J. S. Narasimhan	**Naras.**		M. J. S. Narasimhan	
N. Naoumoff [see N. A. Naumov]				
N. A. Naumov [N. Naoumoff]	**Naumov**		Nikolai Alexandrovich Naumov (1888-1959; see *Bremia graminicola*)	
N. I. Naumov	------		N. I. Naumov [Н. И. НАУМОВ, N. I. Naumov] [data on Russian oomycetes]	
N. N. Naumova	**Naumova**		N. N. Naumova	
A. Naville	**Naville**		A. Naville (see *Dermocystidium ranae*)	
C. G. Nees von Esenbeck	**Nees**		Christian Gottfried Daniel Nees von Esenbeck (1776-1858)	Hawksw. *et al.* (1995)
G. Negrean	**Negrean**		Gavril A. Negrean (1932-)	
A. Negru	**Negru**		Alexander Negru (1914-)	
E. S. Nelen	**Nelen**		E. S. Nelen	
B. Němec	**Němec**		Bohumil Řehoř Němec (1873-1966)	
S. Y. Newell	**S. Y. Newell**		Stephen Y. Newell (1945-)	
G. Nicholas	------		G. Nicholas (see *Pythium marchantiae*)	
G. Niessl	**Niessl**		Gustav Niessl von Mayendorf (1839-1919; see *Peronospora androsaces* 223)	
J. A. Nieuwland	**Nieuwl.**		Julius (Aloysius) Arthur Nieuwland (1878-1936)	
L. Niezabitowski	**Niezab.**		Edward Lubicz Niezabitowski	
M. I. Nikolajeva [Nicolajaeva, Nikolaeva]	**M. I. Nikol.**		M. I. Nikolajeva (see *Peronospora lathyri-pisiformis* - pub. 1970)	
H. Nouvel	**Nouvel**	[ ]	H. Nouvel (see *Thalassomyces boschmai*)	
N. S. Novotel'nova	**Novot.**		N. S. Novotel'nova	
L. Nowakowski	**Nowak.**		Leon Nowakowski (1847-)	
W. Nylander	**Nyl.**		William Nylander (1822-1899)	Hawksw. *et al.* (1995)
C. V. Oescu	**Oescu**		C. V. Oescu (1891-1973; see *Peronospora atriplicis-tataricae*, *P. nitens* see *P. minor*; *P. glechomae* = *P. glechomatis*)	
A. Ogoshi	**Ogoshi**		A. Ogoshi	
A. Ohuchi	------		A. Ohuchi (see *Pythium nelumbii*)	
L. S. Olive	**L. S. Olive**		Lindsay Shepherd Olive (1917-1988)	*Mycologia* **81**: 497-503 (1989)
A. L. B. d' Oliviera	**Oliviera**		Anto'nio Lopes Branquino d'Oliviera (pub. 1940 - *Peronospora coronillicola*)	
F. Onuma	**Onuma**		Fusaji Onuma	
M. Ookubo [M. Ôkubo]	**M. Ôkubo**		Mariko Ookubo (M. Ôkubo)	
L. L. Osipian	**Osipian**		Lia Levonevna Osipian (1930- ; see *Peronospora lathyri-rosei*)	
C. H. Ostenfeld	**Ostenf.**		Carl Emil Hansen Ostenfeld (1873-1931)	
G. Otth	**G. H. Otth**		Gustav Heinrich Otth (1806-1874)	
S. H. Ou	**S. H. Ou**		Shu-Huang Ou	
H. M. Owen	------		H. Malcolm Owen (see *Perkinsus marinus*)	
A. A. Padhye	**A. A. Padhye**		Arvind A. Padhye	
C.-k. Pai	**C. K. Pai**		Chin-kai Pai	
J. Pal	**J. Pal**		J. Pal	

K. P. Pal	**K. P. Pal**	K. P. Pal	
M. Pal	**M. Pal**	Mehendra Pal	
B. T. Palm	**Palm**	Björn Torvald Palm (1887-1956)	
R. Parisi	**Parisi**	Rosa Parisi	
D. Park	**D. Park**	David Park	
A. Pascher	**Pascher**	Adolf (Adolph) A. Pascher (1881-1945)	
I. G. Pascoe	**Pascoe**	Ian G. Pascoe	
G. Passerini	**Pass.**	Giovanni Passerini (1816-1893)	
R. A. Paterson	**R. A. Paterson**	Robert Andrew Paterson (1926-)	
M. K. Patel	**M. K. Patel**	M. K. Patel (pub. 1948 - *Bremia graminicola* var. *indica*)	
M. S. Patil	**M. S. Patil**	M. S. Patil (see *Peronospora canscorina*, *P. satarensis*)	
S. D. Patil	**S. D. Patil**	S. D. Patil	
N. Patouillard	**Pat.**	Narcisse-Théophile Patouillard (1854-1926)	Hawksw. *et al.* (1995)
P. M. Patterson	**P. Patt.**	Paul Morrison Patterson (1902-)	
B. Paul	**B. Paul**	Bernard Paul	
Y. S. Paul	**Y. S. Paul**	Y. S. Paul (pub. 1983 - *Peronospora teucrii*)	
M. S. Pavgi	**Pavgi**	M. S. Pavgi	
M. M. Payak	**Payak**	M. M. Payak	
C. H. Peck	**Peck**	Charles Horton Peck (1833-1917)	Hawksw. *et al.* (199); *Mycologia* **11**: 33-39 (1919)
K. G. Pegg	**Pegg**	K. G. Pegg	
V. Peglion	**Peglion**	Vittorio Peglion (1873- ; see *Peronoplasmopara cannabina*)	
J. J. Pemberton	------	John de Jarnette Pemberton (1887- ; see *Achlya hypogyna*)	
W. R. Pendergrass	**Pend.**	William R. Pendergrass	
O. Penzig	**Penz.**	Albert Julius Otto [Albertus Giulio Ottone] Penzig (1856-1929; see *Gilletia* ≡ *Basidiophora*; *G. spinuligera*)	
J. G. Percy	**Percy**	J. G. Percy	
C. Pérez	**Pérez**	C. Pérez	
R. Pérez-Reyes [Pérez Reyes]	**Pérez Reyes**	Rodolfo Pérez-Reyes	
F. O. Perkins	**F. O. Perkins**	Frank O. Perkins	
E. Perroncito	**Perronc.**	Edourdo Perroncito	
I. Persiel	**Persiel**	Ingetraud Persiel	
C. H. Persoon	**Pers.**	Christiaan Hendrik Persoon (1761-1836)	Hawksw. *et al.* (1995)
H. E. Petersen	**H. E. Petersen**	Henning Eiler Petersen (1877-1946)	
G. H. Pethybridge	**Pethybr.**	George Herbert Pethybridge (1871-1948)	
F. Petrak	**Petr.**	Franz Petrak (1886-1973)	
L. Petri	**Petri**	Lionello Petri (1875-1946)	
W. F. Pfender	------	W. F. Pfender (see *Aphanomyces euteiches* formae)	
E. Pfitzer	**Pfitzer**	Ernst Hugo Heinrich Pfitzer (1846-1906)	
A. J. Pieters	**Pieters**	Adrian John Pieters (1866-1940)	
A. J. van der Plaats-Niterink	**Plaäts-Nit.**	Annie J. van der Plaats-Niterink	
G. L. Po [G. D. Lio-Po ?]	**Po**	Gilda L. Po	
J. Poelt	**Poelt**	Josef Poelt (1924- ; see *Peronospora heliophilae*)	
G. Poirault	**G. Poirault**	(Marie Henri) Georges Poirault (1858-1936 *Peronospora apiospora*)	
A. W. Poitras	**Poitras**	Adrian W. Poitras	
D. Porter	**D. Porter**	David Porter	
R. O. Poyton	**Poyton**	R. O. Poyton	
S. K. Prabhuji	------	S. K. Prabhuji (see *Brevilegnia indica*)	
R. G. Pratt	**R. G. Pratt**	Robert G. Pratt	
G. Preti	**Preti**	G. Preti	
N. Pringsheim	**Pringsh.**	Nathanael Pringsheim (1823-1894)	
G. A. Prowse	**Prowse**	Gerald Albert Prowse (1916-1977)	
C. M. Pueschel	**Pueschel**	Curt M. Pueschel (1950-)	
G. S. Purss	**Purss**	G. S. Purss	
K. A. Pystina	------	[К. А. ПЫСТИНА, K. A. Pystina] [data on Russian peronosporomycetes]	
Y. Qin	**Y. Qin**	Y. Qin	
J. A. Quick	**J. A. Quick**	J. A. Quick	
G. L. Rabenhorst	**Rabenh.**	Gottlob [Gottlieb] Ludwig Rabenhorst (1806-1881)	Hawksw. *et al.* (1995)
M. Raciborski	**Racib.**	Marjan [Maryan, Marian] Raciborski (1863-1917)	Hawksw. *et al.* (1995)
I. M. Rădulescu	**Rădul.**	Ion M. Rădulescu (pub. 1932 - *Peronospora atriplicis-tataricae*, *P. nitens* see *P. minor*; *P. glechomae* = *P. glechomatis*)	
C. S. Rafinesque	**Raf.**	Constantine Samuel Rafinesque (1783-1840)	
S. Raghu Kumar	**Raghuk.**	Seshagiri Raghu Kumar	
J. N. Rai	**J. N. Rai**	J. N. Rai (see *Achlya stellata* var. *multispora*)	
S. Rajgopalan	------	S. Rajgopalan (see *Pythium lobatum* and other spp.)	
K. Ramakrishnan	**K. Ramakr.**	K. Ramakrishnan	
J. Ramsbottom	**Ramsb.**	John Ramsbottom (1885-1974)	*TBMS* **65**: 1-6 (1975)
R. D. Rands	**Rands**	Robert Delafield Rands (1890-1970)	
A. N. S. Rao	**A. N. S. Rao**	A. N. S. Rao	
J. R. Raper	------	John R. Raper (1912-1974; see *Achlya ambisexualis*)	
K. B. Raper	**Raper**	Kenneth Bryan Raper (1908-1987)	*Mycologia* **80**: 599-606(1988)
R. S. Rathore	**Rathore**	R. S. Rathore	

C. Raunkiaer	**Raunk.**		Christen Christiansen Raunkiaer (1860-1938; see *Peronospora stigmaticola*)	
H. W. Ravenel	**Ravenel**		Henry William Ravenel (1814-1887; see *Peronospora*)	
S. M. Ray	**S. M. Ray**		S. M. Ray	
T. Rayss	**Rayss**		Tscharna Rayss (1890-1965)	
E. A. von Regel	**Regel**		Eduard August von Regel (1815-1892)	
P. F. Reinsch	**Reinsch**		Paul Friedrich Reinsch (1836-1914)	
H. S. Reischer	**Reischer**		Helen Simpson Reischer [later Helen Simpson Vishniac]	
J. Renny	**Renny**		J. Renny	
P. Richter	**P. B. Richt.**	[ ]	Paul Boguslav Richter (1853-1911) [fossils]	
P. Richter	**P. G. Richt.**	[ ]	Paul Gerhard Richter (1837-1913; [algae]; see *Rhizophidium oedogonii*)	
W. Richter	-----		W. Richter (see *Saprolegnia richteri*)	
A. Rieth	**Rieth**		Alfred Rieth	
P. Rietschel	**Rietschel**		P. Rietschel	
M. Roberts	**M. Roberts**	[ ]	Margaret Roberts (see *Micromyces oedogonii*)	
R. J. Roberts	**R. J. Roberts**		Ronald John Roberts (1941- ; see *Aphanomyces invadans*)	
G. I. Robertson	**G. I. Robertson**		Graham I. Robertson	
A. L. Rogers	**A. L. Rogers**		Alvin Lee Rogers	
H. Roivainen	**Roiv.**		Heikki Roivainen (1900-1983; see *Peronospora ranunculi-peduncularis*)	
E. F. Roldan	**Roldan**		Emiliano F. Roldan	
L. Rolf	-----		L. Rolf (see *Labyrinthula jeremarina*)	
J. Rosenbaum	**Rosenbaum**		Joseph Rosenbaum (1887-1925)	
J. Rossmann	**Rossmann**		(George Wilhelm) Julius Rossmann (1831-1866; see *Peronospora affinis*)	
C. Rossy-Valderrama	**Rossy-Vald.**		Carmen Rossy-Valderrama	
S. I. Rostovzev[w]	**Rostovzev**		Semen Ivanovich Rostovzev[w]	
E. Rostrup	**Rostr.**		(Frederik Georg) Emil Rostrup (1831-1907; see *Peronospora gentianiae*)	
A. W. Roth	**Roth**		Albrecht Wilhelm Roth (1757-1834)	
H. F. A. de Roussel	**Roussel**		Henri François Anne de Roussel (1747-1812; see *Albugo candida*)	
E. Roze	**Roze**		Ernest Roze (1833-1900)	
F. Rudolphi	**F. Rudolphi**		Friedrich Karl Ludwig Rudolphi (1801-1849)	
W. Rühberg	------		Werner Rühberg (see *Dermocystidium daphniae*)	
R. Sabo	**Sabo**		R. Sabo	
D. Saccardo	**D. Sacc.**		Domenico Saccardo (1872-1952)	
F. Saccardo	**F. Sacc.**		Francesco Saccardo (1869-1896)	
P. A. Saccardo	**Sacc.**		Pier Andrea Saccardo (1845-1920)	Hawksw. *et al.* (1995)
R. Sadebeck	**Sadeb.**		Richard Emil Benjamin Sadebeck (1839-1905)	
K. M. Safeeulla	**Safeeulla**		Kunigal M. Safeeulla (1924-)	
B. C. L. Sah	**Sah**		B. C. L. Sah	
S. Sahtiyanci	**Sahtiy.**		Sehinaz Sahtiyanci	
E. Salmon	**E. S. Salmon**		Ernest Stanley Salmon (1871-1959)	
K. Sampson	**Sampson**		Kathleen Sampson (1892-1980)	*TBMS* 75: 353-354 (1980)
C. Sandu-Ville	**Sandu**		Constantin Sandu-Ville (1897-1969; *Peronospora savulescui*, see *P. erophilae*)	
A. Sarbhoy	**A. Sarbhoy**		Asha Sarbhoy	
A. K. Sarbhoy	**A. K. Sarbhoy**		Ashok Kumar Sarbhoy (1939-)	
J. A. Sarejanni	**Sarej.**		Jean Antoine Sarejanni (1898-1962)	
N. Sarkar	**N. Sarkar**		Nandita Sarkar	
S. C. Sati	**Sati**		S. C. Sati	
C. Sauvageau	**Sauv.**		Camille François Sauvageau (1861-1936)	
D. B. O. Savile	**Savile**		Douglas Barton Osborne Savile (1909-)	
Z. D. Savinceva [Savintseva]	**Savinceva**		Z. D. Savinceva	
A. Săvulescu	**A. Săvul.**		Alice Aronescu Săvulescu (1905-1970)	
O. Săvulescu	**O. Săvul.**		Olga-Constanța Săvulescu (1914-1969)	
T. Săvulescu	**Săvul.**		Traian [Trajan] Săvulescu (1889-1963)	
K. Sawada	**Sawada**		Kaneyoshi (Kenkichi) Sawada (1888-1950)	
J. Schaaschmidt	**Schaaschm.**		Julius Schaaschmidt [later G. Istvanfy] (1860-1930)	
A. Schenk	**Schenk**		Joseph August Schenk (1815-1891)	
A. Schenk	**A. Schenk**	[ ]	Alexander Schenk (1864-1924) [algae]	
A. Scherffel	**Scherff.**		Aladár Scherffel (1865-1939)	
F. Schikora	**Schikora**		Friedrich Schikora (1859-1932)	
K. Schilbersky	**Schilb.**		Károly Schilbersky (1863-1935)	
D. F. L. Schlechtendal	**Schltdl. [Schlecht.]**		Diederich Franz Leonhard von Schlechtendal (1794-1866)	
H. Schmoller	**Schmoller**		H. Schmoller	
J. Schneider	------		Joachim Schneider (see *Thraustochytrium striatum*)	
W. G. Schneider	**W. G. Schneid.**		Wilhelm Gottlieb Schneider (1814-1889)	
E. Scholz	**E. Scholz**		Eduard Scholz (1860-)	
E. Scholz	------		E. Scholz (see *Thraustochytrium pachydermum*)	
C. J. Schröter	**Schröt.**	[ ]	Carl Joseph Schröter (1855-1939)	
J. Schröter [Schroeter]	**J. Schröt.**		Joseph Schröter [Schroeter] (1837-1894)	
E. J. Schwartz	**Schwartz**		Ernest Justus Schwartz (1869-1939)	
L. D. Schweiniz	**Schwein.**		Lewis [Ludwig] David von Schweinitz (1780-1834)	
W. W. Scott	**W. W. Scott**		William Wallace Scott (1920-)	
V. Seethalakshmi	**Seethal.**		V. Seethalakshmi	
I. L. Serbinov	**Serbinow**		Ivan L'rovich Serbinow [И. Л. СЕРБИНОВ, I. L. Serbinov] (1872-1950)	

APPENDIX 2: TAXONOMIC AUTHORITY ABBREVIATIONS

K. S. Sergejeva	**Sergeeva**		Klavdia Spiridonovna Sergejeva [Sergeeva] (1909- ; see *Peronospora fabae*; *P. nonneae*)	
W. A. Setchell	**Setch.**		William Albert Setchell (1864-1943)	
R. L. Seymour	**R. L. Seym.**		Roland Lee Seymour (1939-)	
L. Shanor	**Shanor**		Leland Shanor (1914-1993)	*Mycologia* **87**: 745-748 (1995)
A. D. Sharma	**A. D. Sharma**		A. D. Sharma	
C. G. Shaw	**C. G. Shaw**		Charles Gardner Shaw (1917-1998)	*Mycologia* **90**: 927-929 (1998)
S.-C. Shen	------		S.-C. Shen (see *Rhizophidium parasiticum*)	
H. D. Shew	------		H. David Shew (see *Pythium plurisporium*)	
T. B. Shilova	------		T. B. Shilova [Т. Б. ЙНЛОВА, T. B. Shilova] (general taxonomic data)	
W. A. Shipton	**Shipton**		Warren A. Shipton (1940-)	
M. Shirai	**Shirai**		Mitsutarô ('Kotaro') Shirai (1863-1932)	
W. N. Siang	-----		W. N. Siang (see *Rhizophidium parasiticum*)	
C. P. Sideris	**Sideris**		Christos Plutarchos Sideris (1891-1965)	
A. Sigot	**Sigot**		André Sigot	
H. Siiono	**Siiono**		H. Siiono	
S. A. Simonyan [Simonian]	**Simonyan**		Seda A. Simonyan (1929-)	
J. Simpson	**J. A. Simpson**		Jack A. Simpson (1941-)	
S. D. Singh	**S. D. Singh**		S. D. Singh	
S. L. Singh	**S. L. Singh**		S. L. Singh (see *Aphanomyces brassicae*)	
U. P. Singh	**U. P. Singh**		Udai Pratup Singh (1945-)	
L. C. Singhai	**Singhai**		L. C. Singhai (see *Palaeophthora mohgaonensis* - fossil fungus)	
R. K. Sinha	**R. K. Sinha**		R. K. Sinha	
S. Sinha	**S. Sinha**		S. Sinha (see *Peronospora hypericifoliae*)	
S. K. Sinha	------		S. K. Sinha (recent - see *Brevilegnia indica*)	
B. S. Siradhana	**Siradhana**		B. S. Siradhana	
V. Skalický	**Skalický**		Vladimír Skalický (1930-1994)	
D. I. Skidmore	------		D. I. Skidmore (see *Bremia lactucae* formae)	
W. Siemaszko	**Siemaszko**		Wincenty [Vincent] Siemaszko (1887-1943 *Peronospora ducometii* 98)	
B. W. Skvortzow	**Skvortzov**		Boris Vassilievich Skvortzow (1890-1980)	
M. K. Slifkin	------		Miriam Kresses Slifkin [data on *Olpidiopsis*]	
E. H. Smith	**E. H. Sm.**		Elizabeth Hight Smith (1877-1933)	
R. E. Smith	**R. E. Sm.**		Ralph Elliott (Eliot) Smith (1874-1953)	
W. G. Smith	**Wm. G. Smith**		William Gardner Smith (1866-1928; see *Peronospora exigua*)	
W. G. Smith	**W. G. Sm.**		Worthington George Smith (1835-1917) [fossil fungi]	
H. Solheim	**H. Solheim**		Halvor Solheim	
W. G. Solheim	**Solheim**		Wilhelm Gerhard Solheim (1898-1978)	
H. Sommerstorff	**Sommerst.**		Hermann Sommerstorff (1889-1913)	
N. Sorokin	**Sorokīn**		Nicolai Vasilevitch Sorokīn [Н. СОРОКИН, N. Sorokin] (1846-1909)	
G. R. South	**South**		Graham Robin South (1940-)	
F. K. Sparrow	**Sparrow**		Frederick Kroeber Sparrow (1903-1977)	*Mycologia* **70**: 213-221 (1978)
C. L. Spegazzini	**Speg.**		Carlo Luigi [Carlos Luis] Spegazzini (1858-1926)	
N. N. Speschnew	**Speschnew**		Nicolai Nicolaievich von Speschnew (1844-1907)	
M. C. Srinivasan	**M. C. Sriniv.**		M. C. Srinivasan	
G. C. Srivastava	**G. C. Srivast.**		G. C. Srivastava	
O. N. Srivastava	**O. N. Srivast.**		O. N. Srivastava	
R. C. Srivastava	**R. C. Srivastava**		R. C. Srivastava (see *Aphanomyces pisci*)	
M. E. Stanghellini	**Stangh.**		Michael E. Stanghellini	
N. P. Stenina	**Stenina**		N. P. Stenina (pub. 1968 - *Peronospora lilii* = *P. destructor*)	
F. L. Stevens	**F. Stevens**		Frank Lincoln Stevens (1872-1934)	*Mycologia* **27**: 1-5 (1935)
E. Strasburger	**Strasb.**		Edouard Adolf Strasburger (1844-1912)	
F. C. J. Strauss	**F. Strauss**		Friedrich Karl [Carl] Joseph von Strauss (1787-1855; see *Uredo cubica* = *Cystopus cubicus* = *Albugo tragopogonis*)	
H. J. Su	**H. J. Su**		Horng Jye Su	
L. S. Subramaniam	**Subraman.**		Lekshminarayananpuram Subramania Subramaniam (1890-)	
C. V. Subramanian	**Subram.**		Chirayathumadom Venkatachalier Subramanian (1924-)	
J. N. Suhr	**Suhr**	[ ]	Johannes Nicolous von Suhr (1792-1847; see *Saprolegnia dichotoma*)	
B. M. Sundaram	----- -		B. M. Sundaram (see *Cornumyces catenophlyctidis*)	
N. V. Sundaram	**Sundaram**		Nangavalli Venkatesan Sundaram (1924-)	
W. F. R. Suringar	**Suringar**	[ ]	Willem Frederik Reinier Suringar (1832-1898; see *Chytridium minus*)	
A. P. Swan	------		Allan P. Swan (see *Leptolegnia bandonensis*)	
A. W. Sweeney	**A. W. Sweeney**		A. W. Sweeney	
W. T. Swingle	**Swingle**		Walter Tennyson Swingle (1871-1952)	
H. Sydow	**Syd.**		Hans Sydow (1879-1946)	
P. Sydow	**P. Syd.**		Paul Sydow (1851-1925)	
E. Szijarto	**Szijarto**		E. Szijarto	
R. J. Tabor	**Tabor**		Richard John Tabor (1875-1958)	
F. L. Tai	**F. L. Tai**		Fung [Fang] Lan [L.-y.] Tai (1893-1973)	
M. C. Tai	**M. C. Tai**		Ming Cheih Tai	
J. F. Tao	**J. F. Tao**		Jia Feng Tao (pub. 1926 - *Peronospora capsici*)	
M. Takahashi	**M. Takah.**		M. Takahashi	
K. Takase	-----		K. Takase (see *Tetramyxa elaeagni*)	

S. Takase	**Takase** [*]	S. Takase
S. Takimoto	**S. Takim.**	Seito Takimoto
D.-z. Tang	**D. Z. Tang**	D.-z. Tang (pub. 1984 - *Pseudoperonospora elsholtziae*)
J.-F. [J. F.; J.-f.] Tao	**J. F. Tao**	Jia-Feng Tao (1926-)
H. Tasugi	**Tasugi**	H. Tasugi
P. A. Taylor	**P. A. Taylor**	P. A. Taylor
E. Teiling	**Teiling**	Einar (Johan Sigurd) Teiling (1888-1974)
A. R. Teixeira	**Teixeira**	Alcides Ribiero Teixeira (1918-)
M. Terui	**Terui**	Mutsuo Terui (pub. 1930 - *Peronospora clinopodii*)
J. Thakur	**J. Thakur**	Thakur Ji
R. Thaxter	**Thaxt.**	Roland Thaxter (1858-1932) *Mycologia* **25**: 69-89 (1933)
I. Theohari	**Theoh.**	I. Theohari
J. Therry	**Therry**	Jean Joseph Therry (1833-1888 *Gilletia spinuligera* = *Basidiophora*)
M. J. Thirumalachar	**Thirum.**	Mandayam Jeersannidhi Thirumalachar (1914-)
A. N. Thite	**Thite**	A. N. Thite (pub. 1970 - *Peronospora canscorina*)
A. Thompson	**A. W. Thomps.**	Aldworth William Thompson (1900-1982)
F. Thümen	**Thüm.**	Felix [Karl Albert Ernst Joachim] von Thümen[-Gräfendorf] (1839-1892; see *Peronospora effusa* var. *polygoni*; *P. polygoni*)
G. A. Thuret [394]	**Thur.**	Gustave Adolphe Thuret (1817-1875)
M. Tiesenhausen	**Tiesenh.**	Manfred Tiesenhausen, Baron von Bozen (1875-)
A. Tison	**A. Tison**	Adrien Tison
K. Togashi	**Togashi**	Kogo Togashi
Y. Tokunaga	**Tokun.**	Yosio Tokunaga
A. Tomaschek	**Tomaschek**	Antonín (Anton) Tomaschek (1826-1891)
J. A. Tomlinson	**J. A. Toml.**	J. A. Tomlinson
S. M. Tong	------	Susan M. Tong (see *Developayella elegans*)
J. W. H. Trail	**Trail**	James William Helenus Trail (1851-1919; *Peronospora valerianae*)
R. Tramier	**Tramier**	R. Tramier (see *Peronospora anemonones*)
W. Tranzschel	**Tranzschel**	Woldemar (Andrejovitch) Tranzschel (1868-1942)
G. B. Traverso	**Traverso**	Giovanni Battista Traverso (1878-1955)
H. T. Tribe	**Tribe**	Henry Timothy Tribe (1927-)
J. G. Trog	**Trog**	Jakob Gabriel Trog (1781-1865; see *Peronospora podagrariae* = *Plasmopara podagrariae*)
A. Trotter	**Trotter**	Alessandro Trotter (1874-1967)
A. H. Trow	**Trow**	Albert Howard Trow (1863-1939)
J. H. L. Truscott	**Truscott**	John Henry Lloyd Truscott (1905-)
S. Tsuchiya	**S. Tsuchiya**	S. Tsuchiya
C. M. Tucker	**Tucker**	Clarence Mitchell Tucker (1897-1954)
L. R. Tulasne	**Tul.**	Louis René ['Edmond'] Tulasne (1815-1885; see *Peronospora arenariae*; *P. dipsaci*; *P. effusa*; *P. ficariae*; *P. gangioniformis* = *Bremia lactucae*; *P. papaveris* = *P. arborescens*)
T. V. Tunkina	**Tunkina**	T. V. Tunkina (pub. 1955 - *Peronospora echinopssili*, *Peronospora uljanishchevii*)
G. Ubrizsy	**Ubrizsy**	G. Ubrizsy (1919-1973; see *Bremia zanthii*)
S. Uchida	**S. Uchida**	S. Uchida
E.-R. Uebelmesser	**Uebelm.**	Esther-Ruth Uebelmesser
A. Ulken	**Ulken**	Annemarie Ulken
V. I. Ul'yanishchev	**Uljan.**	V. I. Ul'yanishchev [Uljanishchev]
C. J. Umphlett	**Umphlett**	Clyde J. Umphlett
L. M. Unamuno	**Unamuno**	Luis Mariano Unamuno Yrigoyan (1873-1943)
L. M. Underwood	**Underw.**	Lucien Marcus Underwood (1853-1907; see *Peronospora plantaginis*)
F. Unger	**Unger**	Franz Joseph Andreas Nicolaus Unger (1800-1870)
B. N. Uppal	**Uppal**	Badri Nath Uppal
O. Vaartaja	**Vaartaja**	Otto (Antero) Vaartaja (1912-1975)
W. Vader	-----	Wim Vader (see *Thalassomyces niezabitowskii*)
A. Valkanov	**Valkanov**	Alexander K. Valkanov [A. ВЪЛКАНОВЪ, A. Vülkanov]
H. Vallin	**Vallin**	H. Vallin
J. P. Van Der Meer	**Vanderm. [Van der Meer]**	John P. Van Der Meer (1943-)
S. Vanev	**Vanev**	Simeon (G.) Vanev (pub. 1963 - *Peronospora parietariae*)
K. Vánky	**Vánky**	Kálmán Vánky
L. Vánky	**L. Vánky**	Lia Vánky (1930-)
T. C. Vanterpool	**Vanterp.**	Thomas Clifford Vanterpool (1898-)
G. Vegni	**Vegni**	G. Vegni
A. A. Verhoeven	**Verh.**	Adriana A. Verhoeven (pub. 1972 - *Peronospora parasitica* f. sp. *galligena*)
B. L. Verma	**B. L. Verma**	B. L. Verma
L. Verwoerd	**Verwoerd**	Len Verwoerd (pub. 1924 - *Peronospora mesembryanthemi*)
T. Vestergren	**Vestergr.**	[Jacob] Tycho [Conrad] Vestergren (1875-1930; see *Peronospora*)
P. Viala	**Viala**	Pierre Viala (1859-1936)
A. P. Viégas	**Viégas**	Ahmés Pinto Viégas (1905/6-1986) Hawksw. *et al.* (1995)
G. Viennot-Bourgin	**Vienn.-Bourg.**	Georges Viennot-Bourgin (1906-)
F. Vincens	**Vincens**	François Vincens (1880-1925; see *Peronospora cephalariae*)
H. S. Vishniac	**Vishniac**	Helen Simpson Vishniac [née Reischer]
J. Vörös	**Vörös**	J. Vörös (pub. 1956 - *Bremia zanthii*)

P. Vuillemin	**Vuill.**		Jean Paul Vuillemin (1861-1932)	Hawksw. *et al.* (1995)
S. Wada	-----		Shinpei Wada (see *Halodaphnea awabi*)	
I. Wahl	**Wahl**		I. Wahl	
W. Wahrlich	**Wahrlich**		Woldemar Karlowitsch Wahrlich (1859-1923)	
M. B. Waite	**Waite**		Merton Benway Waite (1865-1945)	
N. Wallich	**Wall.**	[ ]	Nathaniel Wallich (1786-1854; see *Olpidium indicum*)	
R. W. Wallroth	**Wallr.**		Carl [Karl] Friedrich Wilhelm Wallroth (1792-1857)	
J. Walz	**J. Walz**		Jacob Jacoblevič Walz (1840-1904)	
J.-s. Wang	------		Jin-sheng Wang (see *Phytophthora fragariae* var. *oryzyobladis*)	
T. M. Wang	**T. M. Wang**		T. M. Wang (see *Peronospora parasitica-brassicae*, *P. parasitica-raphani* = *P. brassicae*)	
Y. L. Wang	**Y. L. Wang**		Yan Lin Wang	
Y.-x. Wang	**Ying X. Wang**		Ying X. [Ziang] Wang	
M. W. Ward	------		Mary Williams Ward (see *Thraustotheca irregularis*)	
W. M. Ware	**Ware**		William Melville Ware	
A. Wartenweiler	**Wartenw.**		Alfred Wartenweiler	
G. M. Waterhouse	**Waterh.**		Grace Marion Waterhouse (1906-1996)	*Mycol. Res.* **101**: 383-384 (1997)
S. W. Watson	**S. W. Watson**		S. W. Watson	
C. T. Wei	**C. T. Wei**		C. T. Wei (see *Plasmopara satureiae*)	
R. G. Werner	**Werner**		Roger-Guy Werner (1901-1977; see *Peronospora parasitica* var. *biscutellae* Maire in Maire & Werner)	
C. C. Wernham	**C. C. Wernham**		Clifford Charles Wernham (1903-)	
W. H. Weston	**W. Weston**		William Henry Weston (1890-1978)	*Mycologia* **71**: 1103-1106 (1979)
A. J. Whiffen	**Whiffen**		Alma Joslyn Whiffen (1916-1981)	*Mycologia* **74**: 359-362 (1982)
A. J. Whiffen-Barksdale - see A. J. Whiffen				
M. D. Whitehead	**M. D. Whitehead**		Marvin D. Whitehead (pub. 1951)	
R. H. Whittaker	**Whittaker**	[ ]	R. H. Whittaker (see higher taxa)	
A. Whittick	**Whittick**		Alan Whittick	
W. F. Wilcox	------		W. F. Wilcox (see *Phytophthora fragariae* var. *rubi*)	
É. de Wildeman	**De Wild.**		Émil August[e] Joseph de Wildeman	
N. Wille	**Wille**		John Nordal Fischer Wille (1858-1924)	
F. J. Williams	**F. J. Williams**		F. J. Williams (see *Phytophthora*)	
L. G. Willoughby	**Willoughby**		Leonard Guy Willoughby (1926-)	
G. W. Wilson	**G. W. Wilson**		Guy West Wilson (1877-1956)	
B. L. Wing	**Wing**	[ ]	Bruce Larry Wing (1938- ; see *Thalassomyces albatrossi*)	
Ö. Winge	**Winge**		Øjvind Winge (1886-1964)	
G. Winter	**G. Winter**		Heinrich Georg Winter (1848-1887)	Hawksw. *et al.* (1995)
L. Wittmack	**Wittm.**		(Marx Carl) Ludwig [Ludewig] Wittmack (1839-1929)	
C. Wize	**Wize**		Casimir Wize	
F. A. Wolf	**F. A. Wolf**		Frederick Adolph Wolf (1885-1975)	*Mycologia* **68**: 229-232 (1976)
F. T. Wolf	**F. T. Wolf**		Frederick Taylor Wolf (1915-)	
P. C. van der Wolk	**Wolk**		P. C. van der Wolk	
P. T. W. Wong	**P. Wong**		Percy T. W. Wong	
M. Woronin	**Woronin**		Michael Stephanovich Woronin [М. ВОРОНИН, Voronin, M.] (1838-1903)	
Chupp (1934)	______			
E. P. Wright	**E. P. Wright**		Edward Perceval Wright (1834-1910)	
A. Wróblewski	**Wróbl.**		Anton [Antoni] Wróblewski (?-1944; see *Peronospora vistulensis*, *et alia*)	
B. Wrońska	**Wrońska**		B. Wronska	
A. A. Yačeveskie	**Jacz.**		A. A. Yačeveskie [А. А. ЯЧЕВСКИЕ, also spelt Jaczewski, A. A.] (1863-1932)	
P. A. Yačeveskie	**P. A. Jacz.**		P. A. Yačeveskie [П. А. ЯЧЕВСКИЕ, also spelt Jaczewski, P. A.]	
M. Yanagowa	**Yanagowa**		M. Yanagowa	
Z. S. Yang	______		Z. S. Yang	
K. Yendo	**Yendo**	[ ]	Kichisaburo Yendo (1874-1921)	
Y. Yendo	**Y. Yendo**		Y. Yendo (fl. 1932, see above)	
G. Y. Yin	**G. Y. Yin**		Gong Yi Yin (pub. 1912 - *Peronospora microulae*)	
T. Yokochi	**Yokochi**		Toshihiro Yokochi (see *Schizochytrium limacinum*)	
E. L. Young	------		E. L. Young (see *Labyrinthula cienkowskii* var. *zopfii*)	
R. A. Young	**Roy A. Young**		Roy A. Young (pub. 1957)	
Y.-n. [Y. N.] Yu	**Y. N. Yu**		Yong-nian Yu (1923-)	
Zafran	------		Zafran [no other name, Indonesian] (see *Salilagenidium thermophilum*)	
A. Zahlbruchner	**Zahlbr.**		Alexander Zahlbruchner (1860-1938; see *Peronospora*)	
A. Zalewski	**Zalewski**		Aleksander Zalewski (1854-1906 *Cystopus amaranthacearum*)	
N. G. Zaprometov	**Zaprom.**		Nikolai Georgievich Zaprometov (see *Peronospora roemeriae* - pub. 1926)	
L. M. Zelezinskaya	------		L. M. Zelezinskaya [Л. М. ЗЕЛЕЗИНСКАЯ, L. M. Zělězinskaja] (see *Hyphochytrium peniliae*)	
Z.-y. Zhang	**Z. Y. Zhang**		Zhong-yi Zhang	
Zhen Y. Zhao	**Zhen Y. Zhao**		Zhen Y. Zhao	
W.-y. Zhuang	**W. Zhuang**		Wen-ying Zhuang	
A. W. Ziegler	**Ziegler**		Arthur William Ziegler	
M. K. Ziling	**Ziling**		M. K. Ziling (pub. 1929 - *Peronospora anchusae*)	
A. W. P. Zimmerman	**Zimm.**		Albrecht Wilhelm Philip Zimmerman (1860-1931)	
W. Zopf	**Zopf**		Wilhelm Friedrich Zopf (1846-1909)	Hawksw. *et al.* (1995)

GLOSSARY

NOTE:

Many of the definitions given below were written for, and included verbatim in *Ainsworth & Bisby's Dictionary of the Fungi, Eighth Edn* (Hawksworth *et al.*, 1995). Significant deviations from definitions given in Hawksworth *et al.*, (1995) and *Illustrated Glossary of Protoctista* (Margulis, McKhann & Olendzenski, 1992) are noted in parentheses.

AAA pathway: α-aminoadipic acid pathway for lysine synthesis (*DAP pathway*, q.v.).
acronema: extension of flagellum tip containing the two central microtubules but none of the nine peripheral elements (*contra* Margulis *et al.*, 1992).
adhesorium: pad-like adhesive secretion, or part of a cell (see *adhesive disc* - Margulis *et al.*, 1992).
aggregate species: species derived solely from divergence due to mutation in parthenogametic, homokaryotic diplosporic, or heterokaryotic diplosporic species (clonal species). Taxa consistently distinguishable on minor morphological criteria so that each qualifies for recognition as a species.
amoeboid: movement by means of pseudopodia.
anamorph: thallus producing only asexual spores.
anisokont: of flagella that are of different lengths; in mycology normally of flagella with similar morphology; in phycology including flagella of similar length but different, non-straminipilous, morphology (*heterokont, isokont*, q.v.).
antheridium: male gametangium, either formed from a haplophase thallus, or in which meiosis occurs after delimitation.
aplanospore: any naked, amoeboid or non-amoeboid mobile (q.v.) cell; or a sporangiospore (see p. 13).
aplerotic: of an oospore which occupies <60% of the oogonial volume (Shahzad *et al.*, 1992).
apocyte: multinucleate cell in which the multinucleate condition is accidental, transitory or secondary.
apophysis: swelling, particularly of a vegetative axis (hypha or rhizoid) at the place where it subtends a sporangium.
apospory: direct incorporation in a spore of an oogonial or antheridial diploid nucleus with cytoplasm uninfluenced by any meioses at the time of spore wall formation (Dick, 1972).
appressorium: specialized walled cell or hyphal component which functions in the penetration of a host cell wall. In plasmodiophorids, it arises at the end of a short germ tube formed from an encysted zoospore (*contra* Margulis *et al.*, 1992 *re* nutrition).
autecology: ecological studies on a single species and its relationship to the biological and physiochemical aspects of its environment.
automictic sexual reproduction: karyogamy between daughter nuclei of different meioses in the same gametangium (Dick, 1972).
auxiliary zoospore: first-formed zoospore, formed and flagellate within the sporangium, in a species with dimorphic zoospores (Dick, 1973); flagellar insertion apical or sub-apical (contrast *principal zoospore*, q.v.).
axenic: cultures containing only one living species (dual cultures of parasite and host can also be termed axenic if free of all other organisms - Canter-Lund & Lund, 1995: glossary; but see *gnotobiotic* q.v. and compare Hawksworth *et al.*, 1995).
axoneme: main core of a flagellum comprising the two central microtubules plus the nine peripheral double tubules; attached to the *transitional plate* (q.v.).
basal apparatus: accepted synonym of kinetid.
basal body: accepted synonym of kinetosome or blepharoplast (*contra* Hawksworth *et al.* 1995).
bipartite tubular hair: filamentous appendage on the flagellum consisting of a cylindrical shaft and one or more terminal filaments, characteristic of cryptomonads (Andersen *et al.*, 1991).
blastic germination: germination by means of budding (contrast *hyphal germination, polar germination*, q.v.).
boathook hair: characteristic terminally bifid hair, produced intracellularly in principal-form zoospore of *Saprolegnia*, and which ornaments the cyst formed by this zoospore; principal-form cyst or secondary cyst (see Beakes, 1983; Beakes *et al.*, 1995).
bothrosome: accepted synonym of sagenogenetosome (sagenogen).
caducous: of sporangia (esp. *Phytophthora* and *Peronospora*): falling off readily, deciduous.
centriole: short-cylindrical or barrel-shaped cell organelle 0·25 x 4·0 μm diam. (kinetosome lacking an axoneme) associated with microtubule generation (e.g. axoneme, nuclear spindle, cytoskeleton etc.).
CER (chloroplast endoplasmic reticulum): layer of ribosome-studded membrane surrounding the plastid.
chitin: unbranched polymer of ß-1,4 *N*-acetylglucosamine units.
ciliary roots: accepted synonym for flagellar roots.

cilium: alternative term for flagellum. Flagellum is the preferred term in mycology. 'Cilium' is generally used when large numbers of linked kinetosomes subtend axonemata, often when the quasi-sinusoidal wave length is more or less equal to the axoneme length (Dick, 1997*a*) (contrast definitions in Margulis *et al.*, 1992: *flagellum, undulipodium*, q.v.).

cM (centimorgan): relative measure of the linkage distance between genes.

coenocyte: cell or thallus with protoplasm containing many nuclei (cf. apocyte) (*contra* Margulis *et al.*, 1992).

companion cell: contiguous donor gametangium of *Olpidiopsidales*.

conidiosporangiophore: differentiated hypha subtending a conidiosporangium.

conidiosporangium: multinucleate asexual caducous propagule functioning either as a zoosporangium or as a conidium.

crista: tubular, pouch-like or shelf-like inwardly directed fold of the inner membrane of a mitochondrion; site of ATP production during aerobic metabolism (respiration).

cryptic species: see 'sibling species'.

cyst: propagule, especially the walled structure of the encysted zoospore (*Peronosporomycetes*); the spore form associated with sexual reproduction (*Plasmodiophoromycetes*), and resting spores of chrysomonads, etc.

cytoskeleton: intracellular network of protein filaments that is insoluble in non-ionic detergents; normally with two distinct components, microtubules and actin filaments, in different arrays.

cytosorus: group of resting cysts formed from a single plasmodium, fused together or united by exospore secretion.

DAP pathway: α,ϵ-diaminopimelic acid pathway for lysine synthesis (*AAA pathway*, q.v.).

DBV (dense body vesicle): cytoplasmic vesicle, associated with phosphorylated glucan metabolism, found in a variety of TEM morphological states ranging from a single (or several) electron-opaque cores in an amorphous matrix to a highly structured, myelin-like arrangement of alternating electron-opaque and electon-translucent layers (cf. *fingerprint vesicle*). [Sometimes **dense body vacuole**: Margulis *et al.* (1992) define vacuole as: "a small space or cavity in the protoplasm of a cell containing fluid or air and surrounded by a membrane", and vesicle as: "general structural term for a membranous sac. Usu. refers to a cell organelle in protoctists". Earlier definitions for a vacuole include "bounded by a tonoplast membrane" - the tonoplast membrane has distinctive properties. Whether vesicles always have an ER or dictyosomal origin is unclear; whether the tonoplast or DBV membrane has a dictyosomal origin is also unclear; the DBVs are certainly capable of intra-plasmic division and fusion, and therefore may have an independent perpetuation.] (see ms. pp 47)

deciduous: of a sporangium or gemma (rarely an oogonium, as in *Isoachlya*) of *Saprolegniaceae*, which becomes detached before germination.

dehiscence operculum: disk of sporangial wall shed when a circumscissile line of weakness gives way to release the sporangial contents.

dehiscence papilla: morphologically and ultrastructurally distinct protuberance on an undischarged sporangium that becomes converted into the exit tube.

dictyosome: portion of the endomembrane system with few to several flattened cisternae, often stacked in parallel arrays. The Golgi Body is a much more prominent (and differentially stainable in early cytology) stack of such vesicles (*contra* Hawksworth *et al.*, 1995; Margulis *et al.*, 1992).

dimorphic spore: zoospores with two morphologically and ultrastructurally distinct kinds of flagellar insertion; the types of zoospore are formed sequentially, each always followed by an encysted phase, and with the cysts formed from each type of zoospore also morphologically distinct (see ms p. 8) (compare *dimorphic* in Hawksworth *et al.*, 1995, and contrast Margulis *et al.*, 1992).

diplanetism: sequence of two motile flagellate phases with an interspersed mobile aplanosporic phase in the zoosporic part of the life-history; the aplanosporic phase as a walled cyst; motile phases may be monomorphic or dimorphic (see ms p. 10).

diplobiontic: having two assimilative or multinucleate/multicellular phases in the life-cycle.

diplomitotic ploidy cycle: occurrence of two mitotic phases of different ploidy in the life-cycle (Dick, 1987); karyogamy - mitosis - meiosis - mitosis - karyogamy -. Compare *haplomitotic* q.v. Synonymy of *haplomitotic* with *haplobiontic* as given by Margulis *et al.* (1992) is not correct.

endobiotic: assimilative and reproductive phases (except for exit papillae) entirely within the host body, or host cell, or host plasmamembrane.

endospore wall: internal wall of the oospore, laid down by the zygote (or cell containing the two gametic nuclei destined to fuse) and which, in the *Saprolegniomycetidae* and *Peronosporomycetidae*, is resorbed on germination (contrast *endosporium* - the innermost membrane, usually thin, of a spore).

epibiotic: usually confined to monocentric thalli; reproductive structures external to the host surface but with assimilative rhizoids within the host.

epispore wall: initial wall of the oosphere or zygote which defines the shape and size of the oospore and which is not changed on germination.

eucarpic: developing reproductive structures on limited portions of the thallus; residual nucleate protoplasm remaining and capable of further mitotic growth and regeneration.

exit tube: extension of the sporangium, produced prior to or during sporangial discharge, which enables sporangial contents to be released outside the host or substrate.

exospore wall: oospore wall layer(s) deposited outside the epispore wall through the activities or decay of residual nucleated protoplasm in the gametangium (*periplasm*, q.v.).

extra-mural fibrils: fibrils or spines formed perpendicularly and externally by deposition or secretion from host protoplasm around gametangial (rarely sporangial) thalli.

extrusome: membrane-bound structure derived from vesicle of the dictyosome/Golgi system and anchored to the cell membrane by proteinaceous particles; contents extruded in response to stimuli (cf., *ejectosome* of Canter-Lund & Lund, 1995).

facultative parasite: parasite having the power to assimilate nutrients saprotrophically; capable of being grown axenically. Alternatively, growing on both living and dead matter, simultaneously or in successive phases (*obligate parasite*, q.v.).

fertilization tube: tubular extension of the antheridium into the oogonium, developing after contact between the gametangia, and extending through the oogonium to the oosphere.

fertilization hypha: hypha developing from an antheridial gametic cell; passing through the antheridium wall and bridging the gap between non-contiguous gametangia to penetrate the oogonium (see ms p. 51).

fibrillar surface coat: fibrous component attached to the flagellar membrane and covering the entire surface of the flagellum.

flagellar apparatus: complex consisting of one or more basal bodies which may bear flagella; may have microtubular and fibrous roots associated with their bases.

flagellar fibrous roots: roots composed of a bundle of filaments; frequently appearing cross-striated.

flagellar hairs: filamentous appendages usually arranged in one or more rows but not covering the entire surface of a flagellum.

flagellar spines: structure on the paraxonemal body *PAB* (q.v.).

flagellar scales: organic structures of discrete size and shape, often covering the whole surface of the flagellum external to the plasma membrane; usually assembled in the dictyosome.

flagellum: cylindrical extension of a eukaryotic cell, bounded by a plasma membrane and containing an axoneme. Capable of transmitting a quasi-sinusoidal wave generated from the kinetosome at its base and to which the axoneme is attached. (*cilium*, q.v.).

flimmergeissel: flagellum bearing two rows of tripartite tubular hairs.

FPV (fingerprint vesicle): obsolete term for *DBV* (q.v.) with contents in the myelin-like state.

gametangium: cell containing gametes or gametic nuclei; the gametangium may initially be diploid and thus the site of meiosis, or haploid.

gametangiophore: thallus element subtending one or more gametangia.

gamete: naked uninucleate haploid cell with the sole function of fusing with another gamete to produce a zygote.

gametic cell: walled uninucleate cell that functions as a gamete.

gametic nucleus: haploid nucleus that will fuse with another haploid nucleus, without intervening mitosis, but possibly with a time lapse, to form a zygotic nucleus.

genome synteny: Renwick (1971: 86): synteny: two loci so far apart on a single chromosome that linkage cannot be demonstrated, even though presumed to be carried on the same chromosome. Also in relation to identical gene(s) being situated on different chromosomes in similar genomes (see Moore *et al.*, 1995).

germ tube: unbranched hyphal outgrowth resulting from germination of a walled spore; either of very narrow diameter (c.1-3μm), or of hyphal diameter characteristic of the species (see ms pp. 41).

glossoid cell: elongate (tongue-shaped) cell containing an elaborate extrusome (*Haptoglossa*).

gnotobiotic: cultures in which all the living components are known (*axenic*, q.v.)

Golgi body: dictyosome with large numbers of cisternae which may be visible using light microscopy after staining (usually in animals).

gun cell: cell containing the specialized extrusome of *Haptoglossa* and *Plasmodiophora* (cf. *nematocyst, glossoid cell*, q.v.).

haplobiontic: having one assimilative or multinucleate/cellular phase in the life-cycle.

haplomitotic A ploidy cycle: occurrence of one mitotic haploid phase in the life-cycle (Dick, 1987): karyogamy - meiosis - mitosis - karyogamy-. Synonymy with the term *haplobiontic*, q.v. as given by Margulis *et al.* (1992) is not correct.

haplomitotic B ploidy cycle: occurrence of one mitotic diploid phase in the life-cycle (Dick, 1987): karyogamy - mitosis - meiosis - karyogamy-. Synonymy with the term *haplobiontic*, q.v. as given by Margulis *et al.* (1992) is not correct.

haptonema: filamentous appendage (usually coiled) consisting of the plasma membrane, a sheath of endoplasmic reticulum, and a core of several microtubules anchored near the kinetosomes (chromistan algae - *Prymnesiophyceae*).

heterokaryotic (1): having two or more slightly (<5%) genetically different nuclei in common cytoplasm (fungi).

heterokaryotic (2): showing nuclear dimorphism (protists).

heterokaryotic diplospory: incorporation of two or more potentially heterogenetic diploid nuclei in an oospore formed in response to abortive meioses in the oogonium (Dick, 1972).

heterokont (1): condition whereby a flagellum [one of two flagella] possesses two rows of tripartite tubular hairs (heterokont flagellum, syn. flimmergeissel, tinsel flagellum, *straminipilous flagellum*, q.v.).

heterokont (2): of an organism possessing flagella in pairs, the members of which differ in length; type of motion; external appendages. Usually with kinetosomes mutually attached at a wide angle (Manton, 1965). (Obsolete: an organism with flagella of different morphology or length.)

heterothallic: requiring the interaction of two different thalli for sexual reproduction.

histones: lysine, arginine-rich proteins forming complexes with nuclear DNA in chromosomes.

holocarpic: using entire thallus for production of reproductive cells.

homohylic vesicle: sporangial vesicle, the wall of which is continuous with, and of the same material as the wall layer, or one of the wall layers of the sporangium (contrast *plasmamembranic vesicle*, *precipitative vesicle*, q.v.).

homokaryotic: having all nuclei genetically identical.

homokaryotic diplospory: incorporation of one or more homogenetic diploid nuclei in an oospore formed in response to abortive meioses in the oogonium (Dick, 1972).

homothallic: having the ability to reproduce sexually without the interaction of two different thalli.

hypha: constituent of a fungal thallus; tubular, coenocytic, septate or non-septate with apical growth.

hyphal germination: direct germination of a spore to produce a hyphal filamentous germ tube (contrast *blastic germination*, *polar germination*, q.v.).

isokont: condition in which all flagella are of identical length and morphology.

K_2 body: microbody with regularly patterned contents when viewed by TEM; associated with the kinetosome.

karyogamy: fusion of haploid nuclei leading to zygote production.

kb (kilobase pairs): measure of the number of nucleotide base-pairs in a genome.

kinetid: flagellar apparatus including kinetosomes and their associated tubules and fibres.

kinetosome: intracellular, non-membrane bound organelles; microtubular cylinders (0·2 μm diam.) organized in the 9 (A, B, C-microtubules) + 0 pattern and with axoneme extensions in the 9 (A, B- microtubules) + 2 pattern.

laminarin: carbohydrate stored in fucophytes; polymer of glucose and mannitol with ß-1,3- plus some ß-1,6-glycoside linkages.

life-cycle: general pattern of cyclical, combined morphological and nuclear change of an organism. Alternation of Generations only occurs when there are two, *and only two*, bionts (vegetative life forms), the morphology of which may be identical (isomorphic) or different (heteromorphic), *but which differ* in their ploidy levels. *Contra* Hawksworth *et al.* (1995); compare Canter-Lund & Lund (1995).

life-history: the sequence of morphological change in the life forms of an individual or population (contrast *life-cycle*, q.v.).

mastigoneme (obsolete): tripartite tubular hair, also any hair-like ornamentation on a flagellum (Andersen *et al.*, 1991).

Mb (megabase pairs): measure of the number of nucleotide base-pairs in the genome.

microsporangium: secondary sporangium formed from a zoospore cyst, either without any mitoses, or with very few mitoses.

microtubule: hollow structure consisting primarily of α and ß tubulin proteins arranged in a heterodimer; length may vary but diameter is usually 24-25 nm.

mitochondrion: membrane bound intracellular organelle containing enzymes and electron transport chains for oxidative respiration of organic acids and the concomitant production of ATP. Posesses DNA, messenger RNA and small ribosomes and thus capable of protein synthesis.

mobile: capable of passive movement due to extraneous forces (contrast *motile*, q.v.).

monomorphic spores: sporangiospores in only one form; form requires qualification as principal or auxiliary type in Peronosporomycetes (see ms p. 18).

motile: capable of movement as a result of endogenous metabolic activity (contrast *mobile*, q.v.).

mycelium: fungal thallus composed of hyphae.

nematocyst: modified cell with a capsule containing thread-like stinger used for defence, anchoring, capturing prey.

nuclear cycle: obsolete: see ploidy cycle (Dick, 1987, 2000*a*).

obligate parasite: a parasite confined to a living host in the assimilative phase. Alternatively, not known to be capable of culture (*facultative parasite*, q.v.).

oogonium: uninucleate or coenocytic cell producing, and initially containing, female gametes (oospheres).

oosphere: naked or walled female gamete in an oogonium.

ooplast: large membrane bound inclusion of oospores formed by the fusion of DBVs (dense body vesicles) (Howard & Moore, 1970) (characteristic of *Peronosporomycetes*, *contra* Hawksworth *et al.*, 1995).

oospore: walled zygote contained within an oogonium.

opisthokont: cell with a posteriorly directed flagellum originating from a posterior kinetosome.

osmotroph: organism with heterotrophic form of nutrition in which soluble organic compounds are absorbed from the surrounding medium by osmosis, active transport or pinocytosis.

parasitoid: a parasite, usually endobiotic (q.v.) and holocarpic (q.v.), which destroys its host (or host cell) on reproduction.

PAB (paraxonemal body): proteinaceous structure restricted to particular region of the flagellum in chromophytes and euglenoids; also manifest as flagellar spine in male gametes of oogamous fucophytes.

PAR (paraxonemal rod): hollow or solid cylindrical structure extending along most of the length of a flagellum; located between the axoneme and flagellar membrane and usually connected to them. Occurs in dinoflagellates, euglenoids, pedinellids and silicoflagellates but may differ in ultrastructure and biochemical composition.

parthenogamy: the state of an oospore formed with a diploid nucleus resulting from restitution at telophase I or telophase II of meiosis (note that the parthenogametic state cannot be presumed from the absence of an antheridium - *automictic sexual reproduction*, q.v.).

pedicel: the narrow connective (usually *plugged*, q.v.) between the conidiosporangiophore and the conidiosporangium, especially in *Phytophthora* and the *Peronosporales*; possibly homologous with the elongated narrowed constriction between thalloid segments in the *Rhipidiaceae* (Blackwell, 1949).

peitschengeissel: unornamented flagellum.

pellicle: outermost living layer lying below any non-living secreted material; contains the plasma membrane plus underlying epiplasm or other membranes and may show ridges, folds or distinct crests.

phagotroph: organism with heterotrophic form of nutrition in which pseudopods flow over and engulf solid food particles.

planont: any motile (and therefore usually naked) cell (see ms. p. 4; Sparrow, 1943) (compare Hawksworth *et al.*, 1995; Margulis *et al.*, 1992).

plasmamembrane: outer membrane composed of phospholipids and proteins which surrounds a cell and regulates exchange of materials between the cell and its environment.

plasmamembranic vesicle: sporangial vesicle, the membrane of which is the same membrane bounding the protoplasm within the sporangium (contrast *homohylic vesicle*, *precipitative vesicle*, q.v.).

plasmodium: multinucleate mass of cytoplasm bounded by a plasma membrane but lacking a wall; in endobiont parasitism, sometimes inducing additional girdling membranes.

plasmogamy: fusion of two cells or plasmodial cytoplasms without karyogamy (nuclear fusion) or a precursor to karyogamy.

plerotic: strictly, of an oospore which completely fills the oogonial cavity; but also of an oospore which occupies >65% of the oogonium volume (Dick *et al.*, 1992).

ploidy cycle: the sequence of mitosis, meiosis and karyogamy (Dick, 2000a) (*haplomitotic A nuclear cycle*, *haplomitotic B nuclear cycle*, *diplomitotic nuclear cycle*).

plug: occlusion between the hyphal walls separating vegetative and reproductive elements (*septum*, q.v.).

polar germination: germination predestined to produce a vegetative rhizoid and a separate reproductive axis (contrast *blastic germination*, *hyphal germination*).

polymorphic species: species with a series of intergrading morphological features, often resulting from inbreeding or automictic sexual reproduction.

polymorphic spores: sporangiospores in more than one form, either coexisting during sporangiogenesis or showing a sequential progression after release, with or without encysted phases; spores which show amoeboid changes in shape are polymorphic (see ms p. 10).

polyplanetism: sequence of two or more motile flagellate phases with interspersed mobile aplanosporic phases in the zoosporic part of the life-history; the aplanosporic phase may be naked or as a walled cyst; motile phases may be monomorphic or dimorphic (see ms p. 10).

precipitative vesicle: sporangial vesicle, the membrane of which is formed after extrusion of naked sporangial protoplasm, whether partially cleaved into planonts or not, possibly as a precipitation reaction between periprotoplasmic colloids and the environment; amorphous or fibrillar (contrast *homohylic vesicle*, *plasmamembranic vesicle*, q.v.).

primary zoospore: zoospore which germinates directly from a resting cyst (*Plasmodiophoromycetes*).

principal zoospore: first-formed zoospore with laterally inserted flagella in a ventral groove, and which are shed on encystment (contrast *auxiliary zoospore*, q.v.).

quasi-sinusoidal wave: of flagella or cilia; a wave generated from the kinetosome (q.v.), similar to a sine wave but diminishing in amplitude distally.

retraction septum: septum formed when the protoplast retracts from a pre-formed hyphal system (*septum*, q.v.).

retroneme: synonym for a tubular bipartite hair (Andersen *et al.*, 1991).

rhizoid: root-like structure consisting of anucleate filaments for anchorage and absorption (see *polar germination*, q.v.).

rhizoplast (obsolete): synonym of fibrous roots (Andersen *et al.*, 1991).

rohr: extra-cellular infection apparatus of plasmodiophorids (see *schlauch* and *stachel*, q.v.).

rootlet: synonym of flagellar root.

sagenogenetosome: organelle on cell surface from which the ectoplasmic network arises (*Labyrinthista*).

schlauch: open-ended extension of the rohr, orientated toward the cytoplasm of encysted zoospore of plasmodiophorids (see *rohr* and *stachel*).

secondary zoospore: zoospore of sporangial origin (*Plasmodiophoromycetes*); the principal-form zoospore of the peronosporomycetes (oomycetes) as used in much of the earlier literature (e.g. Sparrow, 1960).

septum: a cross wall or partition in a hypha, confluent with the hyphal wall (contrast *plug*, q.v.; see Hawksworth *et al.*, 1995 for expanded definitions).

sexual reproduction: the fusion of two haploid nuclei, each a product from a different meiosis, to produce a diploid nucleus (zygote nucleus) (Dick, 1987); this process may or may not be mediated by separate protoplasts in coenocytes (*automictic sexual reproduction*, q.v.).

sibling species: accepted synonym for cryptic species; species which are genetically distinct and not interbreeding, but are not morphologically distinct.

sporangial vesicle: vesicle produced at the mouth of the sporangium during planont maturation and discharge (*homohylic vesicle, plasmamembranic vesicle, precipitative vesicle*, q.v.).

sporangial plug: non-septal occlusion separating the sporangium from the vegetative system.

sporangiophore: thallus element (usually morphologically differentiated) subtending one or more sporangia.

sporangiospore: walled spore formed within a sporangium.

sporangiosorus: group of spherical sporangia fused together and formed from a single plasmodium; also one or more lobed sporangia formed from a single plasmodium.

sporangium: spore container or vessel in which spores are delimited.

stachel: bullet-like structure in the rohr with its pointed end orientated towards the appressorium and the host cell wall in plasmodiophorids (see *rohr* and *schlauch*, q.v.).

stramenopile: (Patterson, 1989) see *straminipile*.

straminipile: colloquial noun for an organism bearing tubular tripartite hairs (Patterson, 1989) from *stramen, stramini- (s) (gen.)*, a straw, and *pilus*, a hair; *straminipilus* (*s.m.*) (the spelling "(straminopile)" in Hawksworth *et al.*, 1995, is a misprint).

straminipilous: bearing tubular tripartite hairs; applicable to flagella and/or cells, whether uniflagellate, multiflagellate or non-flagellate (the auxiliary cyst of *Saprolegnia* which bears a tuft of tripartite tubular hairs would be a straminipilous cyst).

synaptonemal complex: proteinaceous, longitudinally aligned structure which usually unites homologous chromosomes during the prophase of meiosis.

synecology: ecology of communities.

synzoospore: multinucleate zoospore with many sets of flagella.

teleomorph: thallus bearing structures of sexual reproduction.

thallus: vegetative growth body; bearing, or converting into, reproductive organ.

tinsel flagellum: colloquial term for a flimmergeissel.

tonoplast membrane: membrane bounding the central vacuole.

transitional helix: thick helix (or possibly stacked rings) that lies inside the outer doublets of the axoneme, but apparently not connected to them (Andersen *et al.*, 1991) [note that the biological (Canter-Lund & Lund, 1995) and mathematical definitions of *helix* (a continuous curve formed on a cylinder) and *spiral* (a continuous curve of decreasing radius in two or three dimensions, as in the gasteropod *Helix*) are at variance].

transitional plate: thin plate-like structure oriented perpendicularly to the axoneme, usually filling the lumen of the axoneme and sometimes extending to the flagellar plasmamembrane (Andersen *et al.*, 1991).

TTH (tubular tripartite hair, tripartite tubular hair): filamentous appendage on the flagellum composed of a tapered solid base, a hollow cylindrical shaft and one or more terminal filaments (Andersen *et al.*, 1991).

vesicle (1): single-membrane bound protoplasmic inclusion.

vesicle (2): more or less evanescent membrane enclosing zoospores or protoplasm cleaving into zoospores (see sporangial vesicle).

whiplash flagellum: flagellum with a prolonged narrow extension containing the two central axoneme microtubules (*peitschengeissel*, q.v.).

zoosporangium: sporangium containing zoospores.

zoospore: motile (and flagellate or potentially flagellate) naked cell or planont, normally uninucleate, and usually functioning as a dispersal agent for the same phase of the life-history from whence it was generated (see p. 5).

zygote: cell formed from two haploid gametes, or a uninucleate haploid cell and a donated nucleus, in which karyogamy has occurred (The zygote may be motile as in *Allomyces*; the zygote may be naked or walled; mitochondria may, in theory, be derived from one or both participating gametes.

zygospore: a zygote enclosed in a resistant spore wall (an oospore is a zygote derived from an oosphere).

REFERENCES

AND BIBLIOGRAPHY

REFERENCES AND BIBLIOGRAPHY

NOTES:

In this bibliography I provide the fullest form for journal titles, and where necessary give further annotations, because many of the older and sometimes classic references have been copy-quoted from secondary sources, often with perpetuation and proliferation of typographical errors. The aim has been to provide a source from which abstractions can be made. It is always easier to abbreviate a citation than to search for the expanded form, and very little of the earlier work is in *BIOSIS*.

The bibliography includes ***all*** taxonomic citations for *Pythium sensu lato* to date; ***all*** citations in *The Genus* Achlya (Johnson, 1956*b*); ***all*** citations in *A Monograph of the Genus* Aphanomyces (Scott, 1961); ***all*** citations in *The Genus* Saprolegnia (Seymour, 1970); ***all*** citations in *Predominantly Holocarpic and Eucarpic Simple Biflagellate Phycomycetes* (Karling, 1981*a*) and the citations to biflagellate fungi in *Aquatic Phycomycetes, Second Revised Edition* (Sparrow, 1960). The comprehensive bibliography to the early cytology papers (Dick & Win-Tin, 1973) is also included. Although a substantial bibliography to the *Peronosporales* and *Pythiales* is provided, further citations will be found in Constantinescu (1991*a*). The *Index of Fungi* references given in APPENDIX 1 will enable taxonomic papers additional to these works to be accessed up to 1999.

Entries of authors with a surname preceded by 'Mc' or 'Mac' are listed as if 'MAC'. Treatment of names preceded by 'de' or 'Van' depends on general usage. English double-barrelled surnames without a hyphen are indexed under the first surname and cross-referenced under the second surname.

ABAD, G., SHEW, H. D., GRAND, L. F. & LUCAS, L. T. (1995). A new species of *Pythium* producing multiple oospores isolated from bentgrass in North Carolina. *Mycologia* **87**: 896-901.

ABBEY, G. (1895). The drooping of tomatoes. *Journal of Horticulture, Cottage Gardeners and Country Gentlemen, Series 3* **30**: 360.

ABE, T. (1928). Studies on the rot disease of rice-seedlings. *Nogyo oyobi Engei (Agriculture and Horticulture)* **3**: 259, 381, 511, 643.

ABDELZAHER, H. M. A., ICHITANI, T. & ELNAGHY, M. A. (1994). *Pythiumfluminum* var. *fluminum* from pond water in Osaka. *Mycological Research* **98**: 982-984.

ABRAHAMS, D. & BROWN, W. D. (1977). Toxicity to juvenile European lobster (*Homarus gammarus*) of several antifungal agents used to control *Haliphthoros milfordensis* Vishniac. *Aquaculture* **12**: 31-40.

ACHAR, P. N. (1995). Tissue-culture technique to determine the viability of *Peronospora parasitica* in *Brassica oleracea*. *Journal of Phytopathology* **143**: 647-649.

ADAIR, E. J. & VISHNIAC, H. S. (1958). Marine fungus requiring vitamin B_{12}. *Science, New York* **127**: 147-148.

ADAMS, J. & PETHYBRIDGE, G. H. (1910). A census catalogue of Irish fungi. *Proceedings of the Royal Irish Academy, Section B* **28**: 120-166.

AFANASIEV, M. M. (1948). The relation of six groups of fungi to seedling diseases of sugar beets in Montana. *Phytopathology* **38**: 205-212.

AGARDH, C. A. (1824). *Systema Algarum, Volume I*. Lund, Sweden.

AGERSBORG, H. P. K. (1933). Salient problems in the artificial rearing of salmonoid fishes, with special reference to intestinal fungisitosis and the cause of white spot disease. *Transactions of the American Fisheries Society* **63**: 240-250.

AHRENS, C. (1971). Eine neue *Pythium* - Art aus dem Libanon. *Zeitschrift für Pflanzenkrankheiten und Pflanzenschutz* **78**: 175-180.

AINSWORTH, G. C. (1973). Introduction and keys to higher taxa. *In* G. C. Ainsworth, F. K. Sparrow & A. S. Sussman (Eds), *The Fungi, An Advanced Treatise, Vol. 4B*, pp. 635-648. New York, U.S.A.: Academic Press.

AINSWORTH, G. C. (1976). *Introduction to the History of Mycology*. Cambridge, U.K.: Cambridge University Press.

AIST, J. R. (1974). A freeze-etch study of membranes in *Plasmodiophora*-infected and non-infected cabbage root hairs. *Canadian Journal of Botany* **52**: 1441-1449.

AIST, J. R. & WILLIAMS, P. H. (1971). The cytology and kinetics of cabbage root hair penetration by *Plasmodiophora brassicae*. *Canadian Journal of Botany* **49**: 2023-2034.

ALABI, R. O. (1971*a*). Factors affecting seasonal occurrence of Saprolegniaceae in Nigeria. *Transactions of the British Mycological Society* **56**: 289-299.

ALABI, R. O. (1971*b*). Seasonal periodicity of Saprolegniaceae at Ibadan, Nigeria. *Transactions of the British Mycological Society* **56**: 337-341.

ALABI, R. O. (1972). Thraustothecoid sporangium formation in three Saprolegniaceae from Nigeria. *Transactions of the British Mycological Society* **58**: 519-521.

ALDERMAN, D. J. (1971). Shellfish disease in the United Kingdom 1970-71. *ICESCM* 1971/K: 19.

ALDERMAN, D. J. (1974). Ultrastructure of stages of *Hyalochlorella marina* Poyton. *Veröffentlichungen des Instituts für Meeresforschung in Bremerhaven, Sonderband* **5**: 251-261.

ALDERMAN, D. J. (1976). Fungal disease of marine animals. *In* E. B. G. Jones (Ed.), *Recent Advances in Aquatic Mycology*, pp. 223-261. London, U.K.: Paul Elek.

ALDERMAN, D. J. (1982*a*). Fungal diseases of aquatic animals. *In* R. J. Roberts (Ed.), *Microbial Diseases of Fish*, pp. 189-242. London, U.K.: Academic Press for the Society of General Microbiology, Special Publication 9.

ALDERMAN, D. J. (1982*b*). *In vitro* testing of fisheries chemotheropeutants. *Journal of Fish Diseases* **5**: 113-123.

ALDERMAN, D. J. (1985). Malachite green; a review. *Journal of Fish Diseases* **8**: 289-298.

ALDERMAN, D. J. (1986). Fungi as pathogens of non insect invertebrates. *In* R.A. Samson, J.M. Vlak & D. Peters (Eds), *Fundamental and Applied Aspects of Invertebrate Pathology*, pp. 354-355. Netherlands, Wageningen: Foundation of the Fourth International Colloquium of Invertebrate Pathology, 1986.

ALDERMAN, D. J. (1996). Geographical spread of bacterial and fungal diseases of crustaceans. *Review Scientifique et Technique Office International des Epizooties, 1996* **15**: 603-632.

ALDERMAN, D. J. & GRAS, P. (1969). Gill diseases of Portuguese oysters. *Nature (London)* **224**: 616-617.

ALDERMAN, D. J., HARRISON, J. L., BREMER, G. B. & JONES, E. B. G. (1974). Taxonomic revisions in the marine biflagellate fungi: the ultrastructural evidence. *Marine Biology (Berlin)* **25**: 345-357.

ALDERMAN, D. J., HOLDICH, D. & REEVE, I. (1990). Signal crayfish as vectors in crayfish plague in Britain. *Aquaculture* **86**: 3-6.

ALDERMAN, D. J. & JONES, E. B. G. (1971). Physiological requirements of two marine phycomycetes, *Althornia crouchii* and *Ostracoblabe implexa*. *Transactions of the British Mycological Society* **57**: 213-225.

ALDERMAN, D. J. & POLGLASE, J. L. (1986*a*). Are fungal diseases significant in the marine environment? *In* S. T. Moss (Ed.), *The Biology of Marine Fungi*, pp. 189-198. Cambridge, U.K.: Cambridge University Press.

ALDERMAN, D. J. & POLGLASE, J. L. (1986*b*). *Aphanomyces astaci*: isolation and culture. *Journal of Fish Diseases* **9**: 367-379.

ALDRICH, H. C. & MIMS, C. W. (1970). Synaptonemal complexes and meiosis in Myxomycetes. *American Journal of Botany* **57**: 935-941.

ALEEM, A. A. (1950*a*). A fungus in *Ectocarpus granulosus* C. Agardh near Plymouth. *Nature (London)* **165**: 119.

ALEEM, A. A. (1950*b*). The occurrence of *Eurychasma dicksonii* (Wright) Magnus in England and Sweden. *Meddelanden från Göteborgs Botaniska Trädgård* **18**: 239-245.

ALEEM, A. A. (1950*c*). Phycomycètes marins parasites de Diatomées et d'Algues. *Comptes Rendus Hebdomadaires des Séances de l'Académie des Sciences, Paris, Série D* **231**: 713-715.

ALEEM, A. A. (1950*d*). Phycomycètes marins parasites de Diatomées et d'Algues dans la région de Banyuls-sur-mer (Pyrénées-Orientales). *Vie et Milieu, Paris (Bulletin de la Laboratoire Arago)* **1**: 421-440.

ALEEM, A. A. (1952*a*). Sur l'autécologie d'une spirogyre d'eau saumâtre. *Comptes Rendus Hebdomadaires des Séances de l'Académie des Sciences, Paris*, Série D **234**: 2648-2650.

ALEEM, A. A. (1952*b*). *Olpidiopsis feldmanni* sp. nov., champignon marin parasite d'algues de la famille des Bonnemaisoniacées. *Comptes Rendus Hebdomadaires des Séances de l'Académie des Sciences, Paris, Série D* **235**: 1250-1252.

ALEEM, A. A. (1953). Marine fungi from the west coast of Sweden. *Arkiv för Botanik* **3**: 1-33.

ALEEM, A. A., RUIVO, M. & THÉODORIDÈS, J. (1953). Un cas de maladie à saprolegniale chez une *Atherina* des environs de Salses. *Vie et Milieu (Bulletin de la Laboratoire Arago)* **3**: 44-51.

ALEXOPOULOS, C. J. (1952). *Introductory Mycology*. New York, U.S.A.: Wiley.

ALEXOPOULOS, C. J. (1962). *Introductory Mycology, Second Edition*. New York, U.S.A.: Wiley.

ALFARO, A. A. & MENDOZA, L. (1990). Four cases of equine bone lesions caused by *Pythium insidiosum*. *Equine Veterinary Journal* **22**: 295-297.

ALIAGA, G. R. & ELLZEY, J. T. (1984). Ultrastructural localization of acid phosphatase within oogonia of *Achlya recurva*. *Mycologia* **76**: 85-98.

ALI-SHTAYEH, M. S. (1986). *The Genus* Pythium *in the West Bank and Gaza Strip*. The Research and Documentation Centre, An - Najah University, West Bank, pp. 1-76.

ALI-SHTAYEH, M. S. & DICK, M. W. (1985). Five new species of *Pythium* (Peronosporomycetidae). *Botanical Journal of the Linnean Society* **91**: 297-317.

ALI-SHTAYEH, M. S., LIM-HO, C. L. & DICK, M. W. (1986*a*). An improved method and medium for quantitative estimates of populations of *Pythium* species from soil. *Transactions of the British Mycological Society* **86**: 39-47.

ALI-SHTAYEH, M. S., LIM-HO, C. L. & DICK, M. W. (1986*b*). The phenology of *Pythium* (Peronosporomycetidae) in soil. *Journal of Ecology* **74**: 823-840.

ALLAIN, A. (1934). Cytologie de l'oeuf du *Phytophthora cambivora*. *Compts Rendus de la Société de Biologie de Strasbourg* **117**: 778-780.

ALLEN, R. L., MEEKIN, T. K., PAULEY, G. B. & FUJIHARA, M. P. (1968). Mortality among chinook salmon associated with the fungus *Dermocystidium*. *Journal of the Fisheries Research Board of Canada* **25**: 2467-2475.

ALLEN, R. N. & NEWHOOK, F. J. (1973). Chemotaxis of zoospores of *Phytophthora cinnamomi* to ethanol in capilaries of soil pore dimensions. *Transactions of the British Mycological Society* **61**: 287-302.

ALLISON, L. (1950). Common diseases of fish in Michigan. *Michigan Department of Conservation, Miscellaneous Publication* **5**: 1-27.

AL-REKABI, S. A. W. (1979). Ultrastructural studies of oogonial walls in Saprolegniaceae. PhD Thesis, University of Reading, Reading, U.K.

ALTMANN, K. (1890). *Die Elementaorganismem und ihre Beziehungen zu den Zellen*. Leipzig, Germany: Viet.

AMEMIYA, J. (1982). Granular dermatitis in the horse caused by *Pythium gracile*. *Bulletin of the Faculty of Agriculture, Kagoshima University* **32**: 141-147.

AMEMIYA, J. & NISHIYAMA, S. (1966). On the granular dermatitis in horses which contains a kind of true fungus in the foci. *Bulletin of the Faculty of Agriculture, Kagoshima University* **17**: 215-224.

AMERSON, H. V. & BLAND, C. E. (1973). The occurrence of polycomplexes in the nucleus of encysting spores of *Lagenidium callinectes*, a marine Phycomycete. *Mycologia* **65**: 966-970.

AMLACHER, E. (1965). Pathologische unde histochemische Befunde bei Ichthyosporidiumbefall der Regenbogenforelle (*Salmo gairdneri*) und am 'aquarienfisch *Ichthyophonus*'. *Zeitschrift für Fischerei* **13**: 85-122.

AMON, J. P. (1978*a*). A method for obtaining sporulating *Labyrinthula*. *Mycologia* **70**: 1297-1299.

AMON, J. P. (1978*b*). Thraustochytrids and labyrinthulids of terrestrial, aquatic and hypersaline environments of the Great Salt Lake, U.S.A. *Mycologia* **70**: 1299-1301.

AMON, J. P. & PERKINS, F. O. (1968). Structure of *Labyrinthula* sp. zoospores. *Journal of Protozoology* **15**: 543-546.

ANASTASIOU, C. J. & CHURCHLAND, L. M. (1968). An *Olpidiopsis* parasitic on a marine fungus. *Syesis* **1**: 81-85.

ANDERSEN, R. A. (1989). Absolute orientation of the flagellar apparatus of *Hibberdia magna* comb. nov. (Chrysophyceae). *Nordic Journal of Botany* **8**: 653-669.

ANDERSEN, R. A., BARR, D. J. S., LYNN, D. H., MELKONIAN, M., MOESTRUP, Ø. & SLEIGH, M. A. (1991). Terminology and nomenclature of the cytoskeletal elements associated with the flagellar/ciliary apparatus in protists. *Protoplasma* **164**: 1-8.

ANDREUCCI, (1926). Un nuovo Mixomicete parasita. *Archivio Botanico per la Systematica, Fitogeografia e Genetica* **2**: 19-28.

ANGIOSPERM PHYLOGENY GROUP (APG) (1998). An ordinal classification for the families of flowering plants. *Annals of the Missouri Botanic Garden* **85**: 531-553.

ANN, P. J. & KO, W. H. (1992). Survey of antibiotic resistance and dependence in *Phytophthora*. *Mycologia* **84**: 82-86.

ANONYMOUS (1973). 1973 Public Health Pesticides. *Pest Control* **41**: 17-50.

ANONYMOUS (1979). Data sheet on the biological control agent *Lagenidium giganteum* (Couch 1935) (in relation to mosquito control). Geneva, Switzerland: World Health Organization.

ANONYMOUS (1985*a*). Biological control of vectors. Tropical Disease Research 7th Programme Report, pp. 10/1-10/16. Geneva, Switzerland: World Health Organisation.

ANONYMOUS (1985*b*). Data sheet on the biological control agent *Lagenidium giganteum* (Couch 1935). Geneva, Switzerland: World Health Organization.

ANONYMOUS (1991). Distribution Maps of Plant Diseases, No. 522. Wallingford, U.K.: CAB International.

ANTOINE, R. & RICAUD, C. (1966). Cane Diseases. *Report, Mauritius Sugar Industry Research Institute, 1965*: 71-84.

BIBLIOGRAPHY

APINIS, A. (1929). Untersuchungen über die in Lettland gefunden Saprolegniaceen nebst Bemerkungen über andere Wasserpilze. *Acta Horti Botanici Universitatis Latviensis* **4**: 201-241.

APINIS, A. (1935). Fertilization of oospheres by planogametes in Saprolegniaceae. *Acta Horti Botanici Universitatis Latviensis* **8**: 103-110.

APINIS, A. (1960). Über das Vorkommen niederer Pilze in alluvialen Böden bestimmter Pflanzengesellschaften. *Mitteilungen der Florist-soziologischen Arbeitsgemeinschaft, N. F., Heft* **8**: 110-117.

APINIS, A. (1964). Concerning occurrence of phycomycetes in alluvial soils of certain pastures, marshes and swamps. *Nova Hedwigia* **8**: 104-126.

APSTEIN, C. (1910). *Synchaetophagus balticus*, ein in *Synchaeta* lebender Pilz. *Wissenschaftliche Meeresuntersuchungen der Kommission zur Wissenschaftlichen Untersuchungen der Deutschen Meere, Abteilung Kiel* **12**: 163-166.

ARASAKI, S. (1947). Studies on the rot of *Porphyra tenera* by a *Pythium*. *Nihon Suisan Gakkai-Shi [Bulletin of the Japanese Society of Fisheries]* **13**: 74-90.

ARASAKI, S. (1948). Injury of *Porphyra* and its treatment. *Shokubutsu Boeki [Journal of Plant Protection]* **10**: 243-246.

ARASAKI, S. (1960*a*). A chytridian parasite on the *Porphyra*. *Nihon Suisan Gakkai-Shi [Bulletin of the Japanese Society of Fisheries]* **26**: 543-548.

ARASAKI, S. (1960*b*). The disease of cultured *Porphyra*. *Nihon Suisan Gakkai-Shi [Bulletin of the Japanese Society of Fisheries]* **26**: 1074-1081.

ARASAKI, S. (1962). Studies on the artificial cultures of *Porphyra tenera* Kjellm. III. On the red wasting disease of *Porphyra*, especially on the physiology of the causal fungus *Pythium* sp. nov. *Journal of the Agricultural Laboratory, Akibo, Japan* **3**: 87-93.

ARASAKI, S., AKINO, K. & TOMIYAMA, T. (1968). A comparison of some physiological aspects in a marine *Pythium* on the host and on the arificial medium. *Bulletin of the Misaki Marine Biological Institute, Kyoto University* **12**: 203-206.

ARCHER, W. (1860). On the occurrence of zoospores in the family Desmidiaceae. *Quarterly Journal of Microscopical Science* **8**: 216-239.

ARCHER, W. (1867). On two new species of Saprolegniaceae, referable respectively tothe genus *Saprolegnia* (Nees von Esenb.) and *Achlya* (Nees von Esenb.). *Quarterly Journal of Microscopical Science (N. S.)* **7**: 121-127.

ARCHER, W. (1875). On *Chlamydomyxa labyrinthuloides*. *Quarterly Journal of Microscopical Science* **15**: 107-130.

ARENS, K. (1929). Untersuchungen über Keimung und Zytologie der Oosporen von *Plasmopara viticola* (Berl. et de Toni). *Jahrbücher für wissenschaftliche Botanik* **70**: 57-92.

ARESCHONG, J. E. (1844). Om *Achlya prolifera*, växanda på lefvande fisk. *Öfversigt, Stockholm* **1**: 124-126.

ARESCHOUG, J. E. (1845). Beobachtungen über die merkwürdige Pilze, *Achlya prolifera*... *Archiv skandinavischer Beiträge zur Naturgeschichte Herausgegeben von Christian Friedrich Hornschuch* **1**: 170-172. [The same paper as Areschong (1844) in German; note different surname spelling.]

ARMBRUSTER, B. L. (1982*a*). Sporangiogenesis in three genera of the Saprolegniaceae. I. Pre-sporangium hyphae to early primary spore initial stage. *Mycologia* **74**: 433-459.

ARMBRUSTER, B. L. (1982*b*). Sporangiogenesis in three genera of the Saprolegniaceae. II. Primary spore initial to secondary spore initial stage. *Mycologia* **74**: 975-999.

ARMBRUSTER, B. L. & WEISENSEEL, M. H. (1983). Ionic currents traverse growing hyphae and sporangia of the mycelial water mold *Achlya debaryana*. *Protoplasma* **115**: 65-69.

ARMSTRONG, D. A., BUCHANAN, B. V. & CALDWELL, R. S. (1976). A mycosis caused by *Lagenidium* sp. in laboratory-reared larvae of the Dungeness crab *Cancer magister*, and possible treatment. *Journal of Invertebrate Pathology* **28**: 329-336.

ARMIGER, L. (1964). An occurrence of *Labyrinthula* in New Zealand. *New Zealand Journal of Botany* **2**: 3-9.

ARNAUD, G. (1948). Les Heimerliacées, subdivision des Acrasiales? *Comptes Rendus Hebdomadaires des Séances de l'Académie des Sciences, Paris, Série D* **226**: 1744-1746.

ARNAUD, G. (1949). Les Heimerliacées, subdivision des Acrasiales? *Le Botaniste* **34**: 35-54.

ARNAUD, G. (1952). Mycologie concrète: genera. *Bulletin trimestriel de la Société Mycologique de France* **68**: 181-223.

ARNAUDOW, N. [АРНАУДОВЪ, Н., **Arnaudov', N.**] (1921). Zur morphologie und Biologie von *Zoophagus insidians* Sommerstorff. *Godishiik na Soffiiskiya Universitet (Jahrbuch der Sofianer Universität)* **15-16**: 1-32.

ARNAUDOW, N. (1923*a*). Ein neuer Räderthiere (Rotatoria) fangender Pilz. (*Sommerstorffia spinosa* nov. gen., nov. sp.). *Flora, Jena* **116**: 109-113.

ARNAUDOW, N. (1923*b*). Untersuchungen über *Sommerstorffia spinosa* nov. gen., nov. sp. *Godishiik na Soffiiskiya Universitet (Jahrbuch der Sofianer Universität)* **19**: 161-196.

ARNAUDOW, N. (1925). Untersuchungen über den Tiere fangenden Pilz *Zoophagus insidians* Som. *Flora, Jena* **118**: 1-16.

ARNOLD, D. L. BLAKESLEY, D. & CLARKSON, J. M. (1996). Evidence for the growth of *Plasmodiophora brassicae in vitro*. *Mycological Research* **100**: 535-540.

ARONSON, J. M. (1965). The cell wall. *In* G.C. Ainsworth & A.L. Sussman (Eds), *The Fungi, An Advanced Treatise, Vol. 1*, pp.49-76. New York, U.S.A.: Academic Press.

ARONSON, J. M. & FULLER, M. S. (1969). Cell wall structure of the marine fungus, *Atkinsella dubia*. *Archiv für Mikrobiologie* **68**: 295-305.

ARONSON, J. M. & LIN, C. C. (1978). Hyphal wall chemistry of *Leptomitus lacteus*. *Mycologia* **70**: 363-369.

ARSENAULT, G. P., BIEMAN, K, BARKSDALE, A. W. & McMORRIS, T. C. (1968). The structure of antheridiol, a sex hormone in *Achlya bisexualis*. *Journal of the American Chemical Society* **90**: 5635-5636.

ARTEM[T]CHUK, N. J[Ya]. [АРТЕМЧУК, Н. Я., **Artemchuk, N. Ya.**] (1968). Новости сист. низш. раст. [*Leptolegnia pontica*]. **5**: 77.

ARTEMTCHUK, N. J[Ya]. (1972). The fungi of the White Sea. III. New Phycomycetes, discovered in the Great Salma Strait of the Kandalakshial Bay. *Veröffentlichungen des Instituts für Meeresforschung in Bremerhaven* **13**: 231-237.

ARTEMCHUK, N. Ya. (1974). White Sea fungi. Saprophytic phycomycetes of Kandalasksha Bay in the Great Salma Strait. I. *Mikologiya i Fitopatologiya* [*Микология и Фитопалогия*] **8**: 281-291.

ARTEMCHUK, N. J. (1981). *Микофлора Морей СССР* [Mycoflora of the Seas of the U.S.S.R.] [in Russian]. Moscow, U.S.S.R.: Nauka.

ARTEMCHUK, N. Ya. & ZELEZINSKAYA, L. M. [ЗЕЛЕЗИНСКАЯ, Л. М., **Zèlèzinskaja**] (1969). Морской гриб *Hyphochytrium peniliae* n. sp., поражающий зоопланктонного гачка *Penilia avirostris* (Dana). The sea fungus *Hyphochytrium penilae* n. sp., affecting planktonic crawfish *Penilia avirostris* (Dana). *Mikologiya i Fitopatologiya* [*Микология и Фитопалогия*] **3**: 356-358.

ARYA, H. C. & TIWARI M. M. (1969). Growth of *Sclerospora graminicola* on callus tissue of *Pennisetum typhoides* and in culture. *Indian Phytopathology* **22**: 446-452.

ASCHNER, M. (1958). Isolation of *Labyrinthula macrocystis* from soil. *Bulletin of the Research Council of Israel, Section D* **6**: 174-179.

ASCHNER, M. & KOGAN, S. (1959). Observations on the growth of *Labyrinthula macrocystis*. *Bulletin of the Research Council of Israel, Section D* **8**: 15-24.

ASIEGBU, F. O., LÖNNEBORG, A. & JOHANSSON, M. (1996). Chitin and glucans detected in the cell walls of *Pythium dimorphum* - an oomycetous fungus. *European Journal of Forest Pathology* **26**: 315-321.

ATKINS, D. (1929). On a fungus allied to the Saprolegniaceae found in the pea-crab *Pinnotheres*. *Journal of the Marine Biological Association of the United Kingdom* **16**: 203-219.

ATKINS, D. (1935). *Pythium thalassium* sp. nov. infecting the egg-mass of the pea crab, *Pinnotheres pisum*. *Transactions of the British Mycological Society* **38**: 31-46.

ATKINS, D. (1954*a*). Further notes on a marine member of the Saprolegniaceae, *Leptolegnia marina* n. sp., infecting certain invertebrates. *Journal of the Marine Biological Association of the United Kingdom* **33**: 613-625.

ATKINS, D. (1954*b*). A marine fungus *Plectospira dubia* n. sp. (Saprolegniaceae), infecting crustacean eggs and small crustacea. *Journal of the Marine Biological Association of the United Kingdom* **33**: 721-732.

ATKINSON, G. F. (1895). Damping off. *Bulletin of the Cornell Agricultural Experiment Station* **94**: 233-272

ATKINSON, G. F. (1897). Some fungi from Alabama. *Bulletin of Cornell University* **3**: 1-50.

ATKINSON, G. F. (1909). Some fungus parasites of algae. *Botanical Gazette* **48**: 321-338.

AUSTWICK, P. K. C. & COPLAND, J. W. (1974). Swamp cancer. *Nature (London)* **250**: 84.

AXTELL, R. C. & GUZMAN, D. R. (1987). Encapsulation of the mosquito fungal pathogen *Lagenidium giganteum* (Oomycetes: Lagenidiales) in calcium aginate. *Journal of the American Mosquito Control Association* **3**: 450-459.

AXTELL, R. C., JARONSKI, S. T., MERRIAM, T. L., GRANT, C. D., WASHINO, R. K., LUSK, E. E. & COYKENDALL, R. L. (1983). Efficacy of the mosquito fungal pathogen, *Lagenidium giganteum* (Oomycetes: Lagenidiales). *In* C. D. Grant, R. K. Washino, E. E. Lusk & R. L. Coykendall (Eds), *Proceedings and Papers of the Fiftieth Annual Conference of the California Mosquito and Vector Control Association*, pp. 41-42. Sacramento, U.S.A.: Californian Mosquito and Vector Control Association Press.

AYERS, W. A., & LUMSDEN, R. D. (1977). Mycoparasitism of oospores of *Pythium* and *Aphanomyces* species by *Hyphochytrium catenoides*. *Canadian Journal of Microbiology* **23**: 38-44.

AYSEL, V. & UÇAR, F. (1990). Bazi deniz algerinde *Eurychasma* enfeksionu ile iligili morpholojik ve taksonomik değişiklikler (Infection of some marine alga - changes in algal morphology and taxonomical consequences). *Doğa Türk Botanik Dergisi (Turkish Journal of Botany)* **13**: 502-511.

AZEVEDO, C. (1989). Fine structure of *Perkinsus atlanticus* n. sp. (Apicomplexa, Perkinsea) parasite of the clam *Ruditapes decussatus* from Portugal. *Journal of Parasitology* **75**: 627-635.

AZEVEDO, C. (1990). Virus-like particles in *Perkinsus atlanticus* (Apicomplexa, Perkinsidae). *Diseases of Aquatic Organisms* **9**: 63-65.

AZEVEDO, C., CORRAL, L. & CACHOLA, R. (1990). Fine structure of zoosporulation in *Perkinsus atlanticus* (Apicomplexa, Perkinsea). *Parasitology* **100**: 351-358.

BAHNWEG, G. (1979*a*). Studies on the physiology of the Thraustochytriales. I. Growth requirements and nitrogen nutrition of *Thraustochytrium* spp., *Schizochytrium* sp., *Japonochytrium* sp., *Ulkenia* spp., and *Labyrinthuloides* spp. *Veröffentlichungen des Instituts für Meeresforschung in Bremerhaven* **17**: 245-268.

BAHNWEG, G. (1979*b*). Studies on the physiology of the Thraustochytriales. II. Carbon nutrition of *Thraustochytrium* spp., *Schizochytrium* sp., *Japonochytrium* sp., *Ulkenia* spp., and *Labyrinthuloides* spp. *Veröffentlichungen des Instituts für Meeresforschung in Bremerhaven* **17**: 269-273.

BAHNWEG, G. (1980). Phospholipid and steroid requirements of *Haliphthoros milfordensis*, a parasite of marine crustaceans, and *Phytophthora epistomium*, a facultative parasite of marine fungi. *Botanica Marina, Berlin* **23**: 209-218.

BAHNWEG, G. & BLAND, C. E. (1980). Comparative physiology and nutrition of *Lagenidium callinectes and Haliphthoros milfordensis* fungal parasites of marine crustaceans. *Botanica Marina, Berlin* **23**: 689-698.

BAHNWEG, G. & GOTELLI, D. (1980). Physiology and nutrition of *Lagenidium callinectes* a fungal parasite of the blue crab *Callinectes sapidus*. *Botanica Marina, Berlin* **23**: 219-226.

BAHNWEG, G. & JÄCKLE, I. (1986). A new approach to taxonomy of the Thraustochytriales and Labyrinthulales. *In* S. T. Moss (Ed.), *The Biology of Marine Fungi*, pp. 131-140. Cambridge, U.K.: Cambridge University Press.

BAHNWEG, G. & SPARROW, F. K. (1972). *Aplanochytrium kerguelensis* gen. nov., spec. nov., a new phycomycete from subarctic marine waters. *Archiv für Mikrobiologie* **81**: 45-49.

BAHNWEG. G. & SPARROW, F. K. (1974). Four new species of *Thraustochytrium* from Antarctic regions, with notes on the distribution of zoosporic fungi in the Antarctic marine ecosystems. *American Journal of Botany* **61**: 754-766.

BAIL, T. (1861). Ueber die Identität von *Empusa muscae* Cohn, *Mucor mucedo* Lk., *Achlya prolifera* Nees, und *Hormischium (Cryptococcus) cerevisiae* Bon. *Amtlicher Bericht Versammlung Deutscher Naturforscher und Aertze in Königsberg (1860)* **35**: 252-258.

BAKERSPIGEL, A. (1960). Nuclear structure and division in the vegetative mycelium of the Saprolegniaceae. *American Journal of Botany* **47**: 94-100.

BAKHARIEV, D. (1984). Novo Zabelyavanye po krastavitsite. *Rastitelina Zash* **32**: 13-15 [In Bulgarian, no Latin].

BALAKRISHNAN, M. S. (1948*a*). South Indian Phycomycetes. 1. *Pythium indicum* sp. nov. causing a fruit rot of *Hibiscus esculentus*. *Proceedings of the Indian Academy Science, Section B* **27**: 161-173.

BALAKRISHNAN, M. S. (1948*b*). South Indian Phycomycetes. 2. Some little-known species of *Pythium* occurring in South India. *Proceedings of the Indian Academy of Sciences, Section B* **28**: 27-34.

BALARAMAN, K. & HOTI, S. L. (1986). Factors influencing in vitro zoosporogenesis of mosquito pathogenic fungus *Lagenidium*. *Indian Journal of Medical Research* **83**: 380-383.

BALL, S. L. (1983). Pathogenic variability of downy mildew (*Sclerospora graminicola*) on pearl millet. I. Host cultivar reactions to infection by different pathogen isolates. *Annals of Applied Biology* **102**: 257-264.

BALL, S. L. & PIKE, D. J. (1983). Pathogenic variability of downy mildew (*Sclerospora graminicola*) on pearl millet. II. Statistical techniques for analysis of data. *Annals of Applied Biology* **102**: 265-273.

BALL, S. L. & PIKE, D. J. (1984). Intercontinental variation of *Sclerospora graminicola*. *Annals of Applied Biology* **104**: 41-51.

BANGHAM, R. V. (1933). Parasites of the spotted bass, *Micropterus pseudaplites* Hubbs, and largemouth black bass from Ohio streams. *Transactions of the American Fisheries Society* **63**: 220.

BANIHASHEMI, Z. & MITCHELL, J. E. (1976). Factors affecting oospore germination in *Phytophthora cactorum*, the incitant of apple collar rot. *Phytopathology* **66**: 443-448.

BAO, H., SU, X.-Q., LANG, S.-Y. & ZOU, F.-Z. (1996). Observation on the histological changes of mosquito larvae infected by *Lagenidium giganteum*. *Chun-Kuo Chi Sheng Chung Hsueh Yu Chi Sheng Chung Ping Tsa Chih [Chinese Journal of Parasitology and Parasitic Diseases]* **14**: 303-306 [in Chinese].

BARBIER, J. (1950). Contribution à l'étude de la biologie des phycomycètes aquatiques. *Revue Générale de Botanique* **57**: 23-47.

BARBIER, J. (1969). Le pH et le potentiel d'oxydo-réduction, la reproduction sexuée et asexuée d'une saprolégniée, interaction organism-milieu. *Annales des Sciences Naturelles* **10**: 473-564.

BARKATE, J. A., LARAMORE, C. R., HIRNON, Y. & PERSYN, H. (1974). Some marine organisms related to shrimp diseases. *Proceedings of the World Mariculture Society, 1974*: 267-282.

BARKER, J. (1867). Proceedings of Societies. *Quarterly Journal of Microscopical Science, N. S.* **7**: 232.

BARKER, J. (1868). Proceedings of Societies. *Quarterly Journal of Microscopical Science, N. S.* **8**: 123.

BARKSDALE, A. W. (1960). Interthallic sexual reactions on *Achlya*, a genus of the aquatic fungi. *American Journal of Botany* **47**: 14-23.

BARKSDALE, A. W. (1962*a*). Effect of nutritional deficiency on growth and sexual reproduction of *Achlya ambisexualis*. *American Journal of Botany* **49**: 633-638.

BARKSDALE, A. W. (1962*b*). Concerning the species *Achlya bisexualis*. *Mycologia* **54**: 704-712.

BARKSDALE, A. W. (1963*a*). The uptake of exogenous hormone A by certain strains of *Achlya*. *Mycologia* **55**: 164-171.

BARKSDALE, A. W. (1963*b*). The role of hormone A during sexual conjugation in *Achlya ambisexualis*. *Mycologia* **55**: 627-632.

BARKSDALE, A. W. (1965). *Achlya ambisexualis* and a new cross-conjugating species of *Achlya* (*Achlya heterosexualis*). *Mycologia* **57**: 493-501.

BARKSDALE, A. W. (1966). Segregation of sex in the progeny of a selfed heterozygote of *Achlya bisexualis*. *Mycologia* **58**: 802-804.

BARKSDALE, A. W. (1968). Meiosis in the antheridium of *Achlya ambisexualis* E87. *Journal of the Elisha Mitchell Scientific Society* **84**: 187-194.

BARKSDALE, A. W. (1970). Nutrition and antheridiol-inducing branching in *Achlya ambisexualis*. *Mycologia* **62**: 411-420.

BARKSDALE, A. W. & LASURE, L. L. (1973). Induction of gametangial phenotypes in *Achlya*. *Bulletin of the Torrey Botanical Club* **100**: 199-202.

BARKSDALE, A. W. & LASURE, L. L. (1974). Production of hormone B by *Achlya heterosexualis*. *Appplied Microbiology* **28**: 544-546.

BARKSDALE, A. W., CARLILE, M. J. & MACHLIS, L. (1965). A comparative study of hormone A and sirenin. *Mycologia* **58**: 138-140.

BARKSDALE, A. W., McMORRIS, T. W., SESHADRI. R., ARUNACHALAM, T., EDWARDS, J. A., SUNDEEN, J. & GREEN, D. M. (1974). Response of *Achlya ambisexualis* E87 to the hormone antheridiol and certain other steroids. *Journal of General Microbiology* **82**: 295-299.

BARNES, B. (1954). The classification of the lower fungi. *Proceedings of the Linnean Society of London* **165**: 3.

BARNES, B. & MELVILLE, R. (1932). Notes on British aquatic fungi. *Transactions of the British Mycological Society* **17**: 82-96.

BARR, D. J. S. (1968). A new species of *Olpidium* parasitic on cucumber roots. *Canadian Journal of Botany* **46**: 1087-1091.

BARR, D. J. S. (1970). *Hyphochytrium catenoides*: a morphological and physiological study of North American isolates. *Mycologia* **62**: 492-503.

BARR, D. J. S. (1978). Taxonomy and phylogeny of chytrids. *BioSystems* **10**: 153-165.

BARR, D. J. S. (1979). Morphology and host range of *Polymyxa graminis*, *Polymyxa betae*, and *Ligniera pilorum* from Ontario and some other areas. *Canadian Journal of Plant Pathology* **1**: 85-94.

BARR, D. J. S. (1980*a*). An outline for the reclassification of the Chytridiales, and for a new order, the Spizellomycetales. *Canadian Journal of Botany* **58**: 2380-2394.

BARR, D. J. S. (1980*b*). Fungi Canadenses No. 176. *Olpidium brassicae*. *Fungi Canadenses*. Ottawa, Ontario, Canada: National Mycological Herbarium, Biosystematics Research Centre, Agriculture Canada.

BARR, D. J. S. (1981*a*). Fungi Canadenses No. 199. *Polymyxa graminis*. *Fungi Canadenses*. Ottawa, Ontario, Canada: National Mycological Herbarium, Biosystematics Research Centre, Agriculture Canada.

BARR, D. J. S. (1981*b*). Fungi Canadenses No. 200. *Polymyxa betae*. *Fungi Canadenses*. Ottawa, Ontario, Canada: National Mycological Herbarium, Biosystematics Research Centre, Agriculture Canada.

BARR, D. J. S. (1981*c*). The phylogenetic and taxonomic implications of flagellar rootlet morphology among zoosporic fungi. *BioSystems* **14**: 359-370.

BARR, D. J. S. (1983). The zoosporic grouping of plant pathogens. Entity or non-entity? *In* S. T. Buczacki (Ed.), *Zoosporic Plant Pathogens, a Modern Perspective*, pp.43-83. London, U.K.: Academic Press.

BARR, D. J. S. (1990). Fungi Canadenses No. 329. *Lagena radicicola*. *Fungi Canadenses*. Ottawa, Ontario, Canada: National Mycological Herbarium, Biosystematics Research Centre, Agriculture Canada.

BARR, D. J. S. (1992). Evolution and kingdoms of organisms from the perspective of a mycologist. *Mycologia* **84**: 1-11.

BARR, D. J. S. & ALLAN, P. M. E. (1982). Zoospore ultrastructure of *Polymyxa graminis* (Plasmodiophoromycetes). *Canadian Journal of Botany* **60**: 2496-2504.

BARR, D. J. S. & ALLAN, P. M. E. (1985). A comparison of the flagellar apparatus in *Phytophthora, Saprolegnia, Thraustochytrium,* and *Rhizidiomyces*. *Canadian Journal of Botany* **63**: 138-154.

BARR, D. J. S. & BANDONI, R. (1979). A new species of *Rozella* on a basidiomycete. *Mycologia* **71**: 1261-1264.

BARR, D. J. S. & DÉSAULNIERS, N. L. (1987*a*). *Allochytrium luteum* n. sp.: morphology, physiology and zoospore ultrastructure. *Mycologia* **79**: 193-199.

BARR, D. J. S. & DÉSAULNIERS, N. L. (1987*b*). Ultrastructure of the *Lagena radicicola* zoospore, including a comparison with the primary and secondary *Saprolegnia* zoospores. *Canadian Journal of Botany* **65**: 2161-2176.

BARR, D. J. S. & DÉSAULNIERS, N. L. (1989*a*). The flagellar apparatus of the Oomycetes and Hyphochytriomycetes. *In* J. C. Green, B. S. C. Leadbeater & W. L. Diver (Eds), *The Chromophyte Algae, Problems and Perspectives*, pp. 343-355. Oxford, U.K.: Clarendon Press.

BARR, D. J. S. & DÉSAULNIERS, N. L. (1989*b*). The flagellar apparatus in zoospores of *Phytophthora sojae* f. sp. *glycines* and *P. megasperma*. *Canadian Journal of Botany* **67**: 1916-1926.

BARR, D. J. S. & DÉSAULNIERS, N. L. (1990*a*). The life cycle of *Lagena radicicola*, an oomycetous parasite of wheat roots. *Canadian Journal of Botany* **68**: 813-824.

BARR, D. J. S. & DÉSAULNIERS, N. L. (1990*b*). The flagellar apparatus in the *Phytophthora infestans* zoospore. *Canadian Journal of Botany* **68**: 2112-2118.

BARR, D. J. S. & DÉSAULNIERS, N. L. (1992). The flagellar apparatus in zoospores of *Phytophthora, Pythium* and *Halophytophthora*. *Canadian Journal of Botany* **70**: 2163-2169.

BARR, D. J. S. & HADLAND-HARTMAN, V. E. (1977). Zoospore ultrastructure of *Olpidium cucurbitacearum* (Chytridiales). *Canadian Journal of Botany* **55**: 3063-3074.

BARR, D. J. S. & SLYKHIUS, J. T. (1976). Further observations on zoosporic fungi associated with wheat spindle streak mosaic virus. *Canadian Plant Disease Survey* **56**: 77-81.

BARR, D. J. S., WARWICK, S. I. & DÉSAULNIERS, N. L. (1996). Isozyme variation, morphology, and growth response to temperature in *Pythium ultimum*. *Canadian Journal of Botany* **74**: 753-761.

BARR, K. J. & ASHER, M. J. C. (1992). The host range of *Polymyxa betae* in Britain. *Plant Pathology* **41**: 64-68.

BARR, K. J. & ASHER, M. J. C. (1996). Studies on the life-cycle of *Polymyxa betae* in sugar beet roots. *Mycological Research* **101**: 203-208.

BARRETO, R. W. & DICK, M. W. (1991). Monograph of *Basidiophora* (Oomycetes) with the description of a new species. *Botanical Journal of the Linnean Society* **107**: 313-332.

BARRETT, J. T. (1912*a*). A serious root disease of radish. *Phytopathology* **2**: 96 [abstract].

BARRETT, J. T. (1912*b*). Development and sexuality of some species of *Olpidiopsis*, (Cornu) Fischer. *Annals of Botany, London* **26**: 209-238.

BARRETT, J. T. (1934). A chytridiaceous parasite of *Phytophthora*. *Phytopathology* **24**: 1138.

BARRETT, J. T. (1948). Induced oospore production in the genus *Phytophthora*. *Phytopathology* **38**: 2.

BARRON, G. L. (1973). Nematophagous fungi: a new *Gonimochaete*. *Canadian Journal of Botany* **51**: 2451-2453.

BARRON, G. L. (1976*a*). Nematophagous fungi: a new endoparasite intermediate between *Lagenidium* and *Myzocytium*. *Canadian Journal of Botany* **54**: 1-4.

BARRON, G. L. (1976*b*). Nematophagous fungi: new species of the Lagenidiales endoparasitic on *Rhabditis*. *Antonie van Leeuwenhoek, Journal of Microbiology and Serology* **42**: 131-139.

BARRON, G. L. (1976*c*). Nematophagous fungi: three new species of *Myzocytium*. *Canadian Journal of Microbiology* **22**: 752-762.

BARRON, G. L. (1977*a*). Nematophagous fungi: *Protascus* and its relationship to *Myzocytium*. *Canadian Journal of Botany* **55**: 819-824.

BARRON, G. L. (1977*b*). *The Nematode-Destroying fungi*. Guelph, Ontario, Canada: Canadian Biological Publications.

BARRON, G. L. (1978). Nematophagous fungi: Endoparasites of *Rhabitis terricola*. *Microbial Ecology* **4**: 157-163.

BARRON, G. L. (1980). A new *Haptoglossa* attacking rotifers by rapid injection of an infective sporidium. *Mycologia* **72**: 1186-1194.

BARRON, G. L. (1981). Two new fungal parasites of bdelloid rotifers. *Canadian Journal of Botany* **59**: 1449-1455.

BARRON, G. L. (1985). A new *Gonimochaete* with an oospore state. *Mycologia* **77**: 17-23.

BARRON, G. L. (1987). The gun cell of *Haptoglossa mirabilis*. *Mycologia* **79**: 877-883.

BARRON, G. L. (1989*a*). New species and new records of fungi that attack microscopic animals. *Canadian Journal of Botany* **67**: 267-271.

BARRON, G. L. (1989*b*). Host range studies for *Haptoglossa* and a new species, *Haptoglossa intermedia*. *Canadian Journal of Botany* **67**: 1645-1648.

BARRON, G. L. (1990). A new and unusual species of *Haptoglossa*. *Canadian Journal of Botany* **68**: 435-438.

BARRON, G. L. & PERCY, J. G. (1975). Nematophagous fungi: a new *Myzocytium*. *Canadian Journal of Botany* **53**: 1306-1309.

BARRON, G. L. & SZIJARTO, E. (1986). A new species of *Olpidium* in nematode eggs. *Mycologia* **78**: 972-975.

BARSTOW, W. E., FRESHOUR, G. D. & FULLER, M. S. (1989). The ultrastructure of mitosis during zoosporogenesis in *Rhizidiomyces apophysatus*. *Canadian Journal of Botany* **67**: 3401-3409.

BARTEAUX, M. M. (1950). Aquatic oomycetes of the vicinity of Orono, Maine. *Phytopathology* **40**: 1-2.

BARTLETT, A. W. (1928). *Olpidium radicicolum* de Wildeman and the "hybridisation nodules" of swedes. *Transactions of the British Mycological Society* **13**: 221-238.

BARTNICKI-GARCIA, S. (1966). Chemistry of hyphal walls of *Phytophthora*. *Journal of General Microbiology* **42**: 57-69.

BARTNICKI-GARCIA, S. (1970). Cell wall composition and other biochemical markers in fungal phylogeny. *In* J. G. Harborne (Ed.), *Phytochemical Phylogeny*, pp. 81-103. New York, U.S.A.: Academic Press.

BARTNICKI-GARCIA, S. & HEMMES, D. E. (1976). Some aspects of the form and function of oomycete spores. *In* D. J. Weber & W. M. Hess (Eds), *The Fungal Spore: Form and Function*, pp. 593-641. New York, U.S.A.: John Wiley & Sons.

BARTNICKI-GARCIA, S. & WANG, M. C. (1983). Biochemical aspects of morphogenesis in *Phytophthora*. *In* D. C. Erwin, S. Bartnicki-Garcia & P. H. Tsao (Eds), Phytophthora *Its Biology, Taxonomy, Ecology, and Pathology*, pp. 121-137. St Paul, Minnesota, U.S.A.: American Phytopathological Society.

BARTSCH, A. F. & WOLF, F. T. (1938). Two new saprolegniaceous fungi. *American Journal of Botany* **25**: 392-395.

BARTSCH, G. (1971). Cytologische Beobachtungen an *Labyrinthula coenocystis* Schmoller bei verschiedenen Kulturbedingungen. *Zeitschrift für allgemeine Mikrobiologie* **11**: 79-90.

BATE, C. S. (1888). Report on the Crustacea Macrura collected by H.M.S. Challenger during the years 1873-1876. *Challenger Report* **24**: 1-192.

BATKO, A. (1971*a*). A new species of *Pythiogeton* showing *Achlya*-like diplanetism. *Acta Mycologica* **7**: 241-249.

BATKO, A. (1971*b*). *Nellymyces megaceros* gen. et sp. nov. - a new phycomycete related to *Aqualinderella* and *Araiospora*. *Acta Mycologica, Warszawa* **7**: 251-268.

BATKO, A. (1973). On the new desmids parasite - *Lagenidium podbielkowskii*. *Acta Mycologica, Warszawa* **9**: 105-110.

BATKO, A. (1975). *Zarys Hydromikologie*. Warsaw, Poland: Panstwarwe Whydiawannicto Naukowe.

BATKO, A. (1977). Further observations on *Nellymyces megaceros* and its parasite - *Rozellopsis uliginosa* sp. nov. *Acta Mycologica, Warszawa* **13**: 313-324.

BAUM, B. R. & SAVILE, D. B. O. (1985). Rusts (Uredinales) of Triticeae: evolution and extent of coevolution, a cladistic analysis. *Botanical Journal of the Linnean Society* **91**: 367-394.

BEAKES, G. W. (1980*a*). Electron microscopic study of oospore maturation and germination in an emasculate isolate of *Saprolegnia ferax*. 1. Gross changes. *Canadian Journal of Botany* **58**: 182-194.

BEAKES, G. W. (1980*b*). Electron microscopic study of oospore maturation and germination in an emasculate isolate of *Saprolegnia ferax*. 2. Wall differentiation. *Canadian Journal of Botany* **58**: 195-208.

BEAKES, G. W. (1980*c*). Electron microscopic study of oospore maturation and germination in an emasculate isolate of *Saprolegnia ferax*. 3. Changes in organelle status and associations. *Canadian Journal of Botany* **58**: 209-227.

BEAKES, G. W. (1980*d*). Electron microscopic study of oospore maturation and germination in an emasculate isolate of *Saprolegnia ferax*. 4. Nuclear cytology. *Canadian Journal of Botany* **58**: 228-240.

BEAKES, G. W. (1981*a*). Ultrastructure of the phycomycete nucleus. *In* K. Gull & S. G. Oliver (Eds), *The Fungal Nucleus*, pp. 1-35. British Mycological Society Symposium 5. Cambridge, U.K.: Cambridge University Press.

BEAKES, G. W. (1981*b*). Ultrastructural aspects of oospore differentiation. *In* H. R. Hohl & G. Turian (Eds), *The Fungal Spore: Morphogenetic Controls*, pp. 71-94. London, U.K.: Academic Press.

BEAKES, G. W. (1983). A comparative account of cyst coat ontogeny in saprophytic and fish-lesion (pathogenic) isolates of the *Saprolegnia diclina - parasitica* complex. *Canadian Journal of Botany* **61**: 603-625.

BEAKES, G. W. (1987). Oomycete phylogeny: ultrastructural perspectives. *In* A.D.M. Rayner, C.M. Brasier & D.Moore (Eds), *Evolutionary Biology of the Fungi*, pp. 405-421. Cambridge, U.K.: Cambridge University Press.

BEAKES, G. W. (1989). Oomycete fungi: their phylogeny and relationship to chromophyte algae. *In* J. C. Green, B. S. C. Leadbeater & W. L. Diver (Eds), *The Chromophyte Algae, Problems and Perspectives*, pp. 325-342. Oxford, U.K.: Clarendon Press.

BEAKES, G. W., BURR, A. W., WOOD, S. E. & HARDHAM, A. R. (1995). The application of spore surface features in defining taxonomic versus ecological groupings in oomycete fungi. *Canadian Journal of Botany, Supplement 1, Sections E-H* **73**: S701-S711.

BEAKES, G. W., EL-HAMALAWI, Z. A. & ERWIN, D. C. (1986). Ultrastructure of mature oospores of *Phytophthora megasperma* f. sp. *medicaginis*: preparation protocols and efffects of MTT vital staining and permanganate treatment. *Transactions of the British Mycological Society* **86**: 195-206.

BEAKES, G. & FORD, H. (1983). Esterase isoenzyme variation in the genus *Saprolegnia*, with particular reference to the fish pathogenic *S. diclina - S. parasitica* complex. *Journal of General Microbiology* **129**: 2605-2619.

BEAKES, G. W. & GAY. J. L. (1977). Gametangial nuclear division and fertilization in *Saprolegnia furcata* as observed by light and electron microscopy. *Transactions of the British Mycological Society* **69**: 459-471.

BEAKES, G. W. & GAY. J. L. (1978*a*). Light and electron microscopy of oospore maturation in *Saprolegnia furcata*. 1. Cytoplasmic changes. *Transactions of the British Mycological Society* **71**: 11-24.

BEAKES, G. W. & GAY. J. L. (1978*b*). Light and electron microscopy of oospore maturation in *Saprolegnia furcata*. 2. Wall development. *Transactions of the British Mycological Society* **71**: 25-35.

BEAKES, G. W. & GAY. J. L. (1980). The effects of streptomycin on growth and sporulation of *Saprolegnia* spp. *Journal of General Microbiology* **119**: 361-371.

BEAKES, G. W. & GLOCKLING, S. L. (1998). Injection tube differentiation in gun cells of a *Haptoglossa* species which infects nematodes. *Fungal Genetics and Biology* **24**: 45-68.

BEAN, G. A., PATTERSON, G. W. & MOTTA, J. J. (1972). Sterols and fatty acids of some aquatic phycomycetes. *Comparative Biochemistry and Physiology* **43B**: 935-939.

BEAUCHAMP, P. de (1914). L'évolution et les affinités des protistes du genre *Dermocystidium*. *Comptes Rendus Hebdomadaires des Séances de l'Académie des Sciences, Paris, Série D* **158**: 1359.

BECK, S. J. & ERB, K. (1984). Facultative parasitism of *Spirogyra* sp. (Chlorophyta) by *Saprolegnia asterophora* and *Pythium gracile* (Eumycota, Oomycetes. *Journal of Phycology* **20**: 13-19.

BECKER, E. R. (1926). *Endamoeba citelli* sp. nov. from striped ground squirrel *Citelli tridecemlineatus*, and the life-history of its parasite, *Sphaerita endamoebae* sp. nov. *Biological Bulletin of the Marine Biological Laboratory, Woods Hole* **50**: 444-454.

BECNEL, J. J., GARCIA, J. & JOHNSON, M. (1996). Effects of three larvicides on the production of *Aedes albopictus* based on removal of pupal exuviae. *Journal of the American Mosquito Association* **12**: 499-502.

BEDLAN, G. (1989). Erstmaliger Nachweis von Oosporen von *Pseudoperonospora cubensis* (Berk. et Curt.) Rost. an Gewächshausgurken in Österreich. *Pflanzenschutzberichte* **50**: 119-120.

BEERS, C. D. & SHERWOOD, W. A. (1966). A new species of *Woodruffia*, a zoospore-ingesting ciliate occurring on the water nold *Saprolegnia*. *Transactions of the American Microscopcal Society* **85**; 528-536.

BEECH, P. L., HEIMANN, K. & MELKONIAN, M. (1991). Development of the flagellar apparatus during the cell cycle in unicellular algae. *Protoplasma* **164**: 23-37.

BIBLIOGRAPHY

BEERS, C. D. & SHERWOOD, W. A. (1966). A new species of *Woodruffia*, a zoospore-ingesting ciliate occurring on the water mold *Saprolegnia*. *Transactions of the American Microscopical Society* **85**: 528-536.

BEHRENS, A. (1931). Zytologische Untersuchungen an *Rhypidium europaeum* (Cornu) v. Minden. *"Planta", Archiv für Wissenschaftliche Botanik* **13**: 745-777.

BELCHER, J. H. & SWALE, E. M. F. (1971). The microanatomy of *Phaeaster pascheri* Scherffel (Chrysophyceae). *British Phycological Journal* **6**: 157-169.

BELKHIRI, A., BUCHKO, J. & KLASSEN, G. R. (1992). The 5S ribosomal RNA gene in *Pythium* species: two different genomic locations. *Molecular Biology and Evolution* **9**: 1089-1102.

BELKHIRI, A. & DICK, M. W. (1988). Comparative studies on the DNA of *Pythium* species and some possibly related taxa. *Journal of General Microbiology* **134**: 2673-2683.

BELL, J. V. (1979). Mycopathogens for pest insect suppression. *Journal of the Mississippi Academy of Sciences* **24**: 1-7.

BELL, T. J., LEE, B. & DOMNAS, A. J. (1989). The extracellular acid phosphatase of the mosquito-parasitizing fungus *Lagenidium giganteum*. *Journal of Invertebrate Pathology* **54**: 306-313.

BELSKY, M. M. & GOLDSTEIN, S. (1964). Glucose metabolism of *Thraustochytrium roseum*. *Archiv für Mikrobiologie* **49**: 375-382.

BELSKY, M. M., GOLDSTEIN, S. & SIEGENTHALER, P. A. (1966). Light stimulated phosphate uptake by the non-filamentous marine fungus *Thraustochytrium roseum*. *Bacteriological Proceedings* **66**: 138.

BENEKE, E. S. (1948*a*). A new species of *Achlya* and *Dictyuchus*. *Journal of the Elisha Mitchell Scientific Society* **64**: 261-265.

BENEKE, E. S. (1948*b*). A monographic study of Illinois species of Saprolegniales and Leptomitales. Unpublished PhD Thesis, 166 pp. University of Illinois, Illinois, U.S.A.

BENEKE, E. S. & ROGERS, A. L. (1962). Aquatic phycomycetes isolated in the States of Minas Gerais, São Paulo, and Paraná, Brazil. *Rickia, Brazil* **1**: 181-193.

BENEKE, E. S. & ROGERS, A. L. (1962). Aquatic fungi of Parque Nacional do Itatiaia in the State of Rio de Janeiro. *Rickia, Brazil* **5**: 51-64.

BENEKE, E. S. & SCHMITT, J. A. (1961). Aquatic fungi from South Bass and neighboring islands in western Lake Erie. I. Uniflagellate and biflagellate phycomycetes. *Ohio Journal of Science* **61**: 283-285.

BENEKE, E. S. & SHANOR, L. (1947). Representatives of the Saprolegniales and Leptomitales in Illinois. *American Journal of Botany* **34**: 609.

BENHAMOU, N., CHAMBERLAND, H., NOËL, S. & OUELLETTE, G. B. (1990). Ultrastructural localization of ß-1,4-glucan- containing molecules in the cell walls of some fungi: a comparative study between spore and mycelium. *Canadian Journal of Microbiology* **36**: 149-158.

BENHAMOU, N., CHAMBERLAND, H., OUELLETTE, G. B. & PAUZÉ, F. J. (1987). Ultrastructural localization of ß-1,4-glucans in two pathogenic fungi and in their host tissues by means of an exoglucanase-gold complex. *Canadian Journal of Microbiology* **33**: 404-417.

BENISCH, J. (1940). Kuntslich hervorgerufener *Aphanomyces* Befall bei Wollhandkrabben. *Zeitschrift für Fischerei und Deren Hilfswissenschaften* **38**: 71-80.

BENNETT, H. D. (1972). Cytology of *Phytophthora heveae*. Paper presented to the European Discussion Groups in Plant Pathology, Bari, Italy.

BENSAUDE, M. (1923). A species of *Olpidium* parasitic in the rotsof tomato, tobacco, and cabbage. *Phytopathology* **13**: 451-454.

BERBEE, M. L., CARMEAN, D., WINKLA, K. & ERIKSSON, O. (1998). Phylogenetic resolution and the radiations of the Ascomycota. *Inoculum* **49**: 8.

BERBEE, M. L. & KERWIN, J. L. (1988). Cell surface of *Lagenidium giganteum*, a fungus used for mosquito control. *University of California Mosquito Control Research, Annual Report, 1988*: 23-26.

BERBEE, M. L. & KERWIN, J. L. (1993). Ultrastructural and light microscopic localization of carbohydrates and peroxidase-catalases in *Lagenidium giganteum* zoospores. *Mycologia* **85**: 734-743.

BÉRCZI, L. (1940). Adatok Magyaroszág vízigomba vegetációjához. Vízigombákkal végzett mesterséges feretözési kisérletek (Angeben zur Wasserpilzvegetation ungarns nebst mich Wasserpilen durchgeführten Ansteckungsversuchen). *Geobotany* **3**: 79-99.

BERDAN, H. B. (1938). Revision of the genus *Ancylistes*. *Mycologia* **30**: 396-415.

BERG, L. R. & PATTERSON, G. W. (1983). Effects of triarimol and tridemorph on sterol biosynthesis in 5 species of Oomycetes. *Plant Physiology* **72**: 170.

BERG, L. R. & PATTERSON, G. W. (1986). Phylogenetic implications of sterol biosynthesis in the oomycetes. *Experimental Mycology* **10**: 175-183.

BERGAN, P. (1953). A note on the ellobiopsid genus *Amallocystis* Fage. *Universitet I Bergen Årbok, 1952, Naturvitenskapelig Rekke* **14**: 1-19, plates 1 & 2.

BERGAN, P. (1961). Eiendommelige parasitter. *Fauna, Oslo* **14**: 19-24.

BERKALOFF, C. & ROUSSEAU, B. (1979). Ultrastructure of male gametogenesis in *Fucus serratus*. *Journal of Phycology* **15**: 163-173.

BERKELEY, C. J. A. (1944). Note on the habit and spore formation in a *Saprolegnia*. *Transactions of the British Mycological Society* **27**: 68.

BERKELEY, M. J. (1845). Editorial. *Gardeners' Chronicle and Agricultural Gazette, 1845 (38)*: 640.

BERKELEY, M. J. (1846). Observations, botanical and physiological, on the potato murrain. *Journal of the Horticultural Society of London*, **1**: 9-34.

BERKELEY, M. J. (1851). Observations of the vine mildew. By J. H. Léveillé. Translated from the French, with remarks by the Rev. M. J. Berkeley, M.A., F.L.S. *Journal of the Horticultural Society of London*, **6**: 284-295.

BERKELEY, M. J. (1864). Egg parasites and their relatives. *The Intellectual Observer: Review of Natural History, Microscopic Research, and Recreative Science* **5(3)**: 147-153, 1 plate. [The journal title should not be abbreviated - M. Beasley, pers. comm.]

BERKELEY, M. J. & CURTIS, M. A. (1869). Fungi Cubenses (Hymenomycetes). *Journal of the Linnean Society of London (Botany)*, **10**: 280-392.

BERLESE, A. N. (1898). Ueber die Befruchtung und Entwickelung der Oosphäre bei den Peronosporeen. *Jahrbücher für wissenschaftliche Botanik* **31**: 159-196.

BERLESE, A. N. (1897-1902). Saggio di una monografia delle Peronosporaceae. *Revista di Patologia Vegetale, Padova*, **4**: 1-126, **6**: 78-101, 237-268 (1897); **7**: 19-37 (1898); **9**: 1-126 (1900); **10**: 185-298 (1902).

BERLIN, J. D. & BOWEN, C. C. (1964). Centrioles in the fungus *Albugo candida*. *American Journal of Botany* **51**: 650-652.

BERMÙDEZ, D. (1980). Preliminary experiences in the control of fish diseases in warm water aquaculture operations in Venezuela. *Journal of Fish Diseases* **3**: 355-357.

BEROQUI DE MARTINEZ, M. E. (1969*a*). Una nueva especie de *Achlya* (Saprolegniales, Phycomycetes). *Kurtziana, Córdoba* **5**: 95-99.

BEROQUI DE MARTINEZ, M. E. (1969*b*). Dos especies nuevas del genero *Achlya* (Phycomycetes-Saprolegniales) de la Argentina. *Darwiniana (San Isidro)* **15**: 9-13.

BEROQUI DE MARTINEZ, M. E. (1970). Contribución al estudio de los Phycomycetes acuaticos de la Republica Argentina. I. El genero *Achlya* en aguas de la ciudad de Buenos Aires y Alrededores. *Boletin de la Sociedad Argentina de Botanica* **13**: 109-124.

BERTKE, C. C. & ARONSON, J. M. (1980). Hyphal wall composition in *Apodachlyella completa*. *Current Microbiology* **4**: 235-238.

BERTKE, C. C. & ARONSON, J. M. (1985). Hyphal wall composition of *Mindeniella spinospora* and *Araiospora* sp. *American Journal of Botany* **72**: 467-471.

BERTKE, C. C. & ARONSON, J. M. (1992*a*). Hyphal wall chemistry of *Lagenidium callinectes* and *Lagenidium chthamalophilum*. *Botanica Marina, Berlin* **35**: 147-152.

BERTKE, C. C. & ARONSON, J. M. (1992*b*). Hyphal wall composition of *Lagenidium giganteum*. *Mycologia* **84**: 571-574.

BESCHERELLE, É. (1872). Prodromus bryologiae Mexicanae ou Énumération des mousses de Mexique. *Mémoires de la Société des Sciences Naturelles de Cherbourg* **6**: 145-256.

BESSEY, C. E. (1903). The structure and classification of the Phycomycetes with a revision of the families and a rearrangement of the North American genera. *Transactions of the American Microscopical Society* **24**: 27-54.

BESSEY, C. E. (1907). A synopsis of plant phyla. *University of Nebraska Studies* **7**: 275-373.

BESSEY, E. A. (1942). Some problems in fungus phylogeny. *Mycologia* **34**: 355-376.

BESSEY, E. A. (1950). *Morphology and Taxonomy of Fungi*. Philadelphia, U.S.A.: Blakiston Company.

BEUTE, M. K. & LOCKWOOD, J. L. (1967). Pathogenic variability in *Aphanomyces euteiches*. *Phytopathology* **57**: 57-60.

BEVERWIJK, A. J. van (1948). Observations on submerged water moulds in the Netherlands. *Antonie van Leeuwenhoek, Journal of Microbiology and Serology* **14**: 223-250.

BHALLA, H. S. & MITCHELL, J. E. (1970). A method of obtaining viable, mycelium-free oospores of *Aphanomyces euteiches* using live water snails. *Phytopathology* **60**: 1010-1012.

BHARADWAJ, S. D. & DWIVEDI, R. (1969). A new species of *Synchytrium* from Varanasi. *Proceedings of the 56th Indian Scientific Congress* **3**: 306.

BHARGAVA, K. S. (1944). Fertilisation in *Isoachlya anisospora* (de Bary) Coker var. *indica* Sak. et Bhar. *Current Science, India (Bangalore)* **13**: 236-237.

BHARGAVA, K. S. (1943). Physiological studies of some members of the Saprolegniaceae. I. Enzyme action. *Journal of the Indian Botanical Society* **22**: 85-99.

BHARGAVA, K. S. (1945*a*). Physiological studies of some members of the Saprolegniaceae. II. Sulfur and phosphorus requirements. *Proceedings of the Indian Academy of Sciences, B* **21**: 344-349.

BHARGAVA, K. S. (1945*b*). Physiological studies of some members of the Saprolegniaceae. III. Nitrogen requirements. *Journal of the Indian Botanical Society* **24**: 67-72.

BHARGAVA, K. S. (1945*c*). Physiological studies of some members of the Saprolegniaceae. IV. Carbohydrate requirements. *Lloydia (Cincinnati)* **8**: 60-68.

BHARGAVA, K. S. (1946*a*). Oogenesis and fertilization in *Isoachlya anisospora* var. *indica*. *Transactions of the British Mycological Society* **29**: 101-107.

BHARGAVA, K. S. (1946*b*). Physiological studies on some members of the Saprolegniaceae. V. Growth substances. *Lloydia (Cincinnati)* **9**: 13-23.

BHARGAVA, K. S. (1950). Formation and liberation of zoospores in *Isoachlya anisospora* var. *indica*. *Botanical Gazette* **112**: 24-31.

BHARGAVA, K. S. & SINGH, B. B. (1965). Observations on Indian aquatic fungi III. Some species of Blastocladiales, Monoblepharidales and Peronosporales collected from Gorakhpur. *Proceedings of the National Academy of Sciences, India* **35**: 49-58.

BHARGAVA, K. S. & SRIVASTAVA, G. C. (1966). *Isoachlya luxurians* sp. nov., from India. *Transactions of the British Mycological Society* **49**: 269-273.

BHAT, S. S. & GOWDA, P. S. B. (1985). An efficient method for culturing downy mildew fungi on host callus. *Transactions of the British Mycological Society* **84**: 161-162.

BHAT, S. S., SAFEEULLA, K. M. & SHAW, C. G. (1980). Growth of *Sclerospora graminicola* in host tissue cultures. *Transactions of the British Mycological Society* **75**: 303-309.

BHATTACHARYA, D. & DRUEL, L. D. (1988). Phylogenetic comparison of the small-subunit ribosomal DNA sequence in *Costaria costosa* (Phaeophyta) with those of other algae, vascular plants and oomycetes. *Journal of Phycology* **24**: 539-543.

BHATTACHARYA, D. & EHLTING, J. (1995). Actin coding regions: gene family evolution and use as a phylogenetic marker. *Archiv für Protistenkunde* **145**: 155-164.

BHATTACHARYA, D. & STICKEL, S. K. (1994). Sequence analysis of duplicated actin genes in *Lagenidium giganteum* and *Pythium irregulare* (Oomycota). *Journal of Molecular Evolution* **39**: 56-61.

BHATTACHARYA, D., STICKEL, S. K. & SOGIN, M. L. (1991). Molecular phylogenetic analysis of actin genic regions from *Achlya bisexualis* (Oomycota) and *Costaria costata* (Chromophyta). *Journal of Molecular Evolution* **33**: 525-536.

BIAN, B. Z & EGUSA, S. (1980). *Atkinsiella hamanaensis* sp. nov. isolated from cultivated ova of the mangrove crab, *Scilla serrata* (Forsskål). *Journal of Fish Diseases* **3**: 373-385.

BIAN, B. Z., HATAI, K., PO, G. L. & EGUSA, S. (1979). Studies on the fungal diseases in Crustaceans. I. *Lagenidium scyllae* sp. nov. isolated from cultivated ova and larvae of the mangrove crab (*Scylla serrata*). *Transactions of the Mycological Society of Japan* **20**: 115-124.

BIASOLETTO, B. (1832). *Di Alcune Alghe Microscopiche, Saggio del Dr. B. Biasoletto*. Trieste, Italy: Tipografia Weis.

BIGA, M. L. B. (1955). Riesaminazione delle specie del genere *Albugo* in base alla morfologia dei conidi. *Sydowia* **9**: 339-358.

BILGRAMI, K. S., JAMALUDDIN & RIZVI, M. A. (1979). *Fungi of India, Part I*. New Delhi, India: Today and Tomorrow Publishers.

BIMPONG, C. E. (1975). Changes in metabolic reserves and enzyme activities during zoospore motility and cyst germination in *Phytophthora palmivora*. *Canadian Journal of Botany* **53**: 1411-1416.

BIMPONG, C. E. & HICKMAN, C. J. (1975). Ultrastructural and cytochemical studies of zoospores, cysts, and germinating cysts of *Phytophthora palmivora*. *Canadian Journal of Botany* **53**: 1310-1327.

BIRCHER, U. & HOHL, H. R. (1997). Environmental signalling during induction of appressorium formation in *Phytophthora*. *Mycological Research* **101**: 395-402.

BIRCHFIELD, W. (1960). A new species of *Catenaria* parasitic on nematodes of sugarcane. *Mycopathologia et Mycologia Applicata* **13**: 331-338.

BISHOP, H. (1940). A study of sexuality in *Sapromyces reinschii*. *Mycologia* **32**: 505-529.

BISHOP, M. W. H. (1950). Fungoid infection in *Ectocarpus granulosus*. *Nature (London)* **165**: 937.

BISSONETTE, K. W., SHARP, N. J. H., DYSTRA, M. H., ROBERTSON, I. R., DAVIS, A. A., PADHYE, A. A. & KAUFMAN, L. (1991). Nasal and retrobulbar mass in a cat caused by *Pythium insidiosum*. *Journal of Medical and Veterinary Mycology* **29**: 39-44.

BLACKWELL, E. (1943*a*). The life history of *Phytophthora cactorum* (Leb. and Cohn) Schroet. *Transactions of the British Mycological Society* **26**: 71-89.

BLACKWELL, E. (1943*b*). Presidential address. On germinating the oospores of *Phytophthora cactorum*. *Transactions of the British Mycological Society* **26**: 93-103.

BLACKWELL, E. (1944). Species of *Phytophthora* as water moulds. *Nature (London)* **153**: 496.

BLACKWELL, E. (1953). Haustoria of *Phytophthora infestans* and some other species. *Transactions of the British Mycological Society* **36**: 138-158.

BLACKWELL, E. & WATERHOUSE, G. M. (1941). Watermoulds as a sourse of infection by pathogenic species of *Phytophthora*. *Mycologia* **33**: 449.

BLANC, H. (1887). Notice sur une mortalité exceptionnelle des brochets du Lac Léman en 1887. *Bulletin de la Société Vaudoise des Sciences Naturelles* **23**: 33-37.

BLAND, C. E. & AMERSON, H. V. (1973*a*). Electron microscopy of zoosporogenesis in the marine phycomycete, *Lagenidium callinectes* Couch. *Archiv für Mikrobiologie* **94**: 47-64.

BLAND, C. E. & AMERSON, H. V. (1973*b*). Observations on *Lagenidium callinectes*: isolation and sporangial development. *Mycologia* **65**: 310-320.

BLAND, C. E. & AMERSON, H. V. (1973*c*). Observations on *Lagenidium callinectes* Couch, a fungal parasite of blue crab ova. *Chesapeake Science* **15**: 232-235.

BLANK, I. H. & TIFFNEY, W. N. (1936). The use of ultra-violet irradiated culture media dfor securing bacteria-free cultures of *Saprolegnia*. *Mycologia* **28**: 324-329.

BLATTNY, C. (1935). Príspěveck k póznaní hlenky Bromborové. *Sbornik Vyzhumných ústaru zemědělských RČS [Recuil des instituts de Recherches agronomiques de la République Tchéchoslovaque]* **137**: 21-25.

BLOK, I. & PLAATS-NITERINK, A. J. Van der (1978). *Pythium uncinulatum* sp. nov. and *P. tracheiphilum* pathogenic to lettuce. *Netherlands Journal of Plant Pathology* **84**: 135-147.

BLOMFIELD, J. E. & SCHWARTZ, E. J. (1910). Some observations on the tumors of *Veronica chamaedrys* caused by *Sorosphaera veronicae*. *Annals of Botany, London* **24**: 35-43.

BLUNT, S. J. & ASHER, M. J. C. (1989). *Polymyxa*: the rhizomania vector. *British Sugar Beet Review* **57**: 31-33.

BLUNT, S. J., ASHER, M. J. C. & GILLIGAN C. A. (1991). Infection of sugar beet by *Polymyxa betae* in relation to soil temperature. *Plant Pathology* **40**: 256-267.

BLUNT, S. J., ASHER, M. J. C. & GILLIGAN C. A. (1992). The effect of sowing date on infection of sugar beet by *Polymyxa betae*. *Plant Pathology* **41**: 148-153.

BLY, J. E. & CLEM, L. W. (1991). Temperature-mediated processes in teleost immunity: *in vitro* immunosuppression induced by *in vivo* low temperature in channel catfish. *Veterinary Immunology and Immunopathology* **28**: 365-377.

BLY, J. E., LAWSON, L. A., DALE, D. J., SZALAI, A. J., DURBUROW, R. M. & CLEM, L. W. (1992). Winter saprolegniosis in channel catfish. *Diseases of Aquatic Organisms* **13**: 155-164.

BLY, J. E., LAWSON, L. A., SZALAI, A. J. & CLEM, L. W. (1993). Environmental factors affecting outbreaks of winter saprolegniosis in channel catfish, *Ictalurus punctatus* (Rafinesque). *Journal of Fish Diseases* **16**: 541-549.

BOAG, B. & LOPEZ-LLORCA, L. V. (1989). Nematodes and nematophagous fungi associated with cereal field and permanent pasture in eastern Scotland. *Crop Research (Edinburgh)* **29**: 1-10.

BOCCAS, B. (1972). Contribution à l'étude du cycle chez les *Phytophthora*. Analyse du mode de transmission d'un charactère génétique quantitatif chez une espèce homothallique, le *Phytophthora syringae* Kleb. *Comptes Rendus Hebdomadaires des Séances de l'Academie des Sciences, Paris* **275**: 663-666.

BOCK, C. H., JEGER, M. J., MUGHOGHO, L. K., CARDWELL, K. F. & MTISI, E. (1999). Effect of dew point temperature and conidium age on germination, germ tube growth and infection of maize and sorghum by *Peronosclerospora sorghi*. *Mycological Research* **103**: 859-864.

BOCK, C. H., JEGER, M. J., MUGHOGHO, L. K., CARDWELL, K. F., MTISI, E., KAULA, G. & MUKANSABIMANA, D. (2000). Variability of *Peronosclerospora sorghi* isolates from different geographic locations and hosts in Africa. *Mycological Research* **104**: 61-68.

BOCK, C. H., JEGER, M. J., MUGHOGHO, L. K., MTISI, E. & CARDWELL, K. F. (1998). Production of conidia by *Peronosclerospora sorghi* on sorghum crops in Zimbabwe. *Plant Pathology* **47**: 243-251.

BOCK, K. J. (1952). Zur Ökologie der Ciliaten des marinen Sandgrundes der Kieler Bucht. I. *Kieler Meeresforschungen* **9**: 77.

BOCK, K. J. (1956). Zur Ökologie und Systematik saprophytischer Wasserpilze aus dem Silberse bei Bremerhaven. *Veröffentlichungen des Instituts für Meeresforschung in Bremerhaven* **4**: 25-44.

BOEDJIN, K. (1921). Nieuwe Nederlandsche Saprolegniineae. *Mededeelingen van de Nederlandsche Mycologische Vereeniging* **11**: 16-120.

BOEDJIN, K. (1923). Nieuwe Nederlandische Saprolegniineae II. *Mededeelingen van de Nederlandsche Mycologische Vereeniging* **13**: 84-90.

BOLTOVSKOY, A. (1984). Relacion huesped-parasito entre el quiste de *Peridinium willei* y el oomicete *Aphanomycopsis peridiniella* n. sp. *Limnobios* **2**: 635-645.

BONDE, M. R., PETERSON, G. L. & DUCK, N. B. (1985). Effects of temperatures on sporulation, conidial germination and infection of maize by *Peronosclerospora sorghi* from different geographical regions. *Phytopathology* **75**: 219-222.

BONDE, M. R., PETERSON, G. L., KENNETH, R. G., VERMEULEN, H. D., SUMARTINI & BUSTAMAN, M. (1992). Effect of temperature on conidial germination and systemic infection of maize by *Peronosclerospora* species. *Phytopathology* **82**: 104-109.

BONDE, M. R., SCHMITT, C. G. & DAPPER, R. W. (1978). Effects of dew-period temperature on germination of conidia and systemic infection of maize by *Sclerospora sorghii*. *Phytopathology* **68**: 219-222.

BONING, K. (1932-1933). Das Schwarzwerden der Rettiche. *Praktische Blatter für Pflanzenban und Pflanzenschutz* **10**: 205-219.

BONING, K. (1935-1936). Dungungsmassnahmen zur Bekampfung des Schwarzwerdens der Rettiche. *Praktische Blatter für Pflanzenban und Pflanzenschutz* **13**: 261-269.

BOOSALIS, M. G. & MANKAU, R. (1965). Parasitism and predation of soil microorganisms. *In* K. F. Baker & W. C. Snyder (Eds), *Ecology of Soil-borne Pathogens*, pp. 374-389. Los Angeles, U.S.A.: University of California Press.

BOOTH, T. (1969). Marine fungi from British Columbia: monocentric chytrids and chytridiaceous species from coastal and interior halomorphic soils. *Syesis* **2**: 141-161.

BOOTH, T. (1971*a*). Occurrence and distribution of zoosporic fungi and some Actinomycetales in coastal soils of southwestern British Columbia and the San Juan Islands. *Syesis* **4**: 197-208.

BOOTH, T. (1971*b*). Occurrence and distribution of some zoosporic fungi from soils of Hibben and Moresby, Queen Charlotte Islands. *Canadian Journal of Botany* **49**: 951-965.

BOOTH, T. (1971*c*). Distribution of certain soil inhabiting chytrid and chytridiaceous species related to some physical and chemical factors. *Canadian Journal of Botany* **49**: 1743-1755.

BOOTH, T. (1981). Lignicolous and zoosporic fungi in marine environments of Hudson Bay. *Canadian Journal of Botany* **59**: 1867-1881.

BOOTH, T. & MILLER, C. E. (1968*a*). Comparative morphologic and taxonomic studies in the genus *Thraustochytrium*. *Mycologia* **60**: 480-495.

BOOTH, T. & MILLER, C. E. (1968*b*). Some substrate relationships of the family Thraustochytriaceae. *Veröffentlichungen des Instituts für Meeresforschung in Bremerhaven* **12**: 329-351.

BOOTH, T. & MILLER, C. E. (1969). Morphological development of an isolate of *Schizochytrium aggregatum*. *Canadian Journal of Botany* **47**: 2051-2054.

BOOTSMA, R. (1973). Infections with *Saprolegnia* in pike culture (*Esox lucius* L.). *Aquaculture* **2**: 385-394.

BORGE, O. (1936). Beitrag zur Algenflora von Schweden 5. *Arkiv för Botanik* **28A**: 1-58.

BORNET, E. (1891). Note sur l'*Ostracoblabe implexa* Born. et Flah. *Journal de Botanique* **5**: 397-400.

BORNET, E. & FLAHAULT, C. (1889). Sur quelques plantes vivant dans le test calcaire des mollusques. *Bulletin de la Société Botanique de France, Série 2* **11**: CXLVII-CLXXVI (Congrès de Botanique Tenu à Pars en Août 1889).

BORTNICK, R. N., POWELL, M. J. & BANGERT, T. N. (1985). Zoospore fine structure of the parasite *Olpidiopsis saprolegniae* variety *saprolegniae* (Oomycetes, Lagenidiales). *Mycologia* **77**: 861-879.

BORZÍ, A. (1884). Rhizomixa, *Nuovo ficomycete*. Messina, Italy.

BOSC, M. (1946). Sur la structure des noyaux et la méiose de *Plasmopara viticola* (Berk. et Curt.) Berl. et de Toni. *Comptes Rendus Hebdomadaires des Séances de l'Académie des Sciences, Paris* **223**: 584-586.

BOSCHMA, H. (1948). Sur les organelles d'absorption chez une espèce d'*Amallocystis* (Protozoa, Ellobiopsidae). *Proceedings. Koninklijke nederlandse Akademie van Wetenschappen, Amsterdam* **25**: 281-314.

BOSCHMA, H. (1949). Ellobiopsidae. *Discovery Reports* **25**: 281-314.

BOSCHMA, H. (1959). Ellobiopsidae from tropical West Africa. *Atlantide Report* **5**: 145-175.

BOUCK, G. B. (1971). The structure, origin, isolation, and composition of the tubular mastigonemes of the *Ochromonas* flagellum. *Journal of Cell Biology* **50**: 362-384.

BOUCK, G. B. (1972). Architecture and assemby of mastigonemes. *In* E. J. Du Praw (Ed.), *Advances in Cell and Molecular Biology, Vol. 2*, pp. 237-271. New York, U.S.A.: Academic Press.

BOURRELLY, P. (1968). *Les Algues d'Eau Douce. Inition et Systématique. 2, Les Algues Jaunes et Brunes*. Paris, France: N. Boubée.

BOUSSINGAULT [forename initials not known] (1845). "... extraits suivants d'une lettre de M. le colonel Acosta sur la *maladie des pommes de terre* dans la Nouvelle-Granade". *Comptes Rendus, 2me Semestre* **21** (No. 20) [17 November 1845]: 1114-1115.

BOVEE, E. C. & JAHN, T. L. (1965). Mechanisms of movement in taxonomy of Sarcodina. II. The organization of subclasses and orders in relationship to the classes Autotractea and Hydraulea. *American Midland Naturalist* **73**: 293-298.

BOWER, S. M. (1987*a*). *Labyrinthuloides haliotidis* n. sp. (Protozoa: Labyrinthomorpha), a pathogenic parasite of small juvenile abalone in a British olumbia mariculture facility. *Canadian Journal of Zoology* **65**: 1996-2007.

BOWER, S. M. (1987*b*). Pathogenicity and host specificity of *Labyrinthuloides haliotidis* (Protozoa: Labyrinthomorpha), a parasite of juvenile abalone. *Canadian Journal of Zoology* **65**: 2008-2012.

BOWER, S. M. (1987*c*). Artificial culture of *Labyrinthuloides haliotidis* (Protozoa: Labyrinthomorpha), a pathogenic parasite of abalone. *Canadian Journal of Zoology* **65**: 2013-2020.

BOWER, S. M. (1987*d*). The life cycle and ultrastructure of a new species of thraustochytrid (Protozoa: Labyrinthomorpha) pathogenic to small abalone. *Aquaculture* **67**: 269-272.

BOWER, S. M. (1989). Disinfectants and therapeutic agents for controlling *Labyrinthuloides haliotidis* (Protozoa: Labyrinthomorpha), an abalone pathogen. *Aquaculture* **78**: 207-215.

BOWER, S. M., WHITAKER, D. J. & ELSTON, R. A. (1989). Detection of the abalone parasite *Labyrinthuloides haliotidis* by a direct fluorescent antibody technique. *Journal of Invertebrate Pathology* 53: 281-283.

BOWER, S. M., WHITAKER, D. J. & VOLTOLINA, D. (1989). Resistance to ozone of zoospores of the thraustochytrid abalone parasite, *Labyrinthuloides haliotidis* (Protozoa: Labyrinthomorpha). *Aquaculture* 78: 147-152.

BOWERMAN, L. & DOMNAS, A. (1997). A model for mastigoneme replacement in *Lagenidium* spp. *Inoculum, Newsletter of the Mycological Society of America* 48: Abstracts page 4. [abstract]

BOWERS, J. H. & PARKE, J. L. (1993). Epidemiology of *Pythium* damping-off and *Aphanomyces* root rot of peas after seed treatment with bacterial agents for biological control. *Phytopathology* 83: 1466-1473.

BOWMAN, B. H., TAYLOR, J. W., BROWNLEE, A. G., LEE, J., LU, S. & WHITE, T. J. (1992). Molecular evolution of the fungi: relationship of the Basidiomycetes, Ascomycetes, and Chytridiomycetes. *Molecular Biology and Evolution* 9: 285-296.

BOYD, D. A., HOBMAN, T. C., GRUENKE, S. A. & KLASSEN, G. R. (1984). Evolutionary stability of mitochondrial DNA organization in *Achlya*. *Canadian Journal of Biochemistry and Cell Biology* 62: 571-576.

BRANDT, K. (1881). Untersuchungen an Radiolarien. *Monatsberichte der Preussische Akademie der Wissenschaften der Berlin* 1881: 388-404.

BRASELTON, J. P. (1982). Karyotypic analysis of *Plasmodiophora brassicae* based on serial thin sections of pachytene nuclei. *Canadian Journal of Botany* 60: 403-408.

BRASELTON, J. P. (1983). Karyotype analysis of *Polymyxa betae* (Plasmodiophoromycetes) based on serial thin sections of pachytene nuclei. *Canadian Journal of Botany* 61: 3202-3206.

BRASELTON, J. P. (1984). Karyotypic analysis of *Polymyxa graminis* (Plasmodiophoromycetes) based on serial sections of synaptonemal complexes. *Canadian Journal of Botany* 62: 2414-2416.

BRASELTON, J. P. (1989*a*). Karyotypic analysis of *Ligniera verrucosa* (Plasmodiophoromycetes). *Canadian Journal of Botany* 67: 1216-1218.

BRASELTON, J. P. (1989*b*). Karyotypic analysis of *Membranosorus heterantherae* (Plasmodiophoromycetes). *Canadian Journal of Botany* 67: 1219-1220.

BRASELTON, J. P. (1990). Ultrastructure and karyology of *Tetramyxa parasitica* (Plasmodiophoromycetes). *Canadian Journal of Botany* 68: 594-598.

BRASELTON, J. P. (1992). Ultrastructure and karyology of *Spongospora subterranea* (Plasmodiophoromycetes). *Canadian Journal of Botany* 70: 1228-1233.

BRASELTON, J. P. & DYLEWSKI, D. P. (1986). Karyotypic analysis of *Woronina pythii*. *Mycologia* 78: 511-513.

BRASELTON, J. P. & MILLER, C. E. (1973). Centrioles in *Sorosphaera*. *Mycologia* 65: 220-226.

BRASELTON, J. P. & MILLER, C. E. (1975). Host-parasite interface of *Veronica persica* and *Sorosphaera veronicae* (Plasmodiophoromycetes). *Archives of Microbiology* 104: 97-99.

BRASELTON, J. P. & MILLER, C. E. (1978). Cyst formation in *Sorosphaera veronicae* (Plasmodiophoromycetes). *Micron* 9: 37-38.

BRASELTON, J. P. MILLER, C. E. & PECHAK, D. G. (1975). The ultrastructure of cruciform nuclear division in *Sorosphaera veronicae* (plasmodiophoromycete). *American Journal of Botany* 62: 349-358.

BRASELTON, J. P. & SHORT, F. T. (1985). Karyotypic analysis of *Plasmodiophora diplantherae*. *Mycologia* 77: 940-945.

BRASIER, C. M. (1972). Observations on the sexual mechanism in *Phytophthora palmivora* and related species. *Transactions of the British Mycological Society* 58: 237-251.

BRASIER, C. M. (1975*a*). Stimulation of sex organ formation in *Phytophthora* by antagonistic species of *Trichoderma*. I. The effect *in vitro*. *New Phytologist* 74: 183-194.

BRASIER, C. M. (1975*b*). Stimulation of sex organ formation in *Phytophthora* by antagonistic species of *Trichoderma*. II. Ecological implications. *New Phytologist* 74: 195-198.

BRASIER, C. M. (1992). Evolutionary biology of *Phytophthora* Part I: genetic system, sexuality and the generation of variation. *Annual Review of Plant Pathology* 30: 153-171.

BRASIER, C. M. (1997). Fungal species in practice: identifying species units in fungi. *In* M. F. Claridge, H. A. Dawah & M. R. Wilson (Eds), *Species: The Units of Diversity*, pp.135-170. London, U.K.: Chapman & Hall.

BRASIER, C. M. & BRASIER, C. (1978). Gametic and zygotic nuclei in oospores of *Phytophthora*. *Transactions of the British Mycological Society* 70: 297-302.

BRASIER, C. M., COOKE, D. E. L. & DUNCAN, J. M. (1999). Origin of a new *Phytophthora* pathogen through interspecific hybridization. *Proceedings of the National Academy of Sciences, U.S.A.* 96: 5878-5883.

BRASIER, C. M. & GRIFFIN, M. J. (1979). Taxonomy of '*Phytophthora palmivora*' on cocoa. *Transactions of the British Mycological Society* 72: 111-143.

BRASIER, C. M. & HANSEN, E. M. (1992). Evolutionary biology of *Phytophthora* Part II: phylogeny, speciation, and population structure. *Annual Review of Plant Pathology* 30: 173-200.

BRASIER, C. M. & MEDEIROS, A. G. (1978). The karyotype of *Phytophthora palmivora* morphological form 4. *Transactions of the British Mycological Society* 70: 295-297.

BRASIER, C. M. & SANSOME, E. (1975). Diploidy and gametangial meiosis in *Phytophthora cinnamomi*, *P. infestans*, and *P. drechsleri*. *Transactions of the British Mycological Society* 65: 49-65.

BRAUN, A. (1851). *Betrachtungen über die Erscheinung der Verjüngung in der Natur, insbesondere in der Lebens- und Bildungsgesischte der Pflanze*. Leipzig, Germany.

BRAUN, A. (1853). Reflections on the phenomenon of rejuvenescence in nature, especially in the life and development of plants. *Botanical and Physiological Memoirs of the Ray Society, London*, 341pp. [Same work as Braun (1851) in English.]

BRAUN, A. (1855). Ueber *Chytridium*, eine Gattung einzelner Schmarotzergewächse auf Algen und Infusorien. *Bericht über die zur Bekanntmachung geeigneten verhandlungen der Königlichen Preussischen Akademie der Wissenschaften zu Berlin, 1855* [subsquently (from 1856) *Monatsberichte der Königlichen Preussischen Akademie der Wissenschaften zu Berlin* - cf. Sparrow (1960)]: 378-384.

BRAUN, A. (1856*a*). Über *Chytridium*, eine Gattung einzelliger Schmarotzergewächse auf Algen und Infusorien. *Physicalische Abhandlungen der Königlichen Akademie der Wissenschaften zu Berlin, 1855*: 21-83 + 5 plates.

BRAUN, A. (1856*b*). Über einige neue Arten der Gattung *Chytridium* und die damit verwandte Gattung *Rhizidium*. *Bericht über die zur Bekanntmachung geeigneten verhandlungen der Königlichen Preussischen Akademie der Wissenschaften zu Berlin, 1856* [subsequently (from 1856) *Monatsberichte der Königlichen Preussischen Akademie der Wissenschaften zu Berlin* - cf. Sparrow (1960)]: 587-592.

BRAUN, H. (1924). Geranium stem rot caused by *Pythium complectens* n.sp. Host resistance reactions, significance of *Pythium*-type of sporangial germination. *Journal of Agricultural Research* 29: 399-419.

BRAUN, H. (1925). Comparative studies of *Pythium debaryanum* and two related species from geranium. *Journal of Agricultural Research* 30: 1043-1062.

BREFELD, O. (1881). Zur vergleichenden Morphologie der Pilze. *Botanische Untersuchungen über Schimmelpilze, IV. Heft: 11*: 161-181, Leipzig, Germany: Arthur Felix. [BREFELD, O. (1872-1912). 15 fascicles: *Botanische Untersuchungen über Schimmelpilze*. I: 1872; II: 1874; III: 1877; IV: 1881; *Botanische Untersuchungen über Hefenpilze. Fortsetzung der Schimmelpilze. Untersuchungen aus dem Gesammtgebiet der Mykologie*. V: 1883; *Untersuchungen aus dem Gesammtgebiet der Mykologie. Fortsetzung der Schimmel- und Hefenpilze*. VI: 1884; VII: 1888; VIII: 1889; IX: 18.; X: 189.; XI: 1895; XII: 1895; XIII: 1905; *Untersuchungen aus dem Gesammtgebiet der Mykologie*. XIV: 1908; XV: 1912].

BREHMER, W. von & BÄRNER, J. (1930). Ueber die Viruskrankheiten der Kartoffel. *Arbeiten aus der Biologischen Reichsanstalt für Land- und Forstwirtschaft* 18: 1-54.

BREMER, G. B. (1974). Physiological responses of some thraustochytrid fungi. *Veröffentlichungen des Instituts für Meeresforschung in Bremerhaven, Supplement* 5: 237-250.

BREMER, G. B. (1976). The ecology of marine lower fungi. *In* E. B. G. Jones (Ed.), *Recent Advances in Aquatic Mycology*, pp. 513-542. London, U.K.: Paul Elek.

BRETSNYDER, E. S. (1943). A study of the occurrence and distribution of aquatic phycomycetes in Chrystal Lake. Unpublished MS Thesis, 23 pp. University of Illinois, Champaign, Illinois, U.S.A.

BREY, P. T. (1985). Observations of in-vitro gametangial copulation and oosporogenesis in *Lagenidium giganteum*. *Journal of Invertebrate Pathology* **45**: 276-281.

BREY, P. T. & LEBRUN, R. A. (1987). External morphology of *Lagenidium giganteum* zoospores (Oomycetes: Lagenidiales). *Mycopathologia* **97**: 45-49.

BREY, P. T., LEBRUN, R. A., PAPIEROK, B., OHAYON, H., VENNAVALLI, S. & HAFEZ, J. (1988). Defense reactions by larvae of *Aedes aegypti* during infection by the aquatic fungus *Lagenidium giganteum* (Oomycete). *Cell and Tissue Research* **253**: 245-250.

BREY, P. T., LEBRUN, R. A., PAPIEROK, B., SITA, V. & HAFEZ, J. (1986). Invasive and developmental processes of *Lagenidium giganteum* infecting *Aedes aegypti* larvae. *Conference of the International Society of Developmental and Comparative Immunology on Molecular Aspects of Invertebrate Immunology, 28-30 September, 1986. (Berlin, West Germany)* **10**: 640.

BREY, P. T. & REMAUDIÈRE, G. (1985). Recognition and isolation of *Lagenidium giganteum*. *Bulletin of the Society of Vector Ecologists* **10**: 90-97.

BRIARD, M., DUTERTRE, M., ROUXEL, F. & BRYGOO, Y. (1995). Ribosomal RNA sequence divergence within the Pythiaceae. *Mycological Research* **99**: 1119-1127.

BRIDGES, C. H. & EMMONS, C. W. (1961). A phycomycosis of horses caused by *Hyphomyces destruens*. *Journal of the American Veterinary Medical Association* **138**: 579-589.

BROWN, E. M. (1938). Observations of the aquatic fungi of the Aberystwyth district. *Transactions of the British Mycological Society* **22**: 160-167.

BROWN, J. K. & WASHINO, R. K. (1977). Developments in research with the Clear Lake gnat *Chaoborus astictopus* in relation to the fungus *Lagenidium giganteum*. *In* C. D. Grant (Ed.), *Proceedings and papers of the Forty-fifth Annual Conference of the California Mosquito and Vector Control Association*, p. 106. Visalia, U.S.A.: California Mosquito and Vector Control Association Press.

BROWN, J. K. & WASHINO, R. K. (1979). Evaluating the fungus *Lagenidium giganteum* for the biological control of the Clear Lake gnat, *Chaoborus astictopus*, in an agricultural pond in Lake County, California. *In* C. D. Grant (Ed.), *Proceedings and papers of the Forty-seventh Annual Conference of the California Mosquito and Vector Control Association*, p. 37. Visalia, U.S.A.: California Mosquito and Vector Control Association Press.

BROWN, R. M., FRANKE, W. W., KLEINIG, H. & SITTE, P. (1970). Scale formation in chrysophycean algae. I. Cellulosic and non-cellulosic wall components made by the Golgi apparatus. *Journal of Cell Biology* **45**: 246-271.

BRUG, S. L. (1926). *Nucleophaga intestinalis* n. sp. Parasiet der kern van *Endolimax williamsi* Prow. *Mededelingen van der Dienst der Volksgezond in Nederland Indie* **4**: 250.

BRUMMITT, R. K. (1996). In defence of paraphyletic taxa. *In* L. J. G. van der Maesen, X. M. Van Der Burgt & J. M. Van Medenbach De Rooy (Eds), *The Diversity of African Plants*. pp. 371-384, Dordrecht, The Netherlands: Kluwer Academic Publishers.

BRUMMITT, R. K. & POWELL, C. E. (Eds) (1992). *Authors of Plant Names. A list of authors of scientific names of plants, with recommended standard forms of their names, including abbreviations*. [principal compilers: C. E. Powell; P. M. Kirk (fungi & lichens); P. C. Silva (algae); R. E. Pichi Sermolli; M. R. Crosby (bryophytes); H. T. Clifford (fossils)]. Kew, U.K.: Royal Botanic Gardens.

BRUMPT, E. & LAVIER, G. (1935). Sur une *Nucleophaga* parasite d'*Endolimax nana*. *Annales de Parasitologie Humaine et Comparée* **13**: 439-444.

BRUNCHORST, J. (1886). Über einige Wurzelanschwellungen, besonders diejenigen von *Alnus* und den Elaeagnaceen. *Untersuchungen aus den Botanischen Institut, Tübingen* **2**: 151-177.

BRUNCHORST, J. (1887*a*). Ueber eine sehr verbreite Krankheit der Kartoffelknollen. *Bergens Museums Årsberetning, 1886*: 217-226, plate I.

BRUNCHORST, J. (1887*b*). Zur Bekämpfung der Kohlhernie. *Bergens Museums Årsberetning, 1886*: 227-231.

BRUNO, D. W. & STAMPS, D. J. (1987). Saprolegniasis of Atlantic salmon, *Salmo salar* L., fry. *Journal of Fish Diseases* **10**: 513-517.

BRUNT, S. A. & SILVER, J. C. (1986). Steroid hormone-induced changes in secreted proteins in the filamentous fungus *Achlya*. *Experimental Cell Research* **163**: 22-34.

BRYANT, T. R. & HOWARD, K. L. (1969). Meiosis in the Oomycetes: I. A microspectro-photometric analysis of nuclear deoxyribonucleic acid in *Saprolegnia terrestris*. *American Journal of Botany* **56**: 1075-1083.

BRYNGELSSON, T., GUSTAFSSON, M., GREEN, B. & LIND, C. (1888). Uptake of host DNA by the parasitic fungus *Plasmodiophora brassicae*. *Physiological and Molecular Plant Pathology* **33**: 163-171.

BUCHHOLTZ, W. F. (1944). *Aphanomyces cochlioides* infestation in irrigated sugar-beet-alfalfa rotation plots at Newell, South Dakota. *Phytopathology* **38**: 4 [abstract].

BUCKLEY, J. J. C. & CLAPHAM, P. (1929). The invasion of helminth eggs by chytridiacean fungi. *Journal of Helminthology* **7**: 1-14.

BUCZACKI, S. T. (1977). Root infections from single resting spores of *Plasmodiophora brassicae*. *Transactions of the British Mycological Society* **69**: 328-329.

BUCZACKI, S. T. (1983). *Plasmodiophora* - an interrelationship between biological and practical problems. *In* S. T. Buczacki (Ed.), *Zoosporic Plant Pathogens, A Modern Perspective*, pp. 161-191. London, U.K.: Academic Press.

BUCZACKI, S. T. & CLAY, C. M. (1984). Some observations on secondary zoospore development in *Plasmodiophora brassicae*. *Transactions of the British Mycological Society* **82**: 339-382.

BUCZACKI, S. T. & MOXHAM, S. E. (1980). Karyogamy in *Plasmodiophora brassicae*. *Transactions of the British Mycological Society* **75**: 439-444.

BUCZACKI, S. T. & MOXHAM, S. E. (1983). Structure of the resting spore wall of *Plasmodiophora brassicae* revealed by electron microscopy and chemical digestion. *Transactions of the British mycological Society* **81**: 221-231.

BUDDENHAGEN, I. W. (1958). Induced mutations and variability in *Phytophthora cactorum*. *American Journal of Botany* **45**: 355-365.

BUELL, C. B. & WESTON, W. H. (1947). Application of the mineral oil conservation method to maintaining collections of fungous cultures. *American Journal of Botany* **34**: 555-561.

BUISMAN, C. J. (1927). Root rots caused by Phycomycetes. *Mededeelingen van het Phytopathologisch Laboratorium 'Willie Commelin Scholten'* **11**: 1-51.

BULONE, V., CHANZY, H., GAY, L., GIRARD, V. & FÈVRE, M. (1992). Characterization of chitin and chitin synthase from the cellulosic cell wall fungus *Saprolegnia monoïca*. *Experimental Mycology* **16**: 8-21.

BURKE, D. W., HAGEDORN, D. J. & MITCHELL, J. E. (1969). *Aphanomyces* and *Fusarium* root rot of peas with partial vs. total exposure of roots to infested soil. *Phytopathology* **59**: 1261-1266.

BURKE, D. W., HAGEDORN, D. J. & MITCHELL, J. E. (1970). Soil conditions and distribution of pathogens in relation to pea root rot in Wisconsin soils. *Phytopathology* **60**: 403-406.

BURKE, D. W., MITCHELL, J. E. & HAGEDORN, D. J. (1969). Selective conditions for infection of pea seedlings by *Aphanomyces euteiches* in soil. *Phytopathology* **59**: 1670-1674.

BURRUANO, S., CIOFALO, G. & CONIGLIARO, C. (1994). Studio dell'influenza di alcuni fattori sulla maturazione delle oospore di *Plasmopara viticola*. *Innovazione e prospettive nella difesa fitosanitaria, Ferrara, 24-25 October 1994*: 3-5.

BURRUANO, S., CIOFALO, G. & CONIGLIARO, C. (1995). Ulteriori osservazioni sulla germinazione delle oospore di *Plasmopara viticola* e primo tentativo sperimentale di previsione dell'infezione primaria. *La Difesa delle Piante* **18**: 23-31.

BURRUANO, S., CONIGLIARO, C. & CIOFALO, G. (1997). Nuclear changes in *Plasmopara viticola* during oospore maturation and germination. *Viticulture and Enological Sciences* **3-4**: 138-139.

BURRUANO, S., CONIGLIARO, C. & LAVIOLA, C. (1999). Ulteriori osservazioni sulla evoluzione nucleare nelle strutture gamiche di *Plasmopara viticola*. *Micologia Italiana* (in press).

BURRUANO, S., CONIGLIARO, C., MALTESE, C. E. & LAVIOLA, C. (1992). Prime osservazione sui nuclei di *Plasmopara viticola*. *Phytopatologia Mediterranea* **31**: 107-109.

BURSNALL, L. A. & TRIBE, H. T. (1974). Fungal parasitism in cysts of *Heterodera*. II. Egg parasities of *H. schachtii*. *Transactions of the British Mycological Society* **62**: 595-601.

BURTON, M. G. (1939). *Oogenesis and fertilization in* Pythiopsis intermedia *Coker and Harvey*. Unpublished Thesis, University of North Carolina, Carolina, U.S.A.

BÜSGEN, M. (1882). Die Entwicklung der Phycomycetensporangien. *Jahrbücher für wissenschaftliche Botanik* **13**: 253-285.

BUTLER, E. J. (1907). An account of the genus *Pythium* and some Chytridiaceae. *Memoirs of the Department of Agriculture, India, Botanical Series* **1**: 1-162.

BUTLER, E. J. (1928). Morphology of the chytridiacean fungus *Catenaria anguillulae*, in liver fluke eggs. *Annals of Botany, London* **42**: 813-821.

BUTLER, E. J. & BISBY, G. R. (1932). *The Fungi of India*. Imperial Council for Agricultural Research, Scientific Monograph No. 1. 237pp. Calcutta, India.

BUTLER, J. B. & BUCKLEY, J. J. C. (1927). *Catenaria anguillulae*, as a parasite of the ova of *Fasciola hepatica*. *Scientific Proceedings of the Royal Dublin Society* **18**: 497-512.

BUTLER, J. B. & HUMPHRIES, A. (1932). On the cultivation in artificial media of *Catenaria anguillulae*, a chytridiacean parasite of the ova of the liver fluke, *Fasciola hepatica*. *Scientific Proceedings of the Royal Dublin Society* **20**: 301-324.

BUTTNER, G. (1992). (Nutrient concentration in soil and course of rhizomania infection of sugar beet). *Gesunde-Pflanzen* **44**: 143-145 [only abstract consulted].

BYRT, P. N. & GRANT, B. R. (1979). Some conditions governing zoospore production in axenic cultures of *Phytophthora cinnamomi* Rands. *Australian Journal of Botany* **27**: 103-115.

BYRT, P. N., IRVING, H. R. & GRANT, B. R. (1982). The fffect of cations on zoospores of the fungus *Phytophthora cinnamomi*. *Journal of General Microbiology* **128**: 1189-1198.

CACCIOLA, S. O., MAGNANO DI SAN LIO, G. & BELISARIO, A. (1996). *Phytophthora italica* sp. non. on myrtle. *Phytopatologia Mediterranea* **35**: 177-190.

CAHILL, D. M., COPE, M. & HARDHAM, A. R. (1996). Thrust reversal by tubular mastigonemes: immunological evidence for a role of mastigonemes in forward motion of zoospores of *Phytophthora cinnamomi*. *Protoplasma* **194**: 18-28.

CAMPBELL, W. A. & HENDRIX, F. F. (1967). A new heterothallic *Pythium* from southern United States. *Mycologia* **59**: 274-278.

CAMERON, J. N. & CARLILE, M. J. (1977). Negative geotaxis of zoospores of the fungus *Phytophthora*. *Journal of General Microbiology* **98**: 599-602.

CAMERON, J. N. & CARLILE, M. J. (1980). Negative chemotaxis of zoospores of the fungus *Phytophthora palmivora*. *Journal of General Microbiology* **120**: 347-353.

CAMERON, L. E., HUTSUL, J., THORLACHIUS, L. & LeJOHN, H. B. (1990). Cloning and analysis of β-tubulin gene from a protoctist. *Journal of Biological Chemistry* **265**: 15245-15252.

CAMPOS, V. P. & SILVA, J. F. V. (1990). Isolation of endoparasitic and ectoparasitic fungi associated with nematodes in Brazil and in-vitro tests of parasitic efficacy of *Haptoglossa heterospora* on selected nematodes. *Nematologia* **36**: 337.

CANDOLLE, A. P. de (1828). *Botanicon gallicum seu Synopsis plantarum in Flora gallica descriptarum. Editio secunda. Ex herbariisa et schedio Candollianis propriisque digestum a J. E. Duby*. Paris.

CANNING, E. U., CURRY, A., LACEY, C. J. N. & FENWICK, J. D. (1992). Ultrastructure of *Encephalitozoon* sp. infecting the conjunctival, corneal and nasal epithelia of a patient with AIDS. *European Journal of Protistology* **28**: 226-237.

CANTER, H. M. (1947). On *Myzocytium megastomum* de Wildeman. *Transactions of the British Mycological Society* **31**: 80-85.

CANTER, H. M. (1949*a*). Studies on British chytrids. VI. Aquatic Synchytriaceae. *Transactions of the British Mycological Society* **32**: 69-94.

CANTER, H. M. (1949*b*). On *Aphanomycopsis bacillariacearum* Scherffel, *A. desmidiella* n. sp. and *Ancylistes* spp. in Great Britain. *Transactions of the British Mycological Society* **32**: 162-170.

CANTER, H. M. (1950). Studies on British chytrids IX. *Anisolpidium stigeoclonii* (de Wildeman) n. comb. *Transactions of the British Mycological Society* **33**: 335-344.

CANTER, H. M. (1968). On an unusual fungoid organism, *Sphaerita dinobryoni* n. sp., living in species of *Dinobryon*. *Journal of the Elisha Mitchell Scientific Society* **84**: 56-61.

CANTER, H. M. (1969). Studies on British chytrids. XXIX. A taxonomic revision of certain fungi found on the diatom *Asterionella*. *Botanical Journal of the Linnean Society* **62**: 267-278.

CANTER, H. M. (1979). Fungal and protozoan parasites and their importance in the ecology of the phytoplankton. *Annual Report of the Freshwater Biological Association* **47**: 43-50.

CANTER, H. M. & DICK, M. W. (1994). *Eurychasmopsis multisecunda* gen. et sp. nov., a parasite of the suctorian *Podophrya* sp. (Ciliata). *Mycological Research* **98**: 105-117.

CANTER, H. M. & HEANEY, S. I. (1984). Observations on zoosporic fungi of *Ceratium* spp. in lakes of the English Lake District; importance for phytoplankton population dynamics. *New Phytologist* **97**: 601-612.

CANTER, H. M., HEANEY, S. I. & LUND, J. W. G. (1990). The ecological significance of grazing on planktonic populations of cyanobacteria by the ciliate *Nassula*. *New Phytologist* **114**: 247-263.

CANTER, H. M. & LUND, J. W. G. (1948). Studies on plankton parasites. I. Fluctuations in the numbers of *Asterionella formosa* Hass. in relation to fungal epidemics. *New Phytologist* **47**: 238-261.

CANTER, H. M. & LUND, J. W. G. (1969). The parasitism of planktonic desmids by fungi. *Österreichische Botanische Zeitschrift* **116**: 351-377.

CANTER-LUND, H. [M.] & LUND, J. W. G. (1995). *Freshwater Algae. Their microscopic world explored.* Bristol, U.K.: Biopress Ltd.

CANTINO, E. C. (1949). The growth and nutrition of *Pythiogeton*. *American Journal of Botany* **36**: 747-756.

CANTINO, E. C. (1951). Evidence for an accessory factor involved in fructose utilization by the aquatic fungus *Pythiogeton*. *American Journal of Botany* **38**: 579-585.

CANTINO, E. C. (1955). Physiology and phylogeny in the water molds - a re-evaluation. *Quarterly Review of Biology* **30**: 138-149.

CARLEY, H. E. (1970). Detection of *Aphanomyces euteiches* races usibg a differential bean series. *Plant Disease Reporter* **54**: 943-945.

CARLILE, M. J. (1983). Motility, taxis, and tropism in *Phytophthora*. *In* D. C. Erwin, S. Bartnicki-Garcia & P. H. Tsao (Eds), Phytophthora *Its Biology, Taxonomy, Ecology, and Pathology*, pp. 95-107. St Paul, Minnesota, U.S.A.: American Phytopathological Society.

CANTINO, E. C. & TURIAN, G. F. (1959). Physiology and development of lower fungi (phycomycetes). *Annual Review of Microbiology* **13**: 97-124.

CARLSON, M. C. (1920). Morphological and cytological studies of *Saprolegnia* sp. Unpublished MS Thesis, University of Wisconsin, Madison.

CARLSON, M. C. (1929). Gametogenesis and fertilization in *Achlya racemosa*. *Annals of Botany, London* **43**: 111-117.

CARTER, D. A., ARCHER, S. A., BUCK, K. W., SHAW, D. S. & SHATTOCK, R. C. (1990). Restriction fragment length polymorphisms of mitochondrial DNA of *Phytophthora infestans*. *Mycological Research* **94**: 1123-1128.

CARTER, H. J. (1858). On the spermatology of a n. sp. of *Nais*. *Annals and Magasine of Natural History, III* **2**: 90-104.

CARUS, C. G. (1823). Beitrag zur Geschichte der unter Wasser an verschiedenen Thierkörpern sich erzeugenden Schimmel- oder Algengattungen. *Nova Acta [Novorum Actorum] Physico-medica Academiae Caesareae Leopoldino-carolinae Naturae Curiosorum* **11**: 491-522.

CASSELS, F. & KREBS, C. T. (1983). A comparison of artificial incubation methods using ova of the red crab *Geryon quinquedens* Decapoda Brachyura. *Journal of Crustacean Biology* **3**: 565-574.

CASTANER, D. & O'LEARY, T. J. (1967). Nematode-destroying fungi of Misssouri I. *Transactions of the Missouri Academy of Science* **1**: 3-11.

CASTLEBURY, L. A. & DOMIER, L. L. (1998). Small subunit ribosomal RNA gene phylogeny of *Plasmodiophora brassicae*. *Mycologia* **90**: 102-107.

CASTLEBURY, L. A., MADDOX, J. V. & GLAWE, D. A. (1994). A technique for the extraction and purification of viable *Plasmodiophora brassicae* resting spores from host root tissue. *Mycologia* **86**: 458-460.

CAULLERY, M. (1910). *Ellobiopsis chattoni* n.g., n.sp. Parasite de *Calanus helgolandicus* Claus, appartenant probablement aux Péridiniens. *Bulletin Scientifique de la France et de la Belgique, Paris* **44**: 201-214.

CAVALIER-SMITH, T. (1975). The origin of nuclei and of eukaryotic cells. *Nature (London)* **256**: 463-468.

CAVALIER-SMITH, T. (1978). The evolutionary origin and phylogeny of microtubules, mitotic spindles and eukaryotic flagella. *BioSystems* **10**: 93-114.

CAVALIER-SMITH, T. (1981). Eukaryote kingdoms: seven or nine? *BioSystems* **14**: 461-481.

CAVALIER-SMITH, T. (1986). The Kingdom Chromista: origin and systematics. *In* F. E. Round & D. J. Chapman (Eds), *Progress in Phycological Research*, pp. 309-347. Bristol, U.K.: Biopress.

CAVALIER-SMITH, T. (1987). The origin of fungi and pseudofungi. *In* A. D. M. Rayner, C. M. Brasier & D. Moore (Eds), *Evolutionary Biology of the Fungi*, pp. 339-353. Cambridge U.K.: Cambridge University Press.

CAVALIER-SMITH, T. (1989). The Kingdom Chromista. *In* J. C. Green, B. S. C. Leadbeater & W. L. Diver (Eds), *The Chromophyte Algae, Problems and Perspectives*, pp. 381-407. Oxford, U.K.: Clarendon Press.

CAVALIER-SMITH, T. (1991). The evolution of cells. *In* S. Osawa & T. Honjo (Eds), *Evolution of Life. Fossils, Molecules, and Culture*, pp. 271-304. Tokyo, Japan: Springer-Verlag.

CAVALIER-SMITH, T. (1993*a*). The protozoan phylum Opalozoa. *Journal of Eukaryotic Microbiology* **40**: 609-615.

CAVALIER-SMITH, T. (1993*b*). Kingdom Protozoa and its 18 phyla. *Microbiological Reviews* **57**: 953-994.

CAVALIER-SMITH, T. (1998). A revised six-kingdom system of life. *Biological Reviews of the Cambridge Philosophical Society* **73**: 203-266.

CAVALIER-SMITH, T., CHAO, E. E. & ALLSOPP, M. T. E. P. (1995). Ribosomal RNA evidence for chloroplast loss within Heterokonta: pedinellid relationships and a revised classification of ochristan algae. *Archiv für Protistenkunde* **145**: 209-220.

CAVARA, F. (1897). Malattie crittogamische dei pesci. *Giornale Italiano Pesci e Acquicoltura, Anno I* **1**: 13-18.

*__CEJP, K.__ (1932*a*). Několik poznámek k parasitům Conjugat z Čech. *Rozpravy České Akademie Věd a Umění, Třida II, Praha* **42** *Čislo* **3**: 1-8. (Some remarks to the knowledge of the parasitic phycomycetes of conjugates in Bohemia. *Bulletin International de l'Académie des Sciences de Bohême, class 2, 1932* **2**: 1-7).

CEJP, K. (1932*b*). *Rhipidium americanum* Thaxter a *Apodachlya pirifera* Zopf v Československu. *Rozpravy České Akademie Věd a Umění, Třida II, Praha* **42** *Čislo* **20**: 1-11.

CEJP, K. (1932*c*). První Příspěvek k poznání českých Saprolegnii (Contribution to the knowledge of Saprolegniales of Bohemia. I.). *Zvláštní otisk Časopisu "Mycologia"* **8**: 95-100.

CEJP, K. (1932*d*). *Achlya racemosa. Zvláštní otisk z Vědy přírodní* **13**: 1-5.

CEJP, K. (1933). Further studies on the parasites of conjugates in Bohemia. *Rozpravy České Akademie Věd a Umění, Třida II, Praha* **43** *Čislo* **9**: 1-11. (Further studies on the parasites of conjugates in Bohemia. *Bulletin International de l'Académie des Sciences de Bohême, class 2, 1933*: 1-11.)

CEJP, K. (1934). II. (Contribution to the knowledge of Saprolegniales of Bohemia. II.). *Rozpravy České Akademie Věd a Umění, Třida II, Praha* **44** *Čislo* **26**: 1-13. (*Bulletin International de l'Académie des Sciences de Bohême, class 2, 1934*: 1-13).

CEJP, K. (1935*a*). Parasiti Conjugat z Čech. III. Další příspěvky. *Rozpravy České Akademie Věd a Umění, Třida II, Praha* **45** *Čislo* **3**: 1-12. (The parasites of Conjugates in Bohemia. IIIrd contribution. *Bulletin International de l'Académie des Sciences de Bohême, class 2, 1935*: 1-12).

CEJP, K. (1935*b*). *Sphaerita*, parasit Paramecií. Příspěvek k poznání houbových parasitů Protozoí. (*Sphaerita*, the parasite of the Protozoa). *Spisy přírodovědeckou Fakultou Karlovy University, Praha, No. 141* (Publications de la Faculté des Sciences de l'Université Charles, Prague, No. 141) : 1-7.

CEJP, K. (1936). *Sapromyces androgynus* Thaxter en Europe et étude de l'espèce *Sapromyces reinschii* (Schroeter) Fritsch. *Bulletin de la Société Mycologique de France* **52**: 370-376.

CEJP, K. (1955). Novy parasit perloček z rodu *Pythium* Pringsheim. *Československá Parasitologie* **2**: 15.

CEJP, K. (1957). *Houby I*. Praha, Czechoslovakia: Československé Akademie věd.

CEJP, K. (1959*a*). Contribution to the knowledge of the keratinophilic Phycomycetes. *Academia Republicii Populare Romanî*: 129-138.

CEJP, K. (1959*b*). *Flora ČSR, B Řada Mykologicko-Lichenologická, Svazek 2, Oomycetes I*. Praha, Czechoslovakia: Československé Akademie věd.

CERENIUS, L., OLSON, L. W., LANGE, L. & SÖDERHÄLL, K. (1984). The secondary zoospores of *Aphanomyces astaci* and *A. laevis* (Oomycetes, Saprolegniales). *Nordic Journal of Botany* **4**: 697-706.

CERENIUS, L. & SÖDERHÄLL, K. (1984). Repeated zoospore emergence from isolated spore cysts of *Aphanomyces astaci*. *Experimental Mycology* **8**: 370-377.

CERENIUS, L., SÖDERHÄLL, K., PERSSON, M. & AJAXON, R. (1988). The crayfish plague fungus, *Aphanomyces astaci* - Diagnosis, isolation and pathology. *Freshwater Crayfish* **7**: 131-144.

CESATI, V. (1840). *Stirpes Italicae rariores vel novae descriptionibus, iconibusque illustratae. Accedunt animadversiones in characteres plantarum, pariter tabulis adumbratae. Fasculae 1-3*. Mediolani.

CÉSPEDES, A. E. & CASTILLO, J. (1982). Algunos Chytridiomycetes y Oomycetes aislados de 10 localidades en cuatro estados de la Republica Mexicana (Some Chytridiomycetes and Oomycetes isolated from 10 locations in four states of the Mexican Republic). *Boletin de la Sociedad Mexicana de Micologia, No. 17*: 207-210.

CHADEFAUD, M. (1956). Sur un *Labyrinthula* de Roscoff. *Comptes Rendus Hebdomadaires des Séances de l'Académie des Sciences, Paris, Série D* **243**: 1794-1797.

CHAKRAVARTY, D. K. (1969). Zum Kulturverhalten des marinen parasitischen Pilzes *Lagenisma coscinodisci*. *Veröffentlichungen des Instituts für Meeresforschung in Bremerhaven* **11**: 309-321.

CHAKRAVARTY, D. K. (1974). On the ecology of the infection of the marine diatom *Coscinodiscus granii* by *Lagenisma coscinodisci* in the Weser Estuary. *Veröffentlichungen des Instituts für Meeresforschung in Bremerhaven, Sonderband* **5**: 115-122.

CHAKRAVARTY, D. K. (1979). A critical note on the genus *Thraustochytrium*. *Beihefte zur Nova Hedwigia* **63**: 325-332.

CHAKRAVARTY, D. L. (1979). *Schizochytrium aggregatum*, a marine fungus newly recorded from India. *Kavaka* **6**: 37-40.

CHAMBERLAIN, A. H. L. (1980). Cytochemical and ultrastructural studies on the cell walls of *Thraustochytrium* spp. *Botanica Marina, Berlin* **23**: 669-677.

CHAMBERLAIN, A. H. L. & MOSS, S. T. (1988). The thraustochytrids: a protist group with mixed affinities. *BioSystems* **21**: 341-349.

CHANG-HO, Y. (1970). The effect of pea root exudate on the germination of *Pythium aphanidermatum* zoospore cysts. *Canadian Journal of Botany* **48**: 1501-1514.

CHAPMAN, J. A. & VUJIČIĆ, R. (1965). The fine structure of sporangia of *Phytophthora erythroseptica* Pethyb. *Journal of General Microbiology* **41**: 275-282.

---

*FOOTNOTE: In this journal (and also following publications), held at the Natural History Museum, London, each ***article*** (*Čislo*) is numbered consecutively, but with pages 1- [x]. The texts are in Čzech with or without English subtitles. I have been unable to trace the French-titled journal, which is commonly quoted, at the Natural History Museum, but I have seen some offprints.

CHATTON, E. (1908). Sur la reproduction et les affinités du *Blastulidium poedophthorum* Ch. Pérez. *Comptes Rendus des Séances de la Société Biologie et des Filiales de France* **64**: 34-36.

CHATTON, E. & BRODSKY, A. (1909). Le parasitime d'une Chytridinée du genre *Sphaerita* Dang. chez *Amoeba limax* Dujard. Étude comparative. *Archiv für Protistenkunde* **17**: 1-18.

CHATURVEDI, A. P. & DAYAL, R. (1989). Decomposing leaf litter in ponds of Varanasi (India): lower fungi. *Crop Research (HISAR)* **2**: 30-33.

CHAUDHURI, H. (1942). India water moulds. V. A new genus of the Saprolgniaceae: *Hamidia* gen. nov. *Proceedings of the Indian Academy of Sciences, Section B* **15**: 225-230.

CHAUDHURI, H. & BANERJEE, M. L. (1942). India water moulds. III. *Proceedings of the Indian Academy of Sciences, Section B* **15**: 206-215.

CHAUDHURI, H. & BANERJEE, M. L. (1942). India water moulds. IV. *Proceedings of the Indian Academy of Sciences, Section B* **15**: 216-224.

CHAUDHURI, H. & KOCHHAR, P. L. (1935). Indian water moulds I. *Proceedings of the Indian Academy of Science* **2**: 137-154.

CHAUDHURI, H., KOCHHAR, P. L., LOTUS, S. S., BANERJEE, M. L. & KHAN, A. H. (1947). *A Handbook of Indian Water Moulds, Part I.* Lahore, India: University of the Panjab.

CHAUDHURI, H. & LOTUS, S. S. (1936). India water moulds. II. *Proceedings of the Indian Academy of Sciences, Section B* **3**: 328-333.

CHAUDHURI, M. (1931). On a *Myzocytium* parasite of *Spirogyra affinis*. *Archiv für Protistenkunde* **75**: 462-475.

CHAZE, M. J. (1925). Essais de culture pure d'une Saprolégniacée. *Bulletin de la Société Mycologique de France* **41**: 256-277.

CHEE, H.-Y., KIM, W.-G., CHO, W.-D., JEE, H.-J. & CHOI, Y.-C. (1998). Detection of *Plasmodiophora brassicae* by using polymerase chain reaction. *Korean Journal of Plant Pathology* **14**; 589-593.

CHEN, J. P., RUAN, Y. L., HONG, J. & XU, J. H. (1990). (Serological relationships among three fungus-transmitted viruses of cereals by immunosorbent electron microscopy). *Chinese Journal of Virology* **6**: 239-244 [in Chinese, only abstract consulted].

CHEN, J. P., SWABY, A. G., ADAMS, M. J. & YILI, R. (1991). Barley mild mosaic virus inside its fungal vector, *Polymyxa graminis*. *Annals of Applied Biology* **118**: 615-621.

CHEN, W. (1992). Restriction fragment length polymorphisms in enzymatically amplified ribosomal DNAs of three heterothallic *Pythium* species. *Phytopathology* **82**: 1467-1472.

CHEN, W. (1995). Phylogenetic relationship of *Pythium* with other oomycetes based on small subunit ribosomal DNA sequences. *Phytopathology* **85**: 1118.

CHEN, W. & HOY, J. W. (1993). Molecular and morphological comparison of *Pythium arrrhenomanes* and *P. graminicola*. *Mycological Research* **97**: 1371-1378.

CHEN, W., HOY, J. W. & SCHNEIDER, R. W. (1992). Species-specific polymorphisms in transcribed ribosomal DNA of five *Pythium* species. *Experimental Mycology* **16**: 22-34.

CHEN, W.-H., LIU, M.-C. & CHAO, C.-Y. (1979). The growth of sugar cane downy mildew fungus in tissue culture. *Canadian Journal of Botany* **57**: 528-533.

CHENEY, G. M. (1932). *Pythium* root rot of broad beans in Victoria. *Australian Journal of Experimental Biology and Medical Science* **10**: 143-155.

CHESNICK, J. M., TUXBURY, K., COLEMAN, A., BURGER, G. & LANG, B. F. (1996). Utility of the mitochondrial nad4L gene for algal and protistan phylogenetic analysis. *Journal of Phycology* **32**: 452-456.

CHESTERS, C. G. C. & HICKMAN, C. J. (1944). On *Pythium violae* n.sp. and *P. oligandrum* from cultivated *Viola*. *Transactions of the British Mycological Society* **27**: 55-62.

CHETCHOTISAKD, P., PORNTAVEEVUDHIO, O., MAIRIAG, P., PATJANASOONTORN, B., CHAIPRASERT, A. K., PAIROJKUL, C., SATHAPATAYAVONGS, B., NUNTIROOJ, K., SAEW, O. T. & HASWELL-ELKINS, M. R. (1992). Human pythiosis in Srinagarind Hospital: one year experience. *Journal of the Medical Association, Thailand* **75**: 248-254.

CHIAPPELLI, R. (1933). Indagini sperimentali sulle cause d'infezione di dermatomicosi Saprolegniacea nella carpa. *Giornale di Risicoltura* **23**: 169-173.

CHIDAMBORAM, K. (1942). Fungus disease of Gourami (*Osphromenus goramy* Lacepede) in a pond at Madras. *Current Science, Bangalore, India* **11**: 289, 290.

CHIEN, C.-Y. & LIN, C.-L. (1985). A fungus disease of larval red shrimp caused by *Lagenidium*. *Fish Pathology* [*Gyobyo Kenkyu*] **20**: 263-266.

CHILD, J. J. & HASKINS, R. H. (1971). Induction of sexuality in heterothallic *Pythium* spp. by cholesterol. *Canadian Journal of Botany* **49**: 329-332.

CHILVERS, G. A., LAPEYRIE, F. F. & DOUGLASS, P. A. (1985). A contrast between Oomycetes and other taxa of mycelial fungi in regard to metachromatic granule formation. *New Phytologist* **99**: 203-210.

CHIOU, T.-S. & CHANG, H.-S. (1976). Aquatic phycomycetes of Taiwan II. *Botanical Bulletin of Academia Sinica, Taipei* **17**: 37-53.

CHIOU, T.-S., HSU, W. & CHANG, H.-S. (1975). Aquatic phycomycetes of Taiwan I. *Botanical Bulletin of Academia Sinica, Taipei* **16**: 159-176.

CHMĚLEVSKIJ. V. F. (1888). Къ вопросч о копчляцıи яцерю при половомъ процессѣ у гриовъ. *Эалиски новороссıйскаѕо Оощемов Есмесмвоислбıмамелейаса* **13**: 113-121.

CHMĚLEVSKIJ, V. F. (1889). Zur Frage über die Kopulation der Zellkerne beim Geschlechtsprozesse der Pilze. *Botanisch Centralblatt* **38**: 789-790.

CHO, C. W. & FULLER, M. S. (1989). Ultrastructural organization of freeze-substituted zoospores of *Phytophthora palmivora*. *Canadian Journal of Botany* **67**: 1493-1499.

CHONG, J. & BARR, D. J. S. (1974). Ultrastructure of the zoospores of *Entophlyctis confervae-glomeratae*, *Rhizophydium patellarium*, and *Catenaria anguillulae*. *Canadian Journal of Botany* **52**: 1197-1204.

CHRISTEN, J. & HOHL, H. R. (1972). Growth and ultrastructural differentiation of sporangia in *Phytophthora palmivora*. *Canadian Journal of Microbiology* **18**: 1959-1964.

CHRISTENSEN, J. B., FETTER-LASKO, J. L., WASHINO, R. K., HUSBANDS, R. C. & KAUFFMAN, E. E. (1977). A preliminary field study employing *Lagenidium giganteum*, a fungus, as a possible biological control agent against the pasture mosquito *Aedes nigromaculis*. *In* C. D. Grant (Ed.), *Proceedings and Papers on the Forty-fifth Annual Conference of the Californian Mosquito and Vector Control Association,* p. 105. Visalia, U.S.A.: California Mosquito and Vector Control Association Press.

CHU, F.-L. E. & GREENE, K. H. (1989). Effect of temperature and salinity on in vitro culture of the oyster pathogen *Perkinsus marinus* (Apicomplexa: Perkinsea). *Journal of Invertebrate Pathology* **53**: 260-268.

CHU, F.-L. E., LA PEYRE, J. F. & BURRESON, C. S. (1993). *Perkinsus marinus* infection and potential defense-related activities in eastern oysters, *Crassostrea virginica*: salinity effects. *Journal of Invertebrate Pathology* **62**: 226-232.

CIAFARDINI, G. & MAROTTA, B. (1988). Scanning electron microscopy of the sporosorus in *Polymyxa betae* (Plasmodiophoromycetes). *Canadian Journal of Botany* **66**: 2518-2522.

CIAFARDINI, G. & MAROTTA, B. (1989*a*). Use of the most-probable-number technique to detect *Polymyxa betae* (Plasmodiophoromycetes) in soil. *Applied and Environmental Microbiology* **55**: 1273-1278.

CIAFARDINI, G. & MAROTTA, B. (1989*b*). Localization of fragile areas in walls of resting spores of *Polymyxa betae*. *Canadian Journal of Botany* **67**: 3123-3126.

CICCARONE, A. (1949). Segnalazione di *Ligniera verrucosa Maire e Tison*. *Nuovo Giornale Botanico Italiano* **56**: 289-291.

CIENKOWSKI, L. (1855). Algologische Studien, I. *Botanische Zeitung, Berlin* **13**: 801-806. [N.B. column numbers, not page numbers.]

CIENKOWSKI, L. (1856). Zur genesis eines einzelligen Organismus. *Mélanges Biologiques Tirés du Bulletin Physico-Mathématique de l'Académie Imperiale des Sciences de St.-Petersbourg* **2(4)**: 361-368.

CIENKOWSKI, L. (1858). Über die Pseudogonidien. *Jahrbuch für Wissenschaftliche Botanik* **1**: 371.

CIENKOWSKI, L. (1858). Ueber meinen Beweis für die Generatio primaria. *Melanges Biologiques Tirés du Bulletin Physico-Mathématique de l'Académie Imperiale des Sciences de St.-Petersbourg* **3**: 1-21.

CIENKOWSKI, L. (1859). Ueber meinen Beweis für die *Generatio primaria*. *Bulletin de la Classe Physico-Mathématique de l'Académie Impériale de St.-Pétersbourg (Izvěstiya Imperatorskoï akademiī nauk, St.-Petersburg)* **17**: 81-95.

CIENKOWSKI, L. (1865). Beiträge zur Kenntniss der Monaden. *Archiv für Mikroskopische Anatomie und Entwicklungsmechanik* **1**: 203-232, plates XII-XIV.

CIENKOWSKI, L. (1867). Ueber den Bau und die Entwickelung der Labyrinthuleen. *Archiv für Mikroskopische Anatomie und Entwicklungsmechanik* **3**: 274-310.

CIENKOWSKI, L. (1873). *Guttulina rosea*. Transactions of the Botanical Section of the 4th Meeting of Russian Naturalists at Kazan.

CIENKOWSKI, L. (1876). Ueber einige Rhizopoden und verwandte Organismen. *Archiv für Mikroskopische Anatomie und Entwicklungsmechanik* **12**: 15-50.

CIFERRI, R. (1961). Un chytridiale parasita di Rotiferi termali. *Atti dell'Istituto Botanicao della Universita e Laboratorio Crittogamico, Pavia, Serie 6* **19**: 60-62.

CLARK, G. & DICK, M. W. (1974). Long-term storage and viability of aquatic Oomycetes. *Transactions of the British Mycological Society* **63**: 611-612.

CLARK, J. S. C. & SPENCER-PHILLIPS, P. T. N. (1993). Accumulation of photoassimilate by *Peronospora viciae* (Berk.) Casp. and leaves of *Pisum sativum* L.: evidence for nutrient uptake via intercellular hyphae. *New Phytologist* **124**: 107-119.

CLARK, M. C., LAWRENCE, C. H., PAGE, O. T. & HODGSON, W. A. (1968). Composition of DNA from *Phytophthora infestans*. *Canadian Journal of Microbiology* **14**: 482-483.

CLARKE, D. D. (1983). Potato late blight: a case study. *In* J. A. Callow (Ed.), *Biochemical Plant Pathology*, pp. 3-17. New York, U.S.A.: Wiley.

CLARK-WALKER, G. D. & GLEASON, F. H. (1973). Circular DNA from the water mold *Saprolegnia*. *Archiv für Mikrobiologie* **92**: 209-216.

CLAUSSEN, P. (1908). Über Eintwicklung und Befruchtung bei *Saprolegnia monoica*. *Berichte der Deutschen Botanischen Gesellschaft* **26**: 144-161.

CLAUSZ, J. C. (1968). Factors affecting oogenesis and oospore germination in *Achlya hypogyna*. *Journal of the Elisha Mitchell Scientific Society* **84**: 199-206.

CLAXTON, J. R., POTTER, U. J., BLAKESLEY, D. & CLARKSON, J. M. (1996). An ultrastructural study of the interaction between *Spongospora subterranea* f. sp. *nasturtii* and watercress roots. *Mycological Research* **100**: 1431-1439.

CLAY, C. M. & WALSH, J. A. (1990). Ultrastructure of sporangial development in *Spongospora subterranea* f. sp. *nasturtii* infecting watercress. *Mycological Research* **94**: 463-471.

CLAY, C. M. & WALSH, J. A. (1997). *Spongospora subterranea* f. sp. *nasturtii*, ultrastructure of the plasmodial-host interface, food vacuoles, flagellar apparatus and exit pores. *Mycological Research* **101**: 737-744.

CLAY, R. P., BENHAMOU, N. & FULLER, M. S. (1991). Ultrastructural detection of polysaccharides in the cell walls of two members of the Hyphochytriales. *Mycological Research* **95**: 1057-1064.

CLAYTON, W. D. & RENVOIZE, S. A. (1986). *Genera Graminum, Grasses of the World*. Norwich, UK, HMSO Books.

CLINTON, G. P. (1894). Observations and experiments on *Saprolegnia* infesting fish. *Bulletin of the U.S. Fish Commission* **13**: 163-172.

CLINTON, G. P. (1905). Downy mildew or blight of muskmelons and cucumbers. *Connecticut Agricultural Experiment Station, Annual Report 1904*, 329-362.

CLOKIE, J. (1970). Some substrate relationships of the family Thraustochytriaceae. *Veröffentlichungen des Instituts für Meeresforschung in Bremerhaven* **12**: 329-351.

COCCONI, G. (1894). Ricerche sullo sviluppo evolutivo di due specie nuove di funghi *Lagenidium papillosum* ed *Exoascus flavo-aureus* e sul parassitismo della *Phoma uncinulae* sull' *Uncinula adunca* Lév. *Rendiconti dell'Istituto di Bologna* [*Memorie della Reale Accademia della Scienze dell'Istituto di Bologna - Sezione dell Scienze Naturali*] **4**: 361-372.

COCK, A. W. A. M. de (1986). Marine Pythiaceae from decaying seaweeds in the Netherlands. *Mycotaxon* **25**: 101-110.

COCK, A. W. A. M. de, KARLOVSKY, P., BAHNWEG, G., JAHNKE, K.-D. & BAHNWEG, G. (1996). Similarities in restriction fragment patterns of mitochondrial DNAs of *Phytophthora* species strongly depend on the restriction enzyme used due to heterogeneous base distribution and sequence conservation. *Fungal Genetics and Biology* **20**: 36-42.

COCK, A. W. A. M. de, MENDOZA, L., PADHYE, A. A., AJELLO, L. & KAUFMAN, L. (1987). *Pythium insidiosum* sp. nov., the etiologic agent of pythiosis. *Journal of Clinical Microbiology* **25**: 344-349.

COCK, A. W. A. M. de, NEUVEL, A., BAHNWEG, G., COCK, J. C. J. M. de & PRELL, H. H. (1992). A comparison of morphology, pathogenicity and restriction fragment patterns of mitochondrial DNA among isolates of *Phytophthora porri* Foister. *Tijdschrift over Plantenziekten* [*Netherlands Journal of Plant Pathology*] **98**: 277-289.

COFFEY, M. D. (1975). Ultrastructural features of the haustorial apparatus of the white blister fungus *Albugo candida*. *Canadian Journal of Botany* **53**:1285-1299.

COHEN, Y. & COFFEY, M. D. (1986). Systematic fungicides and the control of oomycetes. *Annual Review of Phytopathology* **24**: 311-318.

COHN, F. (1854). Untersuchungen über die Entwicklungsgesichte der Mikroscopischen Algen und Pilze. *Verhandlungen der Kaiserlichen Leopoldinisch-carolinischen Akademie der Naturforscher* **24**(1): 101-256, plates 15-20 [also as *Novorum Actorum Academiae caesareae Leopoldino-carolinae Naturae Curiosorum*, and as a separate book: *Untersuchungen über die Entwicklungsgesichte der Mikroscopischen Algen und Pilze*. Bonn, Germany: Eduard Weber].

COHN, F. (1856). *Bericht der schlesischen Gesellschaft für 1956*: 24.

COHN, F. (1865). Chytridii speces novae marinae. *Hedwigia* **4**: 169-170.

COHN, F. (1867). Beiträge sur Physiologie der Phycochromaceen und Florideen. *Archiv für Mikroscopische Anatomie und Entwicklungsmechanik* **3**: 1-60.

COHN, F. (1872). Conspectus familiarum cryptogamarum secundum methodum naturalem dispositarum auctore Ferdinand Cohn. *Hedwigia* **11**: 17-20.

COKER, W. C. (1909). *Leptolegnia* from North Carolina. *Mycologia* **1**: 262-264.

COKER, W. C. (1910). Another new *Achlya*. *Botanical Gazette* **50**: 381-383.

COKER, W. C. (1912*a*). *Achlya debaryana* Humphrey and the *prolifera* group. *Mycologia* **4**: 319-324.

COKER, W. C. (1912*b*). *Achlya glomerata* sp. nov. *Mycologia* **4**: 325-326.

COKER, W. C. (1914*a*). Two new species of water molds. *Mycologia* **6**: 285-302.

COKER, W. C. (1914*b*). An *Achlya* of hybrid origin. *Journal of the Elisha Mitchell Scientific Society* **30**: 63.

COKER, W. C. (1923*a*). *The Saprolegniaceae, with Notes on Other Water Molds*. Chapel Hill, North Carolina, U.S.A.: University of North Carolina Press.

COKER, W. C. (1923*b*). A new species of *Thraustotheca*. *Journal of the Elisha Mitchell Scientific Society* **39**: 112-115.

COKER, W. C. (1924). Revision of the genus *Thraustotheca*, with a description of a new species. *Journal of the Elisha Mitchell Scientific Society* **40**: 197-202.

COKER, W. C. (1927). Other water molds from the soil. *Journal of the Elisha Mitchell Scientific Society* **42**: 207-226.

COKER, W. C. & BRAXTON, H. H. (1926). New water molds from the soil. *Journal of the Elisha Mitchell Scientific Society* **42**: 139-149.

COKER, W. C. & COUCH, J. N. (1920). A new species of *Achlya*. *Journal of the Elisha Mitchell Scientific Society* **36**: 100-101.

COKER, W. C. & HYMAN, O. W. (1912). *Thraustotheca clavata*. *Mycologia* **4**: 87-90.

COKER, W. C. & LEITNER, J. (1938). New species of *Achlya* and *Apodachlya*. *Journal of the Elisha Mitchell Scientific Society* **54**: 311-318.

COKER, W. C. & MATTHEWS, V. D. (1937). Blastocladiales, Monoblepharidales, Saprolegniales. *North American Flora* **2**(1): 1-76.

COKER, W. C. & PATTERSON, P. M. (1927). A new species of *Pythium*. *Journal of the Elisha Mitchell Scientific Society* **42**: 247-250.

COKER, W. C. & PEMBERTON, J. D. (1908). A new species of *Achlya*. *Botanical Gazette* **45**: 194-196.

COLHOUN, J. (1958). Club root of crucifers caused by *Plasmodiophora brassicae* Woronin. *Phytopathological Papers* **3**: 1-108.

COLLARD, S. (1966). *Thallasomyces californiensis* sp. n., a parasite of the nervous system of a shrimp, *Pasiphaea emarginata* Rathbun. *Proceedings. Koninklijke nederlandse Akademie van Wetenschappen, Amsterdam* **69**: 37-49.

STRAMINIPILOUS FUNGI

COLLINS, M. I. (1920). Notes on certain variations of the sporocyst in a species of *Saprolegnia*. *Proceedings of the Linnean Society of New South Wales* **45**: 277-284.

COLORNI, A. (1985). A study on the bacterial flora of giant prawn, *Macrobrachium rosenbergii* larvae fed with *Artemia salina nauplii*. *Aquaculture* **49**: 1-10.

COMPS, M., MENU, B. & MOREAU, V. (1993). Massive infections with fungus of the rotifer *Brachionus plicatilis*. *Bulletin of the Europaean Association of Fish Pathology* **13**: 28-29.

CONSTANTINEANU, J. C. (1901). Contributions à la flore mycologique de la Roumanie. *Revue Générale de Botanique* **13**: 369-389.

CONSTANTINESCU, O. (1979). Revision of *Bremiella* (Peronosporales). *Transactions of the British Mycological Society* **72**: 510-515.

CONSTANTINESCU, O. (1985). Notes on *Pseudoperonospora*. *Mycotaxon* **24**: 301-311.

CONSTANTINESCU, O. (1989). *Peronospora* complex on Compositae. *Sydowia* **45**: 79-107.

CONSTANTINESCU, O. (1991*a*). An annotated list of *Peronospora* names. *Thunbergia* **15**: 1-110.

CONSTANTINESCU, O. (1991*b*). *Bremiella sphaerosperma* sp. nov. and *Plasmopara borreriae* comb. nov. *Mycologia* **83**: 473-479.

CONSTANTINESCU, O. (1992). The nomenclature of *Plasmopara* on Umbelliferae. *Mycotaxon* **43**: 471-477.

CONSTANTINESCU, O. (1996*a*). *Paraperonospora apiculata* sp. nov. *Sydowia* **48**: 105-110.

CONSTANTINESCU, O. (1996*b*). *Peronospora* on *Acaena* (Rosaceae). *Mycotaxon* **58**: 313-318.

CONSTANTINESCU, O. (1998). A revision of *Basidiophora* (Chromista, Peronosporales). *Nova Hedwigia* **66**: 251-265.

COOK, H. L. (1971). Fungi parasitic on shrimp. *FAO Aquaculture Bulletin* **3**: 13.

COOK, K. L., HUDSPETH, D. S. S. & HUDSPETH, M. E. S. (2000). A *cox 2* phylogeny of representative marine Peronosporomycetes (Oomycetes). **[in press]**.

COOK, M. T. (1929). Life history of *Ligniera vascularium* (Matz) Cook (formerly known as *Plasmodiophora vascularum*). *Journal of the Department of Agriculture, Puerto Rico* **13**: 19-29.

COOK, M. T. (1937). The organism causing dry rot of sugar cane. *Journal of the Department of Agriculture, Puerto Rico* **21**: 85-96.

COOK, M. T. (1952). Species of *Synchytrium* in Louisiana. VII. A new species on *Urtica chamaedryoides*. *Mycologia* **44**: 827-828.

COOK, P. W. (1963). Host range studies of certain phycomycetes parasitic on desmids. *American Journal of Botany* **50**: 580-588.

COOK, W. R. I. [see IVIMEY COOK, W. R.]

COOKE, D. E. L., KENNEDY, D. M., GUY, D. C., RUSSELL, J., UNKLES, S. E. & DUNCAN, J. M. (1996). Relatedness of group I species of *Phytophthora* as assessed by random amplified polymorphic DNA (RAPDs) and sequences of ribosomal DNA. *Mycological Research* **100**: 297-303.

COOKE, D. E. L., JUNG, T., WILLIAMS, N. A., SCHUBERT, R., BAHNWEG, G., OßWALD, J. M. & DUNCAN, J. M. (1996). Molecular evidence supports *Phytophthora quercina* as a new species. *Mycological Research* **103**: 799-804.

COOKE, D. E. L., WILLIAMS, N. A., WILLIAMSON, B. & DUNCAN, J. M. (2000). Evolution and taxonomy of *Peronospora*: a molecular analysis. *In* P. Spencer-Phillips (Ed.), *Advances in Downy Mildew Research*, Dordrecht, The Netherlands: Kluwer Academic Publishers. **[in press]**

COOKE, M. C. (1871). *Handbook of British Fungi, Volume II*. London, U.K.: Macmillan.

COOKE, R. C. (1968). Relationships between nematode-destroying fungi and soil-borne phytonematodes. *Phytopathology* **58**: 909-913.

COOKE, R. C. & GODFREY, B. E. S. (1968). A key to the nematode-destroying fungi. *Transactions of the British Mycological Society* **47**: 61-74.

COOKE, W. B. (1954). The use of antibiotics in media for the isolation of fungi from polluted water. *Antibiotics and Chemotherapy* **4**: 657-662.

COOKE, W. B. & BARTSCH, A. F. (1959). Aquatic fungi in water with high waste loads. *Sewage and Industrial Wastes* **31**: 1316-1322.

COOKE, W. B. & BARTSCH, A. F. (1960). Aquatic fungi in some Ohio streams. *Ohio Journal of Science* **60**: 144-148.

COOKSON, I. (1937). On *Saprolegnia terrestris* sp. nov., with some preliminary observations on Victorian soil Saprolegniales. *Proceedings of the Royal Society of Victoria (N.S.)* **49**: 235-243.

COONEY, E. W., BARR, D. J. S. & BARSTOW, W. E. (1985). The ultrastructure of the zoospore of *Hyphochytrium catenoides*. *Canadian Journal of Botany* **63**: 497-505.

COOPER, B. A. & ARONSON, J. M. (1967). Cell wall structure of *Pythium debaryanum*. *Mycologia* **59**: 658-670.

COOPER, G. O. (1929*a*). A cytological study of fertilization in *Achlya hypogyna* Coker and Pemberton. *Transactions of the Wisconsin Academy of Science, Arts and Letters* **24**: 303-308.

COOPER, G. O. (1929*b*). Cytological studies on the sporange development and gametogenesis in *Brevilegnia diclina* Harvey. *Transactions of the Wisconsin Acadamy of Science, Arts and Letters* **24**: 309-322.

COPE, M. C. & HARDHAM, A.R. (1994). Synthesis and assembly of flagellar surface antigens during zoosporognesis in *Phytophthora cinnamomi*. *Protoplasma* **180**: 158-168.

COPELAND, H. F. (1956). *The Classification of Lower Organisms*. Palo Alto, U.S.A.: Pacific Books.

COPLAND, J. W. & WILLOUGHBY, L. G. (1982). The pathology of *Saprolegnia* infections of *Anguilla anguilla* L. elvers. *Journal of Fish Diseases* **5**: 421-428.

CORDA, A. J. K. (1837). *Icones Fungorum Hucusque Cognitorum, Vol. 1*. Praha, Czechslovakia.

CORNU, M. (1869). Note sur l'oospore du *Myzocytium* Schenk. *Bulletin de la Société Botanique de France* **16**: 222-223.

CORNU, M. (1870). Note sur une Saprolégniée, parasite d'une nouvelle espèce d'*Oedogonium*. *Bulletin de la Société Botanique de France* **17**: 297-299.

CORNU, M. (1871). Note sur deux genres nouveaux de la famille des Saprolégniées. *Bulletin de la Société Botanique de France* **18**: 58-59.

CORNU, M. (1872*a*). Séance du 22 Janvier: Affinité der Myxomycètes et des Chytridinées; Séance du 9 Février: [response to] "Lettre de M. Émile Bescherelle". *Bulletin de la Société Botanique de France* **19**: 70-71, 75-76.

CORNU, M. (1872*b*). Monographie des Saprolégniées. Étude physiologique et systématique. *Annales des Sciences Naturelles, Botanique, Série V* **15**: 1-198.

CORNU, M. (1877). Remarques sur quelques Saprolégniées nouvelles. *Bulletin de la Société Botanique de France* **24**: 226-228.

COTNER, F. B. (1930). The development of the zoöspores in the Oömycetes at optimum temperatures and the cytology of their active stages. *American Journal of Botany* **17**: 511-546.

COUCH, J. N. (1924). Some observations on spore formation and discharge in *Leptolegnia*, *Achlya*, and *Aphanomyces*. *Journal of the Elisha Mitchell Scientific Society* **40**: 27-42.

COUCH, J. N. (1926*a*). Notes on the genus *Aphanomyces*, wth a description of a new semiparasitic species. *Journal of the Elisha Mitchell Scientific Society* **41**: 213-227.

COUCH, J. N. (1926*b*). Heterothallism in *Dictyuchus*, a genus of the water molds. *Annals of Botany, London* **40**: 849-881.

COUCH, J. N. (1927). Some new water fungi from soil, with observations on spore formation. *Journal of the Elisha Mitchell Scientific Society* **42**: 227-242.

COUCH, J. N. (1931*a*). Observations on some species of water molds connecting *Achlya* and *Dictyuchus*. *Journal of the Elisha Mitchell Scientific Society* **46**: 225-230.

COUCH, J. N. (1931*b*). *Micromyces zygogonii* Dang., parasitic on *Spirogyra*. *Journal of the Elisha Mitchell Scientific Society* **46**: 231-239.

COUCH, J. N. (1932). The development of the sexual organs in *Leptolegnia caudata*. *American Journal of Botany* **19**: 584-99.

COUCH, J. N. (1935*a*). New or little known chytridiales. *Mycologia* **27**: 160-175.

COUCH, J. N. (1935*b*). A new saprophytic species of *Lagenidium*, with notes on other species. *Mycologia* **27**: 376-387.

COUCH, J. N. (1937). Notes on the genus *Micromyces*. *Mycologia* **29**: 592-596.

COUCH, J. N. (1939). Heterothallism in the Chytridiales. *Journal of the Elisha Mitchell Scientific Society* **55**: 409-414.

COUCH, J. N. (1941). The structure and action of the cilia from aquatic Phycomycetes. *American Journal of Botany* **28**: 704-714.

COUCH, J. N. (1942). A new fungus on crab eggs. *Journal of the Elisha Mitchell Scientific Society* **58**: 158-162.

COUCH, J. N. (1945). Observations on the genus *Catenaria*. *Mycologia* **37**: 163-193.

COUCH, J. N., LEITNER, J. & WHIFFEN, A. (1939). A new genus of the Plasmodiophoraceae. *Journal of the Elisha Mitchell Scientific Society* **55**: 399-408.

COUCH, J. N. & ROMNEY, S. V. (1973). Sexual reproduction in *Lagenidium giganteum*. *Mycologia* **65**: 250-252.

COUCH, R. A. (1951). The effect of increasing salinity on the occurrence of aquatic fungi in the lower part of the R. Rheidol. Unpublished BSc dissertation, cited in Wilson (1960).

COUTÉ, A. (1984). Premières observations au M.E.T. et au M.E.B. sur la cytologie de *Ducellieria chodatii* (Ducell.) Teiling (Xanthophyceae, Mischococcales, Chlorobotrydaceae). *Nova Hedwigia* **39**: 651-662.

COUTIÈRE, H. (1911*a*). Sur les *Ellobiopsis* des crevettes bathypélagiques. *Comptes Rendus Hebdomadaires des Séances de l'Académie des Sciences, Paris, Série D* **152**: 409-411.

COUTIÈRE, H. (1911*b*). Les Ellobiopsidæ des crevettes bathypélagiques. *Bulletin Scientifique de la France et de la Belgique, Paris* **45**: 186-206.

COX, B. A. & MACKIN, J. G. (1974). Studies on a new species of *Labyrinthula* (Labyrinthulales) isolated from the marine gastropod *Thais haemastoma floridana*. *Transactions of the American Microscopical Society* **93**: 62-70.

CRAIGIE, J. S. (1984). Disease control in *Chondrus crispus*. *Journal of Phycology, Supplement* **20**: 9.

CRAWFORD, A. R., BASSAM, B. J., DRENTH, A. MacLEAN, D. J. & IRWIN, J. A. G. (1996). Evolutionary relationships among *Phytophthora* species deduced from rDNA sequence analysis. *Mycological Research* **100**: 437-443.

CRISP, L. M. & BLAND, C. E. (1980). Evaluation of 6 strains of *Lagenidium callinectes* via cluster analysis. *Journal of the Elisha Mitchell Scientific Society* **96**: 95.

CRISP, L. M. & BLAND, C. E. (1989). Biosystematics and distribution of *Lagenidium callinectes*, a fungal pathogen of marine Crustacea. *Mycologia* **81**: 709-716.

CRISP, L. M. & BLAND, C. E. (1990). Potential use of ozone to disinfect sea water of fungi causing diseases of cultured marine Crustacea. *Journal of Invertebrate Pathology* **55**: 380-386.

CRISP, L. M., BLAND, C. E. & BAHNWEG, G. (1989). Biosystematics and distribution of *Lagenidium callinectes*, a fungal pathogen of marine Crustacea. *Mycologia* **81**: 709-716.

CROOKS, K. M. (1937). Studies on Australian aquatic Phycomycetes. *Proceedings of the Royal Society of Victoria (New Series)* **49**: 206-232.

CROSS, M. L. & WILLOUGHBY, L. G. (1989). Enhanced vulnerability of rainbow trout (*Salmo gaidneri*) to *Saprolegnia* infection, following treatment of the fish with an androgen. *Mycological Research* **93**: 379-402.

CRUMP, D. H. (1991). Estimation of suppressiveness and isolation of Fungal parasites of cyst nematodes. *Bulletin: SROP* **14**: 18-22.

CRUMP, D. H. & MOORE, H. (1990). Infection of the potato-cyst nematode by the nematophagous fungus, *Nematophthora gynophila*. *Nematologia* **36**: 324-326.

CRUMP, D. H., SAYRE, R. M. & YOUNG, L. D. (1983). Occurrence of nematophagous fungi in cyst nematode populations. *Plant Disease* **67**: 63-64.

CRUMP, E. & BRANTON, D. (1966). Behavior of primary and secondary zoospores of *Saprolegnia* sp. *Canadian Journal of Botany* **44**: 1393-1400.

CRUTE, I. (1981). The host specificity of peronosporaceous fungi and the genetics of the relationship between host and parasite. *In* D. M. Spencer (Ed.), *The Downy Mildews*, pp. 45-56. Academic Press, London, U.K.

CUDA, J. P., HORNBY, J. A., COTTERILL, B. & CATTELL, M. (1997). Evaluation of *Lagenidium giganteum* for biological control of *Mansonia* mosquitoes in Florida (Diptera: Culicidae). *Biological Control* **8**: 124-130.

CUTTER, V. M. (1941). Observations on certain species of *Aphanomyces*. *Mycologia* **33**: 220-240.

CUTTER, V. M. (1943). An undescribed *Lagenidium* parasitic in *Potomogeton*. *Mycologia* **35**: 2-12.

CZECZUGA, B. (1993). Studies on aquatic fungi XXVIII. The presence of predatory fungi in the waters of north-eastern Poland. *Acta Mycologica, Warszawa* **28**: 211-217.

CZECZUGA, B. (1994). Studies of aquatic fungi 17. Aquatic fungi of Lake Háncza in the Suwalki Scenic Park and of some adjacent lakes (northeastern Poland). *Acta Hydrobiologica, Krako'w* **36**: 371-385.

CZECZUGA, B. (1995*a*). Hydromycoflora of thirty-one lakes in Elk Lake District and adjacent waters with reference to the chemistry of the environment. *Acta Mycologica, Warszawa* **30**: 49-63.

CZECZUGA, B. (1995*b*). Hydromycoflora fungi of small forest lakes "Suchary" in the Wigry National Park. *Acta Mycologica, Warszawa* **30**: 167-180.

CZECZUGA, B. (1995*c*). Mycoflora of the Narew River and its tributaries in the strtetch between Tykocin and Ostrołęka. *Acta Mycologica, Warszawa* **30**: 181-191.

CZECZUGA, B., CHOMUTOWSKA, H. & WORONOWICZ, L. (1990). Studies on aquatic fungi. XV. The hydromycoflora of the biosphere sanctuary, Lake Łuknajno. *Acta Mycologica, Warszawa* **26**: 37-44.

CZECZUGA, B. & MUSZYŃSKA, E. (1994). Keratinophilic fungi in various types of water bodies. *Acta Mycologica, Warszawa* **29**: 201-215.

CZECZUGA, B. & MUSZYŃSKA, E. (1997). Aquatic fungi growing on the eggs of some anadromous fish species of the family Clupeidae. *Acta Ichthyologica et Piscatoria* **27**: 83-93.

CZECZUGA, B., MUSZYŃSKA, E. & GODLEWSKA, A. (1998). Aquatic fungi on certain species of crayfish in water from different water bodies. *Polish Journal of Environmental Studies* **7**: 75-82.

CZECZUGA, B. & PRÓBA, D. (1980). The characteristics of the environment of *Sommerstorffia spinosa* (Oomycetes, Saprolegniales), a parasite of certain rotifers. *Mycologia* **72**: 702-707.

CZECZUGA, B., PRÓBA, D. & BRZOZOWSKA, K. (1984). Badania grzybów wodnych. II. Grzyby wodne rzeki Narew na odcinku Suraż-Tykocin oraz u ujścia rzeki Turośłanki i Suprażli na tle zróżnicowango środowiska. *Rocznik AM, Białystok* **29**: 1-17.

CZECZUGA, B. & WORONOWICZ, L. (1991). Studies on aquatic fungi. XXI. The Lake Mamry complex. *Acta Mycologica, Warszawa* **27**: 93-103.

DACKMAN, C., JANSSON, H.-B. & NORDBRING-HERTZ, B. (1992). Nematophagous fungi and their activities in soil. *In* G. Stotzky & J.-M. Bollag (Eds) *Soil Biochemistry, Vol 7. Soils, Plants and the Environment*, pp. 95-130. New York, N.Y., U.S.A.: Marcel Dekker.

DA CUNHA, A. & MUNIZ, J. (1923). Parasitismo de *Trichomonas* por Chytridaceae do genero *Sphaerita* Dang. *Brazil-Medico* **36**: 386.

DAGGETT, S. S., KNIGHTON, J. E. & THERRIEN, C. D. (1995). Polyploidy among isolates of *Phytophthora infestans* from eastern Germany. *Journal of Phytopathology* **143**: 419-422.

DAHL, E. (1951). A note on *Amallocystis capillosus* Fage, an ellobiopsid parasite of *Pasiphaea tarda* Krøyer. *Arkiv för Zoologie, Series 2* **1**: 519-523.

DAHLE, J. (1982). Mykosen bei Krebsen - eine Literaturubersicht. (Mycoses in crayfish, a review.) *Deutsche Tierarztliche Wochenschrift* **89**: 46-50.

DAI, G. H., ANDARY, C., MONDOLOTCOSSON, L. & BOUBALS, D. (1995). Involvement of phenolic compounds in the resistance of grapevine callus to downy mildew (*Plasmopara viticola*). *European Journal of Plant Pathology* **101**: 541-547.

DALE, M. L. & IRWIN, J. A. G. (1991). Stomata as an infection court for *Phytophthora megasperma f. sp. medicaginis* in chickpea and a histological study of infection. *Phytopathology* **81**: 375-379.

D'AMBRA, V. & MUTTO, S. (1975). Ultrastruttura di *Polymyxa betae* Keskin. Plasmodio, sporangio e cistosoro. *Rivista di Patologia Vegetale, Padova, Serie 3* **11**: 115-124.

D'AMBRA, V. & MUTTO, S. (1977). The ultrastructure of *Polymyxa betae* zoospore exit-tube differentiation. *Canadian Journal of Botany* **55**: 831-839.

DAMLE, V. P. (1943). The cytology of *Cystopus platensis* Speg. and its bearing on the identity and phylogeny of the species. *Journal of the Indian Botanical Society* **22**: 137-58.

DANGEARD, P.-A. (1884-85*b*). Note sur le *Catenaria anguillulae* Sorok., *Chytridium zootocum* A. Br. *Bulletin de la Société Linnéenne de Normandie* **9**: 126-136.

DANGEARD, P.-A. (1886*a*). Recherches sur les organisms inférieurs. *Annales des Sciences Naturelles, Botanique, Série VII* **4**: 241-341.

DANGEARD, P.-A. (1886*b*). Sur un nouveau genre de Chytridinées, parasite des rhizopodes et des flagellates. *Bulletin de la Société Botanique de France* **33**: 240-242.

DANGEARD, P.-A. (1888). Les Péridiniens et leurs parasites. *Journal de Botanique* **2**: 126-132.

DANGEARD, P.-A. (1889). Mémoire sur les Chytridinées. *Le Botaniste* **1**: 39-74.

DANGEARD, P.-A. (1890*a*). Contribution à l'étude des organisms inférieurs. *Le Botaniste* **2**: 1-58.

DANGEARD, P.-A. (1890*b*). Recherches histologiques sur les champignons. *Le Botaniste* **2**: 63-149.

DANGEARD, P.-A. (1890*c*). Mémoire sur quelques maladies des algues et des animaux. *Le Botaniste* **2**: 231-268.

DANGEARD, P.-A. (1890*d*). Sur les oospores formés par le concours d'éléments sexuels plurinuclées. *Comptes Rendus Hebdomadaires des Séances de l'Académie des Sciences, Paris, Série D* **111**: 382-384.

DANGEARD, P. A. (1892). Du rôle des noyaux dans la fécondation chez les Oömycètes. *Revue Mycologique, Toulouse* **13**: 53-55.

DANGEARD, P.-A. (1895). Mémoire sur les parasites du noyau et du protoplasma. *Le Botaniste* **4**: 199-248.

DANGEARD, P.-A. (1896). Note sur une nouvelle espèce de Chytridinées. *Le Botaniste* **5**: 21-26.

DANGEARD, P.-A. (1900). Note sur une nouveau parasite des amibes. *Le Botaniste* **7**: 85-87.

DANGEARD, P.-A. (1903). Sur le nouveau genre *Protascus*. *Comptes Rendus Hebdomadaires des Séances de l'Académie des Sciences, Paris. Série D* **136**: 627-628.

DANGEARD, P.-A. (1906). Les ancestres des champignons supérieurs. *Le Botaniste* **9**: 157-303.

DANGEARD, P.-A. (1910). Études sur le développement et la structure des organismes inférieurs. *Le Botaniste* **11**: 1-311, 33 plates.

DANGEARD, P.-A. (1912). Recherches sur quelques algues nouvelle ou peu connues. *Le Botaniste* **12**: 1-21.

DANGEARD, P.-A. (1916). Observations sur le chondriome des *Saprolegnia*, sa nature, son origine, et ses propriétés. *Bulletin de la Société Mycologique de France* **32**: 87-96.

DANGEARD, P.-A. (1931). Mémoire sur la terminologie des éléments cellulares et son application a l'étude des champignons. *Le Botaniste 22 Serie* **6**: 325-493.

DANGEARD, P.-A. (1932). Observations sur la famille des Labyrinthulées et sur quelques autres parasites des *Cladophora*. *Le Botaniste* **24**: 217-258.

DANGEARD, P.-A. (1933). Nouvelles observations sur les parasites des Eugléniens. *Le Botaniste* **25**: 3-46.

DANGEARD, P. (1934). Sur la présence à Roscoff d'une Chytridiale des Ectocarpées, l'*Eurychasma dicksonii* (Wright) Magnus. *Annales Protistologique, Paris* **4**: 69-73.

DANIELS, W. H. (1977). Comparative physiology and chemical control of selected strains of the marine fungus *Lagenidium callinectes* Couch. Thesis, East Carolina University, Greenville, North Carolina, U.S.A.

DARLEY, W. M. & FULLER, M. S. (1970). Cell wall chemistry and taxonomic position of *Schizochytrium*. *American Journal of Botany* **57**: 761 [abstract].

DARLEY, W. M., PORTER, D. & FULLER, M. S. (1973). Cell wall composition and synthesis via Golgi-directed scale formation in the marine eucaryote, *Schizochytrium aggregatum*, with a note on *Thraustochytrium*. *Archiv für Mikrobiologie* **90**: 89-106.

DASGUPTA, S. N. & JOHN, R. (1953). Studies in the Indian aquatic fungi. On the occurrence of Indian Lagenidiales. *Proceedings of the 40th Indian Scientific Congress, Abstracts*: 80-81.

DASGUPTA, S. N. & JOHN, R. (1988). A contribution to our knowledge of the zoosporic fungi. *Bulletin of the Botanical Survey of India* **30**: 1-82.

DASTUR, J. F. (1935). Diseases of pan (*Piper betle*) in the central provinces. *Proceedings of the Indian Academy of Sciences* **1**: 778-815.

DATE, S. G. (1972). A new species of *Achlya* from India. *Journal of the University of Poona* **42**: 79-81.

DAUGHERTY, J., EVANS, T. M., SKILLOM, T., WATSON, L. E. & MONEY, N. P. (1998). Evolution of spore release mechanisms in the Saprolegniaceae (Oomycetes): evidence from a phylogenetic analysis of internal transcribed spacer sequences. *Fungal Genetics and Biology* **24**: 354-363.

DAVIDSON, J. G. N. & BARRON, G. L. (1973). Nematophagous fungi: *Haptoglossa*. *Canadian Journal of Botany* **51**: 1317-1323.

DAVIS, B. M. (1900). The fertilization of *Albugo candida*. *Botanical Gazette* **29**: 297-311.

DAVIS, B. M. (1903). Oogenesis in *Saprolegnia*. *Botanical Gazette* **35**: 233-249, 320-349.

DAVIS, B. M. (1905). Fertilization in the Saprolegniales. *Botanical Gazette* **39**: 61-64.

DAVIS, H. C., LOOSANOFF, V. L., WESTON, W. H. & MARTIN, C. (1954). A fungus disease in clam and oyster larvae. *Science, New York* **120**: 36-38.

DAVIS, H. S. (1947). Studies of the protozoan parasites of fresh-water fish. *United States Department of the Interior. Fish and Wildlife Service, Fishery Bulletin 41* **51**: 1-29.

DAVIS, H. S. & LAZAR, E. C. (1940). A new fungus disease of trout. *Transactions of the American Fisheries Society* **70**: 264-271.

DAVIS, J. J. (1914). A provisional list of the parasitic fungi of Wisconsin. *Transactions of the Wisconsin Academy of Science, Arts, and Letters* **17**: 846-984.

DAVISON, E. M. (1968). Cytochemistry and ultrastructure of hyphae and haustoria of *Peronospora parasitica* (Pers. ex Fr.) Fr. *Annals of Botany, London* **32**: 613-621.

DAWE, V. H. & KUHN, C. W. (1983*a*). Pathogenesis of infection of virus-like particles in the aquatic fungus *Rhizidiomyces*. *Virology* **130**: 10-20.

DAWE, V. H. & KUHN, C. W. (1983*b*). Isolation and characterization of a double-stranded DNA mycovirus infecting the aquatic fungus *Rhizidiomyces*. *Virology* **130**: 21-28.

DAYAL, R. (1958). Some aquatic fungi of Allahabad. *Proceedings of the National Academy of Science, Allahabad, Section B* **28**: 49-57.

DAYAL, R. (1959). Utilization of a mixture of amino acids by some Saprolegniales. *Proceedings of the National Academy of Science, Allahabad. Section B* **29**: 49-52.

DAYAL, R. (1960). Carbon requirements of some members of the family Saprolegniaceae. *Proceedings of the National Academy of Science, Allahabad, Section B* **30**: 340-344.

DAYAL, R. (1961*a*). Utilization of a mixture of sugars by some Saprolegniales. *Proceedings of the National Academy of Science, Allahabad, Section B* **31**: 180-182.

DAYAL, R. (1961*b*). Utilization of oligo- and polysaccharides. *Proceedings of the National Institute of Science, India, B* **27**: 83-90.

DAYAL, R. (1961*c*). Nitrogen requirements of some members of the family Saprolegniaceae. *Proceedings of the National Academy of Science, Allahabad, Section B* **31**: 332-336.

DAYAL, R. (1961*d*). Sulphur requirements of some members of the family Saprolegniaceae. *Proceedings of the National Academy of Science, Allahabad, Section B* **31**: 399-401.

DAYAL, R. (1968). Studies in aquatic fungus of Varanasi. Part 4. New records and notes of taxonomy. *National Academy of Science, Annu 1967* **37**: 108.

DAYAL, R. & TANDON, R. N. (1962). Ecological studies of some aquatic phycomycetes. *Hydrobiologia* **20**: 121-127.

DAYAL, R. & TANDON, R. N. (1963). Ecological studies of some aquatic phycomycetes. II. Fungi in relation to chemical factors of the water. *Hydrobiologia* **22**: 324-330.

DAYAL, R. & THAKUR JI (1965). Studies in aquatic fungi of Varanasi. *Proceedings of the National Academy of Science, Allahabad, Section B* **35**: 309-319.

DAYAL, R. & THAKUR JI (1966). The occurrence and distribution of aquatic fungi in certain ponds of Varanasi. *Hydrobiologia* **27**: 548-558.

DAYAL, R. & THAKUR, J. (1968). Studies in aquatic fungi of Varanasi. III. Observations on some parasitic aquatic phycomycetes. *Sydowia* **22**: 278-283.

DAYAL, R. & THAKUR, J. (1969). Studies on aquatic fungi of Varanasi. II. A new species of *Achlya* with notes on other species. *Mycopathologia et Mycologia Applicata* **38**: 169-174.

DAYAL, R. & USHA KIRAN (1980). Fresh water chytrids from Varanasi (India). III. Some new records. *Hydrobiologia* **76**: 269-273.

DEACON, J. W. & DONALDSON, S. P. (1993). Molecular recognition in the homing responses of zoosporic fungi, with special reference to *Pythium* and *Phytophthora*. *Mycological Research* **97**: 1153-1171.

DEACON, J. W., LAING, S. A. K. & BERRY, L. A. (1990). *Pythium mycoparasiticum* sp. nov., an aggressive mycoparasite from British soils. *Mycotaxon* **42**: 1-8.

DEACON, J. W. & SAXENA, G. (1997). Orientated zoospore attachment and cyst germination in *Catenaria anguillulae*, a facultative endoparasite of nematodes. *Mycological Research* **101**: 513-522.

DEACON, J. W. & SAXENA, G. (1998). Germination triggers of zoospore cysts of *Aphanomyces euteiches* and *Phytophthora parasitica*. *Mycological Research* **102**: 33-41.

DEAN, D. D. & DOMNAS, A. J. (1983*a*). The extracellular proteolytic enzymes of the mosquito-parasitizing fungus *Lagenidium giganteum*. *Experimental Mycology* **7**: 31-39.

DEAN, D. D. & DOMNAS, A. J. (1983*b*). Isolation and partial characterization of collagenolytic enzyme from the mosquito parasitizing fungus *Lagenidium giganteum*. *Archives of Microbiology* **136**: 212-218.

BIBLIOGRAPHY

DEARNALEY, J. D. W. & HARDHAM, A. R. (1994). The Golgi apparatus of *Phytophthora cinnamomi* makes three types of secretory or storage vesicles concurrently. *Protoplasma* **182**: 75-79.

DEARNALEY, J. D. W., MALESZKA, J. & HARDHAM, A. R. (1996). Synthesis of zoospore peripheral vesicles during sporulation of *Phytophthora cinnamomi*. *Mycological Research* **100**: 39-48.

DEASON, T. R. (1971). The origin of flagellar hairs in the xanthophycean alga *Pseudobumilleriopsis pyrenoidosa*. *Transactions of the American Microscopical Society* **90**: 441-448.

DE BARY, A. (1852). Beitrag zur Kenntnis der *Achlya prolifera* Nees. *Botanische Zeitung, 10 Jahrgang* **28**: 473-479, **29**: 489-496, **30**: 505-511. [N.B. column numbers, not page numbers.]

DE BARY, A. (1859). *Die Mycetozoen. Ein Beitrag zur Kenntnis der Niedersten Organismen*, 2nd edition. Leipzig, Germany: Engelmann.

DE BARY, A. (1860). Einige neue Saprolegnieen. *Jahrbücher für wissenschaftliche Botanik* **2**: 169-192.

DE BARY, A. (1863). Recherches sur le développement de quelques champignons parasites. *Annales des Sciences Naturelles, Paris, Série IV* **20**: 5-148.

DE BARY, A. (1866). *Morphologie und Physiologie der Pilze, Flechten und Myxomyceten*. [Volume II of Hofmeister's Handbook, the first edition of de Bary (1887).] Leipzig, Germany.

DE BARY, A. (1876). Researches into the nature of the potato-fungus - *Phytophthora infestans*. *Journal of the Royal Agricultural Society of England, Second Series* **12**: 239-269.

DE BARY, A. (1881*a*). Untersuchungen über die Peronosporeen und Saprolegnieen und die Grundlagen eines natürlichen Systems der Pilze. *Abhandlungen der Senckenbergischen Naturforschenden Gesellschaft* **12**: 225-370.

DE BARY, A. (1881*b*). Zur Kenntniss der Peronosporeen. *Botanische Zeitung, Berlin, 39 Jahrgang* **33**: 521-530, **34**: 537-544, **35**: 553-563, **36**: 569-578, **37**: 585-595, **38**: 601-609, **39**: 617-625. [N.B. column numbers, not page numbers.]

DE BARY, A. (1881*c*). Untersuchungen über die Peronosporeen und Saprolegnieen und der Grundlagen eines natürlichen Systems der Pilze. *In* A. de Bary & M. Woronin, *Beiträge zur Morphologie und Physiologie der Pilze. Vierte Reihe*. Frankfurt, Germany: Christian Winter. [A reprint of the publication: de Bary (1881*a*).]

DE BARY, A. (1883). Zu Pringsheim's neuen Beobachtungen über den Befruchtungsact der Gattungen *Achlya* und *Saprolegnia*. *Botanische Zeitung, 41 Jahrgang* **3**: 38-46, **4**: 54-60. [N.B. column numbers, not page numbers.]

DE BARY, A. (1887). *Comparative Morphology and Biology of the Fungi, Mycetozoa and Bacteria* (The Authorised English Translation by H. E. F. Garnsley revised by I. B. Balfour, of the German original of 1884). London, U.K.: Clarendon Press.

DE BARY, A. (1888). Species der Saprolegnieen. *Botanische Zeitung, 46 Jahrgang* **38**: 597-621, **39**: 629-636, **40**: 645-653. [N.B. column numbers, not page numbers.]

DE BARY, A. & WORONIN, M. (1865). Beitrag zur Kenntniss der Chytridieen. *Berichte der Naturforschenden Gesellschaft zu Freiburg* **3**: 22-61, plates I & II.

DE BARY, A. & WORONIN, M. (1881). [see De Bary, 1881*c*]

DE BRUYN, H. L. D. (1935). Heterothallism in *Peronospora parasitica*. *Phytopathology* **25**: 8.

DE BRUYNE, C. (1890). Monadines et Chytridiacées, parasites des algues du Golfe de Naples. *Archives de Biologie* **10**: 43-104, plates III-V.

DECKENBACH, C. von (1903). *Coenomyces consuens* n.g., n.sp. (ein Beitrag zur phylogenie der Pilze). *Flora, Marburg* **12**: 253-283.

DEHKUIJZEN, H. M. (1979). Electron microscopic studies on the root hairs and cortex of a susceptible and a resistant variety of *Brassica campestris* infected with *Plasmodiophora brassicae*. *Tijdschrift over Plantenziekten* [*Netherlands Journal of Plant Pathology*] **85**: 1-17.

DE-LA-RUE, E. (1869). Note sur l'*Empusa muscae* Cohn et son rapport avec les Saprolégniées. *Byulletin' Moskovskogo Obshchestva Ispytatelei Prirody, Otdel Biologicheskii* **42 (1)**: 468-472.

DE-LA-RUE, E. (1874). Sur un cas de germination des spores des Saprolégniées. *Byulletin' Moskovskogo Obshchestva Ispytatelei Prirody, Otdel Biologicheskii* **46 (2)**: 80-84.

D'ELISCU, P. N. (1977). *Endosphaerium funiculatum*, gen. nov., sp. nov., a new predaceous fungus mutualistic in the gills of freshwater pelecypods. *Journal of Invertebrate Pathology* **30**: 418-421.

DELLA TORRE, B. (1964). Su due *Olpidiopsis* viventi in radici di crucifere. *Rivista di Patologia Vegetale, Padova, Serie 3* **4**: 79-95.

DELWICHE, P. A., GRAU, C. R., HOLUB, E. B. & PERRY, J. B. (1987). Characterization of *Aphanomyces euteiches* isolates recovered from alfalfa in Wisconsin. *Plant Disease* **71**: 155-161.

DENMAN, S. & KNOX-DAVIES, P. S. (1992). Overview of *Pythium* species and diseases recorded in South Africa from 1926 to the end of 1989. *Phytophylactica*. **24**: 79-84.

DENMAN, S., KNOX-DAVIES, P. S. & CALITZ, F. J. (1993). Evaluation of some media and methods for isolating *Pythium* species from lucerne seedlings. *Phytophylactica*. **25**: 149-156.

DENNSTEDT, A. W. (1818). *Schlüssel zum Hortus Indicus Malabaricus, oder dreifaches Register zu diesem Werke*. Weimar, Germany: Landes Industrie Comptour.

de QUEIROZ, K. & GAUTHIER, J. (1994). Toward a phylogenetic system of biological nomenclature. *TREE* **9**: 27-31.

DESHPANDE, K. B. & DESHPANDE, K. S. (1965). A new species of *Pythium*. *Sydowia* **18**. 33-34.

DICK, M. W. (1959). The occurrence and distribution of Saprolegniaceae in certain soils of S. E. England. Unpublished PhD Thesis, Queen Mary College, London, U.K.

DICK, M. W. (1960*a*). *Aplanopsis spinosa* sp. nov. *Transactions of the British Mycological Society* **43**: 60-64.

DICK, M. W. (1960*b*). A critical study of the taxonomy of the *Achlya spinosa, A. stellata, A. cornuta* complex. *Transactions of the British Mycological Society* **43**: 479-488.

DICK, M. W. (1960*c*). *Saprolegnia asterophora*, de Bary (1860). *Transactions of the British Mycological Society* **43**: 597-602.

DICK, M. W. (1961). *Brevilegniella keratinophila* gen. nov., sp. nov. *Papers of the Michigan Academy of Science, Arts, and Letters* **46**: 195-204.

DICK, M. W. (1962). The occurrence and distribution of Saprolegniaceae in certain soils of south-east England. II. Distribution within defined areas. *Journal of Ecology* **50**: 119-127.

DICK, M. W. (1963). The occurrence and distribution of Saprolegniaceae in certain soils of south-east England. III. Distribution in relation to pH and water content. *Journal of Ecology* **51**: 75-81.

DICK, M. W. (1964*a*). *Apodachlyella completa* (Humphrey) Indoh (Peronosporomycetidae: Leptomitales). *Journal of the Linnean Society (Botany)* **50**: 57-62.

DICK, M. W. (1964*b*). A note on the occurrence and host range of *Aphanomyces parasiticus*. *Transactions of the British Mycological Society* **47**: 627-628.

DICK, M. W. (1965). The maintenance of stock cultures of Saprolegniaceae. *Mycologia* **57**: 828-831.

DICK, M. W. (1966). The Saprolegniaceae of the environs of Blelham Tarn: sampling techniques and the estimation of propagule numbers. *Journal of General Microbiology* **42**: 257-282.

DICK, M. W. (1968*a*). *Saprolegnia parasitica* Coker in estuaries. *Nature (London)* **217**: 875.

DICK, M. W. (1968*b*). Considerations of the role of water on the taxonomy and ecology of the filamentous biflagellate fungi in littoral zones. *Veröffentlichen des Instituts für Meeresforschung in Bremerhaven, Sonderband* **3**: 27-38.

DICK, M. W. (1968*c*). A method of preparing cultures of aquatic phycomycetes suitable for mailing. *Bulletin of the British Mycological Society* **2**: 71-72.

DICK, M. W. (1969*a*). The *Scoliolegnia asterophora* aggregate, formerly *Saprolegnia asterophora* de Bary (Oomycetes). *Botanical Journal of the Linnean Society* **62**: 255-266.

DICK, M. W. (1969*b*). Morphology and taxonomy of the Oomycetes, with special reference to Saprolegniaceae, Leptomitaceae and Pythiaceae. I. Sexual reproduction. *New Phytologist* **68**: 751-775.

DICK, M. W. (1969*c*). *Achlya hypogyna* - an aggregate, or a polymorphic species? *Mycologia* **61**: 1002-1008.

DICK, M. W. (1970). Saprolegniaceae on insect exuviae. *Transactions of the British Mycological Society* **55**: 449-458.

DICK, M. W. (1971*a*). Oospore structure in *Aphanomyces*. *Mycologia* **63**: 686-688.

DICK, M. W. (1971*b*). Leptolegniellaceae, fam. nov. *Transactions of the British Mycological Society* **57**: 417-425.

DICK, M. W. (1971*c*). The ecology of the Saprolegniaceae in lentic and littoral muds with a general theory of fungi in the lake ecosystem. *Journal of General Microbiology* **65**: 324-337.

DICK, M. W. (1972). Morphology and taxonomy of the Oomycetes, with special reference to Saprolegniaceae, Leptomitaceae and Pythiaceae. II. Cytogenetic systems. *New Phytologist* **71**: 1151-1159.

DICK, M. W. (1973*a*). Saprolegniales. *In* G. C. Ainsworth, F. K. Sparrow & A. L. Sussman (Eds), *The Fungi, An Advanced Treatise, Volume IVB*, pp. 113-144. New York, U.S.A.: Academic Press.

DICK, M. W. (1973*b*). Leptomitales. *In* G. C. Ainsworth, F. K. Sparrow & A. L. Sussman (Eds), *The Fungi, An Advanced Treatise, Volume IVB*, pp. 145-158. New York, U.S.A.: Academic Press.

DICK, M. W. (1976). The ecology of aquatic phycomycetes. *In* E. B. G. Jones (Ed.), *Recent Advances in Aquatic Mycology*, pp. 513-542. London, U.K.: Elek Press.

DICK, M. W. (1978). Systematics, taxonomy and phylogeny in the Oomycetes. *In* C. V. Subramanian (Ed.), *Taxonomy of Fungi, Part 1*, pp. 82-87. Madras, India: Universiity of Madras.

DICK, M. W. (1981). Resources and regulators of fungal populations. *In* D. T. Wicklow & G. C. Carroll (eds), *The Fungal Community, Its Organization and Role in the Ecosystem*, pp. 355-382. New York, U.S.A.: Marcel Dekker Inc.

DICK, M. W. (1982*a*). Review: J. S. Karling. *Predominantly Holocarpic and Eucarpic Simple Biflagellate Phycomycetes*. 2nd Edition. *Transactions of the British Mycological Society* **79**: 382-383.

DICK, M. W. (1982*b*). Oomycetes. *In* S. Parker (Ed.), *Synopsis and Classification of Living Organisms*, pp. 179-184. New York, U.S.A.: McGraw Hill.

DICK, M. W. (1983). Validation of the class name Hyphochytriomycetes. *In* S. T. Buczacki (Ed.), *Zoosporic Plant Pathogens, A Modern Perspective*, p. 285. London, U.K.: Academic Press.

DICK, M. W. (1985). A technique for teaching: gametangial production in *Mucor*. *Bulletin of the British Mycological Society* **19**: 134-135.

DICK, M. W. (1986). A new family and new genus for two taxa previously assigned to *Apodachlyella completa* (J. E. Humphrey) H. Indoh (Peronosporomycetidae: Leptomitales). *Botanical Journal of the Linnean Society* **93**: 225-229.

DICK, M. W. (1987*a*). Sexual reproduction: nuclear cycles and life-histories with particular reference to lower-eukaryotes. *Biological Journal of the Linnean Society* **30**: 181-192.

DICK, M. W. (1987*b*). The definition and application of the word 'meiosporangium'. *The Mycologist* **1**: 166-167.

DICK, M. W. (1988). Coevolution in the heterokont fungi (with emphasis on the downy mildews and their angiosperm hosts). *In* K. A. Pirozynski & D. L. Hawksworth (Eds), *Coevolution of Fungi with Plants and Animals*, pp. 31-62. London, U.K.: Academic Press.

DICK, M. W. (1989). *Phytophthora undulata* comb. nov. *Mycotaxon* **35**: 449-453.

DICK, M. W. (1990*a*). Phylum Oomycota. *In* L. Margulis, J. O. Corliss, M. Melkonian & D. Chapman (Eds), *Handbook of Protoctista*, chapter 33, pp. 661-685. Boston, U.S.A.: Jones & Bartlett.

DICK, M. W. (1990*b*). The systematic position of *Zoophagus insidians*. *Mycological Research* **94**: 347-354.

DICK, M. W. (1990*c*). *Keys to Pythium*. 64pp. + coloured chart. Reading, U.K.: published by the author. ISBN 0-9516738-0-7.

DICK, M. W. (1992). Patterns of phenology in populations of zoosporic fungi. *In* G. Carroll & D. Wicklow (eds), *The Fungal Community, Its Organization and Role in the Ecosystem, Second Edition*, pp. 355-382. New York, U.S.A.: Marcel Dekker Inc.

DICK, M. W. (1995). Sexual reproduction in the Peronosporomycetes (chromistan fungi). *Canadian Journal of Botany, Supplement 1, Sections E-H* **73**: S712-S724.

DICK, M. W. (1997*a*). Fungi, flagella and phylogeny. *Mycological Research* **101**: 385-394.

DICK, M. W. (1997*b*). The Myzocytiopsidaceae. *Mycological Research* **101**: 878-882.

DICK, M. W. (1998*a*). The species and systematic position of *Crypticola* in the Peronosporomycetes, and new names for the genus *Halocrusticida* and species therein. *Mycological Research* **102**: 1062-1066.

DICK, M. W. (1998*b*). Impact of Molecular Methods on Fungal Systematics. A Joint (SGM Group) Symposium between the Systematics and Evolution Group of the SGM and the BMS, held at the University of Nottingham March 30-31, 1998. *The Mycologist* **12**: 116-117.

DICK, M. W. (1999). Classification of the Peronosporomycetes. *In* R. Robinson, C. Batt & P. Patel (Eds) *Encyclopaedia of Food Microbiology*. London, UK: Academic Press, volume **2**: 871-882.

DICK, M. W. (2000*a*). Towards an understanding of the evolution of the downy mildews. *In* P. Spencer Phillips (Ed.), *Advances in Downy Mildew Research* Volume 1. Dordrecht, The Netherlands: Kluwer Academic Publishers. **[in press]**.

DICK, M. W. (2000*b*). Binomials in the Peronosporales, Sclerosporales and Pythiales. *In* P. Spencer Phillips (Ed.), *Advances in Downy Mildew Research* Volume 1. Dordrecht, The Netherlands: Kluwer Academic Publishers. **[in press]**.

DICK, M. W. (2000*c*). The Peronosporomycetes. *In* D. J. McLaughlin & E. McLaughlin (Eds) *The Mycota, Volume VII, Systematics and Evolution*, pp. 39-72. Springer-Verlag, Germany. [cite as **Dick** (2001).]

DICK, M. W. (2000*d*). Flagellar morphology of *Gracea gracilis*, formerly *Olpidiopsis gracilis*. *Mycological Research* **[accepted]**.

DICK, M. W. (2000*e*). The Peronosporomycetes and other flagellate fungi. *In* D. H. Howard (Ed.) *Fungi Pathogenic for Humans and Animals: Second Edition, Revised and Expanded*. Washington, USA: American Society for Microbiology. Marcel Dekker, Inc. **[with editor]**.

DICK, M. W. & ABRO, H. (1990). Mitochondria of *Lecophagus fasciculatus*. *Mycological Research* **94**: 418-419.

DICK, M. W. & ALI-SHTAYEH, M. S. (1986). Distribution and frequency of *Pythium* species in parkland and farmland soils. *Transactions of the British Mycological Society* **86**: 49-62.

DICK, M. W., CROFT, B. J., MAGAREY, R. C., COCK, A. W. A. M. de, & CLARK, G. (1989). A new genus of the Verrucalvaceae (Oomycetes). *Botanical Journal of the Linnean Society* **99**: 97-113.

DICK, M. W. & GLOCKLING, S. L. (2000). Three new species of the Myzocytiopsidaceae (Peronosporomycetes), with a key to the nematophagous species of the family. *Botanical Journal of the Linnean Society* **[accepted]**.

DICK, M. W. & KOUYEAS, H. (1970). Comments on the abnormal germination of *Phytophthora* oocysts. *Annales de l'Institut Phytopathologique, Benaki, N.S.* **9**: 350-351.

DICK, M. W. & NEWBY, H. V. (1961). The occurrence and distribution of Saprolegniaceae in certain soils of south-east England. I. Occurrence. *Journal of Ecology* **49**: 403-419.

DICK, M. W., VICK, M. C., GIBBINGS, J. G., HEDDERSON, T. A. & LOPEZ LASTRA, C. C. (1999) 18S rDNA for species of *Leptolegnia* and other Peronosporomycetes: justification for the subclass taxa Saprolegniomycetidae and Peronosporomycetidae and division of the Saprolegniaceae *sensu lato* into the families Leptolegniaceae and Saprolegniaceae. *Mycological Research* **103**: 1119-1125.

DICK, M. W. & WIN-TIN (1973). The development of cytological theory in the Oomycetes. *Biological Reviews of the Cambridge Philosophical Society* **48**: 133-158.

DICK, M. W., WONG, P. T. W. & CLARK, G. (1984). The identity of the oomycete causing 'Kikuyu Yellows', with a reclassification of the downy mildews. *Botanical Journal of the Linnean Society* **89**: 171-197.

DIEGUEZ-URIBEONDO, J., CERENIUS, L. & SÖDERHÄLL, K. (1994). Repeated zoospore emergence in *Saprolegnia parasitica*. *Mycological Research* **98**: 810-815.

DIEHL, H. (1935). Beiträge zur Biologie von *Olpidiopsis saprolegniae* Barrett. *Zentralblatt für Bakteriologie, Parasitenkunde, und Infektionskrankheiten, Abt. 2* **92**: 229-249.

DIETRICH, S. M. C. (1973). Carbohydrates from the hyphal walls of some Oomycetes. *Biochimica et Biophysica Acta* **313**: 95-98.

DIETRICH, S. M. C. (1975). Comparative study of hyphal wall components of Oomycetes, Saprolegniaceae and Pythiaceae. *Anais da Academia Brasiliera de Ciências* **47**: 155-162.

DIETRICH, S. M. C. (1976). Presence of polyphosphate of low molecular weight in Zygomycetes. *Journal of Bacteriology* **127**: 1408-1413.

DIETRICH, S. M. C. & CAMPOS, G. M. A. (1978). Effect of polyoxin D on *Achlya radiosa*. *Journal of General Microbiology* **105**: 161-164.

DILLWYN, L. W. (1809). *British Convervae: or Colored Figures and Descriptions of the British Plants referred by Botanists to the Genus* Conferva. London, U.K.: W. Phillips.

DI PIETRO, A., GUT-RELLA, M., PACHLATKO, J. P. & SCHWINN, F. J. (1992). Role of antibiotics produced by *Chaetomium globosum* in biocontrol of *Pythium ultimum*, a causal agent of damping off. *Phytopathology* **82**: 131-135.

DIRIWÄCHTER, G., HERTER, G. & GINDRAT, D. (1979). Observations sur la culture de *Spongospora subterranea* (Plasmodiophoromycetes) en milieu gelose. *Berichte der Schweizerischen Botanischen Gesellschaft* **89**: 105-113.

DIRIWÄCHTER, G. & PARBERY, D. G. (1991). Infection of potato by *Spongospora subterranea*. *Mycological Research* **95**: 762-764.

DISSMANN, E. (1931). Zur Kenntnis einer neuen *Isoachlya*-Art aus dem Erdboden. *Beihefte zum Botanischen Zentralblatt, Cassel, II* **48**: 103-111.

DIXON, G. R. & ROBINSON, D. L. (1986). The susceptibility of *Brassica oleracea* cultivars to *Plasmodiophora brassicae* (club root). *Plant Pathology* **35**: 101-107.

DIXON, P. S. (1963). Studies on marine fungi. I. The occurrence of *Olpidiopsis feldmannii* in the British Isles. *Nova Hedwigia* **5**: 341-345.

DIXON, R. A. & PAIVA, N. L. (1995). Stress-induced phenylproanoid metabolism. *Plant Cell* **7**: 1085-1097.

DOBERSKI, J. W. & TRIBE, H. T. (1978). *Catenaria auxiliaris* (Chytridiomycetes: Blastocladiales) identified in a larva of *Scolytus scolytus* (Coleoptera: Scolytidae). *Journal of Invertebrate Pathology* **32**: 392-393.

DÖBBELER, P. & ITZEROTT, H. (1983). Ein Neufund des muscicolen chytridiomyceten *Pleotrachelus wildemanii*. *Mitteilungen der Botanischen Staatssammlung, Munchen* **19**: 431-434.

DOBSON, R. L. & GABRIELSON, R. L. (1983). Role of primary and secondary zoospores of *Plasmodiophora brassicae* in the development of clubroot in Chinese Cabbage. *Phytopathology* **73**: 559-561.

DOFLEIN, F. (1901). *Die Protozoen als Parasiten und Krankheitserreger nach biologischen Gesichtspunkten dargestellt*. Jena.

DOFLEIN, F. (1907). Fortpflanzungserscheinungen bei Amöben und verwandten Organismen. *Sitzungsberichte der Gesellschaft Morphologie Physiologie, München* **23**: 10-18.

DODGE, J. D. & GREUET, C. (1987). Dinoflagellate ultrastructure and complex organelles. *In* F. J. R. Taylor (Ed.) *The Biology of Dinoflagellates. (Botanical Monographs, Volume 21)*, pp. 92-142. Oxford, U.K.: Blackwell Scientific Publications.

DOGMA, I. J. (1975). Of Philippine mycology and lower fungi. *Kalikasan, Philippine Journal of Biology* **4**: 69-105.

DOGMA, I. J. (1977). Philippine zoosporic fungi: *Olpidium sparrowii*, a new chytridiomycete parasite of rotifer eggs. *Kalikasan, Philippine Journal of Biology* **6**: 9-20.

DOGMA, I. J. (1978). Philippine zoosporic fungi: The algal parasite *Myzocytium megastomum* and its *Rozella* hyperparasite. *Kalikasan, Philippine Journal of Biology* **7**: 47-62.

DOLLFUS, R. P. (1921). Resumé de nos principales connaissances practiques sur les malasies et les ennemis d'huitre. *Notes et Mémoires. Office scientifique et technique des Pêches Maritimes, No. 7* : 46pp.

DOLLFUS, R. P. (1946). *Parasites (Animaux et Végétaux) des Helminths*. Paris, France: Lechevalier.

DOMASHOVA, A. A. [А. А. ДОМАШОВА, **Domashova, A. A.**] (1974). Новый вид грибa из Волги -*Aphanomyces volgensis* Domasch. sp. nov. New water mould species, *Aphanomyces volgensis* Domasch. sp. nov., from the Volga. *Mikologiya I Fitopatologiya* [*Микология и Фитопатология*] **8**: 369-370.

DOMJÁN, A. (1936). "Vízigombás"-Adatok Szeged és Tihany Vidékéröl ("Wasspilz" -daten aus der Umgebung von Szeged und Tihany). *Folia cryptogamica* **2**: 8-59.

DOMNAS, A. J., BISWAS, S. S. & GALLAGHER, P. A. (1994). Squalene metabolism in two species of *Lagenidium*. *Canadian Journal of Microbiology* **40**: 523-531.

DOMNAS, A. J., FAGAN, S. M. & JARONSKI, S. (1982). Factors influencing zoospore production in liquid cultures of *Lagenidium giganteum* (Oomycetes, Lagenidiales). *Mycologia* 74: 820-825.

DOMNAS, A., GIEBEL, P. E. & McINNIS, T. M. (1974). Biochemistry of mosquito infection: preliminary studies of biochemical change in *Culex pipiens quinquifasciatus* following infection with *Lagenidium giganteum*. *Journal of Invertebrate Pathology* **24**: 293-304.

DOMNAS, A., JARONSKI, S. & HANTON, W. K. (1986). The zoospore and flagellar mastigonemes of *Lagenidium giganteum* (Oomycetes, Lagenidiales). *Mycologia* **78**: 810-817.

DOMNAS, A. J., SHIPLEY, R. M. & HICKS, B. F. (1976). Sterol requirement for zoospore production by the mosquito parasite *Lagenidium giganteum*. *Proceedings of the First International Colloquium on Invertebrate Pathology*, p. 427. Kingston, Canada: Queen's University Printing Department.

DOMNAS, A. J., SREBRO, J. P. & HICKS, B. F. (1977). Sterol requirement for zoospore formation in the mosquito-parasitizing fungus *Lagenidium giganteum*. *Mycologia* **69**: 875-886.

DOMNAS, A. J. & WARNER, S. A. (1991). Biochemical activities of entomophagous fungi. *Critical Reviews in Microbiology* **18**: 1-13.

DOMSCH, K. H., GAMS, W. & ANDERSON, T. H. (1980). *Compendium of Soil Fungi* Vol. 1. London, U.K.: Academic Press.

DONALDSON, S. P. & DEACON, J. W. (1992). Role of calcium in adhesion and germination of zoospore cysts of *Pythium*: a model to explain infection in host plants. *Journal of General Microbiology* **138**: 2051-2059.

DONALDSON, S. P. & DEACON, J. W. (1993). Effects of amino acids and sugars on zoospore taxis, encystment and cyst germination in *Pythium aphanidermatum* (Edson) Fitzp., *P. catenulatum* Msatthews and *P. dissotocum* Drechsl. *New Phytologist* **123**: 289-295.

DONOGHUE, M. J. (1985). A critique of the biological species concept and recommendations for a phylogenetic alternative. *The Bryologist* **88**: 172-181.

DOP, P. (1905). Influence de quelques substances sur le dévelopment des Saprolégniées parasites des poissons. *Bulletin de la Société Mycologique de France* **52**: 156-158.

DORRIES, W. & HAASE, L. W. (1930). Über den Einfluss der Wasserstoffionkonzentration auf das Wachstum von *Leptomitus lacteus* in kunstlichen Nahrlosungen. *Kliene Mitteilungen für die Mitglieder des Vereins für Wasser-, Boden- und Lufthygiene. Berlin* **6**: 240-244.

DOWLEY, L. J., BASNNON, L. R., COOKE, L. R., KEANE, T. & O'Sullivan, E. (Eds) (1995). Phytophthora *150*. Dublin, Ireland: Boole Press in conjunction with TEAGASC for the European Association for Potato Research.

DOYLE, J. A., JARDINE, S. & DOERENKAMP, A. (1982). *Afropollis*, a new genus of early angiosperm pollen, with notes on the Cretaceous palynostratigraphy and paleoenvironments of northern Gondwana. *Bulletin Centres Rech Explor Prod Elf Aquitaine* **6**: 39-117.

DOYLE, J. J. (1992). Gene trees and species trees: molecular systematics as one-character taxonomy. *Systematic Botany* **17**: 144-163.

DREBES, G. (1966). Ein parasitischer Phycomycet (Lagenidiales) in *Coscinodiscus*. *Helgoländer wissenschaftliche Meeresuntersuchungen* **13**: 426-435.

DREBES, G. (1968). *Lagenisma coscinodisci*, gen. nov. sp. nov., ein Vertreter der Lagenidiales in der marinen Diatomee *Coscinodiscus*. *Veröffentlichungen des Instituts für Meeresforschung in Bremerhaven* **3**: 67-70.

DREBES, G. (1969). *Lagenisma coscinodisci* (Lagenidiales) - Vegetative Vermehrung in der Kieselalge *Coscinodiscus granii*. *Encyclopaedia Cinematographica* film E. 1398.

DREBES, G. & SCHNEPF, E. (1979). Development and host specificity of *Lagenisma coscinodisci* a marine parasitic oomycete. *Botanica Marina, Berlin* **22**: 406.

DRECHSLER, C. (1927). Two water molds causing tomato rootlet injury. *Journal of Agricultural Research* **34**: 287-296.

DRECHSLER, C. (1928*a*). The occurrence of *Aphanomyces cochlioides* n. sp. on sugar beets in the United States. *Phytopathology* **18**: 149 [abstract].

DRECHSLER, C. (1928*b*). *Pythium arrhenomanes* n.sp., a parasite causing maize root rot. *Phytopathology* **18**: 873-875.

DRECHSLER, C. (1929). The beet water mold and several related root parasites. *Journal of Agricultural Research* 38: 309-361.

DRECHSLER, C. (1930). Some new species of *Pythium*. *Journal of the Washington Academy of Sciences* 20: 398-418.

DRECHSLER, C. (1932). A species of *Pythiogeton* isolated from decaying leaf-sheaths of the common cat-tail. *Journal of the Washington Academy of Science* 22: 421-449.

DRECHSLER, C. (1934*a*). *Pythium scleroteichum* n. sp. causing mottle necrosis of sweet potatoes. *Journal of Agricultural Research* 49: 881-890.

DRECHSLER, C. (1934*b*). Vascular wilt and root rot of pansies due to *Aphanomyces* sp. *Phytopathology* 24: 7-8 [abstract].

DRECHSLER, C. (1935). Occurrence of a species of *Aphanomyces* on roots of spinach and flax. *Phytopathology* 25: 14-15 [abstract].

DRECHSLER, C. (1940). Three fungi destructive to free-living terricolous nematodes. *Journal of the Washington Academy of Sciences* 30: 240-254.

DRECHSLER, C. (1941*a*). Three species of *Pythium* with proliferous sporangia. *Phytopathology* 31: 478-507.

DRECHSLER, C. (1941*b*). Predaceous fungi. *Biological Reviews of the Cambridge Philosophical Society* 16: 265-290.

DRECHSLER, C. (1943). Two species of *Pythium* occurring in southern states. *Phytopathology* 33: 261-299.

DRECHSLER, C. (1946). A nematode-destroying phycomycete forming immotile spores in aerial evacuation tubes. *Bulletin of the Torrey Botanical Club* 73: 1-17.

DRECHSLER, C. (1950). A *Pythium* with stout oogonial spines and coiled antheridial branches. *Bulletin of the Torrey Botanical Club* 77: 442-461.

DRECHSLER, C. (1960). Two root-rot fungi closely related to *Pythium ultimum*. *Sydowia* 14: 106-114.

DRECHSLER, C. (1962). A nematode-capturing phycomycete with distally adhesive branches and proximally imbedded fusiform conidia. *American Journal of Botany* 49: 1089-1095.

DUAN, R. L. (1985). A new species and two new records of *Pythium* in China. *Acta Mycologica Sinica* 4: 1-4 [In Chinese with English abstract].

DUBOSCQ, O. (1921*a*). *Labyrinthomyxa sauvageaui* n. g. n. sp., protéomyxée parasite de *Laminaria lejolisii* Sauvageau. *Compte rendu des Séances de la Société de Biologie, Paris* 84: 27-30.

DUBOSCQ, O. (1921*b*). Les plasmodes de *Labyrinthomyxa sauvageaui*. *Compte rendu des Séances de la Société de Biologie, Paris* 84: 30-33.

DUBY, J. E. (1828) see CANDOLLE, A. P. de (1828).

DUCELLIER, F. (1915). Note sur un nouveau *Coelastrum*. *Bulletin de la Société Botanique de Genève, 2me Série* 7: 73-74.

DUDDINGTON, C. L. (1950). Predacious fungi. *Journal of the Queckett Microscopical Club, Series IV* 3: 67-88.

DUDDINGTON, C. L. (1954). Predacious fungi in agricultural soils. *Nature (London)* 173: 500.

DUDDINGTON, C. L. (1955). Fungi that attack microscopic animals. *Botanical Review* 21: 377-439.

DUDDINGTON, C. L. (1957). *The Friendly Fungi. A New Approach to the Eelworm Problem.* London, U.K.: Faber & Faber.

DUDLER, R. (1990). The single-copy actin gene of *Phytophthora megasperma* encodes a protein considerably diverged from any other known actin. *Plant Molecular Biology* 14: 415-422.

DUDKA, I. A. (1964). Nekotorye biologicheski osobennosti *Saprolegnia parasitica* Coker - vozbuditelya dermatomikoza ryb. (Some specific biological features of *Saprolegnia parasitica* Coker, the pathogen of dermatomycosis). *Pervaya Nauchnaya konferentsiya molodykh biologov (First scientific conference of young biologists) Akad. Nauk SSSR, Kiev*: 84-86.

DUERSCHNER, U. (1983). Communications from the Federal Biological Institute for Agriculture and Forestry No. 217. Fungal endoparasites on mobile nematode stages. *Biologische Bundesanstalt für Land- und Forstwirtschaft, Berlin Dahlem. Mitteilungen.* pp.1-83.

DUNCAN, J. M. (1977). Germination *in vitro* of *Phytophthora fragariae* oospores from infected root tissue. *Transactions of the British Mycological Society* 69: 361-395.

DUNCAN, P. M. (1876). On some unicellular algae parasitic within Silurian and Tertiary coals, with a notice of their presence in *Calceola sandalina* and other fossils. *Quarterly Journal of the Geological Society of London* 32: 205-210.

DUNKERLY, J. S. (1914). *Dermocystidium pusula* Pérez, parasitic in *Trutta fario*. *Zoologischer Anzeiger* 44: 179.

DURVE, V. S. & BAL, D. V. (1960). Shell disease in *Crassostrea gryphoides* Sindermann. *Current Science, Bangalore, India* 29: 489-490.

DYKSTRA, M. J. (1976*a*). Wall and membrane biogenesis in the unusual labyrinthulid-like organism *Sorodiplophrys stercorea*. *Protoplasma* 87: 329-346.

DYKSTRA, M. J. (1976*b*). Nuclear and cell division in *Sorodiplophrys stercorea*, a labyrinthulid-like protist. *Protoplasma* 87: 347-359.

DYKSTRA, M. J., NOGA, E. J., LEVINE, J. F. & MOYE, D. W. (1986). Characterization of the *Aphanomyces* species involved with ulcerative mycosis (UM) in menhaden. *Mycologia* 78: 664-672.

DYKSTRA, M. J. & OLIVE, L. S. (1975). *Sorodiplophrys*: an unusual sorocarp-producing protist. *Mycologia* 67: 873-879.

DYKSTRA, M. J. & PORTER, D. (1984). *Diplophrys marina*, a new scale-forming marine protist with labyrinthulid affinities. *Mycologia* 76: 626-632.

DYLEWSKI, D. P. (1990). Phylum Plasmodiophoromycota. *In* L. Margulis, J. O. Corliss, M. Melkonian & D. J. Chapman (Eds), *Handbook of Protoctista*, pp. 399-416. Boston, U.S.A.: Jones & Bartlett.

DYLEWSKI, D. P., BRASELTON, J. P. & MILLER, C. E. (1978). Cruciform nuclear division in *Sorosphaera veronicae*. *American Journal of Botany* 65: 258-267.

DYLEWSKI, D. P. & MILLER, C. E. (1983*a*). Nuclear elimination during cyst formation in *Woronina pythii*. (Plasmodiophoromycetes). *Mycologia* 75: 271-279.

DYLEWSKI, D. P. & MILLER, C. E. (1983*b*). Cruciform nuclear division in *Woronina pythii* (Plasmodiophoromycetes). *American Journal of Botany* 70: 1325-1339.

DYLEWSKI, D. P. & MILLER, C. E. (1983*c*). Systematic and host range studies of *Woronina pythii* (Plasmodiophoromycetes) and host, *Pythium* species, from axenic culture. *Mycologia* 75: 412-422.

DYLEWSKI, D. P. & MILLER, C. E. (1984*a*). The ultrastructure of mitosis during sporangiogenesis in *Woronina pythii* (Plasmodiophoromycetes). *Protoplasma* 121: 42-53.

DYLEWSKI, D. P. & MILLER, C. E. (1984*b*). The ultrastructure of meiosis in *Woronina pythii* (Plasmodiophoromycetes). *Mycologia* 76: 1075-1088.

DYLEWSKI, D. P., MILLER, C. E. & BRASELTON, J. P. (1978). Sporangial development of *Woronina pythii*. *Micron* 9: 35-36.

EARLE, F. S. (1901). Systematic catalogue of the plants growing without cultivation in Alabama, including descriptions of new, rare, and little-known species. Fungi. *Contributions to the U.S. National Herbarium* 6: 139-263.

EBEL, J. (1998). Oligoglucoside elicitor-mediated activation of plant defense. *BioEssays* 20: 569-576.

EDNEY, K. L. & CHAMBERS, D. A. (1981). Post-harvest treatments for the control of *Phytophthora syringae* storage rot of apples. *Annals of Applied Biology* 97: 237-241.

EDSON, H. A. (1913). Damping-off and root-rot parasites of sugar beets. *Phytopathology* 3: 76 [abstract]

EDSON, H. A. (1915*a*). Seedling diseases of sugar beets and their relation to root-rot and crown-rot. *Journal of Agricultural Research* 4: 135-168.

EDSON, H. A. (1915*b*). *Rheosporangium aphanidermatum*, a new genus and species of fungus parasitic on sugar beets and radishes. *Journal of Agricultural Research* 4: 279-292.

EINARSSON, H. (1945). Euphausiacea. I. North Atlantic species. *Dana-Report, No. 27*: 1-191.

EINHORN, G. & BOCHOW, H. (1990*a*). (The effects of temperature and pH value on the pathogenesis of *Plasmodiophora brassicae* Wor., the causal agent of clubroot). *Archiv für Phytopathologie und Pflanzenschutz* 26: 131-138 [only abstract consulted].

EINHORN, G. & BOCHOW, H. (1990*b*). (Studies on the germination behaviour and survival of *Plasmodiophora brassicae* Wor. resting spores of different ages). *Archiv für Phytopathologie und Pflanzenschutz* 26: 247-255.

ELAD, Y. & CHET, I. (1987). Possible role of competition for nutrients in biocontrol of *Pythium* damping-off by bacteria. *Phytopathology* 77: 190-195.

EL-HAMALAWI, Z. A. & ERWIN, D. C. (1986). Physical enzymic and chemical factors affecting viability and germination of oospores of *Phytophthora megasperma* f. sp. *medicaginis*. *Phytopathology* 76: 503-507.

EL-HISSY, F. (1974). Freshwater fungi in Egypt. *Egyptian Journal of Botany* 17: 187-189.

EL-HISSY, F. (1994*a*). Aquatic phycomycetes from Egyptian soil. *Microbiological Research* **149**: 271-282.

EL-HISSY, F. (1994*b*). Oomycetes and Chytridiomycetes (Mastigomycotina) from water bodies in Tübingen region (Germany). *Journal of Basic Microbiology* **34**: 67-76.

ELIAS, M. K. (1966). Living and fossil algae and fungi, formerly known as structural parts of marine bryozoans. *Palaeobotanist* **14**: 5-18.

ELIASON, E. J. (1928). Comparative virulence of certain strains of *Pythium* in direct inoculation of conifers. *Phytopathology* **18**: 361-367.

ELLENBOGEN, B. B., AARONSON, S., GOLDSTEIN, S. & BELSKY, M. (1969). Polyunsaturated fatty acids of aquatic fungi: possible phylogenetic significance. *Comparative Biochemistry and Physiology* **29**: 805-811.

ELLIOTT, C. G. (1983). Physiology of sexual reproduction in *Phytophthora*. *In* D. C. Erwin, S. Bartnicki-Garcia & P. H. Tsao (Eds), Phytophthora, *Its Biology, Taxonomy, Ecology and Pathology*, pp. 71-80. American Phytopathological Society, St Paul, Minnesota, U.S.A.

ELLIOTT, C. G., HENDRIE, M. R. & KNIGHTS, B. A. (1966). The sterol requirement of *Phytophthora cactorum*. *Journal of General Microbiology* **42**: 425-435.

ELLIOTT, C. G., HENDRIE, M. R., KNIGHTS, B. A. & PARKER, W. (1964). A steroid growth factor requirement in a fungus. *Nature (London)* **203**: 427-428.

ELLIOTT, C. G. & MacINTYRE, D. (1973). Genetical evidence on the life-history of *Phytophthora*. *Transactions of the British Mycological Society* **60**: 311-316.

ELLIOTT, C. G., & SANSOME, E. (1977). The influence of sterols on meiosis in *Phytophthora cactorum*. *Journal of General Microbiology* **98**: 141-145.

ELLIOTT, J. A. (1916). The sweet potato "soil rot" or "pox", a slime mould disease. *Delaware Agriculture Experiment Station Bulletin* **114**: 15.

ELLIOTT, R. F. (1967). Morphological variation in New Zealand Saprolegniaceae. 1. *Achlya caroliniana* Coker and *A. flagellata* Coker. *New Zealand Journal of Botany* **5**: 418-423.

ELLIOTT, R. F. (1968). Morphological variation in New Zealand Saprolegniaceae. 2. *Saprolegnia terrestris* Cookson and *S. australis* sp. nov. *New Zealand Journal of Botany* **6**: 94-105.

ELLIS-EVANS, J. C. (1985). Fungi from maritime Antarctic freshwater environments. *British Antarctic Survey Bulletin* **68**: 37-45.

ELLZEY, J. T. (1974). Ultrastructural observations of meiosis within antheridia of *Achlya ambisexualis*. *Mycologia* **66**: 32-47.

ELLZEY, J. T. & HUIZAR, E. (1977). Synaptonemal complexes in antheridia of *Achlya ambisexualis* E87. *Archives of Microbiology* **112**: 311-313.

EL-NAGDY, M. A. & ABDEL-HAFEZ, S. I. I. (1990). Occurrence of zoosporic and terrestrial fungi in some ponds of Kharga Oases, Egypt. *Journal of Basic Microbiology* **30**: 233-240.

EMERSON, R. (1950). Current trends of experimental research on the aquatic phycomycetes. *Annual Review of Microbiology* **4**: 169-200.

EMERSON, R. & CANTINO, E. C. (1948). The isolation, growth, and metabolism of *Blastocladia* in pure culture. *American Journal of Botany* **35**: 157-171.

EMERSON, R. & HELD, A. A. (1969). *Aqualinderella fermentans* gen. et sp. nov., a phycomycete adapted to stagnant waters. II. Isolation, cultural characteristics, and gase relations. *American Journal of Botany* **56**: 1103-1120.

EMERSON, R. & NATVIG, D. O. (1981). Adaptation of fungi to stagnant waters. *In* D. T. Wicklow & G. C. Carroll (Eds), *The Fungal Community, its Organization and Role in the Ecosystem*, pp. 109-128. Marcel Dekker Inc., New York, USA.

EMERSON, R. & WESTON, W. H. (1967). *Aqualinderella fermentans* gen. et sp. nov., a phycomycete adapted to stagnant waters. I. Morphology and occurrence in nature. *American Journal of Botany* **54**: 702-719.

EMILIANI, C., KRAUS, E. B. & SHOEMAKER, E. M. (1981). Sudden death at the end of the Mesozoic. *Earth Planet. Science Letters* **55**: 317-334.

EMOTO, Y. (1923). On the enzymes of some Saprolegniaceae. *Botanical Magazine, Tokyo* **37**: 13-29.

ENGLISH, P. B. & FROST, A. J. (1984). Phycomycosis in a dog. *Australian Veterinary Journal* **61**: 291-292.

EPSTEIN, G. [ЭПШТЕЙН, Г., Epstein, G.] (1922). К Вопросу о паразитах у простейших. (Über parasitische Infektion bei Darmamöben)]. Архив Русского Протистологиуеского Общества (*Archives de la Société russe de Protistologie*) **1**: 46-81.

ERDOS, G. W. (1971). Ultrastructure and development of *Endemosarca*, a new plasmodial parasite of ciliates. PhD Thesis, University of North Carolina, Chapel Hill, North Carolina, U.S.A.

ERDOS, G. W. (1972*a*). The nuclear cycle and spore discharge in *Endemosarca hypsalyxis*. *Mycologia* **64**: 423-426.

ERDOS, G. W. (1972*b*). Fine structure of the plasmodium of *Endemosarca*. *Cytobiologie* **5**: 83-100.

ERDOS, G. W. (1973). A new species of *Endemosarca* from the Seychelles. *Mycologia* **65**: 229-232.

ERDOS, G. W. & OLIVE, L. S. (1971). *Endemosarca*: a new genus with proteomyxid affinities. *Mycologia* **63**: 877-883.

ERIKSSON, B. & LAANE, M. M. (1982). Examinations of nuclear behaviour during sexual cycle in the diplontic oomycete *Phytophthora palmivora* (Butl.) Butl. *Hereditas* **97**: 73-86.

ÉRSEK, T., HÖLKER, U. & HÖFER, M. (1991). Non-lethal immobilisation of zoospores of *Phytophthora infestans* by Li^+. *Mycological Research* **95**: 970-972.

ERWIN, D. C. & McCORMICK, W. H. (1971) Germination of oospores produced by *Phytophthora megasperma* var. *sojae*. *Mycologia* **63**: 972-977.

ERWIN, D. C. & RIBEIRO, O. K. (1996). Phytophthora *Diseases Worldwide*. American Phytopathological Society, St Paul, Minnesota, USA.

ESSER, R. P. (1976). *Haptoglossa heterospora* Drechsler, a fungus parasite of Florida nematodes. *Nematology Circular, No. 23*: 1-2.

ESSER, R. P. & RIDINGS, W. H. (1973). Pathogenicity of selected nematodes by *Catenaria anguillulae*. *Soil Crop Science Society, Florida* **33**: 60-64.

ESTEY, R. H. & OLTHOF, T. H. A. (1965). The occurring nematophagous fungi in Quebec. *Phyto-protection* **46**: 14-17.

ESTRADA-GARCIA, M. T., CALLOW, J. A. & GREEN, J. R. (1990). Monoclonal antibodies to the adhesive cell coat secreted by *Pythium aphanidermatum* zoospores recognise 200 × 10^3 M, glycoproteins stored within large peripheral vesicles. *Journal of Cell Science* **95**: 199-206.

ESTRADA-GARCIA, T., RAY, T. C., GREEN, J. R., CALLOW, J. A. & KENNEDY, J. F. (1990). Encystment of *Pythium aphanidermatum* zoospores is induced by root mucilage polysaccharides, pectin and a monoclonal antibody to a surface antigen. *Journal of Experimental Biology* **41**: 693-699.

ETTL, H. (1978). Xanthophyceae. *In* H. Ettl, J. Gerloff & H. Heynig (Eds), *Süsswasserflora von Mitteleuropa, 3 (1)*, pp. 1-530. Stuttgart, Germany: G. Fischer.

FAGE, L. (1936). Sur une Ellobiopsidé nouveau, *Amallocystis fasciatus* gen. et sp. nov., parasite des Mysidacés bathypélagiques. *Archives de Zoologie Expérimentale et Générale* **78**: 145-154.

FAGE, L. (1938). A propos d'un nouvel Ellobiopsidé, *Amallocystis capillosus* n. sp., parasite de *Pasiphaea tarda* Krøyer (Crust. Decap.). *Travaux de la Station Zoologique de Wimereaux, Paris* **13**: 287-291.

FAGG, J., WOODS-TOR, A. & MANSFIELD, J. (1991). Comparative study of interactions between *Bremia lactucae* and lettuce cells in suspension culture and in cotyledons. *Physiological and Molecular Plant Pathology* **38**: 105-116.

FAGGIAN, R., BULMAN, S. R., LAWRIE, A. C. & PORTER, I. J. (1999). Specific polymerase chain reaction primers for the detection of *Plasmodiophora brassica* in soil and water. *Phytopathology* **89**: 392-397.

FAMINTZEN, A. (1907). Die Symbiose als mittel der Synthese von Organismen. *Biologische Zeitschrift* **27**: 352-364.

FARLOW, W. G. (1876). On the American grape vine mildew. *Bulletin of the Bussey Institute* **1**: 415-428.

FARLOW, W. G. & SEYMOUR, A. B. (1888). *A Provisional Host Index of the Fungi of the United States*. Cambridge, Massachussetts, U.S.A.: Harvard University Press.

FARO, S. (1972). The role of a cytoplasmic glucan during morphogenesis of sex organs in *Achlya*. *American Journal of Botany* **59**: 919-923.

FARR, D. F., BILLS, G. F., CHAMURIS, G. P. & ROSSMAN, A. Y. (1989). *Fungi on Plants and Plant Products in the United States*. American Phytopathological Society, St Paul, Minnesota, USA.

FEDERICI, B. A. (1981). Mosquito control by the fungi *Culicinomyces, Lagenidium,* and *Coelomomyces*. *In* H. D. Burges (Ed.), *Microbial Control of Pests and Plant Diseases 1970-1980*, pp. 555-572. New York, U.S.A.: Academic Press.

FEDERICI, B. A. (1995). The future of microbial insecticides as vector control agents. *Journal of the American Mosquito Association* 11: 260-268.

FELDMANN, G. (1953). La végétation de l'étang de Salses (riv sud). *Vie et Milieu* 4: 685-700.

FELDMANN, G. (1956). Développement d'une plasmodiophorale marine: *Plasmodiophora bicaudata* J. Feldm., parasite du *Zostera nana* Roth. *Revue Générale de Botanique* 63: 390-421.

FELDMANN, G. (1958). Une nouvelle espèce de plasmodiophorale parasite du *Triglochin maritimum* L.: *Plasmodiophora maritima* nov. sp. *Revue Générale de Botanique* 65: 634-350.

FELDMANN, J. (1936). Les Monocotylédones marines de la Guadaloupe. *Bulletin de la Société de Botanique, France* 83: 604-613.

FELDMANN, J. (1941). Une nouvelle espèce de *Plasmodiophora* (*P. bicaudata*) parasite du *Zostera nana* Roth. *Bulletin de la Société d'Histoire Naturelle de l'Afrique du Nord* 31: 171-177.

FELDMANN, J. (1954). Inventaire de la flore marine de Roscoff. *Supplement 6, Travaux Station Biologique de Roscoff* : 1-152.

FELDMANN, J. & FELDMANN, G. (1940). Note sur une Chytridiale parasite des spores de Céramiacées. *Bulletin de la Société d'Histoire Naturelle de l'Afrique du Nord* 31: 72-75.

FELDMANN, J. & FELDMANN, G. (1956). Observations sur quelques Phycomycètes marins nouveaux ou peu connus. *Revue de Mycologie* 22: 231-251.

FELDMANN, J. & FELDMANN, G. (1967). Le parasitisme d'un *Olpidiopsis* sur le *Radicilingua reptans* (Kylin) Papenfuss. *Le Botaniste* 50: 189-203.

FELL, J. W. & MASTER, I. M. (1975). Phycomycetes (*Phytophthora* spp. nov. and *Pythium* sp. nov.) associated with degrading mangrove (*Rhizophora mangle*) leaves. *Canadian Journal of Botany* 53: 2908-2922.

FENCHEL,T. & PATTERSON, D. J. (1988). *Cafeteria roenbergensis* nov. gen. nov. sp., a heterotrophic microflagellate from marine plankton. *Marine Microbial Food Webs* 3: 9-19.

FERDINANDSEN, C. & WINGE, Ö. (1913). *Plasmodiophora halophilae* sp. n. *Centralblatt für Bakteriologie, Parasitenkunde, und Infektioskrankheiten, Abt. 2* 37: 167.

FERDINANDSEN, C. & WINGE, Ö. (1914). *Ostenfeldiella*, a new genus of Plasmodiophoraceae. *Annals of Botany, London* 28: 643-649.

FERDINANDSEN, C. & WINGE, Ö. (1920). *Clathrosorus*, a new genus of Plasmodiophoraceae. *Annals of Botany, London* 34: 467-469.

FETTER-LASKO, J. L. & WASHINO, R. K. (1977). A three-year study of the ecology of *Lagenidium* infections of *Culex tarsalis* in California. *In* C. D. Grant (Ed.), *Proceedings and Papers on the Forty-fifth Annual Conference of the Californian Mosquito and Vector Control Association*, p. 106. Visalia, U.S.A.: Californian Mosquito and Vector Control Association Press.

FETTER-LASKO, J. & WASHINO, R. K. (1978). Patterns of experimental infection of mosquito larvae by *Lagenidium giganteum*. *In* C. D. Grant (Ed.), *Proceedings and Papers on the Forty-sixth Annual Conference of the Californian Mosquito and Vector Control Association*, p. 90. Visalia, U.S.A.: California Mosquito and Vector Control Association Press.

FETTER-LASKO, J. & WASHINO, R. K. (1979). *Lagenidium giganteum* and mosquitoes associated with irrigated pastures. *In* C. D. Grant (Ed.), *Proceedings and papers of the Forty-seventh Annual Conference of the Californian Mosquito and Vector Control Association*, p. 55. Visalia, U.S.A.: California Mosquito and Vector Control Association Press.

FETTER-LASKO, J. L. & WASHINO, R. K. (1983). In situ studies on seasonality and recycling pattern in California of *Lagenidium giganteum* Couch, an aquatic fungal pathogen of mosquitoes. *Environmental Entomology* 12: 635-640.

FIJAN, N. N. (1969). Systematic mycosis in channel catfish. *Bulletin of the Wildlife Disease Association* 5: 109-110.

FINDLAY, R. H., FELL, J. W., COLEMAN, N. K. & VESTAL, J. R. (1986). Biochemical indicators of the role of fungi and thraustochytrids in mangrove detrital systems. *In* S. T. Moss (Ed.), *The Biology of Marine Fungi*, pp. 91-103. Cambridge, U.K.: Cambridge University Press.

FISCH, C. (1884*a*). Beiträge zur Kenntniss der Chytridiaceen. *Sitzungsberichte der Physicalisch-medicinischen Societät zu Erlangen* 16: 29-66.

FISCH, C. (1884*b*). Über zwei neue Chytridiaceen. *Sitzungsberichte der Physicalisch-medicinischen Societät zu Erlangen* 16: 101-103.

FISCHER, A. (1882). Untersuchungen über die Parasiten der Saprolegnieen. *Jahrbücher für wissenschaftliche Botanik* 13: 286-371.

FISCHER, A. (1892*a*). I Reihe. Archimycetes (Chytridinae). *Die Pilze Deutschlands, Oesterreichs und der Schweiz. Erster Band, IV. Abtheilung: Phycomycetes* pp. 3-160. Leipzig, Germany: Eduard Kummer. [Commonly cited as *Dr. L. Rabenhorst's Kryptogamen-Flora von Deutschland, Oesterreich und der Schweiz, Zweite Auflage, Erster Band: Pilze, IV. Abtheilung: Phycomycetes*.]

FISCHER, A. (1892*b*). III Reihe. Oomycetes. *Die Pilze Deutschlands, Oesterreichs und der Schweiz. Erster Band, IV. Abtheilung: Phycomycetes* pp. 310-490. Leipzig, Germany: Eduard Kummer. [Commonly cited as *Dr. L. Rabenhorst's Kryptogamen-Flora von Deutschland, Oesterreich und der Schweiz, Zweite Auflage, Erster Band: Pilze, IV. Abtheilung: Phycomycetes*.]

FISCHER, J. R., PACE, L. W., TURK, J. R., KREEGER, J. M., MILLER, M. A. & GOSSER, H. S. (1994). Gastrointestinal pythiosis in Missouri dogs: eleven cases. *Veterinary Diagnostic Investigations* 6: 380-382.

FISH, F. W. (1934). A fungus disease of fishes in the Gulf of Maine. *Parasitology* 26: 1-16.

FISHER, F. E. (1950). Entomogenous fungi attacking scale insects and rust mites on citrus fruits in Florida. *Journal of Economic Entomology* 43: 305-309.

FISHER, W. S. (1983). Eggs of *Palaemon macrodactyluis* 3. Infection by the fungus *Lagenidium callinectes*. *Biological Bulletin of the Marine Biological Laboratory, Woods Hole* 164: 214-226.

FISHER, W. S., NILSON, E. H. & SCHLESER, R. A. (1975). Effect of the fungus *Haliphthoros milfordensis* on the juvenile stages of the American lobster *Homarus americanus*. *Journal of Invertebrate Pathology* 26: 41-45.

FITZPATRICK, H. M. (1923). Generic concepts in the Pythiaceae and Blastocladiaceae. *Mycologia* 15: 166-173.

FITZPATRICK, H. M. (1930). *The Lower Fungi: Phycomycetes*. New York, U.S.A.: McGraw Hill.

FLANAGAN, P. W. (1970). Meiosis and mitosis in Saprolegniaceae. *Canadian Journal of Botany* 48: 2069-2076.

FLETCHER, J. (1979). An ultrastructural investigation into the role of calcium in oosphere-initial development in *Saprolegnia diclina*. *Journal of General Microbiology* 113: 315-325.

FLETCHER, R. A. (1987). Plant growth regulating properties of sterol-inhibiting fungicides. *In* S. S. Purchit (Ed.), *Hormonal Regulation of Plant Growth and Development*, pp. 103-114. Martinus Nijhoff, Dordrecht, The Netherlands.

FLETT, S. P. (1983). A technique for detection of *Spongospora subterranea* in soil. *Transactions of the British Mycological Society* 81: 424-425.

FLORA DANICA (1780). Fascicle 14, Tome 5. Hafniae. Copenhagen.

FOIL, C. S. O., SHORT, B. G., FADOK, V. A. & KUNKLE, G. A. (1984). A report of subcutaneous pythiosis in five dogs and a review of the etiological agent *Pythium* spp. *Journal of the American Animal Hospital Association* 20: 959-966.

FOISSNER, I. (1987). Gametangial interaction and oogenesis in the phycomycete *Ciliomyces spectabilis* (Fungi, Lagenidiales). *Archives of Microbiology* 147: 143-151.

FOISSNER, I. & FOISSNER, W. (1986*a*). *Ciliomyces spectabilis*, nov.gen., nov.spec., a zoosporic fungus which parasitizes cysts of the ciliate *Kahliella simplex*. I. Infection, vegetative growth and sexual reproduction. *Zeitschrift für Parasitenkunde* 72: 29-41.

FOISSNER, I. & FOISSNER, W. (1986*b*). *Ciliomyces spectabilis*, nov.gen., nov.spec., a zoosporic fungus which parasitizes cysts of the ciliate *Kahliella simplex*. II. Asexual reproduction, life cycle and systemic account. *Zeitschrift für Parasitenkunde* **72**: 43-55.

FOISSNER, I. & FOISSNER, W. (1995). *Ciliatomyces* nom. nov. for *Ciliomyces* Foissner & Foissner 1986 (*Lagenidiaceae, Oomycota*). *ΦYton (Horn, Austria)* **35**: 115-116.

FOISTER, C. E. (1940). Descriptions of new fungi causing economic diseases in Scotland. *Transactions and Proceedings of the Royal Society of Edinburgh* **33**: 65-68.

FOLEY, M. F. & DEACON, J. W. (1985). Isolation of *Pythium oligandrum* and other necrotrophic mycoparasites from soil. *Transactions of the British Mycological Society* **85**: 631-639.

FOLEY, M. F. & DEACON, J. W. (1986). Physiological differences between mycoparasitic and plant pathogenic *Pythium* species. *Transactions of the British Mycological Society* **86**: 225-231.

FONG, D., RODRIGUEZ, R., KOO, K., SUN, J., SOGIN, M. L., BUSHEK, D., LITTLEWOOD, D. T. J. & FORD, S. E. (1993). Small subunit ribosomal RNA gene sequence of the oyster parasite *Perkinsus marinus*. *Molecular Marine Biology and Biotechnology* **2**: 346-350.

FONTAINE, R. E. & SCHAEFER, C. H. (1978). Integrated pest management research: mosquito control. *California Agriculture* **32**: 29-31.

FORBES, E. J. (1935*a*). Watermoulds of the Manchester district. *Memoirs and Proceedings of the Manchester Literary and Philosophical Society* **79**: 1-11.

FORBES, E. J. (1935*b*). Observations on some British water moulds (Saprolegniales and Blastocladiales). *Transactions of the British Mycological Society* **19**: 221-239.

FÖRSTER, H. & COFFEY, M. D. (1990). Mating behavior of *Phytophthora parasitica*: evidence for sexual recombination in oospores using DNA restriction fragment length polymorphisms as genetic markers. *Experimental Mycology* **14**: 351-359.

FÖRSTER, H. & COFFEY, M. D. (1991). Approaches to the taxonomy of *Phytophthora* using polymorphisms in mitochondrial and nuclear DNA. *In* J. A. Lucas, R. C. Shattock, D. S. Shaw & L. R. Cooke (Eds), Phytophthora, pp. 164-183. Cambridge University Press: Cambridge.

FÖRSTER, H., COFFEY, M. D., ELWOOD, H. & SOGIN, M. L. (1990). Sequence analysis of the small subunit ribosomal RNAs of three zoosporic fungi and implications for fungal evolution. *Mycologia* **82**: 306-312.

FÖRSTER, H., KINSCHERF, T. G., LEONG, S. A. & MAXWELL, D. P. (1988). Estimation of relatedness between *Phytophthora* species by analysis of mitochondrial DNA. *Mycologia* **80**: 466-478.

FÖRSTER, H., KINSCHERF, T. G., LEONG, S. A. & MAXWELL, D. P. (1989). Restriction fragment length polymorphisms of the mitochondrial DNA of *Phytophthora megasperma* isolated from soybean, alfalfa, and fruit trees. *Canadian Journal of Botany* **67**: 529-537.

FÖRSTER, H., LEARN, G. & COFFEY, M. D. (1995). Towards a better understanding of the evolutionary history of the species of the genus *Phytophthora* using isozymes, RFLPs and ribosomal DNA spacers sequences. *In* L. J. Dowley, L. R. Basnnon, L. R. Cooke, T. Keane & E. O'Sullivan (Eds), Phytophthora *150*, pp. 42-54. Dublin, Ireland: Boole Press in conjunction with TEAGASC for the European Association for Potato Research.

FÖRSTER, H., OUDEMANS, P. & COFFEY, M. D. (1990). Mitochondrial and nuclear DNA diversity within six species of *Phytophthora*. *Experimental Mycology* **14**: 18-31.

FÖRSTER, H., RIBEIRO, O. K. & ERWIN, D. C. (1981). Germination of oospores of *Phytophthora megasperma* f. sp. *sojae*. *Phytopathology* **71**: 874 [abstract].

FOTT, B. (1952). Mikroflora moravských rašelin. *Preslia* **24**: 189-209.

FOUST, F. K. (1937). A new species of *Rozella* parasitic on *Allomyces*. *Journal of the Elisha Mitchell Scientific Society* **53**: 197-204.

FOX, N. C. & WOLF, F. T. (1977). Enzymes of *Dictyuchus monosporus*. *Mycologia* **69**: 263-270.

FRANC, A. & ARVY, L. (1969). Sur *Thanatostrea polymorpha* n.g.,n.sp., agent de destruction des branchies et des palpes de l'Huître portugaise. *Comptes Rendus Hebdomadaires des Séances de l'Académie des Sciences, Paris, Série D* **268**: 3189-3190.

FRANCES, S. P. (1991). Pathogenicity, host range and temperature tolerance of *Crypticola clavulifera* (Oomycetes: Lagenidiales). *Journal of the American Mosquito Control Association* **7**: 504-506.

FRANCES, S. P., SWEENEY, A. W. & HUMBER, R. A. (1989). *Crypticola clavulifera* gen. et sp. nov. and *Lagenidium giganteum*: oomycetes pathogenic for dipterans infesting leaf axils in an Australian rain forest. *Journal of Invertebrate Pathology* **54**: 103-111.

FRANCIS, D. M. & MICHELMORE, R. W. (1993). Two classes of chromosome-sized molecules are present in *Bremia lactucae*. *Experimental Mycology* **17**: 284-300.

FRANK, B. (1881). Die Pflanzenkrankheiten. *In* A. Schenck, *Handbuch der Botanik, I*: 327-570.

FRASER, G. C., CALLINAN, R. B. & CALDER, L. M. (1992). *Aphanomyces* species associated with red spot disease, an ulcerative disease of estuarine fish from eastern Australia. *Journal of Fish Diseases* **15**: 173-182.

FRASER, R. S. S. & BUCZACKI, S. T. (1983). Ribosomal RNA molecular weights and the affinities of the Plasmodiophorales. *Transactions of the British Mycological Society* **80**: 107-112.

FRAYMOUTH, J. (1956). Haustoria of the Peronosporales. *Transactions of the British Mycological Society* **39**: 79-107.

FREDERIKSEN, R. A. & RENFRO, B. L. (1977). Global status of maize downy mildew. *Annual Review of Plant Pathology* **15**: 249-275.

FREY, R. (1950). Chitin und Zellulose in Pilzzellwänden. *Bericht der Schweizerischen Botanischen Gesellschaft* **60**: 199-230.

FRIEDMANN, I. (1952). Über neue und wenig bekannte auf Diatomeen parasitierende Phycomyceten. *Österreichische Botanische Zeitschrift* **99**: 173-219.

FRITSCH, A. (1895). Über Parasiten bei Crustaceen und Reederthieren der sussen Gewasser. *Academie des Sciences del'Empereur Francois Joseph I. Bulletin International, Classe II*: 79-85.

FRITSCH, F. E. (1903). Two fungi, parasitic on species of *Tolypothrix* (*Resticularia nodosa*, Dang. and *R. boodlei*, n.sp.). *Annals of Botany, London* **17**: 649-664.

FRITSCH, G. & LYSEK, G. (1984). Nematode destroying endoparasitic fungi from woodland soils in Berlin, West Germany. *Zeitschrift für Mykologie* **49**: 183-194.

FRITSCH, K. (1892). Nomenclatorische Bemerkungen. IV. Der Gattungsname *Naegelia*. *Österreichische Botanische Zeitschrift* **42**: 333-334.

FRITSCH, K. (1893). Nomenclatorische Bemerkungen. VI. *Naegeliella* Schröt. *Österreichische Botanische Zeitschrift* **43**: 420-421.

FRON, G. & GAILLAT, D. F. (1925). Contribution à l'étude du genre *Ligniera*. *Bulletin trimestriel de la Société Mycologique de France* **41**: 388-390.

FRY, W. E. & DOSTER, M. A. (1991). Potato late blight: forecasts and disease suppression. *In* J. A. Lucas, R. C. Shattock, D. S. Shaw & L.R. Cooke (Eds), Phytophthora, pp. 326-336. Cambridge, U.K.: Cambridge University Press.

FUCHS, W. H. (1966). Liberation and behaviour of spores of *Polymyxa betae* Keskin. *In* M. F. Madelin (Ed.), *The Fungus Spore*, pp. 111-112. London, U.K.: Butterworths.

FÜCKEL, L. (1869-1870). Symbolae Mycologicae. Beiträge zur Kenntnis der Rheinischen Pilze. *Jahrbücher des Nassauichen vereins für Naturkunde* **23-24**: 1-459.

FUIJITA, Y. (1990). Diseases of cultivated *Porphyra* in Japan. *In* I. Akatsuka (Ed.). *Introduction to Applied Phycology*, pp. 177-190. The Hague, The Netherlands: SPB Academic Publishing.

FUIJITA, Y. & ZENITANI, B. (1976). Studies on pathogenic *Pythium* of laver red rot in Ariake Sea Farm - I General mycological characteristics. *Bulletin of the Japanese Society of Scientific Fisheries* **42**: 1183-1188.

FUIJITA, Y. & ZENITANI, B. (1977*a*). Studies on pathogenic *Pythium* of laver red rot in Ariake Sea Farm - II Experimental conditions and nutritional requirements for growth. *Nihon Suisan Gakkai-Shi [Bulletin of the Japanese Society of Scientific Fisheries]* **43**: 89-95.

FUIJITA, Y. & ZENITANI, B. (1977*b*). Studies on pathogenic *Pythium* of laver red rot in Ariake Sea Farm - III Environmental and nutritional conditions for production of sexual organs. *Nihon Suisan Gakkai-Shi [Bulletin of the Japanese Society of Scientific Fisheries]* **43**: 921-927.

FUIJITA, Y. & ZENITANI, B. (1977*c*). Studies on pathogenic *Pythium* of laver red rot in Ariake Sea Farm - IV Serological differentiation of pathogenic *Pythium* strains. *Nihon Suisan Gakkai-Shi [Bulletin of the Japanese Society of Scientific Fisheries]* **43**: 1313-1318.

FUKUI, R., KITANO, S., TAKAMATSU, S. & ICHITANI, T. (1987). Taxonomic consideration on two closely related species, *Pythium paddicum* and *P. polypapillatum. Bulletin of the University of Osaka Prefecture, Series B* **39**: 21-31.

FUKUTOMI, M., AKAI, S. & SHIRAISHI, M. (1971). Fine structure of antheridia and oogonia of *Phytophthora macrospora*, the downy mildew fungus of rice plants. *Mycopathologia et Mycologia Applicata* **43**: 249-258.

FULLER, M. S. (1960). Biochemical and microchemical study of the cell walls of *Rhizidiomyces* sp. *American Journal of Botany* **47**: 838-842.

FULLER, M. S. (1962). Growth and development of the water mold *Rhizidiomyces* in pure culture. *American Journal of Botany* **49**: 64-71.

FULLER, M. S. (1966). Structure of the uniflagellate spores of aquatic phycomycetes. *In* M. F. Madelin (Ed.), *The Fungus Spore*, pp. 67-84. London, U.K.: Butterworths.

FULLER, M. S. (1978). *Lower Fungi in the Laboratory: Palfrey Contributions in Botany. No. 1.* Athens, Georgia, U.S.A.: Department of Botany, University of Georgia, Athens, Georgia, U.S.A.

FULLER, M. S. (1990). Phylum Hyphochytriomycota. *In* L. Margulis, J. O. Corliss, M. Melkonian & D. Chapman (Eds), *Handbook of Protoctista*, pp. 380-387. Boston, U.S.A.: Jones & Bartlett.

FULLER, M. S. (1996). The flagellated fungal spore. *In* B. C. Sutton (Ed.), *A Century of Mycology*, pp. 161-192. Cambridge, U.K.: Cambridge University Press.

FULLER, M. S. & BARSHAD, I. (1960). Chitin and cellulose in the cell walls of *Rhizidiomyces* sp. *American Journal of Botany* **47**: 105-109.

FULLER, M. S., FOWLES, B. E. & McLAUGHLIN, D. J. (1964). Isolation and pure culture study of marine phycomycetes. *Mycologia* **56**: 745-756.

FULLER, M. S. & GISI, U. (1985). Comparative studies of the in-vitro activity of the fungicides oxadixyl and metalaxyl. *Mycologia* **77**: 424-432.

FULLER, M. S. & JAWORSKI, A. (Eds) (1987). *Zoosporic Fungi in Teaching & Research.* Athens, Georgia, U.S.A.: Southeastern Publishing Corporation.

FULLER, M. S., LEWIS, B. & COOK, P. (1966). Occurrence of *Pythium* sp. on the marine alga *Porphyra. Mycologia* **58**: 313-318.

FULLER, M. S. & POYTON, R. O. (1964). A new technique for the isolation of aquatic fungi. *Bioscience* **14**: 45-46.

FULLER, M. S. & REICHLE, R. (1965). The zoospore and early development of *Rhizidiomyces apophysatus. Mycologia* **57**: 946-961.

FURTADO, J. S. (1965). A new aquatic fungus, *Achlya benekei*, sp. nov. *Rickia, Brazil* **2**: 121-128.

GABRIELSON, R. L. & ROBAK, J. (1988). Temperature sensitivity of resistance to two pathotypes of *Plasmodiophora brassicae* in *Brassica oleracea. Acta Agrobotanica* **41**: 237-243.

GACUTAN, R. G. & BATICADOS, M. C. L. (1979). Notes on *Lagenidium* from larvae of *Penaeus monodon* Fabricius. I. Isolation and culture in artificial media. *Fisheries Research Journal Philippines* **4**: 24-28.

GAERTNER, A. (1954). Über das Vorkommen niederer Erdphycomyceten in Afrika, Schweden und an einigen mitteleuropäischen Standorten. *Archiv für Mikrobiologie* **21**: 4-56.

GAERTNER, A. (1964). Elektronenmikroskopische Untersuchungen zur Struktur der Geisseln von *Thraustochytrium* spec. *Veröffentlichungen des Instituts für Meeresforschung in Bremerhaven* **9**: 25-28.

GAERTNER, A. (1966). Vorkommen, Physiologie und Verteilung "Marina Niederer Pilze" (Aquatic Phycomycetes). *Veröffentlichungen des Instituts für Meeresforschung in Bremerhaven, Sonderband* **2**: 221-236.

GAERTNER, A. (1967). Ökologische Untersuchungen an einem marinen Pilz aus der Umgebung von Helgoland. *Helgoländer wissenschaftliche Meeresuntersuchungen* **15**: 181-192.

GAERTNER, A. (1968*a*). Die Fluktuationen mariner niederer Pilze in der Deutschen Bucht 1965 und 1966. *Veröffentlichungen des Instituts für Meeresforschung in Bremerhaven, Sonderband* **3**: 105-120.

GAERTNER, A. (1968*b*). Niedere, mit Pollen koderbare marine Pilze diesseits und jenseits des Island-Faroer-Ruckens im Oberflachenwasser und im sediment. *Veröffentlichungen des Instituts für Meeresforschung in Bremerhaven* **11**: 65-82.

GAERTNER, A. (1968*c*). Eine Methode des quantitativen Nachweises niederer, mit Pollen koderbarer Pilze im Meerwasser und im Sediment. *Veröffentlichungen des Instituts für Meeresforschung in Bremerhaven* **3**: 75-92.

GAERTNER, A. (1969). Marine niedere Pilze in Nordsee und Nordatlantik. *Berichte der Deutschen Botanischen Gesellschaft* **82**: 287-306.

GAERTNER, A. (1970*a*). Einiges zur Kultur mariner niederer Pilze. *Helgoländer wissenschaftliche Meeresuntersuchungen* **20**: 29-38.

GAERTNER, A. (1970*b*). Beobachtungen über die sporulation dfer dickwandigen Sporangien von *Thraustochytrium kinnei* Gaertner. *Veröffentlichungen des Instituts für Meeresforschung in Bremerhaven* **12**: 321-327.

GAERTNER, A. (1971). *Thraustochytrium kinnei* (Thraustochytriaceae) Vegetative Entwicklung. *Encyclopaedia Cinematographica* film E. 1664.

GAERTNER, A. (1972). Characters used in the classification of thraustochytriaceous fungi. *Veröffentlichungen des Instituts für Meeresforschung in Bremerhaven* **13**: 183-194.

GAERTNER, A. (1974*a*). Marine niederer Pilze in Nordsee und Nordmeerbecken. *Veröffentlichungen des Instituts für Meeresforschung in Bremerhaven, Supplement* **5**: 123-147.

GAERTNER, A. (1974*b*). Vorkommen und Verteilung mariner niederer Pilze in der Nordsee und im Nordmeerbecken. *Veröffentlichungen des Instituts für Meeresforschung in Bremerhaven, Supplement* **5**: 123-147.

GAERTNER, A. (1974*c*). Aminoverbindungen, Vitamine und Purinkörper als Nährstoffkomponenten einiger Thraustochytriaceae. *Veröffentlichungen des Instituts für Meeresforschung in Bremerhaven, Supplement* **5**: 221-236.

GAERTNER, A. (1977). Revision of the Thraustochytriaceae (lower marine fungi). I. *Ulkenia* nov. gen. with description of three new species. *Veröffentlichungen des Instituts für Meeresforschung in Bremerhaven* **16**: 139-157.

GAERTNER, A. (1979). Some fungal parasites found in the diatom populations of the Rusfjord area (South Norway) during March 1979. *Veröffentlichungen des Instituts für Meeresforschung in Bremerhaven* **18**: 29-33.

GAERTNER, A. (1980). Quantitative studies on the marine phycomycetes, chytrids, and higher mycelial fungi of the upper Tay Estuary. *Proceedings of the Royal Society of Edinburgh, B* **78**: 57-78.

GAERTNER, A. (1981). A new marine phycomycete possessing an unusual type of asexual reproduction. *Veröffentlichungen des Instituts für Meeresforschung in Bremerhaven* **19**: 61-69.

GAERTNER, A. (1982). Lower marine fungi from the Northwest African upwelling areas and from the Atlantic off Portugal. *'Meteor' Forschung-Ergebnisse, Reihe D* **34**: 9-30.

GAERTNER, A. & RAGHU-KUMAR, S. (1980). Ecology of the thraustochytrids (lower marine fungi) in the Fladen Ground and other parts of the North Sea I. *'Meteor' Forschung-Ergebnisse, Reihe A* **22**: 165-185.

GAERTNER, A. & SPARROW, F. K. (1966). A preliminary study of aquatic phycomycetes in the lakes of the Huron Mountains, Michigan. *Veröffentlichungen des Instituts für Meeresforschung in Bremerhaven* **10**: 93-106.

GALAGALI, J. S., BALASUBRAMANIANE, R. & BALARAMAN, K. (1984). Indigenous isolation of F-17 - an oomycetous fungus pathogenic to mosquito larvae. *Indian Journal of Medical Research* **80**: 95-102.

GALINDO, A. J. & GALLEGLY, M. E. (1960). The nature of sexuality in *Phytophthora infestans. Phytopathology* **50**: 123-128.

GALINDO, A. J. & ZENTMYER, G. A. (1967*a*). Genetics and cytology of *Phytophthora. Nature (London)* **214**: 1356.

GALINDO, A. J. & ZENTMYER, G. A. (1967*b*). Genetical and cytological studies of *Phytophthora* strains pathogenic to pepper plants. *Phytopathology* **57**: 1300-1304.

GALLAGHER, P. A., WARNER, S. A. & DOMNAS, A. J. (1994). Presqualene metabolism in two species of *Lagenidium. Canadian Journal of Microbiology* **40**: 858-864.

GALLAUD, I. (1907). Revue des travaux sur les champignons phycomycetès et basidiomycetès. *Revue Général de Botanique* **19**: 392-400.

GALLEGLY, M. E. (1970). Genetics of *Phytophthora. Phytopathology* **60**: 1135-1141.

GALLEGLY, M. E. (1968). Changes of pathogenicity of *Phytophthora infestans. Annual Review of Phytopathology* **6**: 375-396.

GALLEGLY, M. E. & GALINDO, J. A. (1958). Mating types and oospores of *Phytophthora infestans. Phytopathology* **48**: 247-277.

GALLOWAY, T. W. (1891). Notes on the fungus causing damping off and other allied forms. *Transactions of the Massachusetts Horticultural Society, 1891* : 230-239.

GALSTON, A. W. (1983). Polyamines as modulators of plant development. *BioScience* **33**: 382-388.

GALT, J. H. & WHISLER, H. C. (1970). Differentiation of flagellated spores in *Thalassomyces*, ellobiopsid parasite of marine crustaceae. *Archiv für Mikrobiologie* **71**: 295-303.

GALTSOFF, P. S. (1942). Wasting disease causing mortality of sponges in the West Indies and Gulf of Mexico. *Proceedings of the 8th American Science Congress, 1940* **3**: 411-421.

GANAROS, A. E. (1957). Marine fungi infecting eggs and embryos of *Urosalpinx cinerea*. *Science, New York* **125**: 1194.

GARBER, R. C. & AIST, J. R. (1979*a*). The ultrastructure of mitosis in *Plasmodiophora brassicae* (Plasmodiophorales). *Journal of Cell Science* **40**: 89-110.

GARBER, R. C. & AIST, J. R. (1979*b*). The ultrastructure of meiosis in *Plasmodiophora brassicae* (Plasmodiophorales). *Canadian Journal of Botany* **57**: 2509-2518.

GARCIA, R. & SWEENEY, A. W. (1983). The use of pathogens for the medically important vectors. *Pacific Science Congress Proceedings (Australia)* **15**: 79.

GARDNER, I, C. (1965). Observations on the fine structure of the endophyte of the root nodules of *Alnus glutinosa* (L.) Gaertn. *Archiv für Mikrobiologie* **21**: 4-56.

GARRILL, A., CLIPSON, N. J. W. & JENNINGS, D. H. (1992). Preliminary observations on the monovalent cation relations of *Thraustochytrium aureum*, a fungus requiring sodium for growth. *Mycological Research* **96**: 295-304.

GARRILL, A., JACKSON, S. L., LEW, R. R. & HEATH, I. B. (1993). Ion channel activity and tip growth: tip-localized stretch-activated channels generate an essential Ca^{2+} gradient in the oomycete *Saprolegnia ferax*. *European Journal of Cell Biology* **60**: 358-365.

GARRILL, A., LEW, R. R. & HEATH, I. B. (1992). Stretch-activated Ca^{2+} and Ca^{2+}-activated K^+ channels in the hyphal tip plasma membrane of the oomycete *Saprolegnia ferax*. *Journal of Cell Science* **101**: 721-730.

GÄUMANN, E. A. (1918*a*). Über die Formen der *Peronospora parasitica* (Pers.) Fries. *Beihefte Botanische Zentrallblatt* **35**: 395-533.

GÄUMANN, E. A. (1918*b*). Ein Beitrag zu Kenntnis der lappländischen Saprolegnieen. *Botaniska Notiser, 1918-1919*: 151-159.

GÄUMANN, E. [A.] (1923). Beiträge zu einer Monographie der Gattung *Peronospora* Corda. *Beiträge zur Kryptogamenflora der Schweiz* **5 (4)**: 1-360.

GAUTHIER, J. D. & VASTA, G. R. (1993). Continuous *in vitro* culture of the eastern oyster parasite *Perkinsus marinus*. *Journal of Invertebrate Pathology* **62**: 321-323.

GAY, J. L. & GREENWOOD, D. (1966). Structural aspects of zoospore production in *Saprolegnia ferax* with particular reference to the cell and vacuolar membranes. *In* M. F. Madelin (Ed.), *The Fungus Spore*, pp. 95-110. London, U.K.: Butterworths.

GAY, J. L., GREENWOOD, D. & HEATH, I. B. (1971). The formation of vacuoles (vesicles) during oosphere development and zoospore germination in *Saprolegnia*. *Journal of General Microbiology* **65**: 233-241.

GEBER, A., HIGGINS, D. E., WATERS, A. P., BENNETT, J. E. & McCUTCHAN, T. F. (1992). Small subunit ribosomal RNA of *Blastomyces dermatitidis*: sequence and phylogenetic analysis. *Journal of General Microbiology* **138**: 395-399.

GEITLER, L. (1949). Die Differenzierung des Protoplasten des Diatomee *Synedra*. *Österreichische Botanische Zeitschrift* **68**: 625-636.

GERDEMANN, J. W. (1968). Vesicular-arbuscular mycorrhiza and plant growth. *Annual Review of Phytopathology* **6**: 397-418.

GERIK, J. S. (1992). Zoosporic obligate parasites of roots. *In* L. L. Singleton, J. D. Mihail, & C. M. Rush (Eds). *Methods for Research on Soilborne Phytopathogenic Fungi*, pp. 18-24. St Paul, Minnesota, U.S.A.: American Phytopathological society.

GERMAIN, E. (1850). Sur un parasite nouveau. *Compte rendu des Séances de la Société de Biologie, Paris* **2**: 156-158.

GIARD, A. (1888). Sur les *Nephromyces*, genre nouveau de champignons parasites du rein des Molgulidées. *Comptes Rendus Hebdomadaires des Séances de l'Académie des Sciences, Paris* **106**: 1180-1182.

GIBBONS, I. R. & GRIMSTONE, A. V. (1960). On flagellar structure in certain flagellates. *Journal of Biophysical and Biochemical Cytology* **7**: 697-716.

GICKLHORN, J. (1923). *Aphanomyces ovidestruens* nov. spec. - ein Parasit in den Eiern von Diaptomus. *Lotos* **71**: 143-156.

GIEBEL, P. E. & DOMNAS, A. J. (1973). Infection of *Anopheles quadrimaculatus* with *Lagenidium* sp. *Journal of Invertebrate Pathology* **20**: 326-331.

GIEBEL, P. E. & DOMNAS, A. J. (1976). Soluble trehalases from larvae of the mosquito, *Culex pipiens*, and the fungal parasite, *Lagenidium giganteum*. *Insect Biochemistry* **6**: 303-311.

GILL, C. H. (1893). On an endophytic parasite of diatoms. *Journal of the Royal Microscopical Society, London, 1893*: 1-4.

GILMAN, J. C. (1945). *A Manual of Soil Fungi*. Ames, Iowa, U.S.A.: Iowa State College Press

GILMAN, J. C. (1957). *A Manual of Soil Fungi, Second Edition*. Ames, Iowa, U.S.A.: Iowa State College Press.

GILMAN, J. C. & ARCHER, W. A. (1929). The fungi of Iowa parasitic on plants. *Journal of Science of the Iowa State College* **3**: 299-502.

GILPIN, R. H. (1954). Concerning the nutrition of *Apodachlya brachynema*. *Mycologia* **46**: 702-707.

GIL-TURNES, M. S. & FENICAL, W. (1992). Embryos of *Homarus americanus* are protected by epibiotic bacteria. *Biological Bulletin of the Marine Biological Laboratory, Woods Hole* **182**: 105-108.

GIL-TURNES, M. S., HAY, M. E. & FENICAL, W. (1989). Symbiotic marine bacteria chemically defend crustacean embryos. *Science, New York* **246**: 116-118.

GIRARD, V. & FÈVRE, M. (1984). Distribution of (1→3)-β- and (1→4)-β- glucan synthases along the hyphae of *Saprolegnia monoica*. *Journal of General Microbiology* **130**: 1557-1562.

GISI, U. (2000). Chemical control of downy mildews. *In* P. Spencer-Phillips (Ed.), *Advances in Downy Mildew Research*. Dordrecht, The Netherlands: Kluwer Academic Publishers. **[in press]**

GISI, U., HEMMES, D. E. & ZENTMYER, G. A. (1979). Origin and significance of the discharge vesicle in *Phytophthora*. *Experimental Mycology* **3**: 321-339.

GISI, U. & ZENTMYER, G. A. (1980). Mechanism of zoospore release in *Phytophthora* and *Pythium*. *Experimental Mycology* **4**: 362-377.

GIUMA, A. Y. & COOKE, R. C. (1972). Some endozoic fungi parasitic on soil nematodes. *Transactions of the British Mycological Society* **59**: 213-218.

GLARE, T. R. & MILNER, R. J. (1991). Ecology of entomopathic fungi. *In* D. K. Arora, L. Ajello & K. G. Mukerji (Eds). *Handbook of Applied Mycology, Vol. 2. Humans, Animals, and Insects*, pp. 547-612. New York, N.Y., U.S.A.: Marcel Dekker.

GLEASON, F. H. (1968*a*). Respiratory electron transport systems of aquatic fungi I. *Leptomitus lacteus* and *Apodachlya punctata*. *Plant Physiology* **43**: 597-605.

GLEASON, F. H. (1968*b*). Nutritional comparisons in the Leptomitales. *American Journal of Botany* **55**: 1003-1010.

GLEASON, F. (1976). The physiology of the lower freshwater fungi. *In* E. B. G. Jones (Ed.) *Recent Advances in Aquatic Mycology*. pp. 543-572. London, U.K.: Elek Press.

GLEASON, F. H., RUDOLPH, C. R. & PRICE, J. S. (1970). Growth of certain aquatic Oomycetes on amino acids I. *Saprolegnia*, *Achlya*, *Leptolegnia* and *Dictyuchus*. *Physiologia Plantarum* **23**: 513-516.

GLEASON, F. H. & STUART, T. D. (1970). Nitrogen and sulfur requirements in the Leptomitales. *Mycologia* **62**: 1212-1214.

GLEASON, F. H., STUART, T. D., PRICE, J. S. & NELBACH, E. T. (1970). Growth of certain aquatic Oomycetes on amino acids II. *Apodachlya*, *Aphanomyces*, and *Pythium*. *Physiologia Plantarum* **23**: 769-774.

GLEASON, F. H. & UNESTAM, T. (1968). Cytochromes of aquatic fungi. *Journal of Bacteriology* **95**: 1599-1603.

GLENN, F. E. & CHAPMAN, H. C. (1978). A natural epizootic of the aquatic fungus *Lagenidium giganteum* in the mosquito *Culex territans*. *Mosquito News* **38**: 522-524.

GLOCKLING, S. L. (1994). Predacious and parasitoidal fungi in association with herbivore dung in deciduous woodlands. Unpublished PhD Thesis, University of Reading, U.K.

GLOCKLING, S. L. (1997). Isolation of a new species of rotifer-attacking *Olpidium*. *Mycological Research* **101**: 206-208.

GLOCKLING, S. L. & BEAKES, G. W. (2000*a*). Two new species of *Haptoglossa*, *H. erumpens* and *H. dickii*, infecting nematodes in cow manure. *Mycological Research* **104**: 100-106.

GLOCKLING, S. L. & BEAKES, G. W. (2000*b*). A review of the biology and infection strategies of biflagellate zoosporic parasites of nematodes. *Fungal Diversity* **4**: 1-20.

GLOCKLING, S. L. & BEAKES, G. W. (2000*c*). Video-microscopy of spore development in *Haptoglossa heteromorpha*, a new species from cow dung. *Mycologia* **92**: 747-753.

GLOCKLING, S. L. & BEAKES, G. W. (2000*d*). The dimorphic infection cells of *Haptoglossa heteromorpha* illustrate the developmental plasticity of infection apparatus structures in a nematode parasite. *Canadian Journal of Botany* **[in press]**.

GLOCKLING, S. L. & BEAKES, G. W. (2000*e*). Two new species of *Haptoglossa* from N. E. England, *H. northumbrica* and *H. polymorpha*. *Botanical Journal of the Linnean Society* **[in press]**.

GLOCKLING, S. L. & DICK, M. W. (1997). New species of *Chlamydomyzium* from Japan and pure culture of *Myzocytiopsis* species. *Mycological Research* **101**: 883-896.

GOBI, C. [ГОБИ, Хр., Gobi, Khr.] (1899-1900). [Исторія развитія грибка *Pythium tenue* nov. sp. и положеніе его въ системѣ] Entwickelungsgeschichte des *Pythium tenue* nov. sp. *Scriptis Botanicis Horti Universitatis Impericolis Petropolitanae* **15**: 191-226 [191-210 in Russian, 211-226 in German].

GOBI, C. (1900). О Новомъ Паразитномъ Грибкѣ, *Rhizidiomyces ichneumon* nov sp., и Питающемъ его Органнизмѣ-Хозяинѣ, *Chloromonas globulosa* (Perty). Ueber einen neuen parasitischen Pilze, *Rhizidiomyces ichneumon* nov. sp., und seinen Nährorganismus, *Chloromonas globulosa* (Perty). *Scriptis Botanicis Horti Universitatis Impericolis Petropolitanae* **15**: 227-272 [227-250 in Russian, 251-272 in German].

GOEBEL, K. (1884). *Tetramyxa parasitica*. *Flora, Jena* **67**: 517-521.

GOETTEL, M. S., TOOHEY, M. K. & PILLAI, J. S. (1983). Preliminary laboratory infection trials with a Fiji isolate of the mosquito pathogenic fungus *Lagenidium*. *Journal of Invertebrate Pathology* **41**: 1-7.

GOGGIN, C. L. & BARKER, S. C. (1993). Phylogenetic position of the genus *Perkinsus* (Protista, Apicomplexa) based on small subunit ribosomal RNA. *Molecular and Biochemical Parasitology* **60**: 65-70.

GOGGIN, C. L., SEWELL, K. B. & LESTER, R. J. G. (1990). Tolerances of *Perkinsus* spp. (Protista, Apicomplexa) to temperature, chlorine and salinity. *Journal of Shellfish Research* **9**: 145-148.

GOLDIE-SMITH, E. K. (1951). A new species of *Sorodiscus* on *Pythium*. *Journal of the Elisha Mitchell Scientific Society* **67**: 108-121.

GOLDIE-SMITH, E. K. (1952). The sporangial phase of *Pythium undulatum* Petersen. *Journal of the Elisha Mitchell Scientific Society* **68**: 273-292.

GOLDIE-SMITH, E. K. (1954). The position of *Woronina polycystis* in the Plasmodiophoraceae. *American Journal of Botany* **41**: 441-448.

GOLDIE-SMITH, E. K. (1956*a*). The maintenance of stock cultures of aquatic fungi. *Journal of the Elisha Mitchell Scientific Society* **72**: 158-166.

GOLDIE-SMITH, E. K. (1956*b*). A new species of *Woronina*, and *Sorodisus cokeri* emended. *Journal of the Elisha Mitchell Scientific Society* **72**: 348-356.

GOLDSMITH, E. P. (1948). Observations on some water fungi collected in and around Johannesburg. *Journal of South African Botany* **14**: 135-158.

GOLDSTEIN, S. (1963*a*). Morphological variation and nutrition of a new monocentric marine fungus. *Archiv für Mikrobiologie* **45**: 101-110.

GOLDSTEIN, S. (1963*b*). Development and nutrition of new species of *Thraustochytrium*. *American Journal of Botany* **50**: 271-279.

GOLDSTEIN, S. (1963*c*). Studies of a new species of *Thraustochytrium* that displays light stimulated growth. *Mycologia* **55**: 799-811.

GOLDSTEIN, S. (1973). Zoosporic marine fungi (Thraustochytriaceae and Dermatocystidiaceae). *Annual Review of Microbiology* **27**: 13-26.

GOLDSTEIN, S. & BELSKY, M. (1963). B_{12} and B_1 auxotrophy of lower marine phycomycetes. *Archiv für Mikrobiologie* **47**: 161-163.

GOLDSTEIN, S. & BELSKY, M. (1964). Axenic culture of a new marine phycomycete possessing an unusual type of asexual reproduction. *American Journal of Botany* **51**: 72-78.

GOLDSTEIN, S., BELSKY, M & SESNOWITZ-HORN, S. (1980). Endogenous respiration and regulation in a primitive marine fungus. *Archives of Microbiology* **126**: 161-165.

GOLDSTEIN, S. & MORIBER, L. (1966). Biology of a problematic marine fungus, *Dermocystidium* sp. I. Development and cytology. *Archiv für Mikrobiologie* **53**: 1-11.

GOLDSTEIN, S., MORIBER, L. & HERSHENOV, B. (1964). Ultrastructure of *Thraustochytrium aureum*, a biflagellate marine phycomycete. *Mycologia* **56**: 897-904.

GOLKAR, L., LEBRUN, R. A., OHAYON, H., GOUNON, P., PAPIEROK, B. & BREY, P. T. (1993). Variation of larval susceptibility to *Lagenidium giganteum* in three mosquito species. *Journal of Invertebrate Pathology* **62**: 1-8.

GOLUBEVA, O. G. [ГОЛУБЕВА, О. Г., Golubeva, O. G.] (1982). Пресноводные Хитриевые Грибы Ленинградской Области. Fresh-water chytridiae fungi occurring in Leningrad Region. *Mikologiya i Fitopatologiya [Микология и Фитопалогия]* **16**: 385-389.

GOLUEKE, C. G. (1957). Comparative studies of the physiology of *Sapromyces* and related genera. *Journal of Bacteriology* **74**: 337-343.

GÖNNERT, R. (1935). Über Systematik, Morphologie, Entwicklungsgeschichte und Parasiten einiger Dendrosomidae nebst Beschreibung zweier neuer Suktorien. *Archiv für Protistenkunde* **86**: 113-154.

GOODE, P. M. (1956). Infection of strawberry roots by zoospores of *Phytophthora fragariae*. *Transactions of the British Mycological Society* **39**: 367-377.

GOODING, G. V. & LUCAS, G. B. (1959). Factors influencing sporangial formation and zoospore activity in *Phytophthora parasitica* var. *nicotianae*. *Phytopathology* **49**: 277-281.

GOOSEN, P. G. & SACKSTON, W. E. (1968). Transmission and biology of sunflower downy mildew. *Canadian Journal of Botany* **46**: 5-10.

GOTELLI, D. (1969). *Morphology and nutrition of the marine fungus*, Lagenidium callinectes. Ann Arbor, Michigan: University Microfilms 1969.

GOTELLI, D. (1971). *Lagenisma coscinodisci*, a parasite of the marine diatom *Coscinodiscus* occurring in Puget Sound, Washington. *Mycologia* **63**: 171-174.

GOTELLI, D. M. (1974*a*). The morphology of *Lagenidium callinectes*. I. Vegetative development. *Mycologia* **66**: 639-647.

GOTELLI, D. M. (1974*b*). The morphology of *Lagenidium callinectes*. II. Zoosporogenesis. *Mycologia* **66**: 846-858.

GOTELLI, D. (1979). Synaptonemal complexes in the gametangia of *Sapromyces androgynus* Oomycetes. *Mycotaxon* **9**: 90-92.

GOTELLI, D. & HANSON, L. C. (1987). An ultrastructural investigation of the zoospore of *Sapromyces androgynus* (Oomycetes, Leptomitales). *Mycologia* **79**: 745-752.

GOW, N. A. R. (1984). Transhyphal electrical currents in fungi. *Journal of General Microbiology* **130**: 3313-3318.

GOW, N. A. R. & GOODAY, G. W. (1987). Effects of antheridiol on growth, branching and electrical currents of hyphae of *Achlya ambisexualis*. *Journal of General Microbiology* **133**: 3531-3535.

GOW, N. A. R., KROPF, D. L. & HAROLD, F. M. (1984). Growing hyphae of *Achlya bisexualis* generate a longitudinal pH gradient in the surrounding medium. *Journal of General Microbiology* **130**: 2967-2974.

GOWDA, P. S. B. & BHAT, S. S. (1992). Reaction of sorghum lines to downy mildew (*Peronosclerospora sorghi*) in vitro. *Mycological Research* **96**: 889-890.

GRAFF, N. J. & MADELIN, M. F. (1989). Axenic culture of the cyst-nematode parasitizing fungus, *Nematophthora gynophila*. *Journal of Invertebrate Pathology* **53**: 301-306.

GRAFF, P. W. (1928). Contributions to our knowledge of western Montana fungi. II. Phycomycetes. *Mycologia* **20**: 158-179.

GRAHAM, K. M. (1954). Nuclear behavior in *Phytophthora infestans* (Mont.) de Bary. *Phytopathology* **44**: 490 [abstract].

GRAHAME, E. S. (1976). The occurrence of *Lagenisma coscinodisci* in *Palmeria hardmaniana* from Kingston Harbor, Jamaica. *British Phycological Journal* **11**: 57-61.

GRAIN, J., MIGNOT, J.-P. & PUYTORAC, P. de (1988). Ultrastructures and evolutionary modalities of flagellar and ciliary systems in protists. *Biology of the Cell* **63**: 219-237.

GRAN, H. H. (1900). Bemerkungen über einige Plankton-diatomeen. *Nytt Magasin for Naturvidenskapene* **38**: 105-128.

GRANT, B. R., GRANT, J. H. & HARRIS, J. (1992). Inhibition of growth of *Phytophthora infestans* by phosphate and phosphonate in defined media. *Experimental Mycology* **16**: 240-244.

GRANT, B. R., GRIFFITH, J. M. & IRVING, H. R. (1986). A model to explain ion induced differentiation in zoospores of *Phytophthora palmivora*. *Experimental Mycology* **10**: 89-98.

GRANT, B. R., IRVING, H. R. & RADDA, M. (1985). The effect of pectin and related compounds on encystment and germination of *Phytophthora palmivora* zoospores. *Journal of General Microbiology* **131**: 669-676.

GRAU, C. R., MUEHLCHEN, A. M., TOFTE, J. E. & SMITH, R. R. (1991). Variability in virulaence of *Aphanomyces euteiches*. *Plant Disease* **75**: 1153-1156.

GRAVLAND, R., DALE, P. & MITHEN, R. (1992). Gall development in hairy root cultures infected with *Plasmodiophora brassicae*. *Mycological Research* **96**: 225-228.

GRAY, A. B., SACKSTON, W. E. & THAUVETTE, L. (1985). The development of infection structures of *Plasmopara halstedii* in suspensions of sunflower cells. *Canadian Journal of Botany* **63**: 1817-1819.

GRAY, J. (1939). Studies in animal locomotion. VIII. The kinetics of locomotion of *Nereis diversicolor*. *Journal of Experimental Biology* **16**: 9-17.

GRAY, M. W. (1989). Origin and evolution of mitochondrial DNA. *Annual Review of Cell Biology* **5**: 25-50.

GRAY, M. W. (1994). Mitochondrial phylogeny and evolution. Paper read to the Linnean Society of London, meeting held on 14 April 1994.

GRAY, N. F., WYBORN, C. H. E. & LEWIS SMITH, R. I. (1982). Nematophagous fungi from the maritime Antarctic. *Oikos* **38**: 194-201.

GRAY, N. F. (1983). Ecology of nematophagous fungi: distribution and habitat. *Annals of Applied Biology* **102**: 501-510.

GRAY, N. F. & LEWIS SMITH, R. I. (1984). The distribution of nematophagous fungi in the maritime Antarctic. *Mycopathologia* **85**: 81-92.

GRAY, S. F. (1821). *A Natural Arrangement of British Plants....* 1: p. 540. London, U.K.: Baldwin, Crodock & Joy.

GREEN, B. R. & DICK, M. W. (1972). DNA base composition and the taxonomy of the oomycetes. *Canadian Journal of Microbiology* **18**: 963-968.

GREEN, J. (1974). Parasites and epibionts of *Cladocera*. *Transactions of the Zoological Society of London* **32**: 417-515.

GREGORY, C. T. (1912). Spore germination and infection with *Plasmopara viticola*. *Phytopathology* **11**: 235-249.

GRIFFIN, D. H. (1966). Effect of electrolytes on differentiation in *Achlya* sp. *Plant Physiology* **41**: 1254-1256.

GRIFFIN, D. H. & BREUKER, C. (1969). Ribonucleic acid synthesis during differentiation of sporangia in the water mold *Achlya*. *Journal of Bacteriology* **98**: 689-696.

GRIFFIN, D. H., TIMBERLAKE, W. E. & CHENEY, J. C. (1974). Regulation of macromolecular synthesis, colony developmen, and specific growth rate of *Achlya bisexualis* during balanced growth. *Journal of General Microbiology* **80**: 381-388.

GRIFFIN, M. J. & COLEY-SMITH, J. R. (1968). The establishment of hop tissue cultures and their infection by downy mildew *Pseudoperonospora humuli* (Miy. & Tak.) Wilson under aseptic conditions. *Journal of General Microbiology* **53**: 231-236.

GRIFFIN, M. J. & COLEY-SMITH, J. R. (1971). Some effects of streptomycin on *Pseudoperonospora humuli*, the downy mildew of hop. *Journal of General Microbiology* **69**: 117-134.

GRIFFIN, M. J. & COLEY-SMITH, J. R. (1975). Uptake of streptomycin by sporangia of *Pseudoperonospora humuli* and the inhibition of uptake by divalent metal ions. *Transactions of the British Mycological Society* **65**: 265-278.

GRIFFITH, J. M., DAVIS, A. J. & GRANT, B. R. (1992). Target sites of fungicides to control Oomycetes. *In* W. Köllered (Ed.), *Target Sites of Fungicide Action*, pp. 69-100. Boca Raton, Florida, U.S.A.: Chemical Rubber Company Press.

GRIFFITH, J. M., ISER, J. R. & GRANT, B. R. (1998). Calcium control of differentiation in *Phytophthora palmivora*. *Archives of Microbiology* **149**: 565-571.

GROSJEAN, M. C. (1992). Classification et identification des espèces du genre *Pythium*, champignons phytopathogènes du sol, par l'analyse de l'espace inerne transcrit de l'opéron ribosomique. Lyon I University.

GROSSMAN, L. I. & HUDSPETH, M. E. S. (1985). Fungal mitochondrial genomes. *In* J. W. Bennett & L. L. Lasure (Eds), *Gene Manipulations in Fungi*, pp. 65-103. New York, U.S.A.: Academic Press.

GROVE, S. N. (1970). Fine structure of zoospore encystment and germination in *Pythium aphanidermatum*. *American Journal of Botany* **57**: 745.

GROVE, S. N. & BRACKER, C. E. (1978). Protoplasmic changes during zoospore encystment and cyst germination in *Pythium aphanidermatum*. *Experimental Mycology* **2**: 51-98.

GROVE, S. N., BRACKER, C. E. & MORRE, D. J. (1970). An ultrastructural basis for hyphal tip growth in *Pythium ultimum*. *American Journal of Botany* **57**: 245-266.

GRUITHUISEN, F. (1821). Die Branchiennecke und die eine aus ihren Überresten hervorwachsende lebendig-gebährende Conferve. *Nova Acta [Novorum Actorum] Physico-medica Academiae Caesareae Leopoldino-carolinae Naturae Curiosorum* **10**: 437-452.

GRZEMSKI, W. E., LAWRIE, A. C. & GRANT, B. R. (1994). The role of the plasma membrane in differentiation of *Phytophthora palmivora* zoospores. *Mycological Research* **98**: 1051-1058.

GUBLER, F. & HARDHAM, A. R. (1988). Secretion of adhesive material during encystment of *Phytophthora cinnamomi* zoospores, characterized by immunogold labeling with monoclonal antibodies to components of peripheral vesicles. *Journal of Cell Science* **90**: 225-235.

GUBLER, F. & HARDHAM, A. R. (1990). Protein storage in large peripheral vesicles in *Phytophthora* zoospores and its breakdown after cyst germination. *Experimental Mycology* **14**: 393-404.

GUBLER, F., HARDHAM, A. R. & DUNIEC, J. (1989). Characterizing adhesiveness of *Phytophthora cinnamomi* zoospores during encystment. *Protoplasma* **149**: 24-30.

GUILLIERMOND, A. (1922). Observation cytologique sur un *Leptomitus* et en particulier sur la mode de formation et la germination des zoospores. *Comptes Rendus Hebdomadaires des Séances de l'Académie des Sciences, Paris, Série D* **175**: 377-379.

GUILLIERMOND, A. (1931). A propos d'un mémoire récent de M. Dangeard sur la terminologie des éléments cellulaires et son application a l'étude des champignons. *Société Anonyme de l'Imprimerie A. Rey, Lyon. 1931*: 1-48.

GUILLIERMOND, A. (1935). Nouvelles recherches sur la nature et la signification des formations dites de Golgi. *Revue de Cytologie et de Cytophysiologie Végétales* **1**(3): 197-259.

GUILLIERMOND, A. & GAYTHERET, R. (1938). Culture de *Saprolegnia diclina* en milieux additionés de colorants. *Comptes Rendus des Séances de la Société de Biologie* **128**: 493.

GUIMA, A. Y. & COOKE, R. C. (1972). Some endozoic fungi parasitic on soil nematodes. *Transactions of the British mycological Society* **59**: 213-218.

GUNDERSON, J. H., ELWOOD, H., INGOLD, A., KINDLE, K. & SOGIN, M. L. (1987). Phylogenetic relationships between chlorophytes, chrysophytes, and oomycetes. *Proceedings of the National Academy of Sciences, U.S.A.* **84**: 5823-5827.

GUPTA, A. K. & MEHROTRA, R. S. (1992). Some fresh water molds from Kurukshetra - Lagenidiales and Blastocladiales. *India; Advances in Plant Science* **5**: 116-123.

GUPTA, G. D. & HEATH, I. B. (1997). Actin disruption by latrunculin B causes turgor-related changes in tip growth of *Saprolegnia ferax* hyphae. *Fungal Genetics and Biology* **21**: 238-251.

GUPTA, S. C. & SAH, B. C. L. (1964). A new species of *Synchytrium* (*S. kumaonensis* sp. nov.) from Kumaon, India. *Indian Phytopathology* **16**: 377-379.

GUSTAFSSON, M., ARHAMMER, M. & GUSTAFSSON, I. (1983). Linkage between virulence genes, compatibility types and sexual recombination in the Swedish population of *Bremia lactucae*. *Phytopathologische Zeitschrift* **108**: 341-354.

GUSTAVSSON, A. (1959*a*). Studies on Nordic Peronosporas. I. Taxonomic revision. *Opera Botanica* **3** (1): 1-271.

GUSTAVSSON, A. (1959*b*). Studies on Nordic Peronosporas. II. General account. *Opera Botanica* **3** (2): 1-61.

GUYÉNOT, E. & NAVILLE, A. (1922). Un nouveau protiste, du genre *Dermocystidium*, parasite de la grenouille. *Revue Suisse de Zoologie, 1921-1922* **29**: 133-145.

GUZMAN, D. R. & AXTELL, R. C. (1986). Effect of nutrient concentration in culturing three isolates of the mosquito fungal pathogen *Lagenidium giganteum* (Oomycetes: Lagenidiales), on sunflower seed extract. *Journal of the American Mosquito Control Association* **2**: 196-200.

GUZMAN, D. R. & AXTELL, R. C. (1987*a*). Temperature and water quality effects in simulated woodland pools on the infection of *Culex mosquito* larvae by *Lagenidium giganteum* (Oomycetes: Lagenidiales) in North Carolina. *Journal of the American Mosquito Control Association* **3**: 211-218.

GUZMAN, D. R. & AXTELL, R. C. (1987*b*). Population dynamics of *Culex quinquefasciatus* and the fungal pathogen *Lagenidium giganteum* (Oomycetes: Lagenidiales) in stagnant water pools. *Journal of the American Mosquito Control Association* **3**: 442-449.

GWYNNE-VAUGHAN, H. & BARNES, B. (1937). *The Structure and Development of the Fungi.* Cambridge, U.K.: Cambridge University Press.

HAAG, K. H. & BUCKINGHAM, G. R. (1991). Effects of herbicides and microbial insecticides on the insects of aquatic plants. *Journal of Aquatic Plant Management* **29**: 55-57.

HAAN, J. de & HOOGKAMMER, L. J. (1901). Hyphomycosis destruens. *Veeartsenijkundige Bladen voor Nederlandsch-Indie* **13**: 350-374.

HAAN, J. de & HOOGKAMER, L. J. (1903). Hyphomycosis destruens equi. *Archiv für Wissenschaftliche und Praktische Tierheilkunde* **29**: 395-410.

HABER, S., BARR, D. J. S. & PLATFORD, R. G. (1991). Observations on the distribution of flame chlorosis in Manitoba and its association with certain zoosporic fungi and the intensive cultivation of cereals. *Canadian Journal of Plant Pathology* **13**: 241-246.

HAECKEL, E. [H.] (1868). Monographie der Moneren. *Jenaische Zeitschrift für Medicin und Naturwissenschaft herausgegeben von der medicinisch-naturwissenschaftlichen Gesellschaft zu Jena* **4**: 64-137.

HAENSELER, C. M. (1926). Studies on the root-rot of peas (*Pisum sativum*) caused by *Aphanomyces euteiches* Drechsler. *New Jersey Agriculture Experiment Station, Annual Report* **46**: 467,484.

HALL, C. J. J. van (1925). Ziekten en plagen der cultuurgewassen in Nederlandsch-Indie in 1924. *Mededeeling van het Instituut voor Plantenziekten, Buitenzorg* **67**: 1-53.

HALL, G. [S.] (1996). Modern approaches to species concepts in downy mildews. *Plant Pathology* **45**: 1009-1026.

HALLER, R. D. & ROBERTS, R. J. (1980). Dual neoplasia in a specimen of *Sarotherodon spilurus spilurus* (Gunther) (= *Tilapia spilurus*). *Journal of Fish Diseases* **3**: 63-66.

HALLETT, I. C. (1975). Studies on the zoospores of the Oomycetes. PhD Thesis, University of Reading.

HALLETT, I. C. & DICK, M. W. (1976). Intraspecific variation in zoospore cyst size in *Saprolegnia anisospora*. *Transactions of the British Mycological Society* **66**: 179-183.

HALLETT, I. C. & DICK, M. W. (1981). Seasonal and diurnal fluctuations of oomycete propagule numbers in the free water of a freshwater lake. *Journal of Ecology* **69**: 671-692.

HALLETT, I. C. & DICK, M. W. (1986). Fine structure of zoospore cyst ornamentation in the Saprolegniaceae and Pythiaceae. *Transactions of the British Mycological Society* **86**: 457-463.

HALSALL, D. M. (1976). Zoospore chemotaxis in Australian isolates of *Phytophthora* species. *Canadian Journal of Microbiology* **22**:409-422.

HALSALL, D. M. & FORRESTER, R. I. (1977). Effects of certain cations on the formation and infectivity of *Phytophthora* zoospores. I. Effect of calcium, magnesium, potassium, and iron ions. *Canadian Journal of Microbiology* **23**: 994-1001.

HALSALL, D. M. & WILLIAMS, J. D. (1984). Effect of root temperature on the development of *Phytophthora cinnamomi* root rot in *Eucalyptus* seedlings. *Australian Journal of Botany* **32**: 521-528.

HAMASAKI, K. & HATAI, K. (1993*a*). Experimental infection in the eggs and larvae of the swimming crab *Portunus trituberculatus* and the mud crab *Scylla serrata* with seven fungal strains belonging to Lagenidiales. *Nippon Suisan Gakkai-Shi [Bulletin of the Japanese Society of Scientific Fisheries]* **59**: 1059-1066.

HAMASAKI, K. & HATAI, K. (1993*b*). Prevention of fungal infection in the eggs and larvae of the swimming crab *Portunus trituberculatus* and the mud crab *Scylla serrata* by bath treatment with formalin. *Nippon Suisan Gakkai-Shi [Bulletin of the Japanese Society of Scientific Fisheries]* **59**: 1067-1072.

HAMID, A. (1942). Indian water moulds III. *Proceedings of the Indian Academy of Science* **15**: 206-215 [212].

HANNOVER, A. (1842). Fernere Erläuterung der contagiösen Confervenbildung auf fröschen und Wasseralamandern. *In* Müller, *Archiv für Anatomie und Physiologie, Leipzig*: pp. 73-121.

HARA, S. & PILLAY, T. V. R. (1962). Handbook on fish culture in the Indo-Pacific region. *FAO Fish Biology* **14**: 1-204.

HARANT, H. (1931). Contribution à l'histoire naturelle des ascides et de leurs parasites. I Chytridinées. *Annales de l'Institute Oceanographique (Monaco)* **8**: 349-352.

HARBORNE, J. B. (1988). *Introduction to Ecological Biochemistry*, 3rd edition. London, U.K.: Academic Press.

HARDER, R. & GALLWITZ-UEBELMESSER, E. (1959). Über niedere Erdphycomyceten Australiens. *Archiv für Mikrobiologie* **32**: 115-126.

HARDER, R. & PERSIEL, I. (1962). Notiz über das Vorkommen niedere Erdphycomyceten in der Antarktis. *Archiv für Mikrobiologie* **41**: 44-50.

HARDER, R. & UEBELMESSER, E. (1955). Über marine saprophytische Chytridiales und einige andere Pilze vom Meeresboden und Meeresstrand. *Archiv für Mikrobiologie* **22**: 87-114.

HARDHAM, A. R. (1987*a*). Microtubules and the flagellar apparatus in zoospores and cysts of the fungus *Phytophthora cinnamomi*. *Protoplasma* **137**: 109-124.

HARDHAM, A. R. (1987*b*). Ultrastructure and serial section reconstruction of zoospores of the fungus *Phytophthora cinnamomi*. *Experimental Mycology* **11**: 297-306.

HARDHAM, A. R. (1989). Lectin and antibody labeling of surface components of spores of *Phytophthora cinnamomi*. *Australian Journal of Plant Physiology* **16**: 19-32.

HARDHAM, A. R. (1992). Cell biology of pathogenesis. *Annual Review of Plant Physiological Plant Molecular Biology* **43**: 491-526.

HARDHAM, A. R. (1995). Polarity of vesicle distribution in oomycete zoospores: Development of polarity and importance for infection. *Canadian Journal of Botany, Supplement* **73**: S400-S407.

HARDHAM, A. R., CAHILL, D. M., COPE, M., GABOR, B. K., GUBLER, F. & HYDE, G.J. (1994). Cell surface antigens of *Phytophthora* spores: biological and taxonomic characterisation. *Protoplasma* **181**: 213-232.

HARDHAM, A. R. & GUBLER, F. (1990). Polarity of attachment of zoospores of a root pathogen and prealignment of the emerging germ tube. *Cell Biology International Reports* **14**: 947-956.

HARDHAM, A. R., GUBLER, F. & DUNIEC, J. (1990). Ultrastructural and immunological studies of zoospores of *Phytophthora*. *In* J. A. Lucas, R. C. Shattock, D. S. Shaw & L. R. Cooke (Eds), Phytophthora, pp. 50-69. Cambridge, U.K.: Cambridge University Press.

HARDHAM, A. R. & MITCHELL, H. J. (1998). Use of molecular cytology to study the structure and biology of phytopathogenic and mycorrhizal fungi. *Fungal Genetics and Biology* **24**: 252-284.

HARDHAM, A. R. & SUZAKI, E. (1986). Encystment of zoospores of the fungus, *Phytophthora cinnamomi*, is induced by specific lectin and monoclonal antibody binding to the cell surface. *Protoplasma* **133**: 165-173.

HARDHAM, A. R., SUZAKI, E. & PERKIN, J. L. (1985). The detection of monoclonal antibodies specific to surface components on zoospores and cysts of *Phytophthora cinnamomi*. *Experimental Mycology* **9**: 264-268.

HARDHAM, A. R., SUZAKI, E. & PERKIN, J. L. (1986). Monoclonal antibodies to isolate-, species-, and genus-specific components on the surface of zoospores and cysts of the fungus *Phytophthora cinnamomi*. *Canadian Journal of Botany* **64**: 311-321.

HARDMAN, J. M. & DICK, M. W. (1987). Short-term fluctuations in the availability of *Pythium* propagules for isolation from soil. *Transactions of the British Mycological Society* **88**: 29-39.

HARDMAN, J. M. & DICK, M. W. (1989). Comparative resilience to dilution plate procedures shown by zoospore cysts of twenty species of *Pythium* and five species of *Phytophthora*. *Mycological Research* **93**: 383-385.

HARDMAN, J. M., PIKE, D. J. & DICK, M. W. (1989). Short-term retrievability of *Pythium* propagules in simulated soil environments. *Mycological Research* **93**: 199-207.

HARDY, A. A. (1910). Association of alga and fungus in salmon disease. *Proceedings of the Royal Society, Victoria (N. S.)* **23**: 27-32.

HARIOT, P. (1889). Champignons. *Mission Scientifique du Cap Horn, V, Botanique*: 173-200.

HARLAN, J. R. (1976). Diseases as a factor in plant evolution. *Annual Review of Phytopathology* **14**: 31-51.

HARNISH, W. N., BERG, L. A. & LILLY, V. G. (1964). Factors in lima bean and hemp seed required for oospore formation by species of *Phytophthora*. *Phytopathology* **54**: 895.

HAROLD, R. L. & HAROLD, F. M. (1992). Configuration of actin microfilaments during sporangium development in *Achlya bisexualis*: comparison of two staining protocols. *Protoplasma* **171**: 110-116.

HARRIS, S. E., BRASELTON, J. P. & MILLER, C. E. (1980). Chromosomal number of *Sorosphaera veronicae* (Plasmodiophoromycetes) based on ultrastructural analysis of synaptonemal complexes. *Mycologia* **77**: 916-925.

HARRISON, J. G., REES, E. A., BARKER, H. & LOWE, R. (1993). Detection of spore balls of *Spongospora subterranea* on potato tubers by enzyme-linked immunosorbent assay. *Plant Pathology* **42**: 181-186.

HARRISON, J. L. & JONES, E. B. G. (1974*a*). Ultrastructural aspects of the marine fungus *Japonochytrium* sp. *Archiv für Mikrobiologie* **96**: 305-317.

HARRISON, J. L. & JONES, E. B. G. (1974*b*). Zoospore release in *Thraustochytrium striatum* Schneider. *Transactions of the British Mycological Society* **62**: 283-288.

HARRISON, J. L. & JONES, E. B. G. (1974*c*). Ultrastructural observations on the formation of zoospores in *Thraustochytrium kinnei* Gaertner. *Transactions of the Mycological Society of Japan* **15**: 273-288.

HARTIG, R. (1880). Der Buchenkeimlingspilz *Phytophthora (Peronospora) fagi*. *Untersuchungen aus dem forstbotanik Institut zur München* **1**: 33-57.

HARTOG, M. [M.] (1887). On the formation and liberation of the zoospores of the Saprolegniae. *Quarterly Journal of Microscopical Science (New Series)* **27**: 427-438.

HARTOG, M. M. (1888). Recent researches on the Saprolegniae; a critical abstract of Rothert's results. *Annals of Botany, London* **2**: 201-216.

HARTOG, M. [M.] (1889*a*). Technique applicable à l'étude des Saprolegniées. *Bulletin de la Société de Botanique de France* **36**: 208-209.

HARTOG, M. M. (1889*b*). Recherches sur la structure des Saprolegniées. *Comptes Rendus Hebdomadaires des Séances de l'Académie des Sciences, Paris* **108**: 687-689.

HARTOG, M. M. (1892). Some problems of reproduction: a comparative study of gametogeny and protoplasmic senescence and rejuvenescence. *Quarterly Journal of Microscopical Science (N. S.)* **33**: 1-79.

HARTOG, M. [M.] (1895). On the cytology of the vegetative and reproductive organs of the Saprolegnieae. *Transactions of the Royal Irish Academy* **30**: 649-708.

HARTOG, M. [M.] (1896). The cytology of *Saprolegnia*. *Annals of Botany, London* **10**: 98-100.

HARTOG, M. [M.] (1899). The alleged fertilization in the Saprolegnieae. *Annals of Botany, London* **13**: 447-59.

HARVEY, J. V. (1925). A study of the water molds and pythiums occurring in the soils of Chapel Hill. *Journal of the Elisha Mitchell Scientific Society* **41**: 151-164.

HARVEY, J. V. (1927*a*). *Brevilegnia diclina* n. sp. *Journal of the Elisha Mitchell Scientific Society* **42**: 243-246.

HARVEY, J. V. (1927*b*). A survey of the water molds occurring in the soils of Wisconsin as studied during the summer of 1926. *Transactions of the Wisconsin Academy of Science, Arts, and Letters* **23**: 551-562.

HARVEY, J. V. (1928). A survey of the water molds occurring in the soils of North Carolina, Wisconsin, and Oklahoma. *Proceedings of the Oklahoma Academy of Science* **7**: 135.

HARVEY, J. V. (1930). A taxonomic and morphological study of some members of the Saprolegniaceae. *Journal of the Elisha Mitchell Scientific Society* **45**: 319-332.

HARVEY, J. V. (1942). A study of western water molds. *Journal of the Elisha Mitchell Scientific Society* **58**: 16-42.

HARVEY, J. V. (1952). Relationship of aquatic fungi to water pollution. *Sewage and Industrial Wastes* **24**: 1159-1164.

HARZ, C. O. (1906). *Achlya hoferi*, eine neue Saprolegniacee auf lebenden Fischen. *Allgemeine Fischerei-Zeitung* **31**: 365-368.

HASKINS, R. H., BRUSHABER, J. A., CHILD, J. J. & HOLTBY, L. B. (1976). The ultrastructure of sexual reproduction in *Pythium acanthicum*. *Canadian Journal of Botany* **54**: 2193-2203.

HASKINS, R. H., TULLOCH, A. P. & MICETICH, R. G. (1964). Steroids and the stimulation of sexual reproduction of a species of *Pythium*. *Canadian Journal of Microbiology* **10**: 187.

HATAI, K. (1980). Studies on pathogenic agents of saprolegniasis in fresh water fishes. *[Special report of Nagasaki Prefectural Institute of fisheries, No. 8, March 1980]*: 1-95 [in Japanese].

HATAI, K. (1982). On the fungus *Haliphthoros milfordensis* isolated from temporarily held abalone (*Haliotis sieboldii*). *Fish Pathology [Gyobyo Kenkyu]* **17**: 199-204.

HATAI, K. (1986). Fungal diseases of fishes. *Proceedings of the Japanese Association of Mycotoxicology* **24**: 5-8.

HATAI, K., BIAN, B. Z., BATICADOS, M. C. L. & EGUSA, S. (1980). Studies on the fungal diseases in crustaceans. II. *Haliphthoros philippinensis* sp. nov. isolated from cultivated larvae of the jumbo tiger prawn (*Penaeus monodon*). *Nihon Kin Gakkai Kaiho, Japan [Transactions of the Mycological Society of Japan]* **21**: 47-55.

HATAI, K. & EGUSA, S. (1977). Studies on visceral mycosis of salmonid fry. II. Characteristics of fungi isolated from the abdominal cavity of amago salmon fry. *Fish Pathology [Gyobyo Kenkyu]* **11**: 187-193.

HATAI, K., EGUSA, S. & AWAKURA, T. (1977). *Saprolegnia shikotsuensis* sp. nov., isolated from kokanee salmon associated with fish saprolegniasis. *Fish Pathology [Gyobyo Kenkyu]* **12**: 105-110.

HATAI, K., EGUSA, S. & NOMURA, T. (1977). *Saprolegnia australis* Elliott isolated from body surface lesions of rainbow trout fingerlings. *Fish Pathology [Gyobyo Kenkyu]* **11**: 201-206.

HATAI, K., EGUSA, S., TAKAHASHI, S. & OOE, K. (1977). Study on the pathogenic fungus of mycotic granulomatosis I. Isolation and pathogenicity of the fungus from cultured-ayu infected with the desease. *Fish Pathology [Gyobyo Kenkyu]* **12**: 129-133.

HATAI, K. & HOSHIAI, G.-I. (1992*a*). Saprolegniasis in cultured coho salmon (*Oncorhynchus kisutch*). *Fish Pathology [Gyobyo Kenkyu]* **27**: 233-234.

HATAI, K. & HOSHIAI, G. (1992*b*). Mass mortality in cultured coho salmon (*Oncorhynchus kisutch*) due to *Saprolegnia parasitica* Coker. *Journal of Wildlife Diseases* **28**: 532-536.

HATAI, K. & HOSHIAI, G.-i. (1993). Characteristics of two *Saprolegnia* species isolated from Coho salmon with saprolegniosis. *Journal of Aquatic Animal Health* **5**: 115-118.

HATAI, K. & LAWHAVINIT, O.-a. (1988). *Lagenidium myophilum* sp. nov., a new parasite on adult northern shrimp (*Pandalus borealis* Kröyer). *Nihon Kin Gakkai Kaiho, Japan [Transactions of the Mycological Society of Japan]* **29**: 175-184.

HATAI, K., NAKAMURA, K., RHA, S. A., YUASA, K & WADA, S. (1994). *Aphanomyces* infection in Dwarf Gourami (*Colisa lalia*). *Fish Pathology [Gyobyo Kenkyu]* **29**: 95-99.

HATAI, K., RHOOBUNJONGDE, W. & WADA, S. (1992). *Haliphthoros milfordensis* isolated from gills of juvenile kuruma prawn (*Penaeus japonicus*) with black gill disease. *Nihon Kin Gakkai Kaiho [Transactions of the Mycological Society of Japan]* **33**: 185-192.

HATAI, K. & WILLOUGHBY, L. G. (1988). *Saprolegnia parasitica* from rainbow trout inhibited by the bacterium *Pseudomonas fluorescens*. *Bulletin of the European Association of Fish Pathologists* **8**: 27-29.

HATAI, K., WILLOUGHBY, L. G. & BEAKES, G. W. (1990). Some characteristics of *Saprolegnia* obtained from fish hatcheries in Japan. *Mycological Research* **94**: 182-190.

HAUCK, F. (1878). Nitiz über *Rhizophydium dicksonii* Wright. *Österreichische Botanische Zeitschrift* **28**: 321.

HAUSMANN, K. (1978). Extrusive organelles in protists. *International Review of Cytology* **52**: 197-276.

HAWKSWORTH, D. L., SUTTON, B. C. & G. C. AINSWORTH, with the assistance of B. L. BRADY, V. DEMOULIN, P. W. JAMES, P. M. KIRK, J. E. M. MORDUE, S. T. MOSS, D. N. PEGLER, K. A. PIROZYNSKI, B. SPOONER, I. TAVARES and G. M. WATERHOUSE (1983). *Ainsworth & Bisby's Dictionary of the Fungi (Including the Lichens). Seventh Edition.* Kew, Surrey, U.K.: CAB.

HAWKSWORTH, D. L., KIRK, P. M., SUTTON, B. C. & PEGLER, D. N. with the assistance of G. C. AINSWORTH, G. L. BENNY, P. F. CANNON, V. DEMOULIN, M. W. DICK, W. C. ELSIK, R. W. LICHTWARDT, J. E. LONGCORE, J. E. M. MORDUE, F. PANDO, I. L. TAVARES, M. A. J. WILLIAMS and others (1995). *Ainsworth & Bisby's Dictionary of the Fungi. Eighth Edition.* Wallingford, Oxfordshire, U.K.: CAB International.

HÄYRÉN, E. (1903). Verzeichnis einiger in der Nähe von Helsingfors eingesammelter Saprolegnieen. *Meddelanden Societatis pro Fauna et Flora Fennica* **29**: 165-166.

HÄYRÉN, E. (1928). *Saprolegnia asterophora* De Bary. *Memoranda Societatis pro Fauna et Flora Fennica* **4**: 50.

HÄYRÉN, E. (1943). Über das Zusammenballen der Schwärmer des zweiten Schwärmstadiums von *Saprolegnia dioica*. *Memoranda Societatis pro Fauna et Flora Fennica* **18**: 89-93.

HÄYRÉN, E. (1945). Tvenne anteckningar om saprolegniaceer i Helsingforstrakten. *Memoranda Societatis pro Fauna et Flora Fennica* **20**: 148,149.

HÄYRÉN, E. (1954). Der Lebenzyklus von *Saprolegnia dioica* de Bary. *Societas Scientiarum Fennica. Commentationes Biologicae* **15**: 1-45.

HÄYRÉN, E. (1956). Die in Finnland bisher gefundenen Wasserpilze. *Friesia* **5**: 264-265.

HE, S.-H. & SU, X.-Q. (1995). Protoplast formation and regeneration of the mosquito fungal pathogen *Lagenidium giganteum* (Oomycetes: Lagenidiales). *Acta Mycologica Sinica* **14**: 196-201 [in Chinese].

HEATH, I. B. (1974*a*). Mitosis in the fungus *Thraustotheca clavata*. *Journal of Cell Biology* **60**: 204-220.

HEATH, I. B. (1974*b*). Centrioles and mitosis in some oomycetes. *Mycologia* **66**: 354-359.

HEATH, I. B. (1980*a*). Fungal mitoses: the significance of variations on a theme. *Mycologia* **72**: 229-250.

HEATH, I. B. (1980*b*). Variant mitosis in lower eukaryotes: indicators of the evolution of mitosis. *International Review of Cytology* **64**: 1-80.

HEATH, I. B. (1981). An investigation of protistan phylogeny using a numerical taxonomy (cluster) analysis of mitotic systems. *BioSystems* **14**: 261-270.

HEATH, I. B. & GREENWOOD, A. D. (1968). Electron microscope observations of dividing somatic nuclei in *Saprolegnia*. *Journal of General Microbiology* **53**: 287-289.

HEATH, I. B. & GREENWOOD, A. D. (1970). Centriole replication and nuclear division in *Saprolegnia*. *Journal of General Microbiology* **62**: 139-148.

HEATH, I. B. & GREENWOOD, A. D. (1971). Ultrastructural observations on the kinetosomes, and Golgi bodies during the asexual life cycle of *Saprolegnia*. *Zeitschrift für Zellforschung und Mikrosckopische Anatomie* **112**: 371-389.

HEATH, I. B., GREENWOOD, A. D. & GRIFFITHS, H. B. (1970). The origin of flimmer in *Saprolegnia*, *Dictyuchus*, *Synura* and *Cryptomonas*. *Journal of Cell Science* **7**: 445-461.

HEATH, I. B. & HAROLD, R. L. (1992). Actin has multiple roles in the formation and architecture of zoospores of the oomycetes, *Saprolegnia ferax* and *Achlya bisexualis*. *Journal of Cell Science* **102**: 611-627.

HEATH, I. B. & RETHORET, K. (1980). Temporal analysis of the nuclear cycle by serial section electron microscopy of the fungus, *Saprolegnia ferax*. *European Journal of Cell Biology* **21**: 208-213.

HEATH, I. B. & STEINBERG, G. (1999). Mechanisms of hyphal tip growth: tube dwelling amoebae revisited. *Fungal Genetics and Biology* **28**: 79-93.

HEGNAUER, H. & HOHL, H. R. (1978). Cell wall architecture of sporangia, chlamydospores, oogonia, and oospores in *Phytophthora*. *Experimental Mycology* **2**: 216-233.

HEIDT, K. (1937). Beitrag zur Kenntnis der Gattung *Micromyces*. *Berichte der Deutschen Botanischen Gesellschaft* **55**: 204-217.

HEIM, P. (1960). Evolution du *Spongospora*, parasite des racines du cresson. *Revue de Mycologie* **25**: 3-12.

HELD, A. A. (1970). Nutrition and fermentative energy metabolism of the water mold *Aqualinderella fermentans*. *Mycologia* **62**: 339-358.

HELD, A. A. (1972*a*). Host-parasite relations between *Allomyces* and *Rozella*. Parasite penetration depends on a growth response of host cell-wall. *Archiv für Mikrobiologie* **82**: 128-139.

HELD, A. A. (1972*b*). Improved culture methods for *Rozella* and for *Olpidiopsis*. *Mycologia* **64**: 871-886.

HELD, A. A. (1973*a*). Encystment and germination of the parasitic chytrid *Rozella allomycis* on host hyphae. *Canadian Journal of Botany* **51**: 1825-1836.

HELD, A. A. (1973*b*). The Torrey Symposium on current aspects of fungal development III. Development of endoparasitic, zoosporic fungi. *Bulletin of the Torrey Botanical Club* **100**: 203-216.

HELD, A. A. (1974). Attraction and attachment of zoospores of the parasitic chytrid *Rozella allomycis* in response to host-dependent factors. *Archiv für Mikrobiologie* **95**: 97-114.

HELD, A. A. (1975). The zoospore of *Rozella allomycis*: ultrastructure. *Canadian Journal of Botany* **53**: 2212-2232.

HELD, A. A. (1980). Development of *Rozella* in *Allomyces*: a single zoospore produces numerous zoosporangia and resistant sporangia. *Canadian Journal of Botany* **58**: 959-979.

HELD, A. A. (1981). *Rozella* and *Rozellopsis*: naked endoparasitic fungi which dress-up as their hosts. *Botanical Review* **47**: 451-515.

HELD, A. A. & EMERSON, R. (1970). Oogonium production in *Aqualinderella fermentans*. *Mycologia* **62**: 359-364.

HELD, A. A., EMERSON, R., FULLER, M. S. & GLEASON, F. H. (1969). *Blastocladia* and *Aqualinderella*: fermentative water molds with high carbon dioxide optima. *Science, New York* **165**: 706-709.

HELLER, A., ROZYNEK, B. & SPRING, O. (1997). Cytological and physiological reasons for the latent type of infection in sunflower caused by *Plasmopara halstedii*. *Journal of Phytopathology* **145**: 441-445.

HEMMES, D. E. (1983). Cytology of *Phytophthora*. *In* D. C. Erwin, S. Bartnicki-Garcia & P. H. Tsao (Eds), Phytophthora *its Biology, Taxonomy, Ecology, and Pathology*, pp. 9-40. St Paul, Minnesota, U.S.A.: American Phytopathological Society.

HEMMES, D. E. & BARTNICKI-GARCIA, S. (1975). Electron microscopy of gametangial interaction and oospore development in *Phytophthora capsici*. *Archives of Microbiology* **103**: 91-112.

HEMMES, D. E. & HOHL, H. R. (1971). Ultrastructural aspects of encystation and cyst germination in *Phytophthora parasitica*. *Journal of Cell Science* **9**: 175-191.

HEMMES, D. E. & HOHL, H. R. (1973). Mitosis and nuclear degeneration: simultaneous events during germination in *Phytophthora palmivora*. *Canadian Journal of Botany* **51**: 1613-1675.

HEMMES D. E. & RIBIERO, O. K. (1977). Electron microscopy of early gametangial interaction in *Phytophthora megasperma* var. *sojae*. *Canadian Journal of Botany* **55**: 436-447.

HEMMES D. E. & STASZ, T. E. (1984). Ultrastructure of dormant, converted, and germinating oospores of *Pythium ultimum*. *Mycologia* **76**: 924-935.

HEMMI, T. & ABE, T. (1928). An outline of the investigations on the seed- and seedling-rot of rice caused by the water-mould, *Achlya prolifera* Nees. *Japanese Journal of Botany* **4**: 113-123.

HENDRIX, F. F. & CAMPBELL, W. A. (1968). A new heterothallic *Pythium* from the United States and Canada. *Mycologia* **60**: 802-805.

HENDRIX, F. F. & CAMPBELL, W. A. (1969). A new species of *Pythium* with spiny oogonia. *Mycologia* **61**: 387-391.

HENDRIX, F. F. & CAMPBELL, W. A. (1971). A new species of *Pythium* with spiny oogonia and large chlamydospores. *Mycologia* **63**: 978-982.

HENDRIX, F. F. & CAMPBELL, W. A. (1974). Taxonomic value of reproductive cell size in the genus *Pythium*. *Mycologia* **66**: 681-684.

HENDRIX, F. F. & PAPA, K. E. (1974). Taxonomy and genetics of *Pythium*. *Proceedings of the American Phytopathological Society* **1**: 200-207.

HENDRIX, J. W. (1964). Sterols induction of reproduction and stimulation of growth of *Pythium* and *Phytophthora*. *Science, New York* **144**: 1028-1029.

HENDRIX, J. W. (1970). Sterols in growth and reproduction of fungi. *Annual Review of Plant Pathology* **8**: 111-130.

HENDRIX, J. W. (1975). Differential uptake and metabolism of sitosterol and cholesterol by *Achlya*, *Pythium*, and *Phytophthora* species. *Canadian Journal of Microbiology* **21**: 734-737.

HENDRIX, J. W., BENNETT, R. D. & HEFTMANN, E. (1970). Metabolism of cholesterol by *Pythium periplocum*. *Microbios* **5**: 11-15.

HENDRIX, J. W. & GUTTMAN, S.-M. (1968). Annulment of sterol-induced sexual reproduction by estradiol in *Pythium periplocum*. *Science, New York* **161**: 1252.

HERRADO, M. L. & KLEMSDAL, S. S. (1998). Identification of *Pythium aphanidermatum* using the RAPD technique. *Mycological Research* **102**: 136-140.

HERTWIG, R. & LESSER, E. (1874). Ueber Rhizopoden und denselben nahestehende Organismen. *Archiv für Mikroskopische Anatomie und Entwicklungsmechanik (Supplement)* **10**: 139-145.

HESSE, M., KUSEL-FETZMANN, E. & CARNIEL, K. (1989). Life cycle and ultrastructure of *Ducellieria chodati* (Oomycetes). *Plant Systematics and Evolution* **165**: 1-15.

HESSE, R. (1874). Pythium de Baryanum *ein endophytischer Schmarotzer in den Geweben der Keimlinge der Leindotter, der Rüben, des Spergels und einiger anderer landwirthschaftlichen Kulturpflanzen*. Inaugural Dissertation der philosophischen Facultät zu Göttingen. Halle: Hendel.

HEURCK, H. von (1899). *Traité des Diatomées*. Anvers.

HIBBERD, D. J. (1975). Observations on the ultrastructure of the choanoflagellate *Codosiga botrytis* (Ehr.) Saville-Kent with special reference to the flagellar apparatus. *Journal of Cell Science* **17**: 191-207-219.

HIBBETT, D. S. & DONOGHUE, M. J. (1998). Integrating phylogenetic analysis and classification in fungi. *Mycologia* **90**: 347-356.

HIDALGO, T. Z. (1962). The host range and host-parasite relationship of *Dictyomorpha* within the Saprolegniaceae. Unpublished MS Thesis, University of Florida, U.S.A.

HIGHTOWER, R. C. & MEAGHER, R. B. (1986). The molecular evolution of actin. *Genetics* **114**: 315-332.

HIJWEGEN, T. (1988). Coevolution of flowering plants with pathogenic fungi. *In* K. A. Pirozynski & D. L. Hawksworth (Eds), *Coevolution of Fungi with Plants and Animals*. London, UK, Academic Press, pp. 63-77.

HILDEBRAND, A. A. & KOCH, L. W. (1943). Studies on black rot of sugar beet seedlings. *Scientific Agriculture* **23**: 557-567.

HILDEBRAND, F. (1867). Mykologische Beitrage. I. Ueber einige neue Saprolegnieen. *Jahrbücher für wissenschaftliche Botanik* **6**: 249-269.

HILL, A. E., GRAYSON, D. E. & DEACON, J. W. (1998). Suppressed germination and early death of *Phytophthora infestans* sporangia caused by pectin, inorganic phosphate, ion chelators and calcium-modulating treatments. *European Journal of Plant Pathology* **104**: 367-376.

HILL, T. W. & MULLINS, J. T. (1979). Hyphal tip growth in *Achlya*: enzyme activities in mycelium and medium. *Canadian Journal of Botany* **57**: 2145-2149.

HILLER, S., BRUCE, D. M. & JERONIMIDIS, G. (1996). A micro-penetration technique for mechanical testing of plant cell walls. *Journal of Texture Studies* **27**: 559-587.

HINCH, J. M. & CLARKE, A. E. (1980). Adhesion of fungal zoospores to root surfaces is mediated by carbohydrate determinants of the root slime. *Physiological Plant Pathology* **16**: 303-307.

HINCH, J. & WESTE, G. (1979). Behaviour of *Phytophthora cinnamomi* zoospores on roots of Australian forest species. *Australian Journal of Botany* **27**: 679-691.

HINE, F. B. (1878-1879). Observations on several forms of Saprolegnieae. *American Quarterly Microscopical Journal* **1**: 18-28, 136-146.

HIRANE, S. (1960). (Studies on *Pythium* snow blight of wheat and barley with special reference to the taxonomy of the pathogens.) *Transactions of the Mycological Society of Japan* **2**: 82-87.

HIRT, R. P., HEALY, B., VOSSBRINCK, C. R., CANNING, E. U. & EMBLEY, T. M. (1997). A mitochondrial *Hsp 70* orthologue in *Vairimorpha necatrix*: molecular evidence that microsporidia once contained mitochondria. *Current Biology* **7**: 995-998.

HIRUKI, C. (1994). Multiple transmission of plant viruses by *Olpidium brassicae*. *Canadian Journal of Plant Pathology* **16**: 261-265.

HO, H.-H. (1978). Hyphal branching systems in *Phytophthora* and other phycomycetes. *Mycopathologia* **64**: 83-86.

HO, H.-H. (1992). Keys to the species of *Phytophthora* in Taiwan. *Plant Pathology Bulletin* **1**: 104-109.

HO, H. H. & CHANG, H. S. (1992). A re-evaluation of *Phytophthora* species described by K. Sawada in Taiwan. *Mycotaxon* **43**: 297-316.

HO, H. H. & HICKMAN, C. J. (1967*a*). Asexual reproduction and behaviour of zoospores of *Phytophthora megasperma* var. *sojae*. *Canadian Journal of Botany* **45**: 1963-1981.

HO, H. H. & HICKMAN, C. J. (1967*b*). Factors governing zoospore responses of *Phytophthora megasperma var. sojae* to plant roots. *Canadian Journal of Botany* **45**: 1983-1994.

HO, H. H. & JONG, S. C. (1990). *Halophytophthora* gen. nov., a new member of the family Pythiaceae. *Mycotaxon* **36**: 377-382.

HO, H. H., NAKAGIRI, A. & NEWELL, S. Y. (1992). A new species of *Halophytophthora* from Atlantic and Pacific subtropical islands. *Mycologia* **84**: 548-554.

HOCH, H. C. & MITCHELL, J. E. (1972*a*). A continuous flow system for inducing and observing asexual spore formation in *Aphanomyces euteiches*. *Canadian Journal of Botany* **50**: 681-682.

HOCH, H. C. & MITCHELL, J. E. (1972*b*). The ultrastructure of zoospores of *Aphanomyces euteiches* and their encystment and subsequent germination. *Protoplasma* **75**: 113-138.

HOENIGMAN, J. (1954). Novosti s Področja Jadranskega Zooplanktona. (O najbdi elobiopsidov). (Nouveautés concernantes le zooplankton Adriatique. (Sur la trouvaille des Ellobiopsidés)). *Biološki Věstnik, Lyubljana* **3**: 106-116.

HOENIGMAN, J. (1958). Sur la découverte de quelques espèces zooplanctoniques nouvelles pour l'Adriatique, de deux épibiontes nouveaux pour les Mysidacés, et de deux espèces de parasites nouvelles pour le domaine méditerranéen. *Rapport et procès-verbaux des Réunions. Commission Internationale pour l'Exploration scientifique de la Mer Méditerranée* **14**: 263-264.

HOENIGMAN, J. (1960). Faits nouveaux concernant les Mysidacés (Crustacea) et leurs epibiontes dans l'Adriatique. *Rapport et procès-verbaux des Réunions. Commission Internationale pour l'Exploration scientifique de la Mer Méditerranée* **15**: 339-343.

HOFER, B. (1906). *Handbuch der Fischkrankheiten.* Stuttgart, Germany.

HOFFMAN, G. L. (1949). Isolation of *Saprolegnia* and *Achlya* with penicillin-streptomycin, and attempts to infect fish. *Progressive Fish-Culturist* **11**: 171-174.

HOHL, H. R. (1966). The fine structure of the slimeways in *Labyrinthula*. *Journal of Protozoology* **13**: 41-43.

HOHL, H. R. (1983). Nutrition of *Phytophthora*. *In* D. C. Erwin, S. Bartnicky-Garcia & P. H. Tsao (Eds), Phytophthora, *Its Biology, Taxonomy Ecology and Pathology*. American Phytopathological Society, St Paul, Minnesota, USA, pp. 41-54.

HOHL, H. R. & HAMMAMOTO, S. T. (1967). Ultrastructural changes during zoospore formation in *Phytophthora parasitica*. *American Journal of Botany* **54**: 1131-1139.

HÖHNEL, F. von (1903). Mykologische Fragmente. *Annales Mycologici* **1**: 391-414.

HÖHNK, W. (1932). A new parasitic *Pythium*. *Mycologia* **24**: 489-507.

HÖHNK, W. (1933). Polyplanetism and zoospore germination in Saprolegniaceae and *Pythium*. *American Journal of Botany* **20**: 45-62.

HÖHNK, W. (1935*a*). Saprolegniales und Monoblepharidales aus der Umgebung Bremens, mit besonderer Berücksichtigung der Oekologie der Saprolegniaceae. *Abhandlungen herausgegeben vom Naturwissenschaftlichen Verein zu Bremen* **29**: 207-237.

HÖHNK, W. (1935*b*). Zur Cytologie der Oogon- und Eientwicklung bei *Saprolegnia ferax* (Gruith.) Thuret. *Abhandlungen herausgegeben vom Naturwissenschaftlichen Verein zu Bremen* **29**: 308-323.

HÖHNK, W. (1936). On three pythiaceous Oomycetes. *Beihefte zum Botanischen Zentralblatt (Abteilung A)* **55**: 89-99.

HÖHNK, W. (1938). Über heimische Phycomyceten. I. *Abhandlungen herausgegeben vom Naturwissenschaftlichen Verein zu Bremen* **31**: 491-501.

HÖHNK, W. (1939). Ein Beitrag zur Kenntnis der Phycomyceten des Brackwassers. *Kieler Meeresforschungen* **3**: 337-361.

HÖHNK, W. (1952*a*). Die in Nordwestdeutschland gefundenen ufer- und bodenbewohnenden Saprolegniaceae. *Veröffentlichungen des Instituts für Meeresforschung in Bremerhaven* **1**: 52-60.

HÖHNK, W. (1952*b*). Studien zur Brack- und Seewassermykologie. II. Oomycetes: Erster Teil. *Veröffentlichungen des Instituts für Meeresforschung in Bremerhaven* **1**: 247-278.

HÖHNK, W. (1952*c*). Nachtrage zu: Die in Nordwestdeutschland gefundenen ufer- und bodenbewohnenden Saprolegniaceae. *Veröffentlichungen des Instituts für Meeresforschung in Bremerhaven* **1**: 126-128.

HÖHNK, W. (1953). Studien zur Brack- und Seewassermykologie. III. Oomycetes Zweiter Teil. *Veröffentlichungen des Instituts für Meeresforschung in Bremerhaven* **2**: 52-108.

HÖHNK, W. (1956*a*). Mykologische Abwasserstudie I. *Veröffentlichungen des Instituts für Meeresforschung in Bremerhaven* **4**: 67-110.

HÖHNK, W. (1956*b*). Studien zur Brack- und Seewassermykologie. IV. Über die pilzliche Besiedlung verschieden salziger submerser Standorte. *Veröffentlichungen des Instituts für Meeresforschung in Bremerhaven* **4**: 195-213.

HÖHNK, W. (1958). Mykologische Abwasserstudie II. *Veröffentlichungen des Instituts für Meeresforschung in Bremerhaven* **5**: 211-256.

HÖHNK, W. (1960). Mykologische Notizen: II. Phycomyceten von Island und Grönland. III. Phycomyceten von der Insel Strömö (Färöer). *Veröffentlichungen des Instituts für Meeresforschung in Bremerhaven* **7**: 63-69.

HÖHNK, W. (1962). Über die Phycomyceten der Insel Madiera. *Veröffentlichungen des Instituts fürMeeresforschung in Bremerhaven* **8**: 99-108.

HÖHNK, W. & ALEEM, A. A. (1953). Ein Brackwasserpilz: *Olpidium marinum* nov. spec. *Veröffentlichungen des Instituts für Meeresforschung in Bremerhaven* **2**: 224-229.

HÖHNK, W. & BOCK, K. J. (1954). Ein Beitrag zur Ökologie der saprophytischen Wasserpilze. *Veröffentlichungen des Instituts für Meeresforschung in Bremerhaven* **3**: 9-26.

HÖHNK, W. & ULKEN, A. (1979). Pilz aus marinen Schwammen. *Veröffentlichungen des Instituts für Meeresforschung in Bremerhaven* **17**: 199-204.

HÖHNK, W. & VALLIN, S. (1953). Epidermische Absterben von Eurytemora in Bottnischen Meerbusen versacht durch *Leptolegnia baltica* nov. spec. *Veröffentlichungen des Instituts für Meeresforschung in Bremerhaven* **2**: 215-223.

HOLCK, A. R. (1988). Current status of the use of predators pathogens and parasites for the control of mosquitoes. *Florida Entomologist* **71**: 537-546.

HOLLAND, H. D. (1978). *The Chemistry of the Atmosphere and Oceans*. New York, U.S.A.: Wiley.

HOLLAND, H. D. (1984). *The Chemical Evolution of the Atmosphere and Oceans*. Princeton, New Jersey, U.S.A.: Princeton University Press.

HOLLAND, M. (1958). Cave fungi. I. Aquatic Phycomycetes. Unpublished MA Thesis, Virginia Polytechnic Institute, Blacksburg.

HOLLANDE, A. & BALSAC, H. H. de (1942). Parasitism du *Peranema trichophorum* par une chytridinée du genre *Nucleophaga*. *Archives de Zoologie Expérimentale et Générale* **82**: 37-46.

HOLLANDE, A. & CARRUETTE-VALENTIN, J. (1970). Appariement chromosomique et complexes synaptonématiques dans les noyaux en cours de dépolyploïdisation chez *Pyrsonympha flagellata*: le cycle évolutif des Pyrsonymphines symbiontes de *Reticulitermes lucifugus*. *Comptes Rendus Hebdomadaires des Séances de l'Académie des Sciences, Paris, Série D* **270**: 2550-2553. [See also **GRASSÉ, P.-P.** (1970). Remarques sur la note de M A. Hollande et Mme J. Carruette-Valentin: le rôle possible des symptonématiques. *Comptes Rendus Hebdomadaires des Séances de l'Académie des Sciences, Paris, Série D* **270**: 2554-2555.]

HOLLANDE, A. & ENJUMET, M. (1955). Sur l'évolution et la systématique des Labyrinthulidae; étude de *Labyrinthula algeriensis* nov. sp. *Annales des Sciences Naturelles, Zoologie et Biologie Animale, Série 11* **17**: 357-368.

HOLLOWAY, S. A. & HEATH, I. B. (1974). Observations on the mechanism of flagellar retraction in *Saprolegnia terrestris*. *Canadian Journal of Botany* **52**: 939-942.

HOLLOWAY, S. A. & HEATH, I. B. (1977*a*). Morphogenesis and the role of microtubules in synchronous populations of *Saprolegnia* zoospores. *Experimental Mycology* **1**: 9-29.

HOLLOWAY, S. A. & HEATH, I. B. (1977*b*). An ultrastructural analysis of the changes in organelle arrangement and structure between the various spore types of *Saprolegnia*. *Canadian Journal of Botany* **55**: 1328-1339.

HOLUB, E. B., GRAU, C. R. & PARKE, J. L. (1991). Evaluation of the forma specialis concept in *Aphanomyces euteiches*. *Mycological Research* **95**: 147-157.

HOLWILL, M. E. J. (1982). Dynamics of eukaryotic movement. *In* W. B. Amos & J. G. Duckett (Eds), *Prokaryotic and Eukaryotic Flagella*, pp. 289-312. Cambridge, U.K.: Cambridge University Press.

HOLWILL, M. E. J. & PETERS, P. D. (1974). Dynamics of the hispid flagellum of *Ochromonas danica*. *Journal of Cell Biology* **62**: 322-328.

HOLWILL, M. E. J. & SLEIGH, M. A. (1967). Propulsion by hispid flagella. *Journal of Experimental Biology* **47**: 267-276.

HONDA, D., YOKOCHI, T., NAKAHARA,T., ERATA, M. & HIGASHIHARA, T. (1998). *Schizochytrium limacinum* sp. nov., a new thraustochytrid from a mangrove area in the West Pacific Ocean. *Mycological Reasearch* **102**: 439-448.

HONIGBERG, B. M., BALAMUTH, W., BOVEE, E. C., CORLISS, J. O., GOJDICS, M., HALL, R. P., KUDO, R. R., LEVINE, N. D., LOEBLICH, A. R., WEISER, J. & WENRICH, D. H. (1964). A revised classification of the phylum Protozoa. *Journal of Protozoology* **11**: 7-20.

HORD, M. J. & RISTAINO, J. B. (1991). Effects of physical and chemical factors on the germination of oospores of *Phytophthora capsici* in vitro. *Phytopathology* **81**: 1541-1546.

HORGAN, P. A. (1981). The role of the steroid sex pheromone antheridiol in controlling the development of male sex organs in the water mold *Achlya*. *In* D. H. O'Day & P. A. Horgan (Eds), *Sexual Interactions in Eukaryotic Microbes*, pp. 179-188. New York, U.S.A.: Academic Press.

HORI, H., LIM, B.-L. & OSAWA, S. (1985). Evolution of green plants as deduced from 5s rRNA sequences. *Proceedings of the National Academy of Sciences, U.S.A.* **82**: 820-823.

HORI, H. & OSAWA, S. (1987). Origin and evolution of organisms as deduced from 5s ribosomal RNA sequences. *Molecular Biology and Evolution* **4**: 445-472.

HORN, L. (1904). Experimentelle Entwicklungsänderungen bei *Achlya polyandra* de Bary. *Annales Mycologici* **2**: 207-241.

HORNE, A. S. (1930). Nuclear division in the Plasmodiophorales. *Annals of Botany, London* **44**: 199-225.

HOSHINA, T. & OOKUBA, A. (1956). On a fungi-disease of eel. *Journal of the Tokyo University of Fisheries* **42**: 1-13.

HOSHINA, T., SANO, T. & SUNAYAMA, M. (1960). Studies on the saprolegniosis of eel. *Journal of Tokyo University of Fisheries* **47**: 59-79.

HOTI, S. L. & BALARAMAN, K. (1990). A mosquito-pathogenic fungus *Lagenidium*. *Indian Journal of Medical Research, Section A* **91**: 67-69.

HOTI, S. L. & BALARAMAN, K. (1992). Method for storing *Lagenidium* (Oomycetes, Lagenidiales). *Indian Journal of Medical Research, Section A* **95**: 95-98.

HOVASSE, R. (1926). "*Parallobiopsis coutieri*" Collin. Morphologie, cytologie, évolution. Affinités des Ellobiopsidés. *Bulletin Biologique de la France et de la Belgique* **60**: 409-446.

HOVASSE, R. (1951). Contribution à la connaissance biologique des Ellobiopsidæ: la sporulation chez *Ellobiopsis fagei* sp. nov. *Comptes Rendus Hebdomadaires des Séances de l'Académie des Sciences, Paris, Série D* **233**: 980-982.

HOWARD, K. L. (1971). Oospore types in the Saprolegniaceae. *Mycologia* **63**: 697-686.

HOWARD, K. L. & BRYANT, T. R. (1971). Meiosis in the Oömycetes. II. A microspectrophotometric analysis of DNA in *Apodachlya brachynema*. *Mycologia* **63**: 58-68.

HOWARD, K. L. & JOHNSON, T. W. (1969). Aquatic fungi of Iceland. Some filamentous eucarpic, and holocarpic species. *Mycologia* **61**: 496-510.

HOWARD K. L. & MOORE, R. T. (1970). Ultrastructure of oogenesis in *Saprolegnia terrestris*. *Botanical Gazette* **131**: 311-336.

HOWLETT, B. J. (1989). An electrophoretic karyotype for *Phytophthora megasperma*. *Experimental Mycology* **13**: 199-202.

HSIEH, H.-J. & CHANG, H.-S. (1976). Five species of *Pythium*, two species of *Pythiogeton* new for Taiwan and *Pythium afertile*. *Botanical Bulletin of Academia Sinica, Taipei* **17**: 141-150.

HUBSCHMAN, J. H. & SCHMITT, J. A. (1969). Primary mycosis in shrimp larvae. *Journal of Invertebrate Pathology* **13**: 351-357.

HUDSON, C. T. & GOSSE, P. H. (1886, 1889). *The Rotifera: or, Wheel Animalcules, Both British and Foreign*, **I & II**, and Supplement. London, U.K.: Longmans, Green & Co.

HUDSPETH, M. E. S. (1985). Oomycete mitochondrial genomes. *Journal of Cellular Biochemistry S9C* abstract 1608: 187 [abstract].

HUDSPETH, D. S. S., NADLER, S. A. & HUDSPETH, M. E. S. (2000). A *COX2* molecular phylogeny of the Peronosporomycetes. *Mycologia* **92**: 674-684.

HUDSPETH, M. E. S., SHUMARD, D. S., BRADFORD, C. J. R. & GROSSMAN, L. I. (1983). Organization of *Achlya* mtDNA: a population with two orientations and a large inverted repeat containing the rRNA genes. *Proceedings of the National Academy of Sciences, U.S.A.* **80**: 142-146.

HUDSPETH, M. E. S., TIMBERLAKE, W. & GOLDBERG, R. (1977). DNA sequence organization in water mold *Achlya*. *Proceedings of the National Academy of Sciences, U.S.A.* **74**: 4332-4336.

HUGHES, G. C. (1959). The Saprolegniaceae of Clinch County, Georgia. *Bulletin of the Georgia Academy of Sciences* **17**: 88-93.

HUGHES, G. C. (1962). Seasonal periodicity of the Saprolegniaceae in the south-eastern United States. *Transactions of the British Mycological Society* **45**: 519-531.

HUGUENIN, B. & BOCCAS, B. (1970). Étude de la caryocinèse chez le *Phytophthora parasitica* Dastur. *Comptes Rendus Hebdomadaires des Séances de l'Académie des Sciences, Paris* **271**: 660-663.

HUIZAR, H. E. & ARONSON, J. M. (1985). Chitin synthase of *Apodachlya*. *Experimental Mycology* **9**: 302-309.

HUIZAR, H. E. & ARONSON, J. M. (1986). Aspects of cellulin deposition and chitin biosynthesis in the Leptomitaceae. *Mycologia* **78**: 489-492.

HULBERT, S. H., ILOTT, T. W., LEGGE, E. J., LINCOLN, S. E., LANDER, E. S. & MICHELMORE, R. W. (1988). Genetic analysis of the fungus, *Bremia lactucae*, using restriction fragment length polymorphisms. *Genetics* **120**: 947-958.

HUMPHREY, J. E. (1893). The Saprolegniaceae of the United States, with notes on other species. *Transactions of the American Philosophical Society, New Series* **17**: 63-148.

HUNEYCUTT, M. B. (1948). Keratinophilic phycomycetes I. A new genus of the Saprolegniaceae. *Journal of the Elisha Mitchell Scientific Society* **64**: 277-285.

HUNEYCUTT, M. B. (1952). A new water mold on keratinized materials. *Journal of the Elisha Mitchell Scientific Society* **68**: 109-112.

HUXLEY, T. H. (1882). *Saprolegnia* in relation to salmon disease. *Quarterly Journal of Microscopical Science (N. S.)* **22**: 311.

HUYSMANS, E., DAMS, E., VANDENBERGHE, A. & WACHTER, R. de (1983). The nucleotide sequences of the 5s rRNAs of four mushrooms and their use in studying the phylogenetic position of basidiomycetes among the eukaryotes. *Nucleic Acids Research* **11**: 2871-2880.

HWANG, B. K., COCK, A. W. A. M. de, BAHNWEG, G., PRELL, H. H. & HEITEFUSS, R. (1991). Restriction fragment length polymorphisms of mitochondrial DNA among *Phytophthora capsici* isolates from pepper (*Capsicum annuum*). *Systematic and Applied Microbiology* **14**: 111-116.

HYDE, G. J., GUBLER, F. & HARDHAM, A. R. (1991). Ultrastructure of zoosporogenesis in *Phytophthora cinnamomi*. *Mycological Research* **95**: 577-591.

HYDE, G. J. & HARDHAM, A. R. (1992). Confocal microscopy of microtubule arrays in cryosectioned sporangia of *Phytophthora cinnamomi*. *Experimental Mycology* **16**: 207-218.

HYDE, G. J. & HARDHAM, A. R. (1993). Microtubules regulate the generation of polarity in zoospores of *Phytophthora cinnamomi*. *European Journal of Cell Biology* **62**: 75-85.

HYDE, G. J. & HEATH, I. B. (1997). Ca^{2+} gradients in hyphae and branches of *Saprolegnia ferax*. *Fungal Genetics and Biology* **21**: 238-251.

HYDE, G. J., LANCELLE, S. A., HEPLER, A. R. & HARDHAM, A. R. (1991). Freeze substitution reveals a new model for sporangial cleavage in *Phytophthora*, a result with implications for cytokinesis in other eukaryotes. *Journal of Cell Science* **100**: 735-746.

ICHIDA, A. A. & FULLER, M. S. (1968). Ultrastructure of mitosis in the aquatic fungus *Catenaria anguillulae*. *Mycologia* **60**: 141-155.

ICHITANI, T. & AMEMIYA, J. (1980). *Pythium gracile* isolated from the foci of granular dermatitis in the horse (*Equus cabalus*). *Transactions of the Mycological Society of Japan* **21**: 263-265.

ICHITANI, T., KODAMA, T., HORIMOTO, K. & IKEDA, A. (1986). *Aphanomyces iridis* sp. nov. causing aphanomyces basal rot of dutch iris in Japan. *Annals of the Phytopathological Society of Japan* **52**: 590-598.

IDRIS, M. O. & BALL, S.L. (1984). Inter- and intracontinental sexual compatibility in *Sclerospora graminicola*. *Plant Pathology* **33**: 219-223.

IJDENBERG, P., KUMMERT, J. & LEPOIVRE, P. (1986). Complex populations of minor pathogens associated with roots from barley plants infected by barley yellow mosaic virus. *Parasitica* **42**: 137-144.

IKEGAMI, H. (1992). Proliferation and pathogenicity of *Plasmodiophora brassicae* in infected callus tissue (Studies on the clubroot of cruciferous plants X). *Proceedings of the Kansas Plant Protection Society* **34**: 17-28.

ILLINGWORTH, C. A., ANDREWS, J. H., BIBEAU, C. & SOGIN, M. L. (1991). Phylogenetic placement of *Athelia bombacina, Aureobasidium pullulans*, and *Colletotrichum gloeosporioides* inferred from sequence comparisons of small subunit ribosomal RNAs. *Experimental Mycology* **15**: 65-75.

IMWIDTHAYA, P. (1994). Human pythiosis in Thailand. *Journal of Medical and Veterinary Mycology* **32**: 395-399.

INDOH, H. (1937). Aquatic fungi in Senshunyen. *Hakubutsugaku Zasshi (Magazine of Natural History)* **35**: 295-301.

INDOH, H. (1939). Studies on the Japanese aquatic fungi I. On *Apodachlyella completa* sp. nov. with revision of the Leptomitaceae. *Science Reports of the Tokyo Bunrika Daigaku, Section B* **4**: 43-50.

INDOH, H. (1953). Observations on Japanese aquatic fungi. I. *Apodachlya, Sapromyces*, and *Rhipidium*. *Nagaoa* **3**: 26-35.

INGOLD, C. T. (1949). On the genus *Resticularia*. *Annals of Botany, London, N.S.* **13**: 435-443.

INGOLD, C. T. (1952). *Funaria* rhizoids infected with *Pleotrachelus wildermani* [*sic*]. *Transactions of the British Bryological Society* **2**: 53-54.

INGRAM, D. S. (1969*a*). Growth of *Plasmodiophora brassicae* in host callus. *Journal of General Microbiology* **55**: 9-18.

INGRAM, D. S. (1969*b*). The susceptibility of *Brassica* callus to infection by *Peronospora parasitica*. *Journal of General Microbiology* **58**: 391-401.

INGRAM, D. S. (1977). Applications in plant pathology. *In* H. E. Street (Ed.), *Plant Tissue and Cell Culture*, pp 463-500. Oxford, U.K.: Blackwell Scientific Publications.

INGRAM, D. S. (1980). The establishment of dual cultures of downy mildew fungi and their hosts. *In* D. S. Ingram & J. P. Helgeson (Eds), *Tissue Culture Methods for Plant Pathologists*. Blackwell Scientific Publications, Oxford, UK, pp. 139-144.

INGRAM, D. S. & JOACHIM, I. (1971) The growth of *Peronospora farinosa* f. sp. *betae* and sugar beet callus tissues in dual culture. *Journal of General Microbiology* **69**: 211-220.

INGRAM, D. S. & TOMMERUP, I. C. (1972). The life history of *Plasmodiophora brassicae* Woron. *Proceedings of the Royal Society of London, B* **180**: 103-112.

INOUYE, I. (1993). Flagella and flagellar apparatuses of algae. *In* T. Berner (Ed.), *Ultrastructure of Microalgae*, pp. 99-128. Boca Raton: CRC Press.

IRVING, H. R. & GRANT, B. R. (1984). The effect of calcium on zoospore differentiation in *Phytophthora cinnamomi*. *Journal of General Microbiology* **130**: 1569-1576.

IRVING, H. R., GRIFFITH, J. M. & GRANT, B. R. (1984). Calcium efflux associated with encystment of *Phytophthora palmivora* zoospores. *Cell Calcium* **5**: 487-500.

IRWIN, J. A. G. (1976). Observations on the mode of infection of lucerne roots by *Phytophthora megasperma*. *Australian Journal of Botany* **24**: 447-451.

ISER, J. R., GRIFFITH, J. M., BALSO, A. & GRANT, B. R. (1989). Accelerated ion fluxes during differentiation in zoospores of *Phytophthora palmivora*. *Cell Differentiation nd Development* **26**: 29-38.

ISTVÁNFFI, G. VON (1895). Ueber die Rolle derZellkerne bei der Entwickelung der Pilze. *Berichte der Deutschen Botanischen Gesellschaft* **13**: 452-467.

ITO, S. (1913). Kleine Notizen über parasitischer Pilze Japans. *Botanical Magazine, Tokyo* **27**: 217-223.

ITO, S. (1936). *Mycological Flora of Japan. I. Phycomycetes*. 340 pp. Tokyo, Japan.

ITO, S. & NAGAI, M. (1931). On the rot disease of the seeds and seedlings of the rice plant caused by some aquatic fungi. *Journal of the Hokkaido Imperial University* **32**: 45-69.

ITO, S. & TOKUNAGA, Y. (1933). Studies on the rot-disease of rice seedlings caused by *Pythium* species. *Journal of the Faculty of Agriculture, Hokkaido University* **32**: 201-233.

ITO, S. & TOKUNAGA, Y. (1935*a*). Notae mycologiae Asiae orientalis. I. *Transactions of the Sapporo Natural History Society* **14**: 11-33.

ITO, S. & TOKUNAGA, Y. (1935*b*). Notae mycologiae Asiae orientalis. II. *Transactions of the Sapporo Natural History Society* **32**: 45-69.

ITO, S., ICHINOSE, H., YANAGI, C., TANAKA, S., KAMEYA-IWAKI, M. & KISHI, F. (1999). Identification of an *in planta*-induced mRNA of *Plasmodiophora brassicae*. *Journal of Phytopathology [Phytopathologische Zeitschrift]* **147**: 79-82.

ITO, S. [I.], MAEHARA, T., MARUNO, E., TANAKA, S., KAMEYA-IWAKI, M. & KISHI, F. (1999). Development of a PCR-based assay for the detection of *Plasmodiophora brassicae* in soil. *Journal of Phytopathology [Phytopathologische Zeitschrift]* **147**: 83-88.

ITO, S. [I.], MAEHARA, T., TANAKA, S., KAMEYA-IWAKI, M., YANO, S. & KISHI, F. (1997). Cloning of a single copy DNA sequence unique to *Plasmodiophora brassicae*. *Physiological and Molecular Plant Pathology* **50**: 289-300.

ITO, S. [I.], NAKAMURA, T., MATSUMOTO, T., MAEHARA, T., TANAKA, S., KAMEYA-IWAKI, M. & KISHI, F. (1998). Characterization of DNA sequences cloned from resting spores of *Plasmodiophora brassicae*. *Journal of Phytopathology [Phytopathologische Zeitschrift]* **146**: 143-147.

ITO, T. (1942). On some species of Saprolegniaceae found in Kyoto. *Journal of Japanese Botany* **18**: 121-129.

ITO, T. (1944). Some aquatic species of phycomycetes found in Kyoto. *Journal of Japanese Botany* **20**: 51-60.

IVANOVIĆ, M. (1988*a*). (Research on hosts of *Polymyxa betae* Keskin in Yugoslavia). *Zastita Bilja (Yugoslavia)* **39**: 197-202 [only abstract consulted].

IVANOVIĆ, M. (1988*b*). Proučavanje biologije i kcologije *Polymyxa betae* Kesk. na šećernoj repi u Jugoslaviji. (Study of the biology and ecology of *Polymyxa betae* Kesk. on sugar beet in Yugoslavia). *Zastita Bilja* **39**: 259-273.

IVIMEY COOK, W. R. (1926). The genus *Ligniera* Maire and Tison. *Transactions of the British Mycological Society* **11**: 196-213.

IVIMEY COOK, W. R. (1927). The influence of environment on infection by *Ligniera junci*. *Transactions of the British Mycological Society* **12**: 282-290.

IVIMEY COOK, W. R. (1928*a*). Quelques observations sur le genre *Ligniera*. *Bulletin trimestriel de la Société Mycologique de France* **44**: 105-108.

IVIMEY COOK, W. R. (1928*b*). The methods of nuclear division in the Plasmodiophorales. *Annals of Botany, London* **42**: 347-377.

IVIMEY COOK, W. R. (1928*c*). The inter-relationships of the Archimycetes. *New Phytologist* **27**: 230-260, 298-320.

IVIMEY COOK, W. R. (1929). A preliminary account of a new species of Protista. *Archiv für Protistenkunde* **66**: 285-289.

IVIMEY COOK, W. R. (1931). The life history of *Sorodiscus radicicolus*. *Annales Mycologici* **29**: 313-324.

IVIMEY COOK, W. R. (1932*a*). The life-history of *Cystochytrium radicale*, occurring in the roots of *Veronica beccabunga*. *Transactions of the British Mycological Society* **16**: 246-252.

IVIMEY COOK, W. R. (1932*b*). The parasitic slime moulds. *Hong Kong Naturalist, Supplement* **1**: 29-39.

IVIMEY COOK, W. R. (1933). A monograph of the Plasmodiophorales. *Archiv für Protistenkunde* **80**: 179-254.

IVIMEY COOK, W. R. (1935). The genus *Lagenidium* Schenk, with special reference to *L. rabenhorstii* Zopf and *L. entophytum* Zopf. *Archiv für Protistenkunde* **86**: 57-89.

IVIMEY COOK, W. R. & COLLINS, W. B. (1935). The life-history of *Olpidium majus* sp. nov. with notes on *Olpidium radicale* Schwartz & Cook. *Annales Mycologici* **33**: 72-78 and plate 1.

IVIMEY COOK, W. R. & FORBES, E. J. (1933). Investigations on aquatic fungi. *Nature (London)* **132**: 641-642.

IVIMEY COOK, W. R. & MORGAN, E. (1934). Some observations on the Saprolegniaceae of the soils of Wales. *Journal of Botany* **72**: 345-349.

IVIMEY COOK, W. R. & NICHOLSON, W. H. (1933). A contribution to our knowledge of *Woronina polycystis* Cornu. *Annals of Botany, London* **47**: 851-859.

IVIMEY COOK, W. R. & SCHWARTZ, E. J. (1929). The life history of *Sorosphaera radicale*, sp. nov. *Annals of Botany, London* **43**: 81-98.

IVIMEY COOK, W. R. & SCHWARTZ, E. J. (1930). The life-history, cytology, and method of infection of *Plasmodiophora brassicae* Woron., the cause of finger and toe disease of cabbage and other crucifers. *Philosophical Transactions of the Royal Society of London, Series B* **218**: 283-314.

IYENGAR, M. O. P., RAMAKRISHNAN, K. & SUBRAMANIAN, C. V. (1955). A new species of *Sapromyces* from south India. *Journal of the Indian Botanical Society* **34**: 140-145.

IZARD, C., LACHARPAGNE, J. & SCHILTZ, P. (1964). Behaviour of *Peronospora tabacina* in tissue cultures, and role of the leaf epidermis. *Annales de la Direction des Études et de l'Equipment*, section 2, pp 95-99.

IZUMIKAWA, K., ODA, T., YAMANOI, H. & HATAI, K. (1999). Decline of infectivity of Oomycetes in diluted seawater to the larvae of greasyback shrimp *Metapenaeus ensis*. *Nippon Suisan Gakkaishi* **65**: 661-664.

JACKSON, S. L. & HARDHAM, A. R. (1996). A transient rise in cytoplasmic free calcium is required to induce cytokinesis in zoosporangia of *Phytophthora cinnamomi*. *European Journal of Cell Biology* **69**: 180-188.

JACKSON, S. L. & HARDHAM, A. R. (1998). Dynamic rearrangement of the filamentous actin network occurs during zoosporogenesis and encystment in the oomycete *Phytophthora cinnamomi*. *Fungal Genetics and Biology* **24**: 24-33.

JACKSON, S. L. & HEATH, I. B. (1993). Roles of calcium ions in hyphal tip growth. *Microbiological Reviews* **57**: 367-382.

JACOBS, B. F., KINGSTON, J. D. & JACOBS, L. L. (1999). The origin of grass-dominated ecosystems. *Annals of the Missouri Botanical Garden* **86**: 590-643.

JACOBS, R. P. W. M. (1982). The genus *Apodachlya*. *Anthonie van Leeuwenhoek, Journal of Microbiology and Serology* **48**: 389-400.

JACOBSEN, K. R. (1983). Thraustochytriaceae im Institutium ausgewählter Sandwatten des Weserästuars. *Veröffentlichungen des Instituts für Meeresforschung in Bremerhaven* **19**: 145-175.

JACOBSON, D. J., LEFEBVRE, S. M., OJERIO, R. S., BERWALD, N. & HEIKKINEN, E. (1998). Persistent, systemic, asymptomatic infections of *Albugo candida*, an oomycete parasite, detected in three wild crucifer species. *Canadian Journal of Botany* **76**: 739-750.

JACZEWSKI, A. A. (1931) see A. A. YAČEVSKIE & P. A. YAČEVSKIE (1931)

JACZEWSKI, A. A. & JACZEWSKI, P. A. (1931) see A. A. YAČEVSKIE & P. A. YAČEVSKIE (1931)

JAFFEE, B. A. (1985). Parasitism of *Xiphinema americanum* and *Xiphinema rivesi* by certain zoosporic fungi. *Journal of Nematology* **17**: 500.

JAFFEE, B. A. (1986*a*). Parasitism of *Xiphinema rivesi* and *X. americanum* by zoosporic fungi. *Journal of Nematology* **18**: 87-93.

JAFFEE, B. A. (1986*b*). Infection of *Xiphinema* spp. by zoosporic fungi during Paermann funnel extraction. *Journal of Nematology* **18**: 614.

JAHN, T. L. (1933). On certain parasites of *Phacus* and *Euglena*; *Sphaerita phaci*, sp. nova. *Archiv für Protistenkunde* **79**: 349-355.

JAHN, T. L., LANDMAN, M. D. & FONSECA, J. R. (1964). The mechanisms of locomotion of flagellates. II. Function of the mastigonemes of *Ochromonas*. *Journal of Protozoology* **11**: 291-296.

JAHNKE, K.-D & BAHNWEG, G. (1984). Activity and structural characteristics of steroid and non-steroid lipid growth factors for *Haliphthoros milfordensis*, a fungal parasite of marine crustaceans. *Transactions of the British mycological Society* **83**: 447-454.

JAIN, A. C. (1951). Root rot of grape-vines (*Vitis vinifera*) caused by *Pythium debaryanum* var. *viticolum* new var. *Nagpur Agricultural College Magazine* **25**: 7-9.

JANSSEN, H.-B. & THIMAN, L. (1992). A preliminary study of chemotaxis of zoospores of the nematode-parasitic fungus *Catenaria anguillulae*. *Mycologia* **84**: 109-112.

JARONSKI, S. T. (1990). Oomycete fungi for vector control: current status and prospects. *In* D. E. Pinnock (Ed.). *Proceedings and Abstracts, Vth International Colloquium on Invertebrate Pathology and Microbial Control, Adelaide, Australia*, pp. 94-96. Adelaide, Australia: Department of Entomology, University of Adelaide.

JARONSKI, S. T. & AXTELL, R. C. (1982). Effects of organic water pollution on the infectivity of the fungus *Lagenidium giganteum* (Oomycetes: Lagenidiales) for larvae of *Culex quinquefasciatus* (Diptera: Culicidae): field and laboratory evaluation. *Journal of Medical Entomology* **19**: 255-262.

JARONSKI, S. & AXTELL, R. C. (1983*a*). Effects of temperature of infection, growth and zoosporogenesis of *Lagenidium giganteum*, a fungal pathogen of mosquito larvae. *Mosquito News* **43**: 42-45.

JARONSKI, S. & AXTELL, R. C. (1983*b*). Persistence of the mosquito fungal pathogen *Lagenidium giganteum* (Oomycetes: Lagenidales) after introduction into natural habitats. *Mosquito News* **43**: 332-337.

JARONSKI, S. T. & AXTELL, R. C. (1984*a*). Non-susceptibility of *Chaoborus flavicans* (Chaoboridae) to the mosquito pathogen *Lagenidium giganteum* (Oomycetes). *Mosquito News* **44**: 81-82.

JARONSKI, S. T. & AXTELL, R. C. (1984*b*). Simplified production system for the fungus *Lagenidium giganteum* for operational mosquito control. *Mosquito News* **44**: 377-381.

JARONSKI, S., AXTELL, R. C., FAGAN, S. M. & DOMNAS, A. J. (1983). In vitro production of zoospores by the mosquito pathogen *Lagenidium giganteum* (Oomycetes: Lagenidiales) on solid media. *Journal of Invertebrate Pathology* **41**: 305-309.

JEE, H. J., HO, H. H. & CHO, W. D. (2000). *Pythiogeton zeae* sp. nov. causing root and basal stalk rot of corn in Korea. *Mycologia* **92**: 522-527.

JENNEBORG, L.-H. (1977). *Eurychasma* - Infection of marine algae, changes in algal morphology and taxonomical consequences. *Botanica Marina, Berlin* **20**: 499-507.

JENNINGS, D. H. (1986). Fungal growth in the sea. *In* S. T. Moss (Ed.), *The Biology of Marine Fungi*, pp. 1-10. Cambridge, U.K.: Cambridge University Press.

JEPPS, M. W. (1931). Note on a marine *Labyrinthula*. *Journal of the Marine Biological Association of the United Kingdom* **17**: 833-838.

JI, T. (Thakur Ji) see THAKUR, J.

JIANG, J. & ERWIN, D. C. (1990). Morphology, plasmolysis, and tetrazolium bromide stain as criteria for determining viability of *Phytophthora* oospores. *Mycologia* **82**: 107-113.

JIANG, J., STEPHENSON, L. W., ERWIN, D. C. & LEARY, J. V. (1989). Nuclear changes in *Phytophthora* during oospore maturation and germination. *Mycological Research* **92**: 463-469.

JÍROVEC, O. (1939). *Dermocystidium vejdovskyi* n. sp., ein neuer Parasit des Hechtes, nebst einer Bemerkung über *Dermocystidium daphniae*. *Archiv für Protistenkunde* **92**: 137-146.

JÍROVEC, O. (1955). Parasites of our *Cladocera*. *Ceskoslovenska Parasitologie* **2**: 95-98.

JIMENEZ, B. & LOCKWOOD, J. L. (1982). Germination of oospores of *Phytophthora megasperma* f. sp. *glycinea* in the presence of soil. *Phytopathology* **72**: 662-666.

JOHANNES, H. (1949). Zwei neue Arten der Saprolegniaceen. *Archiv für Mikrobiologie* **14**: 594-601.

JOHANNES, H. (1954). Beiträge zur Vitalfärbung von Pilzmycelien. *Protoplasma* **44**: 165-191.

JOHANNES, H. (1955). Die Gattung *Cladolegnia* Joh. gen. nov. (*Isoachlya* Kauffman) der Saprolegniaceae Nees. *Feddes Repertorium.*. **58**: 209-220.

JOHANNES, H. (1957). Zur Ökologie der Saprolegniaceen. *Mitteilungen der Thüringischen Botanischen Gesellschaft* **1**: 90-111.

JOHANSSON, N., SVENSSON, K. M. & FRIDBERG, G. (1982). Studies on the pathology of ulcerative dermal necrosis (U.D.N.) in Swedish salmon, *Salmo salar* L., and sea trout, *Salmo trutta* L., populations. *Journal of Fish Diseases* **5**: 293-308.

JOHN, D. M. (1994). Biodiversity and conservation: an algal perspective. *The Phycologist* **38**: 3-15.

JOHNS, R. M. (1956). Additions to the phycomycete flora of the Douglas Lake region. III. A new species of *Scherffeliomyces*. *Mycologia* **48**: 433-438.

JOHNSON, P. T. (1983). Diseases caused by viruses, rickettsiae, bacteria and fungi. *In* A. J. Provenzano (Ed.) *The Biology of Crustaceae*. New York, U.S.A.: Academic Press.

JOHNSON, T. (1909). *Chrysophlyctis endobiotica* (Schilb.) and other Chytridiaceae. *Scientific Proceedings of the Royal Dublin College of Science, 1909-1910*: 131-144.

JOHNSON, T. W. (1949). A new species of *Achlya* with coiled oogonial stalks. *Mycologia* **41**: 678-685.

JOHNSON, T. W. (1950*a*). A study of an isolate of *Brevilegnia* from New Caledonia. *Mycologia* **42**: 242-252.

JOHNSON, T. W. (1950*b*). A new *Achlya* from Mackinac Island, Michigan, with notes on other species. *Mycologia* **42**: 391-402.

JOHNSON, T. W. (1951*a*). A new *Mindeniella* from submerged, rosaceous fruits. *American Journal of Botany* **38**: 74-78.

JOHNSON, T. W. (1951*b*). An isolate of *Dictyuchus* connecting the false-net and true-net species. *Mycologia* **43**: 365-372.

JOHNSON, T. W. (1952). Notes on the taxonomy of the *Achlya flagellata, Achlya imperfecta* complex. *Mycologia* **44**: 239-244.

JOHNSON, T. W. (1955*a*). A species of *Olpidiopsis* parasitic in *Achlya glomerata*. *Journal of the Elisha Mitchell Scientific Society* **71**: 58-64.

JOHNSON, T. W. (1955*b*). Inoculation studies with a polysporangiate *Rozella* parasitic in *Dictyuchus anomalus*. *American Journal of Botany* **42**: 119-123.

JOHNSON, T. W. (1955*c*). The asexual stage of *Apodachlya minima*. *Transactions of the British Mycological Society* **38**: 415-418.

JOHNSON, T. W. (1956*a*). Notes on fungi from Mississippi I. Aquatic phycomycetes. *American Midland Naturalist* **55**: 184-193.

JOHNSON, T. W. (1956*b*). *The Genus* Achlya*: Morphology and Taxonomy*. Ann Arbor, Michigan, U.S.A.: University of Michigan Press.

JOHNSON, T. W. (1957*a*). Marine fungi. III. Phycomycetes. *Mycologia* **49**: 392-400.

JOHNSON, T. W. (1957*b*). On the marine phycomycete *Thraustochytrium proliferum*. *Transactions of the British Mycological Society* **40**: 292-294.

JOHNSON, T. W. (1957*c*). Resting spore development in the marine phycomycete *Anisolpidium ectocarpii*. *American Journal of Botany* **44**: 875-878.

JOHNSON, T. W. (1958). A fungus parasite in ova of the barnacle *Chthamalus fragilis denticulata*. *Biological Bulletin of the Marine Biological Laboratory, Woods Hole* **114**: 205-214.

JOHNSON, T. W. (1960). Infection potential and growth of *Lagenidium chthamalophilum*. *American Journal of Botany* **47**: 383-385.

JOHNSON, T. W. (1966*a*). A *Lagenidium* in the marine diatom *Coscinodiscus centralis*. *Mycologia* **58**: 131-135.

JOHNSON, T. W. (1966*b*). *Ectrogella* in marine species of *Licmophora*. *Journal of the Elisha Mitchell Scientific Society* **82**: 25-29.

JOHNSON, T. W. (1966*c*). Fungi in planktonic *Synedra* from brackish waters. *Mycologia* **58**: 373-382.

JOHNSON, T. W. (1966*d*). Chytridiomycetes and Oomycetes in marine phytoplankton. *Nova Hedwigia* **10**: 579-588.

JOHNSON, T. W. (1966*e*). A *Rhizophydium* and its hyperparasite on *Polysiphonia*. *American Journal of Botany* **53**: 945-950.

JOHNSON, T. W. (1966*f*). *Rozella marina* in *Chytridium polysiphoniae* from Icelandic waters. *Mycologia* **58**: 490-494.

JOHNSON, T. W. (1966*g*). Monocentric fungi on species of *Rhizosolenia* from saline waters. *Mycopathologia et Mycologia Applicata* **32**: 281-290.

JOHNSON, T. W. (1968). Aquatic fungi of Iceland: Introduction and preliminary account. *Journal of the Elisha Mitchell Scientific Society* **84**: 179-183.

JOHNSON, T. W. (1972). Aquatic fungi of Iceland: *Olpidiopsis* Cornu. *Journal of the Elisha Mitchell Scientific Society* **88**: 83-91.

JOHNSON, T. W. (1973*a*). Aquatic fungi of Iceland: some keratinophilic species. *Norwegian Journal of Botany* **20**: 293-302.

JOHNSON, T. W. (1973*b*). Aquatic fungi of Iceland: *Achlya racemosa* Hildebrand. *Svensk Botanisk Tidskrift* **67**: 438-444.

JOHNSON, T. W. (1974). Aquatic fungi of Iceland: *Brevilegnia bispora* Couch, and some related forms. *Acta Botanica Islandica* **3**: 3-20.

JOHNSON, T. W. (1976). The Phycomycetes: Morphology and taxonomy. *In* E. B. G. Jones (Ed.), *Recent Advances in Aquatic Mycology*, pp. 193-211. London, U.K.: Paul Elek.

JOHNSON, T. W. (1977*a*). Aquatic fungi of Scandinavia: species of *Brevilegnia*. *Mycologia* **77**: 286-298.

JOHNSON, T. W. (1977*b*). Aquatic fungi of Scandinavia: species of *Olpidiopsis*. *Norwegian Journal of Botany* **24**: 255-268.

JOHNSON, T. W. & BONNER, R. R. (1960). *Lagenidium callinectes* Couch in barnacle ova. *Journal of the Elisha Mitchell Scientific Society* **76**: 147-149.

JOHNSON, T. W. & HOWARD, K. (1968). Aquatic fungi of Iceland: species associated with algae. *Journal of the Elisha Mitchell Scientific Society* **84**: 305-311.

JOHNSON, T. W. & PINSCHMIDT, W. C. (1963). *Leptolegnia marina* Atkins in blue crab ova. *Nova Hedwigia* **5**: 413-418.

JOHNSON, T. W. & SEYMOUR, R. L. (1975). Aquatic fungi of Iceland [:] A new species of *Saprolegnia*. *Svensk Botanisk Tidskrift* **69**: 3-20.

JOHNSON, T. W. & SPARROW, F. K. (1961). *Fungi in Oceans and Estuaries*. Weinheim, Germany: J. Cramer.

JOHNSON, T. W. & SURRATT, J. (1955). Taxonomy of the species of *Isochlya* possessing single oospores. *Mycologia* **47**: 122-129.

JOHNSTON, T. H. (1917). Notes on a *Saprolegnia* epidemic amongst Queensland fish. *Proceedings of the Royal Society of Queensland* **29**: 125.

JOHNSTON, T. H. (1921). The fresh-water epidemic in Queensland rivers. *Proceedings of the Royal Society of Queensland* **33**: 205.

JOKL, M. (1916). Eine neue Meereschytridinee: *Pleotrachelus ectocarpii* nov. spec. *Österreichische Botanische Zeitschrift* **66**: 267-272.

JOKL, M. (1918). *Pythium conidiophorum* nov.sp., ein Parasit von *Spirogyra*. *Österreichische Botanische Zeitschrift* **67**: 33-37.

JONES, D. R. & INGRAM, D. S. (1982). Characterization of isolates derived from single resting spores of *Plasmodiophora brassicae* and studies of their interaction. *Plant Pathology* **31**: 239-246.

JONES, E. B. G. & ALDERMAN, D. J. (1971). *Althornia crouchii* gen. et sp. nov., a marine biflagellate fungus. *Nova Hedwigia* **21**: 381-399.

JONES, E. B. G. & HARRISON, J. L. (1976). Physiology of marine phycomycetes. *In* E. B. G. Jones (Ed.), *Recent Advances in Aquatic Mycology*, pp. 193-211. London, U.K.: Paul Elek.

JONES, F. R. & DRECHSLER, C. (1925). Root rot of peas in the United States caused by *Aphanomyces euteiches* (n. sp.). *Journal of Agricultural Research* **30**: 293-325.

JONES, G. M. & O'DOR, R. K. (1983). Ultrastructural observations on a thraustochytrid fungus parasitic in the gills of squid (*Illex illecebrosus* LeSueur). *Journal of Parasitology* **69**: 903-911.

JONES, G. M., O'DOR, R. K., CORNICK, J. W. & ZWICKER, B. M. (1982). Thraustochytrids as pathogens in squid. *Bulletin of the British Mycological Society, Supplement 1* **16**: 5-6.

JONES, J. B., JONES, J. P., STALL, R. E. & ZITTER, T. A. (1991). Compendium of Tomato Diseases. St. Paul, Minnesota, U.S.A.: American Phytopathological Society.

JONES, P. M. (1926*a*). A mycetozoan found on tobacco plants with mosaic-like symptoms. *Phytopathology* **16**: 67.

JONES, P. M. (1926*b*). Structure and cultural history of a mycetozoan found in tobacco plants with mosaic-like symptoms. *Botanical Gazette* **81**: 446-459.

JONES, P. M. (1928). Morphology and cultural history of *Plasmodiophora brassicae*. *Archiv für Protistenkunde* **62**: 313-327.

JONES, P. M. (1930). *Plasmodiophora lewisii* nov. sp. *Ohio Journal of Science* **30**: 296-299.

JONES, S. W., DONALDSON, S. P. & DEACON, J. W. (1991). Behaviour of zoospores and zoospore cysts in relation to root infection by *Pythium aphanidermatum*. *New Phytologist* **117**: 289-301.

JONES, T. W. (1984). A revised classification of the unilocular Nodosariida and Buliminida foraminifera. *Revista Espanola de Micropaleontologia* **[Spain] 16**: 91-160.

JONES, W. (1955). Downy mildew of *Dactylis glomerata* caused by *Sclerophthora cryophila*. *Canadian Journal of Botany* **33**: 350-354.

JONG, S.-C., DAVIS, E. E., McMANUS, C. & KRICHEVSKY, M. I. (1991). Computer coding of strain features of the saprolegnian fungi. *Mycotaxon* **41**: 407-418.

JORSTADT, I. (1923). Hovorledes skal man bekaempe klumproten paa vore Kaalvekster? *Norsk Havetid* **39**: 126-127.

JUDELSON, H. S. (1997). Expression and inheritance of sexual preference and selfing potential in *Phytophthora infestans*. *Fungal Genetics and Biology* **21**: 188-197.

JUDELSON, H. S. (1997). The genetics and biology of *Phytophthora infestans*: modern approaches to a historical challenge. *Fungal Genetics and Biology* **22**: 65-76.

JUDELSON, H. S., COFFEY, M. D., ARREDONDO, F. R. & TYLER, B. M. (1993). Transformation of the oomycete pathogen *Phytophthora megasperma* f. sp. *glycinea* occurs by DNA integration into single or multiple chromosomes. *Current Genetics* **23**: 211-218.

JUDELSON, H. S., SPIELMAN, L. J. & SHATTOCK, R. C. (1995). Genetic mapping and non-mendelian segregation of mating type loci in the oomycete *Phytophthora infestans*. *Genetics* **141**: 503-512.

JUEL, H, O. (1901). *Pyrrhosorus*, eine neue marine Pilzgattung. *Bihang till Kongliga Svenska Vetenskaps-Akademiens Handlingar* **26** (III, 14): 1-16.

KADER, A. A. (1999). Recommendations for maintaining post-harvest quality: cigar end rot [banana finger: *Verticillium theobromae* and or *Trachaesphaera fructigena*]. hHp://postharvest.ucdavis.edu/Produce/ProduceFacts/Fruit/Banana.

KADLUBROWSKA, J. Z. (1970). *Podochytrium clavatum* and *Aphanomycopsis bacillariacearum* Scherffel, a new species in the Polish flora. *Acta Mycologica, Warszawa* **6**: 55-57.

KAJI, S., KANEMATSU, M., TEZUKA, N., FUSHIMI, H. & HATAI, K. (1991). Effects of formalin bath for *Haliphthoros* infection on ova and larvae of the mangrove crab *Scylla serrata*. *Nihon Suisan Gakkai-Shi* [*Bulletin of the Japanese Society of Fisheries*] **57**: 51-55.

KAMINSKYJ, S. G. W., GARRILL, A. & HEATH, I. B. (1992). The relation between turgor and tip growth in *Saprolegnia ferax*: turgor is necessary, but not sufficient to explain apical extension rates. *Experimental Mycology* **16**: 64-75.

KAMOUN, S., HRABER, P., SOBRAL, B., NUSS, D. & GOVERS, F. (1999). Initial assessment of gene diversity for the oomycete pathogen *Phytophthora infestans* based on expressed sequences. *Fungal Genetics and Biology* **28**: 94-106.

KUAN, T. L. & ERWIN, D. C. (1980). Predisposition effect of water saturation of soil on *Phytophthora* root rot of alfalfa. *Phytopathology* **70**: 981-986.

KANE, J. E. (1964). *Thalassomyces marsupii*, a new species of ellobiopsid parasite on the hyperiid amphipod *Parathemisto gaudichaudii* (Guér.). *New Zealand Journal of Science* **7**: 289-303.

KANE, W. D. (1966). A new species of *Leptolegniella*. *Mycologia* **58**: 905-911.

KANOUSE, B. B. (1926). On the distribution of the water molds, with notes on the occurrence in Michigan of members of the Leptomitaceae and Blastocladiaceae. *Papers of the Michigan Academy of Science, Arts, and Letters* **5**: 105-114

KANOUSE, B. B. (1927*a*). A monographic study of special groups of the water molds I. Blastocladiaceae. *American Journal of Botany* **14**: 287-306.

KANOUSE, B. B. (1927*b*). A monographic study of special groups of the water molds II. Leptomitaceae and Pythiomorphaceae. *American Journal of Botany* **14**: 335-357.

KANOUSE, B. B. (1932). A physiological and morphological study of *Saprolegnia parasitica*. *Mycologia* **24**: 431-452.

KANOUSE, B. B. & HUMPHREY, T. (1927). A new species of the genus *Pythium* in the subgenus *Aphragmium*. *Papers of the Michigan Academy of Science, Arts, and Letters* **8**: 129-138.

KAO, C. W. & KO, W. H. (1986). The role of calcium and microorganisms in supression of cucumber damping-off caused by *Pythium splendens* in Hawaiian soil. *Phytopathology* **76**: 221-225.

KAOSIRI, T., ZENTMYER, G. A. & ERWIN, D. C. (1980). Oospore morphology and germination in the *Phytophthora palmivora* complex frm cacao. *Mycologia* **72**: 888-907.

KLISIEWICZ, J. M. (1970). Factors affecting production and germination of oospores of *Phytophthora drechsleri*. *Phytopathology* **60**: 1738-1742.

KARLING, J. S. (1928). Studies in the Chytridiales. III. A parasitic chytrid causing hypertrophy in *Chara*. *American Journal of Botany* **15**: 485-496.

KARLING, J. S. (1934). A saprophytic species of *Catenaria* isolated from roots of *Panicum variagatum*. *Mycologia* **26**: 528-543.

KARLING, J. S. (1937). The structure, development, identity, and relationship of *Endochytrium*. *American Journal of Botany* **24**: 352-364.

KARLING, J. S. (1938). A further study of *Catenaria*. *American Journal of Botany* **25**: 328-335.

KARLING, J. S. (1939). A new fungus with anteriorly uniciliate zoospores: *Hyphochytrium catenoides*. *American Journal of Botany* **26**: 512-519.

KARLING, J. S. (1941*a*). *Cylindrochytrium johnstonii* gen. nov. et sp. nov., and *Nowakowskiella profusum* sp. nov. *Bulletin of the Torrey Botanical Club* **66**: 281-286.

KARLING, J. S. (1941*b*). Texas chytrids. *Torreya* **41**: 105-108.
KARLING, J. S. (1941*c*). Notes on *Endochytrium* du Plessis. *Mycologia* **33**: 356-359.
KARLING, J. S. (1942*a*). *The Simple Holocarpic Biflagellate Phycomycetes*. New York, U.S.A.: Published by the author.
KARLING, J. S. (1942*b*). *The Plasmodiophorales*. New York, U.S.A.: Published by the author.
KARLING, J. S. (1942*c*). A synopsis of *Rozella* and *Rozellopsis*. *Mycologia* **34**: 193-208.
KARLING, J. S. (1942*d*). Parasitism among the chytrids. *American Journal of Botany* **29**: 24-35.
KARLING, J. S. (1943). The life history of *Anisolpidium ectocarpii* gen. nov. et sp. nov., and a synopsis and classification of other fungi with anteriorly uniflagellate zoospores. *American Journal of Botany* **30**: 637-648.
KARLING, J. S. (1944*a*). *Phagomyxa algarum* n. gen., n. sp., an unusual parasite with plasmodiophoralean and proteomyxean characteristics. *American Journal of Botany* **31**: 38-52.
KARLING, J. S. (1944*b*). Brazilian anisochytrids. *American Journal of Botany* **31**: 391-397.
KARLING, J. S. (1944*c*). New lagenidiaceous parasites of rotifers from Brazil. *Lloydia (Cincinnati)* **7**: 328-342.
KARLING, J. S. (1944*d*). Brazilian chytrids IV. Species of *Rozella*. *Mycologia* **36**: 638-647.
KARLING, J. S. (1945). *Rhizidiomyces hirsutus* sp. nov., a hairy anisochytrid from Brazil. *Bulletin of the Torrey Botanical Club* **72**: 47-51.
KARLING, J. S. (1946*a*). Two new chytrid parasites of *Chytriomyces*. *Mycologia* **38**: 103-109.
KARLING, J. S. (1946*b*). Brazilian chytrids. VIII. Additional parasites of rotifers and nematodes. *Lloydia (Cincinnati)* **9**: 1-12.
KARLING, J. S. (1947*a*). Keratinophilic chytrids. II. *Phlyctorhiza variabilis* n. sp. *American Journal of Botany* **34**: 27-32.
KARLING, J. S. (1947*b*). *Lagenidium humanum*, a saprophyte isolated on dead human skin. *Mycologia* **39**: 224-230.
KARLING, J. S. (1948*a*). Chytridiosis of scale insects. *American Journal of Botany* **35**: 246-254.
KARLING, J. S. (1948*b*). An *Olpidium* parasite of *Allomyces*. *American Journal of Botany* **35**: 503-510.
KARLING, J. S. (1949). A new *Olpidiopsis* parasite of *Karlingia rosea* from Maryland. *Mycologia* **41**: 270-276.
KARLING, J. S. (1952). *Sommerstorffia spinosa* Arnaudow. *Mycologia* **44**: 387-412.
KARLING, J. S. (1953*a*). *Micromyces* and *Synchytrium*. *Mycologia* **45**: 276-287.
KARLING, J. S. (1953*b*). *Synchytrium urticae*. *Mycologia* **45**: 613-615.
KARLING, J. S. (1964*a*). Indian Anisochytrids. *Sydowia* **17**: 193-196.
KARLING, J. S. (1964*b*). *Woronina pythii* Goldie-Smith in India. *Bulletin of the Torrey Botanical Club* **91**: 224-227.
KARLING, J. S. (1964*c*). Indian chytrids. II. *Olpidium indicum* sp. nov. *Sydowia* **17**: 302-307.
KARLING, J. S. (1964*d*). Synchytrium. New York, U.S.A.: Academic Press.
KARLING, J. S. (1965). Some zoosporic fungi of New Zealand. I. *Sydowia* **19**: 213-226.
KARLING, J. S. (1966). The chytrids of India with a supplement of other zoosporic fungi. *Sydowia, Beihefte* **6**: 1-125.
KARLING, J. S. (1967*a*). Some zoosporic fungi of New Zealand. II. Synchytriaceae. *Sydowia* **20**: 51-66.
KARLING, J. S. (1967*b*). Some zoosporic fungi of New Zealand. IX. Hyphochytriales or Anisochytridiales. *Sydowia* **20**: 137-142.
KARLING, J. S. (1967*c*). Some zoosporic fungi of New Zealand. X. Blastocladiales. *Sydowia* **20**: 144-150.
KARLING, J. S. (1967*d*). Some zoosporic fungi of New Zealand. XI. Plasmodiophorales. *Sydowia* **20**: 151-156.
KARLING, J. S. (1968*a*). Zoosporic fungi of New Zealand. XII. Olpidiopsidaceae, Sirolpidiaceae, and Lagenidiaceae. *Sydowia* **20**: 190-198.
KARLING, J. S. (1968*b*). Zoosporic fungi of Oceania. I. *Journal of the Elisha Mitchell Scientific Society* **84**: 166-178.
KARLING, J. S. (1968*c*). Zoosporic fungi of Oceania. II. Two saprophytic species of *Aphanomycopsis*. *Mycologia* **60**: 271-284.
KARLING, J. S. (1968*d*). Zoosporic fungi of Oceania. III. Monocentric chytrids. *Archiv für Mikrobiologie* **61**: 112-127.
KARLING, J. S. (1968*e*). *The Plasmodiophorales, Including a Complete Host Index, Bibliography, and a Description of Diseases Caused by Species of this Order*. New York, U.S.A.: Haffner.
KARLING, J. S. (1968*f*). Zoosporic fungi of Oceania. V. Cladochytriaceae, Catenariaceae and Blastocladiaceae. *Nova Hedwigia* **15**: 191-201.
KARLING, J. S. (1969). Zoosporic fungi of Oceania. VI. Species with biflagellate zoospores. *Mycopathologia et Mycologia Applicata* **37**: 161-170.
KARLING, J. S. (1970). Some zoosporic fungi of New Zealand XIV. Additional species. *Archiv für Mikrobiologie* **70**: 266-287.
KARLING, J. S. (1972). The present status of *Sphaerita*, *Pseudosphaerita*, *Morella* and *Nucleophaga*. *Bulletin of the Torrey Botanical Club* **99**: 223-228.
KARLING, J. S. (1976*a*). Additional zoosporic fungi of Africa. *Nova Hedwigia* **27**: 747-771.
KARLING, J. S. (1976*b*). Some zoosporic fungi of Florida. *Nova Hedwigia* **28**: 209-229.
KARLING, J. S. (1977). *Chytridiomycetarum Iconographia*. Vaduz, Lichtenstein: J. Cramer.
KARLING, J. S. (1979*a*). The anisochytrid genus *Elina*. *Mycologia* **71**: 829-830.
KARLING, J. S. (1979*b*). Ross Biological Reserve Fungi I. *Olpidiopsis brevispinosa* Whiffen. *Nova Hedwigia* **31**: 257-264.
KARLING, J. S. (1981*a*). *Predominantly Holocarpic and Eucarpic Simple Biflagellate Phycomycetes*. Vaduz, Lichtenstein: J. Cramer.
KARLING, J. S. (1981*b*). *Rhizidiomyces bullatus* (Sparrow) comb. nov., and other anisochytrids from Venezuela. *Nova Hedwigia* **34**: 669-678.
KARLING, J. S. (1981*c*). *Woronina leptolegniae* n. sp. a plasmodiophorid parasite of *Leptolegnia*. *Nova Hedwigia* **35**: 17-24.
KARLING, J. S. (1981*d*). Some zoosporic fungi from soils of Sri Lanka. *Nova Hedwigia* **35**: 107-116.
KARLING, J. S. (1981*e*). *Rozella irregularis*(?) Butler and some other zosporic fungi from China. *Nova Hedwigia* **34**: 633-644.
KARLING, J. S. (1984). Zoosporic soil fungi of La Guajira, Columbia. *Nova Hedwigia* **40**: 329-340.
KARLING, J. S. (1985). Ross Biological Reserve aquatic fungi II. *Olpidiopsis endogena* n. sp. *Nova Hedwigia* **43**: 7-10.
KARLING, J. S. (1987*a*). Two new species of *Leptolegniella* and other zoosporic fungi in Thailand soils. *Nova Hedwigia* **45**: 433-443.
KARLING, J. S. (1987*b*). Ross Biological Reserve aquatic fungi III. Additional species. *Nova Hedwigia* **45**: 529-535.
KARLING, J. S. (1988). Additional zoosporic fungi from Australian soils. *Nova Hedwigia* **46**: 239-249.
KASANOWSKY, V. (1911). *Aphanomyces laevis* de Bary. I. Entwicklung der Sexualorgane und Befruchtung. *Berichte der Deutschen Botanischen Gesellschaft* **29**: 210-228.
KASSANIS, B. & MACFARLANE, I (1964). Transmission of tobacco necrosis virus to tobacco callus tissues by zoospores of *Olpidium brassicae*. *Nature (London)* **201**: 218-219.
KASTIRR, U. (1991). (Infection of transformed sugarbeet root cultures by *Polymyxa betae* Keskin). *Archiv für Phytopathologie und Pflanzenschutz* **27**: 333-334 [only abstract consulted].
KASTIRR, U., PFEILSTETTER, E. & BURGERMEISTER, W. (1994). Virus content and virulence of *Polymyxa betae* Keskin isolates obtained from different regions in Europe. *Journal of Phytopathology [Phytopathologische Zeitschrift]* **141**: 369-374.
KATO, S., COE, R., NEW, L. & DICK, M. W. (1990). Sensitivities of various Oomycetes to hymexazol and metalaxyl. *Journal of General Microbiology* **136**: 2127-2134.
KATSAROS, C. & GALATIS, E. (1986). Ultrastructural studies on zoosporogenesis of *Halopteris filicina* (Sphacelariales, Phaeophyta). *Phycologia* **25**: 358-370.
KAUFFMAN, C. H. (1906). Unreported Michigan fungi from Petoskey, Detroit, and Ann Arbor for 1905. *Annual Report of the Michigan Academy for Science, Arts, and Letters* **8**: 26-37.
KAUFFMAN, C. H. (1908). A contribution to the physiology of the Saprolegniaceae, with special reference to variations of sexual organs. *Annals of Botany, London* **22**: 361-387.
KAUFFMAN, C. H. (1916). Unreported Michigan fungi for 1911, 1912, 1913, and 1914. *Annual Report of the Michigan Academy for Science, Arts, and Letters* **17**: 194-216.
KAUFFMAN, C. H. (1921). *Isoachlya*, a new genus of the Saprolegniaceae. *American Journal of Botany* **8**: 231-237.
KAWAGUCHI, S. (1942). Investigation on reef-corals. 3. *Kagaku Nauyo (Sci. South Sea)* **5**: 95-106. [In Japanese.]

KAZAMA, F. Y. (1972*a*). Ultrastructure of *Thraustochytrium* sp. zoospores. I. Kinetosome. *Archiv für Mikrobiologie* **83**: 179-188.

KAZAMA, F. Y. (1972*b*). Ultrastructure of *Thraustochytrium* sp. zoospores. II. Striated inclusions. *Journal of Ultrastructural Research* **41**: 60-66.

KAZAMA, F. Y. (1973). Ultrastructure of *Thraustochytrium* sp. zoospores III. Cytolysomes and acid phosphatase distribution. *Archiv für Mikrobiologie* **89**: 95-104.

KAZAMA, F. Y. (1974*a*). Ultrastructure of *Thraustochytrium* sp. zoospores. IV. External morphology with notes on the zoospores of *Schizochytrium* sp. *Mycologia* **66**: 272-280.

KAZAMA, F. Y. (1974*b*). The ultrastructure of nuclear division in *Thraustochytrium* sp. *Protoplasma* **82**: 155-175.

KAZAMA, F. Y. (1975). Cytoplasmic cleavage during zoosporogenesis in *Thraustochytrium* sp.: Ultrastructure and the effects of colchicine and D_2O. *Journal of Cell Science* **17**: 155-170.

KAZAMA, F. Y. (1980). The zoospore of *Schizochytrium aggregatum*. *Canadian Journal of Botany* **58**: 2434-2446.

KAZAMA, F. Y. & SCHORNSTEIN, K. L. (1972). Herpes-like virus particles associated with a fungus. *Science, New York* **177**: 696-697.

KAZAMA, F. Y. & SCHORNSTEIN, K. L. (1973). Ultrastructure of a fungus herpes-type virus. *Virology* **52**: 478-487.

KAZAMA, F. Y., ZACHARY, A. L., & SCHORNSTEIN, K. L. (1975). Observations on *Thraustochytrium* sp.: Development and behavior in culture. *Canadian Journal of Botany* **53**: 360-374.

KEEN, N. T., REDDY, M. N. & WILLIAMS, P. H. (1969). Isolation and properties of *Plasmodiophora brassicae* plasmodia from infected crucifer tissues and from tissue culture callus. *Phytopathology* **59**: 637-644.

KEEN, N. T. & YOSHIKAWA, M. (1983). Physiology of disease and the nature of resistance to *Phytophthora*. *In* D. C. Erwin, S. Bartnicky-Garcia & P. H. Tsao (Eds), Phytophthora, *Its Biology, Taxonomy Ecology and Pathology*. American Phytopathological Society, St Paul, Minnesota, USA, pp. 279-287.

KELMAN, A. (1947). A rare species of *Isoachlya* found in North Carolina. *Journal of the Elisha Mitchell Scientific Society* **63**: 207-211.

KENDRICK, J. B. (1927*a*). Radish black-root caused by *Aphanomyces raphani* n. sp. *Phytopathology* **17**: 43 [abstract].

KENDRICK, J. B. (1927*b*). The black root-rot disease of radish. *Purdue University Agriculture Experiment Station Bulletin, No. 311*: 1-32.

KENNETH, R. G. (1963). Downy mildew of rye grass in Israel caused by *Sclerophthora lolii* sp. nov. *Israel Journal of Botany* **12**: 139.

KENNETH, R. G. (1979). Downy mildews of grasses. *Phytoparasitica* **7**: 50.

KENNETH, R. G. (1981). Downy mildews of graminaceous crops. *In* D. M. Spencer (Ed.), *The Downy Mildews*, pp. 367-394. Academic Press, London, U.K.

KENNETH, R. G., KOLTIN, Y. & WAHL, I. (1964). Barley diseases newly found in Israel. *Bulletin of the Torrey Botanical Club* **91**: 185-193.

KENNETH, R. G. & KRANZ, J. (1973). *Plasmopara penniseti* sp. nov., a downy mildew of pearl millet in Ethiopia. *Transactions of the British Mycological Society* **60**: 590-593.

KERRY, B. R. (1974). A fungus associated with young females of the cereal cyst-nematode, *Heterodera avenae*. *Nematologica* **20**: 259-260.

KERRY, B. R. (1975). Fungi and the decrease of cereal cyst-nematode populations in cereal monoculture. *Bulletin: Organisation Europeenne et Mediterraneenne pour la Protection des Plantes* **5**: 353-361.

KERRY, B. R. (1980). Biocontrol: Fungal parasites of female cyst nematodes. *Journal of Nematology* **12**: 253-259.

KERRY, B. R. (1981). Progress in the use of biocontrol agents for control of nematodes. *In* G. C. Papavizas (Ed.), *Biological Control in Crop Production*, pp.79-90. Totowa, New Jersey, U.S.A.: Allanheld Osmun & Co.

KERRY, B. R. (1988). Fungal parasites of cyst nematodes. *Agriculture, Ecosystems & Environment* **24**: 293-306.

KERRY, B. R. & CRUMP, D. H. (1977). Observations on fungal parasities of females and eggs of the cereal cyst-nematode, *Heterodera avenae* and other cyst nematodes. *Nematologica* **23**: 193-201.

KERRY, B. R. & CRUMP, D. H. (1980). Two fungi parasitic on females of cyst-nematodes (*Heterodera* spp.). *Transactions of the British Mycological Society* **74**: 119-125.

KERRY, B. R., CRUMP, D. H. & MULLEN, L. A. (1980). Parasitic fungi, soil moisture and multiplication of the cereal cyst nematode, *Heterodera avenae*. *Nematologica* **26**: 57-68.

KERRY, B. R., CRUMP, D. H. & MULLEN, L. A. (1982). Studies of the cereal cyst nematode, *Heterodera avenae*, under continuous cereals, 1975-1978. II. Fungal parasitism of nematode females and eggs. *Annals of Applied Biology* **100**: 489-499.

KERWIN, J. L. (1987). Fatty acids and fungal development: structure-activity relationships. *In* G. Fuller & W. D. Nes (Eds) *Ecology and Metabolism of Plant Lipids*, pp. 329-342. Washington, D.C., U.S.A.: American Chemical Society.

KERWIN, J. L. (1995). Glycerophospholipid and eicosanoid regulation of morphogenesis of the mosquito pathogen *Lagenidium giganteum*. *Abstracts of Papers: American Chemical Society* **209**: ANYL 124 [abstract].

KERWIN, J. L., DRITZ, D. A. & WASHINO, R. K. (1988). Nonmammalian safety tests for *Lagenidium giganteum* (Oomycetes: Lagenidiales). *Journal of Economic Entomology* **81**: 158-171.

KERWIN, J. L., DRITZ, D. A. & WASHINO, R. K. (1990). Confirmation of the safety of *Lagenidium giganteum* (Oomycetes: Lagenidiales) to mammals. *Journal of Economic Entomology* **83**: 374-376.

KERWIN, J. L., DRITZ, D. A. & WASHINO, R. K. (1994). Pilot scale production and application in wildlife ponds of *Lagenidium giganteum* (Oomycetes: Lagenidiales). *Journal of the American Mosquito Association* **10**: 451-455.

KERWIN, J. L. & DUDDLES, N. D. (1989). Reassessment of the role of phospholipids in sexual reproduction by sterol-auxotrophic fungi. *Journal of Bacteriology* **171**: 3831-3839.

KERWIN, J. L., DUDDLES, N. D. & WASHINO, R. K. (1991). Effects of exogenous phospholipids on lipid composition and sporulation by three strains of *Lagenidium giganteum*. *Journal of Invertebrate Pathology* **58**: 408-414.

KERWIN, J. L., GRANT, D. F. & BERBEE, M. L. (1991). Specific induction of encystment of *Lagenidium giganteum* by concavalin A and derivatives of chitin and chitosan. *Protoplasma* **161**: 43-51.

KERWIN, J. L., JOHNSON, L. M., WHISLER, H. C. & TUININGA, A. R. (1992). Infection and morphogenesis of *Pythium marinum* in species of *Porphyra* and other red algae. *Canadian Journal of Botany* **70**: 1017-1024.

KERWIN, J. L., McKICHAN, J. K., SEMON, M. J., WIENS, A. M., DeROSE, C. C. & TORVIK, J. J. (1996). Sterol and steryl ester regulation of Phospholipase A2 from the mosquito parasite *Lagenidium giganteum*. *Lipids* **31**: 1179-1188.

KERWIN, J. L., SIMMONS, C. A. & WASHINO, R. K. (1986*a*). Eicosanoid regulation of oosporogenesis by *Lagenidium giganteum*. *Symposium on Lipoxygenase Products and Polyunsaturated Fatty Acids, October 1985. (Limoges, France)* **23**: 173-178.

KERWIN, J. L., SIMMONS, C. A. & WASHINO, R. K. (1986*b*). Oosporogenesis by *Lagenidium giganteum* in liquid culture. *Journal of Invertebrate Pathology* **47**: 258-270.

KERWIN, J. L., TUININGA, A. R., WIENS, A. M., WANG, J. C., TORVIK, J. J., CONRATH, M. L. & MacKICHAN, J. K. (1995). Isoprenoid-mediated changes in the glycerophospholipid molecular species of the sterol auxotrophic fungus *Lagenidium giganteum*. *Microbiology (Reading)* **141**: 399-410.

KERWIN, J. L. & WASHINO, R. K. (1983). Sterol induction of sexual reproduction in *Lagenidium giganteum*. *Experimental Mycology* **7**: 109-115.

KERWIN, J. L. & WASHINO, R. K. (1984). Cyclic nucleotide regulation of oosporogenesis by *Lagenidium giganteum* and related fungi. *Experimental Mycology* **8**: 215-224.

KERWIN, J. L. & WASHINO, R. K. (1985). Efficacy of *Romanomeris culicivorax* and *Lagenidium giganteum* for mosquito control: strategies for use of biological control agents in rice fields of the Central Valley of California. *In* C. D. Grant, J. C. Combs, R. L. Coykendall, E. E. Lusk & R. K. Washino (Eds), *Proceedings and Papers of the Fifty-second Annual Conference of the California Mosquito and Vector Control Association*, pp. 86-92. Sacramento, U.S.A.: California Mosquito and Vector Control Association Press.

KERWIN, J. L. & WASHINO, R. K. (1986*a*). Regulation of oosporogenesis by *Lagenidium giganteum*: promotion of sexual reproduction by unsaturated fatty-acids and sterol availability. *Canadian Journal of Microbiology* **32**: 294-300.

KERWIN, J. L. & WASHINO, R. K. (1986*b*). Ground and aerial application of sexual and asexual stages of *Lagenidium giganteum* (Oomycetes: Lagenidiales) for mosquito control. *Journal of the American Mosquito Control Association* **2**: 182-189.

KERWIN, J. L. & WASHINO, R. K. (1986*c*). Oosporogenesis by *Lagenidium giganteum*: induction and maturation are regulated by calcium and calmodulin. *Canadian Journal of Microbiology* **32**: 663-672.

KERWIN, J. L. & WASHINO, R. K. (1987*a*). Ground and aerial application of the asexual stage of *Lagenidium giganteum* for control of mosquitoes associated with rice cultures in the central valley of California USA. *Journal of the American Mosquito Control Association* **3**: 59-64.

KERWIN, J. L. & WASHINO, R. K. (1987*b*). Artificial culture of the sexual stage of *Lagenidium giganteum* US patent-4687744. August 18 1987. *Official Gazette of the US Patent and Trademark Office Patents, 1081*: 1477.

KERWIN, J. L. & WASHINO, R. K. (1988*a*). Field evaluation of *Lagenidium giganteum* (Oomycetes: Lagenidiales) and description of a natural epizootic involving a new isolate of the fungus. *Journal of Medical Entomology* **25**: 452-460.

KERWIN, J. L. & WASHINO, R. K. (1988*b*). Isolation of a new strain of *Lagenidium giganteum* and implications for control of floodwater and other rapidly developing mosquito species. *Proceedings and Papers of the Fiftysixth Annual Conference of the California Mosquito and Vector Control Association*, pp. 182-183. Sacramento, U.S.A.: California Mosquito and Vector Control Association Press.

KERWIN, J. L., WASHINO, R. K., GRANT, C. D., COMBS, J. C., COYKENDALL, R. L. & LUSK, E. E. (1983*a*). Field evaluation of *Bacillus thuringiensis* var. *israelensis*, *Lagenidium giganteum*, and *Romanomermis culicivorax* in California rice fields. *In* C. D. Grant, J. C. Combs, R. L. Coykendall, E. E. Lusk & R. K. Washino (Eds), *Proceedings and Papers of the Fifty-first Annual Conference of the California Mosquito and Vector Control Association*, pp. 19-25. Sacramento, U.S.A.: California Mosquito and Vector Control Association Press.

KERWIN, J. L., WASHINO, R. K., GRANT, C. D., LUSK, E. E. & COYKENDALL, R. L. (1983*b*). Artificial culture of the sexual and asexual stages of *Lagenidium giganteum*. *In* C. D. Grant, R. K. Washino, E. E. Lusk & R. L. Coykendall (Eds), *Proceedings and Papers of the Fiftieth Annual Conference of the California Mosquito and Vector Control Association*, pp. 43. Sacramento, U.S.A.: California Mosquito and Vector Control Association Press.

KESKIN, B. (1964). *Polymyxa betae* n. sp., ein Parasit in den Wurzeln von *Beta vulgaris* Tournefort, besonders während der Jugendentwicklung der Zuckerrübe. *Archiv für Mikrobiologie* **49**: 348-374.

KESKIN, B. (1971). Beitrag zur Protomitose bei *Polymyxa betae* Keskin. *Archiv für Mikrobiologie* **77**: 344-348.

KESKIN, B. & FUCHS, W. H. (1969). Der Infektions-vorgang bei *Polymyxa betae*. *Archiv für Mikrobiologie* **68**: 218-226.

KEVORKIAN, A. G. (1934). The structure and development of a new aquatic phycomycete. *Mycologia* **26**: 145-152.

KEVORKIAN, A. G. (1935). Studies in the Leptomitaceae. II. Cytology of *Apodachlya brachynema* and *Sapromyces reinschii*. *Mycologia* **27**: 274-285.

KHAN, S. R. (1976). Ultrastructural changes in maturing sporangia of *Albugo candida*. *Annals of Botany, London* **40**: 1285-1292.

KHEW, K. L. & ZENTMYER, G. A. (1973). Chemotactic response of zoospores of five species of *Phytophthora*. *Phytopathology* **63**: 1511-1517.

KHEW, K. L. & ZENTMYER, G. A. (1974). Electrotactic response of zoospores of seven species of *Phytophthora*. *Phytopathology* **64**: 500-507.

KHULBE, R. D. (1980*a*). Occurrence of parasitic watermolds in some lakes of Naini Tal, India. *Hydrobiologia* **70**: 119-122.

KHULBE, R. D. (1980*b*). Occurrence of watermolds in some lakes of Naini Tal, Kumaun Hill, India. *Hydrobiologia* **74**: 77-80.

KHULBE, R. D. (1981). Distribution of aquatic fungi in relation to some ecological factors. *Geobios (Jodhpur)* **8**: 214-216.

KHULBE, R. D. (1983*a*). *Scoliolegnia centrica* sp. nov. - a new watermold from India. *Sydowia, Annales Mycologici* **36**: 150-153.

KHULBE, R. D. (1983*b*). *Protoachlya oryzae* sp. nov. from Nainital. *Current Science, Bangalore, India* **52**: 428.

KHULBE, R. D. & SATI, S. C. (1979). A new record of watermolds from Naini Tal. *Geobios (Jodhpur)* **6**: 229.

KHULBE, R. D. & SATI, S. C. (1983). A new species of *Dictyuchus* and key to the other species. *Bibliotheca Mycologica* **91**: 609-617.

KHULBE, R. D. & VERMA, B. L. (1987). *Saprolegnia bhargavii* sp. nov. from India. *Nova Hedwigia* **44**: 215-217.

KIMURA, T., YOSHIMIZU, M., TAKIJIMA, K. & EZURA, Y. (1980). Disinfection of hatchery water supply by ultraviolet (U.V.) irradiation - II. U.V. susceptibility of some fish pathogenic fungi. *Fish Pathology [Gyobyo Kenkyu]* **14**: 133-137.

KING, C. A. (1903). Observations on the cytology of *Araiospora pulchra* Thaxter. *Proceedings of the Boston Society of Natural History* **31**: 211-245.

KING, T. H. & BISSONETTE, H. (1954). Physiologic specialization in *Aphanomyces euteiches*. *Phytopathology* **44**: 495 [abstract].

KIRAN, U. & DAYAL, R. (1983). Fresh water molds from Varanasi India 6. Some Oomycetes. *Proceedings of the National Academy of Sciences, India, Section B (Biological Sciences)* **53**: 50-52.

KIRAN, U. & DAYAL, R. (1992*a*). A new species of *Woronina* from decomposing plant litter in ponds of Varanasi. *Indian Phytopathology* **45**: 176-178.

KIRAN, U. & DAYAL, R. (1992*b*). Chytrids from leaf litter in ponds of Varanasi VIII. Genus *Olpidium* (Braun) Rabenhorst - new species and records. *Proceedings of the National Academy of Sciences, India, Section B (Biological Sciences)* **62**: 295-298.

KIRBY, H. (1941). Organisms living on and in protozoa. *In* G.N. Calkins & F. M. Summars (Eds), *Protozoa in Biological Research*, pp. 1009-1113. New York, U.S.A.: Columbia University Press.

KIRK, P. M, & ANSELL, A. E. (1992). *Authors of fungal names. A list of authors of scientific names of fungi, with recommended standard forms of their names including abbreviations. Index of Fungi Supplement. 1992*: i-viii + 1-95. Kew, Surrey, U.K.: CAB International.

KIRK, T. W. (1906). Appendix X. - Divisions of biology and horticulture, and publications. *New Zealand Department of Agriculture. The Fourteenth Report of the Department of Agriculture, 1906* **14**: 337-462.

KITANCHAROEN, N. & HATAI, K. (1995). A marine oomycete *Atkinsiella panulirata* sp. nov. from philozoma of spiny lobster, *Panulirus japonicus*. *Mycoscience* **36**: 97-104.

KITANCHAROEN, N. & HATAI, K. (1997). *Aphanomyces frigidophilus* sp. nov. from eggs of Japanese char, *Salvelinus leucomaenis*. *Mycoscience* **38**: 135-140.

KITANCHAROEN, N., HATAI, K., OGIHARA, R. & AYE, D. N. N. (1995). A new record of *Achlya klebsiana* from snakehead, *Channa striatus*, with fungal infection in Myanmar. *Mycoscience* **36**: 235-238.

KITANCHAROEN, N., NAKAMURA, K., WADA, S. & HATAI, K. (1994). *Atkinsiella awabi* sp. nov. isolated from stocked abalone, *Haliotis sieboldii*. *Mycoscience* **35**: 265-270.

KITANCHAROEN, N., ONO, A., YAMAMOTO,A. & HATAI, K. (1997). The fungistatic effect of NaCl on rainbow trout egg saprolegniasis. *Fish Pathology (Gyobyo Kenkyu)* **32**: 159-162.

KITANCHAROEN, N., YAMAMOTO, A. & HATAI, K. (1997). Fungicidal effect of hydrogen peroxide on fungal infection of rainbow trout eggs. *Mycoscience* **38**: 375-378.

KITANCHAROEN, N., YAMAMOTO, A. & HATAI, K. (1998). Effects of sodium chloride, hydrogen peroxide and malachite green on fungal infection in rainbow trout eggs. *Biocontrol Science* **3**: 113-115.

KITANCHAROEN, N., YUASA, K. & HATAI, K. (1995). Morphological aspects of *Saprolegnia diclina* type I isolated from pejerrey, *Odonthetes bonariensis*. *Mycoscience* **36**: 365-368.

KJØLLER, R. & ROSENDAHL, S. (1996). The presence of the arbuscular mycorrhizal fungus *Glomus intraradices* influences enzymatic activity of the root pathogen *Aphanomyces euteiches* in pea roots. *Mycorrhiza* **6**: 487-491.

KJØLLER, R. & ROSENDAHL, S. (1997). Poyacrylamide gel electrophoresis (PAGE) and densitometric measurement of enzymatic activity of the pea root pathogen *Aphanomyces euteiches* in pea roots. *Journal of Phytopathology* **145**: 253-256.

KJØLLER, R. & ROSENDAHL, S. (1998). Enzymatic activity of the mycelium compared with oospore development during infection of pea roots by *Aphanomyces euteiches*. *Phytopathology* **88**: 992-996.

KLASSEN, G. R., BALCERZAK, M., BEDARD, J., SCHURKO, A. & COCK, A. W. A. M. de (1997). Identification of *Pythium* species by analysis of the amplified rDNA intergenic spacer. *Inoculum, Newsletter of the Mycological Society of America* **48**: Abstracts pages 19-20 [abstract].

KLASSEN, G. R., BALCERZAK, M. & COCK, A. W. A. M. de (1996). 5S ribosomal RNA gene spacers as species-specific probes for eight species of *Pythium*. *Phytopathology* **86**: 581-587.

KLASSEN, G. R. & BUCHKO, J. (1990). Subrepeat structure of the intergenic region in the ribosomal DNA of the oomycetous fungus *Pythium ultimum*. *Current Genetics* **18**: 203-205.

KLASSEN, G. R., McNABB, S. A. & DICK, M. W. (1987). Comparison of physical maps of ribosomal DNA repeating units in *Pythium, Phytophthora* and *Apodachlya*. *Journal of General Microbiology* **133**: 2953-2959.

KLASSEN, G. R., McNABB, S. A. WHITMORE, E., GRAY, M. & DICK, M. W. (1987) [abstract]. Restriction sites in ribosomal DNA as characters for the evolutionary study of zoosporic fungi. *Genome* **30** (supplement 1: 16th International Congress of Genetics, Toronto, Abstracts of Papers): 394.

KLEBAHN, H. (1892). Studien über Zygoten. II. *Jahrbücher für Wissenschaftliche Botanik* **24**: 235-267.

KLEBAHN, H. (1905). Eine neue Pilzkrankheit der Syringen. *Centralblatt für Bakteriologie* **15**: 335-336.

KLEBS, G. (1898). Zur Physiologie der Fortpflanzung einiger Pilze. I. *Jahrbücher für wissenschaftliche Botanik* **32**: 1-70.

KLEBS, G. (1899). Zur Physiologie der Fortpflanzung einiger Pilze. II. *Saprolegnia mixta*. *Jahrbücher für wissenschaftliche Botanik* **33**: 513-593.

KLEBS, G. (1900). Zur Physiologie der Fortpflanzung einiger Pilze. III. Allgemeine Betrachtungen. *Jahrbücher für wissenschaftliche Botanik* **35**: 80-203.

KLEINSCHUSTER, S. J. & SWINK, S. L. (1993). A simple method for the in vitro culture of *Perkinsus marinus*. *Nautilus* **107**: 76-78.

KLENK, H.-P., PALM, P. & ZILLIG, W. (1994). DNA-dependent RNA polymerases as phylogenetic marker molecules. *Systematic and Applied Microbiology* **16**: 638-647.

KLENK, H.-P., ZILLIG, W., LANZENDÖRFER, M., GRAMPP, B. & PALM, P. (1995). Location of protist lineages in a phylogenetic tree inferred from sequences of DNA-dependent RNA polymerases. *Archiv für Protistenkunde* **145**: 221-230.

KLIE, H. & MACH, F. (1968). Licht- und elektronmikroscopische Untersuchungen über die Wirkung von *Labyrinthula*-Enzymen auf Bakterien und Hefezellen. *Zeitschrift für allgemeine Mikrobiologie* **8**: 385-395.

KLIMCZAK, L. J. & PRELL, H. H. (1984). Isolation and characterization of mitochondrial DNA of the oomycetous fungus *Phytophthora infestans*. *Current Genetics* **8**: 323-326.

KNIEP, H. (1928). *Die Sexualität der niederen Pflanzen*. Jena, Germany: Gustav Fischer.

KNY, L. (1871). Sitzung am 21. November 1871 [...Entwickelung einer Chytridiee aus der Untergattung *Olpidium*...(93-97)]. *Sitzungs-Bericht der Gesellschaft naturforschender Freunde zu Berlin, 1871*: 89-100.

KNY, L. (1872). *Chytridium (Olpidium) sphacellarum*. *Hedwigia* **11**: 81-87.

KO, W. H., CHANG, H. S., SU, H. J., CHEN, C. C. & LEU, L. S. (1978). Peronophythoraceae, a new family of Peronosporales. *Mycologia* **70**: 380-384.

KOBAYASI, Y. & OOKUBO, M. (1952). Studies on the aquatic fungi of Ozegahara moor. (2). *Journal of Japanese Botany* **27**: 181-188.

KOBAYASHI, Y. & OOKUBO, M. (1953*a*). Studies on the marine Phycomycetes. *Bulletin of the National Science Museum (Tokyo)* **33**: 53-65.

KOBAYAS[H]I, Y. & OOKUBO, M. (1953*b*). On the Labyrinthulales. *Nagaoa* **3**: 41-56. [In Japanese.]

KOBAYAS[H]I, Y. & OOKUBO, M. (1954*a*). Studies on the marine Phycomycetes II. *Bulletin of the National Science Museum (Tokyo), N. S.* **1**[=35]: 62-71.

KOBAYASHI, Y. & ÔKUBO [OOKUBO], M. (1954*b*). Studies on the aquatic fungi of the Ozegahara Moor. *Reports of the Ozegahara General Scientific Survey Commission 1954*: 561-575.

KOBAYAS[H]I, Y. & KONNO, K. (1970). Watermoulds isolated from soil in Tsushima Island (1). *Journal of Japanese Botany* **45**: 325-337.

KOBAYAS[H]I, Y., HIRATSUKA, N., OTANI, Y., TUBAKI, K., UDAGAWA, S.-i., SUGIYAMA, J. & KONNO, K. (1971). Mycological studies of the Angmagssalik region of Greenland. *Bulletin of the National Science Museum (Tokyo) N. S.* **14**: 1-96.

KOCH, W. J. (1951). A study of the motile cells of *Vaucheria*. *Journal of the Elisha Mitchell Scientific Society* **67**: 123-131.

KOCH, W. J. (1966). *Fungi in the Laboratory*. 113 pp. Book Exchange. Chapel Hill, North Carolina, U.S.A.

KOCH, W. J. (1968). Studies of the motile cells of chytrids. V. Flagellar retraction in posteriorly uniflagellate fungi. *American Journal of Botany* **55**: 841-859.

KOCHMAN, J. & MAJEWSKI, T. (1970) *Grzyby (Mycota). [Flora Polska] Tom IV. Glonowce (Phycomycetes) Wroślikowe (Peronosporales)*. Warszawa, Poland: Państwowe Wydawnictwo Naukowe.

KÖLE, A. P. (1954). A contribution to the knowledge of *Spongospora subterranea* (Wallr.) Lagerh., the cause of powdery scab of potatoes. *Tijdschrift over Plantenziekten [Netherlands Journal of Plant Pathology]* **60**: 1-65.

KÖLE, A. P. (1955). Some observations on the zoospores from trhe zoosporangia of *Plasmodiophora brassicae* Woron. *Tijdschrift over Plantenziekten [Netherlands Journal of Plant Pathology]* **61**: 159-162.

KÖLE, A. P. & GIELINK, A. J. (1961). Electron microscope observations on the flagella of the zoosporangial zoospores of *Plasmodiophora brassicae* and *Spongospora subterranea*. *Koninklijk Nederlandse Academie van Wetenschappen - Amsterdam: Proceedings, Series C* **64**: 157-161.

KÖLE, A. P. & GIELINK, A. J. (1962). Electron microscope observations on the resting spore germination of *Plasmodiophora brassicae*. *Koninklijk Nederlandse Academie van Wetenschappen - Amsterdam: Proceedings, Series C* **65**: 117-121.

KÖLE, A. P. & GIELINK, A. J. (1963). The significance of the zoosporangial stage in the life cycle of the Plasmodiophorales. *Tijdschrift over Plantenziekten [Netherlands Journal of Plant Pathology]* **69**: 258-262.

KOLKWITZ, R. (1901). Zur Biologie von *Leptomitus lacteus*. *Berichte der Deutschen Botanischen Gesellschaft* **19**: 288-291.

KOLKWITZ, R. (1903). Über Bau und Leben des Abwasserpilzes *Leptomitus lacteus*. *Berichte der Deutschen Botanischen Gesellschaft* **21**: 147-150.

KOMÁREK, J. & FOTT, B. (1983). Chlorophyceae (Grünalgen); Ordnung: Chlorococcales. *In* G. Heber-Pestalozzi (Ed.), *Die Binnengewässer 16. - Das Phytoplankton de Süsswassers, Teil 7, 1 Hälfte*, pp. 1-1044. Stuttgart, Germany: Schweizerbartsche Verlagsbuchhandlung (Nägele und Obermiller).

KONNO, K. (1969). Studies on Japanese lower aquatic phycomycetes. II. On *Thraustochytrium* collected from a sea weed. *Transactions of the Mycological Society of Japan (Nihon Kin Gakkai Kaiho, Japan)* **10**: 9-13.

KONNO, K. (1972). Studies of Japanese lower aquatic phycomycetes. *Science Reports of the Tokyo Kyoiku Daigaku* **14**: 227-312.

KONNO, K. & TANAKA, J. (1988). *Eurychasma dicksonii*, a parasitic fungus on *Acinetospora crinita* (brown alga) in Japan. *Bulletin of the Natural Science Museum, Tokyo, Series B* **14**: 119-122.

KONSULOFF, S. (1908). Untersuchungen über Rotatorienparasiten. *Archiv für Protistenkunde* **36**: 353-361.

KORAB, J. J. (1929). (Results of a study of the beet nematode *Heterodera schaachtii* at the nematode laboratory of the Belaya Tserkov Research Station) [in Russian with English summary]. *Sbornik Sortovogo Semenovodcheskogo Upravleniya* **8**: 29-67.

KORF, R. P. (1996). Simplified author citations for fungi and some old traps and new complications. *Mycologia* **88**: 146-150.

KORRINGA, P. (1951). Investigations on shell disease in the oyster, *Ostrea edulis*, L. *Rapports et Procès-verbaux, Conseil Permanent International l'Exploration de la Mer* **128**: 50-54.

KORRINGA, P. (1952). Recent advances in oyster biology. *Quarterly Review of Biology* **27**: 257-330, 531-656.

KOSKE, R. E. (1981). *Labyrinthula* inside the spores of a vesicular-arbuscular mycorrhizal fungus. *Mycologia* **72**: 1175-1180.

KOUYEAS, H. & THEOHARI, I. (1977). Notes on the taxonomy of *Pythium vexans* and related species. *Annales de l'Institut Phytopathologique, Benaki, Nouvelle Serie*, **11**: 274-283.

KOUYEAS, V. (1953). On the sexuality of *Phytophthora parasitica* Dastur. *Annales de l'Institut Phytopathologique, Benaki, Année* **7**: 40-53.

KOUYEAS, V. & KOUYEAS, H. (1963). Notes on species of *Pythium*. *Annales de l'Institut Phytopathologique, Benaki, Nouvelle Série* **5**: 207-237.

KRAFT, J. M., ENDO, R. M. & ERWIN, D. C. (1967). Infection of primary roots of bentgrass by zoospores of *Pythium aphanidermatum*. *Phytopathology* **57**: 86-90.

KRAMER, V. I. (1990). Laboratory evaluation of *Lagenidium giganteum* (Oomycetes: Lagenidiales) in water from Contra Costa County, California, mosquito sources. *Journal of the American Mosquito Control Association* **6**: 79-83.

KRAUSE, R. (1960). Untersuchungen über den Einfluss der Assenfactoren auf die Bildung der Oogonien bei *Saprolegnia ferax* (Gruith.) Thuret. *Archiv für Mikrobiologie* **36**: 373-386.

KRENNER, J. A. (1935). Néhány adat hazánk gombaflórájához (Ein. Beiträge zur Pilzflora Ungarns). *Botanikai Közlemenyek* **32**: 201.

KRÖBER, H. (1985). Erfahrungen mit *Phytophthora* de Bary und *Pythium* Pringsheim (Experiences with *Phytophthora* de Bary and *Pythium* Pringsheim) *Mitteilungen aus der Biologischen Bundesanstalt für Land- und Forstwirtschaft Berlin-Dahlem* **225**: 1-175.

KROPF, D. L. (1986). Electrophysiological properties of *Achlya* hyphae: ionic currents studied by intracellular potential recording. *Journal of Cell Biology* **102**: 1209-1216.

KROPF, D. L., CALDWELL, J. H., GOW, N. A. R. & HAROLD, F. M. (1984). Transcellular ion currents in the water mold *Achlya*. *Journal of Cell Biology* **99**: 486-496.

KRUGENS, P., LEE, R. E. & CORLISS, J. O. (1994). Ultrastructure, biogenesis and functions of extrusive organelles in selected non-ciliate protists. *Protoplasma* **181**: 164-190.

KRÜGER, F. (1910). Beitrag zur Kenntnis der Kernerhältnisse von *Albugo candida* und *Peronospora ficariae*. *Zentralblatt für Bakteriologie, Parasitenkunde, und Infektionskrankheiten, Abt. 2* **27**: 186-205.

KUBICEK, Q. B. & KENNETH, R. G. (1984). *Peronosclerospora globosa*, a new downy mildew of Gramineae, attacking cupgrass in Texas. *Phytopathology* **74**: 792 [abstract].

KÜHN, J. (1877). Vorläufiger Bericht über die bisherigen Ergebnisse der seit dem Jahr 1875 im Auftrage des Vereins für Rübenzucker-Industrie ausgeführten Versuche zur Ermittelung der Ursache der Rübenmüdigkeit des Bodens und zur Erforschung der Natur der Nematoden. *Zeitschrift des Vereins für die Rübenzucker-Industrie des Deutschen Reich (ohne Band)*: 452-457.

KÜHN, J. (1881). Die Ergebnisse der Versuche zur Ermittelung der Ursache der Rübenmüdigkeit des Bodens und zur Erforschung der Natur der Nematoden. *Berichte aus dem physiologischen Laboratorium und der Versuchanstalt des landwirthschaftlichen Instituts der Universität Halle* **3**: 1-153.

KUMAR see RAGHU KUMAR

KUNKEL, L. O. (1915). A contribution to the life history of *Spongospora subterranea*. *Journal of Agricultural Research* **4**: 265-278.

KUSABA, T. & TOYAMA, A. (1970). Transmission of soil-born barley mosaic virus. I. Infectivity of diseased root washings. *Nihon Shokubutsu Byori Gakkai-Ho [Annals of the Phytopathological Society of Japan]* **36**: 214-222.

KUSANO, S. (1912). On the life-history and cytology of a new *Olpidium* with special reference to the copulation of motile isogametes. *Journal of the College of Agriculture, Imperial University of Tokyo* **4**: 141-199.

KUSANO, S. (1929). Observations on *Olpidium trifolii* Schroet. *Journal of the College of Agriculture, Imperial University of Tokyo* **10**: 83-99.

KUSEL-FETZMANN, E. & CARNIEL, K. (1984). Beiträge zur Biologie und Ultrastruktur von *Ducellieria chodati*. *Mitteilungsband der Tagung der Deutschen Botanischen Gesellschaft in Wien, Kurzfassung der Beiträge* : 107.

KUSEL-FETZMANN, E. & NOUAK, H. (1981). *Ducellieria chodatii* - Alge oder Pilz? *Plant Systematics and Evolution* **138**: 199-207.

KÜTZING, F. T. (1843). *Phycologia Generalis oder Anatomie, Physiologie und Systemkunde der Tange*. Leipzig (Lipsiae), Germany: F. A. Brockhaus.

KÜTZING, F. T. (1849). *Species Algarum*. Leipzig (Lipsiae), Germany: F. A. Brockhaus.

KUZNETSOV, E. A. (1979). Marine lower fungi of Great Salma Strait of the White Sea USSR. *Biologiya Morya (Soviet Journal of Marine Biology)* **1**: 3-9 (English Translation **5**: 1-6).

KUZNETSOV, E. A. (1981). Anabiosis in lower aquatic fungi. *Mikologiya i Fitopatologiya* [*Микопогия и ФитоПапогия*] **15**: 526-531.

LACEY, L. A. & LACEY, C. M. (1990). The medical importance of riceland mosquitos and their control using alternatives to chemical insecticides. *Journal of the American Mosquito Control Association, Supplement No. 2*, 1-93.

LACEY, L. A. & ORR, B. K. (1994). The role of biological control of mosquitoes in integrated vector control. *American Journal of Tropical Medicine and Hygiene* **50**: 97-115.

LACEY, L. A. & UNDEEN, A. H. (1986). Microbial control of black flies and mosquitoes. *Annual Review of Entomology* 31: 265-296.

LACY, R. C. (1949). Studies in aquatic phycomycetes. *Indian Phytopathology* **2**: 134-141.

LACY, R. C. (1955). Studies in aquatic phycomycetes, II. *Indian Phytopathology* **8**: 208-209.

LAFON, R. & BULIT, J. (1981). Downy mildew of the vine. *In* D. M. Spencer (Ed.), *The Downy Mildews*. Academic Press, London, UK, pp. 601-614.

LAGERHEIM, [N.] G. (1888). Sur un genre nouveau de Chytridiacées parasite des urédospores de certaines Urédinées. *Journal de Botanique* **2**: 432-440, plate X. [This author is inexplicably cited as G. A. Lagerheim (Karling) or A. de Lagerheim (Gwynne-Vaughan]

LAGERHEIM, [N.] G. (1890). *Harpochytrium* and *Achlyella*, zwei neue Chytridiaceen-gattungen. *Hedwigia* **29**: 142-143.

LAGERHEIM, [N.] G. (1891). Remarks on the fungus of potato scab. *Journal of Mycology* **7**: 103-104.

LAGERHEIM, [N.] G. (1900*a*). Om växt- och djurlämningarna i Andrées polarboj. *YMER: Tidskrift utgifven af Svenska Sällskapet för Antropologi och Geografi; Nittonde Årgången 1899*: 425-443.

LAGERHEIM, N. G. (1900*b*). Om växt- och djurlämningarna i Andrées polarboj. *Meddelanden från Stockholms Högskola Botaniska Institut* **3**: 1-19 ["Aftryck ur *Ymer. Årg. 1899*, h4"].

LAHERT, H. & KAVANAGH, J. A. (1985). The fine structure of the cystosorus of *Spongospora subterranea*, the cause of powdery scab of potato. *Canadian Journal of Botany* **63**: 2278-2282.

LAIBACH, F. (1927). Zytologische Untersuchungen über die Monoblepharideen. *Jahrbücher für Wissenschaftliche Botanik* **66**: 596-628.

LAING, S. K. A. & DEACON, J. W. (1990). Aggressiveness and fungal host ranges of mycoparasitic *Pythium* species. *Soil Biology and Biochemistry* **22**: 905-911.

LAING, S. K. A. & DEACON, J. W. (1991). Video microscopical comparison of mycoparasitism by *Pythium oligandrum*, *P. nunn* and an unnamed *Pythium* species. *Mycological Research* **95**: 469-479.

LAIRD, M., WHITTAKER, R. H., MARGULIS, L. & SCHWARTZ, K. V. (1990). Five kingdoms - reflections upon the classification of organisms. *Journal of the American Mosquito Control Association* **6**: 553-556.

LAMBERTI, F. & CIANIO, A. (1992). Biological control of plant parasitic nematodes. *In* E. C. Tjamos, G. C. Papavizas & R. J. Cook (Eds), *Biological Control of Plant Diseases: Progress and Challenges for the Future; NATO Advanced Research Workshop, Athens, Greece, May 19-24, 1991*, pp. 17-20. New York, N.Y., U.S.A.: Plenum Press.

LAMPORT, D. T. A. (1970). Cell wall metabolism. *Annual Review of Plant Physiology* **21**: 235-270.

LANDMANN, J. (1986). The occurrence of endoparasitic nematophagous fungi in rassland soils of Schleswig-Holstein West Germany. *Mitteilungen aus der Biologischen Bundesanstalt fuer land und Forstwirtschaft Berlin-Dahlem (West Germany)*, p. 276.

LANGE, L., EDÉN, U. & OLSON, L. W. (1988). Zoosporogenesis in *Pseudoperonospora cubensis*, the causal agent of cucurbit downy mildew. *Nordic Journal of Botany* **8**: 497-504.

LANGE, L. & INSUNZA, V. (1977). Root-inhabiting *Olpidium* species: the *O. radicale* complex. *Transactions of the British Mycological Society* **69**: 377-384.

LANGE, L. & OLSON, L. W. (1978). The zoospore of *Olpidium radicale*. *Transactions of the British mycological Society* **71**: 43-55.

LANGE, L. & OLSON, L. W. (1979). The uniflagellate phycomycete zoospore. *Dansk Botanisk Arkiv* **33**: 7-95.

LANGE, L. & OLSON, L. W. (1983). The fungal zoospore its structure and biological significance. *In* S. T. Buckzacki (Ed.), *Zoosporic Plant Pathogens, a Modern Perspective*, pp. 1-42. London, U.K.: Academic Press.

LANGE, L., OLSON, L. W. & SAFEEULLA, K. M. (1984). Pearl millet downy mildew (*Sclerospora graminicola*): zoosporogenesis. *Protoplasma* **119**: 178-187.

LANGENBERG, W. G. & VAN DER WAL, D. (1986). Identification of barley yellow mosaic virus by immuno-electron microscopy in barley but not in *Polymyxa graminis* or *Lagena radicicola*. *Tijdschrift over Plantenziekten [Netherlands Journal of Plant Pathology]* **92**: 133-136.

LANGSAM, D. M. (1986). *Achlya bispora*: a taxonomic reassessment of *Brevilegnia bispora*. *Mycologia* **78**: 600-604.

LARGE, E. C. (1940). *The Advance of the Fungi*. Jonathan Cape, London, UK.

LARPENT, J. P. (1962). Action de la methionine de la griséofulvine et de l'histidine sur la mophologie du mycelium d'*Achlya bisexualis* Coker C.B.S. ♀. *Comptes Rendus Hebdomadaires des Séances de l'Academie des Sciences, Paris, Série D* **255**: 151-152.

LARPENT, J. P. (1963). Phénomène de transport et ramification dans le mycelium jeune du *Saprolegnia monoica* Pr. *Comptes Rendus Hebdomadaires des Séances de l'Academie des Sciences, Paris, Série D* **257**: 4006-4008.

LARPENT, J. P. (1964). Transport d'un inhibiteur et ramification du thalle jeune *Saprolegnia monoica* Pr. *Comptes Rendus Hebdomadaires des Séances de l'Academie des Sciences, Paris, Série D* **258**: 686-688.

LARSEN, J., MANSFIELD-GIESE, K. & BØDKER, L. (2000). Quantification of *Aphanomyces euteiches* in pea roots using specific fatty acids. *Mycological Research* **104**: 858-864.

LARSEN, P. (1931). Fungi of Iceland. *In* K. Rosenvinge & W. Warming, *The Botany of Iceland, Volume 2, part 3*. pp. 449-607. Oxford University Press.

LASURE, L. L. & GRIFFIN, D. H. (1974). Evidence for diploidy in *Achlya bisexualis* based on cycloheximide resistance. *Mycologia* **66**: 391-396.

LATHAM, D. H. (1935). *Achlya recurva* from North Carolina soil. *Journal of the Elisha Mitchell Scientific Society* **51**: 183-187.

LAVIER, G. (1935*a*). Action sur la biologie d'une entamibe du parasitisme intranucléaire par une nucleophage. *Compte rendu des Séances de la Société de Biologie, Paris* **118**: 457-458.

LAVIER, G. (1935*b*). Sur une *Nucleophaga*, parasite d'*Entamoeba ranarum*. *Annales de Parasitologie Humaine et Comparée, Paris* **13**: 351.

LAVIOLA, C. (1969). Studies on the genetics of *Phytophthora infestans*. *Dissertation Abstracts* **29**: 3573.

LAVIOLA, C. (1971). Contribution to 'The life history of Oomycetes in relation to cytogenetic studies', a special meeting at the 1st Int. Mycol. Congr. Exeter.

LAVIOLA, C. (1974*a*). On the fertilization tube of *Phytophthora infestans* (Mont.) de Bary. *Phytopatologia Mediterranea* **13**: 55-58.

LAVIOLA, C. (1974*b*). Osservazioni citologiche sulle strutture asessuali di *Phytophthora infestans* (Mont.) de Bary. *Annali della Facolta' di Agraria dell'Universitata' di Bari* **27**: 1-12.

LAVIOLA, C. (1975). On the origin of multinuclearity of sporangia in *Phytophthora infestans* (Mont.) de Bary. (Sull'origine della plurinuclearità degli sporangi in *Phytophthora infestans* (Mont.) de Bary. *Phytopatologia Mediterranea* **14**: 30-31.

LAVIOLA, C. & PORTACCI, M. (1974). Osservazioni citologiche sulle strutture sessuali di *Phytophthora infestans* (Mont.) de Bary. *Phytopatologia Mediterranea* **13**: 87-92.

LEADBEATER, B. S. C. (1989). The phylogenetic significance of flagellar hairs in the Chromophyta. *In* J. C. Green, B. S. C. Leadbeater & W. L. Diver (Eds), *The Chromophyte Algae, Problems and Perspectives*, pp. 145-165. Oxford, U.K.: Clarendon Press.

LEADBEATER, B. S. C. & MORTON, C. (1974). A microscopical study of a marine species of *Codosiga* James-Clark (Choanoflagellata) with special reference to the ingestion of bacteria. *Biological Journal of the Linnean Society* **6**: 337-347.

LEADLAY, E. A. & HEYWOOD, V. H. (1990). The biology and systematics of the genus *Coincya* Porta & Rigo ex Rouy (Cruciferae). *Botanical Journal of the Linnean Society* **102**: 313-398.

LEAL, J. A., FRIEND, J. & HOLLIDAY, P. (1964). A factor controlling sexual reproduction in *Phytophthora*. *Nature (London)* **203**: 545-546.

LEAL, J. A. & GOMEZ-MIRANDA, B. (1965). The effect of light and darkness on the germination of the oospores of certain species of *Phytophthora* on some synthetic media. *Transactions of the British Mycological Society* **48**: 491-494.

LEBEDA, A. & SCHWINN, F. J. (1994). The downy mildews - an overview of recent research progress (Falscher Mehitau - Übersicht über neuere Forschungsresultate). *Zeitschrift für Pflanzenkrankheiten und Pflanzenschutz* **101**: 225-254.

LÉCHER, P. (1978). The synaptonemal complex in the bipartition division of the radiolaria *Aulocantha scolymantha*. *Canadian Journal of Genetics and Cytology* **20**: 85-95.

LECHMERE, A. E. (1910). An investigation of a species of *Saprolegnia*. *New Phytologist* **9**: 305-319.

LECHMERE, A. E. (1911*a*). Further investigations of methods of reproduction in the Saprolegniaceae. *New Phytologist* **10**: 167-203.

LECHMERE, A. E. (1911*b*). Note sur les variations observées dans deux espèces de *Saprolegnia*. *Bulletin de la Musée National d'Histoire Naturelle, Paris* **17**: 376-381.

LECHTOVA-TRNKA, M. (1931). *Étude sur les bactéries des légumineuses et observations sur quelques champignons parasites des nodosites*. These, Paris: 1-488.

LEDINGHAM, G. A. (1934). Zoospore ciliation in the Plasmodiophorales. *Nature (London)* **133**: 534.

LEDINGHAM, G. A. (1935). Occurrence of zoosporangia in *Spongospora subterranea* (Wallroth) Lagerheim. *Nature (London)* **135**: 394.

LEDINGHAM, G. A. (1939). Studies on *Polymyxa graminis*, n. gen., n. sp., a plasmodiophoraceous root parasite of wheat. *Canadian Journal of Research, Section C* **17**: 38-51.

LEE, D. L., VAUGHAN, P. C. & DURSCHNER-PELZ, U. (1992). Ultrastructure of the thallus and secondary spore of the nematophagous fungus *Haptoglossa heterospora* (Oomycetes). *Journal of Invertebrate Pathology* **59**: 33-39.

LEE, H. Y. & ARONSON, J. M. (1975). Composition of cellulin, the unique chitin-glucan granules of the fungus, *Apodachlya* sp. *Archives of Microbiology* **102**: 203-208.

LEE, P. C. (1966). Effects of light and temperature on the formation of sexual structures in the family Saprolegniaceae. *Dissertation Abstracts* **26**: 5676.

LEE, R. E. (1989). *Phycology*, Second Edition. Cambridge, U.K.: Cambridge University Press.

LEE, S. B. & FULLER, M. S. (1992). Ribosomal DNA sequences of *Leptomitus lacteus*, *Apodachlya brachynema*, *Sapromyces elongatus*, *Aqualinderella fermentans* and *Rhipidium* sp. and their evolutionary implications for the oomycete order Leptomitales. *Inoculum, Newsletter of the Mycological Society of America* **43**: 40 [abstract].

LEE, S. B. & TAYLOR, J. W. (1992). Phylogeny of five fungus-like protoctistan *Phytophthora* species, inferred from the internal transcribed spacers of ribosomal DNA. *Molecular Biology and Evolution* **9**: 636-653.

LEEDALE, G. F. (1967). *Euglenoid Flagellates*. Englewood Cliffs, New Jersey, U.S.A.: Prentice-Hall.

LEEDALE, G. F. (1974). How many are the kingdoms of organisms? *Taxon* **23**: 261-270.

LÉGER, L. (1908). Mycétozoaires endoparasites des Insectes. 1. *Sporomyxa scauri* nov. gen. nov. sp. *Archiv für Protistenkunde* **12**: 109-130.

LÉGER, L. (1909). Sur une Mycétozoaire nouveau endoparasite des Insectes. *Comptes Rendus Hebdomadaires des Séances de l'Académie des Sciences, Paris, Série D* **149**: 239-240.

LÉGER, L. (1914). Sur un nouveau protiste du genre *Dermocystidium* parasite de la truite. *Comptes Rendus Hebdomadaires des Séances de l'Académie des Sciences, Paris, Série D* **158**: 807.

LEHNEN, L. P. & POWELL, M. J. (1989). The role of kinetosome-associated organelles in the attachment of encysting secondary zoospores of *Saprolegnia ferax* to substrates. *Protoplasma* **149**: 163-173.

LEHNEN, L. P. & POWELL, M. J. (1991). Formation of K_2-bodies in primary cysts of *Saprolegnia ferax*. *Mycologia* **83**: 163-179.

LEIGY, J. (1850). Entophyta in bodies of animals. *Proceedings of the Academy of Natural Sciences, Philadelphia* **5**: 7-8.

LEIPE, D. D., TONG, S. M., GOGGIN, C. L., SLEMENDA, S. B., PIENIAZEK, N. J. & SOGIN, M. L. (1996). 16S-like rDNA sequences from *Developayella elegans*, *Labyrinthuloides haliotidis*, and *Proteromonas lacertae* confirm that stramenopiles are a primarily heterotrophic group. *European Journal of Protistology* **32**: 449-458.

LEIPE, D. D., WAINRIGHT, P. O., GUNDERSON, J. H., PORTER, D., PATTERSON, D. J., VALOIS, F., HIMMERICH, S. & SOGIN, M. L. (1994). The stramenopiles from a molecular perspective: 16S-like rRNA sequences from *Labyrinthuloides minuta* and *Cafeteria roenbergensis*. *Phycologia* **33**: 369-377.

LEITGEB, H. (1868). Zwei neue Saprolegnieen. *Botanische Zeitung, Berlin, 26 Jahrgang* **31**: 502-503.

LEITGEB, H. (1869). Neue Saprolegnieen. *Jahrbücher für wissenschaftliche Botanik* **7**: 357-389.

LéJOHN, H. B. (1972). Enzyme regulation, lysine pathways and cell wall structures as indicators of major lines of evolution in fungi. *Nature (London)* **231**: 164-168.

LÉJOHN, H. B. (1974). Biochemical parameters of fungal phylogenetics. *In* T. Dobzhansky, M. Hecht & W. C. Steeve (Eds), *Evolutionary Biology* **7**: 79-125.

LENNON, R. E. (1954). Feeding mechanism of the sea lamprey and its effect on host fishes. *U. S. Fish and Wildlife Service, Fisheries Bulletin* **565**: 247-293.

LEONIAN, L. H. (1931). Heterothallism in *Phytophthora*. *Phytopathology* **21**: 941-955.

LEONIAN, L. H. (1934). Identification of *Phytophthora* species. *Agricultural Experiment Station, College of Agriculture, West Virginia University, Bulletin No. 262*: 1-36.

LESEMANN, D. E. & FUCHS, W. H. (1970). Elektronenmikroskopische Untersuchung über die Vorbereitung der Infektion in encystierten Zoosporen von *Olpidium brassicae*. *Archiv für Mikrobiologie* **71**: 9-19.

LESEMANN, D. E. & FUCHS, W. H. (1970). Die Ultrastruktur des Penetrationsvorganges von *Olpidium brassicae* an Kohlrabi-Wurzeln. *Archiv für Mikrobiologie* **71**: 20-30.

LESHEM, Y. Y. (1992). *Plant Membranes, a Biophysical Approach to Structure, Development, and Senescence*. Dodrecht, The Netherlands: Kluver Academic Publishers.

LESTER, R. J. G. (1986). Abalone die-back caused by a protozoan infection? *Australian Fisheries* **45**: 26-27.

LESTER, R. J. G. & DAVIS, G. H. G. (1981). A new *Perkinsus* species (Apicomplexa, Perkinsea) from the Abalone *Haliotis ruber*. *Journal of Invertebrate Pathology* **37**: 181-187.

LEU, L. S. (1973). Effects of temperature on conidial size and sporulation of *Sclerospora sacchari*. *Plant Protection Bulletin, Taiwan* **15**: 106-115.

LÉVEILLÉ, J.-H. (1847). Sur la disposition méthodique des Urédinées. *Annales des Sciences Naturelles, Botanique, Série III* **8**: 369-376.

LÉVESQUE, A. A., BECKENBACH, K., BAILLIE, D. L. & RAHE, J. E. (1993). Pathogenicity and DNA restriction fragment length polymorphisms of isolates of *Pythium* spp. from glyphosate-treated seedlings. *Mycological Research* **97**: 307-312.

LÉVESQUE, A. A., HARLTON, C. E. & COCK, A. W. A. M. de (1998). Identification of some oomycetes by reverse dot blot hybridization. *Phytopathology* **88**: 214-222.

LÉVESQUE, A. A., VRAIN, T. C. & De BOER, S. H. (1994). Development of a species-specific probe for *Pythium ultimum* using amplified ribosomal DNA. *Phytopathology* **84**: 474-478.

LEVINE, N. D. (1978). *Perkinsus* gen. n. and other new taxa in the protozoan phylum Apicomplexa. *Journal of Parasitology* **64**: 549.

LEVINE, N. D. & CORLISS, J. O. (1963). Two new subclasses of sarcodines Labyrinthulia subcl. nov. and Proteomyxidia subcl. nov. *Journal of Protozoology, Supplement* **10**: 27 [abstract].

LEVINE, N. D., CORLISS, J. O., COX, F. E. G., DEROUX, G., GRAIN, J., HONIGBERG, B. M., LEEDALE, G. F., LOEBLICH, A. R., LOM, J., LYNN, D., MERINFELD, E. G., PAGE, F. C., POLJANSKY, G., SPRAGUE, V., VÁVRA, J. & WALLACE, F. G. (1980). A newly revised classification of the protozoa. *Journal of Protozoology* **27**: 37-58.

LEWIS, J. A. (1973). Effect of mineral salts on *Aphanomyces euteiches* and *Aphanomyces* root rot of peas. *Phytopathology* **63**: 989-993.

LEWIS, J. A. & PAPAVIZAS, G. C. (1971). Effect of sulfur-containing volatile compounds and vapors from cabbage decomposition on *Aphanomyces euteiches*. *Phytopathology* **61**: 208-214.

LI, J. & HEATH, I. B. (1994). The behavior of F-actin during the zoosporic phases of the chytridiomycete gut fungi *Neocallimastix* and *Orpinomyces*. *Experimental Mycology* **18**: 57-69.

LIEW, E. C. Y., MacLEAN, D. J. & IRWIN, J. A. G. (1998). Specific PCR based detection of *Phytophthora medicaginis* using the intergenic spacer region of the ribosomal DNA. *Mycological Research* **102**: 73-80.

LIFSHITZ, R., STANGHELLINI, M. E. & BAKER, R. (1984). A new species of *Pythium* isolated from soil in Colorado. *Mycotaxon* **20**: 373-379.

LIGHTNER, D. V. (1981). Fungal diseases of marine crustaceans. *In* E. Davidson (Ed.), *Pathogenesis of Invertebrate Microbial Diseases*, pp. 451-484. New Jersey, U.S.A.: Allenheld, Osmun.

LIGHTNER, D. V. & FONTAINE, C. T. (1973). A new fungus disease of the white shrimp *Penaeus steiferus*. *Journal of Invertebrate Pathology* **22**: 94-99.

LIHNELL, D. (1942). *Tetramyxa rhizophaga* Lihnell n. sp., ein parasite in den Wurzel von *Juniperus communis* L. *Symbolae Botanicae Upsaliensis* **5**: 1-11.

LILIENFELD, F. A. (1911). Przyczynki do znajomości *Haplomitrium hookeri* - Beiträge zur Kenntnis der Art *Haplomitrium Hookeri*. *Bulletin International. Académie Polonaise des Sciences et des Lettres. Classe des Sciences Mathematiques et Naturelles. Série B. Botanique, 1911* **(6B)**: 315-338.

LILLEY, J. H. & ROBERTS, R. J. (1997). Pathogenicity and culture studies comparing the *Aphanomyces* involved in epizootic ulcerative syndrome (EUS) with other similar fungi. *Journal of Fish Diseases* **20**: 135-144.

LIM-HO, C. L. & DICK, M. W. (1986). Comparative estimates of minimum propagule densities of *Aplanopsis* and *Pythium* in soil. *Transactions of the British Mycological Society* **87**: 309-311.

LIN, C. C. & ARONSON, J. M. (1970). Chitin and cellulose in the cell walls of the oomycete, *Apodachlya* sp. *Archiv für Mikrobiologie* **72**: 111-114.

LIN, C. C., SICHER, R. C. & ARONSON, J. M. (1976). Hyphal wall chemistry in *Apodachlya*. *Archives of Microbiology* **108**: 85-91.

LINDER, D. [H.] (1926). A new species of *Araiospora* from British Guiana. *Mycologia* **18**: 172-178.

LINDER, D. H. (1947). Fungi. *In* N. Polunin (Ed.) *Botany of the Canadian Eastern Arctic. II. Thallophyta and Bryophyta*, pp. 234-297. Bulletin No. 97, Biological series No. 26. Canada: National Museum of Canada.

LINDLEY, J. (1845). *Edwards Botanical Register* **31**: 40.

LINDSTEDT, K. (1872). *Synopsis der Saprolegniaceen und Beobachtungen über einige Arten*. Berlin, Germany: R. Friedländer & Sohn.

LINTHICUM, B. & ZIEGLER, A. W. (1954). A summary of physiological studies on *Dictyuchus pseudodictyon*. *Florida State University Studies* **13**: 1-5.

LINFORD, M. B. (1927). Additional hosts of *Aphanomyces euteiches*, the pea root rot fungus. *Phytopathology* **17**: 133-134.

LIO-PO, G. D., BATICADOS, M. C. L., LAVILLA, C. R. & SANCTIVORES, M. E. G. (1985). In vitro effects of fungicides on *Haliphthoros philippinensis*. *Journal of Fish Diseases* **8**: 359-366.

LIO-PO, G. D., LAVILLA, C. R. & TRILLO-LLOBRERA, A. (1978). Toxicity of malachite green to the larvae of *Penaeus monodon* Fabricius (sand shrimp). *Quarterly Research Report, Aquaculture Department, Tigbauan, Philippines* **2**: 3-7.

LIO-PO, G. D. & SANVICTORES, E. G. (1986). Tolerance of *Penaeus monodon* eggs and larvae to fungicides against *Lagenidium* sp. and *Haliphthoros philippinensis*. *Aquaculture* **51**: 161-168.

LIO-PO, G. D., SANVICTORES, M. E. G., BATICADOS, M. C. L. & LAVILLA, C. R. (1982). In vitro effect of fungicides on hyphal growth and sporogenesis of *Lagenidium* spp. isolated from *Penaeus monodon* larvae and *Scylla serrata* eggs. *Journal of Fish Diseases* **5**: 97-112.

LIPKIN, Y, & AVIDOR, A. (1974). *Tetramyxa marina* a new plasmodiophoraceae marine parasite. *Nova Hedwigia* **72**: 799-810.

LIPPS, P. E. (1980). A new species of *Pythium* isolated from wheat beneath snow in Washington. *Mycologia* **72**: 1127-1133.

LIPPMAN, E., ERWIN, D. C. & BARTNICKI-GARCIA, S. (1974). Isolation and chemical composition of oospore-oogonium walls of *Phytophthora megasperma* var. *sojae*. *Journal of General Microbiology* **80**: 131-141.

LITTLEFIELD, L. J., WHALLON, J. H., DOSS, P. J. & HASSAN, Z. M. (1998). Postinfection development of *Polymyxa graminis* in roots of *Triticum aestivum*. *Mycologia* **90**: 869-882.

LLANOS, M. C. & LOCKWOOD, J. L. (1960). Factors affecting zoospore production by *Aphanomyces euteiches*. *Phytopathology* **50**: 826-830.

LOCKWOOD, J. L. (1960). Pea introductions with partial resistance to *Aphanomyces* root rot. *Phytopathology* **50**: 621-624.

LOCKWOOD, J. L. & BALLARD, J. C. (1959). Factors affecting a seedling test for evaluating resistance of pea to *Aphanomyces* root rot. *Phytopathology* **49**: 406-410.

LOHDE, G. (1874). 8. Botanische Section, 23. September Zweite Sitzung (5). Über einige neue parasitische Pilze. *Tageblatt der ... Versammelung Deutscher Naturforscher und Aertze in Breslau* **47**: 203-206.

LOHWAG, H. (1926). Über die Homologie der Sporangien, Oogonien und Antheridien bei den Oomyceten. *Archiv für Protistenkunde* **55**: 1-62.

LONGCORE, J. E. (1996). Chytridiomycete taxonomy since 1960. *Mycotaxon* **60**: 149-174.

LONGCORE, J. E., BROOKS, J. L. & HOMOLA, R. L. (1987). *Apodachlyella* reexamined. *Mycologia* **79**: 621-626.

LOPEZ, S. E. & MacCARTHY, S. (1985). Some parasite phycomycetes on fungi and algae from Argentina. *Darwiniana (Buenos Aires)* **26**: 61-70.

LOPEZ-LLORCA, L. V. & DUNCAN, G. H. (1991). Effects of fungal parasites on cereal cyst nematode (*Heterodera avenae* Woll.) from naturally infested soil - a scanning electron microscopy study. *Canadian Journal of Microbiology* **37**: 218-225.

LORD, J. C. & ROBERTS, D. W. (1985*a*). Solute effects on *Lagenidium giganteum*: zoospore motility and bioassay reproducibility. *Journal of Invertebrate Pathology* **46**: 160-165.

LORD, J. C. & ROBERTS, D. W. (1985*b*). Effects of salinity, *p*H, organic solutes, anaerobic conditions, and the presence of other microbes on production and survival of *Lagenidium giganteum* (Oomycetes: Lagenidiales) zoospores. *Journal of Invertebrate Pathology* **45**: 331-338.

LORD, J. C. & ROBERTS, D. W. (1986). The effects of culture medium quality and host passage on zoosporogenesis, oosporogenesis and infectivity of *Lagenidium giganteum* (Oomycetes: Lagenidiales). *Journal of Invertebrate Pathology* **48**: 355-361.

LORD, J. C. & ROBERTS, D. W. (1987). Host age as a determinant of infection rates with the mosquito pathogen *Lagenidium giganteum* (Oomycetes: Lagenidiales). *Journal of Invertebrate Pathology* **50**: 70-71.

LOTSY, J. P. (1907). *Vorträge über Botanische Stammesgeschichte Gehalten an der Reichsuniversität, Erster Band: Algen und Pilze*. Gustav Fischer, Jena, Germany.

LÖWENTHAL, W. (1905). Weitere Untersuchungen an Chytridiaceen. *Archiv für Protistenkunde* **5**: 221-239.

LOUNSBURY, J. A. (1927). A case of phenomenal zoospore behavior of an apparently sterile *Isoachlya* and a description of the plant. *Transactions of the Wisconsin Academy of Science* **23**: 539-549 [pub. Jan. 1928].

LOUNSBURY, J. A. (1930). Investigations on the nature of *Protoachlya paradoxa* Coker. *Transactions of the Wisconsin Academy of Science* **25**: 215-225.

LOVETT, J. S. & HASELBY, J. A. (1971). Molecular weights of the ribosomal ribonucleic acid of fungi. *Archiv für Mikrobiologie* **80**: 191-204.

LUCAS, J. A. (1994). Plant cell culture and disease resistance. *In* B. V. Charlwood & R. A. Dixon (Eds), *Plant Cell Culture and its Applications*. Sevenoaks, U.K.: Edward Arnold.

LUCAS, J. A., HAYTER, J. B. R. & CRUTE, I. R. (1995). The downy mildews: host specificity and pathogenesis. *In* K. Kohmoto, U. S. Singh & R. P. Singh (Eds), *Pathogenesis and Host Specificity in Plant Diseases*, pp. 217-234. London, U.K.: Pergamon.

LUCAS, J. A., SHATTOCK, R. C., SHAW, D. S. & COOKE, L. R. (Eds) (1991). Phytophthora. Cambridge, U.K.: Cambridge University Press.

LUDVIG-KOHN, H., JAHNKE, K-D. & BAHNWEG, G. (1982). Sterol composition and sterol metabolism of *Haliphthoros milfordensis* and *Atkinsiella dubia*, fungal parasites of marine crustaceans. *Biochimica et Biophysica Acta* **713**: 463-469.

LUDWIG-MULLER, J., BENNETT, R. N., KIDDLE, G., IHMIG, S., RUPPEL, M. & HILGENBERG, W. (1999). The host range of *Plasmodiophora brassicae* and its relationship to endogenous glucosinolate content. *New Phytoplogist* **141**: 443-458.

LUND, A. (1930). A new species of *Pleotrachelus* with remarks on this genus. *Botanisk Tidsskrift* **41**: 240-243.

LUND, A. (1934). Studies on Danish freshwater phycomycetes and notes on their occurrence particularly relative to the hydrogen ion concentration of the water. *Det Kongelige Danske Videnskabernes Selskab Skrifter, naturvidens kabelig og mathematisk Afdeling, 9 Raekke* **6**: 3-97.

LUND, A. (1978). Occurrence of Saprolegniaceae in Danish soils. *Nova Hedwigia* **29**: 577-592.

LUNNEY, C. Z. & BLAND, C. E. (1976). Ultrastructural observations of mature and encysting zoospores of *Pythium proliferum* de Bary. *Protoplasma* **90**: 119-137.

LUTHER, A. (1899). Ueber *Chlorosaccus*, eine neue Gattung der Süsswasseralgen, nebst einigen Bermerkungen zur Systematik verwandter Algen. *Bihang till Kongliga Svenska Vetenskaps-Akademiens Handlingar* **24** (III, 13): 1-22.

LUTHER, H. (1950). Beobachtungen über *Tetramyxa parasitica* Goebel. *Memoranda Societatis pro Fauna et Flora Fennica* **25**: 88-96.

LWOFF, A. (1925). Chytrininées parasites des amibes de l'homme. Possibilité de leur utilisation comme moyen biologique de lutte contre la dysentrie amibienne. *Bulletin de la Société de Pathologie Exotique* **18**: 18.

LYNGBYE, H. C. (1819). *Tentamen hydrophytologie danicae* ... 248 pp. Hafniae, Copenhagen, Denmark.

McCRAY, E. M. (1985). *Lagenidium giganteum* (Fungi). *In* H. C. Chapman (Ed.), *Biological Control of Mosquitoes*, pp. 87-88. Bulletin 6 American Mosquito Control Association.

McCRAY, E. M., UMPHLETT, C. J. & FAY, R. W. (1973). Laboratory studies on a new fungal pathogen of mosquitoes. *Mosquito News* **33**: 54-60.

McCRAY, E. M., WOMELDORF, D. J., HUSBANDS, R. C. & ELIASON, D. A. (1973). Laboratory observations and field tests with *Lagenidium* against California mosquitoes. *In* T. D. Peck & T. D. Mulhern (Eds), *Proceedings and Papers of the Forty-first Annual Conference of the California Mosquito Control Association*, pp. 123-128. Visalia, U.S.A.: California Mosquito and Vector Control Association Press.

McCULLOUGH, J. S. (1977). A survey of nematophagous fungi in Queensland. *Queensland Journal of Agricultural and Animal Science* **34**: 25-34.

MACDONALD, R. (1927). Food and habits of *Meganyctiphanes norvegica*. *Journal of the Marine Biological Association of the United Kingdom* **14**: 753-784.

McDONOUGH, E. S. (1937). The nuclear history of *Sclerospora graminicola*. *Mycologia* **29**: 151-173.

McDONOUGH, E. S. (1946*a*). The cytology of *Sclerospora macrospora* (Sacc.) in relation to its taxonomic position. *American Journal of Botany* **33**: 220.

McDONOUGH, E. S. (1946*b*). A cytological study of the development of the oospore of *Sclerospora macrospora* (Sacc.). *Transactions of the Wisconsin Acadamy of Science, Arts and Letters* **38**: 211-218.

MACFARLANE, I. (1968). Problems in the systematics of the Olpidiaceae. *Veröffentlichungen des Instituts für Meeresforschung in Bremerhaven, Sonderband* **3**: 39-58.

MACFARLANE, I. (1970*a*). Germination of resting spores of *Plasmodiophora brassicae*. *Transactions of the British Mycological Society* **55**: 97-112.

MACFARLANE, I. (1970*b*). *Lagena radicicola* and *Rhizophydium graminis*, two common and neglected fungi. *Transactions of the British Mycological Society* **55**: 113-116.

McGLADDERY, S. E., CAWTHORN, R. J. & BRADFORD, B. C. (1991). *Perkinsus karlssoni* n. sp. (Apicomplexa) in bay scallops *Argopecten irradians*. *Diseases of Aquatic Organisms* **10**: 127-137.

McGILLIVRAY, A. M. & GOW, N. A. R. (1986). Applied electrical fields polarize the growth of mycelial fungi. *Journal of General Microbiology* **132**: 2515-2525.

McINNIS, T. M. (1971). A physiological and biochemical investigation of the aquatic phycomycete *Lagenidium* sp. a facultative parasite of certain mosquito larvae. PhD Dissertation, University of North Carolina, Chapel Hill, North Carolina, U.S.A.

McINNIS, T. M. & DOMNAS, A. J. (1971). A physiological investigation of *Lagenidium* sp., a parasite of mosquito larvae. *First International Mycological Congress, Exeter, England. Abstracts*: 60.

McINNIS, T. M. & DOMNAS, A. J. (1973). The properties of trehalase from the mosquito-parasitizing water mold, *Lagenidium* sp. *Journal of Invertebrate Pathology* **22**: 313-320.

McINNIS, T. M. & DOMNAS, A. J. (1974). An aryl ß-D-glucosidase of the aquatic fungus *Lagenidium giganteum*, a parasite of mosquito larvae. *Archives of Microbiology* **101**: 343-350.

McLEAN, M. A., CAMPBELL, R. N., HAMILTON, R. I. & ROCHON, D. M. (1994). Involvement of the cucumber necrosis virus coat protein in the spwecificity of fungus transmission by *Olpidium bornovanus*. *Virology* **204**: 840-842.

MacKAY, R. M. & DOOLITTLE, W. F. (1982). Two thraustochytrid 5S ribosomal RNAs. *Nucleic Acids Research* **10**: 8307-8310.

McKEEN, C. D. (1952*a*). A species of *Aphanomyces* causing damping-off in peppers and certain other vegetables. *Phytopathology* **42**: 14 [abstract].

McKEEN, C. D. (1952*b*). *Aphanomyces cladogamus* Drech., a cause of damping-off in peppers and certain other vegetables. *Canadian Journal of Botany* **30**: 701-710.

McKEEN, W. E. (1949). A study of sugar beet root rot in southern Ontario. *Canadian Journal of Research, Section C, Botanical Science* **27**: 284-311.

McKEEN, W. E. (1975). Electron microscopy studies of a developing *Pythium* oogonium. *Canadian Journal of Botany* **53**: 2354-2360.

MÄCKEL, H. G. (1928). Zur Cytologie einiger Saprolegniaceen. *Jahrbücher für wissenschaftliche Botanik* **69**: 517-548.

MacKICHAN, J. K., TUININGA, A. R. & KERWIN, J. L. (1994). Preliminary characterization of phospholipase A_2 in *Lagenidium giganteum*. *Experimental Mycology* **18**: 180-192.

MACKIN, J. G. (1962). Oyster disease caused by *Dermocystidium marinum* and other microrganisms in Louisiana. *Publications of the Institute of Marine Sciences, University of Texas* **7**: 132-229.

MACKIN, J. G. & BOSWELL, J. L. (1956). The life cycle and relationships of *Dermocystidium marinum*. *Proceedings of the National Shellfisheries Association* **47**: 112-115.

MACKIN, J. G., OWEN, H. M. & COLLIER, A. (1950). Preliminary note on the occurrence of a new protistan parasite, *Dermocystidium marinum* n. sp., in *Crassostrea virginica* (Gmelin). *Science, New York* **111**: 328-329.

MACKIN, J. G. & RAY, S. M. (1954). Studies on the effect of infection by *Dermocystidium marinum* on ciliary action in oysters (*Crassostrea virginica*). *Proceedings of the National Shellfisheries Association* **45**: 168-181.

MACKIN, J. G. & RAY, S. M. (1966). The taxonomic relationships of *Dermocystidium marinum* Mackin, Owen, and Collier. *Journal of Invertebrate Pathology* **8**: 544-545.

McLARTY, D. A. (1941*a*). Studies in the family Woroninaceae-I. Discussion of a new species including a consideration of the genera *Pseudolpidium* and *Olpidiopsis*. *Bulletin of the Torrey Botanical Club* **68**: 49-66.

McLARTY, D. A. (1941*b*). Studies in the Woroninaceae-II. The cytology of *Olpidiopsis achlyae* sp. nov. (ad int.). *Bulletin of the Torrey Botanical Club* **68**: 75-99.

McLEAN, N. & PORTER, D. (1982). The yellow-spot disease of *Tritonia diomedea* Bergh, 1894 (Mollusca: Gastropoda: Nudibranchia): Encapsulation of the thraustochytriaceous parasite by host amoebocytes. *Journal of Parasitology* **68**: 243-252.

McLEOD, D. M. & MÜLLER-KÖGLER, E. (1970). Insect pathogens: species originally described from their resting spores mostly as *Tarichium* species (Entomophthorales: Entomophthoraceae). *Mycologia* **62**: 33-66.

MacLEOD, H. & HORGAN, P. A. (1979). Germination of the asexual spores of the aquatic fungus, *Achlya bisexualis*. *Journal of Experimental Mycology* **3**: 70-82.

McMEEKIN, D. & MENDOZA, L. (2000). In vitro effect of streptomycin on clinical isolates of *Pythium insidiosum*. *Mycologia* **92**: 371-373.

MACMILLAN, J. R. (1985). Infectious diseases. *In* C. S. Tucker (Ed.), *Channel Catfish Culture*, pp. 458-496. Amsterdam, The Netherlands: Elsevier.

McMORRIS, T. C. (1978). Antheridiol and the oogoniols, steroid hormones which control sexual reproduction in *Achlya. Philosophical Transactions of the Royal Society, London* **284**: 459-470.

McMORRIS, T. C. & BARKSDALE, A. W. (1967). Isolation of a sex hormone from the water mould *Achlya bisexualis. Nature (London)* **215**: 320-321.

McNABB, S. A. (1989). The use of physical maps of mitochondrial and ribosomal DNA to determine evolutionary relationships among zoosporic fungi. MS Thesis, University of Manitoba, Winnipeg, Canada.

McNABB, S. A., BOYD, D. A., BELKHIRI, A., DICK, M. W. & KLASSEN, G. R. (1987). An inverted repeat comprises more than three-quarters of the mitochondrial genome in two species of *Pythium. Current Genetics* **12**: 205-208.

McNABB, S. A., EROS, R. W. & KLASSEN, G. R. (1988). Presence and absence of large inverted repeats in the mitochondrial DNA of Hyphochytriomycetes. *Canadian Journal of Botany* **66**: 2377-2379.

McNABB, S. A. & KLASSEN, G. R. (1988). Uniformity of mitochondrial DNA complexity in Oomycetes and the evolution of the inverted repeat. *Experimental Mycology* **12**: 233-2421.

McNITT, R. (1973). Mitosis in *Phlyctochytrium irregulare. Canadian Journal of Botany* **51**: 2065-2074.

MacQUADE, H. A. & BASSETT, B. (1977). Synaptonemal complexes in premeiotic interphase of pollen mother cells of *Triticum aestivum. Chromosoma* **63**: 153-159.

MAGAREY, R. C. (1996). Microbiological aspects of sugar cane yield decline. *Australian Journal of Agricultural Research* **47**: 307-322.

MAGNUS, P. (1872*a*). Chytrideae. *Jahresbücher: Kommission zur Untersuchungen der Deutschen Meere in Kiel* **2**: 76.

MAGNUS, P. (1872*b*). Sitzung am 19. November 1872 [...über ein *Chytridium*...(87-90)]. *Sitzungs-Bericht der Gesellschaft naturforschender Freunde zu Berlin, 1872*: 79-96.

MAGNUS, P. (1873). Über ein *Chytridium. Hedwigia* **12**: 27-31.

MAGNUS, P. (1875). Die botanischen Ergebnisse der Nordseefahrt von 21. Juli bis 9. Sept. 1872. *Wissenschaft Meeresuntersuchungen der Kommission zur wissenschaftlichen Untersuchung der deutschen Meere, Abteilung Kiel* **2-3**: 59-80.

MAGNUS, P. (1885). Botanische Mitteilungen. *Verhandlungen des Botanischen Vereins der Provinz Brandenburg* **26**: 69-81.

MAGNUS, P. (1905). Über die Gattung, zu der *Rhizophydium dicksonii* Wright gehört. *Hedwigia* **44**: 347-349.

MAIRE, R. (1915). Remarques sur le *Protascus subuliformis* à propos de la communication de M.E. Maupas. *Bulletin de la Société d'Histoire Naturelle de l'Afrique du Nord* **6**: 50-51.

MAIRE, R. & TISON, A. (1909). La cytologie des Plasmodiophoracées et la classe des Phytomyxinae. *Annales Mycologici* **7**: 226-253.

MAIRE, R. & TISON, A. (1911*a*). Nouvelles recherches sur les Plasmodiophoracées. *Annales Mycologici* **9**: 226-246.

MAIRE, R. & TISON, A. (1911*b*). Sur quelques Plasmodiophoracées non hypertrophiantes. *Comptes Rendus Hebdomadaires des Séances de l'Académie des Sciences, Paris, Série D* **152**: 206-208.

MALTESE, C. E., CONIGLIARO, G. & SHAW, D. S. (1995). The development of sporangia of *Phytophthora infestans. Mycological Research* **99**: 1175-1181.

MALVICK, D. K., GRAU, C. R. & PERCICH, J. A. (1998). Characterization of *Aphanomyces euteiches* strains based on pathogenicity tests and random amplified polymorphic DNA analysis. *Mycological Research* **102**: 465-475.

MALVICK, D. K. & PERCICH, J. A. (1998*a*). Variation in pathogenicity and genotype among single-zoospore strains of *Aphanomyces euteiches. Phytopathology* **88**: 52-57.

MALVICK, D. K. & PERCICH, J. A. (1998*b*). Genotypic and pathogenic diversity among pea-infecting strains of *Aphanomyces euteiches* from the Central and Western United States. *Phytopathology* **88**: 915-921.

MANEVAL, W. E. (1937). Parasitic and wood destroying fungi of Boone County Missouri. *University of Missouri Studies* **12**: 65-111.

MANIER, J.-F. (1976). Cycle et ultrastructure de *Blastulidium poedophthorum* Pérez 1903 (phycomycète lagénidiale) parasite des oeufs de *Simocephalus vetulinus* (Mull.) Schoedler (crustacé, Cladocère). *Protistologica* **12**: 225-238.

MANNING, M. A. & MENZIES, S. A. (1984). Pathogenic variability in isolates of *Aphanomyces euteiches* from New Zealand soils. *New Zealand Journal of Agricultural Research* **27**: 569-574.

MANOHARACHARY, C. (1974). A note on some interesting fungi from Hyderabad (India). *Current Science, Bangalore, India* **43**: 129-131.

MANOHARACHARY, C. (1985). Ecological account of some oomycetous fungi. *Indian Botanical Reporter* **4**: 50-51.

MANTON, I. (1956). Plant cilia and associated organelles. *In* D. Rudnick (Ed.), *Cellular Mechanisms in Differentiation and Growth*, pp. 61-71. Princeton, N.J., U.S.A.: Princeton University Press.

MANTON, I. (1959). Observations on the internal structure of the spermatozoid of *Dictyota. Journal of Experimental Botany* **10**: 448-461.

MANTON, I. (1964). The possible significance of some details of flagellar bases in plants. *Journal of the Royal Microscopical Society* **82**: 279-285.

MANTON, I. (1965). Some phyletic implications of flagellar structure in plants. *Advances in Botanical Research* **2**: 1-34.

MANTON, I. (1967). Further observations on the fine structure of *Chrysochromulina chiton* with special reference to the haptonema, "peculiar" Golgi structure and scale production. *Journal of Cell Science* **2**: 265-272.

MANTON, I. & CLARKE, B. (1951). Electron microscope observations on the zoospores of *Pylaiella* and *Laminaria. Journal of Experimental Botany* **2**: 242-246.

MANTON, I., CLARKE, B. & GREENWOOD, A. D. (1951). Observations with the electron microscope on a species of *Saprolegnia. Journal of Experimental Botany* **2**: 321-331.

MANTON, I. CLARKE, B. & GREENWOOD, D. (1953). Further observations with the electron microscope on spermatozoids in the brown algae. *Journal of Experimental Botany* **4**: 319-329.

MANTON, I. & STOSCH, H. A. von (1966). Observations on the fine structure of the male gamete of the marine centric diatom *Lithodesmium undulatum. Journal of the Royal Microscopical Society* **85**: 119-134.

MAO, Y. & TYLER, B. M. (1991). Genome organization of *Phytophthora megasperma* f. sp. *glycinea. Experimental Mycology* **15**: 283-291.

MAO, Y. & TYLER, B. M. (1996). The *Phytophthora sojae* genome contains tandem repeat sequences which vary from strain to strain. *Fungal Genetics and Biology* **20**: 43-51.

MARAITE, H., GOFFART, J. P. & BASTIN, V. (1988). Development of a quantitative method for assessment of *Polymyxa graminis* Led. inoculum potential in soils. *In* R. Cavalloro & K. D. Sunderland (Eds), *Integrated Crop Protection in Cereals. Proceedings of a Meeting of the EC Experts' Group, Littlehampton, 25-27 November 1986*, pp. 259-266. Rotterdam, The Netherlands: A. A. Balkema.

MARCHANT, R. (1968). An ultrastructural study of sexual reproduction in *Pythium ultimum. New Phytologist* **66**: 623-629.

MARCUS, J. H. & WILLOUGHBY, L. G. (1978). Fungi as food for the aquatic invertebrate *Asellus aquaticus. Transactions of the British Mycological Society* **70**: 143-146.

MARGULIS, L., CORLISS, J. O., MELKONIAN, M. & CHAPMAN, D. (Eds) (1990). *Handbook of Protoctista.* Boston, U.S.A.: Jones & Bartlett.

MARGULIS, L., McKHANN, H. I. & OLENDZENSKI, L. (Eds) (1992). *Illustrated Glossary of Protoctista.* Boston, U.S.A.: Jones & Bartlett.

MARKEY, D. R. & BOUCK, G. B. (1977). Mastigoneme attachment in *Ochromonas. Journal of Ultrastructural Research* **59**: 171-177.

MARKEY, D. R. & WILCE, R. T. (1976). The ultrastructure of reproduction in the brown alga *Pylaiella litoralis.* Part II. Zoosporogenesis in the unilocular sporangium. *Protoplasma* **88**: 147-173.

MARKS, G. E. (1965). The cytology of *Phytophthora infestans. Chromosoma (Berlin)* **16**: 681-692.

MARLATT, R. B., LEWIS, R. W. & McKITTRICK, R. T. (1962). Systemic infection of lettuce by *Bremia lactucae. Phytopathology* **52**: 888-890.

MARSHALL WARD, H. (1883*a*). Observations on Saprolegniae. *Quarterly Journal of Microscopical Science* **23**: 272-291.

MARSHALL WARD, H. (1883*b*). Observations on the genus *Pythium*, Pringsh. *Quarterly Journal of Microscopical Science* **23**: 485-515.

MARTIN, F. N. (1989). Maternal inheritance of mitochondrial DNA in sexual crosses of *Pythium sylvaticum. Current Genetics* **16**: 373-374.

MARTIN, F. N. (1990*a*). Taxonomic classification of asexual isolates of *Pythium ultimum* based on cultural characteristics and mitochondrial restriction patterns. *Experimental Mycology* **14**: 47-56.

MARTIN, F. N. (1990*b*). Variation in the ribosomal DNA repeat unit within single-oospore isolates of the genus *Pythium. Genome* **33**: 585-591.

MARTIN, F. N. (1991*a*). Selection of DNA probes useful for isolate identification of two *Pythium*. *Phytopathology* **81**: 742-746.

MARTIN, F. N. (1991*b*). Linear mitochondrial molecules and intraspecific mitochondrial genome stability in a species of *Pythium*. *Genome* **34**: 156-162.

MARTIN, F. N. (1992). *Pythium*. *In* L. L. Singleton, J. D. Mihail & C. M. Rush, *Methods for Research on Soilborne Phytopathogenic Fungi*, pp. 39-49. St Paul, Minnesota, U.S.A.: APS Press.

MARTIN, F. N. (1995*a*). Meiotic instability of *Pythium sylvaticum* as demonstrated by inheritance of nuclear markers and karyotype analysis. *Genetics* **139**: 1233-1246.

MARTIN, F. N. (1995*b*). Electrophoretic karyotype polymorphisms in the genus *Pythium*. *Mycologia* **87**: 333-353.

MARTIN, F. N. (1995*c*). Linear mitochondrial genome organization in vivo in the genus *Pythium*. *Current Genetics* **28**: 225-234.

MARTIN, F. N. (2000). Phylogenetic relationships among some *Pythium* species inferred from sequence analysis of the mitochondrially encoded cytochrome oxidase II gene. *Mycologia* **92**: 711-727.

MARTIN, F. N. & KISTLER, H. C. (1990). Species-specific banding patterns of restriction endonuclease-digested mitochondrial DNA from the genus *Pythium*. *Experimental Mycology* **14**: 32-46.

MARTIN, G. W. (1927). Two unusual water molds belonging to the family Lagenidiaceae. *Mycologia* **19**: 188-190.

MARTIN, R. W. & MILLER, C. E. (1982). Life cycle studies on *Olpidiopsis varians* Olpidiopsidaceae infecting *Aplanopsis terrestris* Saprolegniaceae I. Ultrastructure of sporangial formation. *Ohio Journal of Science* **82**: 13.

MARTIN, R. W. & MILLER, C. E. (1985). Host range and taxonomy of *Olpidiopsis varians*. *Mycotaxon* **24**: 411-418.

MARTIN, R. W. & MILLER, C. E. (1986*a*). Ultrastructure of mitosis in the endoparasite *Olpidiopsis varians*. *Mycologia* **78**: 11-21.

MARTIN, R. W. & MILLER, C. E. (1986*b*). Ultrastructure of zoosporogenesis in the endoparasite *Olpidiopsis varians*. *Mycologia* **78**: 230-241.

MARTIN, R. W. & MILLER, C. E. (1986*c*). Ultrastructure of sexual reproduction in *Olpidiopsis varians*. *Mycologia* **78**: 359-370.

MARTIN, W. W. (1975*a*). A new species of *Catenaria* parasitic in midge eggs. *Mycologia* **67**: 264-272.

MARTIN, W. W. (1975*b*). *Aphanomycopsis sexualis*, a new parasite of midge eggs. *Mycologia* **67**: 923-933.

MARTIN, W. W. (1977). The development and possible relationships of a new *Atkinsiella* parasitic in insect eggs. *American Journal of Botany* **64**: 760-769.

MARTIN, W. W. (1978). Two additional species of *Catenaria* (Chytridiomycetes, Blastocladiales) parasitic in midge eggs. *Mycologia* **70**: 461-467.

MARTIN, W. W. (1981*a*). *Couchia circumplexa*, a water mold parasitic in midge eggs. *Mycologia* **73**: 1143-1157.

MARTIN, W. W. (1981*b*). The natural regulation of midge population by aquatic fungi in Virginia, USA. *Journal of the Elisha Mitchell Scientific Society* **97**: 162-170.

MARTIN, W. W. (1984). The dynamics of aquatic fungi parasitic in a stream population of the midge, *Chironomus attenuatus*. *Journal of Invertebrate Pathology* **44**: 36-45.

MARTIN, W. W. (1991). Egg parasitism by zoosporic fungi in a littoral chironomid community. *Journal of the North American Benthological Society* **10**: 455-462.

MARTIUS, C. F. P. (1842*a*). *Die Kartoffel-Epidemie der letzten Jahre oder die Stockfäule und Ründe der Kartoffeln, Geschildert und in ihren ursachlichen Verhältnissen erörtert.* Munchen.

MARTIUS, C. F. P. (1842*b*). Sur la gangrène des pommes de terre, observée depuis quelques années en Allemagne. *Comptes Rendus Hebdomadaires des Séances de l'Académie des Sciences, Paris* **15**: 310-319.

MASON, P. A. (1973). Studies on the biology of *Bremia lactucae* Regel. PhD thesis, University of Cambridge.

MASSALSKI, A. & LEEDALE, G. F. (1969). Cytology and ultrastructure of the Xanthophyceae. I. Comparative morphology of the zoospores of *Bumilleria sicula* Borzi and *Tribonema vulgare* Pascher. *British Phycological Journal* **4**: 159-180.

MASSEE, G. (1891). *British Fungi. Phycomycetes and Ustilagineae.* London, U.K.

MASSEE, G. (1895). On an orchid disease. *Annals of Botany, London* **9**: 170.

MASSEE, G. (1908). Corky scab of potatoes. *Journal of the Board of Agriculture and Fisheries* **15**: 592.

MASTERS, M. J. (1971). *Chytridium deltanum* n. sp. and other Phycomycetes on *Oocytis* spp. in the Delta Marsh, Manitoba. *Canadian Journal of Botany* **49**: 471-486.

MATSUMOTO, C., KAGEYAMA, K., SUGA, H. & HYAKUMACHI, M. (1999). Phylogenetic relationships of *Pythium* species based on ITS and 5·8S sequences. *Mycoscience* **40**: 321-331.

MATTA, A. (1965). Una malattia della lattuga prodotta da una nuova specie di *Pythium*. *Phytopathologia Mediterranea* **4**: 48-53.

MATTES, O. (1924). Über Chytridineen in Plasma und Kern von *Amoeba terricola*. *Archiv für Protistenkunde* **47**: 413-430.

MATTHEW, J., HAWKE, B. G. & PANKHURST, C. E. (1995). A DNA probe for identification of *Pythium irregulare* in soil. *Mycological Research* **99**: 579-584.

MATTHEWS, V. D. (1928). *Nowakowskiella* and a new species of *Pythium*. *Journal of the Elisha Mitchell Scientific Society* **43**: 229-232.

MATTHEWS, V. D. (1931). *Studies on the Genus* Pythium. Chapel Hill, North Carolina, U.S.A.: University of North Carolina Press.

MATTHEWS, V. D. (1935). Notes on some Oomycetes from the vicinity of Mountain Lake, Virginia. *Journal of the Elisha Mitchell Scientific Society* **51**: 306-310.

MATTHEWS, V. D. (1936). A new species of *Rhipidium* from Mountain Lake, Virginia. *Journal of the Elisha Mitchell Scientific society* **52**: 291-293.

MATTINGLY, P. F. (1972*a*). Mosquito eggs XVII. Further notes on egg parasitization in genus *Armigeres*. *Mosquito Systematics* **4**: 1-8.

MATTINGLY, P. F. (1972*b*). Mosquito eggs XX. Egg parasitism in *Anopheles* with a further note on *Armigeres*. *Mosquito Systematics* **4**: 84-86.

MATZ, J. (1920). A new vascular organism in sugar cane. *Journal of the Puerto Rico Agricultural Experiment Station* **4**: 41-46.

MAUCHLINE, J. (1966). *Thalassomyces fagei*, an ellobiopsid parasite of the euphausiid crustacean, *Thysanoessa raschi*. *Journal of the Marine Biological Association of the United Kingdom* **46**: 531-539.

MAUCH-MANI, B., SCHWINN, F. J. & GUGGENHEIM, R. (1989). Early infection stages of the downy mildew fungi *Sclerospora graminicola* and *Peronosclerospora sorghi* in plants and cell cultures. *Mycological Research* **92**: 455-452.

MAUPAS, E. (1915). Sur un champignon parasite des *Rhabditis*. *Bulletin de la Société d'Histoire Naturelle de l'Afrique du Nord* **6**: 34-49.

MAURIZIO, A. (1894). Zur Entwicklungsgeschichte und Systematik der Saprolegnieen. *Flora, Jena* **79**: 109-158.

MAURIZIO, A. (1895). Zur Kenntniss der schweizerischen Wasserpilze nebst Angaben über eine neue Chytridinee. *Jahresbericht Naturforschung Gesellschaft Grabündens, Chur* **38**: 9-38.

MAURIZIO, A. (1896). Die Sporangiumanlage der Gattung *Saprolegnia*. *Jahrbücher für wissenschaftliche Botanik* **29**: 75-131.

MAURIZIO, A. (1899). Beiträge zur Biologie der Saprolegnieen. *Mittheilungen der Fischereivereine für die Provinz Brandenburg* **7**: 1-66.

MAYTON, H. S., OLIVIER, C., VAUGHN, S. F. & LORIA, R. (1996). Correlation of fungicidal activity of *Brassica* species with allyl isothiocyanate production in macerated leaf tissue. *Phytopathology* **86**: 267-271.

MEHRLICH, F. P. (1932). *Pseudopythium phytophthoron* a synonym of *Phytophthora cinnamomi*. *Mycologia* **24**: 453-454.

MEIER, H. & WEBSTER, J. (1954). An electron microscope study of cysts in the Saprolegniaceae. *Journal of Experimental Botany* **5**: 401-409.

MELHUS, I. E. (1911). Experiments on spore germination and infection in certain species of Oomycetes. *University of Wisconsin, Agricultural Experiment Station, Research Bulletin No 15*: 27-91.

MELKONIAN, M. (1990). Chlorophyte orders of uncertain affinities. Order Pedinomonadales. *In* L. Margulis, J. O. Corliss, M. Melkonian & D. J. Chapman (Eds), *Handbook of Protoctista*, pp. 649-651. Boston, U.S.A.: Jones & Bartlett.

MENDOZA, L., AJELLO, L. & McGINNIS, M. R. (1996). Infections caused by the oomycetous pathogen, *Pythium insidiosum*. *Journal de Mycologie Médicale* **6**: 151-164.

MENDOZA, L. & ALFARO, A. A. (1986). Equine pythiosis in Costa Rica: report of 39 cases. *Mycopathologia* **94**: 123-129.

MENDOZA, L., KAUFMAN, L. & STANDARD, P. (1986). Immunodiffusion test for diagnosing and monitoring pythiosis in horses. *Journal of Clinical Microbiology* **23**: 813-816.

MENDOZA, L., KAUFMAN, L. & STANDARD, P. (1987). Antigenic relationship between the animal and human pathogen *Pythium insidiosum* and nonpathogenic *Pythium* species. *Journal of Clinical Microbiology* **25**: 2159-2162.

MENDOZA, L. & MARIN, G. (1988). Antigenic relationship between *Pythium insidiosum* de Cock *et al.* 1987 and its synonym *Pythium destruens* Shipton 1987. *Mycoses* **32**: 73-77.

MER, G. S. & KHULBE, R. D. (1983). *Pythium vexans* var. *minuta* var. nov. from Kumaun, the Himalayas, India. *Current Science, India (Bangalore)* **52**: 735-736.

MER, G. S., SATI, S. C. & KHULBE, R. D. (1980). Occurrence, distribution ans seasonal periodicity of some aquatic fungi of Sat Tal (Naini Tal), India. *Hydrobiologia* **76**: 201-205.

MERRIAM, T. L. & AXTELL, R. C. (1982*a*). Evaluation of the entomogenous fungi *Culicinomyces clavosporus* and *Lagenidium giganteum* for control of the salt marsh mosquito, *Aedes taeniorhynchus*. *Mosquito News* **42**: 594-602.

MERRIAM, T. L. & AXTELL, R. C. (1982*b*). Salinity tolerance of two isolates of *Lagenidium giganteum* (Oomycetes: Lagenidiales), a fungal pathogen of mosquito larvae. *Journal of Medical Entomology* **19**: 388-393.

MERRIAM, T. L. & AXTELL, R. C. (1983). Relative toxicity of certain pesticides to *Lagenidium giganteum* (Oomycetes: Lagenidiales), a fungal pathogen of mosquito larvae. *Environmental Entomology* **12**: 515-521.

MERZ, U. (1990). Infectivity inoculum density and germination of *Spongospora subterranea* resting spores a solution culture test system. *Bulletin: Organisation Europeenne et Mediterraneenne pour la Protection des Plantes* **19**: 585-592.

MERZ, U. (1992). Observations on swimming pattern and morphology of secondary zoospores of *Spongospora subterranea*. *Plant Pathology* **41**: 490-494.

MEURS, A. (1928). Wortelrot, Veroorzaakt door Schimmels uit de Geslachten *Pythium* Pringsheim en *Aphanomyces* de Bary. *Proefschrift ter Verkrijging van den Graad van Doctor... Rijks-universiteit te Utrecht...*, 1928: 1-94.

MEURS, A. (1934). Parasitic stemburn of Deli tobacco. *Phytopathologische Zeitschrift* **7**: 169-185.

MEYEN, F. J. F. (1839). Von der Fortspflanzung der Pilze. *Neues Systems der Pflanzen-physiologie* **3**: 453.

MEYER, D. (1975). Zur Keimung der Oosporen von *Phytophthora cactorum* (Leb. et Cohn) Schroet. *Zeitschrift für Pflanzenkrankheiten und Pflanzenschutz* **82**: 168-172.

MICHELMORE, R. W. & INGRAM, D. S. (1980). Heterothallism in *Bremia lactucae*. *Transactions of the British mycological Society* **75**: 47-56.

MICHELMORE, R. W. & INGRAM, D. S. (1981). Origin of gametangia in heterothallic isolates of *Bremia lactucae*. *Transactions of the British mycological Society* **76**: 425-432.

MICHELMORE, R. W. & SANSOME, E. (1982). Cytological studies of heterothallism and secondary homothallism in *Bremia lactucae*. *Transactions of the British mycological Society* **79**: 291-297.

MIDDLETON, J. & DICK, M. W. (1989). Retrievability of *Phytophthora cinnamomi* in sand plates. *Mycological Research* **93**: 385-386.

MIDDLETON, J. T. (1943). The taxonomy, host range and geographic distribution of the genus *Pythium*. *Memoirs of the Torrey Botanical Club* **20**: 1-171.

MIDDLETON, J. T. (1952). Generic concepts in the Pythiaceae. *Tijdschrift over Plantenziekten [Netherlands Journal of Plant Pathology]* **58**: 226-235.

MIGULA, W. (1903). Myxomycetes, Phycomycetes, Basidiomycetes, ... *In* Thomé, *Flora Deutschlands, Österreich und der Sweitz, Pilze* **3(1)**: 1-510.

MILANEZ, A. I. (1965*a*). *Myzocytium megastomum* de Wild. in São Paulo, Brazil. *Rickia, Brazil* **2**: 153-158.

MILANEZ, A. I. (1965*b*). *Achlya brasiliensis*, a new species from Brazil. *Rickia, Brazil* **2**: 183-189.

MILANEZ, A. I. (1967). A study of the aquatic "Phycomycetes" of Gull Lake area in Michigan. *Dissertation Abstracts* **28**: no. 6, 2 pp.

MILANEZ, A. I. (1968). Aquatic fungi of the "Cerrado" region of São Paulo State. I. First results. *Rickia, Brazil* **3**: 97-109.

MILANEZ, A. I. (1969*a*). Occurrence of *Achlya radiosa* in the Americas. *Rickia, Brazil* **4**: 41-46.

MILANEZ, A. I. (1969*b*). Occurrence of *Petersenia irregulare* (H.E. Petersen) Sparrow in Brazil. *Rickia, Brazil* **4**: 75-82.

MILANEZ, A. I. (1970). Contributions to the knowledge of aquatic Phycomycetes of São Paulo State. I. Oomycetes from the west region. *Rickia, Brazil* **5**: 23-43.

MILANEZ, A. I. (1984). Fungos zoospóricos do Estado de São Paulo. II. Chytridiomycetes da Região Oeste. *Rickia, Brazil* **11**: 115-127.

MILANEZ, A. I. (1986). Saprolegniaceae no Brasil. *Rickia, Brazil* **13**: 127-131.

MILANEZ, A. I. & BENEKE, E. S. (1966). An unusual *Saprolegnia* from the Gull Lake area. *Papers of the Michigan Academy of Science, Arts, and Letters* **51**: 175-182.

MILANEZ, A. I. & BENEKE, E. S. (1968). New rcords of aquatic Phycomycetes from Michigan. *Papers of the Michigan Academy of Science, Arts, and Letters* **53**: 11-22.

MILANEZ, A. I., PIRES-ZOTTARELLI, C. L. A. & SCHOENLEIN-CRUSIUS, I. H. (1995). Cryptogams of the Parque Estadual das Fontes do Ipiranga: São Paulo, SP. Fungi, 2: Lagenidiales. *Hoehnea* **22**: 115-123.

MILANEZ, A. I. & TRUFEM, S. F. B. (1981). Ficomicetos em frutos submersos, do Parque Estadual das Fontes do Ipiranga, São Paulo. *Rickia, Brazil* **9**: 7-15.

MILANEZ, A. I. & TRUFEM, S. F. B. (1984). Fungos zoospóricos em frutos submersos do Parque Estadual das Fontes do Ipiranga, São Paulo. II. *Rickia, Brazil* **11**: 77-84.

MILANEZ, A. I. & VAL, F. C. (1969). Occurrence of *Petersenia irregulare* (H. E. Petersen) Sparrow in Brazil. *Rickia, Brazil* **4**: 75-82.

MILHOLLAND, R. D. (1975). Pathogenicity and histopathology of *Phytophthora cinnamomi* on highbush and rabbiteye blueberry. *Phytopathology* **65**: 789-793.

MIL'KO, A. A. [А. А. МЦЛЪКО, **A. A. Mil'ko**] (1979). Систематика рода *Saprolegnia*. Taxonomy of the genus *Saprolegnia*. *Mikologiya i Fitopatologiya [Микопогия и ФитоПапогия]* **13**: 288-294.

MIL'KO, A. A. [А. А. МЦЛЪКО, **A. A. Mil'ko**] (1983). Систематика рода *Achlya*. Systematics of *Achlya* genus. *Mikologiya i Fitopatologiya [Микопогия и ФитоПапогия]* **17**: 289-296.

MILLARDET, P. M. A. (1885). Traitment du mildou et du rot. *Journal d'Agriculture Practique* **2**: 513-516 [See Schneiderhan, F. J. (1933) below.]

MILLER, C. E. (1955). *Micromyces grandis*, a new member of the aquatic Synchytriaceae. *Journal of the Elisha Mitchell Scientific Society* **71**: 247-255.

MILLER, C. E. (1958*a*). Morphology and cytology of the zoosporangia and cystosori of *Sorosphaera veronicae*. *Journal of the Elisha Mitchell Scientific Society* **74**: 49-64.

MILLER, C. E. (1958*b*). Zoosporangia of *Sorosphaera veronicae*. *Association of Southeastern Biologists, Bulletin* **5**: 13.

MILLER, C. E. (1959*a*). Studies on *Ligniera verrucosa*, a plasmodiophoraceous parasite. *Association of Southeastern Biologists, Bulletin* **6**: 29.

MILLER, C. E. (1959*b*). Studies on the life cycle and taxonomy of *Ligniera verrucosa*. *American Journal of Botany* **46**: 725-729.

MILLER, C. E. (1961). Some aquatic Phycomycetes from Lake Texoma. *Journal of the Elisha Mitchell Scientific Society* **77**: 293-298.

MILLER, C. E. (1962*a*). A new species of *Petersenia* parasitic on *Pythium*. *Mycologia* **54**: 422-431.

MILLER, C. E. (1962*b*). Some fungal parasites of *Pythium* from Texas soils. *Association of Southeastern Biologists, Bulletin* **9**: 42.

MILLER, C. E. (1965*a*). Observations on some parasitic aquatic Phycomycetes. *Journal of the Elisha Mitchell Scientific Society* **81**: 4-9.

MILLER, C. E. (1965*b*). Annotated list of aquatic phycomycetes from Mountain Lake Biological Station, Virginia. *Virginia Journal of Science* **16**: 219-228.

MILLER, C. E. & DYLEWSKI, D. P. (1980). Life cycle studies on *Woronina pythii* (Plasmodiophoromycetes), an endopasasite of *Pythium* sp. (Oömycetes). *Bulletin de la Société de Botanique, France* **127**: 119-122.

MILLER, C. E. & DYLEWSKI, D. P. (1983). Zoosporic pathogens of lower plants - what can be learned from the likes of *Woronina*? *In* S. T. Buczacki (Ed.), *Zoosporic Plant Pathogens, A Modern Perspective*, pp. 249-283. London, U.K.: Academic Press.

MILLER, C. E., MARTIN, R. W. & DYLEWSKI, D. P. (1985). The ultrastructure of plasmodia, sporangia and cystosori of *Ligniera verrucosa* (Plasmodiophoromycetes). *Canadian Journal of Botany* **63**: 263-273.

MILLER, J. D. & JONES, E. B. G. (1983). Observations on the association of thraustochytrid marine fungi with decaying seaweed. *Botanica Marina, Berlin* **26**: 345-351.

MILLER, J. D. & WHITNEY, N. J. (1981). Fungi of the Bay of Fundy IV. Thraustochytrids. *Nova Hedwigia* **35**: 407-416.

MILLER, R. I. (1983). Investigation into the biology of the three "phycomycotic" agents pathogenic for horses in Australia. *Mycopathologia* **81**: 23-28.

MILLER, R. I. & CAMBELL, R. S. F. (1983). Experimental pythiosis in rabbits. *Sabouraudia* **21**: 331-341.

MILLER, R. I., OLCOTT, B. M. & ARCHER, M. (1985). Cutaneous pythiosis in beef calves. *Journal of the American Medical Association* **186**: 984-986.

MILOVTZOVA, M. O. [М. О. МІЛОВЦОВА, **M. O. Milovtzova**] (1935*a*). Водян грибн [illegible] та Його Околиць (Aquatic phycomycetes from Kharkov and its environs). *Харківський Державний Університет Труди Н.-Д. Інституту Ботаніки* (*Université de Charkov. Travaux de l'Institut Botanique*). 1: 28-37. [In Russian with English summary.]

MILOVTZOVA, M. O. [М. О. МІЛОВЦОВА, **M. O. Milovtzova**] (1935*b*). *Aphanomyces polysporis* sp. nov. *Харківський Державний Університет Труди Н.-Д. Інституту Ботаніки* (*Université de Charkov. Travaux de l'Institut Botanique*). 1: 38-41. [In Russian with English summary.]

MIN, H.-K., HATAI, K. & BAI, S. (1994). Some inhibitory effects of chitosan on fish-pathogenic oomycete, *Saprolegnia parasitica*. *Fish Pathology (Gyobyo Kenkyu)* **29**: 73-77.

MINDEN, M. von (1911-1915). Pilze. [pp. 207-630]. I. Reihe: Chytridiineae [pp. 209-422]. II. Reihe: Ancylistineae [pp. 423-461]. III. Monoblepharidineae [pp. 462-478]. IV. Reihe: Saprolegnineae [pp. 479-606]. *In Kryptogamenflora der Mark Brandenburg und angrenzender Gebiete* (R. Kolkwitz, E. Jahn & M. von Minden) **5**: hefte I (pp. 1-192 published 27 December 1909), hefte II (pp. 193- 352 published 20 April 1911), hefte III (pp. 353-496 published 10 September 1911), hefte IV (pp. 497-608 published 30 January 1912), hefte V (pp. 609-630 published 30 June 1915). Leipzig, Germany: Gebrüder Borntraeger.

MINDEN, M. von (1911*a*). [Chytridiineae, Ancylistineae, Monoblepharidineae, Saprolegniineae] I. Reihe: Chytridiineae [first part, pp. 209-352; second part in Minden (1911*b*)]. *In Kryptogamenflora der Mark Brandenburg und angrenzender Gebiete* **5**: 209-352.

MINDEN, M. von (1911*b*). [Chytridiineae, Ancylistineae, Monoblepharidineae, Saprolegniineae] I. Reihe: Chytridiineae [pp. 353-422]. II. Reihe: Ancylistineae [pp. 423-461]. III. Reihe: Monoblepharidineae [pp. 461-478]. IV. Reihe: Saprolegniineae [first part, pp. 479-496; second part Minden (1912); third part Minden (1915). *In Kryptogamenflora der Mark Brandenburg und angrenzender Gebiete* **5**: 353-496.

MINDEN, M. von (1912). [Chytridiineae, Ancylistineae, Monoblepharidineae, Saprolegniineae]. IV. Reihe: Saprolegniineae [second part; first part Minden (1911*b*); third part Minden (1915)]. *In Kryptogamenflora der Mark Brandenburg und angrenzender Gebiete* **5**: 497-608. 609-612.

MINDEN, M. von (1915). [Chytridiineae, Ancylistineae, Monoblepharidineae, Saprolegniineae]. IV. Reihe: Saprolegniineae [third part; first part Minden (1911*b*); second part Minden (1912)]. *In Kryptogamenflora der Mark Brandenburg und angrenzender Gebiete* **5**: 609-612.

MINDEN, M. von (1916). Beiträge zur Biologie und Systematik einheimischer submerser Phycomyceten. *Mykologische Untersuchungen und Berichte von Dr. Richard Falck* **1** [1923] or **2(2)**: 146-255. [Apparently published in parts or fascicles from 1913 [ie. **2(2)**: 146-255 (1916) as quoted by Sparrow (1960)], but the papers were not collected as a volume [Erster Band] until 1923; the latter date will be most helpful for library retrieval.]

MISHRA, R. C. (1994). Characterization of aquatic moulds from Gujar Lake (Jaunpur). *Indian Journal of Mycology and Plant Pathology* **24**: 209-215.

MITCHELL, D. J. & MITCHELL, J. E. (1973). Oxygen and carbon dioxide concentration effects on the growth and reproduction of *Aphanomyces euteiches* and certain other soil-borne plant pathogens. *Phytopathology* **63**: 1053-1059.

MITCHELL, J. B. (1928). Studies on the life history of a parasite of the Euglenidae. *Transactions of the American Microscopical Society* **47**: 29-35.

MITCHELL, J. E. & HAGEDORN, D. J. (1971). Residual dexon and the persistent effect of soil treatments for control of pea root rot caused by *Aphanomyces euteiches*. *Phytopathology* **61**: 978-983.

MITCHELL, J. E. & YANG, C. Y. (1966). Factors affecting growth and development of *Aphanomyces euteiches*. *Phytopathology* **56**: 917-922.

MITCHELL, R. T. & DEACON, J. W. (1986*a*). Differential (host-specific) accumulation of zoospores of *Pythium* on roots of graminaceous and non-graminaceous plants. *New Phytologist* **102**: 113-122.

MITCHELL, R. T. & DEACON, J. W. (1986*b*). Selective accumulation of zoospores of chytridiomycetes and oomycetes on cellulose and chitin. *Transactions of the British Mycological Society* **86**: 219-223.

MITCHELL, R. T. & DEACON, J. W. (1986*c*). Chemotropism of germ-tubes from zoospore cysts of *Pythium* spp. *Transactions of the British Mycological Society* **86**: 233-237.

MITCHELL, R. T. & DEACON, J. W. (1987). Differential adhesion of zoospore cysts of *Pythium* on roots of graminaceous and non-graminaceous plants. *Transactions of the British Mycological Society* **88**: 401-403.

MITHEN, R. & MAGRATH, R. (1992). A contribution to the life history of *Plasmodiophora brassicae*: secondary plasmodia development in root galls of *Arabidopsis thaliana*. *Mycological Research* **96**: 877-885.

MIURA, K. (1970). Some endozoic fungi parasitic on soil nematodes. *Journal of Japanese Botany* **45**: 233-240.

MIURA, K. (1973). Three entomophthoralean parasites of nematodes collected from Japan. *Kinjin Kenkyusho [Reports of the Tottori Mycological Institute]* **10**: 517-522.

MIX, A. J. (1945). *Aphanomyces* root rot of lettuce, pepper, and eggplant seedlings in northern New Jersey. *Plant Disease Reporter* **29**: 649-650.

MIYABE, K. (1903). *Kawakamia* Miyabe. A new genus belonging to Peronosporaceae. *Botanical Magazine, Tokyo* **17**: 306.

MIYAKE, K. (1901). The fertilization of *Pythium debaryanum*. *Annals of Botany, London* **15**: 653-667.

MOENS, P. B. & PERKINS, F. O. (1969). Chromosome number of a small protist: accurate determination. *Science, New York* **166**: 1289-1291.

MOESTRUP, Ø. (1970). On the fine structure of the spermatozoids of *Vaucheria sescuplicaria* and on the later stages in spermatogenesis. *Journal of the Marine Biological Association of the United Kingdom* **50**: 513-523.

MOESTRUP, Ø. & ANDERSEN, R. A. (1991). Organization of heterotrophic heterokonts. *In* D. J. Patterson & J. Larsen (Eds), *The Biology of Free-living Heterotrophic Flagellates*, pp. 333-360. Oxford, U.K.: Clarendon Press.

MOEWUS, F. (1943). Zur Sexualität der niederen Organismen. *Ergebnisse der Biologie* **19**: 82-142.

MOLISCH, H. (1926). Botanische Beobachtungen in Japan. VI. Mitteilung. *Pseudoplasmodium aurantiacum* n. g. et n. sp. eine neue Acrasiee aus Japan. *Science Reports, Tôhoku Imperial University, Sendai, Series 4* **1**: 119-121.

MOLINA, F. I. (1986). *Petersenia pollagaster* (Oomycetes): an invasive fungal pathogen of *Chondrus crispus* (Rhodophyceae). *In* S. T. Moss (Ed.), *The Biology of Marine Fungi*, pp. 165-175. Cambridge, U.K.: Cambridge University Press.

MOLINA, F. I., HUGHES, G. C. & CRAIGIE, J. S. (1988). Temperature influences infection rates in the *Chondrus crispus - Petersenia pollagaster* pathosystem: a regression analysis. *Marine Biology (Berlin)* **97**: 431-434.

MOLINA, F. I., JONG, S.-C. & MA, G. (1995). Molecular characterization and identification of *Saprolegnia* by restriction analysis of genes coding for ribosomal RNA. *Anthonie van Leeuwenhoek, Journal of Microbiology and Serology* **68**: 65-74.

MÖLLER, E. M., COCK, A. W. A. M. de, & PRELL, H. H. (1993). Mitochondrial and nuclear DNA restriction enzyme analysis of the closely related *Phytophthora* species *P. infestans*, *P. mirabilis*, and *P. phaseoli*. *Journal of Phytopathology* **139**: 309-321.

MÖLLER, H. (1885). *Plasmodiophora alni*. *Berichte der Deutschen Botanischen Gesellschaft* **3**: 102-105.

MÖLLER, H. (1890). Beitrag sur Kenntniss der *Frankia subtilis*. *Berichte der Deutschen Botanischen Gesellschaft* **8**: 215-224.

MØLLER, K. & HOCKENHULL, J. (2001). Biometry of oospores and intraspecific variation of four *Pythium* species. *Mycological Research* **[in press]**.

MOLLIARD, M. (1909). Une nouvelle Plasmodiophorée parasite du *Triglochin palustre* L. *Bulletin de la Société de Botanique, France* **56**: 23-25.

MONEY, N. P. (1998). Why oomycetes have not stopped being fungi. *Mycological Research* **102**: 767-768.

MONEY, N. P. & HILL, T. W. (1997). Correlation between endoglucanase secretion and cell wall strength in oomycete hyphae: implications for growth and morphogeneisis. *Mycologia* **89**: 777-785.

MONEY, N. P. & WEBSTER, J. (1985). Water stress and sporangial emptying in *Achlya intricata*. *Botanical Journal of the Linnean Society* **91**: 319-327.

MONEY, N. P. & WEBSTER, J. (1988). Cell wall permiability and its relationship to spore release in *Achlya intricata*. *Experimental Mycology* **12**: 169-179.

MONEY, N. P., WEBSTER, J. & ENNOS, R. (1988). Dynamics of sporangial emptying in *Achlya intricata*. *Experimental Mycology* **12**: 13-27.

MONEY, N. P., BEAKES, G. W., WEBSTER, J. & WAKLEY, G. (1987). Rudimentary flagella in sporangiospores of *Achlya*. *Transactions of the British mycological Society* **89**: 108-114.

MONSMA, E. Y. (1937). A study of the water molds of the Lydell Stat Fish Hatchery at Comstock Park, Michigan. *Papers of the Michigan Academy of Science, Arts, and Letters* **22**: 165-182.

MONTAGNE, J. F. C. (1845). **[Untitled]**, *L'Institut, Journal Universel des Sciences et des Sociétés Savantes en France et à l'Étranger. 10 Section Sciences mathimatiques, physiques et naturelles, 1845*, **13 (no. 609)**: 312-314.

[*Notes to clarify confusion in recent citations*: date of first publication: 03 Septembre, 1845; diagnosis p. 313; the *Contents* page at the *end* of the volume cites: "Observations sur la maladie des pommes de terre. Montagne..."; reprinted as: *Société Philomatique de Paris. Extraits des procès-verbaux des Séances pendant l'année 1845, Botanique [Séance du 30 Août 1845]*, pp. 98-101 (probably published during 1846). See also *L'Institut, Journal Universel des Sciences et des Sociétés Savantes en France et à l'Étranger. 10 Section Sciences mathimatiques, physiques et naturelles, 1845*, **13 (no. 616)**: (published 22 Octobre 1845): a report of a note presented by M. Montagne at the Séance on 20 October 1845 of the Academy of Sciences (*Compte-rendu hebdomadàire des Séances de l'Académie des Sciences*).

MONTECILLO, C. M., BRACKER, C. E. & POWELL, M. J. (1980). Ultrastructure of *Synchytrium macrosporum* zoospores. *Canadian Journal of Botany* **58**: 1886-1897.

MOORE, C. L. (1909). Some Nova Scotian aquatic fungi. *Proceedings and Transactions of the Nova Scotia Institute of Science* **12**: 217-235.

MOORE, G., DEVOS, K. M., WANG, Z. & GALE, M. D. (1995). Grasses, line up and form a circle. The genomes of six major grass species can be aligned by dissecting the individual chromosomes into segments and rearranging these linkage blocks into highly similar structures. *Current Biology* **5**: 737-739.

MOORE, P. D. (1982). Evolution of photosynthetic pathways in flowerıng plants. *Nature (London)* **295**: 647-648.

MOREAU, F. (1948). Contribution a l'étude de la microflore fongique des eaux douces Normandie. *Bulletin trimestriel de la Société Mycologique de France* **64**: 223-239.

MOREAU, F. & MOREAU, F. (1935*a*). Remarques sur la systématique des Saprolégniacées. *Bulletin de la Société Botanique de France* **82**: 354-358.

MOREAU, F. & MOREAU, F. (1935*b*). Les phénomènes cytologiques du developpement de l'oeuf et de la fécondation chez les Champignons du groupe des Saprolégniées. *Comptes Rendus Hebdomadaires des Séances de l'Académie des Sciences, Paris, Série D* **201**: 1208-1210.

MOREAU, F. & MOREAU, F. (1935*c*). Su des phénomènes d'accrescence des organes de la reproduction sexuelle chez une Saprolégniaciée. *Comptes Rendus des Séances de la Société de Biologie* **119**: 1256.

MOREAU, F. & MOREAU, F. (1936*a*). Action des sucres sur les Saprolégniées. *Comptes Rendus Hebdomadaires des Séances de l'Académie des Sciences, Paris, Série D* **202**: 1086-1087.

MOREAU, F. & MOREAU, F. (1936*b*). Sur la toxicité de quelques cations pour les Saprolégniées. *Comptes Rendus Hebdomadaires des Séances de l'Académie des Sciences, Paris, Série D* **202**: 2175-2177.

MOREAU, F. & MOREAU, F. (1937). Sur l'antagonisme de quelques cations dans les cultures des Saprolégniées. *Comptes Rendus Hebdomadaires des Séances de l'Académie des Sciences, Paris* **203**: 809-811.

MOREAU, F. & MOREAU, F. (1939). Recherches sur les Saprolégniacées. *Annales des Sciences Naturelles, Botanique, Série 11* **1**: 221-332.

MOREAU, F. & MOREAU, F. (1948). Contribution à l'étude de la microflore fongique des eaux douces de Normandie. *Bulletin de la Société Mycologique de France* **64**: 223-237.

MOREL, G. (1944). Le développement du mildiou sur des tissus de vigne cultures in vitro. *Comptes Rendus Hebdomadaires des Séances de l'Academie des Sciences, Paris, Série D* **218**: 50-52.

MOREL, G. (1948). Recherches sur la culture associée de parasites obligatoires et de tissus végétaux. *Annales des Éphiphyties (Série Pathologie Végétale)* **14**: 1-112.

MORGAN, E. (1939). Studies on the Saprolegniaceae of Glamorgan. *Transactions of the British Mycological Society* **23**: 125.

MORINI, F. (1913). Observazioni Micologiche. *Memorie della Reale Accademia delle Scienze dell'Istituto di Bologna, Seria VI* **10**: 297-308.

MORITZI, A. (1913). *Systema Verzeichniss der von H. Zollingewr in den Jahren 1'842-1844 auf Java gesammelten Pflanzen*: 20.

MORRE, D. J., MOLLENHAUER, H. H. & BRACKER, C. E. (1971). Origin and continuity of golgi apparatus. *In* J. Reinert & H. Ursprung (Eds), *Origin and Continuity of Cell Organelles*, pp. 92-126. Berlin, Germany: Springer-Verlag.

MORRIS, B. M. & GOW, N. A. R. (1993). Mechanism of electrotaxis of zoospores of phytopathogenic fungi. *Phytopathology* **83**: 877-882.

MORRIS, B. M., REID, B. & GOW, N. A. R. (1992). Electrotaxis of zoospores of *Phytophthora palmivora* at physiologically relevant field strengths. *Plant, Cell and Environment* **15**: 645-653.

MORUZI, C. & TOMA, N. (1965). Contributii la studiul micoflorei acvatice din complexul lacustru Crapina-Jijila. *Microbiologia* **1**: 407-413.

MORUZI, C. & TOMA, N. (1968). Contribution à la connaissance des Saprolégniées des bassins aquatiques du jardin botanique de Bucarest. *Revue Roumaine de Biologie, Série de Botanique* **13**: 167-173.

MOSS, S. T. (1980). Ultrastructure of the endomembrane-sageno-genetosome-ectoplasmic net complex in *Ulkenia visurgensis* (Thraustochytriales). *Botanica Marina, Berlin* **23**: 73-94.

MOSS, S. T. (1985). An ultrastructural study of taxonomically significant characters of the Thraustochytriales and Labyrinthulales. *Botanical Journal of the Linnean Society* **91**: 329-357.

MOSS, S. T. (Ed.) (1986*a*). *The Biology of Marine Fungi*. Cambridge, U.K.: Cambridge University Press.

MOSS, S. T. (1986*b*). Biology and phylogeny of the Labyrinthulales and Thraustochytriales. *In* S. T. Moss (Ed.), *The Biology of Marine Fungi*, pp. 105-129. Cambridge, U.K.: Cambridge University Press.

MOXHAM, S. E. & BUCZACKI, S. T. (1983). Chemical composition of the resting spore wall of *Plasmodiophora brassicae*. *Transactions of the British mycological Society* **80**: 297-304.

MOXHAM, S. E., FRASER, R. S. S. & BUCZACKI, S. T. (1983). Spore wall proteins of *Plasmodiophora brassicae*. *Transactions of the British Mycological Society* **80**: 497-506.

MÜCKE, M. (1908). Zur Kenntnis der Eintwicklung und Befruchtung von *Achlya polyandra* de Bary. *Berichte der Deutschen Botanischen Gesellschaft* **26**, 367-378.

MUELSTEIN, L. K. & AMON, J. P. (1987*a*). Chemotaxis in zoosporic fungi. *In* M. S. Fuller & A. Jaworski A. (Eds), *Zoosporic Fungi in Teaching and Research*, pp. 284-285. Athens, Georgia, U.S.A.: Southeastern Publishing Corporation.

MUELSTEIN, L. K. & AMON, J. P. (1987*b*). Examining the photoresponse in zoosporic fungi. *In* M. S. Fuller & A. Jaworski A. (Eds), *Zoosporic Fungi in Teaching and Research*, pp. 284-285. Athens, Georgia, U.S.A.: Southeastern Publishing Corporation.

MUEHLSTEIN, L. K., PORTER, D. & SHORT, F. T. (1988). *Labyrinthula* sp., a marine slime mold producing the symptoms of wasting disease in eel grass, *Zostera marina*. *Marine Biology (Berlin)* **99**: 465-472.

MUGNIER, J. & GROSJEAN, M. C. (1995). *PCR Catologue in Plant Pathology*: Pythium. Rhone-Poulenc Agro, Lyon, France.

MUHSIN, T. M. (1994). *Calyptralegnia basraensis* sp. nov. (Saprolegniaceae) from Iraq. *Polskie Archiwum Hydrobiologii* **41**: 415-418.

MÜLLER, D. G. & FALK, H. (1973). Flagellar structure of the gametes of *Ectocarpus siliculosus* (Phaeophyta) as revealed by negative staining. *Archiv für Mikrobiologie* **91**: 313-322.

MULLINS, J. T. (1960*a*). A preliminary host-range study of the chytrid *Pringsheimiella*. *Association of Southeastern Biologists, Bulletin* **8**: 35-36.

MULLINS, J. T. (1960*b*). Studies on the life history and development of *Pringsheimiella dioica*. PhD Thesis, University of North Carolina.

MULLINS, J. T. (1961). The life cycle and development of *Dictyomorpha* gen. nov. (formerly *Pringsheimiella*), a genus of the aquatic fungi. *American Journal of Botany* **48**: 377-387.

MULLINS, J. T. (1965*a*). Parasitism of the chytrid *Dictyomorpha dioica*. *Mycologia* **57**: 352-359.

MULLINS, J. T. (1965*b*). Addendum to "Parasitism of the chytrid *Dictyomorpha dioica*". *Mycologia* **57**: 989.

MULLINS, J. T. (1968). Genetic and hormonal regulation of heterothallism in the water molds. *Journal of the Elisha Mitchell Scientific Society* **84**: 195-198.

MULLINS, J. T. & BARKSDALE, A. W. (1965). Parasitism of the chytrid *Dictyomorpha dioica*. *Mycologia* **57**: 352-359, 989.

MULLINS, J. T. & RAPER, J. R. (1965). Heterothallism in biflagellate aquatic fungi: preliminary genetic analysis. *Science, New York* **150**: 1174-1175.

MULLINS, J. T. & WARREN, C. O. (1975). Nutrition and sexual reproduction in the water mold *Achlya*. *American Journal of Botany* **62**: 770-774.

MUNDKUR, B. B. (1938). *Fungi of India*. Supplement I. Scientific Monograph No 12. Delhi, India: Imperial Council for Agricultural Research.

MUNDKUR, B. D. (1949). Morphology and cytology of development of the sex organs of *Phytophthora himalayensis*. Dastur. *Botanical Gazette* **110**: 475-486.

MUNKVOLD, G. P. & CARLTON, W. M. (1995). Prevalence and distribution of *Aphanomyces euteiches* and *Phytophthora medicaginis* in Iowa farm fields. *Plant Disease* **79**: 1251-1253.

MURDIA, M. S. (1938). Cytological studies of certain members of the family Saprolegniaceae - Part I. *Journal of the Indian Botanical Society* **17**: 301-310.

MURDIA, M. S. (1939). Observations on the development of zoosporangium and liberation of zoospores in *Achlya dubia* Coker. *Proceedings of the National Academy of Sciences, India* **9**: 168-174.

MURDOCH, D. & PARR, D. (1997). *Pythium insidiosum* keratitis. *Australian and New Zealand Journal of Ophthalmology* **25**: 177-179.

MURGOCI, A. A. (1947). Endomycose du sterlet, causée par un champignon Saprolegnicé, (*Saprolegnia exigua* n. sp.). *Académie Roumaine, Bulletin de la Section Scientifique, Bucharest [Académie Românа, Bucharest, Section Stüntifica. Bulletin]* **30**: 151-167.

MURPHY, P. A. (1914). Morphology and cytology of the sexual organs of *Phytophthora erythroseptica* Pethyb. (Preliminary note.). *Annals of Botany* **28**: 735-736.

MURPHY, P. A. (1918). The morphology and cytology of the sexual organs of *Phytophthora erythroseptica*. Pethyb. *Annals of Botany* **32**, 115-53.

MURRAY, G. (1885). Notes on the inoculation of fishes with *Saprolegnia ferax*. *Journal of Botany* **23**: 302-308.

MUSGRAVE, A. & NIEUWENUIS, D. (1975). Metabolism of radioactive antheridiol by *Achlya* species. *Archives of Microbiology* **105**: 313-317.

MUTASA, E. S., CHWARSZCZYNSKA, D. M. & ASHER, M. J. C. (1996). Single-tuber, nested PCR for the diagnosis of *Polymyxa betae* infection in sugar beet roots and colorimetric analysis of amplified products. *Phytopathology* **86**: 493-497.

NABEL, K. (1939). Über die Membran niederer Pilze, besonders von *Rhizidiomyces bivellatus* nov. spez. *Archiv für Mikrobiologie* **16**: 515-541.

NAGAI, M. (1931). Studies on the Japanese Saprolegniaceae. *Journal of the Faculty of Agriculture, Hokkaido Imperial University* **32**: 1-43.

NAGAI, M. & TAKAHASHI, T. (1962). Electron microscope observations on the zoospores of *Saprolegnia diclina*. *Transactions of the Mycological Society of Japan* **3**: 19-23.

NÄGELI, K[arl]. (1842). Botanische Beiträge. *Linnaea, ein Journal für die Botanik in ihrem ganzen Umfange* **16**: 237-285, 1 plate.

NÄGELI, C[arl]. (1844). Zellenkerne, Zellenbildung und Zellenwachsthum bei den Pflanzen. *Zeitschrift für Wissenschaftliche Botanik (Eds: J. Schleiden und C. Nägeli)* **1(1)** [only one volume issued]: 34-133, plates I & II.

NÄGELI, C[arl]. (1846). Zellenkerne, Zellenbildung und Zellenwachsthum bei den Pflanzen. *Zeitschrift für Wissenschaftliche Botanik (Eds: J. Schleiden und C. Nägeli)* **1(3 & 4)** [only one volume issued]: 22-93, plates III & IV.

NÄGELI, C. W. von (1847). *Die neuern Algensysteme und Versuch zur Begründung eines eigener Systems der Algen und Florideen von C. Nägeli* [pp. 1-275, with 10 plates]. Zurich, Switzerland: F. Schulthess.

NÄGLER, K. (1911). Studien über Protozoen aus einen Almtümpel. II. Parasitische Chytridiaceen in *Euglena sanguinea*. *Archiv für Protistenkunde* **23**: 262-268.

NAKAMURA, H. (1965). The uses of tissue cultures in the study of obligate parasites (*Peronospora parasitica*). *Proceedings of the International Conference on Plant Tissue Culture* pp 535-540. Berkeley McCutchen Publishing Co.

NAKAMURA, K. & HATAI, K. (1994). *Atkinsiella parasitica* sp. nov. isolated from a rotifer, *Brachionus plicatilis*. *Mycoscience* **35**: 383-389.

NAKAMURA, K. & HATAI, K. (1995*a*). Three species of Lagenidiales isolated from the eggs and zoeae of the marine crab *Portunus pelagicus*. *Mycoscience* **36**: 87-95.

NAKAMURA, K. & HATAI, K. (1995*b*). *Atkinsiella dubia* and related species. *Mycoscience* **36**: 431-438.

NAKAMURA, K., NAKAMURA, M. & HATAI, K. (1994). *Atkinsiella* infection in the rotifer *Brachionus plicatilis*. *Mycoscience* **35**: 291-294.

NAKAMURA, K., NAKAMURA, M., HATAI, K. & ZAFRAN (1995). *Lagenidium* infection in eggs and larvae of mangrove crab (*Scylla serrata*) produced in Indonesia. *Mycoscience* **36**: 399-404.

NAKAMURA, K., WADA, S., HATAI, K. & SUGIMOTO, T. (1994). *Lagenidium myophilum* infection in the coonstripe shrimp, *Pandalus hypsinotus*. *Mycoscience* **35**: 99-104.

NAKAMURA, K., YUASA, K., SINMUK, S., HATAI, K. & HARA, N. (1995). The ubiquinone system in Oomycetes. *Mycoscience* **36**: 121-123.

NAKATSUJI, N. & BELL, E. (1980). Control by calcium of the contractility of *Labyrinthula* slimeways and of the translocation of *Labyrinthula* cells. *Cell Motility* **1**: 17-19.

NAKATSUJI, N., NAKATSUJI, T. & BELL, E. (1981). Formation of whirling aggregates by *Labrinthula* vegetative cells. *Experimental Cell Research* **131**: 412-416.

NAOUMOFF, N. (1913). Matériaux pour la Flore mycologique de la Russie. *Bulletin de la Société Mycologique de France* **29**: 273-290.

NARISAWA, K., KAGEYAMA, K. & HASHIBA, T. (1996). Efficient root infection with single resting spores of *Plasmodiophora brassicae*. *Mycological Research* **100**: 855-858.

NATVIG, D. O. & GLEASON, F. H. (1983). Oxygen uptake by obligately-fermentative aquatic fungi: absence of a cyanide-sensitive component. *Archives of Microbiology* **134**: 5-8.

NAUMOV, N. I. [Н. И. НАУМОВ, **N. I. Naumov, also N. Naoumoff**] (1925). Болезни огородных растений в Детском Селе за 1923 и 1924 гг. Les maladies des plantes potagères á Detskoe Selo en 1923 et 1924. *La Défense des Plantes. Bulletin du Bureau Permanent des Congrès Entomo-Phytopathologiques de Russie* [Защита Растений от Вредителей. Бюллетень Постоянного Бюро Всероссийских Энтомо-Фитопатологических Съездов] **2**: 242-245.

NEES von ESENBECK, C. G. (1823). Zusatz zu: C. G. Carus, Beitrag zur Geschichte der unter wasser an verwesenden Thierkörpern sich erzeugenden Schimmel- oder Algen-Gattungen. *Nova Acta [Novorum Actorum] Physico-medica Academiae Caesareae Leopoldino-carolinae Naturae Curiosorum* **11**: 507-522.

NEISH, G. A. & GREEN, B. R. (1976). Nuclear and satellite DNA base composition and the taxonomy of *Saprolegnia*. *Journal of General Microbiology* **96**: 215-219.

NEISH, G. A. & HUGHES, G. C. (1980). *Diseases of Fishes, Book 6: Fungal Diseases of Fishes*. Reigate, U.K.: T.F.H. Publications, Ltd.

NELSON, E. H., FISCHER, W. S. & SHLESER, R. A. (1976). A new mycosis of larval lobster (*Homarus americanus*). *Journal of Invertebrate Pathology* **27**: 177-183.

NÉMEC, B. (1911*a*). Über eine Chytridiazee der Zukerrübe. *Berichte der Deutschen Botanischen Gesellschaft* **29**: 48-50.

***NĚMEC, B.** (1911*b*). Příspěvky k poznání nižších hub. I. Nova Chytridiacea *Sorolpidium betae* n. sp. *Rospravy České Akademie Císaře Františka Josefa Pro Vědy, Slovestnost a Umění, Třída II, Praha*, **20** *Číslo* **12**: 1-16[18], plates I & II. ([Zur Kenntnis der niederen Pilze. I. Eine neue Chytridiacee]. *Bulletin International de l'Academie des Sciences de Bohême, class 2, 1911* **16**: 68-87.)

NÉMEC, B. (1911*c*). Příspěvky k poznání nižších hub. III. *Olpidium salicorniue* n. sp. *Rospravy České Akademie Císaře Františka Josefa Pro Vědy, Slovestnost a Umění, Třída II, Praha*, **20** *Číslo* **31**: 1-10, plates I & II. (*Bulletin International de l'Academie des Sciences de Bohême, class 2, 1911* **16**: 136-144.)

NĚMEC, B. (1912*a*). Příspěvky k poznání nižších hub. IV. *Olpidium brassicae* Wor. a dua nové druhy rodu. *Rospravy České Akademie Císaře Františka Josefa Pro Vědy, Slovestnost a Umění, Třída II, Praha*, **21** *Číslo* **8**: 1-11, plates I & II. ([Zur Kenntnis der niederen Pilze. IV. *Olpidium brassicae* und zwei *Entophlyctis*-arten]. *Bulletin International de l'Academie des Sciences de Bohême, class 2, 1911* **16**: 136-144.)

NĚMEC, B. (1912*b* [1913]). Příspěvky k poznání nižších hub. V. *Anisomyxa plantaginis* n. g., np. *Rospravy České Akademie Císaře Františka Josefa Pro Vědy, Slovestnost a Umění, Třída II, Praha*, **21** *Číslo* **35**: 1-16, plates I & II. ([Zur Kenntnis der niederen Pilze. V. Über die Gattung *Anisomyxa plantaginis* n.g, n. sp]. *Bulletin International de l'Academie des Sciences de Bohême, class 2, 1913* **18**: 1-15.)

NĚMEC, B. (1931). *Jaraia salicis*. [ní nižší České Akademie Císaře Františka Josefa Pro Vědy, Slovestnost a Umění, Třída II, Praha (Studies from the Plant Physiological Laboratory of Charles University, Prague)]. *V3stník Královské České Společnosti Nauk, Třída Matematicko-Přírodovědecká, Ročník 1931 (Mémoires de la Société Royale des Sciences de Bohême, Classe des Sciences, Année 1931)*. **4(1)**; *Číslo* **22**: 1-21 [paged as a separate, but see footnote to Němec, 1911*b*].

NES, W. D. (1987). Biosynthesis and requirement for sterols in the growth and reproduction of Oomycetes. *In* G. Fuller & W. D. Nes (Eds), *Ecology and Metabolism of Plant Lipids*, pp. 304-328. Washington, D.C., U.S.A.: American Chemical Society.

NES, W. D. (1990). Stereochemistry, sterol metamorphosis and evolution. *In* P. J. Quinn & J. L. Harwood (Eds), *The Biochemistry, Structure and Utilization of Plant Lipids*, pp. 308-319. Portland, Oregon, U.S.A.: Portland Press.

NES, W. D., LE, P. H., BERG. L. R., PATTERSON, G. W. & KERWIN, J. L. (1986). A comparison of cycloartenol and lanosterol biosynthesis and metabolism by the Oomycetes. *Experientia (Basel)* **42**: 556-558.

NES, W. D., SAUNDERS, G. A. & HEFTMANN, E. (1982). *Lipids (USA)* **17**: 178-183.

---

*FOOTNOTE: In this journal (and also following publications), held at the Natural History Museum, London, each ***article*** (*Číslo*) is numbered consecutively, but paged 1-[x]. The texts are in Čzech without German subtitles. I have been unable to trace the French-titled journal, which is commonly quoted, at the Natural History Museum, and which may be the source of the German titles placed in square brackets.

NES, W. D., NORTON, R. A., GRUMLEY, F. G., MADIGAN, S. J. & KATZ, E. R. (1990). Sterol phylogenesis and algal evolution. *Proceedings of the National Academy of Sciences, U.S.A.* **87**: 7565-7569.

NESTRUD, L. B. & ANDERSON, R. L. (1994). Aquatic safety of *Lagenidium giganteum*: effects on freshwater fish and invertebrates. *Journal of Invertebrate Pathology* **64**: 228-233.

NEWBY, H. V. (1948*a*). Observations on Saprolegniaceae. I. *Saprolegnia anisospora* de Bary. *Transactions of the British Mycological Society* **31**: 254-265.

NEWBY, H. V. (1948*b*). Observations on Saprolegniaceae. II. *Saprolegnia paradoxa* Maurizio. *Transactions of the British Mycological Society* **31**: 266-270.

NEWELL, S. Y. (1976). Mangrove fungi: the succession in the mycoflora of red mangrove (*Rhizophora mangle* L.) seedlings. *In* E. B. G. Jones (Ed.), *Recent Advances in Aquatic Mycology*, pp. 51-91. London, U.K.: Paul Elek.

NEWELL, S. Y., CEFALU, R. & FELL, J. W. (1977). *Myzocytium, Haptoglossa,* and *Gonimochaete* (fungi) in littoral marine nematodes. *Bulletin of Marine Science* **27**: 177-207.

NEWELL, S. Y., MILLER, J. D. & FELL, J. W. (1987). Rapid and pervasive occupation of fallen mangrove leaves by a marine zoosporic fungus. *Applied and Environmental Microbiology* **53**: 2464-2469.

NEWHOOK, F. J., WATERHOUSE, G. M. & STAMPS, D. J. (1978). Tabular key to the species of *Phytophthora* de Bary. *Mycological Papers, Commonwealth Mycological Institute, Kew* **143**: 1-20.

NEWHOOK, F. J., YOUNG, B. R., ALLEN, S. D. & ALLEN, R. N. (1981). Zoospore motility of *Phytophthora cinnamomi* in particulate substrates. *Phytopathologische Zeitschrift* **101**: 202-209.

NEWHOUSE, J. R., TOOLEY, P. W., SMITH, O. P. & FISHEL, R. A. (1992). Characterization of double-stranded RNA in isolates of *Phytophthora infestans* from Mexico, the Netherlands, and Peru. *Phytopathology* **82**: 164-169.

NICHOLAS, G. (1926). Un example nouveau et certain de parasitisme chez les Hépatiques (*Marchantia polymorpha* L.). *Comptes Rendus hebdomadaires des Séances de l'Académie des Sciences, Paris* **182**: 82-83.

NIERE, J. O., GRIFFITH, J. M. & GRANT, B. R. (1990). ^{31}P NMR studies on the effect of phosphite on *Phytophthora palmivora*. *Journal of General Microbiology* **136**: 147-156.

NIEUWLAND, J. A. (1916). Critical notes of new and old genera of plants.-VIII. *American Midland Naturalist* **4**: 379-386.

NIEZABITOWSKI, E. L. (1913). Pasorzyty roślinne morskich raków glebinowych z rodzaju *Pasiphaea*. (Die pflanzlichen Parasiten der Tiefsee-Decapoden- Gattung *Pasiphaea*). *Kosmos, Lwów* **38**: 1563-1572.

NILSON, E. H., FISHER, W. S. & SHLESER, R. A. (1976). A new mycosis of larval lobster *(Homarus americanus)*. *Journal of Invertebrate Pathology* **27**: 177-183.

NISHIMURA, M. (1926). Studies in *Plasmopara halstedii*. *Journal of the College of Agriculture, Hokkaido Imperial University* **17**, 1-61.

NOGA, E. J. & DYKSTRA, M. J. (1986). Oomycete fungi asociated with ulcerative mycosis in menhaden, *Brevoortia tyrannus* (Latrobe). *Journal of Fish Diseases* **9**: 47-53.

NOLAN, R. A. (1966). Nutritional studies with the aquatic phycomycete *Catenaria anguillulae*. *American Journal of Botany* **53**: 624.

NOLAN, R. A. (1970*a*). The phycomycete *Catenaria anguillulae*: growth requirements. *Journal of General Microbiology* **60**: 167-180.

NOLAN, R. A. (1970*b*). Sulfur source and vitamen requirements of the aquatic phycomycete, *Catenaria anguillulae*. *Mycologia* **62**: 568-577.

NOLAN, R. A. (1970*c*). Carbon source and micronutrient requirements of the aquatic phycomycete, *Catenaria anguillulae* Sorokin. *Annals of Botany, London* **34**: 927-939.

NOLARD-TINTIGNER, N. (1973). Étude experimentale sur l'epidmiologie et la pathogenie de la saprolegniose chez *Lebistes reticulatus* Peters et *Xiphophorus helleri* Heckel. *Acta Zoologica et Pathologica Antverpiensia* **57**: 1-127.

NOLARD-TINTIGNER, N. (1974). Contribution a l'étude de la saprolegniose des poissons en region tropicale. *Bulletin de l'Academie Royal de Belgique. Classe des Sciences de naturel medicine (New Series)* **19**: 1-58.

NOLT, B. L., ROMAINE, C. P., SMITH, S. H. & COLE, H. (1981). Further evidence for the association of *Polymyxa graminis* with the transmission of wheat spindle streak mosaic virus. *Phytopathology* **71**: 1269-1272.

NORDBRING-HERTZ, B. (1988). Nematophagous fungi; strategies for nematode exploitation and for survival. *Microbiological Sciences* **5**: 108-116.

NORTON, D. C. (1962). Iowa fungi parasitic on nematodes. *Journal of the Iowa Academy of Science* **69**: 108-117.

NOUVEL, H. (1953). Un Ellobiopsidæ nouveau (*Amallocystis boschmai* n. sp.), parasite d'un Mysidacé en Méditerranée (note préliminaire). *Vie et Milieu, Paris (Bulletin de la Laboratoire Arago)* **4**: 57-58.

NOUVEL, H. & HOENIGMAN, J. (1955). *Amallocystis boschmai* Nouvel 1954. Ellobiopsidé parasite du Mysidacé *Leptomysis gracilis* G. O. Sars. *Vie et Milieu, Supplement No. 6, Paris (Rés. Camp. Pr. Lacaze-Duthiers)* **2**: 7-19.

NOVAES-LEDIEU, M., JIMÉNEZ-MARTÍNEZ, A. & VILLANEUVA, J. R. (1967). Chemical composition of hyphal wall of phycomycetes. *Journal of General Microbiology* **47**: 237-245.

NOWAKOWSKI, L. (1876). Beitrag zur Kenntnis der Chytridiaceen. *Beiträge zur Biologie der Pflanzen* **2**: 73-100.

NYLANDER, W. (1864). Pyrenocarpei quidam Europaei novi. *Flora oder allgemeine botanische Zeitung herausgegeben der königlichen bayerischen botanischen Gesellschaft in Regensburg* [subsequently *Flora, Jena*]: 353-358.

NYBELIN, O. (1931). Undersokningar over kraftpestens orsak. *Ny Svensk Fiskeritidsdrift.*

NYBELIN, O. (1934). Nya undersokningar over kraftpestens orsak. *Ny Svensk Fiskeritidsdrift, No 10*: 110-114.

NYBELIN, O. (1936). Untersuchungen uber die Ursache der in Schweden gegenwartig vorkommenden Krebspest. *Mitteilungen der Anstalt für Binnerfischerei bei Drottningholm, Stockholm, No. 9*: 1-29.

NYHLÉN, L. & UNESTAM, T. (1980). Wound reactions and *Aphanomyces astaci* growth in crayfish cuticle. *Journal of Invertebrate Pathology* **36**: 187-197.

OBEL, P. (1910*a*). Researches on the conditions of the forming of oogonia in *Achlya*. *Annales Mycologici* **8**: 421-443.

OBEL, P. (1910*b*). Undersøgelser over de Forhold, som betinger Forplantningen ved Oosporer hos Saprolegniaceer. *Vidensk Medd Dan Naturhist Foren. Khobenhavn*: 169.

O'BIER, A. H. (1960). A study of the aquatic Phycomycetes associated with diseased fish and fish eggs. Unpublished PhD Thesis, Virginia Polytechnic Institute, Blacksburg, Virginia, U.S.A.

OERTEL, B. & JELKE, E. (1986). Formation of multinucleate zoospores in *Phytophthora infestans* (Mont.) De Bary (*Oomycetes, Pythiaceae*). *Protoplasma* **135**: 173-179.

ÔKUBO, M. see OOKUBO, M.

OLIVE, E. W. (1975). Monograph of the Acrasieae. *Proceedings of the Boston Society for Natural History* **30**: 451-513

OLIVE, L. S. (1975). *The Mycetozoans*. New York, U.S.A.: Academic Press.

OLSON, J. K. (1985). The Riceland Mosquito Management Program (RMMP) - 5 years of progress and product. *Proceedings, 72nd Annual Meeting, New Jersey Mosquito Control Association*, pp. 137-141.

OLSON, L. W. & FULLER, M. S. (1968). Ultrastructural evidence for the biflagellate origin of the uniflagellate fungal zoospore. *Archiv für Mikrobiologie* **62**: 237-250.

OLSON, L. W., LANGE, L. & REICHLE, R. (1978). The zoospore of the aquatic phycomycete *Catenaria anguillulae*. *Protoplasma* **94**: 53-71.

OLSON, L. W. & REICHLE, R. (1978). Meiosis and diploidization in the aquatic phycomycete *Catenaria anguillulae*. *Transactions of the British Mycological Society* **70**: 423-437.

OLSON, R. E., DUNGAN, C. F. & HOLT, R. A. (1991). Water-borne transmission of *Dermocystidium salmonis* in the laboratory. *Diseases of Aquatic Organisms* **12**: 41-48.

OOKUBO, M. (1952). *Lagenidium* sp. on *Aegagrophila*. *Nagaoa* **2**: 124-126.

OOKUBO, M. (1954). Studies on the aquatic fungi collected in the moor and ponds of Hakkôda. *Nagaoa* **4**: 48-60.

OOKUBO, M. & KOBAYASI, Y. (1955). Studies on the water moulds on keratinized materials. *Nagaoa* **5**: 1-10.

OOSAWA, F., KASAE, M., HATANO, S. & ASAKURA, S. (1966). Polymerization of actin and flagellin. *In* G. E. W. Wolstenholme & M. O'Connor (Eds), *Principles of Biomolecular Organization*, pp. 273-307. Boston, Massachusetts, U.S.A.: Little, Brown.

ORDUZ, S. & AXTELL, R. C. (1989). Compatibility of *Bacillus thuringensis israelensis* and *Lagenidium giganteum* for mosquito control. I. Effect of time of application and toxin concentration. *Israel Journal of Entomology* **23**: 99-108.

ORDUZ, S. & AXTELL, R. C. (1991). Compatibility of *Bacillus thuringensis* var. *israelensis* and *Bacillus sphaericus* with the fungal pathogen *Lagenidium giganteum* (Oomycetes, Lagenidiales). *Journal of the American Mosquito Control Association* **7**: 188-193.

ORDUZ, S., ZULUAGA, J. S., DIAZ, T., GOMEZ, C. L. & ROJAS, W. (1992). Effect of antibiotic treatment on mycelial growth of isolates of the mosquito pathogenic fungus *Lagenidium giganteum* (Oomycetes, Lagenidiales) from Australia, Columbia and United States. *Journal of the American Mosquito Control Association* **8**: 427-429.

ORDUZ, S., ZULUAGA, J. S., DIAZ, T. & ROJAS, W. (1992). Five new isolates of the mosquito pathogenic fungus *Lagenidium giganteum* (Oomycetes, Lagenidiales) from Columbia. *Memorias do Instituto Oswaldo Cruz, Rio de Janeiro (Brazil)* **87**: 597-599.

OSBORN, T. G. B. (1911). *Spongospora subterranea* (Wallroth) Johnson. *Annals of Botany, London* **25**: 327-341.

OSTENFELD, C. H. & PETERSEN, H. E. (1930). On a new *Plasmodiophora* found in Canada. *Zeitschrift für Botanik* **23**: 13-18.

O'SULLIVAN, S. M. (1965). The Genus *Isoachlya*. Unpublished PhD Thesis, Queen Mary College, University of London, London, U.K.

OTT, D. W. & BROWN, R. M. (1972). Light and electron microscopical observations on mitosis in *Vaucheria litorea* Hofman ex C. Agardh. *British Phycological Journal* **7**: 361-374.

OTT, D. W. & BROWN, R. M. (1974). Developmental cytology of the genus *Vaucheria* II. Sporogenesis in *V. fontinalis* (L.) Christensen. *British Phycological Journal* **9**: 333-351.

OU, S. H. (1940). Phycomycetes of China. II. *Sinensia* **11**: 427-449.

OVERTON, S. V. & BLAND, C. E. (1981). Infection of *Artemia salina* by *Haliphthoros milfordensis*: a scanning and transmission electron microscope study. *Journal of Invertebrate Pathology* **37**: 249-257.

OVERTON, S. V., THARP, T. P. & BLAND, C. E. (1983). Fine structure of swimming, encysting and germinating spores of *Haliphthoros milfordensis*. *Canadian Journal of Botany* **61**: 1165-1177.

OYARZUN, P. J., DIJST, G., ZOON, F. C. & MAAS, P. W. Th. (1997). Comparison of soil receptivity to *Thielaviopsis basicola*, *Aphanomyces euteiches*, and *Fusarium solani* f.sp. *pisi*. *Phytopathology* **87**: 534-541.

OYARZUN, P. J. & VAN LOON, J. (1989). *Aphanomyces euteiches* as a component of the complex of foot and root rot pathogens of peas in Dutch soils. *Netherlands Journal of Plant Pathology* **95**: 259-264.

PADGETT, D. E. (1978). Salinity tolerance of an isolate of *Saprolegnia australis*. *Mycologia* **70**: 1288-1293.

PADGETT, D. E. (1984). Evidence for extreme salinity tolerance in saprolegniaceous fungi (Oömycetes). *Mycologia* **76**: 372-375.

PADGETT, D. E., KENDRICK, A. S., HEARTH, J. H. & WEBSTER, W.D. (1988). Influence of salinity, temperature, and nutrient availability on the respiration of saprolegniaceous fungi (Oomycetes). *Holarctic Ecology* **11**: 119-126.

PADGETT, D. E. & SEYMOUR, R. L. (1974). Variability of zoospore discharge in species of the genus *Dictyuchus*. *Mycologia* **66**: 615-627.

PAKTITIS, S., GRANT, B. & LAWRIE, A. (1986). Surface changes in *Phytophthora palmivora* zoospores following induced differentiation. *Protoplasma* **135**: 119-129.

PALENIK, B. (1992). Polymerase evolution and organism evolution. *Current Opinions of Genetic Development* **2**: 931-936.

PALLOIX, A., DAUBEZE, A. M. & POCHARD, E. (1988*a*). Time sequences of root infection and resistance expression in an artificial inoculation method of pepper with *Phytophthora capsici*. *Journal of Phytopathology* **123**: 12-24.

PALLOIX, A., DAUBEZE, A. M. & POCHARD, E. (1988*b*). *Phytophthora* root rot of pepper influence of host genotype and pathogen strain on the inoculum density-disease severity relationships. *Journal of Phytopathology* **123**: 25-33.

PALM, B. T. (1918). Sur une Plasmodiophoracée nouvelle, *Ligniera isoetes*. *Svensk Botanisk Tidskrift* **12**: 228-232.

PALM, B. T. & BURK, M. (1933). The taxonomy of the Plasmodiophoraceae. *Archiv für Protistenkunde* **79**: 263-276.

PALTI, J. & KENNETH, R. (1981). The distribution of downy mildew genera over the families and genera of higher plants. *In* D. M. Spencer (Ed.), *The Downy Mildews*. Academic Press, London, UK, pp. 45-56.

PAN, S. Q., YE, X. S. & KUĆ, J. (1992). Induction of chitinases in tobacco plants systemically protected against blue mold by *Peronospora tabacina* or tomato mosaic virus. *Phytopathology* **82**: 119-123.

PANABIÈRES, F., MARAIS, A., TRENTIN, F., BONNET, P. & RICCI, P. (1989). Repetitive DNA polymorphism analysis as a tool for identifying *Phytophthora* species. *Phytopathology* **79**: 1105-1109.

PANABIÈRES, F., PONCHET, M., ALLASIA, V., CARDIN, L. & RICCI, P. (1997). Characterization of border species among Pythiaceae: several *Pythium* isolates produce elicitins, typical proteins from *Phytophthora* spp. *Mycological Research* **101**: 1459-1468.

PAPAVIZAS, G. C. & AYERS, W. A. (1964). Effect of various carbon sources on growth and sexual reproduction of *Aphanomyces euteiches*. *Mycologia* **56**: 816-830.

PAPAVIZAS, G. C. & AYERS, W. A. (1974). *Aphanomyces* species and their root diseases in pea and sugarbeet. *U. S. Department of Agriculture, Agricultural Research Service, Technical Bulletin No. 1485*.

PAPAVIZAS, G. C. & DAVEY, C. B. (1960*a*). Some factors affecting growth of *Aphanomyces euteiches* in synthetic media. *American Journal of Botany* **47**: 758-765.

PAPAVIZAS, G. C. & DAVEY, C. B. (1960*b*). Some factors affecting sexual reproduction of *Aphanomyces euteiches* in synthetic media. *American Journal of Botany* **47**: 884-889.

PAPAVIZAS, G. C. & LEWIS, J. A. (1971). Effect of amendments and fungicides on *Aphanomyces* root rot of peas. *Phytopathology* **61**: 215-220.

PAPAVIZAS, G. C. (1966). Suppression of *Aphanomyces* root rot of peas by cruciferous soil amendments. *Phytopathology* **56**: 1071-1075.

PAPIEROK, B. & BREY, P. T. (1991). A discussion of entomopathogenic fungi and their potential use for the control of insect vectors. *Colloques de l'Institut National de la Recherche Agronomique, France*: 257-270.

PARK, D. (1977). *Pythium fluminum* sp. nov. with one variety and *P. uladhum* sp. nov. from cellulose in fresh-water habitats. *Transactions of the British Mycological Society* **69**: 225-231.

PARKE, J. L. & GRAU, C. R. (1992). *Aphanomyces*. *In* L. L. Singleton, J. D. Mihail & C. M. Rush (Eds), *Methods for Research on Soilborne Phytopathogenic Fungi*, pp. 27-30. St Paul, Minnesota, U.S.A.: American Phytopathological Society.

PARKER, B. C., PRESTON, R. D. & FOGG, G. E. (1963). Studies of the structure and chemical composition of the cell walls of Vaucheriaceae and Saprolegniaceae. *Proceedings of the Royal Society of London, B* **158**: 435-445.

PARKER, S. (Ed.) (1982). *Synopsis and Classification of Living Organisms*. New York, U.S.A.: McGraw Hill.

PARR, D. H., McKINNEY, W. P. & ROGERS, K. (1994). Three unusual isolates from eye infections in New Zealand. *XII Congress of the International Society of Human and Animal Mycology (ISHAM), Adelaide, Australia*: p. D-64.

PARRISH, J. T. (1987). Global palaeogeography and palaeoclimate of the late Cretaceous and early Tertiary. *In* E. M. Friis, W. G. Chaloner & P. R. Crane (Eds), *The Origins of Angiosperms and their Biological Consequences*, pp. 51-73. Cambridge, U.K.: Cambridge University Press.

PARSONS, T. R. (1962). Infection of a marine diatom by *Lagenidium* sp. *Canadian Journal of Botany* **40**: 523.

PASCHER, A. (1931). Systematische Übersicht über die mit Flagellaten in Zusammenhang stehenden Algenreihen und Versuch einer Einrichtung dieser Algenstämme in die Stämme des Pflanzenreiches. *Beihefte zum Botanischen Zentralblatt, Cassel* **48**: 317-332.

PASSERINI, G. (1881). Funghi parmensi enumerati. *Giornale Botanico Italiano* **13**: 267-283.

PATEL, M. K. (1948). *Bremia* sp. on *Arthraxon lancifolius* Hoch in India. *Indian Phytopathology* **1**: 104-106.

PATEL, K. J., RUEDA, L. M. & AXTELL, R. C. (1990). Comparisons of different types and concentrations of alginates for encapsulation of *Lagenidium giganteum* (Oomycetes: Lagenidiales) a fungal pathogen of mosquito larvae. *Journal of the American Mosquito Control Association* **6**: 101-104.

PATEL, K. J., RUEDA, L. M., AXTELL, R. C., STINNER, R. E., SHARPE, J. H. & DeMICHELE, D. W. (1991). Temperature-dependent development of the fungal pathogen *Lagenidium giganteum* (Oomycetes, Lagenidiales) in larvae of *Culex quinquefasciatus* (Diptera: Culicidae). *Journal of Medical Entomology* **28**: 95-100.

PATERSON, R. A. (1958*a*). Parasitic and saprophytic phycomycetes which invade planktonic organisms. I. New taxa and records of chytridiaceous fungi. *Mycologia* **50**: 85-96.

PATERSON, R. A. (1958*b*). Parasitic and saprophytic phycomycetes which invade planktonic organisms. II. A new species of *Dangeardia* with notes on other lacustrine fungi. *Mycologia* **50**: 453-468.

PATERSON, R. A. (1967). Benthic and planktonic Phycomycetes from northern Michigan. *Mycologia* **59**: 405-416.

PATIL, M. S. (1978). Additions to the Synchytria of Maharashtra - II. *Botanique, Nagpur, India* **9**: 203-213.

PATIL, M. S. (1987). Additions to the Synchytria - III. *Indian Phytopathology* **40**: 156-160.

PATIL, S. D. & MAHABALE, T. S. (1964). The genus *Synchytrium* in Maharashtra State. *Journal of the University of Poona [Science and Technology], India* **26**: 67-97.

PATON, F. M. & JENNINGS, D. H. (1989). Evidence that *Thraustochytrium* is unable to synthesize lysine. *Mycological Research* **92**: 470-476.

PATOUILLARD, N. (1918). Quelques champignons de Madagascar. *Bulletin trimestriel de la Société Mycologique de France* **7**: 158-186.

PATTERSON, D. J. (1989). Stramenopiles: chromophytes from a protistan perspective. *In* J. C. Green, B. S. C. Leadbeater & W. L. Diver (Eds), *The Chromophyte Algae, Problems and Perspectives*, pp. 357-379. Oxford, U.K.: Clarendon Press.

PATTERSON, R. T. & RICHARDSON, R. H. (1987). A taxonomic revision on the unilocular Foraminifera. *Journal of Foraminiferal Research* **17**: 212-226.

PATTERSON, P. M. (1927*a*). Fertilization and oogenesis in *Achlya colorata*. *Journal of the Elisha Mitchell Scientific Society* **43**: 108-123.

PATTERSON, P. M. (1927*b*). Oogenesis in *Pythium torulosum*. *Journal of the Elisha Mitchell Scientific Society* **43**: 124-129.

PAU, V. M. & ARONSON, J. M. (1970). Cell wall structure of *Sapromyces elongatus*. *Mycologia* **62**: 531-541.

PAUL, B. (1986*a*). A new species of *Pythium* from Algerian waters. *Hydrobiologia* **131**: 31-38.

PAUL, B. (1986*b*). An aquatic species, *Pythium toruloides* sp. nov., from Algeria. *Transactions of the British Mycological Society* **86**: 330-334.

PAUL, B. (1986*c*). A new non-zoosporic species of *Pythium* from Algeria. *Hydrobiologia* **140**: 233-236.

PAUL, B. (1987*a*). A new species of *Pythium* with filamentous sporangia from Algeria. *Transactions of the British Mycological Society* **89**: 195-198.

PAUL, B. (1987*b*). A new species of *Pythium* with ornamented oogonia from Algeria. *Mycologia* **79**: 797-802.

PAUL, B. (1988*a*). Une nouvelle variété de *Pythium capillosum* dans les sols Algériens. *Cryptogamie, Mycologie* **9**: 35-42.

PAUL, B. (1988*b*). Une nouvelle espèce de *Pythium* isolée d'une saline de l'ouest Algérien. *Cryptogamie, Mycologie* **9**: 325-333.

PAUL, B. (1990). A new aquatic species of *Pythium* from Algeria with filamentous sporangia and thick walled oospores. *Candollea* **45**: 575-579.

PAUL, B. (1991). *Pythium folliculosum*: a new species from the bank of Lake Zürich. *Mycologia Helvetica* **4**: 203-208.

PAUL, B. (1992*a*). *Pythium radiosum*, a new species with ornamented oogonia from France. *Mycologia Helvetica* **5**: 1-8.

PAUL, B. (1992*b*). A new variety of *Pythium* isolated from cultivated soil in France. *Mycologia Helvetica* **5**: 149-156 (1992).

PAUL, B. (1992*c*). *Pythium cylindrosporum* : a new soil fungi from Germany. *International Journal of Mycology and Lichenology* **4**: 337-345.

PAUL, B. (2000). *Pythium contiguanum* nomen novum (syn. *Pythium drechsleri* Paul), its antagonism to *Botrytis cinerea*, ITS1 region of its nuclear ribosomal DNA, and its comparison with related species. *FEMS Microbiology Letters* **183**: 105-110.

PAUL, B., GALLAND, D., BHATNAGAR, T. & DULIEU, H. (1998). A new species of *Pythium* isolated from the Burgundy region in France. *FEMS Microbiology Letters* **158**: 207-213.

PAVILLARD, J. (1914). Observations sur les Diatomées. *Bulletin de la Société Botanique, France* **61**: 164-172.

PAVLETIC, M. M., MILLER, R. I. & TURNWALD, G. H. (1983). Intestinal infraction associated with canine phycomycosis. *Journal of the American Animal Hospital Association* **19**: 913-919.

PAXTON, J. D. (1983). *Phytophthora* root rot and stem rot of soybean: a case study. *In* J. A. Callow (Ed.) *Biochemical Plant Pathology*. Wiley, New York, USA, pp. 19-30.

PAYAK, M. M. & RENFRO, B. L. (1967). A new downy mildew disease of maize. *Phytopathology* **57**: 394-397.

PAYNE, P. A. & ASHER, M. J. C. (1990). The incidence of *Polymyxa betae* and other fungal root parasites of sugar beet in Britain. *Plant Pathology* **39**: 443-451.

PEARSON, J., PIKE, D. J. & DICK, M. W. (1991). A model for short-term fluctuations of *Pythium* propagules in soil. *Mycological Research* **95**: 324-328.

PEDERSÉN, P. M. (1983). Notes on marine, benthic algae from Madeira in nature and in culture. *Bocagiana, Museu Municipal do Funchal No 70*: 1-8.

PELIZON, M. H. & MILANEZ, A. I. (1979). Ficomicetos parasitas de algas da região da grande São Paulo. *Rickia, Brazil* **8**: 51-61.

PEMBERTON, C. M., DAVEY, R. A., WEBSTER, J., DICK, M. W. & CLARK, G. (1990). Infection of *Pythium* and *Phytophthora* species by *Olpidiopsis gracilis* (Oomycetes). *Mycological Research* **94**: 1081-1085.

PENARD, E. (1902). *Faune Rhizopodique du Bassin du Léman*. Geneva, Switzerland: Henry Kündig.

PENDERGRASS, W. R. (1948). A new member of the Plasmodiophoraceae. *Journal of the Elisha Mitchell Scientific Society* **64**: 132-134.

PENDERGRASS, W. R. (1950). Studies on a plasmodiophoraceous parasite, *Octomyxa brevilegniae*. *Mycologia* **42**: 279-289.

PÉREZ, C. (1903). Sur un organisme nouveau, *Blastulidium paedophthorum*, parasite des embryons de *Daphnies*. *Compte rendu des Séances de la Société de Biologie, Paris* **55**: 715-716.

PÉREZ, C. (1905). Nouvelles observations sur le *Blastulidium paedophthorum*. *Compte rendu des Séances de la Société de Biologie, Paris* **58**: 1027-1029.

PÉREZ, C. (1907). *Dermocystis pusula*, organisme nouveau parasite de la peau des tritons. *Compte rendu des Séances de la Société de Biologie, Paris* **63**: 445-446.

PÉREZ, C. (1908). *Dermocystidium pusula*. *Compte rendu des Séances de la Société de Biologie, Paris* **64**: 738.

PÉREZ, C. (1913). *Dermocystidium pusula*, parasite de la peau des tritons. *Archives de Zoologie Expérimentale et Générale* **52**: 343.

PÉREZ REYES, R. (1963). Algunas consideraciones sobre los parasitos de protozoarios incluidos en el genero *Sphaerita*. *Revista de la Sociedad Mexicana de Historia Natural (Mexico)* **24**: 1-6.

PÉREZ-REYES, R., GOMEZ, E. S. (1961). Euglenae del valle de Mexico. IV. Description de algunos endoparasitos. *Revista Latino-americana de Microbiologia (Mexico)* **4**: 53-73.

PÉREZ-REYES, R., MADRAZO-GARIBAY, M. & OCHOTERENA, E. L. (1985). *Pseudosphaerita dryli* sp. nov. parásito de *Phacus acuminata* Stokes (Sarcomastigophora, Euglenidae). *Revista Latino-americana de Microbiologia (Mexico)* **27**: 89-92.

PERKINS, F. O. (1968). Fine structure of zoospores from *Labyrinthulomyxa* sp. parasitizing the clam *Macoma balthica*. *Chesapeake Science* **9**: 198-202.

PERKINS, F. O. (1969). Ultrastructure of vegetative stages in *Labyrinthomyxa marina* (= *Dermocystidium marinum*), a commercially significant oyster pathogen. *Journal of Invertebrate Pathology* **13**: 199-222.

PERKINS, F. O. (1970). Formation of centrioles and centriole-like structures during meiosis and mitosis in *Labyrinthula* sp. (Rhizopodea, Labyrinthulida). *Journal of Cell Science* **6**: 629-653.

PERKINS, F. O. (1972). The ultrastructure of holdfasts, "rhizoids", and "slime tracks" in thraustochytriaceous fungi and *Labyrinthula* spp. *Archiv für Mikrobiologie* **84**: 95-118.

PERKINS, F. O. (1973*a*). A new species of marine labyrinthulid *Labyrinthuloides yorkensis* gen. nov. spec. nov. - Cytology and fine structure. *Archiv für Mikrobiologie* **90**: 1-17.

PERKINS, F. O. (1973*b*). Observations of thraustochytriaceous (Phycomycetes) and labyrinthulid (Rhizopodea) ectoplasmic nets on natural and artificial substrates - an electron microscope study. *Canadian Journal of Botany* **51**: 485-491.

PERKINS, F. O. (1974*a*). Phylogenetic considerations of the problematic thraustochytriaceous-labyrinthulid-*Dermocystidium* complex based on observations of fine structure. *Veröffentlichungen des Instituts für Meeresforschung in Bremerhaven, Supplement* **5**: 45-63.

PERKINS, F. O. (1974*b*). Reassignment of *Labyrinthula minuta* to the genus *Labyrinthuloides*. *Mycologia* **66**: 697-702.

PERKINS, F. O. (1976*a*). Fine structure of lower marine and estuarine fungi. *In* E. B. G. Jones (Ed.), *Recent Advances in Aquatic Fungi*, pp. 279-312. London, U.K.: Paul Elek.

PERKINS, F. O. (1976*b*). Zoospores of the oyster pathogen, *Dermocystidium marinum*. I. Fine structure of the conoid and other sporozoan-like organelles. *Journal of Parasitology* **62**: 959-974.

PERKINS, F. O. & AMON, J. P. (1969). Zoosporulation in *Labyrinthula* sp.; an electron microscope study. *Journal of Protozoology* **16**: 235-257.

PERKINS, F. O. & MENZEL, R. W. (1966). Morphological and cultural studies of a motile stage in the life cycle of *Dermocystidium marinum*. *Proceedings of the National Shellfisheries Association* **56**: 23-30.

PERKINS, F. O. & MENZEL, R. W. (1967). Ultrastructure of sporulation in the oyster pathogen *Dermocystidium marinum*. *Journal of Invertebrate Pathology* **9**: 205-229.

PERRONCITO, E. (1888). *Chytridium elegans*, n. sp., eine parasitäre Chytridinee aus der Classe der Räderthiere. *Centralblatt für Bacteriologie, Parasitenkunde und Infektionskrankheiten* **4**: 295.

PERROTT, P. E. (1960). The ecology of some aquatic Phycomycetes. *Transactions of the British Mycological Society* **43**: 19-30.

PERRY, E. (1967). Host-parasite interactions between *Rozella allomycis* and *Allomyces*. Ph.D. Thesis, University of Indiana, Bloominton, Indiana, U.S.A.

PERSIEL, I. (1960*a*). Über die Verbreitung niederer Phycomyceten in Böden aus verscheidenen Höhenstufen der Alpen und an einigen Standorten subtropischer und tropischer Gebirge. *Archiv für Mikrobiologie* **36**: 257-282.

PERSIEL, I. (1960*b*). Beschreibung neuer Arten der Gattung *Chytriomyces* und einiger seltener niederer Phycomyceten. *Archiv für Mikrobiologie* **36**: 283-305.

PERSIEL, I., SCHOLZ, E. & HARDER, R. (1966). Chytridiales und einige andere iedere Phycomyceten aus Ozeanien. *Archiv für Mikrobiologie* **53**: 173-177.

C. H. PERSOON (1801). *Synopsis Methodica Fungorum* p.223.

PERSSON, M., CERENIUS, L. & SÖDERHÄLL, K. (1987). The influence of haemocyte number on the resistance of the freshwater crayfish, *Pacifastacus leniusculus* Dana, to the parasitic fungus, *Aphanomyces astaci*. *Journal of Fish Diseases* **10**: 471-477.

PERSSON, M. & SÖDERHÄLL, K. (1986). $CaCl_2$-induced germination and peptidase secretion in *Aphanomyces astaci*. *Experimental Mycology* **10**: 205-213.

PETERS, L. (1906). Zur Kenntnis des Wurzelbrandes der Zuckerrube. *Berichte der Deutschen Botanischen Gesellschaft* **24**: 323-329.

PETERS, L. (1911). Über die Erreger des Wurzelbrandes. *Arbeiten aus der Biologischen Reichsanstalt für Land- und Forstwirtschaft* **8**: 211-259.

PETERSEN, A. B. & ROSENDAHL, S. (2000). Phylogeny of the Peronosporomycetes (Oomycetes) based on partial sequences of the large ribosomal subunit (LSU rDNA). *Mycological Research*, xx, xxx.

PETERSEN, E. E., SEMON, M. J., KERWIN, J. L. & BROWER, J. M. (1997). Regulation of attachment, germination, and appressorium formation by zoospores of *Lagenidium giganteum* and related oomycetes by chitin, chitosan, and catecholamines. *Protoplasma* **197**: 96-110.

PETERSEN, H. E. (1905). Contributions à la connaissance des phycomycètes marins (Chytridinae Fischer). *Oversigt over det Kongelige Danske Videnskabernes Seskabs Forhandlinger. Medlemmers Arbeiter, 1905 (no.5)*: 439-488.

PETERSEN, H. E. (1909). Studier over Ferskvands-Phycomyceter. Bidrag til Kundskaben om de Submerse, Phycomyceters Biologi og Systematik, samt om deres Udbredelse i Danmark. *Botanisk Tidsskrift* **29**: 345-440.

PETERSEN, H. E. (1910). An account of Danish freshwater phycomycetes, with biological and systematical remarks. *Annales Mycologici* **8**: 494-560.

PETHYBRIDGE, G. H. (1913). On the rotting of potato tubers by a new species of *Phytophthora* having a method of sexual reproduction hitherto undescribed. *Scientific Proceedings of the Royal Dublin Society* **13**: 529-565.

PEYRONEL, B. (1924). Prime ricerche sulle micorize endtrofiche e sulla micoflora radicicola normale delle fanerogame. *Revista di Biologia* **5-6**: 1-61.

PEYTON, G. A. & BOWEN, C. C. (1963). The host-parasite interface of *Peronospora manshurica* on *Glycine max*. *American Journal of Botany* **50**: 787-797.

PFENDER, W. F. (1984). *Aphanomyces* root rot. *In* D. J. Hagedorn (Ed.), *Compendium of Pea Diseases*, pp. 25-28. St Paul, Minnesota, U.S.A.: American Phytopathological Society.

PFENDER, W. F., DELWICHE, P. A., GRAU, C. R. & HAGEDORN, D. J. (1984). A medium to enhance recovery of *Aphanomyces* from infected plant tissue. *Plant Disease* **68**: 845-847.

PFENDER, W. F. & HAGEDORN, D. J. (1982). *Aphanomyces euteiches* f. sp. *phaseoli*, a causal agent of been root and hypocotyl rot. *Phytopathology* **72**: 303-310.

PFITZER, E. (1869). Physikalische Section. Sitzung vom 20. December. *Sitzungberichte der niederrheinischen Gesellschaft Natur- und Heilkunde, Bonn. 1869*: 218-224.

PFITZER, E. (1872). *Ancylistes closterii*, ein neuer Algen-Parasit aus der Ordnung der Phycomyceten. *Monatsberichte der Königlichen Peussischen Akademie der Wissenschaften zu Berlin*: 379-398.

PICKERING, A. D. & CHRISTIE, P. (1980). Sexual differences in the incidence and severity of ectoparasitic infestation of the brown trout, *Salmo trutta* L. *Journal of Fish Biology* **16**: 669-683.

PICKERING, A. D. & POTTINGER, T. G. (1985). Acclimatation of the brown trout, *Salmo trutta* L., to the stress of daily exposure to malachite green. *Aquaculture* **44**: 145-152.

PICKETT-HEAPS, J. D. (1970). The behavior of the nucleolus during mitosis in plants. *Cytobios* **2**: 69-78.

PIETERS, A. J. (1905). *Annual Report of the Michigan Academy of Science* **8**: 27.

PIETERS, A. J. (1915*a*). The relation between vegetative vigor and reproductiuon in some Saprolegniaceae. *American Journal of Botany* **2**: 529-576.

PIETERS, A. J. (1915*b*). New species of *Achlya* and *Saprolegnia*. *Botanical Gazette* **60**: 483-490.

PIETERS, A. J. (1915*c*). The ferax group of the genus *Saprolegnia*. *Mycologia* **7**: 307-314.

PINE, E. M., HIBBETT, D. S. & DONOGHUE, M. J. (1999). Phylogenetic relationships of cantherelloid and clavariod Homobasidiomycetes based on mitochondrial and nuclear rDNA sequences. *Mycologia* **91**: 944-963.

PIPE, N. D., AZCOITIA, V. & SHAW, D. S. (2000). Self-fertility in *Phytophthora infestans*: heterokaryons segregate several phenotypes. *Mycological Research* **104**: 676-680.

PIROZYNSKI, K. A. (1976*a*). Fossil Fungi. *Annual Review of Phytopathology* **14**: 237-246.

PIROZYNSKI, K. A. (1976*b*). Fungal spores in the fossil record. *Biological Memoirs* **1**: 104-120.

PIROZYNSKI, K. A. & MALLOCH, D. W. (1975). The origin of land plants: a matter of mycotrophism. *BioSystems* **6**: 153-164.

PLAATS-NITERINK, A. J. Van der (1972). The occurrence of *Pythium* in the Netherlands III. *Pythium flevoense* sp.n. *Acta Botanica Neerlandica* **21**: 633-639.

PLAATS-NITERINK, A. J. Van der (1981). Monograph of the genus *Pythium*. *Studies in Mycology, Centraalbureau voor Schimmelcultures, Baarn* **21**: 1-242.

PLAZIAT, J.-C. (1995). Modern and fossil mangroves and mangals: their climatic and biogeographic variability. *In* D. W. J. Bosence & P. A. Allison (Eds), *Marine Palaeoenvironmental Analysis from Fossils*. Geological Society Special Publication No. 83, Geological Society of London, London, UK, pp. 73-96.

PLEHN, M. (1916). Pathogene Schimmelpilze in der Fischniere. *Zeitschrift für Fischerei* **18**: 51-54.

PLESSIS, S. J. du (1933). The life history and morphology of *Olpidiopsis ricciae*, nov. sp., infecting *Riccia* species in South Africa. *Annals of Botany, London* **47**: 755-762.

PODGER, F. D. (1972). *Phytophthora cinnamomi*, a cause of lethal disease in indigenous plant communities in Western Australia. *Phytopathology* **62**: 972-981.

POITRAS, A. W. (1949). A new aquatic species of *Pythium*. *Mycologia*, **41**: 171-176.

POITRAS, A. W. (1955). A preliminary report on the aquatic Phycomyctes for the Pittsburgh region of Western Pennsylvania. *Proceedings of the Pennsylvania Academy of Science* **29**: 127-129.

POKORNY, K. S. (1967). *Labyrinthula*. *Journal of Protozoology* **14**: 697-708.

POKORNY, K. S. (1971). The marine protist *Labyrinthula* Cienkowski: A cytochemical and fine structural study of its morphology and secretion of extracellular material. Ph.D. Thesis, Columbia University, New York.

POKORNY, K. S. (1985). Phylum Labyrinthomorpha. *In* J. J. Lee, S. H. Hutner & E. C. Bobee (Eds), *An Illustrated Guide to the Protozoa*, pp. 318-321. Kansas, U.S.A.: Society of Protozoology.

POLGLASE, J. L. (1980). A preliminary report on the thraustochytrid(s) and labyrinthulid(s) associated with a pathological condition in the lesser octopus *Eledone cirrhosa*. *Botanica Marina, Berlin* **23**: 699-706.

POLGLASE, J. L. (1981). Thraustochytrids as potential pathogens of marine animals. *Bulletin of the British mycological Society, Supplement 1* **16**: 5.

POLGLASE, J. L., ALDERMAN, D. J. & RICHARDS, R. H. (1986). Aspects of the progress of mycotic infections in marine animals. *In* S. T. Moss (Ed.), *The Biology of Marine Fungi*, pp. 155-164. Cambridge, U.K.: Cambridge University Press.

PONGRATZ, E. (1966). De quelques champignons parasites d'organismes planctoniques du Léman. *Schweizerische Zeitschrift für Hydrologie* **28**: 104-132.

PORTER, D. (1969). Ultrastructure of *Labyrinthula*. *Protoplasma* **67**: 1-19.

PORTER, D. (1972). Cell division in the marine slime mold, *Labyrinthula* sp., and the role of the bothrosome in extracellular membrane production. *Protoplasma* **74**: 427-448.

PORTER, D. (1974). Phylogenetic considerations of the Thraustochytriaceae and Labyrinthulaceae. *Veröffentlichungen des Instituts für Meeresforschung in Bremerhaven, Sonderband* **5**: 19-44.

PORTER, D. (1986). Mycoses of marine organisms: an overview of pathogenic fungi. *In* S. T. Moss (Ed.), *The Biology of Marine Fungi*, pp. 141-153. Cambridge, U.K.: Cambridge University Press.

PORTER, D. (1990). Phylum Labyrinthulomycota. *In* L. Margulis, J. O. Corliss, M. Melkonian & D. J. Chapman (Eds), *Handbook of Protoctista*, pp. 388-398. Boston, U.S.A.: Jones & Bartlett.

PORTER, D. & KIRK, P. W. (1987). Marine fungi: taxonomic and ecological considerations. II. Lower fungi. *In* K. G. Mukherji, V. P. Singh & K. L. Garg (Eds), *Frontiers in Applied Microbiology*, **2**, pp.235-256. Lucknow, India: Print House.

PORTER, D. & KOCHERT, G. (1978). Ribosomal RNA molecular weights and the phylogeny of *Labyrinthula*. *Experimental Mycology* **2**: 346-351.

PORTER, D. & LINGLE, W. L. (1992). Endolithic thraustochytrid marine fungi from planted shell fragments. *Mycologia* **84**: 289-299.

PORTER, J. R. & SHAW, D. S. (1978). Aggregation of *Phytophthora drechsleri* zoospores; pattern analysis suggests a taxis. *Transactions of the British Mycological Society* **71**: 515-519.

POST, E. E. & LAUDERMILK, J. D. (1942). A new microchemical reaction for cellulose. *Stain Technology* **17**: 21-24.

POTTER, D., SAUNDERS, G. W. & ANDERSEN, R. A. (1997). Phylogenetic relationships of the Raphidophyceae and Xanthophyceae as inferred from nucleotide sequences of the 18S ribosomal RNA gene. *American Journal of Botany* **84**: 966-972.

POVAH, A. H. (1935). The fungi of Isle Royale, Lake Superior. *Papers of the Michigan Academy of Science, Arts, and Letters* **20**: 113-156.

POWELL, J. R., SCOTT, W. W. & KRIEG, N. R. (1972). Physiological parameters of growth in *Saprolegnia parasitica* Coker. *Mycopathologia et Mycologia Applicata* **47**: 1-40.

POWELL, M. J. (1975). Ultrastructural changes in nuclear membranes and organelle associations during mitosis of the aquatic fungus *Entophlyctis* sp. *Canadian Jovrnal of Botany* **53**: 627-646.

POWELL, M. J. (1980). Mitosis in the aquatic fungus *Rhizophydium sphaerothecea* (Chytridiales). *American Journal of Botany* **67**: 839-853.

POWELL, M. J. (1984). Fine structure of the unwalled thallus of *Rozella polyphagi* in its host *Polyphagus englenae*. *Mycologia* **76**: 1039-1048.

POWELL, M. J. & BLACKWELL, W. H. (1995). Searching for homologous ultrastructural characters in zoosporic fungi. *Canadian Journal of Botany, Supplement 1, Sections E-H* **73**: S693-S700.

POWELL, M. J. & BLACKWELL, W. H. (1998). Phenetic analysis of genera of the Saprolegniaceae (Oomycetes). *Mycotaxon* **68**: 505-516.

POWELL, M. J. & BRACKER, C. E. (1977). Isolation of zoospore organelles from *Phytophthora palmivora*. *In* Proceedings of the 2nd International Mycological Congress, Tampa, Florida, p. 533.

POWELL, M. J., LEHNEN, L. P. & BORTNICK, R. N. (1985). Microbody-like organelles as taxonomic markers among Oomycetes. *BioSystems* **18**: 321-334.

POYTON, R. O. (1970*a*). The characterization of *Hyalochlorella marina* gen. et sp. nov. a new colourless counterpart of *Chlorella*. *Journal of General Microbiology* **62**: 171-188.

POYTON, R. O. (1970*b*). The isolation and occurrence of *Hyalochlorella marina*. *Journal of General Microbiology* **62**: 189-194.

PRASAD, G. & DAYAL, R. (1985). Studies in nematophagous fungi: IX *Myzocytium papillatum* - a new record from India. *Current Science, Bangalore, India* **54**: 286-287.

PRASAD, G. & DAYAL, R. (1986). Studies in nematophagous fungi VIII. Two parasites on *Amphidorylaimus* - sp. *Proceedings of the National Academy of Sciences, India, Section B (Biological Sciences)* **56**: 46-47.

PRASHER, I. B. (1998). *Fungi of Chandigarh*. Dehra Dun, India: Internatiuonal Book Distributors.

PRATT, B. H. SEDGELY, J. H., HEATHER, W. A. & SHEPHERD, C. J. (1972). Oospore production in *Phytophthora cinnamomi* in the presence of *Trichoderma koningii*. *Australian Journal of Biological Sciences* **25**: 861-863.

PRATT, R. G. & MITCHELL, J. E. (1973). A new species of *Pythium* from Wisconsin and Florida isolated from carrots. *Canadian Journal of Botany* **51**: 333-339.

PREISIG, H. R. (1989). The flagellar base ultrastructure and phylogeny of chromophytes. *In* J. C. Green, B. S. C. Leadbeater & W. L. Diver (Eds), *The Chromophyte Algae, Problems and Perspectives*, pp. 167-187. Oxford, U.K.: Clarendon Press.

PREISIG, H. R. (1999). Systematics and evolution of the algae: phylogenetic relationships of taxa within the different groups of algae. *Progress in Botany* **60**: 369-412.

PRETI, G. (1936). Un' infezione de "*Pythium*" su piante di "*Phyllocactus phyllanthoides*". *Rivista di Patologia Vegetale, Padova, Serie* **26**: 331-353.

PRINGSHEIM, N. (1851). Die Entwicklungsgeschichte der *Achlya prolifera*. *Novorum Actorum Academiae caesareae Leopoldino-carolinae Naturae Curiosorum* **23**: 395-460.

PRINGSHEIM, N. (1855). Ueber die Befruchtung der Algen. *Bericht über die ur Bekanntmachung geeigneten verhandlung der Königlichen Preussischen Akademie der Wissenschaften zu Berlin, 1855*: 133-165 [see note on journal title under Braun (1855)].

PRINGSHEIM, N. (1858). Beitrag für Morphologie und Systematik der Algen. [II.] Die Saprolegnieen. *Jahrbuch für Wissenschaftlichen Botanik* **1**: 284-306.

PRINGSHEIM, N. (1860). Beiträge zur Morphologie und Systematik der Algen. IV. Nachträge zur Morphologie der Saprolegnieen. *Jahrbücher für Wissenschaftliche Botanik* **2**: 205-236.

PRINGSHEIM, N. (1873). Weitere Nachtrage zur Morphologie und Systematik der Saprolegnieen. *Jahrbücher für wissenschaftliche Botanik* **9**: 191-234.

PRINGSHEIM, N. (1882). Neue Beobachtungen über den Befruchtungsact der Guttungen *Achlya* und *Saprolegnia*. *Sitzungsberichte der Königlich Preussischen Akademie der Wissenschaften* **2**: 855-890.

PRINGSHEIM, N. (1883*a*). Ueber die vermeintlichen Amöeben in den Schläuchen und Oogonien der Saprolegnieen. *Botanisch Centralblatt* **14**: 378-382.

PRINGSHEIM, N. (1883*b*). Über Cellulinkörner, eine Modification der Cellulose in Körnform. *Berichte der Deutschen Botanischen Gesellschaft* **1**: 288-308.

PRISTOU, R. & GALLEGLY, M. E. (1954). Leaf penetration by *Phytophthora infestans*. *Phytopathology* **44**: 81-86.

PROWSE, G. A. (1951). On *Rozellopsis septigena*. *Transactions of the British Mycological Society* **34**: 400-405.

PROWSE, G. A. (1954*a*). *Aphanomyces daphniae* sp. nov., parasitic on *Daphnia hyalina*. *Transactions of the British Mycological Society* **37**: 22-28.

PROWSE, G. A. (1954*b*). *Sommerstorffia spinosa* and *Zoophagus insidians* predacious on rotifers, and *Rozellopsis inflata* the endoparasite of *Zoophagus*. *Transactions of the British Mycological Society* **37**: 134-150.

PRUETT, C. J. H. & WALLER, J. M. (1989). A report on sugarcane diseases in the Santa Cruz area of Bolivia. *Sugar y Azucar* **84**: 35-37.

PUCKRIDGE, J. T., WALKER, K. F., LANGDON, J. S., DALEY, C. & BEAKES, G. W. (1989). Mycotic dermatitis in a freshwater gizzard shad, the bony bream, *Nematalosa erebi* (Gunther), in the river Murray, South Australia. *Journal of Fish Diseases* **12**: 205-221.

PUEPPKE, S. G. & VAN ETTEN, H. D. (1974). Pisatin accumulation and lesion development in peas infected with *Aphanomyces euteiches*, *Fusarium solani* f. sp., or *Rhizoctonia solani*. *Phytopathology* **64**: 1433-1440.

PUESCHEL, C. M. & VAN DER MEER, J. P. (1985). Ultrastructure of the fungus *Petersenia palmariae* (Oomycetes) parasitic on the alga *Palmaria mollis* (Rhodophyceae). *Canadian Journal of Botany* **63**: 409-418.

PUGH, G. J. F. & JONES, E. B. G. (1986). Antarctic marine fungi: a preliminary account. *In* S. T. Moss (Ed.), *The Biology of Marine Fungi*, pp. 323-330. Cambridge, U.K.: Cambridge University Press.

PUYMALY, A. de (1927). Sur le *Sphaerita endogona*, chytridiacée parasite des Euglènes. *Bulletin de la Société Botanique, France* **74**: 472-476.

PYSTINA, K. A. [ПЫСТИНА, К. А., Pystina, K. A.] (1985). Some data on *Olpidiopsis* Cornu genus. *Mikologiya I Fitopatologiya* [*Микология и Фитопатогия*] [*Ботанический Институт им В. Л. Комарова. Ленинград, СССР*] **19**: 478-281.

PYSTINA, K. A. (1995). *Definitorium Fungorum Rossiae. Classis Oomycetes. Fasc. 1. Ordines Saprolegniales, Leptomitales, Lagenidiales* [Определитель Грибов России. Класс Оомицеты. Вып. 1. Порядки Сапролегниевые, Лептомитовые, Лагенидиевые]. Saint Petersburg, Russia: Nauka.

QUICK, J. A. (1974*a*). A new marine *Labyrinthula* with unusual locomotion. *Transactions of the American Microscopical Society* **93**: 52-61.

QUICK, J. A. (1974*b*). Studies on a new species of *Labyrinthula* (Labyrinthulales) isolated from the marine gastropod *Thais haemostoma floridana*. *Transactions of the American Microscopical Society* **93**: 62-70.

QUICK, J. A. (1974*c*). *Labyrinthuloides schizochytrops* n. sp., a new marine labyrinthula with spheroid "spindle" cells. *Transactions of the American Microscopical Society* **93**: 344-365.

RABENHORST, L. (1843). Über die Knollenkrankheit der Kartoffeln. *Archiv der Pharmazie und Berichte der Deutschen pharmazentischen Gesellschaft* **83**: 300.

RABENHORST, L. (1868). *Flora Europaea Algarum Aquae Dulcis et Submarinae. Sectio III. Algas Chlorophyllophyceas, Melanophyceas et Rhodophyceas Complectens.* Leipzig (Lipsiae), Germany: Eduardum Kummerum.

RACIBORSKI, M. (1886). Roslinne pasorzyty karpi (Vegetative parasites of carp). *Rozprawy i sprawozdania z posiedzeń Wydzialu matematyczno-przyrodniczego Academii umiejetności (Sitsungsbericht der Krakauer Akademie der Wissenschaften)* **14**: 149-168.

RACIBORSKI, M. (1891). *Pythium dictyosporum*, nieznany pasorzyt skretnicy (*Spirogyra*) (*Pythium dictyosporum* ein neuer Parasit der *Spirogyra*). *Anzeiger der Akademie der Wissenschaften in Krakau (Bulletin International de l'Académie des Sciences de Cracovie)* **8**: 283-287 [*also cited as* : RACIBORSKI, M. (1893) *Pythium dictyosporum* nieznany pasorzyt skretnicy (*Spirogyra*) (*Pythium dictyosporum* ein neuer Parasit der *Spirogyra*). *Rozprawy i sprawozdania z posiedzeń Wydzialu matematyczno-przyrodniczego Academii umiejetności* **4**: 25-33.]

RACIBORSKI, M. (1898). Pflanzenpathologisches aus Java. II. *Woroninella psophocarpi* nov. gen. et sp. *Zeitschrift für Pflanzenkrankheiten und Pflanzenschutz* **8**: 195-200.

RADAIS, M. (1898). Sur l'appareil végétatif des Saprolégniées. *Bulletin de la Société Mycologique de France* **14**: 144-148.

RAFIN, C., BRYGOO, Y. & TIRILLY, Y. (1995). Restriction analysis of amplified ribosomal DNA of *Pythium* spp. isolated from soilless culture systems. *Mycological Research* **99**: 277-281.

RAFINESQUE, C. S. (1836). *Flora Telluriana*. Philadelphia, U.S.A.

*__RAGHU KUMAR, C.__ (1978). Physiology of infection of the marine diatom *Licmorpha* by the fungus *Ectrogella perforans*. *Veröffentlichungen des Instituts für Meeresforschung in Bremerhaven* **17**: 1-14.

RAGHU KUMAR, C. (1980*a*). An ultrastructural study of the marine diatom *Licmophora hyalina* and its parasite *Ectrogella perforans*. I. Infection of the host cell. *Canadian Journal of Botany* **58**: 1280-1290.

RAGHU KUMAR, C. (1980*b*). An ultrastructural study of the marine diatom *Licmophora hyalina* and its parasite *Ectrogella perforans*. II. Development of the fungus in its host. *Canadian Journal of Botany* **58**: 2557-2574.

RAGHU KUMAR, C. (1983). Resistance mechanism in the marine diatom *Licmophora* against infection by the fungus *Ectrogella*. *Veröffentlichungen des Instituts für Meeresforschung in Bremerhaven* **19**: 185-190.

RAGHUKUMAR, C. (1986*a*). Fungal parasites of the marine green algae, *Cladophora* and *Rhizoclonium*. *Botanica Marina, Berlin* **29**: 289-297.

RAGHUKUMAR, C. (1986*b*). Thraustochytrid fungi associated with marine algae. *Indian Journal of Marine Science* **15**: 121-122.

RAGHUKUMAR, C. (1987*a*). Fungal parasites of marine algae from Mandapam (South India). *Diseases of Aquatic Organisms* **3**: 137-145.

RAGHUKUMAR, C. (1987*b*). Fungal parasites of the green alga *Chaetomorpha media*. *Diseases of Aquatic Organisms* **3**: 147-150.

RAGHUKUMAR, C., NAGARKAR, S. & RAGHUKUMAR, S. (1992). Association of thraustochytrids and fungi with living marine algae. *Mycological Research* **96**: 542-546.

RAGHU KUMAR, S. (1977*a*). A new species of the genus *Ulkenia* Gaertner (lower marine fungi) from the North Sea. *Veröffentlichungen des Instituts für Meeresforschung in Bremerhaven* **16**: 159-165.

RAGHU KUMAR, S. (1977*b*). Studies on an estuarine isolate of *Catenaria anguillulae* (Phycomycetes) from India. *Veröffentlichungen des Instituts für Meeresforschung in Bremerhaven* **16**: 177-189.

RAGHU-KUMAR, S. (1979). Observations of the life cycle and movement of the thraustochytrid *Ulkenia amoeboidea* (Bahnweg & Sparrow) from the North Sea. *Journal of Protozoology* **26**: 546-566.

RAGHU KUMAR, S. (1980). *Thraustochytrium benthicola* sp. nov.: a new marine fungus from the North Sea. *Transactions of the British Mycological Society* **74**: 607-614.

RAGHU KUMAR, S. (1982*a*). Fine structure of the thraustochytrid *Ulkenia amoeboidea*. I. Vegetative thallus and formation of the amoeboid stage. *Canadian Journal of Botany* **60**: 1092-1102.

RAGHU KUMAR, S. (1982*b*). Fine structure of the thraustochytrid *Ulkenia amoeboidea*. II. The amoeboid stage and formation of zoospores. *Canadian Journal of Botany* **60**: 1103-1114.

RAGHU KUMAR, S. (1985). Enumeration of thraustochytrids (heterotrophic microorganisms) from the Arabian Sea. *Mahasagar, Bulletin of the National Institute of Oceanography* **18**: 429-435.

RAGHUKUMAR, S. (1986). Enumeration of thraustochytrids (heterotrophic microorganisms) from the Arabian Sea. *Mahasagar, Bulletin of the National Institute of Oceanography* **18**: 457-465.

RAGHU-KUMAR, S. (1987). Occurrence of the thraustochytrid, *Corallochytrium limacisporum* gen. et sp. nov. in the coral reef lagoons of the Lakshadweep Islands in the Arabian Sea. *Botanica Marina, Berlin* **30**: 83-89.

RAGHU-KUMAR, S. (1988*a*). *Schizochytrium octosporum* sp. nov. and other thraustochytrids from the North Sea (Rosfjord, Norway). *Transactions of the British Mycological Society* **90**: 273-278.

RAGHU KUMAR, S. (1988*b*). *Schizochytrium mangrovei* sp. nov., a thraustochytrid from mangroves in India. *Transactions of the British Mycological Society* **90**: 627-631.

RAGHU KUMAR, S. & GAERTNER, A. (1978). Ecology of the thraustochytrids (lower marine fungi) in the Fladen Ground and other parts of the north Sea II. *Veröffentlichungen des Instituts für Meeresforschung in Bremerhaven* **18**: 289-308.

RAI, J. N. & MISRA, J. K. (1977). A new variety of *Achlya stellata* de Bary. *Current Science, Bangalore, India* **46**: 28.

RAI, J. N. & MISRA, J. K. (1978). A new species of *Dictyuchus* from alkaline ponds of India. *Current Science, Bangalore, India* **47**: 171-172.

RAI, J. N. & MISRA, J. K. (1981). *Olpidiopsis*: a parasitic fungus from alkaline water. *Geophytology* **11**: 96-98.

RAI, P. V. & STROBEL, G. A. (1966). Chemotaxis of zoospores of *Aphanomyces cochlioides* to sugar beet seedlings. *Phytopathology* **56**: 1365-1369.

RAJAGOPALAN, S. & RAMAKRISHNAN, K. (1971). Phycomycetes in agricultural soils with special reference to the Pythiaceae - II. Species of *Pythium* isolated. *Madras University Journal, Section B* **37-38**: 100-117.

RAMSAY, G. B., SMITH, M. A. & WRIGHT, W. R. (1954). *Peronospora* in radish roots. *Phytopathology* **44**: 384-385.

RAMSBOTTOM, J. (1915). The generic name *Protascus*. *Transactions of the British Mycological Society* **5**: 143-144.

RAMSBOTTOM, J. (1916*a*). A list of the British species of Phycomycetes, etc., with a key to the genera. *Transactions of the British Mycological Society* **5**: 304-317.

RAMSBOTTOM, J. (1916*b*). Notes on the list of British Phycomycetes. *Transactions of the British Mycological Society* **5**: 318-323.

RAMSBOTTOM, J. (1916*c*). Some notes on the history of the classification of the Phycomycetes. *Transactions of the British Mycological Society* **5**: 324-350.

RANDOLPH, L. R. & POWELL, M. J. (1992). Ultrastructure of zoospores of the oomycete *Apodachlya pyrifera*. *Mycologia* **84**: 768-780.

---

*FOOTNOTE: The surnames of Chandralata Raghu Kumar and Seshagiri Raghu Kumar are listed chronologically but are printed as in the citation, whether as one name, or two names with or without a hyphen. Dr C. Raghukumar has told me that neither she nor her husband now use the two word surname.

RANDS, R. D. (1930). Fungi associated with root rots of sugar cane in the southern United States. *Proceedings of the International Society of Sugar Cane Technologists* **3**: 119-131.

RANGASWAMI, G. (1962). *Pythiaceous Fungi* [pp. i-xiii, 1-276]. Government of India Press, Delhi, India.

RAO, A. S. & BRAKKE, M. K. (1969). Relation of soil-borne wheat mosaic virus and its fungal vector *Polymyxa graminis*. *Phytopathology* **59**: 581-587.

RAO, P. N., SALAM, M. A. & THIRUMALACHAR, M. J. (1956). A new genus of downy mildew on sorghum. *Mycologia* **48**: 860-864.

RAFFRAY, J. B. & SEQUEIRA, L. (1971). Dark induction of sporulation in *Bremia lactucae*. *Canadian Journal of Botany* **49**: 237-239.

RAPER, J. R. (1936). Heterothallism and sterility in *Achlya* and observations on the cytology of *Achlya bisexualis*. *Journal of the Elisha Mitchell Scientific Society* **52**: 274-289.

RAPER, J. R. (1937). A method of freeing fungi from bacterial contamination. *Science, New York* **85**: 342.

RAPER, J. R. (1939*a*). Sexual hormones in *Achlya*. I. Indicative evidence for a hormonal coordinating mechanism. *American Journal of Botany* **26**: 639-650.

RAPER, J. R. (1939*b*). Role of hormones in the sexual reaction of heterothallic achlyas. *Science, New York* **89**: 321-322.

RAPER, J. R. (1940*a*). Sexual hormones in *Achlya*. II. Distance reactions, conclusive evidence for a hormonal coordinating mechanism. *American Journal of Botany* **27**: 162-173.

RAPER, J. R. (1940*b*). Sexuality in *Achlya ambisexualis*. *Mycologia* **32**: 710-727.

RAPER, J. R. (1942*a*). Sexual hormones in *Achlya*. III. Hormone A and the initial male reaction. *American Journal of Botany* **29**: 159-166.

RAPER, J. R. (1942*b*). Sexual hormones in *Achlya*. V. Hormone A', a male secreted augmenter or activator of Hormone A. *Proceedings of the National Academy of Sciences, U.S.A.* **28**: 509-516.

RAPER, J. R. (1950*a*). Sexual hormones in *Achlya*. VI. The hormones of the A complex. *Proceedings of the National Academy of Sciences, U.S.A.* **36**: 524-533.

RAPER, J. R. (1950*b*). Sexual hormones in *Achlya*. VII. The hormonal mechanism in homothallic species. *Botanical Gazette* **112**: 1-24.

RAPER, J. R. (1951). Sexual hormones in *Achlya*. *American Scientist* **39**: 110-120. 130.

RAPER, J. R. (1967). The role of specific secretions in the induction and development of sexual organs and in the determination of sexual affinity. *In* W. Ruhland (Ed.) *Handbuch der Pflanzenphysiologie. Volume XVIII*: 214-234.

RAPER, J. R. & HAAGEN SMIT, A. J. (1942). Sexual hormones in *Achlya*. IV. Properties of hormone A of *A. bisexualis*. *Journal of Biological Chemistry* **143**: 311-320.

RAPER, K. B. (1928). Studies on the frequency of the water molds in the soil. *Journal of the Elisha Mitchell Scientific Society* **44**: 133-139.

RATTRAY, J. (1887). Note on *Ectocarpus*. *Transactions of the Royal Society of Edinburgh* **32**: 589-600.

RAYSS, T. (1942). On some lower fungi of Palestine. *Palestine Journal of Botany, Series J* **2**: 247-249.

REDFIELD, G. W. & VINCENT, W. F. (1979). Stages of infection and ecological effects of a fungal epidemic on the eggs of a limnetic copepod. *Freshwater Biology* **9**: 503-510.

REEVES, R. J. & JACKSON, R. M. (1972). Induction of *Phytophthora cinnamomi* oospores in soil by *Trichoderma viride*. *Transactions of the British Mycological Society* **59**: 156-159.

REEVES, R. J. & JACKSON, R. M. (1974). Stimulation of sexual reproduction in *Phytophthora* by damage. *Journal of General Microbiology* **84**: 303-310.

REGEL, E. (1843). Beiträge zur Kenntnis einiger Blattpilze. *Botanische Zeitung, 1 Jahrgang* **39**: 665-667. [N.B. column numbers, not page numbers.]

REGEL, E. A. von (1848). [No title] *Flora, Jena* : 249.

REID, B., MORRIS, B. M. & GOW, N. A. R. (1995). Calcium-dependent, genus specific, autoaggregation of zoospores of phytopathogenic fungi. *Experimental Mycology* **19**: 202-213.

REINKE, J. (1869). Ueber die Geschlechtsverhältnisse von *Saprolegnia monoica*. *Archiv für Mikroskopische Anatomie und Entwicklungsmechanik* **5**: 183-192.

REINSCH, P. F. (1875). *Contributiones ad Algologiam et Fungologiam. Volume I.* Leipzig, Germany.

REINSCH, P. F. (1878). Beobachtungen über einige neue Saprolegnieae, über die Parasiten in Desmidienzellen und über die Stachelkuguln in Achlyaschläuchen. *Jahrbücher für Wissenschaftliche Botanik* **11**: 283-311.

REISCHER, H. S. (1949*a*). Preservation of Saprolegniaceae by the mineral oil method. *Mycologia* **41**: 177-179.

REISCHER, H. S. (1949*b*). A new species of *Achlya*. *Mycologia* **41**: 339-345.

REISCHER, H. S. (1949*c*). The effect of temperature on the papillation of oogonia of *Achlya colorata*. *Mycologia* **41**: 398-402.

REISCHER, H. S. (1951*a*). Growth of Saprolegniaceae in synthetic media. I. Inorganic nutrition. *Mycologia* **43**: 142-155.

REISCHER, H. S. (1951*b*). Growth of Saprolegniaceae in synthetic media. II. Nitrogen requirements and the role of Krebs cycle acids. *Mycologia* **43**: 319-328.

REMY, E. (1948). Über niedere Bodenpycomyceten. *Archiv für Mikrobiologie* **14**: 212-239.

RENFRO, B. L. & BHAT, S. (1981). Role of wild hosts in downy mildew diseases. *In* D. M. Spencer (Ed.) *The Downy Mildews*, pp. 107-120. Academic Press, London, U.K.

RENN, C. E. (1934). Wasting disease of *Zostera* in Americal waters. *Nature (London)* **134**: 416.

RENN, C. E. (1935). A mycetozoan parasite of *Zostera marina*. *Nature (London)* **135**: 544-545.

RENN, C. E. (1936*a*). The wasting disease of *Zostera marina*. *Biological Bulletin of the Marine Biological Laboratory, Woods Hole* **70**: 148-158.

RENN, C. E. (1936*b*). Persistence of the eel-grass disease and parasite on the American Atlantic coast. *Nature (London)* **138**: 507-518.

RENN, C. E. (1937). The eel-grass situation along the middle Atlantic coast. *Ecology* **18**: 323-325.

RENN, C. E. (1942). Demonstration of *Labyrinthula* parasite in eel-grass from the coast of California. *Science, New York* **95**: 122.

RENNERFELT, E. (1936). Untersuchungen über die Entwicklung und Biologie des Krebspestpilzes *Aphanomyces astaci*, Schikora. *Meddelanden från Statens undersöknings-försöksanstalt för sötvattensfisket (Mitteilungen der Anstalt für Binnenfischerei bei Drottningholm, Stockholm) N:r 10*: 1-21.

RENWICK, J. H. (1971). The mapping of human chromosomes. *Annual Review of Genetics* **5**: 81-120.

REYES, R. P. see PÉREZ REYES, R. and PÉREZ-REYES, R.

RIBEIRO, O. K., ERWIN, D. C. & ZENTMYER, G. A. (1975). An improved medium for oospore production and germination of several *Phytophthora* species. *Mycologia* **67**: 1012-1019.*Phytopathology*

RIBEIRO, O. K., ZENTMYER, G. A. & ERWIN, D. C. (1975). Comparative effects of monochromatic radiation on the germination of oospores of three *Phytophthora* species. *Phytopathology* **65**: 904-907.

RICHARDS, R. H. & PICKERING, A. D. (1978). Frequency and distribution patterns of *Saprolegnia* infection in wild and hatchery-reared brown trout *Salmo trutta* L. and *Salvelinus alpinus* (L.). *Journal of Fish Diseases* **1**: 69-82.

RICHARDS, R. H. & PICKERING, A. D. (1979). Changes in serum parameters of *Saprolegnia*-infected brown trout, *Salmo trutta* L. *Journal of Fish Diseases* **2**: 197-206.

RICHTER, W. (1937). Vorarbeiten zu einer Saprolegniaceenflora von Marburg. *Flora, Jena* **131**: 227-262.

RICHTER, W. (1985). Marine Schwamme als Substrat für Thraustochytriaceae (marine niedere Pilze). *Veröffentlichungen des Instituts für Meeresforschung in Bremerhaven* **20**: 141-149.

RIDINGS, W. H. & ZETTLER, F. W. (1973). Aphanomyces blight of Amazon sword plants. *Phytopathology* **63**: 289-295.

RIEHL, R. M., TOFT, D. O., MEYER, M. D., CARLSON, G. L. & McMORRIS, T. C. (1984). Detection of a pheromone-binding protein in the aquatic fungus *Achlya ambisexualis*. *Experimental Cell Research* **153**: 545-549.

RIEMANN, F. & SCHRAGE, M. (1983). On the mass occurrence of a thraustochytroid protist (Fungi or rhizopodan Protozoa) in an Antarctic anaerobic sediment. *Veröffentlichungen des Instituts für Meeresforschung in Bremerhaven* **19**: 191-202.

RIETH, A. (1954). Ein weiterer Beitrag zur Kenntnis algenparasitärer Phycomyceten. *Die Kulturpflanze* **2**: 164-184.

RIETH, A. (1962). Beitrag zur Kenntnis der Phycomyceten IV. *Pleotrachelus wildemani* Petersen neu für Deutschland. *Die Kulturpflanze* **10**: 93-105.

RIETH, A. (1967). Beiträge zur Kenntnis algenparasitärer Phycomyceten. V. Eine Gallbildung verursachende Chytridiale: *Zygorhizidium vaucheriae* nov. spec. *Biologisches Zentralblatt* **86**: 435-448.

RIETHMÜLLER, A., WEISS, M. & OBERWINKLER, F. (1999). Phylogenetic studies of Saprolegniomycetidae and related groups based on nuclear large subunit ribosomal DNA sequences. *Canadian Journal of Botany* **77**: 1790-1800.

RIETSCHEL, P. (1936). Über zwei parasiten von *Tenebrio molitor*. *Archiv für Protistenkunde* **86**: 349-358.

RIOUX, J. A. & ACHARD, F. (1956). Entomophytose mortelle à *Saprolegnia diclina* Humphrey 1892 dans un élevage d'*Aedes berlandi* Séguy 1921. *Vie et Milieu* **7**: 326-335.

RISTAINO, J. B., MADRITCH, M., TROUT, C. L. & PARRA, G. (1998). PCR amplification of ribosomal DNA for species identification in the plant pathogen genus *Phytophthora*. *Applied and Environmental Microbiology* **64**: 948-954.

RITCHIE, D. & SLOVIN, S. F. (1974). *Protoachlya polysporus* rediscovered. *Mycologia* **64**: 362-364.

ROBB, E. J. & BARRON, G. L. (1982). Nature's ballistic missile. *Science, New York* **218**: 1221-1222.

ROBB, J. & LEE, B. (1986*a*). Development sequence of the attack apparatus of *Haptoglossa mirabilis*. *Protoplasma* **135**: 102-111.

ROBB, J. & LEE, B. (1986*b*). Ultrastructure of mature and fired gun cells of *Haptoglossa mirabilis*. *Canadian Journal of Botany* **64**: 1935-1947.

ROBERTS, D. W. (1974). Fungal infections of mosquitoes. *In* A. Aubin, J. P. Bourassa, S. Belloncik, M. Pellissier & E. Lacoursiere (Eds). *Mosquito Control. Proceedings of the International Seminar on Mosquito Control*, pp. 143-193. Montréal, Canada: Presses de l'Université de Québec.

ROBERTS, D. W., LEBRUN, R. A. & SEMEL, M. (1982). Control of the Colorado potato beetle with fungi. *In* J. H. Lashomb & R. Casagrande (Eds). *Advances in Potato Pest Management, Proceedings of a Conference, New Brunswick*, pp. 119-137. Stroudsberg, U.S.A.: Hutchinson Ross.

ROBERTS, D. W. & PANTER, C. (1985). Fungi other than *Coelomomyces* and *Lagenidium*. *Bulletin, American Mosquito Control Association* **6**: 99-109.

ROBERTS, R. E. (1963). A study of the distribution of certain members of the Saprolegniales. *Transactions of the British Mycological Society* **46**: 213-224.

ROBERTS, R. J. (1989). *Fish Pathology*. London, U.K.: Bailliere Tindall.

ROBERTS, R. J. (1993). Mycotic aspects of epizootic ulcerative syndrome (EUS) of Asian fishes. *Journal of Fish Diseases* **16**: 169-183.

ROBERTS, R. J., BALL, H. J., MUNRO, A. L. S. & SHEARER, W. M. (1971). Studies on ulcerative dermal necrosis of salmonids. III. The healing process in fish maintained under experimental conditions. *Journal of Fish Biology* **3**: 221-224.

ROBERTS, R. J., SHEARER, W. M., MUNRO, A. L. S. & ELSON, K. G. R. (1970). On ulcerative dermal necrosis of salmonids. II. The sequential pathology of the lesions. *Journal of Fish Biology* **2**: 373-378.

ROBERTS, R. J., WILLOUGHBY, L. G. & CHINABUT, S. (1993). Mycotic aspects of epizootic ulcerative disease (EUS) of Asian fishes. *Journal of Fish Diseases* **16**: 169-183.

ROBERTSON, G. I. (1979). *Pythium erinaceum* sp. nov. from wheatfields in Canterbury, New Zealand. *New Zealand Journal of Botany* **17**: 283-286.

ROBIN, C. (1853). *Histoire naturelle des végétaux parasites etc., qui croissent sur l'homme et sur les animaux vivants*. Paris, France: J.-B. Baillière.

ROCCUZZO, G. & CIANIO, A. (1991). Notes on nematodes found in irrigation water in southern Italy. *Nematologia Mediterranea* **19**: 105-108.

RODRIGUEZ, F., GODOY, T. & NAVAS, J. I. (1994). Cross-infection with *Perkinsus atlanticus* in *Ruditapes decussatus*, *Ruditapes philippinarum* and *Venerupis pullastra*. *Bulletin of the European Association of Fish Pathologists* **14**: 24-27.

RODRIGUEZ-KABANA, R. (1991). Biological control of plant-parasitic nematodes. *Nematotropica* **21**: 111-122.

RODWAY, R. (1897). Tasmanian fungi. *Papers of the Royal Society of Tasmania*: 144.

ROGERS, A. L. & BENEKE, E. S. (1962). Two new species of *Achlya* in Brazil. *Rickia, Brazil* **1**: 243-249.

ROGERS, A. L., MILANEZ, A. I. & BENEKE, E. S. (1970). Additional aquatic fungi from São Paulo State. *Rickia, Brazil* **5**: 93-110.

ROGERS-TALBERT, R. (1948). The fungus *Lagenidium callinectes* (1942) on eggs of the blue crab in Chesapeake Bay. *Biological Bulletin of the Marine Biological Laboratory, Woods Hole* **95**: 214-228.

ROIVAINEN, H. (1977). Resultados micológicos de la wxpedición a Argentina y Chile en 1969-1970. *Karstenia* **17**: 1-18.

ROLDAN, E. F. (1932). *Pythium* root rot disease of corn in the Philippine Islands. *Philippine Agriculturist* **21**: 165-176.

ROLF, L. (1967). The isolation of *Labyrinthula jeremarina* sp. n. from the marine gastropod *Littorina ziczac*. Masters' Thesis, Texas A. & M., College Station, Texas, U.S.A.

ROSEK, C. E. & TIMBERLAKE, W. E. (1979). Restriction endonuclease mapping by crossed contact hybridization: the ribosomal RNA genes of *Achlya ambisexualis*. *Nucleic Acid Research* **7**: 1567-1578.

ROSENBURG, O. (1903). Ueber die Befruchtung von *Plasmopara alpina* (Johans.). *Bihang till Kongliga Svenska Vetenskaps-Akademiens Handlingar* **28**: 1-20.

ROSENDAHL, S. (1985). Interaction between the vesicular-arbuscular mycorrhizal fungus *Glomus fasciculatum* and *Aphanomyces euteiches*. *Phytopathologische Zeitschrift* **114**: 31-40.

ROSS, I. K. (1979). *Biology of the Fungi*. New York, U.S.A.: McGraw Hill.

ROSSY-VALDERRAMA, C. R. (1955). The taxonomy of some aquatic Phycomycetes from Puerto Rico. Unpublished MS Thesis, Duke University, Durham, North Carolina, U.S.A.

ROSSY-VALDERRAMA, C. R. (1956). Some water molds from Puerto Rico. *Journal of the Elisha Mitchell Scientific Society* **72**: 129-137.

ROSTOWZEW, S. J. (1903). Beiträge zur Kenntnis der Peronosporeen. *Flora, Jena* **92**: 405-430.

ROTH, A. W. (1782, 1783). *Beyträge zu Botanik*, 2 volumes. Bremen, Germany.

ROTH, A. W. (1800). *Catalecta Botanica, Fasciculus Secundus* [*Catalecta Botanica quibus Plantae Novae et Minus Cogntae Describuntur atque Illustrantur 1797-1806*] (volume 1: 1-244, plates 1-8, 1797; volume 2: 1-258, plates 1-9, 1800; volume 3: 1-350, plates 1-12, 1806). Lipsiae, Io. Fr. Gleditschiano.

ROTH, A. W. (1802). *Neue Beyträge zu Botanik*. Frankfurth am Mayn, Germany: Bei F. Wilmans.

ROTHERT, W. (1888). Die Entwickelung der Sporangien bei den Saprolegnieen. *In* Cohn: Beiträge sur Physiologie und Biologie der Pflanzen **5**: 291-349.

ROTHSCHILD, L. J. & HEYWOOD, P. (1987). Protistan phylogeny and chloroplast evolution: conflicts and congruence. *In* D. J. Patterson & J. O. Corliss (Eds). *Progress in Protistology*, **2**, pp. 1-68. Bristol, U.K.: Biopress.

ROUND, F. E. (1981). Some aspects of the origin of diatoms and their subsequent evolution. *BioSystems* **14**: 483-486.

ROUND, F. E. & CRAWFORD, R. M. (1990). Phylum Bacillariophyta. *In* L. Margulis, J. O. Corliss, M. Melconian & D. J. Chapman (Eds), *Handbook of Protoctista*, pp. 574-596. Boston, U.S.A.: Jones & Bartlett.

ROUSSEL, H. F. A. de (1806). *Flore du Calvados, ed.* 2 p. 47.

ROY, A. K. (1982). Effect of Ethoprop on the parasitism of *Catenaria anguillulae* on *Meloidogyne incognita*. *Revue de Nematologie* **5**: 335-336.

ROYLE, D. J. & HICKMAN, C. J. (1964*a*). Analysis of factors governing in vitro accumulation of zoospores of *Pythium aphanidermatum* on roots. I. Behavior of zoospores. *Canadian Journal of Microbiology* **10**: 151-162.

ROYLE, D. J. & HICKMAN, C. J. (1964*b*). Analysis of factors governing in vitro accumulation of zoospores of *Pythium aphanidermatum* on roots. II. Substances causing response. *Canadian Journal of Microbiology* **10**: 201-219.

ROZA, D. & HATAI, K. (1999). *Atkinsiella dubia* infection in the larvae of Japanese mitten crab, *Eriocheir japonicus*. *Mycoscience* **40**: 235-240.

ROZE, E. & CORNU, M. (1869). Sur deux nouveaux types génériques pour les familles des Saprolégniées et Péronosporées. *Annales des Sciences Naturelles, Botanique, Série V* 11: 72-91.

ROZEK, C. E. & TIMBERLAKE, W. E. (1979). Restriction endonuclease mapping by crossed contact hybridization: the ribosomal RNA genes of *Achlya ambisexualis*. *Nucleic Acids Research* **7**: 1567-1578.

ROZSYPAL, J. (1934). [In Czech with German summary - Fungi in cysts of *Heterodera schachtii* Schmidt from Moravian beet soils.] *Věstník Československé Akademie Zemědělské* **10**: 413-422.

RUCH, D. G. (1974). The effects of selected fungicides on growth and development of *Lagenidium callinectes*. Thesis, East Carolina University, Greenville, North Carolina.

RUCH, D. G. & BLAND, C. E. (1979). Ultrastructural changes induced in zoospores of *Lagenidium callinectes* by exposure to captan. *Canadian Journal of Botany* **57**: 2116-2121.

ROZA, D. & HATAI, K. (1999). *Atkinsiella dubia* infection in the larvae of Japanese mitten crab, *Eriocheir japonicus*. *Mycoscience* **40**: 235-240.

RUBEN, D. M. & STANGHELLINI, M. E. (1978). Ultrastructure of oospore germination in *Pythium aphanidermatum*. *American Journal of Botany* **65**: 491-501.

RUDOLPHI, F. (1829). Plantarum vel novarum vel minus cognitarum descriptiones. *Linnaea. Ein Journal für Botanik in ihrem ganzen Umfange* **4**: 114-120.

RUEDA, L. M., PATEL, K. J. & AXTELL, R. C. (1990). Efficacy of encapsulated *Lagenidium giganteum* (Oomycetes: Lagenidiales) against *Culex quinquefasciatus* and *Aedes aegypti* larvae. *Journal of the American Mosquito Control Association* **6**: 694-699.

RUEDA, L. M., PATEL, K. J. & AXTELL, R. C. (1991). Comparison of floating and sinking encapsulated formulations of the fungus *Lagenidium giganteum* (Oomycetes: Lagenidiales) for control of *Anopheles* larvae. *Journal of the American Mosquito Control Association* **7**: 250-254.

RÜHBERG, W. (1933). Über eine Hefeinfektion bei *Daphnia magna*. *Archiv für Protistenkunde* **80**: 72-100.

RUHLAND, W. (1902). Die Befruchtung der *Albugo lepigoni* und einiger Peronosporeen. *Hedwigia* **41**: 179-180.

RUHLAND, W. (1904). Studien über die Befruchtung der *Albugo lepigoni* und einiger Peronosporeen. *Jahrbücher für wissenschaftliche Botanik* **39**: 135-166.

RUMP, A. & KARLOVSKY, P. (1991). Tandem arrangement of tRNA(Asp)-encoding genes in *Phytophthora* spp. *Gene* **102**: 51-56.

SACCARDO, D., BERLESE, A. N. & DE TONI, G. B. (1888). *Sylloge Fungorum, Volume 7 part 1 (Phycomyceteae)*. Padua.

SACCARDO, P. A. (1876). Fungi veneti novi vel critici Serie V, No. 91. *Nuovo Giornale Botanico Italiano e Bolletino della Società Botanica Italiana, Firenze* **8**: 172.

SACCARDO, P. A. (1888). Sylloge Fungorum omnium huscusque cognitorum. Volume 7. Padua, Italy.

SACHAY, D. J., HUDSPETH, D. S. S., NADLER, S. A. & HUDSPETH, M. E. S. (1993). Oomycete mtDNA: *Phytophthora* genes for cytochrome *c* oxidase use an unmodified genetic code and encode proteins most similar to those of plants. *Experimental Mycology* **17**: 7-23.

SACHS, J. (1882). *Text-book of Botany, Morphological and Physiological, 2nd ed.* Oxford, U.K.: Clarendon Press.

SACKSTON, W. E. (1981). Downy mildew of sunflower. *In* D. M. Spencer (Ed.), *The Downy Mildews*. Academic Press, London, UK, pp. 545-575.

SADEBECK, R. (1874). Über *Pythium equiseti*. *Verhandlungen des Botanischen Vereins der Provinz Brandenburg* **16**: 116-124.

SADEBECK, R. (1876). Ueber Infectionen welche *Pythium*-arten bei lebenden Pflanzen hervorbringen. *Tageblatt der ... Versammlung Deutscher Naturforscher und Aerzte in Breslau* **49**: 100.

SADEBECK, R. (1887). Originalberichte gelehrter Gesellschaften. Gesellschaft fur Botanik zu Hamburg. Sitzung vom 25 Februar 1886. *Botanisches Centralblatt* **29**: 318-319.

SADOWSKI, L. A. & POWELL, M. J. (1990). Cytochemical detection of polysaccharides in zoospores of *Aphanomyces euteiches*. *Canadian Journal of Botany* **68**: 1379-1388.

SAFEEULLA, K. M. (1952). Morphological and cytological studies of *Albugo* species on *Ipomoea hederacea*. *Current Science, India (Bangalore)* **21**: 287-288.

SAFEEULLA, K. M. (1976). *Biology and Control of the Downy Mildews of Pearl Millet, Soghum and Finger Millet*. Mysore, India: Downy Mildew Research Laboratory.

SAFEEULLA, K. M. & THIRUMALACHAR, M. J. (1951). Morphological and cytological studies in *Albugo* species on *Ipomoea aquatica* and *Merrimia emarginata*. *La Cellule* **55**: 225-232.

SAFEEULLA, K. M. & THIRUMALACHAR, M. J. (1955). Gametogenesis and oospore formation in *Sclerospora* species on *Sorghum vulgare*. *Mycologia* **47**: 177-184.

SAFEEULLA, K. M. & THIRUMALACHAR, M. J. (1956). A morphological and cytological study of the gametogenesis and oospore formation in *Albugo* species on *Ipomoea hederacea*. *National Institute of Sciences of India, B* **22**: 69-76.

SAFEEULLA, K. M., THIRUMALACHAR, M. J. & SHAW, C. G. (1963). Gametogenesis and oospore formation in *Sclerophthora cryophila* on *Digitaria marginata*. *Mycologia* **55**: 819-23.

SAFFO, M. B. (1981). The enigmatic protist *Nephromyces*. *BioSystems* **14**: 487-490.

SAFFO, M. B. (1982). Distribution of the endosymbiont *Nephromyces* Giard within the ascidian family Molgulidee. *Biological Bulletin of the Marine Biological Laboratory, Woods Hole* **162**: 95-104.

SAFFO, M. B. & DAVIS, W. L. (1982). Modes of infection of the ascidian *Molgula manhattensis* by its endosymbiont *Nephromyces* Giard. *Biological Bulletin of the Marine Biological Laboratory, Woods Hole* **162**: 105-112.

SAFFO, M. B. & FULTZ, S. (1986). Chitin in the symbiotic protist *Nephromyces*. *Canadian Journal of Botany* **64**: 1306-1310.

SAFFO, M. B. & NELSON, R. (1983). The cells of *Nephromyces*: developmental stages of a single life cycle. *Canadian Journal of Botany* **61**: 3230-3239.

SAHTIYANCI, S. (1962). Studien über einige wurzelparasitäre Olpidiaceen. *Archiv für Mikrobiologie* **41**: 187-228.

SAIKAWA, M. & ANAZAWA, T. (1985). Electron microscopy of an endoparasitic fungus *Gonimochaete pyriforme* infecting nematodes. *Canadian Journal of Botany* **63**: 2326-2331.

SAIKAWA, M. & HOSHINO, J. (1986). Electron microscopy on *Sommerstorffia spinosa*, a watermold parasitic on rotifers. *Mycologia* **78**: 554-561.

SAIKAWA, M., OYAMA, M. & YAMAGUCHI, K. (1991). *Harposporum anguillulae* and *Haptoglossa intermedia* parasitizing tardigrades. *Nihon Kin Gakkai Kaiho [Transactions of the Mycological Society of Japan]* **32**: 501-508.

SAKSENA, M. R. K. (1936). Recherches physiologiques et cytologiques sur quelques espèces du genre *Pythium*. *Revue Générale de Botanique* **48**: 156-88, 215-52, 273-313.

SAKSENA, R. K. & BHARGAVA, K. S. (1944). A new variety of *Isoachlya anisospora* (de Bary) Coker. *Current Science, India (Bangalore)* **13**: 236-237.

SALVATORE, M. A., GRAY, F. A. & HINE, R. B. (1973). Enzymatically induced germination of oospores of *Phytophthora megasperma*. *Phytopathology* **63**: 1083-1084.

SALVIN, S. B. (1940). The occurrence of five successive swarming stages in a non-sexual *Achlya*. *Mycologia* **32**: 148-154.

SALVIN, S. B. (1941). Comparative studies on the primary and secondary zoospores of the Saprolegniaceae. I. Influence of temperature. *Mycologia* **33**: 592-600.

SALVIN, S. B. (1942*a*). Variations of specific and varietal character induced in an isolate of *Brevilegnia*. *Mycologia* **34**: 38-51.

SALVIN, S. B. (1942*b*). Factors controlling sporangial type in *Thraustotheca primoachlya* and *Dictyuchus achlyoides*. *American Journal of Botany* **29**: 97-104.

SALVIN, S. B. (1942*c*). Preliminary report on the intergeneric mating of *Thraustotheca clavata* and *Achlya flagellata*. *American Journal of Botany* **29**: 674-676.

SAMOTA, D., IVEZIC, M., MILAKOVIC, Z. & TODOROVIC, M. (1988). (Dynamics of nematofauna and microflora in vineyards as possibilities of biological control of plant parasitic Nematodes). *Zastita Bilja (Yugoslavia)* **39**: 159-170 [only abstract consulted].

SAMSON, K. (1932). Observations on a new species of *Olpidium* occurring in the root hairs of *Agrostis*. *Transactions of the British Mycological Society* **17**: 182-194.

SAMSON, K. (1939). *Olpidium brassicae* (Wor.) Dang. and its connection with *Asterocystis radicis* de Wildeman. *Transactions of the British Mycological Society* **23**: 199-205.

SANDOZ, M. D., ROGERS, R. & NEWCOMBE, C. L. (1944). The fungus infection of eggs of the blue crab *Callinectes sapidus* Rathburn. *Science, New York* **99**: 124-125.

SANSOME, E. (1961). Meiosis in the oogonium and antheridium of *Pythium debaryanum* Hesse. *Nature (London)* **191**: 827-828.

SANSOME, E. R. (1963*a*). Meiosis in *Pythium debaryanum* Hesse and its significance in the life-history of the Biflagellatae. *Transactions of the British Mycological Society* **46**: 63-72.

SANSOME, E. R. (1963*b*). Cytological observations on *Sclerospora sorghi* in relation to the life history of the Oömycetes. *Proceedings of the Science Association of Nigeria* **6**, 48-52.

SANSOME, E. R. (1965). Meiosis in diploid and polyploid sex organs of *Phytophthora* and *Achlya*. *Cytologia* **30**: 103-117.

SANSOME, E. R. (1966). Meiosis in the sex organs of the Oomycetes. *In* C. D. Darlington & K. R. Lewis (Eds), *Chromosomes Today*, vol. **I** pp. 77-83. Edinburgh: Oliver and Boyd.

SANSOME, E. R. (1970). Selfing as a possible cause of disturbed ratios in *Phytophthora* crosses. *Transactions of the British Mycological Society* **54**: 101-107.

SANSOME, E. R. (1976). Gametangial meiosis in *Phytophthora capsici*. *Canadian Journal of Botany* **54**: 1535-1545.

SANSOME, E. R. (1977). Polyploidy and induced gametangial formation in British isolates of *Phytophthora infestans*. *Journal of General Microbiology* **99**: 311-316.

SANSOME, E. R. (1980). Reciprocal translocation heterozygosity in heterothallic species of *Phytophthora* and its significance. *Transactions of the British Mycological Society* **74**: 175-185.

SANSOME, E. (1987). Fungal chromosomes as observed with the light microscope. *In* A. D. M. Rayner, C. M. Brasier & D. Moore (Eds), *Evolutionary Biology of the Fungi*, pp. 97-113. Cambridge, U. K.: Cambridge University Press.

SANSOME, E. & BRASIER, C. M. (1973). Diploidy and chromosomal structural hybridity in *Phytophthora infestans*. *Nature (London)* **241**: 344-345.

SANSOME, E. & BRASIER, C. M. (1974). Polyploidy associated with varietal differentiation in the *megasperma* complex of *Phytophthora*. *Transactions of the British Mycological Society* **63**: 461-467.

SANSOME, E., BRASIER, C. M. & GRIFFIN, M. J. (1979). Chromosome size differences in *Phytophthora palmivora*, a pathogen of cocoa. *Nature (London)* **255**: 704-705.

SANSOME, E., BRASIER, C. M. & SANSOME, F. W. (1979). Further cytological studies on the 'L' and 'S' types of *Phytophthora* from cocoa. *Transactions of the British Mycological Society* **73**: 293-302.

SANSOME, E. R. & HARRIS, B. J. (1962). Use of camphor-induced polyploidy to determine the place of meiosis in fungi. *Nature (London)* **196**: 291-292.

SANSOME, E. & SANSOME, F. W. (1974). Cytology and life-history of *Peronospora parasitica* on *Capsella bursa-pastoris* and of *Albugo candida* on *C. bursa-pastoris* and on *Lunularia annua*. *Transactions of the British Mycological Society* **62**: 323-332.

SARGENT, J. A., INGRAM, D. S. & TOMMERUP, I. C. (1977). Oospore development in *Bremia lactucae* Regel: an ultrastructural study. *Proceedings of the Royal Society of London, B* **198**: 129-138.

SARKAR, N. & DAYAL, R. (1988). A new variety of *Dictyomorpha dioica* (Couch) Mullins. *Proceedings of the National Academy of Sciences, India, Section B (III)* **58**: 403-406.

SASSUCHIN, D., TIFLOV, V. & SCHULZ, R. S. (1934). [Endo- and ectoparasites of *Rhombomys optimus* Licht.] [*Review of Microbiology, Saratow, 1934*]: 335-338.

SASSUCHIN, D. N., TIFLOW, W. E. & SCHULZ, R. S. (1935). Ekto- und enoparasiten der Sandmans *Rhombomys optima* Licht. 3. Mitteilung. *Zeitschrift für Parasitenkunde* **7(5)**: 635-638.

SATHAPATAYAVONGS, O., LECHACHAIKUL, P., PRACHAKTMAN, O., ATICHARTAKARH, V., SRIPHOJANART, T., TRAIRATVORAKUL, P., JIRASIRITHAM, P., NONTASUT, S., EURVILLAICHITA, A. & FLEGEL, T. (1989). Human pythiosis associated with thallassemia hemoglobinopathy syndrome. *Journal of Infectious Diseases* **159**: 274-280.

SATI, S. C. & KHULBE, R. D. (1980). Occurrence of aquatic fungi in soils of Naini Tal Hills. *Geobios (Jodhpur)* **7**: 42-43.

SAVILE, D. B. O. (1968). Possible interrelationships between fungal groups. *In* G. C. Ainsworth & A. L Sussman (Eds), *The Fungi, An Advanced Treatise. Volume III, The Fungal Population*. Academic Press, New York, USA, pp. 649-675.

SAVILE, D. B. O. (1976). Evolution of the rust fungi (Uredinales) as reflected by their ecological problems. *Evolutionary Biology* **9**: 137-207.

SĂVULESCU, O. (1962). A systematic study of the genera *Bremia* Regel and *Bremiella* Wilson. *Revue de Biologie. Academia Republicii Populare Romîne, Bucarest* **7**: 43-62.

SĂVULESCU, O. (1966). Sur quelques principes de systématique des *Péronosporales*. *Feddes Repertorium* **73**: 35-42.

SĂVULESCU, T. & SĂVULESCU, O. (1952). Studiul *Sclerospora, Basidiophora* si *Peronoplasmopara*. *Buletin ştiinţific. Academia Republicii Populare Romîne*, pp. 327-457.

SAWADA, K. (1912*a*). Investigations of the paddy seedling decay in Formosa. *Agricultural Experiment Station, Government of Formosa, Special Bulletin No. 3.* :1-84. [In Japanese]

SAWADA, K. (1912*b*). Studies in the rot disease of rice seedlings. *Special Report. Formosa Agricultural Experiment Station* **3**: 1-93.

SAWADA, K. (1919). Descriptive catalogue of Formosan fungi. *Agricultural Experiment Station, Government of Formosa, Special Bulletin No. 19.*: 1-695. [In Japanese]

SAWADA, K. (1922). Materials for the mycological study of Formosa. *Transactions of the Natural History Society of Formosa*. **62**: 77-84.

SAWADA, K. & CHEN, C. C. (1926). (On the putrefaction disease of *Antirrhinum majus*.) *Journal of the Natural History Society Formosa* **16**: 199-212.

SAXENA, G. & MUKERJI, K. G. (1991). Distribution of nematophagous fungi in Varanasi, India. *Nova Hedwigia* **52**: 487-495.

SAYRE, R. M. & KEELEY, L. S. (1969). Factors influencing *Catenaria anguillulae* infecions in a free-living and a plant-parasitic nematode. *Nematologica* **15**: 492-502.

ŠČERBAK, I. (1910). Ein Beitrage zur Physiologie der Saprolegniaschwärmer. *Mémoires de la Société Nationale, Odessa* **35**: 83.

SCHADE, A. L. (1940). The nutrition of *Leptomitus*. *American Journal of Botany* **27**: 376-384.

SCHADE, A. L. & THIMANN, K. V. (1940). The metabolism of the water-mold, *Leptomitus lacteus*. *American Journal of Botany* **27**: 659-670.

SCHÄPERCLAUS, W. (1927). Krebssterben und Krebskrankheiten in der Mark. *Mitteilungen der Fischereivereine für die Provinz Brandenburg* **19**.

SCHÄPERCLAUS, W. (1935). Die Ursache der Pestartigen Krebssterben. *Zeitschrift für Fischerei* **33**: 343-366.

SCHÄPERCLAUS, W. (1954). *Fischkrankheiten*. 3rd edn. Berlin, Germany: Akademie Verlag.

SCHAREN, A. L. (1960). Germination of oospores of *Aphanomyces euteiches* embedded in plant debris. *Phytopathology* **50**: 274-277.

SCHATZ, S. (1984). Degradation of *Laminaria saccharina* by saprobic fungi. *Mycologia* **76**: 426-432.

SCHENCK, N. C. & NICOLSON, T. H. (1977). A zoosporic fungus occurring on species of *Gigaspora margarita* and other vesicular-arbuscular mycorrhizal fungi. *Mycologia* **69**: 1049-1053.

SCHENK, A. (1858). *Über das Vorkommen contractiler Zellen im Pflanzenreiche*. Würzburg, Germany: F. E. Thein.

SCHENK, A. (1859*a*). *Achlyogeton*, eine neue Gattung der Mycophyceae. *Botanische Zeitung, 17 Jahrgang* **46**: 398-400. [N.B. column numbers, not page numbers.].

SCHENK, A. (1859*b*). Algologische Mittheilungen. V. *Pythium* Pringsh. *Verhandlungen der Physikalisch-medicinische Gesellschaft in Würzburg* **9**: 12-31.

SCHENK, A. (1890). *Handbuch der Botanik, Bande 4*: 508. [A. Schenk, Ed., {*Schenk's*} *Handbuch der Botanik* (4 vols in 5 parts, vol. 3 in 2 parts). Wroclaw (Breslau), Poland: E. Treweundt. Also cited as *Encyklopaedie der Naturwissenschaften*].

SCHERFFEL, A. (1902*a*). Mycologische und algologische Notizen. *Hedwigia* **41**: 105-107.

SCHERFFEL, A. (1902*b*). Néhány adat Magyarhon növény-és állatvilágának ismeretéhez. *Növénytani Közlemények* **1**: 107-111.

SCHERFFEL, A. (1904). Ujabb adatok Magyarország alsorendü szervezeteinek ismeretéhez. (Neuere Beiträge zur Kenntnis der niederen Organismen in Ungarn.) *Növénytani Közlemények* **3**: 116-119.

SCHERFFEL, A. (1914). Kisebb Közlemények a kryptogamok köréböl. *Botanikai Közlemenyek* **13**: 12-17.

SCHERFFEL, A. (1925*a*). Endophytische Phycomyceten-Parasiten der Bacillariaceen und einige neue Monadinen. Ein Beitrag zur Phylogenie der Oomyceten (Schröter). *Archiv für Protistenkunde* **52**: 1-141.

SCHERFFEL, A. (1925*b*). Zur Sexualität der Chytridineen (Der Beiträge zur Kenntnis der Chytridineen, Teil I). *Archiv für Protistenkunde* **53**: 1-58.

SCHERFFEL, A. (1926*a*). Einiges über neue oder ungenügend bekannte Chytridineen (Der Beiträge zur Kenntnis der Chytridineen, Teil II). *Archiv für Protistenkunde* **54**: 167-260.

SCHERFFEL, A. (1926*b*). Beiträge zur Kenntnis der Chytridineen. Teil III. *Archiv für Protistenkunde* **54**: 510-528.

SCHERFFEL, A. (1930). Abstracts of Communications, V International Botanical Congress, Cambridge Cambridge University Press.

SCHERFFEL, A. (1931). Über einige Phycomyceten. *Archiv für Protistenkunde* **73**: 137-146.

SCHIKORA, F. (1903). Über die Krebspest und ihren Erreger. *Fischerei-Zeitung* **6**: 353-355.

SCHIKORA, F. (1906). Die Krebspest. *Fischerei-Zeitung* **9**: 529-532, 549-555, 561-566, 581-583.

SCHIKORA, F. (1922). Über die Krebspest und ihren Erreger *Aphanomyces Magnusi* Schikora. *Verhandlungen der Botanischen Vereins der Provinz Brandenburg* **63**: 87-88.

SCHILBERSZKY, K. (1896). Ein neuer Schorfparasit der Kartoffelknollen. *Berichte der Deutschen Botanischen Gesellschaft* **14**: 36-37.

SCHIMPER, A. F. W. (1883). Ueber die Entwicklung der Chlorophyllkörner und Farbkörper. *Botanische Zeitung, 41 Jahrgang* **7**: 106-112. [N.B. pages, not column numbers, as is the case for many *Botanische Zeitung* references.].

SCHINZ, H. (1920). Myxogasteres (Myxomycetes, Mycetozoa). *Dr. L. Rabenhorst's Kryptogamen-Flora von Deutschland, Österreich und der Schweiz, Zweie Auflage, Erster Band: Pilze, Abtheilüng IV, I Reihe, Archimycetes* **1**(10): 1-472. Leipzig: Kummer.

SCHKORBATOW, L. A. (1923). Nov'e raznovid iz Oomycetes, naïdenn'e v okrestnostíakh g. Khar'kova (New taxa of Oomycetes occurring in the vicinity of Kharkov). *Otdel Sporovykh Rastenií (In Determinations of Sporogenous Plants)*. Botanical Institute, Akademica Nauk SSSR: 34-36.

SCHLÖSSER, L. A. (1929). Geschlechterverteilung und fakultativ Parthenogenese bei Saprolegniaceen. *Planta; Archiv für wissenschaftlich.. Botanik* **8**: 529-570.

SCHMITT, J. A. & BENEKE, E. S. (1962). Aquatic fungi from South Bass and neighboring islands in western Lake Erie. II. Additional biflagellate anbd uniflagellate phycomycetes. *Ohio Journal of Science* **62**: 11-12.

SCHMOLLER, H. (1960). Kultur und Entwicklung von *Labyrinthula coenocystis* n. sp. *Archiv für Mikrobiologie* **36**: 365-372.

SCHMOLLER, H. (1961). Zur Entwickelung der Labyrinthulen. *Archiv für Mikrobiologie* **40**: 224-230.

SCHMOLLER, H. (1966). Beitrag zur Kenntnis der Labyrinthulen-Entwicklung. *Archiv für Protistenkunde* **109**: 226-244.

SCHMOLLER, H. (1971). Die Labyrinthulen und ihre Beziehung zu den Amöben. *Naturwissenschaften* **58**: 142-146.

SCHNEIDER, C. L. (1958). Further studies on the host range of *Aphanomyces cochlioides*. *Phytopathology* **48**: 463-464.

SCHNEIDER, J. (1967). Ein neuer mariner Phycomycet aus der Kieler Bucht (*Thraustochytrium striatum* spec. nov.). *Kieler Meeresforschungen* **23**: 16-20.

SCHNEIDER, J. (1968*a*). Über niedere Phycomyceten der westlichen Ostsee. *Veröffentlichungen des Instituts für Meeresforschung in Bremerhaven, Sonderband* **3**: 93-97.

SCHNEIDER, J. (1968*b*). Über niedere Pilze der westlichen Ostsee. *Berichte der Deutschen Botanischen Gesellschaft* **81**: 375-379.

SCHNEIDER, J. (1969). Zur Taxonomie, Verbreitung un Okologie einiger mariner Phycomyceten. *Kieler Meeresforschung* **25**: 316-327.

SCHNEIDER, J. (1971*a*). Niederer Pilze aus Bodenproben des Brack- und Meerwassers der englischen Westküste. *Kieler Meeresforschung* **27**: 94-101.

SCHNEIDER, J. (1971*b*). Niederer Pilze als Testorganismen für Schadstoffe im Meer- und Brackwasser. Die wirkung von Schwermetallverbindungen und Phenol auf *Thraustochytrium striatum*. *Marine Biology (Berlin)* **16**: 214-225.

SCHNEIDER, J. (1981). Ein ökologischer Vergleich aquatischer niederer Pilze (*Thraustochytrium* sp.) von Meeres- und Binnenlandstandorten. *Botanica Marina, Berlin* **24**: 475-483.

SCHNEIDERHAN, F. J. (1933). *The Discovery of Bordeaux Mixture*. Phytopathological Classics Number 3, American Phytopathological Society, Ithaca, N.Y., USA.

SCHNEPF, E. (1994). A *Phagomyxa*-like endoparasite of the centric marine diatom *Bellerochea malleus* : a phagotrophic plasmodiophoromycete. *Botanica Acta* **107**: 374-382.

SCHNEPF, E., DEICHGRÄBER, G. & DREBES, G. (1978*a*). Development and ultrastructure of the marine, parasitic oomycete, *Lagenisma coscinodisci* Drebes (Lagenidiales). The infection. *Archives of Microbiology* **116**: 133-139.

SCHNEPF, E., DEICHGRÄBER, G. & DREBES, G. (1978*b*). Development and ultrastructure of the marine, parasitic oomycete, *Lagenisma coscinodisci* Drebes. Thallus, zoosporangium, mitosis, and meiosis. *Archives of Microbiology* **116**: 141-150.

SCHNEPF, E., DEICHGRÄBER, G. & DREBES, G. (1978*c*). Development and ultrastructure of the marine, parasitic oomycete, *Lagenisma coscinodisci* Drebes (Lagenidiales): Formation of the primary zoospores and their release. *Protoplasma* **94**: 263-280.

SCHNEPF, E., DEICHGRÄBER, G. & DREBES, G. (1978*d*). Development and ultrastructure of the marine oomycete, *Lagenisma coscinodisci* (Lagenidiales): encystment of primary zoospores. *Canadian Journal of Botany* **56**: 1309-1314.

SCHNEPF, E., DEICHGRÄBER, G. & DREBES, G. (1978*e*). Development and ultrastructure of the marine parasitic oomycete, *Lagenisma coscinodisci* (Lagenidiales): sexual reproduction. *Canadian Journal of Botany* **56**: 1315-1325.

SCHNEPF, E. & DREBES, G. (1977). Über die Entwicklung der marinen parasitischen Phycomyceten, *Lagenisma coscinodisci* (Lagenidiales). *Helgoländer wissenschaftliche Meeresuntersuchungen* **29**: 291-301.

SCHNEPF, E. & HEINZMANN, J. (1980). Nuclear movement tip growth and colchicine effects in *Lagenisma coscinodisci* (Oomycetes, Lagenidiales). *Biochemie und Physiologie der Pflanzen* **175**: 67-76.

SCHNETZLER, J. B. (1887). Infection d'une larve de grenouille par *Saprolegnia ferax*. *Archives des Sciences Physiques et Naturelles. Genève* **18**: 492.

SCHRADER, E. (1937). Die Entwicklung von *Thraustotheca clavata*. *Flora, Jena* **32**: 125-150.

SCHRADER, H. A. (1800-1801). *Journal für die Botanik* **2**: 127.

SCHROETER, J. (1877). Bermerkungen und Beobachtungen über einige Ustilagineen. *Beiträge zur Biologie der Pflanzen* **2**: 249-383.

SCHRÖTER, J. (1879). *Protomyces graminicola* Saccardo. *Hedwigia* **18**: 83-87.

SCHROETER, J. (1885-1889). Die Pilze Schlesiens. *In Kryptogamen-Flora von Schlesien. Namen der Schlesischen Gesellschaft für Vaterländische Cultur herausgegeben von Prof. Dr. Ferdinand Cohn* 3(**1**): 1-814 [published in fascicles: pp. 1-130, 10 April 1885; pp. 131-258, 16 August 1886; pp. 259-386, 27 August 1887; pp. 387-512, 2 June 1888; pp. 513-640, 10 February 1889; pp. 641-814, 15 September 1889].

SCHROETER, J. (1885). Die Pilze Schlesiens. *In Kryptogamen-Flora von Schlesien* **3(1)**: 1-130 [10 April 1885].

SCHROETER, J. (1886*a*). Die Pilze Schlesiens. Phytomyxini. *In Kryptogamen-Flora von Schlesien* **3(1)**: 131-174 [16 August 1886].

SCHROETER, J. (1886*b*). Die Pilze Schlesiens. Chytridiei. *In Kryptogamen-Flora von Schlesien* **3(1)**: 175-224 [16 August 1886].

SCHROETER, J. (1886*c*-1887). Die Pilze Schlesiens. Oomycetes. *In Kryptogamen-Flora von Schlesien* **3(1)**: 225-257 [16 August 1886 - 27 August 1887].

SCHRÖTER, J. (1897). Phytomyxinae. Fungi (Pilze). *In* A. Engler (Ed.), *Die natürlichen Pflanzenfamilien nebst ihren Gattungen und wichtigeren Arten insbesondere den Nutzplanzen, unter Mitwirkung zahlreicher hervorragender Fachgelehrten begründet von A. Engler und K. Prantl. I. Teil, Abteilung I*: 1-. [Published in fascicles, Linnean Society of London accession dates as follows: pp -80, 15 Sept 1892; pp. 81-128, 26 Oct 1893; pp. 129- , 31 Dec 1894.]

SCHRÖTER, J. (1892*a*). Phytomyxinae. *In* A. Engler (Ed.), *Die natürlichen Pflanzenfamilien nebst ihren Gattungen und wichtigeren Arten insbesondere den Nutzplanzen, unter Mitwirkung zahlreicher hervorragender Fachgelehrten begründet von A. Engler und K. Prantl. I. Teil, Abteilung 4*: 5-8.

SCHRÖTER, J. (1892*b*). Angang. Die mit den Myxomyceten nächstverwandten Organismen. *In* A. Engler (Ed.), *Die natürlichen Pflanzenfamilien nebst ihren Gattungen und wichtigeren Arten insbesondere den Nutzplanzen, unter Mitwirkung zahlreicher hervorragender Fachgelehrten begründet von A. Engler und K. Prantl. I. Teil, Abteilung 4*: 36-41.

SCHRÖTER, J. (1892-93 [pp 81-87 1893]). Chytridineae. [1892 includes Olpidiaceae: *Reesia*, *Sphaerita*, *Olpidium*, *Olpidiopsis*, *Pleotrachelus*, *Ectrogella*, *Pleolpidium*; Synchytriaceae: ...*Rozella*, *Woronina* ...; Rhizidiaceae: ...*Rhizidiomyces*. 1893 includes Hyphochytriaceae: *Catenaria*, *Hyphochytrium*...; Oochytriaceae: *Diplophysa*...]. *In* A. Engler (Ed.), *Die natürlichen Pflanzenfamilien nebst ihren Gattungen und wichtigeren Arten insbesondere den Nutzplanzen, unter Mitwirkung zahlreicher hervorragender Fachgelehrten begründet von A. Engler und K. Prantl. I. Teil, Abteilung 4*: 64-87.

SCHRÖTER, J. (1893*a*). Ancylistineae. *In* A. Engler (Ed.), *Die natürlichen Pflanzenfamilien nebst ihren Gattungen und wichtigeren Arten insbesondere den Nutzplanzen, unter Mitwirkung zahlreicher hervorragender Fachgelehrten begründet von A. Engler und K. Prantl. I. Teil, Abteilung 4*: 88-92.

SCHRÖTER, J. (1893*b*). Saprolegniineae. *In* A. Engler (Ed.), *Die natürlichen Pflanzenfamilien nebst ihren Gattungen und wichtigeren Arten insbesondere den Nutzplanzen, unter Mitwirkung zahlreicher hervorragender Fachgelehrten begründet von A. Engler und K. Prantl. I. Teil, Abteilung 4*: 93-105.

SCHRÖTER, J. (1893*c*). Monoblepharidineae. *In* A. Engler (Ed.), *Die natürlichen Pflanzenfamilien nebst ihren Gattungen und wichtigeren Arten insbesondere den Nutzplanzen, unter Mitwirkung zahlreicher hervorragender Fachgelehrten begründet von A. Engler und K. Prantl. I. Teil, Abteilung 4*: 106-107.

SCHRÖTER, J. (1893*d*). Peronosporineae. *In* A. Engler (Ed.), *Die natürlichen Pflanzenfamilien nebst ihren Gattungen und wichtigeren Arten insbesondere den Nutzplanzen, unter Mitwirkung zahlreicher hervorragender Fachgelehrten begründet von A. Engler und K. Prantl. I. Teil, Abteilung 4*: 108-119.

SCHRÖTER, J. & SCHNEIDER, W. G. (1870). Uebersicht der in Schlesien gefundenen Pilze. *Jahresberichte Schlesische Gesellschaft für Vaterländische Kultur* **47**: 140-143.

SCHULZ, P. (1922). Desmidiaceen aus dem Gebiete der Freien Stadt Danzig und dem benachbarten Pomerellen. *Botanisches Archiv* **2(3)**: 113-153.

SCHULZ, P. (1923). Kurze Mitteilungen über Algenparasiten. *Schriften für Süsswasser und Meereskunde* **2**: 178-181.

SCHWAB-STEY, H. & SCHWAB, D. (1974*a*). Hypothetical origin of the particular membrane relationships in *Labyrinthula*. *Protoplasma* **81**: 125-130.

SCHWAB-STEY, H. & SCHWAB, D. (1974*b*). Scanning electron microscopic investigation on the motile stage of *Labyrinthula coenocystis* Schmoller. *Cytobiologia* **8**: 383-394.

SCHWARTZE, C. A. (1922). The method of cleavage in the sporangia of certain fungi. *Mycologia* **14**: 143-172.

SCHWARTZ, E. J. (1910). Parasitic root diseases of the Juncaceae. *Annals of Botany, London* **24**: 511-512.

SCHWARTZ, E. J. (1911). The life-history and cytology of *Sorosphaera graminis*. *Annals of Botany, London* **25**: 791-797.

SCHWARTZ, E. J. (1914). The Plasmodiophoraceae and their relationship to the Mycetozoa and the Chytrideae. *Annals of Botany, London* **28**: 227-240.

SCHWARTZ, E. J. & IVIMEY COOK, W. R . (1928). The life history and cytology of a new species of *Olpidium*: *Olpidium radicale* sp. nov. *Transactions of the British mycological Society* **13**: 205-221.

SCHWINN, F. J. (1983) New developments in chemical control of *Phytophthora*. *In* D. C. Erwin, S. Bartnicki-Garcia & P. H. Tsao (Eds), Phytophthora, *Its Biology, Taxonomy, Ecology and Pathology*, pp. 327-334. American Phytopathological Society, St Paul, Minnesota, U.S.A.

SCOTT, E. S. (1985). Production and characterization of single-spore isolates of *Plasmodiophora brassicae*. *Plant Pathology* **34**: 287-292.

SCOTT, W. W. (1961). A monograph of the genus *Aphanomyces*. *Virginia Agricultural Experiment Station, Technical Bulletin* **151**: 1-95.

SCOTT, W. W. & O'BIER, A. H. (1962). Aquatic fungi associated with diseased fish and fish eggs. *Prog.. Fish Culturist* **24**: 3-15.

SCOTT, W. W., POWELL, J. R. & SEYMOUR, R. L. (1963). Pure culture techniques applied to the growth of *Saprolegnia* spp. on a chemically defined medium. *Virginia Journal of Science* **14**: 42-46.

SCOTT, W. W., SEYMOUR, R. & WARREN, C. (1963). Some new and unusual fungi from Virginia. I. Lower phycomycetes. *Virginia Journal of Science* **14**: 11-15.

SCOTT, W. W. & WARREN, C. O. (1964). Studies of the host range and chemical control of fungi associated with diseased tropical fish. *Virginia Agriculture Experiment Station, Blacksburg, Virginia, Technical Bulletin 171*: 1-24.

SEAMAN, W. L., LARSON, R. H. & WALKER, J. C. (1961). Tinsel-type structure in flagella of zoospores associated with cabbage clubroot tissue. *Nature (London)* **190**: 186-187.

SERBINOV, I. L. [И. Л. СЕРБИНОВ, **I. L. Serbinov**] (1925). К морфологии и биологии грибка *Lagenidium sacculoides* n. sp. Sur la morphologie et la biologie du *Lagenidium sacculoides* n. sp. *La Défense des Plantes. Bulletin du Bureau Permanent des Congrès Entomo-Phytopathologiques de Russie* [Защита Растений от Вредителей. Бюллетень Постоянного Бюро Всероссийских Энтомо-Фнтопатологических Съездов] **2**: 84-87.

SERBINOW, J. L. (1899). Vorläufiger Bericht über die Morphologie und Biologie des *Olpidium ramosum* spec. nov. *Trudy. S.-Peterburg. Obshchestvo estestvoispytatelei* [*Travaux de la Société des Naturalistes de St.-Petersbourg*] **30**: 224-226, 255-256. [The same person as the preceding.]

SERBINOW, J. L. (1907). Kenntniss der Phycomyceten. Organisation und Entwickelungsgesichte einiger Chytridineen Pilze (Chytridineae Schröter). *Scriptis Botanicis Horti Universitatis Impericolis Petropolitanae* **24**: 1-173.

SERBINOW, I. L. (1912). Zur Biologie von *Pythium perniciosum* nov. spec., einem Pilzparasiten der Tabaksämlinge. *Scripta Botanica Horti Universitatis Imperialis Petropolitanae* **28**: 29-30.

SESHADRI, K. & PAYAK, M. M. (1970). Nuclear structure and behaviour in the vegetative hyphae of *Pythium aphanidermatum*. *Mycopathologia et Mycologia Applicata* **40**: 145-153.

SETCHELL, W. A. (1924). Three new fungi. *Mycologia* **16**: 240-244.

SEURAT, L. G. (1920). *Histoire naturelle des Nématodes de la Bérbérie. Travaux du Laboratoire de Zoologie Générale. Faculté des Sciences, Universié d'Alger* pp. 221. [Not seen, not available in U.K.; "available on microfilm at University of California, Berkeley and Arlington National Register of Microform Masters, Virginia, according to OCLC union catalog" (*litt.*).]

SEWARD, A. C. (1898). *Fossil Plants, Volume I.*, pp. 205-222. Cambridge, U.K.: Cambridge University Press.

SEYMOUR, R. L. (1970). The genus *Saprolegnia*. *Nova Hedwigia* **19**: 1-124.

SEYMOUR, R. L. (1984). *Leptolegnia chapmanii*, an oomycete pathogen of mosquito larvae. *Mycologia* **76**: 670-674.

SEYMOUR, R. L. & JOHNSON, T. W. (1973*a*). An unusual oömycete infecting rotifer eggs. *Mycologia* **65**: 941-944.

SEYMOUR, R. L. & JOHNSON, T. W. (1973*b*). Saprolegniaceae: a keratinophilic *Aphanomyces* from soil. *Mycologia* **65**: 1312-1318.

SHAFFER, T. H., PADGETT, D. E. & CELIO, D. A. (1990). Evidence for enhanced salinity tolerance of a suspected fish pathogen of Atlantic menhaden, *Brevoortia tyrannus* Latrobe. *Journal of Fish Diseases* **13**: 335-344.

SHAHZAD, S., COE, R. & DICK, M. W. (1992). Biometry of oospores and oogonia of *Pythium* (Oomycetes): the independent taxonomic value of calculated ratios. *Botanical Journal of the Linnean Society* **108**: 143-165.

SHAHZAD, S. & DICK, M. W. (1990). Effect of different natural baits on zoospore and oospore production by two *Pythium* species. *Pakistan Journal of Botany* **22**: 125-128.

SHANOR, L. (1937). Observations on the development and cytology of the sexual organs of *Thraustotheca clavata* (de Bary) Humph. *Journal of the Elisha Mitchell Scientific Society* **53**: 119-136.

SHANOR, L. (1938). Observations on the development of a new species of *Phytophthora*. *Journal of the Elisha Mitchell Scientific Society* **54**: 154-162.

SHANOR, L. (1939*a*). Studies in the genus *Olpidiopsis*. I. Resting spore germination in a new species. *Journal of the Elisha Mitchell Scientific Society* **55**: 167-177.

SHANOR, L. (1939*b*). Studies in the genus *Olpidiopsis*. II. The relationship of *Pseudolpidium* Fischer and *Olpidiopsis* (Cornu) Fischer. *Journal of the Elisha Mitchell Scientific Society* **55**: 179-195.

SHANOR, L. (1940). Studies in the genus *Olpidiopsis*. III. Some observations on the host range of certain species. *Journal of the Elisha Mitchell Scientific Society* **56**: 165-176.

SHANOR, L. (1942). A new *Rozella* of the polysporangiate series. *Journal of the Elisha Mitchell Scientific Society* **58**: 99-101.

SHANOR, L. (1944). Additional records of aquatic Phycomycetes isolated from Mexican soils. *Journal of the Washington Academy of Science* **34**: 330-333.

SHANOR, L. & CONOVER, R. A. (1943). A new *Protoachlya*. *American Midland Naturalist* **28**: 746-751.

SHANOR, L. & OLIVE, L. S. (1942). Notes on *Araiospora streptandra*. *Mycologia* **34**: 536-542.

SHANOR, L. & SASLOW, H. B. (1944). *Aphanomyces* as a fish parasite. *Mycologia* **36**: 413-415.

SHATLA, M. N., YANG, C. Y. & MITCHELL, J. E. (1966). Cytological and fine-structure studies of *Aphanomyces euteiches*. *Phytopathology* **56**: 923-928.

SHAW, C. G. (1978). *Peronosclerospora* species and other downy mildews of Graminae. *Mycologia* **70**: 594-604.

SHAW, C. G. (1980). *Peronosclerospora noblei*. *Mycologia* **72**: 426.

SHAW, C. G. (1981). Taxonomy and evolution. *In* D. M. Spencer (Ed.), *The Downy Mildews*, pp. 17-29. Academic Press, London, UK.

SHAW, C. G. & WATERHOUSE, G. M. (1980). *Peronosclerospora* (Ito) Shirai & K. Hara antedates *Peronosclerospora* (Ito) C. G. Shaw. *Mycologia* **72**: 425-426.

SHAW, D. S. (1967). A method of obtaining single-oospore cultures of *Phytophthora cactorum* using live water snails. *Phytopathology* **57**: 454.

SHAW, D. S. (1991). Genetics. *In Advances in Plant Pathology* 7. pp. 131-170. New York, U.S.A.: Academic Press.

SHAW, D. S. & ELLIOTT, C. G. (1968). Streptomycin resistance and morphological variation in *Phytophthora cactorum*. *Journal of General Microbiology* **51**: 75-84.

SHAW, D. S. & KHAKI, I. A. (1971). Genetical evidence for diploidy in *Phytophthora*. *Genetics Research* **17**: 165-167.

SHAW, D. S. & SHATTOCK, R. C. (1991). Genetics of *Phytophthora infestans*: the Mendelian approach. *In* J. A. Lucas, R. C. Shattock, D. S. Shaw & L. R. Cooke (Eds), pp. 218-230. Phytophthora. Cambridge, U.K.: Cambridge University Press.

SHEARER, B. L. & HILL, T. C. (1989). Diseases of *Banksia* woodlands on the Bassendean and Spearwood dune systems. *Journal of the Royal Society of Western Australia* **71**: 113-114.

SHEARER, B. L. & TIPPETT, J. T. (1989). Jarrah dieback: the dynamics and management of *Phytophthora cinnamomi* in the Jarrah (*Eucaypsus marginata*) forest of south-western Australia. *Research Bulletin No. 3, Department of Conservation and Land Management, Como, Western Australia*, i-xvii, 1-76.

SHEN, S.-C. & SIANG, W. N. (1948). Studies in the aquatic Phycomycetes of China. *Science Reports of the National Tsing Hua University, Series B: Biology and Psychological Science* **3**: 179-203.

SHEPHARD, A. M. (1956). A short survey of Danish nematophagous fungi. *Friesia* **5**: 396-408.

SHERF, A. F. (1943). A method for maintaining *Phytomonas sepedonica* in culture for long periods without transfer. *Phytopathology* **33**: 330-332.

SHERWOOD, R. T. & HAGEDORN, D. J. (1962). Studies on the biology of *Aphanomyces euteiches*. *Phytopathology* **52**: 150-154.

SHERWOOD, W. A. (1966). Evidence for a sexual hormone in the water mold *Dictyuchus*. *Mycologia* **58**: 215-220.

SHERWOOD, W. A. (1968). Extension of the host range of *Octomyxa brevilegniae*. *Journal of the Elisha Mitchell Scientific Society* **84**: 52-55.

SHERWOOD, W. A. (1969). Sexual reactions between clonal subcultures of a strain of *Dictyuchus monosporus*. *Mycologia* **62**: 251-263.

SHERWOOD, W. A. (1971). Some observations on the sexual behavior of the progeny of six isolates of *Dictyuchus monosporus*. *Mycologia* **63**: 22-30.

SHIBUYA, Y. (1959). On some species of Saprolegniaceae found in Sanshiro Pond in the University of Tokyo. *Japanese Journal of Botany* **34**: 146-151.

SHIELDS, J. D. & KURIS, A. M. (1987). Mechanisms of egg mortality in *Cancer anthonyi*. *American Zoologist* **27**: 15A.

SHILOVA, T. B. [ЙНЛОВА, Т. Б., **Shilova, T. B.**] (1985). Endoparasitic fungi found in the U.S.S.R. in the nematodes' *Gonimochaete* Dreschs., *Meristacrum* Drechs., and *Meria* Drechs. *Сектор Микобиологии Ан Азербайджаиской, ССР. Mikologiya i Fitopatologiya* [*Микопогия и ФитоПапогия*] **19**: 405-407.

SHIPTON, W. A. (1983). Possible relationship of some growth and sporulation responses of *Pythium* to the occurrence of equine phycomycosis. *Transactions of the British Mycological Society* **80**: 13-80.

SHIPTON, W. A. (1987). *Pythium destruens* sp. nov., an agent of equine pythiosis. *Journal of Medical and Veterinary Mycology* **25**: 137-151.

SHIPTON, W. A., MILLER, R. I. & LEA, I. R. (1982). Cell wall, zoospore and morphological characteristics of Australian isolates of a *Pythium* causing equine phycomycosis. *Transactions of the British Mycological Society* **79**: 15-23.

SHIRAI, M. & HARA, K. (1927). *A List of Japanese Fungi Hitherto Known.* Third Ed. Shikuoka, Japan.

SHORT, F. T., MATHIESON, A. C. & NELSON, J. I. (1986). Recurrence of the eelgrass wasting disease at the border of New Hampshire and Maine, U.S.A. *Marine Ecology Progress Series* **29**: 89-92.

SHISHKOFF, N. (1989). Zoospore encystment pattern and germination on onion roots, and the colonization of hypodermal cells by *Pythium coloratum*. *Canadian Journal of Botany* **67**: 258-262.

SHUMARD, D. S., GROSSMAN, L. I. & HUDSPETH, M. E. S. (1986). *Achlya* mitochondrial DNA: gene localization and analysis of inverted repeats. *Molecular and General Genetics* **202**: 16-23.

SHUMARD, D. S. & HUDSPETH, M. E. S. (1986). Mitochondrial DNA in Oomycetes. *Mycological Society of America Newsletter* **37**: 42-43.

SHUMARD-HUDSPETH, D. S. & HUDSPETH, M. E. S. (1990). Genic arrangements in *Phytophthora* mitochondrial DNA. *Current Genetics* **17**: 413-415.

SIANG, W.-N. (1949). Are aquatic phycomycetes present in the air? *Nature (London)* **164**: 1010-1011.

SIDERIS, C. P. (1931). Taxonomic studies in the family Pythiaceae. I. *Nematosporangium*. *Mycologia* **23**: 252-295.

SIDERIS, C. P. (1932). Taxonomic studies in the family Pythiaceae. II. *Pythium*. *Mycologia* **24**: 14-61.

SIEGEL, J. P. & SHADDUCK, J. A. (1987). Safety of the entomopathogenic fungus *Lagenidium giganteum* (Oomycetes: Lagenidiales) to mammals. *Journal of Economic Entomology* **80**: 994-997.

SIEGENTHALER, P. A., BELSKY, M. M. & GOLDSTEIN, S. (1967). Phosphate uptake in an obligately marine fungus: a specific requirement for sodium. *Science, New York* **155**: 93-94.

SIEGENTHALER, P. A., BELSKY, M. M., GOLDSTEIN, S. & MENNA, M. (1967). Phosphate uptake in an obligately marine fungus. II. Role of culture conditions, energy sources and inhibitors. *Journal of Bacteriology* **93**: 1281-1288.

SIETSMA, J. H., EVELEIGH, D. E. & HASKINS, R. H. (1969). Cell wall composition and protoplast formation of some oomycete species. *Biochimica et Biophysica Acta* **184**: 306-317.

SIGOT, A. (1931). Une chytridiacée nouvelle, parasite des oeufs de *Cyclops*: *Blastulidiopsis chattoni* n.g., n. sp. *Comptes Rendus de la Société de Biologie de Strasbourg* **108**: 34-37.

SILBERMAN, J. D., SOGIN, M. L., LEIPE, D. D. & CLARK, C. G. (1996). Human parasite finds taxonomic home. *Nature (London)* **380**: 398.

SILVA, J. F. V. & CAMPOS, V. P. (1991*a*). (Endoparasitic fungi of nematodes found in the south of Minas Gerais State, Brazil 1. Comments on their morphology and the parasitism). *Nematologia Brasileira* **15**: 94-103 [only abstract consulted].

SILVA, J. F. V. & CAMPOS, V. P. (1991*b*). (Endoparasitic fungi of nematodes found in Minas Gerais State, Brazil 2. Field survey report and in-vitro observations). *Nematologia Brasileira* **15**: 104-111 [only abstract consulted].

SILVA, J. F. V. & CAMPOS, V. P. (1992). Effect of the temperature on the parasitic ability of *Haptoglosssa heterospora* in *Panagrellus redivivus*. *Brazilian Phytopathology [Fitopathologica Brasiliera]* **17**: 110-111.

SILVER, J. C. & HORGEN, P. A. (1974). Hormonal regulation of presumptive mRNA in the fungus *Achlya ambisexualis*. *Nature (London)* **249**: 252-254.

SIMMONS, C. A., KERWIN, J. L. & WASHINO, R. K. (1987). Preliminary characterization of lipoxygenase from the entomopathogenic fungus *Lagenidium giganteum*. *In* P. K. Stumpf, J. B. Mudd & W. D. Nes (Eds), *The Metabolism, Structure, and Function of Plant Lipids*, pp. 421-423. U.S.A.: Plenum Publishing Corporation.

SINDERMANN, C. J. (1977). *Disease Diagnosis and Control in North American Marine Aquaculture.* Amsterdam, The Netherlands: Elsevier.

SINGH, K. P. (1989). Artificial culture of *Catenaria anguillulae* from monosporangial zoospores. *Mycological Research* **92**: 107.

SINGH, K. P. & GUPTA, P. (1986). Observations on *Catenaria anguillulae* parasitic on *Heterodera sorgi*. *Advances in Biological Research* **4**: 240-250.

SINGH, K. P., STEPHEN, R. A. & MAKESHKUMAR, T. (1993). Development of *Catenaria anguillulae* in *Heterodera cajani*. *Mycological Research* **97**: 957-960.

SINGH, K. P., STEPHEN, R. A. & VAISH, S. S. (1996). Pathogenicity and development of *Catenaria anguillulae* on some nematodes. *Mycological Research* **100**: 1204-1206.

SINGH, S. L. & PAVGI, M. S. (1977). *Aphanomyces* root rot of cauliflower. *Mycopathologia* **61**: 167-172.

SINGH, U. P. (1973). A new species of *Myzocytium* from India. *Hydrobiologia* **42**: 445-450.

SINGH, U. P., CHAUHAN, V. B., WAGNER, K. G. & KUMAR, A. (1992). Effect of ajoene, a compound derived from garlic (*Allium sativum* on *Phytophthora drechsleri* f. sp. *cajani*. *Mycologia* **84**: 105-108.

SINGH, U. P. & PAVGI, M.S. (1970). Two noteworthy members of Catenariaceae from India. *Mycologia* **62**: 587-590.

SINGH, U. P. & PAVGI, M.S. (1977). *Aphanomyces* root rot of cauliflower. *Mycopathologia* **61**: 167-172.

SINGH, U. S., SHARMA, A. K. & PANDEY, B. K. (1989). Growth of biotrophic plant parasites and symbionts in tissue culture. *In* N. P. Agnihotri, N. Singh, H. S. Chaube, U. S. Singh & T. S. Dwivedi (Eds), *Perspectives in Phytopathology*, pp. 113-139. New Delhi, India: Today and Tomorrow's Printers and Publishers.

SIRADHANA, B. S., DANGE, S. R. S., RATHORE, R. S. & SINGH, S. D. (1980). A new downy mildew on maize in Rajasthan, India. *Current Science, Bangalore, India* **49**: 316-317.

SKALICKÝ, V. (1964). Beitrag zur infraspezifischen Taxonomie der obligat parasitischen Pilze (Prispevek k vnitrodruhove taxonomii nekterych obligatne parasitickych plisni a hub). *Acta Universitatis Carolinae (Praha) 1964*: 25-89.

SKALICKÝ, V. (1966). Taxonomie der Gattungen der Familie Peronosporaceae (Taxonomie rodu celedi Peronosporaceae). *Preslia (Praha)* **38**: 117-129.

SKALICKÝ, V. (1990). Delimitations of the genus *Pseudoperonospora* Rostowz. *Novitates Botanicae Universitatis Carolinae (Praha)* **6**: 7-12.

SKIDMORE, D. I. & INGRAM, D. S. (1985). Conidial morphology and the specialization of *Bremia lactucae* Regel (Peronosporaceae) on hosts in the family Compositae. *Botanical Journal of the Linnean Society* **91**: 503-522.

SKUJA, H. (1934). Beitrag zur Algenflora Lettlands. I. *Acta Horti Botanici Universitatis Latviensis* **7**: 149-177.

SKUJA, H. (1956). Taxonomische und biologische Studien über das Phytoplankton schwedischer Binnengewässer. *Nova Acta Regiae Societatis Scientiarum Upsaliensis. Seria IV* **16**: 1-404.

SKVORTZOW, B. W. (1925). Zur Kenntnis der Phycomyceten aus der Nordmandshurei, China. *Archiv für Protistenkunde* **51**: 428-433.

SKVORTZOW, B. W. (1927). Über einige Phycomyceten aus China. *Archiv für Protistenkunde* **57**: 204-206.

SLEIGH, M. A. (1981). Flagellar beat patterns and their possible evolution. *BioSystems* **14**: 423-431.

SLEIGH, M. A. (1989). *Protozoa and other Protists.* London, U.K.: Edward Arnold.

SLEIGH, M. A. (1991). Mechanisms of flagellar propulsion. A biologist's view of the relation between structure, motion and fluid mechanics. *Protoplasma* **164**: 45-53.

SLIFKIN, M. K. (1961). Parasitism of *Olpidiopsis incrassata* on members of the Saprolegniaceae. I. Host range and effects of light, temperature, and stage of host on infectivity. *Mycologia* **53**: 183-193.

SLIFKIN, M. K. (1962). A new method of purification and preservation of *Olpidiopsis incrassata*. *Mycologia* **54**: 105-106.

SLIFKIN, M. K. (1963). Parasitism of *Olpidiopsis incrassata* on members of the Saprolegniaceae. II. Effect of pH and host nutrition. *Mycologia* **55**: 172-182.

SLIFKIN, M. K. (1967). Nuclear division in *Saprolegnia* as revealed by colchicine. *Mycologia* **59**: 431-445.

SLIFKIN, M. K. (1968). The effect of colchicine on *Olpidiopsis incrassata* and its host, *Saprolegnia delica*. *Journal of the Elisha Mitchell Scientific Society* **84**: 184-186.

SLYKHIUS, J. T. & BARR, D. J. S. (1978). Confirmation of *Polymyxa graminis* as a vector of wheat spindle streak mosaic virus. *Phytopathology* **68**: 639-643.

SMITH, A. (1970). Biometric studies on conidia of *Peronospora tabacina*. *Transactions of the British Mycological Society* **55**: 59-66.

SMITH, C. J. (1996). Tansley Review No. 86. Accumulation of phytoalexins: defence mechanism and stimulus response system. *New Phytologist* **132**: 1-45.

SMITH, E. F. & SALE, W. S. (1992). Regulation of dynein-driven microtubule sliding by the radial spokes in flagella. *Science, New York* **257**: 1557-1559.

SMITH, F. (1884). The pathology of Bursattee. *Veterinary Journal* **19**: 16-17.

SMITH, F. E. V. (1923). On direct nuclear divisions in the vegetative mycelium of *Saprolegnia*. *Annals of Botany, London* **37**: 63-73.

SMITH, R. F. (1973). Considerations on the safety of certain biological agents for arthropod control. *Bulletin of the World Health Organization* **48**: 685-698.

SMITH, R. L. (1940). Studies on two strains of *Aphanomyces laevis* found occurring as wound parasites on crayfish. *Mycologia* **32**: 205-213.

SMITH, S. N., ARMSTRONG, R. A. & RIMMER, J. J. (1984). Influence of environmental factors on zoospores of *Saprolegnia diclina*. *Transactions of the British Mycological Society* **82**: 413-421.

SMITH, V. J. & SÖDERHÄLL, K. (1983). Induction of de-granulation and lysis of haemocytes in the freshwater crayfish, *Astacus astacus* by components of the prophenoloxidase activating system *in vitro*. *Cell Tissue Research* **233**: 295-303.

SMITH, W. G. (1876). The resting spores of the potato fungus. *Gardeners' Chronicle, New Series* **6**: 6-12.

SMITH, W. G. (1878*a*). The salmon disease (*S. ferax*). *Gardeners' Chronicle, New Series* **9**: 560-561.

SMITH, W. G. (1878*b*). The salmon disease (*S. ferax*). *Grevillea* **6**: 152-156.

SMITH, W. G. (1884*a*). *Diseases of Field and Garden Crops Chiefly such as are caused by Fungi*. Macmillan, London, UK.

SMITH, W. G. (1884*b*). Diseases of mushrooms (caused by *Fusisporium mycophytum* and *Saprolegia mucophaga*). *Gardeners' Chronicle* **22**: 245.

SNEH, B., EYE, L. L. & LOCKWOOD, J. L. (1981). Factors affecting germination of oospores of *Phytophthora megasperma* var. *sojae*. *Phytopathologische Zeitschrift* **101**: 314-322.

SNEH, B., HUMBLE, S. J. & LOCKWOOD, J. L. (1977). Parasitism of oospores of *Phytophthora megasperma* var. *sojae*, *P. cactorum*, *Pythium* sp. and *Aphanomyces euteiches* in soil by Oomycetes, Chytridiomycetes, Hyphomycetes, Actinomycetes, and bacteria. *Phytopathology* **67**: 622-628.

SÖDERHÄLL, K. & AJAXON, R. (1982). Effect of quinones and melanin on mycelial growth of *Aphanomyces* spp. and extracellular protease of *Aphanomyces astaci*, a parasite on crayfish. *Journal of Invertebrate Pathology* **39**: 105-109.

SÖDERHÄLL, K. & CERENIUS, L. (1992). Crustacean immunity. *Annual Review of Fish Diseases, 1992*: 3-23.

SÖDERHÄLL, K., DICK, M. W., CLARK, G. FÜRST, M. & CONSTANTINESCU, O. (1991). Isolation of *Saprolegnia parasitica* from the crayfish *Astacus leptodactylus*. *Aquaculture* **92**: 121-125.

SÖDERHÄLL, K. & SMITH, V. J. (1983). Separation of the haemocyte populations of *Carcinus maenas* and other marine decapods, and prophenoloxidase distribution. *Developmental and Comparative Immunolology* **7**: 229-239.

SÖDERHÄLL, K. & SMITH, V. J. (1986). The prophenoloxidase activating system: The biochemistry of its activation and role in arthropod cellular immunity, with special reference to crustaceans. *In* M. Brehélin (Ed.), *Immunity in invertebrates*, pp. 208-223. Berlin & Heidelberg, Germany: Springer-Verlag.

SOGIN, M. L. (1989). Evolution of eukaryotic microorganisms and their small subunit RNAs. *American Zoologist* **29**: 487-499.

SOLBERG, L. (1925). Sygdom paa Erter. *Havedyrkningens Venners Medlemsskr.....* **4**: 1-6.

SOLTIS, D. E., SOLTIS, P. S., CHASE, M. W., ALBACH, D., MORT, M. E., SAVOLAINEN, V. & ZANIS, M. (1998*a*). Molecular phylogenetics of angiosperms: congruent patterns inferred from three genes. Part II. *American Journal of Botany* **85**: 157 [abstract].

SOLTIS, P. S., SOLTIS, D. E., CHASE, M. W., ALBACH, D., MORT, M. E., SAVOLAINEN, V. & ZANIS, M. (1998*a*). Molecular phylogenetics of angiosperms: congruent patterns inferred from three genes. Part I. *American Journal of Botany* **85**: 157-158 [abstract].

SOLTIS, P. S., SOLTIS, D. E. & CHASE, M. W. (1999). Angiosperm phylogeny inferred from multiple genes as a tool for comparative biology. *Nature (London)* **402**: 402-404.

SÖRGEL, G. (1952). Dauerorganbildung bei *Rozella allomycis* Foust, ein Beitrag zur Kenntnis der Sexualität der niederen Phycomyceten. *Archiv für Mikrobiologie* **17**: 247-254.

SOROKIN, N. W. [СОРОКИН, Н., **Sorokin, N.**] (1873). Mykologicheskiia izsliedovaniia [Mycological Researches]. *Trudy Obschchestva Estestvoispytatelei pri Imperatordkom Kasanskom Universitete Kazan* **2** [no.1]: 23.

SOROKIN, N. [W.] [СОРОКИН, Н., **Sorokin, N.**] (1873). [A few words about a new fungus *Synchytrium urticae*]. [*Contributions to the 3rd Meeting of Russian Science, Botany, Kiev*] : 39-42.

SOROKINE, N. (1876*a*). Quelques mots sur le developpement de l'*Aphanomyces stellatus*. *Annales des Sciences Naturelles, Botanique, Série VI* **3**: 46-52.

SOROKIN[E], N. [W.] [СОРОКИН, Н., **Sorokin, N.**] (1876*b*). Les végétaux parasites des Anguillulae. *Annales des Sciences Naturelles, Botanique, Série VI* **4**: 62-71.

SOROKINE, N. (1883). Aperçu systématique des Chytridiacées récoltées en Russie et dans l'Asie Centrale. *Archives Botaniques du Nord de la France* **2**: 1-42.

SOROKINE, N. (1889). Matériaux pour la flore cryptogamique de l'Asie Centrale. *Revue Mycologique, Toulouse* **11**: 69-85; 136-151; 207-208 [illustrations only in Sorokine, 1890].

SOROKINE, N. (1890). Matériaux pour la flore cryptogamique de l'Asie Centrale. *Revue Mycologique, Toulouse* **12**: 3-16; 53-61.

SOUMATI, B. & DICK, M. W. (1989). Biometry of oogonia and oospores of Saprolegniaceae. *Botanical Journal of the Linnean Society* **100**: 219-235.

SPARROW, F. K. (1929). A note on the occurrence of two rotifer-capturing phycomycetes. *Mycologia* **21**: 90-96.

SPARROW, F. K. (1930). The non-sexual stage of *Aphanomyces phycophilus*. *Mycologia* **22**: 118-121.

SPARROW, F. K. (1931*a*). Observations on *Pythium dictyosporum*. *Mycologia* **23**: 191-203.

SPARROW, F. K. (1931*b*). Two new species of *Pythium* parasitic in green algae. *Annals of Botany, London* **45**: 257-277.

SPARROW, F. K. (1932). Observations on the aquatic fungi of Cold Spring Harbor. *Mycologia* **24**: 268-303.

SPARROW, F. K. (1933*a*). New chytridiaceous fungi. *Transactions of the British Mycological Society* **18**: 215-217.

SPARROW, F. K. (1933*b*). Inoperculate chytridiaceous organisms collected in the vicinity of Ithaca, N.Y., with notes on other aquatic fungi. *Mycologia* **25**: 513-535.

SPARROW, F. K. (1934). Observations on marine phycomycetes collected in Denmark. *Dansk Botanisk Arkiv* **8**: 1-24.

SPARROW, F. K. (1936*a*). Biological observations on the marine fungi of Woods Hole waters. *Biological Bulletin of the Marine Biological Laboratory, Woods Hole* **70**: 236-263.

SPARROW, F. K. (1936*b*). A contribution to our knowledge of the aquatic Phycomycetes of Great Britain. *Journal of the Linnean Society (Botany)* **50**: 417-478.

SPARROW, F. K. (1937). Some chytridiaceous inhabitants of submerged insect exuviae. *Proceedings of the American Philosophical Society* **78**: 23-53.

SPARROW, F. K. (1938). Remarks on the genus *Rozella*. *Mycologia* **30**: 375-378.

SPARROW, F. K. (1939*a*). The entomogenous chytrid *Myrophagus* Thaxter. *Mycologia* **31**: 437-444.

SPARROW, F. K. (1939*b*). A new species of *Lagenidium* parasitic in rotifer eggs. *Mycologia* **31**: 527-532.

SPARROW, F. K. (1942). A classification of aquatic Phycomycetes. *Mycologia* **34**: 113-116.

SPARROW, F. K. (1943). *Aquatic Phycomycetes*. Ann Arbor, Michigan, U.S.A.: University of Michigan Press.

SPARROW, F. K. (1950*a*). Some Cuban Phycomycetes. *Journal of the Washington Academy of Sciences* **40**: 50-55.

SPARROW, F. K. (1950*b*). Annual foray, 1948. *Mycologia* **42**: 191-195.

SPARROW, F. K. (1952*a*). A contribution to our knowledge of the phycomycetes of Cuba. Part I. *La Revista de la Sociedad Cubana de Botanica* **9**: 34-40.

SPARROW, F. K. (1952*b*). A contribution to our knowledge of the phycomycetes of Cuba. Part II. *La Revista de la Sociedad Cubana de Botanica* **9**: 68-74.

SPARROW, F. K. (1952*c*). A contribution to our knowledge of the phycomycetes of Cuba. Part III. *La Revista de la Sociedad Cubana de Botanica* **9**: 104-108.

SPARROW, F. K. (1952*d*). Phycomycetes from the Douglas Lake region of northern Michigan. *Mycologia* **44**: 759-772.

SPARROW, F. K. (1957). A further contribution to the phycomycete flora of Great Britain. *Transactions of the British Mycological Society* **40**: 523-535.

SPARROW, F. K. (1958). Interrelationships and phylogeny of the aquatic phycomycetes. *Mycologia* **50**: 797-813.

SPARROW, F. K. (1960). *Aquatic Phycomycetes*, Second Revised Edition. Ann Arbor, U.S.A.: University of Michigan Press.

SPARROW, F. K. (1968). Remarks on the Thraustochytriaceae. *Veröffentlichungen des Instituts für Meeresforschung in Bremerhaven, Sonderband* **3**: 7-18.

SPARROW, F. K. (1969). Zoosporic marine fungi from the Pacific Northwest (U.S.A.). *Archiv für Mikrobiologie* **66**: 129-146.

SPARROW, F. K. (1973*a*). The peculiar marine Phycomycete *Atkinsiella dubia* from crab eggs. *Archiv für Mikrobiologie* **93**: 137-144.

SPARROW, F. K. (1973*b*). Mastigomycotina (zoosporic fungi). *In* G. C. Ainsworth, F. K. Sparrow & A. S. Sussman (Eds), *The Fungi, An Advanced Treatise*, **IVB**, pp. 61-73. New York, U.S.A.: Academic Press.

SPARROW, F. K. (1973*c*). Chytridiomycetes, Hyphochytridiomycetes. *In* G. C. Ainsworth, F. K. Sparrow & A. S. Sussman (Eds.), *The Fungi, An Advanced Treatise* **IVB**, pp. 85-110. New York, U.S.A.: Academic Press.

SPARROW, F. K. (1973*d*). Lagenidiales. *In* G. C. Ainsworth, F. K. Sparrow & A. S. Sussman (Eds), *The Fungi, An Advanced Treatise*, **IVB**, pp. 159-163. New York, U.S.A.: Academic Press.

SPARROW, F. K. (1975). Zoosporic fungi from Tunisia, and the Seychelles Islands (Indian Ocean). *Archives of Microbiology* **104**: 39-49.

SPARROW, F. K. (1976). The present status of classification in biflagellate fungi. *In* E. B. G. Jones (Ed.), *Recent Advances in Aquatic Mycology*, pp. 213-222. London, U.K.: Paul Elek.

SPARROW, F. K. (1977*a*). *Rhizidiomycopsis bullatus*, a new zoosporic fungus. *Mycologia* **69**: 661-666.

SPARROW, F. K. (1977*b*). *Rhizidiomycopsis* on azygospores of *Gigaspora margarita*. *Mycologia* **69**: 1053-1058.

SPARROW, F. K. & BARR, M. E. (1955). Additions to the phycomycete flora of the Douglas Lake region. I. New taxa and records. *Mycologia* **47**: 546-556.

SPARROW, F. K. & CUTTER, V. M. (1941). Observations on *Mindeniella spinosa*. *Mycologia* **33**: 288-293.

SPARROW, F. K. & DOGMA, I. J. (1973). Zoosporic phycomycetes from Hispaniola. *Archiv für Mikrobiologie* **89**: 177-204.

SPARROW, F. K. & ELLISON, B. (1949). *Olpidiopsis schenkiana* and its hyperparasite *Ectrogella besseyi*. *Mycologia* **41**: 28-35.

SPARROW, F. K. & GOTELLI, D. (1969). Is *Atkinsiella* holocarpic? *Mycologia* **61**: 199-201.

SPARROW, F. K. & JOHNSON, T. W. (1961). *Fungi in Oceans and Estuaries*. Weinheim, Germany: J. Cramer.

SPARROW, F. K. & KOCH, T. W. (1959). Additions to the phycomycete flora of the Douglas Lake region. IV. New records and notes for 1956-1957. *Michigan Academy of Science, Arts, and Letters* **44**: 153-159.

SPARROW, F. K., PATERSON, R. A. & JOHNS, R. M. (1965). Additions to the phycomycete flora of the Douglas Lake region. V. New or interesting fungi. *Michigan Academy of Science, Arts, and Letters* **50**: 115-123.

SPAULDING, P. (1961). Foreign diseases of forest trees of the world. *U.S. Agricultural Research Service, Agriculture Handbook no. 197*: 1-361.

SPENCER, D. M. (Ed.) (1981). *The Downy Mildews*. Academic Press, London, UK.

SPENCER, J. A. & COOPER, W. E. (1967). Pathogenesis of cotton by *Pythium* species: zoospore and mycelium attraction and infectivity. *Phytopathology* **57**: 1332-1338.

SPENCER, M. A. & DICK, M. W. (2001). Aspects of graminicolous downy mildew biology: perspectives for plant pathology and peronosporomycete phylogeny. *In* R. Watling (Ed.), *Tropical Mycology, British Mycological Society Meeting, Liverpool, April 2000*. **[in press]**.

SPENCER-PHILLIPS, P. T. N. (1997). Function of fungal haustoria in epiphytic and endophytic infections. *Advances in Botanical Research* **24**: 309-333.

SPESCHNEW, N. N. von (1907). *Die Pilzparasiten des Teestrauches*. Berlin, Germany: Friedländer und Sohn.

SPRAGUE, R. (1950). *Diseases of Cereals and Grasses in North America* New York, U.S.A.: Ronald Press Company.

SPRAGUE, V. (1954). Protozoa. *Fishery Bulletin of the Fish and Wildlife Service, United States Department of the Interior* **89**: 243-256.

SPRAGUE, V. (1965). *Ichthyosporidium* Caullery and Mesnil, 1905, the name of a genus of fungi or a genus of sporozoans? *Systematic Zoology* **14**: 110-114.

SRINIVASAN, M. C., NARASIMHAN, M. J. & THIRUMALACHAR, M. J. (1961). A new *Sclerospora* on *Iseilema laxum*. *Bulletin of the Torrey Botanical Club* **88**: 91-94.

SRIVASTAVA, G. C. (1964). Observations on Indian aquatic fungi. II. Occurrence of *Olpidiopsis varians* Shanor in Gorakhpur. *Indian Phytopathology* **17**: 249-253.

SRIVASTAVA, G. C. (1966). The host range of four species of *Olpidiopsis*. *Transactions of the British Mycological Society* **49**: 69-72.

SRIVASTAVA, G. C. (1967). Some species of *Saprolegnia* collected from Gorakhpur, India. *Hydrobiologia* **30**: 281-292.

SRIVASTAVA, G. C. (1975). *Olpidiopsis indica* sp. nov. from India. *Current Science, Bangalore, India* **44**: 642-644.

SRIVASTAVA, G. C. & BHARGAVA, K. S. (1963). Observations on Indian aquatic fungi. I. Three species of *Olpidiopsis* from Gorakhpur. *Indian Phytopathology* **16**: 271-274.

SRIVASTAVA, G. C. & SRIVASTAVA, R. C. (1975). Fungal infection of the eggs of *Channa punctatus*. *Geobios* **3**: 160.

SRIVASTAVA, G. C. & SRIVASTAVA, R. C. (1977*a*). *Olpidiopsis incrassata* - a new record from India. *Indian Phytopathology* **29**: 83-84.

SRIVASTAVA, G. C. & SRIVASTAVA, R. C. (1977*b*). Host range of *Achlya prolifera* (Nees) de Bary on certain freshwater teleosts. *Mycopathologia* **61**: 61-62.

SRIVASTAVA, O. N. & SINHA, R. K. (1989). *Woronina raii* sp. nov. from Lohardaga, Bihar. *Acta Botanica Indica* **17**: 259-260.

SRIVASTAVA, R. C. (1979). Aphanomycosis - a new threat to fish population. *Mykosen* **22**: 25-29.

SRIVASTAVA, R. C. (1980). Fungal parasites of certain freshwater fishes in India. *Aquaculture* **21**: 387-392.

SRIVASTAVA, R. C. & SRIVASTAVA, G. C. (1978). Fungi associated with the diseases of freshwater fishes. *Mycopathologia* **63**: 121-126.

SRIVASTAVA, S. S. (1986). Distribution of nematophagous fungi in Indian Agricultural Research Institute. Campus soil, New Delhi. *Indian Journal of Nematology* **16**: 128-130.

SRIVASTAVA, S. S. (1987). A simple technique for isolating nematophagous fungi from soil. *Indian Journal of Nematology* **17**: 152-153.

STAMPS, D. J., WATERHOUSE, G. M., NEWHOOK, F. J. & HALL, G. S. (1990). Revised tabular key to the species of *Phytophthora*. *Mycological Papers, Commonwealth Mycological Institute, Kew* **162**: 1-28.

STANGHELLINI, M. E. (1974). Spore germination, growth and survival of *Pythium* in soil. *Proceedings of the Americal Phytopathological Society* **1**: 211-214.

STANGHELLINI, M. E., ADASKAVEG, J. E. & RASMUSSEN, S. L. (1974). Pathogenesis of *Plasmopara lactucae-radicis*, a systemic root pathogen of cultivated lettuce. *Plant Disease* **74**: 173-178.

STANGHELLINI, M. E., KIM, D. H., RASMUSSEN, S. L. & RORABAUGH, P. A. (1996). Control of root rot of peppers caused by *Phytophthora capsici* wth a nonionic surfactant. *Plant Disease* **80**: 1113-1116.

STANGHELLINI, M. E., RASMUSSEN, S. L., KIM, D. H. & RORABAUGH, P. A. (1996). Efficacy of nonionic surfactants in the control of zoospore spread of *Pythium aphanidermatum* in a recirculating hydroponic system. *Plant Disease* **80**: 422-428.

STEBBINS, G. L. (1981). Coevolution of grasses and herbivores. *Annals of the Missouri Botanical Garden* **68**: 75-86.

STECIOW, M. M. (1988). Algunos Oomycetes de ambientes acuaticos de la Provincia de Buenos Aires (Mastigomycotina). *Boletin de la Sociedad Argentina de Botanica* **29**: 211-217.

STECIOW, M. M. (1993). Presencia de Saprolegniales (Mastigomycotina) en Rio Santiage y afluentes (Prov. BS. AS., Argentina). *Boletin de la Sociedad Argentina de Botanica* **29**: 211-217.

STECIOW, M. M. (1998). Aquatic fungi (Chytridiomycota, Oomycota) of the Lake Vitel and its tributaries (Buenos Aires, Argentina). *Darwiniana (San Isidro)* **36**: 101-106.

STEIN, F. [R. von] (1851). Neue Beiträge zur Kenntnis der Entwicklungsgesichte und des feineren Baus der Infusionsthiere. *Zeitschrift für wissenschaftliche Zoologie* **3**: 473-509, plate XVIII.

STEIN, F. [R. von] (1854). *Die Infusionsthiere auf ihre Entwickelungsgesichte*. Leipzig, Germany: Wilhelm Engelmann.

STEIN, F. R. von (1859). *Der Organismus der Infusionsthiere nach eigenen reihenfolge bearbeitet. I Abteilung. Allgemeiner theil und naturgesichte der hypotrichen infusionsthiere* [14 plates]. Leipzig, Germany: Wilhelm Engelmann.

STEIN, F. R. von (1878). *Der Organismus der Infusionsthiere nach eigenen reihenfolge bearbeitet. III Abteilung, I Hälfte. Den noch nicht abgeschlossenen allgemeinen theil nebst erklärung der sämmlichten abbildungen enthaltend*. Leipzig, Germany: Wilhelm Engelmann.

STEUER, A. (1932). Über *Ellobiopsis elongata* n. sp. aus dem Südatlantik. *Note dell'Istituto italo-germanico di Biologia marina di Rovigno d'Istria* **5**: 3-6.

STEVENS, F. L. (1899). The compound oosphere of *Albugo bliti*. *Botanical Gazette* **28**: 149-172, 225-245.

STEVENS, F. L. (1901). Gametogenesis and fertilization in *Albugo*. *Botanical Gazette* **32**: 77-98, 157-169, 238-261.

STEVENS, F. L. (1902). Studies in the fertilization of Phycomycetes. *Botanical Gazette* **34**: 420-425.

STEVENS, F. L. (1904). Oogenesis and fertilization in *Albugo ipomoeae-panduranae*. *Botanical Gazette* **38**: 300-302.

STEVENS, F. L. (1913). *The Fungi Which Cause Plant Disease*. New York, U.S.A.: MacMillan.

STEPHENSON, L. W. & ERWIN, D. C. (1972). Encirclement of the oogonial stalk by the amphigynous antheridium in *Phytophthora capsici*. *Canadian Journal of Botany* **50**: 2439-2441.

STEPHENSON, L. W., ERWIN, D. C. & LEARY, J. V. (1974). Hyphal anastomosis in *Phytophthora capsici*. *Phytopathology* **64**: 149-150.

STEY, H. (1968). Nachweis eines bisher unbekannten Organells bei *Labyrinthula*. *Zeitschrift für Naturforschung B* **23**: 567.

STEY, H. (1969). Elektronmikroskopische Untersuchung an *Labyrinthula coenocystis* Schmoller. *Zeitschrift für Zellforschung und Mikroskopische Anatomie* **102**: 387-418.

STIDD, B. M. & CONSENTINO, K. (1975). *Albugo*-like oogonia from the North American Carboniferous. *Science* **190**: 1092-1093.

STIRLING, A. B. (1877-1878). Notes on the fungus disease affecting salmon. *Proceedings of the Royal Society of Edinburgh* **9**: 726-732.

STIRLING, A. B. (1878-1879). Additional observations on the fungus disease affecting salmon and other fish. *Proceedings of the Royal Society of Edinburgh* **10**: 232, 370.

STIRLING, A. M. & PLATZER, E. G. (1978). *Catenaria anguillulae* in the mermithid nematode *Romanomermis culicivorax*. *Journal of Invertebrate Pathology* **32**: 348-354.

STJERNA-POOTH, I. (1957). *Achlya prolifera* als Abwasserpilz in einem mittelschwedischen Wasserlauf. *Reports of the Institute for Freshwater Research, Drottningholm*: 247-266.

STOBBS, L. W., CROSS, G. W. & MONOCHA, M. S. (1982). Specificity and methods of transmission oof cucumber necrosis virus by *Olpidium radicale* zoospores. *Canadian Journal of Plant Pathology* **4**: 134-142.

STOLL, K. (1936). Saprolegniineen aus der Umgebung von Greifswald. *Mitteilungen Naturwissenschaften Verein Neuvorpommern und Rügen in Greifswald* **63-64**: 20-42.

STORCK, R. & ALEXOPOULOS, C. J. (1970). Deoxyribonucleic acid of fungi. *Bacteriological Reviews* **34**: 126-154.

STÖSSEL, P. & HOHL, H. R. (1981). Sterols in *Phytophthora infestans* and their role in the parasitic interactions with *Solanum tuberosum*. *Bericht der Schweitzerischen Botanischen Gesellschaft* **90**: 118-128.

STPICZŃSKA, E. (1962). Grzby wodne dolów potorfowych kolo wsi Calowanie (Die wasserpilzen in den Torfstichen auf dem Torfmoore Calowanie). *Monographiae Botanicae* **13**: 87-120.

STPICZŃSKA, E. (1963). *Saprolegnia latvica* Apinis W Europie Środkowej (*Saprolegnia latvica* Apinis in Central Europe). *Monographiae Botanicae* **15**: 423-425.

STUART, M. R. & FULLER, H. T. (1968). Mycological aspects of diseased Atlantic salmon. *Nature (London)* **217**: 90-92.

STUEHLING, J. J. & ZIEGLER, A. W. (1958). Weston's leptomitaceous fungus in Florida. *Mycologia* **50**: 947-948.

SU, X.-Q. (1994). The effects of *Lagenidium giganteum* on *Aedes albopictus* and other mosquitoes in Guizhou, China. *Entomologia Sinica* **1**: 283-288 [in Chinese].

SU, X.-Q. & GUO, Q. (1994). A primary study on the effects of some environmental substances on the infection rate of mosquitos by *Lagenidium giganteum* (Oomycetes: Lagenidiales). *Acta Mycologica Sinica* **13**: 111-120 [in Chinese].

SU, X.-Q., GUZMAN, D. R. & AXTELL, R. C. (1986). Factors affecting storage of mycelial cultures of the mosquito fungal pathogen *Lagenidium giganteum* (Oomycetes: Lagenidiales). *Journal of the American Mosquito Control Association* **2**: 350-354.

SU, X.-Q., JIA, S.-H. & XU, H.-M. (1997). A preliminary study on the ability of *Lagenidium giganteum* to control mosquitoes in water accumulations in gardens and tyres. *Mycosystema* **16**: 74-77 [in Chinese].

SUH, C. P. & AXTELL, R. C. (1999). *Lagenidium giganteum* zoospores: effects of concentration, movement, light, and temperature on infection of mosquito larvae. *Biological Control* **15**: 33-38.

SUN, M. H., CHANG, S. C. & TSENG, C. M. (1976). Research advances in sugarcane downy mildew of corn in Taiwan. *Kasertsart Journal* **10**: 89-93.

SUBRAMANIAM, L. S. (1919). A *Pythium* disease of ginger, tobacco and papaya. *Memoirs of the Department of Agriculture, India, Botanical Series* **10**: 181-194.

SUBRAMANIAM, L. S. (1928). Root rot and sclerotial diseases of wheat. *Bulletin of the Agricultural Research Institute, Pusa* **177**: 1-17.

SUBRAMANIAN, C. V. (1974). Two new families of chytrids. *Current Science, Bangalore, India* **43**: 722-723.

SUNDARAM, B. M. (1968). *Petersenia catenophlyctidis* sp. nov. *Journal of the Indian Botanical Society* **47**: 144-149.

SUNDHEIM, L. (1972). Physiologic specialization in *Aphanomyces euteiches*. *Physiological Plant Pathology* **2**: 301-306.

SUZUKI, K., MATSUMIYA, E., UENO, Y. & MIZUTANI, J. (1992). Some properties of germination-stimulating factor from plant for resting spores of *Plasmodiophora brassicae*. *Nihon Shokubutsu Byori Gakkai-Ho [Annals of the Phytopathological Society of Japan]* **58**: 699-705.

SUZUKI, S. (1960). Seasonal variation in the amount of zoospore of aquatic Phycomycetes in Lake Shinseiko. *Botanical Magazine, Tokyo* **73**: 483-486.

SUZUKI, S. (1961*a*). Ecological studies on aquatic fungi in the lakes of volcano Nikko. *Japanese Journal of Ecology* **11**: 1-4.

SUZUKI, S. (1961*b*). Distribution of aquatic Phycomycetes in some inorganic acidotrophic lakes of Japan. *Botanical Magazine, Tokyo* **74**: 317-320.

SUZUKI, S. (1961*c*). A list of the aquatic Phycomycetes in Japan. *Japanese Journal of Botany* **36**: 231-235.

SUZUKI, S. (1961*d*). On the ecological specificity of *Saprolegnia monoica* var. *acidamica*. *Japanese Journal of Botany* **36**: 292-295.

SUZUKI, S. (1962). Ecological studies on aquatic fungi in the Matsubara Lake group. *Japanese Journal of Botany* **37**: 60-64.

SYDOW, H. & BUTLER, E. J. (1907). Fungi Indiae Orientalis. *Annales Mycologici* **5**: 485-515.

SYKES, E. E. & PORTER, D. (1973). Nutritional studies of *Labyrinthula* sp. *Mycologia* **65**: 1302-1311.

SZANISZLO, P. J. (1965). A study of the effect of light and temperature on the formation of oögonia and oöspheres in *Saprolegnia diclina*. *Journal of the Elisha Mitchell Scientific Society* **81**: 10-15.

SZWANKE, A. (1938). Przyczynek do znajomości flory grzybów wodnych okolic Warzsawy (Beitrag zur Kenntnis der Wasserpilze in der Umgebung von Warschau). *Planta Polononica (Contributions à la Flore de la Pologne et des Pays Limatrophes)* **7**: 1-24.

TABOR, R. J. & BUNTING, R. H. (1923). On a disease of cocoa and coffee fruits caused by a fungus hitherto undescribed. *Annals of Botany (London)* **37**: 153-157.

TAKAHASHI, K. (1918). On the rot-disease of rice-seeds. *Byochugai Zasshi* (*Journal of Plant Protection*) **5**: 356,431.

TAKAHASHI, K. (1991). Correlation between fluorescent staining reaction and germination in resting spores of *Plasmodiophora brassicae*. *Nihon Shokubutsu Byori Gakkai-Ho [Annals of the Phytopathological Society of Japan]* **57**: 160-164.

TAKAHASHI, K. & YAMAGUCHI, T. (1988). A method for assessing the pathogenic activity of resting spores of *Plasmodiophora brassicae* by fluorescence microscopy. *Nihon Shokubutsu Byori Gakkai-Ho [Annals of the Phytopathological Society of Japan]* **54**: 466-475.

TAKAHASHI, M. (1954). On the morphology and taxonomy of some species of the genus *Pythium* which causes crop diseases. *Annals of the Phytopathological Society of Japan* **18**: 113-118.

TAKAHASHI, M., ICHITANI, T. & SASAKI, M. (1977). *Pythium porphyrae* sp. nov. causing red rot of marine algae *Porphyra* spp. *Transactions of the Mycological Society of Japan [Nihon Kin Gakkai Kaiho, Japan]* **18**: 279-285.

TAKAHASHI, M. & KAWASE, Y. (1965). Ecologic and taxonomic studies on *Pythium* as pathogenic soil fungi. 5. Several species of *Pythium* causing root rot of strawberry. *Annals of the Phytopathological Society of Japan* **30**: 181-185.

TAKAHASHI, M., OHUCHI, A. & ALICBUSAN, R. V. (1965). Ecologic and taxonomic studies on *Pythium* as pathogenic soil fungi. 6. Some species of *Pythium* causing rhizome rot of Hindu lotus. *Annals of the Phytopathological Society of Japan* **30**: 186-191.

TAKAHASHI, M., TANAKA, Y., ICHITANI, T. & ALICBUSAN, R. V. (1972). Ecologic and taxonomic studies on *Pythium* as pathogenic soil fungi. 10. Several *Pythium* species causing damping-off of sugar beet seedlings. *Annals of the Phytopathological Society of Japan* **38**: 306-312.

TAKENAKA, S. & KAWASAKI, S. (1994). Characterization of alanine-rich, hydroxyproline-containing cell wall proteins and their application for identifying *Pythium* species. *Physiological and Molecular Plant Parthology* **45**: 249-261.

TAKIMOTO, S. (1935). (Phytophthorose of buckwheat.) *Bulletin of the Kyushu Institute of Technology, Science and Technology* **6**: 105-110.

TAKIMOTO, S. (1941). On the *Pythium* causing damping-off of seedling and fruit rot of cucumber. *Annals of the Phytopathological Society of Japan* **11**: 89-91.

TALLEY, M. R., MILLER, C. E. & BRASELTON, J. P. (1978). Notes on the ultrastructure of zoospores of *Sorosphaera veronicae*. *Mycologia* **70**: 1241-1247.

TALUKDAR, M. C. & BARUAH, H. K. (1952). The occurrence of aquatic fungi at Gauhati and their distribution. *Journal of the University of Gauhati* **3**: 117-128.

TAN, I. H. & DRUEHL, L. D. (1993). Phylogeny of the Northeast Pacific brown algal (Phaeophycean) orders as inferred from 18S rRNA gene sequences. *Hydrobiologia* **260/261**: 699-704.

TARLING, D. H. (1980). *Continental Drift and Biological Evolution.* Burlington, North Carolina, U.S.A.: Carolina Biology Readers, Carolina Biological Supply Company.

TASUGI, H. (1953). Studies on the downy mildew of rice plants caused by *Phytophthora macrospora* Ito et Tanaka. *Bulletin of the National Institute of Agricultural Science, Tokyo* C 2: 1-45.

TASUGI, H. & SIIONO, H. (1940). Damping-off of seedlings of China aster and Zinnia. *Annals of the Phytopathological Society of Japan* **10**: 280-295.

TATTERSALL, W. M. (1951). A review of the Mysidacea of the United States National Museum. *United States National Museum, Bulletin* **201**: 1-292.

TAYLOR, F. J. R. (1978). Problems in the development of an explicit hypothetical phylogeny of the lower eukaryotes. *BioSystems* **10**: 67-89.

TAYLOR, G. (1952). Analysis of the swimming of long and narrow animals. *Proceedings of the Royal Society of London* **214A**: 158-183.

TAYLOR, J. W. (1986). Fungal evolutionary biology and mitochondrial DNA. *Experimental Mycology* **10**: 259-269.

TEAKLE, D. S. & NIENHAUS, F. (1974). The effect of plant virus inhibitors on transmission of tobacco necrosis virus by *Olpidium brassicae*. *Phytopathologische Zeitschrift* **80**: 1-8.

TEHLER, A., FARRIS, J. S., LIPSCOMB, D. L. & KÄLLERSJÖ, M. (2000). Phylogenetic analyses of the fungi based on large rDNA sets. *Mycologia* **92**: 459-474.

TEILING, E. (1957). Morphological investigations of asymmetry in desmids. *Botaniska Notiser* **110**: 49-82.

TEMMINK, J. H. M. & CAMPBELL, R. N. (1968). The ultrastructure of *Olpidium brassicae*. I. Formation of sporangia. *Canadian Journal of Botany* **46**: 951-956.

TEMMINK, J. H. M., CAMPELL, R. N. & SMITH, P. R. (1970). Specificity and site of in vitro acquisition of tobacco necrosis virus by zoospores of *Olpidium brassicae*. *Journal of General Virology* **9**: 201-213.

TEMPÈRE, J. (1904). Le *Saprolegnia ferax* et son parasite. *Micrographe Préparateur* **12**: 160.

TeSTRAKE, D. (1958). Estuarine distribution and saline tolerance of some Saprolegniaceae. Unpublished MS Thesis, Duke University, Durham, North Carolina, U.S.A.

TeSTRAKE, D. (1959). Estuarine distribution and saline tolerance of some Saprolegniaceae. *ΦTton, International Journal of Experimental Botany* **12**: 147-152.

TEWARI, J. P. & SKOROPAD, W. P. (1977). Ultrastructure of oospore development in *Albugo candida* on rape seed. *Canadian Journal of Botany* **55**: 2348-2357.

THAKUR JI (1967). Some new records of Indian aquatic fungi of Varanasi. *Indian Phytopathology* **20**: 385-387.

THARP, T. P. & BLAND, C. E. (1977). Biology and host range of *Haliphthoros milfordensis*. *Canadian Journal of Botany* **55**: 2936-2944.

THAXTER, R. (1894). Observations on the genus *Naegelia* of Reinsch. *Botanical Gazette* **19**: 49-55.

THAXTER, R. (1895). New or peculiar fungi. 2. *Gonapodya* Fischer and *Monoblepharis*, nov. gen. *Botanical Gazette* **20**: 477-485.

THAXTER, R. (1896*a*). New or peculiar fungi. 3. *Blastocladia*. *Botanical Gazette* **21**: 45-52.

THAXTER, R. (1896*b*). New or peculiar fungi. 4. *Rhipidium*, *Sapromyces*, and *Araiospora*, nov. gen. *Botanical Gazette* **21**: 317-331.

THERRIEN, C. D. (1966). Microspectrophotometric measurement of nuclear deoxyribonucleic acid content in two myxomycetes. *Canadian Journal of Botany* **44**: 1667-1675.

THIANPRASIT, M. (1986). Fungal infection in Thailand. *Japanese Journal of Dermatology* **96**: 1343-1345.

THIANPRASIT, M. (1990). Human pythiosis. *Tropical Dermatology* **4**: 1-4.

THIRUMALACHAR, M. J. & LACY, R. C. (1951). A leaf spot disease of *Panicum repens* incited by a phycomycetous parasite. *Mycopathologia et Mycologia Applicata* **6**: 72-75.

THIRUMALACHAR, M. J., SHAW, C. G. & NARASIMHAN, M. J. (1953). The sporangial phase of the downy mildew on *Eleusine coracana* with a discussion of the identity of *Sclerospora macrospora* Sacc. *Bulletin of the Torrey Botanical Club* **80**: 299-307.

THIRUMALACHAR, M. J. & WHITEHEAD, M. D. (1952). Sporangial phase of *Sclerospora butleri*. *American Journal of Botany* **39**: 416-418.

THIRUMALACHAR, M. J., WHITEHEAD, M. D. & BOYLE, J. C. (1949). Gametogenesis and oospore formation in *Cystopus (Albugo) evolvuli*. *Botanical Gazette* **110**: 487-491.

THOMAS, D. des S. & McMORRIS, T. C. (1987). Allomonal functions of steroid hormone, antheridiol, in water mold *Achlya*. *Journal of Chemical Ecology* **13**: 1131-1137.

THOMAS, D. des S. & MULLINS, J. T. (1969). Cellulase induction and wall extension in the water mold *Achlya ambisexualis*. *Physiologia Plantarum* **22**: 347-353.

THOMAS, R. & LEWIS, D. T. (1998). Pythiosis in dogs and cats. *Compendium* **20**: 63-75.

THOMPSON, G. E. (1934). Sporangial germination in the genus *Myzocytium*. *Mycologia* **36**: 118-121.

THORNTON, C. R., JARVIS, B. C. & COOKE, R. C. (1991). A chitin assay for the enumeration of *Plasmodiophora brassicae* resting spores in clubroot tissue. *Mycological Research* **95**: 879-882.

THURET, G. (1850). Recherches sur les zoospores des Algues et les Antheridies des Cryptogames. Premier Partie. Zoospores des Algues. *Annales des Sciences Naturelles, Botanique, Série III* **14**: 214-260.

TIEGS, E. & DORRIES, W. (1926). Kann *Leptomitus lacteus* aus anorganischen Stichstoffverbindungen seinen Stichstoffbedarf decken? *Kliene Mitteilungen für die Mitglieder des Vereins für Wasser-, Boden- und Lufthygiene. Berlin* **2**: 105-108.

TIESENHAUSEN, M. von Bozen (1912). Beiträge zur Kenntnis der Wasspilze der Schweiz. *Archiv für Hydrobiologie und Planktonkunde* **7**: 261-308.

TIFFNEY, W. N. (1939*a*). The host range of *Saprolegnia parasitica*. *Mycologia* **31**: 310-321.

TIFFNEY, W. N. (1939*b*). The identity of certain species of Saprolegniaceae parasitic to fish. *Journal of the Elisha Mitchell Scientific Society* **55**: 134-151.

TIFFNEY, W. N. & WOLF, F. T. (1937). *Achlya flagellata* as a fish parasite. *Journal of the Elisha Mitchell Scientific Society* **53**: 298-300.

TIMBERLAKE, W. (1976). Alterations in RNA and protein synthesis associated with steroid hormone-induced sexual morphogenesis in the water mold *Achlya*. *Developmental Biology* **51**: 202-214.

TIMMER, L. W., CASTRO, J., ERWIN, D. C., BELSER, W. L. & ZENTMYER, G. A. (1970). Genetic evidence for zygotic meiosis in *Phytophthora capsici*. *American Journal of Botany* **57**: 1211-1218.

TINGGAL, S. H. & WEBSTER, J. (1981). Technique for single spore infection by *Plasmodiophora brassicae*. *Transactions of the British mycological Society* **76**: 187-190.

TIPPETT, J. T., HOLLAND, A. A., MARKS, G. C. & O'BRIAN, T. P. (1976). Penetration of *Phytophthora cinnamomi* into disease tolerant and susceptible eucalypts. *Archives of Microbiology* **108**: 231-242.

TIWARI, M. M. & ARYA, H. C. (1967). Growth of normal and diseased *Pennisetum typhoides* tissues infected with *Sclerospora graminicola* in tissue culture. *Indian Phytopathology* **20**: 356-368.

TIWARI, M. M. & ARYA, H. C. (1969). *Sclerospora graminicola* - axenic culture. *Science, New York* **163**: 291-293.

TOJO, M., NAKAZONO, E., TSUSHIMA, S., MORIKAWA, T. & MATSUMOTO, N. (1998). Characterization of two morphological groups of isolates of *Pythium ultimum* var. *ultimum* in a vegetable field. *Mycoscience* **39**: 135-144.

TOKUNAGA, J. & BARTNICKI-GARCIA, S. (1971). Structure and differentiation of the cell wall of *Phytophthora palmivora*: cysts, hyphae and sporangia. *Archiv für Mikrobiologie* **79**: 293-310.

TOKUNAGA, Y. (1932). A new species of *Pythium* parasitic on *Aegagropila sauteri* (Nees) Kützing. *Transactions of the Sapporo Natural History Society (Japan)* **12**: 119-123.

TOKUNAGA, Y. (1933*a*). Studies on the aquatic chytrids of Japan. I. Woroninaceae. *Transactions of the Sapporo Natural History Society (Japan)* **13**: 20-28.

TOKUNAGA, Y. (1933*b*). Studies on the aquatic chytrids of Japan. II. Olpidiaceae. *Transactions of the Sapporo Natural History Society (Japan)* **13**: 78-84.

TOKUNAGA, Y. (1934). Notes on the Lagenidiaceae in Japan. *Transactions of the Sapporo Natural History Society (Japan)* **13**: 227-232.

TOMASCHEK, A. (1878-1879). Übere binnenzellen in der grossen Zelle (Antheridiumzelle) des Pollens einiger Coniferen. *Sitzungsberichte der Akademie der Wissenschaften, Mathematische-naturwissenschaftliche Klasse, Wien* **78**: 197-212.

TOMLINSON, J. A. (1958). Crook root of watercress. III. The causal organism *Spongospora subterranea* (Wallr.) Lager. f. sp. *nasturtii* f. sp. nov. *Transactions of the British Mycological Society* **31**: 491-498.

TOMLINSON, J. A., ZITTER, T. A. & CLAY, C. M. (1985). Zoosporic fungi in roots of sweet corn infected with maize white line mosaic virus disease. *National Vegetable Research Station; 35th Annual Report for 1984* pp. 92-93.

TOMMERUP, I. C. (1989). Nuclear behaviour during conidial ontogeny of *Bremia lactucae*. *Mycological Research* **92**: 61-68.

TOMMERUP, I. & INGRAM, D. S. (1971). The life-cycle of *Plasmodiophora brassicae* Woron. in *Brassica* tissue cultures and in intact roots. *New Phytologist* **70**: 327-332.

TOMPKINS, C. M. (1975). World Literature on *Pythium* and *Rhizoctonia*. Contributions of the Reed Herbarium, No. XXIV: 1-169.

TONG, S. M. (1995). *Developayella elegans* nov. gen., nov. spec. a new type of heterotrophic flagellate from marine plankton. *European Journal of Protistology* **31**: 24-31.

TONG, S. M. (1997). Heterotrophic flagellates and other protists from Southampton Water. *Ophelia, International Journal of Marine Biology* **47**: 71-131.

TOOLEY, P. W. & CARRAS, M. M. (1992). Separation of chromosomes of *Phytophthora* species using CHEF gel electrophoresis. *Experimental Mycology* **16**: 188-196.

TOOLEY, P. W. & THERRIEN, C. D. (1987). Cytophotometric determination of the nuclear DNA content of 23 Mexican and 18 non-Mexican isolates of *Phytophthora infestans*. *Experimental Mycology* **11**: 19-26.

TOOLEY, P. W. & THERRIEN, C. D. (1991). Variation in ploidy in *Phytophthora infestans*. *In* J. A. Lucas, R. C. Shattock, D. S. Shaw & L. R. Cooke (Eds), pp. 204-217. Phytophthora. Cambridge, U.K.: Cambridge University Press.

TOTH, R. & MARKEY, D. R. (1973). Synaptonemal complexes in the brown algae. *Nature (London)* **243**: 236-237.

TOYAMA, A. & KUSABA, T. (1970). Transmission of soil-borne barley yellow mosaic virus. 2. *Polymyxa graminis* Led., as vector. *Nihon Shokubutsu Byori Gakkai-Ho* [*Annals of the Phytopathological Society of Japan*] **36**: 223-229.

TRAQUAIR, J. A. & McKEEN, W. E. (1980). Electron microscopy of developing *Aphanomyces* oogonia and oospores. *Mycologia* **72**: 378-394.

TRELEASE, W. (1884). Preliminary list of Wisconsin parasitic fungi. *Transactions of the Wisconsin Academy of Science* **6**: 106-144.

TRIBE, H. T. (1977*a*). A parasite of white cysts of *Heterodera*: *Catenaria auxiliaris*. *Transactions of the British Mycological Society* **69**: 367-376.

TRIBE, H. T. (1977*b*). Pathology of cyst-nematodes. *Biological Reviews of the Cambridge Philosophical Society* **52**: 477-507.

TRIBE, H. T. (1977*c*). On '*Olpidium nematodeae* Skvortzow'. *Transactions of the British Mycological Society* **69**: 509-511.

TRIGIANO, R. N. & SPURR, H. W. (1987). The development of the multinucleate condition of *Peronospora tabacina* sporangia. *Mycologia* **79**: 353-357.

TRIGIANO, R. N., VAN DYKE, C. G., SPURR, H. W. & GRAY, D. J. (1984). Infection and colonization of tobacco callus by *Peronospora tabacina*. *Phytopathology* **74**: 280-285.

TROUTMAN, J. L. & WILLS, W. H. (1964). Electrotaxis of *Phytophthora parasitica* zoospores and its possible role in infection of tobacco by the fungus. *Phytopathology* **54**: 225-228.

TROW, A. H. (1895). The karyology of *Saprolegnia*. *Annals of Botany, London* **9**: 609-652.

TROW, A. H. (1899). Observations on the biology and cytology of a new variety of *Achlya americana*. *Annals of Botany, London* **13**: 131-179.

TROW, A. H. (1901). Observations on the biology and cytology of *Pythium ultimum*, n. sp. *Annals of Botany, London* **15**: 269-312.

TROW, A. H. (1904). On fertilization in the Saprolegnieae. *Annals of Botany, London* **18**: 541-69.

TROW, A. H. (1905). Fertilization in the Saprolegniales. *Botanical Gazette* **139**: 300.

TRUSCOTT, J. H. L. (1933). Observations of *Lagena radicicola*. *Mycologia* **25** : 263-265.

TSANG, K. C. (1929). Recherches cytologiques sur la famille des Péronosporées. Étude spéciale de la reproduction sexuelle. *Botaniste* **21**: 1-128.

TSAO, P. (1970). Selective media for isolation of pathogenic fungi. *Annual Review of Phytopathology* **8**: 157-186.

TSAO, P. & OCANA, G. (1969). Selective isolation of species of *Phytophthora* from natural soils on an improved antibiotic medium. *Nature (London)* **223**: 636-638.

TUCKER, C. M. (1931). Taxonomy of the genus *Phytophthora* de Bary. *University of Missouri, College of Agriculture, Agricultural Experiment Station Research Bulletin* **153**: 1-208.

TULLOCH, J. S. (1934). A new record for *Leptomitus* from Alaska. *Torreya* **34**: 43-44.

TUNLID, A., NIVENS, D. E., JANSSON, H.-B. & WHITE, D. C. (1991). Infrared monitoring of the adhesion of *Catenaria anguillulae* zoospores to solid surfaces. *Experimental Mycology* **125**: 206-214.

TURNER, W. B. (1892). Algae aquae dulcis Indiae Orientalis. The freshwater algae (principally Desmidiae) of East India. *Kongliga Svenska Vetenskaps-Akademiens Handlingar* **25(5)**: 1-187, 23 plates.

TUTIN, T. G. (1938). The autecology of *Zostera marina* in relation to its wasting disease. *New Phytologist* **37**: 50-71.

UEBELMESSER, E.-R. (1956). Über einige neue Chytridineen aus Erdboden (*Olpidium, Rhizophidium, Phlyctochytrium* und *Rhizophlyctis*). *Archiv für Mikrobiologie* **25**: 307-324.

ULKEN, A. (1964). Über einige Thraustochytrien des polyhalinen Brackwassers. *Veröffentlichungen des Instituts für Meeresforschung in Bremerhaven* **9**: 31-41.

ULKEN, A. (1965*a*). Zwei neue Thraustochytrien aus der Aussenwasser. *Veröffentlichungen des Instituts für Meeresforschung in Bremerhaven* **9**: 289-295.

ULKEN, A. (1965*b*). On the occurrence of fungi (*Cephalosporium, Thraustochytrium, Schizochytrium*) in a water sample of the Atlantic ocean near the equator. *Notas Téchnicas. Instituto de Pesquisas da Marinha. Rio de Janeiro* **28**: 1-12.

ULKEN, A. (1966). Zur Physiologie einiger Thraustochytrien. *Veröffentlichungen des Instituts für Meeresforschung in Bremerhaven* **10**: 117-120.

ULKEN, A. (1968*a*). Einige Beobachtungen über das Vorkommen von uniflagellaten Phycomyceten (Chytridiales) in der Wesermündung. *Veröffentlichungen des Instituts für Meeresforschung in Bremerhaven, Sonderband* **3**: 59-66.

ULKEN, A. (1968*b*). Über zwei marine niedere Pilze vom Meeresboden der Nordsee. *Veröffentlichungen des Instituts für Meeresforschung in Bremerhaven, Sonderband* **3**: 74-71.

ULKEN, A. (1968*c*). Morphologische Variationen einiger mariner Thraustochytrien. *Berichte der Deutschen Botanischen Gesellschaft* **81**: 375-379.

ULKEN, A. (1970). Phycomyceten aus der Mangrove bei Cananeia (Sao Paulo, Brasilien). *Veröffentlichungen des Instituts für Meeresforschung in Bremerhaven* **12**: 313-319.

ULKEN, A. (1977). Phycomyceten aus der Laguna de Mandinga, Veracruz, Mexico. *Veröffentlichungen des Instituts für Meeresforschung in Bremerhaven* **16**: 177-189.

ULKEN, A. (1979). Phycomycetenfunde in der Sargossosee. *Veröffentlichungen des Instituts für Meeresforschung in Bremerhaven* **18**: 21-27.

ULKEN, A. (1981). The phycomycete flora of mangrove swamps in the South Pacific. *Veröffentlichungen des Instituts für Meeresforschung in Bremerhaven* **19**: 45-59.

ULKEN, A. (1983*a*). Phycomyceten in Watt des Jadebusens. *Veröffentlichungen des Instituts für Meeresforschung in Bremerhaven* **19**: 177-183.

ULKEN, A. (1983*b*). Distribution of Phycomycetes in mangrove swamps with brackish waters and waters of high salinity. *In* H. J. Teas (Ed.), *Tasks for Vegetation Science* **8**. The Hague, The Netherlands: W. Junk.

ULKEN, A. (1984). The fungi of the mangal ecosystem. *In* F. D. Por & I. Dor (Eds), *Hydrobiology of the Mangal*. The Hague, The Netherlands: W. Junk.

ULKEN, A. (1986). Estimation of thraustochytrid propagules in two mangrove swamps. *Botanica Marina, Berlin* **29**: 85-89.

ULKEN, A. (1990). *Marine Thraustochytrids and Chytridiomycetes in the North Sea Area and in Selected Other Regions*. (Bibliotheca Mycologica, Band 137). Berlin, Germany: J. Cramer.

ULKEN, A., JÄCKLE, I. & BAHNWEG, G. (1985). Morphology, nutrition and taxonomy of an *Aplanochytrium* sp. from the Sargasso Sea. *Marine Biology (Berlin)* **85**: 89-95.

UMPHLETT, C. J. (1973). A note to identify a certain isolate of *Lagenidium* which kills mosquito larvae. *Mycologia* **65**: 970-972.

UMPHLETT, C. J. & HUANG, C. S. (1970). *Lagenidium culicidum* as an agent of biological control of mosquitoes. *A.I.B.S. Bulletin* **17**: 68.

UMPHLETT, C. J. & HUANG, C. S. (1973). Experimental infection of mosquito larvae by a species of the aquatic fungus *Lagenidium*. *Journal of Invertebrate Pathology* **20**: 326-331.

UMPHLETT, C. J. & McCRAY, E. M. (1975). A brief review of the involvements of *Lagenidium*, a aquatic fungus parasite with arthropods. *Marine Fisheries Review* **37**: 61-64.

UNESTAM, T. (1966*a*). Chitinolytic, cellulolytic and pectinolytic activity *in vitro* of some parasitic and saprophytic Oomycetes. *Physiologia plantarum* **19**: 15-30.

UNESTAM, T. (1966*b*). Studies on the crayfish plague fungus *Aphanomyces astaci*. II. Factors affecting zoospores and zoospore production. *Physiologia plantarum* **19**: 1110-1119.

UNESTAM, T. (1968). Some properties of unpurified chitinase from the crayfish plague fungus, *Aphanomyces astaci*. *Physiologia plantarum* **21**: 137-147.

UNESTAM, T. (1969*a*). On the adaptation of *Aphanomyces astaci* as a parasite. *Physiologia plantarum* **22**: 221-235.

UNESTAM, T. (1969*b*). On the physiology of zoospore production in *Aphanomyces astaci*. *Physiologia plantarum* **22**: 236-245.

UNESTAM, T. (1969*c*). Resistance to the crayfish plague in some American, Japanese and European crayfishes. *Institute of Freshwater Research, Drottningholm. Report No. 49, 1969*: 3-8.

UNESTAM, T. (1972). On the host range and origin of the crayfish plague fungus. *Institute of Freshwater Research, Drottningholm. Report No. 52, 1972*: 192-198.

UNESTAM, T. (1981). Fungal diseases of freshwater and terrestrial Crustacea. *In* E. W. Davidson (Ed.), *Pathogenesis of Invertebrate Microbial Diseases*, pp. 484-510. New Jersey, U.S.A.: Allanheld, Osmum & Co.

UNESTAM, T. & GLEASON, F. H. (1968). Comparative physiology of respiration in aquatic fungi. II. The Saprolegniales, especially *Aphanomyces astaci*. *Physiologia plantarum* **21**: 573-588.

UNESTAM, T., NYLUND, J. E. (1972). Blood reactions *in vitro* against a fungal parasite, *Aphanomyces astaci*. *Journal of Invertebrate Patholology* **19**: 94-106.

UNESTAM, T. & SVENSSON, E. (1971). Physiological variation in strains of *Aphanomyces astaci*. *Physiologia plantarum* **25**: 414-416.

UNESTAM, T. & WEISS, D. W. (1970). The host-parasite relationship between freshwater crayfish and the crayfish disease fungus, *Aphanomyces astaci*: Responses to infection by a susceptible and a resistant species. *Journal of General Microbiology* **60**: 77-90.

UNGER, F. (1833). *Die Exantheme der Pflanzen und einige mit diesen verwandte krankheiten der Gewächte pathogenetisch und notographisch dergstellt.* Carl Gerold, Wien [Vienna], Austria.

UNGER, F. (1843). Einiges zur Lebensgeschichte der *Achlya prolifera*. *Linnaea* **17**: 129-152.

UNGER, F. (1844). Recherches sur l'*Achlya prolifera* Nees D'Ebck. *Annales des Sciences Naturelles, Botanique, Série III* **2**: 5-20.

UNGER, F. (1847). Beitrag zur Kenntniss der in der Kartoffelkrankheit vorkommenden Pilze und der Ursache ihres Entstehens. *Botanische Zeitung, 5 Jahrgang* **18**: 305-317, plate 6 [N.B.: column numbers not page numbers].

UNKLES, S. E., MOON, R. P., HAWKINS, A. R., DUNCAN, J. M. & KINGHORN, J. R. (1991). Actin in the oomycetous fungus *Phytophthora infestans* is the product of several genes. *Gene* **100**: 105-112.

UPADHYAYA, G., SHIVANNA, M. B., PRAKASH, H. S. & SHETTY, H. S. (1992). A novel approach to the establishment of dual cultures of pearl millet and *Sclerospora graminicola*. *Plant Cell Tissue and Organ Culture* **31**: 203-206.

VAARTAJA, O. (1965). New *Pythium* species from South Australia. *Mycologia* **57**: 417-430.

VADER, W. (1973*a*). The oldest published record of a *Thalassomyces* species (Ellobiopsidae). *Sarsia* **52**: 171-174.

VADER, W. (1973*b*). A bibliography of the Ellobiopsidae, 1959-1971, with a list of *Thalassomyces* species and their hosts. *Sarsia* **52**: 175-180.

VADER, W. & KANE, J. E. (1968). New hosts and distribution records of *Thalassomyces marsupii* Kane, an ellobiopsid parasite on amphipods. *Sarsia* **33**: 13-20.

VALKANOV, A. [ВЪЛКАНОВЪ, А., Vŭlkanov, A.] (1929*a*). Protistenstudien. 4. Die Natur und die systematische Stellung der Labyrinthuleen. *Archiv für Protistenkunde* **67**: 110-121.

VALKANOV, A. (1929*b*). Protistenstudien. 5. *Hyphochytrium hydrodictii* - ein neuer Algenpilz. *Archiv für Protistenkunde* **67**: 122-127.

VALKANOV, A. (1931*a*). Protistenstudien. 7. Beitrag zur Kenntnis der Süsswasserphycomyceten Bulgariens. *Archiv für Protistenkunde* **73**: 361-366.

VALKANOV, A. (1931*b*). Über Morphologie und Systematik der rotatorienbefallenden Pilze. *Archiv für Protistenkunde* **74**: 5-17.

VALKANOV, A. (1931*c*). Über die Morphologie und Systematik der rotatorienbefallenden Oomyceten. *Godishnik na Sofiiskiya Universitet (Jahrbuch der Sofianer Universität)* **27**: 215-233.

VALKANOV, A. (1932). Nachtrag zu meiner Arbeit über rotatorienbefallende Pilze. *Archiv für Protistenkunde* **78**: 485-496.

VALKANOV, A. (1966). *Dermocystidium cochliopodii* n. sp. Seine Strucktur und Entwicklung. *Nova Hedwigia* **12**: 389-393.

VALKANOV, A. (1969). *Labyrinthodictyon magnificum*, a new rhizoplasmodial unicellular organism. *In* A. A. Strelkov, K. M. Sukhanova & I. B. Raikov (Eds), *Progress in Protozoology. Abstracts of Papers Read at the IIIrd International Congress on Protozoology*, pp. 373-374. Leningrad, USSR: Nauka Publishing House.

VALKANOV, A. (1972*a*). Neue Untersuchungen über *Labyrinthula*. *Archiv für Protistenkunde* **114**: 421-425.

VALKANOV, A. (1972*b*). Untersuchungen über die Struktur und den Entwicklungszyklus von *Labyrinthodictyon magnificum* Valk. *Archiv für Protistenkunde* **114**: 426-443 [posthumous, includes a complete list of publications by A. Valkanov prepared by W. Naidenow & V. Golemansky].

VALLIN, S. (1933). Kraftpestens spridning inom Sverige under 1932-33. *Ny Svensk Fiskeritidskrift* **16**: .

VALLIN, S. (1951). Plankton mortality in the northern Baltic caused by a parasitic water-mould. *Institute of Freshwater Research, Drottningholm, Report No. 32*: 139-148.

VAN DE PEER, Y., VAN DER AUWERA, G. & DE WACHTER, R. (1996). The evolution of stramenopiles and alveolates as derived by "substitution rate calibration" of small ribosomal subunit RNA. *Journal of Molecular Evolution* **42**: 201-210.

VAN DER AUWERA, G., CHAPELLE, S. & DE WACHTER, R. (1994). Structure of the large ribosomal subunit RNA of *Phytophthora megasperma*, and phylogeny of the oomycetes. *FEBS Microbiology Letters* **338**: 133-136.

VAN DER AUWERA, G., DE BAERE, R., VAN DE PEER, Y., DE RIJK, P., VAN DEN BROECK, I. & DE WACHTER, R. (1995). The phylogeny of the Hyphochytriomycota as deduced from ribosomal RNA sequences of *Hyphochytrium catenoides*. *Molecular Biology and Evolution* **12**: 671-678.

VAN DER AUWERA, G. & DE WACHTER, R. (1997). Complete large subunit ribosomal RNA sequences from the heterokont algae *Ochromonas danica*, *Nannochloropsis salina* and *Tribonema aequale*, and phylogenetic analysis. *Journal of Molecular Evolution* **45**: 84-90.

VAN DER AUWERA, G. & DE WACHTER, R. (1998). Structure of the large subunit rDNA from a diatom, and comparison between small and large subunit ribosomal RNA for studying stramenopile evolution. *Journal of Eukaryote Microbiology* **45**: 521-527.

VAN DER GHEYNST, J, S., MAY, B. & KARAGOSIAN, M. (2000). Effect of aeration on the production of *Lagenidium giganteum*. *Abstracts of Papers: American Chemical Society* **219**: BIOT 255 [abstract].

VAN DER LEE, T., DE WITTE, I., DRENTH, A., ALFONSO, C. & GOVERS, F. (1997). AFLP linkage map of the oomycete *Phytophthora infestans*. *Fungal Genetics and Biology* **21**: 278-291.

VAN DER MEER, J. P. & PUESCHEL, C. M. (1985). *Petersenia palmariae* n. sp. (Oomycetes): a pathogenic parasite of the red alga *Palmaria mollis* (Rhodophyceae). *Canadian Journal of Botany* **63**: 404-408.

VAN DER WERFF, A. (1938). A new parasitic organism in *Zostera marina*. *Chronica Botanica* **4**: 498-499.

VANTERPOOL, T. C. (1938). Some species of *Pythium* parasitic on wheat in Canada and England. *Annals of Applied Biology* **25**: 528-543.

VANTERPOOL, T. C. & LEDINGHAM, G. A. (1930). Studies on "browning" rot of cereals. I. The association of *Lagena radicicola* n. gen., n. sp., with root injury of wheat. *Canadian Journal of Research* **2**: 171-194.

VANTERPOOL, T. C. & TRUSCOTT, J. H. L. (1932). Studies on browning root rot of cereals. 2. Some parasitic species of *Pythium* and their relation to the disease. *Canadian Journal of Research, Section C*, **6**: 68-93.

VATAN-DOOST, H., ZAIM, M., MOAZAMI, N., ASMAR, M. & DJAVADIAN, E. (1995). Use of *Lagenidium giganteum* for biological control of mosquito larvae in Iran. *Annals of Tropical Medicine and Parasitology* **89**: 213 [abstract].

VAZIRI-TEHRANI, B. & DICK, M. W. (1980*a*). Amino acid composition of oomycete cell walls. *Transactions of the British Mycological Society* **74**: 225-230.

VAZIRI-TEHRANI, B. & DICK, M. W. (1980*b*). Taxonomic significance of variations between ratios of cell wall amino acids in fungi, with special reference to Oomycetes. *Transactions of the British Mycological Society* **74**: 231-238.

VAZIRI-TEHRANI, B. & DICK, M. W. (1980*c*). Neutral and amino sugars from the cell walls of Oomycetes. *Biochemical Systematics and Ecology* **8**: 105-108.

VERCESI, A., TORNAGHI, R., SANT, S., BURRUANO, S. & FAORO, F. (1999). A cytological and ultrastructural study on the maturation and germination of oospores of *Plasmopara viticola* on overwintering vine leaves. *Mycological Research* **103**: 193-202.

VERMA, P. R. & PETRIE, G. A. (1975). Germination of oospores of *Albugo candida*. *Canadian Journal of Botany* **53**: 836-842.

VERMA, P. R., HARDING, H., PETRIE, G. A. & WILLIAMS, P. H. (1975). Infection and temporal development of mycelium of *Albugo candida* in cotyledons of four *Brassica* species. *Canadian Journal of Botany* **53**: 1016-1020.

VIALA, P. & SAUVAGEAU, C. (1892*a*). Sur la Brunnissure, maladie de la vigne causée par le *Plasmodiophora vitis*. *Comptes Rendus Hebdomadaires des Séances de l'Academie des Sciences, Paris* **114**: 1558-1560.

VIALA, P. & SAUVAGEAU, C. (1892*b*). Sur la maladie de Californie, maladie de la vigne causée par le *Plasmodiophora californica*. *Comptes Rendus Hebdomadaires des Séances de l'Academie des Sciences, Paris* **115**: 67-69.

VIALA, P. & SAUVAGEAU, C. (1892*c*). La Brunnissure et la maladie de Californie. *Journal de Botanique* **6**: 355-363; 378-388 + plate XII.

VIÉGAS, A. P. & TEIXEIRA, A. R. (1943). Alguns fungos do Brasil (Phycomycetos). *Bragantia* **3**: 223-269.

VIENNOT-BOURGIN, G. (1959). Étude de micromycetes parasites récoltés en Guinée. *Annales de l'Institut National de la Recherche Agronomique* **45**: 4-6.

VILLEGAS, C. T. & TI, T. L. (1978). Survival rate of *Penaeus monodon fabricius* larvae fed Chaetoceros sp. and bread yeast. *Quarterly Research Report, Aquaculture Department, Southeast Asian Fisheries Development Center* **2**: 12-15.

VILLOT, A. (1874). Monographie des Dragonneaux. *Archives de Zoologie Expérimentale et Générale* **3**: 181-238.

VIRÁNI, F. (1977). An improved method for detecting systemic infection of sunflower seedlings caused by *Plasmopara halstedii*. *Acta Phytopathologica Academiae Scientiarum Hungariae* **12**: 263-267.

VIRÁNI, F. & SZIRÁKI, I. (1986). Establishment of dual cultures of *Plasmopara halstedii* and sunflower. *Transactions of the British Mycological Society* **87**: 323-325.

VIRGILE, R., PERRY, H. D., PARDANANI, B., SZABO, K., RAHM, E. K., STONE, J., SALKIN, I. & DIXON, D. M. (1993). Human corneal ulcer caused by *Pythium insidiosum*. *Cornea* **12**: 81-83.

VISHNIAC, H. S. (1955*a*). Marine mycology. *Transactions of the New York Academy of Science, Series 2* **17**: 352-0360.

VISHNIAC, H. S. (1955*b*). The nutritional requirements of isolates of *Labyrinthula* spp. *Journal of General Microbiology* **12**: 455-463.

VISHNIAC, H. S. (1955*c*). The activity of steroids as growth factors for a *Labyrinthula* sp. *Journal of General Microbiology* **12**: 464-472.

VISHNIAC, H. S. (1955*d*). The morphology and nutrition of a new species of *Sirolpidium*. *Mycologia* **47**: 633-645.

VISHNIAC, H. S. (1956). On the ecology of the lower marine fungi. *Biological Bulletin of the Marine Biological Laboratory, Woods Hole* **111**: 410-414.

VISHNIAC, H. S. (1957). The occurrence of cholesterol in *Labyrinthula*. *Biochimica et Biophysica Acta* **26**: 430-431.

VISHNIAC, H. S. (1958). A new marine phycomycete. *Mycologia* **50**: 66-79.

VISHNIAC, H. S. & NIGRELLI, R. (1957). The ability of Saprolegniaceae to parasitize platyfish. *Zoologica* **42**: 131-134.

VISHNIAC, H. S. & WATSON, S. W. (1953). The steroid requirements of *Labyrinthula viellina* var. *pacifica*. *Journal of General Microbiology* **8**: 248-255.

VLK, W. (1931). Über die Struktur der Heterokontengeisseln. *Botanisch Centralblatt* **48**: 214-220.

VLK, W. (1939). Über die Geisselstruktur der Saprolegniaceenschwärmer. *Archiv für Protistenkunde* **92**: 157-160.

VOGEL, H. J. (1960). Two modes of lysine synthesis among lower fungi: evolutionary significance. *Biochimica et Biophysica Acta* **41**: 172-173.

VOGEL, H. J. (1961). Lysine synthesis and phylogeny of lower fungi: some chytrids versus *Hyphochytrium*. *Nature (London)* **189**: 1026-1027.

VOGEL, H. J. (1964). Distribution of lysine pathways among fungi: evolutionary implication. *American Naturalist* **98**: 435-446.

VOGLMAYR, H., BONNER, L. & DICK, M. W. (1999). Taxonomy and oogonial ultrastructure of a new aero-aquatic peronosporomycete, *Medusoides* gen. nov. (Pythiogetonaceae fam. nov.). *Mycological Research* **103**: 591-606.

VOGLMAYR, H. & GREILHUBER, J. (1998). Genome size determination in Peronosporales (Oomycota). *Fungal Genetics and Biology* **25**: 181-195.

VON BROEMBSEN, S. L. & DEACON, J. W. (1996). Effects of calcium on germination and further zoospore release from zoospore cysts of *Phytophthora parasitica*. *Mycological Research* **100**: 1498-1504.

VON BROEMBSEN, S. L. & DEACON, J. W. (1997). Calcium interference with zoospore biology and infectivity of *Phytophthora parasitica* in nutrient irrigation systems. *Phytopathology* **87**: 522-528.

VORONICHIN, N. (1920). K florie Phycomycetes Kavkaza (Contributions à la connaissance de Phycomycètes de la Caucasie). *Vestnik Tiflisskogo Botanicheskogo Sada [Moniteur du Jardin Botanique de Tiflis]* **50**: 8-14.

VUILLEMIN, P. (1888). Les tubercules radicaux des légumineuses. *Annales de la Science agronomique française et étrangère* **5**: 121.

VUILLEMIN, P. (1909). Valeur morphologique et biologique des tubercules radicaux des légumineuses. *Bulletin des Séances de la Société des Sciences (et de la Réunion Biologique) de Nancy* **10**: 30-45.

VURBANOV, V. & KROUMOV, K. (1989). (Spread and hosts of *Polymyxa betae*, parasite of sugar beet and vector of rhizomania pathogen). *Rastenievŭdni Nauki* **26**: 80-85.

WADA, S., YUASA, K., RHA, S.-a., NAKAMURA, K. & HATAI, K. (1994). Histopathology of *Aphanomyces* infection in Dwarf Gourami (*Colisa lalia*). *Fish Pathology [Gyobyo Kenkyu]* **29**: 229-237.

WAGER, H. (1889). Observations on the structure of the nuclei in *Peronospora parasitica*, and on their behaviour during the formation of the oöspore. *Annals of Botany* **4**: 127-146.

WAGER, H. (1893). On nuclear division in the Hymenomycetes. *Annals of Botany* **7**: 489-514.

WAGER, H. (1896*a*). Reproduction and fertilization in *Cystopus candidus*. *Annals of Botany* **10**: 89-91.

WAGER, H. (1896b). On the structure and reproduction of *Cystopus candidus* Lév. *Annals of Botany* **10**: 293-342.

WAGER, H. (1899). The sexuality of the fungi. *Annals of Botany* **13**: 575-597.

WAGER, H. (1900). On the fertilization of *Peronospora parasitica*. *Annals of Botany* **14**: 263.

WAGNER, D. T. (1969). A monocentric holocarpic fungus in *Lemna minor*. *Nova Hedwigia* **18**: 203-208.

WAGNER-MERNER, D. T., DUNCAN, W. R. & LAWRENCE, J. M. (1980). Preliminary comparison of Thraustochytriaceae in the guts of a regular and irregular echinoid. *Botanica Marina, Berlin* **23**: 95-97.

WAGNER-MERNER, D. T., TE STRAKE, D. & LAWRENCE, J. M. (1980). Occurrence of fungi (Thraustochytriaceae) in the gut of *Lytechinus variegatus* (Lamark). (Echinodermata, Echinoidea). *Florida Scientist* **43**: 62-63.

WAHRLICH, W. (1887). *Pythium*, n. sp. *Berichte der Deutschen Botanischen Gesellschaft* **5**: 242-246.

WAHRLICH, W. (1889). *Trudy imperatordkom Petrograd Obshchestva Estestvoispytatelei* **19**: 23.

WAKEFIELD, E. M. (1927). The genus *Cystopus* in South Africa. *Bothalia* **2**: 242-246.

WALDHEIM, A. A. von (1877). *Aperçu Systématique des Ustilaginées*. Paris, France.

WALENTOWICZ. (1885). Karpfenfest in Kaniow. *Österreichische Vierteljahresschrift für Wissenschaftliche Veterinärkunde* **64**.

WALLROTH, R. W. (1833). *Flora Cryptogamican Germaniae* **2**: 742.

WALLROTH, R. W. (1842*a*). Die Naturgeschichte der *Erysibe subterranea* Wallr. *[Wallroth's] Beiträge zur Botanik* **1**: 118-123.

WALLROTH, R. W. (1842*b*). Die Knollenbrand der Kartoffel. *Linnaea, Ein Journal für die Botanik in ihrem ganzen Umfange* **16**: 332 only.

WALKER, W. F. & DOOLITTLE, W. F. (1982). Nucleotide sequences of ribosomal RNA from four oomycete and chtrid water moulds. *Nucleic Acids Research* **10**: 5717-5721.

WALSH, J. A. & PHELPS, K. (1991). Development and evaluation of a technique for screening watercress *Rorippa Nasturtium aquaticum* for resistance to watercress yellow spot virus and crook-root fungus *Spongospora subterranea* f. sp. *nasturtii*. *Plant Pathology* **40**: 212-220.

WALZ, J. (1870). Beiträge zur Kenntniss der Saprolegnieen. *Botanische Zeitung, Berlin, 28 Jahrgang* **34**: 537-546; **35**: 553-557. [N.B. column numbers, not page numbers.]

WANG, M. C. & BARTNICKI-GARCIA, S. (1973). Novel phosphoglucans from the cytoplasm of *Phytophthora palmivora* and their selective occurrence in certain life cycle stages. *Journal of Biological Chemistry* **248**: 4112-4118.

WANG, M. C. & BARTNICKI-GARCIA, S. (1974). Mycolaminarans: storage (1-3) -ß-D-glucans from the cytoplasm of the fungus *Phytophthora palmivora*. *Carbohydrate Research* **37**: 331-338.

WARBURTON, A. J. & DEACON, J. W. (1998). Transmembrane Ca^{2+} fluxes associated with zoospore encystment and cyst germination by the phytopathogen *Phytophthora parasitica*. *Fungal Genetics and Biology* **25**: 54-62.

WARD, E. & ADAMS, M. J. (1998). Analysis of ribosomal DNA sequences of *Polymyxa* species and related fungi and the development of genus- and species-specific PCR primers. *Mycological Research* **102**: 965-974.

WARD, H. M. [see MARSHALL WARD, H.]

WARD, M. W. (1939*a*). Observations on a new species of *Thraustotheca*. *Journal of the Elisha Mitchell Scientific Society* **55**: 346-352.

WARD, M. W. (1939*b*). Observations on *Rhizophlyctis rosea*. *Journal of the Elisha Mitchell Scientific Society* **55**: 353-359.

WARNER, S. A., EIERMAN, D. F., DOMNAS, A. J. & DEAN, D. D. (1986). A novel lipoprotein from oomycete fungi. *Experimental Mycology* **10**: 315-322.

WARNER, S. A., GRAHAM, M. S., SOVOCOOL, G. W. & DOMNAS, A. J. (1981). Alkane contamination of lipids extracted from *Lagenidium giganteum* and *Lagenidium callinectes*. *Lipids (USA)* **16**: 628-630.

WARNER, S. A., SOVOCOOL, G. W. & DOMNAS, A. J. (1982). Fungal transformations of triparanol. *Applied and Environmental Microbiology* **44**: 1471-1475.

WARNER, S. A., SOVOCOOL, G. W. & DOMNAS, A. J. (1983*a*). Sterols of selected Oomycetes and Hyphochytridiomycetes. *Mycologia* **75**: 285-291.

WARNER, S. A., SOVOCOOL, G. W. & DOMNAS, A. J. (1983*b*). Comparative utilization of sterols by the Oomycetes *Lagenidium giganteum* and *Lagenidium callinectes*. *Experimental Mycology* 7: 227-232.

WARNER, S. A., SOVOCOOL, G. W., DOMNAS, A. J. & JARONSKI, S. T. (1984). The composition of sterols in mosquito larvae is optimal for zoosporogenesis in *Lagenidium giganteum*. *Journal of Invertebrate Pathology* **43**: 293-296.

WARREN, J. R. (1948). A study of the sugar beet seedling disease in Ohio. *Phytopathology* **38**: 883-892.

WASHBURN, J. O., EGERTER, D. E., ANDERSON, J. R. & SAUNDERS, G. A. (1988). Density reduction in larval mosquito (Diptera: Culicidae) populations by interactions between a parasitic ciliate (Ciliophora: Tetrahymenidae) and an opportunistic fungal (Oomycetes: Pythiaceae) parasite. *Journal of Medical Entomology* **25**: 307-314.

WASHINO, R. K., FETTER, J. L., FUKUSHIMA, C. K. & GONOT, K. (1976). The establishment of *Lagenidium giganteum*, an aquatic fungal parasite of mosquitoes, three years after field introduction. *In* C.D. Grant (Ed.), *Proceedings and Papers of the Forty-fourth Annual Conference of the California Mosquito Control Association*, p. 52. Visalia, U.S.A.: California Mosquito and Vector Control Association Press.

WASHINO, R. K. & FUKUSHIMA, C. K. (1978). Assessment of biological control agents against mosquito larvae in northern California: a progress report. *In* C. D. Grant (Ed.), *Proceedings and Papers of the Forty-sixth Annual Conference of the California Mosquito and Vector Control Association, Inc., January 29-February 1, 1978*, p. 89. Visalia, California, U.S.A.: California Mosquito and Vector Aassociation Press.

WASHINO, R. K. & LAIRD, M. (1983). Biocontrol of mosquitoes associated with California rice fields with special reference to the recycling of *Lagenidium giganteum* Couch and other microbial agents. *Biocontrol of Medical and Veterinary Pests*, pp. 122-139. New York, USA: Praeger Publishers.

WASHINO, R. K., McCRAY, E. M., UMPHLETT, C. J. & LASKO, J. F. (1977). Recovery of *Lagenidium giganteum*, an aquatic fungal parasite of mosquitoes, two years after field introduction. *Proceedings, Sixty-second Annual Meeting, held jointly with Thirty-first Annual Meeting The American Mosquito Control Association*, pp. 125. New Jersey, U.S.A.

WATANABE, T. (1974). Fungi isolated from the underground parts of sugar cane in relation to thepoor ratooning in Taiwan. (2) *Pythium* and *Pythiogeton*. *Transactions of the Mycological Society of Japan* **15**: 343-357.

WATANABE, T. (1983). Formation and deciduousness of sporangia of *Pythium intermedium*. *Transactions of the Mycological Society of Japan* **24**: 25-33.

WATANABE, T. (1987). *Plectospira myriandra*, a rediscovered water mold in Japanese soil. *Mycologia* **79**: 77-81.

WATERHOUSE, G. M. (1940). A chytrid allied to *Pleolpidium inflatum* Butler. *Transactions of the British mycological Society* **24**: 7-19.

WATERHOUSE, G. M. (1942). Some water moulds of the Hogsmill river collected from 1937 to 1939. *Transactions of the British mycological Society* **25**: 315-325.

WATERHOUSE, G. M. (1956). The genus *Phytophthora*. Diagnoses (or descriptions) and figures from the original papers. Diagnoses (or descriptions) and figures from the original papers. *Miscellaneous Publication, Commonwealth Mycological Institute, Kew* **12**: 1-120.

WATERHOUSE, G. M. (1964). The genus *Sclerospora* diagnoses (or descriptions) from the original papers and a key. *Miscellaneous Publication, Commonwealth Mycological Institute, No. 17*: 1-30.

WATERHOUSE, G. M. (1967). Key to *Pythium* Pringsheim. *Mycological Papers, Commonwealth Mycological Institute, Kew* **109**: 1-15.

WATERHOUSE, G. M. (1968*a*). Nomina Conservanda Proposita. Proposal for the conservation of the genus name *Pythium* Pringsheim 1858 (Fungi - Oomycetes: Peronosporales) vs. *Pythium* Nees von Esenbeck in Carus, 1823 (Fungi - Oomycetes: Saprolegniales). *Taxon* **17**: 88.

WATERHOUSE, G. M. (1968*b*). The genus *Pythium* Pringsheim. Diagnoses (or descriptions) and figures from the original papers. *Mycological Papers, Commonwealth Mycological Institute, Kew* **110**: 1-71.

WATERHOUSE, G. M. (1968*c*). *Dicksonomyces* based on a misdetermination. *Mycologia* **60**: 976-978.

WATERHOUSE, G. M. (1970). The genus *Phytophthora* de Bary. Diagnoses (or descriptions) and figures from the original papers. Second Edition. *Mycological Papers, Commonwealth Mycological Institute, Kew* **122**: 1-59.

WATERHOUSE, G. M. (1973). Plasmodiophoromycetes. *In* G. C. Ainsworth, F. K. Sparrow & A. L. Sussman (Eds), *The Fungi, An Advanced Treatise*, **IVB**, pp. 75-82. New York, U.S.A.: Academic Press.

WATERHOUSE, G. M. & BROTHERS, M. P. (1981). The taxonomy of *Pseudoperonospora*. *Mycological Papers, Commonwealth Mycological Institute, Kew* **148**: 1-28.

WATERSTON, J. M. (1946). The fungi of Bermuda. *Bulletin of the Department of Agriculture, Bermuda* **23**: 1-305.

WATSON, A. J. (1971). Foreign bacterial and fungal diseases of food, forage and fiber crops. *U.S. Agricultural Research Service, Agriculture Handbook no. 418*: 1-111.

WATSON, S. W. (1957*a*). Cultural and cytological studies on species of *Labyrinthula*. Ph.D. Thesis, University of Wisconsin, Madison, Wisconsin, U.S.A.

WATSON, S. W. (1957*b*). Techniques for the isolation of *Labyrinthula* and *Thraustochytrium* in pure culture. *Journal of Bacteriology* **73**: 589-590.

WATSON, S. W. & ORDAL, E. J. (1951). Studies on *Labyrinthula*. *University of Washington Oceanography Laboratory, Technical Report* **3**: 1-37.

WATSON, S. W. & ORDAL, E. J. (1957). Techniques for the isolation of *Labyrinthula* and *Thraustochytrium* in pure culture. *Journal of Bacteriology* **73**: 589-590.

WATSON, S. W. & RAPER, K. B. (1957). *Labyrinthula minuta* sp. nov. *Journal of General Microbiology* **17**: 368-377.

WEATHERWAX, P. (1914). *Aphanomyces phycophilus* deBary. *Proceedings of the Indiana Academy of Science, 1913*: 109-111.

WEBB, P. C. R. (1935). The cytology and life-history of *Sorosphaera veronicae*. *Annals of Botany, London* **49**: 41-52.

WEBSTER, M. A. & DIXON, G. R. (1991*a*). Calcium, pH and inoculum concentration influencing colonization by *Plasmodiophora brassicae*. *Mycological Research* **95**: 64-73.

WEBSTER, M. A. & DIXON, G. R. (1991*b*). Boron, pH and inoculum concentration influencing colonization by *Plasmodiophora brassicae*. *Mycological Research* **95**: 74-79.

WEBSTER, R. C. (1943). Influence of nutrition in sporangial formation in *Araiospora streptandra*. *Transactions of the Kentucky Academy of Science* **11**: 31.

WEETE, J. D., FULLER, M. S., HUANG, M. Q. & GANDHI, S. (1989). Fatty acids and sterols of selected Hyphochytriomycetes and Chytridiomycetes. *Experimental Mycology* **13**: 183-195.

WELLS, E. A. (1982). *A developmental study of the aquatic fungus* Hyphochytrium catenoides *Karling*. MS Thesis, University of Georgia, Athens, Georgia, U.S.A.

WELTZIEN, H. C. (1981). Geographical distribution of downy mildews. *In* D. M. Spencer (Ed.) *The Downy Mildews*, pp. 31-43. Academic Press, London, U..

WEMMER, L. (1954). Über die Ruheperiode der Zygosporen von *Saprolegnia*. *Archiv für Mikrobiologie* **21**: 217-229.

WERNHAM, C. C. (1935). A species of *Sorodiscus* on *Heteranthera*. *Mycologia* **27**: 262-273.

WETHERED, J. M. & JENNINGS, D. H. (1985). Major solutes contributing to the solute potential of *Thraustochytrium aureum* and *T. roseum* after growth in media of different salinities. *Transactions of the British mycological Society* **85**: 439-446.

WEST, C. (1916). *Stigeosporium marattiacearum* gen. et sp. nov. *Annals of Botany, London* **30**: 357.

WESTE, G. (1994). Impact of *Phytophthora* species on native vegetation of Australia and Papua New Guinea. *Australian Journal of Plant Pathology* **23**: 190-209.

WESTE, G. & MARKS, G. C. (1987). The biology of *Phytophthora cinnamomi* in Australian forests. *Annual Review of Phytopathology* **25**: 207-229.

WESTON, W. H. (1917). Observations on an *Achlya* lacking sexual reproduction. *American Journal of Botany* **4**: 354-367.

WESTON, W. H. (1919). Repeated zoospore emergence in *Dictyuchus*. *Botanical Gazette* **68**: 287-296.

WESTFALL, J. A., BRADBURY P. C. & TOWNSEND, J. W. (1983). Ultrastructure of the dinoflagellate *Polykrikos*. *Journal of Cell Science* **63**: 245-261.

WETSTEYN, L. P. M. J. & PEPERZAK, L. (1991). Field observations in the Oosterschelde, the Netherlands, on *Coscinodiscus con* (Bacillariophyceae) infected by the marine fungus *Lagenisma coscinodisci* (Oomycetes). *Hydrobiological Bulletin (The Nethen*

WESTON, W. H. (1921). Another conidial *Sclerospora* of Philippine maize. *Journal of Agricultural Research* **20**: 669-684.

WESTON, W. H. (1929). A new *Sclerospora* from Australia. *Phytopathology* **19**: 1107-1115.

WESTON, W. H. (1929). A new *Sclerospora* from Nyasaland. *Phytopathology* **23**: 587-595.

WESTON, W. H. (1938). Heterothallism in *Sapromyces reinschii*. Preliminary note. *Mycologia* **30**: 245-253.

WESTON, W. H. & UPPAL, B. N. (1932). The basis for *Sclerospora sorghi* as a species. *Phytopathology* **22**: 573-586.

WHATLEY, J. M. (1989). Chromophyte chloroplasts - a polyphyletic origin? *In* J. C. Green, B. S. C. Leadbeater & W. L. Diver (Eds), *T and Perspectives* pp. 125-144. Oxford, U.K.: Clarendon Press.

WHIFFEN, A. J. (1938). *Aphanomyces phycophilus* in culture. *American Journal of Botany* **25**: 649-650.

WHIFFEN, A. (1939*a*). The cytology of *Octomyxa achlyae*. Unpublished thesis, University of North Carolina, Chapel Hill, North Carolin

WHIFFEN, A. (1939*b*). The cytology of a new species of the Plasmodiophoraceae. *Journal of the Elisha Mitchell Scientific Society* **55**: 24

WHIFFEN, A. J. (1942). A discussion of some species of *Olpidiopsis* and *Pseudolpidium*. *American Journal of Botany* **29**: 607-611.

WHIFFEN, A. J. (1945). Nutritional studies of representatives of five genera in the Saprolegniaceae. *Journal of the Elisha Mitchell Scienti*

WHIFFEN, A. J. (1946). Two new terricolous phycomycetes belonging to the genera *Lagenidium* and *Blastocladiella*. *Journal of the Elisha Mi*

WHISLER, H. C. (1990). Incertae sedis Ellobiopsida. *In* L. Margulis, J. O. Corliss, M. Melkonian & D. J. Chapman (Eds), *Handbook of P* U.S.A.: Jones & Bartlett.

WHISLER, H. C. & TRAVLAND, L. B. (1973). Mitosis in *Harpochytrium*. *Archiv für Protistenkunde* **115**: 69-74.

WHITFIELD, P. R. & BOTTOMLEY, W. (1983). Organization and structure of chloroplast genes. *Annual Review of Plant Physiology*

WHITE, M. E., POWELL, E. N., WILSON, E. A. & RAY, S. M. (1989). The spatial distribution of *Perkinsus marinus*, a protozoan par (*Crassostrea virginica*) and an ectoparasitic gastropod (*Boonea impressa*). *Journal of the Marine Biological Association of the*

WHITTAKER, R. H. (1969). New concepts of kingdoms of organisms. Evolutionary relations are better represented by new classifications thar *Science, New York* **163**: 150-160.

WHITTICK, A. & SOUTH, G. R. (1971). Remarks on an infection of *Antithamnion floccosum* by a species of *Olpidiopsis*. *British Phyco*

WHITTICK, A. & SOUTH, G. R. (1972). *Olpidiopsis antithamnionis* n. sp. (Oomycetes, Olpidiopsidaceae), a parasite of *Anthithamnion fl* Newfoundland. *Archiv für Mikrobiologie* **82**: 353-360.

WHYTE, S. K., CAWTHORN, R. J. & McGLADDERY, S. E. (1994). Co-infection of bay scallops *Argopecten irradians* with *Perkinsus ka* and an unidentified coccidian parasite. *Diseases of Aquatic Organisms* **18**: 53-62.

WHYTE, S. K., CAWTHORN, R. J. MacMILLAN, R. J. & DESPRES, B. (1993). Isolation and purification of developmental stages of *P* Perkinsea), a parasite affecting bay scallops *Argopecten irradians*. *Diseases of Aquatic Organisms* **15**: 199-205.

WILCE, R. T., SCHNEIDER, C. W., QUINLAN, A. V. & VAN DEN BOSCH, K. (1982). The life history and morphology of free livin (Ectocarpaceae, Ectocarpales) in Nahant Bay, Massachusetts, USA. *Phycologia* **21**: 336-354.

WIGLESWORTH, M. D., NESMITH, W. C., SCHARL, C. D., LI, D. & SIEGEL, M. R. (1994). Use of specific repetitive sequences in *P* detection of the tobacco blue mold pathogen. *Phytopathology* **84**: 425-430.

WILCE, R. T., SCHNEIDER, C. W., VAN DEN BOSCH, K. & QUINLAN, A. V. (1982). Free living *Pilayella littoralis* (L.) Kjellm. in morphology, occurrence and associated species. *Phycologia* **20**: 116.

WILDE, P. (1961). Ein Beitrag zur kenntnis der Variabiität von *Phytophthora infestans* (Mont.) de Bary. *Archiv für Mikrobiologie* **40**: 163

WILDEMAN, É. de (1890*a*). Note sur quelques Saprolegniées parasites des algues. *Bulletin des Séances de la Société Belge de Microscop*

WILDEMAN, É. de (1890*b*). Chytridiacées de Belgique. *Annales de la Société Belge de Microscopie, Mémoires* **14**: 5-28.

WILDEMAN, É. de (1891). Notes sur quelques organismes inférieurs. *Bulletin de la Société Royale de Botanique de Belgique, Bruxelles*

WILDEMAN, É. de (1892). Une espèce nouvelle du genre *Lagenidium* Schenk. *Bulletin de la Société Royale de Botanique de Belgique, E*

WILDEMAN, É. de (1893*a*). Notes mycologiques [premièr fascicule]. *Annales de la Société Belge de Microscopie, Mémoires* **17**: 5-30.

WILDEMAN, É. de (1893*b*). Notes mycologiques [deuxième fascicule]. *Annales de la Société Belge de Microscopie, Mémoires* **17**: 35-63.

WILDEMAN, É. de (1894). Notes mycologiques [troisième fascicule]. *Annales de la Société Belge de Microscopie, Mémoires* **18**: 135-161

WILDEMAN, É. de (1895*a*). Notes mycologiques [quatrième fascicule]. *Annales de la Société Belge de Microscopie, Mémoires* **19**: 59-80.

WILDEMAN, É. de (1895*b*). Notes mycologiques [cinquième fascicule]. *Annales de la Société Belge de Microscopie, Mémoires* **19**: 85-117

WILDEMAN, É. de (1895*c*). Notes mycologiques [sixième fascicule]. *Annales de la Société Belge de Microscopie, Mémoires* **19**: 191-232.

WILDEMAN, É. de (1895*d*). Quelques Chytridiacées nouvelles parasites d'algues. *La Notarisia* **10**: 33-35.

WILDEMAN, É. de (1896*a*). Census Chytridinaearum. *Bulletin de la Société Royale de Botanique de Belgique, Bruxelles* **35**: 7-69.

WILDEMAN, É. de (1896*b*). Notes mycologiques [septième fascicule]. *Annales de la Société Belge de Microscopie, Mémoires* **20**: 21-64.

WILDEMAN, É. de (1896*c*). Notes mycologiques [huitième fascicule]. *Annales de la Société Belge de Microscopie, Mémoires* **20**: 105-136

WILDEMAN, É. de (1897*a*). Notes mycologiques [neuvième fascicule]. *Annales de la Société Belge de Microscopie, Mémoires* **21**: 5-31.

WILDEMAN, É. de (1897*b*). Notes mycologiques [dixième fascicule]. *Annales de la Société Belge de Microscopie, Mémoires* **22**: 125-128

WILDEMAN, É. de (1900*a*). Une nouvelle Chytridinée. *Mémoires de l'Herbier Boissier, 1900* **3**: 1-2.

WILDEMAN, É. de (1900*b*). Observations sur quelques Chytridinées nouvelles ou peu connues. *Mémoires de l'Herbier Boissier, 1900* **15**

WILDEMAN, É. de (1931). Sur quelques Chytridinées parasites d'algues. *Académie Royale de Belgique, Bulletins de la Classe des Scienc*

WILLE, N. (1899). Om nogle Vandsoppe. *Videnskabsselskabets Skrifter. I. Mathematisk-naturvidenskabelig Klasse. 1899. No. 3. Chris*

WILLIAMS, P. H., AIST, S. J. & AIST, J. R. (1971). Response of cabbage root hairs to infection by *Plasmodiophora brassicae*. *Canad*

WILLIAMS, P. H. & McNABOLA, S. (1967). Fine structure of *Plasmodiophora brassicae* in sporogenesis. *Canadian Journal of Botany*

WILLIAMS, P. H. & McNABOLA, S. (1970). Fine structure of the host-parasite interface of *Plasmodiophora brassicae* in cabbage. *Phy*

WILLIAMS, P. H., REDDY, M. N. & STRANDBERG, J. O. (1969). Growth of noninfected and *Plasmodiophora brassicae* infected cab *Journal of Botany* **47**: 1217-1221.

WILLIAMS, P. H. & YUKAWA, Y. B. (1967). Ultrastructural studies on the host-parasite relations of *Plasmodiophora brassicae*. *Phyto*

WILLIAMS, R. J. (1984). Downy mildews of tropical cereals. *In* D. S. Ingram & P. H. Williams (Eds), *Advances in Plant Pathology, Volu* Academic Press.

WILLIAMS, W. T. & WEBSTER, R. K. (1970). Electron microscopy of the sporangium of *Phytophthora capsici*. *Canadian Journal o*

WILLIAMSON, H. R. & YITZCHAK, P. ben (1991). The synaptonemal complexes of *Achlya recurva* (Oomycetes) karyotype analysis and of pachytene nuclei in antheridia and oogonia. *Canadian Journal of Botany* **69**: 1384-1395.

WILLOUGHBY, L. G. (1959). Studies on soil chytrids IV. A chytrid parasite of rotifer eggs. *Transactions of the British mycological Soc*

WILLOUGHBY, L. G. (1962). The occurrence and distribution of reproductive spores of Saprolegniales in fresh water. *Journal of Ecolog*

WILLOUGHBY, L. G. (1969). Pure culture studies on the aquatic phycomycete, *Lagenidium giganteum*. *Transactions of the British myc*

WILLOUGHBY, L. G. (1971). Aquatic fungi from an antarctic island and a tropical lake. *Nova Hedwigia* **22**: 469-488.

WILLOUGHBY, L. G. (1977). An abbreviated life cycle in the salmonid fish *Saprolegnia*. *Transactions of the British mycological Societ*

WILLOUGHBY, L. G. (1978). Saprolegnias of salmonid fish in Windermere: a critical analysis. *Journal of Fish Diseases* **1**: 51-67.

WILLOUGHBY, L. G. (1985). Rapid preliminary screening of *Saprolegnia* on fish. *Journal of Fish Diseases* **8**: 473-476.

WILLOUGHBY, L. G. (1994). *Fungi and Fish Diseases*. Stirling, Scotland, U.K.: Pisces Press.

WILLOUGHBY, L. G. (1998). *Saprolegnia polymorpha* sp nov., a fungal parasite on Koi carp, in the U.K. *Nova Hedwigia* **66**: 507-511.

WILLOUGHBY, L. G. & COPLAND, J. W. (1984). Temperature - growth relationships of *Saprolegnia* pathogenic to fish, especially eels cultivated in warm water. *Nova Hedwigia* **39**: 35-55.

WILLOUGHBY, L. G. & LILLEY, J. H. (1992). The ecology of aquatic fungi in Thailand and the fish disease relationship. *Aquatic Animal Health Research Institute, Bankok, Newsletter 1*: 5-6.

WILLOUGHBY, L. G. & PICKERING, A. D. (1977). Viable Saprolegniaceae spores on the epidermis of the salmonid fish *Salmo trutta* and *Salvelinus alpinus*. *Transactions of the British Mycological Society* **68**: 91-95.

WILLOUGHBY, L. G. & ROBERTS, R. J. (1992*a*). Towards strategic use of fungicides against *Saprolegnia parasitica* in salmonid fish hatcheries. *Journal of Fish Diseases* **15**: 1-13.

WILLOUGHBY, L. G. & ROBERTS, R. J. (1992*b*). The ecology of *Olpidiopsis incrassata, Rozellopsis septigena* and *Woronina polycystis*, biflagellate fungal parasites of *Saprolegnia* fungus in fresh water. *Nova Hedwigia* **55**: 1-9.

WILLOUGHBY, L. G. & ROBERTS, R. J. (1994). Improved methodology for isolation of the *Aphanomyces* fungal pathogen of epizootic ulcerative syndrome (EUS) in Asian fish. *Journal of Fish Diseases* **17**: 541-543.

WILLOUGHBY, L. G., ROBERTS, R. J. & CHINABUT, S. (1995). *Aphanomyces invaderis* sp. nov., the fungal pathogen of freshwater tropical fish affected by epizootic ulcerative syndrome. *Journal of Fish Diseases* **18**: 273-275.

WILSON, G. W. (1907*a*). Studies in North American Peronosporales - I. The genus *Albugo*. *Bulletin of the Torrey Botanical Club* **34**: 61-84.

WILSON, G. W. (1907*b*). Studies in North American Peronosporales - II. Phytophthoreae and Rhysotheceae. *Bulletin of the Torrey Botanical Club* **34**: 387-416.

WILSON, G. W. (1908*a*). Studies in North American Peronosporales - III. New or noteworthy species. *Bulletin of the Torrey Botanical Club* **35**: 361-362.

WILSON, G. W. (1908*b*). Studies in North American Peronosporales - IV. Host index. *Bulletin of the Torrey Botanical Club* **35**: 543.

WILSON, G. W. (1914*a*). Studies in North American Peronosporales - V. A review of the genus *Phytophthora*. *Mycologia* **6**: 54-83.

WILSON, G. W. (1914*b*). Studies in North American Peronosporales - VI. Notes on miscellaneous species. *Mycologia* **6**: 192-210.

WILSON, I. M. (1960). Marine fungi: a review of the present position. *Proceedings of the Linnean Society of London, Session 171, 1958-59*: 53-70.

WING, B. L. (1975). New records of Ellobiopsidae (Protista (*incertae sedis*)) from the North Pacific with a description of *Thalassomyces albatrossi* n. sp., a parasite of the mysid *Stilomysis major*. *Fishery Bulletin* **73**: 169-185.

WINGE, Ö. (1913). Cytological studies in the Plasmodiophoraceae. *Arkiv för Botanik* **12**: 1-39.

WINNER, C. (1966). Untersuchungen über parasitogene Schäden an Wurzeln der Zuckerrübe durch *Aphanomyces*, und über Möglichkeiten ihrer Verhütung. I. Diagnostische Untersuchungen in Feldkulturen. *Phytopathologische Zeitschrift* **57**: 105-120.

WINTER, [H.] G. (1879). Ueber ein natürliches System der Thallophyten ... *Hedwigia* **18**: 1-12.

WINTER, G. (1884). Einleitung. *Die Pilze Deutschlands, Oesterreichs und der Schweitz. Unter Mitwirkung von Professor Dr. A. de Bary und Dr. H. Rehm. Erster Band. Pilze, I. Abteilung: Schizomyceten, Saccharomyceten und Basidiomyceten* pp.1-32. Leipzig, Germany: Eduard Kummer. [Commonly cited as *Dr. L. Rabenhorst's Kryptogamen-Flora von Deutschland, Oesterreich und der Schweiz. Zweite Auflage, Erster Band. Pilze, I. Abteilung: Schizomyceten, Saccharomyceten und Basidiomyceten.*]

WIN-TIN & DICK, M. W. (1975). Cytology of Oomycetes. Evidence for meiosis and multiple chromosome associations in Saprolegniaceae and Pythiaceae, with an introduction to the cytotaxonomy of *Achlya* and *Pythium*. *Archives of Microbiology* **105**: 283-293.

WISSELINGH, C. Van (1898). Mikrochemische Untersuchungen über die Zellwände der Fungi. *Jahrbücher für wissenschaftliche Botanik* **3**: 619-687.

WITMAN, G. B. (1990). Introduction to cilia and flagella. *In* R. A. Bloodgood (Ed.), *Ciliary and Flagellar Membranes*, pp. 1-30. New York, U.S.A.: Plenum.

WITTMACK, L. (1892). *Pythium Sadebeckianum* als ursache oiner krankheit der erbsen. *Mitteilungen des Vereins zur Forderung der Moorkultur* **10**: 83.

WIZE, C. (1904). Choroby komośnika buraczanego (*Cleonus punctiventris*) powodowane przez grzyby owodobójcze, ze szczególnem uwzględnieniem gattunków nowych. (Die durch Pilze hervorgerufenen Krankheiten des Rübenrüssellkäfers (*Cleonus punctiventris* Germ.) mit besonderer Berücksichtigung neuer Arten). (Les maladies du *Cleonus punctiventris* Germ. causées par les champignons entonaophytes en insistant particulièrement sur les espèces nouvelles). *Bulletin International de l'Académie des Sciences de Cracovie. Classe des Sciences Mathématiques et Naturelles* **10**: 713-727 [in Polish; the second title of the journal is in German: *Anzeiger der Akademie der Wissenschaften in Krakou. Mathematisch-Naturwissenschaftliche Classe*; the publisher of the journal (Polska Akademia Umiejętności, Kraków) has sometimes been cited in taxonomic literature].

WOLF, F. T. (1937*a*). Some sporangial variations in *Saprolegnia ferax*. *Mycologia* **29**: 226-228.

WOLF, F. T. (1937*b*). A nutritional study of *Achlya bisexualis* and *Saprolegnia ferax*. *American Journal of Botany* **24**: 119-123.

WOLF, F. T. (1938). Cytological observations on gametogenesis and fertilization in *Achlya flagellata*. *Mycologia* **30**: 456-467.

WOLF, F. T. (1939*a*). Sawada's discovery of *Achlya flagellata* as a parasite of fish. *Mycologia* **31**: 236-237.

WOLF, F. T. (1939*b*). A study of some aquatic Phycomycetes isolated from Mexican soils. *Mycologia* **31**; 376-387.

WOLF, F. T. (1941*a*). A contribution to the life history and geographic distribution of the genus *Allomyces*. *Mycologia* **33**: 158-173.

WOLF, F. T. (1941*b*). A new species of *Achlya* from Costa Rica. *Mycologia* **33**: 274-278.

WOLF, F. T. (1944). *The Aquatic Oomycetes of Wisconsin. Part One*. Madison, Wisconsin, U.S.A.: University of Wisconsin Press.

WOLF, F. T. & WOLF, F. A. (1941). Aquatic phycomycetes from the Everglades region of Florida. *Lloydia* **4**: 270-273.

WOLK, P. C. van der (1913). *Protascus colorans*, a new genus and a new species of the Protoascineae-group: the source of "yellow grains" in rice. *Mycologische Centralblatt* **3**: 153-157.

WOLKE, R. E. (1975). Pathology of bacterial and fungal diseases affecting fish. *In* W. E. Ribelin & G. Migaki (Eds), *The Pathology of Fishes*, pp. 33-166. Madison, Wisconsin, U.S.A.: University of Wisconsin Press.

WOOD, S. E. & WILLOUGHBY, L. G. (1986). Ecological observations on the fungal colonization of fish by Saprolegniaceae in Windermere. *Journal of Applied Ecology* **23**: 737-749.

WOOD, S. E., WILLOUGHBY, L. G. & BEAKES, G. W. (1988). Experimental studies on uptake and interaction of spores of the *Saprolegnia diclina - parasitica* complex with external mucus of brown trout (*Salmo trutta*). *Transactions of the British Mycological Society* **90**: 63-73.

WOODHAM-SMITH, C. (1962). *The Great Hunger, Ireland 1845-9*. Hamish Hamilton, London, UK.

WOODRING, J. L. & KAYA, H. K. (1992). Infectivity of *Lagenidium giganteum* to *Culex tarsalis* (Diptera, Culicidae) in rice field waters, laboratory evaluation. *Environmental Entomology* **21**: 183-190.

WOODRING, J. L., KAYA, H. K. & KERWIN, J. L. (1995). *Lagenidium giganteum* in *Culex tarsalis* larvae: production of infective propagules. *Journal of Invertebrate Pathology* **66**: 25-32.

WORK, T. M., OLSON, J. K., GRANT, C. D., COMBS, J. C., COYKENDALL, R. L., LUSK, E. E. & WASHINO, R. K. (1985). Laboratory studies on the effects of *Lagenidium giganteum* Couch on *Psorophora columbiae* (Dyar & Knab) and *Anopheles quadrimaculatus* Say in Texas. *In* C. D. Grant, J. C. Lombs, R. L. Coykendall, E. E. Lusk & R. K. Washino (Eds), *Proceedings and Papers of the Fifty-second Annual Conference of the California Mosquito and Vector Control Association*, p. 93. Sacramento, U.S.A.: California Mosquito and Vector Control Association Press.

WORONIN, M. (1866). Über die bei der Schwarzerle (*Alnus glutinosa*) und die gewöhnlichen Garten-lupine (*Lupinus mutabilis*) auftretenden Wurzelnschwellungen. *Mémoires de l'Académie Impériale des Sciences de St. Pétersbourg, VIIe Série (Zapiski Imperatorskoĭ Akademii nauk)* **10**: 1-13.

WORONIN, M. [ВОРОНИН, М., **Voronin, M.**] (1877). *Plasmodiophora brassicae*. Организмъ, причиняющiй Капустнымъ растенiямъ болѣзнь, извѣстную подъ названiемъ "Килы" [Organism causing disease known under name 'Hernia' of cabbage plants]. *С.-Петербургъ* [***Trudy Imperatorskago Sankt-Peterburgskago** Obshchestva Estestvoispytatelei*`. *S.-Peterburg*] **81**: 169-201. [The German title often quoted as "Der Organismus der die unter dem Namen Hernie bekannte Krankheit der Kohlpflanzen verusacht" does not appear in the original, but the following references are full translations.]

[**WORONIN, M.** (1878). *Plasmodiophora brassicae*, Urheber der Kohlpflanzen-Hernie. *Jahrbücher für Wissenschaftliche Botanik* **11**: 548-574.

and

CHUPP, C. (1934). *Plasmodiophora brassicae* the cause of cabbage hernia, by Michael Stephanovitch Woronin, 1878, a translation from the German, with a biographical sketch by the translator. *Phytopathological Classics Number 4* pp. 1-31 plus plates. Minnesota, U.S.A.: American Phytopathological Society.]

WORONIN, M. (1878). *Plasmodiophora brassicae*, Urheber der Kohlpflanzen-Hernie. *Jahrbücher für Wissenschaftliche Botanik* **11**: 548-574 [see above].

WRIGHT, E. P. (1879). On a species of *Rhizophydium* parasitic on *Ectocarpus*, with notes on the fructification of the Ectocarpi. *Transactions of the Royal Irish Academy, Dublin (Science)* **26**: 369-380.

WRIGHT, P. J. & EASTON, C. S. (1996). Natural incidence of *Lagenidium giganteum* Couch (Oomycetes: Lagenidiales) infecting the biting midge *Culicoides molestus* (Skuse) (Diptera: Ceratopogonidae). *Australian Journal of Entomology* **35**: 131-134.

WUNDER, W. (1938). Die Ausheilungsvorgänge an der Karpfenkieme bei Kiemenfäule. I. Beschreibung der Verhältnisse bei Betrachtung mit blossem Auge und mit der Lupe. *Zeitschrift für Parasitenkunde* **10**: 280-287.

WYNN, A. R. & EPTON, H. A. S. (1979). Parasitism of oospores of the *Phytophthora erythroseptica* in soil. *Transactions of the British mycological Society* **73**: 255-259.

WYNN, W. K. (1981). Tropic and taxic responses of pathogens to plants. *Annual Review of Phytopathology* **19**: 237-255.

YAČEVSKIE, A. A. & YAČEVSKIE, P. A. [ЯЧЕВСКИЕ, А. А. & ЯЧЕВСКИЕ, П. А. **also spelt Jaczewski, A. A. & Jaczewski, P. A.**] (1931). *Определителъ Грибов. Совершенные Грибы (Диплоидные Стадии). Том **I** - Фикомицеты. С фотографическими снимками и Оригинальными Рисунками, исполненными О. П. Исвевой и Т. Н. Швиндт.* [*Opredelitel' Gribov. Soveršennye griby (diploidnye stadii). Tom I - Fikomicety. S fotografičeski misnimkami i Original'nymi Risunkami, ispolnennymi O. P. Isaevoj i T. I. Švindt*]. pp.294. Leningrad, Russia: Gosudarstvennoe Izdatel'stvo Sel'skozhozyaystvennoi` i Kolkhozno-kooperativnoi` literatury, Moskva.

YANO, S., TANAKA, S., ITO, S. & KAMEYA-IWAKI, M. (1994). Spheroplast isolation and ultrastructure of the cell wall in resting sporangia of *Plasmodiophora brassicae*. *Annals of the Phytopathological society of Japan* **60**: 310-314 [in Japanese].

YANO, S., TANAKA, S. H., ITO, S. I. & KAMEYA-IWAKI, M. (1997). Variations of random amplified polymorphic DNA (RAPD) patterns among field populations of *Plasmodiophora brassicae*. *Annals of the Phytopathological society of Japan* **63**: 179-182.

YANO, S., TANAKA, S. H., KAMEYA-IWAKI, M. & KATUMOTO, K. (1991). (Relation of Ca_2^+ efflux to germination of resting spores of clubroot fungus). *Bulletin of the Faculty of Agriculture, Yamaguchi University, 1991* **39**: 105-112 [only abstract consulted].

YAO, C.-L. (1991). Classification and detection of *Peronosclerospora* species on the basis of DNA southern hybridization and PCR reaction. PhD Thesis, Texaas A. & M. University, Texas, U.S.A.

YENDO, Y. & TAKASE, K. (1932). On the root nodule of *Elaeagnus*. *Bulletin of Sericulture and Silk Industry, Uyeda, Japan* **4**: 4-5.

YARWOOD, C. E. (1977). *Pseudoperonospora cubensis* in rust-infected bean. *Phytopathology* **67**: 1021-1022.

YILI, R., WANHE, Z. & JIANPING, C. (1990). Quantitative assay of resting spores of *Polymyxa graminis* on barley roots. *Annals of Applied Biology* **117**: 343-347.

YMER (1899) [cf. Sparrow (1960: bibliography and text citations - see under [N.] G. Lagerheim (1900*a*)].

YOUNG, E. L. (1937). Notes on the labyrinthulan parasite of eel-grass *Zostera marina*. *Bulletin of the Mount Desert Island Biological Laboratory* pp. 33-35.

YOUNG, E. L. (1938). *Labyrinthula* on Pacific coast eel-grass. *Canadian Journal of Research, Section C* **16**: 115-117.

YOUNG, E. L. (1943). Studies on *Labyrinthula*. The etiologic agent of the wasting disease of eel-grass. *American Journal of Botany* **30**: 586-593.

YU, Y. N. (1973). Five new species of *Pythium*. *Acta Microbiologica Sinica* **13**: 116-123 [in Chinese with Latin diagnoses and English abstract].

YU, Y. N. (1977). *Myzocytium vermicolum*, a nematode-endozoic fungus. *Acta Microbiologica Sinica* **17**: 297-302.

YU, Y.-n. & MA, G.-z. (1989). The genus *Pythium* in China. *Mycosystema* **2**: 1-110.

YU, Y. N. & WANG, Y. L. (1987). Taxonomic studies on the genus *Pythium* from Yunnan Province of China. *Acta Botanica Yunnan* **9**: 129-152 [in Chinese with English abstract].

YUASA, K. & HATAI, K. (1994). An attempt of classification by biological characteristics of some pathogenic water moulds from freshwater fishes. *Nippon Kingakukai Kaiho* **35**: 104-110.

YUASA, K. & HATAI, K. (1995*a*). Drug sensitivity of some pathogenic water moulds isolated from freshwater fishes. *Journal for Antibacterial and Antifungal Agents* **23**: 213-219.

YUASA, K. & HATAI, K. (1995*b*). Relationship between pathogenicity of *Saprolegnia* spp. isolates to Rainbow Trout and their biological characteristics. *Fish Pathology (Gyobyo Kenkyu)* **30**: 101-106.

YUASA, K. & HATAI, K. (1996). Some biochemical characteristics of the genera *Saprolegnia*, *Achlya* and *Aphanomyces* isolated from fishes with fungal infection. *Mycoscience* **37**: 477-479.

YUNG, C. & STENTON, H. (1964). A study of the phycomycetes in the soils of Hong Kong. *Transactions of the British Mycological Society* **47**. 127-139.

ZAKHA, G. A. & VIRÁNI, F. (1991). Axenic culture of the downy mildew fungus *Plasmopara halstedii* in *Agrobacterium rhizogenes*-induced roots of sunflower (*Helianthus annuus*). *Canadian Journal of Botany* **69**: 2709-2715.

ZEBROWSKA, E. (1976). Nikoflora kilku zbiornikow wodnych Puszczy Kampinoskiej. *Acta Mycologica* **12**: 77-89.

ZEBROWSKI, G. (1936). New genera of Cladochytriaceae. *Annals of the Missouri Botanic Garden* **23**: 553-564.

ZENTMYER, G. A. (1961). Chemotaxis of zoospores for root exudates. *Science* **133**: 1595-1596.

ZENTMYER, G. A. (1970). Tactic Responces of zoospores of *Phytophthora*. *In* T. A. Toussoun, R. V. Bege & P. E. Nelson (Eds), *Root Diseases and Soilborne Pathogens*, pp. 109-111. Berkeley, California, U.S.A.: University of California Press.

ZENTMYER, G. A. (1979). Effect of physical factors, host resistance and fungicides on root interaction at the soil-root interface. *In* J. L. Harley & R. Scott Russell (Eds), *The Soil-Root Interface*, pp. 315-328. London, U.K.: Academic Press.

ZENTMYER, G. A. (1980). Phytophthora cinnamomi *and the diseases it causes*. (Monograph 10). St Paul, Minnesota, U.S.A.: American Phytopathological Society.

ZHANG, L. & LANGENBERG, W. G. (1988). Biological characteristics of *Lagena radicicola* culture conditions and light microscope appearance. *Phytopathology* **78**: 1611-1612.

ZHANG, Q., GRIFFITH, J. M. & GRANT, B. R. (1992). Role of phosphatidic acid during differentiation of *Phytophthora palmivora* zoospores. *Journal of General Microbiology* **138**: 451-459.

ZHANG, Q., GRIFFITH, J. M., MOORE, J. G., ISER, J. R. & GRANT, B. R. (1990). The effect of modified pectin, pectin fragments and cations on *Phytophthora palmivora* zoospores. *Phytochemistry* **29**: 695-700.

ZHENG, X. B. & HO, H. H. (2000). The sexual stage of *Phytophthora polygoni* Saw. *Botanical Bulletin of Academia Sinica (Taipei)* **41**: 57-60.

ZHOU, T. & PAULITZ, T. C. (1993). In vitro and in vivo effects of *Pseudomonas spp.* on *Pythium aphanidermatum*: Zoospore behavior in exudates and on the rhizoplane of bacteria-treated cucumber roots. *Phytopathology* **83**: 872-876.

ZICKLER, D. (1973). Fine structure of chromosome pairing in ten ascomycetes. Meiotic and premeiotic (mitotic) synaptonemal complexes. *Chromosoma (Berlin)* **40**: 401-416.

ZIEGLER, A. W. (1948*a*). A comparative study of zygote germination in the Saprolegniaceae. *Journal of the Elisha Mitchell Scientific Society* **64**: 13-40.

ZIEGLER, A. W. (1948*b*). A new species of *Achlya*. *Mycologia* **40**: 336-341.

ZIEGLER, A. W. (1950). A new *Achlya* from Florida. *Mycologia* **42**: 658-662.

ZIEGLER, A. W. (1952). The water molds of Leon County, Florida. *Florida state University Studies* **8**: 1-22.

ZIEGLER, A. W. (1953). Meiosis in the Saprolegniaceae. *American Journal of Botany* **40**: 60-66.

ZIEGLER, A. W. (1958). The Saprolegniaceae of Florida. *Mycologia* **50**: 693-696.

ZIEGLER, A. W. & GILPIN, R. H. (1954). A description of a new species of *Achlya* with some observations on its physiology. *Mycologia* **46**: 647-651.

ZIEGLER, A. W. & LINTHICUM, B. (1950). Some physiological aspects of fertility in *Achlya tuberculata*. *Journal of the Elisha Mitchell Scientific Society* **66**: 182-185.

ZIMMERMAN, A. (1901). *Botanical Microtechnique*. New York, U.S.A.: Henry Holt.

ZOPF, W. (1878). LVIII. Sitzung vom 30. August 1878. Vorsitzender: Herr L. Kny. Ueber einem neuen parasitischen Phycomyceten aus der Abtheilung der Oosporeen. *Verhandlungen der Botanischen Vereins der Provinz Brandenburg* **20**: 77-79.

ZOPF, W. (1884). Zur Kenntniss der Phycomyceten. I. Zur Morphologie und Biologie der Ancylisteen und Chytridiaceen, zugleich ein Beitrag zur Phytopathologie. *Nova Acta der Kaiserlichen Leopoldinisch-carolinischen Akademie der Naturforscher* **47**: 141-236.

ZOPF, W. (1885*a*). Die Pilzthiere oder Schleimpilze...mit...Schnitten. *Separatabdruck aus der Encyklopaedie der Naturwissenschaften* **[3(2)]** i-viii, 1-174. [seen as: **ZOPF, W.** (1887). Die Pilzthiere oder Schleimpilze. Mycetozoa (de Bary). - Myxomycetes Auct. pp. 1-174, *in* [*Schenk's*] *Handbuch der Botanik* [4 vols in 5 parts, vol. 3 in 2 parts] **3(2)**: i-xix, 1-716. Wroclaw (Breslau), Poland: E. Treweundt.]

ZOPF, W. (1885*b*). *Zur Morphologie und Biologie der niederen Pilztiere (Monadinen)*. Leipzig.

ZOPF, W. (1887). Ueber einige den Wasser zu isolieren. *Abhandlungen der Naturforschung Gesellschaft zu Halle* **17**: 77-107.

ZOPF, W. (1888). Zur Kenntnis der Infections-Krankheiten niederer Thiere und Pflanzen. *Nova Acta der Kaiserlichen Leopoldinisch-Carolinischen Deutschen Akademie der Naturforscher* **52**: 313-376.

ZOPF, W. (1890). Die Pilze. *In* (A. Schenk, Ed.), [*Schenk's*] *Handbuch der Botanik*, vol. **4**, pp. 271-755. [Abschnitt VI (Systematik und Entwicklungsgesichte): 552-744]. Wroclaw (Breslau), Poland: E. Trewendt.

ZOPF, W. (1892). Zur Kenntnis der Labyrinthuleen, einer Familie der Mycetozoen. *Beiträge zur Physiologie Morphologie Niederer Organismen* **2**: 36-48.

ZOPF, W. (1893). Ueber eine Saprolegniee mit einer Art von Erysipheen-ähnlicher Fruchtbildung. *Beiträge zur Physiologie Morphologie Niederer Organismen* **3**: 48-59.

ZOPF, W. (1894). Ueber niedere thierische und planzliche Organismen, welche als Krankheitserreger in Algen, Pilzen, niedern Thieren und höheren Pflanzen auftreten. *Beiträge zur Physiologie und Morphologie Niederer Organismen* **4**: 43-68.

ZOU, F.-Z. & SU, X.-Q. (1996). A preliminary study on the effects of six pesticides on *Lagenidium giganteum* (Oomycetes: Lagenidiales). *Chinese Journal of Biological Control* **12**: 123-126 [in Chinese].

INDEXES

INDEX TO TOPICS AND MAJOR GROUPS OF ORGANISMS

INDEX TO GENERA

INDEX TO GENERA

INDEX TO BINOMIALS

NOTE: Names of straminipilous fungi, and other taxa treated nomenclaturally, are given in bold italic. Synonyms, host and substrata taxa are in italic.

INDEX TO BINOMIALS

9780792367802